Theory of Stellar Atmospheres

PRINCETON SERIES IN ASTROPHYSICS
Edited by David N. Spergel

Theory of Rotating Stars, *by Jean-Louis Tassoul*

Theory of Stellar Pulsation, *by John P. Cox*

Galactic Dynamics, Second Edition, *by James Binney and Scott Tremaine*

Dynamical Evolution of Globular Clusters, *by Lyman Spitzer, Jr.*

Supernovae and Nucleosynthesis: An Investigation of the History of Matter, from the Big Bang to the Present, *by David Arnett*

Unsolved Problems in Astrophysics, *edited by John N. Bahcall and Jeremiah P. Ostriker*

Galactic Astronomy, *by James Binney and Michael Merrifield*

Active Galactic Nuclei: From the Central Black Hole to the Galactic Environment, *by Julian H. Krolik*

Plasma Physics for Astrophysics, *by Russell M. Kulsrud*

Electromagnetic Processes, *by Robert J. Gould*

Conversations on Electric and Magnetic Fields in the Cosmos, *by Eugene N. Parker*

High-Energy Astrophysics, *by Fulvio Melia*

Stellar Spectral Classification, *by Richard O. Gray and Christopher J. Corbally*

Exoplanet Atmospheres: Physical Processes, *by Sara Seager*

Physics of the Interstellar and Intergalactic Medium, *by Bruce T. Draine*

The First Galaxies in the Universe, *by Abraham Loeb and Steven R. Furlanetto*

Introduction to Modeling Convection in Stars and Planets: Magnetic Field, Density Stratification, Rotation, *by Gary Glatzmaier*

Dynamics and Evolution of Galactic Nuclei, *by David Merritt*

Theory of Stellar Atmospheres: An Introduction to Astrophysical Non-equilibrium Quantitative Spectroscopic Analysis, *by Ivan Hubeny and Dimitri Mihalas*

Theory of Stellar Atmospheres

An Introduction to Astrophysical Non-equilibrium Quantitative Spectroscopic Analysis

Ivan Hubeny and Dimitri Mihalas

PRINCETON UNIVERSITY PRESS
PRINCETON AND OXFORD

Published by Princeton University Press, 41 William Street, Princeton,
New Jersey 08540
In the United Kingdom: Princeton University Press, 6 Oxford Street,
Woodstock, Oxfordshire OX20 1TW

press.princeton.edu

Library of Congress Cataloging-in-Publication Data

Hubený, I. (Ivan)
Theory of stellar atmospheres : an introduction to astrophysical non-equilibrium quantitative spectroscopic analysis / Ivan Hubeny and Dimitri Mihalas.
pages cm. —(Princeton series in astrophysics)
Includes bibliographical references and index.
ISBN 978-0-691-16328-4 (hardcover : alk. paper) – ISBN 978-0-691-16329-1
(pbk. : alk. paper)
1. Stars–Atmospheres. 2. Stars–Spectra. I. Mihalas, Dimitri, 1939–2013. II. Title. III. Title:
Stellar atmospheres.
QB809.H83 2015
523.8'6—dc23
2014006308

British Library Cataloging-in-Publication Data is available

This book has been composed in LaTeX

Printed on acid-free paper. ∞

Printed in the United States of America

10 9 8 7 6 5 4 3 2 1

Dedication

To my friends and colleagues, Lawrence Auer and David Hummer,
and to the memories of my friends and teachers
William Wilson Morgan, and Walter Stibbs.

D. M.

To the memory of my "scientific mother" and friend Françoise Praderie,
and to the memory of my "scientific father," friend,
and co-author of this book, Dimitri Mihalas, who sadly passed away
just days after the manuscript was submitted to the publisher.

I. H.

Contents

Preface

The purpose of this book is to provide an account of the major advances achieved in making quantitative spectroscopic analyses of the observable outer layers of stars. These analyses can now be done with a much higher degree of realism thanks to a deeper understanding of the interaction between the material and the radiation field in a stellar atmosphere, more accurate and more complete data from space-based and ground-based observations, markedly improved computational algorithms, and the immense increase in the speed of computers. We omit some topics that today are mainly of historical interest and focus instead on providing a solid physical formulation of the problem and methods by which it can be solved.

We have in mind three groups of readers: (1) astronomers who understand the astrophysical context and want to acquire deeper insight into the subject's physical foundations and/or to learn about modern computational techniques for treating non-equilibrium radiative transfer, (2) physicists who know the underlying physics, but are not familiar with the motivation for this work or how to deal with radiative transfer, and (3) students who need to learn both the reasons for, and the methodology of, this discipline. We try to offer a description of both the physics and mathematics that is adequate to do research in the field, rather than a short summary. Several other good books at a more introductory level are available, e.g., [22, 132, 133, 186, 246, 391, 1003, 1006, 1178].

Strands from many diverse topics must be woven together to form the complete picture. For example, the occupation numbers in the material in the outer *observable* layers of a star are determined by kinetic equations containing the rates of radiative and collisional processes. These layers have very low densities, so collision rates are much smaller than radiative rates; hence the latter essentially determine the state of the material. Therefore, the *material's absorption and emission coefficients are set by the local radiation field.* But, in turn, *the local radiation field is the result of not only local photon emission and absorption, but also photons that have penetrated from other (perhaps remote) points in the atmosphere where the physical conditions may be quite different. In short, the radiation field determines the non-equilibrium properties of the material, but those properties in turn determine the radiation field; the two are inextricably coupled.* This is the central problem in computing a theoretical stellar spectrum.

In this book we have concentrated mostly on hot stars, whose atmospheres are dominated by radiative processes and are reasonably well approximated by homogeneous planar or spherical layers. The situation is more complex for cool stars, in which energy is transported by both radiation and convection. The latter is a time-dependent, quasi-turbulent, three-dimensional flow posing problems only now being addressed with some degree of accuracy. We will refer only to the phenomenological *mixing-length picture* of convective energy transport. We make no attempt to discuss magnetically dominated structures in stellar chromospheres and coronae.

The material presented here enables one to calculate radiative transfer in, and to make a quantitative spectroscopic analysis of, any low-density plasma. It applies in the observable layers of any astrophysical object, not just stars. It also provides necessary background for the study of radiation hydrodynamics, in which radiation can drive a flow of the material.

The book is meant to be reasonably self-contained, but the reader should be familiar with the elements of quantum mechanics and special relativity. Also, we point out that the ordering of its chapters is somewhat unconventional compared to other texts on this subject, so we ask serious students of the subject to use an "iterative" approach in reading it. Readers who re-read earlier parts of the book bearing on material they are currently studying will reap solid rewards in their understanding. We believe that a judicious instructor can use this book as a text for beginners or a monograph for advanced students and research scientists. In broad outline, the topics covered are the following.

ASTROPHYSICAL BACKGROUND

Chapter 1 contains a brief sketch of the historical development of the study of stellar atmospheres. In its last section we point out specific areas in which this work makes important contributions to other problems in the larger arena of astrophysics. Chapter 2 describes the wide variety of observational data that both underlie and test the theoretical structures we use to interpret them.

RADIATION

In chapter 3 we present macroscopic, electromagnetic, and quantum mechanical descriptions of a radiation field. These representations are used in all later work. In chapter 4 we use thermodynamics and statistical mechanics to derive the physical properties of matter and radiation in the limit of strict *thermodynamic equilibrium* (TE) and introduce the concept of *local thermodynamic equilibrium* (LTE). Many of the formulae obtained in this chapter are used throughout the remainder the book.

ABSORPTION AND EMISSION OF RADIATION

The quantum mechanics of *photon absorption and emission* in bound-bound, bound-free, and free-transitions is discussed in chapter 5, and the *scattering of continuum radiation* in both the low-energy and relativistic limits is covered in chapter 6. In chapter 7, we sketch the quantum-mechanical calculation of absorption cross sections for some astrophysically important atomic and molecular opacity sources. In chapter 8 we discuss the broadening of spectral lines in frequency resulting from (a) the finite lifetimes of the atomic levels they connect, (b) Doppler shifts produced by the component of an atom's thermal velocity along the line of sight or of a macroscopic velocity of the material, and (c) semi-classical and quantum mechanical descriptions of the interruptions of the wavetrain by collisions with free electrons and/or other atoms.

INTERACTIONS BETWEEN RADIATION AND SPECTRAL LINES

The observable effects of non-equilibrium occupation numbers in the material of a stellar atmosphere are more pronounced in spectral lines than in continua and strongly influence the diagnosis of its physical properties. To understand the interplay between kinetic-equilibrium processes and transport processes, we analyze the physics of spectral line formation in a *given* atmospheric structure in the next four chapters.

In chapter 9 we write the kinetic equations that determine the excitation and ionization occupation numbers of material in a general non-equilibrium (NLTE) radiation field. Here we must account for the fact that photons in a spectral line are not scattered coherently, but in general are redistributed over the line profile. In chapter 10 we describe these phenomena in terms of *redistribution functions* obtained from both a semi-classical approach and a rigorous quantum mechanical analysis. It turns out that the semi-classical and quantum results are essentially the same in the limit of weak radiation fields, which is typically the case in stellar atmospheres.

RADIATION TRANSPORT

In chapter 11 we use the physical information derived in chapters 5–10 to assemble the absorption, emission, and scattering coefficients that appear in the *radiation transport equation*, to give a phenomenological derivation of the transport equation in general coordinates, and to show that it is equivalent to a Boltzmann equation in kinetic theory. We then specialize it to Cartesian, spherical, and cylindrical coordinates, show that its angular moments are dynamical equations for the radiation field, and discuss the closure of the system of moments.

Next we reduce the transport equation to a time-independent *transfer equation*, derive the mathematical operators that connect its moments to the specific intensity of the radiation field, and derive its second-order form, which provides the basis for some of the best computational methods. For physically realistic problems, the solution of the transfer equation cannot be obtained analytically; hence we discretize it (along with constraints discussed in later chapters) to reduce it to a form suitable for numerical computation. Finally, we give a *probabilistic interpretation* of the transfer equation and discuss its very important *asymptotic diffusion limit.*

NUMERICAL SOLUTION OF THE TRANSFER EQUATION

At this point one has the physical theory necessary to compute the transfer of radiation through a stellar atmosphere. In the next two chapters we review current computational methods for solving the transport equation. As mentioned above, the systems of equations that determine the structure of, and spectrum from, a stellar atmosphere are too complex to be addressed analytically. Therefore, we turn to the ever-increasing power and availability of high-speed digital computers to solve them numerically.

In chapter 12 we describe robust *direct methods*, which can give precise solutions of the transfer problems of interest in this discipline. But to fit observed stellar spectra accurately, it may be necessary to account for the non-equilibrium effects of thousands (or more!) of spectral lines and to compute the radiation field at an immense number of frequencies. Unfortunately, direct methods can be too costly to use in such cases. Therefore, in chapter 13 we describe modern, very effective *iterative methods* that allow us to deal with these daunting problems and to compute more physically realistic model atmospheres.

In chapter 14 we make a first study of non-equilibrium spectral line formation. We start with an analysis of the underlying physics of photon diffusion and destruction in, and escape from, an atmosphere, using a model atom consisting of only *two bound levels and a continuum*. Even though it is extremely oversimplified, we gain very useful physical insight from this problem. We then extend the treatment to the physics of photon transport in *multi-level atoms*. In chapter 15 we discuss spectral line formation when there is *partial redistribution* (i.e., noncoherence) of photons scattered by the line.

STRUCTURAL EQUATIONS

In chapter 16 we write the equations that determine the *structure* of stellar atmospheres at different levels of physical reality. We start with general equations of hydrodynamics, specialize them to one-dimensional flow, further specialize to one-dimensional steady flow, and then to the structure of *static* atmospheres in planar and spherical geometry. Next we examine briefly *energy transport by convection* in addition to radiation and, finally, the equations that determine the interior structure of stars.

MODEL ATMOSPHERES

In chapter 17 we describe an efficient method to compute *planar, static, LTE model atmospheres* that satisfy the constraints of *radiative and hydrostatic equilibrium* and account for *line blanketing* by large numbers of spectral lines. The opacity from spectral lines can dominate a star's emergent radiation field and strongly affect its physical structure in its outermost layers. We know today that when one assumes LTE, the results obtained are quantitatively (sometimes even qualitatively) inaccurate. Nevertheless, this problem gives the reader an overview of the issues encountered in constructing models of stellar atmospheres and introduces some of the mathematical techniques used in later work.

Chapter 18 deals with the much more difficult problem of constructing *NLTE model atmospheres*, including line blanketing. The results from such computations yield reliable diagnostic tools to determine the effective temperature, surface gravity, and composition of a wide range of stars. This work has provided valuable insights into the evolutionary histories of stars.

EXTENDED AND EXPANDING ATMOSPHERES

We describe radiative transport in static *spherically symmetric atmospheres* in chapter 19. This geometry is appropriate for highly evolved stars that have very extended envelopes. The spectra of very hot stars often show evidence of rapid mass loss in *stellar winds* for which spherical symmetry may be a reasonable first approximation. Hence we discuss the solution of the radiative transfer equation for material moving radially outward in a stationary *laboratory frame* (also called the *observer's frame*) in which the material is seen to move and in the *comoving frame*, or *fluid frame*, which moves with the material. In the latter case, the *Sobolev approximation* provides a simple method for solving the transfer problem and predicting the emergent spectrum with modest accuracy.

In chapter 20 we study the dynamics of stellar winds. We first briefly describe thermally driven winds. Next, we concentrate on the dynamics of radiatively driven *stellar winds*, which are observed in very hot stars where the rate of outward momentum input from photons into the material exceeds the binding force of gravity. Here great progress has been made with *global model atmospheres* in which the wind is joined self-consistently to the star's photosphere. The results of this work have importance in many other areas of astrophysics.

TYPOGRAPHY

We denote physical scalars with *italic* Roman or Greek letters, vectors with **boldface** letters, and tensors and matrices with bold **sans serif** letters. We denote individual components of vectors, tensors, and matrices with italic letters having the appropriate number of subscripts or superscripts. When relativity enters, we use the summation convention for repeated indices. Sections marked with a ⋆ in the Contents may be omitted on a first reading. As in most technical books, several different physical quantities may be represented by the same symbol because the combined Roman and Greek alphabets do not have enough letters to provide separate characters for each of them.

REFERENCES AND BIBLIOGRAPHY

The complete bibliography of this book is a valuable asset: it lists both the seminal papers at the historical roots of the subject and the large number of papers resulting from its vigorous pursuit today. It has over 4500 references; were it to be printed, it would occupy more than 300 pages, which is excessive. Therefore, in the text we typically give citations for only the initial paper(s) dealing with a topic or indicate that the citations given are only examples with the construction "see, e.g., [199, 824]";

we have taken advantage of modern technology and put the full list on a web site http://press.princeton.edu/titles/10407.html.

This site holds two files: `biblio.ps` and `biblio.pdf` in standard PostScript and PDF format, respectively, which can be viewed on a monitor or printed. It also contains the BibTex file `biblio.bbl`, from which these files were made; it is in plain-text format and can be read with a screen editor and used to make other bibliographies. Readers can then use authors' names and keywords from the titles of the papers cited to search the biblio.bbl file for other relevant references.

Finally, we gratefully acknowledge the help and guidance of many colleagues, too numerous to list here. Although we must leave them anonymous, we offer all of them our most sincere thanks.

Dimitri Mihalas
G.C. McVittie Professor of Astronomy, Emeritus
University of Illinois

Ivan Hubeny
University of Arizona
Tucson, Arizona

Theory of Stellar Atmospheres

Chapter One

Why Study Stellar Atmospheres?

A central objective of astrophysics is to use physical theory to simulate the conditions in astrophysical objects. Today's astrophysicist must, in principle, be familiar with just about all of physics: elementary particle theory and nuclear physics; quantum mechanics and atomic/molecular physics; classical electrodynamics, quantum electrodynamics, and plasma physics; hydrodynamics and magnetohydrodynamics; classical gravitational mechanics, special relativity, and general relativity. The goal of this book is decidedly more modest: (1) to present the physical and mathematical tools needed to make models of stellar atmospheres that are realistic enough to fit closely the observed spectrum of a star, and (2) to show how they can be used to make a reliable quantitative spectroscopic analysis of the physical structure and chemical composition of its outer layers. The problems to be faced in achieving it are developing correct formulations of the physics of spectral line and continuum formation and the transport of radiation.

The complex systems of nonlinear equations that describe cosmic objects can be solved using the extremely fast computers now available. In addition, we now have access to most of the electromagnetic spectrum, ranging from radio wavelengths to gamma rays. These observational data provide a solid basis upon which we can construct theoretical structures to interpret them. It is a reasonable metaphor to say that today our picture of the Universe is developing as rapidly, and as radically, as if Galileo and Newton had lived and worked at the same time.

But given the crowded curriculum faced by astrophysics students one may ask, "Why take time to study the outer layers of stars? Is this work merely a 'cottage industry' of no great relevance to the rest of astrophysics?" This question was put pointedly to one of us (D. M.) nearly 50 years ago by E. Salpeter: "Why in the world would *anyone* want to study stellar atmospheres? They contain only about 10^{-10} of the mass of a star. Surely such a negligible fraction of its mass cannot affect its overall structure and evolution!" His query is reasonable and deserves a good answer.

1.1 A HISTORICAL PRÉCIS

Salpeter's question must be put in the perspective of his own seminal work on stellar structure, stellar evolution, and nucleosynthesis. Relative to those studies, the theory of stellar atmospheres at that time had not produced spectacular results, and did not yet have a sound theoretical basis. To provide some context, we outline below a few milestones in the development of today's observational techniques; the theory of astrophysical radiative transfer and quantitative spectroscopy; and the theory of stellar evolution. Other authors would doubtless select different topics than we have chosen here.

Instrumentation

At the beginning of the 20th century, the most powerful telescopes were long-focus refractors, the largest being the 40″ diameter telescope at Yerkes Observatory, built under the direction of G. E. Hale. They were used for visual observations of the orbits of binary stars (which give data for the determination of stellar masses); photographic observations of *radial velocities* (the component of a star's velocity, relative to the Sun, along the line of sight); *proper motions* (which are proportional to a star's velocity, relative to the Sun, perpendicular to the line of sight); and stellar *parallaxes* (which give a star's distance from the Sun).

After Hale left Yerkes to found the Mt. Wilson Observatory, he directed the construction of the 60″ and 100″ reflectors on Mt. Wilson and the 200″ reflector and 48″ wide-angle Schmidt camera on Palomar Mountain. Each increase in aperture permitted fainter objects to be observed. The instrumentation for these large reflectors was at the cutting edge of technology then available. For example, their spectrographs contained large blazed gratings made with interferometrically controlled grating-ruling engines at the headquarters of the observatory in Pasadena. These had higher efficiency and resolution, and better freedom from ghosts, than any previously made. But until the 1950s astronomical telescopes had little automation other than drives to track the apparent motion of stars across the sky. After about 1950, the sophisticated electronic measurement and control techniques of physics laboratories began to invade the mountaintop redoubts of astronomers.

In the past two to three decades observational astrophysics has enjoyed unparalleled growth. We now have telescopes with apertures of several meters, which have thin mirrors whose optical figures can be adjusted to minimize the effects of turbulence in the Earth's atmosphere and corrected for flexure and other transient defects, in real time, using high-speed computers. Soon, it will be possible to restore images to nearly the diffraction limit of a telescope. With such instruments, we can observe objects at low light levels that were previously inaccessible. The far infrared, ultraviolet, X-ray, and gamma-ray regions of the spectrum can now be observed with space telescopes.

Today's echelle spectrographs can capture spectra at high resolution in many orders simultaneously. Photographic plates have been replaced with CCDs (*charge-coupled devices*), which have very high light detection efficiency and a linear response that permits precise calibration and measurement of images of stars, extended objects, and echelleograms. With them we can now obtain high signal-to-noise spectra of extremely faint objects.

At radio wavelengths, interferometric techniques developed for the Very Large Array (VLA) and the Very Long Baseline Interferometer (VLBI) have produced exquisite pictures showing that many galaxies have massive *black holes* at their centers, which spew immense jets of material racing outward at relativistic speeds. These data imply the existence and continuing modification of an intergalactic medium from which new galaxies form. In short, it is now possible to observe, and begin to model, phenomena totally unknown only a few years ago.

Observations

The two most important observational activities in astrophysics at the start of the 20th century were *photometry* and *spectral classification*. Strenuous efforts were made to set up accurate photometric brightness scales[1] for stars, but they were thwarted by the nonlinear response of photographic emulsions to exposure. Through the 1940s, the apparent brightnesses of stars could not be measured to much better than 10%, and often only to 20%–30%. Brightness measured in different wavelength bands could be used to make estimates of stellar *colors*, which give a low-resolution measure of the distribution of light in their spectra.

The arrival of *photomultipliers* in the early 1950s gave astronomers very sensitive *linear* receivers accurate to 1% and revolutionized stellar photometry. With them, standard photometric systems were established and used for incisive analyses of stellar properties (see § 2.4–§ 2.6). By the mid-1960s, photoelectric *spectrophotometers* calibrated against standard sources having known energy distributions put spectrophotometry on an absolute energy scale (see § 2.3). Today, spectrophotometric measurements of the continua of stars can be matched with high precision to results obtained from physically realistic theoretical model atmospheres that allow for the effects of many thousands of spectral lines. With arrays of CCDs containing thousands of individual detectors one can measure simultaneously the apparent brightness of huge numbers of stars in large fields of the sky.

The state of stellar spectroscopy was better. In the early 1900s, it was found that stellar spectra could be arranged in a sequence that correlates closely with a star's *effective temperature*.[2] This classification scheme was quickly supplemented with additional criteria based on subtle effects that correlate with the average density in a star's atmosphere. These phenomena could be interpreted using theoretical work in statistical mechanics [344]. By 1914, E. Hertzsprung and H. Russell [488, 915] showed that stars fall in definite sequences in the *Hertzsprung-Russell diagram* (or *HR diagram*), a plot of their *luminosity* versus effective temperature. This discovery had profound implications for the development of a theory of stellar structure and evolution.

Measurement of stellar spectra progressed from mere eye estimates of "line intensities" in the early 1900s to photographic measurements of *line profiles* and *equivalent widths* in the 1930s. In contrast to the absolute photometry needed to determine stellar magnitudes, these measurements require only *relative* photometry, i.e., comparing the light at several wavelengths in a line to the local continuum. Hence the results were more accurate. Today, with linear receivers such as CCDs, very precise measurements can be made simultaneously for a large range of wavelengths.

[1] Measured in *magnitudes*; see § 2.4 for their definition.

[2] A representative temperature of the material in its atmosphere; see equation (2.3) for its formal definition.

A breakthrough came in 1944 when W. Baade [73, 74] (see especially the excellent review [938]) made the seminal discovery of two *stellar populations* in the Galaxy, whose properties were determined by, and give information about, the formation and subsequent development of the Galaxy. His work integrated our observational picture of the stellar content of galaxies and also provided critical guidance to stellar evolution theory. He found that the distributions of the two populations of stars in the HR diagram are distinctively different. HR diagrams for *Population I* and *Population II* stars are shown in figures 2.6 and 2.7, respectively.

Population I objects in our Galaxy are typified by (1) *galactic clusters* (loose clusters containing $\sim 10^2 - 10^3$ stars), which include (a) a *main sequence* extending from massive, hot, very luminous *blue dwarfs* to cool, less massive, much less luminous *red dwarfs* (see table 2.4); (b) very luminous, cool, red *supergiants* in young clusters; and (c) *subgiants* and *red giants* in older clusters; (2) *Cepheid pulsating variables*; and (3) *interstellar material*. These objects are located near the central plane of the disk of the Galaxy. In other spiral galaxies, they are found in bright *spiral arms* bordered by dark *dust lanes*.

The great majority of stars in the solar neighborhood belong to Population I and are on the main sequence. In their cores, these stars are converting hydrogen to helium in thermonuclear reactions. This process releases the largest amount of energy per reaction, and hydrogen is the most abundant element in stellar material. Therefore, a star spends more time "burning" hydrogen to helium than in any other stage in its evolutionary history, so most stars will be found in this phase. Population I stars have relatively small velocities with respect to the Sun. They move on high angular momentum, nearly circular orbits around the Galactic center. They have near-solar abundances of "*metals*" (astrophysical jargon for elements with $Z \geq 6$; Z being the atomic number) [379, 472, 473, 474, 475, 476, 1129].

Typical Population II objects in our Galaxy are (1) *globular clusters* (gravitationally bound spherical systems containing $\sim 10^5 - 10^6$ stars); (2) individual *halo stars*, weak-lined *high-velocity stars*, and *subdwarfs* in the solar neighborhood; (3) *RR Lyrae pulsating variables*; and (4) *planetary nebulae* and their *nuclei*, which are old stars in late stages of their life. These objects are not strongly concentrated in the Galactic plane. Indeed, the distribution of globular clusters is roughly isotropic around the center of the Galaxy. Population II stars make up the central bulge in other spiral galaxies and most elliptical galaxies. They have low angular momentum and low velocities around the center of the Galaxy; hence they have high velocities relative to the Sun. The most extreme of these stars, and globular clusters, move on almost radial "plunging" orbits with respect to the Galactic center [120], [chapter 10]. This kinematic behavior gives clues about the formation of the Galaxy [305]. In 1951, the extreme subdwarfs HD 19445 and HD 140283 were found [211] to have "metal" abundances smaller by at least a factor of 25 to 40 than the Sun (a much too conservative estimate; see the discussion in [938, p. 433]). Since then, spectroscopic analyses of many Population II stars have been made, which give metal abundances factors of 10 to 10^5(!) smaller than solar; see, e.g., [85, 206, 258, 273, 347, 635, 802, 1022, 1024, 1036, 1117, 1130, 1131, 1132, 1133, 1136]. These stars presumably represent a primeval population.

At the opposite extreme, some (quite young?) *super-metal-rich* (SMR) Population I stars have a higher metal abundance than the Sun; see, e.g., [204, 293, 327, 1020, 1070, 1134]. It is clear that as a function of time there has been a progressive enrichment of the metal abundance of the material from which stars form.

The conventional notation used to indicate a star's metal abundance relative to the Sun is

$$[\mathrm{Fe/H}] \equiv \log\left[(\mathrm{Fe}_*/\mathrm{H}_*) \big/ (\mathrm{Fe}_\odot/\mathrm{H}_\odot) \right]. \tag{1.1}$$

Here iron is used as a proxy for all elements with $Z \geq 6$. This notation is oversimplified because there are variations from star to star in the ratio of the abundance of any chosen element to that of iron. Using criteria based on their distribution, kinematics, and abundances, the idea of stellar populations has been elaborated into a picture having several groups intermediate between the extremes represented by the most metal-poor globular clusters and youngest galactic clusters [800, 805].

The earliest generation of stars was composed of about 90% hydrogen by number; about 10% helium by number; and very small amounts of some isotopes of Li and Be, which were formed from primeval hydrogen in the *Big Bang*. Modern work shows that the He/H ratio is about the same in the interiors of both Population I and Population II stars. This fact implies that essentially all He was formed in the Big Bang. On the other hand, the existence of young and very old stars, having high and low metal abundances, respectively, shows that *there has been a progressive enrichment of elements with* $Z \geq 6$ *in the interstellar material from which stars form.* The heavy elements in the interstellar medium are created by *nucleosynthesis* [160, 1137], i.e., by thermonuclear processing of material in the cores of highly evolved, massive stars. This material is deposited into the interstellar medium by *supernovae*.

Stellar Structure and Evolution

The earliest models of the internal structure of stars were based on Ritter and Emden's theory of polytropic gas spheres with self gravity [318]. By assuming that the gas pressure $p_{\mathrm{gas}} = K\rho^{(n+1)/n}$, where ρ is the mass density [gm/cm^3], the equations for hydrostatic pressure balance can be combined with Poisson's equation for the gravitational field to get a single second-order differential equation for $\rho(r)$, the variation of density with radius in the star. It can be solved analytically for $n = 0$, 1, and 5 and by numerical integration for other values of n.

Beginning in the late 19th century, and through the 1930s, intellectual giants such as S. Chandrasekhar, A. Eddington, R. Emden, R. Fowler, E. Hopf, H. Lane, E. Milne, A. Ritter, A. Schuster, and W. Thomson (Lord Kelvin) developed and refined the theory of polytropes. With the assumption that the material is a perfect gas in which radiation pressure is unimportant, their work yielded realistic lower bounds on a star's central and average pressure: $p_c > 1.3 \times 10^{15}(\mathsf{M}^2/\mathsf{R}^4)$ and $\langle p \rangle \geq 5.4 \times 10^{14}(\mathsf{M}^2/\mathsf{R}^4)$ dynes/cm^2, and its average temperature: $\langle T \rangle \geq 4.6 \times 10^6 \mu(\mathsf{M}/\mathsf{R})$ K. In these inequalities M and R are a star's total mass and its radius in units of the solar mass and radius and μ is the "effective mean molecular weight" of the gas. They also showed that the ratio of radiation pressure to total pressure increases monotonically toward a star's center, and with increasing mass. For the Sun this ratio is < 0.03. A rigorous discussion of the reasoning leading to these bounds is given in Chandrasekhar's classic book *An Introduction to the Study of Stellar Structure* [223].

The publication of Eddington's book *The Internal Constitution of the Stars* in 1926 [302] was a milestone in studies of stellar structure. Eddington showed that (1) stars have central energy sources; (2) for massive stars, energy is transported to the surface by radiation; (3) and for these stars, radiation pressure produces the dominant force supporting them against gravity. His book contains good discussions of the thermodynamics of radiation, then-current quantum theory, ionization, diffusion, his pioneering work on the opacity of stellar material, and speculations about stars' energy source, which he could not identify but suspected must come from subatomic processes. He developed a "standard model" that assumes that the average rate of energy release per unit mass, $[L(r)/\mathsf{L}]/[M(r)/\mathsf{M}]$, within a volume of radius r, times the opacity coefficient k, is a constant. This assumption is ad hoc, but it allowed him to derive a mass-luminosity relation for main-sequence stars in fair agreement with observation and to derive the basic relation $P\sqrt{\langle\rho\rangle} =$ constant for Cepheid variables; here P is the star's pulsation period and $\langle\rho\rangle$ is its average density.

Also in 1926, Fowler [341] realized that *white dwarfs*, being $\sim 10^{-4}$ times as bright as normal stars of the same effective temperature, must have radii about 10^{-2} times smaller and are such compact, dense objects that electrons in the material are *degenerate*, i.e., obey Fermi-Dirac statistics. Further analysis [33, 212, 213, 1055, 1056] showed that degenerate material behaves like a polytrope with $p \propto \rho^{5/3}$ if the electrons' speeds are non-relativistic and $p \propto \rho^{4/3}$ if they are relativistic.[3] In the early 1930s Chandrasekhar [212, 213, 215] wrote a number of fundamental papers in which he represented white dwarfs in terms of polytropes, using the exact formula for the transition from non-relativistic to relativistic degeneracy with increasing mass. He found the remarkable result[4] that if a star's mass exceeds $\mathsf{M}_{\rm lim} = (5.84/\mu_{\rm e}^2)\,\mathsf{M}_\odot$, the *Chandrasekhar limit*, relativistic degeneracy pressure is insufficient to support the star, and it will collapse to "zero" radius, becoming either a *neutron star* or a black hole. Here μ_e is the number of atomic mass units per free electron in the gas, ≈ 2 for material with $Z \geq 2$, which is appropriate for material that has been processed in thermonuclear reactions. Thus the mass of a stable white dwarf at the end of nuclear burning must be $\lesssim 1.46\,\mathsf{M}_\odot$.

A persistent question from the beginning of studies of stellar structure was "What is the energy source that supports a star's luminosity?" The classical answer was that it is the release of *gravitational potential energy*. The gravitational potential energy between two particles with masses m_1 and m_2 separated by a distance r is

$$\mathsf{V} = -\frac{Gm_1m_2}{r}. \tag{1.2}$$

[3] That is, with a polytropic index $n = 3/2$ if non-relativistic and $n = 3$ if relativistic.

[4] For which he received the Nobel Prize in 1983.

Here, G is the gravitational constant. Thus the total potential energy of a star of radius R with a radial mass distribution $M(r)$ is

$$\mathsf{V} = -\int_0^{\mathsf{R}} \frac{GM(r)}{r}\, dM(r). \tag{1.3}$$

v and V are negative because particles are more tightly bound gravitationally as they approach one another.

At each position, particles in non-degenerate material have a Maxwellian velocity distribution at the local temperature. Let T be the total *thermal energy* in the star, i.e., the sum of the kinetic energies of all the particles. Consider the star to be evolving so slowly that it is in quasi-static equilibrium at any instant. Then we can apply the *virial theorem*, which states that

$$2\mathsf{T} + \mathsf{V} = 0 \quad \text{or} \quad \Delta\mathsf{T} = -\tfrac{1}{2}\Delta\mathsf{V}. \tag{1.4}$$

From (1.4) we see that if a star contracts as a whole, so that V becomes more negative, gravitational binding energy is released; half of it is converted into thermal energy (heats the material), and the other half is radiated away.[5] To get an order of magnitude estimate of the timescale of this process for the Sun, suppose we compute V by taking $M(r) \sim \frac{4}{3}\pi\,\overline{\rho}\,r^3$, and $dM(r) \sim 4\pi\,\overline{\rho}\,r^2\,dr$, where $\overline{\rho} \approx 1.4$ gm/cm^3 is the Sun's average density. With these approximations

$$\mathsf{V} = -\frac{16\pi^2 G\overline{\rho}^2}{3}\int_0^{\mathsf{R}} r^4\,dr = -\frac{3}{5}\frac{G\mathsf{M}^2}{\mathsf{R}} \sim -2.3\times 10^{48}\ \text{erg}. \tag{1.5}$$

The Sun's luminosity is 4×10^{33} erg/sec $\approx 1.2\times 10^{41}$ erg/year. So the maximum length of time that luminosity can be supported by release of the Sun's entire potential energy, the *Kelvin-Helmholtz timescale*, is $\tau_{\mathrm{KH}} \equiv -\frac{1}{2}\mathsf{V}/\mathsf{L}_\odot \approx 10^7$ years. In the absence of any evidence to the contrary, this result was regarded as reasonable. But it soon became clear that τ_{KH} is about 100 times smaller than the geological timescale derived from careful analyses of rock strata at many sites around the world.

The source of energy production in stars could not be understood until G. Gamow showed [360, 361, 413] that quantum mechanical tunneling allows a positively charged particle to penetrate a nucleus even though its thermal energy is smaller than the Coulomb barrier between the two particles. This theory was applied by R. Atkinson [43, 44] to a three-step *proton-proton,* or *p-p, reaction* in which four protons are synthesized into a He^4 nucleus. There are three branches for this reaction. The dominant one starts with a proton penetrating the Coulomb barrier between it and another proton, fusing them into a deuteron (H^2) plus a positron and neutrino; followed by the deuteron fusing with another proton to make a He^3 nucleus and a

[5] This result shows that gravitationally bound systems effectively have a *negative specific heat* in the sense that when energy is removed from the system by being radiated away, the material left behind becomes *hotter*!

gamma ray; finally, the He^3 nucleus fuses with another He^3 to make He^4 plus two protons, with a release of 26.2 MeV of energy:

$$\begin{aligned} &p + p \rightarrow H^2 + \beta^+ + \nu, \\ &H^2 + p \rightarrow He^3 + \gamma, \\ &He^3 + He^3 \rightarrow He^4 + 2p. \end{aligned} \tag{1.6}$$

He showed that at the conditions that apply in its core ($T \sim 10^7$ K and $\rho \sim 100$ gm/cm^3) this process would produce the Sun's luminosity.

In 1939, H. Bethe [115] found another path to generate thermonuclear energy: the fusion of four protons into a He^4 nucleus in the *CNO cycle*, a chain of six reactions with isotopes of carbon, nitrogen, and oxygen, which releases 25.0 MeV of energy.[6] The dominant form of this reaction is

$$\begin{aligned} &C^{12} + p \rightarrow N^{13} + \gamma, \quad && N^{13} \rightarrow C^{13} + \beta^+ + \nu, \\ &C^{13} + p \rightarrow N^{14} + \gamma, && N^{14} + p \rightarrow O^{15} + \gamma, \\ &O^{15} \rightarrow N^{15} + \beta^+ + \nu, && N^{15} + p \rightarrow C^{12} + He^4. \end{aligned} \tag{1.7}$$

In this process the CNO isotopes act as *catalysts* and recover their initial abundances by the end of the cycle.

To estimate the Sun's *nuclear timescale*, assume it is composed of 100% H, which is all converted to He. $m_H = 1.67 \times 10^{-24}$ g and $M_\odot = 2 \times 10^{33}$ g, so it contains $\sim 1.2 \times 10^{57}$ protons. Each cycle of both reaction chains "consumes" 4 protons, so there can be $\sim 3 \times 10^{56}$ reactions, each with a yield $\sim 4 \times 10^{-5}$ erg. Hence the Sun's nuclear energy reservoir is $\sim 1.2 \times 10^{52}$ erg. With these assumptions, the solar luminosity of 1.2×10^{41} erg/year could be sustained for $\sim 10^{11}$ years, about 10^4 times longer than τ_{KH}, and 10 times longer than the currently estimated lifetime of the Universe. Modern stellar evolution calculations show that only about 10% of the Sun's hydrogen is converted to helium, so the lifetime of a solar-type star is actually $\sim 10^{10}$ years, in agreement with the ages estimated for the oldest globular clusters.

Near $T \approx 1.5 \times 10^7$ K, the rate of energy release in the p-p chains can be fit with a power law: $\epsilon_{pp} \approx \epsilon_0 \rho X_H^2 T_6^4$. Here X_H is the fraction, by mass, of the material that is hydrogen; ρ is its density, and $T_6 \equiv T/10^6$. Near the same temperature the rate of energy release in the CNO cycle can be fit with the power law $\epsilon_{CNO} \approx \epsilon_0' \rho X_H X_{CNO} T_6^{20}$, where X_{CNO} is the total mass fraction of the material that is carbon, nitrogen, and oxygen. Low-mass stars have lower central temperatures than more massive stars, so the p-p reaction is active in the cores of stars with masses $\lesssim 1.2\,M_\odot$; at higher masses the CNO cycle is dominant in the core.

Because of the relatively weak temperature dependence of the p-p reaction, the temperature gradient in the cores of low-mass stars is small enough for them to be in *radiative equilibrium* and the released energy is initially transported outward by *absorption and emission of photons*. In this process, high-energy photons are systematically degraded to lower energies, but at each position the *total amount*

[6] This reaction was also noticed by C. von Weizsäcker [1146]; see [678] for historical commentary.

of radiant energy emitted is exactly balanced by the total amount absorbed. The envelopes of these stars generally have zones in which hydrogen ionizes, or helium becomes once or twice ionized, so the opacity of the material is very large. Then radiative energy transfer is inefficient, and energy is transported outward by a *turbulent convective flow* in which hot, less dense "parcels" of hotter material rise, expand, release energy to the surroundings, cool, compress back to a higher density, and sink. The material is then in *convective equilibrium.*[7] In some cases (e.g., a red giant) practically the entire outer envelope of a star is convective. In contrast, because the carbon cycle has such a steep temperature dependence, stars more massive than about $1.2\mathsf{M}_\odot$ have convective cores, whereas their envelopes are in radiative equilibrium.

The physical structure of a star is determined by its mass M, age, and chemical composition. Stars on the *zero-age main sequence* (ZAMS), where fusion of hydrogen into helium in their cores has just begun, have the same chemical composition throughout, i.e., are *chemically homogeneous.* As thermonuclear fusion converts one element ("fuel") to another ("ash"), a star's chemical composition begins to vary with depth. As hydrogen is converted to helium, the molecular weight of the gas rises, and to sustain the pressure needed to support the overlying hydrogen envelope, the material in the core is compressed to a higher density.

Eventually a star completely exhausts all the hydrogen at its center. It develops an inert, isothermal, helium core surrounded by a hydrogen-burning shell. In 1938, E. Öpik [818] showed that the core undergoes contraction, and in accordance with the virial theorem, half of the released gravitational potential energy goes into thermal energy that heats the core, and half raises the star's luminosity. The heating of the core causes the shell to generate additional ("excess") energy, which is absorbed in the star's outer envelope, causing it to expand, and lowers the surface temperature. Thus in the HR diagram, a star moves upward, toward higher luminosity, and to the right, toward lower T_{eff}. A star having about a solar mass evolves up the subgiant branch toward becoming a red giant. The higher the luminosity of a star, the more rapidly it consumes its hydrogen. This means that the more massive (hence more luminous) a star is, the faster it departs from the main sequence. Consequently, there is a characteristic *turnoff point* at the upper end of the main sequence in the HR diagram of a cluster, where stars move toward the domain of red giants (see figure 2.7).

In 1942, M. Schönberg and Chandrasekhar showed that one cannot construct simple core-envelope models if the core mass exceeds about 10% of the star's total mass [968] because then the star begins to contract on a Kelvin-Helmholtz timescale. The contraction could be halted if the core material becomes degenerate. But in this phase the hydrogen shell source generally cannot support the star's luminosity, so the core continues to contract and heats further. In 1957, Salpeter [937] showed that when a star's central temperature rises to $\sim 10^8$ K, the *triple-alpha reaction,* in which three He^4 nuclei are fused to make C^{12}, comes into operation. The energy released in this reaction halts the contraction, leaving the star with a hydrogen-burning shell and a helium-burning core. The triple-alpha reaction is less robust than the p-p reaction

[7] The criteria separating radiative and convective equilibria are given in chapter 16.

or CNO cycle because the beryllium isotope created in the intermediate reaction $\alpha + \alpha \rightarrow \mathrm{Be}^{8*}$ is in an excited state (hence that part of the reaction is endothermic) and is unstable. Thus an equilibrium with sufficient numbers of this isotope must be reached for the final reaction, $\mathrm{Be}^{8*} + \alpha \rightarrow \mathrm{C}^{12}$, to proceed, with a net release of about 7.4 MeV of energy.[8]

In the 1930s, B. Strömgren [1060] computed improved opacities applicable at the temperatures and densities in stellar interiors. After World War II better values became available from work at U.S. government laboratories. Currently, the best results come from the Opacity Project (OP) [1192–1232]; its successor, the Iron Project (IP) [1233–1253]; and the Livermore Opacity Library (OPAL), [1254–1271].

Mathematically, computations of stellar structures pose two-point boundary value problems for a system of four nonlinear first-order differential equations. In the early 1950s, models could be made for chemically homogeneous stars and stars with a single change in composition between core and envelope. The solution was obtained by integrating a family of (dimensionless) trial solutions from the center toward the surface, and another from the surface toward the center. The value and slope of a solution in one set were fitted to a solution in the other set at some intermediate point. A discussion of this technique is given in [982, § 13 and § 14]. In 1952, calculations by A. Sandage and M. Schwarzschild [940] showed that after core hydrogen exhaustion, a star evolves quite rapidly from a main-sequence dwarf into a red giant. Their results were consistent with observed *color-magnitude diagrams* for globular clusters and indicated that those clusters must be of the order of 3 billion years old (current work gives ages of about 14 billion years).

As a star ages, it can develop a very complex internal structure. Starting with hydrogen, it burns a nuclear fuel in its core, leaving behind the product of the reaction (in this case helium). Eventually all of that fuel in the core is exhausted, and energy generation shifts to a *shell source* surrounding an inert core (e.g., a hydrogen-burning shell surrounding a helium core). As the star ages further, the shell source becomes thinner (less massive) and eventually is unable to support the luminosity of the star. At that point the star contracts, which releases gravitational potential energy, and its central density and temperature rise until the next possible exoergic thermonuclear reaction can proceed, in this case the triple-alpha reaction, converting helium to carbon. The same series of events takes place until the new core fuel is exhausted and starts burning in another shell source around an inert core. For example, the star may have an unburned hydrogen outer envelope and then a hydrogen shell source, surrounding an inert helium shell, below which there is a helium-burning shell, and an inert carbon core, etc. If a star is massive enough, this sequence may be repeated several times, and the star develops an "*onion skin*" structure with a number of shell sources burning simultaneously. In addition, different zones can be in either radiative or convective equilibrium.

The computational procedure outlined above works when a star's interior structure is relatively simple. But it becomes difficult to use, or fails altogether, for stars that

[8] Note that helium burning produces less energy per reaction than hydrogen burning.

have several shells with different compositions, types of energy release, or modes of energy transport. When fast digital computers arrived in the 1960s, L. Henyey and his coworkers [480, 481] made a gigantic step forward by developing an efficient two-point boundary-value linearization method. This strongly convergent method can handle models of great structural complexity given a reasonable starting model, and it was immediately adopted by all workers in the field. Over the past 50 years, comprehensive calculations have been carried out tracing a star's evolution from its formation in the interstellar medium to its ultimate fate as either a supernova or a white dwarf.

For most stars, the rate of mass loss is negligible during the time they are near the main sequence, so its mass M can be used as a basic input parameter for a model of their interior structure. But for the most massive upper main-sequence stars, the mass-loss rate is large enough that it must be taken into account from the time a star is formed. As a massive star evolves away from the main sequence and becomes a red supergiant, its outermost layers becomes less strongly bound gravitationally, and the star may develop a very strong *stellar wind*, driven by momentum deposited in the material by photons emerging from the star. The mass-loss rate in the wind may be large enough to strip away much, or even most, of the star's envelope, thus exposing material that has undergone nuclear burning.

Stars of near-solar mass develop similar complex structures and winds in the very late phases (red giant and beyond) of their evolution. We can track this process with quantitative spectroscopic analysis of their atmospheres because we begin to see successive layers in which the relative abundances of elements have been altered in the "burned" material. Indeed, in the most extreme cases known, we can observe a bare stellar core; see [1155, 1157].

Calculation of the most advanced stages of stellar evolution is at best very difficult. A number of good books and review articles about the subject are available, e.g., [233, 234, 558, 559, 619, 725, 773, 949, 952].

Precise models of stellar structure can also yield very important, unexpected results. For example, for some time the rate of neutrino emission from the Sun calculated with even the most refined models differed by an order of magnitude from observations made at neutrino observatories. This discrepancy ultimately led to the conclusion that neutrinos are *not* massless, as previously supposed in "standard" elementary particle theory, but having a finite mass, they can oscillate from one type to another (electron, τ, and μ neutrinos) in the time it takes them to travel from the Sun to the Earth.

Stellar Atmospheres

The theory of *radiative transfer* started with the work of A. Schuster [979], who formulated the *transfer equation*[9] for radiation passing through a *scattering* medium. His work was extended by K. Schwarzschild [980, 981], who made the distinction

[9] The terms "radiative transfer" and "transfer equation" are used when the material is static. The terms "radiation transport" and "transport equation" are used in problems in which the material is in motion or is time-dependent. See [1002] for an interesting review of the development of radiative transfer theory.

between scattering and *thermal absorption and emission* processes,[10] developed an elegant integral equation formalism to describe the transfer of radiation; showed that the variation of the brightness of the solar continuum from the center to the edge of the Sun's disk is consistent with radiative, not convective equilibrium; and demonstrated that the frequency variation of the depths of the H and K Fraunhofer lines of ionized calcium are consistent with them being formed by scattering.

In the 1920s, work by Fowler, E. Milne, A. Pannekoek, and M. Saha; see, e.g., [343, 344, 757, 760, 836, 932], showed that excitation and ionization of material in *thermodynamic equilibrium* can be described with statistical mechanics. Using this theoretical framework, they made the first correct interpretation of the observed spectral sequence. Cecilia Payne showed in her Ph.D. thesis, later published as the monograph *Stellar Atmospheres* [849], that hydrogen is the most abundant element in stellar atmospheres, a result confirmed in an analysis of solar hydrogen line profiles by A. Unsöld [1099].

Milne [753, 754] derived approximate results for the temperature distribution in, and emergent radiation from, a *gray* (i.e., the opacity of the stellar material is independent of frequency) plane parallel atmosphere in radiative equilibrium and *local thermodynamic equilibrium* (LTE).[11] E. Hopf, M. Bronstein, and C. Mark found the exact solution to this problem; see [150, 151, 511, 711]. It provides a valuable benchmark for evaluating approximate methods of solving transfer equations. S. Rosseland showed that radiative transfer in a stellar interior can be described as a diffusion process with an average opacity known as the *Rosseland mean* [907]. And in a prescient paper B. Gerasimovic opened the question of departures from LTE in stellar atmospheres [372].

In the 1930s, W. McCrea [719] constructed the first model stellar atmosphere by numerical integration of the structural equations, allowing for the depth dependence of pressure, temperature, ionization, and opacity. In this era radiative transfer in spectral lines was treated with schematic models based on the paradigm of pure absorption and *coherent* scattering processes; see, e.g., [301, 756, 837, 1071, 1175]. This work showed that these two mechanisms result in very different emergent spectra. In reality, the scattering of photons in lines is *not* coherent. Some simplified studies of line formation allowing for noncoherent scattering were made by several authors. But at a more basic level, the picture of "absorption" and "scattering" processes is oversimplified. In reality more complicated chains of transitions, which cause *interlocking* of spectral lines [1174], are implicit in the equations of *kinetic equilibrium*; they did not receive satisfactory treatment until the 1970s. Observational and theoretical studies of stars showing strong emission lines had also started [97, 98, 373].

Despite the disruption of research by World War II, important advances were made in the 1940s. Chandrasekhar developed the *discrete-ordinate method* to solve the

[10] This distinction is extremely important. Its physical and mathematical implications are described in chapters 5, 6, 9–11, 13–15.

[11] LTE hypothesizes that gradients of physical conditions in stellar material are so small that its properties can be computed using thermodynamic equilibrium formulae at the local value of the temperature and density. This approximation is generally invalid in the observable outer layers of stars.

transfer equation subject to the constraint of radiative equilibrium in gray material; see, e.g., [218]. An outcome of this work was the conclusive identification of H^- as the dominant opacity source in the solar atmosphere. This method is a powerful and flexible numerical tool that has been applied in most astrophysical radiation transport problems.

An outstanding breakthrough in this era was made by V. Sobolev, working in near isolation in the Soviet Union. Through extraordinary physical insight, he recognized that the velocity gradient in the rapidly expanding envelopes of *Wolf-Rayet stars* and *novae* actually *simplifies* the problem of spectral line formation, even under extreme non-equilibrium conditions. The reason is that a steep velocity gradient limits the geometric size of the *interaction sphere* within which a line photon emitted at one point in the atmosphere can interact with material at another point before it is Doppler-shifted out of the line's bandwidth. This insight allowed him to derive an analytical expression for a photon's escape probability in terms of the *local* properties of the gas and the value of the velocity gradient. With it, he was able to compute realistic emission line spectra for expanding stellar envelopes and nebulae. His work was first published only in Soviet journals that were unavailable in the West during wartime, and it remained almost unnoticed until the 1960s when it was translated into English [1028, 1029]. The "Sobolev approximation" is still a very useful tool in astrophysics.

After World War II ended, Unsöld and his coworkers pushed forward detailed studies of stellar spectra and atmospheric structure, based on the hypothesis of LTE; see, e.g., [1100–1103]. But in the 1950s, astrophysicists began to take a more iconoclastic view of the assumption of LTE and the classical picture of line formation and to examine critically the effect of abandoning these a priori assumptions. In 1957, R. Thomas made his classic study of line formation for a simplified model atom having two bound levels plus continuum, a good first caricature for strong resonance lines [1075]. His work showed that photons emitted in the core of a line are redistributed so efficiently over the line's absorption profile that transfer in the line cannot be treated as coherent. A much better approximation is the limit of *complete redistribution*, i.e., where the frequency of the photon emerging from the scattering process has no correlation with that of the incoming photon.

The important physical consequence of this difference is that line-core photons absorbed and emitted at physical depths where they would be trapped by the high opacity at line center can be redistributed into a line's wings, where the opacity is smaller, and escape to the surface. This leakage allows the source term in the transport process to differ (by orders of magnitude) from its LTE value at the surface. A dark absorption line is produced even in an isothermal atmosphere, whereas no absorption feature would exist in LTE. Subsequently, in a series of seminal papers, J. Jefferies and Thomas studied non-LTE (NLTE) effects in the formation of resonance lines in a medium having a "chromosphere" (i.e., an outward rise in temperature) and delineated the behaviors of *collision dominated* and *photoionization dominated* lines. These papers [579–581] demolished earlier arguments that LTE is valid for line formation calculations and set the stage for the developments of the next 40 years.

As was the case for stellar evolution, work on stellar atmospheres changed radically in the 1960s with the arrival of fast computers that had compilers able to convert a mathematics-like language into machine code. It became possible to use more realistic descriptions of radiative transfer (which is strongly nonlinear), and theoretical work morphed into a combination of analysis and computation. A flood of new results followed. Algorithms to solve the central problem of finding the temperature distribution for a nongray atmosphere in radiative equilibrium were developed by E. Avrett, M. Krook, and L. Lucy [67, 703]. A major step forward was the development of an efficient algorithm for solving very general transfer equations by P. Feautrier [323–325], who generalized Schuster's second-order differential equation form of the transfer equation to allow for an angle- and frequency-dependent radiation field, a depth-dependent opacity, and arbitrary scattering functions. It did for stellar atmospheres modeling what Henyey's method did for stellar evolution. Large grids of LTE models were calculated for various classes of stars and used to estimate element abundances. Computations were made of the effects of "blanketing" by H lines in the visible [728] and by strong lines in the UV of hot stars; see, e.g., [141, 238, 747, 865]. S. Strom and R. Kurucz showed how to treat blanketing by millions of spectral lines by constructing *opacity distribution functions*, ODFs [1059]. In parallel, other investigators, e.g., [595, 866], developed direct *opacity sampling* techniques. Both methods are powerful tools.

Enormous progress was also made in the physical theory of spectral line formation. Accurate computations of Stark broadening of the H and He lines (used to determine the effective temperature, surface gravity, and helium abundance of a stellar atmosphere) became available; see, e.g., [80, 90, 91, 376, 399, 404, 1120, 1121, 1122]. And accurate descriptions of noncoherent scattering in spectral lines were worked out; see, e.g., [250, 251, 519, 530, 544, 546, 547, 699, 817, 1186]. This work put the physics of absorption and emission of line photons on a sound physical footing for the first time.

In the past 35 years great progress in modeling stellar atmospheres has resulted from (1) incisive physical formulations of radiative transfer; (2) calculation of high-quality cross sections for radiative and collisional processes in plasmas; (3) efficient algorithms to calculate the radiation field and material properties in a stellar atmosphere using simultaneous self-consistent solutions of the radiative transfer and kinetic equilibrium equations that allow for thousands of spectral lines. Stellar atmosphere modeling has also benefited greatly from the immense increases in computer speed and capacity.

In the area of algorithm development, the extreme difficulty of handling all of the *physical* coupling between radiative transfer and multi-level model atoms was solved by L. Auer and D. Mihalas in 1969 by the complete linearization (CL) method [53]. By the mid-1970s, calculations had been made for a number of NLTE models of O-stars. It was found that in NLTE, the strengths of the hydrogen and helium lines were considerably larger than those predicted by LTE, in agreement with observations for the first time [55, 56, 734]. But as originally formulated, the method was computationally too expensive to treat the complicated transition arrays of multielectron atoms and ions.

In the mid-1980s, several authors, e.g., [176, 918, 955, 957, 958, 959, 960], developed fast, approximate methods (which can be iterated to self-consistency) for solving transfer problems with large scattering terms. Generically, they are referred to as *Accelerated Lambda Iteration*(ALI) methods. In 1986, L. Auer, R. Buchler, and G. Olson [815] made a breakthrough by showing that inversion of only the *diagonal* of the matrix representing the depth-coupling in a transfer problem yields a convergent iterative method, which is *immensely* more efficient than direct solutions. It opened the door to constructing model atmospheres having huge numbers of spectral lines in their spectra. In the late 1980s, L. Anderson realized that we can group physically similar atomic levels into *superlevels* and represent entire transition arrays with a small number of *superlines* [29–32] between the superlevels. The combination of these three ideas (CL/ALI/superlevels) led to very powerful, fast model atmosphere codes; see, e.g., [432, 458, 523, 532, 533, 1148, 1149, 1150], that can handle atoms/ions having thousands of energy levels and millions of spectral lines, in the full non-equilibrium regime. It is now possible to compute theoretical spectra that fit observational data precisely; see, e.g., [290, 388, 427, 452, 453, 682, 683]. Indeed, they are reliable enough to use as absolute calibration standards for spectrophotometric observations; see, e.g., [124, 126, 679].

The effectiveness of photon momentum deposition in ultraviolet resonance lines to produce stellar winds from O-type and Wolf-Rayet stars was demonstrated conclusively by L. Lucy and P. Solomon in 1970 [705]. Soon after, J. Castor, D. Abbott, and R. Klein [199], using the Sobolev approximation to solve the transfer problem, and accounting for the huge numbers of lines present in the spectrum, made models that could give the massive flows actually observed. By the late 1980s, a number of groups in Germany produced models, e.g., [354, 355, 358, 430, 431, 433, 435, 436, 450], in which the photon momentum deposition in spectral lines in the expanding envelope of a star is treated self-consistently with the hydrodynamic flow it produces. ALI methods again had great impact on such computations. At present, we can even begin to analyze the exploding outer layers of novae and supernovae; see, e.g., [455, 459, 1004, 1065].

- *The essential problem faced in modeling stellar interiors is determining the depth variation of all physical variables in a star as a function of mass, age, and composition, in the face of having only two observational checks: the star's luminosity and radius.*
- *The essential problems faced in modeling stellar atmospheres are that the radiation field is formed in a non-equilibrium boundary layer and that the entire emergent spectrum must be computed as a function of depth, angle, and frequency, with a heavy burden of non-equilibrium physics.*

The tasks in both disciplines are very challenging!

1.2 THE BOTTOM LINE

Although stellar atmospheres theory originally lagged behind the theory of stellar structure and evolution, the progress made in this discipline in the past 40 years has put it on a firm physical foundation, and it now makes predictions in close agreement with observable data. The fact is, *the study of stellar atmospheres provides a perfect arena for the development of diagnostic tools that can be used in the analysis of radiation from all kinds of astrophysical and laboratory sources.*

The answers we can now make to Salpeter's question are more compelling today than they could have been 45 years ago:

1. The atmosphere of a star is what we can *see*. Once the gas in a stellar envelope becomes opaque, we cannot obtain *direct* information about conditions inside the star. In the case of the Sun, nonradial oscillation modes are observed on its disk, and some information about its internal structure can be inferred. But other stars are seen as unresolved points, so we get information only from their outer layers, and it must be used to the fullest.
2. Virtually everything known about *all* astrophysical objects is derived by analysis of their emitted radiation. The methods developed for stellar spectra can be applied to other objects, e.g., H II regions, planetary nebulae, neutron stars, and active galactic nuclei. Radiation is also an active ingredient in determining the structure and dynamics of some of these objects.
3. With a dependable theory of radiative transfer in a star's atmosphere that explicitly accounts for the coupling of the radiation field with the atomic/ionic occupation numbers of the material, one can now make accurate calculations for the emergent continuum and line spectra including the effects of millions of spectral lines. These allow reliable computation of the transformations between an observer's *absolute magnitudes* M_V and *colors*, e.g., *UBVRIJHK*, and a theoretician's *bolometric magnitude*, effective-temperature, and surface-gravity scales. As a result, the outputs of stellar structure calculations can be connected reliably with observed color-magnitude diagrams and thus provide critical tests of stellar evolution theory. Without a trustworthy theory of stellar atmospheres, such a connection would be extremely difficult. Further, these models provide good estimates of the ionizing fluxes that determine the physical state of the interstellar medium and gaseous nebulae.
4. Using accurate computations of spectral line strengths, we can make sound quantitative chemical analyses of stellar compositions for stars of different masses and ages. We are now able to perform such analyses even in the rapidly expanding, non-equilibrium atmospheres of very highly evolved stars. This work provides information about the internal structure and evolution of these stars and insight into the past history of the material in its interior that has undergone nucleosynthesis in thermonuclear reactions.
5. With line-blanketed NLTE model atmospheres and accurate stellar evolution tracks we can study stellar populations in galaxies by computing synthetic spectra for an assumed mixture of stars and comparing them with observed galaxy spectra [713]. When a good fit is achieved, one can infer information about their ages and chemical evolution. Here, very hot and massive stars have special importance because they are so luminous we can observe their spectra even in remote galaxies. With these results it is possible to map the change of

relative abundances of chemical elements in the Universe as a function of time with confidence. This work has provided strong observational support to, and tests of, the Big Bang picture.

6. Comparison of physically realistic computations of stellar spectra with high-quality data can yield two constraints between the three fundamental parameters L, M, and R. If one of them can be found by an independent method, we can solve for all three. The results provide guidance to theories of stellar structure and evolution at its most fundamental level.
7. A star's mass M and luminosity L are well defined in the sense that they can, in principle, be measured by a remote observer. However, as will be shown later, if a star has an *extended envelope*, its radius R is not, even in principle, uniquely defined. Yet stars in the most sensitive stages of their evolution (giants, supergiants) have such envelopes! In stellar structure calculations, the outermost layers of a star have long been treated using a severely oversimplified model of radiation transport, which is invalid near the surface (i.e., several photon mean free paths into the material). As a result, the calculated values of R or $T_{\rm eff}$ may have significant uncertainties. Furthermore, the state of the material in such extended envelopes is far from LTE [87, 453, 1005]. *For giants and supergiants, it is an absolute necessity to take into account NLTE effects and stellar winds to model their outer envelopes, to interpret their observed radiation field (colors, spectral energy distribution, and total luminosity), and to provide realistic outer boundary conditions for the calculation of their internal structure.*
8. Stars in all parts of the HR diagram have been analyzed. The main sequence has been examined from massive, hot, very luminous stars, down to faint substellar objects that have very low surface temperatures and such small masses that their central temperatures are too low for thermonuclear fusion to generate enough energy to support their luminosity. The study of the post main-sequence evolution of stars has greatly profited from a very close interplay of stellar evolution calculations that tell us about the internal structure of a star and model atmosphere calculations that are now sufficiently accurate to make reliable direct connections to observed data.
9. Despite its small fractional mass, a star's atmosphere *can* affect its evolution profoundly through mass loss in stellar winds. We now know the following:
   a) Highly evolved massive stars have Wolf-Rayet spectra showing intense emission lines that indicate very rapid mass loss from their envelopes. Evolutionary calculations of the massive upper main-sequence stars and Wolf-Rayet stars *must* take mass loss in winds into account in order to get realistic evolution tracks in the HR diagram. Impressive self-consistent stellar structure and model atmosphere calculations can now be made; see, e.g., [233, 387, 725, 773, 947, 948, 951, 1039]. The fate of some of these stars will be to undergo collapse of a degenerate iron core, from which thermonuclear energy can no longer be extracted, and become supernovae. The remnant from this gigantic explosion may be a neutron star or a black hole.

b) Very old stars of about solar mass found on the red giant branch (RGB) of the HR diagram have a core in which hydrogen has been converted to helium, surrounded by a hydrogen-burning shell source, inside a hydrogen envelope [438, 439, 983, 987]. While the outer envelope expands, the core contracts; eventually it reaches temperatures at which the helium ignites and begins to produce a helium/carbon core. The star moves abruptly to the foot of the *asymptotic giant branch* (AGB). When the helium in the core is depleted, the star has both hydrogen-burning and helium-burning shells surrounding an inert carbon core. Instabilities occur in this *double shell-source* phase [440, 984], and hydrogen may be mixed into the helium-burning shell by helium-shell flashes [985]. In addition, AGB stars may develop extensive convective atmospheres that can *dredge up* material from deeper layers that has been processed through multiple stages of thermonuclear burning. Such episodes may explain the origin of spectral types C and S.

Copious mass loss strongly influences the evolution of these stars as they ascend the AGB branch into post-AGB phases that include the extremely hot PG 1159 stars; then to very hot *planetary nebula nuclei* [441]; and ultimately to white dwarfs, which have consumed all their thermonuclear fuel, are radiating away the heat in their interiors, and ultimately slide down the white-dwarf sequence in the HR diagram into a dark oblivion. Stellar atmospheres theory is now able to make fairly reliable calculations of these critical phenomena; see, e.g., [113, 483, 895, 896, 1152, 1153].

10. For theoretical reasons alone, a correct description of photon transfer in the outer layers of a star is a fascinating challenge, because here radiation emerges from the near-perfect thermodynamic-equilibrium stellar interior into the blackness of space. The observable atmospheric layers are a severely non-equilibrium environment and require detailed analysis of the role of radiation in determining the excitation/ionization state of the material, which, in turn, determines the opacity and emissivity of the material, and hence the radiation field. Attaining a consistent treatment of these tightly interlocked processes has been a significant scientific achievement.
11. Most of the physical insights and mathematical methods developed in the study of stellar atmospheres are applicable to the diagnosis of spectra from laboratory plasmas.
12. Finally, the ability to make quantitative spectroscopic analyses of cosmic objects can have profound implications. For example, current models of the structure of the Universe attribute only about 30% of its mass-energy to baryonic matter. The other 70% is given the name *Dark Energy*; its origin and nature are not understood within the current framework of physics. At present, astrophysicists are trying to interpret small departures of the observed redshift-distance relation (*Hubble diagram*) for Type Ia supernovae from a standard model for the expansion of the Universe. These may imply a time variation

of the dynamical effects of Dark Energy. The redshifts of the supernovae can be measured directly, but their distance must be deduced from their maximum luminosity. It is very important to continue to refine the models of the dynamic atmospheres of these objects in order to establish a precise relationship between the maximum brightness observed in their light curves and their absolute luminosities. It is possible that future modeling of these objects may modify some of the profound conclusions mentioned above.

On more general grounds, one must recognize that astrophysics is an intricate construct assembled using many different lines of reasoning. It is all connected: the activities in its different subfields must be viewed as *cooperative*, not competitive. For example, who would have guessed in the 1950s that the then "old-fashioned" discipline of celestial mechanics would re-emerge into the spotlight in the 1960s, when it was needed to plan the orbits of space vehicles? And who would have guessed that this magnificent 10- to 12-digit formalism developed by great mathematicians such as Brown, Euler, Gauss, Hill, Lagrange, Laplace, Newcomb, and Poincaré, coupled with precision measurement of distance by radar and radio astronomy, and precise timing using atomic clocks, would provide definitive proof of the correctness of general relativity?

Chapter Two

Observational Foundations

In laboratory physics, an investigator can perform controlled *experiments* designed to examine a physical effect or test theoretical predictions. In contrast, the empirical foundation of astrophysics is *observational*. There is no way to alter the conditions where the phenomena being studied occur: they are literally beyond our reach, and the scales of space, time, mass, and energy at play are far too large for human intervention. Furthermore, interpretation of the data can be difficult because what is observed is generally the result of nonlinear interactions between several diverse physical phenomena.

Despite the relatively passive role to which they have been relegated, astronomers have taken advantage of a kind of "astrophysical ergodic theorem"; that is, they can observe the behavior of many examples of a given type of object, e.g., a star, which have differing initial conditions and different ages, and use that information to infer the time history of a single member of the ensemble. In the hands of masters, this small bit of leverage has been sufficient to construct a fairly comprehensive picture of the Universe.

In this chapter we discuss the observations upon which the theoretical framework used to analyze the physical structure of a star's atmosphere and interior rests. We emphasize direct measurements of stellar properties and ignore secondary schemes astrophysicists have devised to make (often quite rough) estimates of a quantity needed when direct data are unavailable. We also introduce some of the specialized astronomical vocabulary used to describe the data and the objects to which they pertain; additional discussion of this information can be found in [132], [133, chapters 1–3], and [139, chapter 3].

To a research scientist or student who has had exposure to it, the material in this chapter may seem elementary. But from an epistemological viewpoint it is essential, *because every physical theory must explain the experimental and/or observational facts upon which it is based*. These facts can validate a theory when it fits the data and guide its growth by showing where it fails. Without secure knowledge of the observational information described in this chapter, there would be no theory of stellar atmospheres, stellar structure, or stellar evolution.

2.1 WHAT IS A STELLAR ATMOSPHERE?

A stellar atmosphere comprises those layers of a star from which photons escape freely into space, and can be measured by an outside observer. The surface from which an emitted photon has about a 50% probability of escape into space is called the star's *photosphere*. Photons emitted in deeper layers, its *envelope* and *interior*, are reabsorbed by overlying material before they can escape, so spectroscopic diagnostics do not yield much information about physical conditions in subphotospheric layers.

The material in stellar atmospheres is a rarefied gas. Its density in their observable layers is so low that it behaves mechanically like an ideal gas even when the dissociation, excitation, and ionization balance in the material is far from equilibrium.[1] If we assume the atmosphere is *static*, i.e., there are no macroscopic material motions, the material will be in hydrostatic and radiative equilibrium; see chapter 16.

The most fundamental properties of a star are its *mass* M [gm]; its *chemical composition* (which in general is a function of depth because it can be altered by the thermonuclear reactions that produce the radiation emerging from its surface); and its *age*. Using the equations of stellar evolution, a star's *luminosity* L [erg/sec] (i.e., the rate it emits radiation, integrated over all frequencies) and *radius* R [cm] can be computed as a function of time.

In favorable cases, accurate values of M, L, and R for a real star can be derived from analysis of its observed radiation field. Knowledge of these quantities provides important checks on the theory of stellar structure and evolution. Those results may be uncertain if the material in the star's atmosphere have been affected by mass exchange with other stars in a binary or multiple system or was "contaminated" by material in a stellar wind from a neighboring star.

The meanings of M and L are unambiguous in the sense that in principle both of them can be determined by "suitable" experiments. L can be found using the methods described in § 2.4 and § 2.5. Its measurement may be complicated by practical problems in determining a star's distance, calibrating a receiver, and correcting for absorption by interstellar material. But these can often be overcome with other supporting analyses. M can be found from a star's dynamical behavior when it interacts gravitationally with another star in a binary; see § 2.5. The main point is that *in principle* both M and L can be found by a remote observer, independently of the internal structure of the star. Methods for measuring and interpreting these data are well developed and reliable.

In contrast, the meaning of R is not always unambiguous. For stars with a *compact atmosphere* (e.g., main-sequence stars), the depth range Δr over which the transition from transparent to opaque material occurs is much smaller than the radius of the star's photosphere. For example, the radius of the Sun's photosphere is $\mathsf{R}_\odot \approx 7 \times 10^{10}$ cm, whereas the density scale height in the photosphere is only $\Delta r \sim 10^7$ cm. In such cases, R is well defined, and if $\Delta r \ll \mathsf{R}$, the atmosphere can be treated as *planar*.

But for *giants* and *supergiants* the concept of a definite radius fails because the depth in their extended envelopes from which photons emerge depends strongly on wavelength; i.e., the size of their photosphere is larger at wavelengths where

[1] In white dwarfs, the ideal gas law holds in a thin, but opaque, atmospheric surface layer; degeneracy comes into play only at depths we cannot observe directly.

the material is opaque than at wavelengths where it is transparent; see [966]. This ambiguity makes the meaning of R uncertain for giants and almost meaningless for supergiants.

Worse, the envelopes of supergiants are observed to be extremely inhomogeneous and highly structured [237, 1169, 1170, 1171]. Even if inhomogeneities are ignored, generalizations of the standard stellar structure equations (chapter 16) would be needed to assign an appropriate radius, *surface gravity*, and *effective temperature* to such stars. One should keep these difficulties in mind when reading § 2.6.

For a given chemical composition, the structure of a compact atmosphere is determined by two parameters: its surface gravity g,

$$g \equiv \mathsf{GM}/\mathsf{R}^2 \quad [\,\mathrm{cm/sec}^2\,], \tag{2.1}$$

which controls its density stratification, and its effective temperature T_{eff} [K], defined such that the total radiation flux emitted per unit area of the stellar surface is

$$F = \sigma_R T_{\mathrm{eff}}^4 \quad [\,\mathrm{erg/cm^2/sec}\,], \tag{2.2}$$

so its total luminosity is

$$\mathsf{L} = 4\pi \mathsf{R}^2 \sigma_R T_{\mathrm{eff}}^4 \quad [\,\mathrm{erg/sec}\,]. \tag{2.3}$$

In (2.2) and (2.3) σ_R is the Stefan-Boltzmann constant [cf. equation (4.69)]. Had a star radiated as a black body at temperature T, which is never actually true, then $T_{\mathrm{eff}} \equiv T$. In reality, T_{eff} *is* approximately equal to the kinetic temperature T of the material at the depth where photons freely escape into space, and it has heuristic value because it immediately tells an astronomer what features to expect in a star's spectrum.

A star's surface gravity can be determined from measurements of its mass and radius. In many cases, it can be inferred more easily by detailed analysis of the *profiles* of spectral lines that are sensitive to pressure broadening (i.e., the density in the star's atmosphere). A star's effective temperature can be found observationally by integrating its absolute energy distribution (see § 2.3) over all wavelengths to obtain its total flux f [$\mathrm{erg/cm^2/s}$] at the Earth, and then using its angular diameter to convert f to the star's flux at its photosphere; see (2.29). It can also be found theoretically by matching the star's observed line and continuum spectra to a high-quality NLTE line-blanketed model atmosphere having T_{eff} as an input parameter. Ideally, the two results should agree.

Most quantitative analyses of stellar spectra assume that photospheres are static, although ad hoc "microturbulent" or "macroturbulent" motions have been invoked in order to fit observed spectral line profiles; see § 17.5. These motions are not turbulence in a hydrodynamic sense (the inferred "turbulent" velocities may be supersonic), but are likely to be small-scale motions produced by the overshoot of convective motions from a deeper convectively unstable layer into the photosphere, and/or the result of high-order non-radial pulsation modes of the star.

The atmospheres of some stars are in motion on the largest scales, e.g., as a *wind* flowing from the star into interstellar space. Stellar winds can be driven by *radiative forces* or by local heating resulting from mechanical and/or electromagnetic *energy deposition* into the material. In the latter case (see [669]), sound and magnetohydrodynamic waves created in a convection zone deeper in a star's envelope dissipate and heat the material as they propagate upward. We confine our discussion in chapter 20 to radiatively driven winds and only briefly mention other types of winds.

Because stars are gaseous, they do not rotate as solid bodies, but *differentially*, i.e., at a rate that depends on latitude. Hence there are shearing motions in the ionized plasma. In solar-type stars, the coupling of convective motions to differential rotation produces a *dynamo*, which generates magnetic fields [841, 888, 1054]. On the Sun, very intense magnetic fields are observed in *active regions* near *sunspots*, and in thin *flux tubes*, which are ubiquitous in the quiet photosphere. Such stars have *chromospheres*, highly structured regions where violent motions are driven by magnetohydrodynamic forces. For the Sun, *spicules* (magnetically confined supersonic jets of material) are found in the chromosphere, and *flares*, which release gigantic amounts of energy on extremely short timescales, occur in the photosphere.

Yet higher in its atmosphere, a star may be surrounded by a *corona*, a highly ionized, low-density region that also is dominated by magnetohydrodynamic forces. The corona is the source of the stellar wind for a star like the Sun. Stellar coronae are typically transparent except in strong spectral lines of highly ionized elements. Successful modeling of these phenomena poses great challenges. The paradigm of an "atmosphere" used in this book for the discussion of stellar quantitative spectroscopy is totally inadequate to describe these parts of a stellar atmosphere.

2.2 SPECTROSCOPY

Spectral Classification

Spectral lines, produced by transitions between bound states of molecules, atoms, and ions, appear in absorption or emission in a star's spectrum. In the 1890s, E. C. Pickering and Williamina Fleming separated stellar spectra empirically into an arbitrary set of *spectral types* denoted with capital letters. About 1910, Annie Jump Cannon used advances in laboratory spectroscopy to rearrange these spectral types into the sequence O, B, A, F, G, K, M, ordered in decreasing effective temperature. Characteristic features of spectral types are defined precisely in [769] and summarized briefly in table 2.1; see also [1003, pp. 274–279].

In astrophysical parlance, types O–F are called *early-type*, type G *solar-type*, and types K and M *late-type* stars. (These names came from erroneous contemporary ideas about stellar evolution; but they are permanently embedded in astrophysical usage.) Individual types are subdivided by a numerical suffix, again ordered in decreasing temperature: O3, ..., O9, B0, ..., B9, A0, ..., A9, etc. Over the next two decades Cannon classified more than 200,000 (!) objective prism spectra of stars to create the *Henry Draper Catalog* and its extensions. Stars in these catalogs are referred to by the prefix "HD" followed by a six digit number. The original HD types have been supplemented by spectral types C, S, L, T, Y, and D, which show features not embraced in the HD scheme.

Table 2.1 Characteristic Features of Spectral Types

Type	Major Spectral Features
O	Strong UV continuum; He II lines; He I weaker than type B; lines from C III, N III, O III, Si IV
B	He II lines absent; maximum of He I lines at B2; H lines stronger at later subtypes; lines from C II, O II, Si III
A	H lines reach maximum strength at A0, weaken at later subtypes; Ca II lines weak; Mg II and Si II strong
F	H lines weaken; Ca II lines stronger; lines of first ions of metals prominent; lines of neutral metals appear
G	Solar-type spectrum; very strong Ca II lines; strong neutral, and weaker ionic, metal lines; strong G band; H lines weaker
K	Blue continuum weak; neutral metal lines dominate; Ca I strong; H lines very weak; CH and CN molecular bands
M	Red continuum; Ca I very strong; strong TiO bands
C	Carbon stars; strong bands of C_2, CN, and CO; no TiO; $T_{\rm eff}$ in same range as types K and M
S	Heavy-element stars; TiO absent; bands of ZrO, YO, LaO; neutral atoms strong as in K and M; $T_{\rm eff}$ like K and M
L	Substellar mass; infrared continuum; CO, H_2O, CaH, FeH, and CrH bands; strong lines of easily ionized atoms: Li, Na, K, Rb, Cs
T	Substellar mass; strong infrared continuum; CH_4, H_2O, and NH_3 bands
Y	Substellar mass; strong mid-infrared continuum; CH_4, H_2O, and NH_3 bands
D	White dwarfs; degenerate stars; DO – He II lines; no H or He I lines DB – only He I lines; no H or metals lines DA – only H lines; DZ – metal lines only; no H or He lines DC – continuum only; no lines deeper than 5%

The effective temperatures of stars differ by a factor $\sim$ 15 in the range from O to M; hence the spectra of different spectral types are dominated by distinctive sets of lines. In O stars, the atmospheric gas is a strongly ionized plasma, as indicated by prominent lines from He II. With decreasing atmospheric temperature, He^+ recombines, becoming mainly neutral in B stars, where He I lines are prominent. In A stars, hydrogen recombines, the Balmer lines are prominent in the visible, and the Balmer continuum dominates in the near ultraviolet. Hydrogen is almost completely neutral in solar-like stars (types G and K), in which elements with $Z \geq 6$ are either mainly neutral (e.g., C, N, O) or mainly ionized (e.g., the alkalis and alkali earths). In K stars, almost all elements are neutral, and molecule formation begins. In the coolest normal stars (type M), hydrogen is mainly molecular, and other diatomic and triatomic molecules are prominent in the spectrum.

Spectral types L, T, and Y, the *brown dwarfs* and similar objects that orbit a main-sequence star called *extrasolar giant planets*, are *substellar objects*. Their masses are too small for their central temperatures to be high enough to support their luminosities by thermonuclear fusion of H $\rightarrow$ He in their cores. Fusion of the more easily burned hydrogen isotope deuterium and the element lithium can contribute initially to the luminosity of early L types.

In 1897, Antonia Maury found that some stellar spectra could be classified into a *two-dimensional* scheme, in which stars having the same spectral type fall into three *luminosity classes*, *a*, *b*, and *c*. For a given spectral type, *c* stars show a higher average degree of ionization of metals and narrower profiles for lines that are sensitive to pressure broadening (e.g., hydrogen) than in *a* stars. Hertzsprung [488] showed that *c* stars are very distant, and hence have high luminosity. Higher luminosity at a given spectral type (temperature) implies a larger radius and thus a lower surface gravity and density in the star's atmosphere. These stars are called *supergiants* because of their huge radii. After Saha derived his ionization equation [931, 932], later extended by Fowler and Milne, see, e.g., [340, 756], it was realized that the observed phenomena implied a much lower density of ions and electrons in the atmospheres of the most luminous stars.

The standard two-dimensional scheme, the *MK system*, was developed by Morgan and Keenan [768, 769]. It has played a key role in understanding stellar atmospheres, stellar evolution and galactic structure. The MK system is an autonomous empirical system defined in terms of standard stars on the sky. It can be reproduced by any observer with spectra having about the same spectroscopic dispersion as the defining standards. The significance of this point is discussed in [363, 364, 612]. A panoramic view of the diversity of, and vast information content in, stellar spectra is best seen in spectral atlases, e.g., [6, 512, 659, 766, 912, 1088, 1128, 1185].

In the MK system a star's spectrum is assigned both a spectral type and a *luminosity class*, which together define its *MK spectral class*. Luminosity classes are denoted with roman numerals: I for supergiants; II for bright giants; III for giants; IV for subgiants; and V for main-sequence stars; see § 2.4–§ 2.6. The MK spectral class of the Sun is G2 V. Class I stars may be further subdivided (from most to least luminous) using the suffixes 0, a, ab, b, bc, and c.

The classification criteria for luminosity classes are based on strengths of lines from atoms or ions having different excitation and ionization potentials, which happen to lie close together in wavelength. Luminosity criteria are described in detail in the references cited above. Some of them are summarized in table 2.2. The reclassfication of all Henry Draper Catalog stars in a two-dimensional system based on MK standards was carried out by Nancy Houk and her coworkers [512, 513].

A number of other types have been recognized in addition to the spectral classes described above. For example, some spectra, e.g., those of luminous O stars, novae, and supernovae, show large-scale, sometimes explosive, expansion of the star's atmosphere; see, e.g., [3, 98, 434, 978, 1128, 1168]. At the hot end of the main sequence *Wolf-Rayet* (WR) stars are undergoing rapid mass loss; they have spectra exhibiting very broad, intense, C or N emission lines. Of stars are O stars with spectra showing emission lines from He II at λ4686 Å; N III at $\lambda\lambda$ 4634–4642 Å;

Table 2.2 Luminosity Class Indicators

Type	Lines Used for Luminosity Classification
O6	Hγ weaker; N IV stronger; He II λ4686 and N III $\lambda\lambda$ 4634–41 emission at higher luminosity
O9	H lines weaker and narrower; He II λ4686 weaker; certain C III, N III, Si IV lines stronger at higher luminosity
B0	H lines weaker and narrower; certain C III, O II, Si IV lines stronger at higher luminosity
B2	H lines weaker and narrower; He I lines weaker; certain N II, O II, Si III lines stronger at higher luminosity
A0	H lines weaker and narrower; certain Si II, Fe II lines stronger at higher luminosity
A2	H lines weaker and narrower at higher luminosity; certain Ti II, Fe II lines stronger
F0	H lines weaker and narrower; certain Ti II, Fe I, Fe II lines stronger at higher luminosity
F2	Some Ti II, Fe I, Fe II lines stronger at higher luminosity
F5	CN band; certain Fe I and II, Ti II, Sr II lines stronger at higher luminosity
G0	CN band; certain Ti II, Fe I, Sr II lines stronger at higher luminosity
G5 – K2	H lines; G band; CN band; Sr II λ4215 stronger; ratio Fe I λ4144/Sr II λ4077 weaker at higher luminosity
K2 – M5	H, Sr II lines, MgH band stronger; Ca I λ4226, TiO weaker at higher luminosity

and Si IV at $\lambda\lambda$ 4089 and 4116 Å, which provide indicators of their luminosity [1127]. In order of increasing luminosity, O(ff) stars have weak N III emission and strong He II absorption lines; O(f) stars have stronger N III emission, and He II in weak absorption, or "neutralized"; Of stars have He II and N III emission; and in Of$^+$ stars, He II, N III, and Si IV are all in emission. C- and S-type stars are low-mass objects in very late phases of their evolution that show mixing of nuclear fusion products from their interiors into their surface layers. The spectra of barium stars and some S stars result from mass transfer in binary systems.

At lower effective temperatures among main-sequence A stars a group of *chemically peculiar* stars (CP stars) are found [887]. They are separated into four main subclasses on the basis of differences between their spectra and those of normal A stars:

1. CP1, *metallic line*, or Am stars;
2. CP2, *peculiar*, or Ap stars;
3. CP3, *mercury–manganese*, or Hg–Mn stars;
4. CP4, *helium-weak* stars.

The Am stars have effective temperatures in the range 7000–10,000 K, weak lines of singly ionized Ca and/or Sc, but enhanced abundances of heavier metals. They tend to be slow rotators. The Ap stars have effective temperatures in the range 8000–15,000 K and are characterized by strong magnetic fields. They have enhanced abundances of Si, Cr, Sr, and Eu, and are generally slow rotators. The Hg–Mn stars have effective temperatures in the range 10,000–15,000 K and show increased abundances of singly ionized Hg and Mn. They have sometimes been placed in the Ap category, but they do not show the strong magnetic fields characteristic of classical Ap stars. They are very slow rotators. The He-weak stars show weaker He lines than would be expected from their observed Johnson UBV colors.

It has been suggested that the observed peculiar surface compositions of these main-sequence stars result from processes that cause some elements, e.g., He, N, and O, to diffuse and settle into deeper layers of the atmosphere, while other elements, e.g., Mn, Sr, Y, Zr, are radiatively "levitated" toward the surface [726]. The bulk chemical composition of the entire star is thought to remain normal, reflecting the composition of the interstellar material from which it formed. If these very slow processes are to be effective, the atmosphere would have to be exceptionally stable against convective mixing. It has been proposed (but not fully demonstrated) that unusually large magnetic fields in CP stars may provide a mechanism leading to such stability.

The MK system, used as the framework for the discussion above, was developed primarily for Population I stars (see § 2.6) whose ages are $\lesssim$ the Sun's and that have [Fe/H] ≈ 0. It is worth a theoretician's effort to become familiar with the criteria used in stellar classification; they can guide stellar atmospheric modeling by highlighting spectral features that are particularly sensitive to, say, the temperature and/or density structure of an atmosphere.

A star's MK spectral class (or photometric colors) correlates with its mass, radius, luminosity, surface gravity, and effective temperature; see § 2.5. These calibrations are secondary and should be used only for first estimates. On the other hand, given a star's MK spectral class, an astronomer immediately has an idea what its physical properties are.

High Dispersion Spectroscopy

Stellar spectral classification uses qualitative estimates of line strengths or line-strength ratios. To get quantitative results from stellar spectroscopy, one must measure line profiles on spectra with high spectral resolution. From these we obtain its *absorption depth*

$$A_\lambda \equiv 1 - [\,F(\lambda)/F_c(\lambda)\,], \tag{2.4}$$

or *residual flux* $R_\lambda \equiv F(\lambda)/F_c(\lambda) = 1 - A_\lambda$ in the line. Here $F(\lambda)$ is the flux measured at wavelength λ in the line, and $F_c(\lambda)$ is the interpolated *continuum flux* at that wavelength. Note that A_λ is negative when a line is in emission above the continuum ($F_\lambda > F_c$). Line profiles from high dispersion spectroscopy provide the most detailed information about a star's spectrum.

For stars, we can observe only the flux integrated over the disk. But for the Sun, we can observe line profiles at each point on the disk, which gives

$$a_\lambda(\mu) \equiv 1 - [\, I_\lambda(\mu)/I_c(\mu)\,], \tag{2.5}$$

where μ is the cosine of the heliocentric angle of a position from the center of the Sun's disk. This information about the *center-to-limb variation* of line profiles, see, e.g., [128, 1072, 1126, 1160], and the continuum, see, e.g., [489, 763, 764, 868], is valuable because it gives some information about the depth variation of the basic physical properties of an atmosphere, and hence provides additional constraints on the theory. The existence of such information for the Sun is a reason why its spectrum has been a critical testing ground for theories of spectral line formation. At some time in the future it may be possible to get similar information for stars from space interferometers.

In the past, astronomical spectrographs using photographic plates were too inefficient to allow accurate measurement of line profiles for any but the brightest stars. For faint stars the entrance slit of the spectrograph had to be opened wide enough to admit as much light as possible from seeing-degraded stellar images, thus reducing the effective spectral resolution and smearing the true line profiles. Nevertheless, a slit passes all wavelengths equally well, so its convolution with the stellar spectrum still conserves energy. Hence astronomers often used the *equivalent width* of a spectral line:

$$W_\lambda \equiv \int A_\lambda \, d\lambda = \int [\, 1 - (F_\lambda/F_c)\,]\, d\lambda, \tag{2.6}$$

which measures the total amount of energy removed from the spectrum in units of the local continuum. Such data are usually analyzed with a *curve of growth* (see § 17.5), which relates the increase of the equivalent width of a line to the number of absorbers along the line of sight.

Measurement of both line profiles and equivalent widths requires only *relative intensities*; they depend on the *linearity* of the detector but not its absolute calibration. Compensation for the inherent nonlinearity between exposure and density on a photographic plate can be made with (laborious) experimental calibrations (see [391] and [494, chapters 2, 4, and 13]) so equivalent widths of acceptable accuracy can be measured photographically; see, e.g., [406, 407, 914]. Today, with spectrographs equipped with linear electronic detectors, and data reduction by computers, most of the past limitations have been overcome; it is possible to measure line profiles for faint stars and to determine equivalent widths with an accuracy estimated to be about ± 2 mÅ. Until recently, relative element abundances were known only to $\sim\ \pm 0.3$ dex (i.e., a factor of 2) because of errors in the observed equivalent widths, errors in atomic data (absorption cross sections), and systematic errors in the methods used for theoretical analysis of line spectra (e.g., the assumption of LTE). With modern techniques it is estimated that we may now attain an accuracy of ± 0.05 dex in favorable cases [202]. Equivalent widths may be adequate for the interpretation of the relatively weak lines of metals. But line profiles are necessary for interpretation of H, He I, and He II lines and those showing the effects of stellar winds. Achieving a detailed fit of computed and observed line profiles, equivalent widths, and the continuum energy distribution in stellar spectra is the primary goal of stellar atmospheres theory.

In short, analyses of stellar spectra furnish the most sensitive and comprehensive set of diagnostics we have to determine the composition, physical conditions, and motions in the outer layers of stars. But the results obtained will be reliable only if the underlying theoretical structure provides an accurate representation of spectrum formation.

2.3 SPECTROPHOTOMETRY

To compare stellar spectra with model atmosphere predictions, we need to know a star's continuum flux in physical units [ergs/cm^2/s/Å] as a function of wavelength, in addition to spectral line profiles. Absolute stellar energy distributions can be obtained using a *spectrophotometer* (comprising a telescope, mirrors, spectrograph, order-separating filters, and photoelectric detector) that has been calibrated by measuring the ratio of its response to a star's flux in several wavelength bands (a few Å wide) and its response, in the same bands, to the flux from a laboratory source whose energy distribution is known, using exactly the same instrument.

Compared to high dispersion spectroscopy, spectrophotometric measurements have the next lower level of spectral resolution. Early spectrophotometric work in the visible spectral region using photographic methods, e.g., [201, 617, 1167], was not reliable because of difficulties in making absolute calibrations, the nonlinear response of photographic plates, and, in some cases, poor atmospheric conditions at the observing sites. Current work uses photoelectric detectors at high-altitude sites having excellent atmospheric conditions. Modern calibration procedures and results in the near ultraviolet, visible, and near infrared parts of the spectrum are described in [240, 461, 463, 809, 813].

Standard Sources

Only two radiation processes have flux distributions that can be predicted in absolute units by theory: *thermal radiation* and *synchrotron radiation*. Thermal radiation is emitted through a small aperture in the side of a *hohlraum*, an enclosure in equilibrium at a precisely known absolute temperature, such as the freezing point of molten gold, copper, or platinum. Radiation from tungsten filament lamps calibrated against standard hohlraums may also be used. These sources are much cooler (1350 K to 2800 K) than the radiation temperatures of hot stars; they give very low count rates, and hence have large statistical errors, in the ultraviolet. Synchrotron radiation is rich in the ultraviolet. Thus radiation from a 240-MeV synchrotron storage ring source (which is similar to the continuum from a B2 V star) at 3000 Å is over 600 times, and at 1500 Å about 10^8 times, brighter than that from a tungsten filament lamp having the same brightness near 5556 Å [241].

Atmospheric Extinction

Measurements from both stars and the laboratory source must be corrected for *atmospheric extinction*. In a planar atmosphere, the air mass along the line of sight to a star is proportional to $\sec z$, where z is star's zenith distance. Hence the *vertical extinction* $X_{\text{vert}}(\lambda, z) \propto \exp(-k_\lambda \sec z)$. Empirical corrections for X_{vert} in each wavelength band are made by observing a star's flux at several zenith distances, plotting its brightness $\log[\,\ell(\lambda, \sec z)\,]$ vs. $\sec z$, and extrapolating a straight-line fit to those data back to "$\sec z = 0$" (i.e., to zero air mass) to derive the incident flux outside the atmosphere.

Data from the laboratory source must be corrected for *horizontal extinction* along the line of sight between it and the telescope. A complication here is that different extinction components (Rayleigh scattering, absorption by water vapor and ozone, aerosols, air pollution) have different scale heights in the Earth's atmosphere; hence they make different contributions per unit air mass in the vertical and horizontal directions. It is not adequate simply to scale the empirically measured vertical extinction data by the ratio of the horizontal to vertical air mass. Fortunately, the small horizontal extinction corrections can be calculated accurately from laboratory data; see [462].

Calibration in the Visual and Infrared

The goal is to convert a measured instrumental reading into an absolute flux. Suppressing technical details, the procedure is as follows: (1) Let $L(\lambda)$ be the known flux distribution of the laboratory source at wavelength λ in physical units. Let $R_{\text{lab}}(\lambda)$ be the response, corrected for horizontal extinction, of the spectrometer to $L(\lambda)$ in some convenient output units (e.g., photon numbers, DC current). Use these data to generate a *calibration factor*

$$C(\lambda) \equiv L(\lambda)/R_{\text{lab}}(\lambda) \tag{2.7}$$

for a prechosen set of wavelength bands. This is the most difficult part of the procedure. (2) Make high signal-to-noise measurements of the spectrophotometer's responses $R_{\text{std}}(\lambda)$, corrected for vertical extinction, to the flux from a *standard star*, using the same instrument, in the same wavelength bands. (3) Then the flux from the standard star in physical energy units is

$$f_{\text{std}}(\lambda) = C(\lambda)R_{\text{std}}(\lambda) \equiv \left[\frac{R_{\text{std}}(\lambda)}{R_{\text{lab}}(\lambda)}\right]L(\lambda). \tag{2.8}$$

In effect, this procedure creates a standard source having a known absolute flux distribution directly on the sky. (4) Thus $f_*(\lambda)$ for any other star is

$$f_*(\lambda) = \left[\frac{R_*(\lambda)}{R_{\text{std}}(\lambda)}\right]f_{\text{std}}(\lambda). \tag{2.9}$$

(5) Normalize both $f_{\rm std}(\lambda)$ and $f_*(\lambda)$ at a standard wavelength (usually $1/\lambda = 1.8\ \mu^{-1} \Rightarrow \lambda = 5556$ Å) to derive the star's *relative absolute flux distribution*:

$$\frac{f_*(\lambda)}{f_*(5556\,\text{Å})} = \left[\frac{R_*(\lambda)}{R_*(5556\,\text{Å})}\right]\left[\frac{R_{\rm std}(5556\,\text{Å})}{R_{\rm std}(\lambda)}\right]\left[\frac{f_{\rm std}(\lambda)}{f_{\rm std}(5556\,\text{Å})}\right]. \tag{2.10}$$

Each of the ratios in the first two brackets in (2.10) is determined directly from observation; they are independent of the calibration of the standard star. No new observations are required if that calibration changes; we need update only the ratio in the last bracket.

The best calibrated standard star is α Lyrae (Vega), MK class A0 V. It is favorable because it is bright, and it crosses the meridian near the zenith for northern hemisphere sites. Taking Vega as a primary standard, secondary standards have been established at other positions of the sky [812], for fainter stars [811], and for stars of special interest, e.g., white dwarfs [810]. In the visible and near ultraviolet, the absolute calibration is accurate to $\approx 1\%$ and agrees well with modern model stellar atmospheres calculations. Calibrations in the infrared are more difficult because of strong water vapor absorption in the Earth's atmosphere, which can be highly variable. Absolute calibrations of Vega in selected infrared windows have been reported in [35, 126, 463, 993, 994].

Absolute Fluxes

To convert relative absolute fluxes into physical units, we must, of course, know $f_{\rm std}(5556\,\text{Å})$ for the standard star. This measurement is difficult experimentally. Years of work by many observers has determined $f_{\rm Vega}(5556\ \text{Å}) = 3.44 \pm 0.05$ erg/cm^2/s/Å at the top of the Earth's atmosphere. The absolute flux at 5556 Å of any other star is given by this number and the observed ratio $R_*(5556\ \text{Å})/R_{\rm Vega}(5556\ \text{Å})$. The absolute energy distribution of the Sun is also of importance in developing a theory of stellar atmospheres. It has been intensively studied; see, e.g., [693, 710, 867, 893]. It is difficult to measure because of the Sun's immense apparent brightness. That problem can partially be avoided by measuring the energy distribution of "solar twins," i.e., stars that are close analogs to the Sun; see, e.g., [203, 245].

Calibration in the Ultraviolet

Instruments for space observations in the range 912 Å $< \lambda <$ 3300 Å can be calibrated pre-flight. But such calibrations are difficult, and unfortunately they may change once the instrument is in space. Therefore, it is important to have standard stars in the sky whose energy distribution is known in this spectral region. Initially, extremely hot halo stars like BD $+28°4211$ (see figure 2.1), which has a featureless spectrum and a flux $\propto \nu^2$ in the visible (appropriate for a black body as $T \to \infty$), were used.

But it has been found [124, 125, 126] that an even better absolute calibration from 1150 Å to 1 μ can be obtained by fitting absolute fluxes from NLTE model

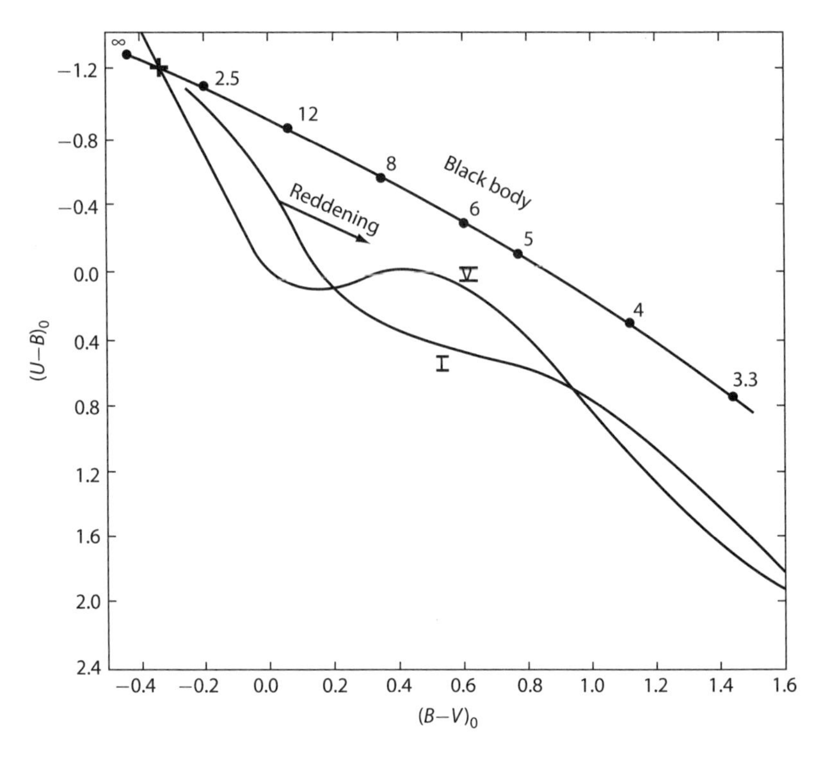

Figure 2.1 Intrinsic color-color diagram for stars of luminosity classes I and V, showing a reddening line. The upper curve shows theoretical colors of black bodies; labels give $T/10^3 K$. The cross marks colors of BD +28°4211. Adapted from [37, 495].

atmospheres to the observed data for hot DA white dwarfs. The models are reliable because the physics of their atmospheres can be treated accurately by current theory and computational technique. The final calibration appears to be good to $\approx 1\%$ in the range $0.5 - 0.8\,\mu$ and better than $\approx 4\%$ in the larger range $0.12 - 3\,\mu$. It may seem surprising that a good calibration can be obtained with relative ease in the UV and IR regions of the spectrum; this is so only because we can make physically realistic model atmospheres for these stars.

2.4 PHOTOMETRY

At first, photographic plates were used to accumulate light from faint stars by means of long exposures. Astronomers measure stellar brightness in terms of *magnitudes*. If ℓ_1 and ℓ_2 are the energy (corrected for extinction) received in a wavelength range $\Delta\lambda$ from stars "1" and "2," their *apparent magnitudes* m_1 and m_2 are defined by

$$m_1 - m_2 \equiv -2.5 \log(\ell_1/\ell_2) = -1.086 \ln(\ell_1/\ell_2), \tag{2.11}$$

or

$$\ell_1/\ell_2 = 10^{0.4(m_2 - m_1)}. \tag{2.12}$$

The brighter a star, the smaller its magnitude: $\Delta m = 0.01 \Rightarrow$ the stars differ by 1% in brightness; $\Delta m = 0.1 \Rightarrow 10\%$; $\Delta m = 1.0 \Rightarrow$ a factor of 2.5; $\Delta m = 2.5 \Rightarrow$ a factor of 10. Because photographic plates have a nonlinear response, photographic magnitudes were accurate to about ± 0.1 or ± 0.2 mag. (i.e., $\pm 10\%$ or $\pm 20\%$ in brightness). The introduction of photoelectric photometers in the early 1950s revolutionized astrophysical photometry. They have much greater sensitivity and accuracy, ± 0.01 mag (i.e., $\pm 1\%$), and a linear response over a huge range of apparent brightness.

Apparent Magnitude

Let $f_*(\lambda)$ be the stellar flux at wavelength λ incident at the top of the Earth's atmosphere; $X(\lambda$, air mass, pollution, clouds) the fraction of it transmitted by the atmosphere; $R(\lambda$, reflectivity, dirt) the telescope's optical transmission function; $T(\lambda)$ the filter transmission function; $S(\lambda)$ the sensitivity function of the detector; and $\mathcal{S}(\lambda, z)$ the transmission function of the total system. Then the light measured in a wavelength band $\Delta\lambda$ is

$$\ell(\Delta\lambda, z) = \int_{\Delta\lambda} f_*(\lambda) X(\lambda, z) R(\lambda) T(\lambda) S(\lambda)\, d\lambda \equiv \int_{\Delta\lambda} f_*(\lambda) \mathcal{S}(\lambda, z)\, d\lambda. \quad (2.13)$$

In this integral, $X(\lambda, z)$ tells us to observe near the zenith, in clean air, on a clear night; $R(\lambda)$ is controlled by efficient design and clean optics; $T(\lambda)$ isolates a prechosen wavelength band $\Delta\lambda$ in the spectrum; and, ideally, $S(\lambda)$ is linear, so the receiver's response is directly proportional to the input light.

As mentioned in § 2.3, to a first approximation $X(\lambda, z) \propto \exp(-k_\lambda \sec z)$ in a planar atmosphere. This result holds only if the absorption in $\Delta\lambda$ grows linearly with the number of atoms along the line of sight; see [715]. Saturation can be detected if it occurs in the narrow wavelength bands used in spectrophotometry, but in broad band photometry, it causes $f_*(\lambda)$ to be weighted differently at different wavelengths, and hence to contribute differently in (2.13) at different zenith distances.

Because $X(\lambda, z)$ varies from night to night at a given location, $R(\lambda)$ is different from observatory to observatory, $S(\lambda)$ is different for different kinds of receivers, and $T(\lambda)$ can vary for different sets of filters at a given observatory, it is necessary to transform measured magnitudes, after correction for atmospheric extinction, to a set of standards adopted by the International Astronomical Union; see, e.g., [592, 594, 597, 767]. The zero point of a magnitude scale is arbitrary; for the visual band V, it is chosen so that the bright star Vega (A0 V) has $V_{\text{Vega}} = 0.03$.

Photometric Systems and Colors

Magnitude differences (corrected for atmospheric extinction)

$$(m_b - m_a) = -2.5 \log \left[\frac{\int_{\Delta\lambda_b} f_*(\lambda) R_b(\lambda) T_b(\lambda) S_b(\lambda) d\lambda}{\int_{\Delta\lambda_a} f_*(\lambda) R_a(\lambda) T_a(\lambda) S_a(\lambda) d\lambda} \right] + C_{ab} \quad (2.14)$$

Table 2.3 Extended Johnson-Morgan System

Band	λ_{eff}	$\Delta\lambda$	Band	λ_{eff}	$\Delta\lambda$
U	3650 Å	680 Å	J	1.25 μ	0.3 μ
B	4400 Å	980 Å	H	1.65 μ	0.4 μ
V	5500 Å	890 Å	K	2.2 μ	0.5 μ
R	0.70 μ	0.22 μ	L	3.4 μ	0.7 μ
I	0.88 μ	0.24 μ	M	5.0 μ	1.2 μ

measured in two wavelength bands generate a system of *color indexes*. In setting up an observational system, the constants C_{ab} can be chosen at will; they allow for the arbitrary zero point of the magnitude scale in each spectral band. If comparisons are to be made with colors computed using absolute fluxes, or model atmospheres, these constants must be calibrated.

Many photometric systems have been devised. The most commonly used is the broad band Johnson-Morgan *UBV* system [767] and its infrared extension [592, 594]; table 2.3 shows the parameters defining this system.

The data given by *UBV* measurements are the V magnitude, and two colors: $(B-V)$ and $(U-B)$. The constants in (2.14) are adjusted such that all colors for an average A0 V star are 0.00. Then color indices for the Sun are $(U - B) = +0.20$, $(B - V) = +0.66$, $(V - R) = +0.54$, $(V - I) = +0.88$, and $(V - K) = +1.59$. Colors give a measure of the flux ratio in the continuum in two spectral bands. This ratio gives information about a star's temperature: stars in which the radiation in the shorter wavelength band exceeds that in the longer wavelength band are hotter than stars where the reverse is true. For example, the U filter measures stellar flux at wavelengths shorter than the threshold for continuous absorption from the $n = 2$ state of neutral hydrogen, near 3650 Å. The B filter measures flux at wavelengths larger than this threshold. Thus for hot stars $(U-B)$ gives information about the size of the *Balmer jump* in the continuum, whereas $(B-V)$ gives information about the slope of the *Paschen continuum*, which in hot stars is dominated by photoionization from, and recombination to, the $n = 3$ state of hydrogen. The filters R through M give information about the flux distribution emitted by cool stars.

Colors give the coarsest information about stellar spectra. But because photometric systems use large bandwidths, colors can be obtained much more quickly than spectra for bright stars and can be measured accurately even for very faint stars for which spectroscopy is impractical.

Interstellar Absorption and Reddening

A practical problem encountered in interpreting photometric measurements is the existence of an interstellar absorbing medium consisting of "dust" and gas. It both dims starlight and reddens it (i.e., absorbs shorter wavelengths more efficiently). It is concentrated mostly near the plane of the Galaxy and strongly affects measurements of magnitudes and colors of Population I stars at large distances. Its effect is smaller for stars at high galactic latitudes, in particular for Population II stars in globular clusters.

In the presence of pure extinction (no re-emission), a star that has an *intrinsic intensity* $I_0(\lambda_i)$ has an *observed intensity* $I(\lambda_i) = I_0(\lambda_i)10^{-k(\lambda_i)D}$ at distance D; see equation (11.98), where $k(\lambda_i)$ measures the absorptivity of the material at λ_i.

In terms of magnitudes, see (2.11), $m(\lambda_i) = m_0(\lambda_i) + 2.5k(\lambda_i)D \equiv m_0(\lambda_i) + A(\lambda_i)$. Thus interstellar absorption produces an additive shift in a star's observed magnitude that is linear in distance, and hence linear in the total absorption along the line of sight. Similarly, at any other wavelength λ_j, $m(\lambda_j) = m_0(\lambda_j) + A(\lambda_j)$. Thus if a star's *intrinsic color index* is $C^0_{ij} \equiv m_0(\lambda_i) - m_0(\lambda_j)$, then its *observed color index* C_{ij} is

$$C_{ij} = C^0_{ij} + [\,A(\lambda_i) - A(\lambda_j)\,] \equiv C^0_{ij} + E_{ij}. \tag{2.15}$$

E_{ij} is called the *color excess*. In the *UBV* system, $E(B-V) \equiv (B-V) - (B-V)_0$ and $E(U-B) \equiv (U-B) - (U-B)_0$. In the visible and near ultraviolet regions of the spectrum, $A(\lambda_i) > A(\lambda_j)$ for $\lambda_i < \lambda_j$; hence in the *UBV* system color excesses are positive (see figure 2.1).

Color-Color Diagram

Fortunately, interstellar reddening and absorption can sometimes be determined directly from observation. A star's spectral class is unaffected by interstellar absorption and reddening; but its observed energy distributions can be changed by different amounts of reddening. A large sample of O–B stars having known spectral types and *UBV* magnitudes was studied in [495, 597]. They are ideal because (1) they are luminous and can be seen to large distances, where they are strongly reddened; (2) their spectrum is rich in blue and UV light, both reddening sensitive; and (3) they have distinctive spectra, and hence are easily classified.

One can plot the observed colors of the stars in their sample in a *two-color diagram*, $(U-B)$ vs. $(B-V)$. By assuming the bluest stars for each spectral type are unreddened, one can construct the *intrinsic two-color curve* $(U-B)_0$ vs. $(B-V)_0$, shown in figure 2.1. At very high temperature $(B-V)$ and $(U-B)$ approach limiting values because as $T \to \infty$, the Planck function (describing the equilibrium, or hohlraum, radiation—see chapter 4) in the visible reduces to its Rayleigh-Jeans form (see equation 4.55) $B_\nu(T_{\rm rad}) \propto T_{\rm rad}/\nu^2$, where $T_{\rm rad}$ is a characteristic *radiation temperature*. In this limit, the flux ratio in any two bands is independent of temperature. For the extremely hot halo O star BD +28°4211, $(U-B)_0 = -1.26$ and $(B-V)_0 = -0.34$; these are close to infinite-temperature blackbody values. The rapid drop of the actual two-color curve away from this limit point is caused by lower photospheric temperatures, which imply both less ultraviolet radiation relative to blue and the blocking of continuum radiation by many spectral lines in the ultraviolet. The dip in the two-color curve for main-sequence stars near $(B-V)_0 \approx 0.0$ is caused by strong absorption both in the hydrogen Balmer lines as they converge to the Balmer limit and in the Balmer continuum itself, which are seen in the U band.

This feature is much weaker for supergiants, in which hydrogen is much more strongly ionized, and its lines are less pressure broadened.

For O–B9 stars, all stars of the same MK spectral class are found to lie along nearly parallel *reddening lines* in the two-color diagram, having slopes

$$E(U-B)/E(B-V) = 0.72 + 0.05E(B-V). \tag{2.16}$$

If we neglect the small color term in (2.16), the quantity

$$Q \equiv (U-B) - [\,E(U-B)/E(B-V)\,](B-V) \tag{2.17}$$

is reddening-independent for stars on this spectral range. Thus a measurement of Q gives both the intrinsic colors and spectral type of a star. This method works only for early-type stars because for later types, the reddening lines are nearly parallel to the two-color curve itself. In practice, in a star cluster we estimate the reddening and absorption for early-type stars and use these values for all other stars in the same cluster.

Using data in infrared bands, where interstellar absorption effects are small, to fix the zero point, absorption in the V band is found to be

$$A_V = [\,3.3 + 0.28(B-V)_0 + 0.04E(B-V)\,]E(B-V). \tag{2.18}$$

Knowledge of a star's color excess gives the amount its light has been dimmed, in magnitudes, by interstellar absorption and allows us to correct m_V to m_{0V}. As shown in figures 2.2 and 2.3, A_λ is enormous in the ultraviolet.

The interstellar medium (ISM) is essentially opaque for $\lambda \lesssim 912$ Å because of absorption of photons by interstellar hydrogen in its ground state. It becomes more transparent in the extreme ultraviolet (EUV) and soft X-ray regions of the spectrum. Tabulations of unreddened $(B-V)$, $(U-B)$, $(V-R)$, $(V-K)$, $(J-H)$, $(H-K)$, $(K-L)$, and $(K-M)$ colors versus M_V and MK spectral type from O5 to M8, for luminosity classes I, III, and V are given in [18, pp. 204 and 206], [261, tables 7.6–7.8 and 15.3.1] along with estimated effective temperatures for some of these stars.

Absolute Magnitude

The apparent magnitude of the Sun is $V_\odot = -26.75 \pm 0.06$ [461, 1040], whereas the faintest stars measurable in dark sky have $V \approx +25$, which implies an apparent brightness range of 10^{20}! This immense range of apparent brightness[2] is not intrinsic to stars, but results from the inverse-square falloff of the flux from a star (seen as a point source) at a distance D (see equation 3.72). If we are to infer the luminosities of stars from observation, we must find a way to compensate for the effect of its

[2] Such a range cannot be handled by a single instrument, so we use telescopes of different sizes to measure overlapping sets of magnitude standards at fainter and fainter light levels.

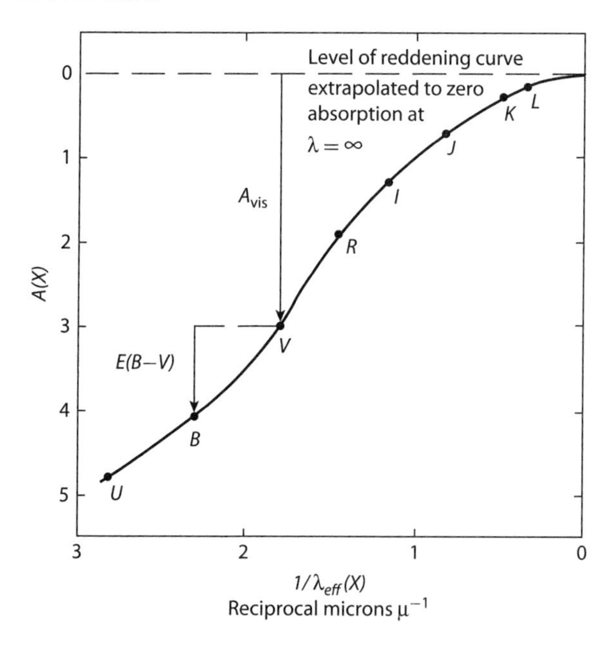

Figure 2.2 Interstellar absorption at different wavelengths, normalized to $E(B-V) = 1.0$. $X \equiv 1/\lambda_{\text{eff}}$ in reciprocal microns (μ^{-1}). Adapted from [593].

distance. To this end, we define the *absolute magnitude* M_V of a star to be the magnitude it would have if it were seen at a standard distance D_0. Then from (2.14)

$$M_V = m_V - 2.5 \log(D/D_0)^2 = m_V - 5 \log(D/D_0). \tag{2.19}$$

Now the question is: how are we to obtain D? The fundamental way of determining a star's distance is to use the techniques of *astrometry* to measure its parallax from very high precision measurements of the positions of stars in the sky.

Parallax

A star's parallax is the very small shift in its position as seen from the Earth when it is on opposite sides of its orbit around the Sun. If R is the radius of the Earth's orbit, measured in *astronomical units* [1 au] $= 1.495979 \times 10^8$ km, and the star's distance is D in the same units, then its parallax in radians is $\pi = R/D$ rad. In observational practice π is measured in *arc-seconds* [as], i.e., $\pi'' = 206,265 \times \pi$ rad, and D is measured in *parsecs* [pc] $\equiv$ 206,265 au $= 3.0857 \times 10^{13}$ km ≈ 3.262 light years. Then D [pc] $= 1/\pi''$. Stellar parallaxes are *very small*. The largest known is 0.76 as, for the star α Centauri C, corresponding to a distance of 1.32 light years.

In (2.19) the standard distance D_0 is taken to be 10 parsecs, so

$$M_V = m_V + 5 - 5 \log D. \tag{2.20}$$

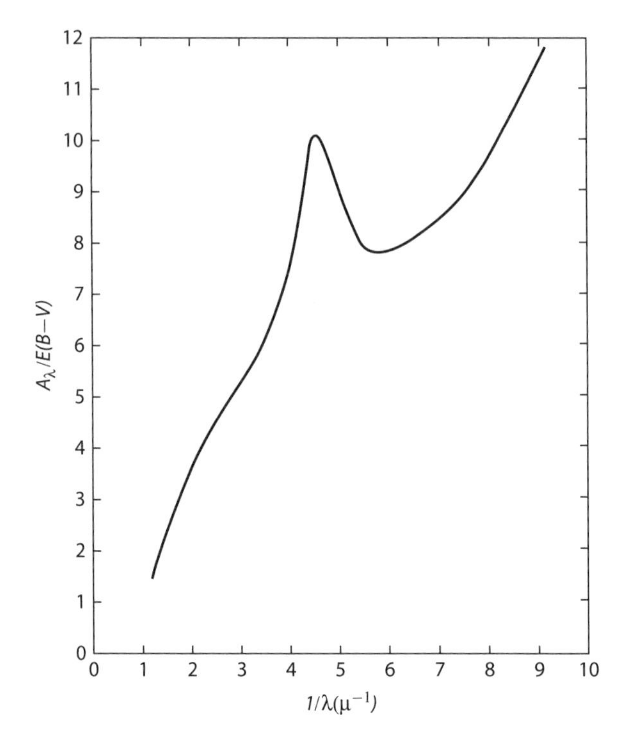

Figure 2.3 Ratio of A_λ in the ultraviolet to $E(B-V)$. $X \equiv 1/\lambda$ in reciprocal microns (μ^{-1}). Adapted from [495, 593].

$(m_V - M_V) = 5\log(D) - 5$ is called the *distance modulus* of an object. For the Sun, $D = 1$ au $= 1/206,265$ pc; hence $M_V(\odot) = +4.82$. For stars, errors in D cause the greatest uncertainty in absolute magnitudes. If there is significant interstellar absorption, (2.20) must be replaced by

$$M_{0V} = m_V - 5\log(D/10) - A_V. \tag{2.21}$$

It should be noted that unlike photometric measurements, spectral classification is unaffected by interstellar absorption and reddening.

It is not trivial to obtain reliable parallaxes! To begin with, the Earth is not an inertial frame:

- The barycenter of the Earth-Moon system moves on an elliptical orbit around the Sun.
- The Earth moves on an elliptical orbit around that barycenter.
- The Earth's axis of rotation is inclined to its orbital plane. Hence gravitational torques exerted by the Sun and Moon on the (nonspherical) Earth force its rotation axis to revolve slowly around the normal to its orbital plane—the *ecliptic plane*.
- Forces exerted by the Moon and planets make additional perturbations on the direction of its rotation axis.

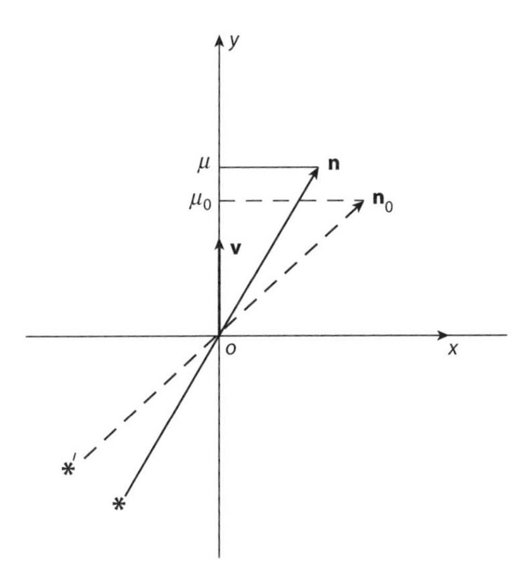

Figure 2.4 Change in apparent position of a star produced by motion across line of sight to the star.

Collectively, these effects lead to a rotation of the coordinate system called *precession* at a rate of 50.25″/year, which is very much larger than a stellar parallax. It is subtle work to make observations that simultaneously determine the moving coordinate system and measure the small changes in position that reveal a star's distance. Good discussions of traditional methods for acquiring the necessary data are given in [393, 787, 1015].

Aberration

The next largest effect to be taken into account is *aberration*, which is the change in the apparent direction to a star when the observer moves perpendicular to the line of sight to the star. It was discovered by Bradley in 1728 from observations of the star γ Dra.

In figure 2.4, let light from the star $*$ travel along the propagation vector $\mathbf{n} = [(1-\mu^2)^{1/2}, \mu]$ to a point $\mathbf{O}$, which is at rest relative to $*$. An observer $\mathbf{O}'$ moving with velocity $\mathbf{v} = (0, v)$ along the y axis with respect to the rest frame sees the star in a direction $\mathbf{n}_0 = [(1-\mu_0^2)^{1/2}, \mu_0]$, which is slightly displaced from $\mathbf{n}$.

Lorentz transformation from the rest frame $\mathbf{O}$ to the moving observer's frame $\mathbf{O}'$ gives [see equation (B.6) in Appendix B]

$$\mu_0 = (\mu - \beta)/(1 - \beta\mu), \tag{2.22}$$

where $\beta \equiv v/c$, and c is the speed of light. The Earth's speed of revolution around the Sun is $v_\oplus \approx 29.8$ km/s, and $c \approx 3 \times 10^5$ km/s; hence $\beta \approx 10^{-4}$. Upon expanding (2.20) to first order in β, one finds $\mu_0 - \mu = \beta(\mu^2 - 1)$. The maximum difference occurs when $\mu = 0$, i.e., $\mathbf{n}$ is perpendicular to $\mathbf{v}$; then $\mu_0 = -\beta$. The apparent position of $*$ seen by the moving observer $\mathbf{O}'$ is displaced from its position seen by an observer at rest at $\mathbf{O}$ by $\beta \approx 10^{-4}$ radians $\approx 20.5'' \equiv \kappa$, the *aberration constant*, in the same direction as the observer's motion. Fortunately, aberration effects can be eliminated from parallax measurements because they can be calculated precisely from the known elements of the Earth's orbit.

Ground-based optical astrometry used two methods to measure parallaxes. "Fundamental" positions were measured using meridian circles, which had short focal lengths, and hence small scale. Differential positions were measured on photographic plates (CCDs now) obtained with long-focus telescopes.

Both methods suffer from serious problems:

- Atmospheric seeing, the blurring and motion of images caused by turbulence in the Earth's atmosphere.
- Seasonal changes, e.g., summer-to-winter temperature changes, that produce optical and/or mechanical changes in the telescope.
- Refraction in the Earth's atmosphere, which systematically changes the observed position of a star as it approaches the horizon.
- Lack of parallaxes for the reference stars used in differential position measurements.

Therefore, the accuracy of parallaxes from ground-based observations is seriously compromised by the effects listed above.

Radio Astrometry

Thanks to radio and space observations, astrometry has undergone a stunning renaissance, now yielding vast amounts of data of previously unachievable accuracy [600, 634, 686, 900]. Use of radio interferometers revolutionized position measurement; see [393, chapter 15]. Observations at radio wavelengths are essentially unaffected by atmospheric seeing, and very high spatial resolution is achieved by means of interferometry. The VLBI (Very Long Baseline Interferometer) can routinely measure positions of very distant compact radio sources (e.g., quasars), to better than $0.001'' = 1$ mas (milli-arc-seconds), far surpassing the accuracy of ground-based optical data. These data have been used to define an *International Celestial Reference Frame* (ICRF). All of these sources are so distant that any motion they may have through space produces no observable changes in their positions. Thus the ICRF constitutes a genuine kinematic inertial frame.

Space Astrometry

A similar advance at optical wavelengths was obtained with the ESA satellite HIPPARCOS. In space, images are small and stable, with point spread functions at the diffraction limit of the telescope. In addition, the whole sky can be observed in a single day, during which changes from parallax, aberration, and proper motion (the angular motion produced by the component of a star's space velocity across the line of sight) are negligible. Hence we can *close* a set of observations over the entire sky in a single day and construct an absolute coordinate grid by requiring that the relative positions of all stars in the dataset be consistent with one another. The optical counterparts of the radio sources that define the ICRF are too faint to be observed by HIPPARCOS. A minute correction was needed to remove a small spurious global rotation of the HIPPARCOS system and align it with the ICRF [600].

The HIPPARCOS Catalog contains ~120,000 stars, whose positions are known to about 0.7 mas [634]. The median precision of its parallaxes is 0.97 mas, a factor of 10 better than for most ground-based data, giving $\approx$ 21,000 stars that have distances determined to better than 10% and $\approx$ 50,000 stars having distances determined to better than 20%. This is a huge advance in our knowledge of reliable stellar distances. The data from HIPPARCOS have had a profound impact on studies of stellar structure, stellar evolution, galactic dynamics, and the cosmic distance scale [686, 900]. Astrometric space satellites planned for the future are expected to yield data with accuracies measured in μas, i.e., *micro*-arc-seconds!

Precise HIPPARCOS parallaxes permit one to make *color-magnitude diagrams* (see below) with markedly improved accuracy [374, 613, 822, 880, 900]. In table 19.1 of [261], one sees that the 100 stars nearest the Sun are either main-sequence dwarfs cooler than the Sun or white dwarfs. Exceptions are α CMa (A1 V), α CMi (F5 IV–V), and α Aql (A7 IV–V). All of these stars have HIPPARCOS distances accurate to 1% or better, implying errors in their absolute magnitudes $\lesssim \pm 0.02$ mag. Most apparently bright stars are very luminous and so distant that in the past their measured parallaxes had large fractional errors. But among the 100 apparently brightest stars [261, table 19.2] we find a good number with luminosity classes from III to Ia that have HIPPARCOS parallaxes good to $1\% \leq \epsilon_\pi/\pi \leq 5\%$. For other spectral classes, the main method used to calibrate absolute magnitudes is color-magnitude diagram fitting.

Color-Magnitude Diagrams

The original form of the HR diagram was a plot of the absolute magnitude M_V of a star versus its spectral class. Unfortunately, the discrete nature of spectral types causes stars to appear in artificial clumps in this diagram, which reduces its usefulness for quantitative work.

Variants of the diagram using proxies for the ordinate or abscissa, or both, overcome this difficulty. For example, one can plot V_0 or M_{V0}, corrected for interstellar absorption, versus an unreddened color index, say $(B - V)_0$, which correlates with T_{eff}. The result is a *color-magnitude diagram* (or *CM diagram*). CM diagrams have the advantage that the photometric quantities plotted can be measured with high precision. Alternatively, we might plot M_{V0} vs. $\log T_{\text{eff}}$, or $\log \mathsf{L}$ vs. $\log T_{\text{eff}}$ (the "theoretician's HR diagram"); but theory is required to convert M_{V0} and $(B - V)_0$ to $\log \mathsf{L}$ and $\log T_{\text{eff}}$. So for the present, we consider CM diagrams, which employ only directly observable information.

Population I

CM diagrams are of great importance because they can be used to calibrate the absolute magnitudes of types of stars for which fundamental data do not exist.

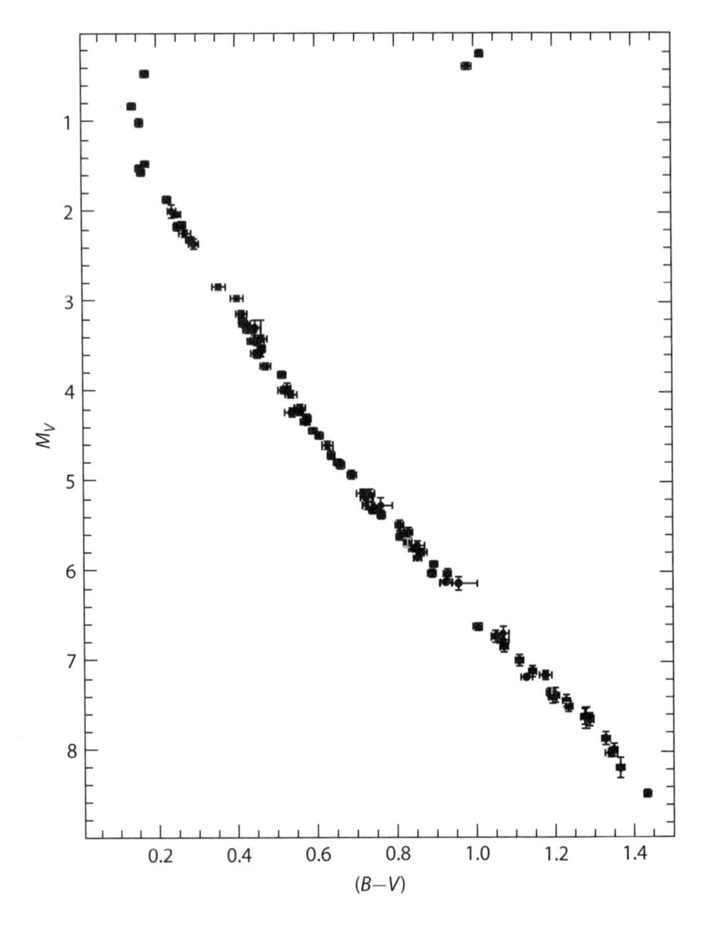

Figure 2.5 Plot of absolute visual magnitude of Hyades stars versus $(B-V)$. Adapted from [156].

A very accurate CM diagram for the Hyades cluster is shown in figure 2.5. The colors and apparent magnitudes of Hyades stars are known with high precision. The cluster is so close ($\approx$ 40 pc) that the effects of interstellar absorption and reddening are small and easily corrected. Using HIPPARCOS parallax data, their apparent magnitudes are converted to absolute magnitudes. Notice the very small scatter of the data points around a smooth curve. This CM diagram defines a *fundamental main sequence*. The three stars at the very top of the diagram are evolved red giants, the bluest one being right at the turnoff point, which marks the transition from when hydrogen in a star's core is exhausted and its luminosity is produced in a hydrogen-burning shell surrounding an inert helium core.

Given a fundamental sequence, the *main-sequence fitting method* can be used to produce a composite Population I CM diagram such as that shown in figure 2.6. Magnitudes and colors, corrected for reddening, are measured for stars in several

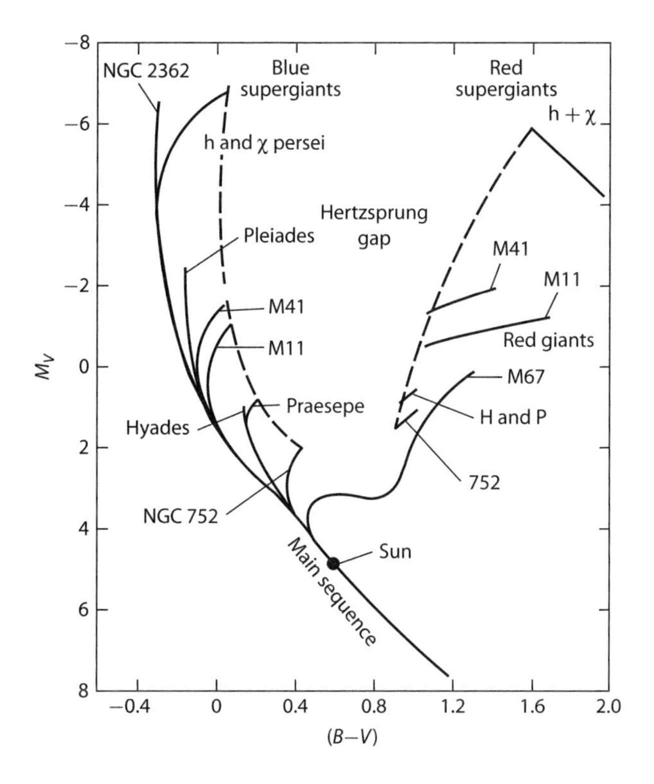

Figure 2.6 Schematic composite color-magnitude diagram for Population I clusters.

clusters. The CM diagram of a cluster (e.g., NGC 752 and Praesepe) is shifted vertically at constant $(B-V)_0$ to match the lowest portion of its main sequence to the Hyades sequence to minimize evolutionary effects. This procedure is repeated for each cluster in the figure, each time fitting to the cumulative main sequence of the preceding plot, thereby fixing the zero point of its absolute-magnitude scales. By this method we determine the *zero-age main sequence*, or ZAMS. By successive fits the main sequence is established for increasingly luminous stars, up to $M_{V0} \approx -4$. When this process is applied to a cluster containing supergiants (e.g., h and χ Persei), the absolute magnitudes of those stars are also determined. From such studies we can calibrate the rarer types of MK spectral classes. These results can be used to establish average relations between spectral type, colors, and absolute magnitudes (§ 2.5). They allow us to make approximate estimates of the properties of stars, but may not be accurate for an individual star; in the end there is no substitute for a detailed spectral analysis.

Population II

As illustrated in figure 2.7, the CM diagram for a globular cluster, which is a group of coeval Population II stars, is distinctively different from the Population I diagram.

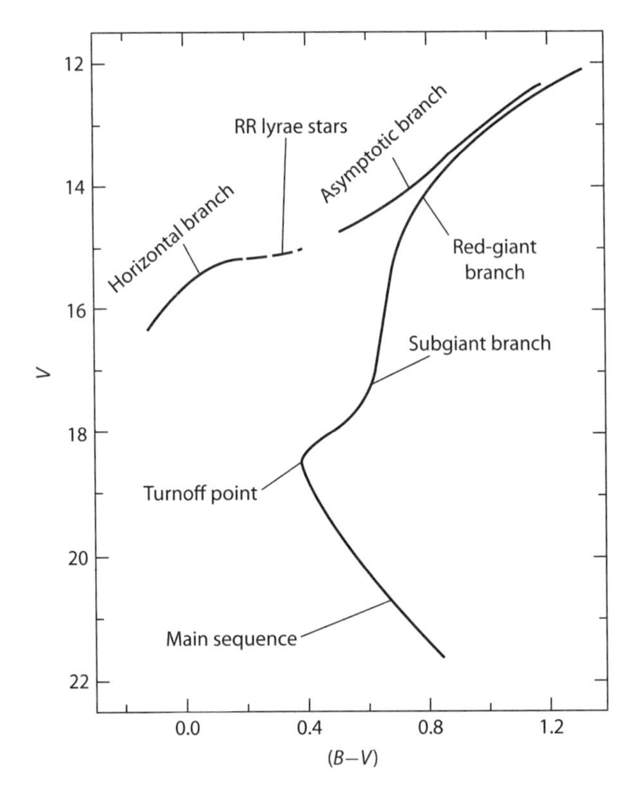

Figure 2.7 Schematic color-magnitude diagram for a metal-poor globular cluster. Adapted from [96, 599].

These stars are metal-poor and are much older, so their CM diagram contains only the faint end of the main sequence up to a turnoff point at about one solar mass.

A number of accurate CM diagrams for globular clusters have been measured; see, e.g., [96, 599, 938, 941]. After the turnoff point, the diagram continues toward higher luminosities and cooler temperatures along a *subgiant branch*, and then to a well-developed *red-giant branch*. A second sequence of giants lies along the *asymptotic giant branch*, which starts near the tip of the red-giant branch and runs back toward higher temperatures and more moderate luminosities. On the *horizontal branch*, one finds RR Lyrae-type pulsating variable stars.

Subdwarfs are *weak-line* and/or *high-velocity* Population II stars in the solar neighborhood. Because they are near the Sun, their parallaxes, and hence absolute magnitudes can be determined reliably, and the position of their main sequences can be established as a function of [Fe/H]. But the sample is insufficient to define their complete CM diagram. In contrast, globular clusters are rich in stars, but none of them are close enough to have its parallax measured directly, so the zero points of their absolute-magnitude scales are unknown. By careful fitting of the lowest part of the main sequence in a cluster CM diagram to that for nearby subdwarfs having the same [Fe/H], one could both calibrate the zero point of the cluster's absolute-magnitude scale and get a comprehensive view of the CM diagrams of metal-poor stars. Another method to set the zero point of a globular cluster CM diagram would be to use the absolute magnitude of white dwarfs in the cluster, which can now be observed with the Hubble Space Telescope.

Synthetic Colors and Bolometric Correction

An estimate of a star's properties can be made by comparing its observed unreddened colors with *synthetic colors* computed by folding sensitivity curves of filters with the absolute energy distributions of standard stars, or with fluxes from line-blanketed model atmospheres. Likewise, we can compute synthetic colors for black body or synchrotron radiation sources. Arp [37] calculated colors of black bodies, and Mathews and Sandage [715] computed them for both black body and synchrotron radiation sources. (See also [1158].) For a filter $x = (u, b, v)$ (denoting U, B, or V), define the theoretical magnitude

$$x_n \equiv -2.5 \log\left[\int_0^\infty \mathcal{S}_n^x(\lambda) F(\lambda)\, d\lambda \Big/ \int_0^\infty \mathcal{S}_n^x(\lambda)\, d\lambda \right]. \qquad (2.23)$$

Here $\mathcal{S}_n^x$ is the system's total sensitivity curve in band x, allowing for transmission by the filter function x, two reflections from aluminum-coated mirrors, and extinction by n air masses. Both studies account for atmospheric extinction, so a direct comparison can be made with observed data. The theoretical system has an arbitrary zero point and is normalized to the observational system by requiring $(B - V)_0 \equiv 0.0$ for an average A0 star. Synthetic colors in other photometric systems have been computed for model atmospheres over wide ranges of $T_{\rm eff}$, $\log g$, and [Fe/H]; see, e.g., [24, 105, 107, 108, 169, 191, 378, 420, 657, 652, 694, 770, 862, 903, 909]. These results can provide good first estimates of the atmospheric parameters of actual stars.

Results from [715] for the colors of black bodies are shown in a two-color diagram in figure 2.1. Points on the black body curve are labeled with $T/10^3$. The cross at the upper left corner represents the star BD +28°4211; this star is so hot that in the visible its continuum closely resembles that of an infinite-temperature black body. The case BD +28°4211 shows that for a very hot star $(U - B)$ and $(B - V)$ colors lose their sensitivity to $T_{\rm eff}$. The same is true for very cool stars (see below).

A star's total luminosity is often expressed in terms of its *absolute bolometric magnitude*, $M_{\rm bol*}$, defined such that

$$\log(\mathsf{L}_*/\mathsf{L}_\odot) = 0.4(M_{\rm bol\odot} - M_{\rm bol*}). \qquad (2.24)$$

$M_{\rm bol*}$ is usually written as the sum of the star's absolute visual magnitude M_{V*} in the V band and a *bolometric correction*, $B.C.(*)$, which converts the fraction of the star's flux that is measured in the V band to its total flux integrated over wavelength:

$$M_{\rm bol*} = M_{V*} + B.C.(*). \qquad (2.25)$$

The terminology is historical; early attempts to measure the total flux from the Sun were made using *bolometers*, devices that had nearly unit sensitivity over the wavelength range where the Earth's atmosphere is transparent.

If a star's absolute energy distribution $f_*(\lambda)$ above the Earth's atmosphere is known in physical units for its entire spectrum, the bolometric correction is

$$B.C.(*) = 2.5 \log \left[\int_{\Delta\lambda} f_*(\lambda) \mathcal{S}_V(\lambda) d\lambda \bigg/ \int_0^\infty f_*(\lambda) d\lambda \right] + \text{constant}. \qquad (2.26)$$

Here $\mathcal{S}_V$ is the receiver's absolute response function in the V band, and the constant is to normalize this theoretical scale to the observational V magnitude scale. Other photometric bands may be used instead of V.

Because M_{V*} measures only a fraction of the total flux, $M_{\text{bol}*}$ must be $\leq M_{V*}$. For solar-type stars, the Earth's atmosphere is fairly transparent to radiation emitted in the visible range, so their bolometric corrections are small. The traditional estimate for $B.C.(\odot)$ is -0.07 mag. Bolometric corrections can now be calculated from grids of line-blanketed models. The zero point of the scale can be chosen so that no bolometric correction is positive; the result is $B.C.(\odot) \approx -0.19$ and $M_{\text{bol}\odot} \approx 4.63$ [169, 191]. The zero point is immaterial because when (2.24) is used to compute L_*/L, it is the same for both $B.C.(*)$ and $B.C.(\odot)$ and cancels out.

For other kinds of stars the bolometric correction can be large. Hot stars emit most of their radiation at wavelengths where the Earth's atmosphere is opaque; hence space observations are necessary. Model atmosphere calculations show that a B5 star emits only 20% of its radiation in the V band; B1 and B2 stars emit $\sim$ 90% of their flux in the range 912 Å $< \lambda <$ 3300 Å, which is accessible to space vehicles. But an O6 star emits $\sim$ 50% of its flux at $\lambda <$ 912 Å, where the interstellar medium is opaque and no direct measurements are possible. For these stars accurate bolometric corrections can be obtained only by fitting a line-blanketed, NLTE model atmosphere to the observable part of their spectrum and using the model to calculate the fraction of the energy in the unobservable region.

Stars of types A5–K5 emit $\sim$90% of their radiation in the visible and infrared range, 3300 Å $< \lambda$. But only $\sim$22% of the radiation from an M0 star is in the V band. For stars with $T_{\text{eff}} \lesssim 4500$ K, we must use data from the infrared J, H, K, and L bands to get a good calibration between T_{eff} and color; see figure 2 in [593]. There are extensive calculations of $B.C.$ from model atmospheres in the literature; see, e.g., [25, 78, 114, 169, 170, 185, 191, 242, 270, 337, 657, 682, 683, 772, 909].

2.5 MASS, LUMINOSITY, AND RADIUS

Mass

Mass of the Sun

Fundamental stellar masses are obtained from Kepler's third law

$$\mathsf{M}_1 + \mathsf{M}_2 = 4\pi^2 a^3 / G P^2 \qquad (2.27)$$

for the orbital motion of two gravitationally bound bodies. Here M_1 and M_2 are their masses, P their orbital period, a the semi-major axis of their elliptical orbit, and G is the Newtonian constant of gravitation. The Earth's mass is obtained by applying (2.27) to orbits of artificial satellites. Their masses are so small that they can be ignored in (2.27). One finds $M_\oplus = 5.9742 \times 10^{27}$ gm; its accuracy is limited by uncertainty in the value of G. The mass of the Moon can be found using data from artificial satellites in orbit around it. The ratio of these two masses (which is known to seven digits) agrees with astrometric measurements of the Earth's motion around the Earth-Moon barycenter. Hence the sum of the Earth's and Moon's masses in gm is known.

The sum of the masses of the Sun and the Earth-Moon system also follows from (2.27). The length of the *sidereal year* (measured relative to distant stars) is known to eight digits in terms of SI seconds. The distance from the Earth to other planets in astronomical units is known precisely from accurate theories of celestial mechanics. The scale of the Earth's orbit in kilometers can be set by radar ranging to suitable planets. The resulting value of the semi-major axis a of the Earth's orbit in km is known to 11 digits! Subtracting the known mass of the Earth-Moon system gives the mass of the Sun. The final result is $M_\odot = 1.9891 \times 10^{33}$ gm; again, its accuracy is limited by uncertainty in the value of G.

Visual Binaries

To measure stellar masses, (2.27) is applied to visual binary stars:

$$M_1 + M_2 = a^3/P^2, \tag{2.28}$$

where $M_1 + M_2$ is measured in solar masses, a in astronomical units, and P in years. Observations give the separation a'' and the *position angle* (the angle between the line connecting the stars and the direction to the north celestial pole) of the fainter star 2 relative to the brighter star 1.

From classical mechanics we know that the path of an object moving in a known force field is determined by six initial conditions: three components of the vector of its initial position $\mathbf{r}$ and three components of the vector of velocity $\dot{\mathbf{r}}$. When the observed (projected) elliptical orbit of a star is well defined, we can derive six equivalent *orbital elements*: its period; true (deprojected) semi-major axis; true ellipticity e; and three angles that fix its orientation in space: the inclination i between the orbital plane and the plane of the sky; the angle of periastron passage (measured in the plane of the orbit); and the direction of the line of nodes (where the orbital plane intersects the plane of the sky) [393, chapter 19]. Conversion of a'' to astronomical units is made by multiplying by $1/\pi''$.

Reliable distances to many visual binaries are known from HIPPARCOS observations. Separations for closely spaced binaries can now be measured down to $\approx 0.015''$ by interferometric methods, particularly *speckle interferometry* [375, 665, 666, 716, 717].

Astrometric Binaries and Spectroscopic-Visual Binaries

From the above analysis, we obtain only the sum of M_1 and M_2. To obtain individual masses we must know the ratio M_2/M_1. It can be determined in two ways:

(1) In a few cases it is possible to measure astrometrically the periodic wobble in the path of the stars relative to the straight path of their barycenter. These data yield a_1/a_2; hence $M_2/M_1 \equiv a_1/a_2$. Such stars are called *astrometric binaries*. Ground-based observations give accurate results for only a few astrometric binaries. New solutions have been obtained for several systems with separation a'' on the sky $\gtrsim 0.08''$ and periods $10 \leq P \leq 55$ years by combining data from ground-based catalogs with HIPPARCOS observations [381].

(2) A visual binary may also be a *spectroscopic binary* for which radial velocities $v \sin i$, in km/s, of both components can be measured. Because $\sin i$ can be determined in the solution of the visual binary orbit, the measured velocities can be deprojected, yielding a_1 and a_2 in km, independent of knowledge of the stars' distance. The ratio of the velocity amplitudes $|v_1|_{\max}/|v_2|_{\max} = a_1/a_2$, from which M_2/M_1 follows. Unfortunately, spectroscopic-visual binaries are rare: if $i \approx 90°$ so that $\sin i \approx 1$ the velocity amplitudes are large, but the orbit is seen nearly edge-on, so it is difficult to determine its orbital elements accurately; and if $i \approx 0°$, the orbit is easy to see, but the velocity amplitudes are small.

Spectroscopic Binaries

A spectroscopic binary is an unresolved pair of stars whose spectral lines show periodic Doppler shifts resulting from their orbital motions. For stars of nearly equal brightness, both spectra are observable (double-line binaries). If one component is much brighter than the other, only one spectrum is observable (single-line binaries). Double-line binaries yield more complete results. For a double-line binary, we can derive [393, chapter 19] the orbital-velocity amplitudes $v_1 \sin i$ and $v_2 \sin i$ in km/s from an analysis of the radial velocity curves; hence, from the known period, $a_1 \sin i$ and $a_2 \sin i$ in km. The ratio $a_1 \sin i/a_2 \sin i$ gives M_2/M_1. But $\sin i$ is still unknown, so (2.27) gives only $(M_2 + M_1) \sin^3 i$.

For an *eclipsing spectroscopic* binary the coplanar orbits must be nearly perpendicular to the plane of the sky for eclipses to occur, so $\sin i \approx 1$. In the optimum case, we can deduce the elements of the relative orbit from spectroscopic data alone, obtaining a_1 and a_2, and hence the mass ratio from a_1/a_2, the sum of the masses from Kepler's law, and then the mass of each star.

The weak point of this method is ignorance of the value of $\sin i$. Light curves are now analyzed with elaborate computer codes. Inputs to these programs are the ratio of the radii of the stellar disks, their relative brightness, the center-to-limb intensity variation for each star, spectroscopic orbital elements, and an estimate of $\sin i$. In favorable cases, $\sin i$ can be determined by iterative improvement of the solution. Then from eclipse durations and orbital velocities we can find the radii of the two components in km in addition to their masses.

Results

Authoritative discussions of the present state of stellar mass determinations appear in [28, 879, 880]. Reliable results exist for a number of eclipsing-spectroscopic, spectroscopic-visual, and astrometric binaries. Almost all of these systems are of luminosity class V. Tabulations of $\log(\mathsf{M}/\mathsf{M}_\odot)$ versus MK spectral type from O5 to M8 for luminosity classes I, III, and V are given in [18, p. 209], [261, table 15.8]. Many of these data are obtained by indirect methods; they have lower accuracy.

Luminosity

Solar Luminosity

Careful reduction of existing space data to the Space Absolute Radiometer Reference [352] yields a solar flux of $1.366 \pm 0.1\% \times 10^6$ erg/cm^2 at the Earth-Sun mean distance. From knowing the astronomical unit R in physical units (§ 2.2), one finds the Sun's luminosity is $\mathsf{L}_\odot = 3.842 \times 10^{33}$ erg/s, and from the solar radius quoted below, the flux at the Sun's surface is $F_\odot = 6.31 \times 10^{10}$ erg/cm^2.

Stellar Luminosities

To determine the luminosity of a star we need (1) its absolute visual magnitude from its apparent magnitude and parallax, or its position in a color-magnitude diagram with known zero point; (2) its bolometric correction; (3) and the luminosity of the Sun. Items (1) and (2) yield the star's absolute bolometric magnitude, and we get L_* from $\mathsf{L}_\odot$ using (2.24).

Mass-Luminosity Relation

The mass-luminosity relation is derived from data for stars having reliable fundamental masses and luminosities. It can be represented by the formulae $\log(\mathsf{M}/\mathsf{M}_\odot) = 0.48 - 0.105 M_{\mathrm{bol}}$ for $-8 \le M_{\mathrm{bol}} < 10.5$, or $\log(\mathsf{L}/\mathsf{L}_\odot) = 3.8 \log(\mathsf{M}/\mathsf{M}_\odot)$ for $\mathsf{M} > 0.2\mathsf{M}_\odot$ [18, p. 209], [261, p. 382]. These relations apply for main-sequence stars; as a star evolves away from the main sequence its luminosity increases, markedly in the case of lower main-sequence stars ascending the red-giant branch.

Radius

The Solar Radius

Given the angular diameter α_* (in radians) of a star and its distance d, then $\mathsf{R}_* = d \sin(\alpha_*/2)$. For example, the Sun's angular radius seen from the Earth is $\alpha_\odot = 959.63'' = 4.6524 \times 10^{-3}$ rad. From the value of 1 au quoted above, we get $\mathsf{R}_\odot = 6.9599 \times 10^{10}$ cm. As mentioned earlier, the Sun's atmospheric density scale height is 10^7 cm, so the photosphere appears to have an angular thickness

$\sim 0.1''$, which is far below the resolving power of the eye, explaining why the solar disk appears to have a crisp edge.

Interferometric Methods

Interferometric methods for measuring stellar angular diameters are the following:

(1) Michelson interferometry uses phase interference between two coherent beams of light from a star. Early efforts [727, 857, 858] gave the angular diameters of some red supergiants. Modern phase-interference instruments now yield reliable data about the angular separations of close binary stars, angular diameters of their components, and even maps of the brightness over the surface of individual stars [157].

(2) Intensity interferometry, developed by Brown and Twiss [153, 154, 271], uses correlations in the fluctuations in the intensity signal from a star in the Rayleigh-Jeans part of its spectrum. It works best for hot stars.

(3) Speckle interferometry uses the fact that the "seeing disk" of a star, blurred by variations in the index of refraction in, and motions of, many atmospheric cells over the surface of a telescope mirror, consists of numerous bright granules, each having a size at the resolution limit of the full aperture. The angular sizes of these granules can be recovered by Fourier deconvolution of the blurred image [716].

To convert a star's angular size to its radius, we must know its distance. Errors in its distance translate into errors in its inferred radius. But a star's angular diameter can be used to convert its flux at the Earth to the flux emergent at its surface, independent of its distance. Then direct comparisons with model atmospheres can be made. See discussion of "effective temperature" below.

Eclipsing Binaries

Accurate stellar radii can be obtained from the radial-velocity and light curves of spectroscopic eclipsing binary stars. Using the velocity data, one determines the relative orbit of the two components, excepting the projection factor $\sin i$. Because the stars eclipse one another, $\sin i \approx 1$. More detailed modeling of the light curve can discriminate between total and partial eclipses, in favorable cases giving information about i. By combining velocity data with the time between onset and maximum depth of the eclipses one determines the radius of each star in km.

The difficulty of assigning a radius to a giant or supergiant with an extended envelope remains (§ 2.1). An extreme example is the binary ζ Aurigae (K3 Ib + B8). When the smaller, brighter, B star passes behind the K supergiant, it shines through the supergiant's diffuse envelope, and its spectrum contains clusters of spectral lines, which show the existence of "... discrete and rather dense condensations whose nature is unexplained" [1170]. In other words, the envelope of the supergiant appears to be a random collection of separate blobs of gas.

Results

Tabulations of $\log(\mathsf{R}/\mathsf{R}_\odot)$ versus MK spectral type in the range O5 to M8 for luminosity classes I, III, and V can be found in [18, p. 209] and [261, table 15.8].

The mass-radius relation for stars on the ZAMS is

$$\log(\mathsf{R}/\mathsf{R}_\odot) = 0.640 \log(\mathsf{M}/\mathsf{M}_\odot) + 0.011$$

for $0.12 < \log(\mathsf{M}/\mathsf{M}_\odot) < 1.3$, and

$$\log(\mathsf{R}/\mathsf{R}_\odot) = 0.917 \log(\mathsf{M}/\mathsf{M}_\odot) - 0.020$$

for $-1.0 < \log(\mathsf{M}/\mathsf{M}_\odot) < 0.12$ [261, p. 382].

Data for luminosity classes V, and probably III, are reliable. The supergiant ζ Aurigae gives a sobering caveat for class I.

Surface Gravity

The surface gravity of a star of known mass and radius is $\log(g/g_\odot) = \log(\mathsf{M}/\mathsf{M}_\odot) - 2\log(\mathsf{R}/\mathsf{R}_\odot)$. Tabulations of $\log g$ versus MK spectral type for luminosity classes I, III, and V are given in [18, p. 209] and [261, table 15.8]. The mass and radius of the Sun give $g_\odot = 2.65 \times 10^4$ cm/s^2, or $\log g = 4.44$. For main-sequence stars $g \approx 10^4$; for early-type giants $g \approx 2 \times 10^3$; for late-type giants $20 \lesssim \log g \lesssim 10^3$; for early-type supergiants $g \approx 10^2$ to 2×10^3; for late-type supergiants $g \approx 1$ to 30; and for white dwarfs $g \approx 10^8$.

Effective Temperature

The effective temperature of a star follows from (2.3). The radius of the Sun is precisely known, and almost all its radiant energy can be observed from the ground or by space vehicles. One finds $T_{\rm eff}(\odot) = 5777$ K. A star's effective temperature can be estimated by fitting its absolute energy distribution to blanketed NLTE model atmospheres. It may be possible to observe only a fraction of a star's total flux, so the accuracy of the result depends on the quality of a model's predictions in unobservable regions of the spectrum.

An empirical approach is to combine a star's integrated flux f_* obtained from its absolute energy distribution with its measured angular diameter α [242]. Then from equation (3.72)

$$\sigma_R T_{\rm eff}^4(*) = F_* = 4f_*/\alpha^2. \tag{2.29}$$

The paper cited gives reliable fundamental determinations of $T_{\rm eff}$. Note again that effective temperature is not a fundamental property, being *defined* by (2.3). It is meaningful for compact atmospheres and can be determined observationally or with theoretical models.

Relations Among Stellar Properties

Main Sequence

Table 2.4 summarizes correlations for masses, luminosities, radii, surface gravities, effective temperature, absolute visual magnitude, $(B - V)_0$ and $(U - B)_0$ colors,

Table 2.4 Correlation of Physical and Photometric Properties of Main-Sequence Stars with Spectral Type

Type	log M	log L	log R	log g	T_{eff}	M_V	$(B-V)_0$	$(U-B)_0$	$B.C.(V)$
O5 V	1.65	5.67	1.08	3.9	41500	−5.6	−0.33	−1.19	−3.94
O7 V	1.50	5.28	1.05	3.9	35500	−5.1	−0.32	−1.15	−3.48
O9 V	1.28	4.86	0.95	3.8	31500	−4.4	−0.31	−1.12	−3.13
B0 V	1.24	4.58	0.86	4.0	30000	−3.9	−0.30	−1.08	−2.92
B2 V	0.99	3.69	0.71	4.0	22000	−2.3	−0.24	−0.84	−2.30
B5 V	0.77	2.87	0.59	4.0	15500	−1.1	−0.17	−0.58	−1.42
A0 V	0.46	1.63	0.43	4.0	9500	0.8	−0.02	−0.02	−0.29
A2 V	0.38	1.37	0.31	4.1	9000	1.4	0.05	0.05	−0.19
A5 V	0.30	1.11	0.20	4.1	8200	2.0	0.15	0.10	−0.14
F0 V	0.20	0.76	0.18	4.2	7200	2.8	0.30	0.03	−0.08
F2 V	0.15	0.65	0.16	4.2	6900	3.0	0.35	0.00	−0.10
F5 V	0.11	0.47	0.13	4.3	6500	3.6	0.44	−0.02	−0.13
G0 V	0.02	0.12	0.02	4.4	6000	4.5	0.58	0.06	−0.17
G2 V	0.00	0.00	0.00	4.44	5750	4.8	0.63	0.12	−0.19
G5 V	−0.04	−0.15	−0.04	4.5	5650	5.2	0.68	0.20	−0.20
K0 V	−0.10	−0.43	−0.07	4.5	5100	6.0	0.81	0.46	−0.30
K2 V	−0.13	−0.62	−0.12	4.6	4800	6.6	0.91	0.64	−0.41
K5 V	−0.17	−0.86	−0.17	4.6	4250	7.5	1.15	1.09	−0.71
M0 V	−0.29	−1.27	−0.24	4.7	3750	8.9	1.40	1.25	−1.37
M2 V	−0.40	−1.50	−0.30	4.7	3450	10.0	1.49	1.18	−1.61
M5 V	−0.68	−2.02	−0.57	4.9	3100	12.4	1.64	1.24	−2.72

and bolometric corrections of main-sequence stars versus spectral type. M, L, and R are in solar units. The entry for spectral type G2 represents the Sun.

For O stars, $(B-V)$ and $(U-B)$ do not discriminate spectral types well; ultraviolet colors from space vehicles should be used. For M stars $(J-H)$ and $(H-K)$ colors are more suitable than $(B-V)$ and $(U-B)$. Their luminosities should be derived from M_J and $B.C.(J)$. The accuracy of these data has greatly improved thanks to the following:

(1) extensive grids of line-blanketed NLTE model atmospheres for hot [60, 61, 682, 683, 714] and cool [451–454, 620] stars;

(2) broad band colors from line-blanketed models [114, 169, 191, 770];

(3) accurate ultraviolet observations from space vehicles [125, 126, 501, 679, 1087] and sensitive receivers in the far infrared.

Note that no data are given for giants or supergiants. Other references, e.g., [18, 261, 677], tabulate results for luminosity classes III and I; the photometric data, masses, and luminosities can be as reliable as for class V. But the radii, gravity, and effective temperatures, especially for class I, are not secure.

Table 2.5 Physical and Photometric Properties of Substellar Objects

Type	log M	log L	log R	$\log g$	$T_{\rm eff}$	M_J	$(J-H)_0$	$(H-K)_0$	$B.C.(J)$
L0	−1.35	−3.6	−0.96	∼5	2200	11.8	0.73	0.52	−1.8
L2	−1.36	−3.8	−0.97	∼5	1950	12.4	0.87	0.60	−1.7
L5	−1.42	−4.2	−0.99	∼5	1600	13.7	0.99	0.64	−1.4
T0	−1.40	−4.7	−.96	∼5	1200	13.8	0.75	0.46	−2.6
T2	−1.40	−4.8	−.97	∼5	1100	14.1	0.53	0.29	−2.5
T5	−1.41	−5.0	−.98	∼5	1000	14.8	−0.22	−0.12	−2.3
T8	−1.42	−5.4	−.99	∼5	800	16.0	−0.38	−0.13	−2.1

Substellar Objects

Substellar objects types L, T, and Y have been studied very actively. These objects are very challenging to analyze theoretically because molecules such as CO, CH_4, N_2, NH_3, and H_2O that exist at the low temperatures in their atmospheres produce extremely heavy line blanketing. In addition, condensates from refractory elements such as Mg, Al, Si, Ca, Ti, V, and Fe form compounds that can condense into clouds. The best fits to the data are obtained with models that have clouds. Some results from these models are listed in table 2.5.

Analysis of the spectra of these objects gives $T_{\rm eff}$ and $\log g$ directly; $\log g$ is not very well determined, but nearly always lies between 4.5 and 5.5 so we take $\log g = 5$ as a representative value. In the table we have used (a) M_J (from photometry) and $B.C.(J)$ (from models) to estimate L; (b) that L and $T_{\rm eff}$ (from models) to estimate R; and (c) that R and $\log g$ (from models) to estimate M. The masses quoted may be uncertain by a factor of 3 because of the uncertainty in $\log g$. In the table, M, L, and R are in solar units. Only classes L and T are considered here.

2.6 INTERPRETATION OF COLOR-MAGNITUDE DIAGRAMS

The discussions in § 2.4 and § 2.5 show how the observational information needed to make a connection between color-magnitude diagrams and theoretical stellar evolution tracks in the HR diagram is obtained. An immense amount of work has been done in stellar evolution studies using high-fidelity physics and powerful numerical algorithms. Here we can only briefly summarize some of the most basic results that have been found. Detailed authoritative discussions of this huge body of work can be found in [483, 557–559, 619, 1104, 1153].

Population I

The color-magnitude diagram for a group of coeval Population I stars (a galactic cluster) in figure 2.6 shows a well-populated main sequence extending from low-mass and low-luminosity to high-mass and high-luminosity stars. The main sequence is prominent in the cluster's HR diagram because the stars spend most of their lifetime "burning" hydrogen into helium. In both the p-p and CNO cycles, four protons, each with a mass 1.673×10^{-24} gm, are fused into a helium nucleus, which has a mass of 6.645×10^{-24} gm. In the process, the binding energy of the He nucleus, equivalent to $\Delta m \approx 4.6 \times 10^{-26}$ gm of matter, is released. The energy released per transmuted proton is $\frac{1}{4}\Delta mc^2 = 1.15 \times 10^{-26} \times 9 \times 10^{20} = 1.035 \times 10^{-5}$ erg/proton.

Consider the case of the Sun: if it were pure hydrogen it would contain $2 \times 10^{33}\text{gm}/1.673 \times 10^{-24}$gm/proton $\sim 1.2 \times 10^{57}$ protons, so the energy available for hydrogen burning is $\sim 1.24 \times 10^{52}$ erg. If *all* of the Sun's hydrogen were converted to helium, its current luminosity, 4×10^{33} erg/sec, could be supported for $\sim 10^{11}$ years, an order of magnitude greater than its present age.[3] Thus stars spend most of their lives on or near the hydrogen-burning locus in the HR diagram, and hence are most likely to be found there. Indeed, the Sun lies nearly on the main sequence.

The amount of fuel available for thermonuclear reactions in a star scales as its mass M, and the rate at which this fuel is consumed scales as its luminosity L. On the main sequence, the mass-luminosity relation in § 2.5 implies that $\mathsf{L} \sim \mathsf{M}^{3.8}$. Therefore, the time to exhaust hydrogen at a star's center scales roughly as $\tau_{\rm H} \sim \mathsf{M}/\mathsf{L} \propto \mathsf{M}^{-2.8}$ or $\tau_{\rm H} \propto \mathsf{L}^{-0.74}$. Thus the brightest, most massive stars in a cluster exhaust their hydrogen fuel soonest. Therefore, the youngest clusters have the brightest turnoff points. Hence the progression from NGC 2362 to M 67 indicates increasing age. Stars with $\mathsf{M} \lesssim 0.9\mathsf{M}_\odot$ cannot exhaust their hydrogen in the age of the Universe.

Shortly before complete hydrogen exhaustion in its core, a star with mass $\mathsf{M} \gtrsim 1.5\mathsf{M}_\odot$ undergoes a brief structural adjustment in which energy generation shifts from the core to H $\rightarrow$ He burning in a shell around the core. The star contracts slightly, moving briefly to a slightly higher effective temperature and luminosity. Afterward the temperature decreases monotonically along its evolution track until it star reaches the red-giant branch.

As the star ages, the hydrogen-burning shell becomes thinner, and at some point it cannot produce the energy being radiated away. The star contracts, its central temperature rises, and helium burning He $\rightarrow$ C begins in the core. In a massive star, this sequence of events is repeated for a succession of nuclear fuels with ever-increasing Z, creating an "onion-skin" structure that may have several shell sources burning at the same time. The shell in each nuclear-burning episode leaves behind layers of the "nuclear ash" from its energy-release process. In this way, the α *elements* He, C, O, Ne, Mg, Si, Ca are created by successive additions of α particles (He^4 nuclei) in the chain $\text{He}^4 + \alpha \rightarrow \text{Be}^{8*} + \alpha \rightarrow \text{C}^{12} + \alpha \rightarrow \text{O}^{16} + \alpha \rightarrow \text{Ne}^{20} + \alpha \rightarrow \text{Mg}^{24} + \alpha \rightarrow \text{Si}^{28} \cdots + \alpha \rightarrow \text{Ca}^{40}$ and so on. As the star's evolution proceeds, each new fuel "ignites" at progressively higher temperatures so the thermal energy in the core of the star increases. To satisfy the virial theorem, when the temperature of the core rises, its gravitational binding energy must also increase, so it contracts. At the same time, the star's envelope outside its outermost shell source expands by a large factor, and its outer layers cool.

[3] In reality only about 10% of the hydrogen is burned.

If the stars at the turnoff point are massive enough, they evolve rapidly across the sparsely populated *Hertzsprung gap*, the region bounded schematically by dashed lines in figure 2.6, toward the region of red supergiants. In these stages, a star moves in a complicated series of wide loops between high and low effective temperatures in the HR diagram [557]. Even the large-scale features of these loops depend sensitively on the physical models implemented in the computer code. Given the ambiguity of the meaning of the stellar "radius" for extended atmospheres, the positions of these highly evolved stars in the HR diagram may be only qualitative.

While crossing the Hertzsprung gap, massive stars may enter the *Cepheid instability strip* and pulsate, exhibiting large, regular variations in luminosity. These pulsations are driven by a thermodynamic engine resulting from the behavior of the opacity in the $He^{+} \rightarrow He^{2+}$ ionization zone. The opacity in this zone increases when the material in it is compressed, so radiation is trapped. The material becomes overheated, creating an overpressure that causes the star's envelope to expand. As it expands, the opacity in the envelope decreases and the trapped radiation is released. The overexpanded envelope then falls back on the star, and the cycle starts anew.

Eventually the star becomes a red supergiant, and the "burnable" nuclear fuel in its core is exhausted. The core then consists of iron, which has the most tightly bound nucleons of any element; hence no *exoergic* thermonuclear reaction is possible, and this is the end point of the nuclear-burning chain. The inert iron core contracts until its material becomes degenerate. At the same time, the star's envelope expands in a dense stellar wind that drains off most of its mass, and the star moves rapidly across the HR diagram to the white-dwarf sequence.

As described in § 1.1, if the core's mass is below the Chandrasekhar limit of about $1.4\mathsf{M}_\odot$, the degenerate electron gas in it is non-relativistic and can support the weight of the overlying layers. But having exhausted its nuclear fuel, the core cannot provide an energy source to sustain the star's luminosity. Moreover, it cannot contract to release gravitational energy because the degenerate material is too "stiff." Hence the star simply cools; starting at the hot end of the white-dwarf sequence, it slides down it, becomes fainter and redder, and ultimately fades to invisibility.

But if the core's mass is above the Chandrasekhar limit, the degenerate electrons in it become relativistic and it can no longer support the weight of the overlying layers. The core continues to contract until the central temperature rises to the point where iron photodisintegrates into α particles in an *endoergic* reaction. This demand on the internal energy of the gas causes the star to collapse on an extremely short timescale to become a neutron star, at essentially nuclear density. The star's outer layers are no longer supported by gas pressure and fall inward, smashing into the rigid neutron core, which produces a huge shock that ultimately blows away the star's envelope in a supernova explosion, a mechanism first suggested by Baade and Zwicky in 1934 to explain Type II supernovae [75].

The CM diagram of an old galactic cluster (e.g., M 67 or NGC 188) is quite different. In a very old cluster, the turnoff point occurs at little more than $1\mathsf{M}_\odot$. Such clusters have no Hertzsprung gap and no supergiants, but instead well-developed subgiant and red-giant branches [598]. The transition from core to shell-source H burning is made on the subgiant branch. Shell-source burning of both H and He takes over on the red-giant branch. The subsequent evolution of these stars is qualitatively similar to that described below for Population II stars.

Mapping to the Theoretical HR Diagram

To convert an observational color-magnitude diagram to the theoretical HR diagram (log L vs. $T_{\rm eff}$), one needs (1) accurate distances, or precise CM-fitting to the fundamental Hyades sequence, to convert V to M_V [869, 870]; (2) reliable bolometric corrections, to convert M_V to $M_{\rm bol}$, and hence L; and (3) absolute energy distributions, to convert observed color indices to the integrated flux F received from a star. Items (2) and (3) can be obtained either empirically or from line-blanketed model atmospheres.

Population I stars are abundant in the solar neighborhood, so many have reliable distances. Some are bright enough to be observed with infrared detectors and in the UV from space observatories; hence L and F can be determined quite well. Also, Population I stars near the main sequence have compact atmospheres, which can be interpreted reliably. Thus they can accurately be mapped into the HR diagram and compared with theoretical stellar evolution. The accuracy of the mapping process is lower for giants and supergiants because of the difficulty in determining the "radius" of an extended envelope. Some current stellar evolution models use stellar atmospheres (which may allow mass loss through radiation-driven winds in massive stars) to provide accurate outer boundary conditions; see, e.g., [950, 951, 953, 954].

Age Estimates

The estimates for the hydrogen-depletion time $\tau_{\rm H}$ in stars' cores made above are only semi-quantitatively correct. Modern stellar evolution models provide much more accurate calculations of $\tau_{\rm H}$ and also detailed tracks in the HR diagram. Hence age dating of galactic clusters can now be done with good confidence. Age estimates range from very young ($\sim 10^6$ years for NGC 2369) to ages comparable to those of globular clusters: $\sim 6.5 \times 10^9$ years for NGC 188 and $\sim 8 \times 10^9$ years for Berkeley 17.

Population II

Population II stars are characterized by low metal abundances and high space velocities relative to the Sun. The best color-magnitude diagrams for Population II stars are obtained from globular clusters because they contain huge numbers of stars. A schematic color-magnitude diagram for Population II stars in figure 2.7 shows that they are very old because the turnoff point from the main sequence occurs at masses $\lesssim 1\mathsf{M}_\odot$. In modern work, values of [α/Fe] are measured spectroscopically in addition to [Fe/H].

On the Population II main sequence, stars have a hydrogen-burning core. As the hydrogen in its core is exhausted, a star moves up the subgiant branch with an inert helium core and an active hydrogen-burning shell. It continues up the red-giant branch, which is terminated when a star's core becomes hot enough to ignite helium burning in an almost explosive *helium flash*. The star as a whole contracts, and it shoots over to the lower end of the asymptotic giant branch in the HR diagram, at a luminosity about 2 mag above the cluster's turnoff point and a surface temperature on the order of 10^4 K (comparable to an A star).

As the helium core continues to burn, the star becomes more luminous until it reaches the horizontal branch about 3.5 mag above the turnoff point. On the horizontal branch stars pass through the RR Lyrae and W Virginis instability strips and undergo pulsations powered by the same mechanism as Cepheid pulsation. As its core burns helium to exhaustion, a star climbs up the asymptotic giant branch, along which it has an inert C/O core and both a helium-burning and a hydrogen-burning shell [558, 559]. At some point, the two nuclear-burning shells approach each other, and the star becomes unstable against double shell-source *relaxation oscillations*, which lead to variations in the star's luminosity and results in mixing of core material into its envelope. Convection zones develop that may connect zones in which different nuclear burning processes have occurred. These lead to episodes of *dredge-up* of core material, such as carbon and "exotic" isotopes of other elements resulting from extensive processing by thermonuclear reactions, which drastically alter the composition of the envelope.

Eventually the temperature in the core becomes sufficiently high that carbon burning, $C^{12} \rightarrow O^{16}$, ignites. In the absence of mass loss, the highest luminosity a star reaches at the upper tip of the asymptotic giant branch is set either by its total mass if $M < 1.4M_{\odot}$ (the Chandrasekhar limit) or by the onset of carbon burning, whichever comes first. At the tip of the asymptotic giant branch, stars suffer copious mass loss in stellar winds. Owing to the high carbon content in their atmospheres, these stars are seen spectroscopically as carbon stars (class C). Complex molecules and grains form in their winds, and the star is enveloped in a *cocoon* that is essentially impenetrable to visible light. We can detect the mass loss from large Doppler shifts and breadths of spectral lines observed in infrared radiation, which penetrates the cocoon.

These stars are not massive enough to undergo core collapse and supernova explosion. Rather, their mass loss makes them evolve to the upper-left corner of the HR diagram as very hot, dense planetary nebula nuclei. They are extremely luminous, but not very massive; hence their envelopes are not strongly gravitationally bound. Consequently, they can blow large quantities of matter into space in radiatively driven winds and become surrounded by an extended nebula. Some of them are extremely hot and emit EUV and soft X-ray radiation; indeed, the object H $1504 + 65$ appears to be a bare stellar core [1157] that has just arrived at the upper end of the white-dwarf sequence.

Mapping to the Theoretical HR Diagram

To convert a globular cluster color-magnitude diagram to a theoretical HR diagram, we need the same kinds of information as for Population I, namely, reliable bolometric corrections [25] and relations between energy distributions or colors and effective temperatures [26]; these are now available from a combination of good model atmosphere calculations [106, 174, 192, 421, 708, 1117] and observations in the red and infrared [131, 173, 236, 239, 317, 596]. Attempts to achieve this mapping have a long history. The outstanding problem in these studies has always been in setting the zero point of globular cluster stars' absolute magnitudes. As mentioned above, no globular cluster is near enough to have a directly measured parallax. Only recently has it been possible to fit cluster CM diagrams below their turnoff points reliably to the Population II main sequence defined by subdwarfs in the solar neighborhood [159, 899, 901].

Population II stars in the solar neighborhood are mainly subdwarfs that have solar, or lower, effective temperatures and are found, at a given $(B - V)$, to lie ~1 mag below the Population I main sequence in the color-magnitude diagram. These stars also have an *ultraviolet excess*; at a given $(B - V)$ they are brighter in the ultraviolet than Population I stars and thus lie above the Population I curve in the two-color diagram. The reason for an ultraviolet excess is easily understood. The spectroscopic analyses cited in the bibliography all show that subdwarfs are metal-poor: $[\,\mathrm{Fe/H}\,] \sim -2$ to -3 in extreme cases. The absorption lines of metals are strongest in the ultraviolet spectral region. They are much weaker in the spectrum of a star with a low metals abundance; hence they do not block as much of the emergent UV continuum flux as in a Population I star.

The question of whether subdwarfs lie below the Population I main sequence in the HR diagram, not just in the color-magnitude diagram, is more significant. The earliest study [986], allowing approximately for differential line-blocking effects, concluded that subdwarfs have lower luminosities than Population I main-sequence stars of the same effective temperature. Stellar structure calculations [902] then available showed that very metal-poor stars lie the observed amount below the Population I sequence if they also have about the same helium abundance as the Sun. The important implication is that helium is formed from hydrogen mainly in the primordial Big Bang, and the heavier elements are formed later by nucleosynthesis (in agreement with the present-day picture).

An alternative attempt to answer this question was made using broad band (U, B, V) data [1164] to correct for line blocking. Surprisingly, this work indicated [939] that subdwarfs lie below the Population I main sequence in the color-magnitude diagram because of line blocking in the photometric bands, and that adjustments for this blocking moved their positions in the HR diagram on to the Hyades main sequence. This mapping of extreme subdwarfs to the Hyades main sequence would imply that they have N(He)/N(H) $\simeq 0.0$ [306].

But a careful analysis [205] of the location of subdwarfs in the HR diagram using (G, R, I) data from the Stebbins-Whitford six-color system, blocking-insensitive red and infrared *VRIJK* colors from the Johnson system, intermediate-band vby colors from the Strömgren system, all of which are less affected by line blocking than *UBV* colors, combined with critically evaluated parallax data, showed that extreme subdwarfs, with masses low enough to be unevolved, lie below the Population I main sequence by 0.7 ± 0.3 mag, so their He abundance must be nearly the same as Population I.

Subsequent six-color photometric measurements including red and infrared bands led to the same conclusion [989]. It was shown that in a plot of M_V vs. $(G - I)$ (a color index little affected by line blocking) the cooler subdwarfs fall ~0.5 to

0.75 mag below the Hyades main sequence, and a reanalysis of the broad band color data [304] using line-blocking insensitive $(R-I)$ colors showed that the earlier UBV result that subdwarfs lie on the Hyades main sequence was incorrect. The error arose because somewhat evolved stars were included in the subdwarf sample. The cooler, less evolved stars do lie about 1 mag below the Population I old-disk main sequence.

More recent studies, e.g., [184], using spectrophotometry and model atmospheres to deduce T_{eff}, more precise trigonometric parallaxes, and reliable values for $B.C.$ [185], show clearly that subdwarfs are about 0.6 mag below the Hyades main sequence. Analysis of the HR diagram for Population II stars [207] made using bolometric fluxes obtained from *UBVRIJHK* photometry [25], effective temperatures [26] from the bolometric flux, reliable model atmospheres [658], and HIPPARCOS data giving distances accurate to $\pm 5\%$ makes this result definitive.

Fits of the Population II diagram using these data yield a helium/hydrogen ratio a bit smaller than for Population I. Thus it appears that most cosmic helium was formed during the Big Bang, and there has been a small enrichment since by nucleosynthesis in stars.

Age Estimates

Once the zero point of the absolute-magnitude scale of a globular cluster's color-magnitude diagram has been determined by fitting it to the main sequence of nearby Population II subdwarfs having the same [Fe/H] [236], or to a white-dwarf sequence, we can use stellar evolution calculations to estimate its age from its color-magnitude diagram. This approach may possibly be used to set lower limits on the age of the Universe [210, 390, 878, 899]. But caution is needed because, unlike the large effect of age on the turnoff points in young galactic clusters, globular clusters are so old ($\sim 10^{10}$ years) that significant differences in age result in relatively small changes in their HR diagrams. Thus, small errors in the fitting procedure can produce large errors in the inferred ages. And the mapping from an observed CM diagram to a theoretical $\log T_{\text{eff}}, \log \mathsf{L}$ diagram is less certain for the giant, horizontal, and asymptotic giant branches. These are problems for future work.

Outlook

The HR diagram gives us insight into the structure and evolution of stars of differing chemical composition. A great accomplishment of theories of stellar structure and evolution, and stellar atmospheres, is that they provide a meaningful interpretation of the HR diagram. It is now possible:

(1) to set accurate zero points on the absolute-magnitude scales in both galactic and globular cluster CM diagrams;

(2) to use spectrophotometric absolute energy distributions and/or stellar atmospheres theory to generate reliable bolometric corrections to convert M_V to $\log \mathsf{L}$ (at least for stars with compact atmospheres);

(3) to make reliable conversions from observed colors, including ultraviolet and far infrared bands, to log $T_{\rm eff}$ (for stars with compact atmospheres); and

(4) to make comparison with theoretical stellar evolution tracks (in regions where the "interiors" specialist's notion of the meaning of $T_{\rm eff}$ makes sense) to obtain reliable ages for galactic, and in some cases, globular clusters. In the future, such work may even lead to results of cosmological significance.

Chapter Three

Radiation

The radiation we receive from a star contains an enormous wealth of information about the structure and composition of its atmosphere. The goal of the theory of stellar atmospheres is to develop methods that can recover it. In this chapter we describe radiation using three different, but equivalent, formalisms: (1) macroscopic quantities, with which we can make a direct connection with observables; (2) photons; and (3) electromagnetic fields. Here we consider only radiation itself; radiation-matter interactions and the transfer of radiation will be discussed in later chapters.

3.1 SPECIFIC INTENSITY

Macroscopic Definition

The *specific intensity* $I(\mathbf{x}, t; \mathbf{n}, \nu)$ [erg/cm^2/sr/Hz/s] of radiation at a position $\mathbf{x}$ [cm], time t [s], traveling in direction $\mathbf{n}$, into a solid angle $d\Omega$ [sr],[1] having a frequency in the range $(\nu, \nu + d\nu)$ [Hz], is defined such that the amount of energy passing through an *oriented surface element* $d\mathbf{S} \equiv dS\hat{\mathbf{s}}$ (see figure 3.1), in a time interval dt, is

$$\delta\mathcal{E} \equiv I(\mathbf{x}, t; \mathbf{n}, \nu)\, \mathbf{n} \cdot d\mathbf{S}\, d\Omega\, d\nu\, dt \quad [\,\text{erg}\,]. \tag{3.1}$$

$I(\mathbf{x}, t; \mathbf{n}, \nu)$ is a *seven-dimensional distribution function*, defined on a *four-dimensional spacetime manifold* $(\mathbf{x}, t)$, and an *independent set of three-dimensional angle-frequency spaces* $(\mathbf{n}, \nu)$, also called *tangent spaces*, attached to each spacetime event $(\mathbf{x}, t)$. Note that only two components of $\mathbf{n}$ are independent because it is a unit vector. The specific intensity provides a complete macroscopic description of unpolarized radiation. To simplify the notation, we will generally suppress explicit mention of $(\mathbf{x}, t)$ and use a subscript ν to indicate frequency dependence.

Photon Number Density

Radiation consists of an ensemble of *photons*, which are massless relativistic particles that travel (in vacuum) at the same speed c in all frames. In inertial frames they propagate on straight lines with constant frequency until they interact with material. In describing a radiation field, a useful alternative to the specific intensity is the *photon number density* $\varphi(\mathbf{n}, \nu)$, which is defined to be the number of photons

[1] "sr" stands for *steradian*, a unit of solid angle.

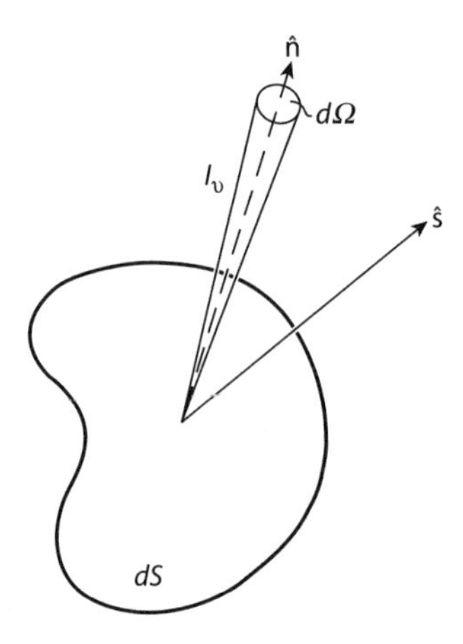

Figure 3.1 Specific intensity $I(\mathbf{x}, t; \mathbf{n}, \nu)$.

with frequencies in the range $(\nu, \nu + d\nu)$, traveling in direction $\mathbf{n}$, into solid angle $d\Omega$, through an oriented surface area element $d\mathbf{S}$, in a time interval dt. Each photon carries an energy $h\nu$ [erg/photon], so the energy passing through $d\mathbf{S}$ is

$$\delta\mathcal{E} \equiv I_\nu(\mathbf{n})\,\mathbf{n}\cdot d\mathbf{S}\,d\Omega\,d\nu\,dt = ch\nu\,\varphi_\nu(\mathbf{n})\mathbf{n}\cdot d\mathbf{S}\,d\Omega\,d\nu\,dt; \tag{3.2}$$

hence

$$\varphi_\nu(\mathbf{n}) = I_\nu(\mathbf{n})/ch\nu \quad [\,\text{photons/cm}^3/\text{sr/Hz}\,]. \tag{3.3}$$

The total photon number density is

$$\varphi(\mathbf{x}, t) = \int_0^\infty d\nu \oint \varphi(\mathbf{x}, t; \mathbf{n}, \nu)\,d\Omega \quad [\,\text{photons/cm}^3\,]. \tag{3.4}$$

The reader should not confuse the photon number density in coordinate space, as defined here, with the radiation distribution function, or the photon occupation number in phase space, defined immediately below.

Radiation Distribution Function

The three-momentum of a photon of energy $h\nu$ moving in direction $\mathbf{n}$ is

$$\mathbf{p} = (h\nu/c)\,\mathbf{n} \quad [\,\text{erg s/cm/photon}\,]. \tag{3.5}$$

In statistical mechanics a *particle distribution function* gives the number of particles having spin α per unit volume of *phase space*; see [673, § 10]. The *photon distribution function* f_α is defined such that $f_\alpha(\mathbf{x}, t; \mathbf{p})\,d^3x\,d^3p$ is the number of photons in polarization state α, in a coordinate-space volume d^3x, having momenta in a momentum-space volume d^3p. The *radiation distribution function* is the sum over all spin states:

$$f_R(\mathbf{x}, t; \mathbf{p})\, d^3x\, d^3p \equiv \sum_\alpha f_\alpha(\mathbf{x}, t; \mathbf{p})\, d^3x\, d^3p. \tag{3.6}$$

For photons, spin corresponds to polarization. Photons are bosons (see § 4.4) that have spin 1. In principle they could have spin projections $m_s = -1, 0, 1$ relative to a space axis. But experimentally we find they have only two polarization states. This is because photons travel at the speed of light in all frames, and hence *have no rest frame*, and one can determine only whether a photon's spin is aligned along, or opposite to, its propagation direction $\mathbf{n}$. Thus a spin projection $m_s = 0$ is meaningless, so the sum extends only over $\alpha = 1$ and 2. For unpolarized radiation, both spin states are populated equally, so $f_1 = f_2$ and $f_R = 2f_\alpha$.

An alternative expression for a momentum-space volume element is

$$d^3p = p^2 dp\, d\Omega = (h^3\nu^2/c^3)\, d\Omega\, d\nu. \tag{3.7}$$

Changing to angle-frequency variables we have

$$f_R(\mathbf{p})\, d^3x\, d^3p \rightarrow (h^3\nu^2/c^3) f_R(\mathbf{n}, \nu)\, d^3x\, d\Omega\, d\nu. \tag{3.8}$$

Then $(h^3\nu^2/c^3) f_R(\mathbf{x}, t; \mathbf{n}, \nu)\mathbf{n}\cdot d\mathbf{S}\, d\Omega\, d\nu\, dt$ is the number of unpolarized photons at position $\mathbf{x}$ and time t, with frequencies in the range $(\nu, \nu + d\nu)$, propagating along $\mathbf{n}$, through an oriented surface element $d\mathbf{S}$, into solid angle $d\Omega$ in a time interval dt. The energy they transport through $d\mathbf{S}$ is

$$\begin{aligned} \delta\mathcal{E} &= (h\nu)(h^3\nu^2/c^3) f_R(\mathbf{n}, \nu)\, \mathbf{n}\cdot d\mathbf{S}\, d\Omega\, d\nu\, cdt \\ &\equiv I(\mathbf{n}, \nu)\, \mathbf{n}\cdot d\mathbf{S}\, d\Omega\, d\nu\, dt. \end{aligned} \tag{3.9}$$

Therefore,

$$f_R(\mathbf{x}, t; \mathbf{n}, \nu) = (c^2/h^4\nu^3) I(\mathbf{x}, t; \mathbf{n}, \nu) \quad [\,\text{photon/erg}^3/\text{s}^3/\text{sr}\,]. \tag{3.10}$$

We show in Appendix B that f_R is a relativistic invariant.

Photon Occupation Number

An important concept from quantum statistical mechanics is the *photon occupation number* in a unit volume of phase space. The basic unit of phase space is h^3, so the occupation number for photons in polarization state α is

$$n_\alpha \equiv h^3 f_\alpha. \tag{3.11}$$

Hence for unpolarized radiation,

$$I_\nu = (h^4\nu^3/c^2) f_R = (2h^4\nu^3/c^2) f_\alpha \equiv (2h\nu^3/c^2)\, n_\alpha, \tag{3.12}$$

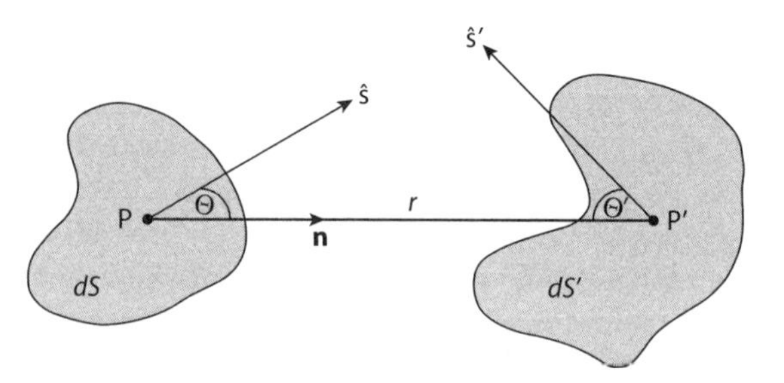

Figure 3.2 Geometry used in proof of invariance of specific intensity.

or

$$n_\alpha(\mathbf{n}, \nu) = \left(c^2/2h\nu^3\right)I(\mathbf{n}, \nu) \quad [\text{photon/sr}]. \tag{3.13}$$

In § 5.2 and § 5.4 we will show that $n_\alpha(\mathbf{n}, \nu)$ is connected with the process of *induced* (or stimulated) *emission* (see equations 5.14 and 5.103).

Invariance Properties

If there is no emission or absorption of radiation along the line of sight, the specific intensity from a resolved source is *invariant*. Consider a pencil of rays moving in direction **n**, which passes through both the projected surface area $\mathbf{n}\cdot d\mathbf{S}$ at point P *and* the projected surface area $\mathbf{n}\cdot d\mathbf{S}'$ at point P', which are at rest and separated by a distance r. The amount of energy $\delta\mathcal{E}_\nu$ with frequencies $(\nu, \nu + d\nu)$ passing through both areas in a time interval dt must be the same. Writing $\cos\Theta \equiv \mathbf{n}\cdot\hat{\mathbf{s}}$ and $\cos\Theta' \equiv \mathbf{n}\cdot\hat{s}'$, we find

$$\delta\mathcal{E}_\nu = I_\nu\, dS\, \cos\Theta\, d\Omega\, d\nu\, dt \equiv \delta\mathcal{E}_\nu{}' = I_\nu{}' dS'\cos\Theta' d\Omega' d\nu\, dt. \tag{3.14}$$

In figure 3.2 we see that the solid angle subtended by dS' as seen from P is $d\Omega = dS'\cos\Theta'/r^2$, and the solid angle subtended by dS as seen from P' is $d\Omega' = dS\,\cos\Theta/r^2$. Using these identities in (3.14) we find $I_\nu{}' \equiv I_\nu$. This result is important because it means that if we know the solid angle $d\Omega$ subtended by a remote source, then the intensity of radiation I_ν measured per unit time in a frequency band $d\nu$ by a receiver of known collecting area and detection efficiency at position P of the observer gives the intensity $I_\nu{}'$ *in absolute units at the source itself.* This method is limited to objects that are *spatially resolved*, e.g., the Sun, nebulae, galaxies.

If a resolved source has spherical symmetry, the angle cosine $\mu \equiv \cos\Theta$ of a radial ray relative to the local normal can be computed. Hence the observed variation of I_ν at different positions on the object's projected disk yields the angular variation $I_\nu(\mu)$ at the surface of the object. To a first approximation, the radiation emergent at some point on the surface is produced mainly by the emission material at about the same *geometric depth measured along the ray*; call it D. The *radial depth* Δr corresponding to D for a ray emerging at angle cosine μ is $\Delta r = \mu D$. Hence measurements of $I_\nu(\mu)$ across an object's (e.g., the Sun's) disk made at a remote location outside the object can give some information about the depth variation of radiant source terms inside it.

This discussion shows only that specific intensity has *geometric, not relativistic*, invariance and applies only to radiation received from a source fixed with respect to the observer. If there is relative motion, relativistic effects must be taken into account; see chapter 19 and Appendix B.

Symmetry

In this book we will consider mainly *one-dimensional, gravitationally stratified* stellar atmospheres that have *planar or spherical symmetry*, i.e., are composed either of homogeneous plane-parallel layers or homogeneous spherical shells. In a planar atmosphere, all physical variables are functions of depth z, measured upward opposite to the direction of gravity $\mathbf{g}$. Thus the specific intensity I at a given frequency has no dependence on (x, y), is a function of z only, and depends only on the polar angle Θ between the outward normal $\hat{z}$ and its propagation direction $\mathbf{n}$. In a spherically symmetric atmosphere, all physical variables are functions of the radius r only, are measured outward, and are independent of the positional angular coordinates θ and ϕ. In particular, the specific intensity is a function of only the radius r and the polar angle Θ between the local normal $\hat{\mathbf{r}}$ and the direction $\mathbf{n}$ of a ray and is independent of the azimuthal angle Φ around $\mathbf{n}$. In short, $I \equiv I(r, t; \Theta, \nu)$.

3.2 MEAN INTENSITY AND ENERGY DENSITY

Macroscopic Description

Eddington defined the *mean intensity* as the average of the specific intensity over all solid angles:

$$J_\nu \equiv \frac{1}{4\pi} \oint I_\nu(\mathbf{n})\, d\Omega = \frac{1}{4\pi} \int_0^{2\pi} d\Phi \int_{-1}^{1} I_\nu(\Theta, \Phi)\, d\mu. \tag{3.15}$$

Here we used the identity $d\Omega = \sin\Theta\, d\Theta\, d\Phi = -d\mu\, d\Phi$. The mean intensity has dimensions erg/cm^2/Hz/s; it is a scalar.

To calculate the *energy density* in a radiation field in the frequency range $(\nu, \nu + d\nu)$, consider a small spherical volume V through which energy flows from all solid angles. The amount of energy flowing into it in a time dt from a solid angle $d\Omega$ through an element of surface area dS is

$$\delta\mathcal{E}_\nu = I_\nu(\mathbf{n})\, d\nu\, dt\, (dS\, \cos\Theta)\, d\Omega. \tag{3.16}$$

Include only those photons in flight within V. If their path length across V is ℓ, then the time they are contained within V is $dt = \ell/c$. Further, $\ell\, dS \cos\Theta = dV$, the differential volume element through which they sweep. Hence the energy in dV coming from $d\Omega$ is

$$\delta\mathcal{E}_\nu = I_\nu(\mathbf{n}) d\nu\, d\Omega\, dV/c. \tag{3.17}$$

Integrating over solid angle and the entire volume, the total energy contained in V is

$$\mathcal{E}_\nu d\nu = \frac{1}{c}\int_V dV \oint I_\nu(\mathbf{n})\, d\nu\, d\Omega. \tag{3.18}$$

In the limit of infinitesimal V, the intensity I is independent of position, and the integrations over V and Ω can be carried out separately. Thus the *monochromatic radiation energy density* E_ν is

$$E_\nu = \frac{1}{c}\oint I_\nu(\mathbf{n})\, d\Omega \equiv \frac{4\pi}{c}J_\nu \quad [\,\text{erg/cm}^3/\text{Hz}\,]. \tag{3.19}$$

The *total radiation energy density* is the integral of (3.19) over frequency.

$$E = \frac{4\pi}{c}\int_0^\infty J_\nu\, d\nu \equiv \frac{4\pi}{c}J \quad [\,\text{erg/cm}^3\,]. \tag{3.20}$$

Photon Picture

These expressions for radiation energy density are consistent with the photon description of radiation and are easier to understand in terms of photons. The total monochromatic radiation energy density is the photon density, multiplied by the energy per photon, summed over all solid angles:

$$E_\nu = h\nu \oint \varphi(\mathbf{n}, \nu)\, d\Omega\,. \tag{3.21}$$

But from (3.3), $(c\, h\nu)\, \varphi_\nu = I_\nu$; hence (3.21) is identical to (3.19).

Electromagnetic Description

Maxwell's Equations

In electromagnetic theory, Maxwell's equations predict the existence of a radiation field; see, e.g., [575, chapter 6] and [838, chapter 9]. Here we show there is a one-to-one correspondence between the electromagnetic and macroscopic pictures of radiation.

Let $\mathbf{E}$ be the *electric field*; $\mathbf{D} \equiv \mathbf{E} + 4\pi\mathbf{P} = \varepsilon\mathbf{E}$ the *electric displacement*; $\mathbf{H}$ the *magnetic field*; $\mathbf{B} \equiv \mathbf{H} + 4\pi\mathbf{M} = \mu\mathbf{H}$ the *magnetic induction*; ϱ the *charge density*; and $\mathbf{j} = \varrho\mathbf{v}$ the *current density*. All of these variables are functions of $(\mathbf{x}, t)$. In the these definitions ε is the *dielectric constant* and μ is the *magnetic permeability* of the material in which the fields exist, and $\mathbf{P}$ and $\mathbf{M}$ are the *polarization* and *magnetization* induced in the material by $\mathbf{E}$ and $\mathbf{H}$.

In Gaussian units, Maxwell's equations are

$$\nabla\cdot\mathbf{D} = 4\pi\varrho, \tag{3.22a}$$

$$\nabla\cdot\mathbf{B} = 0, \tag{3.22b}$$

$$(\nabla\times\mathbf{E}) = -\frac{1}{c}\frac{\partial\mathbf{B}}{\partial t}, \tag{3.22c}$$

$$(\nabla\times\mathbf{H}) = \frac{1}{c}\frac{\partial\mathbf{D}}{\partial t} + \frac{4\pi}{c}\mathbf{j}. \tag{3.22d}$$

The divergence of (3.22d), with $\nabla\cdot\mathbf{D}$ eliminated via (3.22a), gives an equation for *conservation of charge density*

$$\frac{\partial\varrho}{\partial t} = -\nabla\cdot\mathbf{j}. \tag{3.23}$$

In addition, the *Lorentz force* acting on a particle with charge q is[2]

$$\mathbf{F} = q\left(\mathbf{E} + \frac{\mathbf{v}}{c}\times\mathbf{E}\right). \tag{3.24}$$

For electromagnetic fields in vacuum, $\varrho = \mathbf{j} = 0, \mathbf{P} = \mathbf{M} = 0$, and $\varepsilon = \mu = 1$, so $\mathbf{E} \equiv \mathbf{D}$, and $\mathbf{H} \equiv \mathbf{B}$. Then

$$\nabla\cdot\mathbf{E} = 0, \tag{3.25a}$$

$$\nabla\cdot\mathbf{H} = 0, \tag{3.25b}$$

$$(\nabla\times\mathbf{E}) = -\frac{1}{c}\frac{\partial\mathbf{H}}{\partial t}, \tag{3.25c}$$

$$(\nabla\times\mathbf{H}) = \frac{1}{c}\frac{\partial\mathbf{E}}{\partial t}. \tag{3.25d}$$

Plane wave solutions

Take the time derivative of (3.25d), substitute from (3.25c) for $(\partial\mathbf{H}/\partial t)$, and taking note of (3.25a), use the vector identity

$$\nabla\times(\nabla\times\mathbf{V}) = \nabla(\nabla\cdot\mathbf{V}) - \nabla^2\mathbf{V} \tag{3.26}$$

to eliminate $\nabla\times(\nabla\times\mathbf{E})$. The result is the *wave equation*

$$\frac{1}{c^2}\frac{\partial^2\mathbf{E}}{\partial t^2} - \nabla^2\mathbf{E} = 0. \tag{3.27}$$

By similar manipulations, the same equation is found for $\mathbf{H}$.

[2] In both electromagnetic and quantum mechanical formulae the symbol q will denote an arbitrary charge of either sign, whereas the symbol e will denote the absolute value of the electronic charge.

Maxwell's equations were developed from practical experiments using circuits, so $\mathbf{E}$ and $\mathbf{H}$ here refer to the *total* field, with no reference to its frequency spectrum. Even so, they have solutions for traveling *monochromatic plane waves in vacuum*, propagating along a pathlength $\mathbf{r}$ with speed c:

$$\mathbf{E_k}(\mathbf{x}, t) = \mathbf{E_{0k}} e^{i(\mathbf{k}\cdot\mathbf{r}-\omega t)}, \tag{3.28}$$

and

$$\mathbf{H_k}(\mathbf{x}, t) = \mathbf{H_{0k}} e^{i(\mathbf{k}\cdot\mathbf{r}-\omega t)}. \tag{3.29}$$

Here $\mathbf{k} \equiv k\,\mathbf{n}_0$ is the wave's *propagation vector* with $k \equiv 2\pi/\lambda \equiv \omega/c$ [cm^{-1}], where λ is its *wavelength* [cm] and $\omega \equiv 2\pi c/\lambda = 2\pi\nu$ is its *circular frequency*. Taking the curl of (3.28) and (3.29) we find

$$\nabla \times \mathbf{E_k} = i\mathbf{k} \times \mathbf{E_{0k}}, \tag{3.30}$$

and

$$\nabla \times \mathbf{H_k} = i\mathbf{k} \times \mathbf{H_{0k}}. \tag{3.31}$$

Using (3.28) and (3.29) in (3.25c) we find

$$i\mathbf{k} \times \mathbf{E_{0k}} = i(\omega/c)\mathbf{H_{0k}} \quad \text{or} \quad \mathbf{H_{0k}} = \mathbf{n}_0 \times \mathbf{E_{0k}}. \tag{3.32}$$

Similarly, use of (3.28) and (3.31) in (3.25d) yields

$$\mathbf{E_{0k}} = -\mathbf{n}_0 \times \mathbf{H_{0k}}. \tag{3.33}$$

$\mathbf{E_{0k}}$ and $\mathbf{H_{0k}}$ have the same magnitude, i.e., $|\mathbf{E_{0k}}| = |\mathbf{H_{0k}}|$; and $(\mathbf{E_{0k}}, \mathbf{H_{0k}}, \mathbf{n}_0)$ form a right-handed orthogonal triad.

Electromagnetic Energy Density and Energy Flux

Use of the general vector identity

$$\mathbf{H}\cdot(\nabla \times \mathbf{E}) - \mathbf{E}\cdot(\nabla \times \mathbf{H}) = \nabla\cdot(\mathbf{E} \times \mathbf{H}) \tag{3.34}$$

in (3.25c) and (3.25d) yields

$$\frac{1}{8\pi}\frac{\partial}{\partial t}(E^2 + H^2) + \frac{c}{4\pi}\nabla\cdot(\mathbf{E} \times \mathbf{H}) = 0. \tag{3.35}$$

By integrating (3.35) over a volume V and making use of the divergence theorem we obtain *Poynting's theorem*

$$\frac{1}{8\pi}\frac{\partial}{\partial t}\int_V (E^2 + H^2)\,dV = -\frac{c}{4\pi}\int_S (\mathbf{E} \times \mathbf{H})\cdot d\mathbf{S}, \tag{3.36}$$

which is of the general form of an equation of continuity for the electromagnetic energy in the field. Thus we identify the first term in (3.36) as the rate of change of the *electromagnetic energy density* in V,

$$\epsilon = \frac{1}{8\pi}(E^2 + H^2) \qquad [\,\mathrm{erg/cm^3}\,], \tag{3.37}$$

and the second as the outward normal component of the *Poynting vector*, i.e., the *electromagnetic energy flux*, integrated over the surface S enclosing V:

$$\mathbf{S} \equiv \frac{c}{4\pi}(\mathbf{E} \times \mathbf{H}) \qquad [\,\mathrm{erg/cm^2/s}\,]. \tag{3.38}$$

For harmonic time variation of the fields, the real parts of $\mathbf{E}$ and $\mathbf{H}$ are used in (3.37) and (3.38). From (3.28) and (3.29), the instantaneous energy density in a plane monochromatic electromagnetic wave is

$$\epsilon_{\mathbf{k}}(t) = \frac{1}{8\pi}\left\{[\,\Re\mathrm{e}(E)\,]^2 + [\,\Re\mathrm{e}(H)\,]^2\right\} = \frac{1}{4\pi}E_{0\mathbf{k}}^2 \cos^2 \omega t. \tag{3.39}$$

The time average of the energy density over a cycle is

$$\langle \epsilon_{\mathbf{k}} \rangle_{\mathrm{T}} = \frac{1}{4\pi}\langle E_0^2 \cos^2 \omega t \rangle_{\mathrm{T}} = \frac{E_{0\mathbf{k}}^2}{8\pi} \tag{3.40}$$

because $\langle \cos^2 \omega t \rangle_{\mathrm{T}} = \frac{1}{2}$. Similarly, the time average of the electromagnetic flux over a cycle is

$$\langle \mathbf{S}_{\mathbf{k}} \rangle_{\mathrm{T}} = \frac{c}{4\pi}\langle\, |\mathbf{E} \times \mathbf{H}|\, \rangle_{\mathrm{T}} = \frac{c}{8\pi}E_{0\mathbf{k}}^2 \mathbf{n}_0. \tag{3.41}$$

Equations (3.40) and (3.41) provide the basis on which the identity of the macroscopic and electromagnetic pictures of a radiation field can be established.

Scalar and Vector Potential

We define a vector potential $\mathbf{A}$ and a scalar potential ϕ from which the magnetic and electric fields can be derived as

$$\mathbf{H} = \nabla \times \mathbf{A}, \tag{3.42}$$

which satisfies (3.25b), and

$$\mathbf{E} = -\nabla\phi - \frac{1}{c}\frac{\partial \mathbf{A}}{\partial t}, \tag{3.43}$$

which satisfies (3.25c) because the curl of a gradient is zero.

Gauge

As defined by (3.42), $\mathbf{H}$ is unaltered if the gradient of a scalar function is added to $\mathbf{A}$. Then (3.33) shows that $\mathbf{E}$ is also unchanged. But if $\mathbf{E}$ is to be unaltered, (3.43) implies there must be a change in ϕ. This lack of uniqueness means that given a choice of $\mathbf{A}$ and ϕ that satisfy equations (3.25a)–(3.25d), one may make a *gauge transformation* to

$$\mathbf{A}' \equiv \mathbf{A} + \nabla\xi \quad \text{and} \quad \phi' \equiv \phi - \frac{1}{c}\frac{\partial \xi}{\partial t}, \tag{3.44}$$

where $\xi(\mathbf{x}, t)$ is arbitrary, and yet leave the physically measurable variables $\mathbf{E}$ and $\mathbf{H}$ in (3.42) and (3.43) unchanged. That is, Maxwell's equations are *gauge invariant.*

One convenient choice is the *Lorentz gauge*:

$$\nabla\cdot\mathbf{A} = -\frac{1}{c}\frac{\partial \phi}{\partial t}. \tag{3.45}$$

Substituting (3.43) and (3.45) into (3.25a), we find

$$\nabla^2\phi - \frac{1}{c^2}\frac{\partial^2\phi}{\partial t^2} = -4\pi\rho. \tag{3.46}$$

In (3.25d), replacing $\mathbf{H}$ with (3.42), and using (3.34), (3.43), and (3.45), we get

$$\nabla^2\mathbf{A} - \frac{1}{c^2}\frac{\partial^2\mathbf{A}}{\partial t^2} = -\frac{4\pi}{c}\mathbf{j}. \tag{3.47}$$

Equations (3.46) and (3.47) are wave equations for ϕ and $\mathbf{A}$.

Liénard–Wiechert Potentials

Formal solutions for ϕ and $\mathbf{A}$ measured by a distant observer at position $\mathbf{r}$ and time t, resulting from a charge distribution ϱ that has a velocity distribution $\mathbf{v}$, are given by the *Liénard–Wiechert potentials* [575, chapter 14], [838, chapter 19]:

$$\phi(\mathbf{r}, t) = \int_{V'} \frac{\varrho(\mathbf{r}', t')}{|\mathbf{r}' - \mathbf{r}|}\, dV' \tag{3.48}$$

and

$$\mathbf{A}(\mathbf{r}, t) = \frac{1}{c}\int_{V'} \frac{\varrho(\mathbf{r}', t')\mathbf{v}(\mathbf{r}', t')}{|\mathbf{r}' - \mathbf{r}|}\, dV'. \tag{3.49}$$

Here ϱ and $\mathbf{v}$ at the position $\mathbf{r}'$ of the charge are to be evaluated at the *retarded time* $t' = t - |\mathbf{r}' - \mathbf{r}|/c$ in order to take into account the finite propagation speed of the electromagnetic field from $\mathbf{r}'$ to $\mathbf{r}$. If $\varrho(\mathbf{r}', t')$ and $\mathbf{v}(\mathbf{r}', t')$ are known continuous functions of $\mathbf{r}'$ and t', we can calculate ϕ and $\mathbf{A}$ from (3.48) and (3.49).[3] Once ϕ and $\mathbf{A}$ are known we can derive $\mathbf{E}$ and $\mathbf{H}$ from (3.43) and (3.42).

For the important case where the charge is concentrated into a moving point source, e.g., an electron, $\varrho(\mathbf{r}', t')$ is a delta function. Equations (3.48) and (3.49) still apply, but care must be taken in their evaluation. From a fully covariant analysis [575, Eq. 14.8], one finds

$$\phi(\mathbf{r}, t) = \left. \frac{q}{R(1 - \boldsymbol{\beta}\cdot\mathbf{n})} \right|_{t'} \tag{3.50}$$

and

$$\mathbf{A}(\mathbf{r}, t) = \left. \frac{q\boldsymbol{\beta}}{R(1 - \boldsymbol{\beta}\cdot\mathbf{n})} \right|_{t'}, \tag{3.51}$$

where $R \equiv |\mathbf{r} - \mathbf{r}'(t')|$, $\mathbf{n}$ is a unit vector in the direction of $\mathbf{r} - \mathbf{r}'(t')$, and $\boldsymbol{\beta} \equiv \mathbf{v}(t')/c$. In (3.50) and (3.51) the notation $|_{t'}$ means that all quantities in the expression are to be evaluated at the retarded time.

The calculation of the electric and magnetic fields from (3.50) and (3.51) becomes more complicated because derivatives in the observer's frame must be connected with those in the moving frame of the charge. The result is [575, Eq. 14.14]:

$$\mathbf{E}(\mathbf{r}, t) = q \left. \frac{(1 - \beta^2)(\mathbf{n} - \boldsymbol{\beta})}{(1 - \boldsymbol{\beta}\cdot\mathbf{n})^3 R^2} \right|_{t'} + \frac{q}{c} \left. \frac{\mathbf{n} \times [\,(\mathbf{n} - \boldsymbol{\beta}) \times \dot{\boldsymbol{\beta}}\,]}{(1 - \boldsymbol{\beta}\cdot\mathbf{n})^3 R} \right|_{t'} \tag{3.52}$$

and

$$\mathbf{H} = \mathbf{n} \times \mathbf{E}. \tag{3.53}$$

The first term in (3.52) is the *static near field*, which falls off as R^{-2}, and the second term is the *radiation far field*, which falls off as R^{-1}.

For uniform motion of a point charge, the first term in (3.52) can be obtained by Lorentz transformation of the antisymmetric electromagnetic field tensor. To obtain both the second term and the first term for nonuniform motion, a more detailed procedure of connecting derivatives in the two frames is needed; see [575, chapter 14], [838, chapter 19], [905, § 40].

Symmetry, Isotropy, and Plane Waves

In planar symmetry, I is independent of Φ; hence

$$J_\nu(z) = \frac{1}{4\pi} \int_0^{2\pi} d\Phi \int_{-1}^{1} I_\nu(z, \mu)\, d\mu = \frac{1}{2} \int_{-1}^{1} I_\nu(z, \mu)\, d\mu. \tag{3.54}$$

For radial symmetry, replace z in (3.54) with r.

[3] In vacuum, ϱ and $\mathbf{j}$ are zero, so these expressions are not useful. This fact has no impact on the existence of monochromatic wave solutions for $\mathbf{E}$ and $\mathbf{H}$ because we can assume that $\mathbf{E}_{\mathbf{k}0}$ and $\mathbf{H}_{\mathbf{k}0}$, propagating at the point of measurement, were set at a distant source, e.g., a star.

In a one-dimensional medium it is useful to employ *angular moments* (i.e., angular averages) in both the physical and mathematical description of a radiation field. The angular moments are defined as

$$\mathsf{M}_\nu^n \equiv \frac{1}{2}\int_{-1}^{1} I_\nu(\mu)\,\mu^n\,d\mu. \tag{3.55}$$

The *mean intensity J is the angular moment of order zero* of the specific intensity. If the radiation field is *isotropic*, then $I_\nu \equiv J_\nu$.

In the macroscopic picture, a *plane wave* with frequency ν_0 propagating in direction $\mathbf{n}_0$ set by the angles (Θ_0, Φ_0) has a specific intensity

$$I(\mu, \Phi, \nu) = I_{0\nu}\,\delta(\mu - \mu_0)\,\delta(\Phi - \Phi_0)\,\delta(\nu - \nu_0). \tag{3.56}$$

Substitution of (3.56) into (3.19) yields a radiation energy density $E_\nu = I_{0\nu}/c$, which is clearly correct for a plane wave propagating with speed c. Comparing this result with (3.40) from electromagnetic theory, we see that correspondence between the two descriptions is obtained by identifying[4]

$$I_{0\nu} = \frac{cE_{0k}^2}{8\pi}, \tag{3.57}$$

which also gives consistent relations between the Poynting vector and Maxwell stress tensor for a plane wave and their macroscopic counterparts.

Although the preceding discussion deals only with monochromatic plane waves, Maxwell's equations are linear, so an arbitrary angle and frequency distribution of the radiation field can be constructed by summing a superposition of suitably chosen elementary plane waves.

3.3 RADIATION FLUX

Macroscopic Description

The *flux* of radiation $\mathbf{F}_\nu$ is a vector defined such that $\mathbf{F}_\nu \cdot d\mathbf{S}$ gives the *net rate of energy flow* of radiation in the frequency range $(\nu, \nu + d\nu)$ across an arbitrarily oriented surface $d\mathbf{S}$ per unit time and frequency interval. From (3.1), the energy transported in a pencil of radiation moving in direction $\mathbf{n}$ is $d\mathcal{E}_\nu = I_\nu \mathbf{n} \cdot d\mathbf{S}$. Integrating over all solid angles we get the *monochromatic radiation flux*

$$\mathbf{F}_\nu = \oint I_\nu(\mathbf{n})\,\mathbf{n}\,d\Omega \quad [\,\text{erg/cm}^2/\text{Hz/s}\,]. \tag{3.58}$$

Integration over frequency gives the *total radiation flux*

$$\mathbf{F} = \int_0^\infty d\nu \oint I_\nu(\mathbf{n})\,\mathbf{n}\,d\Omega \quad [\,\text{erg/cm}^2/\text{s}\,], \tag{3.59}$$

[4] Note the unfortunate, but customary, double usage of the letter E here.

or in component form,

$$F^i = \int_0^\infty d\nu \oint I_\nu(\mathbf{n})\, n^i \, d\Omega. \tag{3.60}$$

The dyadic form (3.59) emphasizes the vectorial nature of $\mathbf{F}$, but the component form in (3.60) is more useful in writing geometrically and relativistically covariant equations.

In cartesian coordinates, $\mathbf{n}$ is resolved into components (n_x, n_y, n_z) along the right-handed set of unit vectors $\mathbf{i}$, $\mathbf{j}$, and $\mathbf{k}$ on the x, y and z axes. Then

$$(F_x, F_y, F_z) = \left(\oint I\, n_x \, d\Omega, \oint I\, n_y \, d\Omega, \oint I\, n_z \, d\Omega \right), \tag{3.61}$$

where

$$n_x = \sin\Theta\cos\Phi, \quad n_y = \sin\Theta\sin\Phi, \quad \text{and} \quad n_z = \cos\Theta \equiv \mu. \tag{3.62}$$

Here Θ is the polar angle between $\mathbf{n}$ and $\mathbf{k}$, and Φ is the azimuthal angle of the projection of $\mathbf{n}$ onto the (x, y) plane measured counterclockwise from $\mathbf{i}$.

In spherical coordinates, resolving $\mathbf{F}$ along $\hat{\boldsymbol{r}}$, $\hat{\boldsymbol{\theta}}$, and $\hat{\boldsymbol{\phi}}$ we find

$$\begin{pmatrix} F_r \\ F_\theta \\ F_\phi \end{pmatrix} = \begin{pmatrix} \sin\theta\cos\phi & \sin\theta\sin\phi & \cos\theta \\ \cos\theta\cos\phi & \cos\theta\sin\phi & -\sin\theta \\ -\sin\phi & \cos\phi & 0 \end{pmatrix} \begin{pmatrix} F_x \\ F_y \\ F_z \end{pmatrix}. \tag{3.63}$$

The matrix is orthogonal, so

$$\begin{pmatrix} F_x \\ F_y \\ F_z \end{pmatrix} = \begin{pmatrix} \sin\theta\cos\phi & \cos\theta\cos\phi & -\sin\phi \\ \sin\theta\sin\phi & \cos\theta\sin\phi & \cos\phi \\ \cos\theta & -\sin\theta & 0 \end{pmatrix} \begin{pmatrix} F_r \\ F_\theta \\ F_\phi \end{pmatrix}. \tag{3.64}$$

A similar procedure gives the connection between F_x, F_y, F_z and F_r, F_ϕ, F_z in cylindrical coordinates.

Photon Energy and Photon Momentum Flux

We get the same results by describing radiation in terms of photons. The net *photon number flux*, i.e., the net number of photons passing with speed c through a unit area oriented at angle Θ to the beam, per unit time, is

$$N_{\text{photon}}(\nu) = c \oint \varphi(\mathbf{n}, \nu) \cos\Theta \, d\Omega \quad [\,\text{photon/cm}^2\text{/s/Hz}\,]. \tag{3.65}$$

Each photon has energy $h\nu$, so the net *photon energy flux* is

$$\mathbf{F}_{\text{photon}}(\nu) = c\, h\nu \oint \varphi(\mathbf{n}, \nu)\, \mathbf{n} \, d\Omega \quad [\,\text{erg/cm}^2\text{/s/Hz}\,]. \tag{3.66}$$

In view of (3.3), (3.66) is identical to (3.58). Photons of energy $h\nu$ propagating in direction $\mathbf{n}$ have momentum $(h\nu/c)\,\mathbf{n}$. Hence the *photon momentum flux* is

$$\mathbf{M}_{\text{photon}}(\nu) = h\nu \oint \varphi(\mathbf{n}, \nu)\,\mathbf{n}\,d\Omega = \mathbf{F}_{\text{photon}}(\nu)/c \quad [\,\text{erg/cm}^3/\text{Hz}\,]. \tag{3.67}$$

Symmetry, Isotropy, and Plane Waves

As noted in § 3.1, in a one-dimensional atmosphere $I(\mathbf{x}; \nu, \mathbf{n})$ does not depend on the azimuthal angle Φ, so there is a ray-by-ray cancellation of the net energy transport in a plane perpendicular to the outward axis, i.e., $\mathbf{k}$ or $\hat{\mathbf{r}}$. Thus in a planar atmosphere, $F(x; \nu) \equiv 0$ and $F(y; \nu) \equiv 0$, so we have only

$$F_\nu = F(z; \nu) = 2\pi \int_{-1}^{1} I(z; \mu, \nu)\mu\,d\mu. \tag{3.68}$$

Similarly, in a spherically symmetric atmosphere, only F_r is nonzero. For cylindrical symmetry, only $F_\phi \equiv 0$, and both F_r and F_z can, in general, be nonzero. Because only one component of the flux is needed in planar or spherical atmospheres, it is often referred to as "the flux."

In parallel with (3.54) for J_ν, the *Eddington flux* is defined as

$$H_\nu \equiv \left(F_\nu/4\pi\right) = \frac{1}{2}\int_{-1}^{1} I_\nu\,\mu\,d\mu, \tag{3.69}$$

which is the angular moment of order one of the specific intensity. Remembering that $\mathbf{F}$ is a vector, we can extend (3.69) to

$$\mathbf{H}_\nu \equiv \mathbf{F}_\nu/4\pi. \tag{3.70}$$

In terms of macroscopic quantities, the flux associated with the plane wave (3.56) is

$$\mathbf{F}_{0\nu} = \oint I\,\mathbf{n}\,d\Omega = \oint I_{0\nu}\,\delta(\mathbf{n} - \mathbf{n}_0)\,\mathbf{n}\,d\Omega = I_{0\nu}\,\mathbf{n}_0. \tag{3.71}$$

From (3.57), we see that $\mathbf{F}_0$ in (3.71) is identical to $\langle\,\mathbf{S}\,\rangle_{\text{T}}$ given by (3.41) for a plane wave. Again, this result can be generalized to an arbitrary angle and frequency distribution of the radiation field.

Observational Significance

The energy received at the Earth from a star is proportional to the flux F_ν emitted at the star's surface. Assume that the distance D between the star and observer is much larger than the stellar radius r_*, so that all rays received by the observer are essentially parallel. The energy per unit detector area (normal to the line of sight) received from a differential area on the stellar disk seen projected on the sky is

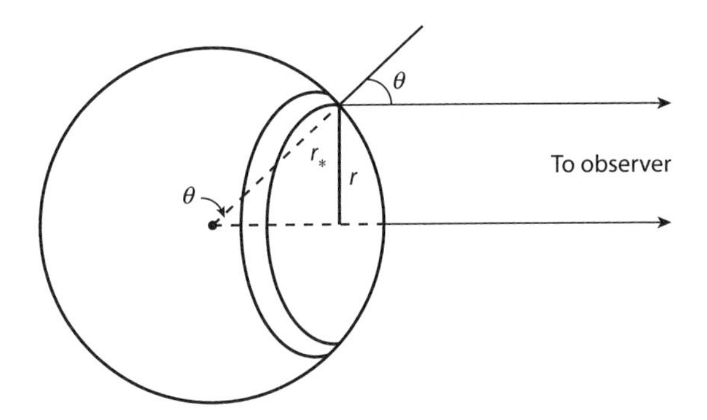

Figure 3.3 Geometry of measurement of stellar flux by a distant observer.

$df_{\rm obs}(\nu) = I_\nu(r_*, \mu)\, d\Omega$, where $d\Omega$ is the solid angle subtended by that projected area, and $I_\nu(r_*, \mu)$ is the specific intensity at the stellar surface.

From the geometry shown in figure 3.3, $r = r_* \sin\theta$, so that the area of a differential annulus on the projected stellar disk is $dS = 2\pi r\, dr = 2\pi r_*^2\, \mu\, d\mu$; therefore, $d\Omega = 2\pi (r_*/D)^2\, \mu\, d\mu$. Integration over the projected disk gives

$$f_{\rm obs}(\nu) = 2\pi \left(\frac{r_*}{D}\right)^2 \int_0^1 I_\nu(r_*, \mu)\, \mu\, d\mu = \left(\frac{r_*}{D}\right)^2 F_\nu(r_*) = \tfrac{1}{4}\, \alpha_*^2 F_\nu(r_*), \qquad (3.72)$$

where α_* is the *angular diameter* of the star. Here we assumed that no radiation is incident upon the surface of the star; i.e., $I_\nu(r_*, -\mu) = 0$.

For *unresolved* objects (e.g., stars), we can measure only the flux. The energy received falls off as the inverse square of the distance because the area of a sphere through which the fixed amount of energy passes increases as D^2. But if a star's angular diameter has been measured interferometrically, then the absolute energy flux at the Earth can be converted to the absolute flux at the star, which allows direct comparison of observed stellar spectral energy distributions with fluxes predicted by model atmospheres.

3.4 RADIATION PRESSURE TENSOR

Macroscopic Description

The *radiation stress tensor*, or *radiation pressure tensor*, $\mathbf{P}_\nu$, is most easily defined in terms of photon momentum flux. Let P_ν^{ij} be the net rate of transport of the jth component of photon momentum of frequency ν through a unit area perpendicular to the ith coordinate axis, per unit time. The number of photons moving with angle cosine n^i crossing a unit area perpendicular to the ith axis is $\varphi_\nu(\mathbf{n})\, c\, n^i$. Each photon has momentum $(h\nu/c)n^j$ in the jth direction. Integrating over solid angles we find

$$P_\nu^{ij} = \oint \varphi_\nu(\mathbf{n})(c\, n^i)(h\nu/c)n^j\, d\Omega, \qquad (3.73)$$

or, from (3.3),

$$P_\nu^{ij} = \frac{1}{c}\oint I_\nu(\mathbf{n})\, n^i\, n^j\, d\Omega \quad [\,\text{erg/cm}^3/\text{Hz}\,]. \tag{3.74}$$

In cartesian coordinates,

$$P_\nu^{xx} = \frac{1}{c}\int_0^{2\pi} d\Phi \int_{-1}^{1} d\mu\, (1-\mu^2)\, \cos^2\Phi\, I_\nu(\mu,\, \Phi), \tag{3.75a}$$

$$P_\nu^{yy} = \frac{1}{c}\int_0^{2\pi} d\Phi \int_{-1}^{1} d\mu\, (1-\mu^2)\, \sin^2\Phi\, I_\nu(\mu,\, \Phi), \tag{3.75b}$$

$$P_\nu^{zz} = \frac{1}{c}\int_0^{2\pi} d\Phi \int_{-1}^{1} d\mu\, \mu^2\, I_\nu(\mu,\, \Phi), \tag{3.75c}$$

$$P_\nu^{xy} = \frac{1}{c}\int_0^{2\pi} d\Phi \int_{-1}^{1} d\mu\, (1-\mu^2)\, \sin\Phi\cos\Phi\, I_\nu(\mu,\, \Phi), \tag{3.75d}$$

$$P_\nu^{xz} = \frac{1}{c}\int_0^{2\pi} d\Phi \int_{-1}^{1} d\mu\, \mu\, (1-\mu^2)^{\frac{1}{2}}\, \cos\Phi\, I_\nu(\mu,\, \Phi), \tag{3.75e}$$

$$P_\nu^{yz} = \frac{1}{c}\int_0^{2\pi} d\Phi \int_{-1}^{1} d\mu\, \mu\, (1-\mu^2)^{\frac{1}{2}}\, \sin\Phi\, I_\nu(\mu,\, \Phi). \tag{3.75f}$$

In a spherically symmetric medium, equations (3.75) apply with z in the superscripts replaced by r; x replaced by θ; and y replaced by ϕ. Components of the stress tensor in cylindrical coordinates are most easily obtained by use of the standard transformation rules of tensor analysis.

P_ν^{ij} is symmetric in i and j, so in 1D, 2D, and 3D media, there are one, three, and six nonredundant components, respectively. Note that the trace of P_ν^{ij}, i.e., the sum of its diagonal elements $\sum_i P_\nu^{ii} \equiv E_\nu$, is a scalar; hence it is independent of the coordinate system. In dyadic notation,

$$\mathsf{P}_\nu = \frac{1}{c}\oint I_\nu(\mathbf{n})\, \mathbf{n}\,\mathbf{n}\, d\Omega. \tag{3.76}$$

Integrating over frequency, we get the total radiation stress

$$\mathsf{P} = \frac{1}{c}\int_0^{\infty} d\nu \oint I_\nu(\mathbf{n})\, \mathbf{n}\,\mathbf{n}\, d\Omega \quad [\,\text{erg/cm}^3\,]. \tag{3.77}$$

Again, the dyadic form in (3.76) and (3.77) emphasizes the tensorial nature of P, but the component form in (3.75) is more useful in writing covariant equations.

Symmetry, Isotropy, and Plane Waves

In Eddington's notation, the *angular moment of order two* over angle of the specific intensity in a planar atmosphere is

$$K_\nu \equiv \frac{1}{2}\int_{-1}^{1} I_\nu\, \mu^2\, d\mu = \frac{c}{4\pi} P_\nu. \tag{3.78}$$

In parallel with (3.54) and (3.70), we define

$$\mathbf{K}_\nu = \frac{1}{4\pi} \oint I_\nu(\mathbf{n})\, \mathbf{n}\,\mathbf{n}\, d\Omega. \tag{3.79}$$

or

$$K_\nu^{ij} = \frac{1}{4\pi} \oint I_\nu(\mathbf{n})\, n^i\, n^j\, d\Omega = \frac{c}{4\pi} P_\nu^{ij}. \tag{3.80}$$

In one-dimensional symmetry, I is independent of Φ, so from (3.77) we have

$$\mathbf{P} = \begin{pmatrix} P^{xx} & 0 & 0 \\ 0 & P^{yy} & 0 \\ 0 & 0 & P^{zz} \end{pmatrix} = \begin{pmatrix} \frac{1}{2}(E-P) & 0 & 0 \\ 0 & \frac{1}{2}(E-P) & 0 \\ 0 & 0 & P \end{pmatrix}$$

$$\equiv \begin{pmatrix} P & 0 & 0 \\ 0 & P & 0 \\ 0 & 0 & P \end{pmatrix} - \frac{1}{2} \begin{pmatrix} 3P-E & 0 & 0 \\ 0 & 3P-E & 0 \\ 0 & 0 & 0 \end{pmatrix}. \tag{3.81}$$

For isotropic radiation, say deep in a stellar interior, $K = \frac{1}{3}J$, or in terms of dynamical quantities, $P \equiv \frac{1}{3}E$; hence by (3.81) the radiation pressure tensor is diagonal and isotropic:

$$\mathbf{P} = \begin{pmatrix} P & 0 & 0 \\ 0 & P & 0 \\ 0 & 0 & P \end{pmatrix}. \tag{3.82}$$

Thus in the isotropic limit we can replace $\mathbf{P}$ by a single *scalar* hydrostatic pressure, p_{rad}, often called "the" radiation pressure.

Near the surface of a star, radiation propagates preferentially outward, so the integral (3.78) over μ for K weights $I(\mu \approx 1)$ more heavily than it does in the integral (3.15) for J; hence $K > \frac{1}{3}J$. Thus at the surface P^{zz} is greater than either P^{xx} or P^{yy}. In the extreme limit of *outward streaming*, i.e., for an outgoing plane wave with $I(\mu) = I_0\,\delta(\mu - 1)$, $J \equiv H \equiv K$. Then $P \equiv E$ so that from (3.81) $P_0^{xx} = P_0^{yy} = 0$, and $P_0^{zz} = E$.

More generally, using (3.57) for a plane wave traveling in direction $\mathbf{n}_0$,

$$P_{0\nu}^{ij} = \frac{1}{c} \oint I_{0\nu}\, \delta\,(\mathbf{n} - \mathbf{n}_0)\, n^i\, n^j\, d\Omega = \frac{I_{0\nu}\, n_0^i\, n_0^j}{c} = \frac{E_{0k}^2\, n_0^i\, n_0^j}{8\pi}. \tag{3.83}$$

Electromagnetic Description: Maxwell Stress Tensor

In electromagnetic theory, radiation stress is described by the *Maxwell stress tensor* [575, p. 239], defined as

$$P_{\mathrm{M}}^{ij} \equiv \frac{1}{4\pi} \left[\tfrac{1}{2}\delta_{ij}(E^2 + H^2) - E_i E_j - H_i H_j \right] \tag{3.84}$$

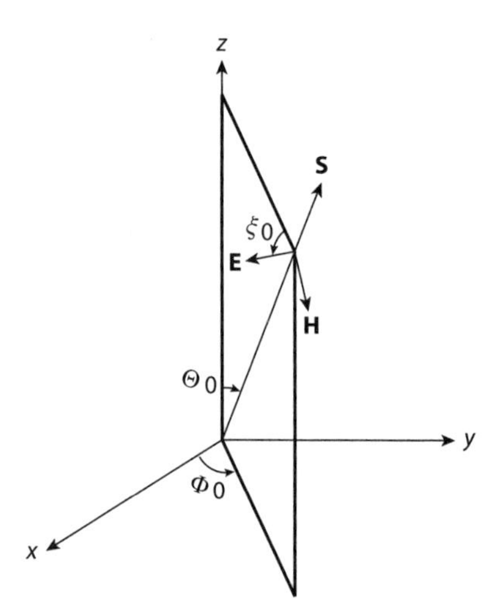

Figure 3.4 A plane electromagnetic wave (**E**, **H**) propagates along the Poynting vector **S**; ξ_0 is the angle of rotation of **E** around **S**, out of the plane defined by $\mathbf{n}_0$ and $\hat{\mathbf{k}}$.

where δ_{ij} is the Kronecker delta function. This expression for P_{M}^{ij} must be identical to the stress tensor given by macroscopic theory for a plane wave, (3.83). Let the electromagnetic field be such that the Poynting vector **S** lies along $\mathbf{n}_0$ in the direction (Θ_0, Φ_0). The polarization of the wave is set by ξ_0, the angle of rotation of **E** around **S** from the plane containing $\mathbf{n}_0$ and $\hat{\mathbf{k}}$; see figure 3.4. Then

$$E_x = E_0 (\sin\xi_0 \sin\Phi_0 - \cos\xi_0 \cos\Phi_0 \cos\Theta_0), \tag{3.85a}$$

$$E_y = -E_0 (\sin\xi_0 \cos\Phi_0 + \cos\xi_0 \sin\Phi_0 \cos\Theta_0), \tag{3.85b}$$

$$E_z = E_0 \cos\xi_0 \sin\Theta_0. \tag{3.85c}$$

Substitution of equations (3.85) and the corresponding equations for **H** into (3.84) gives the components of $\mathbf{P}_{\mathrm{M}}$. For example,

$$P_{\mathrm{M}}^{zz} = \left[\tfrac{1}{2} (E^2 + H^2) - E_z^2 - H_z^2 \right] / 4\pi \tag{3.86}$$

$$= E_0^2 (1 - \sin^2\Theta_0)/4\pi = E_0^2 \cos^2\Theta_0/4\pi. \tag{3.87}$$

Averaging over a cycle, $\langle P_{\mathrm{M}}^{zz} \rangle_{\mathrm{T}} = (E_0 \cos\Theta_0)^2/8\pi$, which equals P_0^{zz} given by macroscopic theory in (3.83). Note that the result is independent of ξ_0.

3.5 ★ TRANSFORMATION PROPERTIES OF I, E, F, P

Specific Intensity

In § 3.1 we discussed the *spatial* invariance of the specific intensity, in the absence of material, in a fixed frame. Complications arise when the material that produces (and interacts with) the radiation field moves with a general velocity field $\mathbf{v}(\mathbf{x}, t)$ relative to that frame. Then we need transformation laws between the *laboratory frame* and the set of noninertial *fluid frames* embedded in the moving material. *Lorentz transformations* can be used to account for the aberration in direction and Doppler shift in energy of individual photons between moving frames. (One invokes conservation of photon four-momentum under Lorentz transformation.) However, as we emphasized in § 3.1, Lorentz transformation does not apply to the macroscopic specific intensity $I(\mathbf{x}, t; \mathbf{n}, \nu)$ because it is a *distribution function* in four-dimensional spacetime and a set of three-dimensional phase spaces.

We also mentioned there, but did not prove, that the photon distribution function $f_R(\mathbf{x}, t; \mathbf{p})$ is a relativistic invariant, so it remains unchanged in all frames. In Appendix B (see also [746, p. 153], [877, p. 270]) we use *Thomas transformations* [1074] to prove the invariance of f_R between two frames in relative motion and to account for all velocity effects (classically: aberration, advection, and Doppler shifts) on both the radiation field and the material properties (absorption, emission, scattering coefficients) that determine it.

Radiation Energy Density, Flux, and Stress Tensor

The frequency-dependent radiation energy density E_ν and mean intensity J_ν are scalars (hence invariants) in three-dimensional space. Likewise, in three-space, the frequency-dependent flux $\mathbf{F}_\nu$ is a vector and the frequency-dependent radiation pressure P_ν is a tensor. They can be transformed from one set of coordinates to another by the standard rules of tensor analysis. Their spatial derivatives in curvilinear coordinates can be computed with covariant derivatives appropriate for the metric of the coordinate system [1068]. But none of these quantities are *world scalars* in four-dimensional spacetime, because they contain residual information about the distribution of photons in phase space, i.e., their distribution over frequency.

On the other hand, in four-dimensional spacetime the angle-frequency-integrated radiation energy density E and mean intensity J *are* world scalars. They can be combined with the frequency-integrated flux vector $\mathbf{F}$ and radiation pressure tensor P to form the *four-dimensional radiation stress-energy tensor*, which is Lorentz transformable.

Radiation Stress–Energy Tensor

In a given frame, the four-dimensional radiation stress–energy tensor used in radiation hydrodynamics is

$$\mathsf{R} = \begin{pmatrix} E & c^{-1}\mathbf{F}^{\mathsf{T}} \\ c^{-1}\mathbf{F} & \mathsf{P} \end{pmatrix}. \tag{3.88}$$

In this 4×4 matrix the indices of $R^{\alpha\beta}$ run over $(0, 1, 2, 3)$; 0 is a *time-like* component, and $(1, 2, 3)$ are *space-like* components. R^{00} is the radiation energy density. The flux

F occupies rows (1, 2, 3) of the 0th column, and its transpose occupies columns (1, 2, 3) of the 0th row. The radiation pressure tensor **P** fills the 3×3 matrix R^{ij} for $(i, j = 1, 2, 3)$. **R** is covariant for three-space coordinate transformations and Lorentz transformations in spacetime.

3.6 QUANTUM THEORY OF RADIATION IN VACUUM

Hamiltonian

In § 3.2 we saw that the classical electromagnetic field can be represented by an ensemble of plane waves, each propagating in a direction **k** as a harmonic oscillation having a polarization α and circular frequency $\omega_k = c\,|\mathbf{k}|$. Dirac realized [280, 286] that a direct route to quantization of the radiation field is to associate the energy in each electromagnetic mode with an eigenstate of a set of *quantized harmonic oscillators*, each of which has an integer number of photons. The Hamiltonian operator for a quantized one-dimensional harmonic oscillator is

$$\hat{H}_{\mathbf{k}} = \tfrac{1}{2}(\hat{p}_{\mathbf{k}}^2 + \omega_k^2 \hat{q}_{\mathbf{k}}^2). \tag{3.89}$$

Here $\hat{q}_{\mathbf{k}}$ is the *position operator*, and $\hat{p}_{\mathbf{k}} \equiv -i\hbar(\partial/\partial q_{\mathbf{k}})$ is the *momentum operator*. They obey the commutation rule

$$[\,\hat{q}_{\mathbf{k}}, \hat{p}_{\mathbf{k}}\,] \equiv \hat{q}_{\mathbf{k}}\hat{p}_{\mathbf{k}} - \hat{p}_{\mathbf{k}}\hat{q}_{\mathbf{k}} = i\hbar. \tag{3.90}$$

Inasmuch as the commutator is nonzero, $\hat{p}$ and $\hat{q}$ cannot be specified simultaneously [275, 428, 946].

Creation and Destruction Operators

Define the *creation operator* $\hat{a}_{\mathbf{k}}^{\dagger}$ (or *raising operator*) and the *destruction operator* $\hat{a}_{\mathbf{k}}$ (or *lowering operator*)[5] as

$$\hat{a}_{\mathbf{k}}^{\dagger} \equiv (\hat{p}_{\mathbf{k}} + i\omega_k \hat{q}_{\mathbf{k}})/(2\hbar\omega_k)^{1/2}, \tag{3.91}$$

and

$$\hat{a}_{\mathbf{k}} \equiv (\hat{p}_{\mathbf{k}} - i\omega_k \hat{q}_{\mathbf{k}})/(2\hbar\omega_k)^{1/2}. \tag{3.92}$$

The inverse relation between $(\hat{q}_{\mathbf{k}}, \hat{p}_{\mathbf{k}})$ and $(\hat{a}_{\mathbf{k}}^{\dagger}, \hat{a}_{\mathbf{k}})$ is

$$\hat{q}_{\mathbf{k}} \equiv (\hbar\omega_k/2)^{\frac{1}{2}}(\hat{a}_{\mathbf{k}} + \hat{a}_{\mathbf{k}}^{\dagger}) \tag{3.93}$$

and

$$\hat{p}_{\mathbf{k}} \equiv -i(\hbar/2\omega_k)^{\frac{1}{2}}(\hat{a}_{\mathbf{k}} - \hat{a}_{\mathbf{k}}^{\dagger}). \tag{3.94}$$

[5] The appropriateness of these names will become obvious below.

The commutator of $\hat{a}_{\mathbf{k}}$ and $\hat{a}_{\mathbf{k}}^{\dagger}$ is

$$[\,\hat{a}_{\mathbf{k}},\ \hat{a}_{\mathbf{k}}^{\dagger}\,] \equiv \hat{a}_{\mathbf{k}}\hat{a}_{\mathbf{k}}^{\dagger} - \hat{a}_{\mathbf{k}}^{\dagger}\hat{a}_{\mathbf{k}} = 1, \tag{3.95}$$

and by induction

$$[\,\hat{a}_{\mathbf{k}}, \hat{a}_{\mathbf{k}}^{\dagger n}] = \hat{a}_{\mathbf{k}}\hat{a}_{\mathbf{k}}^{\dagger n} - \hat{a}_{\mathbf{k}}^{\dagger n}\hat{a}_{\mathbf{k}} = n\hat{a}^{\dagger n-1}. \tag{3.96}$$

Using (3.89)–(3.96) it is easy to see that

$$\hat{a}_{\mathbf{k}}\hat{a}_{\mathbf{k}}^{\dagger} = (\hat{H}_{\mathbf{k}} + \tfrac{1}{2}\hbar\omega_k)/(\hbar\omega_k) \tag{3.97}$$

and

$$\hat{a}_{\mathbf{k}}^{\dagger}\hat{a}_{\mathbf{k}} = (\hat{H}_{\mathbf{k}} - \tfrac{1}{2}\hbar\omega_k)/(\hbar\omega_k). \tag{3.98}$$

By adding (3.97) and (3.98) we obtain the *Hamiltonian operator*

$$\hat{H}_{\mathbf{k}} = \tfrac{1}{2}\hbar\omega_k(\hat{a}_{\mathbf{k}}^{\dagger}\hat{a}_{\mathbf{k}} + \hat{a}_{\mathbf{k}}\hat{a}_{\mathbf{k}}^{\dagger}). \tag{3.99}$$

Clearly, $\hat{H}_{\mathbf{k}}$ is self-adjoint, i.e., $\hat{H} \equiv \hat{H}^{\dagger}$, and hence *Hermitian*. Hermitian operators have real eigenvalues and correspond to measurable dynamical quantities.

Eigenkets and Eigenvalues

If $|\,n_{\mathbf{k}}\rangle$ is an *eigenfunction* of $\hat{H}_{\mathbf{k}}$, whose energy *eigenvalue* is E_{nk}, then

$$\hat{H}_{\mathbf{k}}|\,n_{\mathbf{k}}\rangle = E_{nk}|\,n_{\mathbf{k}}\rangle. \tag{3.100}$$

From (3.98),

$$\hat{H}|\,n_{\mathbf{k}}\rangle = \hbar\omega_k(\hat{a}_{\mathbf{k}}^{\dagger}\hat{a}_{\mathbf{k}} + \tfrac{1}{2})|\,n_{\mathbf{k}}\rangle = E_{nk}|\,n_{\mathbf{k}}\rangle. \tag{3.101}$$

Multiplying (3.101) by $\hat{a}_{\mathbf{k}}^{\dagger}$ from the left, we have

$$\hbar\omega_k(\hat{a}_{\mathbf{k}}^{\dagger}\hat{a}_{\mathbf{k}}^{\dagger}\hat{a}_{\mathbf{k}} + \tfrac{1}{2}\hat{a}_{\mathbf{k}}^{\dagger})|\,n_{\mathbf{k}}\rangle = E_{nk}\hat{a}_{\mathbf{k}}^{\dagger}|\,n_{\mathbf{k}}\rangle, \tag{3.102}$$

which, using (3.95), becomes

$$\hbar\omega_k(\hat{a}_{\mathbf{k}}^{\dagger}\hat{a}_{\mathbf{k}}\hat{a}_{\mathbf{k}}^{\dagger} - \hat{a}_{\mathbf{k}}^{\dagger} + \tfrac{1}{2}\hat{a}_{\mathbf{k}}^{\dagger})|\,n_{\mathbf{k}}\rangle = E_{nk}\hat{a}_{\mathbf{k}}^{\dagger}|\,n_{\mathbf{k}}\rangle. \tag{3.103}$$

Transfer the second term on the left-hand side to the right; then

$$\hbar\omega_k(\hat{a}_{\mathbf{k}}^{\dagger}\hat{a}_{\mathbf{k}} + \tfrac{1}{2})\hat{\mathbf{a}}_{\mathbf{k}}^{\dagger}|\,n_{\mathbf{k}}\rangle = (E_{nk} + \hbar\omega_k)\hat{a}_{\mathbf{k}}^{\dagger}|\,n_{\mathbf{k}}\rangle, \tag{3.104}$$

or from (3.98),

$$\hat{H}_{\mathbf{k}}\hat{a}_{\mathbf{k}}^{\dagger}|\,n_{\mathbf{k}}\rangle = (E_{nk} + \hbar\omega_k)\hat{a}_{\mathbf{k}}^{\dagger}|\,n_{\mathbf{k}}\rangle. \tag{3.105}$$

Equation (3.105) says that when the creation operator $\hat{a}^{\dagger}_{\mathbf{k}}$ acts on the ket $|\,n_{\mathbf{k}}\rangle$ *it produces a new ket*

$$\hat{a}^{\dagger}_{\mathbf{k}}|\,n_{\mathbf{k}}\rangle \equiv |\,n_{\mathbf{k}}+1\rangle, \tag{3.106}$$

which has an energy eigenvalue $E_{nk}+\hbar\omega_k$. That is, it raises the oscillator to its next higher energy level or *creates a photon* of energy $\hbar\omega_k$. Repetition of this process produces an infinite ladder of kets with energies separated by $\hbar\omega_k$.

By a similar analysis,

$$\hat{H}_{\mathbf{k}}\hat{a}_{\mathbf{k}}|\,n_{\mathbf{k}}\rangle = (E_{nk}-\hbar\omega_k)\hat{a}_{\mathbf{k}}|\,n_{\mathbf{k}}\rangle. \tag{3.107}$$

Thus when the destruction operator $\hat{a}_{\mathbf{k}}$ acts on the ket $|\,n_{\mathbf{k}}\rangle$ *it produces a new ket*

$$\hat{a}_{\mathbf{k}}|\,n_{\mathbf{k}}\rangle \equiv |\,n_{\mathbf{k}}-1\rangle, \tag{3.108}$$

which has an energy eigenvalue $E_{nk}-\hbar\omega_k$. That is, it *destroys a photon* of energy $\hbar\omega_k$ in the radiation field.

Standard Ket

Repeated application of the destruction operator on any eigenket produces a series of kets that terminates with a ket $|\,0\rangle$, which contains no photons. This ket is called the *standard ket* because any other ket can be expressed as a power of $\hat{a}^{\dagger}$ times $|\,0\rangle$. For brevity we omit the mode indicator **k** on operators and kets in equations (3.109) through (3.126).

Using the Hamiltonian operator in the form of (3.101),

$$\hat{H}|\,0\rangle \equiv E_0|\,0\rangle = (\hat{a}^{\dagger}\hat{a}+\tfrac{1}{2}\hbar\omega)|\,0\rangle = \tfrac{1}{2}\hbar\omega|\,0\rangle \tag{3.109}$$

because $\hat{a}|\,0\rangle \equiv 0$. Starting from the standard ket, we can form a sequence of orthogonal kets

$$|\,0\rangle, \qquad \hat{a}^{\dagger}|\,0\rangle, \qquad \hat{a}^{\dagger 2}|\,0\rangle, \qquad \hat{a}^{\dagger 3}|\,0\rangle, \ \ldots, \tag{3.110}$$

or

$$|\,0\rangle, \qquad |\,1\rangle, \qquad |\,2\rangle, \qquad |\,3\rangle, \ \ldots, \tag{3.111}$$

which have eigenvalues

$$\tfrac{1}{2}\hbar\omega, \qquad \tfrac{3}{2}\hbar\omega, \qquad \tfrac{5}{2}\hbar\omega, \qquad \tfrac{7}{2}\hbar\omega, \ \ldots. \tag{3.112}$$

That is,

$$E_n = (n+\tfrac{1}{2})\hbar\omega, \tag{3.113}$$

$n = 0, 1, \ldots$, a result first derived by Schrödinger [972], [973, eq. 25′] by a different analysis.

$E_0 \equiv \frac{1}{2}\hbar\omega$ is called the *zero-point energy* of a quantized oscillator. Its existence is an unavoidable consequence of Heisenberg's Uncertainty Principle. E_0 could be zero only if the expectation values $\langle q^2 \rangle$ and $\langle p^2 \rangle$ were simultaneously zero, in violation of the uncertainty principle. In contrast to a classical oscillator, a quantized oscillator in its lowest energy level is not at rest.

Number Operator

From (3.101) and (3.113),

$$\hat{\mathbf{n}} \big| n \big\rangle \equiv \hat{\boldsymbol{a}}^{\dagger}\hat{\boldsymbol{a}} \big| n \big\rangle = n \big| n \big\rangle. \tag{3.114}$$

The operator $\hat{\mathbf{n}}$ is called the *number operator*; it gives the occupation number of photons in level n. Using (3.101) one finds that $\hat{\boldsymbol{H}}$ and $\hat{\mathbf{n}}$ commute; hence $\big| n \big\rangle$ is simultaneously an eigenket of both operators. In contrast, $\hat{\boldsymbol{a}}$ and $\hat{\boldsymbol{a}}^{\dagger}$ do not commute, and neither commutes with $\hat{\boldsymbol{H}}$; *those operators, acting by themselves, do not produce observable quantities.*

Orthogonality and Normalization

Eigenkets with different eigenvalues are orthogonal. $\hat{\boldsymbol{H}}$ is Hermitian, so

$$\big\langle E' \,\big|\, \hat{\boldsymbol{H}} E \big\rangle \equiv \big\langle \hat{\boldsymbol{H}} E' \,\big|\, E \big\rangle. \tag{3.115}$$

or

$$E \big\langle E' \,\big|\, E \big\rangle = E' \big\langle E' \,\big|\, E \big\rangle. \tag{3.116}$$

Therefore,

$$(E - E')\big\langle E' \,\big|\, E \big\rangle = 0, \tag{3.117}$$

so if $E' \neq E$, $\big\langle E' \,\big|\, E \big\rangle \equiv 0$, i.e., $\big| E \big\rangle$ and $\big| E' \big\rangle$ are orthogonal.

The kets in (3.106) and (3.108) are unnormalized. To normalize them such that

$$\big\langle n+1 \,\big|\, n+1 \big\rangle = \big\langle n \,\big|\, n \big\rangle = \big\langle n-1 \,\big|\, n-1 \big\rangle = 1, \tag{3.118}$$

introduce a normalization constant such that

$$\hat{\boldsymbol{a}} \big| n \big\rangle = C_n \big| n-1 \big\rangle. \tag{3.119}$$

Take the inner product of (3.119) with its Hermitian conjugate and use (3.114); then

$$\big\langle n \,\big|\, \hat{\boldsymbol{a}}^{\dagger}\hat{\boldsymbol{a}} \big| n \big\rangle \equiv n = \big\langle n-1 \,\big|\, C_n^* C_n \big| n-1 \big\rangle, \tag{3.120}$$

which shows that $|C_n| = n^{1/2}$, and

$$\hat{\boldsymbol{a}} \big| n \big\rangle = n^{1/2} \big| n-1 \big\rangle. \tag{3.121}$$

Similarly,

$$\hat{a}^{\dagger}\big|\,n\big\rangle = (n+1)^{1/2}\big|\,n+1\big\rangle. \tag{3.122}$$

Equations (3.121) and (3.122) will be of great importance later.

Alternatively, the ladder of kets $|\,1\rangle$, $|\,2\rangle$, $|\,3\rangle \ldots$ can be normalized. From (3.96) and (3.109) it follows that

$$\hat{a}\hat{a}^{\dagger n}\big|\,0\big\rangle = n\hat{a}^{\dagger n-1}\big|\,0\big\rangle, \tag{3.123}$$

which, using (3.96), implies that the square of the length of $\hat{a}^{\dagger n}|\,0\rangle$ is

$$\begin{aligned}\big\langle 0\,\big|\,\hat{a}^{n}\hat{a}^{\dagger n}\big|\,0\big\rangle &= \big\langle 0\,\big|\,\hat{a}^{n-1}(\hat{a}\hat{a}^{\dagger n})\big|\,0\big\rangle \\ &= \big\langle 0\,\big|\,\hat{a}^{n-1}(\hat{a}^{\dagger n}\hat{a} + n\hat{a}^{n-1})\big|\,0\big\rangle = n\big\langle 0\,\big|\,\hat{a}^{n-1}\hat{a}^{\dagger n-1}\big|\,0\big\rangle\end{aligned} \tag{3.124}$$

because $\hat{a}|\,0\rangle \equiv 0$. Then, by induction,

$$\big\langle 0\,\big|\,\hat{a}^{n}\hat{a}^{\dagger\, n}\big|\,0\big\rangle = n! \tag{3.125}$$

and

$$\big|\,n\big\rangle = (n!)^{-1/2}\hat{a}^{\dagger\, n}\big|\,0\big\rangle \tag{3.126}$$

will be normalized if $|\,0\rangle$ is. The factor $(n!)^{-1/2}$ accounts for the $n!$ possible permutations of the n identical photons in state $|\,n\rangle$.

Ensemble of Photons

The ability, mentioned in § 3.2, to superpose classical electromagnetic fields to construct an arbitrary radiation field implies that photons do not interact.[6] Then the state vector for the set of all photons in mode **k** can be written as the product of their individual kets:

$$\big|\,\{n_{\mathbf{k}}\}\big\rangle \equiv \prod_{n=1}^{\infty}\big|\,n_{\mathbf{k}}\big\rangle. \tag{3.127}$$

Each ket $|\,n_{\mathbf{k}}\rangle$ has an energy $E_{nk} = (n_{\mathbf{k}} + \frac{1}{2})\hbar\omega_k$. The state vector for the entire radiation field can also be written as the product of the state vectors of all modes, i.e., as $\prod_{\mathbf{k}}|\,\{n_{\mathbf{k}}\}\rangle$. If the state vectors of the individual modes are normalized, the state vector of the ensemble is normalized. Making use of commutation of the

[6] Actually, quantum electrodynamics shows there is a very small cross section for two photons to create an electron-positron pair, which are virtual particles for photon energies that sum to $\lesssim 1$ MeV and can be real particles at higher photon energies. They immediately annihilate one another, releasing two different photons. That high-energy process is outside the domain of this book; see [109, § 127].

number and Hamiltonian operators, and normalization and orthogonality of the eigenfunctions, the total energy in the system is

$$\prod_{\mathbf{k}} \left\langle \{n_{\mathbf{k}}\} \middle| \sum_{\mathbf{k},n} \hat{\boldsymbol{H}}_{n_{\mathbf{k}}} \middle| \prod_{\mathbf{k}} \{n_{\mathbf{k}}\} \right\rangle = \sum_{\mathbf{k}} \sum_{n=1}^{\infty} \langle n_{\mathbf{k}} \,|\, \hat{\boldsymbol{H}}_{n_{\mathbf{k}}} \,|\, n_{\mathbf{k}} \rangle$$

$$= \sum_{\mathbf{k}} \sum_{n=1}^{\infty} \langle n_{\mathbf{k}} \,|(\hat{\mathbf{n}}_{n_{\mathbf{k}}} + \tfrac{1}{2})|\, n_{\mathbf{k}} \rangle = \sum_{\mathbf{k}} \sum_{n=1}^{\infty} (n_{\mathbf{k}} + \tfrac{1}{2})\hbar\omega_k. \quad (3.128)$$

Equation (3.128) applies to a general radiation field. But it leaves the occupation numbers $n_{\mathbf{k}}$ of each mode unspecified. These occupation numbers are determined by the interaction of the radiation field with material, i.e., by absorption, emission, and scattering of photons. Use of creation and destruction operators allowed us to determine the eigenvalues of the eigenfunctions of the radiation field, but not the numbers of photons in those states.

Only in the special case of radiation in strict thermodynamic equilibrium in a hohlraum is there a unique set of occupation numbers; see chapter 4. But in a stellar atmosphere, which has an open boundary, the occupation numbers of photons of different frequencies are the result of *non-equilibrium of transfer* of radiation through the atmosphere and must be calculated.

Chapter Four

Statistical Mechanics of Matter and Radiation

The physical conditions in the atmospheres of stars vary greatly. On the main sequence, the effective temperatures of stellar atmospheres range over a factor of about 25, and about an order of magnitude more if we include *bare stellar cores*, which are the remains of highly evolved stars that have had almost all of their envelopes stripped away in dense stellar winds, leaving very hot remnants that emit soft X-rays. The particle number density in main-sequence photospheres typically range between 10^{10} and 10^{17} cm^{-3}. Stellar spectra show features characteristic of a strongly ionized plasma in early-type stars, mostly neutral gas in the Sun, and diatomic and triatomic molecules (even "dust grains") in late spectral types.

In the limit of LTE, the state of material is determined by collisional processes; radiation is trapped in an opaque region having almost constant material properties. In this limit one can apply statistical mechanics to describe the properties of both material and radiation. But the outer, observable layers of stellar atmospheres are far from being in thermodynamic equilibrium: the properties of the material can change greatly over a photon mean free path, and radiation at frequencies with low opacity can escape freely. Therefore, estimates of the physical conditions in the visible layers of stars based on the assumption of LTE can be seriously in error, and a more general methodology is required.

To obtain physically meaningful results one must solve transfer equations (chapter 11), which determine the radiation field produced by the material's absorption and emission coefficients (§ 5.1), simultaneously with non-equilibrium kinetic equations (chapter 9), which determine the state of the material, consistent with that radiation field. The development of such techniques occupies the remainder of this book. Statistical mechanics provides many fundamental relationships needed in that work. In addition, it applies in the deeper layers of stellar envelopes and in the interiors of stars, where the overlying material is, in fact, opaque, and photons are trapped in layers in which changes in temperature and density over a photon mean free path are small. It also can be applied to many laboratory experiments with plasmas.

4.1 THERMODYNAMIC EQUILIBRIUM

Detailed Balance

Both kinetic theory and statistical mechanics show that *thermodynamic equilibrium* (TE) in a gaseous system is produced by random collisions among the particles in the gas [232, 696]. A basic requirement for thermodynamic equilibrium is that *all processes are balanced exactly by their inverses.* A system composed of $\mathcal{N}$ identical material particles plus radiation, which has a total energy $\mathcal{E}$, is confined in a fixed volume $\mathcal{V}$, and is completely isolated from the external environment (i.e., in a hohlraum), eventually evolves toward thermodynamic equilibrium. In that condition, the phase-space distribution functions of both particles and photons are unique and constant in space and time. They can be driven away from their equilibrium distributions when detailed balance in all transitions is not achieved. Einstein [309, 310] used this powerful principle to establish the connection between upward and downward radiative transition probabilities in spectral lines and the existence of the phenomenon of induced (or stimulated) emission; see chapter 5. We will point out its use in other contexts.

Thermodynamic Probability

An isolated system in TE is in its most probable state; if it were not, some process would act to lead it to a more probable state. Put differently, the system is in a state of maximum disorder; otherwise, a process driven by some order in the system would move it to a more disordered state.

Assume the particles are weakly interacting (e.g., do not bind into molecules) and experience the same background potential $V(\mathbf{x})$, that each has a single-particle ket $|\,i\rangle$ with energy ϵ_i, and that they are distributed into *cells* that contain $n_i(\epsilon_i)$ particles that have energies in the ranges $\epsilon_1 \pm \Delta\epsilon_1$; $\epsilon_2 \pm \Delta\epsilon_2$; ... ; $\epsilon_N \pm \Delta\epsilon_N$. The energies ϵ_i are measured from a ground-state energy ϵ_0. Each cell may contain a number of *degenerate levels* (i.e., those having very nearly the same energy), which are grouped into a single state and given a *statistical weight* g_i (the number of degenerate levels). A specific set of the cell populations n_i is called a *macrostate* of the system; a specific distribution of particles over the energy levels within each cell is called a *microstate*.

The *thermodynamic probability* W of a macrostate is proportional to the number of microstates that produce it. We make the assumption that all microstates of a given macrostate are equally likely to occur. The material evolves from one microstate to another by collisions among the particles. We will see below that there is an enormous number of macrostates that give the correct number of particles and correct total energy in the system. The goal is to find the *most probable macrostate* that satisfies the constraints imposed by the given values of $\mathcal{N}$, $\mathcal{E}$, and $\mathcal{V}$.

Relation of Statistical Mechanics to Thermodynamics

In macroscopic thermodynamics the disorder of a system is measured by its *entropy* S. From the discussion above, thermodynamic equilibrium is a state of maximum entropy. The connection between S and W is given by *Boltzmann's law*:

$$S = k \ln W, \tag{4.1}$$

where k is *Boltzmann's constant.* It would take us too far afield to justify (4.1); suffice it to say that it leads to an internally consistent theoretical structure that agrees closely with experiment; see [515, § 7.2 and § 7.3], [771, chapter 17].

4.2 BOLTZMANN STATISTICS

Boltzmann statistics applies to *classical* (i.e., *distinguishable*) particles. With it we can deduce equilibrium distribution functions for particle velocities; atomic excitation and ionization equilibria; and molecular dissociation. It is a limiting case of quantum statistics, valid when the number of particles is much smaller than the number of phase-space cells available to them. In this theory it is assumed that the particles suffer random *elastic collisions* with one another, leading to *equipartition of energy* among them; in other words, they are *thermalized.*

Boltzmann Excitation Equation

Maximum Entropy Distribution Function

Suppose the material is composed of a fixed number $\mathcal{N}$ of identical classical particles. For now assume that the particles have no internal structure (we relax this assumption below). Then the eigenstate of the complete system can be written as the product of single-particle eigenkets:

$$\big|\, 1\, 2 \ldots i \ldots \mathcal{N} \big\rangle = \big|\, \mathbf{x}_1\, \epsilon_1 \big\rangle \big|\, \mathbf{x}_2\, \epsilon_2 \big\rangle \ldots \big|\, \mathbf{x}_i\, \epsilon_i \big\rangle \ldots \big|\, \mathbf{x}_{\mathcal{N}}\, \epsilon_{\mathcal{N}} \big\rangle. \tag{4.2}$$

The first label in the single-particle kets denotes the space and spin coordinates of a particle, and the second denotes its energy. As described above, we divide the total energy range of the system into cells, each having a narrow width $\pm\Delta\epsilon_i$, much smaller than the total energy $\mathcal{E}$, but large enough to include a very large number g_i of states with energy $\approx \epsilon_i$. The number of particles in cell i is its *occupation number* n_i. Acceptable sets $\{n_i\}$ must yield the correct number of particles and total energy:

$$\mathcal{N} = \sum_i n_i \tag{4.3}$$

and

$$\mathcal{E} = \sum_i n_i \epsilon_i. \tag{4.4}$$

To determine the thermodynamic probability of the system, count the number of physically distinct microstates associated with this macrostate, i.e., the number of ways $\mathcal{N}$ distinguishable particles can be distributed into the set of cells. First, there are $\mathcal{N}!$ ways a list of the particles can be ordered. Next, take sets of n_i particles from this list sequentially for all of the cells. Any of the n_i particles in the ith cell can be in any of the g_i states in that cell; hence there are $(g_i)^{n_i}$ possible ways to distribute them into those states. But each of the $n_i!$ permutations of the order in which they are put into those states produces the same microstate. Therefore, the thermodynamic probability $W(\{n_i\})$, i.e., the number of physically distinct microstates associated with this macrostate, i.e., the number of sets $\{n_i\}$ of particles in cell i, is

$$W(\{n_i\}) = \mathcal{N}! \prod_i \left(g_i^{n_i}/n_i!\right). \tag{4.5}$$

Note that $W(\{n_i\})$ is not affected by empty cells because $0! = 1$, and $g_i^0 \equiv 1$.

The assumption that the particles are distinguishable is artificial; quantum mechanically they are not. Therefore, the order in which particles are chosen in the process of sorting them into cells actually does not matter (i.e., cannot be detected experimentally); hence the factor of $\mathcal{N}!$ in the numerator must be discarded.[1] Then the thermodynamic probability of this macrostate with *correct Boltzmann counting* is

$$W(\{n_i\}) = \prod_i \left(g_i^{n_i}/n_i!\right). \tag{4.6}$$

The total thermodynamic probability of the state of the system is the product of the thermodynamic probabilities of the macrostates associated with it; thus

$$\ln W = \sum_i \ln W(\{n_i\}). \tag{4.7}$$

In principle, one could evaluate the sum in (4.7); but that leads to algebraic complications. However, for very large $\mathcal{N}$, one term, $W_{\max}$, dominates all others in the sum, so analysis of it alone will yield the correct result (see [342] for a rigorous justification of this assertion). Hence in (4.7) we replace W with the largest term given by (4.6):

$$W = \left[\prod_i \left(g_i^{n_i}/n_i!\right)\right]_{\max}. \tag{4.8}$$

Then using *Stirling's formula* $\ln x! \approx x \ln x - x + \frac{1}{2}\ln x + \frac{1}{2}\ln 2\pi \cdots \to x \ln x$ (valid for $x \gg 1$) in (4.8) we obtain

$$\ln W^* \approx \ln W_{\max} = \sum_i \left(n_i^* \ln g_i - \ln n_i^*!\right) \approx \sum_i \left[\, n_i^* - n_i^* \ln\left(n_i^*/g_i\right)\right]. \tag{4.9}$$

The superscript * denotes a thermodynamic equilibrium value.

In equilibrium, W^* is at its maximum, so its variation caused by small fluctuations δn_i^* of the occupation numbers is zero. Hence for fixed $\mathcal{N}^*$,

$$\delta W^*/W^* = -\sum_i \delta n_i^* \ln\left(n_i^*/g_i\right) = 0. \tag{4.10}$$

[1] The entropy of a system composed of two different gases originally separated into volumes V_1 and V_2 increases when the two gases are allowed to mix and fill the full volume $V_1 + V_2$. If the correction just mentioned were not made, then (4.5) would predict an increase in entropy even if the gases in the two initial volumes are the same. This absurd result, known as the *Gibbs paradox*, is corrected when the counting procedure is modified as in (4.6); see [499, § 7.6].

The fluctuations (4.10) cannot be arbitrary, but must also be subject to the constraints $\sum_i \delta n_i^* \equiv 0$ (particle conservation) and $\delta \mathcal{E}^* = \sum_i \epsilon_i \delta n_i^* \equiv 0$ (energy conservation). All three requirements can be met by solving for the distribution that satisfies an arbitrary linear combination of them. That is,

$$\sum_i \left[\ln \alpha - \beta \epsilon_i - \ln \left(n_i^* / g_i \right) \right] \delta n_i^* \equiv 0, \tag{4.11}$$

where α and β are yet to be determined.

The δn_i^*'s in (4.11) can now be considered to be independent. The solution of (4.11) for arbitrary δn_i^* is

$$n_i^* = \alpha g_i \exp(-\beta \epsilon_i), \tag{4.12}$$

where α is a normalization coefficient. Using (4.12) in (4.3) we have

$$\mathcal{N}^* = \alpha \sum_i g_i \exp(-\beta \epsilon_i) \equiv \alpha\, \mathcal{U}^*, \tag{4.13}$$

where the *partition function* is defined as

$$\mathcal{U}^* \equiv \sum_i g_i \exp(-\beta \epsilon_i). \tag{4.14}$$

From (4.13) $\alpha \equiv \mathcal{N}^* / \mathcal{U}^*$, so (4.12) becomes

$$n_i^* / \mathcal{N}^* = g_i \exp(-\beta \epsilon_i) \,/\, \mathcal{U}^*, \tag{4.15}$$

and (4.4) becomes

$$\mathcal{E}^* / \mathcal{N}^* = \sum_i \epsilon_i g_i \exp(-\beta \epsilon_i) / \mathcal{U}^*. \tag{4.16}$$

Then using (4.12–4.16) in (4.9), we find

$$\ln W^* = \mathcal{N}^* \left[1 + \ln \left(\mathcal{U}^* / \mathcal{N}^* \right) \right] + \beta \mathcal{E}^*, \tag{4.17}$$

so from (4.1)

$$\mathcal{S}^* = \mathcal{N}^* k \left[1 + \ln \left(\mathcal{U}^* / \mathcal{N}^* \right) \right] + \beta k \mathcal{E}^*. \tag{4.18}$$

A direct connection of (4.12–4.18) can now be made with thermodynamics. Using the general expansion

$$d\mathcal{S} = \left(\frac{\partial \mathcal{S}}{\partial \mathcal{E}} \right)_{\mathcal{V}} d\mathcal{E} + \left(\frac{\partial \mathcal{S}}{\partial \mathcal{V}} \right)_{\mathcal{E}} d\mathcal{V} \tag{4.19}$$

in the first law of thermodynamics $T\, d\mathcal{S} = d\mathcal{E} + p\, d\mathcal{V}$, one sees that $(\partial \mathcal{S} / \partial \mathcal{E})_{\mathcal{V}} \equiv 1/T$. On the other hand, for fixed $\mathcal{N}^*$ the expression $(\partial \mathcal{S} / \partial \mathcal{E})^*_{\mathcal{V}} = k\beta$ can be obtained by direct differentiation of (4.18) with respect to $\mathcal{E}^*$. Thus $\beta \equiv 1/kT$, where T is the *absolute thermodynamic temperature of the material.* This expression for β is to be substituted into (4.12) through (4.18).

Mixture of Materials

To generalize these formulae to a gas having a mixture of chemical elements each having multiple ionization and excitation states, let n^*_{ijl} be the number of particles in excitation state i of ionization stage j of chemical element l, which has an excitation potential ϵ_{ijl} relative to the ground state of that ion and statistical weight g_{ijl}. Let $N^*_{jl} \equiv \sum_i n^*_{ijl}$ be the number of particles of chemical element l in all excitation states i of ionization stage j, and let $N^*_l \equiv \sum_j N^*_{jl} = \sum_j \sum_i n^*_{ijl}$ be the total number of particles of element l. Then (4.15) becomes

$$\left(n^*_{ijl}/N^*_{jl}\right) = g_{ijl} \exp\left(-\epsilon_{ijl}/kT\right) / U^*_{jl}(T), \tag{4.20a}$$

from which we see that

$$\left(n^*_{i'jl}/n^*_{ijl}\right) = \left(g_{i'jl}/g_{ijl}\right) \exp\left[-\left(\epsilon_{i'jl} - \epsilon_{ijl}\right)/kT\right]. \tag{4.20b}$$

Further,

$$E^*_{jl}(T) = \left[N^*_{jl}/U^*_{jl}(T)\right] \sum_i \epsilon_{ijl} g_{ijl} \exp\left(-\epsilon_{ijl}/kT\right) \tag{4.20c}$$

and

$$S^*_{jl}(T) = N^*_{jl} k \left\{1 + \ln\left[U^*_{jl}(T)/N^*_{jl}\right]\right\} + \left[E^*_{jl}(T)/T\right]. \tag{4.20d}$$

Equation (4.20b) is the usual form for the *Boltzmann excitation equation*. In equations (4.20), n and N may denote either total particle numbers or particle densities [cm^{-3}].

The Coulomb potential of an atom's nucleus has infinite range; therefore, an atom or ion nominally can have an infinite number of bound levels, and the sum in the partition function U would diverge because its smallest terms are finite: $\exp(-\epsilon_I/kT)$, where ϵ_I is the atom's ionization potential. In reality, a loosely bound electron in a high level can be so strongly perturbed by neighboring particles that it can become unbound, a processes called *pressure ionization*. To get a rough estimate of the number of levels remaining bound to an atom or ion, let the number density of perturbers be $\mathfrak{n}$. Each perturber occupies an average volume $\frac{4}{3}\pi r_0^3 \sim 1/\mathfrak{n}$. The atomic "radius" of a hydrogenic atom or ion with charge Z in a state with principal quantum number n is $r_n \approx n^2 a_0/Z$, where the *Bohr radius* $a_0 = 5.3 \times 10^{-9}$ cm. Consider an electron to be bound if r_n is no larger than the distance to the nearest perturber, i.e., $r_n \leq r_0$. Then if we set $r_0 = r_n$ we find $n^2 = Z/(\frac{4}{3}\pi\mathfrak{n})^{1/3} a_0$. Taking $\mathfrak{n} \sim 10^{16}$ as a typical photospheric particle density, we find $n_{\max} \approx 24\, Z^{1/2}$. With this phenomenological cutoff, the number of terms in the sum clearly remains finite.

Maxwellian Velocity Distribution

Maxwell used kinetic theory to derive the velocity distribution of free particles in TE; it can also be obtained using Boltzmann statistics. Consider particles moving with velocities $(\mathbf{v}, \mathbf{v} + d\mathbf{v})$ in a phase-space volume element $dp_x dp_y dp_z dV = m^3 dv_x dv_y dv_z dV = 4\pi m^3 v^2 dv\, dV$. The unit of phase space is h^3, so there are

$$g_{\text{trans}} = 4\pi m^3 v^2 dv\, dV/h^3 \tag{4.21}$$

states in this phase volume. The kinetic energy of the particles in that volume is $\frac{1}{2}mv^2$. Applying (4.15) we have

$$d^6 N^*(\mathbf{x}, v)/N^* = \left(4\pi m^3/h^3 U^*_{\text{trans}}\right) \exp(-mv^2/2kT)\, v^2 dv\, dx\, dy\, dz. \tag{4.22}$$

Integrating (4.22) over space we obtain

$$d^3 N^*(v)/N^* = \left(4\pi m^3 V/h^3 U^*_{\text{trans}}\right) \exp\left(-mv^2/2kT\right) v^2 dv; \tag{4.23}$$

then integrating over velocity we find

$$\left(m^3 V/h^3 U^*_{\text{trans}}\right) (2\pi kT/m)^{3/2} = 1. \tag{4.24}$$

Thus the partition function for translational motion is

$$U^*_{\text{trans}} = \left(2\pi mkT/h^2\right)^{3/2} V, \tag{4.25}$$

so from (4.22) the normalized distribution function for particle speeds is

$$f(v)dv = (m/2\pi kT)^{3/2} \exp\left(-mv^2/2kT\right) 4\pi v^2 dv. \tag{4.26}$$

For this distribution function: (1) The *most probable speed* is $\overline{v} = (2kT/m)^{1/2} = 11.85(T/10^4 A)^{1/2}$ km/s, where A is the particle's atomic weight. (2) The *root mean square speed* is $\langle v^2 \rangle^{1/2} = (3kT/m)^{1/2}$; thus *in TE all particles have the same kinetic energy*: K.E. $= \frac{1}{2}m\langle v^2 \rangle = \frac{3}{2}kT$. (3) The *root mean square velocity component along a line of sight* is $\langle v_x^2 \rangle^{1/2} = (kT/m)^{1/2}$, so if a spectrum line were a δ function in the atom's frame, in the laboratory frame it would be broadened into a normalized Gaussian profile

$$\phi_{\text{D}}(x) = \exp(-x^2)/\sqrt{\pi} \quad \Rightarrow \quad \int_{-\infty}^{\infty} \phi_{\text{D}}(x)\, dx = 1, \tag{4.27}$$

where $x \equiv \Delta\lambda/\Delta\lambda_{\text{D}}$, and the *Doppler width* is $\Delta\lambda_{\text{D}}/\lambda = (2kT/m)^{1/2}/c$.

Particles with Internal and Translational Energy

The material in stellar atmospheres is composed of atoms and ions that (except for fully stripped ions) have internal excitation states with statistical weights g_i, lying at energies ϵ_i above the ground level. Let the translational energy of an atom (or ion) whose speed lies in a phase-space cell t with statistical weight g_t be ϵ_t. Then the total energy of an atom in excitation state i moving in a phase-space element t is $\epsilon = \epsilon_i + \epsilon_t$, and its statistical weight is $g = g_i\, g_t$. The partition function for such an atom is

$$U^*(T) = \sum_i \sum_t g_i\, g_t\, e^{-(\epsilon_i+\epsilon_t)/kT}. \tag{4.28}$$

In dilute gases the distribution functions for translational and internal excitation states are statistically independent, because an atom with an electron in excitation state i can move at any speed, and in equilibrium it must have the same Maxwellian speed distribution as atoms in all other excitation states. Therefore, for each term in the sum over i, there is a complete sum over t, and the total partition function can be factored as

$$U^*(T) = \sum_i g_i\, e^{-\epsilon_i/kT} \sum_t g_t\, e^{-\epsilon_t/kT} = U^*_{\text{trans}}\, U^*_{\text{exc}}. \tag{4.29}$$

Saha Ionization Equation

Above the discrete eigenstates of an atom there exists a *continuum* of states in which an electron has become unbound and has a nonzero kinetic energy, leaving behind an ion in its ground state. The energy above the ground level at which this continuum begins is the atom's *ionization potential* (or *binding energy*) ϵ_I. The relative numbers of atoms and ions in successive stages of ionization are given by the *Saha ionization equation*, which is an extension of Boltzmann's excitation equation to the continuum of free electron states.

Consider a process in which an ion of chemical element l, with net charge j ($j = 0$ for a neutral atom), in its ground state with number density n_{0jl}, is ionized to the ground level of next highest ion, plus a free electron moving in the plasma with kinetic energy $\frac{1}{2}m_e v^2$. The energy required by this process is $\epsilon_{Ijl} + \frac{1}{2}m_e v^2$, where ϵ_{Ijl} is the ionization potential of ion j of element l. Let g_{0jl} and $g_{0,j+1,l}$ be the statistical weights of the ground state of ions l and $l+1$, respectively. The statistical weight of the final state of ion $l+1$ plus an electron with speed in the range $(v, v+dv)$ is $g(v)\,dv \equiv g_{0,j+1,l}\, g_e(v)\,dv$, where $g_e(v)\,dv = 2V \times 4\pi p^2\, dp/h^3 = (8\pi m_e^3 v^2\, dv/h^3)/n_e$ is the statistical weight of the free electron. The factor of 2 accounts for two possible orientations of the electron's spin, and V has been chosen to contain one free electron.

Writing $n_{0,j+1,l}(v)$ for the number density of ions in their ground states from which a free electron moving with speed v has been stripped, and applying (4.20b) to the initial and final states of this process, we have

$$\frac{n^*_{0,j+1,l}(v)\,dv}{n^*_{0jl}} = \frac{8\pi m_e}{h^3 n_e}\left(\frac{g_{0,j+1,l}}{g_{0jl}}\right) \exp\left[\frac{-\left(\epsilon_{Ijl} + \frac{1}{2}m_e v^2\right)}{kT}\right] v^2 dv. \tag{4.30}$$

Summing over all final states by integrating over electron speed, we find the basic form of *Saha's ionization equation*:

$$\frac{n^*_{0,j+1,l}}{n^*_{0jl}} = \frac{2}{n_e}\left(\frac{2\pi m_e kT}{h^2}\right)^{\frac{3}{2}}\left(\frac{g_{0,j+1,l}}{g_{0jk}}\right)\exp\left(\frac{-\epsilon_{Ijl}}{kT}\right). \tag{4.31}$$

Using (4.20b) in (4.31), the occupation number of excitation state i of ion stage j in terms of temperature, electron density, and the ground-state population of ion $j+1$ is

$$n^*_{ijl} = C_I\left(\frac{n_e n^*_{0,j+1,l}}{T^{3/2}}\right)\left(\frac{g_{ijl}}{g_{0,j+1,l}}\right)\exp\left(\frac{\epsilon_{Ijl} - \epsilon_{ijl}}{kT}\right), \tag{4.32}$$

where $C_I \equiv \frac{1}{2}(h^2/2\pi m_e k)^{3/2} = 2.07 \times 10^{-16}$ in cgs units.

Further, from (4.20a), $n^*_{0,j+1,l}/g_{0,j+1,l} = N^*_{j+1,l}/U^*_{j+1,l}$, so n^*_{ijl} in terms of the total number density $N^*_{j+1,l}$ of ion $j+1$ is

$$n^*_{ijl} \equiv n_e N^*_{j+1,l}\Phi_{ijl}(T), \tag{4.33}$$

where

$$\Phi_{ijl}(T) \equiv \frac{C_I g_{ijl}}{T^{3/2}U^*_{j+1,l}}\exp\left(\frac{\epsilon_{Ijl} - \epsilon_{ijl}}{kT}\right). \tag{4.34}$$

Equation (4.33) is the most useful form of Saha's equation for our work. We will use it to compute induced emission corrections to opacities.

Finally, summing (4.32) over all bound levels of both the lower and upper ionization states, we get an expression for the total number of ions in each successive stage of ionization:

$$N^*_{jl} = C_I\, n_e N^*_{j+1,l}\left(\frac{U^*_{jl}}{U^*_{j+1,l}}\right)\frac{\exp\left(\epsilon_{Ijl}/kT\right)}{T^{3/2}} \equiv n_e N^*_{j+1,l}\Phi_{jl}(T). \tag{4.35}$$

Equipped with the expressions developed above, we can write an explicit formula for the fraction f_{jl} of element l in ionization state j. Let Z_l be the charge of the last ion of element l that is considered. Define $r_{jl} \equiv N^*_{jl}/N^*_{j+1,l} = n_e\Phi_{jl}(T)$, and set $r_{Z_l,l} \equiv 1$. Then f_{jl} for $(j = 0, \ldots, Z_l)$ is

$$f_{jl}(n_e, T) \equiv \left(N^*_{jl}/N^*_l\right) = \left(N^*_{jl}/N^*_{Z_l}\right)\Big/\left(N^*_l/N_{Z^*_l}\right) \tag{4.36a}$$

$$= \frac{r_{jl}\, r_{j+1,l}\, \cdots\, r_{Z_l-1,l}}{1 + r_{Z_l-1,l} + r_{Z_l-1,l}\, r_{Z_l-2,l} + \cdots + r_{Z_l-1,l} \cdots r_{0,l}} \tag{4.36b}$$

$$= \prod_{m=j}^{Z_l-1} n_e\Phi_{ml}(T)\Big/\sum_{m=0}^{Z_l}\prod_{j=m}^{Z_l} n_e\Phi_{jl}(T) \tag{4.36c}$$

$$\equiv P_{jl}(n_e, T)\big/S_l(n_e, T), \quad (j = 1, \ldots Z_l). \tag{4.36d}$$

Here we have tacitly assumed that the electron density n_e is known so that we can calculate $N^*_{j+1,k}/N^*_{jk}$, etc. But as shown at the beginning of this section, the total particle density N is more fundamental. In §17.2 we will see that by using (4.36) and demanding *charge conservation* between electrons, and all ions N^*_{jl} for all j and l, we can invert the calculation in order to obtain $n_e(N, T)$. In short, when the material is in TE, if T, and n_e or N, are known, we can compute the fraction of any chemical element l in any ionization state j or in a specific excitation state i of that ionization state.

Molecular Dissociation

When they are separated at large distances, the atoms in a molecule are effectively isolated (i.e., noninteracting) and the potential of the system is constant. As the atoms approach each other, their electrons "feel" the attraction of the positively charged nuclei in the system. Their individual wave functions become distorted such that their collective wave function has an increased probability of negative charge density between the nuclei, which partially shields the repulsive force between the positive nuclear charges. The nuclei are, in turn, attracted to this concentration of negative charge, so the potential of the system can become negative, and it is bound. As the atoms approach yet more closely, the repulsive interaction of their positively charged nuclei dominates, so the system's potential becomes large and positive, thus limiting the atoms' minimum separation. The resulting potential well has a finite depth, and hence a finite *dissociation energy* D_0.

A molecule's internal partition function is more complicated than an atom's. The electrons in a molecule can be in quantized excited states. In addition, the molecule can vibrate and rotate. Hence the entire system has multiple series of quantized rotation and vibration states. Fortunately, the analysis can generally be simplified because the energy spacing associated with these different modes differs greatly: $\Delta E_{\rm rot} \ll \Delta E_{\rm vib} \ll \Delta E_{\rm exc}$.[2]

Dissociation Formula

Saha's formula (4.30) can be adapted to molecular dissociation. For the dissociation of a molecule AB into atoms A and B we can write an equation having the general form

$$\frac{N_A N_B}{N_{AB}} = \frac{U_A U_B}{U_{AB}}\, e^{-D_0/kT}. \tag{4.37}$$

Here U_{AB}, U_A, and U_B are, respectively, the *total* partition functions of the molecule AB and its atoms A and B. As in (4.29) we can approximate each of them as a product of their translational and internal partition functions:

$$U_{\rm tot} \equiv U_{\rm trans}\, U_{\rm int}. \tag{4.38}$$

[2] These inequalities will be justified in § 7.3.

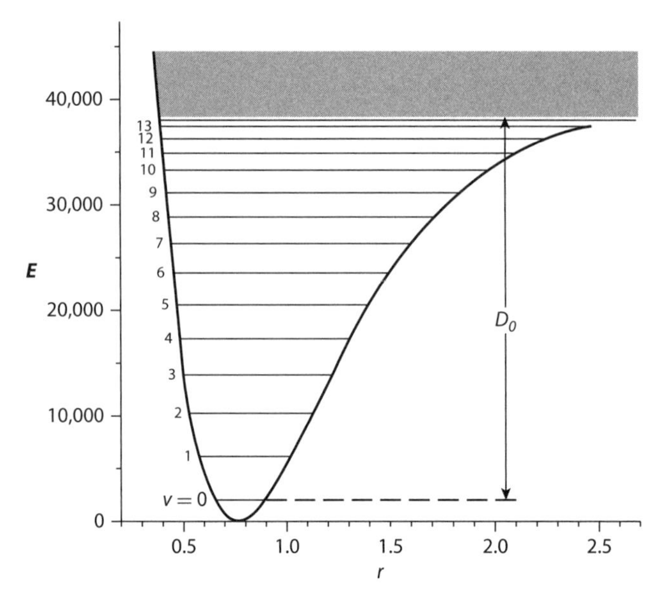

Figure 4.1 Potential well for ground electronic state of H_2 showing vibrational levels through $v = 13$ and continuum above the dissociation energy D_0. *Abscissa*: separation of the nuclei in units of 10^{-8} cm. *Ordinate*: energy in units of cm^{-1}.

As shown in figure 4.1, the dissociation energy of a molecule is measured from its lowest vibrational level.

For some molecules, accurate internal partition functions can be computed from the empirically known energies of their individual rotation, vibration, and excitation states. In particular, a large body of data exists for the important molecules H_2 and H_2^+, e.g., [76, 135, 256, 487, 556, 695, 786, 1190]. But in many cases, for lack of such data we assume that the three modes of excitation are independent and approximate a molecule's internal partition function by their product [661, 945, 1069]:

$$U_{\text{int}} = U_{\text{exc}}\, U_{\text{rot}}\, U_{\text{vib}}. \tag{4.39}$$

This factorization is not as accurate for molecules as it would be for atoms because rapid rotation distorts a molecule's shape and changes its vibrational potential well, and hence its vibrational levels. At the same time vibration changes its moments of inertia, and hence its rotational energy levels. If the vibrational plus rotational energy of a molecule exceeds its dissociation energy, in effect it is torn apart, and its constituent atoms become unbound.

The translational partition function for each component of a molecule has the form $U_{\text{trans}} = (2\pi mkT/h^2)^{3/2}$, where $m = m_A, m_B$, and m_{AB} for the atoms and bound molecule, respectively. Hence (4.37) has a general form

$$\frac{N_A N_B}{N_{AB}} = C_I (2\pi \mathsf{m} kT/h^2)^{\frac{3}{2}} \left(\frac{U_A U_B}{U_{AB}}\right)_{\text{int}} e^{-D_0/kT}, \tag{4.40}$$

where now m is the *reduced mass* $\mathsf{m} \equiv m_A m_B/(m_A + m_B)$ of the molecule.

Rotation and Vibration

For a quantized rigid rotator, Schrödinger [973] showed that

$$E_{\rm rot} = \tfrac{1}{2}\hbar^2 J(J+1)/I, \tag{4.41}$$

where I is the rotator's moment of inertia, and J is a non-negative integer. For a given J there are $g_J = (2J+1)$ degenerate states. Hence

$$U_{\rm rot} = \sum_{J=0}^{\infty} (2J+1)\exp[-J(J+1)\Theta_{\rm rot}/T], \tag{4.42}$$

where

$$\Theta_{\rm rot} \equiv \hbar^2/2kI \tag{4.43}$$

is a characteristic *rotation temperature*, only a few K. Because the spacing between rotational levels is so small, we can pass to the continuum limit. Write $x \equiv J(J+1)$; then

$$U_{\rm rot} \to \int_0^\infty \exp[-x\Theta_{\rm rot}/T]\,dx = T/\Theta_{\rm rot} = 8\pi^2 IkT/h^2. \tag{4.44}$$

This result is correct for heteronuclear molecules; for homonuclear molecules, $U_{\rm rot}$ in (4.44) must be divided by 2. For vibration, we use equation (3.113) for a quantized oscillator:

$$E_{\rm vib} = \left(n + \tfrac{1}{2}\right)\hbar\omega_0. \tag{4.45}$$

Ignore the zero-point energy because energies are measured relative to the lowest vibrational state.[3] Then, defining $\Theta_v \equiv \hbar\omega_0/k$,

$$U_{\rm vib} = \sum_{n=0}^{\infty} e^{-n\hbar\omega_0/kT} = \frac{1}{1-\exp(-\hbar\omega_0/kT)} \equiv \frac{1}{1-\exp(-\Theta_v/T)}. \tag{4.46}$$

In contrast to atomic/ionic partition functions, a molecule's electronic partition function does not diverge because its potential well has finite depth and width, and hence only a finite number of excited states. Both $\Theta_{\rm rot}$ and Θ_v are determined empirically.

The most important molecules in stellar atmospheres are usually diatomic. The structure of an atmosphere is most affected when H is converted to H_2. But H_2 does not have a permanent dipole moment; hence its interaction with photons is weak, and normally its influence on the total opacity is small. In very cool stars, the opacity of H_2 can be greatly increased by the process of *collision-induced absorption*; see § 7.3. Prominent features in the visible and infrared spectra of cool stars are produced by HCN, H_2O, CH, C_2, CN, CO, NH, N_2, NO, OH, TiO, YO, ZrO, and LaO.

[3] Note: the vibrational energy levels of H_2 are *not* equally spaced as predicted by (4.45).

Ideal Gas Equation of State

We showed in (4.26) that the particles of all species (atoms, ions, and free electrons) in an equilibrium gas are thermalized and have an isotropic velocity distribution, and each has a kinetic energy $\frac{3}{2}kT$. The *pressure* $p_{\rm gas}$ in a gas is the rate at which momentum, perpendicular to a unit surface area, is deposited by these particles. When a particle with momentum $m_i\mathbf{v}_i$ strikes an immovable elastic wall,[4] its speed v and the tangential component of its momentum are unchanged. But the normal component of its momentum is reversed. So the impulse it delivers to the wall is $2m_iv_i\cos\theta$, where θ is the angle between the normal $\mathbf{n}$ to the surface and the velocity vector $\mathbf{v}_i$. Thus the momentum flux delivered to the wall by particles with speeds $(v_i, v+dv_i)$ coming from a solid angle $d\Omega$ is

$$2(m_iv_i\cos\theta)N(v_i)\,v_i\,dv_i\,(d\Omega/4\pi) = m_iN(v_i)\,v_i^2\,dv_i\,\sin\theta\cos^2\theta\,d\theta\,d\phi/2\pi, \quad (4.47)$$

where ϕ is the azimuthal angle around $\mathbf{n}$. Summing over all solid angles and types of particles, recalling that $\frac{1}{2}m_i\langle v_i^2\rangle \equiv \frac{3}{2}kT$ for all particle species, we find

$$\begin{aligned} P &= \sum_i m_i \int_0^\infty N(v_i)\,v_i^2\,dv_i \int_0^{2\pi}\frac{d\phi}{2\pi}\int_0^{\pi/2}\sin\theta\cos^2\theta\,d\theta \\ &= \tfrac{1}{3}\sum_i N_im_i\langle v_i^2\rangle = \tfrac{1}{3}\,(3kT)\sum_i N_i = NkT, \end{aligned} \quad (4.48)$$

which is the equation of state for an ideal gas.

4.3 THERMAL RADIATION

The *Planck function*, the distribution function for *thermal radiation*, was derived by Planck by combining the Rayleigh–Jeans calculation of the number of electromagnetic modes per unit frequency interval with a computation of the average energy of all modes at a given frequency obtained using Boltzmann statistics, and the additional hypothesis that each mode contains an *integral number of quanta of energy*.

Experiments and Thermodynamics

At the end of the 19th century, a central question in physics was "What is the frequency distribution of equilibrium radiation field in a hohlraum?" It was addressed with both experiments and theory. With astute "gedanken experiments," Kirchhoff,

[4] Calculating the exchange of momentum with a wall simplifies the argument. In § 4.4 we calculate pressure in terms of the stress tensor.

Wien, and others showed that radiation in a hohlraum at temperature T has the following properties [758]:

- The intensity $I_\nu^*(T)$ in a vacuum cavity is everywhere the same because the walls of the cavity equilibrate to unique temperature, and I is invariant.
- It is isotropic, i.e., $I_\nu^*(T) \equiv J_\nu^*(T)$. If it were not, beams with unequal intensities traveling in opposite directions would produce a directional transport of energy, from which work could be extracted, in violation of the second law of thermodynamics for an isolated adiabatic enclosure.
- The system is in a *steady state*. Therefore, the same amount of energy is emitted from the walls as is absorbed. This statement must be true frequency by frequency, or energy could be extracted from the system to do external work, in violation of the second law of thermodynamics. Hence in equilibrium the material and radiation obey *Kirchhoff's law*

$$\eta^*_{\text{thermal}}(\nu, T) = \kappa^*_{\text{thermal}}(\nu, T) I_\nu^*(T), \tag{4.49}$$

  where $\kappa^*_{\text{thermal}}$ [cm^{-1}] is the material's TE opacity; η^*_{thermal} [erg/cm^3/s/sr] is its TE emissivity; and $I_\nu^*(T)$ is the specific intensity of the TE radiation field. Here κ^* includes a correction for induced emission; see § 5.2 and § 5.4. Scattering terms have been ignored in both κ and η. In a hohlraum, (4.49) still holds if scattering is included. They are excluded here so we can apply (4.49) to the case of local thermodynamic equilibrium; see § 4.5.
- For isotropic thermal radiation both the electromagnetic and macroscopic descriptions give $P_\nu^*(T) = \frac{1}{3}E_\nu^*(T)$ for radiation pressure and $F_\nu^*(T) = \frac{1}{4}cE_\nu^*(T)$ for the flux emitted by a wall. Integrating over frequency, we get $P^*(T) = \frac{1}{3}E^*(T)$ and $F^*(T) = \frac{1}{4}cE^*(T)$.
- From thermodynamics, in an adiabatic change dV of the cavity's volume, $d(E^*V) + P^*dV = 3VdP^* + 4P^*dV = 0$, or $P^*V^{4/3} = \text{constant}$. Hence equilibrium radiation acts like a polytrope with $\gamma = \frac{4}{3}$.
- The frequencies of photons in a cavity with perfectly reflecting walls are altered by Doppler shifts when the cavity expands or compresses adiabatically. From its polytropic behavior, the final temperature T_f of the radiation in the cavity is a unique function of its initial temperature T_i. Wien showed from thermodynamic arguments that the frequencies of the maximum of the radiation distribution function are connected to the cavity's temperature by the *displacement law* $\nu_i/T_i = \nu_f/T_f$. The energy densities in the two states are related by $(dE_\nu)_i/(dE_\nu)_f = (T_i/T_f)^4$, where $dE_{\nu_i} \equiv (4\pi/c)J^*(\nu_i, T_i)\,d\nu_i$ and $dE_{\nu_f} \equiv (4\pi/c)J^*(\nu_f, T_f)\,d\nu_f$. Thus

$$\begin{aligned} J^*(\nu_i, T_i) &= (T_i/T_f)^4 J^*(\nu_f, T_f)(d\nu_f/d\nu_i) \\ &= (T_i/T_f)^3 J^*[(\nu_i T_f/T_i), T_f] = (T_i/T_f)^3 f(\nu_i/T_i). \end{aligned} \tag{4.50}$$

That is, the *shape* of the distribution function of equilibrium radiation, $J^*(\nu, T)$ versus ν/T, is the same for all temperatures.

- Experiments showed that the formula $J^*_{\text{exp}}(\nu) = c_1\nu^3/[\,e^{c_2\nu/kT} - 1\,]$ gives an accurate fit to the data. The challenge to theory is to show why.

Classical Analysis: Rayleigh–Jeans Formula

Rayleigh and Jeans attempted to use electromagnetic theory to find the spectral distribution of thermal radiation. Suppose that a field at frequency ν in a rectangular hohlraum of dimensions (X, Y, Z) is a superposition of oscillatory *standing waves* (modes), having zero amplitude on each boundary surface of the cavity. Then the amplitude of each mode is proportional to $f_{\mathbf{k}}(x, y, z) = \sin(k_x x)\sin(k_y y)\sin(k_z z)$, where the *wave vector*

$$\mathbf{k} \equiv (k_x, k_y, k_z) = (n_x\pi/X,\ n_y\pi/Y,\ n_z\pi/Z).$$

Here n_x, n_y, and n_z are positive integers. The magnitude of $\mathbf{k}$ is $k = 2\pi\nu/c$, so

$$\nu = \tfrac{1}{2}c\,[\,(n_x/X)^2 + (n_y/Y)^2 + (n_z/Z)^2\,]^{1/2}. \tag{4.51}$$

For a given frequency, (n_x, n_y, n_z) for all modes lie on the surface of the ellipsoid

$$(n_x/a_x)^2 + (n_y/a_y)^2 + (n_z/a_z)^2 = 1, \tag{4.52}$$

where $(a_x, a_y, a_z) = (2\nu/c)(X, Y, Z)$. Thus $\mathfrak{N}(\nu)$, the total number of modes with $\nu' \leq \nu$ equals twice (two polarizations) the number of points having integer coordinates contained inside one octant (all n's ≥ 0) of the ellipsoid for frequency ν. The volume of an ellipsoid is $(4\pi a_x a_y a_z/3)$; hence

$$\mathfrak{N}(\nu) = 2 \times (1/8) \times (4\pi/3)(8XYZ\nu^3/c^3) = (8\pi\nu^3 V/3c^3), \tag{4.53}$$

where $V = XYZ$ is the volume of the hohlraum. The *density of modes* per unit frequency is

$$\mathfrak{n}(\nu) \equiv \frac{d\mathfrak{N}(\nu)}{d\nu} = \frac{8\pi\nu^2 V}{c^3}. \tag{4.54}$$

Further, assume each mode is in energy equipartition at temperature T and has energy kT. Thus the Rayleigh–Jeans law for thermal equilibrium radiation is

$$J^*_{\text{RJ}}(\nu, T) = \frac{c}{4\pi}E^*_{\text{RJ}}(\nu) = \frac{c}{4\pi V}\left(\frac{8\pi\nu^2 V}{c^3}\right)kT = \frac{2kT\nu^2}{c^2}. \tag{4.55}$$

Note that (4.55) does not contain Planck's constant, so it is classical.

Semi–Classical Derivation of Planck's Law

The Rayleigh–Jeans formula fits experimental data at low frequencies, but fails to show the maximum observed experimentally. Worse, the total energy density $\int_0^\infty E^*_{\rm RJ}(\nu)\,d\nu$ is infinite: this is called the "*ultraviolet catastrophe*." This problem arises because each mode (frequency) is assumed to have the equipartition energy kT, and the range of frequencies is infinite.

In contrast, the empirical formula mentioned above has a maximum and yields a finite total energy density. The Rayleigh–Jeans formula departs radically from the data at high frequencies. In 1901, Planck [874] solved the problem, using a different (quantum) set of assumptions. We merely outline it here; for details see [576, chapter 1], [875]. Planck hypothesized the following:

- The radiation field interacts with monochromatic, one-dimensional, charged harmonic oscillators (a caricature of atoms in the walls).
- In thermal equilibrium, these oscillators are distributed in energy according to Boltzmann statistics (§ 4.2).
- Each oscillator can absorb energy from the radiation field continuously, in accordance with classical electromagnetic theory.
- But an oscillator radiates only when its energy is an integral multiple of a unit (*quantum*) of energy; when it does, it emits all of it.
- The energies the oscillators can radiate are $E_n = nh\nu$, where $h = 6.626 \times 10^{-27}$ erg s is *Planck's constant*.

In this picture, radiation is part of a system (radiation plus matter), and its equilibrium distribution function is determined by the interaction of its two constituents. From the second hypothesis above, the relative probability of an oscillator with energy $nh\nu$ in a hohlraum at temperature T is

$$\pi_n = e^{-nh\nu/kT} \Big/ \sum_{n=0}^{\infty} e^{-nh\nu/kT}. \tag{4.56}$$

The average energy of all modes at frequency ν is

$$< h\nu > = \sum_{n=0}^{\infty} nh\nu\, e^{-nh\nu/kT} \Big/ \sum_{n=0}^{\infty} e^{-nh\nu/kT}. \tag{4.57}$$

Defining $x \equiv e^{-h\nu/kT}$, rewrite the denominator of equation (4.57) as

$$1 + x + x^2 + \ldots = \frac{1}{1-x} \tag{4.58}$$

and its numerator as

$$h\nu(x + 2x^2 + 3x^3 + \ldots) = (h\nu x)\frac{d}{dx}(1 + x + x^2 + \ldots)$$
$$= (h\nu x)\frac{d}{dx}\left(\frac{1}{1-x}\right) = \frac{h\nu x}{(1-x)^2}. \quad (4.59)$$

The radiation energy density in the hohlraum is the number of modes per unit volume times the average energy per mode; combining equations (4.54) and (4.57–4.59) we find

$$E_\nu^*(T) = \frac{8\pi h\nu^3}{c^3}\frac{1}{e^{h\nu/kT} - 1} \quad [\,\mathrm{erg/cm^3/Hz}\,]. \quad (4.60)$$

But $I^* = J^*$ and $J^* = cE^*/4\pi$, so the specific intensity of radiation in a hohlraum in thermal equilibrium at absolute temperature T is the *Planck function*:

$$B_\nu(T) \equiv \frac{2h\nu^3}{c^2}\frac{1}{e^{h\nu/kT} - 1} \quad [\,\mathrm{erg/cm^2/s/Hz/sr}\,]. \quad (4.61)$$

Thus in thermodynamic equilibrium Kirchhoff's law (4.49) becomes

$$\eta^*(\nu) = \kappa^*(\nu)B_\nu[\,T(\mathbf{x}, t)\,], \quad (4.62)$$

the *Kirchhoff–Planck* relation.

For $h\nu/kT \ll 1$, the Planck function recovers the Rayleigh–Jeans formula (4.55), and the high-frequency tail of the spectrum fits *Wien's law*

$$J^*_{\mathrm{Wien}}(\nu, T) = \frac{2h\nu^3}{c^2}e^{-h\nu/kT} \quad \text{for} \quad (h\nu/kT) \gg 1. \quad (4.63)$$

Equation (4.61) gave spectacular agreement with all available experimental data and ushered in the new discipline of quantum theory.

Integrated Planck Function and Stefan's Law

The frequency-integrated Planck function, $B(T)$, is given by

$$B(T) \equiv \int_0^\infty B_\nu(T)\,d\nu = \frac{2k^4T^4}{c^2h^3}\int_0^\infty x^3e^{-x}(1 - e^{-x})^{-1}\,dx \quad (4.64)$$
$$= \frac{2k^4T^4}{c^2h^3}\int_0^\infty x^3e^{-x}(1 + e^{-x} + e^{-2x} + \ldots)\,dx \quad (4.65)$$
$$= \frac{12k^4T^4}{c^2h^3} \times (1 + 2^{-4} + 3^{-4} + \ldots). \quad (4.66)$$

The sum in the last line equals $\zeta(4) = \pi^4/90$, where ζ is the *Riemann zeta function* [5, p. 807]. Thus the total energy density of thermal radiation is given by *Stefan's law* [223, p. 55]:

$$E^*(T) = \frac{4\pi}{c} B(T) = a_R T^4. \tag{4.67}$$

The *radiation constant* $a_R = (8\pi^5 k^4/15c^3h^3) = 7.56 \times 10^{-15}$ erg/cm^3/K^4.

Stefan's law holds in the deeper, opaque layers of a star's atmosphere and in its interior, where both the escape probability of photons from the surface and the thermal gradient over a photon mean free path are very small. When these conditions hold, the radiation field becomes essentially isotropic and can thermalize to the equilibrium distribution. It does not apply near a star's surface, where there is a large temperature gradient and the overlying material begins to become transparent in some frequency range. In such regions the radiation field is strongly anisotropic and has a non-Planckian energy distribution.

The Planck function is isotropic; hence $\mathbf{F} \equiv 0$ in a hohlraum's interior. A hohlraum is also called a *black body* because if there is a small aperture in one of its walls, radiation that falls there enters the cavity, is absorbed by an interior wall, and is thermalized. By the same token, we can observe the equilibrium radiation that escapes from the interior through the aperture. The emergent monochromatic flux is

$$F_{\mathrm{BB}}(\nu) = 2\pi \int_0^1 B_\nu(T)\,\mu d\mu = \pi B_\nu(T). \tag{4.68}$$

The frequency-integrated flux from a black body is

$$F_{\mathrm{BB}}(T) = \pi \int_0^\infty B_\nu(T)\,d\nu = \pi B(T) = \sigma_R T^4 \quad [\,\mathrm{erg/cm^2/s}\,], \tag{4.69}$$

where $\sigma_R \equiv a_R c/4 = (2\pi^5 k^4/15c^2h^3) = 5.67 \times 10^{-5}$ erg/cm^2/s/K^4 is the *Stefan–Boltzmann constant.*

4.4 QUANTUM STATISTICS

Indistinguishable Particles

In quantum mechanics, two *identical particles*, say electrons or photons, are indistinguishable. Interchanging them produces *no observable effect*. So the Hamiltonian for an ensemble of identical particles must be a *symmetric function of its independent variables*; indeed, any measurable property of the ensemble must be a symmetric function of these variables. The overall symmetry of the state vector of the ensemble, once established, is *invariant*.

As in Boltzmann statistics, write the state vector of a system of $\mathcal{N}$ particles as

$$\big| abc \ldots \mathcal{N} \big\rangle = \big| \mathbf{x}_1\, a \big\rangle \big| \mathbf{x}_2\, b \big\rangle \big| \mathbf{x}_3\, c \big\rangle \ldots \big| \mathbf{x}_{\mathcal{N}}\, N \big\rangle. \tag{4.70}$$

The first label on the single-particle kets denotes both the space and spin coordinates of a particle, and the second the eigenstate it is in. The ket (4.70) is representative. A general ket for an ensemble has the form

$$\big|\, abc \ldots \mathcal{N}\big\rangle = \big|\, a\rangle\big|\, b\rangle\big|\, c\rangle \ldots \big|\, N\big\rangle + \big|\, b\rangle\big|\, a\rangle\big|\, c\rangle \ldots \big|\, N\big\rangle$$
$$+ \ldots \text{all permutations of the } \mathcal{N} \text{ identical particles.} \quad (4.71)$$

The particle coordinates have been suppressed in (4.71) because they are irrelevant for indistinguishable particles. An interchange of particles can be expressed in terms of a *permutation operator* $\hat{\boldsymbol{P}}_{rs}$,

$$\hat{\boldsymbol{P}}_{rs}\big|\, abc \ldots r\, s \ldots \mathcal{N}\big\rangle \longrightarrow \big|\, abc \ldots s\, r \ldots \mathcal{N}\big\rangle, \quad (4.72)$$

which can be applied repeatedly to get any ordering of the particles. In the process of interchanging particles, a ket may (or may not) change sign, so that in going from an initial list $\{a_i, b_i, \text{etc.}\}$ to a final list $\{a_f, b_f, \text{etc.}\}$,

$$\hat{\boldsymbol{P}}_{if}\big|\, a_i b_i c_i \ldots \mathcal{N}_i\big\rangle = \pm\big|\, a_f b_f c_f \ldots \mathcal{N}_f\big\rangle. \quad (4.73)$$

This seemingly minor change leads to very different kinds of statistics. A ket for an ensemble is called *symmetric* if its sign is unchanged by *any* permutation of the particles and *antisymmetric* if its sign is *unchanged by an even permutation* of the particles, but is *changed by an odd permutation*.[5]

Bose–Einstein Statistics

As mentioned in § 3.1, photons are *bosons* and obey Bose–Einstein (BE) *statistics*; see [138,312,313],[515, chapters 11 and 12],[771, chapters 25 and 26], which gives the equilibrium distribution function of particles with integral spin (e.g., α particles, neutral helium atoms, and, specifically, photons). BE statistics is symmetric in the sense stated above. When the operation in (4.73) is symmetric, we can say that certain kets are occupied, without being able to say which particle is in which state. Two or more kets may be the same; hence two or more particles can be in the same state.

Maximum Entropy Distribution Function

Suppose the ensemble has two particles, 1 and 2, and two kets, $\big|\, a\rangle$ and $\big|\, b\rangle$, which have the same probability of being occupied. In classical statistics with distinguishable particles there are four possible states: $\big|\, a1\rangle\big|\, a2\rangle$ with probability $\frac{1}{4}$; $\big|\, b1\rangle\big|\, b2\rangle$ with probability $\frac{1}{4}$; $\big|\, a1\rangle\big|\, b2\rangle$ with probability $\frac{1}{4}$; and $\big|\, a2\rangle\big|\, b1\rangle$ with probability $\frac{1}{4}$. But in BE statistics the particles are indistinguishable. Therefore, there are three

[5] A permutation is called *even* when it has an even number of particle interchanges and *odd* when it has an odd number of interchanges.

symmetric states: $|a1\rangle|a1\rangle$ with probability $\frac{1}{3}$; $|b1\rangle|b1\rangle$ with probability $\frac{1}{3}$; and $|a1\rangle|b2\rangle + |a2\rangle|b1\rangle$ with probability $\frac{1}{3}$. *Hence in BE statistics there is a greater probability of both particles occupying the same state than in classical statistics.*

In deriving the general distribution function, let the number of bosons in cell i be n_i. In BE statistics each cell may contain an unlimited number of bosons. To divide any given cell into g_i states, $g_i - 1$ *partitions* are required. The total number of distinct ways a list of bosons and partitions can be made is $(n_i + g_i - 1)\,!$. The bosons are indistinguishable; hence in BE statistics, any permutation of bosons within a cell produces physically indistinguishable results. Likewise, we can relabel the partitions in any order. Hence the number of arrangements of photons and partitions that produce a distinct macrostate within cell i is

$$W_i(\{n_i\}) = \prod_i \frac{(n_i + g_i - 1)\,!}{n_i\,!(g_i - 1)\,!}. \tag{4.74}$$

As in Boltzmann statistics, the thermodynamic probability of the ensemble's macrostate is $\sum_i W_i$. But one macrostate dominates, so for that state,

$$\ln W = \sum_i [\ln(n_i + g_i - 1)\,! - \ln n_i\,! - \ln(g_i - 1)\,!]. \tag{4.75}$$

From Stirling's formula,

$$\ln W \approx \sum_i [\,(n_i + g_i - 1)\ln(n_i + g_i - 1) - n_i \ln n_i - (g_i - 1)\ln(g_i - 1)\,]. \tag{4.76}$$

At maximum entropy, the first variation of W with respect to changes in the occupation numbers is zero. Noting that $n_i \gg 1$, we have

$$\delta \ln W = \sum_i \left[\,\ln(n_i + g_i) - \ln n_i\,\right]\delta n_i = 0. \tag{4.77}$$

Again, the set $\{\delta n_i\}$ is subject to the constraints $\delta\mathcal{E} = \sum_i \epsilon_i \delta n_i \equiv 0$ (energy conservation) and $\sum_i \delta n_i \equiv 0$ (number conservation). As in Boltzmann statistics, all three requirements can be met by solving for the distribution that satisfies an arbitrary linear combination of them. Thus

$$\sum_i \left[\,\ln(n_i + g_i) - \ln n_i - \beta(\epsilon_i - \mu)\,\right]\delta n_i = 0. \tag{4.78}$$

By the same analysis as used above for Boltzmann statistics, $\beta = 1/kT$. Hence the BE distribution function is

$$n_i = \frac{g_i}{e^{(\epsilon_i - \mu)/kT} - 1}. \tag{4.79}$$

In this formulation, μ, called the *chemical potential*, is related to the number of particles. Using (4.79) one can derive expressions for the pressure and energy density of the gas; note that for it to make physical sense, the energy of the lowest state in the distribution must have $\epsilon_i > \mu$.

For the particular case of photons, *number conservation does not apply*. For example, suppose an atom in a bound state a absorbs a photon, causing an electron to jump to a higher state c, and then decays $a \to c \to b \to a$. Two photons have emerged from the interaction when *one* was put in. Discarding number conservation makes the chemical potential in (4.79) zero for photons.

Planck Function Revisited

Planck's discovery of the correct distribution law for thermal radiation led to a decisive break with classical physics on the atomic scale and to the creation of quantum mechanics. But some of the assumptions made in his derivation of the formula need revision in the light of quantum statistics:

(1) Planck assumed that thermal radiation results from interaction of a *classical radiation field* with *quantized material harmonic oscillators*. But Einstein's analysis [307] of the photoelectric effect showed that *it is the radiation field that is quantized into photons*.

(2) Planck used classical Boltzmann statistics to derive (4.57) for the average energy of his oscillators. He got the correct expression for the equilibrium photon distribution function, but as mentioned earlier, the assumption that photons are distinguishable is incorrect.

Instead, the Planck function must be derived using BE statistics. In (4.79) write $\epsilon_i \equiv h\nu$, and rather than using the Rayleigh–Jeans formula for the number of electromagnetic modes in a cavity, use the relativistic expression $p^2 = (h\nu/c)^2$ for a massless particle and an elementary phase volume h^3 to compute the number of phase-space elements in a frequency interval $d\nu$:

$$dx\,dy\,dz\,dp_x\,dp_y\,dp_z/h^3 = dV\,4\pi\,p^2 dp/h^3 = dV\,4\pi\,\nu^2 d\nu/c^3. \tag{4.80}$$

Multiplying (4.80) by (4.79) and 2 (to account for two polarizations) $\times\, h\nu$, we get the photon energy density in the frequency interval $d\nu$:

$$E_\nu d\nu = \frac{8\pi h\nu^3}{c^3}\frac{d\nu}{e^{h\nu/kT}-1}. \tag{4.81}$$

Multiplying (4.81) by $c/4\pi$ to convert E_ν to specific intensity, we recover the Planck function (4.61).

Fermi–Dirac Statistics

Maximum Entropy Distribution Function

Particles with half-integral spin (electrons, protons, ..., neutrinos) obey *Fermi–Dirac (FD) statistics*; see [284, 285], [286, chapter 9], [328, 329]. In FD statistics at most *one* particle can occupy an elementary phase-space volume (a cell; now meaning a distinct space-momentum volume *and* a value for the spin) because of the *Pauli exclusion principle* [845]. This can be assured if the eigenstate of the system is an *antisymmetric* combination of single-particle kets in the sense of (4.73).

A mathematical expression that meets these requirements is the *Slater determinant* [1010]:

$$\big|\, abc \ldots \mathcal{N}\big\rangle = \frac{1}{\sqrt{\mathcal{N}\,!}} \begin{vmatrix} \big|\mathbf{x}_1\, a\big\rangle & \big|\mathbf{x}_1\, b\big\rangle & \ldots & \big|\mathbf{x}_1\, \mathcal{N}\big\rangle \\ \big|\mathbf{x}_2\, a\big\rangle & \big|\mathbf{x}_2\, b\big\rangle & \ldots & \big|\mathbf{x}_2\, \mathcal{N}\big\rangle \\ \vdots & \vdots & & \vdots \\ \big|\mathbf{x}_{\mathcal{N}}\, a\big\rangle & \big|\mathbf{x}_{\mathcal{N}}\, b\big\rangle & \ldots & \big|\mathbf{x}_{\mathcal{N}}\, \mathcal{N}\big\rangle \end{vmatrix}. \tag{4.82}$$

From the definition of a determinant, $\big|\, abc \ldots \mathcal{N}\big\rangle$ changes sign if any two particles [i.e., any two columns in (4.82)], or if any two single-particle kets [i.e., any two rows in (4.82)], are interchanged and is zero if any two particles are in the same ket.

Suppose we have n_i particles to be distributed into $g_i \geq n_i$ states and, momentarily, that both the particles and the states are distinguishable. There are $n_i\,!$ ways we can order the choice of particles. We can choose from g_i states for the first particle, $g_i - 1$ for the second, down to $g_i - n_i$ for the last one, which gives us a total of $n_i\,!/(g_i - n_i)\,!$ unique arrangements. But the particles are actually indistinguishable, so the order in which they are chosen is immaterial, and the number of different microstates is $g_i\,!/n_i\,!(g_i - n_i)\,!$ Thus the thermodynamic probability of this macrostate is

$$W(n_i) = \prod_i \frac{g_i\,!}{n_i\,!\,(g_i - n_i)\,!}. \tag{4.83}$$

Applying Stirling's formula, and maximizing the probability, we find

$$-\ln n_i + \ln(g_i - n_i) - \alpha - \beta(\epsilon_i - \mu) = 0. \tag{4.84}$$

Hence $n_i = 1\big/[e^{(\epsilon_i - \mu)/kT} + 1]$, or in the conventional notation,

$$n_i = \frac{1}{e^{(\epsilon_i/kT - \psi)} + 1}, \tag{4.85}$$

where ϵ is the energy of the particle, and ψ is called the *degeneracy parameter* of the gas. For *any* choice of ϵ or ψ, this distribution function allows *at most one particle per phase-space volume element*, so it satisfies the Pauli exclusion principle. Comparing (4.85) with (4.79) we also see that when $\epsilon_i/kT \ll 1, n_i(\mathrm{BE})/n_i(\mathrm{FD}) \gg 1$, which shows that *Bose condensation* can occur.

To get a bit of insight into the difference between BE and FD statistics, consider a system with two identical particles interacting only with each other, and whose spins are decoupled from outside forces [690, § 7.2]. The Hamiltonian of this system must be invariant for (1) exchange of both phase-space and spin coordinates of both particles, (2) exchange of only their phase-space coordinates, or (3) exchange of only their spin coordinates. If we say that the spins are unaffected by outside forces, the wave function for each of the particles is a product function:

$$\Psi_i = \Phi_i(\mathbf{q}_i, \mathbf{p}_i)\; \Xi_i(\sigma_i). \tag{4.86}$$

Both $\Phi_i(\mathbf{q}_i, \mathbf{p}_i)$ and $\Xi_i(\boldsymbol{\sigma}_i)$ must be symmetric or antisymmetric functions under the exchanges. Normalized wave functions with these properties are of the form

$$\Phi_i(\mathbf{q}_i, \mathbf{p}_i) = \frac{1}{\sqrt{2}}[\,\phi_\alpha(1)\phi_\beta(2) \pm \phi_\alpha(2)\phi_\beta(1)\,] \tag{4.87}$$

and

$$\Xi(\boldsymbol{\sigma}_i) = \frac{1}{\sqrt{2}}[\,\xi_\alpha(1)\xi_\beta(2) \pm \xi_\alpha(2)\xi_\beta(1)\,]. \tag{4.88}$$

The plus signs in (4.87) and (4.88) apply for bosons, and the minus signs apply for fermions. The expectation value of the square of the distance between the two particles is

$$\langle(\mathbf{r}_2 - \mathbf{r}_1)^2\rangle = \tfrac{1}{2}\iiint [\phi_\alpha^*(1)\phi_\beta^*(2) \pm \phi_\alpha^*(2)\phi_\beta^*(1)] \times (r_1^2 - 2\mathbf{r}_1\cdot\mathbf{r}_2 + r_2^2)$$
$$\times\, [\phi_\alpha(1)\phi_\beta(2) \pm \phi_\alpha(2)\phi_\beta(1)]\, d^3\mathbf{x}_1\, d^3\mathbf{x}_2. \tag{4.89}$$

Expanding this expression, and using the orthogonality of the single-particle eigenfunctions, we find

$$\langle(\mathbf{r}_2 - \mathbf{r}_1)^2\rangle = \langle r^2\rangle_\alpha + \langle r^2\rangle_\beta - 2\langle\mathbf{r}\rangle_\alpha\cdot\langle\mathbf{r}\rangle_\beta \mp 2\left|\langle\alpha|\mathbf{r}|\beta\rangle\right|^2. \tag{4.90}$$

Here

$$\langle r^2\rangle_n \equiv \int \phi_n^* r^2 \phi_n\, d^3\mathbf{x},\quad \langle\mathbf{r}\rangle_n \equiv \int \phi_n^*\, \mathbf{r}\, \phi_n\, d^3\mathbf{x},\quad \langle\alpha|\mathbf{r}|\beta\rangle \equiv \int \phi_\alpha^*\, \mathbf{r}\, \phi_\beta\, d^3\mathbf{x}. \tag{4.91}$$

The first three terms on the right in (4.90) are what would be found for distinguishable particles. The fourth term reflects the different signs in (4.87) and (4.88) for exchange of bosons and fermions.

The upper sign (−) applies for bosons and the lower (+) for fermions. Thus because of *exchange interactions*, the expected distance between bosons is less than or equal to that for distinguishable particles, and the opposite is true for fermions. In this sense, bosons tend to “attract one another,” and fermions tend to “repel one another.”[6]

Degenerate Matter Equation of State

Although FD statistics has its widest application in describing solid state materials, say a crystal lattice, the material in the cores of stars can be so dense that essentially all available phase-space elements are occupied by a particle, and the plasma

[6] The discussion above is very oversimplified. In [847], Pauli writes, “we conclude for the relativistically invariant equation for free particles: From postulate (I) [requiring that] energy must be positive, the necessity of Fermi–Dirac statistics for particles with arbitrary half-integral spin; from postulate (II) [requiring that] observables on different space-time points with a space-like distance are commutable, the necessity of Bose–Einstein statistics for particles with arbitrary integral spin. ... the connection between spin and statistics is one of the most important applications of the special relativity theory.”

becomes degenerate. For example, in the cores of white dwarfs, the density of material is on the order of $10^5 \lesssim \rho \lesssim 10^8$ gm/cm^3. At such densities, pressure ionization strips essentially all electrons from atoms and ions, and one has a dense plasma of free electrons and bare nuclei. We will see below that the free electrons are the degenerate component in the plasma; a density of $\rho \sim 10^{14}$ would be required for the nuclei to become degenerate.

To satisfy the Pauli exclusion principle, the number density of electrons $n_e(p)dp$ with momenta in the range $(p, p+dp)$ is limited to[7]

$$n_e(p)dp \leq 8\pi p^2 dp/h^3. \tag{4.92}$$

Electrons obey Fermi statistics, and (4.85) gives a distribution function

$$n(p)dp = \frac{8\pi p^2 dp}{h^3} \frac{1}{e^{(\epsilon_i/kT - \psi)} + 1}. \tag{4.93}$$

For our purposes it is sufficient to consider only the case of *complete degeneracy* in which every phase-space cell, starting with $p = 0$, is filled.[8] Comparing (4.92) and (4.93), we see that for complete degeneracy $\psi \to \infty$,

$$n_e(p)dp = (8\pi p^2 dp/h^3) \quad \text{for} \quad p \leq p_F$$

and (4.94)

$$n_e(p)dp = 0 \quad \text{for} \quad p > p_F,$$

where p_F is called the *Fermi momentum*. In this limit,

$$n_e = \int_0^{p_F} \frac{8\pi p^2 dp}{h^3} = \frac{8\pi}{3h^3} p_F^3, \tag{4.95}$$

which defines the Fermi momentum and shows that $p_F \propto n_e^{1/3}$. Further, if the electrons are non-relativistic, the *Fermi energy* $E_F \equiv p_F^2/2m_e \propto n_e^{2/3}$.

To obtain the equation of state for a fully degenerate electron gas we can use the distribution function (4.94) in a version of (4.48) re-expressed in terms of the momentum of the particles:

$$P_e = \int_{4\pi} d\Omega \int_0^{p_F} f(p)\, v(p)\, p \cos^2\theta \, dp = \frac{8\pi}{3h^3} \int_0^{p_F} p^3 v(p)\, dp. \tag{4.96}$$

From equation (A.30) we know $p = \gamma m_e v$, which can be inverted to give

$$\frac{v}{c} = \frac{p/(m_e c)}{[1 + (p/m_e c)^2]^{1/2}}. \tag{4.97}$$

[7] In the distribution function we use **p** instead of **v** because (a) there may be particles moving at relativistic speeds and (b) in Hamiltonian mechanics, momentum **p** is the canonical coordinate to position **x**; see, e.g., [380, chapters 6 and 7].

[8] Discussion of the six possible regimes having non-degenerate, partially degenerate, and completely degenerate distribution functions, for both relativistic and non-relativistic electrons, can be found in sections § 24.1–§ 24.9 of [262].

Using (4.95) and (4.97) in (4.96), we have

$$P_e(x) = \frac{8\pi c}{3h^3}\int_0^{p_F} \frac{p^4/(m_e c)}{[1+p^2/(m_e c^2)]^{1/2}}dp \;=\; \frac{8\pi m_e^4 c^5}{3h^3}\int_0^x \frac{\xi^4 d\xi}{(1+\xi^2)^{1/2}}, \quad (4.98)$$

where $\xi \equiv p/m_e c$ and $x \equiv p_F/m_e c$. Note that $P_e(x)$ does not depend at all on temperature. The integral in (4.98) can be evaluated analytically [262, p. 822] and [619, p. 121], yielding

$$f(x) = \tfrac{1}{8}[x(2x^2-3)(x^2+1)^{1/2} + 3\sinh^{-1} x], \quad (4.99)$$

hence $P_e(x) = (\pi m_e^4 c^5/3h^3) f(x)$, and, from (4.95), $n_e(x) = (8\pi m_e^3 c^3/3h^3)x^3$. Analysis of (4.99) shows that for small x (non-relativistic limit), $P_e \propto f(x) \to cx^5$. The material is fully ionized, so its mass density ρ is proportional to $n_e(x) \propto x^3$. Therefore, in this limit $P \propto \rho^{5/3}$. In contrast, for large x (relativistic limit), $P_e \propto f(x) \to cx^4$; hence $P \propto \rho^{4/3}$.

Astrophysical Significance

At first sight, one might think that the small change in the polytropic index of the density-pressure relation for the cases of non-relativistic versus relativistic degeneracy might not be of great importance. Actually, it may have catastrophic consequences for the stability of a star. One can understand why by means of a simple dimensional analysis. Consider a star of a given mass M with an initial radius R. Its average density scales as $\rho \propto M/R^3$. If the material near its center is non-relativistically degenerate, the gas pressure scales as $P \propto \rho^{5/3} \propto M^{5/3}/R^5$. If that material is relativistically degenerate, the gas pressure scales as $P \propto \rho^{4/3} \propto M^{4/3}/R^4$. The gravitational binding force scales as $\rho GM/r^2 \propto M^2/R^5$. For hydrostatic equilibrium it must be balanced by the pressure gradient. The average pressure gradient would scale as $dP/dr \propto M^{5/3}/R^6$ if the material were non-relativistically degenerate and as $dP/dr \propto M^{4/3}/R^5$ if relativistically degenerate.

Assuming that the mass of the star remains unchanged, compare the pressure gradient with the gravitational force. In the non-relativistic case, those two forces scale with a different power of the radius; therefore, the star can bring the two into balance by adjusting its radius. But in the relativistic case, the two forces scale with the *same* power of the radius. Therefore, a star cannot bring the two forces into balance by adjusting its radius. As described in § 2.6, massive stars that have exhausted their nuclear fuel sources have relativistically degenerate cores. If those cores have masses above the Chandrasekhar limit, they collapse and produce Type II supernovae.

4.5 LOCAL THERMODYNAMIC EQUILIBRIUM

Basic Concept

In the limit of strict thermodynamic equilibrium, the temperature and a particle number density (e.g., the electron number density) are sufficient to determine the distribution functions for both the particles and radiation in a hohlraum. But TE is a *theoretical abstraction* that can be only approximated, even in carefully designed laboratory experiments. Strictly speaking, it applies only in a hohlraum where all the material has a perfectly constant temperature and density. Of course, a stellar atmosphere is *not* a hohlraum; it has an open boundary through which photons escape freely and regions where there are large temperature and density gradients. We emphasize that the formulae in § 4.2–§ 4.4 hold *only* in TE and *cannot be used to describe the radiation field or the excitation and ionization state of the material in stellar atmospheres.*

Constructing a model atmosphere under such circumstances is difficult. Before fast computers, and the algorithms needed to solve the non-equilibrium problem, were available, it was argued that if the temperature and density gradients in the material are "small enough," and the density is "large enough" for collisions to drive the material's distribution functions nearly to equilibrium, then the material might be homogeneous enough over a particle mean free path that its thermodynamic properties (including opacity and emissivity) could have "essentially" their TE values at the *local* values of T and ρ, i.e., be in *local thermodynamic equilibrium, LTE.* Then the radiative transfer equations for the radiation field's angle-frequency dependence could be solved using those material properties. However, even if the material were homogeneous over a *particle* mean free path, *photon* mean free paths are always much larger, so the radiation field can respond to markedly different conditions in distant regions of the atmosphere, in particular, to the existence of an open boundary. In reality, the occupation numbers of the material are determined by balance between the rates of excitation and de-excitation, and ionization and recombination, by both radiative and collisional processes.

At the low densities in the observable layers of stellar atmospheres, the radiative rates generally dominate the collisional rates. Hence it is inevitable that the character of the radiation field, and the formation of both continua and spectral lines in the observable upper layers of an atmosphere, are out of thermodynamic equilibrium. If we are to obtain a self-consistent description of the state of the material and the radiation field in stellar photospheres, the transfer equation and non-equilibrium kinetic equations must be solved simultaneously.

For decades the question of whether LTE could be a valid approximation to use in analyzing stellar atmospheres was debated without a conclusive answer; a good example of an argument supporting its use can be found in [129]. But today the theoretical framework is solid; the necessary atomic and molecular data are available; and the computers and algorithms are fast enough that we can now make calculations with sufficient accuracy to decide how important "departures from LTE" are in a wide range of models. The result is that *although LTE might give a fair representation of some of the features in some stars' spectra, it generally fails badly*. The good news is that the answer to the question above in a specific case is no longer a matter of conjecture; it can be settled by computation.

Electron Velocity Distribution

An exception where LTE does apply, even in very low density material, is the velocity distribution of free electrons. While they are in the continuum, electrons suffer elastic collisions with other electrons, and *inelastic collisions* with atoms or ions, which can excite or ionize them. The elastic collisions thermalize the electrons and lead to equipartition, and hence a Maxwellian velocity distribution in a *relaxation time* t_c [1037, chapter 5]. One finds that the rate of inelastic collisions for conditions in stellar atmospheres is low compared to t_c, so they do not hinder thermalization. Indeed, this result holds even at the extremely low densities in planetary nebulae [127]. Another question is whether the atoms and ions also have Maxwellian distributions, with equal kinetic temperatures, i.e., $T_{\text{ion}} \approx T_e$. An analysis [118] for a hydrogen plasma comprising atoms, ions, electrons, and radiation in steady state shows that if $n_e > 10^{10}$ and $5 \times 10^3 < T_e < 10^5$, then $|T_e - T_{\text{ion}}| \lesssim 10^{-3} T_e$. Thus it is reasonable to assume that the particles do indeed have a single kinetic temperature almost everywhere in a static stellar atmosphere.

Chapter Five

Absorption and Emission of Radiation

Analysis of a star's spectrum is our most effective tool for deducing the structure and composition of its outer layers and yields data that can support or test stellar evolution theory. To compute the theoretical spectrum of a star we solve transfer equations that describe how radiation is transported through its atmosphere, which are formulated in terms of material coefficients that set the rates of *absorption, emission*, and *scattering* of photons. These depend on (1) the nature of the *quantum mechanical interactions* that couple radiation to the material; (2) their *cross sections* as a function of wavelength; and (3) the *number density* of atoms[1] that can interact with radiation at a given wavelength. We discuss thermal absorption and emission in spectral lines and continua in this chapter and in chapter 7, continuum scattering in chapter 6, and scattering in spectral lines in chapter 10. We assume the material is *isotropic on the microscale.*

In § 5.1 we give qualitative examples of thermal absorption and emission and contrast them with scattering. In § 5.2 we summarize Einstein and Milne's *detailed-balance* analyses, which show that when a photon is emitted *spontaneously* in a spectral line or continuum, thermodynamic equilibrium in a hohlraum cannot be attained unless it *induces* (or *stimulates*) additional identical photons proportional to the photon number density in the ambient radiation field.

In § 5.3 we describe a semi-classical calculation by Slater [1009] of the absorption probability for a bound-bound transition. But the bound-bound emission probability cannot be computed this way. In § 5.4 we present Dirac's quantum mechanical calculation [280] of both bound-bound photon absorption and emission probabilities. His work also showed that (1) Einstein and Milne had obtained the correct relative probabilities for absorption, induced emission, and spontaneous emission; (2) an induced photon is identical in both frequency and direction to the photon emitted spontaneously; and (3) the number of induced photons emitted equals the phase-space density of that type of photon in the radiation field. We sketch the computation of photoionization cross sections in § 5.5 and give a semi-classical treatment of free-free transitions § 5.6.

We consider only *allowed transitions*, which result from the interaction of an atom's electric dipole moment with the radiation's electric field. They are orders of magnitude stronger than those produced by interactions of higher-order multipoles of the atom or of the atomic dipole moment with the magnetic component of the radiation field. In-depth discussions of these subjects are given in [286, 444, 471, 723, 961, 1025] and the references listed at the beginning of chapter 7.

[1] In this chapter, we suppress mention of the dependence of physical variables on $(\mathbf{x}, t)$ unless necessary and use the word "atom" to mean atom, ion, or molecule.

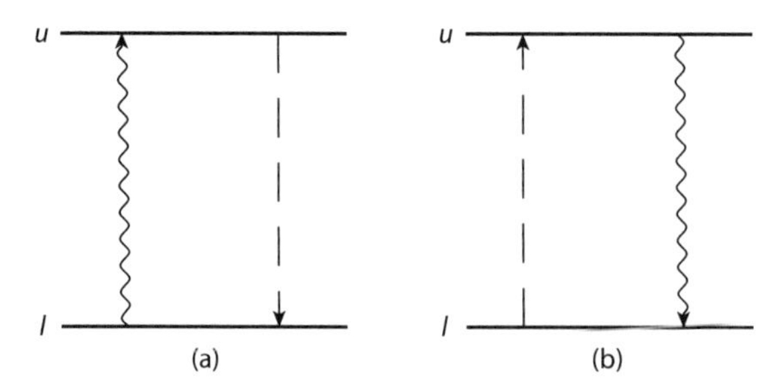

Figure 5.1 (a) Photon destruction. (b) Photon creation. *Wavy line*: radiative transition; *dashed line*: collisional transition.

5.1 ABSORPTION AND THERMAL EMISSION

When a photon is *absorbed*, it is *destroyed*. In the particular process shown in figure 5.1(a), a photon excites an electron from a lower bound state l of an atom to a higher bound state u. The photon's energy has been removed from the radiation field and deposited into the internal energy of the atom. If the excited atom is then de-excited by an inelastic collision with a free electron, which returns the bound electron to its initial lower level, the kinetic energy of the collision partner is increased. It, in turn, collides with other particles, and its excess energy is deposited into the internal energy of the gas. Thus absorption processes feed radiative energy into the *thermal pool* of the gas. In the inverse process, shown in figure 5.1(b), an electron in bound state l of an atom is excited to a higher bound state u by an inelastic collision with a particle. The excited atom decays radiatively back to the original lower level by emitting a photon. The net result is *thermal emission*, in which a photon is *created* from energy in the thermal pool. Likewise, a bound electron of an atom (or ion) may be *photoionized* by absorbing a photon. The photon is destroyed, and the excess of its energy over the electron's binding energy goes into the electron's kinetic energy and ultimately into the thermal pool when the electron is thermalized by elastic collisions with other particles. The inverse process is *radiative recombination*, in which a free electron is captured by the positive charge of an ion, and a photon with energy equal to the sum of the electron's kinetic and binding energy is created. Thus the combined actions of bound-bound *thermal* absorption and emission processes, and of photoionization and radiative recombination, tend to couple the radiation field to the local thermal properties of the gas and to lead to local *thermal equilibrium*[2] between radiation and material.

In a *scattering* process, a photon excites an atom from a bound lower state l to a higher bound state u. The excited electron jumps back to level l, and a photon is emitted. In general, the upper and lower levels of a line are not perfectly sharp because they are perturbed by nearby particles, and/or have finite lifetimes (cf. chapter 8). Thus a scattering process may not be perfectly *coherent* because

[2] Note: local *thermal* equilibrium, not local *thermodynamic* equilibrium.

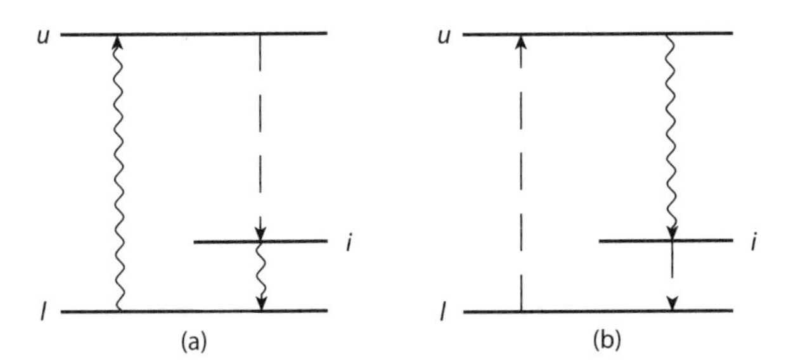

Figure 5.2 Ambiguous cases of photon creation/destruction. *Wavy line*: radiative transition; *dashed line*: collisional transition.

the frequencies of the incoming and outgoing photons scattered in a spectral line may differ (cf. chapter 10). Photons may also scatter by interactions with free electrons (electron scattering), or by a non-resonant interaction with atoms or molecules (Rayleigh scattering); see chapter 6.

When sequences of transitions among several states occur, ambiguities arise, so that a given transition cannot rigorously be described as an "absorption" or "scattering" event. Indeed, in view of the complicated couplings shown in the kinetic equilibrium equations (cf. chapter 9), it would not be useful to try to do so. For example:

- Suppose an atom has a lower level l, an upper level u, and an intermediate level i. An electron in l is excited by a photon to u. It then undergoes a collisional de-excitation $u \rightarrow i$, followed by emission of a photon $i \rightarrow l$, as shown in figure 5.2(a). In terms of the descriptions given above, the original photon has not been scattered because the frequencies of the incoming and outgoing photons are quite different. But was the outgoing photon "emitted thermally" inasmuch as some of its energy was derived from the original radiative transition?
- In the process shown in figure 5.2(b), an electron in level l is excited to level u by an inelastic collision. A photon is emitted in the transition $u \rightarrow i$, and then the electron undergoes a collisional de-excitation $i \rightarrow l$. The outgoing photon was created at the expense of the thermal pool. But was it emitted "thermally" when only part of the collision partner's energy returns to the thermal pool?
- Suppose an electron is excited from level l to level u by a photon and then decays radiatively through an intermediate state i' to level l.[3] In this process of *fluorescence*, a single energetic photon with energy $h\nu_{lu} = E_u - E_l$ can produce two or more photons whose energies sum to $E_u - E_l$. The original photon was not "absorbed" because its energy makes no direct contribution to the thermal energy of the gas. And the emergent photons were not directly created at the expense of the thermal pool, so they are not "thermally emitted." Yet the original photon was not "scattered" because the frequencies of the outgoing photons may not be close

[3] In *LS* coupling, this process cannot work with a single intermediate state because of selection rules.

to the incoming photon's, and they may have very different escape probabilities through the atmosphere's boundary surface from that of the original photon, so their *nonlocal* behavior is different.

5.2 DETAILED BALANCE

Einstein Relations for Bound–Bound Transitions

In two seminal papers [309, 310], Einstein showed from detailed balancing arguments that absorption and emission of photons by an atom in thermodynamic equilibrium with a radiation field require the existence of *three* different processes. Consider transitions between a lower level l with statistical weight g_l, and an upper level u with statistical weight g_u, in a material at rest. The three processes are as follows.

(1) A photon is absorbed by an atom in the lower level, causing a bound electron to jump to the upper level. In terms of *Einstein's absorption probability* B_{lu}, the direct rate $r_{\mathrm{lu}}(\mathbf{n}, \nu)$ of photon absorption in the transition $l \rightarrow u$ from a beam with specific intensity $I(\mathbf{n}, \nu)$ in the frequency range $(\nu, \nu + d\nu)$ traveling in direction $\mathbf{n}$, out of solid angle $d\Omega$, is[4]

$$r_{\mathrm{lu}}(\mathbf{n}, \nu)\, d\nu\, (d\Omega/4\pi) \equiv n_l\, \phi_{lu}(\nu) B_{lu} I(\mathbf{n}, \nu)\, d\nu\, (d\Omega/4\pi). \tag{5.1}$$

If the material is at rest in the laboratory frame, and atoms are not oriented preferentially in the material, then the *absorption* profile $\phi_{lu}(\nu)$ in (5.1) is isotropic.[5]

(2) Photons are emitted when electrons in the upper level spontaneously drop to the lower level. In terms of *Einstein's spontaneous emission probability* A_{ul}, the rate $r^{\mathrm{s}}_{\mathrm{ul}}(\mathbf{n}, \nu)$ of spontaneous emission in the transition $u \rightarrow l$ in the frequency range $(\nu, \nu + d\nu)$, traveling in direction $\mathbf{n}$, into solid angle $d\Omega$, is

$$r^{\mathrm{s}}_{\mathrm{ul}}(\mathbf{n}, \nu)\, d\nu\, (d\Omega/4\pi) \equiv n_u \psi_{ul}(\mathbf{n}, \nu) A_{ul}\, d\nu\, (d\Omega/4\pi). \tag{5.2}$$

Here $\psi_{ul}(\mathbf{n}, \nu)$ is the *emission profile* seen in the laboratory frame, normalized such that $\oint (d\Omega/4\pi) \int_0^\infty \psi_{ul}(\mathbf{n}, \nu)\, d\nu = 1$. In general, the emission profile $\psi_{ul}(\nu_0)$ is not the same as $\phi_{ul}(\nu_0)$, even in the comoving frame, because before being emitted

[4] Einstein defined his probabilities in terms of radiation energy density $E_\nu = 4\pi J_\nu/c$. Milne used specific intensity. In a hohlraum, $I_\nu \equiv J_\nu$, and the two treatments differ only by the conversion factor $4\pi/c$. But Milne's choice, which we use, allows for anisotropy of the radiation field and its departure from thermodynamic equilibrium. Both of these aspects are critical in describing the physics of a stellar atmosphere.

[5] If the material were in motion, the line profile would become $\phi_{lu}(\mathbf{n}, \nu)$, which is *anisotropic* in the laboratory frame, because identical photons with the same laboratory-frame frequency ν, coming from different directions, are Doppler-shifted to different frequencies $\nu_0 = \nu - \mathbf{n}\cdot\mathbf{v}/c$ in the *comoving frame* (a set of frames, each of which moves with a fluid element) and thus to different values $\phi_{lu}(\nu_0)$. $\phi_{lu}(\mathbf{n}, \nu)$ is still normalized such that $\oint (d\Omega/4\pi) \int_0^\infty \phi_{lu}(\mathbf{n}, \nu)\, d\nu = 1$. Because the characteristic frequency width of a bound-bound transition is small, even relatively small material velocities can result in large changes in both absorption and emission profiles.

a photon may be *redistributed* to another frequency while the excited atom is in the upper level (see chapter 10). Spontaneous emission at a given frequency ν_0 is isotropic in the comoving frame, but not in the laboratory frame if the material is moving.

(3) Photons are also emitted when an excited electron is *induced* to drop to a lower level by the presence of other photons having the same frequency, direction, and polarization in the comoving frame as a photon emitted. The rate $r^{\mathrm{i}}_{\mathrm{ul}}(\mathbf{n}, \nu)$ at which these transitions occur is

$$r^{\mathrm{i}}_{\mathrm{ul}}(\mathbf{n}, \nu)\, d\nu\, (d\Omega/4\pi) \equiv n_u \psi_{ul}(\mathbf{n}, \nu) B_{ul} I(\mathbf{n}, \nu)\, d\nu\, (d\Omega/4\pi), \tag{5.3}$$

where B_{ul} is the *Einstein induced emission probability*.

In an atom's rest frame, *spontaneous emission is isotropic*, whereas even in that frame *induced emission may be anisotropic* because it is proportional to the photon occupation number, and hence specific intensity, of the ambient radiation field, which may be anisotropic.

To attain equilibrium, the number of upward transitions must balance, frequency by frequency, angle by angle, the total number of downward transitions, i.e.,

$$r_{\mathrm{lu}}(\mathbf{n}, \nu) \equiv r^{\mathrm{s}}_{\mathrm{ul}}(\mathbf{n}, \nu) + r^{\mathrm{i}}_{\mathrm{ul}}(\mathbf{n}, \nu). \tag{5.4}$$

In thermodynamic equilibrium the material is at rest; hence $\phi_{lu}(\mathbf{n}, \nu) \to \phi(\nu)$ and $\psi_{ul}(\mathbf{n}, \nu) \to \psi_{ul}(\nu)$; and the radiation field is isotropic; hence $I(\mathbf{n}, \nu) \to I^*(\nu)$. Using these identities in (5.1)–(5.4) we can write

$$n_l^* B_{lu} I_\nu^* \phi_\nu \equiv n_u^* (A_{ul} + B_{ul} I_\nu^*) \psi_\nu. \tag{5.5}$$

Here the superscript $*$ means a variable has its thermodynamic equilibrium value. Solving for I^* we find

$$I_\nu^* = \frac{(A_{ul}/B_{ul})}{\left(n_l^* B_{lu} \phi_\nu / n_u^* B_{ul} \psi_\nu\right) - 1}. \tag{5.6}$$

Then, by (1) evaluating (5.6) at the line-center frequency $\nu = \nu_{lu}$, (2) taking $\psi_\nu \equiv \phi_\nu$ to get exact detailed balance, (3) setting I_ν^* to its equilibrium value, the Planck function (4.61), and (4) using Boltzmann's formula (4.20b) for the equilibrium occupation numbers n_l^* and n_u^*, we find

$$g_l B_{lu} = g_u B_{ul} \tag{5.7}$$

and

$$A_{ul} = (2h\nu_{lu}^3/c^2) B_{ul}. \tag{5.8}$$

Although we invoked thermodynamic equilibrium in this analysis, we show in § 5.3 and § 5.4 that the Einstein coefficients are determined only by quantum mechanical properties of an atom; *hence their ratios in* (5.7) *and* (5.8) *hold in general*. The actual values of these coefficients cannot be determined from detailed-balancing arguments alone, but can be measured experimentally and computed quantum mechanically.

By multiplying (5.1)–(5.3) by $h\nu$ we get the energy absorption and emission rate. Absorption and induced emission rates are both proportional to $I(\mathbf{n}, \nu)$. Therefore, we subtract the latter from the former to obtain the *net absorption rate corrected for induced emission.* Thus the *macroscopic line absorption coefficient* is

$$\kappa_{lu}(\mathbf{n}, \nu) \equiv [\, n_l B_{lu} \phi_{lu}(\nu) - n_u B_{ul} \psi_{ul}(\mathbf{n}, \nu)\,](h\nu/4\pi), \tag{5.9}$$

and the *macroscopic line emission coefficient* is

$$\eta_{ul}(\mathbf{n}, \nu) \equiv n_u A_{ul} \psi_{ul}(\mathbf{n}, \nu)(h\nu/4\pi). \tag{5.10}$$

In thermodynamic equilibrium, the ambient radiation field is the Planck function, so the ratio of induced emission to spontaneous emission is

$$\frac{r^{\mathrm{i}}_{\mathrm{ul}}}{r^{\mathrm{s}}_{\mathrm{ul}}} = \frac{B_{ul}}{A_{ul}} B(\nu_{lu}, T) = \frac{c^2}{2h\nu^3_{lu}} \frac{2h\nu^3_{lu}/c^2}{(e^{h\nu_{lu}/kT} - 1)} = \frac{1}{(e^{h\nu_{lu}/kT} - 1)}. \tag{5.11}$$

In the low-frequency limit, $(h\nu_{lu}/kT) \ll 1$; so $\exp(h\nu_{lu}/kT) \approx 1 + h\nu_{lu}/kT$; hence

$$(r^{\mathrm{i}}_{\mathrm{ul}}/r^{\mathrm{s}}_{\mathrm{ul}}) \approx (kT/h\nu_{lu}) \gg 1. \tag{5.12}$$

Thus in thermodynamic equilibrium *induced emission dominates spontaneous emission in transitions for which* $h\nu_{lu}/kT \ll 1$, which explains why it is easier to make masers than lasers and easier to make red lasers than blue.

Further, detailed balance for a radiation field $I(\mathbf{n}, \nu)$ requires that

$$n_l B_{lu} I(\mathbf{n}, \nu_{lu}) = n_u [\, A_{ul} + B_{ul} I(\mathbf{n}, \nu_{lu})\,], \tag{5.13}$$

so the ratio of the total rate of emission to the spontaneous rate is

$$\frac{r^{\mathrm{t}}_{\mathrm{ul}}}{r^{\mathrm{s}}_{\mathrm{ul}}} = 1 + \frac{B_{ul}}{A_{ul}} I(\mathbf{n}, \nu_{lu}) = 1 + \frac{c^2}{2h\nu^3_{lu}} I(\mathbf{n}, \nu_{lu}) = 1 + n_\alpha(\mathbf{n}, \nu_{lu}), \tag{5.14}$$

where $n_\alpha(\nu_{lu})$ is the photon occupation number defined in equation 3.13.[6]

[6] As shown in § 5.4, for Fermi–Dirac statistics the factor $1 + n_\alpha$ would be replaced by $1 - n_\alpha$, because emission of a fermion is forbidden by Pauli's exclusion principle if another fermion already occupies the same phase-space volume.

Milne Relations for Bound–Free Transitions

Milne [755] extended Einstein's analysis to photoionization and radiative recombination. Consider the unit process: (atom $+ h\nu$) $\leftrightarrow$ (ion + free electron with kinetic energy $\frac{1}{2}m_e v^2$). Let

- n_0 = number density of atoms in the ground state; n_1 = number density of ions in the ground state; $n_e(v)dv$ = number density of electrons with speeds in the range $(v, v + dv)$, all in cm^{-3}.
- p_ν [cm^2/erg] = the photoionization probability of atoms by radiation with specific intensity I_ν in the range $(\nu, \nu + d\nu)$.
- $F(v)$ = the *spontaneous radiative recombination probability* in the reaction [ion + electron with speed $(v, v + dv) \rightarrow$ atom].
- $G(v)$ = the *induced radiative recombination probability* in the reaction [ion + electron with speed $(v, v+dv) + h\nu \rightarrow$ atom] in the presence of radiation with specific intensity I_ν.

Then the photoionization rate is

$$r_{\mathrm{bf}}(\mathbf{n}, \nu) = n_0 p_\nu I(\mathbf{n}, \nu)\, d\nu, \tag{5.15}$$

and the total recombination rate is

$$r^{\mathrm{t}}_{\mathrm{fb}}(\mathbf{n}, \nu) = n_1 n_e(v)[\, F(v) + G(v) I(\mathbf{n}, \nu)\,] v\, dv. \tag{5.16}$$

Because continua are normally smooth over large frequency intervals, Doppler shifts resulting from (non-relativistic) material motions do not have much effect on these rates.

In thermodynamic equilibrium $I_\nu = I^*_\nu \equiv B_\nu, n_0 = n^*_0, n_1 = n^*_1$, and the photoionization and recombination rates must be in detailed balance. When a photon ionizes an atom, $h\nu = \chi_{\mathrm{ion}} + \frac{1}{2}m_e v^2$; hence $h\, d\nu = m_e v\, dv$. Thus in thermodynamic equilibrium

$$n^*_0 p_\nu B_\nu = n^*_1 n^*_e(v)[\, F(v) + G(v) B_\nu\,](h/m_e), \tag{5.17}$$

or

$$B_\nu = \frac{[\, F(v)/G(v)]}{(m_e/h)[n^*_0 p_\nu / n^*_1 n^*_e(v) G(v)] - 1}, \tag{5.18}$$

which, from equation (4.61), implies that

$$F(v) = (2h\nu^3/c^2) G(v) \tag{5.19}$$

and

$$G(v) = (m_e/h) \left[\, n^*_0 / n^*_1 n^*_e(v)\, \right] p_\nu e^{-h\nu/kT}. \tag{5.20}$$

As in the case of bound-bound transitions, (5.19) and (5.20) depend only on the structure of the atom and ion. Thus they are *general*, independent of the assumption of thermodynamic equilibrium, which was invoked here just to derive them easily.

By expressing $n_e^*(v) = n_e f(v)$ and using (4.26) for $f(v)$, expressing $(n_1/n_0)^*$ through the Saha equation, (4.31), and employing the energy balance equation $\frac{1}{2}m_e v^2 + \chi_{\rm ion} = h\nu$, one obtains

$$p_\nu = (8\pi m_e v^2 g_1/h^2 g_0)G(v) = (4\pi c^2 m_e v^2 g_1/h^3 g_0 \nu^3)F(v), \tag{5.21}$$

where the second equality follows from (5.19).

The energy absorption and emission rates are given by (5.15) and (5.16) times $h\nu$. Then using (5.19) and (5.20) we find that the macroscopic spontaneous emission coefficient is

$$\begin{aligned}
\eta_{\rm fb}^{\rm s}(\nu)d\nu &= n_1\, n_e(v)\, F(v)\, v\, dv\, h\nu \\
&= n_1 n_e(v) \left(\frac{n_0}{n_1 n_e(v)}\right)^* \left(\frac{m_e}{h}\right)\left(\frac{h}{m_e}\right)\left(\frac{2h\nu^3}{c^2}\right) e^{-h\nu/kT}\, p_\nu\, h\nu\, d\nu \\
&= n_0^*\, \alpha_{\rm bf}(\nu) \left(\frac{2h\nu^3}{c^2}\right) e^{-h\nu/kT}\, d\nu \\
&= n_0^*\, \alpha_{\rm bf}(\nu)(1 - e^{-h\nu/kT}) B_\nu(T)\, d\nu,
\end{aligned} \tag{5.22}$$

where the photoionization cross section $\alpha_{\rm bf}$ is defined by

$$\alpha_{\rm bf}(\nu) \equiv h\nu\, p_\nu. \tag{5.23}$$

In (5.22), we denoted

$$n_0^* \equiv n_1 n_{\rm e}(v) \left(\frac{n_0}{n_1 n_{\rm e}(v)}\right)^*, \tag{5.24}$$

which has the meaning of the LTE value of n_0 computed from the Saha equation using *actual* values of n_1 and $n_{\rm e}(v)$ (but remember that $n_{\rm e}^*(v) \equiv n_{\rm e}(v)$ because electrons posses an equilibrium velocity distribution). We will use this definition throughout this book, keeping in mind that this value is generally different from n_0^* considered in (5.15)–(5.20) that represents an "absolute" LTE value, i.e., that computed from the Saha equation with the LTE value of $n_1 = n_1^*$. Equation (5.22) also shows that because the emission process is collisional, the emission rate is proportional to the LTE population of the bound state to which recombination occurs.

Analogously, we find the macroscopic continuum absorption coefficient corrected for induced emission is

$$\kappa_{\rm bf}(\nu) \equiv \left(n_0 - n_0^* e^{-h\nu/kT}\right) \alpha_{\rm bf}(\nu). \tag{5.25}$$

Moreover, using (5.19), and (5.20) we find

$$\frac{r_{\rm fb}^{\rm t}}{r_{\rm fb}^{\rm s}} = 1 + \frac{G(v) I(\mathbf{n}, \nu)}{F(v)} = 1 + \frac{c^2}{2h\nu^3} I(\mathbf{n}, \nu) \equiv 1 + n_\alpha(\nu), \tag{5.26}$$

where n_α is the photon occupation number in phase space, the same result as (5.14) for bound-bound transitions, and the total emission rate is again the spontaneous rate multiplied by $(1 + n_\alpha)$.

To extend Milne's analysis to free-free transitions, one would replace p_ν with an absorption probability depending on both the photon's frequency and the speed of the emitted electron and $F(v)$ and $G(v)$ with probabilities depending on both the initial and final speeds of the electron. It is noteworthy that the Einstein–Milne analyses were made before the advent of quantum mechanics, yet their results are quantum mechanically correct!

5.3 BOUND–BOUND ABSORPTION PROBABILITY

The most probable interactions between radiation and material are when a photon is absorbed (or emitted) in an *electric dipole transition*.[7] In an absorption process a radiation ket changes from $|n_\mathbf{k}\rangle$ to $|n_\mathbf{k} - 1\rangle$, and a bound electron jumps to a higher energy level. The absorption probability B_{lu} for a transition from level l to level u can be calculated semi-classically, by treating the atom quantum mechanically and the radiation field with classical electrodynamics.

Hamiltonian

From equations (A.47), (A.7), and (A.8) the relativistic Hamiltonian for a free particle with rest mass m_0 and four-momentum $\mathfrak{p} \equiv \gamma m_0 (d\mathbf{x}/dt)$ is

$$\mathcal{H}_0 = \sqrt{m_0^2 c^4 + \mathfrak{p}^2 c^2} = m_0 c^2 \sqrt{1 + \gamma^2 \beta^2}. \tag{5.27}$$

The Hamiltonian for a particle having charge q in an electromagnetic field with a vector potential $\mathbf{A}$ and a scalar potential ϕ is [575, p. 575]

$$\mathcal{H} = \sqrt{m_0^2 c^4 + (c\mathfrak{p} - q\mathbf{A})^2} + q\phi. \tag{5.28}$$

Expanding (5.28) to first order (i.e., letting $\gamma \to 1$), we have

$$\mathcal{H} = m_0 c^2 \sqrt{1 + (\mathfrak{p} - q\mathbf{A}/c)^2 / m_0^2 c^2} + q\phi$$

$$\approx m_0 c^2 + (\mathbf{p} - q\mathbf{A}/c)^2 / 2m_0 + q\phi. \tag{5.29}$$

Then dropping the constant rest-mass energy $m_0 c^2$, the Hamiltonian for an atom with N electrons, each with mass m_e and charge $-e$, and its interaction with an electromagnetic field is

$$\mathcal{H} = \sum_{j=1}^{N} \left(\frac{p_j^2}{2m_e} + V_j + \frac{e}{m_e c} \mathbf{A}_j \cdot \mathbf{p}_j + \frac{e^2}{2m_e c^2} \mathbf{A}_j \cdot \mathbf{A}_j \right). \tag{5.30}$$

[7] Electric quadrupole and magnetic dipole transition probabilities are orders of magnitude smaller.

Here $q\phi$ in (5.29) has been generalized to a potential energy V_j that accounts for the mutual electrostatic interactions among the N atomic electrons, each with charge $-e$, and with a nucleus with charge $+Ze$:

$$V_j \equiv -\frac{Ze^2}{|\mathbf{x}_j|} + \sum_{i \neq j}^{N} \frac{e^2}{|\mathbf{x}_j - \mathbf{x}_i|}. \tag{5.31}$$

The vector potential $\mathbf{A}_j$ comes from the radiation's electromagnetic field. Recalling that $\hat{\boldsymbol{p}}_{\mathbf{k}} \equiv -i\hbar(\partial/\partial q_{\mathbf{k}})$, we can write (5.30) in operator form as

$$\hat{\boldsymbol{H}}_A + \hat{\boldsymbol{H}}_I \equiv \sum_{j=1}^{N} \left(-\frac{\hbar^2}{2m_e} \nabla_j^2 + V_j \right) + \frac{e}{m_e c} \sum_{j=1}^{N} \hat{\mathbf{A}} \cdot \hat{\boldsymbol{p}}_j + \frac{e^2}{2m_e c^2} \hat{\mathbf{A}} \cdot \hat{\mathbf{A}}. \tag{5.32}$$

In (5.32) we have set $\mathbf{A}_j$ to $\mathbf{A}$ at the atom's nucleus because the wavelength of visible and ultraviolet radiation is much larger than atomic dimensions.

The term linear in $\hat{\boldsymbol{A}}$ results in emission (creation) or absorption (destruction) of a single photon. The quadratic term describes the following:

(a) Scattering, in which an incoming photon is converted to an outgoing photon with different properties (frequency and/or direction) (see chapters 6 and 10).

(b) Processes in which two photons are emitted or absorbed in a single transition, e.g., the forbidden transition $2s\,^2S \rightarrow 1s\,^2S$ in hydrogen, which produces an (observed) continuum starting at Lyα and extending to longer wavelengths; see [821, p. 89], [1038, 1183, 1184]. This continuum can be strong enough to affect the interpretation of the spectra of planetary nebulae.

As written, the term containing $\hat{\boldsymbol{A}} \cdot \hat{\boldsymbol{p}}$ implies $\hat{\boldsymbol{p}}$ and $\hat{\boldsymbol{A}}$ commute, which is justifiable for this problem (see below). In order to use (5.32) to compute transition probabilities, we need a quantum mechanical representation of $\hat{\boldsymbol{A}}$.

Atomic Eigenkets

The atom's Hamiltonian operator $\hat{\boldsymbol{H}}_A$ does not depend explicitly on time. Its bound levels are time-independent kets $|k\rangle$ that are solutions of the time-independent Schrödinger's equation for energies E_k:

$$\hat{\boldsymbol{H}}_A |k\rangle = E_k |k\rangle. \tag{5.33}$$

The kets must satisfy boundary conditions, which cannot be done for arbitrary choices of E_k. Thus (5.33) poses an eigenvalue problem. It must be solved for a spectrum of time-independent eigenkets and eigenvalues. For simplicity, we assume the eigenkets have been normalized and are *non-degenerate*, and hence *orthogonal*:

$$\langle i | j \rangle = \delta_{ij}. \tag{5.34}$$

The time-dependent form of Schrödinger's equation,

$$\hat{\boldsymbol{H}}_A \big| kt\big\rangle = i\hbar \frac{\partial \big| kt\big\rangle}{\partial t}, \tag{5.35}$$

yields time-dependent kets $\big| kt\big\rangle$. They are also solutions of (5.33):

$$\hat{\boldsymbol{H}}_A \big| k\, t\big\rangle = E_k \big| k\, t\big\rangle. \tag{5.36}$$

Thus the space and time dependence of $\big| k\, t\big\rangle$ is separable. Specifically,

$$\big| k\, t\big\rangle = e^{-iE_k t/\hbar} \big| k\, 0\big\rangle. \tag{5.37}$$

A general state $\big| \mathbf{x}\, 0\big\rangle$ of the atom at time $t = 0$ can be expanded in terms of the complete set of eigenkets $\big| k\, 0\big\rangle$ by writing

$$\big| \mathbf{x}\, 0\big\rangle = \sum_k a_k \big| k\, 0\big\rangle, \tag{5.38}$$

where $a_k \equiv \big\langle k\, 0 \,\big|\, \mathbf{x}\, 0\big\rangle$. At a later time t this general state is

$$\big| \mathbf{x}\, t\big\rangle = \sum_k a_k \big| k\, t\big\rangle = \sum_k a_k e^{-iE_k t/\hbar} \big| k\, 0\big\rangle. \tag{5.39}$$

The probability that a measurement finds the system in state k is $|a_k|^2$.

Vector Potential

As in § 3.2, the classical magnetic field $\mathbf{H} \equiv \nabla \times \mathbf{A}$, so $\nabla \cdot \mathbf{A}$ can be specified arbitrarily. To quantize $\mathbf{A}$, instead of using the Lorentz gauge (3.45), it is better to work in the *Coulomb gauge* (also called the *radiation gauge*), in which

$$\nabla \cdot \mathbf{A} = 0. \tag{5.40}$$

Then using equations (3.25a) and (5.40) in the divergence of equation (3.43) we obtain

$$-\nabla^2 \phi = \nabla \cdot \mathbf{E} \equiv 0, \tag{5.41}$$

so that ϕ is a solution of Poisson's equation.[8] A valid solution in vacuum is $\phi = 0$, which can be considered as part of the gauge condition.

Note that for a differentiable function f, $\nabla \cdot (\mathbf{A} f) = (\nabla \cdot \mathbf{A}) f + \mathbf{A} \cdot (\nabla f)$. But in the Coulomb gauge $\nabla \cdot \mathbf{A} \equiv 0$, so $\mathbf{A} \cdot (\nabla f) - \nabla \cdot (\mathbf{A} f) = 0$. Hence $\hat{\boldsymbol{p}} = i\hbar \nabla$ and $\hat{\mathbf{A}}$ commute, so their order in a matrix element $\big\langle b \,\big| \hat{\mathbf{A}} \cdot \hat{\boldsymbol{p}} \big|\, a\big\rangle$ is immaterial, as stated above in connection with (5.32).

[8] If ϕ is derived from Poisson's equation instead of the Liénard–Wiechert potential (3.48), time retardation is omitted in its calculation. But relativistic causality requires that information not propagate faster than c. The resolution of the problem is that only $\mathbf{E}$ and $\mathbf{H}$ are observable quantities, whereas ϕ is not; see [147], [702, p. 125].

Standing Waves

As before, using equations (3.42) and (3.43) in equation (3.25d), we find

$$\nabla(\nabla\cdot\mathbf{A}) - \nabla^2\mathbf{A} = -\frac{1}{c}\frac{\partial}{\partial t}\left(\nabla\phi + \frac{1}{c}\frac{\partial \mathbf{A}}{\partial t}\right). \tag{5.42}$$

In view of the gauge conditions, (5.42) simplifies to

$$\frac{1}{c^2}\frac{\partial^2\mathbf{A}}{\partial t^2} - \nabla^2\mathbf{A} = 0, \tag{5.43}$$

which is the same as equation (3.27) for **E** and also has traveling-wave solutions.

Its general solution is a linear combination of monochromatic waves $\mathbf{A_k}(\mathbf{x}, t)$, subject to appropriate boundary conditions. The monochromatic wave for circular frequency ω_k and wave vector **k** is

$$\mathbf{A_k} = \mathfrak{a}_\mathbf{k} e^{i(\omega_k t - \mathbf{k}\cdot\mathbf{x})} + \mathfrak{a}_\mathbf{k}^* e^{-i(\omega_k t - \mathbf{k}\cdot\mathbf{x})} \equiv \mathbf{a_k}(t)e^{-i\mathbf{k}\cdot\mathbf{x}} + \mathbf{a}_\mathbf{k}^*(t)e^{i\mathbf{k}\cdot\mathbf{x}}. \tag{5.44}$$

Here $\mathfrak{a}_\mathbf{k}$ and its complex conjugate $\mathfrak{a}_\mathbf{k}^*$ are constant amplitudes, and the time dependence $e^{\pm i\omega_k t}$ has been absorbed into the coefficients $\mathbf{a_k}(t)$ and $\mathbf{a}_\mathbf{k}^*(t)$.

To obtain a field of standing waves, Fourier analyze **A** in a large, but finite, vacuum cavity of size $(L \times L \times L)$, with periodic boundary conditions at its surface:

$$\mathbf{A_k}(L, y, z) = \mathbf{A_k}(0, y, z) \tag{5.45a}$$

$$\mathbf{A_k}(x, L, z) = \mathbf{A_k}(x, 0, z) \tag{5.45b}$$

$$\mathbf{A_k}(x, y, L) = \mathbf{A_k}(x, y, 0). \tag{5.45c}$$

These requirements fix the components of **k**:

$$(k_x, k_y, k_z) = 2\pi(n_x, n_y, n_z)/L, \tag{5.46}$$

where

$$(n_x, n_y, n_z) = 0, \pm1, \pm2, \ldots. \tag{5.47}$$

With these values for (n_x, n_y, n_z), the electric and magnetic fields have *nodes*, i.e., zero amplitude, on the surface of the cavity. Thus we have a finite series of closely spaced, but discrete, modes.

To account for polarization, resolve **A** along two orthogonal unit vectors $\mathbf{e}_{\mathbf{k}\alpha}$ and, using (5.44), write the general spacetime behavior of **A** as

$$\mathbf{A} = \sum_\mathbf{k}\sum_{\alpha=1}^{2} \mathbf{e}_{\mathbf{k}\alpha}\left[a_{\mathbf{k}\alpha}(t)e^{-i\mathbf{k}\cdot\mathbf{x}} + a_{\mathbf{k}\alpha}^*(t)e^{i\mathbf{k}\cdot\mathbf{x}}\right]. \tag{5.48}$$

Equation (5.43) is homogeneous; hence each mode may be scaled by an arbitrary constant. For later convenience, rewrite (5.48) as

$$\mathbf{A} = \sum_{\mathbf{k}\alpha}(2\pi\hbar c/k)^{1/2}\mathbf{e}_{\mathbf{k}\alpha}\left[a_{\mathbf{k}\alpha}(t)e^{-i\mathbf{k}\cdot\mathbf{x}} + a_{\mathbf{k}\alpha}^*(t)e^{i\mathbf{k}\cdot\mathbf{x}}\right]. \tag{5.49}$$

Now, $a_{\mathbf{k}\alpha}(t)$ is a function of t only, $\mathbf{e}_{\mathbf{k}\alpha}$ is fixed in space, and $\nabla\boldsymbol{\cdot}\mathbf{A} = 0$, so

$$\nabla\boldsymbol{\cdot}\mathbf{A}_{\mathbf{k}\alpha} = \mathbf{k}\boldsymbol{\cdot}\mathbf{A}_{\mathbf{k}\alpha} = 0, \tag{5.50}$$

and the same for the complex conjugate. Hence $\mathbf{A}$ is *transverse*, lying in the plane perpendicular to its direction of propagation, $\mathbf{k}$.

Electric and Magnetic Fields

Using (5.49) in equations (3.43) and (3.42) in the Coulomb gauge, we find

$$\mathbf{E}_{\mathbf{k}\alpha} = -i(2\pi\hbar ck)^{1/2}\,\mathbf{e}_{\mathbf{k}\alpha}\left[a_{\mathbf{k}\alpha}(t)e^{-i\mathbf{k}\boldsymbol{\cdot}\mathbf{x}} - a^{*}_{\mathbf{k}\alpha}(t)e^{i\mathbf{k}\boldsymbol{\cdot}\mathbf{x}}\right] \tag{5.51}$$

and

$$\mathbf{H}_{\mathbf{k}\alpha} = -i(2\pi\hbar c/k)^{1/2}(\mathbf{k}\times\mathbf{e}_{\mathbf{k}\alpha})\left[a_{\mathbf{k}\alpha}(t)e^{-i\mathbf{k}\boldsymbol{\cdot}\mathbf{x}} - a^{*}_{\mathbf{k}\alpha}(t)e^{i\mathbf{k}\boldsymbol{\cdot}\mathbf{x}}\right]. \tag{5.52}$$

The square bracket in (5.49) is the sum of a quantity and its complex conjugate; hence $\mathbf{A}$ is real. The brackets in (5.51) and (5.52) contain a quantity minus its complex conjugate, so they are imaginary; the leading factor of i makes both $\mathbf{E}_{\mathbf{k}\alpha}$ and $\mathbf{H}_{\mathbf{k}\alpha}$ real. Also, recalling that $\omega_k = kc$, we see that (5.51) and (5.52) are consistent with equations (3.32) and (3.33).

From equation (3.40) the time-averaged energy density in each mode $\mathbf{k}$ is $\langle e_{\mathbf{k}}\rangle_{\mathrm{T}} = \langle E^2_{\mathbf{k}}\rangle_{\mathrm{T}}/4\pi$. For unpolarized radiation use (5.51), sum over polarization vectors $\mathbf{e}_{\mathbf{k}\alpha}$, and write $\xi \equiv i(\mathbf{k}\boldsymbol{\cdot}\mathbf{x} - \omega_k t)$. Then, from (5.51),

$$E^2_{\mathbf{k}} = (2\pi\hbar kc)\left[\mathfrak{a}^{*}_{\mathbf{k}}\mathfrak{a}_{\mathbf{k}} + \mathfrak{a}_{\mathbf{k}}\mathfrak{a}^{*}_{\mathbf{k}} - \mathfrak{a}_{\mathbf{k}}\mathfrak{a}_{\mathbf{k}}e^{-2\xi} - \mathfrak{a}^{*}_{\mathbf{k}}\mathfrak{a}^{*}_{\mathbf{k}}e^{2\xi}\right]. \tag{5.53}$$

In (5.53), both $\mathfrak{a}_{\mathbf{k}}$ and $\mathfrak{a}^{*}_{\mathbf{k}}$ are independent of time. The exponentials containing ξ average to zero over a cycle, so the energy density in mode $\mathbf{k}$ is

$$\langle e_{\mathbf{k}}\rangle_{\mathrm{T}} = \langle E^2_{\mathbf{k}}\rangle_{\mathrm{T}}/4\pi = \tfrac{1}{2}\hbar\omega_k\left(\mathfrak{a}^{*}_{\mathbf{k}}\mathfrak{a}_{\mathbf{k}} + \mathfrak{a}_{\mathbf{k}}\mathfrak{a}^{*}_{\mathbf{k}}\right). \tag{5.54}$$

For a classical field, both $\mathfrak{a}_{\mathbf{k}}$ and $\mathfrak{a}^{*}_{\mathbf{k}}$ are complex numbers, so their order in (5.54) is immaterial. But when these amplitudes are interpreted as quantum operators, their order relative to one another must be preserved.

Conversion to Quantum Description

Equation (5.54) gives the classical radiation-field Hamiltonian H_R in (5.27) for photons of type $\mathbf{k}$. *It is identical to the Hamiltonian for a quantized harmonic oscillator in equation* (3.99)

$$\hat{\boldsymbol{H}}_{\mathbf{k}} = \tfrac{1}{2}\hbar\omega_k\left(\hat{\boldsymbol{a}}^{\dagger}_{\mathbf{k}}\hat{\boldsymbol{a}}_{\mathbf{k}} + \hat{\boldsymbol{a}}_{\mathbf{k}}\hat{\boldsymbol{a}}^{\dagger}_{\mathbf{k}}\right) \tag{5.55}$$

if $\mathfrak{a}_{\mathbf{k}}^*$ *is identified with the creation operator* $\hat{\boldsymbol{a}}_{\mathbf{k}}^{\dagger}$, *and* $\mathfrak{a}_{\mathbf{k}}$ *with the destruction operator* $\hat{\boldsymbol{a}}_{\mathbf{k}}$. It was to achieve this correspondence that we introduced the scaling used in (5.49). Thus the quantized version of (5.49) for **A** is

$$\hat{\mathbf{A}} = \sum_{\mathbf{k}\alpha} (2\pi\hbar c/k)^{1/2} \mathbf{e}_{\mathbf{k}\alpha} \left(\hat{\boldsymbol{a}}_{\mathbf{k}\alpha} e^{-i\mathbf{k}\cdot\mathbf{x}} + \hat{\boldsymbol{a}}_{\mathbf{k}\alpha}^{\dagger} e^{i\mathbf{k}\cdot\mathbf{x}} \right), \tag{5.56}$$

and similarly for **E** and **H** in (5.51) and (5.52).

Time Development

If the atom is unperturbed, i.e., only the Hamiltonian $\hat{\boldsymbol{H}}_A$ in (5.32) is in operation, then the a_k's in (5.39) are constant. But if there is an additional *perturbing potential* $\hat{\boldsymbol{V}}(t)$, e.g., the interaction Hamiltonian $\hat{\boldsymbol{H}}_I$ in (5.32), Schrödinger's equation is

$$\hat{\boldsymbol{H}} \big| \mathbf{x}\, t \big\rangle = [\, \hat{\boldsymbol{H}}_A + \hat{\boldsymbol{V}}(t)\,] \big| \mathbf{x}\, t \big\rangle = i\hbar \frac{\partial \big| \mathbf{x}\, t \big\rangle}{\partial t}, \tag{5.57}$$

and the expansion parameters become slowly varying functions of time, i.e.,

$$\big| \mathbf{x}\, t \big\rangle = \sum_k a_k(t) e^{-iE_k t/\hbar} \big| \, k\, 0 \big\rangle. \tag{5.58}$$

The time variation of the a_k's is interpreted as the atom making *transitions* from one level to another. This problem can be treated by the standard techniques of quantum mechanical perturbation theory.

Substitution of (5.58) into (5.57) gives

$$\Big[\hat{\boldsymbol{H}}_A + \hat{\boldsymbol{V}}(t) \Big] \sum_k a_k(t) e^{-iE_k t/\hbar} \big| \, k\, 0 \big\rangle = i\hbar \frac{\partial}{\partial t} \sum_k a_k(t) e^{-iE_k t/\hbar} \big| \, k\, 0 \big\rangle \tag{5.59}$$

or

$$\sum_k a_k(t) [\, E_k + \hat{\boldsymbol{V}}(t)\,] e^{-iE_k t/\hbar} \big| \, k\, 0 \big\rangle = i\hbar \sum_k \left[\frac{da_k}{dt} + \frac{a_k(t) E_k}{i\hbar} \right] e^{-iE_k t/\hbar} \big| \, k\, 0 \big\rangle; \tag{5.60}$$

hence

$$\sum_k \frac{da_k}{dt} e^{-iE_k t/\hbar} \big| \, k\, 0 \big\rangle = (i\hbar)^{-1} \sum_k a_k(t) \hat{\boldsymbol{V}}(t) e^{-iE_k t/\hbar} \big| \, k\, 0 \big\rangle. \tag{5.61}$$

This system is complicated. The information we seek can be obtained by computing the behavior of a specific level u. Thus, if we multiply (5.61) by the bra $\langle u\, 0 | e^{iE_u t/\hbar}$, integrate over space, and use (5.34), we obtain

$$\frac{da_u}{dt} = (i\hbar)^{-1} \sum_k a_k(t) \langle u\, 0 | \hat{V}(t) | k\, 0 \rangle e^{i(E_u - E_k)t/\hbar}. \tag{5.62}$$

Here it was assumed that any shift of the atomic energy levels that might result from the small perturbing potential can be ignored.

For brevity write $\langle u | \hat{V}(t) | k \rangle \equiv V_{uk}(t)$, and write $\hat{p}_j = m(\partial \hat{r}_j / \partial t)$. Then the term linear in $\hat{A}$ in (5.32) becomes

$$V_{uk}(t) = \frac{e}{m_e c} \hat{A} \cdot \sum_j \langle u | \hat{p}_j | k \rangle = \frac{e}{c} \hat{A} \cdot \sum_j \left\langle u \left| \frac{\partial \hat{r}_j}{\partial t} \right| k \right\rangle. \tag{5.63}$$

The matrix element in (5.63) can be evaluated by switching from the *Schrödinger representation*, in which operators are time-independent and their eigenkets $| k\, t \rangle$ time-dependent, to the *Heisenberg representation*, in which the eigenkets are time-independent and operators time-dependent. They are connected by a *unitary transformation* defined such that in the Schrödinger representation,

$$| k\, t \rangle_S \equiv \mathsf{U}_t | k\, 0 \rangle_S. \tag{5.64}$$

Here the subscript "S" denotes "Schrödinger." Then from (5.46),

$$i\hbar \frac{\partial}{\partial t} \mathsf{U}_t | k\, 0 \rangle_S = \hat{H} \mathsf{U}_t | k\, 0 \rangle_S, \tag{5.65}$$

which implies that

$$i\hbar \frac{\partial \mathsf{U}_t}{\partial t} = \hat{H} \mathsf{U}_t, \tag{5.66}$$

or

$$\mathsf{U}_t = \exp(-iE_k t/\hbar), \tag{5.67}$$

which shows that U_t is the *time development operator* in the Schrödinger representation.

Define $| k\, t \rangle_H = \mathsf{U}_t^{-1} | k\, t \rangle_S$; the subscript "H" denotes "Heisenberg." Then

$$| k\, t \rangle_H = \mathsf{U}_t^{-1} \mathsf{U}_t | k\, t \rangle_S \equiv | k\, 0 \rangle_S, \tag{5.68}$$

which verifies that $| k \rangle_H$ is indeed independent of time.

Because a matrix element must have the same value in both representations, for an operator $\hat{O}_S$ in the Schrödinger representation, the corresponding operator in the Heisenberg representation is

$$\begin{aligned} \langle k\, t |_S \hat{O}_S | k\, t \rangle_S &\equiv \langle k\, 0 |_S \mathsf{U}_t^{-1} \hat{O}_S \mathsf{U}_t | k\, 0 \rangle_S \equiv \langle k |_H \hat{O}_H(t) | k \rangle_H \\ &\Rightarrow \hat{O}_H(t) = \mathsf{U}_t^{-1} \hat{O}_S \mathsf{U}_t = e^{iE_k t/\hbar} \hat{O}_S e^{-iE_k t/\hbar}. \end{aligned} \tag{5.69}$$

Differentiating the second line of (5.69) with respect to time, we find

$$i\hbar\frac{d\hat{\boldsymbol{O}}_{\mathrm{H}}(t)}{dt} = \hat{\boldsymbol{O}}_{\mathrm{H}}\hat{\boldsymbol{H}} - \hat{\boldsymbol{H}}\hat{\boldsymbol{O}}_{\mathrm{H}}. \tag{5.70}$$

Then, taking $\hat{\boldsymbol{O}}_H(t) = \hat{\boldsymbol{r}}_j(t)$, we see that

$$\begin{aligned}\left\langle u \left| \sum_j e\frac{\partial \hat{\boldsymbol{r}}_j}{\partial t} \right| k \right\rangle &= [\, i(E_u - E_k)/\hbar\,]\langle u \,|\, \sum_j e\,\hat{\boldsymbol{r}}_j \,|\, k\rangle \\ &\equiv [\, i(E_u - E_k)/\hbar\,]\langle u \,|\, \hat{\boldsymbol{d}} \,|\, k\rangle,\end{aligned} \tag{5.71}$$

where $\hat{\boldsymbol{d}} = \sum_j e\hat{\boldsymbol{r}}_j$ is the atom's *total electric dipole* moment.[9]

Define $\langle u \,|\hat{\boldsymbol{d}}|\, k\rangle \equiv \mathbf{d}_{ku}$ and $\nu_{lu} \equiv \nu_u - \nu_l = (E_u - E_l)/h$. Using (5.63) and (5.71) we have

$$V_{uk}(t) = 2\pi i\nu_{ku}\mathbf{A}\cdot\mathbf{d}_{ku}/c. \tag{5.72}$$

Equation (5.72) shows that $V_{kk}(t) \equiv 0$, i.e., transitions out of a given level back into the same level have no net effect. Hence (5.62) becomes

$$\frac{da_u}{dt} = (i\hbar)^{-1}\sum_{k\neq u} a_k(t)V_{uk}(t)e^{2\pi i\nu_{ku}t}. \tag{5.73}$$

Transition Rate

Now assume that at $t = 0$, the atom is in a definite eigenstate l, so that $a_l(0) = 1$ and $a_k(0) = 0$ for $k \neq l$. Furthermore, take a time interval T short enough that $a_l(t) \approx 1$ for $t \leq T$. Then the sum in (5.73) reduces to the single term

$$\frac{da_u}{dt} = -\frac{4\pi^2\nu_{lu}}{ch}\mathbf{A}\cdot\mathbf{d}_{lu}e^{2\pi i\nu_{lu}t}. \tag{5.74}$$

Write the vector potential $\mathbf{A}$ as

$$\mathbf{A} = \mathbf{A}_\nu\cos(2\pi\nu t + \zeta_\nu). \tag{5.75}$$

The phases ζ_ν are random for incoherent radiation and can be ignored. Express $\cos x$ in (5.75) in terms of complex exponentials; then (5.74) becomes

$$\frac{da_u}{dt} = -\frac{2\pi^2\nu_{lu}}{ch}(\mathbf{A}_\nu\cdot\mathbf{d}_{lu})\left[e^{2\pi i(\nu+\nu_{lu})t} + e^{-2\pi i(\nu-\nu_{lu})t}\right], \tag{5.76}$$

[9] The equivalence of (5.63) and (5.71) shows that *exact* eigenkets $|k\rangle$ must give the same transition probability using the *velocity operator* in (5.63) *or* the *length operator* in (5.71). How closely the results actually agree gives an indication of the accuracy of a set of approximate wave functions.

which, when integrated with respect to time, yields

$$a_u(t) = \frac{i\pi\nu_{lu}}{ch}(\mathbf{A}_\nu \cdot \mathbf{d}_{lu})\left[\frac{e^{2\pi i(\nu+\nu_{lu})t}-1}{\nu+\nu_{lu}} + \frac{e^{-2\pi i(\nu-\nu_{lu})t}-1}{\nu-\nu_{lu}}\right], \tag{5.77}$$

In calculating $a_u^*(t)a_u(t)$, note the following:

- In an absorption process $E_u > E_l$, so $\nu_{lu} > 0$. The dominant contribution to $a_u(t)$ comes from the resonance at $\nu \approx \nu_{lu}$ in the denominator of the second term of (5.77). Thus radiation near the frequency of the spectral line joining levels l and u is the most effective in producing transitions $l \to u$. Therefore, the first term in (5.77) is negligible compared to the second and can be dropped.
- For typical astrophysical conditions the amplitude of $\mathbf{A}_\nu$ is essentially constant over the frequency width of the transition. For example, the specific intensity of thermal radiation at 5000 K (about the surface temperature of the Sun) varies little over a frequency interval $\Delta \sim (0.01kT/h) \sim 10^{12}$. A typical transition time is $t \sim 10^{-8}$ s; hence $\delta\nu \sim 10^8 \ll \Delta$.

Now write $\mathbf{A}_\nu \equiv A_0\,\hat{\mathbf{a}}$, where $\hat{\mathbf{a}}$ is a unit vector pointing along $\mathbf{A}$, and $\mathbf{d}_{ul} \equiv d_{ul}\,\hat{\mathbf{d}}$, where $\hat{\mathbf{d}}$ is a unit vector pointing along $\mathbf{d}_{ul}$, so $\mathbf{A}_\nu \cdot \mathbf{d}_{ul} \to A_0\,d_{ul}\,\hat{\mathbf{a}}\cdot\hat{\mathbf{d}} \equiv A_0\,d_{ul}\cos\theta$. Keeping only the resonant term in (5.77) and using equation 404.12 from [294] we get

$$\begin{aligned} |a_u(t)|^2 &= \frac{\pi^2\nu_{lu}^2}{c^2h^2}(A_0 d_{lu}\cos\theta)^2\left[\frac{2-2\cos[\,2\pi(\nu-\nu_{lu})t\,]}{(\nu-\nu_{lu})^2}\right] \\ &= \frac{\nu_{lu}^2}{c^2\hbar^2}(A_0 d_{lu}\cos\theta)^2\,\frac{\sin^2[\,\pi(\nu-\nu_{lu})t\,]}{(\nu-\nu_{lu})^2}. \end{aligned} \tag{5.78}$$

Equation (5.78) can be interpreted as the probability, per atom in level l, that an $l \to u$ transition is driven by radiation of frequency ν in a time t. To obtain the total probability, integrate (5.78) over a frequency range $\nu - \nu_{lu} = \pm\delta$ containing the entire line profile. Thus, set $x \equiv (\nu - \nu_{lu})$. Then the probability of a transition $l \to u$ is

$$P_{lu} = \frac{\nu_{lu}^2}{c^2\hbar^2}(A_0 d_{lu}\cos\theta)^2 \int_{-\delta}^{\delta} \frac{\sin^2(\pi x t)}{x^2}\,dx. \tag{5.79}$$

The resonance at $x = 0$ is so sharp that the limits on the integral can formally be extended to $\pm\infty$.

Make the additional transformation $y = \pi x t$; then

$$P_{lu} = \frac{\pi\nu_{lu}^2 t}{c^2\hbar^2}(A_0 d_{lu}\cos\theta)^2 \int_{-\infty}^{\infty} \frac{\sin^2 y}{y^2}\,dy. \tag{5.80}$$

From equation 858.652 in [294] the value of this definite integral is π. And (5.80) shows that the total probability that an atom jumps from l to u is linear in time.

Because the particles in the material are oriented at random, the angular average of $< \cos^2\theta >$ over solid angle is $\frac{1}{3}$. The final result is

$$P_{lu} = \frac{\pi^2 \nu_{lu}^2 A_0^2}{3c^2\hbar^2} \left|\langle u \left|\mathbf{d}\right| l\rangle\right|^2 t. \tag{5.81}$$

Einstein's Absorption Probability B_{lu}

By using (5.75) for $\mathbf{A}$ in the general relation $\mathbf{E} \equiv -c^{-1}(\partial\mathbf{A}/\partial t)$, taking time averages, and equating the macroscopic expression for the radiation energy density in (3.19) to the electromagnetic expression in (3.41), we find

$$\langle A_0^2\rangle_{\mathrm{T}} = \frac{2c^2}{\pi\nu_{lu}^2}\frac{\langle E_0^2\rangle_{\mathrm{T}}}{8\pi} = \frac{2c^2}{\pi\nu_{lu}^2}\frac{4\pi J(\nu_{lu})}{c} = \frac{8cJ(\nu_{lu})}{\nu_{lu}^2}. \tag{5.82}$$

Thus

$$P_{lu} = \frac{8\pi^2}{3c\hbar^2}\left|\langle u \left|\mathbf{d}\right| l\rangle\right|^2 J(\nu_{lu})\, t. \tag{5.83}$$

In terms of Einstein's absorption probability, $P_{lu} = B_{lu}J(\nu_{lu})\,t$, so

$$B_{lu} = \frac{8\pi^2}{3c\hbar^2}\left|\langle u \left|\mathbf{d}\right| l\rangle\right|^2 = \frac{32\pi^4}{3ch^2}\left|\langle u \left|\mathbf{d}\right| l\rangle\right|^2. \tag{5.84}$$

If $E_l > E_u$, (5.76)–(5.81) can be used with trivial modifications to compute B_{ul}. Given B_{ul}, the Einstein relations (5.7) and (5.8) could be used to compute A_{ul}. But more fundamentally, we show below that it can be derived directly from a full quantum mechanical computation.

5.4 BOUND–BOUND EMISSION PROBABILITY

The semi-classical theory used in § 5.3 to calculate bound-bound absorption probabilities is incomplete because it accounts only for transitions resulting from the action of a *given* radiation field on the atom. But it omits the back reaction of transitions in the atom (i.e., emission of photons) on the radiation field; hence it cannot be used to compute spontaneous emission probabilities. In a major step forward, Dirac [280] extended the theory to account for the *mutual* interaction between a quantized atom and a quantized radiation field.

Quantized Interaction Hamiltonian

Using (5.56) for $\hat{A}$, write the quantized interaction Hamiltonian in (5.32) as $\hat{H}_I = \hat{H}_1 + \hat{H}_2$, where $\hat{H}_1$ represents $\hat{A}\cdot\hat{p}$ and $\hat{H}_2$ represents $\hat{A}\cdot\hat{A}$. Then

$$\hat{H}_1 = (e/m_e c) \sum_{\mathbf{k}\alpha} \sqrt{2\pi\hbar c/k}\ \mathbf{e}_{\mathbf{k}\alpha} \cdot \hat{p} \left(\hat{a}_{\mathbf{k}\alpha} e^{-i\mathbf{k}\cdot\mathbf{x}} + \hat{a}^{\dagger}_{\mathbf{k}\alpha} e^{i\mathbf{k}\cdot\mathbf{x}} \right) \tag{5.85}$$

$$\begin{aligned} \hat{H}_2 = (e^2/2m_e c^2) \sum_{\mathbf{k}'\alpha'} \sum_{\mathbf{k}\alpha} (2\pi\hbar c/\sqrt{kk'}) \mathbf{e}_{\mathbf{k}'\alpha'} \cdot \mathbf{e}_{\mathbf{k}\alpha} \\ \times \Big[\hat{a}_{\mathbf{k}\alpha}\hat{a}_{\mathbf{k}'\alpha'} e^{-i(\mathbf{k}+\mathbf{k}')\cdot\mathbf{x}} + \hat{a}_{\mathbf{k}\alpha}\hat{a}^{\dagger}_{\mathbf{k}'\alpha'} e^{-i(\mathbf{k}-\mathbf{k}')\cdot\mathbf{x}} \\ + \hat{a}^{\dagger}_{\mathbf{k}\alpha}\hat{a}_{\mathbf{k}'\alpha'} e^{i(\mathbf{k}-\mathbf{k}')\cdot\mathbf{x}} + \hat{a}^{\dagger}_{\mathbf{k}\alpha}\hat{a}^{\dagger}_{\mathbf{k}'\alpha'} e^{i(\mathbf{k}+\mathbf{k}')\cdot\mathbf{x}} \Big]. \end{aligned} \tag{5.86}$$

Effect of the Interaction Hamiltonian on Radiation Eigenkets

Before calculating the emission rate from a bound level, it is instructive to examine the effects of $\hat{H}_1$ and $\hat{H}_2$ on radiation kets to understand better the connection between spontaneous and induced emission.

Emission

The unperturbed Hamiltonian $\hat{H} = \hat{H}_A + \hat{H}_R$ for a system containing matter and radiation has state vectors of the form

$$\begin{aligned} |\,\text{atom} + \text{radiation}\rangle &= |\,a\rangle_{\text{atom}} |\,n_1\rangle |\,n_2\rangle \ldots |\,n_{\mathbf{k}}\rangle \ldots \rangle_{\text{rad}} \\ &\equiv |\,a\rangle_{\text{atom}} |\,n_1 \ldots n_{\mathbf{k}} \ldots \rangle_{\text{rad}}. \end{aligned} \tag{5.87}$$

As in § 3.6, the first form above indicates that photons do not interact with one another; the second is a shorthand notation.

Suppose an electron initially in an upper level $|\,u\rangle$ makes a transition to a lower level $|\,l\rangle$, emitting a photon of type $|n_{\mathbf{k}}\rangle|$. The initial and final states are

$$|\,i\rangle = |\,u\rangle_{\text{atom}} |\ldots n_{\mathbf{k}} \ldots \rangle_{\text{rad}} \tag{5.88}$$

and

$$|f\rangle = |\,l\rangle_{\text{atom}} |\ldots n_{\mathbf{k}} + 1 \ldots \rangle_{\text{rad}}. \tag{5.89}$$

The transition $i \to f$ results from the action of the second term in the interaction Hamiltonian $\hat{H}_1$ in (5.85), which we can represent as an operator $\hat{\mathcal{O}}_{\mathbf{k}}$ containing numerical factors, the electron's momentum, and the photon's wave vector and polarization and the creation operator $\hat{a}^{\dagger}_{\mathbf{k}}$, acting on the initial atomic ket $|\,u\rangle$.

The discussion in § 3.6 of normalization of the radiation field's kets showed that $\hat{a}^\dagger|n\rangle = (n+1)^{1/2}|n+1\rangle$ and $\hat{a}|n\rangle = n^{1/2}|n-1\rangle$. Hence

$$\langle f|\hat{H}_1|i\rangle = \langle l|\hat{O}_\mathbf{k}|u\rangle\langle\ldots n_\mathbf{k}+1\ldots|\hat{a}_k^\dagger|\ldots n_\mathbf{k}\ldots\rangle$$
$$= \langle l|\hat{O}_\mathbf{k}|u\rangle(n_\mathbf{k}+1)^{1/2}. \qquad (5.90)$$

The transition rate $u \to l$ is proportional to

$$r_{\mathrm{ul}} \propto |\langle f|\hat{H}_1|i\rangle|^2 = |\langle l|\hat{O}_\mathbf{k}|u\rangle|^2(n_\mathbf{k}+1)\,. \qquad (5.91)$$

The "1" in the factor $(n_\mathbf{k}+1)$ accounts for spontaneous emission; the factor of $n_\mathbf{k}$ accounts for induced emissions.

Absorption

For absorption, reverse the direction of the process. Now the system is operated on by the first term of $\hat{H}_1$ in (5.85), which contains the product of an operator $\hat{O}_\mathbf{k}$ and the destruction operator $\hat{a}_\mathbf{k}$. In this case

$$|i\rangle = |l\rangle_{\mathrm{atom}}|\ldots n_\mathbf{k}\ldots\rangle_{\mathrm{rad}} \qquad (5.92\mathrm{a})$$

$$|f\rangle = |u\rangle_{\mathrm{atom}}|\ldots n_\mathbf{k}-1\ldots\rangle_{\mathrm{rad}} \qquad (5.92\mathrm{b})$$

so

$$\langle f|\hat{H}_1|i\rangle = \langle u|\hat{O}_\mathbf{k}|l\rangle\langle\ldots n_\mathbf{k}-1\ldots|\hat{a}_k|\ldots n_\mathbf{k}\ldots\rangle = \langle u|\hat{O}_\mathbf{k}|l\rangle\, n_\mathbf{k}^{1/2} \qquad (5.92\mathrm{c})$$

and

$$r_{\mathrm{lu}} \propto |\langle f|\hat{H}_1|i\rangle|^2 = |\langle u|\hat{O}_\mathbf{k}|l\rangle|^2 n_\mathbf{k}. \qquad (5.92\mathrm{d})$$

The transition rate $l \to u$ is proportional to the number of photons present.

Fermi's Golden Rule

The second term in (5.85) represents photon emission. As seen in (5.91) and (5.92d), the creation and destruction operators affect only the radiation field, so only the matrix element connecting the atom's upper level to the lower level needs to be evaluated. Let the initial and final states be $|i\rangle = |u\rangle|n\rangle$ and $|f\rangle = |l\rangle|n+1\rangle$. $E_i \equiv E_u$ and $E_f = E_l + h\nu_{lu}$ are the total energies of the initial and final states, where E_u and E_l are the energies of the upper and lower atomic levels, and $h\nu_{lu}$ is the energy of the emitted photon.

Consider a transition driven by a perturbation $\hat{V}$ that starts "instantaneously" and then remains constant. Assume that the atom initially is entirely in the upper level, whose population is not significantly altered as it decays to the lower level. Then from (5.62),

$$\frac{da_f}{dt} = (i\hbar)^{-1}\langle f|\hat{\mathbf{V}}|i\rangle \exp(i\omega t), \tag{5.93}$$

where $\omega = (E_f - E_i)/\hbar$. Integrating over time,

$$a_f = \langle f|\hat{\mathbf{V}}|i\rangle \frac{1-\exp(i\omega t)}{E_f - E_i}, \tag{5.94}$$

so

$$|a_f|^2 = 2|\langle f|\hat{\mathbf{V}}|i\rangle|^2 \frac{1-\cos\omega t}{(E_f - E_i)^2}. \tag{5.95}$$

The transition probability becomes large when $E_i - E_f$ is close to zero, i.e., when $h\nu_{lu}$, the energy of the emitted photon, very nearly equals $E_u - E_l \equiv E_0$, so that *the total energy of the atom + radiation system is (very nearly) conserved in the transition.*

Both the upper and lower levels of the atom have a finite spread over energy, so this criterion can be satisfied by a continuum of states of the radiation field. Let $\tilde{\rho}(E)$ be the density of such states; then the total number of transitions in a time t is

$$N_{if} = \int_{E_0-\epsilon}^{E_0+\epsilon} \tilde{\rho}(E)|a_f|^2 dE = 2|\langle f|\hat{\mathbf{V}}|i\rangle|^2 \int_{E_0-\epsilon}^{E_0+\epsilon} \tilde{\rho}(E)\left[\frac{1-\cos\omega t}{E_f - E_i}\right]^2 dE.$$

Make the transformation $x \equiv \omega t = (E - E_0)t/\hbar$. Then

$$N_{if} = \frac{2t}{\hbar}|\langle f|\hat{\mathbf{V}}|i\rangle|^2 \int_{-\epsilon t/\hbar}^{\epsilon t/\hbar} \tilde{\rho}\left(E_0 + \frac{\hbar x}{t}\right)\frac{(1-\cos x)^2}{x^2}\,dx, \tag{5.96}$$

which increases linearly with time. As in (5.79), the integrand is essentially a delta function at $x = 0$, so we can take the density of states $\tilde{\rho}(E_0)$ outside the integral, set its limits to $\pm\infty$, and note that $\int_{-\infty}^{\infty}(1-\cos x)^2/x^2 = \pi$. The result is *Fermi's Golden Rule* for the transition probability per unit time:

$$r_{\text{if}} = \frac{2\pi}{\hbar}\tilde{\rho}(E_0)\langle f|\hat{\mathbf{V}}|i\rangle^2. \tag{5.97}$$

This formula greatly simplifies many quantum mechanical calculations.

Einstein's Spontaneous and Induced Emission Probabilities A_{ul} and B_{ul}

Emission is the result of action by the term in (5.85) containing the creation operator. For a transition $u \to l$,

$$\langle f|\hat{\boldsymbol{H}}_1|i\rangle = \frac{e}{m_e c}\left(\frac{2\pi\hbar c}{k}\right)^{\frac{1}{2}} \langle f|\hat{\boldsymbol{a}}_{\mathbf{k}}^{\dagger} e^{i\mathbf{k}\cdot\mathbf{x}}\hat{\boldsymbol{p}}|i\rangle \cdot \mathbf{e}_{\mathbf{k}\alpha}\,. \tag{5.98}$$

As noted in § 5.4, the wavelength of visible and ultraviolet radiation is much larger than the size of an atom, so $e^{i\mathbf{k}\cdot\mathbf{x}} \to 1$. Then, applying the creation operator to the radiation kets contained in $|i\rangle$ and $|f\rangle$, we have

$$\langle f\,|\hat{\boldsymbol{H}}_1|\,i\rangle = \frac{e}{m_e c}\left(\frac{2\pi\hbar c}{k}\right)^{\frac{1}{2}}\langle u\,|\hat{\boldsymbol{p}}|\,l\rangle\cdot\mathbf{e}_{\mathbf{k}\alpha}(n_{\mathbf{k}}+1)^{\frac{1}{2}}. \tag{5.99}$$

From (5.71),

$$\frac{e}{m_e c}\langle u\,|\hat{\boldsymbol{p}}|\,l\rangle\cdot\mathbf{e}_{\mathbf{k}\alpha} = \frac{2\pi i\nu_{lu}}{c}\langle l\,|\mathbf{d}|\,u\rangle\cdot\mathbf{e}_{\mathbf{k}\alpha}. \tag{5.100}$$

Substituting (5.100) into (5.99), and averaging $\mathbf{d}\cdot\mathbf{e}_\alpha$ over solid angle for random orientations of the atom's dipole moment, we obtain

$$\left|\langle f\,|\hat{\boldsymbol{H}}_1|\,i\rangle\right|^2 = \frac{2\pi h\nu_{lu}}{3}\left|\langle u\,|\mathbf{d}|\,l\rangle\right|^2[\,n(\nu_{lu})+1\,]. \tag{5.101}$$

From equation (4.54) or, more correctly, (4.80), the density of states $\tilde{\rho}(E)$ is

$$\tilde{\rho}(E)dE = \mathfrak{n}(\nu)d\nu = \frac{8\pi\nu^2}{c^3}d\nu \Rightarrow \tilde{\rho}(E) = \frac{8\pi\nu^2}{hc^3}. \tag{5.102}$$

Substitution of (5.101) and (5.102) into Fermi's Golden Rule (5.97) gives the total emission rate

$$\begin{aligned} r_{\mathrm{ul}} &= \left(\frac{4\pi^2}{h}\right)\left(\frac{8\pi\nu_{lu}^2}{hc^3}\right)\left(\frac{2\pi h\nu_{lu}}{3}\right)\left|\langle u\,|\mathbf{d}|\,l\rangle\right|^2\left[n(\nu_{lu})+1\right] \\ &= \left(64\pi^4\nu_{lu}^3/3c^3h\right)\left|\langle u\,|\mathbf{d}|\,l\rangle\right|^2\left[n(\nu_{lu})+1\right]. \end{aligned} \tag{5.103}$$

Spontaneous emission comes from the "1" in the square bracket, i.e.,

$$A_{ul} = \frac{64\pi^4\nu_{lu}^3}{3c^3h}\left|\mathbf{d}_{lu}\right|^2, \tag{5.104}$$

where $|d_{lu}| = e\left|\langle\,l\,|\mathbf{r}|\,u\,\rangle\right|$ is the dipole matrix element connecting l and u.

Dividing (5.104) by $(2h\nu_{lu}^3/c^2)$, we get $(32\pi^4/3ch^2)\left|\langle l\,|\mathbf{d}|\,u\rangle\right|^2 \equiv B_{lu}$ in (5.84), as expected from Einstein's detailed-balance result (5.8). The above derivation of A_{ul} yields the correct result. But it is not relativistically invariant, and in the Hamiltonian of multi-electron atoms it omits terms that account for the Coulomb repulsion among pairs of electrons and spin-orbit terms arising from coupling of the orbital and spin angular momenta of the electron making the transition. We give a brief description of these phenomena in chapter 7; for details see [882, § 2.13–§ 2.16].

Oscillator Strength and Line Strength

If we were to associate a classical oscillator with a transition between two bound levels in an atom, there would be a single cross section $\sigma_{cl} = (\pi e^2/mc)$ for all transitions (see § 6.1). But the absorption probability B_{ij} for a transition between levels i and j actually depends on the quantum mechanical properties of those levels. Hence it is convenient to write an absorption probability in terms of a dimensionless *oscillator strength*, or *f-value*, f_{ij} defined such that

$$\sigma_{ij} = f_{ij}\frac{\pi e^2}{mc} \equiv \frac{h\nu_{ij}}{4\pi}B_{ij}. \tag{5.105}$$

This parameterization is useful because $f_{ij} \sim 1$ for strong lines, and $f_{ij} \ll 1$ for very weak lines.

In our derivation of B_{lu} and A_{ul} we tacitly assumed l and u are single states so that $g_l = g_u = 1$. If the upper and/or lower states are degenerate, the total energy emitted by spontaneous transitions $u \to l$ is obtained by summing over all their substates u' and l':

$$E_{ul} = \left(\frac{64\pi^4\nu_{lu}^4}{3c^3}\right)\sum_{l',u'}|\mathbf{d}_{l'u'}|^2. \tag{5.106}$$

The sum in (5.106) is called the *line strength*

$$S(l,u) = \sum_{l',u'}\left|\mathbf{d}_{l'u'}\right|^2, \tag{5.107}$$

which is symmetric in l and u. Then

$$g_uA_{ul} = (64\pi^4\nu_{lu}^3/3c^3h)S(l,u) = (2h\nu_{lu}^3/c^2)g_uB_{ul}; \tag{5.108a}$$

hence

$$g_lB_{lu} = g_uB_{ul} = (32\pi^4/3ch^2)S(l,u), \tag{5.108b}$$

and from (5.105)

$$g_lf_{lu} = (8\pi^2m_e\nu/3e^2h)S(l,u). \tag{5.108c}$$

Or, writing $g_l \equiv \sum_{l'} g_{l'}$, the total oscillator strength of the transition is

$$f_{lu} = \sum_{l',u'} g_{l'}f_{l'u'}/g_l. \tag{5.109}$$

Effects of Induced Emission

Induced emission powers lasers and masers. In a resonant cavity, the atoms or molecules in a sample of gas can be driven into a *metastable* level (i.e., unable to decay by spontaneous emission) via permitted transitions to higher levels, followed by cascades into the metastable level. The result is a huge overpopulation in the metastable level, which has a long lifetime, and hence a narrow frequency width. If the gas is then illuminated with radiation at the frequency of the transition between the metastable and a lower state, the electrons in the metastable level can decay by induced emission. The radiation emitted is trapped and amplified in the cavity, resulting in an intense burst of coherent radiation in a narrow frequency range. The amplification process is efficient because if a photon with direction and frequency $(\mathbf{n}, \nu)$ in the laboratory frame is Lorentz transformed to $(\mathbf{n}', \nu')$ in an atom's frame and induces emission of other $(\mathbf{n}', \nu)'$ photons, they go through the inverse of that transformation and emerge back into the laboratory frame as $(\mathbf{n}, \nu)$ photons.

5.5 PHOTOIONIZATION

Density of Free States

Consider now the process of *photoionization* of an atom in bound state $|b\rangle$ by absorption of a photon with energy $\hbar\omega$ and momentum $\hbar\mathbf{k}$ that results in an electron being liberated into a *continuum* of states having momenta $\mathbf{p} = \hbar\boldsymbol{\kappa}$ and energies $E_\kappa = \hbar\omega - \chi_{\rm ion} = p^2/2m_e \equiv \hbar^2\kappa^2/2m_e$. The initial and final states of the atom + radiation system are

$$|i\rangle = |b\rangle|\ldots n_{\mathbf{k}}\ldots\rangle, \tag{5.110}$$

and

$$|f\rangle = |\kappa\rangle|\ldots n_{\mathbf{k}} - 1\ldots\rangle. \tag{5.111}$$

Assume that as $r \to \infty$ the free electron's wave function can be taken to be periodic in a box with sides L, so that $k_x = 2\pi n_x/L$, etc. Then the density of free states per unit wavenumber interval is

$$\tilde{\rho}(\kappa)d\kappa = (L/2\pi)^3\, d^3\kappa = (L/2\pi)^3\kappa^2\, d\kappa\, d\Omega, \tag{5.112}$$

where $d\Omega$ is an element of solid angle. As before, we will choose the volume to contain one electron, and set $L = 1$. The density of free states per unit energy interval is $\tilde{\rho}(E_\kappa) = \tilde{\rho}(\kappa)(d\kappa/dE_\kappa)$, where $dE_\kappa = p\,dp/m_e = \hbar^2\kappa\, d\kappa/m_e$, so

$$\tilde{\rho}(E_\kappa) = (\kappa^2\, d\kappa\, d\Omega/8\pi^3) \times (m_e/\hbar^2\kappa\, d\kappa) = m_e\kappa\, d\Omega/8\pi^3\hbar^2. \tag{5.113}$$

Differential Cross Section

Using (5.113) and the first term of (5.85) in Fermi's Golden Rule (5.97) we find that the probability per unit time that electrons in a bound state $|b\rangle$ are removed by photoionization $(b \to \kappa)$ into a free state $|\kappa\rangle$ and ejected into solid angle $d\Omega$ is

$$\begin{aligned} r_{\rm bf} &= \frac{2\pi}{\hbar}\frac{e^2}{m_e^2c^2}\frac{m_e\kappa}{8\pi^3\hbar^2}\frac{2\pi\hbar c^2}{\omega_k} n_{\mathbf{k}\alpha}\left|\langle\kappa\,|e^{i\mathbf{k}\cdot\mathbf{r}}\mathbf{e}_\alpha\cdot\hat{\boldsymbol{p}}|\,b\rangle\right|^2 \sin\theta\, d\theta\, d\phi \\ &= \frac{e^2\kappa}{2\pi\hbar^2 m_e\omega_k} n_{\mathbf{k}\alpha}\left|\langle\kappa\,|e^{i\mathbf{k}\cdot\mathbf{r}}\mathbf{e}_\alpha\cdot\hat{\boldsymbol{p}}|\,b\rangle\right|^2 d\Omega. \end{aligned} \tag{5.114}$$

By dividing r_{bf} by the photon flux $cn_{\mathbf{k}\alpha}$ and the exit solid angle $d\Omega$, we get the differential cross section for photoionization:

$$\frac{d\sigma_{\text{bf}}}{d\Omega} = \frac{e^2\kappa}{2\pi c\hbar^2 m_e \omega_k}\left|\left\langle\kappa\left|e^{i\mathbf{k}\cdot\mathbf{r}}\mathbf{e}_\alpha\cdot\hat{\boldsymbol{p}}\right|b\right\rangle\right|^2. \tag{5.115}$$

In view of (5.71) we can rewrite (5.115) as

$$\frac{d\sigma_{\text{bf}}}{d\Omega} = \frac{m_e e^2\kappa k}{2\pi\hbar^2}\left|\left\langle\kappa\left|e^{i\mathbf{k}\cdot\mathbf{r}}\mathbf{e}_\alpha\cdot\hat{\boldsymbol{r}}\right|b\right\rangle\right|^2, \tag{5.116}$$

which is general in the non-relativistic regime. When the wavelength of the ionizing radiation is much larger than atomic dimensions, we can set $e^{i\mathbf{k}\cdot\mathbf{r}} = 1$ (the *dipole approximation*).

To obtain actual photoionization cross sections, we need to know accurate wave functions for both the bound state $|b\rangle$ of the (usually multi-electron) atom and the final state $|\kappa\rangle$ of the free electron. The former can be written analytically only for an atom with a single bound electron; see § 7.1. For multi-electron systems $|b\rangle$ has to be determined by numerical computation; see § 7.2. In the limit $\hbar\omega \gg \chi_{ion}$, to lowest order the wave function of the emergent electron is not affected by the structure of the atom and can be represented by a plane wave (the *Born approximation*): $|\kappa\rangle \sim e^{i\boldsymbol{\kappa}\cdot\mathbf{x}}$. More accurate representations of the electron's final state $|\kappa\rangle$ are discussed in, e.g., [961, § 38] and [1006, pp. 240–242].

5.6 FREE-FREE TRANSITIONS

In a collision of an electron with an ion, sketched in figure 5.3, the two particles form a *time-varying dipole moment* that can emit or absorb photons in free-free transitions. If a photon is absorbed (emitted) in this process, the electron's kinetic energy is increased (decreased), and, classically speaking, it moves off on a different "orbit" relative to the ion. As it does, it exchanges energy collisionally with other particles, and hence tends to heat (cool) the material. Both particles are unbound before and after the collision.

Panel (a) in figure 5.3 shows a collision with high magnification near the instant of impact. The electron's path O is represented as a straight line with *impact parameter b*. In panel (b), at lower magnification, we see that O is actually a hyperbola with a large radius of curvature.

If the system emits a photon ($h\nu \le \frac{1}{2}mv^2$) during the collision, then the electron has less energy at infinity, so its hyperbolic path E curves inward away from O. If the system absorbs a photon during the collision, then the electron has higher energy, and its hyperbolic path A curves outward away from O.

For the low-density conditions in stellar atmospheres a classical treatment based on electromagnetic theory is often adequate, because in a rarefied medium b is usually much larger than an atomic dimension. In this section we calculate the free-free emission (*bremsstrahlung*) coefficient (see, e.g., [927, 1007]) and then

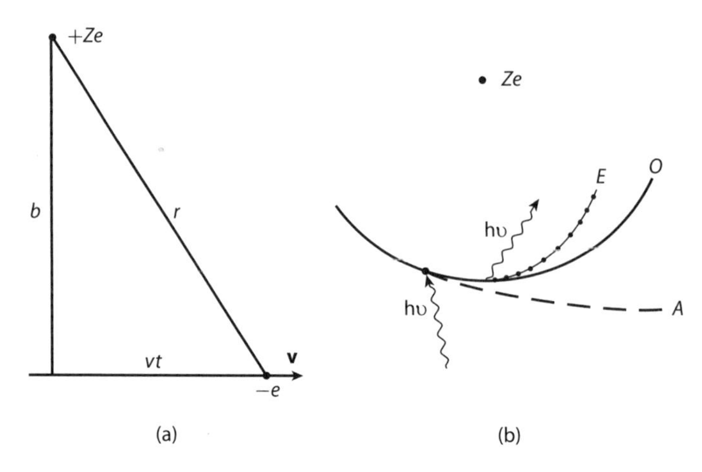

Figure 5.3 Collision geometry.

convert it to an absorption coefficient by using the Kirchhoff–Planck relation $\eta^*_{\rm ff} = \kappa^*_{\rm ff} B_\nu(T)$.

In § 7.1 we will sketch a quantum mechanical calculation based on a generalization of the formalism for bound-bound transitions in hydrogenic ions to free-free transitions. See also [444, pp. 56–60] for a quantum electrodynamics calculation.

Electromagnetic Flux Radiated in an Electron–Ion Collision

The frequency-integrated electric and magnetic fields $\mathbf{E}(t)$ and $\mathbf{H}(t)$ generated by a moving point charge are given by equations (3.52) and (3.53). We assume that the distance R between the observer and the electron is very much greater than the impact parameter b so that we need only the radiation far-field part of those expressions, that the electron moves with a speed much less than the speed of light so relativistic terms can be ignored, and that time retardation can be ignored for this application because the time delay required for the radiated electromagnetic field to reach the observer is irrelevant. Then equations (3.52) and (3.53) simplify to

$$\mathbf{E}(t) = \frac{-e \sin\theta(t)}{c^2 R} \ddot{\mathbf{r}}(t)\, \hat{\mathbf{e}} \tag{5.117}$$

and

$$\mathbf{H}(t) = \frac{-e \sin\theta(t)}{c^2 R} \ddot{\mathbf{r}}(t)\, \hat{\mathbf{h}}. \tag{5.118}$$

The frequency-integrated electromagnetic energy flux through a unit area is the Poynting vector

$$\mathbf{S}(t) = \frac{c}{4\pi} \mathbf{E}(t) \times \mathbf{H}(t). \tag{5.119}$$

$\mathbf{E}(t)$ and $\mathbf{H}(t)$ are orthogonal and $|\mathbf{E}| = |\mathbf{H}|$; hence the instantaneous flux per unit area is

$$\mathbf{S}(t) = \frac{e^2\, \ddot{r}(t)^2 \sin^2\theta(t)}{4\pi\, c^3\, R^2}\mathbf{n} \quad [\,\mathrm{erg/cm^2/sec}\,]. \tag{5.120}$$

In (5.118)–(5.120), $\hat{\mathbf{e}}$, $\hat{\mathbf{h}}$, and $\mathbf{n}$ form a right-handed orthonormal triad; $r(t)$ is the distance between the ion and electron; $\mathbf{v}(t)$ the electron's velocity vector; R the distance from the ion to the observer; $\mathbf{n}$ the unit vector from the ion toward the observer; and $\theta(t)$ the angle between $\ddot{\mathbf{r}}(t) \equiv \dot{\mathbf{v}}(t)$ and $\mathbf{n}$.

The total (frequency- and time-integrated) flux per unit solid angle radiated in the collision is

$$\mathcal{F} = \int_{-\infty}^{\infty} |\mathbf{S}(t)|\, dt = \frac{c}{4\pi}\int_{-\infty}^{\infty} |\mathbf{E}(t)|^2\, dt \quad [\,\mathrm{erg/sr}\,]. \tag{5.121}$$

Time–Frequency Reciprocity

The *frequency spectrum* of $\mathbf{E}$(t) is its *Fourier transform*:

$$\mathbf{E}(\omega) = \frac{1}{\sqrt{2\pi}}\int_{-\infty}^{\infty} \mathbf{E}(t)e^{i\omega t}\, dt, \tag{5.122}$$

which has the inverse transform

$$\mathbf{E}(t) = \frac{1}{\sqrt{2\pi}}\int_{-\infty}^{\infty} \mathbf{E}(\omega)e^{-i\omega t}\, d\omega. \tag{5.123}$$

Physically, we require that $\mathbf{E}(t)$ be real; hence (5.122) can be rewritten as

$$\mathbf{E}^*(\omega) \equiv \frac{1}{\sqrt{2\pi}}\int_{-\infty}^{\infty} \mathbf{E}(t)e^{-i\omega t}\, dt \equiv \mathbf{E}(-\omega). \tag{5.124}$$

Here $*$ denotes complex conjugate. Thus the unphysical negative frequencies appearing formally in (5.123) can be eliminated, and *Parseval's theorem* becomes

$$\int_{-\infty}^{\infty} |\mathbf{E}(t)|^2\, dt = \int_{-\infty}^{\infty} |\mathbf{E}(\omega)|^2\, d\omega = 2\int_0^{\infty} |\mathbf{E}(\omega)|^2\, d\omega, \tag{5.125}$$

so we can rewrite (5.121) as

$$\mathcal{F} = \int_0^{\infty} |\mathbf{S}(\omega)|\, d\omega = \frac{c}{2\pi}\int_0^{\infty} |\mathbf{E}(\omega)|^2\, d\omega \quad [\,\mathrm{erg/sr}\,]. \tag{5.126}$$

Therefore, the flux per unit solid angle and circular frequency interval is

$$S(\omega) \equiv \frac{c}{2\pi}|\mathbf{E}(\omega)|^2. \tag{5.127}$$

Total Radiated Flux

To evaluate $\mathcal{F}$, start with the definition of the Fourier transform in (5.123)

$$\mathbf{r}(t) = \frac{1}{\sqrt{2\pi}} \int_{-\infty}^{\infty} \mathbf{r}(\omega) e^{-i\omega t} \, d\omega, \tag{5.128}$$

which defines $\mathbf{r}(\omega)$. Differentiating (5.128) twice with respect to time gives

$$\ddot{\mathbf{r}}(t) = \frac{-1}{\sqrt{2\pi}} \int_{-\infty}^{\infty} \omega^2 \mathbf{r}(\omega) e^{-i\omega t} \, d\omega. \tag{5.129}$$

Combining (5.123) with (5.120) and (5.129), we can write

$$\frac{1}{\sqrt{2\pi}} \int_{-\infty}^{\infty} \mathbf{E}(\omega) e^{-i\omega t} \, d\omega \equiv \mathbf{E}(t) \equiv \mathcal{K} \, \ddot{\mathbf{r}}(t) = \frac{-\mathcal{K}}{\sqrt{2\pi}} \int_{-\infty}^{\infty} \omega^2 \mathbf{r}(\omega) e^{-i\omega t} \, d\Omega \tag{5.130}$$

and hence

$$\mathbf{E}(\omega) = -\mathcal{K} \, \omega^2 \mathbf{r}(\omega) \equiv -(e\omega^2/c^2 R) \mathbf{r}(\omega) \sin\theta. \tag{5.131}$$

Using (5.131) in (5.127), we find that the energy flux per unit solid angle and circular frequency interval is

$$(d^2\mathcal{F}/d\omega \, d\Omega) = (c/2\pi)(e^2\omega^4/c^4 R) |\mathbf{r}(\omega)|^2 \sin^2\theta. \tag{5.132}$$

A differential solid angle around the electron is $d\Omega = d\mu \, d\phi$, so a differential area on a sphere of radius R surrounding the charge is $dA = R^2 \, d\mu \, d\phi$. Writing $\sin^2\theta = (1-\mu^2)$ and integrating (5.132) over area we get a factor of $8\pi R^2/3$, so the total energy flux radiated in all directions by a collision of an electron with an ion is

$$\frac{d\mathcal{F}}{d\omega} = \frac{4e^2\omega^4}{3c^3} |\mathbf{r}(\omega)|^2. \tag{5.133}$$

Collision Integral

Impact Approximation

We must now evaluate $\mathbf{r}(\omega)$. The Coulomb force acting on the electron is large only for a very short time interval, $|t| \lesssim$ the *collision time* τ, which we estimate as $\tau \approx b/v$, where b is the impact parameter of the collision. Therefore, we make the *impact approximation* and consider $\mathbf{E}(t)$ to be an abrupt pulse that is large when $|t| \lesssim \tau \rightarrow 0$ and the exponential in (5.123) is essentially unity, but then drops abruptly to zero (and hence contributes little to the integral) for $|t| \gtrsim \tau$. Then from (5.131),

$$\omega^2 \mathbf{r}(\omega) = -\frac{1}{\sqrt{2\pi}} \int_{-\infty}^{\infty} \ddot{\mathbf{r}}(t) e^{i\omega t} \, dt \approx -\frac{1}{\sqrt{2\pi}} \int_{-\tau}^{\tau} \dot{\mathbf{v}}(t) \, dt \equiv \frac{\Delta|\mathbf{v}(b, v)|}{\sqrt{2\pi}}. \tag{5.134}$$

Here $\Delta\mathbf{v}(b, v)$ is the change in the electron's speed resulting from the collision. Using (5.134) in (5.133) we find

$$\frac{d\mathcal{F}(b,v)}{d\omega} = \frac{2e^2}{3\pi c^3}|\Delta\mathbf{v}|^2. \tag{5.135}$$

$|\Delta\mathbf{v}(b, v)|$ is determined by the dynamics of the electron-ion collision.

For small-angle scattering, the electron's (symmetric) path around the ion is almost linear, so $|\Delta\mathbf{v}(b, v)|$ is the time integral of the component of acceleration perpendicular to its path, i.e., $m_e|\mathbf{a}_\perp| = (Ze^2/r^2)\sin\delta$, where $r^2 = (b^2 + v^2t^2)$ and $\sin\delta = b/r$. Thus

$$|\Delta\mathbf{v}(b,v)| = \frac{Ze^2}{m_e}\int_{-\infty}^{\infty}\frac{b\,dt}{(b^2+v^2t^2)^{3/2}} = \frac{2Ze^2}{m_e bv}. \tag{5.136}$$

Hence for small-angle scattering in the impact approximation, the flux from a collision between an electron having a given b and v with an ion is

$$\frac{d\mathcal{F}(b,v)}{d\omega} \equiv \frac{8\,e^6\,Z^2}{3\pi c^3\,m_e^2\,b^2\,v^2}. \tag{5.137}$$

Cutoff Procedure

In a plasma with ion density n_{ion} and electron density n_e, the rate per unit volume of collisions of an electron moving with speed v with an ion is $n_{ion}n_e(v)v\,2\pi b\,db$, so the omnidirectional emission rate per unit volume summed over all impact parameters is

$$\mathcal{E}(v) = \frac{16\,e^6\,Z^2}{3\,c^3\,m_e^2\,v}\,n_{\rm ion}\,n_e(v)\,\mathfrak{L}(v), \tag{5.138}$$

where

$$\mathfrak{L}(v) \equiv \int_{b_{\rm min}(v)}^{b_{\rm max}} db/b = \ln\left[\frac{b_{\rm max}}{b_{\rm min}(v)}\right]. \tag{5.139}$$

Values for $b_{\rm min}$ and $b_{\rm max}$ are needed to keep (5.139) finite.

For the classical path approximation to be valid, $b_{\rm min}$ must be $\gtrsim$ the electron's *de Broglie wavelength* $\lambda_{\rm dB} \equiv h/m_e v$. To set $b_{\rm max}$, note that an ion Z attracts a cloud of electrons around it, which partially shields its charge, reducing the probability W_e that an electron will be found near the ion. If there were no interaction, that probability would be $n_e dV$. But if $\phi > 0$ at position $\mathbf{r}$, electrons will migrate toward it; if $\phi < 0$, they migrate away from it.

Ecker suggested [297–300] using a linearized approximation to a Boltzmann factor $\exp(-q\phi/kT)$ in calculating W, taking $q = -e$ for electrons and $q = +Ze$ for ions. Define $\psi \equiv (e\phi/kT)$; then for $\psi \ll 1$,

$$n_e W_e\,dV = n_e\exp(\psi)\,dV \approx n_e(1+\psi)\,dV \tag{5.140}$$

and

$$n_i W_i \, dV = n_i \exp(-Z_i \psi) \, dV \approx n_i (1 - Z_i \psi) \, dV. \qquad (5.141)$$

To calculate an effective potential around an ion, smear the electron cloud into a charge density ϱ and use Poisson's equation $\nabla^2 \phi = -4\pi \varrho$ to determine ϕ. In a large volume a plasma must be electrically neutral,

$$n_e = \sum_i Z_i n_i. \qquad (5.142)$$

Using (5.140)–(5.142), we can write the charge density near an ion as

$$\varrho = -e n_e W_e + e \sum_i n_i Z_i W_i = -e\psi \Big(n_e + \sum_i Z_i^2 n_i \Big) \qquad (5.143)$$

and rewrite Poisson's equation as $\nabla^2 \phi = -\phi / \lambda_{\rm D}^2$, where

$$\lambda_{\rm D} \equiv (kT/4\pi e^2)^{1/2} \Big(n_e + \sum_i Z_i^2 n_i \Big)^{-1/2} \qquad (5.144)$$

is the *Debye length* in the plasma. The solution of Poisson's equation with constant $\lambda_{\rm D}$ is $\phi = r^{-1}(Ae^{-r/\lambda_{\rm D}} + Be^{r/\lambda_{\rm D}})$. Neutrality of the plasma requires that $\phi \to 0$ as $r \to \infty$, so we set $B \equiv 0$. Thus beyond $\lambda_{\rm D}$, an ion's field is effectively shielded, so $\lambda_{\rm D}$ sets an upper limit on (a) the distance over which two charged particles can interact, (b) the size of the region in which appreciable departures from charge neutrality can occur, and (c) the wavelength of electromagnetic radiation that can propagate through the plasma without strong dissipation. Therefore, we choose $b_{\rm max} = \lambda_D$. Neither $b_{\rm min}$ nor $b_{\rm max}$ are determined precisely, but they enter (5.139) only logarithmically, so the result is correct to within a small factor.

Emission from Thermal Electrons

Equation (5.132) gives the total energy emitted in all directions by an electron moving with speed v. Averaging it over the Maxwellian velocity distribution we get the total emission from thermal electrons. In computing that average we note that to emit a photon of energy $h\nu$, an electron must have an energy $\frac{1}{2} m_{\rm e} v^2 \geq \frac{1}{2} m_{\rm e} v_{\rm min}^2 = h\nu$. Thus

$$\mathcal{E}(\omega, T) = \frac{16\, e^6\, Z^2}{3\, c^3\, m_e^2} \left(\frac{m_e}{2\pi kT} \right)^{3/2} 4\pi n_e\, n_{\rm ion} \int_{v_{\rm min}}^{\infty} \mathfrak{L}(v)\, e^{-m_{\rm e} v^2/2kT}\, v\, dv. \qquad (5.145)$$

Evaluating the integral and converting from circular to ordinary frequency units, we get the total rate of emission in all directions per unit volume per (ordinary) frequency interval:

$$\mathcal{E}_{\rm ff}(\nu, T) = \frac{32\pi e^6\, Z^2}{3c^3\, m_e^2} \left(\frac{2\, m_e}{\pi k} \right)^{1/2} \frac{n_e n_{\rm ion}}{T^{1/2}} e^{-h\nu/kT} \langle \mathfrak{L} \rangle \;\; [\,{\rm erg/cm^3/s/Hz}\,], \qquad (5.146)$$

where

$$\langle \mathfrak{L} \rangle \equiv \frac{m_e}{kT} \int_{v_{\min}}^{\infty} \mathfrak{L}(v)\, e^{-m_e(v^2 - v_{\min}^2)/2kT} v\, dv = \frac{m_e}{kT} e^{h\nu/kT} \int_{v_{\min}}^{\infty} \mathfrak{L}(v)\, e^{-m_e v^2/2kT} v\, dv. \tag{5.147}$$

In the astrophysical literature, (5.146) is conventionally written in terms of the velocity-averaged *Gaunt factor*, a quantum mechanical correction factor introduced by Gaunt [365, 366] to be applied to the semiclassical absorption coefficient of hydrogenic ions derived by Kramers [636] (see also § 7.1) using *Bohr's correspondence principle* and the Kirchhoff–Planck relation (4.62). In the present notation,

$$\overline{g}_{\rm ff}(\nu, T) \equiv \sqrt{3}\, \langle \mathfrak{L} \rangle / \pi. \tag{5.148}$$

Remember that in astrophysical work, the emission coefficient is measured per steradian, i.e., $\eta_{\rm ff}(\nu, T) = \mathcal{E}_{\rm ff}(\nu, T)/4\pi$.

Absorption Coefficient

Free-free absorption and emission are collisional processes; hence they occur at LTE rates and are connected by the Kirchhoff–Planck relation. Thus the absorption coefficient corrected for induced emission is

$$\kappa_{\rm ff}(\nu, T) = \frac{\sqrt{32\pi}\, Z^2 e^6\, \overline{g}_{\rm ff}(\nu, T)}{3\sqrt{3}\, ch\, (k m_e^3)^{\frac{1}{2}}} \frac{n_e n_{\rm ion}}{T^{\frac{1}{2}} \nu^3} (1 - e^{-h\nu/kT}) \quad [\,{\rm cm}^{-1}\,]. \tag{5.149}$$

Chapter Six

Continuum Scattering

A photon is *scattered* when it interacts with a *scattering center* and moves away in a different direction and perhaps with a different frequency. Unlike absorption and emission processes, which create and destroy photons in first-order transitions between well-defined quantum states, photon scattering is the result of higher-order quantum interactions with free charged particles and resonances with bound electrons in molecules and atoms. If the energy of an incoming photon is much less than the rest energy of the scattering center, so little energy is delivered to the particle that the internal (excitation, ionization, kinetic) energy of the particles in the gas is essentially unaltered. In this case, the rate at which the radiation field is changed by its interaction with the material is set primarily by the local value of the radiation field itself, and only secondarily by the thermodynamic properties of the gas. An important point is that *in all scattering processes photon numbers are conserved*; i.e., they are neither created nor destroyed, as in absorption and emission processes. Scattering processes can also *polarize* radiation. We will not address these phenomena; see, e.g., [225, 573, 574, 789, 1085].

In hot stellar atmospheres, hydrogen and helium are strongly ionized; hence the main continuum scattering process affecting the radiation field is *Thomson scattering* by the abundant free electrons.[1] Although we describe other continuum scattering processes in this chapter, we consider mainly scattering by free electrons in the remainder of this book.

In § 6.1 we use electromagnetic theory to compute the energy scattered by an electron having a hypothetical "natural oscillation frequency" ω_0 (e.g., because it is bound in an atom or molecule) that is driven into motion by radiation having frequency ω. By equating the energy in the incident and scattered radiation we obtain the classical value of the Thomson scattering cross section. We find the same cross section by a quantum mechanical calculation in § 6.2.

In § 6.3 we discuss *Rayleigh scattering*, which is a special case of *Raman scattering*, of photons by atoms (mainly hydrogen) or molecules (mainly H_2). These are the main scattering processes of continuum radiation in cool stellar atmospheres (solar and later spectral type) and in planetary atmospheres (e.g., the Earth). They are characterized by very large cross sections at frequencies near resonances with permitted transitions in the bound system, plus a long tail extending to lower frequencies.

Thomson scattering by electrons ceases to be valid in extremely hot material, such as that found in very high energy laboratory experiments and some

[1] Usually called "electron scattering" in stellar atmospheres work.

astrophysical sources. In such environments, photons with energies approaching $m_e c^2$ experience *Compton scattering* on electrons moving at relativistic speeds. In § 6.4 we treat this interaction as a relativistic collision of a photon and electron in which the electron recoils, so the scattering process is noncoherent, and the change in four-momentum of both the photon and electron is typically large. Hence the laboratory-frame frequency of a Compton-scattered photon can be changed drastically. The exact cross section for this process was derived in a quantum mechanical calculation by Klein and Nishina [621]. We merely quote it and then outline its convolution with a relativistic electron velocity distribution.

In § 6.5 we outline some aspects of the effects of Compton scattering on the state of the early Universe. The analysis by Kompaneets [629] shows that Compton scattering plays an essential role in establishing thermal equilibrium between matter and radiation in a closed system.

6.1 THOMSON SCATTERING: CLASSICAL ANALYSIS

Before the development of quantum mechanics, quantitative calculations of absorption and emission cross sections could not be made. However, classical electrodynamics could be used to calculate the energy scattered by an electron driven into harmonic oscillation by the electric field of an incident light wave, allowing for the decay of its amplitude at the rate set by the energy lost into the scattered radiation field. The discussion here will parallel some of that in § 5.6, with the difference that there we dealt with radiation from the transient dipole between an ion and electron during a collision of finite duration, whereas here the radiation + electron system may reach a steady state.

Radiation from a Classical Oscillator

Let ω_0 be the "characteristic frequency" of an electron bound to a nucleus, oscillating with a circular frequency ω, and $\mathbf{r}$ be the vector from the electron to a remote observer. If the electron's velocity relative to the nucleus is $v \ll c$, and its oscillation amplitude $|\,\mathbf{d}(t) \ll \mathbf{r}\,|$, relativistic effects and time retardation in the radiated electromagnetic field can both be neglected. Then equations (3.52) and (3.53) for the radiation far-field part of the vector electric and magnetic fields produced by the electron's motion reduce to

$$\mathbf{E}(\mathbf{r}, t) = \frac{-e\ddot{\mathbf{d}}(t)}{c^2 r} \sin\theta\, \hat{\boldsymbol{\theta}} \tag{6.1}$$

and

$$\mathbf{H}(\mathbf{r}, t) = \frac{-e\ddot{\mathbf{d}}(t)}{c^2 r} \sin\theta\, \hat{\boldsymbol{\phi}}. \tag{6.2}$$

Here we have taken $\ddot{\mathbf{d}}(t)$ to be along the z axis. The unit vectors $\hat{\boldsymbol{r}}$, $\hat{\boldsymbol{\theta}}$, and $\hat{\boldsymbol{\phi}}$ are a right-handed orthonormal triad in spherical coordinates; see figure 6.1. Note that $\mathbf{E}$ and $\mathbf{H}$ are determined by the electron's *acceleration*.

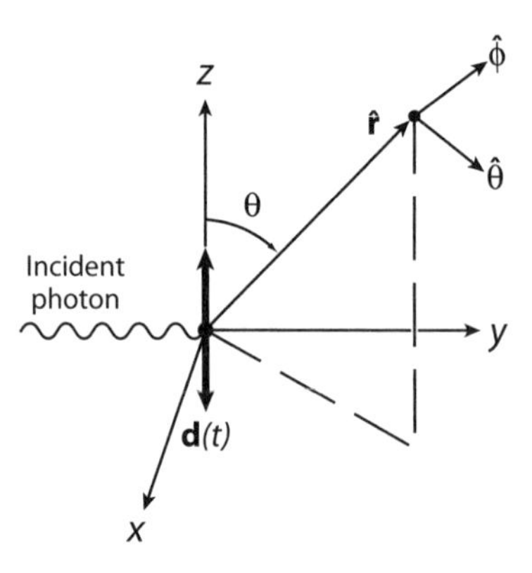

Figure 6.1 Radiation propagating in the $\hat{\mathbf{r}}$ direction from an electron oscillating along the $\hat{\mathbf{z}}$ axis.

Let the amplitude of the electron's motion be $\mathbf{d}(t) = \mathbf{d}_0 \sin \omega_0 t$; then its velocity is $\mathbf{v}(t) = \dot{\mathbf{d}}(t) = \omega_0 \, \mathbf{d}_0 \cos \omega_0 t$, and its acceleration is $\dot{\mathbf{v}}(t) = \ddot{\mathbf{d}}(t) = -\omega_0^2 \mathbf{d}_0 \sin \omega_0 t = -\omega_0^2 \mathbf{d}(t)$, all along the z axis. Using (6.1) and (6.2) in equation (3.38), we find that the Poynting vector along $\hat{\boldsymbol{r}}$ is

$$\mathbf{S} = \frac{c}{4\pi}(\mathbf{E} \times \mathbf{H}) = \frac{e^2 \ddot{d}^2(t)}{4\pi c^3 r^2} \sin^2 \theta \, \hat{\boldsymbol{r}} \quad [\mathrm{erg/cm^2/s/sr}]. \tag{6.3}$$

$\mathbf{E}$ and $\mathbf{H}$ are orthogonal and $|\mathbf{E}| = |\mathbf{H}|$. Hence $|S(t)| = c|E(t)|^2/4\pi$, in agreement with equation (3.38).

An ensemble of randomly oriented oscillators in an isotropic radiation field produces Poynting vectors uniformly distributed in all directions. Thus integrating (6.3) over a sphere of radius r by forming $\mathbf{S} \cdot d\mathbf{A}$, where $d\mathbf{A} = r^2 \, d\Omega \, \hat{\mathbf{r}} = (r^2 d\mu \, d\phi) \, \hat{\mathbf{r}}$, and writing $\sin^2 \theta = (1 - \mu^2)$, we find that the total power scattered in all directions is

$$P(t) = \frac{e^2 \ddot{d}^2}{4\pi c^3} \int_0^{2\pi} d\phi \int_{-1}^{1} (1 - \mu^2) \, d\mu = \frac{2e^2 \ddot{d}^2}{3c^3}. \tag{6.4}$$

Substituting the harmonic oscillator expression for $\ddot{d}^2$ and averaging over a cycle with period T, noting that $\langle \cos^2 \omega_0 t \rangle_\mathrm{T} = \frac{1}{2}$, we obtain

$$\langle P(\omega_0) \rangle_\mathrm{T} = \frac{e^2 d_0^2 \omega_0^4}{3c^3}. \tag{6.5}$$

Driven and Damped Classical Oscillators

Because the oscillators radiate away energy, their amplitude will eventually decay. For simplicity we will describe this loss of energy in terms of a hypothetical "viscous damping force" exerted on the moving electron. To calculate the effective damping force $\mathbf{F}_{\rm rad}$, we equate the average rate of energy loss by the oscillator to the rate of work by done such a force. Then from (6.4),

$$\mathbf{F}_{\rm rad}\boldsymbol{\cdot}\mathbf{v} = -\frac{2e^2\dot{v}^2}{3c^3}. \tag{6.6}$$

Integrating by parts over one period T of the oscillation we have

$$\int_0^{\rm T} \mathbf{F}_{\rm rad}\boldsymbol{\cdot}\mathbf{v}\,dt = -\frac{2e^2}{3c^3}\left(\Big|_0^{\rm T} \mathbf{v}\boldsymbol{\cdot}\dot{\mathbf{v}} - \int_0^{\rm T} \ddot{\mathbf{v}}\boldsymbol{\cdot}\mathbf{v}\,dt\right). \tag{6.7}$$

The first term on the right-hand side vanishes for a periodic oscillation, so we identify $\mathbf{F}_{\rm rad}$ as

$$\mathbf{F}_{\rm rad} = \frac{2e^2\ddot{\mathbf{v}}}{3c^3}. \tag{6.8}$$

The rate of damping is small, so to a good approximation $\ddot{\mathbf{v}} = -\omega_0^2\mathbf{v}$, and the damping force can be represented as $\mathbf{F}_{\rm rad} \equiv -m\,\gamma\,\mathbf{v}$, where the *classical damping constant* γ is

$$\gamma \equiv \frac{2e^2\omega_0^2}{3mc^3}. \tag{6.9}$$

This picture is only a rough caricature of the physics; nevertheless, it yields useful results.

Suppose now that an oscillator is *driven* by radiation that has an electric field $\mathbf{E}_0(\omega)$ with frequency $\omega \neq \omega_0$. The equation of motion of a particle of mass m and charge e in this field, allowing for radiative damping, is

$$m(\ddot{\mathbf{d}} + \omega_0^2) = e\mathbf{E}_0(\omega) - m\,\gamma\,\dot{\mathbf{d}}. \tag{6.10}$$

Taking a trial solution $\mathbf{d} \propto \exp(i\omega t)$ we find that the steady-state solution is

$$\mathbf{d} = -\frac{e}{m}\Re\mathrm{e}\left[\frac{\mathbf{E}_0(\omega)e^{i\omega t}}{(\omega^2 - \omega_0^2) + i\,\gamma\,\omega}\right]; \tag{6.11}$$

hence

$$\ddot{\mathbf{d}} = \dot{\mathbf{v}} = \frac{e\,\omega^2}{m}\Re\mathrm{e}\left[\frac{\mathbf{E}_0(\omega)e^{i\omega t}}{(\omega^2 - \omega_0^2) + i\,\gamma\,\omega}\right]. \tag{6.12}$$

Using (6.12) in (6.4) and averaging over a cycle, we obtain

$$\langle P(\omega)\rangle_{\rm T} = \left(\frac{e^4\omega^4}{3m^2c^3}\right)\frac{E_0^2}{(\omega^2 - \omega_0^2)^2 + \gamma^2\omega^2}. \tag{6.13}$$

To make a correspondence with the macroscopic description of radiation in chapter 3, use equation (3.57) for an isotropic field, $I_0 \equiv cE_0^2/8\pi$, and write the scattering cross section as $\sigma(\omega)$, so that $\langle P(\omega)\rangle_T = \oint \sigma(\omega) I_0 \, d\Omega = (cE_0^2/8\pi)\sigma(\omega)$. Then from (6.13) we have

$$\sigma(\omega) = \frac{(8\pi e^4\omega^4/3m^2c^4)}{[(\omega^2-\omega_0^2)^2+\gamma^2\omega^2]}. \tag{6.14}$$

In wavelength units, γ corresponds to $\Delta\lambda_{cl} = 1.2 \times 10^{-4}$ Å, which is much smaller than the resolution typically obtained in stellar spectra, and in frequency units $\gamma/\omega = \Delta\lambda_{cl}/\lambda \ll 1$. Hence to a good approximation

$$(\omega^2-\omega_0^2) = (\omega-\omega_0)(\omega+\omega_0) \approx 2\omega_0(\omega-\omega_0), \tag{6.15}$$

and using (6.9), (6.14) can be rewritten as

$$\sigma(\omega) = \left(\frac{\pi e^2}{mc}\right)\left[\frac{\gamma}{(\omega-\omega_0)^2+(\gamma/2)^2}\right]. \tag{6.16}$$

Defining $x \equiv 4\pi(\nu-\nu_0)/\gamma$, and noting that $-4\pi\nu_0/\gamma$ is essentially $-\infty$, we obtain the *total classical cross section*

$$\sigma_{cl} = \frac{\pi e^2}{mc}\int_0^\infty \frac{(\gamma/4\pi^2)\, d\nu}{(\nu-\nu_0)^2+(\gamma/4\pi)^2} \rightarrow \frac{\pi e^2}{mc}\frac{1}{\pi}\int_{-\infty}^{\infty}\frac{dx}{x^2+1} = \frac{\pi e^2}{mc}. \tag{6.17}$$

We can rewrite the integrand in (6.17) as

$$\sigma(x) = \frac{\pi e^2}{mc}\,\phi_L(x), \tag{6.18}$$

where the normalized profile function

$$\phi_L(x) = \frac{1}{\pi(x^2+1)} \tag{6.19}$$

is a *Lorentz profile*, also called a *damping profile*. $\phi_L(x)$ has a narrow peak at the resonance frequency ω_0 and broader wings falling off as $1/x^2 \propto 1/\Delta\omega^2$. We see that σ has half its peak value at $x = 1$; thus the full half-intensity width of the scattered radiation is $\Delta\omega = \gamma$. In some cases quantum mechanical descriptions of spectral line broadening also yield Lorentz profiles, but we stress that the classical analysis presented here is *not* a physically meaningful description of those processes.

Thomson Cross Section

At $T = 10^4$ K, the most probable speed of an electron given by equation (4.26) is $v \approx 5.4 \times 10^7$cm/s, so its de Broglie wavelength is $\lambda_{dB} = h/m_e v \approx 12$ Å, at least

100 times smaller than the wavelengths of visible and ultraviolet photons. Therefore, quantum effects in the scattering of photons at these wavelengths by free electrons are small, and to a first approximation it is reasonable to treat the process by classical electrodynamics.

For a free electron, $\omega_0 \equiv 0$, and recalling that $\gamma \ll \omega$, we see from (6.14) that $\sigma(\omega)$ is independent of frequency:

$$\sigma(\omega) \to \sigma_T \equiv \frac{8\pi e^4}{3m_e^2 c^4} = 6.65 \times 10^{-25}\ \text{cm}^2. \tag{6.20}$$

σ_T is the *Thomson scattering cross section*, usually called the *electron scattering cross section* in stellar atmospheres work. It is sometimes written as $\sigma_T = \frac{8}{3}\pi r_0^2$, where $r_0 = e^2/m_e c^2$ is called the *"classical electron radius."*

Angular Phase Function

From (6.1)–(6.3) and figure 6.1, we see that the **E** field for unpolarized radiation propagating along the y axis has two equal components with amplitudes $|E_x| = E_0$ and $|E_z| = E_0$ along the x axis and the z axis. A distant observer who views the scattered radiation along a line of sight at an angle $\xi \equiv 90° - \theta$ between the incoming and scattered radiation sees these components with amplitudes $|E_x| = E_0$ and $|E_z| = E_0 \cos\xi$. The intensity in each component is proportional to the square of its amplitude; hence the intensity of scattered radiation has an angular dependence

$$g(\xi) = K(1 + \cos^2\xi). \tag{6.21}$$

K is a normalization constant chosen to conserve photon number when integrated over solid angle $d\omega \equiv \sin\xi\, d\xi\, d\phi$. Thus

$$\frac{K}{4\pi}\int_0^{2\pi} d\phi \int_{-\pi/2}^{\pi/2} (1 + \cos^2\xi)\, \sin\xi\, d\xi = \tfrac{4}{3}K \equiv 1, \tag{6.22}$$

so $K = \frac{3}{4}$, and

$$g(\xi) = \tfrac{3}{4}(1 + \cos^2\xi), \tag{6.23}$$

which is the *Rayleigh* (or *dipole*) *phase function*. Thus for unpolarized radiation,

$$n_e\sigma_e(\xi) = \tfrac{3}{4}(1 + \cos^2\xi) n_e\sigma_T. \tag{6.24}$$

Equation (6.23) shows that a photon has an equal probability of being *forward scattered* ($\xi = 0$) and *backscattered* ($\xi = \pi$), and half that probability of being scattered 90° away from the direction of the incoming photon. In one-dimensional (planar or spherical) media, the radiation field is independent of the azimuthal angle Φ, so (6.24) holds for all Φ at a given ξ.

In deriving (6.23) and (6.24) we assumed the geometry shown in figure 6.1. For a more general geometry one would write

$$g(\mathbf{n}', \mathbf{n}) = \tfrac{3}{4}[1 + (\mathbf{n}' \cdot \mathbf{n})^2], \tag{6.25}$$

where $\mathbf{n}'$ and $\mathbf{n}$ are unit vectors along the directions of the incoming and outgoing photons. The invariance of (6.25) under interchange of $\mathbf{n}$ and $\mathbf{n}'$ is expected because the process is symmetric under time reversal.

6.2 THOMSON SCATTERING: QUANTUM MECHANICAL ANALYSIS

The Scattering Process

It is instructive to derive the Thomson scattering cross section quantum mechanically. A free electron has no dipole moment; hence it cannot absorb or emit a photon through a first-order process contained in $\hat{\boldsymbol{H}}_1$ in the interaction Hamiltonian (5.85) without violating energy and momentum conservation. But it can scatter a photon through higher-order interactions contained in $\hat{\boldsymbol{H}}_2$ in equation (5.86).

Suppose that in a scattering event an electron is converted from an initial ket $|p_i\rangle$ with momentum $\hbar\,\mathbf{p}_i$ to a final ket $|p_f\rangle$ with momentum $\hbar\,\mathbf{p}_f$, while an incident photon with momentum $\hbar\,\mathbf{k}_i$ in direction $\boldsymbol{\alpha}_i$ is converted to one with momentum $\hbar\,\mathbf{k}_f$ in direction $\boldsymbol{\alpha}_f$. The initial and final states of the electron + radiation system are

$$|\,i\rangle = |p_i\rangle|\ldots n_{k_i\alpha_i}\ldots n_{k_f\alpha_f}\ldots\rangle \tag{6.26}$$

and

$$|f\rangle = |p_f\rangle|\ldots n_{k_i\alpha_i} - 1\ldots n_{k_f\alpha_f} + 1\ldots\rangle. \tag{6.27}$$

The electron kets are the solutions of

$$\hat{\boldsymbol{H}}_A|p\rangle = -\frac{\hbar^2}{2m_e}\nabla^2|p\rangle + \phi(\mathbf{x})|p\rangle. \tag{6.28}$$

For a free electron the electric potential $\phi(\mathbf{x}) \equiv 0$; hence $|p\rangle = e^{i\mathbf{p}\cdot\mathbf{x}}$, with eigenvalue $E_p = \hbar^2 p^2/2m_e$. These eigenkets are normalized such that

$$\iiint\limits_V \langle p_f\,|\,p_i\rangle\, d^3x = \delta(\mathbf{p}_i - \mathbf{p}_f). \tag{6.29}$$

The volume V for the spatial integration is chosen to contain one free electron in the initial and final states.

The transformation from (6.26) to (6.27) is the result of the action of the third term in the interaction operator $\hat{\boldsymbol{H}}_2$ in (5.83):

$$\begin{aligned}\langle f\,|\hat{\boldsymbol{H}}_2|\,i\rangle &= \langle p_f|\hat{\mathcal{O}}_{k_ik_f}|\,p_i\rangle\langle\ldots n_{k_i} - 1\; n_{k_f} + 1\ldots|\hat{\boldsymbol{a}}^\dagger_{k_f}\hat{\boldsymbol{a}}_{k_i}|\ldots n_{k_i}n_{k_f}\ldots\rangle \\ &= \langle p_f|\hat{\mathcal{O}}_{k_ik_f}|\,p_i\rangle\, n_{k_i}^{1/2}\,(n_{k_f}+1)^{1/2}.\end{aligned} \tag{6.30}$$

Here, to simplify the notation we have written n_{k_i} for $n_{k_{i\alpha_i}}$ and $\hat{\boldsymbol{a}}_{k_i}$ for $\hat{\boldsymbol{a}}_{k_i\alpha_i}$, and similarly for n_{k_f} and $\hat{\boldsymbol{a}}^{\dagger}_{k_f\alpha_f}$. The scattering rate is

$$R_{k_ik_f} \propto \left| \langle \mathbf{p}_f \left| \hat{\mathcal{O}}_{k_ik_f} \right| \mathbf{p}_i \rangle \right|^2 n_{k_i} (n_{k_f} + 1). \tag{6.31}$$

As in chapter 5, $\hat{\mathcal{O}}$ is a function of the initial and final states of the electron, the wave vectors and polarization of the incoming and outgoing photons, and a destruction and a creation operator.

In (6.31) n_{k_i} is the number of incident photons. As shown by Pauli [844] the factor $(n_{k_f} + 1)$ accounts for the number of direct and induced photons emerging from the scattering process.[2]

Cross Section

To compute the rate of transitions $i \rightarrow f$ we evaluate the result of the second-order term of the interaction Hamiltonian in equation (5.85) using Fermi's Golden Rule. In Thomson scattering, for a given initial state the final state that satisfies the requirement of energy conservation is unique, so the density of states in (5.97) can be replaced by a delta function that selects that particular state. Then

$$R_{if} = \frac{2\pi}{\hbar} \left(\frac{e^2}{2m_e c^2} \right)^2 (2\pi \hbar c^2)^2 \sum_{i,f} n_i (n_f + 1)\, 4 \frac{|\mathbf{e}_i \cdot \mathbf{e}_f|^2}{\omega_i \omega_f} \left| \langle \mathbf{p}_f \left| e^{i(\mathbf{k}_i - \mathbf{k}_f)\cdot \mathbf{x}} \right| \mathbf{p}_i \rangle \right|^2$$
$$\delta \left[\hbar \left(\omega_i - \omega_f \right) + \left(\hbar^2 / 2m_e \right) \left(p_i^2 - p_f^2 \right) \right]. \tag{6.32}$$

Because the total momentum and energy of the photon and scattering center are conserved (see below), the reverse process $f \rightarrow i$ is simply the time reversal of $i \rightarrow f$; hence the second term in $\hat{\boldsymbol{H}}_2$ makes an equal contribution to the matrix element, which introduces the numerical factor of 4 in (6.32). Further, for the purpose of computing only the scattering cross section, we assume the incident photons are scattered into vacuum and set $n_f = 0$.

In evaluating the sum over the final states f, recall that we imposed periodic boundary conditions on $\mathbf{A}$ in equations (5.45)–(5.47). In the continuum limit $(L \rightarrow \infty)$, the sum over all values of (k_x, k_y, k_z) for the electron can be replaced by an integral, and from (5.113) we have

$$\sum_{k_f} \longrightarrow \frac{V}{8\pi^3} \int d^3 k_f. \tag{6.33}$$

[2] Pauli's recognition of induced scattering preceded quantum mechanics. He derived the correct result using detailed-balanced arguments.

Choose the volume V to contain a single electron. Then (6.32) becomes

$$R_{if} = n_i \frac{e^4 \hbar}{4\pi c^2 m_e^2} \sum_{\alpha_f} \sum_{\mathbf{p}_f} |\mathbf{e}_i \cdot \mathbf{e}_f|^2 \int \frac{d^3 k_f}{k_i k_f} \left| \langle \mathbf{p}_f \left| e^{i(\mathbf{k}_i - \mathbf{k}_f) \cdot \mathbf{x}} \right| \mathbf{p}_i \rangle \right|^2 \times \\ \delta \left[\hbar \left(\omega_i - \omega_f \right) + \left(\hbar^2 / 2m_e \right) \left(p_i^2 - p_f^2 \right) \right]. \tag{6.34}$$

In view of (6.29), the matrix element in (6.34) is

$$\langle \mathbf{p}_f \left| e^{i(\mathbf{k}_i - \mathbf{k}_f) \cdot \mathbf{x}} \right| \mathbf{p}_i \rangle = \iiint_V e^{i(\mathbf{p}_i + \mathbf{k}_i - \mathbf{p}_f - \mathbf{k}_f) \cdot \mathbf{x}} \, d^3 x \\ = \delta \left(\mathbf{p}_i + \mathbf{k}_i - \mathbf{p}_f - \mathbf{k}_f \right). \tag{6.35}$$

The delta function vanishes unless the total initial momentum $\hbar \left(\mathbf{p}_i + \mathbf{k}_i \right)$ of the electron and photon equals their total final momentum $\hbar \left(\mathbf{p}_f + \mathbf{k}_f \right)$. Thus *both momentum and energy are conserved in the scattering process*. Then writing $d^3 k_f$ as $4\pi k_f^2 dk_f$, and inserting (6.35) into (6.34), we have

$$R_{if} = n_i \frac{e^4 \hbar}{c^2 m_e^2} \sum_{\alpha_i} \sum_{\mathbf{p}_f \alpha_f} \left\{ \int k_f^2 \, dk_f \frac{|\mathbf{e}_i \cdot \mathbf{e}_f|^2}{k_i k_f} \right. \\ \left. \times \, \delta \left(\mathbf{p}_i + \mathbf{k}_i - \mathbf{p}_f - \mathbf{k}_f \right) \delta \left[\hbar (\omega_i - \omega_f) + \frac{\hbar^2}{2m_e} \left(p_i^2 - p_f^2 \right) \right] \right\} \\ = n_i \frac{e^4 \hbar}{c^2 m_e^2} \sum_{\alpha_i} \sum_{\alpha_f} |\mathbf{e}_i \cdot \mathbf{e}_f|^2 \int \frac{k_f}{k_i} \, dk_f \\ \times \, \delta \left[\hbar c (k_i - k_f) + \frac{\hbar^2}{2m_e} (p_i^2 - |\mathbf{p}_i + \mathbf{k}_i - \mathbf{k}_f|^2) \right]. \tag{6.36}$$

The second term in the delta function in (6.36) corresponds to the Compton shift of the incoming photon's wavelength (see § 6.4); it is about a factor of $1/c$ smaller than the first, and for the present purpose it can be ignored. Then the delta function reduces to $\delta[\hbar c (k_i - k_f)] \equiv \delta(k_i - k_f)/\hbar c$, which shows that Thomson scattering is coherent, and (6.36) collapses to

$$R_{if} = \left(e^4 / m_e^2 c^3 \right) n_i \sum_{\alpha_i} \sum_{\alpha_f} |\mathbf{e}_i \cdot \mathbf{e}_f|^2 \int \frac{k_f}{k_i} \delta(k_i - k_f) \, dk_f \\ = \left(e^4 / m_e^2 c^3 \right) n_i (1 + \cos^2 \xi). \tag{6.37}$$

To obtain the second equality, (a) integrate over the delta function, which is nonzero only if $k_f \equiv k_i$, i.e., the scattering process is coherent; and note that (b) the incoming photons travel in direction $\mathbf{e}_i = \hat{j}$, and the outgoing photons travel in direction $\mathbf{e}_f = \cos\xi\,\hat{j} + \sin\xi\,\hat{k}$, so $|\mathbf{e}_i{\cdot}\mathbf{e}_i|^2 = 1$, $|\mathbf{e}_f{\cdot}\mathbf{e}_f|^2 = 1$, and $|\mathbf{e}_i{\cdot}\mathbf{e}_f|^2 = |\mathbf{e}_f{\cdot}\mathbf{e}_i|^2 = \cos^2\xi$; so that $\sum_{\alpha_i}\sum_{\alpha_f} |\mathbf{e}_i{\cdot}\mathbf{e}_f|^2 = 2(1+\cos\xi^2)$.

Finally, identify $R_{if} = n_i\, c\, \sigma_{\mathrm{T}}$ [photons/s/sr] as the rate at which photons are scattered into a unit solid angle, average over the initial polarizations, and integrate over solid angle. Then in view of (6.22), we have

$$\sigma_{\mathrm{T}} = \left(\frac{e^2}{m_e c^2}\right)^2 \int_0^{2\pi} d\phi \int_{-\pi/2}^{\pi/2} (1+\cos^2\xi)\cos\xi\, d\xi = \frac{8\pi e^4}{3m_e^2 c^4}. \tag{6.38}$$

The classical and quantum mechanical analyses both yield the same cross section and show that Thomson scattering is (1) *coherent* and (2) *anisotropic*.

6.3 ⋆ RAYLEIGH AND RAMAN SCATTERING

Kramers-Heisenberg Dispersion Formula

Rayleigh scattering and Raman scattering by atoms and molecules affect radiation over a wide range of frequencies. These processes are very efficient when the frequency of an incident photon is close to resonance with a bound-bound transition in the scattering center. In Rayleigh scattering, an incident photon raises a bound electron from an eigenstate of the scattering center to a "virtual state," i.e., an intermediate state *that is not an eigenstate of the system*, followed by the direct return of the electron to the original eigenstate with the release of a photon of the same frequency as the input photon. See figure 6.2a. Like Thomson scattering, Rayleigh scattering is coherent and has an anisotropic phase function.

In Raman scattering, an incident photon raises an electron from an eigenstate of the scattering center to an intermediate state (again *not* an eigenstate of the system), followed by a jump of the electron to a *different* eigenstate of the system, along with the release of a photon having a frequency different from that of the input photon; see figure 6.2b. Hence we have the following: (1) Unlike Thomson and Rayleigh scattering, Raman scattering is noncoherent. (2) When there is a downward cascade of the electron from the second eigenstate, *photon number is not conserved*. And (3) Raman scattering is anisotropic.

If one thinks of these processes as a "collision" of a photon with the scattering center, Rayleigh scattering could be viewed as an "elastic collision" and Raman scattering as an "inelastic collision."

Cross sections for both processes are obtained from the second and third terms of the operator $\hat{\boldsymbol{H}}_2$ in equation (5.82). The calculation is more complicated than those for Thomson or Compton scattering because the scattering centers are not free particles, but are in bound states. The exact expression for the differential cross section for Rayleigh and Raman scattering is the *Kramers-Heisenberg dispersion formula* [637], which they obtained before the development of quantum mechanics by applying Bohr's correspondence principle to the classical formula. The same

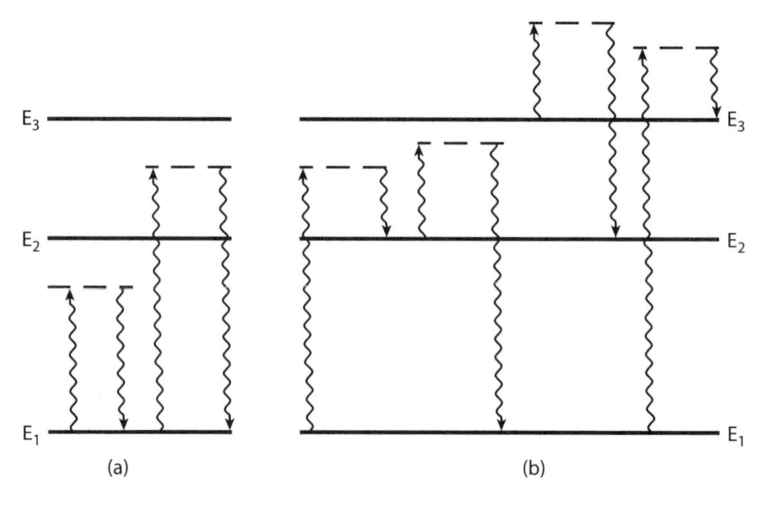

Figure 6.2 Examples of (a) Rayleigh scattering and (b) Raman scattering. Solid lines are energy eigenstates of the scattering center. Dashed lines are intermediate states that are not eigenstates of the system.

formula was later derived quantum mechanically by Dirac [281] and Schrödinger [976]; details of more modern calculations can be found in [243, 244].

Differential Cross Section

The differential scattering cross section for an incident photon with energy $\hbar\omega$ and polarization vector $\mathbf{e}$ raising a bound electron of the scattering center from an initial state $|A\rangle$ through an intermediate state $|I\rangle$ to a final state $|B\rangle$, with the emission of a photon with energy $\hbar\omega'$ and polarization vector $\mathbf{e}'$ is

$$(d\sigma/d\Omega) = r_0^2(\omega'/\omega)|\mathfrak{M}_{AB}|^2, \tag{6.39}$$

where $r_0 = e^2/mc^2$ is the classical electron radius and $|\mathfrak{M}_{AB}|$ is the matrix element obtained from the second and third terms in the operator $\hat{\boldsymbol{H}}_2$ in equation (5.83) acting on a system with an infinite number of bound states and continuum:

$$\mathfrak{M}_{AB} = (\mathbf{e}_A \cdot \mathbf{e}'_B)\delta_{AB}$$
$$- \frac{1}{m}\mathsf{S}_I \left[\frac{\langle B|\mathbf{p}\cdot\mathbf{e}'|I\rangle\langle I|\mathbf{p}\cdot\mathbf{e}|A\rangle}{E_I - E_A - \hbar\omega} + \frac{\langle B|\mathbf{p}\cdot\mathbf{e}|I\rangle\langle I|\mathbf{p}\cdot\mathbf{e}'|A\rangle}{E_I - E_A + \hbar\omega'} \right]; \tag{6.40}$$

see, e.g., [367], [471, p. 192], [944]. Here $\mathbf{p}$ is the electron's momentum operator, and S_I denotes a sum over all bound excited states and an integral over the continuum.

Using the commutation relation $\mathbf{p} = m[\mathbf{x}, \hat{\boldsymbol{H}}]/i\hbar$, which implies that[3] $\langle u\,|\mathbf{p}|\,l\rangle = im\omega_{lu}\langle u\,|\mathbf{x}|\,l\rangle$, and the completeness relation one finds

$$\frac{1}{i\hbar}\sum_{I}\left[\langle A\,|\mathbf{p}\cdot\mathbf{e}'|\,I\rangle\langle I\,|\mathbf{x}\cdot\mathbf{e}|\,B\rangle - \langle A\,|\mathbf{x}\cdot\mathbf{e}|\,I\rangle\langle I\,|\mathbf{p}\cdot\mathbf{e}'|\,A\rangle\right] = \mathbf{e}_A\cdot\mathbf{e}'_B\delta_{AB}. \tag{6.41}$$

Alternatively, using matrix elements of the position operator $\mathbf{x}$, one has

$$\frac{d\sigma}{d\Omega} = r_0^2 m^2 \omega\,\omega'^3 \left| \mathsf{S}_I \left[\frac{\langle B\,|\mathbf{x}\cdot\mathbf{e}'|\,I\rangle\langle I\,|\mathbf{x}\cdot\mathbf{e}|\,A\rangle}{E_I - E_A - \hbar\omega} + \frac{\langle B\,|\mathbf{x}\cdot\mathbf{e}|\,I\rangle\langle I\,|\mathbf{x}\cdot\mathbf{e}'|\,A\rangle}{E_I - E_A + \hbar\omega'} \right] \right|^2. \tag{6.42}$$

This form is useful because dipole matrix elements of hydrogen are readily available. Another form for Raman scattering from an initial state $A = |\,ns\rangle$ to a final state $B = |\,n'l'\rangle$, written in terms of oscillator strengths, is

$$\sigma(\omega,\omega') = \frac{\sigma_{\mathrm{T}}}{16}\,\omega\,\omega'^3 \left| \sum_I \sqrt{\frac{(gf)_{AI}}{\omega_{AI}}}\sqrt{\frac{(gf)_{BI}}{\omega_{BI}}}\left(\frac{1}{\omega_{BI}-\omega} + \frac{1}{\omega_{AI}+\omega'}\right) \right.$$
$$\left. + \int_0^\infty dI'\,\sqrt{\frac{(gf)_{AI'}}{\omega_{AI'}}}\sqrt{\frac{(gf)_{BI'}}{\omega_{BI'}}}\left(\frac{1}{\omega_{BI'}-\omega} + \frac{1}{\omega_{AI'}+\omega'}\right) \right|^2. \tag{6.43}$$

A typical example is a transition from the ground state $|1s\rangle\,^2S$ through an intermediate state $|I'p\rangle\,^2P$ to the metastable level $|2s\rangle\,^2S$, and the reverse.

Rayleigh Scattering by Atomic Hydrogen

Rayleigh scattering by atomic hydrogen can be important for metal-poor G- and K-type stars, whose atmospheric hydrogen is mostly neutral and in the ground state. The low abundance of easily ionized elements implies a paucity of free electrons; hence the opacity of H^- is greatly diminished (see § 7.2). Thus photons in the visible part of the spectrum can interact with the extended wings of spectral lines in the hydrogen Lyman series $(1 \rightarrow j)$, which lie far in the ultraviolet.

For Rayleigh scattering, the initial and final states $|A\rangle$ and $|B\rangle$ in (6.40), (6.42), or (6.43) are the same. The cross section $\sigma_{\mathrm{Ray}}(\omega)$ for Rayleigh scattering from the $|\,1s\rangle\,^2S$ ground state of atomic hydrogen including resonances with all bound $|np\rangle\,^2P$ states and states in the $|\kappa p\rangle\,^2P$ continuum has been evaluated accurately in [367]. Figure 6.3 shows the result of computations including all bound states with $n \leq 1000$

[3] Recall equation (5.68).

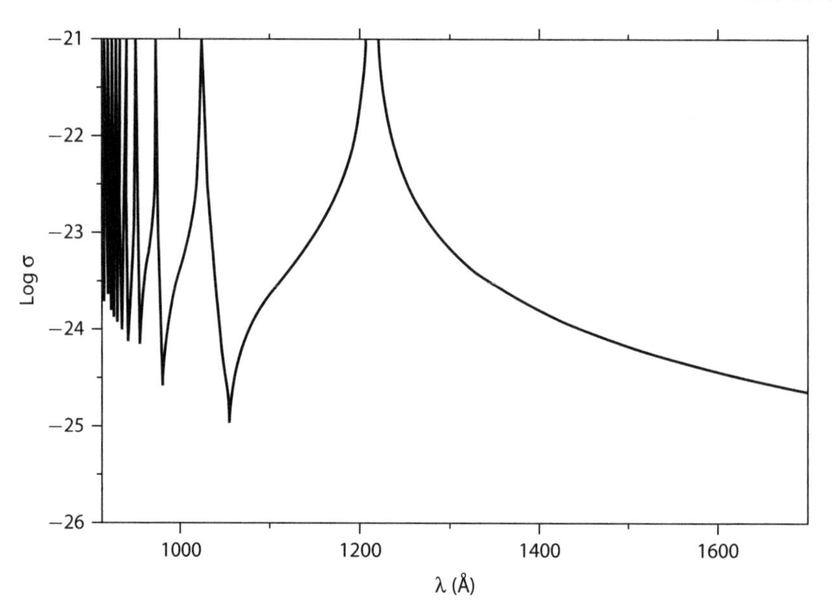

Figure 6.3 Cross section for Rayleigh scattering from the ground state of atomic hydrogen, showing resonances near wavelengths of Lyman lines. *Ordinate:* log σ (cm^2). Adapted from [689].

and the continuum. Accurate fitting formulae for computing $\sigma_{\text{Ray}}(\omega)$ at frequencies $\omega < \omega(\text{Ly}\alpha)$ are given in [688, 689]. If the cross section is computed using upper-state eigenfunctions given by the time-independent Schrödinger equation, the resonances at the positions of the Lyman lines are singular. This artifact is eliminated by accounting for the breadths of the upper states resulting from radiative decay to lower eigenstates.

One can get an approximate cross section for Rayleigh scattering from the classical analysis leading to equation (6.14) in the limit that γ is ignored, and the frequency ω of the scattered radiation is much smaller that ω_{ij}, the frequency of a Lyman line. Then (6.14) simplifies to

$$\sigma_{\text{Ray}}(\omega) \propto \frac{\sigma_T \omega^4}{(\omega_{ij}^2 - \omega^2)^2} \to \sigma_T \left(\frac{\omega}{\omega_{ij}}\right)^4 \quad [\text{cm}^2]. \tag{6.44}$$

At long wavelengths $\sigma_{\text{Ray}}(\omega)$ varies as λ^{-4}, so there is a strong color dependence of the scattered radiation. For example, the blue color of the sky results from more efficient scattering of the shorter wavelengths of sunlight by N_2 and O_2 molecules in the Earth's atmosphere.

In the limit $\omega \ll \omega_{1j}$ the electromagnetic field of the incident radiation varies slowly enough that inertial effects in the driven oscillation can be neglected. Then the size of the induced dipole moment will be directly proportional to the applied field, so for a transition $(1 \to j)$ with oscillator strength f_{1j}, we take $d \propto f_{1j} E_0$ in equation (6.11). Summing over all Lyman lines we rewrite (6.44) as

$$\sigma_{\text{Ray}}(\omega) \approx \sigma_{\text{T}} \left(\sum_j \frac{f_{1j}\omega^2}{\omega_{1j}^2 - \omega^2} \right)^2 \ [\text{cm}^2]. \tag{6.45}$$

Again the cross section has a "singularity" near the frequency of a Lyman line.

The Rayleigh scattering cross section of a gas is also related to its index of refraction n [1102, § 49]:

$$\sigma_{\text{Ray}}(\omega) = \frac{(n^2 - 1)^2}{6\pi c^4 N} \omega^4, \tag{6.46}$$

where N is the number of scattering centers per cm^{-3}. The quantity $(n^2 - 1)$ can be measured experimentally and computed theoretically; see, e.g., [266, 267, 876]. For elements like H and He, precise wave functions are known so the theoretical results are likely more accurate than the experimental measurements. Numerical fitting formulae for $(n^2 - 1)$ are given in [267] for H and He. In cool stars (types K and M) Rayleigh scattering by molecular hydrogen can be important. Accurate results from a quantum mechanical calculation using the methods in [267] are given in [269] along with a numerical fitting formula.

Raman Scattering by Atomic Hydrogen

The Raman scattering cross section for a transition from the hydrogen ground state $|1s\rangle\,^2S$ to its metastable $|2s\rangle\,^2S$ states via an intermediate state $|I\,'p\rangle\,^2P$ has also been calculated accurately, see, e.g., [687, 804, 944], and is shown in figure 6.4. Notice the first resonance occurs near the wavelength of Lyβ, corresponding to the transition $|1s\rangle\,^2S \to |3p\rangle\,^2P \to |2s\rangle\,^2S$. Note also that the Raman scattering cross sections are an order of magnitude smaller than the Rayleigh scattering cross sections shown in figure 6.3 and fall off much more quickly with increasing λ.

Raman scattering in hydrogen can have direct observational effects despite its smaller cross section compared to Rayleigh scattering and that its first resonance near Lyβ lies farther out in the ultraviolet than the first Rayleigh scattering resonance near Lyα. For example, an ultraviolet photon trapped in a very opaque hydrogen cloud by coherent Rayleigh scattering in a Lyman resonance line may have to random walk a huge number of times before it can escape without being destroyed. But at each scattering the cumulative probability for it being Raman scattered can finally become large enough that the photon takes that route and escapes.

In addition, there are numerous spectral lines of ions of He, C, O, and S close to Lyβ; see, e.g., [804, 963, 964]. If any of these lines are in emission, their radiation may raise an electron from the $|1s\rangle\,^2S$ ground level of hydrogen to an intermediate state $|I\rangle = |n\,'p\rangle\,^2P$ near its $|3p\rangle\,^2P$ eigenstate, which then decays to the $|2s\rangle\,^2S$ level in a permitted transition. A particular example is that strong emission in the O VI resonance doublet $2s\,^2S_{1/2} \to 2p\,^2P^o_{3/2}$ at 1032 Å and $2s\,^2S_{1/2} \to 2p\,^2P^o_{1/2}$ at 1038 Å in the spectra of "symbiotic stars" is responsible for the previously unexplained broad emission bands at 6827 Å and 7084 Å. See figure 6.5. Such Raman-scattered

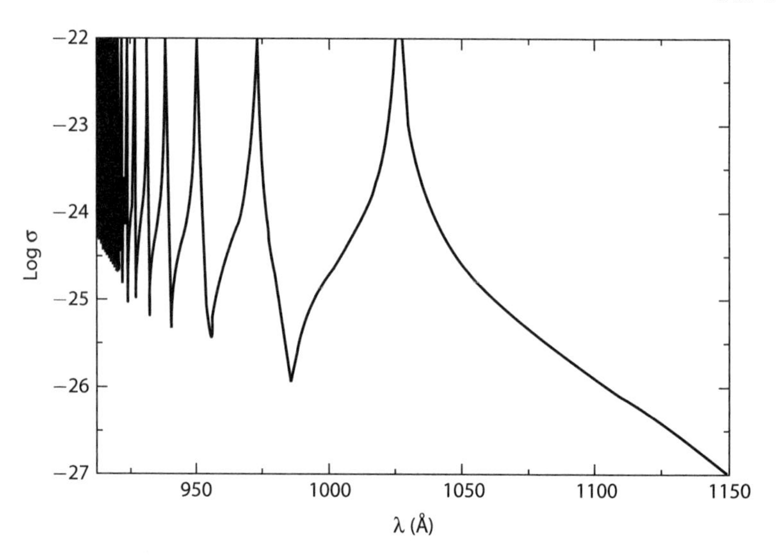

Figure 6.4 Raman scattering cross section for atomic hydrogen transition $1s\ ^2S \rightarrow 2s\ ^2S$ via intermediate bound $np\ ^2P$ and continuous $\kappa p\ ^2P$ states. *Ordinate:* $\log \sigma$ (cm^2). Adapted from [687].

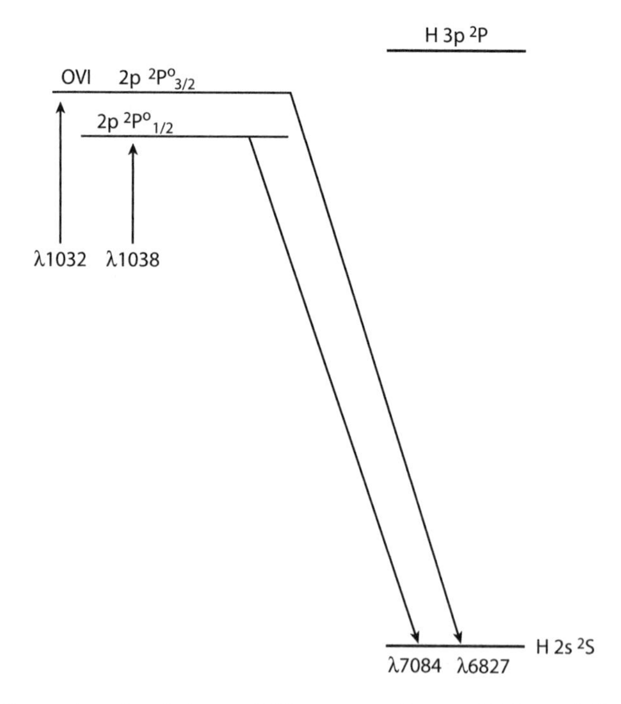

Figure 6.5 Conversion of emission in O VI resonance lines at 1032 and 1038 Å into emission bands at 6827 and 7084 Å by Raman scattering in hydrogen in the process $1s\,^2S \rightarrow$ an intermediate state near $3p\,^2P \rightarrow 2s\,^2S$. Adapted from [963].

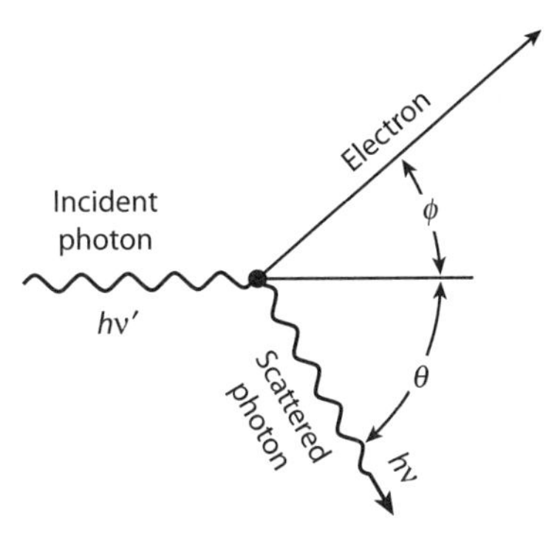

Figure 6.6 Compton scattering by an electron of an incoming photon with energy $h\nu'$ into an outgoing photon with energy $h\nu$.

features can provide sensitive diagnostics of the physical conditions in the extended envelopes of stars and in gaseous nebulae.

6.4 COMPTON SCATTERING

When the energy of an incident photon is a significant fraction of m_ec^2, Thomson scattering is replaced by *Compton scattering*, which was discovered in studies of scattering of hard X-rays in solids. In Compton's analysis [247] the scattering process was treated as a relativistic collision between a photon and a classical material particle (electron).

Kinematics

Suppose that an incoming photon of energy $h\nu'$ moving along the x axis scatters from an electron as an outgoing photon with energy $h\nu$ in the direction $\mathbf{n}$ at an angle θ with respect to $\mathbf{n}'$, and that the electron recoils with momentum $\mathfrak{p}$ at an angle ϕ with respect to $\mathbf{n}'$; see figure 6.6.

In the collision, each component of the sum of the four-momenta of the photon and particle is conserved. Thus

$$(h\nu/c)\cos\theta + \mathfrak{p}\cos\phi = h\nu'/c, \tag{6.47a}$$

$$(h\nu/c)\sin\theta - \mathfrak{p}\sin\phi = 0, \tag{6.47b}$$

$$h\nu + E = h\nu' + m_ec^2. \tag{6.47c}$$

Here $E = (\mathfrak{p}^2c^2 + m_e^2c^4)^{1/2}$ is the postcollision total energy of the electron, m_e is its proper mass, and $\mathfrak{p} = \gamma m_e v$ is the magnitude of its relativistic momentum; see equation (A.48). This system appears to be underdetermined because it has three equations with four unknowns $(\nu, \mathfrak{p}, \theta, \phi)$.

Eliminate ϕ by rearranging (6.47a) and (6.47b), squaring, and adding:

$$
\begin{aligned}
& h^2\nu^2 - 2h^2\nu\nu'\cos\theta + h^2\nu'^2 = \mathfrak{p}^2c^2 \\
\Rightarrow\ & h^2(\nu' - \nu)^2 + 2h^2\nu\nu'(1-\cos\theta) = \mathfrak{p}^2c^2.
\end{aligned}
\tag{6.48}
$$

Likewise, rearranging (6.47c) and squaring, we have

$$h^2(\nu' - \nu)^2 + 2h(\nu' - \nu)m_ec^2 + m_e^2c^4 = E^2. \tag{6.49}$$

Subtracting (6.48) from (6.49), we find

$$(\nu' - \nu)\, m_ec^2 = h\nu\nu'(1-\cos\theta), \tag{6.50}$$

which, rewritten, becomes

$$c\,(\nu' - \nu)/(\nu'\nu) = \lambda - \lambda' \equiv \Delta\lambda = (h/m_ec)(1-\cos\theta), \tag{6.51}$$

or

$$\nu = \nu'/[1 + (h\nu'/m_ec^2)(1-\cos\theta)]. \tag{6.52}$$

Hence for a given ν', we can compute the frequency ν of the scattered photon as a function of its scattering angle θ. Knowing ν', ν, and θ, we can solve the system for ϕ, $\mathfrak{p}$, and E. The *Compton wavelength of an electron* is

$$\lambda_C \equiv (h/m_ec) = 0.02426\ \text{Å}, \tag{6.53}$$

and the energy corresponding to λ_C is an electron's rest energy

$$h\nu_C = m_ec^2. \tag{6.54}$$

If the incoming photon is forward scattered, $\cos\theta \equiv 1$, and (6.52) shows that $\nu \equiv \nu'$ independent of the size of $h\nu'$. In this event, (6.47) implies $\mathfrak{p} \equiv 0$, so the electron remains at rest. More generally, if the energy of the incoming photon $h\nu' \ll m_ec^2$, (6.52) shows that $\nu \to \nu'$ for all values of $\cos\theta$, and we recover Thomson scattering. Again, $\mathfrak{p} \to 0$.

If the photon is backscattered, $\cos\theta \equiv -1$, and (6.52) shows it is redshifted an amount $\Delta\lambda_{\max} \equiv 2\lambda_C$. In this case (6.47) implies that $\sin\phi = 0$, $\cos\phi = 1$, and the electron has forward momentum $\mathfrak{p} = h(\nu' - \nu)/c$. In the case $h\nu' \gg m_ec^2$, (6.52) shows that for a backscattered photon $h\nu \to \frac{1}{2}m_ec^2$, and (6.47) shows that the electron's recoil momentum is $\mathfrak{p}c = h\nu' - \frac{1}{2}m_ec^2$. The scattering process is symmetric around $\mathbf{n}'$, so that for a given θ an equal number of photons are scattered at all angles ϕ.

Cross Section

Klein and Nishina [621, 796] derived the exact quantum mechanical expression for the differential cross section of Compton scattering *in the electron's frame*; see, e.g., [398, pp. 174–185] and [471, § 22]. For scattering of unpolarized radiation by an electron at rest, into solid angle $d\Omega$,

$$\sigma_{\mathrm{KN}}(\theta) = \frac{3}{4}\sigma_{\mathrm{T}} \left(\frac{\nu}{\nu'}\right)^2 \left(\frac{\nu'}{\nu} + \frac{\nu}{\nu'} - \sin^2\theta\right). \tag{6.55}$$

Notice that when $m_e c^2 \gg h\nu'$, $\nu' \to \nu$, and (6.55) reduces to the Thomson cross section.

Using (6.52) in (6.55) one can show that the Klein-Nishina cross section integrated over solid angle and all outgoing frequencies is

$$\frac{4}{3}\frac{\sigma_{\mathrm{KN}}(x)}{\sigma_{\mathrm{T}}} = \frac{1+x}{x^3}\left[\frac{2x(1+x)}{1+2x} - \ln(1+2x)\right] + \frac{\ln(1+2x)}{2x} - \frac{1+3x}{(1+2x)^2}, \tag{6.56}$$

where $x \equiv h\nu'/m_e c^2$. For $h\nu' \ll m_e c^2$, $\sigma_{\mathrm{KN}}(x) \to (1-2x)\sigma_{\mathrm{T}}$, and for $h\nu' \gg m_e c^2$, $\sigma_{\mathrm{KN}}(x) \to \frac{3}{8}\sigma_{\mathrm{T}}\left(\ln 2x + \frac{1}{2}\right)/x$. The ratio of $\sigma_{\mathrm{KN}}(x)$ to σ_{T} as a function of $x \equiv h\nu'/m_e c^2$ of the incoming photon is shown in figure 6.7. It begins to differ from unity for $x \sim 10^{-2}$, i.e., for $h\nu' \sim 5$ keV, which corresponds to $T \sim 5 \times 10^7$ K in a thermal

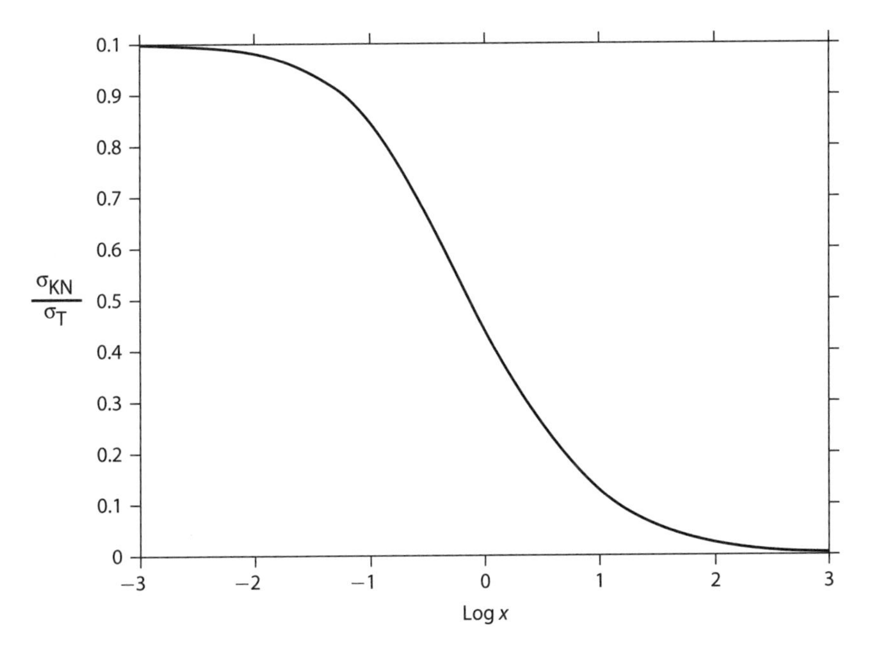

Figure 6.7 Ratio of angle-integrated Klein-Nishina cross section to Thomson cross section. *Ordinate:* $\sigma_{\mathrm{KN}}(x)/\sigma_{\mathrm{T}}(x)$. *Abscissa:* $\log x$, where $x \equiv h\nu'/m_e c^2$ for the incoming photon. Adapted from [621].

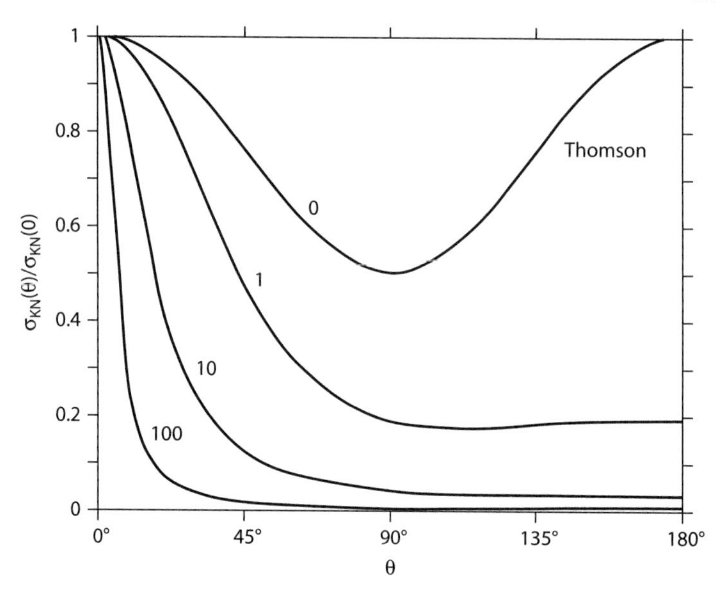

Figure 6.8 Angular distribution of Compton-scattered radiation. Adapted from [621].

distribution. The angular distribution of Compton-scattered radiation is shown in figure 6.8. The curves are labeled with the value of $x \equiv h\nu'/m_e c^2$ of the incoming photon. Note the very strong forward peaking of the distribution for $x \gtrsim 1$. The discussion above assumes the electron is free. But even an electron bound to a nucleus is effectively free for an extremely energetic incoming photon when $h\nu \gg$ than the electron's binding energy. In that case the Klein-Nishina formula can become inapplicable because an incoming photon in the strong electric field of a nucleus may produce an electron-positron pair, see, e.g., [398, pp. 203–207], [471, § 26], whose total kinetic energy equals $h\nu' - 2m_e c^2$. The positron is immediately annihilated by an electron in the plasma, and a 1MeV gamma-ray photon is emitted.

Convolution of Cross Section with Relativistic Velocity Distribution

The shift $\Delta\lambda_C$, measured in the rest frame of the electron, is always positive and quite small. But for Compton scattering in very hot plasmas, in which the electrons have high thermal velocities, an incoming photon's frequency in the laboratory frame can be changed radically by Doppler shifts between an electron's frame and the laboratory frame. The Doppler shifts of the scattered photons' frequencies in the laboratory frame results in a *diffusion of photons in frequency space*. If an incident photon's energy is greater than the kinetic energy of the electron, the photon is generally *downscattered* to a lower frequency ("normal Compton scattering"), and the plasma is heated. If the incident photon's energy is less than the kinetic energy of the electron, the photon is generally *upscattered* to a higher frequency ("inverse Compton scattering"), and the gas is cooled. In fact, this process alone can lead to equilibration between radiation and material in a hohlraum [629].

In very hot plasmas a beam of radiation is Compton scattered by an ensemble of electrons that have a relativistic Maxwellian (or Jüttner [605]) velocity distribution

$$f(v) = m_0\gamma^5 \exp(-\gamma m_0 c^2/kT)\big/4\pi ckTK_2(m_0c^2/kT), \tag{6.57}$$

which is normalized such that

$$4\pi \int_0^c f(v)v^2 dv = 1, \tag{6.58}$$

where c is the speed of light. Here $K_2(x)$ is the second-order modified Bessel function of the second kind. [See equations (A.75) and (A.76).]

The convolution of the Compton scattering cross section with an isotropic relativistic electron velocity distribution is complicated. In one approach: (1) We are given an incoming photon's momentum $(h\nu\,\mathbf{n}/c)$, which specifies both its propagation direction $\mathbf{n}$ and its energy $h\nu$ as seen in the laboratory frame, and (2) the temperature of the electrons. (3) The photon's momentum is Lorentz transformed from the laboratory frame to an electron's frame, where (4) it interacts with the Klein-Nishina scattering cross section. (5) The energy and direction of the scattered photon are calculated in that electron's frame. (6) This result is averaged over the electron velocity distribution, and (7) the resulting distribution of photons is transformed back to the laboratory frame. Alternatively, we can transform the electron-frame scattering cross section into the laboratory frame and evaluate the convolution in that frame.

Thus the number of photons $\mathfrak{N}$, as seen in the laboratory frame, scattered from a material element that contains n electrons is

$$\mathfrak{N} = n\sigma(\nu \to \nu'; \mathbf{n} \to \mathbf{n}')\frac{I(\mathbf{n},\nu)}{h\nu}\left[1 + \frac{c^2}{2h\nu'^3}I(\mathbf{n}',\nu')\right] d\nu\, d\Omega\, d\nu' d\Omega' d\mathbf{r}\, dt. \tag{6.59}$$

This must be the same *number* of photons scattered as seen in the frame of the moving material. Using the Thomas transformations in Appendix B, we have the following Lorentz invariants:

$$d\mathbf{r}\, dt = d\mathbf{r}_e dt_e, \tag{6.60a}$$

$$\nu d\nu d\Omega = \nu_e d\nu_e d\Omega_e, \tag{6.60b}$$

and

$$I(\mathbf{n},\nu)/\nu^3 = I_e(\mathbf{n}_e,\nu_e)/\nu_e^3. \tag{6.60c}$$

The subscript "e" denotes a quantity measured in an electron's frame, and the unadorned symbols denote quantities measured in the laboratory frame.

We can then rewrite (6.59) in invariant form as

$$\mathfrak{N} = n\frac{\nu}{\nu'}\sigma(\nu\to\nu';\mathbf{n}\to\mathbf{n}')\frac{I(\mathbf{n},\nu)}{\nu^3}(\nu\, d\nu\, d\Omega)(\nu' d\nu' d\Omega')d\mathbf{r}dt \equiv \mathfrak{N}_e$$

$$= n_e\frac{\nu_e}{\nu'_e}\sigma_e(\nu_e\to\nu'_e;\mathbf{n}_e\to\mathbf{n}'_e)\frac{I_e(\mathbf{n}_e,\nu_e)}{\nu_e^3}(\nu_e d\nu_e d\Omega_e)(\nu'_e d\nu'_e d\Omega'_e)d\mathbf{r}_e dt_e. \quad (6.61)$$

Because they contain only constants and the invariant I/ν^3, the square brackets in (6.59) have the same value in both frames and have been canceled. And in view of (6.60), (6.61) collapses to

$$\sigma(\nu\to\nu';\mathbf{n}\to\mathbf{n}') = (n_e\nu_e\nu'/n\nu\nu'_e)\sigma_e(\nu_e\to\nu'_e;\mathbf{n}_e\to\mathbf{n}'_e)$$
$$= (D/\gamma D')\sigma_e(\nu_e\to\nu'_e;\mathbf{n}_e\to\mathbf{n}'_e). \quad (6.62)$$

Here $D \equiv (1-\mathbf{n}\cdot\boldsymbol{\beta})$ and $D' \equiv (1-\mathbf{n}'\cdot\boldsymbol{\beta})$ [877], and $n = \gamma n_e$. Then, averaging over the isotropic electron velocity distribution function, we obtain

$$d\sigma(\nu\to\nu';\mathbf{n}\to\mathbf{n}') = 4\pi\int_0^c (D/\gamma D')\sigma_e(\nu_e\to\nu'_e;\mathbf{n}_e\to\mathbf{n}'_e)f(v)v^2 dv. \quad (6.63)$$

The Klein-Nishina formula averaged over polarization states in the electron's frame can be written [38, eq. 7.110], [877]

$$\sigma_e(\nu_e\to\nu'_e,\cos\xi_e) = \frac{r_0^2}{2x_e\nu_e}\left[1+\cos^2\xi_e + x_e x'_e(1-\cos\xi_e)^2\right]$$
$$\times\,\delta\left(\cos\xi_e - 1 + \frac{1}{x'_e} - \frac{1}{x_e}\right), \quad (6.64)$$

where $\cos\xi_e \equiv \mathbf{n}_e\cdot\mathbf{n}'_e$, $x_e \equiv h\nu_e/m_e c^2$ and $x'_e \equiv h\nu'_e/m_e c^2$. As in (6.37) for Thomson scattering, the delta function enforces conservation of energy and momentum.

To perform the integration in (6.63), all electron-frame angle and frequency variables in (6.64) must be expressed in the laboratory frame. Using the transformations (B.7)–(B.9) one obtains finally

$$\sigma(\nu\to\nu',\mathbf{n}\to\mathbf{n}') = \frac{2\pi r_0^2}{x\nu}\int_0^c \frac{1}{\gamma}\left\{1+\left[1-\frac{(1-\cos\xi)}{\gamma^2 DD'}\right]^2 + \frac{xx'(1-\cos\xi)^2}{\gamma^2 DD'}\right\}$$
$$\times\,\delta\left[\cos\xi - 1 + (\gamma D/x') - (\gamma D'/x)\right]v^2 f(v)\,dv. \quad (6.65)$$

Efficient algorithms for this complicated computation are available in the literature; see, e.g., [615, 791, 886, 996, 1163].

6.5 COMPTON SCATTERING IN THE EARLY UNIVERSE

Compton scattering plays a major role in establishing the conditions in the early Universe. According to the Big Bang theory, the Universe began as a hot ($> 10^{12}$ K), dense volume of energy called the *primeval fireball*. This infant Universe is thought to have been about as dense as the nucleus of an atom. It progressed through a series of stages in an extremely rapid expansion, during which photons collided to form particles of matter, and the forces that bind matter separated from one another.

The Matter Era

The greatest changes in the energy and matter in the fireball occurred during the first few seconds. A good discussion of these phenomena is given in [1188]. The first stage was the *heavy-particle era*, which lasted from about 10^{-40} to 10^{-4} seconds. During this stage, pairs of particles, matter and antimatter, were created from colliding photons. The Universe then might be described as a soup of high-energy, light and massive, elementary particles. Neither atoms nor the nuclei of atoms existed. Next came the *light-particle era*, from about 10^{-4} to 10 seconds. By then most of the energy of the infant Universe had already been converted to matter, and enough remained to produce only lighter particles, such as electrons and positrons. Protons and electrons interacted to form neutrons. The Universe then entered a period of nuclear interactions, which lasted from about 10 seconds to about 10^6 years. Neutrons and protons reacted to form the nuclei of light elements, including deuterium, helium, and tritium. After about a million years, the temperature fell to about 3000 K, and nuclei began to capture electrons, forming the first atoms. Almost all of these atoms were either hydrogen or helium.

Radiation Era

After about a thousand years the matter became transparent to radiation, so photons were no longer trapped. Matter, now freed from its coupling with radiation, clumped together to form the first stars and galaxies. The radiation left over cooled to 2.7 K, which we can still observe in all directions as the faint *cosmic microwave background* (CMB). The second remnant of the initial cosmic explosion is the *universal redshift*. We observe galaxies moving away from us in all directions. Their redshifts tell us how fast they are moving. To know how long they have been moving, and thus the age of the Universe, we need to know their distances. This information is obtained from the apparent magnitudes of their *Cepheid variables*, whose luminosities vary in cycles, the period of which depends on their absolute magnitudes. The relation between period and luminosity can be calibrated from observations of star clusters in our Galaxy. Thus by measuring the period and apparent magnitude of a remote Cepheid we can infer its distance. This information combined with their rate of recession yields their age. The data indicate that our Galaxy formed at about 5 billion years after the Big Bang, and our solar system was formed at about 10 billion years.

Establishment of Thermal Equilibrium Between Photons and Electrons

Kompaneets [629] first studied how photons and electrons reached thermal equilibrium in the very early Universe. Suppose we have pure radiation in a closed system such as a vacuum container that has perfectly reflecting walls. He recognized that such a system cannot reach thermodynamic equilibrium because the equations of electrodynamics are linear; hence the exchange of energy between photons of different frequencies, polarization, and direction of propagation cannot occur. To achieve equilibrium, one must add a material particle that can *absorb* and *emit* (in the sense discussed in chapter 4) radiation of all frequencies. Over a "sufficiently long time" the radiation field will evolve to the equilibrium (Planck) distribution. On the other hand, if the particle is a free electron, it can only *scatter* radiation, not absorb or emit it. This means that the occupation number n of photons of any frequency remains unchanged in the process, and we do not have the phenomenon of induced emission, which would result in $n + 1$ photons. In this situation the distribution function for photons of different frequencies would be the same as the energy levels in an ideal gas, namely, $n \propto e^{-h\nu/kT_e}$, i.e., Wien's law. To obtain Planck's law, induced emission is necessary.

Comptonization

Suppose now the material consists of completely ionized atoms. Then large numbers of low-energy photons are emitted and absorbed by the free-free mechanism (*bremsstrahlung*), which is more probable the lower the photon's frequency [384]. Thus at sufficiently low frequencies, thermal equilibrium with an average energy close to $0.5kT_e$ can be established. But at high frequencies, the most probable interaction is Compton scattering by energetic electrons, which quickly leads to a Wien distribution with an average energy $3kT_e$; this process is called *comptonization*. The final photon distribution is the result of the joint operation of free-free processes, which produce photons, and Compton scattering, which redistributes them into higher-energy parts of the spectrum.

As shown in [560], comptonization plays a decisive role in low-density, high-temperature plasmas, i.e., $N_e < 10^{-11}T_e^{4.5}$, that have a large optical thickness in Thomson scattering, i.e., $\tau_e > (m_e c^2/kT_e)^{\frac{1}{2}}$. The radiation field then evolves through a series of Bose-Einstein distributions, and this *Bose condensation* process ultimately reaches an equilibrium Planck distribution. In the hot early Universe, this mechanism determines the shape of the primordial energy spectrum [561].

Kompaneets analyzed this process in detail for the distribution function n of photons, first accounting for Compton interactions only, with the kinetic equation (which bears his name)

$$\left(\frac{\partial n}{\partial t}\right)_{\mathrm{C}} = -\int d\tau \int [n(1+n')N(\epsilon) - n'(1+n)N(\epsilon + h\nu - h\nu')]\, dW. \quad (6.66)$$

Here $N(\epsilon)$ is the distribution function for free electrons of energy ϵ, $d\tau$ is the phase-space element occupied by the electrons, and dW is the differential probability of transition from a given state to another, compatible with the laws of conservation of energy and momentum. He notes that the electrons in the plasma can arrive at statistical equilibrium very quickly and independently of the radiation field, so that $N(\epsilon)$ can be considered to be Maxwellian. Then if n and n' are assumed to be a Planckian distribution, i.e., $n = 1/(e^{h\nu/kT} - 1)$, both the left and right sides of (6.66) vanish exactly, as required.

The non-relativistic approximation for the conservation of energy and momentum laws in the collision are

$$h\nu + (p^2/2m_e) = h\nu' + (p'^2/2m_e) \tag{6.67}$$

and

$$(h\nu/c)\mathbf{n} + \mathbf{p} = (h\nu'/c)\mathbf{n}' + \mathbf{p}'. \tag{6.68}$$

Here $h\nu$ and $h\nu'$ are the photon's energy and $\mathbf{n}$ and $\mathbf{n}'$ its direction before and after the collision, and $\mathbf{p}$ and $\mathbf{p}'$ are the electron's momentum before and after the collision. By eliminating $\mathbf{p}'$ from these equations one obtains an equation for $h\nu'$ as a function of $h\nu$, $\mathbf{p}$, and the angles of scattering, which can then be rewritten in terms of $\Delta \equiv h\nu' - h\nu \ll h\nu$. Limiting to the case that $n \ll 1$, and to terms only linear in Δ, Kompaneets shows that (6.66) becomes

$$\left(\frac{\partial n}{\partial y}\right)_{\mathrm{C}} = \frac{1}{x^2}\frac{\partial}{\partial x}\left[x^4\left(\frac{\partial n}{\partial x} + n\right)\right]. \tag{6.69}$$

Here $x \equiv h\nu/kT$, and y is a dimensionless time $y \equiv (mc^2/kT)(l/c)$, where l is the Compton range, which is set by the total cross section $(8\pi/3)(e^2/mc^2)$.

Multiplying both sides of (6.69) by x^3 and integrating over x, one finds

$$\frac{d}{dy}\int_0^\infty nx^3dx = 4\int_0^\infty nx^3dx - \int_0^\infty nx^4dx. \tag{6.70}$$

While $h\nu$ is still small in comparison with kT, the second integral on the right-hand side is negligible, and we find $x = x_{\mathrm{init}}e^{4y}$. Thus the time required for the energy of a photon to increase by a factor of e as a result of the Compton effect is

$$\tau_{\mathrm{C}} = (mc^2/4kT)(l/c). \tag{6.71}$$

The Bremsstrahlung Spectrum

In the Born approximation, which is appropriate for light elements at high temperatures ($\geq$ tens of keV), and in the non-relativistic limit, the bremsstrahlung cross section is [629, eq. 16]

$$d\sigma = \frac{3Z^2e^6}{8mhc^3}\frac{1}{\epsilon}\ln\left[\frac{\left(\sqrt{\epsilon} + \sqrt{\epsilon h\nu}\right)^2}{h\nu}\right]\frac{d\nu}{\nu}, \tag{6.72}$$

where Z is the charge on the ion and ϵ is the electron's energy. To obtain the number of photons of a given frequency emitted per second, multiply the cross section by $LN(\epsilon)v d\tau$, where L is the number of nuclei per unit volume and v is the electron's velocity; integrate over all electron states; and divide by the number of photon states of a given frequency per unit volume. The result is

$$\frac{1}{\tau_{\mathrm{B}}^{0}} = 3\sqrt{\frac{\pi^3}{8m^3kT}}\frac{Z^3L^2e^6}{h\nu^3}e^{-h\nu/2kT}K_2\left(\frac{h\nu}{2kT}\right) \equiv B\frac{e^{-x/2}K_2(x/2)}{h\nu^3}, \tag{6.73}$$

where K_2 is the modified Bessel function of the second kind of order two (see Appendix A). Here τ_{B}^{0} is the characteristic bremsstrahlung time. Accounting for the emission (plus induced emission) of photons, we obtain

$$\left(\frac{\partial n}{\partial t}\right)_{\mathrm{B}} = \frac{1}{\tau_{\mathrm{B}}^{0}}\left[(1+n) - ne^{h\nu/kT}\right]. \tag{6.74}$$

As desired, this derivative vanishes if n is given by the Planck distribution. Setting $n(0)=0$ and integrating (6.74) over time, one finds

$$n(t) = \frac{1-\exp[-\left(t/\tau_{\mathrm{B}}^{0}\right)(e^{h\nu/kT}-1)]}{(e^{h\nu/kT}-1)}. \tag{6.75}$$

From (6.75) we see that the relaxation time for thermal equilibrium of the photons by pure bremsstrahlung is

$$n(t) = \tau_{\mathrm{B}}^{0}/(e^{h\nu/kT}-1). \tag{6.76}$$

Thus the total kinetic equation, accounting for both Compton scattering and bremsstrahlung, is

$$\begin{aligned}\left(\frac{\partial n}{\partial t}\right)_{\mathrm{C}} + \left(\frac{\partial n}{\partial t}\right)_{\mathrm{B}} &\equiv \left(\frac{\partial n}{\partial t}\right)\\ &= \left(\frac{kT}{mc^2}\frac{c}{l}\right)\frac{1}{x^2}\frac{\partial}{\partial x}\left[x^4\left(\frac{\partial n}{\partial x}+n+n^2\right)\right] + \frac{1}{\tau_{\mathrm{B}}^{0}}\left[(1+n)-ne^{x}\right].\end{aligned} \tag{6.77}$$

Or, in terms of the dimensionless time y,

$$\left(\frac{\partial n}{\partial y}\right) = \frac{1}{x^2}\frac{\partial}{\partial x}\left[x^4\left(\frac{\partial n}{\partial x}+n+n^2\right)\right] + \frac{4\tau_{\mathrm{C}}}{\tau_{\mathrm{B}}^{0}}\left[(1+n)-ne^{x}\right]. \tag{6.78}$$

The braking time τ_{B}^{0} decreases with decreasing frequency; hence photons at sufficiently low frequencies go into statistical equilibrium by emission and absorption processes. Photons of higher frequency increase in frequency by means of Compton scattering, eventually approaching the Wien distribution. The transition occurs at some $h\omega_0$ at which $\tau_C/\tau_B \approx 1$; it is important when $h\omega_0 \ll kT$. The total energy per unit volume emitted by bremsstrahlung is

$$\left(\frac{dE}{dt}\right)_B = \int \frac{1}{\tau_B}\left(\frac{h\nu^3}{\pi^2 c^3}\right) d\nu = \frac{2BkT}{\pi^2 c^3 h}. \tag{6.79}$$

Photons with energies $> \nu_0$ transfer energy to the electrons by Compton scattering; on the average their energy approaches $3kT$, so

$$\left(\frac{dE}{dt}\right)_C = 3kT \int_{\nu_0}^{\infty} \frac{\nu^2}{\pi^2 c^3 \tau_B}\, d\nu. \tag{6.80}$$

To avoid divergence of the integral, we replace ∞ with $\nu = kT/h$; then one finds [629, eq. 28]

$$\left(\frac{dE}{dt}\right)_C \bigg/ \left(\frac{dE}{dt}\right)_B = \frac{3}{4}\left(\ln \frac{4kT}{\gamma h\nu_0}\right)^2, \tag{6.81}$$

where $\ln \gamma = 0.577$. This ratio can be of the order of a few tens. The final result is that the mean frequency of any photon approaches $3kT/h$, independent of its initial frequency.

As pointed out by Rybicki [921], the Kompaneets equation incorporates several important physical effects:

1. Conservation of photon number
2. Detailed balance (leads to correct equilibrium solution)
3. Frequency spreading by the thermal Doppler effect
4. Frequency shift by thermal Doppler effect (i.e., inverse Compton effect)
5. Frequency shift by electron recoil (Compton effect)
6. Stimulated scattering.

It treats the first two of these exactly, but the remaining effects are treated only to the lowest order in $h\nu/mc^2$ and kT/mc^2. Its major shortcoming is that it is unable to handle situations where the radiation field changes significantly as a function of frequency on the scale of Compton frequency shifts, i.e., in the neighborhood of spectral lines. This fact led Rybicki to develop an improved kinetic equation for Compton scattering in [921], subsequently replaced by a Fokker-Planck equation [922] that overcomes these problems, handles stimulated scattering correctly, and satisfies detailed balance.

Chapter Seven

Atomic and Molecular Absorption Cross Sections

In chapters 5 and 6 we described the interaction of radiation and material in bound-bound, bound-free, and free-free absorption-emission processes and by scattering, using both detailed-balancing arguments and simplified quantum mechanical analyses that contain only symbolic atomic eigenkets. Here we examine the interaction of radiation with specific atomic structures and molecules more closely.

There is a vast literature dealing with these subjects; we refer the reader to [116, 117, 244, 248, 286, 303, 444, 471, 485, 486, 542, 566, 702, 723, 850, 851, 853, 854, 855, 882, 961, 1011, 1012, 1013, 1021, 1025]. Such a long list of references implies that these topics (1) have been studied thoroughly, and (2) can be complicated. To avoid going too far afield, we restrict attention to a small set of problems having the greatest importance in modeling stellar atmospheres.

Hydrogen and helium are the most abundant elements in the Universe. In A-, B-, and O- star atmospheres, hydrogen absorption is the main continuum opacity source for $\lambda \geq 912$ Å. Absorption by the ground state of H dominates the ultraviolet in the range 504 Å $\leq \lambda \leq$ 912 Å; the ground state of He dominates for 228 Å $\leq \lambda \leq$ 504 Å; and the ground state of He^+ dominates for $\lambda \leq$ 228 Å in O-stars. The spectral line profiles and equivalent widths of H, He I, and He II lines provide sensitive indicators of the effective temperatures and surface gravities of early-type stars, and also the He/H abundance ratio.

Thus even the simplest problem of a single electron moving in a spherically symmetric nuclear potential field, discussed in § 7.1, is of great importance in astrophysics. Exact solutions of both the non-relativistic Schrödinger equation and the relativistic Dirac equation can be obtained for it, yielding analytical expressions for interaction cross sections with radiation. Neutral helium with its two bound electrons provides the simplest example of a multi-electron atomic structure, discussed in § 7.2. For an atom, the task is to find the wave functions of its electrons. Electrons are fermions; hence they collectively must be in configurations that are consistent with the Pauli exclusion principle. Electrons in bound states move on quantized "orbits."[1] Each bound state has a quantized *binding energy*, *orbital angular momentum*, and *spin angular momentum*. Molecules, discussed in § 7.3, have additional quantized *vibration* and *rotation* states of the molecule as a whole.

[1] This misleading terminology implies that electrons move on definite paths around the nucleus. In fact, we can determine only the *probability distribution* of an electron's position.

7.1 HYDROGEN AND HYDROGENIC IONS

Wave Equation and Wave Function

The time-dependent Schrödinger equation for an atom consisting of a nucleus with mass m_N at position (x_N, y_N, z_N) and a single electron with rest mass m_e at position (x_e, y_e, z_e), which has potential energy V, is

$$i\hbar\frac{\partial\Psi}{\partial t} = \left[-\frac{\hbar^2}{2m_N}\nabla_N^2 - \frac{\hbar^2}{2m_e}\nabla_e^2 + V(x_N, y_N, z_N, x_e, y_e, z_e)\right]\Psi, \quad (7.1)$$

where $\Psi = \Psi(x_N, y_N, z_N, x_e, y_e, z_e, t)$. In a *central potential*, the potential energy depends only on the distance between the particles, so we can use *center of mass coordinates* (X, Y, Z), defined as

$$\mathsf{M}X \equiv m_N x_N + m_e x_e, \quad \mathsf{M}Y \equiv m_N y_N + m_e y_e, \quad \mathsf{M}Z \equiv m_N z_N + m_e z_e, \quad (7.2a)$$

where $\mathsf{M} = m_N + m_e$ is the total mass of the system, and *relative coordinates* (x, y, z), are defined as

$$x \equiv x_e - x_N, \quad y \equiv y_e - y_N, \quad z \equiv z_e - z_N. \quad (7.3)$$

Then (7.1) becomes

$$i\hbar\frac{\partial\Psi}{\partial t} = \left[-\frac{\hbar^2}{2\mathsf{M}}\left(\frac{\partial^2}{\partial X^2} + \frac{\partial^2}{\partial Y^2} + \frac{\partial^2}{\partial Z^2}\right) - \frac{\hbar^2}{2\mathsf{m}}\left(\frac{\partial^2}{\partial x^2} + \frac{\partial^2}{\partial y^2} + \frac{\partial^2}{\partial z^2}\right) + V(x, y, z)\right]\Psi, \quad (7.4)$$

where m is the *reduced mass* of the system

$$\frac{1}{\mathsf{m}} \equiv \frac{1}{m_N} + \frac{1}{m_e} = \frac{m_N + m_e}{m_N m_e}. \quad (7.5)$$

Ψ can now be written as the product of functions that depend on the center of mass and relative coordinates only:

$$\Psi(x, y, z, X, Y, Z, t) = \chi(X, Y, Z)\, \psi(x, y, z)\, e^{-i(E+\epsilon)t/\hbar}. \quad (7.6)$$

The Schrödinger equation then separates into

$$-\frac{\hbar^2}{2\mathsf{M}}\tilde{\nabla}^2\chi = \epsilon\,\chi \quad (7.7a)$$

and

$$-\frac{\hbar^2}{2\mathsf{m}}\nabla^2\psi + V(x, y, z)\psi = E\,\psi. \quad (7.7b)$$

The solution $\chi(X, Y, Z)$ of (7.7a) describes the rectilinear motion of the center of mass of the nucleus + electron system as a free particle with energy ϵ. The solution $\psi(x, y, z)$ of (7.7b) describes the motion of an electron with a total (potential + kinetic) energy E relative to the nucleus.

Stationary States

Stationary state (i.e., time-independent) solutions of (7.7b) for ψ must be everywhere bounded, single-valued, continuous, and differentiable. Consider an electron that has a potential energy $-Ze^2/r$ in the *spherically symmetric* central potential $+Ze/r$ of a nucleus with charge $+Ze$. Then we can write (7.7b) in polar coordinates, where r is the radial distance between the electron and nucleus, θ is the polar angle (i.e., colatitude) between them, and ϕ is the azimuthal angle of the electron, measured around the z axis:

$$-\frac{\hbar^2}{2\mathrm{m}}\nabla^2\psi - \frac{Ze^2}{r}\psi = E\,\psi, \tag{7.8}$$

or

$$-\frac{\hbar^2}{2\mathrm{m}}\frac{1}{r^2}\left[\frac{\partial}{\partial r}\left(r^2\frac{\partial\psi}{\partial r}\right)+\frac{1}{\sin\theta}\frac{\partial}{\partial\theta}\left(\sin\theta\frac{\partial\psi}{\partial\theta}\right)+\frac{1}{\sin^2\theta}\frac{\partial^2\psi}{\partial\phi^2}\right]-\frac{Ze^2}{r}\psi = E\,\psi. \tag{7.9}$$

The solution of (7.9) can be factored as

$$\psi(r,\theta,\phi) = R(r)\,\Theta(\theta)\,\Phi(\phi). \tag{7.10}$$

By substituting (7.10) into (7.9), we obtain three equations, each depending on one of the independent variables. They are consistent only if each equation equals a *separation constant* depending on three integers (n, l, m) such that

$$\frac{d^2\Phi}{d\phi^2} + m^2\Phi = 0, \tag{7.11a}$$

$$\frac{1}{\sin\theta}\frac{d}{d\theta}\left(\sin\theta\frac{d\Theta}{d\theta}\right) + \left[l(l+1) - \frac{m^2}{\sin^2\theta}\right]\Theta = 0, \tag{7.11b}$$

and

$$\frac{1}{r^2}\frac{d}{dr}\left(r^2\frac{dR}{dr}\right) - \frac{l(l+1)}{r^2}R + \frac{2\mathrm{m}}{\hbar^2}\left(E + \frac{Ze^2}{r}\right)R = 0. \tag{7.11c}$$

Azimuthal Equation

Bound states (i.e., $E < 0$) have specific values for the electron's angular momentum, set by its *orbital angular momentum quantum number* l and *magnetic quantum number* m, which measures the projection of l onto the z axis. Solutions of the azimuthal-angle equation (7.11) are

$$\Phi_m(\phi) = \frac{1}{\sqrt{2\pi}}e^{im\phi}, \tag{7.12}$$

which are normalized such that

$$\int_0^{2\pi} \Phi_m^*(\phi)\Phi_{m'}(\phi)\, d\phi = \delta_{mm'}. \tag{7.13}$$

In a spherically symmetric potential, orbital angular momentum is conserved, so its z component is quantized with values $m\hbar$, $m = 0, \pm 1, \pm 2, \ldots, \pm l$.

Orbital Equation

Solutions of the orbital equation (7.11b) are *associated Legendre functions* $P_{l|m|}(\theta)$, where $l = 0, \pm 1, \ldots, \pm(n-1)$, and $|m| \leq |l|$; see [5, p. 338]. These are orthogonal and are normalized such that

$$\int_0^{\pi} P_{l|m|}(\mu)P_{l'|m|}(\mu)\, d\mu = \frac{2(l+|m|)!}{(2l+1)(l-|m|)!}\delta_{ll'}. \tag{7.14}$$

Here $\mu \equiv \cos\theta$. In view of (7.14), normalized solutions of (7.11b) are

$$\Theta_{l|m|}(\mu) = \left[\frac{(2l+1)(l-|m|)!}{2(l+|m|)!}\right]^{\frac{1}{2}} P_{l|m|}(\mu). \tag{7.15}$$

$\Theta_{lm}(\mu)$ is real, and

$$\int_0^{\pi} \left[\Theta_{l|m|}(\mu)\, \Theta_{l'|m|}(\mu)\right]^2\, d\mu = \delta_{ll'}. \tag{7.16}$$

By combining (7.12) and (7.15) we obtain the angular part of the complete wave function as a *spherical harmonic*:

$$Y_{lm}(\theta,\phi) \equiv (-1)^{\frac{1}{2}(m+|m|)} \left[\frac{(2l+1)(l-|m|)!}{4\pi(l+|m|)!}\right]^{\frac{1}{2}} P_{l|m|}(\mu)e^{im\phi}. \tag{7.17}$$

In view of (7.13) and (7.14), these spherical harmonics are normalized:

$$\int_0^{2\pi}\int_0^{\pi} Y_{lm}^*(\theta,\phi)Y_{l'm'}(\theta,\phi)\sin\theta\, d\theta\, d\phi = \delta_{ll'}\delta_{mm'}. \tag{7.18}$$

Note that both the azimuthal and orbital equations are independent of the analytical form of the potential V and the total energy E. Hence these solutions apply in any central potential.

Radial Equation

As $r \to \infty$, $R(r)$ must $\to 0$ fast enough that the integral $\int_0^\infty r^2 R(r) dr$ is finite. For bound states this requirement is met only if n is an integer, the *principal quantum number* ($n = 1, 2, \ldots$), which determines the binding energy of a state. Solutions of the radial equation (7.11c) are *associated Laguerre polynomials* $L_b^a(x)$, which are a special case of the confluent hypergeometric function; see [5, chapters 13 and 22].

Define

$$\rho \equiv (2Z/na_0)r, \tag{7.19}$$

where

$$a_0 \equiv (\hbar^2/\mathrm{m}e^2) \tag{7.20}$$

is the *Bohr radius*. Then

$$R_{nl}(r) = \left[\left(\frac{2Z}{na_0}\right)^3 \frac{(n-l-1)!}{2n[(n+l)!]^3}\right]^{\frac{1}{2}} \rho^l e^{-\rho/2} L_{n+l}^{2l+1}(\rho), \tag{7.21}$$

where $l < n$. These functions are orthogonal and normalized such that

$$\int_0^\infty R_{nl}(r)R_{n'l}(r)\, r^2\, dr = \delta_{nn'}. \tag{7.22}$$

Another useful function is $P_{nl}(r) \equiv rR_{nl}(r)$, for which $\int_0^\infty P_{nl}(r)P_{n'l}(r)\, dr = \delta_{nn'}$. Both $R_{nl}(r)$ and $P_{nl}(r)$ are real.

Energy Eigenvalues

The bound states have energies

$$E_n^0 = -(Z^2\mathcal{R}/n^2), \tag{7.23}$$

where

$$\mathcal{R} \equiv (e^4\mathrm{m}/2\hbar^2) = 2.1799 \times 10^{-11}\ \mathrm{erg} \tag{7.24}$$

is the *Rydberg constant* in energy units. In contrast, when $E > 0$, (7.11c) yields a continuum of unbound states for any value of E and l.

The frequency $\nu_{n'n}$ of an emission line transition $n' \to n$ ($n' > n$) is

$$h\nu_{n'\to n} = Z^2\mathcal{R}\left(\frac{1}{n^2} - \frac{1}{n'^2}\right). \tag{7.25}$$

At this level of approximation, the energy eigenvalues of hydrogen are *degenerate* in as much as they depend only on n. The number of states in the nth energy eigenstate (its *statistical weight*)[2] is

$$2\sum_{l=0}^{n-1}(2l+1) = 2n^2. \tag{7.26}$$

This degeneracy is resolved when one accounts for the interaction of the orbital angular momentum and spin of the electron with the nuclear potential.

[2] There are two electron spin states for each choice of (n, l, m); see below.

Normalized Wave Functions

Normalized wave functions for the first three (n-degenerate) levels of hydrogen and hydrogenic ions are

$$\begin{aligned}
\psi_{100} &= \left(Z^3/\pi a_0^3\right)^{\frac{1}{2}} e^{-(Zr/a_0)} \\
\psi_{200} &= \left(Z^3/32\pi a_0^3\right)^{\frac{1}{2}} [2-(Zr/a_0)]e^{-(Zr/2a_0)} \\
\psi_{210} &= \left(Z^5/32\pi a_0^5\right)^{\frac{1}{2}} re^{-(Zr/2a_0)} \cos\theta \\
\psi_{21\pm1} &= \left(Z^5/64\pi a_0^5\right)^{\frac{1}{2}} re^{-(Zr/2a_0)} \sin\theta e^{\pm i\phi} \\
\psi_{300} &= \tfrac{1}{81}\left(Z^3/3\pi a_0^3\right)^{\frac{1}{2}} \left[27-18(Zr/a_0)+2(Zr/a_0)^2\right] e^{-(Zr/3a_0)} \\
\psi_{310} &= \tfrac{1}{81}\left(2Z^5/\pi a_0^5\right)^{\frac{1}{2}} \left[6r-(Zr^2/a_0)\right] e^{-(Zr/3a_0)} \cos\theta \\
\psi_{31\pm1} &= \tfrac{1}{81}\left(Z^5/\pi a_0^5\right)^{\frac{1}{2}} \left[6r-(Zr^2/a_0)\right] e^{-(Zr/3a_0)} \sin\theta e^{\pm i\phi} \\
\psi_{320} &= \tfrac{1}{81}\left(Z^7/6\pi a_0^7\right)^{\frac{1}{2}} r^2 e^{-(Zr/3a_0)} (3\cos^2\theta-1) \\
\psi_{32\pm1} &= \tfrac{1}{81}\left(Z^7/\pi a_0^7\right)^{\frac{1}{2}} r^2 e^{-(Zr/3a_0)} \sin\theta\cos\theta\, e^{\pm i\phi} \\
\psi_{32\pm2} &= \tfrac{1}{162}\left(Z^7/\pi a_0^7\right)^{\frac{1}{2}} r^2 e^{-(Zr/3a_0)} \sin^2\theta\, e^{\pm 2i\phi}.
\end{aligned} \tag{7.27}$$

The subscript on each denotes the values of (n, l, m).

Time-Dependent Eigenfunctions

Time-dependent eigenfunctions for a one-electron atom/ion are

$$\psi_{nlm}(r, \mu, \phi, t) = R_{nl}(r)\Theta_{lm}(\mu)\Phi_m(\phi)e^{-iE_nt/\hbar}, \tag{7.28}$$

and the wave function of an arbitrary state of the system can be written

$$\Psi(r, \mu, \phi, t) = \sum_{n=1}^{\infty}\sum_{l=0}^{n-1}\sum_{m=-l}^{l} a_{nlm}R_{nl}(r)\Theta_{lm}(\mu)\Phi_m(\phi)e^{-iE_nt/\hbar}. \tag{7.29}$$

$|a_{nlm}|^2$ is the initial probability of state (n, l, m) and $\sum_{nlm} |a_{nlm}|^2 \equiv 1$.

Orbital Angular Momentum

Classically, the angular momentum of a single particle with linear momentum $\mathbf{p}$ at radius vector $\mathbf{r}$ from an origin is $\mathbf{L} = \mathbf{r} \times \mathbf{p}$. The quantum mechanical operator for linear momentum is $\hat{p}_k \equiv -i\hbar(\partial/\partial q_k)$; hence we take the *angular momentum operator* to be [148]

$$\hat{l} = -i\hbar\,\mathbf{r}\times\nabla. \tag{7.30}$$

Writing (7.30) in Cartesian coordinates, we have

$$\hat{l}_x = yp_z - zp_y = -i\hbar\left(y\frac{\partial}{\partial z} - z\frac{\partial}{\partial y}\right),$$

$$\hat{l}_y = zp_x - xp_z = -i\hbar\left(z\frac{\partial}{\partial x} - x\frac{\partial}{\partial z}\right), \tag{7.31}$$

and

$$\hat{l}_z = xp_y - yp_x = -i\hbar\left(x\frac{\partial}{\partial y} - y\frac{\partial}{\partial x}\right).$$

In this form it is easy to find the commutators of the components of $\hat{\boldsymbol{l}}$:

$$[\hat{l}_x, \hat{l}_y] \equiv i\hbar\hat{l}_z, \qquad [\hat{l}_y, \hat{l}_z] \equiv i\hbar\hat{l}_x, \qquad [\hat{l}_z, \hat{l}_x] \equiv i\hbar\hat{l}_y \tag{7.32}$$

or, equivalently,

$$\hat{\boldsymbol{l}} \times \hat{\boldsymbol{l}} = i\hat{\boldsymbol{l}}. \tag{7.33}$$

Converting to spherical coordinates, we find

$$\hat{l}_x = i\hbar\left(\sin\phi\frac{\partial}{\partial\theta} + \cot\theta\cos\phi\frac{\partial}{\partial\phi}\right),$$

$$\hat{l}_y = -i\hbar\left(\cos\phi\frac{\partial}{\partial\theta} - \cot\theta\sin\phi\frac{\partial}{\partial\phi}\right), \tag{7.34}$$

and

$$\hat{l}_z = -i\hbar\frac{\partial}{\partial\phi},$$

and the total orbital angular momentum operator is

$$\hat{\boldsymbol{l}}\cdot\hat{\boldsymbol{l}} = \hat{l}_x^2 + \hat{l}_y^2 + \hat{l}_z^2 = -\hbar^2\left[\frac{1}{\sin\theta}\frac{\partial}{\partial\theta}\left(\sin\theta\frac{\partial}{\partial\theta}\right) + \frac{1}{\sin^2\theta}\frac{\partial^2}{\partial\phi^2}\right]. \tag{7.35}$$

Note that $\hat{\boldsymbol{l}}\cdot\hat{\boldsymbol{l}}$ and $\hat{l}_z$ commute, so the eigenvalues of these operators can be determined simultaneously. The operator in (7.35) the same as appeared in (7.11b), so Y_{lm} is an eigenfunction of (7.35). It has eigenvalues

$$\hat{\boldsymbol{l}}\cdot\hat{\boldsymbol{l}}\, Y_{lm}(\theta,\phi) = l(l+1)\,\hbar^2\, Y_{lm}(\theta,\phi) \tag{7.36}$$

and

$$\hat{l}_z Y_{lm}(\theta,\phi) = m\,\hbar\, Y_{lm}(\theta,\phi). \tag{7.37}$$

For the central field problem, angular momentum is a constant of the motion in both quantum and classical mechanics.

Relativistic Variation of the Electron Mass

The energy levels of hydrogen and hydrogenic ions actually do not depend on n only, as stated in (7.23), but are slightly shifted (i.e., have *fine structure*) because (a) relativistic effects make the laboratory-frame mass of a moving electron differ from its rest mass and (b) its magnetic moment (resulting from its motion around the nucleus) interacts with the nuclear potential. Consider first the relativistic mass-variation effects.

From equation (5.25) the energy of an electron with rest mass m_e, charge $-e$, and linear momentum $\mathfrak{p} = \gamma m_e \mathbf{v}$ in an electric potential $\phi(r) = +Ze/r$ is $\mathcal{H} = (m_e^2 c^4 + \mathfrak{p}^2 c^2)^{1/2} - Ze^2/r$. Expanding $\mathcal{H}$ to first order in the limit that $\gamma \to 1$ and subtracting the electron's rest energy, we have

$$E = \mathcal{H} - m_e c^2 \approx \frac{p^2}{2m_e} - \frac{p^4}{8m_e^3 c^2} - \frac{Ze^2}{r}. \tag{7.38}$$

The non-relativistic energy of an electron with principal quantum number n is

$$E_n^0 = \frac{p^2}{2m_e} - \frac{Ze^2}{r} = -\frac{Z^2 \mathcal{R}}{n^2}. \tag{7.39}$$

Thus the energy shift produced by the relativistic increase of the electron's laboratory-frame mass is

$$\Delta E_{nl} = -\frac{p^4}{8m_e^3 c^2} = -\frac{1}{2m_e c^2}\left(\frac{p^2}{2m_e}\right)^2 = -\frac{1}{2m_e c^2}\left(E_n^0 + \frac{Ze^2}{r}\right)^2, \tag{7.40}$$

i.e., it becomes more tightly bound. Averaging over an orbit, we have

$$\langle \Delta E_{nl} \rangle = -\frac{1}{2m_e c^2}\left[(E_n^0)^2 + 2E_n^0 Ze^2 \langle r^{-1} \rangle_{nl} + Z^2 e^4 \langle r^{-2} \rangle_{nl}\right]. \tag{7.41}$$

The averages in (7.41) can be calculated from the radial functions R_{nl} in (7.21), yielding $\langle r^{-1} \rangle_{nl} = Z/n^2 a_0$ and $\langle r^{-2} \rangle_{nl} = Z^2/n^3 (l + \frac{1}{2}) a_0^2$ [1025, p. 7]. Using these expressions and the definitions of $\mathcal{R}$ and a_0, the energy shift is

$$\langle \Delta E_{nl} \rangle = \alpha^2 \left(\frac{3}{4n} - \frac{1}{l + \frac{1}{2}}\right) \frac{Z^4}{n^3} \quad [\,\text{Ryd}\,], \tag{7.42}$$

where

$$\alpha \equiv e^2/\hbar c = 1/137.036 \tag{7.43}$$

is the *fine-structure constant*. Note that ΔE_{nl} is always negative and that levels with the smallest l lie lowest (are most tightly bound).

Spin-Orbit Interaction

The Stern-Gerlach experiment (1922) showed that neutral atoms[3] sent through a magnetic field that has an intensity gradient along its path have a *magnetic dipole moment* produced by the orbital and intrinsic angular momentum of their electrons. If the field through which they moved were homogeneous, the forces on opposite ends of the dipole would cancel, and their trajectories would be unaffected. But when the field has a spatial gradient, the force on one end of the dipole is greater than that on the other end, so there is a torque on the dipole, which produces a net force that deflects the particle's trajectory. If atoms were classical objects, their angular momentum vectors would be oriented randomly, so each particle would be deflected by a different amount and their distribution on the detector would be smooth. In reality they are deflected up or down by a specific amount, which shows that their magnetic moments have quantized values.

Orbital Magnetic Field

An electron in motion around an atom's nucleus both (1) experiences an electric field, i.e., attraction by the positively charged nucleus, and (2) as a reflection of its own motion, "sees" the nucleus moving around it, which is, in effect, an electric current that, by Ampere's law, generates a magnetic field at the electron's position.

We can use equation (3.52) to calculate these fields. For simplicity, assume circular motion of the electron. Take the electron's path around the nucleus to be $\mathbf{r}(t) = r\,(\cos\omega t\,\mathbf{i} + \sin\omega t\,\mathbf{j}) \equiv r\mathbf{n}(t)$, where $r \equiv |\mathbf{r}|$ is constant and $\mathbf{n}$ is a unit vector along $\mathbf{r}$. Then $\boldsymbol{\beta} \equiv \dot{\mathbf{r}}/c = -(\omega r/c)(\sin\omega t\,\mathbf{i} - \cos\omega t\,\mathbf{j})$, which is perpendicular to $\mathbf{n}$, so that $\boldsymbol{\beta}\cdot\mathbf{n} \equiv 0$ in the denominators of both terms in (3.52). Further, $\dot{\boldsymbol{\beta}} = \ddot{\mathbf{r}}/c = -(\omega^2 r/c)\mathbf{n}$; therefore, the second term in (3.52) vanishes identically, leaving, to $O(v/c)$, an electric field $\mathbf{E} = Z\,e\,(\mathbf{n} - \boldsymbol{\beta})/r^2$. Hence, *in the frame of the electron*,

$$\mathbf{H}_{\text{orb}} = (\mathbf{n}\times\mathbf{E}) = -\frac{Z\,e}{c\,r^3}(\mathbf{r}\times\mathbf{v}) = \frac{\hbar}{c\,e\,m_e}\left(\frac{1}{r}\frac{\partial V}{\partial r}\right)\hat{\boldsymbol{l}}. \tag{7.44}$$

Here we replaced the hydrogenic Coulomb potential with a generalized potential $V(r)$, and the classical angular momentum ($\mathbf{r} \times m_e\mathbf{v}$) with the quantum expression $\hbar\hat{\boldsymbol{l}}$.

But as shown by L. H. Thomas [1073], if the coordinate system is fixed in the atom's frame (or lab frame) instead of the accelerating electron's frame, a relativistic kinematical effect of the "precession" of the electron's spin axis in its "orbit" around the nucleus reduces the magnetic field calculated above by a factor of two[4] so that

$$\mathbf{H}_{\text{orb}} = \frac{\hbar}{2m_e c\,e}\left(\frac{1}{r}\frac{\partial V}{\partial r}\right)\hat{\boldsymbol{l}}. \tag{7.45}$$

The ratio $\mu_0 \equiv e\,\hbar/2m_e c$ is called the *Bohr magneton.*

[3] Neutral atoms are used to avoid the large deflections that charged particles would have in the magnetic field, allowing internal atomic effects to dominate.

[4] See also [575, § 11.8 and § 11.11].

Intrinsic Electron Spin

The mechanism described above does not fully explain spectroscopic observations of fine-structure splitting in many line transitions. This deficiency led Goudsmit and Uhlenbeck [383, 1090] to postulate in 1926 that electrons have an additional quantized angular momentum called *spin*, which has an associated magnetic moment. They showed that with this hypothesis one could obtain a satisfactory explanation of the observed Zeeman-effect splitting of atomic spectral lines in external magnetic fields.

Although he did not give it the name "spin," it was Pauli who first proposed the existence of a fourth quantum number as a necessary ingredient of quantum mechanics. Unlike mass and charge, there is no classical analog to spin. He asserted that Thomas's image of a "spinning sphere" was too descriptive and that spin must instead be viewed as an *intrinsic* property, like mass and charge, associated with specific types of particles. In 1927 he worked out a mathematical theory for spin in terms of *spinors* [846]. Electron spin appears naturally in, and is an essential part of, the relativistic quantum mechanics formulated by Dirac [282, 284, 286, § 70]. In modern terminology, spin can be viewed as a type of angular momentum by considering it, consistent with Noether's theorem, to be a *generator of rotations*.

An electron is assigned a spin momentum $\hat{s}$ of magnitude

$$|\,\hat{s}\,| = \sqrt{s\,(s+1)}\hbar, \tag{7.46}$$

with $s = 1/2$. The projection of $\hat{s}$ on the z axis is

$$s_z \equiv m_s\,\hbar = \pm\tfrac{1}{2}\,\hbar. \tag{7.47}$$

The exact solution of Dirac's relativistic theory shows that an electron has a magnetic moment

$$\hat{\boldsymbol{\mu}}_e = -g_e \frac{e\hbar}{2m_e c}\hat{s} = -2\mu_0 \hat{s} \tag{7.48}$$

i.e., $g_e = 2$. This is twice the value it would have if the electron were considered to be a classical charge.[5] Subsequent work with precision experiments [806], and using quantum electrodynamics to correct for vacuum fluctuations ("virtual photons"), gives $g_e = 2.0023\,19304362$, with any uncertainty confined to the last digit.

Fine Structure

An electron's magnetic dipole moment $\hat{\boldsymbol{\mu}}_e$ with its axis oriented at an angle θ relative to the magnetic field $\mathbf{H}_{\text{orb}}$ experiences a torque $\mathbf{T}(\theta) = -|\mu_e\, H_{\text{orb}}|\sin\theta$, which acts to align the dipole moment to its stable direction along the field $\mathbf{H}_{\text{orb}}$. The alignment process would require an amount of work

$$\Delta E_{\text{orb}} = -\int |\mu_e\, H_{\text{orb}}|\sin\theta\, d\theta = -\mu_e \hat{\mathbf{s}}\cdot\mathbf{H}_{\text{orb}} \tag{7.49}$$

[5] See [575, § 11.8].

to be done against the torque. Combining (7.45) and (7.49), we find that the energy shift of a level resulting from *spin-orbit interaction* is

$$\Delta E_{so} = \frac{\hbar^2}{2m_e^2c^2}\left(\frac{1}{r}\frac{\partial V}{\partial r}\right)\hat{\boldsymbol{l}}\cdot\hat{\boldsymbol{s}}. \tag{7.50}$$

An electron's *total angular momentum* is $\hat{\boldsymbol{j}} \equiv \hat{\boldsymbol{l}} + \hat{\boldsymbol{s}}$. Thus

$$\hat{\boldsymbol{j}}\cdot\hat{\boldsymbol{j}} = \hat{\boldsymbol{l}}\cdot\hat{\boldsymbol{l}} + 2\hat{\boldsymbol{l}}\cdot\hat{\boldsymbol{s}} + \hat{\boldsymbol{s}}\cdot\hat{\boldsymbol{s}}, \tag{7.51}$$

so from (7.36) and (7.46) we have

$$\hat{\boldsymbol{l}}\cdot\hat{\boldsymbol{s}} = \tfrac{1}{2}[j(j+1) - l(l+1) - s(s+1)]\hbar^2. \tag{7.52}$$

Using (7.52) in (7.50), taking $r^{-1}(\partial V/\partial r) = Ze^2r^{-3}$ for a hydrogenic ion, and averaging over an orbit with $\langle r^{-3}\rangle_{nl} = Z^3/[a_0^3n^3l(l+\frac{1}{2})(l+1)]$ [1025, p. 7], one finds (for $l > 0$)

$$\langle\Delta E_{so}\rangle = \alpha^2\left[\frac{j(j+1) - l(l+1) - \frac{3}{4}}{2l(l+\frac{1}{2})(l+1)}\right]\frac{Z^4}{n^3} \quad [\,\text{Ryd}\,]. \tag{7.53}$$

Finally, combining (7.42) and (7.53), we find that for the two possible cases, $j = l + \frac{1}{2}$ and $j = l - \frac{1}{2}$, the energy shift is

$$\langle\Delta E_{nlj}\rangle = \langle\Delta E_{nl}\rangle + \langle\Delta E_{so}\rangle = \alpha^2\left(\frac{3}{4n} - \frac{1}{j+\frac{1}{2}}\right)\frac{Z^4}{n^3} \quad [\,\text{Ryd}\,]. \tag{7.54}$$

That is, the energy levels of hydrogenic ions are split into *doublets*, and the state with the largest j lies highest in energy. In table 7.1 we list the separation $\langle\Delta E_{nlj}\rangle$ of the $2p_{1/2}$, $2p_{3/2}$ levels from spectroscopic data tables for elements of astrophysical interest. It is very small for hydrogen, but becomes important for large Z. A numerical check verifies that these separations do in fact scale as Z^4.

Table 7.1 Splitting of $2p_{1/2}$ and $2p_{3/2}$ Levels of Hydrogenic Ions

Ion	Z	$\langle\Delta E\rangle$ eV	Ion	Z	$\langle\Delta E\rangle$ eV
H	1	4.54×10^{-5}	Ne X	10	0.455
He II	2	7.26×10^{-4}	Mg XII	12	0.946
Li III	3	3.68×10^{-3}	Si XIV	14	1.755
C VI	6	5.89×10^{-2}	Ar XVIII	18	4.816
N VII	7	1.09×10^{-1}	Ca XX	20	7.359
O VIII	8	1.86×10^{-1}	Fe XXVI	26	21.214

Hyperfine Structure

In addition to the fine structure discussed above, there is a *hyperfine* splitting of the $1s$ ground state of atomic hydrogen produced by a change in the relative orientation of the spins of the proton and electron. In the higher-energy state the proton and electron have parallel spins and anti-parallel magnetic moments owing to their opposite charge. If the spin of the nucleus is $\hbar\hat{\boldsymbol{I}}$, its magnetic moment is

$$\hat{\boldsymbol{M}} = g_N \frac{Ze\hbar}{2m_N c}\hat{\boldsymbol{I}}, \tag{7.55}$$

which is a factor of m_e/M_N smaller than $\hat{\boldsymbol{\mu}}_e$. The *spin flip* transition between these levels via the *spin-spin interaction* releases radiation about 5×10^{-6} eV in energy, at 1420.40575177 MHz, or a vacuum wavelength of 21.10611405413 cm, commonly called the *21 cm line*. This transition has the extremely small probability of $2.9\times10^{-15}\mathrm{sec}^{-1}$, which means an isolated atom of neutral hydrogen will undergo this transition only once in about 10^7 years. But the total number of atoms of neutral hydrogen in the interstellar medium is very large, so this emission line is easily observed with radio telescopes. Using these data it has been possible to map out the distribution of neutral hydrogen in the plane of the Galaxy.

Dipole-Transition Selection Rules

We saw in chapter 5 that radiation can interact strongly with the electric dipole moment of an atom (which is the only component of the field for hydrogen and hydrogenic ions). A transition between states nlm and $n'l'm'$ has nonzero probability if it satisfies the *selection rules*:

$$\Delta l = l' - l = \pm 1 \quad \text{and} \quad \Delta m = m_l' - m_l = 0, \text{ or } \pm 1, \tag{7.56}$$

and the parity of the wave function changes (i.e., the wave function changes sign on reflection through the origin) between the initial and final states. If these restrictions are not satisfied, the transition is "*forbidden*." It may still occur for a multi-electron atom that has an electric quadrupole or magnetic dipole moment; in such cases the transition probability is typically smaller than a factor of 10^5 or more than for a permitted electric dipole transition.

We can re-state the selection rules in terms of nj, namely, $\Delta j = 0, \pm1$, except $(j=0 \leftrightarrow j'=0)$. Transitions of the form $nlj \leftrightarrow n'l'j'$ are *multiplets*. For hydrogenic ions there are two possible cases $j = l \pm \frac{1}{2}$, and (7.54) reduces to

$$\Delta E_{jj'} = \alpha^2 Z^4 / n^3 l(l+1) \quad [\,\mathrm{Ryd}\,]. \tag{7.57}$$

For hydrogen, levels $j=\frac{1}{2}$ and $j=\frac{3}{2}$ are split by a mere 0.36, 0.12, and 0.044 cm^{-1} for n = 2, 3, and 4, respectively.

Hydrogenic Bound-Bound Matrix Elements and f-Values

Given the above expressions for $\psi(r,\mu,\phi)$, we can compute the absorption probability B_{lu} from (5.81) or the oscillator strength f_{lu} from (5.102) for hydrogenic ions, analytically. The calculation below provides a specific example of the general theory described in § 5.3 and § 5.4 with the advantage of having analytical wave functions.

Write $\mathbf{r}=x\,\mathbf{i}+y\,\mathbf{j}+z\,\mathbf{k}$; then for the transition $(n',l',m')\to(n,l,m)$ we have

$$\left|\langle nlm\|e\mathbf{r}\|n'l'm'\rangle\right|^2=\left|\langle nlm\|ex\|n'l'm'\rangle\right|^2+\left|\langle nlm\|ey\|n'l'm'\rangle\right|^2 +\left|\langle nlm\|ez\|n'l'm'\rangle\right|^2. \tag{7.58}$$

In terms of polar coordinates,

$$x=r\sin\theta\cos\phi\equiv r(1-\mu^2)^{\frac{1}{2}}\,(e^{i\phi}+e^{-i\phi})/2, \tag{7.59a}$$

$$y=r\sin\theta\sin\phi\equiv r(1-\mu^2)^{\frac{1}{2}}\,(e^{i\phi}-e^{-i\phi})/2i, \tag{7.59b}$$

and

$$z=r\cos\theta\equiv r\mu. \tag{7.59c}$$

Consider first the matrix element

$$\left|\langle nlm\|ez\|n'l'm'\rangle\right|= \frac{e}{2\pi}\int_0^\infty P_{nl}\,r\,P_{n'l'}\,dr\int_{-1}^{1}\Theta_{lm}\Theta_{l'm'}\,\mu\,d\mu\int_0^{2\pi}e^{i(-m+m')\phi}\,d\phi. \tag{7.60}$$

To non-dimensionalize the radial integral in (7.21), divide out a factor (a_0/Z), and for brevity write

$$\sigma(n',l';n,l)\equiv\int_0^\infty P_{nl}\,r\,P_{n'l'}\,dr. \tag{7.61}$$

The ϕ integral (7.13) shows that $\left|\langle nlm\|ez\|n'l'm'\rangle\right|\equiv 0$ unless $m=m'$, so this matrix element reduces to

$$\left|\langle nlm'\|ez\|n'l'm'\rangle\right|^2=\left(\frac{a_0^2e^2}{Z^2}\right)\sigma^2(n',l';n,l)\left(\int_{-1}^{1}\Theta_{lm'}\Theta_{l'm'}\,\mu\,d\mu\right)^2. \tag{7.62}$$

To evaluate (7.62) we use the expansion formula [5, pp. 333–334]

$$\mu\Theta_{l|m|}(\mu)=\left[\frac{(l-|m|)(l+|m|)}{(2l-1)(2l+1)}\right]^{\frac{1}{2}}\Theta_{(l-1)|m|}(\mu) +\left[\frac{(l-|m|+1)(l+|m|+1)}{(2l+1)(2l+3)}\right]^{\frac{1}{2}}\Theta_{(l+1)|m|}(\mu). \tag{7.63}$$

From (7.13) and (7.16) we see that $|\langle nlm' \| ez \| n'l'm' \rangle|$ is nonzero only for $l = l' \pm 1$. Therefore,

$$|\langle n(l'-1)|m'| \, \| ez \| n'l'|m'| \rangle|^2 = \left(\frac{a_0^2 e^2}{Z^2} \right) \sigma^2(n', l'; n, l'-1) \frac{(l' - |m'|)(l' + |m'|)}{(2l'-1)(2l'+1)} \quad (7.64)$$

and

$$|\langle n(l'+1)|m'| \, \| ez \| n'l'|m'| \rangle|^2 = \left(\frac{a_0^2 e^2}{Z^2} \right) \sigma^2(n', l'; n, l'+1) \frac{(l' + 1 - |m'|)(l' + 1 + |m'|)}{(2l'+1)(2l'+3)}. \quad (7.65)$$

Consider now the matrix element

$$|\langle nlm \| ex \| n'l'm' \rangle| = \left(\frac{a_0 e}{4\pi Z} \right) \sigma(n', l'; n, l) \times \int_{-1}^{1} \Theta_{l'm'} \Theta_{lm} (1 - \mu^2)^{\frac{1}{2}} \, d\mu \int_0^{2\pi} [e^{-i(m'-m+1)\phi} + e^{-i(m'-m-1)\phi}] \, d\phi. \quad (7.66)$$

The ϕ integral shows that $|\langle nlm \| ex \| n'l'm' \rangle| \equiv 0$ unless $m = m' \pm 1$, so this matrix element reduces to

$$|\langle n\, l\, m'-1 \| ex \| n'l'm' \rangle| = \left(\frac{a_0 e}{2Z} \right) \sigma(n', l'; n, l) \int_{-1}^{1} \Theta_{l(m'-1)} \Theta_{l'm'} (1 - \mu^2)^{\frac{1}{2}} \, d\mu \quad (7.67a)$$

and

$$|\langle n\, l\, m'+1 \| ex \| n'l'm' \rangle| = \left(\frac{a_0 e}{2Z} \right) \sigma(n', l'; n, l) \int_{-1}^{1} \Theta_{l(m'+1)} \Theta_{l'm'} (1 - \mu^2)^{\frac{1}{2}} \, d\mu. \quad (7.67b)$$

To evaluate (7.67) we use the expansion formula

$$(1 - \mu^2)^{\frac{1}{2}} \Theta_{l|m|}(\mu) = \left[\frac{(l + |m| + 1)(l + |m| + 2)}{(2l+1)(2l+3)} \right]^{\frac{1}{2}} \Theta_{(l+1)\,(|m|+1)}(\mu) - \left[\frac{(l - |m| - 1)(l - |m|)}{(2l-1)(2l+1)} \right]^{\frac{1}{2}} \Theta_{(l-1)\,(|m|+1)}(\mu). \quad (7.68)$$

This matrix element is nonzero only for $l = l' \pm 1$.

The *selection rules* $\Delta m = 0$ or $\Delta m = \pm 1$ and $l \equiv l' \pm 1$ apply generally for electric dipole transitions and are not restricted to hydrogen. An electric dipole has *odd parity*. Hence *the initial and final eigenstates must have opposite parity so the product of their eigenfunctions and the dipole moment will have even parity and not integrate to zero.*

Using the above results in (7.58), we obtain

$$\left|\left\langle n\, l'-1\ |m'|-1 \left\| ex \right\| n'l'|m'|\right\rangle\right|^2 = \left(\frac{a_0^2 e^2}{4Z^2}\right)\sigma^2(n',l';n,l'-1)\frac{(l'+|m'|-1)(l'+|m'|)}{(2l'-1)(2l'+1)}, \quad (7.69a)$$

$$\left|\left\langle n\, l'-1\ |m'|+1 \left\| ex \right\| n'l'|m'|\right\rangle\right|^2 = \left(\frac{a_0^2 e^2}{4Z^2}\right)\sigma^2(n',l';n,l'-1)\frac{(l'-|m'|-1)(l'-|m'|)}{(2l'-1)(2l'+1)}, \quad (7.69b)$$

$$\left|\left\langle n\, l'+1\ |m'|-1 \left\| ex \right\| n'l'|m'|\right\rangle\right|^2 = \left(\frac{a_0^2 e^2}{4Z^2}\right)\sigma^2(n',l';n,l'+1)\frac{(l'-|m'|+1)(l'-|m'|+2)}{(2l'+1)(2l'+3)}, \quad (7.69c)$$

and

$$\left|\left\langle n\, l'+1\ |m'|+1 \left\| ex \right\| n'l'|m'|\right\rangle\right|^2 = \left(\frac{a_0^2 e^2}{4Z^2}\right)\sigma^2(n',l';n,l'+1)\frac{(l'+|m'|+1)(l'+|m'|+2)}{(2l'+1)(2l'+3)}. \quad (7.69d)$$

All the matrix elements $\left|\left\langle nlm \left\| ey \right\| n'l'm'\right\rangle\right|^2$ are the same as $\left|\left\langle nlm \left\| ex \right\| n'l'm'\right\rangle\right|^2$.

To find the oscillator strength $f(n',l';n,l)$ we sum the matrix elements above over m and m'. Thus for $l = l'+1$, sum over m, which yields

$$\begin{aligned}
\sum_m \left|\left\langle n\, l'+1\ m \left\| e\mathbf{r} \right\| n'l'm'\right\rangle\right|^2 &= \left|\left\langle n\, l'+1\ m' \left\| ez \right\| n'l'm'\right\rangle\right|^2 \\
&+ \left|\left\langle n\, l'+1\ m'-1 \left\| ex \right\| n'l'm'\right\rangle\right|^2 + \left|\left\langle n\, l'+1\ m'+1 \left\| ex \right\| n'l'm'\right\rangle\right|^2 \\
&+ \left|\left\langle n\, l'+1\ m'-1 \left\| ey \right\| n'l'm'\right\rangle\right|^2 + \left|\left\langle n\, l'+1\ m'+1 \left\| ey \right\| n'l'm'\right\rangle\right|^2 \\
&= \left(\frac{a_0^2 e^2}{Z^2}\right)\frac{\sigma^2(n',l';n,l'+1)}{(2l'+1)(2l'+3)}\Big[\tfrac{1}{2}(l'-|m'|+1)(l'-|m'|+2) \\
&+ \tfrac{1}{2}(l'+|m'|+1)(l'+|m'|+2) + (l'-|m'|+1)(l'+|m'|+1)\Big] \quad (7.70)
\end{aligned}$$

or

$$\sum_m |\langle n\, l'+1\, m\|e\mathbf{r}\|n'l'm'\rangle|^2 = \left(\frac{a_0^2e^2}{Z^2}\right)\frac{\sigma^2(n',l';n,l'+1)\,(l'+1)}{2l'+1}. \tag{7.71}$$

This result is independent of m'. By a similar calculation one finds

$$\sum_m |\langle n\, l'-1\, m\|e\mathbf{r}\|n'l'm'\rangle|^2 = \left(\frac{a_0^2e^2}{Z^2}\right)\frac{\sigma^2(n',l';n,l'-1)\,l'}{2l'+1}, \tag{7.72}$$

which is also independent of m'. In short, (7.71) and (7.72) show that

$$\sum_m |\langle nlm\|e\mathbf{r}\|n'l'm'\rangle|^2 = \left(\frac{a_0^2e^2}{Z^2}\right)\frac{\sigma^2(n',l';n,l)\max(l',l)}{2l'+1}, \tag{7.73}$$

where $l = l' \pm 1$. Now sum over m'. Because both (7.63) and (7.64) are independent of m', the sum over the $(2l'+1)$ values $m' = (-l', \ldots, 0, \ldots, l')$ merely puts a factor of $2l'+1$ into the numerator of (7.61). Finally, allowing for the two spin states of the electron, we obtain for $l = l' \pm 1$

$$\sum_s\sum_{m'}\sum_m |\langle nlm\|e\mathbf{r}\|n'l'm'\rangle|^2 = \left(\frac{2a_0^2e^2}{Z^2}\right)\sigma^2(n',l';n,l)\max(l',l). \tag{7.74}$$

Thus from (5.103) the line strength for the transition $n'l' \rightarrow nl$ is

$$S(n',l';n,l) = \left(\frac{2a_0^2e^2}{Z^2}\right)\sigma^2(n',l';n,l)\max(l',l), \tag{7.75}$$

where $l = l' \pm 1$.

Then using (5.105c) and the definitions of a_0, $\mathcal{R}$, and σ^2, we find that the corresponding oscillator strength is

$$\begin{aligned} f_{\text{bb}}(n',l';n,l) &= \frac{8\pi^2 m_e\nu}{3e^2h}\,\frac{2a_0^2e^2\sigma^2(n',l';n,l)\max(l',l)}{2(2l'+1)} \\ &= \frac{\max\,(l',l)}{3(2l'+1)}\left(\frac{1}{n'^2}-\frac{1}{n^2}\right)\left(\int_0^\infty P_{n'l'}rP_{nl}\,dr\right)^2. \end{aligned} \tag{7.76}$$

In (7.76) a factor of 2 in $g_{n',l'}$ is included to account for electron spin. Extensive tables for $S(n',l';n,l)$ and $f(n',l';n,l)$ are contained in [392].

For any transition $(n',l' \rightarrow n,l)$ in a hydrogenic ion of nuclear charge Z, (7.75) shows that $S_Z(n',l';n,l) = S_H(n',l';n,l)/Z^2$. From (7.76) we have the practical result that $f_Z(n', l'; n, l) = f_H(n', l'; n, l)$; i.e., aside from relativistic effects, for a hydrogenic ion, *the oscillator strength in a given transition is independent of Z.* In contrast, from (5.105b), and the Z-scaling of S, we find $B_{lu}(Z) = B_{lu}(H)/Z^2$, and the same for B_{ul}. Equations (5.105a) and (7.67) show that $A_{ul}(Z) = Z^4 A_{ul}(H)$.

Total Bound-Bound f-Values

For most work it is more useful to have the total oscillator strength summed over all allowed transitions between substates of n' and n. An analytical expression for $\sigma^2(n', l'; n, l)$ was derived by Gordon [382]. By summing Gordon's formula over l' and l, Menzel and Pekeris [722] derived a general formula for $f(n', n)$:

$$f(n', n) = \frac{64D}{3g_{n'}} \left| \frac{\left[(n - n')(n + n')\right]^{2(n+n')}}{n'^2 n^2 (1/n'^2 - 1/n^2)^3} \frac{\Delta(n', n)}{(n - n')} \right|, \tag{7.77}$$

where

$$\Delta(n', n) \equiv$$

$$\left[F\left(-n + 1, -n', 1, \frac{-4n'n}{(n - n')^2}\right)\right]^2 - \left[F\left(-n' + 1, -n, 1, \frac{-4n'n}{(n - n')^2}\right)\right]^2. \tag{7.78}$$

Here F is the hypergeometric function (Kummer's formula) [5, eq. 13.1.2]:

$$F(\alpha, \beta, \gamma, x) \equiv 1 + \frac{\alpha\beta x}{\gamma} + \frac{\alpha(\alpha + 1)\beta(\beta + 1)x^2}{2!\gamma(\gamma + 1)} \dots. \tag{7.79}$$

In (7.77), D is a normalizing factor, which is unity for bound-bound transitions, and the absolute value sign signifies the modulus of complex numbers, which is needed because imaginary quantum numbers will be used to describe bound-free and free-free transitions.

For bound-bound transitions, n' and n are real and some of the arguments of F in (7.79) are negative, so the series truncates in a finite number of terms. Analytical expressions for $\Delta(n, n')$ for $n' = 1 \leq 7$ are given in [722]. With a computer, (7.78) and (7.79), and hence (7.77), are easily evaluated numerically. The total absorption coefficient per atom in state n' is given by (5.102).

Historically, (7.77) was often written as

$$f(n', n) = g_{\text{bb}}(n', n) f_{\text{K}}(n, n), \tag{7.80}$$

where

$$f_{\text{K}}(n', n) \equiv \frac{32}{3\pi\sqrt{3}} \left(\frac{1}{n'^2} - \frac{1}{n^2}\right)^{-3} \left(\frac{1}{n'^5 n^3}\right) \tag{7.81}$$

is the semiclassical formula derived by Kramers [636] using the Bohr correspondence principle before the advent of quantum mechanics, and $g_{\text{bb}}(n', n)$ is the quantum mechanical *Gaunt factor* [392] for bound-bound transitions.

Photoionization Cross Section

Gordon [382] pointed out that by replacing the discrete real quantum number n for the upper state with a continuous pure imaginary quantum number ik ($0 \leq k \leq \infty$), matrix elements for bound-bound transitions could be extended to bound-free transitions. By analytic continuation into the complex plane Menzel and Pekeris generalized (7.69) to apply to any bound-free or free-free transition mediated by the absorption of a photon.

In the case of photoionization, a photon that has energy $h\nu \geq$ the ionization potential $\chi_{\text{ion}}(n')$ of an atom/ion in level n' can be absorbed in a continuum of bound-free transitions in which the bound electron is liberated with kinetic energy

$$\tfrac{1}{2}m_e v^2 = h\nu - Z^2\mathcal{R}/n'^2. \tag{7.82}$$

If we represent the free state with an imaginary quantum number ik in the Rydberg formula (7.25), then we can write

$$h\nu_{n'k} \equiv Z^2\mathcal{R}\left(1/n'^2 + 1/k^2\right) = (Z^2\mathcal{R}/n'^2) + \tfrac{1}{2}m_e v^2. \tag{7.83}$$

In (7.25) we see that $h\nu_{n'n} \to \chi_{\text{ion}}(n')$ when $n \to \infty$. In (7.83) we see that $h\nu_{n'k}$ goes to the same limit and $\frac{1}{2}mv^2 \to 0$ when $k \to \infty$. Thus at the ionization threshold, both the bound and free states join continuously. As k decreases, the continuum state lies at ever higher energies above the bound state, and both $h\nu_{n'k}$ and $\frac{1}{2}mv^2 \to \infty$ as $k \to 0$.

As was described in § 5.5, in astrophysics the bound-free absorption coefficient was written as the product of the semiclassical absorption coefficient for hydrogenic ions derived by Kramers [636] and a bound-free Gaunt factor [365] g_{bf} that corrects it to the quantum mechanical result. Then

$$\alpha_{\text{bf}}(n', \nu) = \frac{64\pi^4 Z^4 m_e e^{10} g_{\text{bf}}(n', \nu)}{3\sqrt{3}ch^6 n'^5 \nu^3} \equiv \mathcal{K}\frac{Z^4}{n'^5\nu^3} g_{\text{bf}}(n', \nu), \tag{7.84}$$

where $\mathcal{K} = 2.815 \times 10^{29}$.

To obtain the quantum mechanical expression for the absorption cross section $\alpha_{\text{bf}}(n', \nu)$ per atom in state n', consider a narrow band dk of free states of the transition $n' \to ik$, and write the mean value of the oscillator strength $f_{\text{bf}}(n', k)$ as $\bar{f}_{\text{bf}}(n', k)$. Then in the frequency interval $d\nu$ corresponding to dk we can write the absorption coefficient as

$$\alpha_{\text{bf}}(n', \nu)d\nu = \frac{\pi e^2}{m_e c}\bar{f}_{\text{bf}}(n', k)\left(\frac{dk}{d\nu}\right)d\nu. \tag{7.85}$$

Menzel and Pekeris showed that for the transition $n' \to ik$, (7.69) becomes

$$f_{\text{bf}}(n', k) = \frac{32}{3n'^2}\frac{e^{-4k\tan^{-1}(n'/k)}\left|\Delta(n', ik)\right|}{n'^3k^3\left(1/n'^2 + 1/k^2\right)^{7/2}(1 - e^{-2\pi k})}. \tag{7.86}$$

If we define

$$g_{\rm bf}(n',\nu) \equiv \frac{\sqrt{3}\pi n' k\, e^{-4k\tan^{-1}(n'/k)}\,\left|\Delta(n',ik)\right|}{(k^2+n'^2)^{\frac{1}{2}}(1-e^{-2\pi k})}, \tag{7.87}$$

we can write

$$f_{\rm bf}(n',k) = \frac{32}{3\sqrt{3}\pi}\,\frac{g_{\rm bf}(n',\nu)}{n'^5k^3\left(1/n'^2+1/k^2\right)^3}, \tag{7.88}$$

and from (7.83)

$$\frac{dk}{d\nu} = -\frac{hk^3}{2Z^2\mathcal{R}}. \tag{7.89}$$

Substitution of (7.88) and (7.89) into (7.85) gives

$$\alpha_{\rm bf}(n',\nu) = \frac{\pi e^2}{m_e c}\left(\frac{32}{3\sqrt{3}\pi}\right)\frac{1}{n'^5k^3}\,\frac{g_{\rm bf}(n',\nu)}{(h\nu/Z^2\mathcal{R})^3}\left(\frac{hk^3}{2Z^2\mathcal{R}}\right) \tag{7.90}$$

for $\nu \geq \nu_{n'}$. Or, using (7.24), we have

$$\alpha_{\rm bf}(n',\nu) = \frac{64\pi^4 Z^4 e^{10} m_e}{3\sqrt{3}ch^6}\,\frac{g_{\rm bf}(n',\nu)}{n'^5\nu^3}, \tag{7.91}$$

which is identical to Kramers's formula, but now with an exact expression for $g_{\rm bf}(n',\nu)$ from (7.7). Using the method of steepest descents to evaluate contour integrals in the complex plane, Menzel and Pekeris derived approximate formulae for $g_{\rm bf}(n',\nu)$ in the limits $k/n' \gg 1$, i.e., near the ionization threshold, and $n'/k \gg 1$, i.e., as $\nu \to \infty$. Today, accurate values of $g_{\rm bf}(n,\nu)$ for all n' and ν are easily calculated with computers that can perform complex arithmetic. Extensive calculations of $g_{\rm bf}(n',\nu)$ for $1 \leq n' \leq 15$ are contained in [611]. There tables of $g_{\rm bf}(n',\nu)$ for $1 \leq n' \leq 15$ are given as a function of $h\nu/Z^2\mathcal{R}$, which allows the correct Z-scaling to hydrogenic ions.

Z-Scaling at Threshold and Continuum Oscillator Strength

At first sight (7.91) appears to say that the bound-free absorption coefficient from level n' of a hydrogenic ion scales as Z^4. But using the definition of the threshold ionization frequency $h\nu_{n'} = Z^2\mathcal{R}/n'^2$ and (7.24) for $\mathcal{R}$ we find

$$\begin{aligned}\alpha_{\rm bf}(n',\nu) &= \frac{8h^3n'}{3\sqrt{3}\pi^2 c\, e^2 m_e^2 Z^2}\left(\frac{\nu_{n'}}{\nu}\right)^3 g_{\rm bf}(n',\nu)\\ &= 7.904\times10^{-18}\left(\frac{n'}{Z^2}\right)\left(\frac{\nu_{n'}}{\nu}\right)^3 g_{\rm bf}(n',\nu) \quad [\mathrm{cm}^{-2}],\end{aligned} \tag{7.92}$$

so the threshold cross section actually scales as Z^{-2}. The total oscillator strength for the bound-free continuum $n' \to k$ is defined as

$$f_c(n') \equiv \frac{m_e c^2}{\pi e^2} \int_{\nu_0}^{\infty} \alpha_{\rm bf}(n', \nu)\, d\nu = \frac{16\mathcal{R}^2 Z^4\, \overline{g}_{\rm bf}(n')}{3\pi\sqrt{3}h^2 n'^5} \int_{\nu_0}^{\infty} \frac{d\nu}{\nu^3} = \frac{8}{3\pi\sqrt{3}} \frac{\overline{g}_{\rm bf}(n')}{n'}, \tag{7.93}$$

where $\overline{g}_{\rm bf}(n')$ is the Gaunt factor appropriately averaged over frequency. For example, in the hydrogen Lyman continuum, $f_c(n'=1)=0.4089$, and in the Balmer continuum $f_c(n'=2)=0.2375$. In comparison, in the Lyman series, $f\,(\mathrm{Ly}\alpha) \equiv f_{\rm bb}(1,2)=0.4162$, and in the Balmer series, $f\,(\mathrm{H}\alpha) \equiv f_{\rm bb}(2,3)=0.6407$. It is noteworthy that a single strong spectral line can have a larger oscillator strength than the entire integrated continuum.

Free-Free Cross Section

Free-free absorption can be treated with the same formalism used for bound-free transitions by replacing the lower state's discrete quantum number n' with a second continuous pure imaginary number ik', where $(0 \le k' \le \infty)$ and $k' > k$. Then the kinetic energy of an electron with initial velocity v is

$$Z^2\mathcal{R}/k'^2 = \tfrac{1}{2} m v^2, \tag{7.94a}$$

and the energy of the absorbed photon is

$$h\nu_{k'k} = Z^2\mathcal{R}(1/k^2 - 1/k'^2). \tag{7.94b}$$

Assume that an absorption goes from a band of states dk' into a band of states $dk=(dk/d\nu)d\nu$. Then in parallel with (7.85) write

$$\alpha_{\rm ff}(k', k) = \frac{\pi e^2}{m_e c} \overline{f}(k', k) dk' \frac{dk}{d\nu}. \tag{7.95}$$

For a free-free transition $(ik' \to ik)$, Menzel and Pekeris wrote

$$f_{\rm ff}(k', k) = \frac{64}{3\sqrt{3}\pi} \frac{g_{\rm ff}(k', k)}{g_e\, k'^3\, k^3 \left(1/k^2 - 1/k'^2\right)^3}. \tag{7.96}$$

And again using analytic continuation, they showed

$$g_{\rm ff}(k', k) \equiv \frac{\sqrt{3}\pi\, k'k\, e^{-2\pi k} |\Delta(ik', ik)|}{(k'-k)(1-e^{-2\pi k'})(1-e^{-2\pi k})}. \tag{7.97}$$

From equations (4.22) and (7.94a) we have

$$g_e = \frac{8\pi m_e^3 v'^2\, dv'}{h^3} = \frac{16\pi m_e^2 Z^2 \mathcal{R} v'}{h^3 k'^3} dk'. \tag{7.98}$$

Using (7.24), (7.60), (7.97), and (7.98) in (7.95), we find that the absorption cross section per ion and electron with speed $(v, v+dv)$ is

$$\alpha_{\text{ff}}(\nu, v') = \frac{4\pi}{3\sqrt{3}} \frac{Z^2 e^6 g_{\text{ff}}(\nu, v')}{chm_e^2 \nu^3} \frac{1}{v'}. \tag{7.99}$$

Assume the electrons have a Maxwellian velocity distribution. Then integrating over v' and using the result $\langle 1/v \rangle = (2m_e/\pi kT)^{1/2}$, the free-free absorption coefficient per electron and ion is

$$\alpha_{\text{ff}}(\nu) = \frac{\sqrt{32\pi}\, Z^2 e^6\, \overline{g}_{\text{ff}}(\nu, T)}{3\sqrt{3}\, ch(km_e^3 T)^{1/2}\, \nu^3}. \tag{7.100}$$

This expression is the same as the semiclassical result (5.144), but now contains the exact quantum mechanical Gaunt factor.

The notion of an impact parameter is no longer relevant because an electron is not treated as a classical point charge, but as a wave function. Hence it is unnecessary to use its de Broglie wavelength to prevent divergence of the calculation. On the other hand, it is assumed here that the electron and ion are alone in space, so no plasma cutoff is included. Because stellar atmospheres are rarefied, collisions with impact parameters near λ_D are weak, so only very low energy photons are affected.

Calculations of $\overline{g}_{\text{ff}}$ eliminating the approximations above and including relativistic effects and degeneracy for large ranges of temperatures, densities, and frequencies are given in [110, 187, 231, 268, 389, 549, 567, 568, 569, 570, 611, 646, 790, 852, 943, 1066].

Hydrogenic Spectra

The spectrum of a hydrogenic ion consists of (1) a set of photoionization continua whose cross sections (ignoring Gaunt factors) fall off as $1/\nu^3$ toward higher frequencies from the threshold ionization frequency of each bound level, (2) an overlapping free-free continuum, and (3) series of spectral lines coupling each lower level n' to all higher levels n. Vacuum wavelengths of H lines for transitions $n' \to n$ are listed in table 7.2 and for He II transitions in table 7.3. The last row, $n = \infty$, gives the wavelength at which the continuum from level n' starts. The sets of hydrogen lines coming from levels $n' = 1$, 2, 3, and 4, respectively, are the Lyman, Balmer, Paschen, and Brackett series. In stellar spectra, the Lyman lines, in the ultraviolet, can be observed only from space vehicles. The entire Balmer series can be observed with terrestrial telescopes. The Paschen lines approaching their series limit can be observed from the Earth's surface with special infrared detectors.

He II lines from levels with the same values of n' have much shorter wavelengths because the binding energy of electrons in the He^+ helium nucleus potential is four times as large (scales as Z^2). Many of the conspicuous He II lines in the visible part of the spectrum come from $n' = 4$. Those having even values of n are overlapped by the (generally stronger) hydrogen (Balmer) lines. Prominent, readily observed

Table 7.2 Vacuum Wavelengths (Å) of Hydrogen Lines

	n'			
n	1	2	3	4
2	1215.68			
3	1025.73	6564.70		
4	972.55	4862.74	18756.27	
5	949.75	4341.73	12821.67	40522.82
6	937.81	4102.94	10941.16	26258.78
7	930.76	3971.24	10052.19	21661.29
8	926.24	3890.19	9548.65	19450.95
9	923.16	3836.51	9231.60	18179.16
10	920.97	3799.01	9017.44	17366.92
∞	911.76	3647.05	8205.87	14588.21

Table 7.3 Vacuum Wavelengths (Å) of Ionized Helium Lines

	n'					
n	1	2	3	4	5	6
2	303.80					
3	256.33	1640.51				
4	243.04	1215.19	4687.16			
5	237.34	1084.99	3204.11	10126.58		
6	234.36	1025.32	2734.18	6562.02	18642.11	
7	232.59	992.40	2512.02	5413.12	11629.74	30917.22
8	231.46	972.15	2386.19	4860.76	9347.61	18748.64
9	230.70	958.74	2306.96	4542.94	8239.15	14764.55
10	230.15	949.37	2253.44	4339.96	7594.93	12816.45
∞	227.85	911.39	2050.63	3645.57	5696.20	8202.53

He II lines not overlapped by hydrogen lines are located at $\lambda\lambda$ 1640 ($2 \rightarrow 3$), 4686 ($3 \rightarrow 4$), 3203 ($3 \rightarrow 5$), 10124 ($4 \rightarrow 5$), 5412 ($4 \rightarrow 7$), and 4542 ($4 \rightarrow 9$).

Figure 7.1 shows the sum of bound-free and free-free absorption cross sections of hydrogen in LTE at two temperatures. The photoionization edges are labeled with the quantum numbers of their initial level. Except for the hottest stars, most of the hydrogen is in its ground state, so the absorption edge at λ 912 Å is extremely strong. For 912 Å $\leq \lambda \leq$ 3647 Å, absorption from the ground state cannot occur, and the main opacity source is photoionization from the $n = 2$ level (Balmer continuum). For 3647 Å $\leq \lambda \leq$ 8206 Å, neither $n = 1$ nor $n = 2$ can absorb, so the dominant source is from $n = 3$ (Paschen continuum). The picture changes markedly when we account for the overlapping of each continuum edge by series of lines starting from the same level.

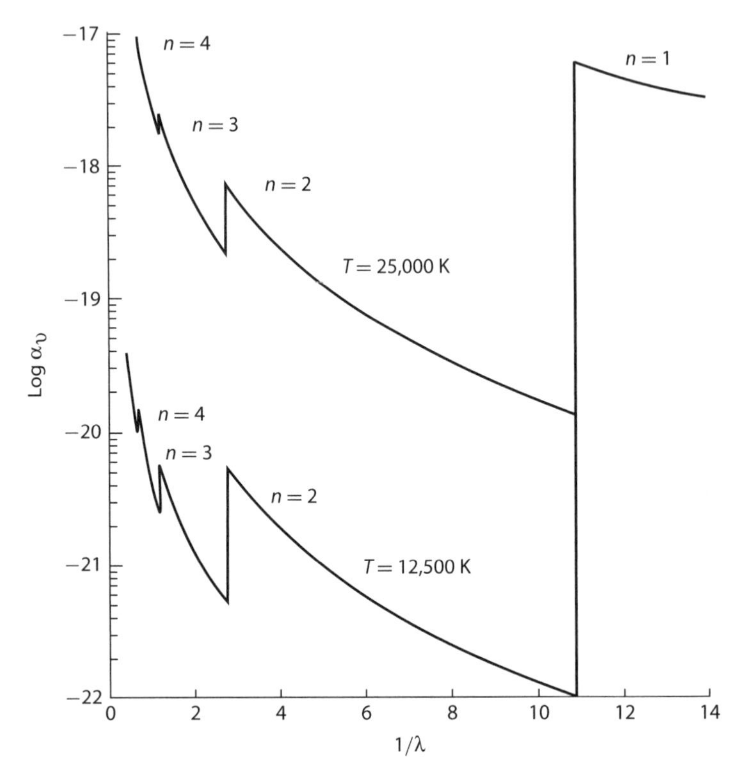

Figure 7.1 Sum of bound-free and free-free absorption cross sections of hydrogen [cm^2/atom], in LTE at $T = 15{,}000$ K and $T = 25{,}000$ K.

7.2 MULTI-ELECTRON ATOMS

In classical gravitational dynamics, we can obtain exact solutions for the two-body problem in the Newtonian limit and, in certain cases, the special, and even general, relativistic limits. But no closed solutions are known for even the three-body problem. Similarly, in quantum mechanics we have an exact solution[6] for a single electron moving in the field of a massive nucleus. But an exact solution is not possible for multi-electron systems, because of the many ways (electromagnetic, spin, exchange) the electrons can interact with the nucleus and each other. Their wave functions (and cross sections) must be obtained numerically.

Atomic Shell Structure

The overall picture of the shell structure of atoms is the direct result of, and follows directly from, the Pauli exclusion principle. Good discussions of the empirical (i.e., spectroscopic) basis of atomic structure can be found in [1000, chapter 2]

[6] Ignoring some of the effects of quantum electrodynamics.

and [1025, chapters 2–5]. To each electron in an atom one assigns a principal, azimuthal, magnetic, and spin quantum number (n, l, m_l, s). All electrons with a given n belong to a *shell.* Following traditional spectroscopic notation, electrons with $l = 0, 1, 2, 3, 4, \ldots$ are labeled $s, p, d, f, g, \ldots$, respectively. The "exponent" on these letters gives the number of electrons in that shell. According to the exclusion principle, *no two electrons in an atom can have the same values for all of these four quantum numbers.*

The $n = 1$ shell has two s electrons with $l = 0$ and $m_l = 0$; to satisfy the exclusion principle they must have $m_s = \pm 1/2$, i.e., spin up or down. The $n = 2$ shell has two s electrons and as many as six p electrons with $l = 1$, having $m_l = -1, 0, 1$ and two spin states, for a total of up to eight electrons in the shell. The $n = 3$ shell has two s electrons, at most six p electrons, and up to ten d electrons with $l = 2$ and $m_l = -2, -1, 0, 1, 2$ and two spin states, for a total of up to 18 electrons, and so on. A *closed shell* contains all electrons of a given n that can satisfy the exclusion principle. In a closed shell, an electron with a given (m_l, m_s) has a counterpart with $(-m_l, -m_s)$, so both the total orbital and spin angular momenta of such an atom are zero. An element with charge $Z + 1$ builds on the shell structure of element Z by adding an electron into a partially filled shell of element Z or into the next shell of $Z + 1$ having the lowest energy.

The ground-state configurations and ionization potentials (in eV) of astrophysically important elements through iron ($Z = 26$) are shown in table 7.4. When Z is ≤ 18 (argon), electrons in a given n shell are added in order of increasing

Table 7.4 Ground-State Configurations of Selected Elements

Element	Z	Ground Configuration	Ionization Potential (eV)	Ground Terms
H	1	$1s$	13.598	$^2S_{1/2}$
He	2	$1s^2$	24.587	$^1S_0$
C	6	$1s^2 2s^2 2p^2$	11.260	$^3P_0\ ^1D_2\ ^1S_0$
N	7	$1s^2 2s^2 2p^3$	14.534	$^4S_{3/2}\ ^2D_{5/2}\ ^2P_{1/2}$
O	8	$1s^2 2s^2 2p^4$	13.618	$^3P_2\ ^1D_2\ ^1S_0$
Ne	10	$1s^2 2s^2 2p^6$	21.564	$^1S_0$
Na	11	$1s^2 2s^2 2p^6 3s$	5.139	$^2S_{1/2}$
Mg	12	$1s^2 2s^2 2p^6 3s^2$	7.646	$^1S_0$
Si	14	$1s^2 2s^2 2p^6 3s^2 3p^2$	8.151	$^3P_0\ ^1D_2\ ^1S_0$
S	16	$1s^2 2s^2 2p^6 3s^2 3p^4$	10.360	$^3P_2\ ^1D_2\ ^1S_0$
Ar	18	$1s^2 2s^2 2p^6 3s^2 3p^6$	15.759	$^1S_0$
K	19	$1s^2 2s^2 2p^6 3s^2 3p^6 4s$	4.341	$^2S_{1/2}$
Ca	20	$1s^2 2s^2 2p^6 3s^2 3p^6 4s^2$	5.113	$^1S_0$
Fe	26	$1s^2 2s^2 2p^6 3s^2 3p^6 3d^6 4s^2$	7.870	$^5D_4$
Cs	55	$1s^2 2s^2 2p^6 3s^2 3p^6 3d^{10}$ $4s^2 4p^6 4d^{10} 5s^2 5p^6 6s$	3.894	$^2S_{1/2}$

l (i.e., $s, p, d, \ldots$) because orbitals with the smallest angular momenta have probability distributions that are larger near the nucleus, and hence are more tightly bound than the more extended high-l states. But at $Z = 19$ (potassium) and $Z = 20$ (calcium), $4s$ electrons become more tightly bound than $3d$ electrons. The $3d$ shell doesn't start filling until $Z = 21$ (Sc) and is not completed until $Z = 29$ (Cu). The usual order in which shells are filled as inferred from spectroscopic data is

$$1s, 2s, 2p, 3s, 3p, [\,4s, 3d\,], 4p, [\,5s, 4d\,], 5p, [\,6s, 4f, 5d\,], 6p, [\,7s, 5f, 6d\,];$$

however, the shells listed in square brackets have very nearly the same binding energy, so they are not always filled in the order listed.

The *noble gases* (He, Ne, Ar) have closed outer shells; they are nonflammable and chemically inactive. They have ionization potentials significantly larger than elements with nearly the same Z. He has the highest ionization potential of all elements, because the charge of the atom's nucleus is only partially shielded by the unexcited $1s$ electron. Electrons outside of closed shells are *valence electrons*, which act in chemical reactions. The *alkali metals* (Li, Na, K, Rb, Cs) have one very loosely bound valence electron. Cs has the lowest ionization potential of all elements. The *halogens* (F, Cl, Br) lack an electron to close their last shell. Both types are chemically very active.

Angular Momentum

Total Orbital and Spin Angular Momentum

The angular and spin momenta, $\hat{\boldsymbol{l}}_i$ and $\hat{\boldsymbol{s}}_i$, of an atom's electrons couple *vectorially* into a total orbital angular momentum $\mathbf{L}$ and total spin $\mathbf{S}$:

$$\mathbf{L} = \sum_i \hat{\boldsymbol{l}}_i \quad \text{and} \quad \mathbf{S} = \sum_i \hat{\boldsymbol{s}}_i. \tag{7.101}$$

The dynamical variables of different particles commute, so

$$[\,\hat{\boldsymbol{l}}_i, \hat{\boldsymbol{l}}_j\,] = \hat{\boldsymbol{l}}_i \cdot \hat{\boldsymbol{l}}_j - \hat{\boldsymbol{l}}_j \cdot \hat{\boldsymbol{l}}_i \equiv 0, \quad i \neq j, \tag{7.102}$$

and

$$\mathbf{L} \times \mathbf{L} = i\hbar \mathbf{L}. \tag{7.103}$$

Spin has no classical counterpart and cannot be expressed in terms of canonically conjugate variables. Its coordinate space is independent of the space coordinates in $\mathbf{L}$, so we postulate that $\mathbf{S}$ has the same commutation rule:

$$\mathbf{S} \times \mathbf{S} = i\hbar \mathbf{S}. \tag{7.104}$$

L–S Coupling

Electrons in multi-electron atoms experience non-central forces by interacting with other electrons, so $\hat{\boldsymbol{l}}_i$ of an individual electron is not conserved. Ignoring small terms in the Hamiltonian, and assuming no external torque on an atom, the electrostatic forces between pairs of electrons lie along the lines joining them, so **L** *is* conserved. In this approximation we also ignore external effects on an electron's spin because it is an intrinsic property of each electron; hence **S** is conserved. Thus the *total angular momentum* $\mathbf{J} = \mathbf{L} + \mathbf{S}$ is conserved, and as in (7.51), we can write

$$\mathbf{L}\cdot\mathbf{S} = \tfrac{1}{2}\left(|\mathbf{J}|^2 - |\mathbf{L}|^2 - |\mathbf{S}|^2\right) = \tfrac{1}{2}[J(J+1) - L(L+1) - S(S+1)]. \quad (7.105)$$

In this limit we have *L–S* or *Russell-Saunders coupling* [916], which applies for small-to-moderate Z (aside from some exceptional cases). For higher-Z atoms it breaks down, and other coupling schemes are used; they are discussed in detail in § 3.2–3.5 and § 4.1–4.2 of [1025]. In L–S coupling, we label terms with the notation ${}^{M}L_{J}$, where $L \equiv |\mathbf{L}|$ and $J \equiv |\mathbf{J}|$ for the lowest level in the term. The *multiplicity* M of a term is the number of possible values for M_S, i.e., $M = (2S+1)$, where $S \equiv |\mathbf{S}|$. Terms with $S = 0, \frac{1}{2}, 1, \frac{3}{2} \ldots$ are called *singlets, doublets, triplets, quartets, . . .*, etc.

Assuming $L-S$ coupling, we can understand qualitatively how the energy associated with a given configuration of electrons gets split into *systems* of spectroscopic terms, which are further split into *multiplets*, and finally into individual *levels* as sketched in figure 7.2 for two non-equivalent p electrons (i.e., having the same p but different n.)

In L–S coupling, there are three useful empirical rules from spectroscopic data that we can understand by qualitative theoretical arguments:

Hund's Rule #1: For given values of (n, L), *the terms with the largest multiplicity have the lowest energy.*

Electrons are fermions and obey Fermi-Dirac statistics. The largest splitting of the energies in a spherically symmetric potential field is produced by *exchange forces* of the type discussed in chapter 4, which result from the requirement of *antisymmetry of the total wave function.* For definiteness, consider the case of neutral helium, with one electron in the ground state, 1s, and the other in the lowest excited state, 2s. For two electrons we can have $m_s = \frac{1}{2} - \frac{1}{2} = 0$ or $m_s = \frac{1}{2} + \frac{1}{2} = 1$,

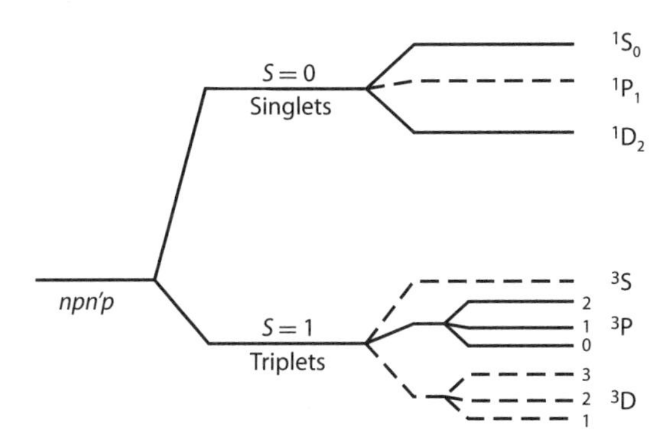

Figure 7.2 Splitting of a central-field $(np, n'p)$ orbital into systems, terms, and levels. If the electrons are *equivalent*, i.e., $n = n'$, the levels and components shown with dotted lines are absent.

and hence two systems of spectroscopic terms: singlets ($S=0$) and triplets ($S=1$). The separation of these two systems can be thought of as follows:

1. The largest multiplicity corresponds to parallel spins, so the spin part of the total wave function is symmetric for an exchange of the two electrons. But the total wave function must be antisymmetric because electrons are fermions, so the space part of the wave function must be antisymmetric.
2. For an antisymmetric spatial wave function the electrons have a larger average separation (Pauli principle) than if it were symmetric. Also, an antisymmetric spatial wave function $\rightarrow 0$ as $r \rightarrow 0$. Both effects imply that a small separation of the electrons has a lower probability than would be the case for a symmetric wave function.
3. Because the electrons are on the average farther apart, their mutual electrostatic repulsion, which tends to raise the energy of the excited electron in the nuclear potential well, is smaller, and also the nuclear charge is less well shielded by the ground-state electron. Therefore, the excited electron is more tightly bound and lies at a lower energy in the potential well.

Hund's Rule #2: *For terms with a given multiplicity, the state having the highest angular momentum (largest L) lies lowest (is more tightly bound).*

Consider the configuration $np\, n'p$ shown in figure 7.2. If the electrons "orbit" in the same direction, they "meet" less often than if they "orbit" in opposite directions. Hence on the average, the mutual repulsive potential they experience is less, so they can lie deeper in the potential well. Their energies are determined mainly by non-central electrostatic forces between all electrons.

Hund's Rule #3: When a shell is less than half-filled (e.g., p^2), the level with the lowest value of J lies lowest in energy.

In L–S coupling the spin-orbit term (4) in (7.125) splits the levels of a given term into fine-structure components, separated in energy by $\Delta E_{JLS} = K[J(J+1) - L(L+1) - S(S+1)]$. Both the magnitude and sign of K depend on L, S, and the electron configuration. The two possible signs produce *normal* and *inverted* ordering of the components in a multiplet.

J–J Coupling

L–S coupling breaks down when the energy of the spin-orbit term exceeds the inter-electron electrostatic term. Then the most important quantum number physically is $\hat{j} = \hat{l} + \hat{s}$. In this case the energy of a central-field orbital is split first by spin-orbit interactions into terms that depend on the combined $\hat{j}$ value of the valence electrons. Electrostatic interactions and exchange interactions in the antisymmetrization process split these into levels labeled with a total J. Some atoms, e.g., neutral neon, don't obey either scheme and are in *intermediate coupling*. More detailed discussions of coupling schemes are given in [248], [882, § 2.5–2.9], [1000, chapters 6 and 8], and [1025, § 2.2, 2.3, 3.2–3.5, chapter 4].

Zeeman Effect

In § 7.1 we mentioned that the concept of electron spin was introduced to explain the observed Zeeman splitting of spectral lines emitted by atoms in an externally imposed magnetic field. The total magnetic moment $\boldsymbol{\mu}$ of a multi-electron atom is the sum of the orbital and spin magnetic moments of its individual electrons:

$$\boldsymbol{\mu} = -\mu_0 \sum_i (\mathbf{l}_i + 2\mathbf{s}_i) = -\mu_0(\mathbf{L} + 2\mathbf{S}), \tag{7.106}$$

which clearly is not aligned with its total angular momentum $\mathbf{J} = \mathbf{L} + \mathbf{S}$.

In $L-S$ coupling, $\mathbf{L}$, $\mathbf{S}$, and $\boldsymbol{\mu}$ "precess" around $\mathbf{J}$ with a frequency $\Delta E_{so}/h$. In the absence of an external magnetic field $\mathbf{H}$, $\mathbf{J}$ is constant. But the torque of a magnetic field on the atom's magnetic moment causes $\mathbf{J}$ to "precess" slowly around $\mathbf{H}$. In a weak field, the rate of that precession is much slower than the spin-orbit precession, so the time-averaged component of $\boldsymbol{\mu}$ is essentially equal to its component along $\mathbf{J}$ times the component of $\mathbf{J}$ along the field, i.e.,

$$\Delta E_Z = \mu_0[\,(\mathbf{L} + 2\mathbf{S})\cdot\mathbf{J}\,](\mathbf{J}\cdot\mathbf{H})/|J|^2. \tag{7.107}$$

The term $|J|^2$ in the denominator converts the two appearances of $\mathbf{J}$ in the numerator into unit vectors. Taking $\mathbf{H}$ to be along the z axis, $\mathbf{J}\cdot\mathbf{H} = J_z|\mathbf{H}| = M_J|\mathbf{J}||\mathbf{H}|$; and noting that $\mathbf{L} = \mathbf{J} - \mathbf{S}$, we see that $2\mathbf{J}\cdot\mathbf{S} = \mathbf{J}\cdot\mathbf{J} + \mathbf{S}\cdot\mathbf{S} - \mathbf{L}\cdot\mathbf{L}$; hence

$$\Delta E_Z = \mu_0 M_J|\mathbf{H}| \left[\, |\mathbf{J}|^2 + \tfrac{1}{2}(|\mathbf{J}|^2 + |\mathbf{S}|^2 - |\mathbf{L}|^2) \right] \Big/ |\mathbf{J}|^2 \equiv \mu_0 g_L M_J|\mathbf{H}|, \tag{7.108}$$

where

$$g_L \equiv 1 + [J(J+1) + S(S+1) - L(L+1)]/2J(J+1) \tag{7.109}$$

is the *Lande′ g-factor* [675, 676].

Selection Rules

The selection rules for electric dipole transitions in $L-S$ coupling are the following:

1. change in parity
2. $\Delta l = \pm 1$ for the electron making the transition
3. $\Delta S = 0$, i.e., no allowed intersystem lines
4. $\Delta M_S = 0$
5. $\Delta L = 0, \pm 1$
6. $\Delta M_L = 0$
7. $\Delta J = 0, \pm 1$ except $J = 0 \leftrightarrow J = 0$ is forbidden
8. $\Delta M_J = 0, \pm 1$ except $M_J = 0 \leftrightarrow M_J = 0$ is forbidden if $\Delta J = 0$

These selection rules, specifically rule 3, can be violated in magnetic dipole and electric quadrupole transitions.

Hamiltonian

The Hamiltonian for multi-electron atoms is complicated because it must account for several different kinds of interparticle interactions. Thus it is not possible to derive closed analytical solutions for their structure or their optical properties (transition probabilities and photoionization cross sections). The electric potential for such systems is no longer a spherically symmetric Coulomb field, but has higher multipole moments that can strongly affect the "orbits" of the bound electrons and how they can interact with an imposed radiation field.

These interactions can lead to nonzero probabilities for transitions that would be forbidden in a pure Coulomb field. In the past, bound-bound transition probabilities and photoionization cross sections for atoms and ions of multi-electron elements were often calculated with approximate theories, e.g., [162,163,856,991]; they have at best modest accuracy. But thanks to important advances in the physical theory of atomic processes, and the availability of very high-speed computers, a large body of accurate results has now been derived computationally, even for high-Z atomic systems, using the R-matrix method in the Opacity Project (OP) [1198–1232] and its successor, the IRON Project (IP) [1234–1253], and with a somewhat lower accuracy in the Livermore Opacity Library (OPAL) [1254–1271]. In this text we can sketch only some aspects of the underlying theory for, and the results from, this work.

Central Field Approximation

In the *central field approximation*, developed by Hartree [445, 446], one calculates a preliminary wave function for each atomic electron by assuming it moves in a spherically symmetric potential central field of the nucleus and a smeared-out electron-cloud representation of the other electrons. One then uses these wave functions to correct the potential field for each electron for the effects of all the other electrons.

Suppose we have a nucleus with charge Z surrounded by N electrons. We approximate the Hamiltonian of an electron in the system as

$$H_i = H_{cf} + H_{ee} + H_{so}. \tag{7.110}$$

Here H_{cf} represents the interaction of electron i with the potential of the nucleus,

$$H_{cf}(r_i) = -\frac{\hbar^2}{2m_e}\nabla_i^2 - \frac{Ze^2}{r_i}, \tag{7.111}$$

H_{ee} represents the repulsion between electron i and all other electrons in the atom,

$$H_{ee}(r_i) = +\sum_{j\neq i}\int |u_j(\mathbf{r}_j)|^2 \frac{e^2}{|\mathbf{r}_j - \mathbf{r}_i|}\, d^3 r_j, \tag{7.112}$$

and H_{so} is the spin-orbit interaction of electron i with the spherically averaged potential $V(r)$ of the whole system

$$H_{so}(r_i) = \frac{\hbar^2}{2m_e^2c^2}\frac{1}{r_i}\left.\frac{\partial V(r)}{\partial r}\right|_{r_i} \mathbf{l}_i \cdot \mathbf{s}_i. \quad (7.113)$$

The iteration is started by assuming a potential $V(r)$, evaluating a first estimate of H_{cf}, H_{ee}, and H_{so}, and using the results to compute new wave functions that are used to estimate new values of these terms. We can constrain the first estimate of $V(r)$ by demanding (1) $V(r) \le 0$, (2) $V(r) \to 0$ as $r \to \infty$, and (3) $V(r) \to -Ze^2/r$ as $r \to 0$. The process is continued until the potentials are self-consistent to a high accuracy.

Angular momentum is conserved for motion in a spherically symmetric potential; hence each stationary state is defined by a principal quantum number n and orbital angular momentum quantum number l. The radial part $R_{nl}(r)$ satisfies

$$\frac{1}{r^2}\frac{d}{dr}\left(r^2\frac{dR_{nl}}{dr}\right) + \frac{2\mathrm{m}}{\hbar^2}[E_{nl} - U_{nl}(r)]R_{nl}(r) = 0. \quad (7.114)$$

The effective potential in (7.114) for an electron with principal quantum number n and orbital quantum number l is

$$U_{nl}(r) = V_n(r) + \frac{\hbar^2}{2\mathrm{m}}\frac{l(l+1)}{r^2}, \quad (7.115)$$

which depends on l but not m. Thus an (n, l) energy level contains $(2l + 1)$ states with the same energy.

The angular part of the solutions of (7.110) remain spherical harmonics $Y_{lm}(\theta, \phi)$. Coupling these with the radial part $R_{nl}(r)$ yields a set of *orbitals* for each choice of n and l. The assumption of spherical symmetry for the potential is accurate for electrons in a closed shell (i.e., given n and all l states $0 \le l \le n - 1$) because then the complete set of spherical harmonics adds to a spherically symmetric charge distribution.

Hartree-Fock Method

In the central field approximation, the wave function for the set of all electrons is the product of the one-electron orbitals:

$$\psi(\mathbf{r}_1, \mathbf{r}_2, \ldots, \mathbf{r}_N) = \psi_1(\mathbf{r}_1)\psi_2(\mathbf{r}_2)\ldots\psi_N(\mathbf{r}_N). \quad (7.116)$$

This wave function is not fully satisfactory because it ignores correlations among the electrons and is not antisymmetric as required by the Pauli principle. These deficiencies are overcome in the Hartree-Fock method [338], [882, § 2.11], [1012, chapter 17] by expressing the wave function for all electrons in terms of a Slater determinant such as shown in equation (4.82) in § 4.4.

Neutral Helium

Neutral helium, consisting of an α-particle nucleus and two electrons, is the simplest multi-electron atom. Extensive calculations have been made of the energy levels, bound-bound transition probabilities, and photoionization cross sections for He and He-like ions [712, 1138, 1203]. Theoretical and experimental studies have been made of the line profiles of both permitted and forbidden bound-bound transitions in He I; see, e.g., [278, 376, 401, 405, 631, 1208]. These have been of great value in interpreting the observed spectrum of visible and ultraviolet He I lines in the spectra of B- and O-stars.

Here we merely outline a simple variational method for estimating the binding energy of the ground state $1s^2$ of He I. Ignoring spin-orbit and other higher-order terms, the Hamiltonian of the system is

$$H = -\frac{\hbar^2}{2m_e}(\nabla_1^2 + \nabla_2^2) - Ze^2\left(\frac{1}{r_1} + \frac{1}{r_2}\right) + \frac{e^2}{|\mathbf{r}_1 - \mathbf{r}_2|}. \tag{7.117}$$

Neglecting the interaction between the electrons, the wave function would be the product of the hydrogenic wave functions for the two electrons:

$$\psi_0(r_1, r_2) = \psi_{100}(r_1)\psi_{100}(r_2), \tag{7.118}$$

where from (7.27)

$$\psi_{100}(r) = \left(Z^3/\pi a_0^3\right)^{\frac{1}{2}} e^{-(Zr/a_0)}, \tag{7.119}$$

and the binding energy would be $E_0 = -2(Z^2e/2a_0) = -108.2$ eV.

The first-order correction to E_0 is

$$\Delta E_0 = \left\langle \frac{e^2}{\mathbf{r}_1 - \mathbf{r}_2} \right\rangle = e^2 \int\int \frac{\psi_{100}^2(\mathbf{r}_1)\psi_{100}^2(\mathbf{r}_2)}{|\mathbf{r}_1 - \mathbf{r}_2|}\, d^3\mathbf{r}_1\, d^3\mathbf{r}_2. \tag{7.120}$$

To evaluate this integral, we use the expression from potential theory

$$\frac{1}{|\mathbf{r}_1 - \mathbf{r}_2|} = \frac{1}{r_>}\sum_{l=0}^{\infty}\left(\frac{r_<}{r_>}\right)^l P_l(\cos\theta), \tag{7.121}$$

where $r_> \equiv \max(r_1, r_2)$, $r_< \equiv \min(r_1, r_2)$, and $\cos\theta \equiv \mathbf{r}_1 \cdot \mathbf{r}_2$. In as much as ψ_{100}^2 is spherically symmetric, only the $l = 0$ term in (7.121) survives. The final result is $\Delta E_0 = \frac{5}{8}Ze^2/a_0$, so the first-order value for the binding energy is

$$E_0 = -\frac{Z^2e^2}{a_0} + \frac{5}{8}\frac{Ze^2}{a_0} = -74.4 \text{ eV}. \tag{7.122}$$

We can improve this estimate with a variational technique. Instead of taking (7.119) for both electrons, we use it for electron 1 and replace Z with Z' for electron 2. Then we find

$$E_0(Z') = -(e^2/a_0)\left(2ZZ' - Z'^2 - \tfrac{5}{8}Z'\right). \tag{7.123}$$

Differentiating with respect to Z' and setting the derivative to zero, we find $Z' = Z - \frac{5}{16}$; hence

$$E_0 = -(e^2/a_0)\left(Z - \tfrac{5}{16}\right)^2 = -77.1 \text{ eV}, \tag{7.124}$$

which is close to the experimental value: -78.6 eV.

Breit-Pauli Hamiltonian

As the charge Z on the atom's nucleus increases, relativistic effects become more and more important. As mentioned earlier, for hydrogen with its single electron, an *exact* relativistic Hamiltonian can be written and solved in closed form. This is not possible for multi-electron systems. Nevertheless, they can be treated to order $(Z\alpha)^2(Z^2)$ Ryd in energy using the *Breit-Pauli Hamiltonian*, which, in atomic units (see below) for an N-electron atom with nuclear charge Z, is [117, §38 and 39], [377, 819], [882, § 2.17]

$$\begin{aligned}\hat{\boldsymbol{H}}_{\text{BP}} = \sum_{i=1}^{N}\Bigg\{ &-\tfrac{1}{2}\nabla_i^2 - \frac{Z}{r_i} + \sum_{j>i}^{N}\frac{1}{r_{ij}} \\ &+\alpha^2\Bigg[\tfrac{1}{2}Z\frac{\mathbf{l}_i\cdot\mathbf{s}_i}{r_i^3} + \sum_{j>i}\frac{1}{r_{ij}^3}\left(\mathbf{s}_i\cdot\mathbf{s}_j - \frac{3(\mathbf{s}_i\cdot\mathbf{r}_{ij})(\mathbf{s}_j\cdot\mathbf{r}_{ij})}{r_{ij}^2}\right) \\ &-\tfrac{1}{2}\sum_{j\neq i}\left(\frac{\mathbf{r}_{ij}}{r_{ij}^3}\times\mathbf{p}_i\right)\cdot(\mathbf{s}_i + 2\mathbf{s}_j) - \tfrac{1}{2}\sum_{j>i}\left(\frac{\mathbf{p}_i\cdot\mathbf{p}_j}{r_{ij}} + \frac{(\mathbf{r}_{ij}\cdot\mathbf{p}_i)(\mathbf{r}_{ij}\cdot\mathbf{p}_j)}{r_{ij}^3}\right) \\ &-\tfrac{1}{8}\nabla_i^4 - \tfrac{1}{8}Z\nabla_i^2\left(\frac{1}{r_i}\right) + \tfrac{1}{4}\sum_{j>i}\nabla_i^2\left(\frac{1}{r_{ij}}\right) - \tfrac{8}{3}\pi\sum_{j>i}(\mathbf{s}_i\cdot\mathbf{s}_j)\delta(r_{ij})\Bigg]\Bigg\}. \end{aligned} \tag{7.125}$$

The first line of (7.125) contains non-relativistic terms: (1) the electrons' *kinetic energies*, (2) their *potential energy* in the central field of the nucleus, and (3) the energy of their *mutual electrostatic interactions*. The remaining terms in (7.125) are relativistic in origin. In the second line we have (4) the sum of the energies of the spin-orbit interaction of each electron's magnetic moment with the magnetic field produced by its motion in the Coulomb field of the nucleus and (5) the energy of the *spin-spin* interactions between the magnetic moments of all pairs of electrons. The third line represents (6) the sum over all pairs of electrons of their *spin-other-orbit* energies, which consist of two terms, the interaction of the spin of electron i with the Coulomb field of electron j, and the interaction of the spin magnetic moment of electron j with the orbital current of electron i; and (7) the sum over all electron pairs of the energy of the *orbit-orbit* interaction between the magnetic moments resulting from their motions around the nucleus. The fourth line contains (8) terms for the electrons' *relativistic mass variation*, (9) and (10) the one-body and two-body *Darwin interaction* terms, which are relativistic corrections to the potential energies of the electrons, and (11) the *spin-contact* term, which becomes nonzero when two electrons try to occupy the same position. Terms (4) and (8) are multi-electron generalizations of the interactions discussed in § 7.1 for ions with a single electron. In (7.125), $\mathbf{p}_i$ is the linear momentum of electron i. This Hamiltonian accounts for all physical interactions that have a major effect on atomic structure.

Configuration Interaction

In addition, for atoms of charge Z with several valence electrons, there may be spectroscopic terms with approximately the same energies arising from several different configurations of the electrons in the "parent" ion of charge $Z-1$. For example, for Fe I, numerous spectroscopic terms resulting from the configurations $3d^7 4s$ and $3d^6 4s\, 4p$ are interspersed among those from the ground configuration $3d^6 4s^2$. Because the three systems have relatively small energy separations, they can perturb each other and become coupled, which is described as *configuration interaction*. For excited states, configuration interaction occurs frequently. These interactions can be taken into account by means of *fractional parentage coefficients, Clebsch-Gordon coefficients, Wigner 3j, 6j, and 9j symbols*, or *Racah coefficients*. Detailed discussions of these matters are given in [248], [1000, chapters 6 and 9], [1025, chapter 4].

Modern Computer Codes

Clearly, many factors enter into the solution of Schrödinger's equation for a multi-electron atom/ion. One of the most successful methods for computing atomic structure, oscillator strengths, photoionization cross sections, and collisional excitation and ionization cross sections uses the *close-coupling approximation* [166], [882, § 3.4], as implemented in the *R-matrix method* [111, 164, 165, 167], [882, § 3.5], [988]. In this method one considers a system consisting of an electron in the field of an N-electron atom/ion *target*.

In computational work it is convenient to use a system of *atomic units* defined by setting $e = m_e = \hbar = 1$. Then the atomic unit of length is the Bohr radius, $a_0 = \hbar^2/me^2 = 5.2918 \times 10^{-9}$ cm; of energy, $E_0 = e^2/a_0 = 2[\,\text{Ryd}\,] = 4.3597 \times 10^{-11}$ erg $= 27.2114$ eV; and of time, $\tau_0 = \hbar^3/m_e e^4 = 2.4189 \times 10^{-17}$ sec. The speed of light in atomic units is the inverse of the fine-structure constant, $c = 1/\alpha = 137.036$.

The wave function for a single target electron in a central field has the form

$$\psi(\mathbf{x}_k) = Y_{lm}(\hat{\boldsymbol{r}}_k)\chi(\boldsymbol{\sigma}_k, m_{sk})P_{nl}(r_k)/r_k, \tag{7.126}$$

where r_k is the distance of electron k from the nucleus, $\hat{\boldsymbol{r}}_k$ specifies the two angles of the radius vector $\mathbf{r}_k$ in a spherical coordinate system, $\boldsymbol{\sigma}_k$ is its spin with z component quantum number m_{sk}, and $\mathbf{x}_k \equiv (\mathbf{r}_k, \boldsymbol{\sigma}_k)$, $k = 1, \ldots, N$ gives its combined space-spin coordinates. The orbitals are antisymmetrized as in equation (4.82) to obtain single-configuration wave functions

$$\phi_j(\mathbf{x}_1,\ldots,\mathbf{x}_N) = \phi_j(1s^{n_1}\,2s^{n_2}\,2p^{n_3},\ldots,\alpha_j\,L_j\,S_j\,M_{L_j}\,M_{S_j}|\mathbf{x}_1,\ldots,\mathbf{x}_N). \quad (7.127)$$

Here n_λ is the occupation number of atomic shell λ, where $\sum_\lambda n_\lambda = N$. The target states are described by multi-configurational wave functions

$$\Phi_i(\mathbf{x}_1,\mathbf{x}_2,\ldots,\mathbf{x}_N) = \sum_j a_{ij}\phi_j(\mathbf{x}_1,\mathbf{x}_2,\ldots,\mathbf{x}_N). \quad (7.128)$$

The orbital and spin angular momenta of the N electrons are coupled vectorially to form $\mathbf{L}_j\mathbf{S}_j\mathbf{M}_{\mathbf{L}_j}\mathbf{M}_{\mathbf{S}_j}$; α_j gives the contribution of the coupling of open shells to $\mathbf{L}_j$ and $\mathbf{S}_j$. All the ϕ_j in (7.127) have the same $\mathbf{L}_j\mathbf{S}_j\mathbf{M}_{\mathbf{L}_j}\mathbf{M}_{\mathbf{S}_j}$, but may differ in their occupation numbers $m_{\lambda j}$ and coupling parameters α_j.

The spin-orbit, spin-other-orbit, and spin-spin terms in (7.125) commute with L^2, S^2, J^2, and J_z and can therefore be treated in a $|\alpha LSJM_J\rangle$ representation. All the other terms commute with L^2, S^2, L_z, and S_z (hence also with J^2 and J_z) and can therefore be considered in a $|\alpha LSM_LM_S\rangle$ representation. To represent both types of terms one uses a representation that is diagonal in J^2 and J_z. Accurate representations of the target states allowing for configuration interaction can be calculated using the powerful CIV3 [491], MCHF [351], and SUPERSTRUCTURE [316] codes.

The $(N+1)$-electron wave function for the target + electron collision is expanded as

$$\Psi_k(\mathbf{x}_1,\ldots,\mathbf{x}_{N+1}) = \mathcal{A}\sum_{i,j} c_{ijk}\overline{\Phi}_i(\mathbf{x}_i,\ldots,\mathbf{x}_N,\hat{\mathbf{r}}_{N+1},\boldsymbol{\sigma}_{N+1})u_{ij}(\mathbf{r}_{N+1})$$
$$+\sum_j d_{ij}\phi_j(\mathbf{x}_1,\ldots,\mathbf{x}_N). \quad (7.129)$$

Here the operator $\mathcal{A}$ antisymmetrizes the scattered electron's coordinates against the coordinates of the N atomic/ionic electrons. The functions $\overline{\Phi}$ are formed by coupling the multi-configurational functions Φ_i in (7.128) with spin-angle wave functions u_{ij} for the scattered electron to give eigenstates with total orbital angular momentum $\mathbf{L}$, total spin $\mathbf{S}$, and parity π of the combined system. The u_{ij} are basis functions for the free electron.

The second sum contains wave functions ϕ_j for *correlation functions* (or *pseudostates*) that are used to improve the representation of short-range correlations in the collision. An astute choice of the pseudostates ϕ_j can greatly improve the accuracy of the results. For example, calculated cross sections for collisional excitation of $1s^2S \rightarrow 2s^2S$ and $1s^2S \rightarrow 2p^2P$ in hydrogen agreed much better with experiment when $\overline{3s}$ and $\overline{3p}$ pseudostates were included as correlation functions in the computation [167].

In calculations of oscillator strengths and continuum cross sections for C II [1200, 1201], the configurations included are (C III+e), where the core ion C III is represented by the states $2s^2(^1S)$; $2s2p(^3P,^1P)$ and $2p^2(^3P,^1D,^1S)$; and C II correlation configurations of the type $(2s^2, 2s2p, 2p^2)nl$, where $n=3$ and 4, e.g., $2s^23d$, $2s2p3s$, $2p^23s$, and so on. This work produced a large amount of accurate data for C II energy levels, oscillator strengths, and photoionization cross sections. The coefficients c_{ij} and d_{ij} in (7.129) are determined by diagonalizing the $(N + 1)$-electron Hamiltonian. The total energy of the system is $E = E_i + \epsilon_i$, where E_i and ϵ_i are, respectively, the energy of the target state and the colliding electron. For $\epsilon_i \geq 0$, we have an *open channel* and the electron is free after the collision. For $\epsilon_i < 0$, we have a *closed channel* and the electron is left in a bound or quasi-bound (resonance) state.

Oscillator Strengths and Photoionization Cross Sections

The web sites listed at [1272] contain collections of these and other data. Here we only mention a few interesting results. A comparison of gf_L and gf_V values from dipole-length and dipole-velocity calculations [1200] for C II is shown in figure 7.3.

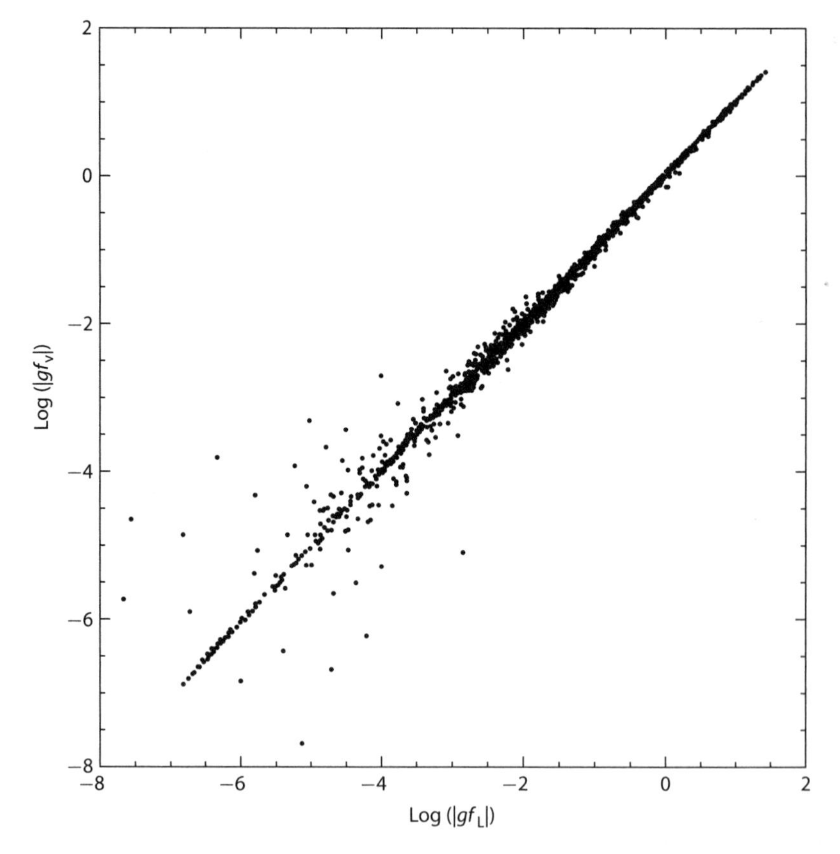

Figure 7.3 $\log(|f_V|)$ versus $\log(|f_L|)$ for bound-bound transitions of C II. The larger scatter for $gf \lesssim 10^{-3}$ is caused by numerical cancellations in the wave functions for these transitions. Adapted from [1200].

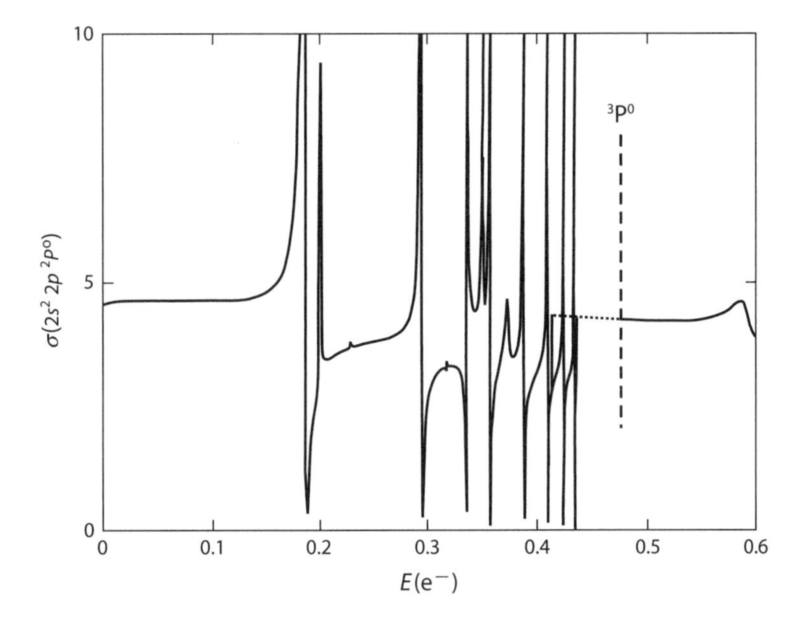

Figure 7.4 Photoionization cross section of the $1s^2\,2s^2\,2p\,^2P^\circ$ state of C II versus ejected electron energy [Ryd]. Adapted from [1201].

Photoionization cross sections for multi-electron atoms/ions often show *resonances*. At threshold, a photon has just enough energy to raise an electron from the ground configuration of the original system to the ground state of the next ion. At higher energies, the incident photon may have enough energy not only to liberate an electron from the original system but also to simultaneously raise one of the valence electrons of the next ion to an excited state. These two states can interfere quantum mechanically, resulting in a much larger, rapidly varying cross section near this critical energy.

The photoionization cross section of the $1s^2\,2s^2\,2p\,^2P^\circ$ state of C II is shown in figure 7.4 from the first ionization threshold $1s^2\,2s^2\,^1S$ to the second threshold $1s^2\,2s\,2p\,^3P^\circ$. In this process there are three channels allowed by dipole selection rules: ^2S, ^2P, and ^2D. Only the ^2S and ^2D channels are open between the ^1S and $^3P^\circ$ thresholds, giving free-electron $(^1S)ks\,^2S$ and $(^1S)kd\,^2D$ states and a set of $(^3P^\circ)np\,^2P$ bound states. Above the $^3P^\circ$ threshold, these become $(^3P^\circ)kp\,^2P$ states. The open channels have resonances with $(^3P^\circ)ns\,^2S$, $(^3P^\circ)np\,^2D$, and $(^3P^\circ)nf\,^2D$ bound states [1201].

In that condition the system can *autoionize*, leaving the next ion in an excited state, which then decays to the ground state of that ion. The inverse of this process is *dielectronic recombination*. Here a collision between an electron and an ion produces a doubly excited state of the next lower ionization state, which then stabilizes to a singly excited state by emitting a photon; see also § 9.3.

A dramatic example of these phenomena is shown in figure 7.5. We see that the cross sections for photoionization from the $4f\,^2F^\circ$ states of He II and C II agree

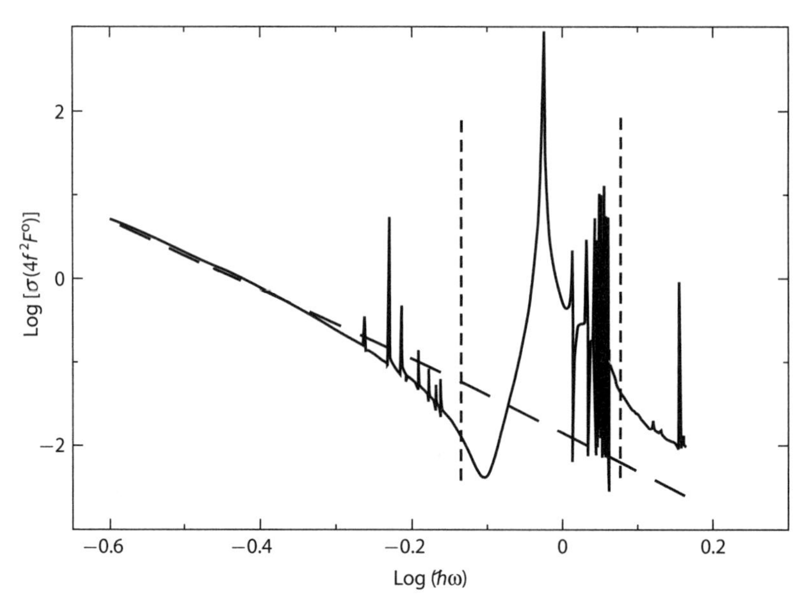

Figure 7.5 Comparison between photoionization cross section of 4f state of He II (dashed curve) and 4f state of C II (solid curve). *Ordinate*: log σ. *Abscissa*: log photon energy $\hbar\omega$ [Ryd]. Adapted from [1201].

closely at near-threshold energies. At higher energies, we see resonances in the C II cross section below the ionization potentials to the $^3P^\circ$ and $^1P^\circ$ states of C III. The dramatic peak near one Rydberg is a *PEC* (photoexcitation of the core) *resonance*, in which

$$2s^2(^1S)4f + \hbar\omega \rightarrow 2s2p(^1P^\circ)4f \rightarrow 2s^2\ ^1S + e(v)$$

or

$$2s^2(^1S)4f + \hbar\omega \rightarrow 2s2p(^1P^\circ)4f \rightarrow 2s2p\ ^3P^\circ + e(v').$$

Negative Hydrogen Ion

Some elements can form a stable *negative ion* in which an extra electron is attached to a neutral atom. Because the polarizability of hydrogen is large, it can form the ion H^-, consisting of a proton and two electrons. In a sense it is a special kind of multi-electron system. H^- can absorb/emit photons via both bound-free (*photodissociation*)

$$\mathrm{H}^- + h\nu \leftrightarrow \mathrm{H} + e(v)$$

and free-free (collision of an electron with a neutral H atom)

$$\mathrm{H} + e(v) + h\nu \leftrightarrow \mathrm{H} + e(v');\ \tfrac{1}{2}mv'^2 = \tfrac{1}{2}mv^2 + h\nu$$

processes.

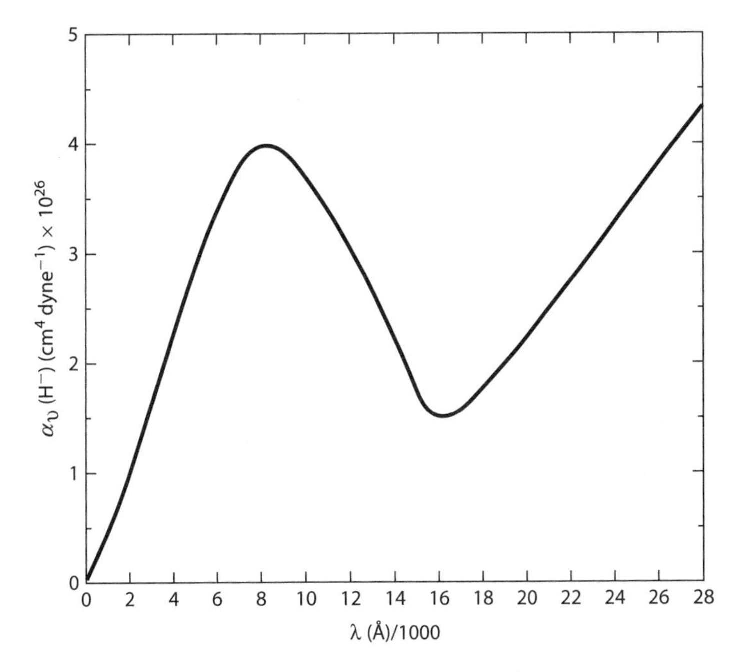

Figure 7.6 Bound-free plus free-free absorption opacity of H^- per neutral H atom and per unit electron pressure p_e. Ordinate: α_ν [cm^4/dyne]$\times 10^{26}$, in LTE at $T = 6300$ K.

The potential importance of free-free absorption by H^- in the Sun's atmosphere was suggested in 1930 by Pannekoek [837, p. 166]. Wildt later pointed out [1159, 1165] that the electron affinity of hydrogen is large enough for a second electron to be bound. Its binding energy is 0.754 eV, which corresponds to a bound-free threshold at $\lambda \approx 1.65\ \mu$. Because the binding energy of H^- is small, it is destroyed by dissociation in hot stellar atmospheres, so it is important for solar-type and cooler stars. The opacity of H^- is proportional to the number densities of neutral hydrogen atoms and free electrons; in cooler stars the electrons come from "metals" that have low ionization potentials. Thus H^- absorption is much less important in stars with very low metal abundances.

The maximum bound-free cross section is $\approx 4 \times 10^{-17}$ cm^2/H^-ion at $\lambda \approx$ 8500 Å and decreases toward shorter wavelengths. The free-free cross section is about equal to the bound-free at $\lambda \approx 1.5\ \mu$ and increases toward longer wavelengths. The summed absorption coefficient has a minimum at about 1.6 μ; see figure 7.6. In LTE, it is customary to use the Saha equation to write $n^*(H^-) = n(H) p_e \Phi(T)$, where $\Phi(T)$ contains the temperature dependence of the ionization equilibrium, and to express the absorption coefficient in units cm^4/dyne, so that $n(H) p_e \alpha_\nu$ has the correct units of cm^{-1}.

Numerous studies have been devoted to deriving an accurate quantum mechanical model of H^-, and the calculation of its continuum cross sections, from the early work of Chandrasekhar and his colleagues, see, e.g., [219, 221, 222, 224, 226–229], to more modern work using more elaborate models, culminating in definitive R-matrix calculations; see, e.g., [12, 103, 588, 762, 1053]. A critical assessment [588] shows that the calculations in [103] are accurate to about 1% for wavelengths >5000 Å over a temperature range from 1400 K to 10,080 K. Calculations of cross sections for other negative ions are contained in, e.g., [104, 585–587, 591].

More generally, if an electron collides with a neutral atom, the radiation produced is considered free-free emission by the atom's negative ion, see, e.g., [103, 288, 371, 583, 588, 807, 808], the most important of which astrophysically is H^-. Calculations have also been made for free-free emission from He^-, Ne^-, and Ar^-, e.g., [584, 590, 591]. Similarly, a collision with a molecule results in free-free emission by a *negative molecular ion*, e.g., [102, 104, 586, 589].

7.3 MOLECULES

In the spectra of K-, M-, S-, and C-stars, molecular absorption produces prominent *bands* composed of thousands of individual lines; see, e.g., [482, 672, 724, 769]. Molecules also make minor contributions to the continuous opacity in the atmospheres of the Sun and solar-type stars [588]. They play a pivotal role in modeling the atmospheric structure of substellar objects, whose spectra are totally dominated by molecular bands.

In this section we outline some aspects of molecular structure and molecular spectroscopy to provide an overview of the essentials. We consider only diatomic molecules. More complete discussions of these subjects are given in, e.g., [116, 484–487, 848, 1006, 1013].

Born-Oppenheimer Approximation

We saw in § 7.2 that the Breit-Pauli Hamiltonian $\hat{\boldsymbol{H}}_{\mathrm{BP}}$ used to compute the properties of atoms with moderate Z is intricate because it includes numerous relativistic terms. Different kinds of complications in the Hamiltonian of a molecule result from additional degrees of freedom in the system: in addition to states determined by the kinetic and excitation energies of the electrons, the nuclei in the molecule can vibrate with respect to one another, and the molecule as a whole can rotate about one or more axes. Diatomic molecules vibrate along the axis joining the two nuclei. The nuclei in triatomic molecules define a plane; their vibrations have components along two perpendicular axes. In general, the vibrations of polyatomic molecules have components along all three coordinate axes. In addition, molecules can rotate. Diatomic molecules rotate about an axis perpendicular to the line joining the nuclei; triatomic molecules can rotate around two axes; and polyatomic molecules can tumble freely in space. The energies associated with these vibrations and rotations are all quantized.

In principle, we want to find the eigenfunctions and eigenvalues of a system consisting of all the nuclei and their attendant electrons. In a fundamental paper Born and Oppenheimer [136] pointed out that the problem can be simplified greatly by taking advantage of the fact that an electron's mass is orders of magnitude smaller than a nucleon's mass; hence its speed is orders of magnitude faster than a that of a nucleus. In the absence of external forces, the position

$$\mathbf{X}_0 \equiv \sum_n M_n \mathbf{X}_n + m_e \sum_l \mathbf{x}_l \tag{7.130}$$

of a molecule's center of mass translates with a uniform velocity. There is no dynamical effect of this motion, so $\mathbf{X}_0$ can be taken as the origin of coordinates and at rest. In (7.130) M_n is the mass of a nucleus of type n, located at position $\mathbf{X}_n$, and m_e is the mass of an electron, located at position $\mathbf{x}_l$. Setting $d\mathbf{X}_0/dt = 0$, and taking averages $\langle F \rangle \equiv \int \psi^* F \, \psi \, d^3x$, we have

$$\sum_n M_n \langle \mathbf{V} \rangle_n + m_e \sum_l \langle \mathbf{v} \rangle_l = 0. \tag{7.131}$$

The particles are fermions, and hence indistinguishable, so on the average the electrons all have the same speed, as do all nuclei of a given species. Then (7.131) implies that for a typical nucleus $M\,|\langle \mathbf{V} \rangle| \approx Zm_e |\langle \mathbf{v} \rangle| \Rightarrow |\langle \mathbf{V} \rangle| \approx 5 \times 10^{-4} (Z/A) |\langle \mathbf{v} \rangle|$, where A is its atomic number and Z is its charge. Note that whereas an electron's average speed is much greater than a nucleon's, (7.131) shows they have comparable momenta. Therefore, the ratio of the kinetic energy of a nucleon is also $\mathrm{O}(m_e/M_n)$ that of an electron.

The disparity in their masses means that the electrons in a molecule move much faster relative to the center of mass than the nuclei. *Therefore, they can assume a quasi-stationary configuration almost instantaneously compared to the much slower nuclei.* This is the physical essence of the *Born-Oppenheimer approximation.*

Ignoring at present the spins of the particles, the Hamiltonian of the total system is

$$\hat{\boldsymbol{H}} = \hat{\boldsymbol{H}}_{\mathrm{n}} + \hat{\boldsymbol{H}}_{\mathrm{e}}. \tag{7.132}$$

$\hat{\boldsymbol{H}}_{\mathrm{n}}$ represents the kinetic energy of the nuclei and $\hat{\boldsymbol{H}}_{\mathrm{e}}$ the kinetic energy of the electrons plus the total potential energy from the electrostatic interactions among all the particles. The time-independent Schrödinger equation for the molecule is

$$\hat{\boldsymbol{H}} \psi_{\nu\lambda}(\mathbf{X}, \mathbf{x}) = \mathcal{E}_{\nu\lambda} \psi_{\nu\lambda}(\mathbf{X}, \mathbf{x}). \tag{7.133}$$

In (7.133), $\mathbf{X}$ denotes a set of all nuclear coordinates ($\{\mathbf{X}_n\}, n = 1, \ldots, N$); $\mathbf{x}$ a set of all electronic coordinates ($\{\mathbf{x}_l\}, l = 1, \ldots, L$); $\psi_{\nu\lambda}(\mathbf{X}, \mathbf{x})$ is the wave function for all the particles; $\mathcal{E}_{\nu\lambda}$ is the total energy of the system; ν represents the quantum numbers associated with the motions of the nuclei; and λ represents those associated with the motions of the electrons.

The nuclear contribution to $\hat{\boldsymbol{H}}$ is

$$\hat{\boldsymbol{H}}_{\mathrm{n}} = -\frac{\hbar^2}{2} \sum_{n=1}^{N} \frac{\nabla_n^2}{M_n}, \tag{7.134}$$

and the electronic contribution is

$$\hat{\boldsymbol{H}}_{\mathrm{e}} = -\frac{\hbar^2}{2m_e}\sum_{l=1}^{L}\nabla_l^2 + V(\mathbf{X},\mathbf{x}), \tag{7.135}$$

where V is the potential energy

$$V(\mathbf{X},\mathbf{x}) = \sum_{n>m}^{N}\sum_{m=1}^{N-1}\frac{Z_m Z_n e^2}{r_{mn}} + \sum_{l>k}^{L}\sum_{k=1}^{L-1}\frac{e^2}{r_{kl}} - \sum_{n=1}^{N}\sum_{l=1}^{L}\frac{Z_n e^2}{r_{ln}}. \tag{7.136}$$

Here $r_{mn} \equiv |\mathbf{X}_m - \mathbf{X}_n|$; $r_{kl} \equiv |\mathbf{x}_k - \mathbf{x}_l|$; $r_{ln} \equiv |\mathbf{x}_l - \mathbf{X}_n|$; ∇_n^2 contains derivatives with respect to the nuclear coordinates $\mathbf{X}_n$; and ∇_l^2 contains derivatives with respect the electronic coordinates $\mathbf{x}_l$. If we could solve the system (7.133)–(7.136), we would obtain an exact wave function $\psi_{\upsilon\lambda}(\mathbf{X},\mathbf{x})$. Such a calculation is difficult.

In light of Born and Oppenheimer's astute observation, we first solve (7.133) for an approximate wave function for the electrons, *assuming that the configuration of the nuclei is frozen* so their contribution to the potential can be treated *parametrically*. That is, for a *fixed* configuration $\boldsymbol{\xi}$ of the nuclei we calculate a wave function $\phi_\lambda(\mathbf{x}|\boldsymbol{\xi})$ and its energy eigenvalue $E_\lambda(\boldsymbol{\xi})$, for the electrons alone, using the equation

$$\left[-\frac{\hbar^2}{2m_e}\sum_{l=1}^{L}\nabla_l^2 + V(\boldsymbol{\xi},\mathbf{x})\right]\phi_\lambda(\mathbf{x}|\boldsymbol{\xi}) = E_\lambda(\boldsymbol{\xi})\phi_\lambda(\mathbf{x}|\boldsymbol{\xi}). \tag{7.137}$$

We must compute $\phi_\lambda(\mathbf{x}|\boldsymbol{\xi})$ and $E_\lambda(\boldsymbol{\xi})$ for all possible arrangements of the nuclei in order to obtain $\phi_\lambda(\mathbf{X},\mathbf{x})$ and $E_\lambda(\mathbf{X})$ for all sets $\mathbf{X}$.[7] Then we can rewrite (7.137) as

$$\left[-\frac{\hbar^2}{2m_e}\sum_{l=1}^{L}\nabla_l^2 + V(\mathbf{X},\mathbf{x})\right]\phi_\lambda(\mathbf{X},\mathbf{x}) = E_\lambda(\mathbf{X})\phi_\lambda(\mathbf{X},\mathbf{x}), \tag{7.138}$$

which yields the electronic wave function $\phi_\lambda(\mathbf{X},\mathbf{x})$.

In some cases it is possible to determine $E_\lambda(\mathbf{X})$ from experiment. For diatomic molecules $E_\lambda(\mathbf{X})$ may be fairly well represented by an analytical *Morse potential*

$$E_{\mathrm{Morse}}(R) = E_0\left[e^{-a(R-R_0)} - 2e^{-b(R-R_0)}\right]. \tag{7.139}$$

In (7.139), a, b and E_0 are all positive and $a > b$. The constants in the formula may be optimized by fitting to real data. The first term in $E(R)$ is positive; it is the result of internuclear repulsion at small separations $R < R_0$. The second term is negative, and hence attractive. The two terms cancel most strongly at $R = R_0$, yielding a binding energy $E(R_0) = -E_0$. At large separations, the nuclei are shielded by their atomic

[7] An example of $E_\lambda(\mathbf{X})$ for H_2 in its ground electronic state is shown in figure 4.1. Because H_2 is a diatomic molecule, $\mathbf{X}$ is changed only by the separation R between the nuclei; hence $E_\lambda(\mathbf{X}) \to E_\lambda(R)$.

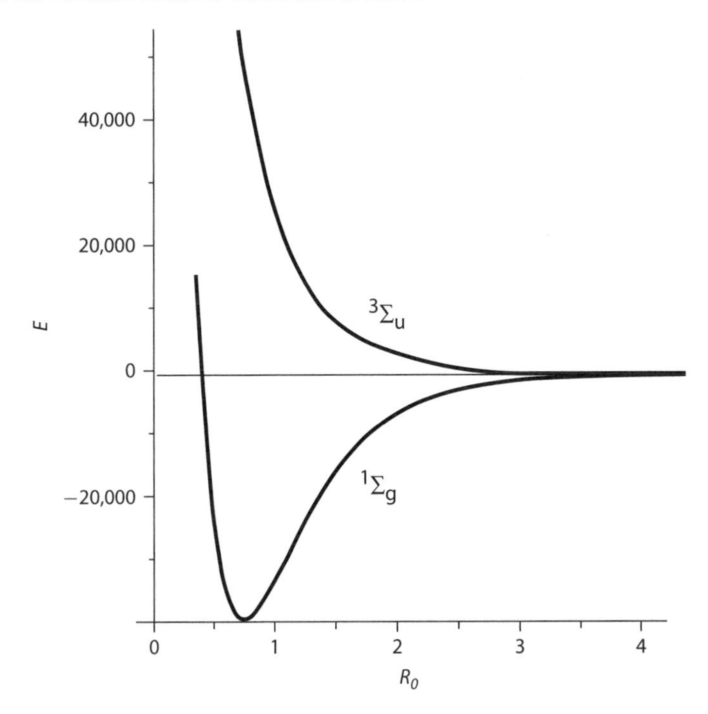

Figure 7.7 The energies of the $^1\Sigma_g$ and $^3\Sigma_u$ states of H_2 as a function of internuclear separation. *Ordinate:* Energy in cm^{-1}. *Abscissa:* Internuclear separation in 10^{-8} cm.

electrons, so $E(R) \to 0$. Physically this is because at large enough separation, the nuclei and their electrons are simply two distinct atoms, and they no longer interact.

Thus far, we have ignored the effect of electron spin. But it can radically alter the qualitative discussion of the process of molecular binding given in chapter 4. For example, if the electron spins of two hydrogen atoms are aligned, then as they approach one another, their collective wave function is not distorted in such a way as to increase the density of negative charge between the nuclei because the electrons are forbidden to occupy the same phase space by the exclusion principle. Therefore, the internuclear repulsive force will always prevail. On the other hand, if the spins are opposed, then the description of the process in chapter 4 is qualitatively correct. As shown in figure 7.7, the $^1\Sigma_g$ state of H_2, in which the electron spins are opposed, is bound, and hence stable, whereas the $^3\Sigma_u$ state in which the electron spins are aligned is never bound and is unstable.

Finally, given $\phi_\lambda(\mathbf{X}, \mathbf{x})$ we calculate an approximate wave function $\Phi_{\nu\lambda}(\mathbf{X})$ for the motion of the nuclei by using $E_\lambda(\mathbf{X})$ as the potential function in

$$\left[-\sum_{n=1}^{N} \frac{\hbar^2}{2M_n} \nabla_n^2 + E_\lambda(\mathbf{X}) \right] \Phi_{\nu\lambda}(\mathbf{X}) = \mathcal{E}_{\nu\lambda} \Phi_{\nu\lambda}(\mathbf{X}). \tag{7.140}$$

Feynman [330] showed that $E_\lambda(\mathbf{X})$ is a good potential to use in calculating $\Phi(\mathbf{X})$ in as much as its derivative with respect to any X_i gives the same force component F_i

as one obtains from the gradient of the total electrostatic potential V of the system, i.e., $(\partial E/\partial X_i) = \int \phi^*(\mathbf{X},\mathbf{x})(\partial V/\partial X_i)\phi(\mathbf{X},\mathbf{x})\, d^3x$.

In essence we have tried to make a factorization such that the product $\phi_\lambda(\mathbf{X},\mathbf{x})\Phi_{\nu\lambda}(\mathbf{X})$ is a good first approximation to the wave function $\psi_{\nu\lambda}(\mathbf{X},\mathbf{x})$ of the whole system. The factorization cannot be exact because $\psi_{\nu\lambda}$ depends *simultaneously* on two sets of variables in the *same* coordinate space. Thus unlike, say, space and spin coordinates, $\mathbf{X}$ and $\mathbf{x}$ are not physically independent. Now the question is: "How accurate is the result of this process?" If we multiply (7.138) on the right by $\Phi_{\nu\lambda}(\mathbf{X})$, (7.140) on the left by $\phi_\lambda(\mathbf{X},\mathbf{x})$, and add, we get

$$\left[-\frac{\hbar^2}{2m_e}\sum_{l=1}^{L}\nabla_l^2 + V(\mathbf{X},\mathbf{x})\right]\phi_\lambda(\mathbf{X},\mathbf{x})\Phi_{\nu\lambda}(\mathbf{X}) - \phi_\lambda(\mathbf{X},\mathbf{x})\sum_{n=1}^{N}\frac{\hbar^2}{2M_n}\nabla_n^2\Phi_{\nu\lambda}(\mathbf{X}) = \mathcal{E}_{\nu\lambda}\phi_\lambda(\mathbf{X},\mathbf{x})\Phi_{\nu\lambda}(\mathbf{X}). \quad (7.141)$$

By rearranging the second term on the left-hand side, we find

$$\left[-\frac{\hbar^2}{2}\left(\sum_{n=1}^{N}\frac{\nabla_n^2}{M_n} + \sum_{l=1}^{L}\frac{\nabla_l^2}{m_e}\right) + V\right]\phi_\lambda\Phi_{\nu\lambda} = \mathcal{E}_{\nu\lambda}\phi_\lambda\Phi_{\nu\lambda} - \sum_{n=1}^{N}\frac{\hbar^2}{2M_n}\left(2\nabla_n\phi_\lambda\cdot\nabla_n\Phi_{\nu\lambda} + \Phi_{\nu\lambda}\nabla_n^2\phi_\lambda\right). \quad (7.142)$$

The extra terms on the right-hand side prevent the factorization of $\phi_\lambda\Phi_{\nu\lambda}$ to be exact. We show below that they are small, so the product function is actually a good first approximation. They cannot always be discarded because they determine the transition rate between electronic levels.

Application to Diatomic Molecules

In what follows we examine the specific case of diatomic molecules, many of which, e.g., H_2, C_2, CH, CN, CO, OH, and TiO, have considerable astrophysical importance. For such systems the analysis is greatly simplified, but still provides a fairly comprehensive description of the major features in molecular spectra.

Consider a molecule composed of two nuclei, A and B, with charges $Z_A e$ and $Z_B e$, located at positions $\mathbf{X}_A$ and $\mathbf{X}_B$, respectively, and two valence electrons, a and b, located at $\mathbf{x}_a$ and $\mathbf{x}_b$. An example is H_2. The system is portrayed in figure 7.8. From (7.134) and (7.135) the nuclear contribution to the Hamiltonian $\hat{\boldsymbol{H}}$ is

$$\hat{\boldsymbol{H}}_{\rm n} = -\frac{\hbar^2}{2}\left(\frac{\nabla_A^2}{M_A} + \frac{\nabla_B^2}{M_B}\right), \quad (7.143)$$

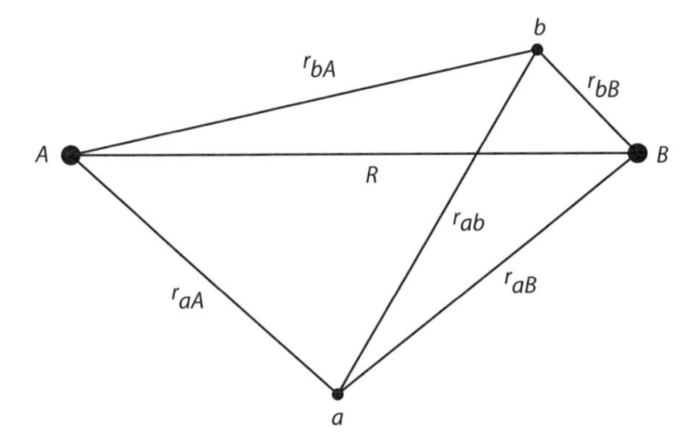

Figure 7.8 Schematic diatomic molecule.

and the electronic contribution is

$$\hat{\boldsymbol{H}}_{\mathrm{e}} = -\frac{\hbar^2}{2m_e}\left(\nabla_a^2 + \nabla_b^2\right) + V, \tag{7.144}$$

where V is the potential energy

$$V(\mathbf{X}, \mathbf{x}) = -\frac{Z_A e^2}{r_{aA}} - \frac{Z_B e^2}{r_{aB}} - \frac{Z_A e^2}{r_{bA}} - \frac{Z_B e^2}{r_{bB}} + \frac{Z_A Z_B e^2}{R} + \frac{e^2}{r_{ab}}. \tag{7.145}$$

Here ∇_A^2, ∇_B^2, ∇_a^2, and ∇_b^2 contain derivatives with respect to $\mathbf{X}_A$, $\mathbf{X}_B$, $\mathbf{x}_a$, and $\mathbf{x}_b$, respectively; $r_{ab} \equiv |\mathbf{x}_a - \mathbf{x}_b|$, $r_{aA} \equiv |\mathbf{x}_a - \mathbf{X}_A|$, $r_{aB} \equiv |\mathbf{x}_a - \mathbf{X}_B|$, etc.; and $R \equiv |\mathbf{X}_A - \mathbf{X}_B|$ is the *internuclear separation.*

Applying the Born-Oppenheimer procedure to this specific case, we factor the wave function ψ for the total system into the product of an electronic wave function ϕ_{e} and a nuclear wave function Φ_{n}:

$$\psi(\mathbf{X}, \mathbf{x}) \approx \phi_{\mathrm{e}}(\mathbf{x}_a, \mathbf{x}_b, R)\Phi_{\mathrm{n}}(\mathbf{X}_A, \mathbf{X}_B). \tag{7.146}$$

Again, the factorization in (7.146) is not exact because the electronic configuration still depends on the *scalar* internuclear separation R. In this approximation, Schrödinger's equation (7.142) for the system's total wave function ψ (using abbreviated notation) is

$$\left[-\frac{\hbar^2}{2}\left(\frac{\nabla_A^2}{M_A} + \frac{\nabla_B^2}{M_B}\right) - \frac{\hbar^2}{2m_e}(\nabla_a^2 + \nabla_b^2) + V\right]\phi\Phi = \mathcal{E}\phi\Phi$$
$$-\left(\frac{\hbar^2}{M_A}\nabla_A\phi \cdot \nabla_A\Phi + \frac{\hbar^2}{M_B}\nabla_B\phi \cdot \nabla_B\Phi\right) - \frac{\hbar^2\Phi}{2}\left(\frac{\nabla_A^2}{M_A} + \frac{\nabla_B^2}{M_B}\right)\phi. \tag{7.147}$$

In the second term on the right-hand side, $\nabla_A \phi = (\mathbf{R}/R)(\partial\phi/\partial R) = -\nabla_B \phi$, so it becomes

$$\hbar^2 \frac{\partial \phi}{\partial R}\frac{\mathbf{R}}{R}\left(\frac{1}{M_B}\nabla_B - \frac{1}{M_A}\nabla_A\right)\Phi. \tag{7.148}$$

The parentheses contains the quantum mechanical operator for the *relative velocity* of the two nuclei, $\mathbf{v}_B - \mathbf{v}_A$. *In a bound molecule, it averages to zero*, so we drop this term. In the last term on the right-hand side, ∇_A^2 essentially operates on *differences* $(\mathbf{x}_l - \mathbf{X}_A)$ in spatial coordinates so it is of the same order as ∇_a^2 on the left-hand side. Thus the last term is a factor m_e/M smaller than the middle term on the left, so we drop it as well.

Separation of Individual Nuclear Degrees of Freedom

To separate the individual modes of motion of the nuclei, we follow the procedure used earlier for atoms. The center of mass $\mathbf{X}_0$ of the two nuclei is

$$\mathbf{X}_0 \equiv \frac{M_A \mathbf{x}_A + M_B \mathbf{x}_B}{M_A + M_B}, \tag{7.149}$$

and the molecule's reduced mass is

$$\mu \equiv \frac{M_A M_B}{M_A + M_B}. \tag{7.150}$$

As we did for the hydrogen atom, write the operator $\hat{\boldsymbol{H}}_{\rm n}$ in (7.143) as the sum of operators for (1) the kinetic energy in the *translational motion* of the center of mass as a free particle and (2) the kinetic energies of the *internal motions* of the nuclei relative to the center of mass:

$$\hat{\boldsymbol{H}}_{\rm n} \equiv -\frac{\hbar^2}{2M}\nabla_0^2 - \frac{\hbar^2}{\mu}\nabla_R^2, \tag{7.151}$$

where $M \equiv M_A + M_B$. ∇_0 contains derivatives with respect to $\mathbf{X}_0$ only, and ∇_R with respect to R only. Because each of these operators depends on only one of the two independent sets of coordinates $\mathbf{X}_0$ and R, the wave function can be factored exactly into the product

$$\psi_{\rm n} = \psi_{\rm trans}(\mathbf{X}_0)\psi_{\rm int}(R). \tag{7.152}$$

Substituting (7.152) into (7.140), and dividing both sides by $\psi_{\rm trans}\psi_{\rm int}$, we obtain

$$-\frac{1}{\psi_{\rm trans}}\frac{\hbar^2}{2M}\nabla_0^2\psi_{\rm trans} = \frac{1}{\psi_{\rm int}}\frac{\hbar^2}{2\mu}\nabla_R^2\psi_{\rm int} - E_{\rm e}(R) + \mathcal{E}, \tag{7.153}$$

where $\mathcal{E}$ is the total energy of the system.

Because the left-hand side depends only on $\mathbf{X}_0$ and the right-hand side only on the internuclear separation R, both sides must equal a constant. We take it to be the *translational energy* E_{trans} of the system. Then the left-hand side of (7.153) gives us

$$\hat{\boldsymbol{H}}_{\text{trans}}\psi_{\text{trans}} \equiv -\frac{\hbar^2}{2M}\nabla_0^2\psi_{\text{trans}} = E_{\text{trans}}\psi_{\text{trans}}. \tag{7.154}$$

The translational wave function for a free particle has the form $\psi_{\text{trans}} \propto \exp(i\mathbf{k}\cdot\mathbf{X}_0)$. If we assume that $\mathbf{k}$ is unchanged during a transition between internal energy states of the molecule, i.e., photons do not couple with the translational degrees of freedom, then we may work in the frame of the molecule's center of mass and set $\mathbf{k}=0$, $\psi_{\text{trans}} \equiv 1$, and $E_{\text{trans}} =$ constant. Henceforth we will ignore the translational motion.

The right-hand side of (7.153) gives us

$$\hat{\boldsymbol{H}}_{\text{int}}\psi_{\text{int}} \equiv -\left[\frac{\hbar^2}{2\mu}\nabla_R^2 + E_{\text{e}}(R)\right]\psi_{\text{int}} \equiv E_{\text{int}}\psi_{\text{int}}, \tag{7.155}$$

where the *internal energy* associated with the nuclei is

$$E_{\text{int}} \equiv \mathcal{E} - E_{\text{trans}}. \tag{7.156}$$

Energy Levels

The nuclei in the molecule can vibrate, and the molecule as a whole can rotate. The energies of each of these modes of motion are quantized. Thus a molecule's internal energy can be written

$$E_{\text{int}} \equiv E_{\text{e}} + E_{\text{vib}} + E_{\text{rot}}, \tag{7.157}$$

where E_{e}, E_{vib}, E_{rot} are, respectively, contributions from the electrons' kinetic energies and their interaction with the electrostatic potential V [see (7.138)], the molecule's vibrational energy, and its rotational energy.

Explicit expressions can be obtained for E_{rot} and E_{vib}. Write the position vector $\mathbf{R}$ of the equivalent particle having reduced mass μ in spherical coordinates (R, θ, ϕ). In these coordinates, the Laplacian ∇_R^2 is

$$\nabla_R^2 = \frac{1}{R^2}\left[\frac{\partial}{\partial R}\left(R^2\frac{\partial}{\partial R}\right) - \frac{L^2}{\hbar^2}\right], \tag{7.158}$$

where $\mathbf{L}$ is the angular momentum operator

$$\mathbf{L} \equiv \mathbf{R} \times (-i\hbar\nabla_R)]. \tag{7.159}$$

Because E_{e} depends only on the internuclear separation R, (7.155) can be separated in the same way as for the hydrogen atom into a radial (*vibrational*) part and an angular (*rotational*) part, which is a spherical harmonic. We can then write

$$\psi_{\text{int}}(\mathbf{R}) = \frac{1}{R}W_{\text{vib}}(R)\,Y_{Jm}(\theta, \phi). \tag{7.160}$$

Rotational Levels

Consider first the angular part of (7.160). The quantum numbers J and m are analogous to the quantum numbers l and m in atoms. As in (7.36) and (7.37)

$$|\hat{\boldsymbol{L}}\cdot\hat{\boldsymbol{L}}\, Y_{Jm}| = J(J+1)\hbar^2\, Y_{Jm} \tag{7.161}$$

and

$$L_z Y_{Jm} = m\hbar Y_{Jm}. \tag{7.162}$$

As mentioned in § 4.2, Schrödinger [973] showed that the energy eigenvalues for a quantized rigid rotator with moment of inertia I are

$$E_{\text{rot}} = J(J+1)\frac{\hbar^2}{2I} = J(J+1)\frac{\hbar^2}{2\mu R_0^2} \equiv J(J+1)B. \tag{7.163}$$

B is the *rotation constant* of the molecule, J is a non-negative integer, μ is the molecule's effective mass, and R_0 is the equilibrium position of the vibrating nuclei.

Vibrational Levels

By substituting (7.160), (7.161), and (7.163) into (7.155), we can eliminate the contributions from rotation and are left with

$$\hat{\boldsymbol{H}}_{\text{vib}} W_{\text{vib}} \equiv -\frac{\hbar^2}{2\mu}\frac{d^2 W_{\text{vib}}}{dR^2} + E_{\text{e}}(R) W_{\text{vib}} = E_{\text{vib}} W_{\text{vib}} \tag{7.164}$$

for the vibrational part of the wave function. Although (7.164) has the same general form as the Schrödinger equation for a harmonic oscillator, $E_{\text{e}}(R)$ on the left-hand side depends on R, so it actually represents an *anharmonic oscillator*. Its eigenfunctions would be difficult to obtain analytically for a general $E_{\text{e}}(R)$. To obtain a useful approximate solution we expand $E_{\text{e}}(R)$ in a Taylor series to transform (7.164) into an equation for a pure harmonic oscillator. Thus write

$$E_{\text{e}}(R) = E_{\text{e}}(R_0) + E_{\text{e}}''(R_0)(R-R_0)^2 + \ldots, \tag{7.165}$$

where R_0 is the equilibrium position of the nuclei, i.e., the point at which the potential energy has its minimum, and also the first derivative of $E_{\text{e}}(R)$ vanishes. For notational convenience write $x \equiv R - R_0$, and

$$E_{\text{e}}''(R_0) \equiv \mu\omega_0^2. \tag{7.166}$$

Here ω_0 can be considered as the *natural frequency of vibration* of the pair of nuclei. Then (7.164) becomes

$$-\frac{\hbar^2}{2\mu}\frac{d^2 W_{\text{vib}}}{dx^2} + \tfrac{1}{2}\mu\omega_0^2 x^2 W_{\text{vib}} = E_{\text{vib}} W_{\text{vib}}. \tag{7.167}$$

The vibrational energy E_{vib} of the nuclei can now be written

$$E_{\text{vib}} = E_{\text{int}} - E_{\text{el}}(R_0) - J(J+1)\frac{\hbar^2}{2\mu R_0^2}. \tag{7.168}$$

The eigenfunctions of (7.164) for W_{vib} are Hermite polynomials times an exponential [723, chapter 5]. As was shown by a different method in § 3.6, their eigenvalues are

$$E_{\text{vib}} = \left(v + \tfrac{1}{2}\right)\hbar\omega_0, \qquad v = 0, 1, 2, \ldots, \tag{7.169}$$

where v is the *vibrational quantum number.*

In summary, each electronic state of a molecule has a ladder of vibrational states separated an energy $\hbar\omega_0$. Each of these vibrational states has a set of rotational states separated by much smaller energy differences $J\hbar^2/\mu R_0^2$ between levels with rotational quantum number J and $J+1$.

Electronic Levels

To understand the electronic energy states of a molecule, consider first their similarities to, and differences from, electronic states in an atom. In an atom, electrons in a valence shell move in the (partially shielded) spherically symmetric electrostatic potential of the nucleus and the non-symmetric mutual electrostatic potential of other electrons outside of closed shells. The orbital momentum of an individual electron is not conserved, but the *total* orbital angular momentum and spin of all electrons outside of closed shells are both conserved. All other terms (e.g., spin-orbit coupling) in the potential are of order α^2 and to first order can be neglected.

In a molecule, the potential is not at all spherically symmetric because there are two or more (possibly quite strongly) charged nuclei present. (For a diatomic molecule it is at least axially symmetric about the internuclear axis.) Consequently, only the *component* of the electronic orbital angular momentum along the internuclear axis is conserved. The vector representing the total electronic orbital angular momentum *precesses* around this axis. Quantum mechanically, the component along the axis is $\hbar M_L$, where M_L can have only the integer values $(L, L-1, L-2, \ldots, 0, \ldots, -L)$.

It is customary to write the unsigned component of $\mathbf{L}$ along the internuclear axis as $\Lambda = |M_L|$. For a given value of L, Λ can have the values $\Lambda = 0, 1, 2, \ldots, L$, so there are $L+1$ states with distinct energy. In parallel with atomic spectroscopic notation, states with $\Lambda = 0, 1, 2, 3, \ldots$ are denoted as $\Sigma, \Pi, \Delta, \Phi, \ldots$ states. The energy of a state does not depend on the sign of M_L, so states with the same $|M_L|$ are degenerate. Thus the Σ states are degenerate; the $\Pi, \Delta, \Phi, \ldots$ states are doubly degenerate because M_L can have two values, $+\Lambda$ and $-\Lambda$.

The individual electrons contribute to the total spin $\mathbf{S}$ and the corresponding quantum number S. If $\Lambda = 0$ (Σ states), the spin angular momentum is constant (unless there is an external magnetic field). If $\Lambda \neq 0$ (i.e., $\Pi, \Delta, \ldots$ states), there is an internal magnetic field along the internuclear axis produced by the motion of the electrons, which leads to a precession of the vector $\mathbf{S}$ about this axis. As with orbital angular momentum, the component along the internuclear axis is described quantum mechanically as $\hbar M_S$, where M_S is an integer in the range $(S, S-1, S-2, \ldots, 0, \ldots, -S)$. M_S is, unfortunately, usually denoted as Σ (which is not to be confused with a state with $\Lambda = 0$).

The component of the total electronic angular momentum along the internuclear axis, Ω, is obtained by adding the orbital and spin contributions. Unlike the case for atoms, where vector addition is used, here the *projections* along the internuclear axis are added algebraically, i.e., $\Omega = |\Lambda + \Sigma|$.

Transitions

Order of Magnitude Transition Energies

In what follows, it is useful to have order of magnitude estimates of the energies of electronic, vibrational, and rotational transitions.

The dissociation energies D_0 of the astrophysically important molecules listed in table 2.1 are about 4–7 eV, i.e., about $\frac{1}{3}$ to $\frac{1}{2}$ of a Rydberg $\mathcal{R} \approx e^4 m_e / 2\hbar^2$. The excitation energies of the electronic states in a molecule are of the same order of magnitude as D_0, so

$$D_0 \sim E_0 \sim e^4 m_e / \hbar^2 \sim e^2 / a_0 \sim \hbar^2 / m_e a_0^2. \tag{7.170}$$

Here we used atomic units to eliminate e^2 in favor of the Bohr radius a_0. Thus the dissociation energy, at which a molecule has its minimum internuclear potential and stable molecular bonds can form, occurs when the nuclei are separated by a distance comparable to a_0.

To order of magnitude, the vibrational energies of a molecule are comparable to a fraction of the depth of the potential well. If the nuclei are displaced from their equilibrium separation ($R_0 \approx a_0$), they oscillate about the equilibrium position with a characteristic circular frequency $\omega_{\rm vib}$. Hence the vibrational energy is $M \omega_{\rm vib}^2 a_0^2 \approx D_0 \sim \hbar^2 / m_e a_0^2$. If a photon is emitted during a transition $(v+1 \rightarrow v)$ it has an energy $E_{\rm vib} = \hbar \omega_{\rm vib}$:

$$E_{\rm vib} \sim \hbar \frac{\hbar}{a_0^2 (m_e M)^{1/2}} = \left(\frac{m_e}{M}\right)^{1/2} E_{\rm el}. \tag{7.171}$$

To estimate the rotational energy of a molecule, we assume the equilibrium separation of nuclei is $\approx a_0$, and the molecule rotates with a characteristic angular frequency $\omega_{\rm rot}$. Then its angular momentum is $\sim M a_0^2 \omega_{\rm rot}$. Quantum mechanically, this angular momentum has values that are integer multiples of $\hbar$, i.e., $M a_0^2 \omega_{\rm rot} \sim \hbar$; hence the energy of photons emitted when the rotational quantum number changes by one unit is

$$E_{\rm rot} = \hbar \omega_{\rm rot} \sim \hbar^2 / M a_0^2 \approx (m_e / M) E_{\rm el}. \tag{7.172}$$

Thus to order of magnitude, the relative energies of photons emitted in electronic, vibrational, and rotational transitions are

$$E_{\rm rot} : E_{\rm vib} : E_{\rm el} \sim (m_e / M) : (m_e / M)^{\frac{1}{2}} : 1. \tag{7.173}$$

Typical electronic energies are a few eV; therefore, electronic transitions typically lie in the optical or the ultraviolet region. Because $m_e/M \approx 5 \times 10^{-4}$, vibrational transitions typically lie in the near- or mid-infrared, and rotational transitions in the far-infrared or millimeter-wave regions.

Electric Dipole Transitions

The interaction Hamiltonian for dipole transitions is

$$\hat{H}_{\mathrm{D}} = -\mathbf{E} \cdot \mathbf{d}, \tag{7.174}$$

where $\mathbf{E}$ is the electric field, and $\mathbf{d}$ is the electric dipole moment. The latter is the vector sum of the electronic and nuclear dipole moments:

$$\mathbf{d} = \mathbf{d}_{\mathrm{el}} + \mathbf{d}_{\mathrm{nuc}}. \tag{7.175}$$

The electronic dipole moment is

$$\mathbf{d}_{\mathrm{e}} = \mathbf{d}_a + \mathbf{d}_b = -e\,(\mathbf{X}_a + \mathbf{X}_b), \tag{7.176}$$

where $\mathbf{d}_a$ and $\mathbf{d}_b$ are the dipole moments of the individual electrons. The nuclear dipole moment is

$$\mathbf{d}_{\mathrm{nuc}} = Z_A e \mathbf{X}_A + Z_B e \mathbf{X}_B, \tag{7.177}$$

where $\mathbf{X}_A = (M_B/M)\mathbf{R}$ and $\mathbf{X}_B = -(M_A/M)\mathbf{R}$.[8] The dipole matrix element for a transition is

$$\langle \Psi_f \,|\hat{H}_{\mathrm{D}}|\, \Psi_i \rangle = -\langle \Psi_f \,|\mathbf{E}\cdot(\mathbf{d}_{\mathrm{e}} + \mathbf{d}_{\mathrm{nuc}})|\, \Psi_i \rangle. \tag{7.178}$$

Consider now three different types of transitions with increasing complexity, i.e., differing by how many quantum numbers change during the transition.

Pure Rotational Transitions

The simplest transitions are those in which the initial and final electronic and vibrational states are unchanged; recall that they lie in the microwave domain. The dipole transition matrix element (7.178) is

$$\langle \Psi_f |\hat{H}_{\mathrm{D}}|\Psi_i \rangle = \int_0^{2\pi} d\phi \int_0^{\pi} d\theta \, \sin\theta \, Y^*_{Jm}(\theta,\phi) \langle W^f_{\mathrm{vib}} |\hat{H}_{\mathrm{vib}}| W^i_{\mathrm{vib}} \rangle Y_{J'm'}(\theta,\phi), \tag{7.179}$$

[8] In a *homonuclear molecule* (such as H_2), the two nuclei are identical, $A \equiv B$; hence the nuclear dipole moment $\mathbf{d}_{\mathrm{nuc}} \equiv 0$. This is easy to understand physically because in such a molecule the center of mass coincides with the center of nuclear charge.

where primes indicate the quantum numbers of the initial state. Take $\mathbf{E}$ in (7.178) to be a harmonic electric field (i.e., a photon) of frequency ω. Then, recalling that the vibrational state does not change in a pure rotational transition,

$$\left\langle W^f_{\text{vib}} \middle| \hat{\boldsymbol{H}}_{\text{vib}} \middle| W^i_{\text{vib}} \right\rangle \equiv -\int_0^\infty \mathbf{E}_\omega \cdot \mathbf{D}(R,\theta,\phi) \left| W_{\text{vib}}(R) \right|^2 dR, \tag{7.180}$$

where

$$\mathbf{D}(R,\theta,\phi) \equiv \int (\mathbf{d}_\text{e} + \mathbf{d}_\text{n}) \psi_\text{e}(\mathbf{X}_1, \mathbf{X}_2, R)\, d^3X_1\, d^3X_2 \tag{7.181}$$

is the expectation value of the electric dipole moment integrated over all electronic coordinates in the center of mass frame. (More precisely, this procedure also includes a summation over spin, but we will not write this explicitly.)

$\mathbf{D}$ is the *permanent dipole moment* of the molecule; (7.180) shows that pure rotational transitions have a nonzero probability only if the permanent dipole moment is nonzero. This is different from allowed electronic transitions which require only that the dipole moment must change during the transition: $\langle \psi_f \left| \mathbf{d} \right| \psi_i \rangle \neq 0$. As mentioned above, in a homonuclear molecule $\mathbf{D}_\text{nuc} \equiv 0$. Further, in this case $\mathbf{D}_\text{e} \equiv 0$ as well because the system's center of mass also coincides with the center of electronic charge, which leads to a zero value for the electronic dipole moment. Thus, although H_2 is by far the dominant molecule in cool media (interstellar matter, the coolest stars, and substellar-mass objects), it cannot provide much opacity in the far infrared and microwave regions because it is homonuclear.[9]

To calculate an explicit result for the rotational electric dipole transition probability, suppose $\mathbf{E}_\omega$ is at an angle θ with respect to $\mathbf{R}$; then

$$\mathbf{E}_\omega \cdot \mathbf{D} = |\mathbf{E}_\omega|\, D(R) \cos\theta. \tag{7.182}$$

One usually invokes the *rigid rotator approximation*, in which D is evaluated at the equilibrium separation of the nuclei:

$$D(R) \approx D(R_0) \equiv D_0. \tag{7.183}$$

D_0 is the *reduced dipole moment*. Then the dipole matrix element (7.179) is

$$\left\langle \Psi_f \middle| \hat{\boldsymbol{H}}_\text{D} \middle| \Psi_i \right\rangle = -|E_\omega| D_0 \int_0^{2\pi} d\phi \int_0^\pi d\theta\, \sin\theta \cos\theta\, Y^*_{Jm}(\theta,\phi)\, Y_{J'm'}(\theta,\phi). \tag{7.184}$$

[9] As discussed below, H_2 may acquire a transient dipole moment during an encounter with a nearby particle (such as another H_2 molecule) and thus contribute to the opacity in the infrared region. This phenomenon is called *collision-induced absorption* (CIA). For other diatomic molecules the permanent dipole is nonzero. The most important among them is CO, because C and O are the most abundant elements after hydrogen and helium (which does not form a molecule). Furthermore, C and O atoms have strong affinity for each other.

From (7.184) and the properties of the spherical harmonics we obtain the dipole selection rules for purely rotational transitions:

$$\Delta J = \pm 1 \qquad \text{and} \qquad \Delta m = 0, \tag{7.185}$$

along with the requirement that $D_0 \neq 0$.

For example, consider a transition from a rotational level with $J' = J + 1$ to level J, accompanied by spontaneous emission of a photon with energy $\hbar\omega = 2(J + 1)B$. From any sublevel m' of level J', the total transition probability to all sublevels m of level J that satisfy the selection rules is obtained by summing over all allowed values of m. Then the total transition dipole moment $D_{J+1,J}$ is

$$|D_{J+1,J}|^2 \equiv \sum_{m=-J}^{J} D_0^2 \left| \oint \sin\theta \cos\theta \; Y_{Jm}^*(\theta, \phi) \; Y_{J+1,m'}(\theta, \phi) \, d\Omega \right|^2 = \frac{J+1}{2J+3} D_0^2. \tag{7.186}$$

Using (7.186) in (5.101) we find that the Einstein coefficient for spontaneous emission from level $J + 1$ to J is

$$A_{J+1,J} = \frac{4\omega^3}{3\hbar c^3} |D_{J+1,J}|^2 = \frac{4(J+1)}{3(2J+3)} \frac{\omega^3}{\hbar c^3} D_0^2. \tag{7.187}$$

The frequency of an allowed rotational transition $(J + 1) \to J$ given by (7.164) is

$$\omega_{J+1,J} = \frac{E_{J+1} - E_J}{\hbar} = 2(J+1)\frac{B}{\hbar} = (J+1)\frac{\hbar}{\mu R_0^2}, \tag{7.188}$$

which shows that the frequencies of allowed rotational transitions increase linearly with the rotational quantum number J.

Vibrational-Rotational Transitions

Now consider transitions in which both the vibrational and rotational quantum numbers may change, but the electronic state remains unchanged. Then the dipole matrix element (7.179) in the harmonic oscillator approximation is

$$\langle \Psi_f | \hat{\boldsymbol{H}}_\text{D} | \Psi_i \rangle = -|\mathbf{E}_\omega| \int_0^{2\pi} d\phi \int_0^{\pi} d\theta \; \sin\theta \cos\theta \; Y_{Jm}^*(\theta, \phi) \langle v | D | v' \rangle \, Y_{J'm'}(\theta, \phi), \tag{7.189}$$

where the dipole matrix element for transitions between levels with vibrational quantum numbers v' and v is given by

$$\langle v | D | v' \rangle \equiv -n_v n_{v'} \int_{-\infty}^{\infty} H_v(x/x_0) \, D(R) \, H_{v'}(x/x_0) \, e^{-(x/x_0)^2} \, dx. \tag{7.190}$$

Here $x \equiv R - R_0$; $x_0 \equiv (\hbar/\mu\omega_0)^{\frac{1}{2}}$ is the harmonic oscillator length scale, H_v is the Hermite polynomial of order v, and n_v is its associated normalization constant.

As in (7.165), expand $D(R)$ in a Taylor series about R_0:

$$D(R) = D_0 + \left(\frac{dD}{dR}\right)_0 x + \cdots . \tag{7.191}$$

Hermite polynomials of different order are orthogonal; hence the term D_0 gives a nonzero result in (7.190) only if $(v' \equiv v)$; these correspond to pure rotational transitions. For vibrational transitions $v' \neq v$. Take the linear term in (7.191) as a perturbation. Then, using the recursion relation

$$H_v(u) = uH_{v-1}(u) + \tfrac{1}{2}H_{v+1}(u) \tag{7.192}$$

for Hermite polynomials, and writing $u \equiv x/x_0$, we find that the dipole matrix element is

$$\langle v| D |v'\rangle = x_0^2 n_v n_{v'} \left(\frac{dD}{dR}\right)_0 \int_{-\infty}^{\infty} \left[uH_{v-1}(u) + \tfrac{1}{2}H_{v+1}(u)\right] H_{v'}(u) e^{-u^2} du. \tag{7.193}$$

Equation (7.193) shows that $\langle v| D |v'\rangle$ is nonzero only if $v' = v \pm 1$. Thus the selection rules for vibrational transitions are

$$\Delta v = \pm 1 \qquad \text{and} \qquad (dD/dR)_0 \neq 0. \tag{7.194}$$

This selection rule applies to the linear term in (7.191). Such transitions define the *fundamental mode*. The higher-order terms in the expansion (7.191), proportional to x^2, x^3, etc., lead respectively to transitions with $\Delta v = \pm 2$, which are *first overtone* transitions, $\Delta v = \pm 3$, *second overtone* transitions, etc. The ratio of two successive terms in the expansion (7.191) scales as $(x/R) \sim x_0/R_0$, so the ratio of transition probabilities (given by the square of the matrix elements) goes as $x_0^2/R_0^2 = \hbar/(\mu\omega_0 R_0^2) \propto (m_e/M)^{1/2}$. Thus the strength of transitions in higher overtones rapidly decreases with order of the overtone.

In the harmonic oscillator approximation, the energy of level v is $E_v = (v + \frac{1}{2})\hbar\omega_0$. Thus all vibrational transitions in the fundamental mode with $\Delta v = \pm 1$ have energy $\hbar\omega_0$. Vibrational transitions in the first overtone have $\Delta v = \pm 2$ and energy $2\hbar\omega_0$, and so on for higher overtones. In reality, this uniform spacing is not quite accurate, first, because of departures from the harmonic oscillator approximation and, second, because vibrational transitions typically are also accompanied by changes in the rotational states. Adopting both the harmonic-oscillator and rigid-rotator approximations, the energy of a combined vibration-rotation level is

$$E_{v,J} = (v + \tfrac{1}{2})\hbar\omega_0 + J(J+1)B. \tag{7.195}$$

At first sight it may seem that the selection rules derived for rotational transitions, namely, $\Delta J = \pm 1$, $\Delta m = 0$, apply for vibrational-rotational transitions as well. However, it turns out that this is true only for transitions within the electronic state with $\Lambda = 0$. For $\Lambda \neq 0$, there is an additional possibility for a dipole-allowed transition, namely, that with $\Delta J = 0$. This is because a rotational state with $\Lambda > 0$ corresponds to two states (which is called "Λ doubling"), with an angular momentum parallel or antiparallel to the internuclear axis **A**. These states have opposite parities. So the selection rule $\Delta J = \pm 1$ that applies for transitions within the same electronic state because a parity change is required in a dipole-allowed transition can be extended to include also $\Delta J = 0$ if $\Lambda \neq 0$. That is, dipole-allowed vibrational transitions can occur with $\Delta J = 0$ without violating the requirement that parity must change in a transition.

Among vibrational levels, three types of transitions occur. Transitions with $v' = v \pm 1$ and $J' = J - 1$ have frequencies

$$\omega = \omega_0 - 2JB. \tag{7.196}$$

These transitions form the *P-branch* of the vibration-rotation complex. Transitions with $v' = v \pm 1$ and $J' = J + 1$ have frequencies

$$\omega = \omega_0 + 2(J + 1)B/\hbar. \tag{7.197}$$

These transitions form the *R-branch* of the complex. For transitions within an $\Lambda \neq 0$ electronic state, there is another branch, called the *Q-branch*, which corresponds to $\Delta J = 0$.

The resulting spectrum is a *band* of lines equally spaced around the frequency ω_0 of the vibrational transition in the fundamental mode. There are analogous weaker bands for the second, etc., overtones around frequencies $2\omega_0$, etc. In reality, the spacing of rotational transitions is not precisely equidistant because of departures from the rigid-rotator approximation. A detailed discussion of vibration-rotation spectral features is given in [485, 486].

Electronic-Vibrational-Rotational Transitions

The selection rules [485, 486] for dipole-allowed transitions of a diatomic molecule are

$$\Delta\Lambda = 0, \pm 1 \qquad \Delta S = 0. \tag{7.198}$$

Like atoms, molecules possess different excited states, analogous to atomic states with different principal quantum numbers n. For H_2, the ground electronic state is labeled X, and the excited states are labeled alphabetically, $B, C, D, \ldots$. Transitions between the ground electronic states with different vibrational and rotational levels, and the first and second excited electronic states, B and C, produce called the *Lyman* and *Werner bands*, respectively. Because the first excited electronic states of H_2 are $\gtrsim 10$ eV above the ground state, these transitions occur in the far ultraviolet, $\lambda = 900$–1100 Å. A dramatic demonstration of their importance are the numerous H_2 interstellar lines in the spectra obtained from the *Far Ultraviolet Spectroscopic Explorer* (FUSE) spectrograph; see, e.g., [1033].

From the discussion above it is clear that the appearance of a molecular spectrum is quite different from that of an atomic spectrum. For a transition between electronic states $n' \to n$, there are many possible transitions with different vibrational quantum numbers, and for each such possibility $v' \to v$, there is still finer rotational structure with all possible $\Delta J = 0, \pm 1$ combinations. Therefore, instead of a single line, or a multiplet composed of a few lines, which we have in atomic spectroscopy, molecular spectra are typically composed of bands containing a large number of vibrational-rotational components.

Within a band, the individual lines (in the three branches P, Q, and R) are spaced almost equidistantly because of the simple relations between energies of individual rotational and vibrational levels. In reality there are departures from the harmonic oscillator and the rigid-rotator approximations on which the equidistant spacing depends. Hence the spacing is not strictly equidistant, but it is highly regular.

Transient Dipoles and Collision-Induced Absorption (CIA)

So far, we have dealt with stable molecules. The dipole moments responsible for photon absorption and emission are set by intra-atomic or intra-molecular dynamics. We ignored interactions (collisions) with other particles during the time a radiative transition occurs and derived transition probabilities that are independent of such interactions.

In chapter 8 we show that relatively weak collisions can perturb the energies of the upper and lower states in a transition, which leads to pressure broadening of spectral lines. As noted above, close encounters of an atom or a molecule with another atom or molecule can also give rise to a transient or *interaction-induced dipole moment*. The resulting transient system can interact with radiation. In this case, collisions with neighboring particles are not mere perturbations of an electric dipole transition, but are the *source* of the dipole (CIA). These close encounters must be numerous enough to produce observable effects; therefore, this phenomenon is more pronounced at higher densities. An excellent textbook on this topic is [353], to which an interested reader is referred for a detailed theoretical and experimental analysis of the process.

The most important example is H_2, which is by far the most abundant molecule, but which does not posses a permanent dipole moment and thus cannot contribute directly to the opacity at optical and infrared wavelengths, where most of the radiation of cool stars is transported. However, a transient dipole moment can be produced during a close encounter of an H_2 molecule with another H_2 molecule (or with a He atom, the second most abundant element in the cool atmospheres). Then H_2 is able to contribute significantly to the opacity in the infrared. The effects are largest in low-temperature (so H_2 is abundant), high-density (so the collision frequency is large) atmospheres that are very deficient in elements with $Z \geq 6$ (so opacity from molecular species such as H_2O, CH, C_2, CN, CO, and TiO is small). Results from model atmospheres [137] with $T_{\rm eff} = 2800$ K, $\log g = 4.0$, and $Z = 10^{-3} Z_\odot$ that include (curve with highest peak) and omit CIA opacity from $H_2 + H_2$ pairs are shown in figure 7.9. Because CIA opacity is largest at long wavelengths, the emergent flux is shifted to shorter wavelengths.

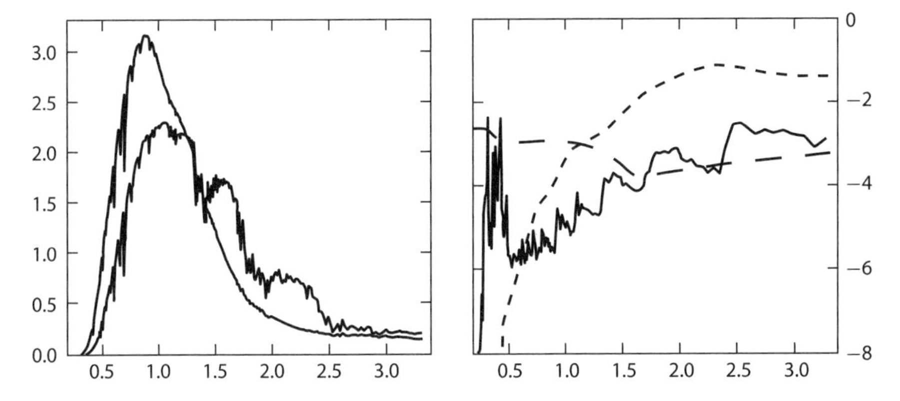

Figure 7.9 *Abscissa:* Wavelength in microns. *Ordinate:* Left panel: Emergent flux (relative units). Right panel: log opacity [cm^2/gm] at depth of continuum formation. *Dotted curve:* Opacity from CIA. *Dashed curve:* Sum of opacities from all other continuum sources. *Solid curve:* Molecular opacity. Adapted from [137].

A different, but related, type of opacity is the *quasi-molecular satellites* of hydrogen Lyman lines. This mechanism produces observable features in the spectra of medium cool ($T_{\rm eff} \approx$ 10,000 to 20,000 K) pure-hydrogen (DA) white dwarfs; see, e.g., [16, 17, 465, 466, 626]. It is also important in the spectra of very metal-poor A-type stars. The features again result from a dipole moment produced in a close encounter of a neutral hydrogen atom with a proton, which creates a transient H_2^+ molecular hydrogen ion (see below), or with another hydrogen atom to form a transient H_2 molecule.

The quasi-molecular satellites are transitions between electronic states of these transient molecules. This mechanism cannot be viewed simply as a broadening of a Lyman line, because it can lead to discrete features quite distant from the line. Calculations and observations both show that the two most important features are at 1400 and 1600 Å, which are produced by collisions of H with H^+ and H with H, respectively. See figure 7.10.

Astrophysically Interesting Molecules

One may look at the importance of a molecule from two different points of view. First, molecules can be reservoirs where atoms of a chemical element are sequestered and may therefore be important for the equation of state of an atmosphere. Second, molecules can be sources of opacity. It might seem that the most abundant molecules would be important sources of opacity, but this is not always true. As mentioned above, H_2, usually the dominant molecule in a cool atmosphere, produces very little opacity at visible or infrared wavelengths; its strong absorption features lie in the ultraviolet, where there is little radiation in a cool atmosphere. On the other hand, molecules like titanium oxide, TiO, which are essentially negligible in the overall chemical balance, can be very important sources of opacity in cool stars despite their low abundances.

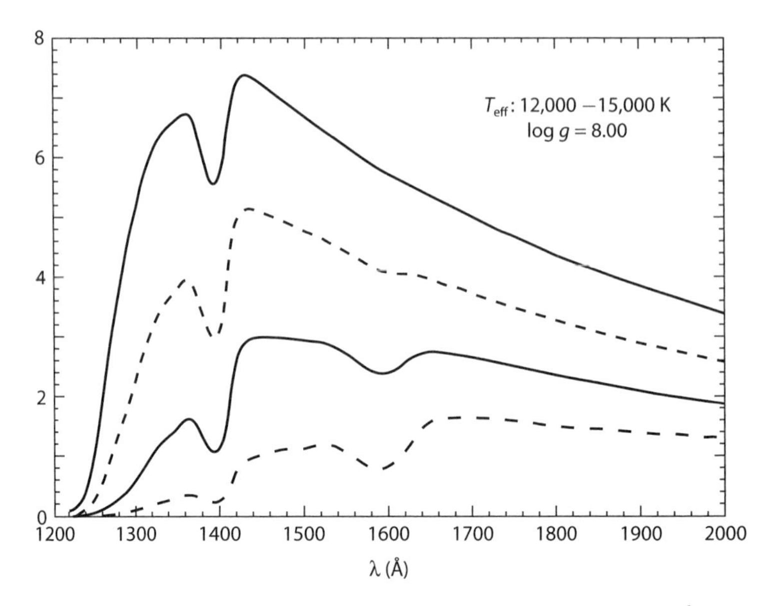

Figure 7.10 *Ordinate:* Emergent flux (relative units). *Abscissa:* Wavelength (Å). Curves are for 12,000 K (lowest) to 15,000 K. Adapted from [16].

When carbon is bound in a molecule, it is sequestered almost exclusively in CO at temperatures $\gtrsim$1500 K (depending on pressure). At cooler temperatures it is mainly in CH_4 (methane). Similarly, at higher temperatures nitrogen is sequestered in N_2, but it is optically inactive because it is a homonuclear molecule. At lower temperatures it is mainly in NH_3 (ammonia), which has strong absorption features. The dividing line between the two states is around 1000 K. Oxygen is tied up in CO and H_2O at higher temperatures and is almost entirely in H_2O at cooler temperatures. Together with H_2, these six molecules are dominant species for chemical balance. At solar atmospheric temperatures, other important molecules are CN, CH, C_2, OH, and MgH. Non-homonuclear or polyatomic molecules that have appreciable abundances are often an important source of opacity. All the molecules listed above, except H_2 and N_2, are important opacity sources under conditions where they are abundant. For example, water has a very rich line spectrum (over 10^9 individual lines) because it is a triatomic molecule and has many more modes of motion than diatomic molecules. Some diatomic molecules such as TiO, or the even less abundant vanadium oxide (VO), are important opacity sources because they possess numerous strong transitions. These molecules are composed of two nuclei with very different masses; their vibrational motions are much more vigorous than those of molecules with nuclei of equal mass.

Molecular Ions

The molecular hydrogen ion H_2^+ is an important source of opacity at the highest temperatures (around 7000 K) at which significant numbers of molecules exist in stellar atmospheres. It is the simplest molecular ion, composed of two protons and one electron. H_2^+ has bound states and can absorb photons in both photodissociation reactions and free-free transitions:

$$H_2^+ + h\nu \rightarrow H^+ + H \tag{7.199}$$

and

$$H_2^+ + e^-(v) + h\nu \rightarrow H_2^+ + e^-(v'). \tag{7.200}$$

Useful expressions for the absorption cross sections of these processes are given in [94, 158]. The number density of H_2^+ is proportional to $n(H)n_p$, so H_2^+ contributes significantly to the total opacity only in the temperature/density domain where H is about half ionized. This range is characteristic of A-stars, where H_2^+ makes about a 10% contribution to the total opacity at visible wavelengths (the H_2^+ absorption peak at $\lambda 1100\,\text{Å}$ is swamped by the hydrogen Balmer continuum).

Another molecular ion of importance is H_2^-, which like H^- has an extra electron, three in all. It is the most important molecular contributor to the continuous infrared opacity in the atmospheres of late-type stars (e.g., M-stars). H_2^- has no stable bound states, so its opacity is entirely free-free, caused by the transient dipole moment produced in a collision of an electron with an H_2 molecule; see, e.g., [102, 104, 588, 1032]. The H_2^- continuum tends to fill in the H^- opacity minimum at $1.6\,\mu$.

Chapter Eight

Spectral Line Broadening

The analysis of spectral lines in a star's spectrum is our most effective diagnostic tool for determining the conditions in, and the composition and structure of, a stellar atmosphere. Quantitative analysis of spectral lines requires knowledge of (1) the number of atoms that can absorb radiation at each frequency in a line, (2) the wavelength distribution of the line's opacity, i.e., its *absorption profile*, and (3) how that profile depends on the temperature and density in the plasma.[1]

The absorption profiles of spectral lines are broadened in frequency by *radiation damping*, set by the finite lifetimes of both the upper and lower levels of the transition; *Doppler shifts* resulting from the random thermal motions of the absorbing atoms along the line of sight; and *pressure broadening* induced by perturbations of the radiating atom by other particles in the plasma. A number of excellent books about spectral line broadening are now available, e.g., [95, 400, 403, 404].

8.1 NATURAL DAMPING PROFILE

In chapter 5 we found that the loss of energy into the electromagnetic field radiated by a charged classical oscillator leads to a decay of its amplitude in time. The Fourier transform of this time-dependent radiation field yields a frequency spectrum having a Lorentz profile with a damping width set by its rate of decay. The quantized atomic levels of real atoms have finite lifetimes because of photon emission and absorption in lines that couple a given level to other levels in the transition array. Hence in accordance with the Heisenberg Uncertainty Principle they are broadened in frequency by *radiation damping*, which acts in even a solitary, isolated atom.

Some Mathematical Tools

Energy Spectrum

We saw in equation (5.124) that the energy spectrum radiated by an accelerated charge per unit solid angle and unit circular frequency interval is

$$\mathcal{F}(\omega) \equiv \frac{c}{2\pi} |\mathbf{E}(\omega)|^2 . \tag{8.1}$$

[1] In this chapter the term "atom" may sometimes be used loosely to mean an atom, ion, or molecule, and the term "absorb" to mean any interaction that removes a photon from a specific ray of radiation, whether it is "absorbed" or "scattered" in the sense used in chapter 5.

Power Spectrum

For some applications one does not use the energy spectrum, but instead the energy delivered per unit time, i.e., the *power spectrum*, defined as

$$\phi(\omega) \equiv \lim_{T\to\infty} \frac{1}{2\pi T} \left| \int_{-T/2}^{T/2} f(t) e^{-i\omega t} dt \right|^2 . \tag{8.2}$$

Note that for oscillations of finite duration or with, say, an exponential decay, the power spectrum is zero because when averaged over an infinite time interval, the finite energy emitted yields zero power. In such a case we use the energy spectrum itself, postulating that we observe an ensemble of oscillators created at a constant rate with random phases, which would produce finite power having a frequency distribution proportional to the energy spectrum of an individual oscillator.

Autocorrelation Function

In some cases neither the energy spectrum nor the power spectrum can be computed directly using (8.1) and (8.2). In such cases we can use the *autocorrelation function*:

$$\Phi(s) \equiv \lim_{T\to\infty} \frac{1}{T} \int_{-T/2}^{T/2} f^*(t) f(t+s)\, dt, \tag{8.3}$$

from which the power spectrum is obtained as

$$\phi(\omega) \equiv \frac{1}{2\pi} \int_{-\infty}^{\infty} \Phi(s) e^{-i\omega s}\, ds. \tag{8.4}$$

This result may be verified by using a limiting procedure in the integration over s. The autocorrelation function is useful for calculating power spectra from a radiating atom perturbed by collisions.

Quantum Mechanical Oscillator

The first quantum mechanical treatments of natural broadening were made by V. Weisskopf and E. Wigner in the 1930s [1140–1145]. They showed that the probability of finding an atom initially in an excited state j making transitions to the ground state 1 at time t is

$$P_j(t) = \psi_j^*(\mathbf{r}) \psi_j(\mathbf{r}) e^{-\Gamma_j t}, \tag{8.5}$$

where $\Gamma_j = A_{j1}$, the spontaneous emission rate from level j to the ground state. The time development of the wave function for this state is

$$\psi_j(\mathbf{r}, t) e^{-\Gamma_j t/2} = u_j(\mathbf{r}) e^{-(iE_j t/h)} e^{-\Gamma_j t/2} = u_j(\mathbf{r}) e^{-(i\omega_j + \Gamma_j/2)t}. \tag{8.6}$$

Consistent with the uncertainty principle, we see that the decaying state (with a characteristic lifetime $\Delta t_j \approx 2/\Gamma_j$) no longer has a perfectly defined energy E_j, but is a superposition of states with energies spread around E_j over a width $\Delta E_j \sim h/\Delta t_j$. From the fundamental reciprocity relations of quantum mechanics, the amplitude of the energy distribution is given by the Fourier transform of the time dependence of the wave function, and the probability distribution of energy states is given by the square of this amplitude. By a calculation similar to that leading to equation (6.19) we find this transition has a profile

$$\phi_{1j}(\omega) = \frac{\Gamma_j/2\pi}{[(\omega - \omega_{1j})^2 + (\Gamma_j/2)^2]}. \tag{8.7}$$

We see from (8.5) that Γ_j is to be interpreted as the reciprocal of the mean lifetime of the upper state. If there are several permitted transitions out of the upper state j, then

$$\Gamma_j = \sum_{i<j} A_{ji}. \tag{8.8}$$

This expression accounts only for *spontaneous* transitions $j \to i$ and is strictly correct only in the absence of a radiation field. But the radiation field in a stellar atmosphere can be intense, so the initial state can also be depopulated by induced emissions to lower bound states and absorptions to higher bound states. In this case a better estimate of the width of state j would be

$$\Gamma_j = \sum_{i<j} \left[A_{ji} + B_{ji} I(\nu_{ij})\right] + \sum_{k>j} B_{jk} I(\nu_{jk}). \tag{8.9}$$

We can make a rough estimate of the importance of these additional processes by assuming that the radiation field at each frequency is approximately Planckian. Then using the relations among the Einstein coefficients given in § 5.1, we find

$$\Gamma_j \approx \sum_{i<j} \frac{A_{ji} e^{-h\nu_{ij}/kT}}{1 - e^{-h\nu_{ij}/kT}} + \sum_{k>j} \frac{(n_k/n_j)^* A_{kj}}{1 - e^{-h\nu_{jk}/kT}}, \tag{8.10}$$

which suggests that the additional terms in (8.9) may be of importance if there are strong transitions out of or into level j for which $h\nu \ll kT$.

In deriving (8.7) and (8.8) we assumed that the upper level is broadened, and ground level has an "infinite" lifetime, and hence is sharp, which is appropriate for principal series lines in an atomic spectrum. But subordinate lines are transitions between two broadened excited levels, and the lower level has a width Γ_ℓ. Hence in general a line transition $\ell \leftrightarrow u$ reflects the width of both levels.

Assume each level ($s = \ell$ or u) has a Lorentz distribution of sublevels around its nominal energy, with $\gamma_s \equiv \Gamma_s/2$. Let $x_s \equiv (E_s - E_{s0})/h$ be the frequency displacement of a sublevel from the nominal energy. Further, assume that the probability of ending in a particular sublevel x_u of the upper level is independent of the sublevel

x_ℓ of the lower level from which the transition originated. Then the joint probability of starting from x_ℓ and ending in x_u is

$$p(x_\ell, x_u) = \frac{\gamma_\ell \gamma_u}{\pi^2[(x_\ell^2 + \gamma_\ell^2)(x_u^2 + \gamma_u^2)]}. \tag{8.11}$$

If we restrict attention to transitions producing radiation of a particular frequency ν, then $\nu = \nu_0 + x_u - x_\ell$, where ν_0 is the line-center frequency. Or, if writing $x_0 \equiv \nu_0 - \nu$, then $x_u \equiv x_\ell - x_0$. The total intensity at ν is obtained by summing over all upper states x_u:

$$\phi_{\ell u}(\omega) = \int_{-\infty}^{\infty} p(x_\ell, x_\ell - x_0)\, dx_\ell = \frac{\gamma_\ell \gamma_u}{\pi^2} \int_{-\infty}^{\infty} \frac{dx_\ell}{(x_\ell^2 + \gamma_\ell^2)[(x_\ell - x_0)^2 + \gamma_u^2]}. \tag{8.12}$$

This integral can be evaluated by contour integration using the residue theorem, noting the poles at $z = \pm i\gamma_\ell$ and $z = x_0 \pm i\gamma_u$. We again find (8.7), now with

$$\Gamma_j \equiv \gamma_\ell + \gamma_u, \tag{8.13}$$

i.e., the profile is Lorentzian with a half-intensity width equal to the sum of the half-intensity widths of both levels; see [927, p. 290], [1178, p. 114]. The Lorentz profiles calculated above are emission profiles. By the principle of detailed balancing, the absorption profile has the same form.

To convert to an absorption cross section per atom, recall from equation (5.102) that in terms of oscillator strengths,

$$\int_{-\infty}^{\infty} \alpha_{ij}(\nu)\, d\nu = \frac{\pi e^2}{mc} f_{ij}. \tag{8.14}$$

Using the profile function (8.7) with Γ_j given by (8.13) and converting to ordinary frequency units, we find that the absorption cross section is

$$\alpha_{ij}(\nu) = \frac{\pi e^2}{mc} f_{ij} \frac{\Gamma_{ij}/4\pi^2}{[(\nu - \nu_0)^2 + (\Gamma_{ij}/4\pi)^2]}. \tag{8.15}$$

Radiation damping is primarily of importance for strong lines in very low density media, e.g., Lyman α in the interstellar medium. However, in most cases of interest in stellar atmospheres a line is formed in regions where the density of perturbing particles is large enough that pressure broadening is significant or dominant.

8.2 DOPPLER BROADENING: VOIGT FUNCTION

The lines seen in a stellar spectrum or laboratory plasma are produced by the absorption or emission of all atoms along the line of sight into the material. Each atom has a velocity component ξ (measured positive away from the observer) along that path, so as seen by a fixed observer, the line profile of that atom is Doppler-shifted in frequency. If the material has a kinetic temperature T, and the velocity distribution of the atoms is Maxwellian, the probability that an atom has a velocity in the range $(\xi, \xi + d\xi)$ is

$$W(\xi)\, d\xi = (1/\sqrt{\pi}) \exp(-\xi^2/\xi_0^2)(d\xi/\xi_0). \tag{8.16}$$

Here $\xi_0 \equiv (2kT/m)^{1/2} = 12.85\,(T/10^4 A)^{1/2}$ km/sec and A is the atom's atomic weight.

Radiation observed at frequency ν from an atom moving with velocity ξ was absorbed from or emitted at frequency $\nu[1-(\xi/c)]$ in the atom's rest frame. The observed line profile for a transition $i \to j$ is the convolution of intrinsic profile α_{ij} of those atoms with their velocity distribution. Hence the material's absorption coefficient at frequency ν is

$$\alpha_{ij}(\nu) = \int_{-\infty}^{\infty} \alpha_{ij}[\nu(1-\xi/c)] W(\xi)\, d\xi. \tag{8.17}$$

Equation (8.17) can be applied to any absorption profile to account for the effects of Doppler broadening.

Consider the case of a Lorentzian intrinsic profile. Let the line center be at frequency ν_0 in the atom's frame. The *Doppler width* of the line is

$$\Delta\nu_D \equiv \xi_0 \nu_0 / c. \tag{8.18}$$

Define

$$a \equiv \Gamma/4\pi\,\Delta\nu_D, \tag{8.19}$$

$$v \equiv (\nu - \nu_0)/\Delta\nu_D, \tag{8.20}$$

and

$$y \equiv \xi/\xi_0. \tag{8.21}$$

Then using (8.16) in (8.17) we find

$$\alpha_{ij}(\nu) = \frac{\sqrt{\pi} e^2}{mc\Delta\nu_D} f_{ij} H(a, v), \tag{8.22}$$

where

$$H(a, v) \equiv \frac{a}{\pi} \int_{-\infty}^{\infty} \frac{e^{-y^2} dy}{(v-y)^2 + a^2} \tag{8.23}$$

is known as the *Voigt function*. One can show that $\int_{-\infty}^{\infty} H(a, v) dv = \sqrt{\pi}$; thus in practice it is often preferable to use the normalized profile function $U(a, v) \equiv H(a, v)/\sqrt{\pi}$.

Extensive tabulations of the Voigt function can be found in [335, 443, 500], and especially [545]. Now that fast computers are readily available, several algorithms for calculating it on line exist; see, e.g., [155, 254, 543, 625, 651, 999, 1089, 1147]. A useful schematic representation of the Voigt function is

$$H(a, v) \approx e^{-v^2} + a/(\pi^{1/2} v^2). \tag{8.24}$$

The first term applies in the line core, $v \leq v^*$, and the second in the line wind, $v \geq v^*$, with the quantity v^* being chosen such that the two terms are equal. The line core is clearly dominated by Doppler broadening, while the line wings are dominated by the damping profile.

8.3 SEMICLASSICAL IMPACT THEORY

Natural broadening of spectral lines accounts for the finite widths of the upper and lower levels of the transition set by their radiative decay lifetimes. If the radiating atom is embedded in a plasma, its profile is also affected by pressure broadening resulting from collisions with neutral or charged particles. Classically, pressure broadening is described in terms of two limiting approximate theories.

In classical *impact theory* [1141], the radiating atom is considered to be an oscillator that suffers an interruption of the emitted wavetrain by a collision with a perturber, which generates an instantaneous phase shift or induces a transition in the line or to another atomic level. In this picture the radiating oscillator is forced to "start" and "stop" in finite time intervals, leading to both a frequency spread in the radiated wavetrain and a shift of the line from its unperturbed wavelength. An alternative view is the *statistical theory* in which the radiating atom is viewed as embedded in plasma of slowly moving charged perturbers. The field at the position of the atom fluctuates around some average value, so the position of the spectral line is shifted slightly. The intensity of the emitted radiation as a function of wavelength is taken to be proportional to the probability that a perturbation of strength is adequate to shift line center by a given amount. The classical theories of line broadening are limited by their inability to account for the actual structure of the radiating atoms and for the rate of transitions in the atom triggered by the perturbing field. Both of these limitations are overcome in the quantum theory of pressure broadening, which yields results in very good agreement with experiment.

Weisskopf Approximation

In classical impact theory [1141], one assumes that a radiating atom is unperturbed between collisions. Suppose the time between collisions is T, and that in that interval the radiator emits a monochromatic wavetrain $f(t) = \exp(i\omega_0 t)$. Its Fourier transform is

$$F(\omega, T) = \int_0^T e^{i(\omega-\omega_0)t}\, dt = \frac{\exp[i(\omega - \omega_0)T - 1]}{i(\omega - \omega_0)}. \tag{8.25}$$

Using this expression in (8.1), we obtain the energy spectrum $E(\omega, T)$ of this wavetrain.

In a rarefied medium such as a stellar atmosphere, the collisions occur randomly. Because the particle density is low, the collisions can be considered to be distinct, and if the mean time between collisions is τ_c, the probability of two successive collisions in the time interval $(T, T + dT)$ is

$$W(T)dT = e^{-T/\tau} dT/\tau. \tag{8.26}$$

Averaging over all collision intervals gives the mean energy spectrum

$$E(\omega) \equiv \langle E(\omega, T) \rangle_T = \frac{1}{2\pi} \int_0^\infty F^*(\omega, T) F(\omega, T) W(T)\, dT. \tag{8.27}$$

Evaluation and normalization of (8.27) yields

$$E(\omega) = \frac{1/\pi\tau_c}{(\omega - \omega_0)^2 + (1/\tau_c)^2}. \tag{8.28}$$

Thus in the Weisskopf approximation we again find a Lorentz profile (the result of assuming the collisions are distinct) with $\Gamma \equiv 2/\tau_c$. As was done for a radiation-damped oscillator, we assume that the profile of an ensemble of continuously created, randomly phased oscillators is proportional to the energy spectrum of a single oscillator [averaged over all times, as in (8.28)]. If both radiation and collision damping occur, with widths Γ_R and Γ_C, respectively, and are assumed to be completely uncorrelated, then the profile is again Lorentzian with a total width $\Gamma = \Gamma_R + \Gamma_C$. Doppler broadening is accounted for as in §10.2, which yields a Voigt profile with damping width Γ.

To complete the theory we need an estimate of τ_c. If the radiating atoms and perturbing particles have atomic weights A_r and A_c, respectively, and both have a Maxwellian velocity distribution at temperature T, then their average relative velocity is

$$\overline{v} = \langle v^2 \rangle^{1/2} = \left[\left(8kT/\pi m_H \right) \left(A_r^{-1} + A_c^{-1} \right) \right]^{1/2}. \tag{8.29}$$

Assuming that the effective impact parameter of the collisions responsible for the broadening is ρ_0, we have

$$\tau_c^{-1} = \pi \rho_0^2 N \overline{v} \tag{8.30}$$

and

$$\Gamma_C = 2\pi \rho_0^2 N \overline{v}, \tag{8.31}$$

where N is the perturber density. We must now determine ρ_0.

Following Weisskopf [1141], we assume that (i) the perturber is a *classical particle*, (ii) the perturber moves with constant velocity past the atom on a *straight-line path* with velocity $v = \overline{v}$ and *impact parameter* ρ, (iii) the *interaction* between atom and perturber is described approximately by

$$\Delta\omega = C_p / r^p, \tag{8.32}$$

where $r(t) = (\rho^2 + v^2t^2)^{\frac{1}{2}}$ and setting $t = 0$ at the point of closest approach, and finally (iv) *no transitions* in the atom are produced by the action of the perturber. The validity of these assumptions will be examined later. The form of the interaction in equation (8.32) is only approximate but holds over a fairly wide range of distances. The value of the exponent p depends upon the nature of the interaction. The values of astrophysical interest and the interaction they represent are as follows:

$p = 2$, *linear Stark effect* (hydrogen + charged particle);
$p = 3$, *resonance broadening* (atom A + atom A);
$p = 4$, *quadratic Stark effect* (non-hydrogenic atom + charged particle);
$p = 6$, *van der Waals force* (atom A + atom B).

The interaction constant C_p must be calculated from quantum theory or measured by experiment.

The *phase shift* induced by the perturbation is

$$\eta(t) = \int_{-\infty}^{t} \Delta\omega(t')dt' = C_P \int_{-\infty}^{t} (\rho^2 + v^2t'^2)^{-p/2} dt'. \tag{8.33}$$

The *total phase shift* introduced by the perturbation is $\eta(\rho) \equiv \eta(t = \infty)$ is found directly to be

$$\eta(\rho) = C_P \psi_P / \overline{v} \rho^{p-1}, \tag{8.34}$$

where

$$\psi_P = \pi^{\frac{1}{2}} \Gamma[(p-1)/2] / \Gamma(p/2). \tag{8.35}$$

Here Γ denotes the usual gamma function; for $p = (2, 3, 4, 6)$ one finds $\psi_p = (\pi, 2, \pi/2, 3\pi/8)$.

We now assume that only those collisions that produce a total phase shift greater than some critical value of η_0 are effective in broadening the line. The effective impact parameter for such collisions is thus

$$\rho_0 = (C_p \psi_p / \eta_0 \overline{v})^{1/(p-1)}, \tag{8.36}$$

and the corresponding value for the damping constant is

$$\Gamma = 2\pi N \overline{v} (C_p \psi_p / \eta_0 \overline{v})^{2/(p-1)}. \tag{8.37}$$

Weisskopf arbitrarily adopted $\eta_0 = 1$ as the critical phase shift; with this choice we obtain the *Weisskopf radius* ρ_W from equation (8.36) and the Weisskopf damping parameter Γ_w from equation (8.37).

If C_p is given, the theory described above yields a definite value for Γ, and the results are found to be of the right order of magnitude. Yet there remain serious defects in it: (i) The choice $\eta_0 = 1$ is arbitrary, and there is no means of determining a priori the correct value of η_0 to be used. (ii) The theory does not account for the collisions that produce small phase shifts, even though a number of such collisions increases as ρ^2. (iii) The theory fails to predict the existence of a *line shift*; as will be shown below, this failure arises from the omission of weak collisions, as mentioned in (ii).

Lindholm Approximation

A significant improvement in the classical impact theory was made by Lindholm [697, 698] and Foley [339]. In this approach we consider the radiator to have an *instantaneous frequency* $\omega(t)$, which, because of perturbations, differs from the nominal frequency ω_0 by an amount $\Delta\omega(t)$. Then we write

$$f(t) = \exp\left[i\omega_0 t + i\int_{-\infty}^{t} \Delta\omega(t')\,dt'\right] \equiv e^{i[\omega t+\eta(t)]}, \tag{8.38}$$

where $\eta(t)$ is the *instantaneous phase* of the oscillator. To obtain the line profile, we calculate the autocorrelation function $\Phi(s)$ defined by equation (8.3). Let $\phi(s)$ be defined by $\phi(s) \equiv e^{-i\omega_0 s}\Phi(s)$, which eliminates the unperturbed oscillations. Then from equation (8.3),

$$\begin{aligned}\phi(s) &= \lim_{T\to\infty} T^{-1}\int_{-T/2}^{T/2} e^{-i\omega_0 s}\, e^{-i[\omega_0 t+\eta(t)]}\, e^{i[\omega_0(t+s)+\eta(t+s)]}\,dt \\ &= \lim_{T\to\infty} T^{-1}\int_{-T/2}^{T/2} e^{i[\eta(t+s)-\eta(t)]}\,dt.\end{aligned} \tag{8.39}$$

The quantity $\phi(s)$ is then interpreted as the time-averaged value of the *additional phase shift* occurring in the time interval s. For brevity, write

$$\eta(t, s) \equiv \eta(t+s) - \eta(t). \tag{8.40}$$

Then

$$\phi(s) = \langle e^{i\eta(t,s)} \rangle_T. \tag{8.41}$$

Further, writing $d\phi(s) = \phi(s+ds) - \phi(s)$, we have

$$d\phi(s) = \langle e^{i\eta(t,s)}(e^{i\eta'} - 1)\rangle_T, \tag{8.42}$$

where η' denotes the *change in phase* occurring in the time interval ds as a result of collisions that take place in this interval. The phase *change* cannot be correlated with the *current value* of the phase if the collisions occur at random. Thus the average of the product can be replaced by the product of the averages, i.e.,

$$d\phi(s) = \langle e^{i\eta(t,s)}\rangle_T \langle e^{i\eta'} - 1\rangle_T = \phi(s)\langle e^{i\eta'} - 1\rangle_T. \tag{8.43}$$

If we can calculate the average of $e^{\eta'}$, we obtain a differential equation for $\phi(s)$.

By forming an average over a sufficiently long time interval T, the randomly occurring collisions will happen at all values of ρ with an appropriate statistical frequency. We then invoke the *ergodic hypothesis* to replace the average over time by the appropriate sum over impact parameters. The number of impacts that occur in the range $(\rho, \rho + d\rho)$ in time ds is just $(2\pi\rho\, d\rho)N\overline{v}\, ds$; hence

$$\langle e^{i\eta'} - 1\rangle_T \rightarrow \langle e^{i\eta'(\rho)} - 1\rangle_\rho = 2\pi N\overline{v}\, ds \int_0^\infty \left[e^{i\eta'(\rho)} - 1\right] \rho\, d\rho. \tag{8.44}$$

The integral in equation (8.44) has both a real and imaginary part, so we write

$$\langle e^{i\eta'} - 1\rangle_\rho = -N\overline{v}\, ds(\sigma_R - i\sigma_I), \tag{8.45}$$

where

$$\sigma_R \equiv 2\pi \int_0^\infty [1 - \cos\eta(\rho)]\, \rho\, d\rho = 4\pi \int_0^\infty \sin^2[\eta(\rho)/2]\, \rho\, d\rho \tag{8.46}$$

and

$$\sigma_I \equiv 2\pi \int_0^\infty \sin\eta(\rho)\, \rho\, d\rho. \tag{8.47}$$

Combining equations (8.42), (8.43), and (8.44) and solving the resulting differential equation with the initial condition $\phi(0) = 1$, we find

$$\phi(s) = \exp[-N\overline{v}(\sigma_R|s| - i\sigma_I s)]. \tag{8.48}$$

Finally, calculating the intensity from equation (8.4) and normalizing the profile, we obtain

$$I(\omega) = \frac{(N\overline{v}\sigma_R/\pi)}{(\omega - \omega_0 - N\overline{v}\sigma_I)^2 + (N\overline{v}\sigma_R)^2}. \tag{8.49}$$

Thus Lindholm theory yields a Lorentz profile with a damping *width*

$$\Gamma = 2N\overline{v}\sigma_R \tag{8.50}$$

and a *line shift*

$$\Delta\omega_0 = N\overline{v}\sigma_I. \tag{8.51}$$

The prediction of a shift is in agreement with experiment for cases that such shifts are observed. Quantum theory yields profile of the same form as equation (8.49) and gives explicit expressions for Γ and $\Delta\omega_0$ in terms of matrix elements of the perturbing potential and transitions within the atom. As we shall see below, Lindholm theory yields a unique value of $\Gamma/\Delta\omega_0$ for each choice of p; quantum theory shows that this ratio actually varies over a moderate range as T and N vary and is different for each line. The effects of Doppler broadening are taken into account using a Voigt profile with the appropriate damping parameter and shifted from its rest frequency by an amount $\Delta\omega_0$.

The dominant contributions to σ_R and σ_I come from quite different ranges of impact parameter ρ. From equation (8.34) we note that $\eta(\rho) \propto \rho^{-(p-1)}$. Thus for $(\rho/\rho_W) > 1$, the integrand of σ_R rapidly drops to zero [1083, p. 16] and [1102, p. 305], and the dominant contribution to the line broadening comes from (strong) collisions inside the Weisskopf radius—i.e., $(\rho/\rho_w) < 1$. In contrast, for σ_I, the integrand for $(\rho/\rho_w) < 1$ fluctuates in sign and averages to nearly zero. Thus the dominant contribution to the line shift comes from (weak) collisions outside the Weisskopf radius. It is easy to understand physically why the shift arises. The very weak collisions ($\eta \ll 1$, $\rho \gg \rho_W$) are extremely numerous and occur at an essentially constant rate, yielding an average phase shift per unit time

$$\overline{\eta} = 2\pi N \overline{v} \int_{\rho^*}^{\infty} \eta(\rho)\rho \, d\rho, \tag{8.52}$$

where η^* is chosen to assure that $\eta(\rho^*) \ll 1$. But as can be seen from equation (8.38), the rate of change of phase is by definition a change $\Delta\omega$ for the oscillator's frequency.

Specific Cases

Lindholm theory has been most widely applied in astrophysical work for the cases $p=3, 4$, and 6. (For the case $p=2$, sophisticated quantum mechanical approaches have provided much more accurate results.) For these cases the integrals in equations (8.46) and (8.47) can be evaluated explicitly to yield the values listed in table 8.1; see [1083, p. 14] for details. The last line of the table gives the value of η_0 that when inserted into the Weisskopf formula, equation (8.37), gives the Lindholm Γ. As η_0 is always less than unity, it can be seen that the Weisskopf formula always leads to too small a value of Γ.

Resonance broadening, $p=3$, is of importance mainly for collisions of hydrogen atom with one another. As the atmosphere must be hot enough that hydrogen is excited to the $n=2$ state (to produce the observable Balmer lines) but cool enough that it is not dominantly ionized, resonance broadening effects are of interest for solar-type stars. (For resonance broadening of Lyman lines, the atmosphere must also be hot enough, not because of excitation to $n=2$, but because otherwise there

Table 8.1 Results of Lindholm Theory

p	3	4	6
Γ	$2\pi^2 C_3 N$	$11.37\, C_4^{\frac{2}{3}} \overline{v}^{\frac{1}{3}} N$	$8.08\, C_4^{\frac{2}{5}} \overline{v}^{\frac{3}{5}} N$
$\Delta\omega_0$		$9.85\, C_4^{\frac{2}{3}} \overline{v}^{\frac{1}{3}} N$	$2.94\, C_4^{\frac{2}{5}} \overline{v}^{\frac{3}{5}} N$
$\Gamma/\Delta\omega_0$		1.16	2.75
η_0	0.64	0.64	0.61

is essentially no observable flux in the Lyman lines region.) The interaction constant C_3 in equation (8.32) for level n is

$$C_3 = e^2 f_{1n}/(2m\omega_{1n}) \tag{8.53}$$

(see [1075], [144, p. 231]). A quantum mechanical calculation gives a Γ slightly different from the Lindholm value, namely,

$$\Gamma_3 = 16\pi n_1 C_3/3 = 4n_1 e^2 f_{1n}/(3m\nu_{1n}). \tag{8.54}$$

For the hydrogen lines, there is no shift $\Delta\omega_0$ because individual Stark components split symmetrically about line center (see §8.4) and the shift is identically zero. Resonance broadening is most significant for the lowest members of a series where Stark broadening is the smallest. The effects of resonance broadening have been shown to be important in the solar $H\alpha$ line, but negligible for higher series members [208].

Quadratic Stark effect, $p=4$, is important for the broadening of lines of non-hydrogenic atoms and ions by impacts with charged particles (electrons), and is the dominant pressure-broadening mechanism for these lines in the atmospheres of early-type stars. In applications of the classical Lindholm theory the interaction constant C_4 was typically estimated from experimental measurements of line shifts in electric fields or from time-independent perturbation theory for the quadratic Stark effect (see [1083, pp. 319–320], [1102, pp. 326–328] for examples of this procedure). The resulting damping widths are usually much too small, however, because the Lindholm approximation assumes that the collisions are *adiabatic* (i.e., do not cause transitions in the radiating atom); this assumption is frequently poor, and accurate quantum mechanical calculations including non-adiabatic effects (cf.§8.5) yield much larger line widths.

Van der Waals interactions, $p=6$, of non-hydrogenic atoms in collisions with neutral hydrogen (and, to a smaller extent, helium) atoms are the major source of pressure broadening in solar-type and cooler stars. The usual classical treatment accounts for the dipole-dipole interaction term in the potential, and yields [1083, pp. 91–97], [1102, pp. 331–334]

$$\Delta\omega = C_6/r^6 = e^2\alpha a_0^2[\overline{R}_u^2 - \overline{R}_l^2]/(hr^6), \tag{8.55}$$

where α is the polarizability of hydrogen and $\overline{R}^2$ is the mean square radius of the two levels. Quantum mechanical results are sometimes available for $\overline{R}^2$; if not, hydrogenic estimates are used. Using C_6 determined in this manner, Γ can be computed from Lindholm theory. When this is done (e.g., for lines of Fe I), it is found that the predicted values are much too small, by factors of 5 to 30 [664]. Quantum mechanical calculations that again employ only the dipole-dipole term do not lead to large increases in Γ, e.g., [400, p. 96], which points to the breakdown of the dipole-dipole approximation rather than other theoretical problems; see also [496,497,911]. Calculations using the more realistic Lennard-Jones potential [498] lead to significantly larger widths. Another approach is based on the proposal that the dominant interaction leading to the line broadening is between the perturber and the valence electron [910]. One can then use a Smirnov potential [1016] and obtain an expression for the damping parameter. This approach has been used to produce extensive tables [274], giving the parameters α and β in the formula $\Gamma = N\alpha T^{\beta}$, as functions of the effective quantum numbers n^* of the lower and upper levels, for *s-p*, *p-d*, and *d-f* transitions.

Validity Criteria

The classical impact theory is based on four major approximations that are applicable if the following four criteria are met:

(1) An effective impact time τ_s can be defined such that τ_s times the peak value of $\Delta\omega$ for a collision at the effective impact parameter, namely, $C_p \rho_0^{-p}$, yields the total phase shift given by equation (8.34). This gives

$$\tau_s = \psi_p \rho_0 / \overline{v}. \tag{8.56}$$

For impact theory to be valid, we demand that one collision at a time occur, so that $\tau_s < \tau = 1/(N\pi\rho_0^2\overline{v})$. Writing $N = 3/(4\pi r_0^3)$, where r_0 is the mean interparticle distance, we find $\frac{3}{4}\psi_p(\rho_0/r_0)^3 < 1$. Thus impact theory will be valid only when the density of particles is so low that the *Weisskopf radius is small compared to the interparticle distance.*

(2) It is clear that, as $\rho \to \infty$, the effective impact time τ_s becomes larger and larger, and eventually exceeds the mean time between collisions, τ, so that collisions will overlap. Instead of having well-separated collision events, one begins to approach the opposite limit, namely, that a radiator is subjected to essentially *continuous* perturbation, and here we expect statistical theory (§8.4) to begin to be valid. We saw earlier that weak collisions produce the line *shift*, just as would be given by a steady perturbation, but the impact theory fails to treat line *broadening* properly, because it does not consider weak collisions in a logically consistent way.

(3) Impact theory fails for large frequency displacements $\Delta\omega$ from the line center, and statistical theory (see §8.4) becomes valid. In impact theory, it follows from the general properties of Fourier transforms that the characteristic interruption time τ corresponding to the frequency displacement $\Delta\omega$ is $\tau \sim 1/\Delta\omega$. For sufficiently large $\Delta\omega$ we will eventually have $\tau \ll \tau_s$, and impact theory breaks down. These values correspond to large phase shifts (i.e., $\Delta\omega\tau_s \gg 1$); and hence to the strong collisions that occur *inside* the Weisskopf radius. It is difficult to construct a theory that makes the transition from impact to quasi-static theory. A useful conceptualization is to suppose that there is a "boundary" frequency $\Delta\omega_g$ inside of which impact theory holds and beyond which quasi-static theory is valid. To a fair approximation, see [502, 1101],

$\Delta\omega_g \approx \Delta\omega_W$, where $\Delta\omega_W$ denotes the frequency shift produced by a perturber at the Weisskopf radius ρ_W, i.e.,

$$\Delta\omega_W = (\overline{v}^p / C_p \psi_p^p)^{1/(p-1)}. \tag{8.57}$$

Note that $\Delta\omega_W$ corresponds to a phase shift of unity. We shall see in §8.4 that equation (8.57) implies that the broadening of the hydrogen lines by *ions* follows the *quasi-static* theory, while *electron* broadening is given by *impact theory*.

(4) Classical impact theory assumes that the collisions are *adiabatic*, i.e., transitions are not induced in the atom. A collision occurring in an impact time τ_s will have Fourier components of frequencies up to $\Delta\omega_s \sim 1/\tau_s$. To guarantee that the collision is adiabatic, $\Delta\omega_s$ should be much smaller than any characteristic transition frequency ω_{ij}; i.e.,

$$\Delta\omega_s = 1/\tau_s \ll \omega_{ij} = |E_i - E_j|/\hbar. \tag{8.58}$$

For *non-degenerate levels*, the energy separation is often large enough that the condition stated above will be met. But for *degenerate levels* (e.g., for hydrogen), the energy separation between levels will be proportional to the perturbing field itself; i.e., $|E_i - E_j| \approx \hbar C_p/\rho^p = \hbar\Delta\omega(\rho)$. Then the condition of adiabaticity implies that $\Delta\omega(\rho)\tau_s = \eta(\rho) \gg 1$; that is, *only collisions inside the Weisskopf radius will be adiabatic*. In the case of hydrogen, $\Delta\omega_W$ for ions is very small, and for virtually the entire profile the statistical theory is valid, and the collisions causing the broadening occur inside the Weisskopf radius. Thus the ion broadening will be adiabatic. Precisely the opposite is true for electrons. Here $\Delta\omega_W$ will be large, and almost the whole profile is described by impact theory with the relevant collisions occurring outside the Weisskopf radius. The electron broadening is strongly non-adiabatic, and hence must be described by quantum theory. When the adiabatic assumption breaks down, much larger damping parameters than those predicted by classical theory are found; for this reason the modern quantum mechanical results are often drastically different from earlier classical work.

8.4 STATISTICAL THEORY: QUASI-STATIC APPROXIMATION

This approach is essentially an approximation opposite to the impact theory described above. The basic picture is that the atom is viewed as radiating in a statistically fluctuating field produced by randomly distributed perturbers. Most importantly, *the motions of the perturbers are ignored*, hence the name *quasi-static approximation*. As we shall see later, this approximation is appropriate for slowly moving ions, e.g., protons, in the plasma. A specific distribution of perturbers produces a definite field; the relative probability of fields of different strengths is thus determined by the statistical frequency with which particle distributions producing the appropriate strengths are realized. For a given value of the field, the oscillation frequency of the radiator is shifted by a definite $\Delta\omega$. The intensity of the radiation at this frequency displacement $\Delta\omega$ is assumed to be proportional to the statistical frequency of the occurrence of the appropriate field. The central problem of this theory is thus to determine the probability distribution of the perturbing fields. Once this is known, line profiles can be computed. The applications in this section will be restricted to quasi-static broadening of hydrogen lines by linear Stark-effect interactions with protons, although the theory is relevant in other cases as well.

Nearest-Neighbor Approximation

As a first approximation, we assume that the main effect on the radiator results from the *strongest* perturbation acting at any given instant, namely, that from the *nearest neighbor*, and that the effects of all other particles are neglected. Then if $W(r)dr$ is the probability that the nearest neighbor is located in the range $(r, r+dr)$ from the radiator, the frequency spectrum is

$$I(\Delta\omega)d(\Delta\omega) \propto W(r)[dr/d(\Delta\omega)]\, d(\Delta\omega), \tag{8.59}$$

where it is assumed that $\Delta\omega$ is given by equation (8.32); i.e., $\Delta\omega = C_p/r^p$.

To find $W(r)$ we calculate the probability that a particle is located in the range $(r, r+dr)$, *and* that none is at distance less than r. Then, assuming a uniform particle density N, the probability $W(r)$ is given by

$$W(r)\, dr = \left[1 - \int_0^r W(x)dx\right](4\pi r^2 N)\, dr, \tag{8.60}$$

where the factor $(4\pi r^2 N)\, dr$ is the relative probability of a particle lying in the shell $(r, r+dr)$, while the term in the square brackets is the probability that no particle lies inside this shell. By differentiation we find

$$\frac{d}{dr}\left[\frac{W(r)}{4\pi r^2 N}\right] = -(4\pi r^2 N)\left[\frac{W(r)}{4\pi r^2 N}\right], \tag{8.61}$$

and thus by integration and normalization

$$W(r) = 4\pi r^2 N \exp\left(-\tfrac{4}{3}\pi r^3 N\right). \tag{8.62}$$

It is customary to adopt the *mean interparticle distance* $r_0 = (4\pi N/3)^{-\frac{1}{3}}$ as the *reference distance* at which a perturber produces the *normal frequency shift* $\Delta\omega_0 = C_p/r_0^p$. Then

$$(\Delta\omega/\Delta\omega_0) = (r_0/r)^p, \tag{8.63}$$

and equation (8.62) can be rewritten as

$$W(r) = \exp[-(\Delta\omega_0/\Delta\omega)^{3/p}]\, d(\Delta\omega_0/\Delta\omega)^{3/p}. \tag{8.64}$$

For the case of the *linear Stark effect*, the perturbing field is $F = (e/r^2)$. If we define the *normal field strength* to be

$$F_0 = (e/r_0^2) = e\left(\tfrac{4}{3}\pi N\right)^{\frac{2}{3}} = 2.5985\, eN^{\frac{2}{3}}, \tag{8.65}$$

and measure F in units of F_0, i.e., $\beta \equiv F/F_0$, then the nearest-neighbor theory yields

$$W(\beta)\, d\beta = \tfrac{3}{2}\beta^{-\frac{3}{2}} \exp(\beta^{-\frac{3}{2}})\, d\beta. \tag{8.66}$$

The asymptotic value at large β

$$W(\beta) \to \tfrac{3}{2}\beta^{-\frac{3}{2}} \quad \text{for} \quad \beta \to \infty; \tag{8.67}$$

hence statistical theory predicts that, in the wings of a line broadened by the linear Stark effect, the profile falls off as $\Delta\omega^{-\frac{3}{2}}$, in contrast with the prediction $\Delta\omega^{-2}$ given by the impact theory.

The basic failing of this theory is that the profile is, of course, the result of perturbations by *all* particles, not just the nearest neighbor. To obtain accurate results, a more elaborate theory must be constructed.

Holtsmark Theory

The effect of an *ensemble* of particles upon a radiator was first determined by Holtsmark [503], who calculated the net vector field, at the position of the radiating atom, from the superposition of the field vectors of all perturbers. An elegant treatment of the problem was given in [217]; this paper should be consulted for the derivation of the results quoted here.

For an interaction of the form $F = C_p/r^p$, the analysis yields

$$W(\beta) = (2\beta/\pi) \int_0^\infty \exp(-y^{3/p})\, y \sin(\beta y)\, dy. \tag{8.68}$$

Here $\beta = F/F_0$, the normal field strength F_0 now being defined as

$$F_0 = \gamma C_p N^{p/3}, \tag{8.69}$$

where

$$\gamma = \{(2\pi^2/p)[3(p+3)\Gamma(3/p)\sin(3\pi/2p)]\}^{p/3}. \tag{8.70}$$

In particular, for the linear Stark effect, $p = 2$, $C_p = e$, and $\gamma = 2.6031$ so that $F = 2.6031\, eN^{\frac{2}{3}}$, which differs only inconsequentially from the normal field strength given by the nearest-neighbor theory.

The integral in equation (8.68) cannot be evaluated exactly for $p = 2$, and $W(\beta)$ must be expressed in a series expansion. For small β,

$$W(\beta) = \left(\frac{4}{3\pi}\right) \sum_{l=0}^{\infty} (-1)^l \Gamma\left(\frac{4l+6}{3}\right) \frac{\beta^{2l+2}}{(2l+1)!}, \tag{8.71}$$

while for $\beta \gg 1$, one finds an asymptotic expansion

$$W(\beta) = 1.496\, \beta^{-\frac{3}{2}} \left(1 + 5.107\, \beta^{-\frac{3}{2}} + 14.43\, \beta^{-3} + \cdots\right), \tag{8.72}$$

the leading term of which is essentially the same as that given by the nearest-neighbor theory. Tabulations of $W(\beta)$ are given in [217] and [1083, p. 28].

Debye Shielding

In deriving the probability distributions described above, *interactions among the perturbers* were ignored. In reality, the probability that a particle is found in a volume dV is not just $N\, dV$, but depends also upon the *electrostatic potential* ϕ in dV. For example, if at some point $\phi > 0$, electrons will tend to migrate toward it while ions will tend to migrate away, and vice versa if $\phi < 0$. Following Ecker [297–300], we may account for these effects schematically by introducing a Boltzmann factor depending on $\psi \equiv (e\phi/kT)$. Thus for electrons and ions, respectively, we write

$$n_{\mathrm{e}} W_{\mathrm{e}} dV = n_{\mathrm{e}} e^{\psi} dV \approx n_{\mathrm{e}}(1 + \psi)\, dV \tag{8.73}$$

and

$$n_{\mathrm{i}} W_{\mathrm{i}} dV = n_{\mathrm{i}} e^{-Z_i \psi} dV \approx n_{\mathrm{i}}(1 - Z_i \psi)\, dV, \tag{8.74}$$

where n_{e} and n_{i} are the electron and ion densities, Z_i is the ionic charge, and it is assumed that $\psi \ll 1$. As the plasma is *electrically neutral* over sufficiently large volumes,

$$n_{\mathrm{e}} = \sum_i Z_i n_i. \tag{8.75}$$

We now calculate the potential around a particular ion under the simplifying assumption that all particles can be smeared out into an equivalent charge density. We then may use *Poisson's equation*

$$\nabla^2 \psi = -4\pi e \rho \tag{8.76}$$

to determine ψ, where ρ is given by

$$e\rho = -e n_{\mathrm{e}} W_{\mathrm{e}} + e \sum_i Z_i n_i W_i. \tag{8.77}$$

In view of equations (8.73) through (8.75), this expression reduces to

$$e\rho = -e\psi\left(n_e + \sum_i Z_i^2 n_i\right). \tag{8.78}$$

Substituting equation (8.78) into (8.76) we may rewrite Poisson's equation as $\nabla^2\phi = (\phi/D^2)$, where

$$D \equiv (kT/4\pi e^2)^{\frac{1}{2}}\left[n_e + \sum_i Z_i^2 n_i\right]^{-\frac{1}{2}} \tag{8.79}$$

is the *Debye length*. Solving for ϕ, we find $\phi = r^{-1}[Ae^{-r/D} + Be^{r/D}]$. Demanding that $\phi \to 0$ as $r \to \infty$, we set $B \equiv 0$. Further, to recover the potential of the ion itself as $r \to 0$, we set $A = Z_i e$. Thus the potential produced by an ion imbedded in a plasma is

$$\phi(r) = [Z_i e \exp(-r/D)]/r. \tag{8.80}$$

As is clear from equation (8.80), beyond the Debye length the field of an ion is strongly shielded and rapidly vanishes. Physically, this occurs because a charged particle tends to polarize the plasma in its vicinity, and the oppositely charged particles that cluster around it shield the field of the original particle at large distances. Thus the Debye length sets an upper limit on (i) the distance over which two charged particles can effectively interact, (ii) the size of a region in which appreciable departures from charge neutrality can occur, and (iii) the wavelength of electromagnetic radiation that can propagate through plasma without dissipating.

In most astrophysical applications we can assume a practically pure hydrogen plasma; then $Z_i = 1$, and $n_i = n_e$, and inserting numerical constants into equation (8.79), we find

$$D = 4.8\,(T/n_e)^{\frac{1}{2}}\ \text{cm}. \tag{8.81}$$

To calculate the effect of shielding on $W(\beta)$ one can make the very simple assumptions [299, 300] that the field of the perturber is unchanged for $r \le D$, but is identically zero for $r > D$. One then finds

$$W(\beta,\delta) = (2\beta\delta^{\frac{4}{3}}/\pi)\int_0^\infty e^{-\delta g(y)} y \sin(\delta^{\frac{1}{2}}\beta y)\,dy, \tag{8.82}$$

where

$$g(y) = \tfrac{3}{2}y^{\frac{3}{2}}\int_y^\infty (1 - z^{-1}\sin z)z^{-\frac{5}{2}}\,dz \tag{8.83}$$

and

$$\delta = \tfrac{4}{3}\pi D^3 N, \tag{8.84}$$

which is the number of particles contained in the Debye sphere. As $\delta \to \infty$, one expects to recover the Holtsmark distribution; i.e., $W(\beta, \infty) = W_H(\beta)$. From equation (8.82) one can indeed show that for $\delta \to \infty$, $W(\beta, \delta) = W_H(\beta) + \delta^{-\frac{1}{3}} F(\beta)$, where $F(\beta)$ is a bounded definite integral. Recovery of the Holtsmark distribution for large δ can also be seen from the asymptotic expansion

$$W(\beta, \delta) \approx 1.496\, \beta^{-\frac{3}{2}} \left(1 + 5.107\, \beta^{-\frac{3}{2}} + 6.12\, \delta^{-\frac{1}{3}} \beta^{-2} + \cdots\right), \tag{8.85}$$

which may be compared to equation (8.72). In principle, for small δ the theory should merge continuously into the nearest-neighbor approximation, but in practice, when $\delta \leq 5$, the assumptions employed (particularly that a smeared-out charge distribution may be used) break down.

The treatment of perturber interactions given above is somewhat oversimplified. Very precise calculations of the perturber field strength distributions have been made using cluster-expansion methods [84, 774, 863] and numerical integrations using Monte Carlo techniques [504–506]. Most modern treatments of hydrogen-line broadening use these refined distribution functions. In practice, the effects of shielding are often quite important in laboratory plasmas, while in stellar atmospheres the densities are so low that the number of particles in the Debye sphere is large ($\delta \gtrsim 100$), and hence the departures from the Holtsmark distribution are not large.

The presence of nearby charged particles neutralizes the effects of the nuclear charge upon an orbital electron and thereby weakens the potential well in which the electron is bound. The reduction of binding energy can be calculated using the Debye potential in equation (8.80), as

$$\Delta E = (Ze^2/r)[\exp(-r/D) - 1] \approx -Ze^2/D \tag{8.86}$$

for $r \ll D$. An electron in a state that, in an unperturbed atom, has an energy $\Delta\chi$ below the ionization limit, can be considered unbound if $\Delta\chi \lesssim \Delta E$. That is, the *ionization potential of the atom is decreased* by an amount

$$\Delta\chi = Ze^2/D = (27.2\, Za_0/D)\ \mathrm{eV} = 3 \times 10^{-8} Z n_e^{\frac{1}{2}} T^{-\frac{1}{2}}\ \mathrm{eV}, \tag{8.87}$$

where use has been made of equation (8.81). This calculation is only schematic. A much better treatment of these effects is using the *occupation probabilities* of the individual levels, which will be described in chapter 9.

Quasi-static Hydrogen-Line Broadening

In the absence of a perturbing field, each level of hydrogen is degenerate with $2n^2$ sublevels. When an electric field is applied, these sublevels separate, and, because hydrogen has a permanent dipole moment, the energy shift is directly proportional to the applied field strength F, the so-called *linear Stark effect* [320]. If no other broadening mechanisms are operative, the line profile will consist of a number of *Stark components*, arising from transitions between the sublevels of the lower and upper states. Each Stark component has a characteristic relative intensity I_k [975] and will be displaced from line center by a characteristic shift

$$\Delta\lambda_k = (3h\lambda^2 n_k / 8\pi^2 cmeZ)F \equiv C_k F, \tag{8.88}$$

where Z is the charge of the atom ($=1$ for hydrogen) and n_k is an integer. The resulting line will be a superposition of these components, weighted by their relative intensities and the probability to be shifted to the appropriate wavelength position.

The *Stark pattern* of a hydrogen line is symmetric about the line center with identical components at $\pm k$ with $I_{-k}=I_k$, $C_{-k}=C_k$; see, e.g., [1102, p. 328]; and [1083, p. 73]. Assuming that the intensities are normalized such that $\sum_k I_k = 1$, where the sum extends over all components, the line profile will be

$$\begin{aligned} I(\Delta\lambda)\, d(\Delta\lambda) &= \sum_k I_k W(F/F_0)\,(dF/F_0) \\ &= \sum_k I_k W(\Delta\lambda/C_k F_0) d(\Delta\lambda)/(C_k F_0). \end{aligned} \tag{8.89}$$

It is customary to define the parameter α as

$$\alpha \equiv \Delta\lambda / F_0, \tag{8.90}$$

where F_0 is the normal field strength, $F_0 = 2.6\, eN^{\frac{2}{3}}$. Then the line profile is given by a function $S(\alpha)$,

$$S(\alpha)\, d\alpha = \sum_k I_k W(\alpha/C_k)(d\alpha/C_k), \tag{8.91}$$

which is normalized in the range $(-\infty, \infty)$ for α. The absorption cross section per atom can be written as

$$\kappa_\nu(\Delta\lambda) = (\pi e^2/mc) f S(\Delta\lambda/F_0)(\lambda^2/cF_0). \tag{8.92}$$

Extensive tables of C_k, I_k, and $S(\alpha)$ for numerous hydrogen lines are given in [1094]. The largest C_k for a line of upper quantum number n increases as n^2, and recalling that for $\beta \gg 1$, $W(\beta) \propto \beta^{-\frac{5}{2}}$, we see from equation (8.90) or (8.92) that the Stark widths of lines rise rapidly up a series, roughly as n^3.

Let us now consider when the quasi-static profiles are applicable. Let $\overline{n}_k = \sum I_k n_k / \sum I_k$, the sum being taken over positive values only. Then, writing $\overline{\Delta\omega} = \overline{C}_2/r^2$ and $F = e/r^2$, it follows from equation (8.88) that

$$\overline{C}_2 = (3h\overline{n}_k/4\pi m) = 1.738\, \overline{n}_k, \tag{8.93}$$

with $\overline{n}_k \approx \frac{1}{2}n(n-1)$ for $n \gg 1$. From equation (8.57) the wavelength shift delimiting the transition between the impact and statistical theories is, for $p=2$,

$$\Delta\lambda_W = (\overline{v}^2\lambda^2/2\pi^3 c\overline{C}_2). \tag{8.94}$$

Table 8.2 Division Wavelength $\Delta\lambda_W$ (Å) between Impact and Quasi-static Broadening of Hydrogen Lines

Line	Perturber	$T = 10^4$ K	$T = 2.5 \times 10^4$ K
$H\alpha$	Electrons	230.0	580.0
	Protons	0.25	0.63
$H\beta$	Electrons	48.0	120.0
	Protons	0.05	0.13
$H\gamma$	Electrons	19.0	48.0
	Protons	0.02	0.05
$H\delta$	Electrons	13.0	32.0
	Protons	0.01	0.03

Notice that $\Delta\lambda_W \propto \overline{v}^2$, and thus $\Delta\lambda_W$ (electron) $\approx 10^3 \Delta\lambda_W$ (proton). Using equations (8.93) and (8.29), we obtain the results listed in table 8.2. It is obvious that the ion broadening is very well described by the quasi-static theory, especially when we note that Doppler motions will dominate in the core. The electrons, however, are in the impact-broadening regime except at very large displacement from line center and, as mentioned in §8.3, are non-adiabatic. The complete profile consists of the effects of both ions and electrons, and, as we shall see in §8.5, the latter increase the line broadening markedly. The functions $S(\alpha)$ for ions alone seriously underestimate the hydrogen-line widths, and a satisfactory theoretical description of stellar hydrogen-line profiles became possible only after the development of the quantum mechanical line broadening theory.

8.5 ★ QUANTUM THEORY OF LINE BROADENING

Quantum mechanical calculations yield much more accurate profiles for pressure-broadened lines than those provided by semiclassical theories. The development of this theory brought about a major improvement in one of the most important (and difficult) applications of atomic physics in the analysis of stellar spectra. The field was pioneered by Baranger [80–82], and excellent discussion of the general theory can be found in in several reviews and textbooks, e.g., [95, chapter 13], [249], [400, chapter 4], [403], [404, chapter 4]; only a brief outline will be presented here.

General Theory

As was shown in chapter 5, the power radiated by an *isolated atom* in a transition from an upper state j to a lower state i is [cf. equation (5.101), with $P \equiv h\nu_{ij}A_{ji}$]

$$P = (4\omega^4/3c^3)\left|\langle i|\mathbf{d}|j\rangle\right|^2, \qquad (8.95)$$

where $\mathbf{d}$ is the atomic dipole moment, $\mathbf{d} = -e \cdot \mathbf{r}$, e is electron charge, and $\mathbf{r}$ is the position vector of the radiating electron measured from the nucleus (or of the sum of such vectors if more electrons are involved). The total emission, summed over all possible substates contributing to a line, and weighted by the probability ρ_j of finding the system in the initial state j is

$$P(\omega) = (4\omega^4/3c^3) \sum_{i,j} \rho_j \delta(\omega - \omega_{ij}) \left| \langle i | \mathbf{d} | j \rangle \right|^2. \tag{8.96}$$

Here, the δ function accounts for the energy conservation, $\hbar\omega_{ij} = E_j - E_i$, where E_i and E_j are the energies of the stationary states i and j.

The crucial point of the theory of line broadening is that unlike describing an emission from an isolated atom, the states i and j refer to the *complete quantum mechanical system*, consisting of a radiating atom *and* all perturbers. In most applications it is assumed that the contribution from the individual radiators add incoherently (which is usually well satisfied in astrophysical conditions, but may easily break down in gas lasers) and that the radiating atoms do not interact with each other (which is well satisfied except the case of resonance broadening). We also assume that the factor ω^4 varies slowly across the frequency range of a line. It is thus convenient to introduce a line shape (or line profile) function as

$$L(\omega) = \sum_{i,j} \rho_j \delta(\omega - \omega_{ij}) \left| \langle i | \mathbf{d} | j \rangle \right|^2, \tag{8.97}$$

where ρ_j represents a probability of a particular state j of the system that is represented by a radiating atom and in principle all perturbers. In quantum statistical mechanics, this probability is described through the *density matrix* of the system. In the following, we shall refer to ρ as a density matrix.

It is important to realize that in most applications it is not necessary to consider the corresponding formulae for absorption and induced emission because they are related to each other by Kirchhoff's law. The fundamental condition for the validity of this assumption is that the probability of the initial state, ρ_j, does *not* depend on radiation field in the line of interest. If it does, the profiles of absorption and emission and emission may differ. In other words, it happens if a photon is emitted in a line immediately after a previous absorption of a photon within the same line, that is, when one deals specifically with the problem of radiation *scattering* in a line. This approach is usually referred to in astrophysical applications as the *line redistribution* or *partial redistribution* problem and will be considered in detail in detail in chapter 10.

While one can use equation (8.97) directly, most studies employ a modified formalism that is better amenable to theoretical analysis. One introduces a Fourier transform $C(s)$ of the line shape function $L(\omega)$,

$$\begin{aligned} C(s) &= \int_{-\infty}^{\infty} \exp(i\omega s) L(\omega) d\omega \\ &= \sum_{ij} \exp(i\omega_{ij} s) \left| \langle i | \mathbf{d} | j \rangle \right|^2 \rho_j, \end{aligned} \tag{8.98}$$

which is analogous to the classical *autocorrelation function*; see [95, p. 498]. The line profile function then follows from the inverse Fourier transform,

$$L(\omega) = (1/\pi)\,\mathrm{Re}\int_0^\infty \exp(-i\omega s)C(s)\,ds; \tag{8.99}$$

because the autocorrelation function satisfies $C(-s) = [C(s)]^*$, only positive values of s are needed. Equation (8.98) can be rewritten as

$$\begin{aligned} C(s) &= \sum_{ij} \rho_j e^{(E_j - E_i)s/\hbar} \left|\langle i|\mathbf{d}|j\rangle\right|^2 \\ &= \sum_{ij} \rho_j \langle j|\mathbf{d}|i\rangle e^{-E_i s/\hbar} \langle i|\mathbf{d}|j\rangle e^{E_j s/\hbar} \\ &= \sum_{ij} \rho_j \langle j|\mathbf{d}T(s,0)|i\rangle\langle i|\mathbf{d}T^\dagger(s,0)|j\rangle. \end{aligned} \tag{8.100}$$

The last equality follows from an introduction of the *time-development operator*, $T(t,0)$, defined such that a state of the system at time t is related to the state of the system at time $t = 0$ by

$$|\alpha, t\rangle = T(t,0)|\alpha, 0\rangle. \tag{8.101}$$

As $|\alpha, t\rangle$ satisfies the Schrödinger equation

$$H|\alpha, t\rangle = i\hbar(d|\alpha, t\rangle/dt), \tag{8.102}$$

one obtains by substituting equation (8.101) into (8.102) a Schrödinger equation for $T(t,0)$,

$$HT(t,0) = i\hbar[dT(t,0)/dt], \tag{8.103}$$

which has the solution

$$T(t,0) = e^{-iHt/\hbar}, \tag{8.104}$$

where the exponential has to be understood as an *operator*. Finally, using the *expansion rule*, relative to a *complete* set of states $|\gamma\rangle$,

$$\langle\alpha|\beta\rangle = \sum_\gamma \langle\alpha|\gamma\rangle\langle\gamma|\beta\rangle, \tag{8.105}$$

equation (8.100) can be rewritten as

$$C(s) = \sum_j \rho_j \langle j|\mathbf{d}T\,\mathbf{d}T^\dagger|j\rangle = \mathrm{Tr}(\rho\,\mathbf{d}T\,\mathbf{d}T^\dagger). \tag{8.106}$$

The trace includes both the atomic and perturber states. This expression is quite general; the detailed form of $C(s)$ depends upon the form of T and on the nature of perturbers and their interaction with the radiator.

Classical Path Approximation

To make progress, let us examine the time-development operator T in more detail. Let ψ be the wave function that provided the solution of the complete system of radiating atom plus a single perturber,

$$i\hbar(d\psi/dt) = (H_{\rm A} + H_{\rm P} + V)\psi, \tag{8.107}$$

where the atomic (radiator) Hamiltonian $H_{\rm A}$ is independent of perturber coordinates, $H_{\rm P}$ is independent of atomic coordinates, and V depends on both. The formalism is based on the two main assumptions: (i) Wave function of the system of atom plus perturber is assumed to be *separable*, i.e.,

$$\psi(t) = \alpha(t)\pi(t), \tag{8.108}$$

where $\alpha(t)$ and $\pi(t)$ are the atomic and perturber wave functions, respectively. (ii) The perturber bath is independent of the state of the atom and is fixed. In this way the effect of the perturber on the atom is taken into account, but the back-reaction of the atom on the perturber is ignored. This is a good approximation as long as the energy gained or lost by the perturbed (of order $\hbar\Gamma$, where Γ is the line width) is much less than its kinetic energy, kT, a condition that is almost always satisfied in the typical conditions in stellar atmospheres. A few collisions in which large energy exchanges take place will always occur, but these will not invalidate the assumption if they do not dominate the broadening.

Under these assumptions, the perturber wave function satisfies a simple Schrödinger equation

$$i\hbar[d\pi(t)/dt] = H_{\rm P}\pi(t), \tag{8.109}$$

and the time-evolution operator for the perturber is

$$T_{\rm P}(t,0) = e^{-iH_{\rm P}t/\hbar}. \tag{8.110}$$

The Schrödinger equation for the atom is derived by multiplying equation (8.107) by π^*, integrating over perturber coordinates, and using the normalization condition for π, $\int \pi^*(t)\pi(t)d\tau_{\rm P} = 1$; we obtain

$$i\hbar[d\alpha(t)/dt] = \left(H_{\rm A} + \int \pi^* V\pi\, d\tau_{\rm P}\right)\alpha(t). \tag{8.111}$$

If the perturber wave packets are narrow enough that the perturbers can be considered to be classical particles on classical paths, then we can make the identification $\int \pi^* V\pi\, d\tau_{\rm P} \to V_{cl}(t)$, where $V_{cl}(t)$ is a *classical interaction potential.* The Schrödinger equation for the atom's time-development operator then becomes

$$i\hbar[dT_{\rm A}(t,0)/dt] = [H_{\rm A} + V_{cl}(t)]T_{\rm A}(t,0), \tag{8.112}$$

and the time-development operator for the complete system is

$$T(t,0) = T_{\rm A}(t,0)T_{\rm P}(t,0) = T_{\rm A}(t,0)e^{-iH_{\rm P}t/\hbar}. \tag{8.113}$$

Finally, we assume that the density matrix ρ is separable as well, $\rho = \rho_{\rm A}\rho_{\rm P}$, where $\rho_{\rm A}$ refers to atomic states only, and $\rho_{\rm P}$ refers to perturber states only and is diagonal in the perturber coordinates. Again, this may be done if the back-reaction of the atom on the perturber can be neglected. Substituting equations (8.108) and (8.113) and the expressions for ρ into equation (8.106), we find

$$C(s) = {\rm Tr}\{\rho_{\rm A}\mathbf{d}T_{\rm A}\,\mathbf{d}T_{\rm A}^{\dagger}\}_{\rm Av}, \tag{8.114}$$

where $\{\cdots\}_{\rm Av}$ refers to a *thermal average* over the perturbers, while the trace is carried out over the *atomic states* only. This important result was first obtained by Baranger [80].

Impact Approximation

In the classical path approximation, the general program for line-broadening calculations is as follows. First, one has to solve the Schrödinger equation (8.112) for all potentials $V_{cl}(t)$ produced by perturbers passing by with various impact parameters, velocities, and times of closest approach. These solutions are then substituted into the expression for the correlation function (8.114), which requires the calculation of statistical average over the perturber trajectories. Finally, the line shape function is obtained by the Fourier transform of the correlation function, according to equation (8.99).

To calculate the correlation function by equation (8.114) we assume that both the initial state a and the final state b consist of several substates, denoted by α and β, respectively, and that the dipole transitions exist only between substates of a and b, but the radiative transitions among substates of a or b can be ignored. On the other hand, the collision-induced transitions between states a and b are ignored, and we assume that collisions can result only in transitions among substates of a or b. Consequently, we have $\langle\alpha|\mathbf{d}|\alpha'\rangle = 0$, $\langle\beta|\mathbf{d}|\beta'\rangle = 0$, and $\langle\alpha|T_{a,b}|\beta\rangle = 0$. Then, writing out the trace in equation (8.114), we find

$$C(s) = \rho_a \sum_{\alpha,\alpha',\beta,\beta'} \langle\alpha|\mathbf{d}|\beta\rangle\langle\beta'|\mathbf{d}|\alpha'\rangle\langle\alpha|\langle\beta|\{T_b T_a^*\}_{\rm Av}|\alpha'\rangle|\beta'\rangle. \tag{8.115}$$

This expression takes into account that the transpose of the Hermitian conjugate of $T(t,0)$ is simply its complex conjugate and assumes that the density matrix ρ_a is constant among substates of a and b. Note that only the time-development operators are affected by the statistical averages. The last expression in equation (8.115) is often written using a "double-index" notation,

$$C(s) = \rho_a \sum_{\alpha,\alpha',\beta,\beta'} \mathbf{d}_{\alpha\beta}\langle\langle\alpha\beta|\{T_b T_a^*\}_{\rm Av}|\alpha'\beta'\rangle\rangle\mathbf{d}^*_{\alpha'\beta'}. \tag{8.116}$$

It is convenient to write the time-development operator in the *interaction representation*, defined as

$$U(t,0) \equiv e^{iH_{\mathrm{A}}t/\hbar}T(t,0). \tag{8.117}$$

This definition factors out the time behavior of the eigenstates and isolates the effects of perturbers; indeed, for an unperturbed eigenstate $U(t,0) \equiv 1$.

In view of equation (8.112), U obeys the Schrödinger equation

$$i\hbar[dU(t,0)/dt] = e^{iH_{\mathrm{A}}t/\hbar}V_{cl}e^{-iH_{\mathrm{A}}t/\hbar}U(t,0) \equiv V'_{cl}(t)U(t,0). \tag{8.118}$$

According to the standard quantum mechanical procedure, this equation can be solved by iteration, starting with $U(t,0) = 1$ as a first approximation:

$$U(t,0) = 1 + (i\hbar)^{-1}\int_0^t dt_1 V'_{cl}(t_1) + (i\hbar)^{-2}\int_0^t dt_2 V'_{cl}(t_2)\int_0^{t_2} dt_1 V'_{cl}(t_1) + \cdots . \tag{8.119}$$

By substituting equation (8.117) into (8.116) we have

$$C(t) = \rho_a \sum_{\alpha,\alpha',\beta,\beta'} e^{i(E_a-E_b)t/\hbar}\mathbf{d}_{\alpha\beta}\langle\langle\alpha\beta\big|\{U_b(t,0)U_a^*(t,0)\}_{\mathrm{Av}}\big|\alpha'\beta'\rangle\rangle\mathbf{d}^*_{\alpha'\beta'}. \tag{8.120}$$

To calculate the statistical average $\{U_bU_a^*\}$, we consider the change in a time interval δt (we skip the subscript Av):

$$\begin{aligned}\Delta\{U_b(t,0)U_a^*(t,0)\} &\equiv U_b(t+\Delta t,0)U_a^*(t+\Delta t,0) - U_b(t,0)U_a^*(t,0)\}\\ &= U_b(t+\Delta t,t)U_a^*(t+\Delta t,t) - 1\}\{U_b(t,0)U_a^*(t,0)\}.\end{aligned}$$

That is, we consider the change in $\{U_bU_a^*\}$ caused by a specific collision (treated as an *impact*) in a time interval $(t, t+\Delta t)$ to be statistically independent of the current values so that we can replace the average of the product as a product of the averages. Using equation (8.119) on the interval $(t, t+\Delta t)$, we obtain

$$\{U_b(t+\Delta t)U_a^*(t+\Delta t) - 1\} = e^{i(H_b-H_a)t/\hbar}(\Phi_{ab}\Delta t)e^{-i(H_b-H_a)t/\hbar} \tag{8.121}$$

and we obtain a differential equation for $\{U_bU_a^*\}$,

$$\frac{d}{dt}\{U_bU_a^*\} = e^{i(H_b-H_a)t/\hbar}\Phi_{ab}e^{-i(H_b-H_a)t/\hbar}\{U_bU_a^*\}, \tag{8.122}$$

which has a solution

$$\{U_b(t,0)U_a^*(t,0)\} = e^{i(H_b-H_a)t/\hbar}\exp[-i(H_b-H_a)t/\hbar + \Phi_{ab}t]. \tag{8.123}$$

Here, all the physics of the interaction of an atom and perturbers is contained in the quantity Φ_{ab}, which is given by

$$\begin{aligned}\Phi_{ab} = (1/\Delta t)\Bigg\{&(i\hbar)^{-1}\int_0^{\Delta t} dt_1\left[V_b'(t_1) - V_a'^*(t_1)\right]\\ &+(i\hbar)^{-2}\Bigg[\int_0^{\Delta t} dt_1 V_b'(t_1)\int_0^{t_1} dt_2 V_b'(t_2)\\ &+\int_0^{\Delta t} dt_1 V_a'^*(t_1)\int_0^{t_1} dt_2 V_a'^*(t_2)\\ &-\int_0^{\Delta t} dt_1 V_b'(t_1)\int_0^{\Delta t} dt_2 V_a'^*(t_2)\Bigg]\cdots\Bigg\}. \end{aligned} \tag{8.124}$$

Now, substituting equation (8.123) into (8.120), we obtain for the correlation function

$$C(t) = \rho_a \sum_{\alpha,\alpha',\beta,\beta'} \mathbf{d}_{\alpha\beta}\langle\langle\alpha\beta|e^{-i(H_b-H_a)t/\hbar+\Phi_{ab}t}|\alpha'\beta'\rangle\rangle\mathbf{d}^*_{\alpha'\beta'}. \tag{8.125}$$

Finally, using the inverse Fourier transformation, equation (8.99), we obtain the line shape function

$$L(\omega) = \frac{\rho_a}{\pi}\,\mathrm{Re}\sum_{\alpha,\alpha',\beta,\beta'} \mathbf{d}_{\alpha\beta}\langle\langle\alpha\beta|[i\omega + i(H_b - H_a)/\hbar - \Phi_{ab}]^{-1}|\alpha'\beta'\rangle\rangle\mathbf{d}^*_{\alpha'\beta'}. \tag{8.126}$$

The result is valid if (1) the interval Δt in equation (8.121) can be chosen to include a complete collision, (2) when the collisions overlap, they are weak enough that their contributions to the iterative solution for U are simply additive, and (3) the perturbers can be treated as classical particles. These validity criteria must be checked in each case.

Quasi-static Approximation

As pointed out above, in many circumstances the emitting atoms are under the influence of electric fields produced by relatively rapidly moving electrons and slowly moving ions. The following model is therefore appropriate: The ion motions are neglected and the impact approximation is used to calculate the electron broadening in the line corresponding to a transition between states of the Hamiltonian $H_{\mathrm{A}}(F)$, which now depends on the electric field strength produced by perturbing ions, i.e., describes the usual static Stark effect. Equation (8.97) is then modified to read

$$L(\omega) = \int \sum_{i,j} \rho_j \delta[\omega - \omega_{ij}(F)]\,|\langle i|\mathbf{d}|j\rangle|^2 W(F)\,dF, \tag{8.127}$$

where $W(F)$ is the ion field distribution, considered in detail in §8.4. It is assumed that the density matrix of the radiator does not depend on the ion field F. The formalism is then the same as outlined above, and the final line profile is obtained by taking the appropriate average of equation (8.125) over the ion field strength,

$$L(\omega) = \frac{\rho_a}{\pi} \int dF\, W(F)$$
$$\mathrm{Re} \sum_{\alpha,\alpha',\beta,\beta'} \mathbf{d}_{\alpha\beta} \langle\langle \alpha\beta | \left[i\omega + i[H_b(F) - H_a(F)]/\hbar - \Phi_{ab}(F) \right]^{-1} | \alpha'\beta' \rangle\rangle \mathbf{d}^*_{\alpha'\beta'}. \tag{8.128}$$

Equation (8.128) is quite general and has been used in most quantum mechanical calculations of Stark-broadened line profiles. Although this approach is called in the literature the "quasi-static" approximation, it should be kept in mind that the quasi-static approximation is being made *only* with respect to *ion* broadening, while the electron broadening is treated within the impact approximation.

It should be pointed out that for very high densities, a separation of the radiator and perturber parts of the density matrix is no longer possible, and one has to consider a molecular dynamics simulation of clusters of particles (a radiator plus one or more perturbers).

Unified Stark Broadening Theory

Assuming the quasi-static approximation is valid for ion broadening, equation (8.128) still breaks down for large distances from the line center where the impact approximation for electrons ceases to be applicable. A formalism that is able to treat a general case of electron broadening, including the range beyond the validity of impact approximation, is usually called the *unified theory* [1018, 1120, 1121]. We omit here the derivation because it is quite involved; for details the reader is referred to the original references and the textbook [404].

The line shape is given by

$$L(\omega) = -(\hbar/\pi)\,\mathrm{Im}\,\mathrm{Tr} D[\hbar\omega - H_{\mathrm{A}} - \mathcal{L}(\omega)]^{-1}, \tag{8.129}$$

where D is defined by

$$\langle i | D | i' \rangle = \sum_f \langle i | \mathbf{d} | f \rangle \langle f | \mathbf{d} | i' \rangle, \tag{8.130}$$

which may be called a "doubled" dipole operator. All the details of interaction of the radiator with perturber bath is contained in an operator $\mathcal{L}$, called the *line shape operator*.

If $\mathcal{L}$ can be approximated by a constant operator, and if off-diagonal matrix elements of this operator are not important, then the line profile is actually Lorentzian, with width and shift given by

$$\hbar\Gamma = -\mathrm{Im}\,\langle j | \mathcal{L}(\omega_0) | j \rangle \tag{8.131}$$

and

$$\hbar\,\Delta\omega_0 = \mathrm{Re}\,\langle j|\mathcal{L}(\omega_0)|j\rangle. \tag{8.132}$$

Notice that in the usual impact approximation this is indeed the case, and in this case the collisional operator Φ is given by $\Phi = -i\mathcal{L}/\hbar$. Otherwise, one has to evaluate matrix elements of $\mathcal{L}$ using appropriate averages of the time-evolution operator, as described in [1018, 1120, 1121]. Unified theory was applied mainly for hydrogen (and also for He II; see below).

For hydrogen, although the exact expressions are very complicated, it turned out that one can write down useful approximate expression that can be used in many astrophysical applications [1120] (see also [251]). The line profile is written formally as a Lorentz profile, but with a *frequency-dependent damping parameter*, given as, for a transition between lower level i and upper level j,

$$I_{ij}(\Delta\omega) = \frac{\Gamma_{ij}(\Delta\omega)/\pi}{\Delta\omega^2 + \Gamma_{ij}(\Delta\omega)^2}, \tag{8.133}$$

where

$$\Gamma_{ij}(\Delta\omega) = \gamma_j(\Delta\omega) + \Gamma_{ij}^{\mathrm{rad}}/2, \tag{8.134}$$

and where $\Gamma_{ij}^{\mathrm{rad}}$ is the usual radiative damping parameter

$$\Gamma_{ij}^{\mathrm{rad}} = \sum_{\ell<i} A_{i\ell} + \sum_{\ell<j} A_{j\ell}, \tag{8.135}$$

and the "collisional damping" parameters are given by, for the first few levels of hydrogen (for a general expression, refer to [1120]),

$$\gamma_{2p} = q(2)/3 \tag{8.136}$$

$$\gamma_{3s} = 2q(6)/3 \tag{8.137}$$

$$\gamma_{3p} = [q(6) + 2q(3)]/3 \tag{8.138}$$

$$\gamma_{3d} = [q(6) + 6q(3)]/15, \tag{8.139}$$

where $q(j)$ is the sum of the electron (e) and ion (i) contributions,

$$q(j) = q_e(j) + q_i(j). \tag{8.140}$$

Index j is the product of the principal quantum number and the parabolic quantum number [1120]. The quantities $q_e(j)$ and $q_i(j)$ are constant in the *impact regime*, which is bounded by the respective plasma frequencies $\Delta\omega_{pe}$ and $\Delta\omega_{pi}$, which are given by

$$\Delta\omega_{pe} = (4\pi e^2 n_{\mathrm{e}}/m_{\mathrm{e}})^{1/2} \tag{8.141}$$

$$\Delta\omega_{pi} = (m_{\mathrm{e}}/m_{\mathrm{i}})^{1/2}\,\Delta\omega_{pe}, \tag{8.142}$$

where n_e is the electron density, m_e its mass, and m_i the mass of the perturbing ions (protons). In the impact regime we have for electrons

$$q_e(j) = (3j\hbar/2m_e)^2(8\pi m_e/kT)^{1/2}n_e[0.27 - \ln(4C^2)] \quad \text{for} \quad \Delta\omega < \Delta\omega_{pe}, \tag{8.143}$$

and for ions

$$q_i(j) = \mu(3j\hbar/2m_e)^2(8\pi m_e/kT)^{1/2}n_e[0.27 - \ln(4C^2\mu^2)] \quad \text{for} \quad \Delta\omega < \Delta\omega_{pi}, \tag{8.144}$$

where $\mu = (m_r/m_e)^{1/2}$, with m_r being the reduced mass of the radiator-perturber pair, $\mu = 30.2$ for hydrogen, and

$$4C^2 = (2\pi n_e/m_e)\,(3j\hbar e/kT)^2. \tag{8.145}$$

For frequency separations larger than Weisskopf frequency, the broadening enters the quasi-static regime. The appropriate Weisskopf frequencies are

$$\Delta\omega_{We}(j) = 4kT/(3j\hbar) \tag{8.146}$$

and

$$\Delta\omega_{Wi}(j) = \Delta\omega_{We}(j)/\mu^2, \tag{8.147}$$

for electrons and ions, respectively. Then for ions in the quasi-static regime,

$$q_i(\Delta\omega, j) = \pi^2 n_e(3\hbar j/2m_e)^{3/2}(\Delta\omega)^{-1/2} \quad \text{for} \quad \Delta\omega > \Delta\omega_{Wi}(j), \tag{8.148}$$

and in the electron quasi-static regime,

$$q_e(\Delta\omega, j) = q_i(\Delta\omega, j) \quad \text{for} \quad \Delta\omega > \Delta\omega_{We}(j). \tag{8.149}$$

Between the impact and the quasi-static regime, i.e., between the appropriate $\Delta\omega_p$ and $\Delta\omega_W(j)$, it is adequate to interpolate linearly in $\log(\Delta\omega)$.

Model Microfield Method

The approaches described so far neglect the effects of ion motions in the ionic broadening and assume *static* ions. A fairly successful model that is able to treat ion dynamics effects is the Model Microfield Method (MMM) developed in [149]. In this model, the dynamics of the microfield fluctuations, both ionic and electronic, are treated as a stochastic process where the microfields are assumed to be constant most of the time, with sudden jumps to another constant value at random times. The jumping times follow Poisson statistics, with a field-dependent frequency $\nu(F)$. The actual calculation of line profile requires the knowledge of the electronic and ionic field distribution functions, for which Hooper's distribution functions [504–506] are used. Another free parameter of the theory is the jumping frequency $\nu(F)$, which is typically chosen so as to reproduce the field autocorrelation function correctly [149, 1044]. Although not unambiguous, this procedure ensures that the line profiles obtained are correct in the impact and quasi-static limits.

Quantum Mechanical One-Perturber Approach

For accurate calculations of opacity in strong lines, in particular hydrogen lines, it is important to supplement detailed profile calculations that are typically limited to moderate distances from the line center with asymptotic formulas for the extended line wings. For this purpose, it is usually assumed [403] that both ion and electron effects could be estimated using the nearest-neighbor, quasi-static approximation. However, in order to produce a large frequency shift $\Delta\omega$, the radiator-perturber separation $r \propto 1/\Delta\omega$ becomes so small that one can no longer treat the system as radiating atom plus a perturber, but instead as a transient multi-particle system, for instance, H^- in the case of electron broadening of hydrogen or H_2^+ in the case of proton broadening. In other words, a radiator plus one perturber is treated as a single quantum mechanical system [322, 992, 1042, 1082, 1209].

An application to Lyman-α broadening by electrons [322, 1081] and by protons [616] revealed that the line profile is asymmetric with various discrete features that arise due to resonances in the H^- and H_2^+ systems, respectively. We have previously mentioned this effect in §7.3 in the context of molecular absorption. As pointed out there, a distinction between separated processes of "line broadening" and "continuum absorption" is blurred and, in fact, loses its usual meaning.

8.6 APPLICATIONS

The theory outlined in the previous sections, in particular the quantum theory, is very complex and involves complicated calculations. On the other hand, most astrophysical applications need just the final results of such calculations, together with an estimate of their accuracy. We will summarize the most important applications and practical results below.

Hydrogen

One of the most important applications of the theories outlined above has been to the calculation of hydrogen line broadening. The impact theory outlined above has been intensively applied by Griem and collaborators, summarized in [400, 403], essentially using equation (8.129), i.e., assuming impact approximation for electrons and quasi-static approximation for protons. Although the evaluation of Φ_{ab} is straightforward in principle, it is complicated in practice, for it entails cutoffs at both small and large impact parameters; the former arising due to strong collisions inside the Weisskopf radius, which are no longer correctly described by the series expansion of the time-development operator U, and the latter to account for Debye shielding effects. Further cutoff procedures are required to allow for the transition of the electrons from the impact to quasi-static regimes.

Using the unified theory outlined above that automatically accounts for the transition of electrons from the impact to quasi-static regime, extensive tables have been published [1122] for the first members of the Lyman and Balmer series in temperature-density ranges appropriate to stellar atmospheres. These tables include convolutions of the Stark profile with a thermal velocity distribution of the hydrogen atoms. These calculations were later extended in [692] for the Lyman, Balmer, Paschen, and Brackett series of hydrogen. These tables cover the series up to $n = 22$ for an extended grid of temperatures and electron densities, $2500 \le T \le 1.6 \times 10^5$ K; $10^{10} \le n_e \le 10^{18}$ cm^{-3}.

Applying the Model Microfield Method, [1043] presented tables of Stark broadening for the Lyman, Balmer, and Paschen series, restricted, however, to a rather narrow range of temperature and density. Subsequently, [1045] calculated extended tables for $2500 \le T \le 1.3 \times 10^5$ K; $10^{10} \le n_e \le 10^{19}$ cm^{-3}.

Finally, [1084] produced tables that are based on the unified theory and take into account effects of level dissolution (see §9.4); hence they are preferable to the previous work for high densities occurring, e.g., in white-dwarf atmospheres.

In astrophysical applications, the hydrogen lines are significantly affected by broadening mechanisms other than the Stark effect. The core of the line is dominated by Doppler broadening. The effect of radiation and resonance broadening may be important in the wings at low electron densities. Assuming that these mechanisms are all uncorrelated, we may account for their combined effect by a convolution procedure. Folding the Doppler profile with the Lorentz profiles from radiation and resonance damping gives a Voigt profile $H(a, v)$, where $a = (\Gamma_{\rm rad} + \Gamma_{\rm res})/4\pi\,\Delta\nu_D$, and $v = (\nu - \nu_0)/\Delta\nu_D$. This Voigt profile is then convolved (numerically) with the Stark profile $S(\alpha)$, yielding the cross section

$$\alpha_\nu(\Delta\nu) = (\pi^{\frac{1}{2}} e^2/mc) \int_{-\infty}^{\infty} S^*(\Delta\nu + v\Delta\nu_D) H(a, v)\, dv, \tag{8.150}$$

where S^* is $S(\alpha)$ converted to frequency units. However, many published tables of hydrogen-line Stark profiles mentioned above already contain a convolution with Doppler and natural broadening as described by equation (8.150). One can also use approximate expressions such as that suggested in [531]. They are based on the observation the Voigt function $H(a, v)$ exhibits a sharp peak around $v = 0$, essentially a Gaussian profile, while function S^* is smooth and slowly varying function in the wings, so that $\alpha(\Delta\nu) \approx S^*(\Delta\nu)$ in the line wings, while the line center is dominated by the Doppler core of the Voigt function, $\alpha(\Delta\nu) \propto H(a, x) \approx e^{-x^2}$, with $x = \Delta\nu/\Delta\nu_D$.

Hydrogenic Ions

Hydrogenic ions, of charge Z, have Stark patterns identical to those of hydrogen, though energies are, of course, different. The profile from ion broadening alone is

$$S(\alpha) = Z^5 S_{QS}(Z^5 \alpha), \tag{8.151}$$

where S_{QS} is the quasi-static hydrogen profile given, e.g., in [1094]. Note that on a wavelength scale the ion lines are narrower by a factor of $1/Z^5$. Indeed, because the Stark widths decrease for ions while radiative transition probabilities increase, there comes a point where Stark broadening can be neglected compared to Γ_{rad}.

The effects of electron broadening for hydrogenic ions are similar to those for hydrogen, though the expressions for Φ_{ab} and $\mathcal{L}$ change because now the perturber moves on a *hyperbola* around the positively charged ion instead of a *straight-line path*. Early calculations of the broadening of lines of He II $\lambda\lambda$ 256, 304, 1085, 1216, 1640, 4686, and 3203 are given in [614]. Unified theory of [1018] was modified so as to include the effects of the ionic charge on the paths of the perturbing electrons [394, 396], and actual calculations for He II λ304 are given in [395]. Finally, [970] adapted the unified theory and produced extensive tables [971] for He II λ1640 (transition 2–3), Paschen series [transitions 3–4 (λ4686) to 3–8 (λ2386)], and Pickering series [transitions 4–5 (λ10125) to 4–12 (λ4101)], for temperatures ranging from 1×10^4 K to 3.2×10^5 K, and electron densities from 3.162×10^{10} to 10^{16} cm^{-3}. These tables provide a very useful tool for analyzing spectra of hot stars.

Neutral Helium

The lines of He I are prominent in the spectra of B-stars. Here the electron impact broadening and quasi-static ion broadening act by quadratic Stark effect and for isolated lines yield profile of the form

$$I(\delta\omega) = \frac{w}{\pi} \int_0^\infty \frac{W(F)\, dF}{(\Delta\omega - d - C_4 F^2/e^2)^2 + w^2}, \tag{8.152}$$

where $W(F)$ is the probability of a field of strength F, w is the electron impact width of the line, and d the line shift. Note that because F enters as F^2, the ionic field always skews the line components in one direction and thus leads to an *asymmetric* profile.

Explicit expressions for w and d are given in [253, 400, 403, 405]. For *isolated* lines, tables allowing the calculation of the profile in terms of convenient dimensionless units are given in [405] together with detailed numerical results for w and d for several lines. Improved results are given in [88, 89, 253]. Comprehensive tables for electron and proton impact line broadening for lines originating from upper levels with $4 \leq n \leq 10$ have been published in [276, 277], and improved calculations in [279].

A much more difficult case is presented by the diffuse series lines (2^3P–n^3D, 2^1P–n^1D) for $n \geq 4$. As was first recognized by Struve [1062, 1063], the He I λ4471 line (2^3P–4^3D) shows a *"forbidden component"* (2^3P–4^3F) at λ4470. The other diffuse series lines also show these components, which arise from the mixing of the (^3D, ^3F) or (^1D, ^1F) states in the presence of the electric fields in the plasma. As the diffuse-series lines are among the best observed in stellar spectra, a reliable theory for them is of great interest. The first attempts at constructing such theories [89, 401] were not too successful, for they gave "forbidden" components that were too narrow and too intense. A comprehensive better theory was then developed for He I λ4471 [90] and λ4921 (2^1P–4^1D) [91] lines; predictions based on this theory are in excellent agreement with observed stellar spectra [735, 736].

Other Light Elements

Electron collisions broaden the lines of other elements observed in stellar spectra. Widths and shifts for these lines may be calculated using techniques similar to those employed for He I; see [403, §§II.3c, II.3.d]. Extensive results for neutral atoms are given in [400, pp. 454–527] and [403, Appendix IV]. For charged ions, the Coulomb interaction between the radiator and perturber implies a hyperbolic path for the latter [142, 143], and line widths allowing for this are substantially larger than those calculated with a straight-line path. Extensive results of detailed calculations for ions are presented in [403, Appendix V]; [143, 933, 934, 936]. Convenient approximate formulae for estimating Stark widths are given in [260, 402, 936].

Recently, a comprehensive effort was started to collect, maintain, and gradually upgrade available theoretical Stark broadening data within the international project Virtual Atomic and Molecular Data Center (VAMDC), called STARK-B [935]. It is designed to provide a database for line broadening by collisions with charged particles. The data are calculated using three basic approximations: (1) impact approximation (essentially, assuming that perturbing interactions are separated in time, and the atom interacts with one perturber only at a time), (2) complete collision approximation (the atom does not emit or absorb a photon during a collision), and (3) isolated lines. The database contains a large number of data for astrophysically interesting atoms and ions.

Chapter Nine

Kinetic Equilibrium Equations

Stellar atmospheres are regions of high temperature and low density. Therefore, the gas is composed mainly of single atoms, ions, and free electrons. The basic characteristics of the medium are a relatively low density and the presence of strong gradients in structural parameters. The latter arise because of the nature of a stellar atmosphere, which is bounded by the stellar interior on one side and by a cool, essentially empty space on the other. Therefore, the microscopic state of the gas cannot generally be described by equilibrium thermodynamics; instead, one has to use *non-equilibrium statistical mechanics* to describe its microscopic state.

9.1 LTE VERSUS NON-LTE

Local thermodynamic equilibrium (LTE) is characterized by the following three equilibrium distributions:

- Boltzmann excitation equation,

$$(n_i/N_I) = (g_i/U_I)\, e^{-E_i/kT}, \tag{9.1}$$

  where n_i is the population of level i, g_i is its statistical weight, and E_i is the level energy, measured from the ground state; N_I and U_I are the total number density and the partition function of the ionization stage I to which level i belongs, respectively.

- Saha ionization equation,

$$(N_I/N_{I+1}) = n_{\mathrm{e}}\,(U_I/U_{I+1})\,(h^2/2\pi m_{\mathrm{e}}kT)^{\frac{3}{2}}\, e^{\chi_I/kT}, \tag{9.2}$$

  where χ_I is the ionization potential of ion I.

- Maxwellian velocity distribution of particles, written here in terms of distribution for the absolute values of velocity,

$$f(v)\,dv = (m/2\pi kT)^{3/2}\, \exp(-mv^2/2kT)\, 4\pi v^2\, dv, \tag{9.3}$$

  where m is the particle mass and k the Boltzmann constant.

Combining equations (9.1) and (9.2), we obtain a useful expression for LTE level populations with respect to the ground state of the next higher ion. As is customary, we denote an LTE population with an asterisk.

$$n_i^* = n_e n_1^+ \, \Phi_i(T), \tag{9.4}$$

where n_1^+ is the population of the ground state of the next higher ion, and

$$\Phi_i(T) = (g_i/g_1^+)\,(h^2/2\pi m_e kT)^{\frac{3}{2}}\, e^{(\chi_I - E_i)/kT} = C(g_i/g_1^+) T^{-\frac{3}{2}}\, e^{(\chi_I - E_i)/kT} \tag{9.5}$$

is the Saha-Boltzmann factor. The constant has the value $C = 2.07 \times 10^{-16}$ in cgs units. The advantage of equation (9.5) is that it expresses LTE level populations without using a partition function.

Equations (9.1) though (9.3) define the state of LTE from the macroscopic point of view. Microscopically, LTE holds if all atomic processes are in *detailed balance*, i.e., if the number of processes $A \rightarrow B$ is exactly balanced by the number of inverse processes $B \rightarrow A$. By A and B we mean any particle states between which there exists a physically reasonable transition. For instance, A is an atom in an excited state i, and B the same atom in another state j (either of the same ion as i, in which case the process is an excitation/de-excitation, or of a higher or lower ion, in which case the term is an ionization/recombination). An illuminating discussion is presented in [834, chapter 1].

In contrast, by the term non-LTE (or NLTE), we describe any state that departs from LTE. In practice, one usually means that populations of some selected energy levels of some selected atoms/ions are allowed to depart from their LTE value, while the velocity distributions of all particles are assumed to be Maxwellian, (9.3), moreover, with the same kinetic temperature T.

To understand why and where we may expect departures from LTE, let us turn to the microscopic definition of LTE. It is clear that LTE breaks down if the detailed balance in at least one transition $A \rightarrow B$ fails. We distinguish between *collisional transitions* (arising due to interactions between two or more massive particles) and *radiative transitions* (interactions involving particles and photons)—see § 9.3. Under stellar atmospheric conditions, collisions between massive particles tend to maintain the local equilibrium (since velocities are Maxwellian). Therefore, the validity of LTE hinges on whether the radiative transitions are in detailed balance or not.

The very fact that the radiation escapes from a star implies that LTE must eventually break down at a certain point in the atmosphere. Essentially, this is because detailed balance in radiative transitions generally breaks down at a certain point near the surface. Since photons escape (and more so from the uppermost layers), there must be a lack of them there. Consequently, the number of photoexcitations (or any atomic transition induced by absorbing a photon) becomes lower than the number of inverse spontaneous processes, as their number does not depend on the number of available photons.

Thus we may expect departures from LTE if the following two conditions are met: (i) radiative rates in some important atomic transition dominate over the collisional rates, and (ii) radiation is not in equilibrium, i.e., the intensity does not have the Planckian distribution. Collisional rates are proportional to the particle density; it is therefore clear that for low densities departures from LTE will be significant or even crucial. Therefore, departures from LTE are expected to be largest in the upper layers of the atmosphere. In contrast, deep in the atmosphere, photons do not escape, and so the intensity is close to the equilibrium value. Departures from LTE are therefore small, even if the radiative rates dominate over the collisional rates.

9.2 GENERAL FORMULATION

Classical View

Under non-equilibrium conditions, the natural starting point to describe the microscopic state of the matter is to introduce the *distribution function*, f_i, for a particle of type i, where i refers not only to the general type of a particle (e.g., an individual atomic species) but also to the internal degree of freedom, i.e., the ionization and excitation state, of the particle. The physical meaning of f_i is such that $f_i(t, \mathbf{r}, \mathbf{p})\, d^3\mathbf{r}\, d^3\mathbf{p}$ represents a number of particles of type i at time instant t, located in an elementary volume in phase space $d^3\mathbf{r}\, d^3\mathbf{p}$ around position $\mathbf{r}$ and momentum $\mathbf{p}$.

The behavior and time evolution of the distribution function is determined by the *Boltzmann equation*, or *kinetic equation*, which is a phase-space conservation equation for particles of type i:

$$(\partial f_i/\partial t) + (\mathbf{p}/m_i)\cdot\nabla f_i + \mathbf{F}\cdot\nabla_{\mathbf{p}} f_i = \{\delta f_i/\delta t\}_{\rm coll}, \tag{9.6}$$

where m_i is the mass the particle, $\mathbf{p} = m_i\mathbf{u}$ for non-relativistic particles with $\mathbf{u}$ being the velocity of the particle, and $\mathbf{F}$ is the external force acting on the particles of the gas. Here the left-hand side represents the time change and the streaming term—a flow of particles through the volume of the phase space—while the right-hand side is the *collisional term*, which represents the net rate of removing the particle of type i from the elementary volume $d^3\mathbf{r}\, d^3\mathbf{p}$ by *interactions with other particles* and photons. The interactions, or *collisions*, are of two kinds. First, those that remove the particle from the elementary volume $d^3\mathbf{p}$ around $\mathbf{p}$, but do not change the type of particle (its internal degree of freedom). In other words, they change only the velocity of the particle, not its internal state—*elastic collisions*. The second kind is collisions that change the internal degree of freedom of a particle, possibly together with its velocity—*inelastic collisions*. In stellar atmospheres, we can safely neglect another possible kinds of interactions that may change even the type of a particle, such as thermonuclear reactions, pair production and annihilations, etc., which require much higher energies than those met in the typical conditions in stellar atmospheres.

Equation (9.6) describes an evolution of the distribution function f_i in a seven-dimensional space and provides a full description of f_i. Although it is used in its general form for some applications (e.g., in the solar transition region between the chromosphere and the corona), it is unnecessarily complex for most applications. The usual procedure adopted in statistical mechanics is to introduce *moments* of the distribution function with respect to the momentum, the simplest one being the *number density* of the particles of type i, or the zero-order moment, defined by

$$n_i(t,\mathbf{r}) = \int f_i(t,\mathbf{r})\, d^3\mathbf{p}. \tag{9.7}$$

We also introduce the mean (macroscopic) velocity of the particles of type i,

$$\mathbf{v}_i(t,\mathbf{r}) = \int (\mathbf{p}/m_i) f_i(t,\mathbf{r})\, d^3\mathbf{p}. \tag{9.8}$$

In principle, different types of particles may possess different mean velocities. However, in most circumstances, all the mean velocities are the same, $\mathbf{v}_i = \mathbf{v}$. If the medium is globally at rest, $\mathbf{v} = 0$, and $\mathbf{u}_i$ describes the thermal motion of the particles.

Integrating equation (9.6) over all velocities, we obtain the *conservation equation for the number density*,

$$(\partial n_i/\partial t) + \nabla \cdot (n_i \mathbf{v}_i) = \{\delta n_i/\delta t\}_{\rm coll}, \tag{9.9}$$

where the right-hand side represents the collisional term averaged over momentum,

$$\{\delta n_i/\delta_t\}_{\rm coll} = \int \{\delta f_i/\delta t\}_{\rm coll}\, d^3\mathbf{p}. \tag{9.10}$$

The term $\{\delta n_i/\delta_t\}_{\rm coll}$ contains contributions only from the inelastic collisions, because by integrating over all velocities the elastic collision contributions cancel [a negative contribution to $f_i(\mathbf{p})$ due to an elastic collision $\mathbf{p} \to \mathbf{p}'$ will appear as the positive contribution to $f_i(\mathbf{p}')$]. The collision term may then be written as

$$\{\delta n_i/\delta_t\}_{\rm coll} = -n_i \sum_{j\neq i} P_{ij} + \sum_{j\neq i} (n_j P_{ji}), \tag{9.11}$$

where P_{ij} denotes the *total rate* of transition from state i to state j by *all* processes, i.e., those involving photons as well as massive particles. We stress that the summation extends over all excitation and ionization states of the species. The general form of the *kinetic equilibrium equation* is then written as

$$(\partial n_i/\partial t) + \nabla \cdot (n_i \mathbf{v}) = \sum_{j\neq i} (n_j P_{ji}) - n_i \sum_{j\neq i} P_{ij}. \tag{9.12}$$

Note that by summing over all states of the given species and writing $N = \sum_i n_i$, one arrives at the *continuity equation* for this species,

$$(\partial N/\partial t) + \nabla \cdot (N\mathbf{v}) = 0, \tag{9.13}$$

because, again, by summing over all states of the species, the inelastic collision contributions cancel. By multiplying equation (9.13) by m, the mass of the species, and summing over all species, we obtain the standard hydrodynamical *continuity equation*,

$$(\partial\rho/\partial t) + \nabla \cdot (\rho \mathbf{v}) = 0. \tag{9.14}$$

For a steady state ($\partial/\partial t \equiv 0$), equation (9.12) simplifies to

$$n_i \sum_{j\neq i} P_{ij} - \sum_{j\neq i} (n_j P_{ji}) = -\nabla \cdot (n_i \mathbf{v}), \tag{9.15}$$

and for a *static* atmosphere ($\mathbf{v} = 0$), we have

$$n_i \sum_{j\neq i} P_{ij} - \sum_{j\neq i} (n_j P_{ji}) = 0. \tag{9.16}$$

As we consider only static media through chapter 18, we shall deal almost exclusively with equation (9.16), which is called the *kinetic equilibrium equation.*

The total rate P_{ij} is written as

$$P_{ij} = R_{ij} + C_{ij}, \tag{9.17}$$

and the kinetic equilibrium, or the *rate* equations, are written as

$$n_i \sum_{j\neq i} (R_{ij} + C_{ij}) = \sum_{j\neq i} n_j (R_{ji} + C_{ji}), \tag{9.18}$$

where R_{ij} is the *radiative rate*, which describes transition $i \to j$ caused by an interaction with photons or a spontaneous emission that results in an emission of a photon, and C_{ij} is the *collisional rate*, which corresponds to a transition caused by an interaction with particles. We shall consider these rates in detail in § 9.3.

Quantum Mechanical View

The formalism presented above is based on an implicit assumption that an atom is always in a definite state i and, due to interactions with neighboring photons and material particles (mainly electrons), makes an instantaneous transition to a different state j.

From the point of view of quantum mechanics, this is the case only for isolated atoms, in which case the probability of finding an atom in state $|i\rangle$ is given by $p_i = |\psi_i|^2$, where ψ_i is the wave function. However, an ensemble of identical atoms immersed in a bath of perturbing particles cannot be fully described by a wave function; instead, one introduces the concept of *density matrix* (or *density operator*) to describe such an ensemble. The density operator is defined by

$$\rho \equiv \sum_i |i\rangle p_i \langle i|, \tag{9.19}$$

where p_i is the probability of finding the system in state i. The evolution of the density matrix is described by the *master equation*,

$$d\rho(t)/dt = -(i/\hbar)[H, \rho(t)] \equiv -iL\rho(t), \tag{9.20}$$

where the *Liouville operator* L denotes the commutator with the Hamiltonian for the entire system, including atoms, perturbers, radiation field, and their mutual interactions. As follows from equation (9.19), the population of level i is proportional to the diagonal matrix element of the density matrix, $n_i \propto \langle i|\rho|i\rangle \equiv \rho_{ii}$. Therefore, a quantum mechanical analog of the kinetic equilibrium equation can be found by obtaining a stationary solution of the master equation (9.20). Importantly, due to interaction with other particles and photons, such a solution will contain not only diagonal elements ρ_{ii}, i.e., populations n_i of all levels, but also off-diagonal elements ρ_{ij}, called *resonances* [250]. Consequently, the classical kinetic equation (9.18) that contains only the populations n_j is generally incomplete from the quantum mechanical point of view. Nevertheless, as shown in [250], the off-diagonal density matrix elements are negligible in most astrophysical applications, and the diagonal elements are determined by an equation analogous to (9.18), so one may safely use the classical form of the kinetic equilibrium. An artifact of the quantum mechanical nature of the problem is some subtleties and conceptual problems when taking the classical description too literally. An example is a treatment of two-step processes, such as dielectronic recombination or resonance line scattering (cf. chapter 10), which cannot generally be split into two simple separate processes when writing down the kinetic equilibrium equation. But such complications are minor, so throughout this book we will adopt the classical formulation of the rate equations.

9.3 TRANSITION RATES

Overview of Transitions

We shall first summarize all possible types of transitions that may occur under typical stellar atmospheric conditions.

Photon-Induced Processes

First, there are radiative transitions that do not change the ionization state, namely, *radiative excitation* and *radiative de-excitation*,

$$X_i^Z + h\nu \rightleftarrows X_j^Z, \tag{9.21}$$

where X^Z denotes an ion of charge Z ($Z = 0$ for neutral atoms) and i and j are the particular excitation states of the same ion, with $E_i < E_j$. The process in the rightward direction, $\rightarrow$, is *radiative excitation*, or *photo-excitation*, an absorption of a photon with energy $h\nu$ with a simultaneous transition of an atom/ion from state i to a higher state j. An inverse process, $\leftarrow$, is a *spontaneous radiative de-excitation*, an emission of a photon $h\nu$ accompanied with a transition $j \to i$. As we pointed out in chapter 5, any spontaneous emission process is accompanied by a corresponding *stimulated emission process*,

$$X_j^Z + h\nu \to X_i^Z + h\nu + h\nu, \tag{9.22}$$

i.e., a process in which a photon $h\nu$ stimulates the transition $j \to i$ of an atom/ion with a simultaneous emission of another, identical photon.

Radiative transitions that involve a change of the ionization state of the atom/ion,

$$X_i^Z + h\nu \rightleftarrows X_j^{Z+1} + e, \tag{9.23}$$

are called *photoionization* ($\rightarrow$) and *radiative recombination* ($\leftarrow$). Here, e denotes a free electron. The state j of the higher ion is usually its ground state, $j = 1$, although this is not necessary, and ionizations to, or recombinations from, an excited state of an ion are possible and sometimes important. Again, the radiative recombination has its stimulated counterpart,

$$X_j^{Z+1} + e + h\nu \to X_i^Z + h\nu + h\nu. \tag{9.24}$$

We note that very high energy photons (such as those in the X-ray region) may remove more electrons from an atom/ion, an *Auger ionization*,

$$X_i^Z + h\nu \to X_j^{Z+k} + ke, \tag{9.25}$$

with $k \geq 2$. Usually, but not necessarily always, $k = 2$.

Particle-Induced Processes

These are collisional transitions, caused by interaction with neighboring particles. One usually considers collisions with free electrons. The reason is that charged particles dominate the interactions owing to the long-range nature of Coulomb interactions. Moreover, the collision frequency is proportional to the flux of impinging particles, and hence to their velocity. We can estimate the ratio of the velocity of free electrons and ions using thermal equilibrium, where $\overline{v}_e/\overline{v}_i = (m_{\rm H}A/m_{\rm e})^{\frac{1}{2}} \approx 43\,A^{\frac{1}{2}}$, with A being the atomic mass. Consequently, in a fully or partially ionized medium, the collisions with free electrons largely dominate over collisions with ions. For cool stars where there are few free electrons present, collisions with other particles, for instance, neutral hydrogen atoms, are important. In the following, we consider only collisions with free electrons.

The collisional bound-bound processes are analogous to the radiative ones.

$$X_i^Z + e \leftrightarrows X_j^Z + e' \tag{9.26}$$

are *collisional excitation* and *de-excitation*; here e' denotes an electron with energy lower than that of electron e. The ionization/recombination processes are more complicated. First, there are direct processes,

$$X_i^Z + e \leftrightarrows X_j^{Z+1} + e' + e'', \tag{9.27}$$

where e, e', and e'' are electrons with different energies. It turns out that an efficient channel for ionization/recombination is provided by a two-step process that involves a *doubly excited* state of an ion. The doubly excited state is that in which not only the valence electron is excited, but also one of the inner-shell electrons is excited. The energy of such system may exceed the ionization energy of the ion, so the ion exists in an excited state with an energy higher than the binding energy of the valence electron. Autoionization and dielectronic recombination are two-step processes that involve such a doubly exited state.

An autoionization process is defined as

$$X_j^{Z-1} + e \rightarrow X_{j^{**}}^{Z-1} + e' \rightarrow X_i^Z + e' + e'', \tag{9.28}$$

where j^{**} denotes a doubly excited state with an energy above the ionization limit. The inverse process is a two-step collisional transition that involves an electron-impact excitation from state i to the doubly excited state j^{**}, followed by a transition to the ordinary level j of ion Z.

A dielectronic recombination is a similar process:

$$X_j^Z + e \rightarrow X_{j^{**}}^{Z-1} \rightarrow X_i^{Z-1} + h\nu. \tag{9.29}$$

The first step is a reverse of the last step of an autoionization process, while the second step is a transition to an ordinary state of ion $Z-1$ with a simultaneous emission of a photon $h\nu$. The final result is the same as in the case of radiative recombination, namely, a capture of a free electron by an ion, resulting in an ion with a lower charge plus a photon. But because dielectronic recombination involves doubly excited states with specific energies, photons with certain discrete frequencies are created in the process, in contrast with usual radiative recombination, where a *continuous* spectrum of photons is generated.

At first sight it might seem that it would be simpler to treat both steps of dielectronic recombination separately, i.e., to describe the first step as a regular collisional excitation from j to j^{**} (of the lower ion), and the second part of a bound-bound transition—a spectral line—from state j^{**} to i (of the same ion). However, the fundamentally quantum mechanical nature of the process leads to *resonance* phenomena, which prevents making such a separation of the two steps of the process. From a practical point of view, the majority of the available data for dielectronic recombination rates are obtained for the complete two-step process because the rates are calculated by a quantum mechanical approach that does not make an artificial separation of the two steps. In fact, some calculations yield the total recombination rate, radiative plus dielectronic, without even separating them.

An inverse process to dielectronic recombination resembles an ordinary photoionization, but with discrete features. It is treated as *resonances* in the photoionization cross section, as we mentioned in chapter 7.

Finally, there is a process of *charge transfer*, which is an inelastic collision between an ion and a neutral particle (atom or molecule), in which m electrons, originally bound to neutral species, are transferred to the ion,

$$X^Z + Y^0 \rightleftarrows X^{Z-m} + Y^{+m}. \tag{9.30}$$

For instance, charge transfer with neutral hydrogen, where obviously $m = 1$, is

$$X^Z + \mathrm{H} \rightleftarrows X^{Z-1} + \mathrm{H}^+. \tag{9.31}$$

Charge transfer with He or, in cold atmospheres, with H_2 can have $m = 1$ or 2. For heavier atoms, such as oxygen, reactions even with $m > 2$ are possible, but under most circumstances, the single-electron exchange ($m = 1$) usually dominates. The process in the right direction, $\rightarrow$, is called *charge transfer recombination*, while the process in the opposite direction is *charge transfer ionization*.

Background Theory

Modern theoretical approaches [1198, 1199, 1234] treat essentially all these processes in an *ab initio* manner using the *close coupling* approximation and the R-matrix method [165, 1198, 1199]. Many such calculations were performed with the *Opacity Project* (OP) [1192–1232] and the *Iron Project* (IP) [1233–1253] international collaborative efforts.

Here we briefly outline the essential points of the theoretical approach. One describes the system that consists of an ion with N electrons, called the "target ion," plus the interacting $(N+1)$th, free electron. The wave function of the total system of $N+1$ electrons is written as

$$\psi_E(e + ion) = A \sum_i^N \chi_i(ion)\theta_i + \sum_j c_j \Phi_j(e + ion), \tag{9.32}$$

where χ_i is the target ion wave function in a specific state i and θ_i is the wave function of the interacting electron with a specific incident kinetic energy. In the second term, Φ_j is the correlation function of the electron + ion system that accounts for short-range correlation interactions. The target wave functions are obtained by the appropriate atomic structure calculations, as for instance was done in the Opacity and Iron Projects by using the code SUPERSTRUCTURE [316]. The wave function ψ_E is obtained by solving the Schrödinger equation with a relativistic Hamiltonian in the Breit-Pauli approximation:

$$H_{N+1}^{\mathrm{BP}} = H_{N+1}^{\mathrm{NR}} + H_{N+1}^{\mathrm{corr}}, \tag{9.33}$$

where H_{N+1}^{NR} is the non-relativistic Hamiltonian given by

$$H_{N+1}^{\mathrm{NR}} = \sum_{i=1}^{N+1} \left[-\nabla_i^2 - \frac{2Z}{r_i} + \sum_{j>i}^{N+1} \frac{2}{r_{ij}} \right], \tag{9.34}$$

and H_{N+1}^{corr} is a correction term; for its explicit form, see [165, 1198]. Here, r_i is the distance from the ith electron from the nucleus of charge Z, r_{ij} is the distance between electrons i and j, and ∇_i is the ∇ operator with respect to the coordinates of the ith electron. The Schrödinger equation is written as

$$H_{N+1}^{\rm BP}\psi_E = E\psi_E. \tag{9.35}$$

Substitution of the wave function expansion (9.32) into (9.35) produces a set of coupled equations that are solved using the R-matrix method. The resulting wave functions, with *positive energies*, $E > 0$, called *continuum wave functions*, ψ_F, describe the scattering process with a free electron interacting with a target ion at positive energies, whereas at *negative* energies ($E < 0$), the solutions correspond to pure bound states ψ_B.

Cross sections for particular processes are obtained, as explained in chapter 7, by computing the elements of the *transition matrix*, which for the individual processes are schematically given by

$\langle\psi_B||\mathbf{d}||\psi_{B'}\rangle$ for radiative bound-bound processes,
$\langle\psi_B||\mathbf{d}||\psi_F\rangle$ for radiative bound-free processes,
$\langle\psi_F||H(e+ion)||\psi_{F'}\rangle$ for collisional bound-bound processes,

where $\mathbf{d}$ is the dipole operator, $\mathbf{d} = e\sum_i \mathbf{r}_i$.

Radiative Rates

Bound-Bound Transitions

Consider the transition from bound level i to a higher bound level j. The number of transitions produced by incident specific intensity of radiation I_ν in the frequency interval $d\nu$ and solid angle $d\Omega$ is $n_i B_{ij} I_\nu \phi_{ij}(\nu,\mathbf{n})\,d\nu(d\Omega/4\pi)$, where B_{ij} is the Einstein coefficient for absorption and $\phi_{ij}(\nu)$ is the *absorption profile coefficient*, or simply *absorption profile*, for the transition $i\to j$. Specifically, $\phi_{ij}(\nu,\mathbf{n})\,d\nu\,d\Omega$ represents a *conditional probability* that *if* a photon is absorbed in transition $i\to j$, it is absorbed with frequency in the range $(\nu,\nu+d\nu)$ and in a solid angle $d\Omega$ around direction $\mathbf{n}$. As a consequence, the absorption profile is *normalized*, i.e., $\int \phi_{ij}(\nu,\mathbf{n})\,d\nu(d\Omega/4\pi) = 1$. The total number of transitions is obtained by integrating over all frequencies and angles. Since the transition *rate* is defined as a number of transitions divided by the number density of the initial state, the radiative rate is given by

$$R_{ij} = B_{ij}\int_0^\infty \oint I(\nu,\mathbf{n})\phi_{ij}(\nu,\mathbf{n})\,(d\Omega/4\pi)\,d\nu. \tag{9.36}$$

In a static medium, ϕ is isotropic, so we have

$$R_{ij} = B_{ij}\int_0^\infty \phi_{ij}(\nu)J_\nu\,d\nu \equiv B_{ij}\overline{J}_{ij}, \tag{9.37}$$

where $\overline{J}_{ij}$ is called the *frequency-averaged mean intensity*.

It is advantageous to introduce a *cross section* for the transition $i\to j$ as

$$\alpha_{ij}(\nu) \equiv (h\nu/4\pi)B_{ij}\phi_{ij}(\nu), \tag{9.38}$$

so that the radiative rate can also be written as

$$R_{ij} = 4\pi \int_0^\infty \alpha_{ij}(\nu)(h\nu)^{-1} J_\nu \, d\nu \approx (4\pi/h\nu_{ij}) \int_0^\infty \alpha_{ij}(\nu) J_\nu \, d\nu, \qquad (9.39)$$

where in the second approximate equality we used the fact that the width of the absorption profile is much smaller than the frequency of the transition. In moving media (chapter 19), we may consider either the *co-moving (Lagrangian) frame*, in which case equation (9.39) remains valid if J_ν is the mean intensity as measured in this frame (i.e., by an observer at rest with respect to the moving fluid), or the *observer's (Eulerian) frame*, in which case one uses equation (9.36), and a double integration over frequencies and angles must be carried out explicitly. Similar remarks apply to other radiative rates given below.

Similarly, the rate of transitions $j \to i$ is given by

$$R_{ji} = A_{ji} + B_{ji} \int_0^\infty \oint I(\nu, \mathbf{n}) \psi_{ji}(\nu, \mathbf{n}) \, (d\Omega/4\pi) \, d\nu. \qquad (9.40)$$

This is a general expression that takes into account the fact that the profile of emission coefficient, or the *emission profile*, may generally be different than the absorption profile, essentially because the probability of emission may be influenced by properties of a previously absorbed photon in the scattering process $i \to j \to i$. We shall consider this topic in detail in chapter 10. Here, we shall assume *complete redistribution*, in which case both profiles are identical, $\psi_{ji}(\nu) = \phi_{ij}(\nu)$. The downward radiative rate can then be written as

$$R_{ji} = A_{ji} + B_{ji} \overline{J}_{ij} = (g_i/g_j) B_{ij} \left[(2h\nu_{ij}^3/c^2) + \overline{J}_{ij} \right], \qquad (9.41)$$

where we used the relations between the Einstein coefficients; see equations (5.7) and (5.8). Using the cross section defined above, the rate may also be written as

$$R_{ji} \approx (4\pi/h\nu_{ij})(g_i/g_j) \int_0^\infty \alpha_{ij}(\nu) \left[(2h\nu^3/c^2) + J_\nu \right] d\nu. \qquad (9.42)$$

In order to be able to calculate the radiative rates needed to solve the kinetic equilibrium equations, we need to know (i) the Einstein coefficients, which are calculated as described in chapter 7, or, in some instances, one can take measured values; (ii) the absorption profile coefficient ϕ_{ij}, whose evaluation is described in detail in chapter 8; and (iii) the mean intensity of radiation in the line. While the first two ingredients are provided by atomic physics and are either given constants (Einstein coefficients) or are prescribed functions of temperature and electron density (the absorption profile), the radiation intensity has to be obtained by solving the transfer equation. The opacities and emissivities needed to solve the transfer equations depend, in turn, on level populations, so one is faced with a complicated coupled problem of a simultaneous solution of the transfer equation together with the kinetic equilibrium equations. Chapter 14 is devoted specifically to this problem.

Bound-Free Transitions

In analogy to the expressions derived above, the radiative rate of the transition from a bound state i to state k of the next higher ion is given by

$$R_{ik} = 4\pi \int_{\nu_0}^{\infty} \alpha_{ik}(\nu)(h\nu)^{-1} J_\nu \, d\nu, \tag{9.43}$$

where $\alpha_{ik}(\nu)$ is the photoionization cross section.

We calculate the rate of *spontaneous recombination* by use of a detailed-balancing argument. In *thermodynamic equilibrium*, the total number of spontaneous recombinations must be equal to the number of photoionizations given by $n_i R_{ik}$ when (i) J_ν has its equilibrium value, i.e., $J_\nu = B_\nu$; (ii) we correct for stimulated emission at the thermodynamic equilibrium value by multiplying by a factor $(1 - e^{-h\nu/kT})$ (cf. §5.2); and (iii) all level populations have their equilibrium values, given by the Boltzmann distribution, $n_j = n_j^*$. Thus,

$$(n_k R_{ki})^*_{\text{spon}} = n_i^* \, 4\pi \int_{\nu_0}^{\infty} \alpha_{ik}(\nu)(h\nu)^{-1} B_\nu (1 - e^{-h\nu/kT}) \, d\nu. \tag{9.44}$$

The recombination process is a collisional process involving electrons and ions, and therefore its rate is proportional to $n_e \cdot n_k$. For a given electron velocity and given T, which by definition describes the electron velocity distribution, the rate just calculated must still apply *per ion*, even out of thermodynamic equilibrium. Hence, to obtain the general, non-LTE, spontaneous recombination rate we need to correct equation (9.44) only by using the actual number density n_k. Then

$$\begin{aligned}(n_k R_{ki})_{\text{spon}} &= n_k (n_i/n_k)^* \, 4\pi \int_{\nu_0}^{\infty} \alpha_{ik}(\nu)(h\nu)^{-1} B_\nu (1 - e^{-h\nu/kT}) \, d\nu \\ &= n_k (n_i/n_k)^* \, 4\pi \int_{\nu_0}^{\infty} \alpha_{ik}(\nu)(h\nu)^{-1} \, (2h\nu^3/c^2) e^{-h\nu/kT} \, d\nu. \end{aligned} \tag{9.45}$$

The number of stimulated recombinations is calculated using the same procedure; in thermodynamic equilibrium,

$$(n_k R_{ki})^*_{\text{stim}} = n_i^* \, 4\pi \int_{\nu_0}^{\infty} \alpha_{ik}(\nu)(h\nu)^{-1} B_\nu \, e^{-h\nu/kT} \, d\nu. \tag{9.46}$$

To generalize the result to the non-LTE case, we replace the equilibrium radiation field B_ν by the actual value and use the actual number density n_k:

$$(n_k R_{ki})_{\text{stim}} = n_k (n_i/n_k)^* \, 4\pi \int_{\nu_0}^{\infty} \alpha_{ik}(\nu)(h\nu)^{-1} J_\nu \, e^{-h\nu/kT} \, d\nu. \tag{9.47}$$

The total number of recombinations is, therefore,

$$n_k R_{ki} = n_k \left(\frac{n_i}{n_k}\right)^* 4\pi \int_{\nu_0}^{\infty} \frac{\alpha_{ik}(\nu)}{h\nu} \left(\frac{2h\nu^3}{c^2} + J_\nu\right) e^{-h\nu/kT} \, d\nu. \tag{9.48}$$

The number of recombinations is sometimes expressed in terms of a *recombination coefficient* $\alpha_{\rm RR}(T)$ defined such that the total recombination rate as given by equation (9.48) is $n_k n_{\rm e} \alpha_{\rm RR}(T)$.

It is often advantageous to use a synoptic notation in which both bound-bound and bound-free radiative rates are given by the same expression. For all *upward* transitions $i \to j$, for j belonging to the same ion or not (i.e., bound or free), we write the total number of transitions as $n_i R_{ij}$, where the rate is given by

$$R_{ij} = 4\pi \int_0^\infty \alpha_{ij}(\nu)(h\nu)^{-1} J_\nu \, d\nu, \tag{9.49}$$

and it is understood that for a photoionization process $\alpha_{ij}(\nu) = 0$ for $\nu < \nu_0$. For all *downward* radiative transitions $j \to i$, the total number of transitions is written in terms of a *modified downward rate* as $n_j (n_i/n_j)^* R'_{ji}$, where

$$R'_{ji} = 4\pi \int_0^\infty \alpha_{ij}(\nu)(h\nu)^{-1} [(2h\nu^3/c^2) + J_\nu] \, e^{-h\nu/kT} \, d\nu. \tag{9.50}$$

For radiative recombinations, equation (9.50) follows directly from (9.48). For de-excitations, one applies equation (9.42) and expresses g_i/g_j using the Boltzmann excitation formula, equation (9.1), $(n_j/n_i)^* = (g_j/g_i) e^{-h\nu_{ij}/kT}$, and noticing that $e^{-h\nu_{ij}/kT}$ may be replaced by $e^{-h\nu/kT}$ and put into the integral over frequencies because the frequency varies negligibly over the range where the cross section $\alpha_{ij}(\nu)$ have non-vanishing values. Note that in equilibrium $R^*_{ij} = (R')^*_{ji}$.

So far, the recombination rate given by equation (9.48) or (9.50) can be understood as the *radiative recombination* rate, because it is expressed through the cross section for radiative ionization (photoionization). However, as mentioned above, modern approaches that compute the photoionization cross sections, such as those based on the close-coupling R-matrix approach, produce cross sections that exhibit a number of discrete, line-like features that arise due to resonances with doubly excited states above the ionization threshold. Therefore, by calculating the total recombination rate by exact integration of a detailed photoionization cross section over all frequencies, one in fact automatically includes dielectronic recombination. This is a consequence of the power of the R-matrix approach that does not have to make an artificial distinction between the radiative and dielectronic recombination.

Separating these two parts, and a necessity of having an extra *dielectronic recombination rate*, was an issue in previous astrophysical treatments that worked either in terms of total recombination rates, given by simplified atomic physics calculations, or in terms of smooth photoionization cross sections without resonances, in which case an extra dielectronic recombination rate was needed.

Collisional Rates

As explained above, collisions with free electrons usually dominate over collisions with heavier particles, so we first consider the electron collisions. Let $\sigma_{ij}(v)$ be the cross section for producing the transition $i \to j$ (where j may belong to the same ion as i or not) by collisions with electrons of velocity v (relative to the atom). Then the total number of transitions is

$$n_i C_{ij} = n_i n_e \int_{v_0}^{\infty} \sigma_{ij}(v) f(v) v \, dv \equiv n_i n_e q_{ij}(T), \tag{9.51}$$

where v_0 is the velocity corresponding to E_{ij}, the threshold energy of the process, i.e., $\frac{1}{2} m v_0^2 = E_{ij}$. The downward rate ($j \rightarrow i$) can be obtained immediately on the basis of detailed-balancing arguments, for the electron velocity distribution is the equilibrium, Maxwellian, distribution; hence we must have

$$n_i^* C_{ij} = n_j^* C_{ji}, \tag{9.52}$$

from which it follows that the number of downward transitions is

$$n_j C_{ji} = n_j (n_i/n_j)^* C_{ij} = n_j (n_i/n_j)^* n_e q_{ij}(T). \tag{9.53}$$

The actual cross sections σ_{ij} required to compute rates are found either experimentally or by rather involved quantum mechanical calculations. It would take us too far afield to describe these methods here; instead, we refer the interested reader to [883] and references therein. These calculations typically provide the cross section as a function of energy of the colliding electron. It is often expressed in units of πa_0^2, where a_0 is the Bohr radius, i.e., we write $\sigma_{ij} = \pi a_0^2 Q_{ij}(E)$, with $E = \frac{1}{2} m v^2$. Substituting (9.3) into (9.51), we find

$$q_{ij}(T) = C_0 T^{\frac{1}{2}} \int_{u_0}^{\infty} Q_{ij}(ukT) u \, e^{-u} \, du, \tag{9.54}$$

where $u \equiv E/kT$, $u_0 \equiv E_{ij}/kT$, and $C_0 = \pi a_0^2 (8k/m\pi)^{\frac{1}{2}} = 5.465 \times 10^{-11}$. Performing the integration in equation (9.54), we obtain

$$q_{ij}(T) = C_0 T^{\frac{1}{2}} \exp(-E_{ij}/kT) \, \Gamma_{ij}(T), \tag{9.55}$$

where

$$\Gamma_{ij}(T) \equiv \int_0^{\infty} Q_{ij}(E_{ij} + xkT)(x + u_0) e^{-x} \, dx. \tag{9.56}$$

Equation (9.55) is useful if atomic physics calculations provide numerical values of Γ_{ij} or detailed cross section Q_{ij}. On the other hand, the calculations performed within the Opacity and Iron Projects calculated a quantity analogous to Q_{ij} directly from the scattering matrix of the system, where the incident electron energies were measured in Rydbergs. The collisional excitation rate is then given by

$$q_{ij}(T) = 8.63 \times 10^{-6} \, T^{-\frac{1}{2}} g_i^{-1} \exp(-E_{ij}/kT) \, \gamma_{ij}(T), \tag{9.57}$$

where γ_{ij} is a quantity analogous to Γ_{ij}, which is obtained by integrating the energy (velocity) of the incident electron over the Maxwellian velocity distribution but expressed in Rydbergs. Therefore, the factor $C_0 T^{\frac{1}{2}}$ that enters equation (9.54) has to be multiplied by $E_H/kT = 157885/T$, with E_H being the ionization energy of hydrogen. The new factor g_i^{-1} follows from the scattering matrix and was absorbed in Γ_{ij} when writing down equation (9.54).

Equations (9.54) or (9.57) yield quite accurate results for the transition rate q_{ij}, provided that the original cross section Q_{ij} is either measured in the laboratory (as it is in a few cases, mostly transitions from a ground state) or computed theoretically. A characteristic difficulty for astrophysical work is that for many transitions of interest, $kT \ll E_{ij}$, so that the rate depends very sensitively upon values of Q_{ij} near the threshold. Unfortunately, for $E \sim E_{ij}$ a great computational effort is required to obtain accurate cross sections because several simplifying approximations that are valid for $E \gg E_{ij}$ break down, and because complicated variations of Q_{ij} result from various *resonances* in the collision process.

In cases where accurate calculations or experimental values are not available, one has to resort to approximate expressions. One of the most popular and often-used approximation is the *Van Regemorter formula* [1107], which expresses the collisional excitation rate for *dipole-permitted* transitions in terms of the oscillator strength f_{ij} as

$$q_{ij}(T) = C_0 T^{\frac{1}{2}} \, [14.5 f_{ij} (E_H/E_{ij})^2] \, u_0 \, e^{-u_0} \, \Gamma_e(u_0), \tag{9.58}$$

where, *for ions*,

$$\Gamma_e(u_0) \equiv \max[\overline{g}, 0.276 \, e^{u_0} \, E_1(u_0)], \tag{9.59}$$

where E_1 is the first exponential integral; see equation (11.107). The parameter $\overline{g}$ is about 0.7 for transitions of the form $nl \rightarrow nl'$ and about 0.2 for transitions of the form $nl \rightarrow n'l'$, $n' \neq n$ [122]. For neutral atoms, Γ_e has a different form (e.g., [57] for approximations of q_{ij} for Ne I). It is worth stressing that as the collisional cross sections do not depend on the dipole moment, collisional transitions are *not restricted* by the dipole selection rules $\Delta l = \pm 1$, and cross sections for other values of Δl (dipole-forbidden transitions) may be as large as those for $\Delta l = \pm 1$ despite f_{ij} being essentially zero in the dipole approximation. Therefore, while equations (9.58) and (9.59) provide an acceptable approximation for dipole-permitted transitions, they fail completely for dipole-forbidden transitions. For those, one typically uses equation (9.57) with an empirical estimate (or, very often, an educated guess) of γ_{ij}.

For *collisional ionizations*, if no detailed calculations are available, one typically uses an approximate formula, known as *Seaton's formula* [95, p. 374], which yields the rate [578, p. 121]

$$q_{ik} = 1.55 \times 10^{13} \, T^{-\frac{1}{2}} \overline{g}_i \alpha_{ik}(\nu_0) \, e^{-u_0}/u_0, \tag{9.60}$$

where $\alpha_{ik}(\nu_0)$ is the threshold photoionization cross section and $\overline{g}_i$ is of order 0.1, 0.2, and 0.3 for $Z = 1$, 2, and > 2, respectively, where Z is the charge of the ion. There are several other empirical formulas available. For instance, [701] suggested a formula for the cross section, which yields the rate

$$q_{ik} = C_0 T^{\frac{1}{2}} [2.5\zeta (E_H/E_0)^2]\, u_0 \,[E_1(u_0) - b e^c u_0 E_1(u_1)/u_1], \tag{9.61}$$

where ζ, b, and c are empirical quantities fitted to individual atoms and $u_1 = u_0 + c$.

Collisions with Hydrogen

As mentioned above, in solar- and later-type stars where the ionization of the medium, and hence the electron density, become low, inelastic collisions with a neutral hydrogen atom become the most important mechanism for collisional bound-bound and bound-free processes.

An approach most widely used in astrophysical literature is to employ the *Drawin formula* [289], which is essentially a modified classical Thomson formula. It provides merely an order of magnitude estimate. The original study dealt with collisions between two atoms of the same species; a generalization presented in [1041] considers collisions between hydrogen and other atomic species. The corresponding rate for transition $i \to j$ in atom A is given by

$$q_{ij}(T) = 16\pi a_0^2 \left(\frac{2kT}{\pi\mu}\right)^{\frac{1}{2}} \left(\frac{E_H}{E_{ij}}\right)^2 f_{ij} \,\frac{m_{\rm A}}{m_{\rm H}} \,\frac{m_{\rm e}}{m_{\rm H} + m_{\rm e}} \Psi(u_0), \tag{9.62}$$

where

$$\Psi(u) = e^{-u}/(1 + 2/u), \tag{9.63}$$

and where $m_{\rm A}$ and $m_{\rm H}$ are the mass of atom A and hydrogen, respectively, and μ is the reduced mass of the atom-hydrogen system, $\mu = m_{\rm A} m_{\rm H}/(m_{\rm A} + m_{\rm H})$. As for the Van Regemorter formula, equation (9.58), the Drawin formula (9.62) also breaks down completely for dipole-forbidden transitions. In those cases, and for more accurate values of the collisional rates for allowed transition, one has to employ results of detailed quantum mechanical calculations, whenever available. An overview of some quantum mechanical results, and a discussion of accuracy of Drawin formula, is presented in [86].

Charge Transfer Reactions

Detailed quantum mechanical calculations for this process are relatively difficult and rare. However, one can obtain reasonably accurate results by use of the so-called Landau-Zener approximation; see, e.g., [171]. A comprehensive study that performs more extended calculations [618] employs an analytic fit to the theoretical results, as well as available experimental data, by using an expression

$$\sigma_{\rm rec}(T) = a\, T_4^b\, (1 + e^{d\, T_4}) \tag{9.64}$$

and provides fitting coefficients a, b, c, d for the first 30 elements. Here $T_4 = T/10^4$. The subscript "rec" refers to the process of *charge transfer recombination*, which is the process $\mathrm{A}^Z + \mathrm{H} \rightarrow \mathrm{A}^{Z-1} + \mathrm{H}^+$. The inverse process is called *charge transfer ionization*; its rate follows from the usual detailed balancing arguments.

9.4 LEVEL DISSOLUTION AND OCCUPATION PROBABILITIES

The formalism presented above is not complete. It takes into account some aspects of the interaction of the atom/ion under study (called "radiator") with neighboring particles (perturbers) and photons, namely, the induced transitions between atomic levels (discussed in this chapter), as well as broadening of spectral lines (chapter 8). However, it does not consider another aspect, namely, an influence of perturbers on the electrostatic potential of the atomic nucleus and consequent distortion of atomic eigenstates. One particular aspect of the latter effect is known as *Debye shielding* (cf. § 8.4), accompanied with a lowering of the ionization potential, χ, by an amount $\Delta\chi$, which is a specified function of temperature and ion density; see equation (8.87).

A classical description of this effect assumes that the energy levels (eigenstates) with energies below $\chi - \Delta\chi$ (measured from the ground state) do exist and are unperturbed, while levels with energies higher than $\chi - \Delta\chi$ do not exist, and an electron with energy in this range is viewed as free. The kinetic equilibrium equations derived above would then apply only for such "existing levels."

In reality, an interaction with neighboring particles influences all energy levels, so a much improved description is based on introducing the *occupation probability*, w_i, for each level i (that is, any eigenstate of the atom), defined such that the LTE population of level i is given by

$$(n_i/n_I)^* = w_i\,(g_i/U_I)\,e^{-E_i/kT}, \tag{9.65}$$

which is a direct generalization of equation (9.1). Physically, w_i is the probability that the atom in question is in a state i relative to that in a similar ensemble of non-interacting atoms. Correspondingly, $(1 - w_i)$ is the probability that the state i is dissolved, i.e., it lies in the continuum. In other words, the corresponding electron is free, and the atom/ion with charge Z in a dissolved state i is counted among ions with charge $Z + 1$. Another advantage of this formalism is that the partition function is defined as

$$U = \sum_{i=1}^{\infty} w_i\, g_i\, e^{-E_i/kT}. \tag{9.66}$$

Since w_i monotonically decreases with increasing i and quickly approaches zero when i exceeds some value depending on the local conditions, this expression naturally avoids divergence of the classical partition function, as well as artificial cutoff procedures.

Clearly, in the classical treatment $w_i = 1$ for levels with $E_i < \chi_I - \Delta\chi_I$, and $w_i = 0$ for levels with $E_i \geq \chi_I - \Delta\chi_I$. An improved theory should provide a definite w_i, with $0 < w_i \le 1$, for each i. Such a theory was developed by Hummer and Mihalas in [1192], and later updated in [792] and [531], to which the interested reader is referred for details. Here we present only a brief outline of the most important results.

Evaluation of the Occupation Probabilities

A treatment of occupation probabilities depends on whether the perturbers are neutral or charged particles. For *neutral* perturbers, one can use a simple hard-sphere model [1192] to obtain

$$w_i^{\rm neut} = \exp[-(4\pi/3)\, n_p (r_i + r_{p1})^3], \tag{9.67}$$

where n_p is the perturber number density, r_i is the orbital radius associated with state i, and r_{p1} is the orbital radius associated with the perturber in the ground state. Equation (9.67) assumes that there is only one kind of perturber and that interaction with perturbers residing in excited states is negligible—the so-called *low-excitation approximation*. As shown in [1192], both of these approximations may be easily relaxed.

Perturbations with *charged particles* are usually more important for astrophysical applications (an exception being, for instance, the cool white dwarfs). Here, the effect of charged particles on an atom or ion is represented by a fluctuating microfield, exactly as in Holtsmark theory of quasi-static line broadening [cf. § 8.4]. Consider a hydrogenic state with the principal quantum number i. The basic idea is that for each bound state i of an unperturbed atom/ion, there is a critical value of electric field $F_c(i)$ such that the state in question cannot exist if the field exceeds the critical value. Hence the probability that a given state *does* exist is simply the probability that the field strength is less than $F_c(i)$, i.e.,

$$w_i = \int_0^{F_c(i)} W(F)\, dF = \int_0^{\beta_c(i)} W(\beta)\, d\beta, \tag{9.68}$$

where $\beta = F/F_0$ is the field strength expressed in units of normal field strength, F_0, with F_0 given by equation (8.65). As shown in [1192], the critical field strength is given by

$$\beta_c(i) = 8.3 \times 10^{14}\, n_e^{-\frac{2}{3}}\, Z^3\, k_i\, i^{-4}, \tag{9.69}$$

where Z is the charge of perturbing ions, and

$$k_i = \begin{cases} 1, & \text{for} \quad i \le 3, \\ (16/3)\, i\, (i+1)^{-2}, & \text{for} \quad i > 3. \end{cases} \tag{9.70}$$

Neglecting correlations between perturbers, the microfield distribution function is given by the well-known Holtsmark distribution (cf. § 8.4). When plasma correlation effects are important, the microfield distribution functions are given, e.g., by

Hooper's distribution function [504]. Using this distribution, one can numerically integrate equation (9.68) and fit it by an approximate analytical formula [531, Appendix A], [792]. One obtains

$$w_i = f/(1+f), \tag{9.71}$$

where

$$f = \frac{c_1(x + 4Z_r a^3)\beta_c^3}{1 + c_2 x \beta_c^{3/2}}, \tag{9.72}$$

where

$$a = 0.09\, n_e^{\frac{1}{6}} T^{-\frac{1}{2}} \tag{9.73}$$

is the correlation parameter, Z_r is the radiator charge (=0 for neutral atoms), and $c_1 = 0.1402$, $c_2 = 0.1285$, and $x = (1+a)^{3.15}$.

Rate Equations with Occupation Probabilities

We follow here an analysis presented in [531], to which the reader is referred for additional details.

The population of state i represents the number density of atoms in a true, *undissolved* state i. The transition probabilities, considered above, are unchanged, provided that they are understood as *conditional probabilities*, given that state i is undissolved. The crucial point to realize is that any transition from a certain state i to another state j, $j > i$, may leave an atom either in the bound (undissolved) state j or in an unbound, ionized state at the same energy. Thus, the total rate of transitions out of state i is given by

$$P_i^{\rm out} = \sum_{j\neq i} w_j P_{ij} + \sum_{j\neq i} (1 - w_j) P_{ij} + P_{ik}, \tag{9.74}$$

where the first term describes transitions $i \to j$ to the undissolved fraction of state j, the second term represents an ionization into the dissolved fraction of this state, and the third term describes an ordinary ionization from state i in the absence of level dissolution. The second and the third term are naturally combined to form an effective, total ionization rate

$$P_{ik}^{\rm tot} = P_{ik} + \sum_{j\neq i} (1 - w_j) P_{ij}. \tag{9.75}$$

This equation also applies for both radiative and collisional rates individually. The total collisional ionization rate is obtained by performing a summation in equation (9.75) explicitly, truncating a sum at a level i where $(1 - w_i)$ is sufficiently small.

The case of total photoionization rate is more complicated. As in the case of ordinary photoionization, we need to express the rate as an integral over frequencies of the cross section times the mean intensity of radiation. As shown in [531, 1197], the corresponding cross section can be written, to a good approximation, as

$$\alpha_{ik}^{\rm tot}(\nu) = D_i(\nu)\,\alpha_{ij}^{\rm ext}(\nu), \tag{9.76}$$

where $\alpha_{ik}^{\rm ext}(\nu) = \alpha_{ik}(\nu)$ for $\nu \geq \nu_{ik}$, while for lower frequencies it represents an *extrapolated cross section*, and

$$D_i(\nu) = \begin{cases} 1, & \text{if} \quad \nu \geq \nu_{ik}, \\ 1 - w_{m_i^*}(\nu), & \text{if} \quad \nu < \nu_{ik}. \end{cases} \tag{9.77}$$

Here ν_{ik} is the ionization frequency from level i, and $m_i^* = [i^{-2} - (\nu/\nu_{ik})]^{-\frac{1}{2}}$ is an effective quantum number of the highest state that can be reached from state i by the absorption of a photon with frequency ν (in the hydrogenic approximation). It does not have to be an integer, and its occupation probability is evaluated by analytic expressions (9.71)–(9.73) that allow for non-integer values of m. The quantity $D_i(\nu)$ is called the *dissolved fraction*. When the cross section given by (9.76) and (9.77) is employed for computing the opacity, the part corresponding to $\nu < \nu_{ik}$ is called the *pseudo-continuum opacity*.

The total rate of transitions *into* state i is

$$P_i^{\rm in} = w_i\left(\sum_{j\neq i} n_j P_{ji} + n_k P_{ki}^{\rm tot}\right), \tag{9.78}$$

because now i is the final state of transitions $j \to i$, and therefore the probability that it is not dissolved has to be taken into account. Obviously, the remaining part of transitions into the dissolved fraction of state i does not have to be dealt with because those transitions are in fact collisions between ions and therefore do not contribute to the population of true, undissolved atoms in state i.

The downward rates are obtained by using the usual detailed balancing arguments. For instance, for collisional bound-bound processes, the detailed balance for transition $i \leftrightarrow j$ stipulates that in equilibrium $n_i^* w_j C_{ij} = n_j^* w_i C_{ij}$, and hence $C_{ji}/C_{ij} = (g_i/g_j)\exp[(E_j - E_i)/kT]$, exactly as in the case of no dissolution, as it should be. By the same reasoning, $C_{ki} = (n_i/n_k)^* C_{ik}$, and for the downward radiative bound-free rate one can use equation (9.50), replacing α_{ik} with $\alpha_{ik}^{\rm tot}$.

The kinetic equilibrium equation for level i is then

$$n_i\left[\sum_{j\neq i} w_j(R_{ij} + C_{ij}) + R_{ik}^{\rm tot} + C_{ik}^{\rm tot}\right]$$
$$= w_i\left[\sum_{j\neq i} n_j(R_{ji} + C_{ji}) + (n_i/n_k)^*(R_{ki}^{\prime\,\rm tot} + C_{ik}^{\rm tot})\right]. \tag{9.79}$$

The rate equations can be written in the identical form as the original equations (9.18), provided one multiplies any transition rate by the *occupation probability of the final state*, $P_{ij} \to w_j P_{ij}$. In the subsequent text we will make no notational distinction between the original and modified transition rates.

9.5 COMPLETE RATE EQUATIONS

Closure Relation

The kinetic equilibrium equation for level i is

$$n_i\Big(\sum_{j\neq i} R_{ij} + C_{ij}\Big) = \sum_{j\neq i} n_j\Big(R_{ji} + C_{ji}\Big). \tag{9.80}$$

We stress that equation (9.80) is written for the levels of all ionization stages of the given atom. The total number of levels is $NL = \sum_I NL_I$, where NL_I is the number of levels of ionization stage I.

If (9.80) were to be written for all levels of the atom, the resulting equations would be a linearly dependent system because the last equation is a linear combination of the remaining equations because all the transitions into and out of the last level were already accounted for in the equations for the previous levels.

Therefore, one of the equations has to be replaced by some other relation. Typically, one uses the expression for the total number density of the atom, sometimes called the *abundance definition equation*,

$$\sum_{j=1}^{NL} n_j = n_{\rm atom}. \tag{9.81}$$

Here $n_{\rm atom}$ is the total number density of the given atomic species in all excitation and ionization states, which can also be written as $n_{\rm atom} = A_{\rm atom} n_H$, where A is the *chemical abundance* of that atom relative to hydrogen, and n_H is the *total number density of hydrogen atoms and ions*. There is still freedom as to which equation is to be replaced by the abundance definition condition. Two possibilities are used in the literature. The common one is to replace the equation for the last level, $i = NL$. Another choice, suggested in [200], sometimes more advantageous in terms of numerical stability, is to replace the rate equation for the level with the largest population.

Equation (9.81) applies, strictly speaking, only if *all* levels of the atomic species that have a non-negligible population are included in the set of kinetic equilibrium equations. In practice, the number of such levels may be enormous, and therefore the following strategy is adopted: the lowest NL_I levels of the ionization stage I of the given atom are being considered in detail. The remaining, higher, levels are not treated in detail, but their contribution to the total number density of the species is taken into account, assuming that the populations of these levels are in LTE with respect to the ground state of the next ion. The total population of such high states of ion I can then be written as

$$n_I^{\rm up} = n_{\rm e} n_1^{I+1}\, \Sigma_I, \tag{9.82}$$

where

$$\Sigma_I \equiv \sum_{i=j_{\rm min}}^{j_{\rm max}} w_i \Phi_i(T), \tag{9.83}$$

where w_i is the occupation probability of level i, Φ_i is its Saha-Boltzmann factor defined by (9.5), $j_{\rm min}$ is the index of the lowest level not being treated explicitly, and $j_{\rm max}$ is an index of highest level that contributes to the total upper level populations. The latter is not a well-defined quantity. A better approach is to use the partition function and to write [521]

$$\Sigma_I = (h^2/2\pi m_e kT)^{\frac{3}{2}}\, e^{\chi_I/kT} \left(U_I/g_1^{I+1}\right) - \sum_{i=1}^{NL_I} w_i \Phi_i(T). \tag{9.84}$$

The first term of equation (9.84) multiplied by $n_e n_1^+$ is the LTE population of the ground state multiplied by the partition function, i.e., the total LTE number density of the ion. The second term multiplied by $n_e n_1^+$ is the total LTE population of the levels treated explicitly, so their difference is just the exact total LTE population of higher, non-explicit levels.

Equation (9.81) is then modified to read

$$\sum_{j=1}^{NL} n_j(1 + S_j) = n_{\rm atom}, \tag{9.85}$$

where

$$S_i = \begin{cases} 0 & \text{if } i \text{ is not the ground state of an ion,} \\ n_e \Sigma_I & \text{if } i \text{ is the ground state of ion } I+1. \end{cases} \tag{9.86}$$

Matrix Form

The set of equations (9.80) and (9.85) can be written in matrix notation as

$$\mathsf{A} \cdot \mathbf{n} = \mathbf{b}, \tag{9.87}$$

where $\mathbf{n}$ is a vector of level populations, $\mathbf{n} = (n_1, n_2, \ldots, n_{NL})^T$, and $\mathbf{b}$ is a right-hand side vector, which, in the case of the abundance definition equation replacing the NLth equation, is $\mathbf{b} = (0, 0, \ldots, 0, n_{\rm atom})^T$. The matrix A, often called the *rate matrix*, has elements

$$A_{ii} = \sum_{j \neq i} (R_{ij} + C_{ij}), \tag{9.88a}$$

$$A_{ij} = -(R_{ji} + C_{ji}), \text{ for } j \neq i \text{ and } i \neq i_{\rm ref}, \tag{9.88b}$$

$$A_{i_{\rm ref},j} = 1 + S_j \tag{9.88c}$$

where $i_{\rm ref}$ is the index of the "reference level," namely, the level for which the rate equation is replaced by the abundance definition equation, and j_1^+ are the indices of the ground states of all ions except the lowest ion considered. In this notation, the elements of vector $\mathbf{b}$ are given by $b_i = \delta_{i,i_{\rm ref}} n_{\rm atom}$, where δ is the Kronecker symbol.

One sometimes uses an alternative form of the kinetic equation that uses modified downward rates defined by equation (9.50), where one splits the contributions of bound-bound processes, and where one assumes, for simplicity, that all ionizations from the bound states of ion I go to the ground state only of ion $I+1$ (generalization is easy but complicates the formalism). We then write the kinetic equation for level i as

$$-\sum_{\ell<i} n_\ell (R_{\ell i} + C_{\ell i}) + n_i \left[\sum_{\ell<i} (n_\ell/n_i)^* (R'_{i\ell} + C_{\ell i}) + \sum_{u>i}^{k} (R_{iu} + C_{iu}) \right] - \sum_{u>i}^{k} n_u (n_i/n_u)^* (R'_{ui} + C_{iu}) = 0. \quad (9.89)$$

Here, the first term represents the total number of transitions (radiative + collisional) into i from lower bound states; the second term represents the total number of transitions from i to lower bound states; the third term represents the total number of transitions from i to higher states, including the continuum (bound-free transitions); and the last term represents the total number of transitions from higher states (including continuum) into i.

Limiting Cases

We consider here several limiting cases that apply mostly in very rarefied media, such as nebulae, stellar coronae, or the interstellar medium. It was realized by the early 20th century that departures from LTE play a crucial role for these objects. However, due to the lack of computer power, it was necessary to resort to approximate approaches. We describe here some of those approximations, not only for their historical interest, but also because they provide valuable physical insight into the results of otherwise purely numerical simulations.

One-Level Ions

We consider first an atom consisting of a single bound level that ionizes into its continuum. We neglect stimulated emission and rewrite equation (9.89) by expressing explicitly the ionization and recombination rates using equations (9.49), (9.50), and (9.52):

$$(n_1/n_1^*) = \left[\int_{\nu_0}^{\infty} (\alpha_\nu B_\nu / h\nu)\, d\nu + n_e q_{1k} \right] \Big/ \left[\int_{\nu_0}^{\infty} (\alpha_\nu J_\nu / h\nu)\, d\nu + n_e q_{1k} \right]. \quad (9.90)$$

We consider two limiting forms. First, for *high densities* the collisional rates dominate over the radiative rates; then

$$(n_1/n_1^*) \approx (n_e q_{1k} / n_e q_{1k}) = 1, \quad (9.91)$$

i.e., LTE is recovered, as it should be. Further, at large depths, $J_\nu \to B_\nu$, and then also $(n_1/n_1^*) \to 1$; i.e., if the radiation field is perfectly Planckian, we recover LTE, as expected.

For the opposite limit, at *low densities*, the radiation rates dominate, and we obtain from equation (9.90)

$$(n_1/n_1^*) \approx \int_{\nu_0}^{\infty} (\alpha_\nu B_\nu/h\nu)\, d\nu \Big/ \int_{\nu_0}^{\infty} (\alpha_\nu J_\nu/h\nu)\, d\nu. \tag{9.92}$$

This expression states that if the recombination rate exceeds the photoionization rate, the level is *overpopulated* with respect to LTE, and it is underpopulated if the reverse is true.

Equation (9.92) can be rewritten in another useful form by expressing $J_\nu = WB_\nu(T_R)$, with W being the *dilution factor* and T_R the *radiation temperature*, as

$$(n_1/n_1^*) \approx \int_{\nu_0}^{\infty} (\alpha_\nu B_\nu(T_e)/h\nu)\, d\nu \Big/ \left[W \int_{\nu_0}^{\infty} (\alpha_\nu B_\nu(T_R)/h\nu)\, d\nu \right]. \tag{9.93}$$

This expression is useful, for instance, in estimating departures from LTE in nebular conditions.

As a zero-order estimate of the effect, we note that $\alpha(\nu)$ is often a rapidly decreasing function of frequency with the maximum at the threshold (recall that the hydrogenic cross section varies as ν^{-3}), so that the integrals can be replaced by the values at the threshold,

$$(n_1/n_1^*) \approx B_{\nu_0}(T)/J_{\nu_0} \approx B_{\nu_0}(T)/[WB_{\nu_0}(T_R)]. \tag{9.94}$$

Note that $B_{\nu_0}(T)$ depends on the local temperature, while J_{ν_0} is the radiation field that may reflect physical conditions of the medium far away. If a nebula is optically thin in frequency ν_0, and if the nebular material is cold and is irradiated by a strong radiation, $B_{\nu_0} < J_{\nu_0}$, and the level is *underpopulated*. It is easily understood on physical grounds, because in this case there are more ionizations (caused by strong incident radiation) than recombinations (that reflect the local temperature).

It is sometimes useful to use equation (9.94) to express an approximate ionization balance. To this end, write $n_1 = g_1 N_I/U_I$, where N_I is the total number density of the ion I of which the level 1 is the ground state, and U_I is its partition function. Similarly, $n_1^* = n_e n_1^+ \Phi_1(T)$, with $n_1^+ = g_1^+ N_{I+1}/U_{I+1}$. Using equation (9.5) for the Saha-Boltzmann factor, equation (9.94) becomes

$$\frac{N_I}{N_{I+1}} \approx CT^{-\frac{3}{2}} \frac{n_e}{W} \frac{B_{\nu_0}(T)}{B_{\nu_0}(T_R)} \frac{U_I}{U_{I+1}} \exp(\chi_I/kT). \tag{9.95}$$

In the uppermost layers of a stellar atmosphere, where equation (9.92) holds, we may estimate the mean intensity by using the Eddington-Barbier relation (see § 11.4), namely, $J_{\nu_0}(\tau_0 \to 0) \approx \frac{1}{2} B_{\nu_0}(\tau_0 = \frac{1}{2})$, where τ_0 is the optical depth at frequency ν_0. In other words, $W = \frac{1}{2}$ and $T_R = T(\tau_0 = \frac{1}{2})$ in this case. If ν_0 is larger than the frequency corresponding to the maximum of the Planck function at the local temperature, $B_\nu(T)$ decreases exponentially with T (the Wien tail). If the temperature in the atmosphere decreases outward (which is usually the case;

see chapters 17 and 18), then the decrease of the Planck function from $\tau_0 \sim \frac{1}{2}$ to $\tau_0 \sim 0$ may vastly dominate over the dilution factor $W \sim \frac{1}{2}$ coming from the Eddington-Barbier relation, so that $B_{\nu_0}(\tau_0 \sim 0) \ll \frac{1}{2} B_{\nu_0}(\tau_0 = \frac{1}{2})$, and one obtains $(n_1/n_1)^* \ll 1$, i.e., a strong *underpopulation* of level 1.

Coronal Ionization Equilibrium

This is a variant of the previous case. Here we have T ($\sim 10^6$ K) $\gg T_R(\sim 6000$ K), which implies that the collisional ionizations exceed radiative. In contrast, recombinations (radiative plus dielectronic), with a total rate denoted α_R, which proceed at a rate specified by T, exceed collisional recombinations. Then

$$n_1 n_e q_{1k} = n_k n_e \alpha_R, \tag{9.96}$$

so that

$$(n_k/n_1) = q_{1k}/\alpha_R = f(T). \tag{9.97}$$

That is, the coronal ionization balance depends only on temperature and is independent of the electron density, a fact that vastly simplifies analysis of the corona. Both the coronal and nebular situations represent extreme departures from LTE.

Optically Thin Cascades

This is a simplified non-LTE multilevel problem for hydrogen, applicable in nebulae. Take a volume of gas illuminated by a very diluted radiation field. Virtually all of the hydrogen will be in its ground state, so we assume that the resonance lines (Lyman lines) are completely opaque, and hence in detailed radiative balance. Further, we assume that, after an atom is photoionized from the ground state, recombinations occur to *all* states, but the populations of the upper states are so small and the incident radiation field so diluted that (i) photoionizations from these states can be ignored, and (ii) electrons in any state cascade downward at rates determined by the Einstein coefficients A_{ji} without reabsorption upward; i.e., the subordinate lines are *transparent*. We further assume that densities are so low that collisions may be neglected altogether. Then the kinetic equilibrium equation for the continuum state k (i.e., protons) reads

$$n_1 R_{1k} = n_k n_e \sum_{i=1}^{NL} \alpha_R(i, T) = n_e^2 \sum_{i=1}^{NL} \alpha_R(i, T), \tag{9.98}$$

because $n_k = n_e$ in a pure hydrogen gas where free electrons are generated only by ionization of hydrogen. The particle conservation equation is

$$\sum_{i=1}^{NL} n_i + n_e = n_H, \tag{9.99}$$

where $n_{\rm H}$ is the (given) total hydrogen number density. The photoionization rate R_{1k} is assumed to be given in terms of $J_\nu = WB_\nu(T_{\rm R})$. For any subordinate state we can calculate the population in terms of the *branching ratios*, $a_{ji} \equiv A_{ji}/\sum_{\ell<j} A_{j\ell}$, and the *cascade probabilities*, p_{ji}, which are defined recursively as

$$p_{i+1,i} = a_{i+1,i},$$
$$p_{ji} = a_{ji} + \sum_{\ell=i+1}^{j-1} p_{j\ell} a_{\ell i}, \quad (j \ge i+2). \tag{9.100}$$

Then for level i we find

$$n_i \sum_{\ell<i} A_{i\ell} = n_{\rm e}^2 \alpha_{\rm R}(i,T) + \sum_{j>i}^{NL} A_{ji} n_j = n_{\rm e}^2 \left[\alpha_{\rm R}(i,T) + \sum_{j>i}^{NL} p_{ji} \alpha_{\rm R}(j,T) \right]. \tag{9.101}$$

From equation (9.101) we may estimate ratios of occupation numbers, and hence ratios of line intensities along a series. For example, we can compute the relative intensities of the Balmer lines—the *Balmer decrement*—as

$$I({\rm H}_j)/I({\rm H}_i) = (n_j A_{j2} h\nu_{j2})/(n_i A_{i2} h\nu_{i2}) \tag{9.102}$$

and compare the theoretical results with observations. The approach outlined in equations (9.98)–(9.102), with extensive elaboration and refinement, forms the basis of the analysis of nebulae; see [23, chapter 4], [720, pp. 40–110], and [821, chapter 4].

Rosseland Theorem of Cycles

Consider an atom that consists of three states $(1, 2, 3)$ in order of increasing energy in rarefied medium (neglect collisions) and a dilute radiation field. A famous result regarding such a system is *Rosseland's theorem of cycles*, which states that the number of radiative transitions in the direction $1 \to 3 \to 2 \to 1$ exceeds the number of transitions in the inverse direction $1 \to 2 \to 3 \to 1$. A consequence of this result is that energetic photons are systematically *degraded* from high energies (say, far ultraviolet) to low (optical or infrared). For example, in a nebula, Lyman continuum photons are degraded, e.g., into Balmer continuum photons plus Lα photons. In this case, state $1 = 1s$, state $2 = 2p$, and state $3 =$ continuum. We may calculate the ratio $R_{1\to3\to2\to1}/R_{1\to2\to3\to1}$ quite easily. The number of excitations $1 \to 3$ is $n_1 B_{13} WB(\nu_{13})$. Of the excited atoms in state 3, a fraction $A_{32}/(A_{32} + A_{31})$ decays to state 2, and of the atoms in state 2, a fraction $A_{21}/[A_{21} + B_{23} WB(\nu_{23})]$ decays to state 1. Here we neglected stimulated emission. Thus

$$n_1 R_{1\to3\to2\to1} = \frac{n_1 B_{13} WB(\nu_{13}) A_{32} A_{21}}{(A_{32} + A_{31})[A_{21} + B_{23} WB(\nu_{23})]}. \tag{9.103}$$

By similar reasoning,

$$n_1 R_{1\to2\to3\to1} = \frac{n_1 B_{12} W B(\nu_{12}) B_{23} W B(\nu_{23}) A_{31}}{(A_{32}+A_{31})[A_{21}+B_{23} W B(\nu_{23})]}, \tag{9.104}$$

so that

$$R_{1\to2\to3\to1}/R_{1\to3\to2\to1} = W[B_{12}B(\nu_{12})/A_{21}][B_{23}B(\nu_{23})/A_{32}][A_{31}/B_{13}B(\nu_{13})]. \tag{9.105}$$

By using relations among Einstein coefficients and writing B in the Wien approximations ($h\nu/kT \gg 1$), one finds $[B_{ij}B(\nu_{ij})/A_{ji}] = (n_j/n_i)^*$, so equation (9.105) reduces to $R_{1\to2\to3\to1}/R_{1\to3\to2\to1} = W < 1$, which proves the theorem. This result follows from the fact that in the cycle $1 \to 3 \to 2 \to 1$ the dilution factor enters only *once*, while in the inverse process it enters *twice*. In stellar atmospheres, Rosseland's theorem is relevant because at certain depths one may have resonance lines that are opaque (i.e., $W \sim 1$) exciting atoms to upper states, from which the subordinate lines are transparent; in such cases we anticipate systematic photon degradation.

Fine-Structure Levels in the Interstellar Medium

Here we have a medium with extremely low density and temperature. Therefore, the atoms and ions are found essentially in the ground level. For atoms/ions for which the ground level is not a singlet state the question arises of how the total population of the ground state is distributed among the individual J-states. Since these levels are lower levels of the components of resonance doublets or triplets (or even higher multiplets) that are observed, one needs an estimate of their population.

A simple description, developed in [77], uses the following physical model: The only processes that occur between the individual J-states are collisional excitation and de-excitation and spontaneous emission from the upper states. All other processes are neglected. Thus, for atoms/ions with a doublet-state fine structure, i.e., states with $J = \frac{1}{2}$ and $J = \frac{3}{2}$ (for instance, C II, N III, Al I, Si II, Si IV) one deals with a simple two-level atom, with kinetic equilibrium equation

$$n_1 C_{12} = n_2(A_{21} + C_{21}) \;\;\Rightarrow\;\; (n_2/n_1) = C_{12}/(A_{21} + C_{21}). \tag{9.106}$$

This equation is supplemented by the condition $n_1 + n_2 = n_I$, n_I being the total number density of the ion, assuming that the populations of the higher levels are negligible. For atoms/ions with a ground-state triplet, $J = 0, 1, 2$ (e.g., C I, N II, O I, O III, Si I, S I, S III), we have analogous equations for a simple three-level atom

$$\begin{aligned} n_1(C_{12} + C_{13}) &= n_2(A_{21} + C_{21}) + n_3 C_{31} \\ n_2(A_{21} + C_{21} + C_{23}) &= n_1 C_{12} + n_3(A_{32} + C_{32}). \end{aligned} \tag{9.107}$$

Note that we did not include the term with A_{31} because this transition is not permitted by dipole selection rules. Solution of equations (9.107) is easily found to be

$$\frac{n_2}{n_1} = \frac{C_{12}P_3 + C_{13}(A_{32} + C_{32})}{(A_{21} + C_{21})P_3 + C_{23}C_{31}}$$
$$\frac{n_3}{n_1} = \frac{C_{12} + C_{13}}{C_{31}} - \frac{A_{21} + C_{21}}{C_{31}}\frac{n_2}{n_1}, \tag{9.108}$$

where $P_3 \equiv A_{32} + C_{32} + C_{31}$ is the total rate of transitions out of level 3.

It should be stressed that C_{ij} has to contain contributions from collisions with free electrons as well as neutral hydrogen atoms.

Chapter Ten

Scattering of Radiation in Spectral Lines

In this chapter we analyze the scattering of radiation in spectral lines. In § 10.1, we consider a semiclassical approach based on work by Weisskopf, Wigner, and Woolley [1140, 1143, 1173–1177] in the 1930s, and elucidated in astrophysics in the distinguished text by Woolley and Stibbs [1178] in 1953. The basic concept in this formalism is that a broadened atomic level can be decomposed into *sublevels*, and that transitions between sublevels of a pair of atomic levels can be treated statistically to obtain a *redistribution function*, which gives the probability that, measured in the atom's frame, an incoming photon of frequency ξ' is scattered as an outgoing photon of frequency ξ.

A rigorous quantum mechanical derivation of the redistribution function is outlined in § 10.2. Perhaps surprisingly, the semiclassical and quantum results are essentially the same in the limit of weak radiation fields, which often applies in stellar atmospheres. So for most astrophysical applications the semiclassical picture is sufficiently accurate.

In § 10.3 we derive a system of basic redistribution functions based on the semiclassical picture. These can be used as building blocks in formulating physically realistic redistribution functions for certain processes, in particular, resonance scattering. We first consider atom's-frame functions, then their laboratory-frame counterparts, then angle-averaged redistribution functions, and, finally, useful approximate forms of these functions.

In § 10.4 we discuss more complicated redistribution functions, such as a generalized function for resonance Raman scattering, redistribution functions for hydrogen, and a general form of a redistribution function allowing for collisions. In § 10.5 we write expressions for the line emission coefficient and line source function, which are the most important quantities in actual applications.

Throughout this chapter, we use a consistent notation that distinguishes between atom's-frame and observer's-frame quantities. Specifically we denote an atom's-frame redistribution function with a lowercase letter r, and a laboratory-frame (i.e., observer's-frame) redistribution function with an uppercase letter R. Similarly, the line absorption profile in the atom's frame is written as φ, and the line absorption profile in the laboratory frame as ϕ. We follow the convention that primed quantities apply to the frequency and direction of an incoming photon, and unprimed quantities to the frequency and direction of an emitted photon. Frequencies measured in the atom's frame are written as ξ' and ξ, and frequencies measured in the laboratory frame as ν' and ν. Likewise, the directions of an incoming and outgoing photon are $\mathbf{n}'$ and $\mathbf{n}$.

For the non-relativistic flows considered in this book the transformations between the atom's frame and the observer's frame for all of these quantities contain terms that are $O(v/c)$. We can ignore them for the direction vectors because $v/c \ll 1$. But they must be taken into account in the transformation of frequencies between frames because line profiles vary swiftly with frequency; for example, in a Doppler profile, the frame-dependent terms are in effect amplified to $O(v/v_{\text{thermal}})$.

One should bear in mind that although the essential physics of the redistribution of scattered light is determined by the form of the atom's-frame redistribution function r, the most important quantity in application is the laboratory-frame redistribution function R.

10.1 SEMICLASSICAL (WEISSKOPF–WOOLLEY) PICTURE

The semiclassical description of scattering of radiation in spectral lines is a somewhat intuitive picture based on a combination of quantum mechanical and classical concepts. It is remarkably successful in explaining (and even predicting) many of the features of line scattering in astrophysical conditions, specifically the *weak radiation limit*; see equation (10.12). An internally consistent formulation of the early work cited above and its generalization to multilevel atoms was achieved in [539, 540], and summarized in chapter 6 and Appendix B of [834]. Despite its intuitive appeal and ability to predict the general form of the redistribution function and the line emission profile, the semiclassical picture is not a self-consistent theory and does not obviate the necessity of a quantum mechanical analysis. Yet it helps one get a better understanding of the quantum mechanical results. That is, whatever the limitations of the semiclassical picture in complex situations may be, it serves as a good reference point for the physical interpretation of more exact quantum mechanical calculations.

Basic Postulates

The semiclassical picture is based on the following postulates:

(1) The energy levels of an atom are determined using quantum mechanics as described in chapter 7. However, instead of a single energy E_n (an eigenvalue of the atomic Hamiltonian) for level n, each level is viewed as a continuous distribution of *sublevels* with energies close to E_n. Two types of processes that broaden an energy level into sublevels are *lifetime broadening* and *pressure* (or *collision*) *broadening*, discussed in chapter 8.

(a) The energy-time Heisenberg Uncertainty Principle

$$\Delta E_n \Delta t_n \approx \hbar \tag{10.1}$$

explains lifetime broadening. If level n has a lifetime Δt_n, it must have a spread in energy about $\Delta E_n \approx \hbar/\Delta t_n$. The lifetime of a level is set by the sum of all transition rates out of it. Usually that sum is dominated by spontaneous emission rates, so the lifetime is about $\Delta t_n \approx 1/\sum_{m<n} A_{nm}$; hence the energy width of the level is about $\Delta E_n \approx \hbar \sum_{m<n} A_{nm}$. Lifetime broadening applies to all energy levels in all physical regimes. And, as we saw in

equations (5.38), (5.39), and (5.58), an atom is always in a superposition of states, so its probability of occupying a particular state varies in time even in the absence of an external radiation field and/or interaction with other particles.

(b) Collisional broadening results from perturbations of the dipole moment in an atomic transition by its interactions with neighboring particles. These produce small shifts in the positions of the lower and upper levels, and hence a spread in frequency of the corresponding spectral line; see chapter 8. We normally view this broadening as resulting from *elastic collisions*, even if, strictly speaking, energy is not exactly conserved. This is an example of why one cannot take the semiclassical picture literally.

From the quantum mechanical point of view, elastic collisions with neighboring particles change the *phase* of the wave function but not its energy state. But because the energy exchange in these collisions is very small, much less than the separation between the atomic levels, they can usually be ignored when computing the material internal energy density in the gas, so we can consider them to be elastic. Collisional broadening may be unimportant compared to lifetime broadening in very rarefied media, but it dominates in the deeper layers of a stellar atmosphere. The breadth of a spectral line is set by the combined effects of both lifetime and collisional broadening of the levels connected by that line transition.

(2) The semiclassical picture postulates that the atom is in a definite sublevel of some level. And a transition in a spectral line is considered to be an *instantaneous* transition between a *definite* sublevel of an initial level to a *definite* sublevel of a final level. *The premises of definite sublevels and instantaneous transitions are, of course, both incompatible with quantum mechanics.* In reality one can specify only *probabilities*, not certainties.

(3) Consider transitions among three atomic levels: an *initial* level i, with energy eigenvalue E_i; an *excited* (or *intermediate*) level e, with energy eigenvalue E_e, ($E_i < E_e$); and a *final* level f, with energy eigenvalue E_f, ($E_f < E_e$). The final and initial levels may be the same. Three types of transitions among these levels play a role in the formation of spectral lines:

(a) An instantaneous transition from a definite sublevel of the initial level i to a definite sublevel of the excited level e.

(b) An instantaneous transition between sublevels of the excited level e, produced by elastic collisions.

(c) An instantaneous transition from a definite sublevel of the intermediate level e to a definite sublevel of the final level f (which could also be the initial level i).

In general, the final level f is different from the initial level i. But in the case of *resonance scattering*, the usual scattering process for absorption and emission in the same line, the initial and final levels are the same. When the final and initial levels are different, the same formalism still applies, and the process is called *resonant Raman scattering*. The influence of resonant Raman scattering can sometimes be important in the interpretation of observed spectral line profiles. For example, the process $2s \rightarrow 3p \rightarrow 1s$ has significant effects on the formation of the solar $L\beta$ profile [535].

(4) Denote the energies (expressed in frequency units, i.e., divided by $\hbar$) of sublevels relative to that of their parent levels with the labels χ_i, χ'_e, and χ_f. The frequency of an incoming photon that produces a transition from sublevel χ_i of level i to sublevel χ'_e of level e is

$$\xi' = (E_e - E_i)/h + (\chi'_e - \chi_i) \equiv \xi_{ie} + (\chi'_e - \chi_i), \tag{10.2}$$

which implies

$$\chi'_e = \xi' - \xi_{ie} + \chi_i. \tag{10.3}$$

See figure 10.1. Note that *the sublevel χ'_e of the excited level e is uniquely determined by the frequency ξ' of the incoming photon and the sublevel χ_i of the initial level i.*

The frequency ξ of an outgoing photon produced by a transition from sublevel χ'_e of level e to sublevel χ_f of level f is

$$\xi = (E_e - E_f)/h + (\chi'_e - \chi_f) \equiv \xi_{ef} + (\chi'_e - \chi_f), \tag{10.4}$$

which, in view of (10.3), implies

$$\chi_f = \chi'_e + \xi_{ef} - \xi = \chi_i + \xi' - \xi - \xi_{if}. \tag{10.5}$$

Again, *the sublevel χ_f of the final state f is uniquely determined by the sublevel χ'_e of the excited level from which the transition originates and the frequency ξ of the outgoing photon.*

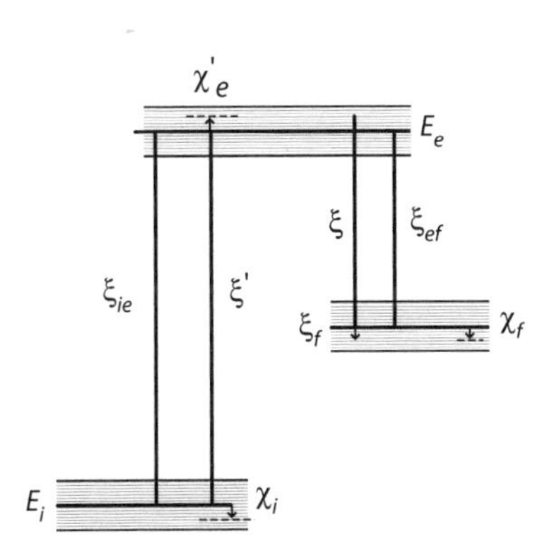

Figure 10.1 Schematic illustration of a division of levels into sublevels in the semiclassical picture. Adapted from [1178].

(5) The semiclassical picture postulates that in the process of photon scattering, an atom stays in a given sublevel unless there are elastic collisions that cause a transition to a different sublevel. Thus if there are no elastic collisions, the next transition starts from the same sublevel where the previous transition ended. This fact poses a serious conceptual problem for the semiclassical picture. *In the absence of elastic collisions, an atom "remembers" the frequency of the photon that excited it (or possibly, de-excited it in a stimulated emission) to its present state.* Hence it will also "remember" what happened one, two, three, . . ., transitions before. Therefore, we must ask where to put the "beginning" of such a correlated chain of transitions. Even in the idealized case of a two-level atom, there are correlated chains: not only $1 \to 2 \to 1$ (normal resonance scattering), but also $1 \to 2 \to 1 \to 2 \to 1$, and $1 \to 2 \to 1 \to 2 \to 1 \to 2 \to 1$, etc. So things start to look very complex and even ill-defined.

(6) To cope with the problem outlined above, one introduces the notion of the *natural population* of a level. A level n is said to be naturally populated if the probability of emitting a photon $(\xi, \mathbf{n})$ in a transition to a lower level, when averaged over an ensemble of identical atoms, is independent of the previous history of the ensemble, and hence of the manner in which that level has been populated. A level populated by collisions (excitations from lower energy levels and de-excitations from higher energy levels), by both radiative and three-body (collisional) recombinations, and by spontaneous emissions from a higher state is naturally populated. In contrast, a level is generally *not* naturally populated by radiative processes in which there is a correlation with properties of previously absorbed/emitted photons (i.e., absorption or stimulated emission from other levels).[1]

Returning to the example above of ever-increasing chains of correlated transitions, we see that if the stimulated emission contribution is small (weak-radiation limit) one has to consider only the $1 \to 2 \to 1$ chain, not the more complicated ones, because most of the other $2 \to 1$ transitions are spontaneous emissions that lead to natural populations and thus destroy "memory" of what happened before.

Lorentz Distribution

The distribution of individual sublevels of a naturally populated level n at E_n is usually (but not necessarily) taken to be a Lorentz profile[2] $L(\chi, \gamma)$:

$$L(\chi, \gamma) \equiv \frac{\gamma}{\pi(\chi^2 + \gamma^2)}. \tag{10.6}$$

[1] Induced radiative transitions could also produce a natural population if the intensity of radiation is constant over the line width (the *white-light approximation*). In astrophysical problems, this is the case when the optical depth in a line is large.

[2] See also equations (6.19) and (8.7), where different notation is used.

The Lorentz profile, or Lorentzian, is an even function of χ, i.e., $L(\chi, \gamma) = L(-\chi, \gamma)$, a property we use below. As above, χ is the energy between the center of level n and one of its sublevels expressed in frequency units, i.e., $\chi = (E - E_n)/h$. And γ_n is the total half-width corresponding to the total lifetime of level n [cf. equation (8.7)]:

$$2\gamma_n = \sum_{m<n} A_{nm} + \sum_{m\neq n} B_{nm} \overline{J}_{nm} + R_{n\kappa} + \sum_{m\neq n} C_{nm} + C_{n\kappa} + Q_n. \tag{10.7}$$

Here the A's and B's are the bound-bound Einstein coefficients; the R's are photoionization rates; the C's are collisional rates (both bound-bound and bound-free); and Q_n is the effective rate of elastic collisions in level n. $\overline{J}_{nm}$ is a frequency-averaged mean intensity of radiation for the transition $n \rightarrow m$, given by

$$\overline{J}_{nm} = \int_0^\infty \oint I(\nu, \mathbf{n}) \phi_{nm}(\nu, \mathbf{n}) \, (d\Omega/4\pi) \, d\nu. \tag{10.8}$$

This form applies, strictly speaking, only for transitions from n to a higher level m. The summation over induced radiative transitions in equation (10.7) must also include transitions from level n to lower states via stimulated emissions, which are described in terms of *emission* profile coefficients ψ_{nm}. This is another conceptual problem with the semiclassical picture. We discuss this problem later on; for the moment it suffices to say that we assume that the stimulated emission rates are much smaller than the corresponding spontaneous emission rates, so we may consider the absorption profile coefficient to apply in evaluating both the induced absorption and emission rates.

The atom's-frame absorption profile for a transition between a broadened initial level i and a broadened excited level e is given by the convolution

$$\varphi_{ie}(\xi') = \int_{-\infty}^{\infty} L(\chi_i, \gamma_i) \, L(\chi_e', \gamma_e) \, d\chi_i = \int_{-\infty}^{\infty} L(\chi_i, \gamma_i) \, L(\xi' - \xi_{ie} + \chi_i, \gamma_e) \, d\chi_i, \tag{10.9}$$

which is interpreted as follows:

(a) $L(\chi_i, \gamma_i) \, d\chi_i$ is the natural occupation probability of sublevels on the range $(\chi_i, \chi_i + d\chi_i)$ of the initial level i.

(b) When an atom in this sublevel absorbs a photon of frequency ξ', according to (10.3) it ends up at sublevel $\chi_e' = (\xi' - \xi_{ie} + \chi_i)$ of the excited level, where it has an occupation probability $L(\xi' - \xi_{ie} + \chi_i, \gamma_e)$.

(c) The total probability of making the transition is the product of these two distributions summed over all initial sublevels χ_i.

(d) Equation (10.9) can be evaluated explicitly using Cauchy's residue theorem; see, e.g., [1178, p. 113]. One finds

$$\varphi_{ij}(\xi') = L(\xi' - \xi_{ij}, \gamma_{ij}) = \frac{\gamma_{ij}}{\pi \left[(\xi' - \xi_{ij})^2 + \gamma_{ij}^2 \right]}, \tag{10.10}$$

where

$$\gamma_{ij} \equiv \gamma_i + \gamma_j. \tag{10.11}$$

This result also follows from Fourier transform theory. In short, the convolution of two Lorentz profiles yields another Lorentz profile with half-width equal to the sum of the individual half-widths. Note that $\varphi_{ij}(\xi')$ is normalized: $\int_{-\infty}^{\infty} \varphi_{ij}(\xi')d\xi' = 1$.

Validity of the Semiclassical Picture

The conditions for validity of the semiclassical picture are the following:

- The radiation is "weak" enough that spontaneous emissions outweigh induced emissions:

$$A_{ji} \gg B_{ji} \int_0^{\infty} \oint I(\nu, \mathbf{n})\psi_{ji}(\nu, \mathbf{n})\, (d\Omega/4\pi)\, d\nu, \tag{10.12}$$

  for all level pairs i and j ($j > i$) having strong transitions. When we use the semiclassical picture of redistribution, transitions for which i is the ground state or one of the lowest excited levels are of primary interest.

  In transitions between close-lying high-energy levels, (10.12) may break down. We can use an order of magnitude estimate to see why. Assume the specific intensity is given by the Planck function, and assume $g_i/g_j = 1$. Then the inequality above reduces to $B_{ji}B_\nu \ll A_{ji}$, or $(B_{ji}/A_{ji})B_\nu \ll 1$, or

$$(c^2/2h\nu^3)(2h\nu^3/c^2)[\exp(h\nu/kT) - 1]^{-1} = [\exp(h\nu/kT) - 1]^{-1} \ll 1,$$

  where the first factor comes from the Einstein relation for B_{ji}/A_{ji}. For low-frequency transitions, $h\nu/kT < 1$, so the relation above may indeed break down. We shall see below that in such transitions, we may use the alternative limit of *complete frequency redistribution*.
- Because we have assumed a Lorentzian distribution of sublevels within each level, the line profile is also Lorentzian. Therefore, from the theory of line broadening in chapter 8 we would conclude that, strictly speaking, the semiclassical picture is valid only when the impact approximation applies, i.e., in the limit that the duration of any elastic collisions contributing to line broadening is much shorter than the average time between collisions.
- The spectral line is isolated; i.e., the energy difference between any two levels that participate in a redistribution process of interest is much larger than the energy widths of the levels. This requirement is necessary because there are quantum mechanical interference effects between two close-lying levels that the semiclassical picture cannot treat properly.

Note that all of the above assumptions break down for a hydrogen atom. First, its levels are l-degenerate. Second, quasi-static line broadening, resulting from the interaction of the radiating atom with a sea of slowly moving perturbers, plays an important role, and the requirements for its validity are opposite to those of the impact approximation. Yet we will see later that even for hydrogen one can use an appropriately modified semiclassical picture of line scattering.

Stimulated Emission

As stated above, the semiclassical picture applies only if stimulated emission is "small" in comparison with spontaneous emission. But that does not mean it is completely neglected. We summarize below what is, and is not, neglected in practice, and what parts of the treatment are approximate.

- The process of stimulated emission itself is *not* neglected, even if its contribution may be small in specific applications. The effects of stimulated emission are taken into account in the line's *emission coefficient.*
- Stimulated emission is not allowed to produce a non-natural population of the final state. This is an approximation, but it avoids having complex multi-photon chains of correlated transitions.
- When calculating *transition rates* in the kinetic equilibrium equations, stimulated emission is assumed to have the line absorption profile. Although one could compute them with the emission profile, see [539], that leads to unnecessary complications: for instance, having branching ratios for coherent scattering that are dependent on radiation intensity, etc.
- The physical motivation for the above approximation is discussed in [252]. There it is shown that treating stimulated emission by using the correct emission profile coefficient in the transfer equation, while replacing it with the absorption profile in the rate equations, yields consistent first-order terms in the corresponding perturbation expansions; see [250], and [252].

We need not go into such subtleties here; suffice it to say that using the absorption profile in calculating the stimulated emission rate is not only convenient, but also consistent with the degree of approximation that underlies the whole semiclassical picture.

Redistribution Function

The *redistribution function* describes any correlation between an absorbed and emitted photon during the scattering in a spectral line. It is defined such that $r(\xi', \mathbf{n}'; \xi, \mathbf{n})\, d\xi' d\xi\, d\Omega' d\Omega$ *is the joint probability density that a photon in the frequency range* $(\xi', \xi' + d\xi')$ *and propagating in an elementary solid angle* $d\Omega'$ *around direction* $\boldsymbol{n}'$ *is absorbed, and subsequently a photon in the frequency range* $(\xi, \xi + d\xi)$ *and propagating in an elementary solid angle* $d\Omega$ *around direction* $\boldsymbol{n}$ *is emitted spontaneously in the transitions* $i \rightarrow e \rightarrow f$. The concept of redistribution function may also be used for continuum scattering, although in practice it is rarely done (except for the Compton scattering).

We specifically consider only *spontaneous* re-emission of a photon in a scattering process for two reasons. First, we avoid conceptual problems connected with stimulated emission. Had the re-emission probability included a stimulated emission contribution, the redistribution function would depend explicitly on a radiation field, which is not useful for practical applications. Second, an emission in a given transition is the result of many different processes, some correlated with a previous photon (such as in a resonance scattering), and some that are not (e.g., a thermal emission—an emission after a collisional transition to an upper level). The total emission coefficient for a given transition is a superposition of all possible processes. Regardless of the exact nature of the processes, we know from quantum mechanics that the emission coefficient corresponding to the stimulated emission is given as $(c^2/2h\nu^3)\,I_\nu$ times the coefficient for spontaneous emission. It is therefore sufficient to consider only spontaneous emissions for describing the individual emission contributions.

As noted in § 6.2, in the non-relativistic limit (i.e., except for Compton scattering) we can usually factor the frequency and angular parts of the full redistribution function into the product of a *frequency redistribution function* r^f and an *angular phase function* g:

$$r(\xi', \mathbf{n}'; \xi, \mathbf{n}) = r^f(\xi', \xi)\, g(\mathbf{n}', \mathbf{n}). \tag{10.13}$$

The two most commonly used phase functions are

$$\begin{aligned} g(\mathbf{n}', \mathbf{n}) &= 1, \quad \text{for isotropic scattering,} \\ g(\mathbf{n}', \mathbf{n}) &= \tfrac{3}{4}[1 + (\mathbf{n}' \cdot \mathbf{n})^2], \quad \text{for dipole scattering.} \end{aligned} \tag{10.14}$$

In some texts, the redistribution function is factored into a product of the probability of absorption (i.e., the absorption profile) and a probability of re-emission $r(\xi', \xi)$. In this book, the atom's-frame redistribution function is defined such that $r(\xi', \xi)\, d\xi' d\xi$ is the *joint probability density that a photon in the frequency range $\xi', \xi' + d\xi'$ is absorbed, and subsequently a photon in the frequency range $\xi, \xi + d\xi$ is emitted, in the transitions $i \to e \to f$*. This similarity of notation is unfortunate; the reader should be careful to notice which definition is being used.

Unlike most treatments, which start with the simplest cases of basic redistribution functions, we proceed in the opposite way, deriving a general semiclassical redistribution function and then considering limiting cases, because this approach is more natural from the viewpoint of the semiclassical picture.

Completely Correlated Scattering

Consider first the case of *completely correlated scattering* for a resonance transition $i \to e \to f$ in which an atom does not experience elastic collisions while it is in the excited level, and in accordance with Postulate 5, there is no reshuffling of photons to other sublevels within level e. We call this redistribution function r_{ief}^{corr}, which is given by

$$r_{ief}^{\text{corr}}(\xi', \xi) = \int_0^\infty L(\chi_i, \gamma_i)\, L(\chi_e, \gamma_e)\, L(\chi_f, \gamma_f)\, d\chi_i. \tag{10.15}$$

Here $L(\chi_i, \gamma_i)$ is the occupation probability at χ_i, the sublevel of level i where the scattering process begins; χ_e is the sublevel in level e to which the atom is excited (which remains *unchanged* during the scattering process), having an occupation probability $L(\chi_e, \gamma_e)$; and χ_f is the sublevel of level f to which the atom returns, which has an occupation probability $L(\chi_f, \gamma_f)$.

The redistribution function for an ensemble of identical atoms is obtained by integrating over the initial sublevel energy χ_i, with the constraints on the excited sublevel χ'_e and the final sublevel χ_f given by (10.3) and (10.5):

$$r_{ief}^{\text{corr}}(\xi', \xi) = \int_{-\infty}^{\infty} L(\chi_i, \gamma_i)\, L(\xi' - \xi_{ie} + \chi_i, \gamma_e)\, L(\xi' - \xi - \xi_{if} + \chi_i, \gamma_f)\, d\chi_i. \tag{10.16}$$

This is the *general redistribution function for correlated scattering*. Unlike the integral over two Lorentzians, which yields another Lorentzian, this integral over three Lorentzians is much more complicated. An explicit formula for it was first derived by Henyey [479]; see also [1178, p. 166].[3] Henyey showed that

$$\begin{aligned}
&\frac{\gamma_1\gamma_2\gamma_3}{\pi^3}\int_{-\infty}^{\infty}\frac{dx}{[(x-x_1)^2+\gamma_1^2][(x-x_2)^2+\gamma_2^2][(x-x_3)^2+\gamma_3^2]}\\
&= \frac{4\gamma_1\gamma_2\gamma_3(\gamma_1+\gamma_2+\gamma_3)/\pi^2}{[(x_1-x_2)^2+(\gamma_1+\gamma_2)^2][(x_2-x_3)^2+(\gamma_2+\gamma_3)^2][(x_3-x_1)^2+(\gamma_1+\gamma_3)^2]}\\
&\quad + \frac{\gamma_1\gamma_2/\pi^2}{[(x_2-x_3)^2+(\gamma_2+\gamma_3)^2][(x_3-x_1)^2+(\gamma_1+\gamma_3)^2]}\\
&\quad + \frac{\gamma_2\gamma_3/\pi^2}{[(x_3-x_1)^2+(\gamma_1+\gamma_3)^2][(x_1-x_2)^2+(\gamma_1+\gamma_2)^2]}\\
&\quad + \frac{\gamma_1\gamma_3/\pi^2}{[(x_1-x_2)^2+(\gamma_1+\gamma_2)^2][(x_2-x_3)^2+(\gamma_2+\gamma_3)^2]}.
\end{aligned} \tag{10.17}$$

For the case of *completely correlated scattering*, set $x = \chi_i$, $x_1 = 0$, $x_2 = \xi_{ie} - \xi'$, $x_3 = \xi' - \xi - \xi_{if}$, and $(\gamma_1, \gamma_2, \gamma_3) = (\gamma_i, \gamma_e, \gamma_f)$. Then

$$\begin{aligned}
&r_{ief}^{\text{corr}}(\xi', \xi) =\\
&\frac{4\gamma_i\gamma_e\gamma_f(\gamma_i+\gamma_e+\gamma_f)/\pi^2}{[(\xi'-\xi_{ie})^2+(\gamma_i+\gamma_e)^2][(\xi-\xi_{ef})^2+(\gamma_e+\gamma_f)^2][(\xi'-\xi-\xi_{if})^2+(\gamma_i+\gamma_f)^2]}\\
&\quad + \frac{\gamma_i\gamma_e/\pi^2}{[(\xi-\xi_{ef})^2+(\gamma_e+\gamma_f)^2][(\xi'-\xi-\xi_{if})^2+(\gamma_i+\gamma_f)^2]}\\
&\quad + \frac{\gamma_e\gamma_f/\pi^2}{[(\xi'-\xi-\xi_{if})^2+(\gamma_i+\gamma_f)^2][(\xi'-\xi_{ie})^2+(\gamma_i+\gamma_e)^2]}\\
&\quad + \frac{\gamma_i\gamma_f/\pi^2}{[(\xi'-\xi_{ie})^2+(\gamma_i+\gamma_e)^2][(\xi-\xi_{ef})^2+(\gamma_e+\gamma_f)^2]}.
\end{aligned} \tag{10.18}$$

[3] Unfortunately, Henyey's formula as printed in [479] contains a typographical error: each γ in the denominator on the left-hand side of his equation (76) should be γ^2. The correct form is given in [1178].

This is the most general expression for the redistribution function in the semiclassical picture. For resonance scattering, $i \equiv f$, we have $\xi_{if} = 0$. We shall consider special cases in § 10.3 and 10.4.

Completely Uncorrelated Scattering

Consider now the opposite case of complete decorrelation of the absorbed and emitted photon; this limit is referred to as *complete frequency redistribution* (CFR), or simply *complete redistribution*. In the semiclassical picture, it corresponds to complete reshuffling of the excited electrons over all the sublevels of level e, so that re-emission of a photon from level e is proportional to the number of sublevels at each value of χ_e. The redistribution function is now given by

$$r_{iei}^{\text{ncorr}}(\xi', \xi) = \int_{-\infty}^{\infty} \int_{-\infty}^{\infty} L(\chi_i', \gamma_i)\, L(\chi_e', \gamma_e)\, L(\chi_e, \gamma_e)\, L(\chi_i, \gamma_i)\, d\chi_i'\, d\chi_e$$
$$= L(\xi_{ie} - \xi', \gamma_{ie})\, L(\xi_{ie} - \xi, \gamma_{ie}) = \varphi_{ie}(\xi')\, \varphi_{ie}(\xi). \tag{10.19}$$

The notation here is the same as in (10.10) and (10.11).

As in fully correlated redistribution, the atom starts at sublevel χ_i' of level i, is excited to sublevel χ_e' of level e by absorption of a photon of frequency ξ', is reshuffled randomly by elastic collisions to another sublevel χ_e of level e, and then makes a transition to sublevel χ_i accompanied by an emission of a photon of frequency ξ. We must integrate the sublevel population distributions $L(\chi_i', \gamma_i)$, $L(\chi_e', \gamma_e)$, $L(\chi_e, \gamma_e)$, and $L(\chi_i, \gamma_i)$, over all initial sublevel population energies, χ_i', and over all reshuffled sublevel energies, χ_e. These integrations are independent because there is no correlation between χ_e' and χ_e. The second and third equalities in (10.19) follow from (10.9) and (10.10). In the CFR limit the frequency distribution of the emitted photons is the same as the absorption profile.

The derivation given here illustrates that, like correlated scattering, complete redistribution is the result of a sequence of three distinct processes. In fact, the redistribution function r^{corr} can also be derived by replacing the third Lorentzian in (10.19), $L(\chi_e, \gamma_e)$, which accounts for the reshuffling of the upper-state sublevels, by the δ-function $\delta(\chi_e' - \chi_e)$.

General Redistribution Function

The general redistribution function for a mixture of some reshuffling and some no-reshuffling in the excited state is referred to in astrophysical literature as *partial frequency redistribution* (PFR). In practice, the label PFR may refer to *any* kind of departure from complete redistribution, including the case of full coherence. Within the semiclassical picture, the general redistribution function can be viewed as follows: The probability that a photon is spontaneously re-emitted at all in a transition $j \to i$ is

$$p_{ji}^{\text{reem}} = A_{ji}/P_j, \tag{10.20}$$

where P_j is the total rate of transitions out of level j given by (10.7) with elastic collisions omitted:

$$P_j = \sum_{\ell<j}(A_{j\ell} + B_{j\ell}\overline{J}_{j\ell} + C_{j\ell}) + \sum_{u>j}(B_{ju}\overline{J}_{ju} + C_{ju}) + R_{j\kappa} + C_{j\kappa}. \tag{10.21}$$

Thus the probability that a spontaneous emission takes place before an elastic collision, or any other interaction that causes a transition from a given sublevel either to a different sublevel or to a different level entirely, so that the photon is re-emitted *coherently*, is given by

$$p_{ji}^{\mathrm{coh}} = A_{ji}/(P_j + Q_j), \tag{10.22}$$

where Q_j is the rate of elastic collisions, i.e., transitions to another sublevel of the same level. Because we deal with normalized redistribution functions, the conditional probability of correlated re-emission, i.e., the probability that a photon is re-emitted from the same sublevel (before an elastic collision), provided that it is re-emitted at all, is given by

$$p_j^{\mathrm{corr}} = \frac{p_{ji}^{\mathrm{coh}}}{p_{ji}^{\mathrm{reem}}} = \frac{A_{ji}}{P_j + Q_j}\frac{P_j}{A_{ji}} = \frac{P_j}{P_j + Q_j}\,. \tag{10.23}$$

Although this conclusion may seem to be counterintuitive (because it says that the probability of correlated re-emission is proportional to the *total* rate of transitions out of level j, not just by the spontaneous rate A_{ji}), it results from the definition of the redistribution function as a *normalized* function; i.e., as the probability of (absorption + re-emission), provided that the re-emission occurs at all. The quantity p_j^{corr} is also called the *coherence fraction.*

In short, the general redistribution function is written as

$$r_{iji}(\xi', \xi) = p_j^{\mathrm{corr}} r_{iji}^{\mathrm{corr}}(\xi', \xi) + (1 - p_j^{\mathrm{corr}}) r_{iji}^{\mathrm{ncorr}}(\xi', \xi). \tag{10.24}$$

10.2 ⋆QUANTUM MECHANICAL DERIVATION OF REDISTRIBUTION FUNCTIONS

We outline here a quantum mechanical derivation of a general redistribution function for the correlated transition

$$|i\rangle \overset{\text{photon 1}}{\longrightarrow} |e\rangle \overset{\text{photon 2}}{\longrightarrow} |f\rangle.$$

Here photon 1 is absorbed in the transition $i \to e$ (with $E_i < E_e$), and photon 2 is emitted in the transition $e \to f$.

Let n_1 and n_2 be occupation numbers of photons in modes 1 and 2, respectively. Then following Dirac, the combined states for the system (atom + radiation field) are

$$|I\rangle \equiv |i, n_1, n_2\rangle, \tag{10.25}$$

$$|E\rangle \equiv |e, n_1 - 1, n_2\rangle, \tag{10.26}$$

$$|F\rangle \equiv |f, n_1 - 1, n_2 + 1\rangle. \tag{10.27}$$

The amplitude for an interaction between an atom and the radiation field to occur during a time interval t is found by taking the matrix element of the operator representing the process with the time-evolution operator

$$U(t) = \exp[-i(H + H_r + V)t/\hbar], \tag{10.28}$$

where H is the Hamiltonian of the particle system (atom + perturbers), H_r is the Hamiltonian of the radiation field, and V is the Hamiltonian of the interaction between the atom and the radiation field (the interaction of perturbers with the radiation field is neglected).

In the dipole approximation, the interaction Hamiltonian is given by

$$V = -\mathbf{d} \cdot \mathbf{E}(\mathbf{R}) = -i\mathbf{d} \cdot \sum_{\mathbf{k}\alpha} (2\pi\hbar ck)^{1/2} \mathbf{e}_{\mathbf{k}\alpha} \left(\hat{a}_{\mathbf{k}\alpha} e^{-i\mathbf{k}\cdot\mathbf{R}} - \hat{a}^{\dagger}_{\mathbf{k}\alpha} e^{i\mathbf{k}\cdot\mathbf{R}} \right), \tag{10.29}$$

where $\mathbf{d}$ is the electric dipole operator of the radiating atom and $\mathbf{R}$ is its position. The sums extend over all possible values of the wave vector $\mathbf{k}$ and over two independent polarizations α; $\mathbf{e}_{\mathbf{k}\alpha}$ is the polarization vector, and $\hat{a}_{\mathbf{k}\alpha}$ and $\hat{a}^{\dagger}_{\mathbf{k}\alpha}$ are, respectively, the photon annihilation and creation operators; see § 5.3, equations (5.48), and (5.53).

Unlike the description of a photon absorption or emission by a single atom discussed in chapter 5, here we have an absorption and emission of a photon by a radiating atom in the presence of its stochastic interactions with neighboring particles. Such an analysis is a generalization of the quantum mechanical treatment of line broadening, discussed in chapter 8.

The rate of the process $|I\rangle \longrightarrow |F\rangle$ is given by

$$G = \lim_{t\to\infty} \frac{1}{t} \left[\langle F|U(t)|I\rangle \right]^2_{\text{av}}, \tag{10.30}$$

where $[\dots]_{\text{av}}$ denotes a summation over the final states and an average over the initial states of the system (atom + perturbers), described in terms of the particle density matrix, ρ.

In astrophysical applications, it is sufficient to consider the case of low intensities, for which a two-photon process is described by the second-order term in the expansion of $U(t)$ in powers of the interaction V. The matrix element $A(t) = \langle F|U(t)|I\rangle$ is given by

$$A(t) = C \int_0^t dt_2 \int_0^{t_2} dt_1\, e^{i\omega_2(t_2-t)}\, e^{-i\omega_1 t_1} \times \langle f| e^{-iH(t-t_2)/\hbar} D_2\, e^{-iH(t_2-t_1)/\hbar} D_1 e^{-iHt_1/\hbar} |i\rangle, \tag{10.31}$$

where

$$C = (2\pi/\hbar)[\omega_1 \omega_2 n_1 (n_2 + 1)]^{1/2}, \tag{10.32}$$

$\omega_1 = ck_1$ and $\omega_2 = ck_2$ are the circular frequencies of photons 1 and 2, respectively, and n_1 and n_2 their occupation numbers. Here

$$D_1 \equiv d_{ei}\epsilon_1 e^{-i\mathbf{k}\cdot\mathbf{R}} \qquad \text{and} \qquad D_2 \equiv d_{fe}\epsilon_2^* e^{i\mathbf{k}\cdot\mathbf{R}}, \tag{10.33}$$

where $d_{ei} \equiv \langle e|\mathbf{d}|i\rangle$ is the matrix element of the atomic dipole moment between states i and e; and ϵ_1 is the polarization vector of the photon 1; and similarly for $d_{fe}\epsilon_2$.

We evaluate $|A(t)|^2$ using

$$|A(t)|^2 = \langle F|U^\dagger(t)|I\rangle\langle I|U(t)|F\rangle, \tag{10.34}$$

and the averaging is done as

$$\{|A(t)|^2\}_{\text{av}} = \text{Tr}\left[\langle F|U(t)^\dagger|I\rangle\rho\langle I|U(t)|F\rangle\right], \tag{10.35}$$

where ρ is the density matrix of the system atom + perturbers.

The transition rate is then written as

$$G = C^2 F(\omega_1, \omega_2), \tag{10.36}$$

where

$$F(\omega_1,\omega_2) = \lim_{t\to\infty}\frac{1}{t}\int_0^t dt_4 \int_0^{t_4} dt_3 \int_0^t dt_2 \int_0^{t_2} dt_1\, e^{-i\omega_1(t_1-t_3)}\, e^{i\omega_2(t_2-t_4)} \times$$
$$\text{Tr}\Big[e^{-iH(t-t_2)/\hbar} D_2^\dagger e^{-iH(t_2-t_1)/\hbar} D_1 e^{-iHt_1/\hbar}$$
$$\rho\; e^{iHt_3/\hbar}\, D_1^\dagger\, e^{iH(t_4-t_3)/\hbar}\, D_2\, e^{iH(t-t_4)/\hbar}\Big]. \tag{10.37}$$

As shown in [336, 795, 817] the contribution to (10.37) from the integration domain $t_4 > t_2$ is the complex conjugate of the contribution from $t_4 < t_2$, so that

$$\int_0^t dt_4 \int_0^{t_4} dt_3 \int_0^t dt_2 \int_0^{t_2} dt_1 \cdots = 2\,\text{Re}\int_0^t dt_4 \int_0^{t_4} dt_3 \int_0^{t_4} dt_2 \int_0^{t_2} dt_1 \cdots . \tag{10.38}$$

Following these references, one recognizes three possible time sequences in the integral in (10.38): (i) $t_1 < t_2 < t_3 < t_4$, (ii) $t_1 < t_3 < t_2 < t_4$, and (iii) $t_3 < t_1 < t_2 < t_4$.

In each of the three cases the four times t_1, t_2, t_3, t_4 embrace three successive time intervals denoted τ_1, τ_2, τ_3. The corresponding contributions to F are

$$\begin{aligned} F(\omega_1,\omega_2) = 2\mathrm{Re} \int_0^\infty \int_0^\infty \int_0^\infty d\tau_1\, d\tau_2\, d\tau_3 \times \\ \Big(e^{i\omega_1(\tau_1+\tau_2)}\, e^{-i\omega_2(\tau_2+\tau_3)}\, \mathrm{Tr}\, D_2^\dagger e^{-iL\tau_3} \Big\{ e^{-iL\tau_2} \Big[D_2 e^{-iL\tau_1} (D_1\rho) \Big] D_1^+ \Big\} \\ + e^{i\omega_1\tau_1} e^{-i\omega_2\tau_3}\, \mathrm{Tr}\, D_2^\dagger e^{-iL\tau_3} \Big\{ D_2 e^{-iL\tau_2} \Big[e^{-iL\tau_1} (D_1\rho) D_1^+ \Big] \Big\} \\ + e^{-i\omega_1\tau_1} e^{-i\omega_2\tau_3}\, \mathrm{Tr}\, D_2^\dagger e^{-iL\tau_3} \Big\{ D_2 e^{-iL\tau_2} \Big[D_1 \rho e^{-iL\tau_1} D_1^+ \Big] \Big\} \Big) \\ = F_{(\mathrm{i})} + F_{(\mathrm{ii})} + F_{(\mathrm{iii})}, \end{aligned} \tag{10.39}$$

where "Tr" denotes the trace, and the Liouville operator L denotes the commutator with the particle Hamiltonian,

$$L\rho \equiv [H, \rho]/\hbar. \tag{10.40}$$

A formalism that employs the Liouville operator is advantageous because for any operator $\hat{\boldsymbol{d}}$ the following identity applies (see, e.g., [817]):

$$e^{iHt}\hat{\boldsymbol{d}}\, e^{-iHt} = e^{-iLt}\hat{\boldsymbol{d}}. \tag{10.41}$$

The Liouville operator describes the evolution of the particle system resulting from interactions between the radiator and elastic and inelastic collisions by the perturbers. The effect of spontaneous transitions on the evolution of the particle system may be incorporated by adding a damping term to the Liouville operator (see, e.g., [817]),

$$-iL\rho_{ab} = -(i/\hbar)\, [H, \rho_{ab}] - \tfrac{1}{2}(\Gamma_a + \Gamma_b)\, \rho_{ab}; \;\; a, b = i, e, f, \tag{10.42}$$

with Γ_a being the radiative decay rate from state $|a\rangle$.

The complex physics of photon absorption and emission under the influence of radiator-perturber interactions is simplified here using two approximations. These were already encountered when describing spectral line broadening. In fact, a treatment of line scattering (redistribution—a two-photon process) is a relatively straightforward generalization of the treatment of line broadening (absorption or emission of a photon—a one-photon process), treated in chapter 8. The two simplifications are the following:

(1) The *factorization approximation*, which essentially states that the radiator and perturbers are uncorrelated:

$$\rho \rightarrow \rho_a\, \rho_p\,, \quad e^{-iL\tau_j} \rightarrow \rho_p \mathrm{Tr}_p\, e^{-iL\tau_j}, \quad j = 1, 2, \tag{10.43}$$

where ρ_a is the initial density matrix of the radiator, ρ_p is the density matrix of the perturbers, and Tr_p indicates a trace over the perturber states. The trace is expressed through the *correlation function* [795, 817],

$$\mathrm{Tr}_p e^{-iLt}(A_{ab})\rho_p = C_{ab}(t)\, A_{ab}, \quad a, b = i, e, f, \tag{10.44}$$

where C_{ab} is the correlation function, and A_{ab} is an arbitrary operator between states $|a\rangle$ and $|b\rangle$ of the radiator.

(2) The *impact approximation*, which was discussed in chapter 8. Using the present formalism, it can be expressed as

$$C_{ab}(t) = \exp(-i\omega_{ab}t - i\Delta_{ab}t - \gamma_{ab}t), \quad a, b = i, e, f, \tag{10.45}$$

where

$$\omega_{ab} = (E_a - E_b)/\hbar \equiv -\omega_{ba} \tag{10.46}$$

is the frequency of the transition between states $|a\rangle$ and $|b\rangle$, and

$$\gamma_{ab} = \tfrac{1}{2}(\Gamma_a + \Gamma_b) + \gamma_{ab}^c \tag{10.47}$$

represents the collisional width and Δ_{ab} the collisional shift.

On evaluating (10.39), one obtains

$$F_{(i)} = 2|D_1|^2|D_2|^2 \rho_{ii} \times \mathrm{Re}\left[\frac{1}{i(\widetilde{\omega}_{ei} - \omega_1) + \gamma_{ei}}\, \frac{1}{i(\widetilde{\omega}_{fe} + \omega_2) + \gamma_{fe}}\, \frac{1}{i(\widetilde{\omega}_{fi} - \omega_1 + \omega_2) + \gamma_{fi}}\right], \tag{10.48}$$

and

$$F_{(ii)} + F_{(iii)} = 2|D_1|^2|D_2|^2 \frac{2\rho_{ii}}{\gamma_{ee}} \frac{\gamma_{ei}}{(\omega_1 - \widetilde{\omega}_{ei})^2 + \gamma_{ei}^2} \frac{\gamma_{fe}}{(\omega_2 - \widetilde{\omega}_{fe})^2 + \gamma_{fe}^2}, \tag{10.49}$$

where $\widetilde{\omega}_{ab} \equiv \omega_{ab} + \Delta_{ab}$ is a shifted atomic frequency, and the diagonal element of the density matrix ρ_{ii} represents the fraction of atoms (population) in the initial state $|i\rangle$.

To simplify the notation, write

$$\Delta\omega_1 \equiv \omega_1 - \widetilde{\omega}_{ei}, \quad \Delta\omega_2 \equiv \widetilde{\omega}_{ef} - \omega_2, \quad \text{and } \Delta\omega_{12} \equiv \Delta\omega_1 + \Delta\omega_2. \tag{10.50}$$

Upon evaluation of the real part, (10.48) and (10.49) become

$$F_{(i)} = A \frac{1}{\Delta\omega_1^2 + \gamma_{ei}^2} \frac{1}{\Delta\omega_2^2 + \gamma_{fe}^2} \frac{1}{\Delta\omega_{12}^2 + \gamma_{fi}^2} \times \left(\gamma_{ei}\gamma_{fe}\gamma_{fi} + \gamma_{ei}\Delta\omega_2\Delta\omega_{12} - \gamma_{fe}\Delta\omega_1\Delta\omega_{12} + \gamma_{fi}\Delta\omega_1\Delta\omega_2\right), \tag{10.51}$$

and

$$F_{(ii)} + F_{(iii)} = \frac{2A}{\gamma_{ee}} \frac{\gamma_{ei}}{\Delta\omega_1^2 + \gamma_{ei}^2} \frac{\gamma_{ee}}{\Delta\omega_2^2 + \gamma_{fe}^2}, \tag{10.52}$$

where

$$A = 2|D_1|^2|D_2|^2\rho_{ii} \tag{10.53}$$

is a constant.

The next step consists of a judicious rewriting of equation (10.51). The derivation is lengthy, so we outline only the basic steps. The first term represents a product of three Lorentzian-type expressions [the frequency dependence is contained only in the form $(\Delta\omega^2+\gamma^2)^{-1}$], while the next three terms contain mixed expressions where the terms $\Delta\omega$ appear also in the numerator. To show how they can be eliminated, consider the first such term,

$$\begin{aligned}\gamma_{ei}\Delta\omega_2\Delta\omega_{12} &= \gamma_{ei}\Delta\omega_2(\Delta\omega_1 + \Delta\omega_2) = \gamma_{ei}(\Delta\omega_2^2 + \Delta\omega_1\Delta\omega_2)\\ &= \gamma_{ei}(\Delta\omega_2^2 + \gamma_{fe}^2) - \gamma_{ei}\gamma_{fe}^2 + \gamma_{ei}\Delta\omega_1\Delta\omega_2.\end{aligned} \tag{10.54}$$

The term splits into three parts. The first part contains a factor $(\Delta\omega^2 + \gamma^2)$ that cancels with a corresponding Lorentz-type factor in equation (10.51); the second part does not contain any frequency dependence; and the third part contains a factor proportional to $\Delta\omega_1\Delta\omega_2$. A similar procedure is applied to the next term of equation (10.54), which again splits into three parts: the first is proportional to $(\Delta\omega_1^2+\gamma_{ei}^2)$; the second part is frequency-independent; and the third term is proportional to $\Delta\omega_1\Delta\omega_2$. The last term of (10.54) contains only $\Delta\omega_1\Delta\omega_2$. The combined contribution of these factors is dealt with by writing $\Delta\omega_1\Delta\omega_2 = \frac{1}{2}(\Delta\omega_{12}^2 - \Delta\omega_1^2 - \Delta\omega_2^2)$. Upon adding and subtracting corresponding γ-terms, one can cancel the $(\Delta\omega^2 + \gamma^2)$ terms with the appropriate factors in the denominator.

Now, defining

$$\mathcal{L}_1 \equiv 1/(\Delta\omega_1^2 + \gamma_{ei}^2), \quad \mathcal{L}_2 \equiv 1/(\Delta\omega_2^2 + \gamma_{fe}^2), \quad \mathcal{L}_{12} \equiv 1/(\Delta\omega_{12}^2 + \gamma_{fi}^2), \tag{10.55}$$

and using the algebraic manipulations outlined above, we obtain the final expression

$$\begin{aligned}F \equiv F_{(i)} + F_{(ii)} + F_{(iii)} = \tfrac{1}{2}A\Big\{&\mathcal{L}_1\mathcal{L}_2\,[(4/\gamma_{ee}) - \gamma_{ei} - \gamma_{fe} - \gamma_{fi}]\\ &+ \mathcal{L}_1\mathcal{L}_{12}\,(\gamma_{fi} - \gamma_{ei} + \gamma_{fe}) + \mathcal{L}_2\mathcal{L}_{12}\,(\gamma_{fi} + \gamma_{ei} - \gamma_{fe})\\ &+ \mathcal{L}_1\mathcal{L}_2\mathcal{L}_{12}\Big[-\gamma_{fi}(\gamma_{ei}^2 + \gamma_{fe}^2) + \gamma_{ei}(\gamma_{fe}^2 + \gamma_{fi}^2) + \gamma_{fe}(\gamma_{ei}^2 + \gamma_{fi}^2)\\ &+(\gamma_{fi}^3 - \gamma_{ei}^3 - \gamma_{fe}^3 + 2\gamma_{ei}\gamma_{fe}\gamma_{fi})\Big]\Big\}.\end{aligned} \tag{10.56}$$

In the case of pure radiative damping, all the collisional damping terms γ^c and shifts Δ vanish. Then

$$\gamma_{ei} \equiv \tfrac{1}{2}(\Gamma_e + \Gamma_i), \; \gamma_{fe} \equiv \tfrac{1}{2}(\Gamma_f + \Gamma_e), \; \gamma_{fi} \equiv \tfrac{1}{2}(\Gamma_f + \Gamma_i), \text{ and } \gamma_{ee} \equiv \Gamma_e, \tag{10.57}$$

so (10.56) may be rewritten in the symmetric form

$$F = (A/2\Gamma_e)\Big\{\mathcal{L}_1\mathcal{L}_2\Gamma_i\Gamma_f + \mathcal{L}_1\mathcal{L}_{12}\Gamma_e\Gamma_f + \mathcal{L}_2\mathcal{L}_{12}\Gamma_i\Gamma_e + \mathcal{L}_1\mathcal{L}_2\mathcal{L}_{12}\left[\Gamma_i\Gamma_e\Gamma_f(\Gamma_i + \Gamma_e + \Gamma_f)\right]\Big\}. \tag{10.58}$$

Using Henyey's identity [equation (10.17) or (10.18)], equation (10.58) may also be written as

$$F = C\int_{-\infty}^{\infty} L\left(\omega - \omega_1, \tfrac{1}{2}\Gamma_i\right) L\left(\omega - \omega_{ei}, \tfrac{1}{2}\Gamma_e\right) L\left(\omega - \omega_2 - \omega_{fi}, \tfrac{1}{2}\Gamma_f\right) d\omega, \tag{10.59}$$

where $C \equiv \frac{1}{2}A\pi^2/\Gamma_e$, which agrees with the semiclassical expression (10.15) for the correlated redistribution function. Note that because the astrophysical formalism deals with *normalized* redistribution functions, the constant C is unimportant. With some further algebraic manipulations (see [469, 516]) one finally obtains for the general case

$$F = (A/2\Gamma_e)\left[\alpha\int_{-\infty}^{\infty} L(\omega - \omega_1, \Gamma_I)\, L(\omega - \omega_{ei}, \Gamma_E)\, L(\omega - \omega_2 - \omega_{fi}, \Gamma_F)\, d\omega + \beta\, L(\omega - \omega_1, \gamma_{ei}) L(\omega - \omega_2, \gamma_f)\right], \tag{10.60}$$

where

$$\gamma_I = \gamma_{ei} - \gamma_{fe} + \gamma_{fi}, \tag{10.61a}$$

$$\gamma_E = \gamma_{ei} + \gamma_{fe} - \gamma_{fi}, \tag{10.61b}$$

$$\gamma_F = -\gamma_{ei} + \gamma_{fe} + \gamma_{fi}, \tag{10.61c}$$

and the coefficients α and β are given by

$$\alpha = \Gamma_e/(\gamma_{ei} + \gamma_{fe} - \gamma_{fi}), \tag{10.62}$$

and

$$\beta = \alpha(\gamma_{ei} + \gamma_{fe} - \gamma_{fi} - \gamma_{ee})/\gamma_{ee}. \tag{10.63}$$

Equation (10.60) shows that the general redistribution function is indeed a linear combination of the redistribution function for completely correlated scattering and completely uncorrelated scattering, as suggested by the semiclassical picture in (10.24).

To show that the coefficients α and β agree with the semiclassical picture, we must express the damping parameters γ_{ab}^c, $a, b = i, e, f$ [see (10.47)] in terms of elastic and inelastic collisional rates. According to [83, equation 83], in the isolated line approximation they are given by

$$\gamma_{ab}^c = \tfrac{1}{2}\left(Q_a^{\text{inel}} + Q_b^{\text{inel}} + Q_{ab}^{\text{el}}\right), \tag{10.64}$$

where

$$Q_a^{\text{inel}} = \sum_{n \neq a} C_{an}, \tag{10.65}$$

and C_{an} is the rate of inelastic collisions $a \to n$. The elastic rate Q_{ab}^{el} cannot, in general, be expressed as a sum of contributions from level a and b. However, neglecting quantum interference effects, the elastic rate can be separated into the contributions from those two states,

$$Q_{ab}^{\text{el}} = Q_a^{\text{el}} + Q_b^{\text{el}}. \tag{10.66}$$

Then

$$\alpha = \Gamma_e / (\Gamma_e + Q_e^{\text{inel}} + Q_e^{\text{el}}), \tag{10.67}$$

and

$$\beta = \frac{\Gamma_e}{\Gamma_e + Q_e^{\text{inel}}} \left(\frac{Q_e^{\text{el}}}{\Gamma_e + Q_e^{\text{inel}} + Q_e^{\text{el}}} \right). \tag{10.68}$$

The redistribution function in equation (10.60) is not normalized because a photon may be destroyed by an elastic collision with the intermediate state e. One recognizes that $\Gamma_e + Q_e^{\text{inel}} \equiv P_e$ is the total rate of transitions out of state e. When normalization of the redistribution function is enforced, i.e., one multiplies (10.60) by $(\Gamma_e + Q_e^{\text{inel}}) / \Gamma_e$, the coefficients α and β become

$$\alpha = P_e / (P_e + Q_e^{\text{el}}), \quad \beta = 1 - \alpha, \tag{10.69}$$

which agrees with the semiclassical expression (10.23).

10.3 BASIC REDISTRIBUTION FUNCTIONS

Atom's-Frame Redistribution Functions

A system of basic redistribution functions (also called "elementary redistribution functions" in the literature) was first introduced by Hummer in [544] as a set of more or less empirical functions. As outlined in § 10.5, it became clear that these basic redistribution functions do not necessarily provide an actual redistribution function for a transition in a realistic atom, but the actual redistribution functions are usually linear combinations of the basic functions. Therefore, it is important to summarize their mathematical form and physical meaning.

In his original study, Hummer recognized four cases, denoted by Roman numerals I to IV. His notation became universal and was later extended. Here, $i \equiv f$, and we call the upper state j instead of e as in previous sections. Also, we now write the line center frequency $\xi_0 \equiv \xi_{ij}$.

Case I is a limiting case of a correlated "redistribution function" between two sharp levels, i.e., $\gamma_i, \gamma_j \to 0$. Then the Lorentzians in (10.15) and (10.16) are replaced by δ-functions, and we find

$$
\begin{aligned}
r_{\rm I}(\xi', \xi) = \lim_{\gamma_i, \gamma_j \to 0} r_{iji}^{\rm corr}(\xi', \xi) &= \int_{-\infty}^{\infty} \delta(x - \xi')\, \delta(x - \xi_0)\, \delta(x - \xi)\, dx \\
&= \delta(\xi' - \xi_0)\, \delta(\xi' - \xi),
\end{aligned}
\tag{10.70}
$$

which can be understood as a product of the absorption profile coefficient, $\delta(\xi' - \xi_0)$ and the re-emission probability $\delta(\xi' - \xi)$. Note that $r_{\rm I}$ can also be viewed as a limiting form of the non-correlated redistribution function $r^{\rm ncorr}$,

$$
r_{\rm I}(\xi', \xi) = \lim_{\gamma_i, \gamma_j \to 0} r_{iji}^{\rm ncorr}(\xi', \xi) = \delta(\xi' - \xi_0)\, \delta(\xi - \xi_0) = \delta(\xi' - \xi_0)\, \delta(\xi' - \xi).
\tag{10.71}
$$

This result is not surprising because *any* "redistribution" over two perfectly sharp levels is necessarily "complete redistribution" over those levels. Consequently, line transfer solutions using the corresponding laboratory-frame redistribution functions show that the global scaling of the radiation field for redistribution with $R_{\rm I}$ and with complete redistribution is very similar; see § 15.6.

Case II is a limiting case of the correlated redistribution function with the lower level sharp and the upper level broadened,

$$
\begin{aligned}
r_{\rm II}(\xi', \xi) = \lim_{\gamma_i \to 0} r^{\rm corr}(\xi', \xi) &= \int_{-\infty}^{\infty} \delta(x - \xi') L(x - \xi_0, \gamma_j) \delta(x - \xi)\, dx \\
= L_j(\xi' - \xi_0)\, \delta(\xi' - \xi) &= \varphi_{ij}(\xi')\, \delta(\xi' - \xi),
\end{aligned}
\tag{10.72}
$$

which is again interpreted as a product of the absorption profile and the re-emission probability.

Case III is the case of non-correlated redistribution, considered earlier:

$$
r_{\rm III}(\xi', \xi) = r_{iji}^{\rm ncorr}(\xi', \xi) = L(\xi' - \xi_0, \gamma_{ij})\, L(\xi - \xi_0, \gamma_{ij}) = \varphi_{ij}(\xi')\varphi_{ij}(\xi).
\tag{10.73}
$$

Case IV was meant to describe correlated redistribution between two broadened levels; i.e., the function given by (10.16), which, as mentioned above, was first derived by Woolley and Stibbs [1178]. However, at the time of developing the set of elementary redistribution functions it seemed that a different (but incorrect) redistribution function, suggested in [471], was more appropriate. Therefore, the following function was adopted:

$$
r_{\rm IV}(\xi', \xi) = L(\xi - \xi', \gamma_i)\, L(\xi - \xi_0, \gamma_j).
\tag{10.74}
$$

To gain more insight into this function, note that it can be rewritten as

$$
r_{\rm IV}(\xi', \xi) = \int_{-\infty}^{\infty} L(x - \xi', \gamma_i)\, L(x - \xi_0, \gamma_j)\, \delta(x - \xi)\, dx,
\tag{10.75}
$$

which is the limiting case of $r^{\rm corr} = \int L(x - \xi', \gamma_i)\, L(x - \xi_0, \gamma_j)\, L(x - \xi, \gamma_i)\, dx$, where the first Lorentzian $L(x, \gamma_i)$ is kept, while the second one is replaced by a δ-function. In other words, the lower level is treated as broadened in the absorption part of the scattering process (when the level acts as an initial state), but as a sharp one when it acts as the final state, which is inconsistent.

Case V After the first rigorous quantum mechanical derivation of the redistribution fuction for two broadened levels in [817], it became generally accepted that the original function $r_{\rm IV}$ is inconsistent. Some authors began calling the correct function $r_{\rm IV}$ also, which led to some confusion. Therefore, it was suggested in [468] to call it $r_{\rm V}$, thus extending Hummer's original system. This function is given by (10.16),

$$r_{\rm V}(\xi', \xi) = \int_{-\infty}^{\infty} L(x - \xi', \gamma_i)\, L(x - \xi_0, \gamma_j)\, L(x - \xi, \gamma_i)\, dx, \tag{10.76}$$

which can be evaluated as shown in (10.18). For resonance scattering $i \to e \to i$, (10.18) may be rewritten as

$$\begin{aligned} r_{\rm V}(\xi', \xi) = & \left[2\pi \gamma_i \gamma_j (\gamma_i + \gamma_{ij})/\gamma_{ij}^2\right] L(\xi' - \xi, 2\gamma_i)\, L(\xi' - \xi_0, \gamma_{ij})\, L(\xi - \xi_0, \gamma_{ij}) \\ & + \left(\gamma_j/2\gamma_{ij}\right) L(\xi - \xi_0, \gamma_{ij})\, L(\xi - \xi', 2\gamma_i) \\ & + \left(\gamma_j/2\gamma_{ij}\right) L(\xi' - \xi_0, \gamma_{ij})\, L(\xi - \xi', 2\gamma_i) \\ & + \left(\gamma_i^2/\gamma_{ij}^2\right) L(\xi' - \xi_0, \gamma_{ij})\, L(\xi - \xi_0, \gamma_{ij}). \end{aligned} \tag{10.77}$$

However, this form is not very advantageous for practical applications. We will see in the next section that the form of the corresponding laboratory-frame redistribution function is more easily derived from the original integral form. One can also easily verify that $r_{\rm V}$ can be expressed as a two-dimensional convolution of $r_{\rm II}$ with two Lorentzians [468],

$$r_{\rm V}(\xi', \xi) = \int_{-\infty}^{\infty} \int_{-\infty}^{\infty} L(t, \gamma_i)\, L(t', \gamma_i)\, r_{\rm II}(\xi' - t', \xi - t)\, dt'\, dt. \tag{10.78}$$

Observer's-Frame Redistribution Functions

The previous redistribution functions were evaluated in the rest frame of the radiating atom. In order to be able to obtain formulae useful in a radiative transfer problem, we need to average the atom's-frame functions over the velocity distribution of the radiating atoms.

The large majority of astrophysical studies of frequency redistribution make the following three assumptions:

(i) The atoms in all energy levels have a Maxwellian velocity distribution, with a kinetic temperature equal to the electron temperature.

(ii) An atom's velocity is unchanged during the scattering process. This is usually a good approximation. A correct treatment of this effect, and a discussion of some subtle points, not discussed here, are given in [528].

(iii) We neglect the aberration of photon directions in transforming from the atom's frame to the laboratory frame and use the lowest-order form of the Doppler effect. The corresponding atom's-frame frequencies for absorption and emission are

$$\xi' = \nu' - \nu_0(\mathbf{v}\cdot\mathbf{n}')/c, \quad \xi = \nu - \nu_0(\mathbf{v}\cdot\mathbf{n})/c. \tag{10.79}$$

Velocities are expressed in dimensionless thermal units,

$$\mathbf{u} = \mathbf{v}/v_{\text{th}} = (m_A/2kT)^{1/2}\mathbf{v}, \tag{10.80}$$

where m_A is the mass of an atom. In this notation, the Maxwellian velocity distribution is

$$f(u_1,u_2,u_3)\,du_1\,du_2\,du_3 = \pi^{-1/2}\exp(-u_1^2-u_2^2-u_3^2)\,du_1\,du_2\,du_3. \tag{10.81}$$

We also use dimensionless frequency displacements from the line center,

$$x' \equiv (\nu'-\nu_0)/w, \qquad x \equiv (\nu-\nu_0)/w. \tag{10.82}$$

Here, $w \equiv (\nu_0/c)(2kT/m_A)^{1/2}$ is the Doppler width (denoted as w, not as $\Delta\nu_D$ as in the customary notation).

We then have

$$R_v(\nu',\mathbf{n}';\nu,\mathbf{n}) = r(\nu'-\nu_0\mathbf{v}\cdot\mathbf{n}'/c, \nu-\nu_0\mathbf{v}\cdot\mathbf{n})\,g(\mathbf{n}',\mathbf{n}), \tag{10.83}$$

where the subscript v indicates redistribution produced by an atom moving with velocity $\mathbf{v}$.

The redistribution function for an ensemble of atoms is obtained by averaging over the Maxwellian velocity distribution. To perform this average, choose orthogonal basis vectors $(\mathbf{n}_1,\mathbf{n}_2,\mathbf{n}_3)$, with $\mathbf{n}_1$ and $\mathbf{n}_2$ taken to be coplanar with $\mathbf{n}'$ and $\mathbf{n}$, and with $\mathbf{n}_1$ bisecting the angle Θ between them, $\cos\Theta = \mathbf{n}'\cdot\mathbf{n}$. Then

$$\mathbf{n}' = \alpha\mathbf{n}_1 + \beta\mathbf{n}_2, \qquad \mathbf{n} = \alpha\mathbf{n}_1 - \beta\mathbf{n}_2, \tag{10.84}$$

where

$$\alpha = \cos(\Theta/2), \qquad \beta = \sin(\Theta/2). \tag{10.85}$$

Averaging (10.83) over the velocity distribution (10.81) we have

$$\begin{aligned} R(\nu',\mathbf{n}';\nu,\mathbf{n}) &= \int_{-\infty}^{\infty} du_1 \int_{-\infty}^{\infty} du_2 \int_{-\infty}^{\infty} du_3 f(u_1,u_2,u_3) R_u(\nu',\mathbf{n}';\nu,\mathbf{n}) \\ &= \frac{g(\mathbf{n}',\mathbf{n})}{\pi}\int_{-\infty}^{\infty} du_1\, e^{-u_1^2} \int_{-\infty}^{\infty} du_2\, e^{-u_2^2}\, r[\nu'-w(\alpha u_1+\beta u_2), \nu-w(\alpha u_1-\beta u_2)], \end{aligned} \tag{10.86}$$

where the integration over u_3 has been carried out explicitly.

If the re-emission process is coherent, i.e., if r contains $\delta(\xi' - \xi)$, one has

$$\delta(\xi' - \xi) = \delta(\nu - \nu' + 2w\beta u_2), \tag{10.87}$$

which simplifies the double integral.

We can now derive results for the individual cases, obtained by substituting the individual basic redistribution functions into (10.86).

Case I: We obtain

$$R_{\mathrm{I}}(\nu', \mathbf{n}'; \nu, \mathbf{n}) = \frac{g(\mathbf{n}', \mathbf{n})}{2\pi\alpha\beta w^2} \exp\left[\frac{-(\nu' - \nu)^2}{4\beta^2 w^2}\right] \exp\left[\frac{-(\nu + \nu' - 2\nu_0)^2}{4\alpha^2 w^2}\right]. \tag{10.88}$$

Noting that $2\alpha\beta = 2\sin(\Theta/2)\cos(\Theta/2) = \sin\Theta$, and $\alpha^2 + \beta^2 = 1$, and transforming to dimensionless frequency units, we have

$$R_{\mathrm{I}}(x', \mathbf{n}'; x, \mathbf{n}) = \frac{g(\mathbf{n}', \mathbf{n})}{\pi \sin\Theta} \exp\left[-x^2 - (x' - x\cos\Theta)^2 \csc^2\Theta\right]. \tag{10.89}$$

Case II: The averaging procedure gives

$$R_{\mathrm{II}}(\nu', \mathbf{n}'; \nu, \mathbf{n}) = \frac{g(\mathbf{n}', \mathbf{n})}{2\pi\alpha\beta w^2} \exp\left[\frac{-(\nu' - \nu)^2}{4\beta^2 w^2}\right]$$
$$\times \left(\frac{\gamma}{\pi\alpha w}\right) \int_{-\infty}^{\infty} \left[\left(\frac{\nu' + \nu - 2\nu_0}{2\alpha w} - u\right)^2 + \left(\frac{\gamma}{\alpha w}\right)^2\right]^{-1} e^{-u^2} du. \tag{10.90}$$

Transforming to dimensionless frequencies using (10.82), and recalling the definition of the Voigt function $H(a, v)$ [equation (8.23)], we obtain

$$R_{\mathrm{II}}(x', \mathbf{n}'; x, \mathbf{n}) = \frac{g(\mathbf{n}', \mathbf{n})}{\pi \sin\Theta} \exp\left[-\tfrac{1}{4}(x - x')^2 \csc^2(\Theta/2)\right]$$
$$\times H[a\sec(\Theta/2), \tfrac{1}{2}(x + x')\sec(\Theta/2)], \tag{10.91}$$

where $a \equiv \gamma/w$. Although this result is relatively complicated, efficient methods to evaluate $H(a, v)$ exist, and R_{II} can be evaluated fairly easily. Notice that for $\Theta \to 0$, R_{II} approaches the δ-function, $R_{\mathrm{II}}(\nu', \nu; 0) = g(\Theta = 0)\delta(\nu' - \nu)$. This result is easily understood on physical grounds, because for $\Theta = 0$ we have forward scattering, so that there is no change of projected atomic velocity during the scattering event; therefore, the laboratory-frame function is equal to the atom's-frame function.

Case III: Performing the integrations and converting to dimensionless frequency units, we find

$$R_{\mathrm{III}}(x', \mathbf{n}'; x, \mathbf{n}) =$$
$$\frac{g(\mathbf{n}', \mathbf{n})}{\pi^2} a\csc\Theta \int_{-\infty}^{\infty} \frac{\exp(-u^2) H(a\csc\Theta, x\csc\Theta - u\cot\Theta)}{(x' - u)^2 + a^2} du. \tag{10.92}$$

This result is no longer expressible in terms of simple functions and must be evaluated by numerical integration, for which efficient numerical methods have been developed in [898] and [468].

Cases IV and V: Because r_{IV} is a special case of r_V, and because the redistribution function r_{IV} is inappropriate for describing redistribution between two broadened levels, we consider here only case V. One can use the form (10.77) and average over velocities (as was done in [544]), but the resulting expressions are cumbersome. A more elegant procedure was developed in [468], which applied the convolution theorem for Fourier transforms (see, e.g., [140]), denoted by $\mathcal{F}$, to (10.76) to write

$$\begin{aligned}\mathcal{F}_{r,s}[r_V(\xi',\xi)] &= \exp[-\gamma_i|\,r\,| - \gamma_i|\,s\,|\,] \times \mathcal{F}_{r,s}[r_{II}(\xi',\xi)] \\ &= \exp[-\gamma_i|\,r\,| - \gamma_i|\,s\,| - \gamma_j|\,r+s\,| - i\xi_0(r+s)]. \end{aligned} \tag{10.93}$$

The inverse Fourier transform of (10.93) is

$$r_V(\xi',\xi) = \frac{1}{4\pi^2}\iint_{-\infty}^{\infty} \exp[-\gamma_i|\,r\,| - \gamma_i|\,s\,| - \gamma_j|\,r+s\,| + itr + it's]\,dr\,ds, \tag{10.94}$$

where $t \equiv \xi - \xi_0$, and $t' \equiv \xi' - \xi_0$. Using the notation

$$\big\langle f(u_1,u_2)\big\rangle \equiv \frac{1}{4\pi^2}\iint_{-\infty}^{\infty} e^{-(u_1^2+u_2^2)} f(u_1,u_2)\,du_1\,du_2 \tag{10.95}$$

for the velocity averaging, the laboratory-frame redistribution function R_V may be written as

$$R_V = \frac{g(\mathbf{n}',\mathbf{n})}{4\pi^2}\iint_{-\infty}^{\infty} \exp[\,-\gamma_i|\,r\,| - \gamma_i|\,s\,| - \gamma_j|\,r+s\,|\,]\big\langle e^{itr+it's}\big\rangle\,dr\,ds, \tag{10.96}$$

because only the frequency differences t and t' contain velocity components from the Doppler shifts. After some algebra (the interested reader is referred to [468]), one obtains

$$R_V(x',\mathbf{n}';x,\mathbf{n}) = \frac{g(\mathbf{n}',\mathbf{n})}{4\pi^2\sin\Theta}\Big\{H\left[a_j\sec(\Theta/2),\,C/2\right] \times H\,[a_i\csc(\Theta/2),\,D/2] + E_V(x',x,\Theta)\Big\}, \tag{10.97}$$

where

$$C \equiv (x+x')\sec(\Theta/2), \qquad D \equiv (x-x')\csc(\Theta/2). \tag{10.98}$$

and the "correction term" is given by

$$E_V(x', x, \Theta) = \frac{4}{\pi} \int_0^\infty dv\, e^{-[u^2+2a_j \sec(\Theta/2)u]} \cos(Cu) \times \int_{v \cot(\Theta/2)}^\infty du\, e^{-[v^2-2a_i \sec(\Theta/2)\cot(\Theta/2)v]} \cos(Dv) \tag{10.99}$$

This function has to be evaluated numerically; an efficient algorithm was developed in [468].

Symmetry Properties

First, rewrite the general result (10.86) in terms x and x'. Introducing $\tilde{r}(\nu' - \nu_0, \nu - \nu_0) \equiv r(\nu', \nu)$, one finds

$$R(x', \mathbf{n}'; x, \mathbf{n}) = \frac{w^2 g(\mathbf{n}', \mathbf{n})}{\pi} \int_{-\infty}^{\infty} du_1\, e^{-u_1^2} \int_{-\infty}^{\infty} du_2\, e^{-u_2^2}\, \tilde{r}[w(x' - \alpha u_1 - \beta u_2), w(x - \alpha u_1 + \beta u_2)]. \tag{10.100}$$

The basic redistribution function is symmetric for cases I, II, III, and V (but is not symmetric for case IV), i.e.,

$$\tilde{r}_i(\xi', \xi) = \tilde{r}_i(\xi, \xi'), \qquad i = \text{I}, \text{II}, \text{III}, \text{V}. \tag{10.101}$$

Let us first examine the symmetry of the laboratory-frame redistribution function over an exchange of the absorption and emission frequency. From (10.100) and (10.101),

$$\begin{aligned} R(x, \mathbf{n}'; x', \mathbf{n}) &= \frac{w^2 g(\mathbf{n}', \mathbf{n})}{\pi} \int_{-\infty}^{\infty} du_1 \int_{-\infty}^{\infty} du_2\, e^{-(u_1^2+u_2^2)} \tilde{r}[w(x - \alpha u_1 - \beta u_2), w(x' - \alpha u_1 + \beta u_2)] \\ &= \frac{w^2 g(\mathbf{n}', \mathbf{n})}{\pi} \int_{-\infty}^{\infty} du_1 \int_{-\infty}^{\infty} du_2\, e^{-(u_1^2+u_2^2)} \tilde{r}[w(x' - \alpha u_1 + \beta u_2), w(x - \alpha u_1 - \beta u_2)] \\ &= \frac{w^2 g(\mathbf{n}', \mathbf{n})}{\pi} \int_{-\infty}^{\infty} du_1 \int_{-\infty}^{\infty} du_2\, e^{-(u_1^2+u_2^2)} \tilde{r}[w(x' - \alpha u_1 - \beta u_2), w(x - \alpha u_1 + \beta u_2)] \\ &= R(x', \mathbf{n}'; x, \mathbf{n}), \end{aligned} \tag{10.102}$$

where the second equality follows from the symmetry of the atom's-frame functions, (10.101), and the third equality follows from replacing u_2 with $-u_2$, which is possible because the integral spans the whole range $(-\infty, \infty)$, and $\exp(-u_2^2)$ is a symmetric function of u_2. Thus

$$R_i(x', \mathbf{n}'; x, \mathbf{n}) = R_i(x, \mathbf{n}'; x', \mathbf{n}), \qquad i = \text{I, II, III, V}. \tag{10.103}$$

Similarly, one can show (see [468] for a detailed proof) that the function R_V is unchanged when we replace $x' \to -x'$ and $x \to -x$, and inasmuch as cases I, II, and III are limiting cases of R_V,

$$R_i(-x', \mathbf{n}'; -x, \mathbf{n}) = R_i(x', \mathbf{n}'; x, \mathbf{n}), \qquad i = \text{I, II, III, V}. \tag{10.104}$$

Finally, a transformation from $\mathbf{n}' \to -\mathbf{n}'$ changes the sign of α and β in the expressions for absorbed photons; hence if $g(\mathbf{n}', \mathbf{n})$ is an even function of $\mathbf{n}' \cdot \mathbf{n}$, then

$$\begin{aligned} R(-x', -\mathbf{n}'; x, \mathbf{n}) &= \frac{w^2 g(\mathbf{n}', \mathbf{n})}{\pi} \int_{-\infty}^{\infty} du_1 \int_{-\infty}^{\infty} du_2\, e^{-(u_1^2+u_2^2)} \tilde{r}[w(-x'+\alpha u_1+\beta u_2), w(x-\alpha u_1+\beta u_2)] \\ &= R(x', \mathbf{n}'; x, \mathbf{n}) \end{aligned} \tag{10.105}$$

for cases I and III.

Angle-Averaged Redistribution Functions

We shall see in § 10.5 that the basic quantity that enters the emission coefficient is the *redistribution integral*, defined as

$$\overline{R}(\nu, \mathbf{n}) \equiv \int_0^{\infty} d\nu' \oint (d\Omega'/4\pi)\, I(\nu', \mathbf{n}') R(\nu', \mathbf{n}'; \nu, \mathbf{n}). \tag{10.106}$$

Thus the redistribution integral represents a contribution to the emission coefficient from emissions after previous absorption of a photon in the same line. If the radiation field is isotropic, we can write

$$\overline{R}(\nu) = \int_0^{\infty} J(\nu') R(\nu', \nu)\, d\nu', \tag{10.107}$$

where $J(\nu')$ is the mean intensity of radiation and $R(\nu', \nu)$ is the *angle-averaged redistribution function*

$$R(\nu', \nu) \equiv \oint R(\nu', \mathbf{n}'; \nu, \mathbf{n})\, (d\Omega'/4\pi). \tag{10.108}$$

Notice that although we average over one direction ($\mathbf{n}'$) only, the resulting angle-averaged function does not depend on direction $\mathbf{n}$ because the original redistribution function $R(\nu', \mathbf{n}'; \nu, \mathbf{n})$ does not depend on the individual directions $\mathbf{n}'$ and $\mathbf{n}$, but only on the angle Θ between them. The angle-averaged redistribution function may also be written as

$$R(\nu', \nu) = 2\pi \int_0^{\pi} R(\nu', \mathbf{n}', \nu, \mathbf{n}) \sin \Theta \, d\Theta. \tag{10.109}$$

In astrophysical radiative transfer an approximation is often made that (10.107) can be used for a general radiation field, not only one that is isotropic. The rationale is as follows: At any given frequency the radiation field departs significantly from isotropy only at points whose optical depths τ_ν from the surface are of order of unity or less; at depths τ_ν greater than $\simeq 1$, the radiation field is essentially isotropic. We will see in chapters 14 and 15 that a characteristic of NLTE line formation is that the surface value of the source function is determined by photons contributed over *an entire destruction length*, and in virtually all of this region the radiation field will, in fact, be isotropic. Therefore, we may expect that even at the surface where $I(\mu, \tau)$ shows departures from isotropy, S_ν will still have a value *already* fixed by processes occurring at depths where the anisotropy is negligible; hence the value of $I(\tau_\nu = 0, \mu, \nu)$ computed from this S_ν should be quite accurate. Thus use of the angle-averaged redistribution function accounts fully for the critical aspects of redistribution in frequency and sacrifices information only of secondary importance.

The function defined by (10.108) is normalized such that

$$\int_0^{\infty} \int_0^{\infty} R(\nu', \nu) \, d\nu' \, d\nu = 1. \tag{10.110}$$

Integration over all emitted photons yields the absorption profile coefficient

$$\int_0^{\infty} R(\nu', \nu) \, d\nu = \phi(\nu'), \tag{10.111}$$

and integration over all absorptions yields what is called the *natural-excitation* emission profile coefficient, which is equal to the absorption profile coefficient for all physically consistent basic redistribution functions, i.e., cases I, II, III, and V, and thus is consistent with the semiclassical picture, i.e.,

$$\int_0^{\infty} R_i(\nu', \nu) \, d\nu' \equiv \psi^*(\nu) = \phi(\nu), \qquad i = \mathrm{I, II, III, V}. \tag{10.112}$$

An evaluation of the angle-averaged basic redistribution functions is straightforward but lengthy. Although one can apply the definition (10.108), it is rather complicated. It is simpler to perform the angle average for atom's-frame redistribution functions $r_i(\xi', \xi)$ first, for each velocity, and then to average over velocities. Only the final expressions will be given here; the interested reader is referred to [544].

We will simplify the notation introduced in [544], which adds an additional subscript A for the case of isotropic phase function $g(\mathbf{n}', \mathbf{n}) = 1$ and a subscript B for a dipole phase function. We will also consider, as was done throughout the history of astrophysical line transfer, only the case of an isotropic phase function. Moreover, if one intends to study details of angular redistribution and the effects of the dipole phase function, it is more appropriate to employ angle-dependent redistribution functions.

Case I:

$$R_{\rm I}(x', x) = \tfrac{1}{2}\,{\rm erfc}[\max(|x'|, |x|)]\,, \tag{10.113}$$

where the complimentary error function is defined as

$$\mathrm{erfc}(x) \equiv \frac{2}{\sqrt{\pi}} \int_x^\infty e^{-z^2}\, dz. \tag{10.114}$$

The redistribution function is easy to compute from well-known approximation formulas for $\mathrm{erfc}(x)$; see [5, p. 99].

Case II:

$$R_{\rm II}(x', x) = \frac{1}{\pi^{3/2}} \int_{u_{\min}}^\infty e^{-u^2} \left[\arctan\left(\frac{\underline{x}+u}{a}\right) - \arctan\left(\frac{\overline{x}-u}{a}\right)\right] du, \tag{10.115}$$

where $u_{\min} \equiv \frac{1}{2}(\overline{x} - \underline{x})$, and

$$\overline{x} \equiv \max(|x|, |x'|), \qquad \underline{x} \equiv \min(|x|, |x'|). \tag{10.116}$$

An accurate method for the evaluation of $R_{\rm II}$ is given in [7].

The redistribution function $R_{\rm II}$ is of great interest. By itself, it describes the important case of scattering by a resonance line that is broadened by radiation (natural) damping, but as we shall see later on, it also describes a corresponding part of the redistribution corresponding to an emission before an elastic collision in the case of general broadening. Because the exact evaluation of $R_{\rm II}$ is somewhat costly, several approximations have been introduced that will be discussed below.

Case III:

$$R_{\rm III}(x', x) = \frac{1}{\pi^{3/2}} \int_0^\infty e^{-u^2} \left[\arctan\left(\frac{x'+u}{a}\right) - \arctan\left(\frac{x'-u}{a}\right)\right] \times \left[\arctan\left(\frac{x+u}{a}\right) - \arctan\left(\frac{x-u}{a}\right)\right] du. \tag{10.117}$$

On occasion it has been argued on intuitive grounds that if the redistribution process is completely noncoherent in the atom's frame, and if it is combined with random Doppler motions, then photon redistribution should be completely noncoherent in the observer's frame as well. However, this conclusion is false. Doppler motions introduce a correlation between incoming and outgoing frequencies (but recall that this is only if we assume no velocity-changing collisions during the scattering process), and the deviations from complete frequency redistribution in the laboratory frame may be significant.

However, despite these deviations, it turns out (see § 15.6) that the assumption $R(x', x) = \phi(x')\phi(x)$ produces line profiles quite similar to those obtained from the exact $R_{\rm III}(x', x)$, and in practice, the case of complete noncoherence in the atom's frame may be treated as complete noncoherence in the laboratory frame without serious errors.

Case V: There are no useful analytical expressions for $R_{\rm V}$. In practice, this function has to be evaluated by numerical integration of the angle-dependent function, equations (10.97)–(10.98).

Approximate Forms of Redistribution Functions

The basic redistribution functions are complicated and costly to evaluate. Therefore, many attempts have been made to develop approximate forms that are easier to evaluate, yet capture the essential physical features of the solution of the transfer problem.

It is important to bear in mind that a redistribution function is merely a *tool* for formulating the emission coefficient. As we shall see in § 10.5, the quantity of interest is the scattering or redistribution *integral*. Therefore, in approximating the redistribution functions we do not necessarily strive to approximate well the redistribution function, but rather the redistribution integral. Because the redistribution integral depends on the radiation field, the approximate form of the redistribution function may also depend, to some extent, on expected properties of the radiation field, which would be absurd if we aim only to approximate mathematical functions.

Actually, the only important basic redistribution functions are $R_{\rm II}$, $R_{\rm III}$, and $R_{\rm V}$ (because $R_{\rm I}$ is simple and $R_{\rm IV}$ is physically inconsistent). Approximations were developed for angle-averaged functions because there is no point to solve numerically a complex problem of frequency and angular redistribution when using an approximate form of the redistribution function. Moreover, angle-dependent redistribution functions are mathematically simpler than their angle-averaged counterparts, and solutions of transfer problems with angular dependence are more time-consuming than evaluation of the redistribution functions, so approximations here are unnecessary, and even counterproductive.

Case II: In figure 10.2 we see the basic features. $R_{\rm II}$ behaves like complete redistribution for an incoming photon absorbed in the line core, whereas it has a narrow peak about one Doppler width wide for a photon absorbed in the wing. Because the radiation field in the line wing is a relatively smooth function of frequency, variations across one Doppler width are rather small. It is thus reasonable to replace $R_{\rm II}$ with a δ-function, $\delta(x' - x)$, for large x'. This is essentially what was suggested in [582], namely,

$$R_{\rm II}(x', x) \approx [1 - a(x)]\, \phi(x')\phi(x) + a(x)\phi(x')\delta(x' - x). \tag{10.118}$$

This approximation is often referred to as the *Jefferies–White* approximation. There is some freedom in the choice of the function $a(x)$. In the original paper, a was chosen to be a unit step function, $a(x) = S(x, 3)$, where $S(x, y) = 1$ for $x > y$, and $S(x, y) = 0$ for $x \leq y$. But as pointed out in [624], such a function is neither

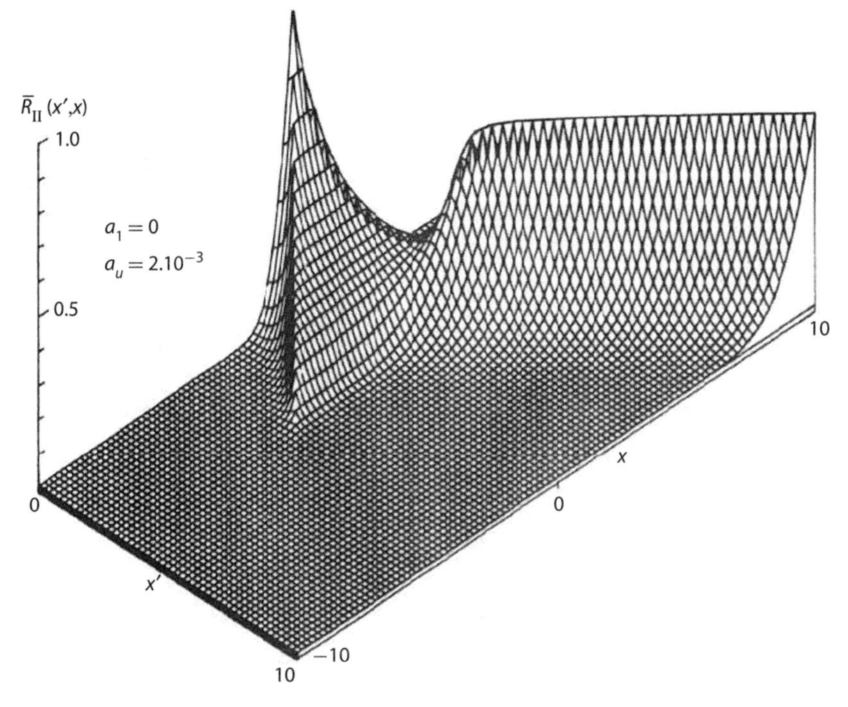

Figure 10.2 Probability of re-emission at frequency x, per absorption at frequency x', for the angle-averaged redistribution function $R_{\rm II}(x',x)$. The damping parameter of the upper level is $a_u = 2 \times 10^{-3}$. The ordinate is $\overline{R}_{\rm II}(x',x) \equiv R_{\rm II}(x',x)/\phi(x')$. From [470].

normalized nor symmetric. The modification suggested was to replace this $a(x)$ by the symmetric function

$$a(x',x) = a(x,x') = \tilde{a}(x), \quad \text{for} \quad x > x', \tag{10.119}$$

and

$$a(x',x') = \int_0^\infty a(x',x)\phi(x)\,dx, \tag{10.120}$$

and to approximate $R_{\rm II}$ by

$$R_{\rm II}(x',x) \approx [1 - a(x',x)]\phi(x')\phi(x) + a(x',x)\phi(x')\delta(x'-x). \tag{10.121}$$

The function $\tilde{a}$ can be chosen based on experience; [624] suggested the form

$$\tilde{a}(x) = 1 - \exp\left[-(x-x_0)^2/\Delta^2\right], \quad \text{for} \quad x \geq x_0, \tag{10.122}$$

and $\tilde{a}(x) = 0$ for $x < x_0$, with $x_0 = 2$ and $\Delta = 2$ recommended as the best choice.

The approximation defined by (10.118) or (10.121) is often referred to as *partial coherent scattering approximation*. It provides good results for weak lines, and optically thin or only moderately thick situations, but, as we shall discuss in more detail in § 15.2, it neglects the important phenomenon called "Doppler diffusion." The point is that the peak of $R_{\rm II}(x', x)$ is not symmetric around x for large x'; there is a bias for re-emission at $x < x'$ rather than at $x > x'$, as can be seen from (10.91). An averaged photon frequency is thus gradually shifted toward the line center, while when using the partial coherent scattering approximation (10.118) or (10.121), the photons are trapped in the line wing. The large-scale behavior of the radiation field, and the thermalization length, are thus different for exact $R_{\rm II}$ redistribution and for the partial coherent scattering approximation.

There are several possible ways to cope with this problem, discussed in § 15.2. One is the Fokker–Planck-type approach developed in [442]; the other simple yet efficient method was suggested in [517]. The latter approach consists of adopting the Jefferies-White approximation, but taking the division frequency to be depth-dependent, i.e.,

$$a(x) = 0, \quad \text{for} \quad x < x_D(\tau), \qquad a(x) = 1, \quad \text{for} \quad x \geq x_D(\tau), \tag{10.123}$$

where $x_D(\tau) \propto \tau^{1/3}$. We return to this method, as well as to explaining the physical meaning of the division frequency, in § 15.2. Here we simply stress that this is an example of the fact that an approximate redistribution function must yield a good approximation of the redistribution integral.

Nevertheless, there are several useful approximations of $R_{\rm II}$ in the mathematical sense. They essentially follow from an expansion of the Voigt function [545],

$$H(a, x) \approx e^{-x^2} + \frac{a}{\sqrt{\pi}\, x^2} \left[1 + \left(\tfrac{3}{2} - a^3 \right) / x^2 + \ldots \right], \tag{10.124}$$

which allows the integral in (10.108) to be evaluated analytically using (10.91). We thus obtain

$$R_{\rm II}(x', x) \approx \tfrac{1}{2} \operatorname{erfc}(|r| + |s|) + \left(1 - \frac{a^2}{s^2} + \frac{a^4}{s^4} \right) \frac{a}{\pi s^2} \operatorname{ierfc}(|r|) + \ldots, \tag{10.125}$$

where

$$r \equiv (x' - x)/2, \qquad s \equiv (x' + x)/2, \tag{10.126}$$

and the *integrated error function* ierfc is defined by

$$\operatorname{ierfc}(z) \equiv \int_z^\infty \operatorname{erfc}(t)\, dt = \frac{1}{\sqrt{\pi}} e^{-z^2} - z \operatorname{erfc}(z). \tag{10.127}$$

We recognize the first term as the function $R_{\rm I}$. Taking the leading term of the second term of (10.125), we have

$$R_{\rm II}(x', x) \approx R_{\rm I}(x', x) + \frac{a}{\pi s^2} \operatorname{ierfc}(|r|). \tag{10.128}$$

In the far wings, the contribution from $R_{\rm I}$ is negligible; one is left with the limiting form of $R_{\rm II}$ first derived in [1098], which is sometimes called the *Unno approximation*. A more complete form of the expansion of (10.124) is given in [7], which can also be used for an evaluation of $R_{\rm II}$ in certain frequency regimes.

Another approximation, called sometimes the Ayres approximation [72], is a simple yet efficient generalization of (10.128), namely,

$$R_{\rm II}(x',x) \approx R_{\rm I}(x',x) + \frac{a}{\pi \widetilde{x}^2} \mathrm{ierfc}(|r|) = R_{\rm I}(x',x) + \frac{a}{\pi \widetilde{x}^2} \mathrm{ierfc}\left(\frac{|x'-x|}{2}\right), \tag{10.129}$$

where $\widetilde{x} = \max(s,\alpha) = \max[(x+x')/2,\alpha]$, and α is an empirical parameter of the order of unity. Ayres in [72] showed that $\alpha = 2$ is a good choice for all interesting values of the damping parameter a; for large values of $a > 0.1$, the choice $\alpha = 1$ gives better results.

Case III: A convenient approximation, commonly used in numerical work, is to replace $R_{\rm III}$ by complete redistribution in the laboratory frame:

$$R_{\rm III}(x',x) \approx \phi(x')\,\phi(x). \tag{10.130}$$

Case V: As can be seen in figure 10.3 a natural approximation for $R_{\rm V}$, demonstrated numerically in [530], is

$$R_{\rm V}(x',x) \approx \frac{a_\ell}{a_u + a_\ell}\,\phi(x')\phi(x) + \frac{a_u}{a_u + a_\ell} R_{\rm II}(x',x), \tag{10.131}$$

i.e., as a linear combination of complete redistribution and $R_{\rm II}$. If $a_\ell \to 0$, then $R_{\rm V}(x',x) \to R_{\rm II}(x',x)$, as it should be. One can further apply the Jefferies-White approximation for $R_{\rm II}$, so that the approximate $R_{\rm V}$ reads

$$R_{\rm V}(x',x) \approx \frac{a_\ell}{a_u + a_\ell}\,\phi(x')\phi(x) + \frac{a_u}{a_u + a_\ell}\,\delta(x'-x)\,S(x',3)\,, \tag{10.132}$$

where $S(x,y)$ is again a unit step function.

10.4 MORE COMPLEX REDISTRIBUTION FUNCTIONS

Generalized Redistribution Functions

So far, we have considered the case of absorption and re-emission of a photon in the same line, i.e., the case of resonance scattering, the correlated transition $i \to j \to i$. But the picture developed above is not limited to these transitions. One can extend the formalism to any kind of radiative (induced) transition from an initial level i to an intermediate level e, and a subsequent radiative (induced) transition from the

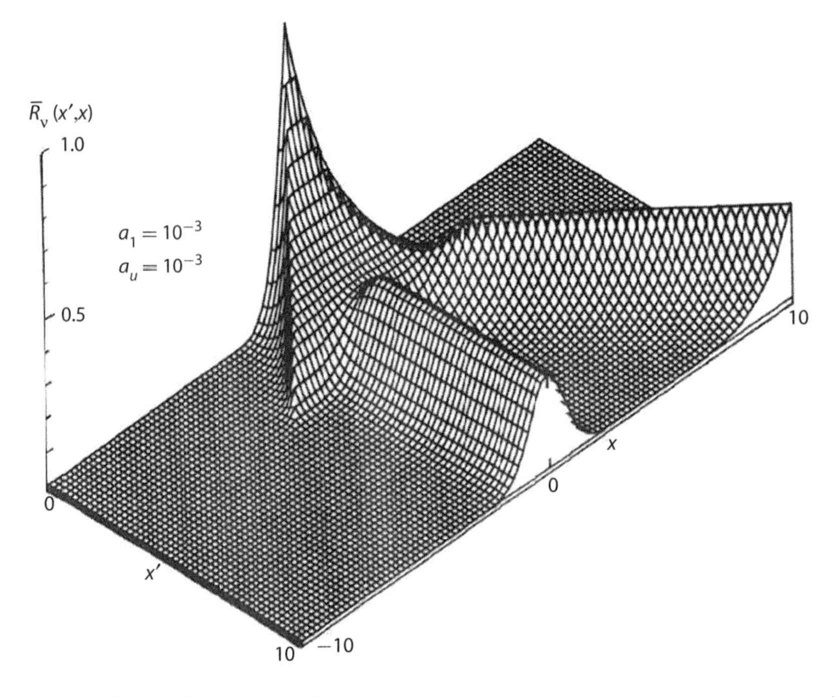

Figure 10.3 Probability of re-emission at frequency x, per absorption at frequency x', for the angle-averaged redistribution function $R_V(x',x)$. The damping parameters are $a_l = a_u = 10^{-3}$. The ordinate is $\overline{R}_V(x',x) \equiv R_V(x',x)/\phi(x')$. From [470].

intermediate state e to the final state f. In principle there is no restriction on the relative energies of the these levels. There are four possible cases:

$E_i < E_e, E_f < E_e$	(resonance Raman scattering)
$E_i < E_e, E_f > E_e$	(correlated two-proton absorption)
$E_i > E_e, E_f < E_e$	(correlated two-photon emission)
$E_i > E_e, E_f > E_e$	(inverse resonance Raman scattering)

The latter two processes that start with an emission of a photon must start with an induced emission; otherwise, the spontaneous emission leads to a natural population of the intermediate level, and there will be no correlation with its frequency and the frequency of the photon involved in the subsequent transition $e \rightarrow f$. In view of our assumption of the weak radiation field, these processes are negligible, and we will not consider them in the following discussion.

The two-photon absorption may be of importance in some special cases, but it would lead to a significant complication of the formulation of the absorption coefficient, which would then also contain correlated terms. Fortunately, in most astrophysical situations, this process is negligible. The only potentially significant contribution might come from an absorption from the ground state to a low excited state (in a resonance line), followed by another absorption in a subordinate line.

However, the optical depths of the resonance and subordinate lines are typically quite different. At layers where the optical depth in the subordinate becomes compared to unity (and where the correlation effects may be important), the optical depth in the resonance line is still large, and thus the radiation field is constant over the line profile, and therefore the absorption is a process that leads to the natural population of level e. In the layers where the resonance line becomes transparent, and thus the absorption $i \to e$ has the potential to populate level e non-naturally, the optical thickness in the subordinate line is small, and the radiation field does not change any longer; hence the two-photon absorption is negligible

The only interesting process left is resonance Raman scattering. Thus we consider the probability density that a photon in the range $(\xi', \xi' + d\xi')$ is absorbed in the transition $i \to e$, and subsequently a photon in the range $(\xi, \xi + d\xi)$ is emitted in the transition $e \to f$. These functions were treated generally in [516], where the term *generalized redistribution function* was suggested, and a system of functions parallel to the (extended) Hummer's system was developed. Earlier studies (e.g., [752]) called one special function of this kind a "cross-redistribution function." Here we will continue to call them generalized redistribution functions and label them as r_{ief}.

The form of a generalized redistribution is derived in § 10.1 from the semi-classical point of view and in § 10.2 from the quantum point of view, namely [cf. equation (10.24)],

$$r_{ief}(\xi', \xi) = p_e^{\rm corr} r_{ief}^{\rm corr}(\xi', \xi) + (1 - p_e^{\rm corr}) r_{ief}^{\rm ncorr}(\xi', \xi), \tag{10.133}$$

where the "correlated" function is given by (10.16) or (10.18) and the "non-correlated" one by (10.19) and the coherence fraction, which in this case is better called a "correlation fraction," is [cf. equation (10.23)]

$$p_e^{\rm corr} = \frac{P_e}{P_e + Q_e}. \tag{10.134}$$

The system of basic generalized redistribution functions was derived in [516]; the functions were denoted by p, with Roman numeral subscripts that indicate the correspondence with Hummer's system. The derivation is essentially analogous to that given above for ordinary redistribution functions; thus we present only the resulting expressions.

Case I: All three levels sharp:

$$p_{\rm I}(\xi', \xi) = \delta(\xi' - \xi_{ie})\, \delta(\xi' - \xi - \xi_{ie} + \xi_{ef}). \tag{10.135}$$

We see that the coherent re-emission probability $\delta(\xi' - \xi)$ that applies in the case of ordinary redistribution functions is replaced by $\delta[(\xi' - \xi_{ie}) - (\xi - \xi_{ef})]$, i.e., we have a kind of "coherence" not for the absolute values of frequencies, but for frequency differences from the line centers.

Case II: Initial and final levels sharp, the intermediate level broadened:

$$p_{\rm II}(\xi', \xi) = L(\xi' - \xi_{ie}, \gamma_e)\, \delta(\xi' - \xi - \xi_{ie} + \xi_{ef}). \tag{10.136}$$

Case III: Non-correlated case:

$$p_{\mathrm{III}}(\xi', \xi) = L(\xi' - \xi_{ie}, \gamma_{ie})\, L(\xi - \xi_{ef}, \gamma_{ef}). \tag{10.137}$$

Case IV: Levels i and e broadened; level f sharp. Here we see that function analogous to r_{IV} has a well-defined meaning because the initial and final states are different; hence one can be broadened and the other sharp. The redistribution function is given by

$$p_{\mathrm{IV}}(\xi', \xi) = L(\xi_{ie} - \xi_{ef} - \xi' + \xi, \gamma_i)\, L(\xi' - \xi_{ef}, \gamma_e), \tag{10.138}$$

Case $\overline{\mathrm{IV}}$: This is case is similar to case IV, but now with the initial level sharp, and the intermediate and final levels broadened. In the original paper it was suggested that it be called $\overline{\mathrm{IV}}$ in order to reserve case V for the actual analog to case V of the ordinary functions:

$$\overline{p_{\mathrm{IV}}}(\xi', \xi) = L(\xi' - \xi_{ie}, \gamma_e)\, L(\xi' - \xi - \xi_{ie} + \xi_{ef}, \gamma_f). \tag{10.139}$$

Case V: All levels broadened; correlated scattering:

$$p_{\mathrm{V}}(\xi', \xi) = \int_{-\infty}^{\infty} L(x - \xi', \gamma_i)\, L(x - \xi_{ie}, \gamma_e)\, L(x - \xi_{ie} + \xi_{ef} - \xi, \gamma_f)\, dx. \tag{10.140}$$

The corresponding laboratory-frame and angle-averaged redistribution functions are given in [516], to which the interested reader is referred. Here we write only the form of the function P_{II} (called R_X in [752]), the only one that was used in actual applications [535, 997, 998].

$$\begin{aligned} P_{\mathrm{II}}(x', \mathbf{n}'; x, \mathbf{n}) = & \\ & \frac{1}{\pi \sin\Theta} H\left[\frac{\gamma_e w_3}{w_1 w_2 \sin\Theta}, \frac{x'(w_2 - w_1\cos\Theta) + x(w_1 - w_2\cos\Theta)}{w_3 \sin\Theta}\right] \\ & \times \exp\left[-\left(\frac{w_1 x' - w_2 x}{w_3 \sin\Theta}\right)^2\right], \end{aligned} \tag{10.141}$$

where

$$w_3 \equiv (w_1^2 + w_2^2 - 2 w_1 w_2 \cos\Theta)^{1/2}, \tag{10.142}$$

and w_1 and w_2 are the Doppler widths for the transitions $i \to e$ and $e \to f$, respectively. The dimensionless frequencies are defined for each transition separately:

$$x' \equiv (\nu' - \nu_{ie})/w_1, \qquad x \equiv (\nu - \nu_{ef})/w_2. \tag{10.143}$$

The angle-averaged function is given by

$$P_{\mathrm{II}}(x', x) = \frac{1}{\pi} \int_0^{\pi} \frac{1}{\sin\Theta} H(\gamma_e w_2/a_1, b_1 x' + b_2 x)\, e^{-(a_1 x' - a_2 x)^2}\, d\Theta, \tag{10.144}$$

where

$$a_1 \equiv w_1/(w_3 \sin\Theta), \qquad a_2 \equiv w_2/(w_3 \sin\Theta), \tag{10.145}$$

and

$$b_1 \equiv a_2 - a_1 \cos\Theta, \qquad b_2 = a_1 - a_2 \cos\Theta. \tag{10.146}$$

Finally, one can derive expressions for "redistribution" functions for three or more photons by an extension of the semiclassical derivation in § 10.2. Expressions for three-photon functions are given in [537,538], but they have hardly any practical use.

Redistribution in Hydrogen

We have mentioned in § 10.1 that the basic assumptions of the semiclassical picture, namely, the validity of the impact approximation, and isolated lines, break down for hydrogen. At the same time, the hydrogen lines are among the most important lines in astrophysics, and departures from complete redistribution are particularly important for the $L\alpha$ line. Fortunately, it is possible to extend the semiclassical picture to treat the hydrogen lines.

A rigorous quantum mechanical treatment of redistribution in hydrogen lines was developed in [1186] and [250]; [251] gave detailed expressions useful in astrophysical applications. This formalism was used in [535] to calculate profiles for the $L\alpha$, $L\beta$, and $H\alpha$ lines originating in the solar chromosphere. We will not follow a detailed quantum mechanical treatment here, as it is quite complicated. Instead, we will use results obtained in the references above to provide an intuitive explanation within the framework of an extended semiclassical picture.

First, we note that the problem of redistribution in the hydrogen atom was often treated incorrectly. Although it is sufficient for the purposes of kinetic equilibrium to lump all l-dependent states (e.g., $2s$ and $2p$, or $3s$, $3p$, and $3d$), into one state characterized by the principal quantum number (i.e., $n=2$ or $n=3$), it is not permissible to use such an approach for the purposes of treating line scattering.

Consider a model hydrogen atom with levels up to $n=3$, resolved into (n, l) states; see figure 10.4. The higher states need not be considered in detail because,

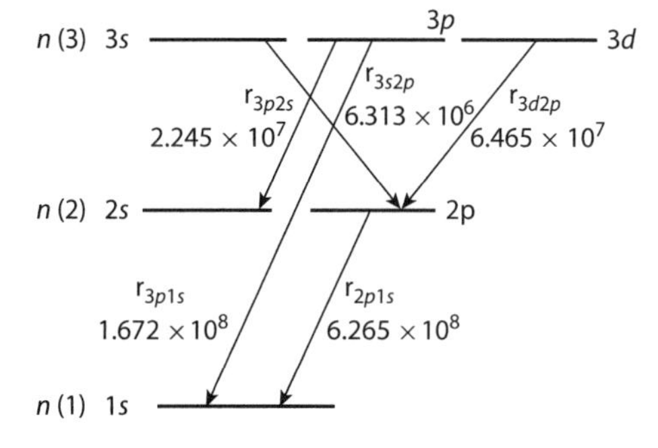

Figure 10.4 Permitted transitions among the $n = 1, 2, 3$ states of hydrogen. From [251].

in the presence of numerous interaction chains, they are populated almost naturally. Therefore, the correlated two-photon processes are

$1s \to 2p \to 1s$	(resonance scattering in $L\alpha$)
$1s \to 3p \to 1s$	(resonance scattering in $L\beta$)
$2s \to 3p \to 1s$	(resonance Raman scattering $H\alpha \to L\beta$)
$1s \to 3p \to 2s$	(resonance Raman scattering $L\beta \to H\alpha$)

These are the only dipole-permitted transitions.

In describing the transition probabilities, we cannot use averaged coefficients such as A_{21}. Rather, l-dependent coefficients, such as $A_{2p,1s}$ are required. For level populations we may still use equilibrium populations (e.g., $n_{2s} = \frac{1}{4}n_2$, $n_{2p} = \frac{3}{4}n_2$). The correct formulation of the problem in terms of l-dependent states leads to different branching ratios than those obtained when we lump together all states with the same main quantum numbers.

Second, recall that all redistribution functions introduced so far are formulated through Lorentzians (or their limiting cases), which is the consequence of adopting the impact approximation. But we know that for hydrogen lines the impact approximation is valid only very close to the line center, but in the line wings (where as we shall see shortly, the departures from CFR are most important), the line broadening approaches the quasi-static approximation. Hence the frequency dependence of the absorption profile coefficient switches from $\propto \xi^{-2}$ (an asymptotic form of a Lorentzian far from the line center) to $\propto \xi^{-3/2}$, which applies in the quasi-static limit.

A natural extension of the original semiclassical picture still describes the atom's-frame absorption profile coefficient in terms of a Lorentzian, but with a frequency-dependent broadening parameter γ,

$$\varphi_{ij}(\xi) = L[\xi, \gamma_{ij}(\xi)]. \tag{10.147}$$

Actual expressions are given in, for instance, [251]. It is also natural to extend the semiclassical picture by considering all the redistribution functions to be formulated with frequency-dependent $\gamma(\xi)$ instead of constant γ. For instance, the function $R_{\rm II}$ modified for hydrogen is given by

$$\widetilde{r}_{\rm II}(\xi', \xi) = L[\xi', \gamma_j(\xi')]\, \delta(\xi' - \xi), \tag{10.148}$$

and similarly for a generalized redistribution function $p_{\rm II}(\xi', \xi)$ for describing resonance Raman scattering $2s \to 3p \to 1s$. It turns out that detailed quantum mechanical calculations [251, 1186] yield the same result.

A complication can in principle arise when averaging over the velocity distribution of the radiators, i.e., when evaluating the laboratory-frame redistribution function. Strictly speaking, the additional frequency-dependent term $\gamma(\xi)$ makes the above averaging procedure invalid. However, $\gamma(\xi)$ is only a slowly varying function of ξ, with a scale of variation much larger than one Doppler width, which, on the other hand, is the bandwidth of the function with which one performs the velocity average. Therefore, one can neglect frequency variations of γ in the velocity averaging. This approach then gives [251]

$$\widetilde{R}_{\rm II}(x', x) = R_{\rm II}[x', x, \gamma(x')]. \tag{10.149}$$

Collisional Redistribution

The redistribution function for a general case of line scattering in the presence of elastic collisions was already given in (10.24), which was derived within the framework of the semiclassical picture. The general form is a linear combination of the correlated and non-correlated redistribution functions. In the case of sharp initial state, the correlated part is described with type II basic redistribution functions ($r_{\rm II}$ in the atom's frame, $R_{\rm II}$ in the laboratory frame for resonance scattering, and $p_{\rm II}$ and $P_{\rm II}$ for resonance Raman scattering). The branching ratio is given by (10.23).

As we found in § 10.2, one obtains the same result using a rigorous quantum mechanical derivation, which was first done in the important paper by Omont, Smith, and Cooper [817]. In the physics-oriented literature, this case is usually referred to as "collisional redistribution."

In the case of a two-level atom, we can express the re-emission probability with the actual redistribution function given by (10.24):

$$\begin{aligned} R_{121}(\nu', \nu) &= \gamma R_{\rm II}(\nu', \nu) + (1-\gamma) R_{\rm III}(\nu', \nu) \\ &\approx \gamma R_{\rm II}(\nu', \nu) + (1-\gamma)\phi_{12}(\nu')\phi_{12}(\nu), \end{aligned} \tag{10.150}$$

where the coherence fraction is given by

$$\gamma = \frac{P_2}{P_2 + Q_2} = \frac{A_{21} + C_{21}}{A_{21} + C_{21} + Q_2}. \tag{10.151}$$

10.5 EMISSION COEFFICIENT

The emission coefficient is the most important part of the theoretical description of the line-scattering process. A redistribution function is merely a *tool* to express the emission profile coefficient for the transition of interest. We have already mentioned that in the case of complete frequency redistribution (CFR), the absorption and emission profile coefficients are equal, while they are generally different if photon correlation phenomena such as "partial frequency redistribution" (PFR) are present.

Let us introduce the following notation:

$$\overline{J}_{ij} \equiv \int_0^\infty \oint I(\nu', \mathbf{n}')\phi_{ij}(\nu')\,(d\Omega'/4\pi)\,d\nu' = \int_0^\infty J(\nu')\phi_{ij}(\nu')\,d\nu', \tag{10.152}$$

which is usually called the "frequency-averaged mean intensity" for the transition $i \rightarrow j$. Further,

$$\overline{R}_{ief}(\nu, \mathbf{n}) \equiv \int_0^\infty \oint R_{ief}(\nu', \mathbf{n}'; \nu, \mathbf{n})\, I(\nu', \mathbf{n}')\,(d\Omega'/4\pi)\,d\nu', \tag{10.153}$$

which is called the *scattering integral*, and

$$\mathcal{R}_{ief}(\nu, \mathbf{n}) = \overline{R}_{ief}(\nu, \mathbf{n}) \Big/ \overline{J}_{ie}, \tag{10.154}$$

which is called the *normalized scattering integral*, or the *re-emission probability*, i.e., the conditional probability that a photon in the frequency range $(\nu, \nu + d\nu)$ is emitted in the transition $e \to f$ *if* it was previously been absorbed in the transition $i \to e$.

When we make the approximation of using angle-averaged redistribution functions, the above expressions simplify to

$$\overline{R}_{ief}(\nu) = \int_0^\infty R_{ief}(\nu', \nu) J(\nu')\, d\nu', \tag{10.155}$$

and

$$\mathcal{R}_{ief}(\nu) = \overline{R}_{ief}(\nu) / \overline{J}_{ie}. \tag{10.156}$$

Recall that in CFR, $R_{ief}(\nu', \nu) = \phi_{ie}(\nu')\phi_{ef}(\nu)$. Thus $\overline{R}_{ief}(\nu) = \overline{J}_{ie}\phi_{ef}(\nu)$ and $\mathcal{R}_{ief}(\nu) = \phi_{ef}(\nu)$.

To formulate the emission coefficient for a transition from an upper level j to a lower level i, we must distinguish two types of processes that populate level j: those that lead to the *natural population* of level j and those that do not. That is, those that involve a correlation with previously absorbed (or, more generally, emitted) photons. We denote the probability of natural population as $p(\to j^*)$. When dealing with processes that involve a correlation with previous photon(s), we consider all such processes separately because they depend on individual redistribution (i.e., correlation) functions. We follow derivation presented in [540]. We consider two cases: a strictly two-level atom, which is the simplest case that allows us to explain the basic features, and a multi-level atom.

Two-Level Atom

For an atom with two levels, we denote the ground state as "1" and the excited state as "2." Let us assume that stimulated emission is negligible. In this case there are only two processes that populate the upper level: radiative transitions $1 \to 2$ and collisional excitation $1 \to 2$. The probability $p(\to 2^*)$ is then

$$p(\to 2^*) = \frac{n_1 C_{12}}{n_1 (B_{12} \overline{J}_{12} + C_{12})}, \tag{10.157}$$

where the denominator represents the total number of transitions into level 2 per unit time, and $n_1 C_{12}$ is the number of the transitions into level 2 that lead to its natural population.

We denote the remaining probability, i.e., the probability of starting from a naturally populated level 1 and making a transition to level 2 that does *not* lead to natural population, as $p(1^* \to 2)$. That is,

$$p(1^* \to 2) = \frac{n_1 B_{12} \overline{J}_{12}}{n_1 (B_{12} \overline{J}_{12} + C_{12})} = 1 - p(\to 2^*). \tag{10.158}$$

In the case of a two-level atom (and if stimulated emission is neglected), level 1 is always populated naturally, so the asterisk seems unnecessary; we have used this notation to prepare for the more general case of multi-level atoms.

The emission profile coefficient for the transition $2 \rightarrow 1$ is then given by

$$\psi_{21}(\nu, \mathbf{n}) = p(\rightarrow 2^*)\,\phi_{12}(\nu) + p(1^* \rightarrow 2)\mathcal{R}_{121}(\nu, \mathbf{n}), \tag{10.159}$$

where the first term corresponds to an ensemble of emissions after natural population of level 2, and the second term describes an ensemble of emissions after previous absorption of a photon in a transition $1 \rightarrow 2$. We have to use the conditional re-emission probability $\mathcal{R}_{121}$ because the probability of absorption $1 \rightarrow 2$ is already contained in $p(1^* \rightarrow 2)$. This expression may be rewritten in a different form by noting that due to kinetic equilibrium, the total number of transitions out of level 2 may also be written as

$$n_1(B_{12}\overline{J}_{12} + C_{12}) = n_2(A_{21} + C_{21}) \equiv n_2 P_2. \tag{10.160}$$

Here we wrote the total rate of transitions out of level 2 as P_2. Therefore,

$$\begin{aligned}\psi_{21}(\nu, \mathbf{n}) &= \frac{n_1 C_{12}}{n_2 P_2}\phi_{12}(\nu) + \frac{n_1 B_{12}\overline{J}_{12}}{n_2 P_2}\mathcal{R}_{121}(\nu, \mathbf{n}) \\ &= \phi_{12}(\nu) + \frac{n_1 B_{12}\overline{J}_{12}}{n_2 P_2}\left[\mathcal{R}_{121}(\nu, \mathbf{n}) - \phi_{12}(\nu)\right],\end{aligned} \tag{10.161}$$

where we used the kinetic equilibrium equation to write

$$n_1(C_{12} + B_{12}\overline{J}_{12})/(n_2 P_2) = 1. \tag{10.162}$$

The second equality in (10.161) expresses the emission profile coefficient as the absorption profile coefficient plus a correction that depends on the re-emission probability. It is clear that in complete redistribution, where $\mathcal{R}_{121} = \phi_{12}(\nu)$, the correction term is zero.

We can now write the emission coefficient $\eta_{21}(\nu, \mathbf{n}) = n_2 A_{21}\psi_{21}(\nu, \mathbf{n})$ in a more revealing form. Using (10.161), we have

$$\begin{aligned}\eta_{21}(\nu, \mathbf{n}) &= \frac{A_{21}}{A_{21} + C_{21}}\left[n_1 C_{12}\phi_{12}(\nu) + n_1 B_{12}\overline{J}_{12}\mathcal{R}_{121}(\nu, \mathbf{n})\right] \\ &= n_2 A_{21}\phi_{12}(\nu) + \frac{A_{21}}{P_2}\, n_1 B_{12}\overline{J}_{12}\left[\mathcal{R}_{121}(\nu, \mathbf{n}) - \phi_{12}(\nu)\right].\end{aligned} \tag{10.163}$$

In the first equality, the term $A_{21}/(A_{21} + C_{21})$ is the probability that the atom in level 2 is de-excited radiatively by a photon emission; the first term in the square bracket corresponds to an excitation of level 2 by a collisional transition, and the second term describes a resonance scattering $1 \rightarrow 2 \rightarrow 1$. In the last equality, we can also write

$$
\begin{aligned}
&\frac{1}{\phi_{12}(\nu)} \overline{J}_{12} \left[\mathcal{R}_{121}(\nu, \mathbf{n}) - \phi_{12}(\nu) \right] \\
&= \int_0^\infty \oint \left[\frac{R_{121}(\nu', \mathbf{n}'; \nu, \mathbf{n})}{\phi_{12}(\nu)} - \phi_{12}(\nu') \right] I(\nu', \mathbf{n}') \, (d\Omega'/4\pi) \, d\nu' \equiv \overline{J}^R_{121}(\nu),
\end{aligned}
\tag{10.164}
$$

which is sometimes called the *redistribution integral.* From (10.164) one sees that $\overline{J}^R \equiv 0$ in the case of complete redistribution.

Finally, the source function, defined by $S_{21}(\nu, \mathbf{n}) \equiv \eta_{21}(\nu, \mathbf{n}) / \kappa_{12}(\nu, \mathbf{n})$; see § 11.4, is written (again neglecting stimulated emission)

$$
S_{21}(\nu, \mathbf{n}) = \frac{n_2 A_{21}}{n_1 B_{12}} + \frac{A_{21}}{P_2} \overline{J}^R_{121}(\nu, \mathbf{n}) = S^{\mathrm{CFR}} + \frac{A_{21}}{P_2} \overline{J}^R_{121}(\nu, \mathbf{n}),
\tag{10.165}
$$

where S^{CFR} is the frequency-independent source function in the case of complete redistribution.

Stimulated Emission

When stimulated emission is taken into account, we have to replace the total rate of transition out of level 2 with

$$
P_2 = A_{21} + B_{21} \overline{J}_{12} + C_{21}.
\tag{10.166}
$$

As explained in § 10.1, it is satisfactory, and in fact more consistent, to describe the stimulated emission rate using $\overline{J}_{12}$ instead of $\overline{J}_{21}$, which would contain the emission profile coefficient ψ_{21} and thus would involve a nonlinear dependence on the radiation field.

There are several possibilities to formulate the emission coefficient and the source function in the presence of stimulated emission.

(a) The first, and often used, approach is to treat stimulated emission as a negative absorption. In this case the emission coefficient is a spontaneous emission coefficient given by (10.163), and the source function is written as

$$
\begin{aligned}
S_{21}(\nu, \mathbf{n}) &= \frac{n_2 A_{21} \psi_{21}(\nu, \mathbf{n})}{n_1 B_{12} \phi(\nu) - n_2 B_{21} \psi_{21}(\nu, \mathbf{n})} \\
&= \frac{n_2 A_{21} \psi_{21}(\nu, \mathbf{n})}{n_1 B_{12} \phi(\nu)} \cdot \left[1 - \frac{n_2 B_{21}}{n_1 B_{12}} \frac{\psi_{21}(\nu, \mathbf{n})}{\phi_{12}(\nu)} \right]^{-1} \\
&= \left[\frac{n_2 A_{21}}{n_1 B_{12}} + \frac{A_{21}}{P_2} \overline{J}^R_{121}(\nu, \mathbf{n}) \right] \cdot \left[1 - \frac{n_2 g_1}{n_1 g_2} \frac{\psi_{21}(\nu, \mathbf{n})}{\phi_{12}(\nu)} \right]^{-1},
\end{aligned}
\tag{10.167}
$$

where the correction for stimulated emission $[1 - (n_2 g_1/n_1 g_2)(\psi/\phi)]^{-1}$ is treated iteratively, as shown in § 15.5.

(b) Another possibility is to denote the emission coefficient given by equation (10.163) as $\eta_{21}^{\rm spont}$ and to write the total emission coefficient as

$$\eta_{21}(\nu, \mathbf{n}) = \eta_{21}^{\rm spont}(\nu, \mathbf{n}) \left[1 + \frac{c^2}{2h\nu^3} I(\nu, \mathbf{n}) \right]. \tag{10.168}$$

The redistribution term would lead to a nonlinear dependence of the emission coefficient on the radiation field. This would be cumbersome (though not impossible) to handle in numerical work. It is rarely necessary in actual applications because partial redistribution effects are important only for strong resonance lines, which typically have large frequencies, and the stimulated emission term is negligible.

(c) A reasonable approximation of (10.168) for practical applications is to take the emission coefficient as

$$\eta_{21}(\nu, \mathbf{n}) = n_2 A_{21} \phi(\nu) \left[1 + \frac{c^2}{2h\nu^3} I(\nu, \mathbf{n}) \right] +$$

$$(A_{21}/P_2)\, n_1 B_{12} \overline{J}_{12} \left[\mathcal{R}_{121}(\nu, \mathbf{n}) - \phi_{12}(\nu) \right] \left[1 - \exp(-h\nu/kT) \right]^{-1}, \tag{10.169}$$

i.e., to take the exact stimulated emission expression only for the first, thermal term, while approximating the specific intensity by the Planck function in the redistribution term. In practice, the exponential term $\exp(-h\nu/kT)$ is typically neglected, and the stimulated emission contribution in the thermal part of the emission coefficient is treated as negative absorption, so the net effect of stimulated emission is taken into account by using the total rate P_2 as given by equation (10.166) instead of (10.160).

Multi–Level Atom

Generalizing the procedure for a two-level atom, we can write the emission profile coefficient for the transition $j \to i$, neglecting stimulated emission, as

$$\psi_{ji}(\nu, \mathbf{n}) = p(\to j^*)\, \phi_{ij}(\nu) + p(\to i^* \to j)\, \mathcal{R}_{iji}(\nu, \mathbf{n}) + \sum_{\ell<j, \ell \neq i} p(\to \ell^* \to j) \mathcal{R}_{\ell ji}(\nu, \mathbf{n}). \tag{10.170}$$

Here the first term corresponds to the emission after natural population, the second term to resonance scattering $i \to j \to i$, and the new third term to all possible processes of the resonant Raman scattering $\ell \to j \to i$.

The corresponding probabilities are given by given by

$$p(\to j^*) = \left[\sum_{\ell<j} n_\ell C_{\ell j} + \sum_{u>j} n_u (C_{uj} + A_{uj}) + n_\kappa (C_{\kappa j} + R_{\kappa j}) \right] \Big/ n_j P_j, \tag{10.171}$$

where the denominator expresses the total number of transitions *into* level j, which, to attain kinetic equilibrium, must be equal to the total number of transitions $n_j P_j$ *out* of level j. Similarly,

$$p(\to i^* \to j) = p(\to i^*)\, n_i B_{ij} \bar{J}_{ij} \big/ n_j P_j. \tag{10.172}$$

The emission profile coefficient for the transition $j \to i$ is given by

$$\begin{aligned}\psi_{ji}(\nu, \mathbf{n}) = \phi_{ij}(\nu) &+ \frac{n_i B_{ij} \bar{J}_{ij}}{n_j P_j} \left[p_i^* \mathcal{R}_{iji}(\nu, \mathbf{n}) - \phi_{ij}(\nu)\right] \\ &+ \sum_{\ell<j, \ell\neq i} \frac{n_\ell B_{\ell j} \bar{J}_{\ell j}}{n_j P_j} \left[p_\ell^* \mathcal{R}_{\ell ji}(\nu, \mathbf{n}) - \phi_{ij}(\nu)\right],\end{aligned} \tag{10.173}$$

where we used a shorthand notation $p_i^* \equiv p(\to i^*)$. Bear in mind here that while $p_1^* = 1$ because the ground state is assumed to be naturally populated, we have in general $p_i^* < 1$.

The source function is given by (neglecting stimulated emission)

$$S_{ji}(\nu, \mathbf{n}) = \frac{n_j A_{ji} \psi_{ji}(\nu, \mathbf{n})}{n_i B_{ij} \phi_{ij}(\nu)} = \frac{n_j A_{ji}}{n_i B_{ij}} + \sum_{\ell<j} \frac{n_\ell B_{\ell j}}{n_i B_{ij}} \frac{A_{ji}}{P_j} \bar{J}^R_{\ell ji}(\nu, \mathbf{n}), \tag{10.174}$$

where we again recognize the first term on the right-hand side as S^{CFR}, the source function for complete redistribution. Equation (10.164) is generalized to

$$\bar{J}^R_{\ell ji}(\nu, \mathbf{n}) \equiv \int_0^\infty \oint \left[p_\ell^* \frac{R_{\ell ji}(\nu', \mathbf{n}'; \nu, \mathbf{n})}{\phi_{ij}(\nu)} - \phi_{\ell j}(\nu')\right] I(\nu', \mathbf{n}')\, (d\Omega'/4\pi)\, d\nu'. \tag{10.175}$$

We do not make any formal distinction between resonance scattering ($\ell = i$) and Raman scattering ($\ell < j$). When the stimulated emission is taken account one has, similarly to (10.167),

$$S_{ji}(\nu, \mathbf{n}) = \left[\frac{n_j A_{ji}}{n_i B_{ij}} + \sum_{\ell<j} \frac{n_\ell B_{\ell j}}{n_i B_{ij}} \frac{A_{ji}}{P_j} \bar{J}^R_{\ell ji}(\nu, \mathbf{n})\right] \cdot \left[1 - \frac{n_j g_i}{n_i g_j} \frac{\psi_{ji}(\nu, \mathbf{n})}{\phi_{ij}(\nu)}\right]^{-1}. \tag{10.176}$$

Since in most cases of interest the population of the ground state dominates over the populations of the excited states, we may write the natural-excitation probability as

$$p_j^* = \frac{n_j P_j - n_1 B_{1j} \bar{J}_{1j}}{n_j P_j}. \tag{10.177}$$

This expression shows that besides the ground state, which is always populated naturally (in the weak-field limit), other states that are populated naturally, or almost naturally, are the states that are not connected to the ground state by a dipole-allowed transition. This is an important point. Consider, for instance, resonance Raman scattering $2 \to 3 \to 1$ in the hydrogen atom, i.e., a correlated absorption of a photon in $H\alpha$ followed by an emission of a photon in $L\beta$. If we view the transition chain as $2 \to 3 \to 1$, we may conclude at first sight that the process is unimportant because the probability of natural population of the initial state, p_2^*, may be small because the dominant source of population of this state may be radiative transitions $1 \to 2$. However, the process is, in fact, the chain $2s \to 3p \to 1s$; other components (e.g., $2p \to 3s \to 1s$) are dipole forbidden. The state $2s$ is not connected to the ground state $1s$ by a dipole-allowed transition. Therefore, the $2s$ state is populated naturally; hence it may serve as a start of the chain for the Raman scattering. One can thus feed photons from the wings of $H\alpha$ to the wings of $L\beta$; this effect was shown to be important in the solar chromosphere [535]; cf. § 15.6.

Expressions with Collisional Redistribution

Here we write expressions that consider the redistribution function as a superposition of coherent scattering and complete redistribution, i.e., those describing a general case of collisional redistribution. Substituting (10.151) and (10.150)—the second, approximate equality there—into (10.154), we obtain (for a two-level atom; expressions for a multi-level atom are analogous)

$$\mathcal{R}_{121}(\nu, \mathbf{n}) = \frac{\gamma}{\overline{J}_{12}} \int_0^\infty \oint R_{\text{II}}(\nu', \mathbf{n}'; \nu, \mathbf{n}) I(\nu', \mathbf{n}') \frac{d\Omega'}{4\pi} d\nu' + \frac{1-\gamma}{\overline{J}_{12}} \overline{J}_{12} \phi_{12}(\nu) \tag{10.178}$$

and hence

$$\mathcal{R}_{121}(\nu, \mathbf{n}) - \phi_{12}(\nu) = \gamma[\mathcal{R}_{\text{II}}(\nu, \mathbf{n}) - \phi_{12}(\nu)]. \tag{10.179}$$

We can then transform emission profile (10.161) to a useful form:

$$\psi_{21}(\nu, \mathbf{n}) = \phi_{12}(\nu) + \frac{n_1 B_{12} \overline{J}_{12}}{n_2(P_2 + Q_2)} [\mathcal{R}_{\text{II}}(\nu, \mathbf{n}) - \phi_{12}(\nu)]. \tag{10.180}$$

Finally, the source function is given by

$$\begin{aligned} S_{21}(\nu, \mathbf{n}) &= \frac{n_2 A_{21} \psi_{21}(\nu, \mathbf{n})}{n_1 B_{12} \phi_{12}(\nu)} \\ &= \frac{n_2 A_{21}}{n_1 B_{12}} + \frac{A_{21}}{P_2 + Q_2} \frac{\overline{J}_{12}}{\phi_{12}(\nu)} [\mathcal{R}_{\text{II}}(\nu, \mathbf{n}) - \phi_{12}(\nu)] \\ &= S^{\text{CFR}} + \frac{A_{21}}{P_2 + Q_2} \int_0^\infty \oint \left[\frac{R_{\text{II}}(\nu', \mathbf{n}'; \nu, \mathbf{n})}{\phi_{12}(\nu)} - \phi_{12}(\nu') \right] I(\nu', \mathbf{n}') \frac{d\Omega'}{4\pi} d\nu'. \end{aligned} \tag{10.181}$$

This expression is analogous to the general expression (10.165), where a general redistribution function R_{iji} has been replaced by R_{II}, and the branching ratio changed appropriately.

Chapter Eleven

Radiative Transfer Equation

The flow of radiation through material is governed by the *radiative transfer equation*. If the material is *static*, i.e., motionless, photons are seen to move on straight paths at constant frequency. If the material is in nonuniform motion, either in space or in time, one must distinguish between two frames: the *laboratory frame*, a fixed frame in which the material is seen to move, and the *comoving frame*, in which the coordinate system at each position moves with the material. When the material is moving, the transfer equation's differential operators are simple in the laboratory frame. But the material's opacity and emissivity become anisotropic because photons experience Doppler shifts, aberration, and advection that vary from point to point in the flow. In the comoving frame, the material properties are independent of its motion, but the differential operators become complicated because they depend on the velocity field. In this chapter we discuss the transfer equation *in static material only.*

In § 11.1 we write expressions for the macroscopic absorption, emission, and scattering coefficients of the material. The transfer equation is derived in § 11.2, first for a set general coordinates $\mathbf{q}$ and then in planar, spherical, and cylindrical geometry. In cartesian coordinates, $\mathbf{q} = (x, y, z)$; in spherical coordinates, $\mathbf{q} = (r, \theta, \phi)$; in cylindrical coordinates, $\mathbf{q} = (r, z, \phi)$. The angles Θ and Φ specify the direction $\mathbf{n}$ of a photon relative to an orthonormal triad appropriate to the coordinate system, and ν is a photon's frequency. As before, to simplify the notation, for any function $f(\mathbf{q}, t; \mathbf{n}, \nu)$, we may suppress reference to $\mathbf{q}$ and t and write $f_{\mathbf{n}\nu}$.

A solution of the transfer equation gives the angle-frequency distribution of the radiation field throughout the entire atmosphere. In § 11.3 we integrate the transfer equation over solid angle and frequency, which yields *moment equations* that connect the *radiation energy density*, the *radiation energy flux*, and the *radiation stress tensor*. In § 11.4 we define the concepts of *optical depth* and *source function*, state the transfer equation's boundary conditions, and solve it for some simple problems.

We discuss Schwarzschild's and Milne's integral formulae for calculating angular moments of the specific intensity in § 11.5. In § 11.6 we derive Schuster's second-order transfer equation, which provides an efficient and stable method for solving very general transfer problems, and show how it is discretized for numerical computation in § 11.7. We give a probabilistic interpretation of the transfer equation in § 11.8 and discuss its asymptotic *diffusion limit* in § 11.9.

11.1 ABSORPTION, EMISSION, AND SCATTERING COEFFICIENTS

Photons passing through a stellar atmosphere are repeatedly absorbed, emitted, and scattered. They are removed from the radiation field and destroyed by thermal *absorption* processes. They are created and added to the radiation field by thermal *emission*. They can also be removed from a specific beam of radiation by *outscattering*, or put into it by *inscattering*. In any scattering process the *number* of photons is conserved. All of these processes are described with macroscopic material coefficients.

In static material the thermal absorption coefficient κ_ν and thermal emission coefficient η_ν are isotropic. However, even the simplest form of scattering (pure Thomson scattering), which has a frequency-independent scattering coefficient σ_{T}, is anisotropic. Rayleigh, Raman, and Compton scattering vary in frequency and are also anisotropic; thus in general the scattering coefficient becomes $\sigma_{\mathbf{n}\nu}$. We define the material's *extinction coefficient* as the sum of the absorption and scattering coefficients, $\chi_{\mathbf{n}\nu} \equiv \kappa_\nu + \sigma_{\mathbf{n}\nu}$.

K. Schwarzschild pointed out [981] that on a *global scale* photons interact differently with material when absorbed and emitted, or when scattered. A statistical ensemble of photons carries a "signature" (e.g., the frequency distribution of the Planck function) of the conditions in the material where they were created. A photon travels a *mean free path* $\lambda \approx 1/\chi$ before being absorbed or scattered. At each interaction with the material a photon has a *single-scattering destruction probability* $\epsilon \equiv \kappa/\chi$ (also called the *thermal coupling parameter*) that it is absorbed instead of scattered. This is the probability that its energy is deposited into the thermal pool. If there is *no scattering,* ($\epsilon = 1$), a photon emitted at a given position is not directly linked to the properties of any photon previously absorbed there, but reflects only the local physical conditions, i.e., temperature, density, and occupation numbers of bound and free states of the material, at the position where it is created.

In the presence of scattering, properties of the radiation field at one point can be communicated to other points in the material. The probability that a photon is *not* absorbed when it interacts with the material is its *single-scattering albedo* $1 - \epsilon$. When $\epsilon \ll 1$, it may travel a distance much larger than its single-flight mean free path by repeated scattering. *Thus if photons have a large scattering albedo, they can transport information about the physical conditions where they were created over large distances in the atmosphere.* In that case, the radiation field at a given point may have little to do with the local physical properties of the material.

Thermal Absorption and Emission Coefficients

Bound-Bound Transitions

From equation (5.9), the *absorption coefficient* of the bound-bound transition $l \to u$, corrected for induced emission, is

$$\kappa_{\mathrm{bb}}(\mathbf{n}, \nu) = n_l B_{lu} \left(\frac{h\nu}{4\pi}\right) \left[\phi_{lu}(\nu) - \left(\frac{g_l n_u}{g_u n_l}\right) \psi_{ul}(\mathbf{n}, \nu)\right]. \tag{11.1}$$

Here B_{lu} is the Einstein absorption probability. Departures from LTE can change κ_{bb} by altering the occupation numbers of *either* bound state. In stationary material the line absorption profile ϕ_{lu} is isotropic, but the emission profile ψ_{ul} is, in general, given through an angle-frequency-dependent *redistribution function* (see chapter 10) and may be anisotropic even in stationary material.

In the limit of *complete redistribution*, in which the excited electrons are randomly redistributed (e.g., by collisions) over the substates of the upper level before emission, $\psi_\nu \equiv \phi_\nu$ and (11.1) simplifies to

$$\kappa_{\text{bb}}(\nu) = n_l B_{lu} \left(\frac{h\nu}{4\pi}\right) \left[1 - \left(\frac{g_l n_u}{g_u n_l}\right)\right] \phi_{lu}(\nu). \tag{11.2}$$

In the special case of LTE *only*,

$$\left(\frac{g_l n_u}{g_u n_l}\right) = e^{-h\nu_{lu}/kT}, \tag{11.3}$$

and (11.2) becomes[1]

$$\kappa^*_{\text{bb}}(\nu) = n^*_l \left(\frac{B_{lu} h\nu_{lu}}{4\pi}\right) \left(1 - e^{-h\nu_{lu}/kT}\right) \phi_{lu}(\nu). \tag{11.4}$$

From equation (5.10) the spontaneous[2] bound-bound emission coefficient is

$$\eta_{\text{bb}}(\mathbf{n}, \nu) = n_u A_{ul} \left(\frac{h\nu}{4\pi}\right) \psi_{ul}(\mathbf{n}, \nu). \tag{11.5}$$

Bound-Free Transitions

From equation (5.25), the absorption coefficient, corrected for induced emission, for the bound-free transition $l \to f$ is

$$\kappa_{\text{bf}}(\nu) = \left(n_l - n^*_l e^{-h\nu/kT}\right) \alpha_{\text{bf}}(\nu). \tag{11.6}$$

Note that departures from LTE change κ_{bf} only by affecting the occupation number of the bound state. In the special case of LTE *only*,

$$\kappa^*_{\text{bf}}(\nu) = n^*_l \, \alpha_{\text{bf}}(\nu) \left(1 - e^{-h\nu/kT}\right). \tag{11.7}$$

From equation (5.22), the rate of emission by spontaneous radiative recombinations $f \to l$ is

$$\eta^*_{\text{fb}}(\nu) = n^*_l \, \alpha_{\text{bf}}(\nu) \left(1 - e^{-h\nu/kT}\right) B_\nu(T), \tag{11.8}$$

which is proportional to the LTE population of the recombined ion and is isotropic in the fluid frame.

[1] In (11.4)–(11.9), the factor $(1 - e^{-h\nu/kT})$ is commonly called "the correction for induced emission"; this terminology is correct *only* in the limit of LTE.

[2] Remember that induced emission is treated as a correction to the absorption coefficient as in (11.1).

Free-Free Transitions

Free-free transitions are collisional processes; hence both absorption and induced emission take place at the LTE rate. The free-free absorption coefficient corrected for induced emission is; see equation (5.149),

$$\kappa_{\rm ff}^{*}(\nu, T) = n_{\rm ion} n_e \, \alpha_{\rm ff}(\nu, T) \left(1 - e^{-h\nu/kT}\right), \tag{11.9}$$

where $n_{\rm ion}$ and n_e are the ion and electron densities. The rate of spontaneous emission in free-free transitions is

$$\begin{aligned} \eta_{\rm ff}^{*}(\nu, T) &= n_{\rm ion} n_e \, \alpha_{\rm ff}(\nu, T)(2h\nu^3/c^2) e^{-h\nu/kT} \\ &= n_{\rm ion} n_e \, \alpha_{\rm ff}(\nu, T) \left(1 - e^{-h\nu/kT}\right) B_\nu(T). \end{aligned} \tag{11.10}$$

Note that $\kappa_{\rm bf}^{*}$ and $\eta_{\rm bf}^{*}$, and $\kappa_{\rm ff}^{*}$ and $\eta_{\rm ff}^{*}$, obey the Kirchhoff–Planck relation.

Total Thermal Absorption and Emission Coefficients

The total absorption coefficient is a sum over the occupation numbers of all levels i, of all ions j, of all elements k, times nonzero bound-bound, bound-free, and free-free cross sections at frequency ν. These processes are assumed to be independent and to add linearly. For bound-bound transitions the lower level is labeled l and the upper level u.

$$\begin{aligned} &\kappa_{\rm thermal}(\mathbf{n}, \nu) = \\ &\sum_k \sum_j \left\{ \sum_{u>l} \sum_l \left[n_{ljk} \phi_{lu,jk}(\mathbf{n}, \nu) - (g_l/g_u) n_{ujk} \psi_{ul,jk}(\mathbf{n}, \nu) \right] (B_{lu,jk} h\nu/4\pi) \right. \\ &\left. + \sum_i \left(n_{ijk} - n_{ijk}^{*} e^{-h\nu/kT} \right) \alpha_{ijk}^{\rm bf}(\nu) + \, n_e \, n_{jk} \left(1 - e^{-h\nu/kT}\right) \alpha_{jk}^{\rm ff}(\nu, T) \right\}. \end{aligned} \tag{11.11}$$

The total thermal emission coefficient is

$$\begin{aligned} \eta_{\rm thermal}(\mathbf{n}, \nu) = \sum_k \sum_j &\left\{ \sum_{u>l} \sum_l n_{ujk} \psi_{ul,jk}(\mathbf{n}, \nu)(A_{ul,jk} h\nu/4\pi) \right. \\ &\left. + (2h\nu^3/c^2) e^{-h\nu/kT} \left[\sum_i n_{ijk} \alpha_{ijk}^{\rm bf}(\nu) + n_e \, n_{jk} \alpha_{jk}^{\rm ff}(\nu, T) \right] \right\}. \end{aligned} \tag{11.12}$$

$\kappa(\nu)$ has units of cm^{-1}, and $\eta(\nu)$ has units of ergs/cm^3/sec/Hz/sr.

Continuum Scattering

In chapter 6 we focused on the scattering of individual photons. Consider now the scattering of photons from an angle-frequency-dependent specific intensity. In a hot stellar atmosphere, say $T_{\text{eff}} \gtrsim 3 \times 10^4$ K, the material is highly ionized, and the dominant scattering mechanism for photons with energies $h\nu \ll m_e c^2$ is Thomson scattering by free electrons, which has a frequency-independent cross section σ_T. At such temperatures we can ignore relativistic effects in their velocity distribution function.

Using a notation similar to that in equation (6.59), the energy per unit solid angle *outscattered* by n_e electrons from a beam of incoming $(\mathbf{n}, \nu)$ photons into outgoing $(\mathbf{n}', \nu')$ photons is

$$E_{\text{out}}(\mathbf{n}, \nu) = n_e \int_0^\infty d\nu' \oint_{4\pi} \frac{d\Omega'}{4\pi}\, \sigma(\nu \to \nu'; \mathbf{n} \to \mathbf{n}')\, I(\mathbf{n}, \nu) \left[1 + \left(\frac{c^2}{2h\nu'^3} \right) I(\mathbf{n}', \nu') \right]. \quad (11.13)$$

The energy per unit solid angle *inscattered* by n_e electrons from beams of incoming $(\mathbf{n}', \nu')$ photons into outgoing $(\mathbf{n}, \nu)$ photons is

$$E_{\text{in}}(\mathbf{n}, \nu) = n_e \int_0^\infty d\nu' \oint_{4\pi} \frac{d\Omega'}{4\pi} \left(\frac{\nu}{\nu'} \right) \sigma(\nu' \to \nu; \mathbf{n}' \to \mathbf{n})\, I(\mathbf{n}', \nu') \left[1 + \left(\frac{c^2}{2h\nu^3} \right) I(\mathbf{n}, \nu) \right]. \quad (11.14)$$

In (11.14) the factor $(1/\nu')$ converts $I(\mathbf{n}', \nu')$ to the *number* of incoming $(\mathbf{n}', \nu')$ photons, and the factor ν converts that number to the *energy* of $(\mathbf{n}, \nu)$ photons.[3]

These expressions have some interesting properties:

- For both inscattering and outscattering, the induced emission factor depends on the direction and frequency of the *outgoing* photon.[4]
- These scattering source/sink terms are *explicitly* nonlinear in I. In contrast, in NLTE transfer, the source terms are *implicitly nonlinear* because the occupation numbers of the material, and hence its opacity and emissivity, which *determine* the radiation field, are *determined by* the radiation field.
- For Thomson scattering these expressions simplify because (a) it is *coherent*, i.e., $\nu' \equiv \nu$, and (b) its phase function $g(\mathbf{n}', \mathbf{n}) = \frac{3}{4}[1 + (\mathbf{n}' \cdot \mathbf{n})^2]$ is *symmetric* in $\mathbf{n}$ and $\mathbf{n}'$. Hence $\sigma(\nu \to \nu'; \mathbf{n} \to \mathbf{n}') \equiv \sigma(\nu' \to \nu; \mathbf{n}' \to \mathbf{n}) = \sigma_T\, g(\mathbf{n}', \mathbf{n})\, \delta(\nu - \nu')$. Thus (11.13) and (11.14) collapse to

$$E_{\text{out}}(\mathbf{n}, \nu) = \frac{\sigma_T n_e}{4\pi} \oint d\Omega'\, g(\mathbf{n}, \mathbf{n}')\, I(\mathbf{n}, \nu) \left[1 + \left(\frac{c^2}{2h\nu^3} \right) I(\mathbf{n}', \nu) \right] \quad (11.15)$$

[3] Remember that in scattering processes the *number* of photons is conserved.

[4] Induced scattering can sometimes be ignored in stellar atmospheres work because photon occupation numbers at visible and ultraviolet wavelengths are small at stellar surface temperatures.

and

$$E_{\text{in}}(\mathbf{n}, \nu) = \frac{\sigma_{\text{T}} n_e}{4\pi} \oint d\Omega' \, g(\mathbf{n}', \mathbf{n}) \, I(\mathbf{n}', \nu) \left[1 + \left(\frac{c^2}{2h\nu^3} \right) I(\mathbf{n}, \nu) \right]. \tag{11.16}$$

- The induced emission terms in the integrands of (11.15) and (11.16) are identical and cancel exactly in the net energy exchange $E_{\text{in}} - E_{\text{out}}$; hence they can be discarded.
- Then the energy per unit solid angle outscattered from a beam $I(\mathbf{n}, \nu)$ is

$$E_{\text{out}}(\mathbf{n}, \nu) = n_e \, \sigma_{\text{T}} I(\mathbf{n}, \nu) \frac{1}{4\pi} \oint g(\mathbf{n}, \mathbf{n}') \, d\Omega' = n_e \, \sigma_{\text{T}} I(\mathbf{n}, \nu). \tag{11.17}$$

  The angular redistribution effects of the phase function have vanished in the integration over outgoing directions.
- But the energy inscattered into a beam of $I(\mathbf{n}, \nu)$ radiation is different:

$$E_{\text{in}}(\mathbf{n}, \nu) = \frac{n_e \sigma_{\text{T}}}{4\pi} \oint g(\mathbf{n}', \mathbf{n}) I(\mathbf{n}', \nu) \, d\Omega' \equiv n_e \sigma_{\text{T}} \, \mathfrak{J}(\mathbf{n}, \nu). \tag{11.18}$$

  That is, in general it is a *function* of the angular distribution of the incoming photons. In classical stellar atmospheres calculations, the anisotropy of the scattering process was usually ignored and (11.18) replaced with

$$E_{\text{in}}(\nu) = \frac{n_e \sigma_{\text{T}}}{4\pi} \oint I(\mathbf{n}', \nu) \, d\Omega' = n_e \sigma_{\text{T}} J_\nu. \tag{11.19}$$

  We may use (11.19) for expository convenience in the next few chapters. We will show how the transfer equation is solved using the anisotropic source term (11.18) in § 12.5 and estimate the size of the error made if (11.19) were to be used instead.

11.2 FORMULATION

Let $\mathbf{q}$ denote a set of spatial coordinates in stationary material. The change in the energy in a beam of radiation that has specific intensity $I(\mathbf{q}, t; \mathbf{n}, \nu)$ in frequency interval $d\nu$, that enters a material element of length ds and cross section dS oriented perpendicular to $\mathbf{n}$, from solid angle $d\Omega$, at position $\mathbf{q}$ and time t, and emerges at position $\mathbf{q} + \Delta\mathbf{q}$ at time $t + \Delta t$ (see figure 11.1) equals the amount of energy $\eta_{\mathbf{n}\nu}$ emitted thermally and inscattered by the material minus the amount of energy χ_ν it absorbs thermally and outscatters. The cross sections for thermal absorption and, as shown in (11.17), for Thomson outscattering are isotropic; therefore, χ_ν is isotropic. Thermal emission is also isotropic, but as shown in (11.18), Thomson inscattering is not; hence in general $\eta_{\mathbf{n}\nu}$ may be anisotropic. Thus

$$\begin{aligned} &[I(\mathbf{q} + \Delta\mathbf{q}, t + \Delta t; \mathbf{n}, \nu) - I(\mathbf{q}, t; \mathbf{n}, \nu)] \, dS \, d\Omega \, d\nu \, dt \\ &\quad = [\eta_{\mathbf{n}\nu}(\mathbf{q}, t) - \chi_\nu(\mathbf{q}, t) I(\mathbf{q}, t; \mathbf{n}, \nu)] \, dS \, ds \, d\Omega \, d\nu \, dt \end{aligned} \tag{11.20}$$

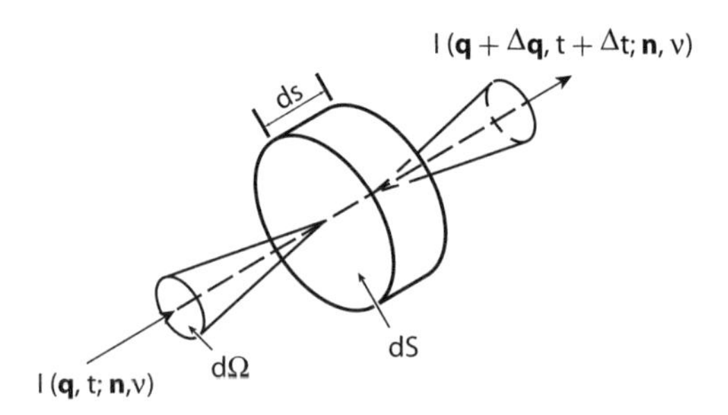

Figure 11.1 Beam of radiation passing through an element of absorbing, emitting, and scattering material.

or

$$\frac{\delta I_{\mathbf{n}\nu}(\mathbf{q},t)}{\delta s} = \eta_{\mathbf{n}\nu}(\mathbf{q},t) - \chi_\nu(\mathbf{q},t)I_{\mathbf{n}\nu}(\mathbf{q},t), \tag{11.21}$$

which is the radiation transport equation in arbitrary coordinates.

In (11.21) the notation $(\delta I/\delta s)$ denotes the *intrinsic derivative* (or *absolute derivative*) of I with respect to path length s along a ray, which is a *geodesic* in spacetime. This derivative takes into account the variation of I with respect to all variables, i.e., space coordinates, angles, frequency, and time, on which it depends. Writing out (11.21), we have

$$\frac{\delta I_{\mathbf{n}\nu}}{\delta s} = \frac{\partial I}{\partial t}\frac{\partial t}{\partial s} + \sum_{i=1}^{3}\frac{\partial I}{\partial q^i}\frac{\partial q^i}{\partial s} + \frac{\partial I}{\partial \Theta}\frac{\partial \Theta}{\partial s} + \frac{\partial I}{\partial \Phi}\frac{\partial \Phi}{\partial s} + \frac{\partial I}{\partial \nu}\frac{\partial \nu}{\partial s} = \eta_{\mathbf{n}\nu} - \chi_\nu I_{\mathbf{n}\nu}. \tag{11.22}$$

Equation (11.22) is a general form of the transfer equation, valid in any inertial frame, in both cartesian and curvilinear coordinates, and in static or moving media. In the absence of any special symmetries, this equation is seven-dimensional.

At first sight, knowledge of $\eta_{\mathbf{n}\nu}$ and χ_ν appears to provide a complete *macroscopic* description of the rate radiation is put into and removed from a specific beam of radiation by its interaction with material. However, this "completeness" is illusory because the occupation numbers of the material's bound and free states that determine these coefficients *are in turn determined by* processes such as photoexcitation, photoionization, and radiative recombination.

Kinetic Theory Derivation—The Transfer Equation as a Boltzmann Equation

The basic equation describing particle transport in kinetic theory is the *Boltzmann equation*. The transfer equation (11.22) is a Boltzmann equation for photons. Consider a distribution function $f(\mathbf{q},\mathbf{p},t)$ that gives the density of particles in a phase volume element $(\mathbf{q},\mathbf{q}+\mathbf{dq})$, $(\mathbf{p},\mathbf{p}+\mathbf{dp})$. Follow the evolution of f within that phase-space element for a time interval dt, in which $\mathbf{q}_0 \rightarrow \mathbf{q} = \mathbf{q}_0 + \mathbf{v}dt$ and

$\mathbf{p}_0 \to \mathbf{p} = \mathbf{p}_0 + \mathbf{F}\,dt$, where $\mathbf{F}$ is any externally imposed force acting on the particles. Then the evolution in size and shape of the phase-space volume element is given by

$$d^3q_0\,d^3p_0 \longrightarrow d^3q\,d^3p = \mathrm{J}\left(\frac{q^1,\,q^2,\,q^3,\,p^1,\,p^2,\,p^3}{q_0^1,\,q_0^2,\,q_0^3,\,p_0^1,\,p_0^2,\,p_0^3}\right) d^3q_0\,d^3p_0, \tag{11.23}$$

where J is the Jacobian of the transformation: the 6×6 determinant

$$\mathrm{J} = \begin{vmatrix} (\partial q^1/\partial q_0^1) & (\partial q^1/\partial q_0^2) & (\partial q^1/\partial q_0^3) & (\partial q^1/\partial p_0^1) & (\partial q^1/\partial p_0^2) & (\partial q^1/\partial p_0^3) \\ (\partial q^2/\partial q_0^1) & (\partial q^2/\partial q_0^2) & (\partial q^2/\partial q_0^3) & (\partial q^2/\partial p_0^1) & (\partial q^2/\partial p_0^2) & (\partial q^2/\partial p_0^3) \\ (\partial q^3/\partial q_0^1) & (\partial q^3/\partial q_0^2) & (\partial q^3/\partial q_0^3) & (\partial q^3/\partial p_0^1) & (\partial q^3/\partial p_0^2) & (\partial q^3/\partial p_0^3) \\ (\partial p^1/\partial q_0^1) & (\partial p^1/\partial q_0^2) & (\partial p^1/\partial q_0^3) & (\partial p^1/\partial p_0^1) & (\partial p^1/\partial p_0^2) & (\partial p^1/\partial p_0^3) \\ (\partial p^2/\partial q_0^1) & (\partial p^2/\partial q_0^2) & (\partial p^2/\partial q_0^3) & (\partial p^2/\partial p_0^1) & (\partial p^2/\partial p_0^2) & (\partial p^2/\partial p_0^3) \\ (\partial p^3/\partial q_0^1) & (\partial p^3/\partial q_0^2) & (\partial p^3/\partial q_0^3) & (\partial p^3/\partial p_0^1) & (\partial p^3/\partial p_0^2) & (\partial p^3/\partial p_0^3) \end{vmatrix}. \tag{11.24}$$

The elements of J are straightforward to compute. For the upper-left 3×3 partition with $(i=1\ \ldots\ 3; j=1\ \ldots\ 3)$, $\mathrm{J}_{ij} = \delta_{ij} + O(dt)$; for the upper-right partition with $(i=1\ \ldots\ 3; j=4\ \ldots\ 6)$, $\mathrm{J}_{ij} = O(dt)$. Here δ_{ij} is the Kronecker delta, and $O(dt)$ denotes a term of the form (constant $\times\ dt$). Likewise, for the lower-left 3×3 partition with $(i=4\ \ldots\ 6;\ j=1\ \ldots\ 3)$, $\mathrm{J}_{ij} = O(dt)$; for the lower-right partition with $(i=4\ \ldots\ 6; j=4\ \ldots\ 6)$, $\mathrm{J}_{ij} = \delta_{ij}$, unless the forces depend on velocity (e.g., electromagnetic forces), in which case they are $\mathrm{J}_{ij} = \delta_{ij} + O(dt)$. Thus the value of the determinant is $\mathrm{J} = 1 + O(\delta t)$.

Therefore, in the limit as $dt \to 0$, $\mathrm{J} \to 1$. Thus although a volume element may be *deformed*, its phase volume is *unchanged*. If all external forces $\mathbf{F}$ are continuous, then the deformation of a phase-space element is continuous, and all particles originally within the volume remain there. Because a phase volume is unchanged, the particle density is unchanged.

But if, in addition, collisions among the particles occur, individual particles may be reshuffled from one element of phase space to another "discontinuously"; i.e., their neighbors in phase space may be unaffected during the same time interval. Hence any change in the particle number density within a phase-space element must equal the net number introduced into the element by collisions. For example, in cartesian coordinates,

$$\begin{aligned} &\frac{\partial f}{\partial t} + \left(\frac{\partial x}{\partial t}\right)\left(\frac{\partial f}{\partial x}\right) + \left(\frac{\partial y}{\partial t}\right)\left(\frac{\partial f}{\partial y}\right) + \left(\frac{\partial z}{\partial t}\right)\left(\frac{\partial f}{\partial z}\right) \\ &\quad + F_x \frac{\partial f}{\partial p_x} + F_y \frac{\partial f}{\partial p_y} + F_z \frac{\partial f}{\partial p_z} = \left(\frac{Df}{Dt}\right)_{\text{coll}}, \end{aligned} \tag{11.25}$$

or, in more compact notation,

$$\frac{\partial f}{\partial t} + (\mathbf{v}\cdot\nabla)f + (\mathbf{F}\cdot\nabla_p)f = \left(\frac{Df}{Dt}\right)_{\text{coll}}. \tag{11.26}$$

Here the operator D/Dt denotes the *Lagrangian time derivative*, defined as $D/Dt \equiv (\partial/\partial t) + \mathbf{v}\cdot\nabla$, which is the time derivative taken while following a definite element of fluid.

For photons, $\mathbf{F} \equiv 0$ in the absence of general relativistic effects; thus photons move on straight lines, with velocity $\mathbf{v} = c\,\mathbf{n}$ and constant frequency in any inertial frame. The distribution function f_R is related to the specific intensity by equation (3.10). The analog of "collisions" are interactions of photons with the material (i.e., absorption, emission, and scattering). The net number of photons put into a volume element equals the energy thermally emitted and inscattered by the material, minus the energy absorbed and outscattered, divided by the energy per photon. That is,

$$\frac{1}{ch\nu}\left(\frac{\partial I_{\mathbf{n}\nu}}{\partial t} + c\mathbf{n}\cdot\nabla I_{\mathbf{n}\nu}\right) = \frac{\eta_{\mathbf{n}\nu} - \chi_\nu I_{\mathbf{n}\nu}}{h\nu}, \tag{11.27}$$

which is identical to the transfer equation (11.22).

To summarize, the transfer equation is a *linear Boltzmann equation* for a relativistic gas that responds to no external forces, but interacts strongly with the material. Actually, it is only *quasi-linear* because the material properties are determined by the radiation field that is consistent with the requirements of kinetic equilibrium and energy balance. These constraints imply a complicated and highly nonlinear coupling between the material and radiation. A physically correct solution for this system may require simultaneous solution of transfer equations in scores of continua and thousands of spectral lines, coupled to kinetic equilibrium equations for dozens of ionization stages of several elements, each represented by hundreds of discrete atomic levels.

Cartesian Coordinates

The transfer equation in Cartesian coordinates is relatively simple. In the laboratory frame photons are seen to move with constant direction and constant frequency. Let Θ be the polar angle between a ray and the z axis, and Φ be the azimuthal angle between its projection on the x-y plane measured counterclockwise from the x axis; see figure 11.2. The propagation vector is

$$\mathbf{n} = n_x\,\mathbf{i} + n_y\,\mathbf{j} + n_z\,\mathbf{k} = \sin\Theta\cos\Phi\,\mathbf{i} + \sin\Theta\sin\Phi\,\mathbf{j} + \cos\Theta\,\mathbf{k} \tag{11.28}$$

and hence

$$\frac{\partial t}{\partial s} = \frac{1}{c},\quad \frac{\partial x}{\partial s} = n_x,\quad \frac{\partial y}{\partial s} = n_y,\quad \frac{\partial z}{\partial s} = n_z,\quad \frac{\partial\Theta}{\partial s} = 0,\quad \frac{\partial\Phi}{\partial s} = 0,\quad \frac{\partial\nu}{\partial s} = 0. \tag{11.29}$$

Then the transfer equation (11.21) is

$$\begin{aligned}\frac{\delta I_{\mathbf{n}\nu}}{\delta s} &= \frac{1}{c}\frac{\partial I_{\mathbf{n}\nu}}{\partial t} + \sin\Theta\cos\Phi\frac{\partial I_{\mathbf{n}\nu}}{\partial x} + \sin\Theta\sin\Phi\frac{\partial I_{\mathbf{n}\nu}}{\partial y} + \cos\Theta\frac{\partial I_{\mathbf{n}\nu}}{\partial z}\\ &= \eta_{\mathbf{n}\nu} - \chi_\nu I_{\mathbf{n}\nu}.\end{aligned} \tag{11.30}$$

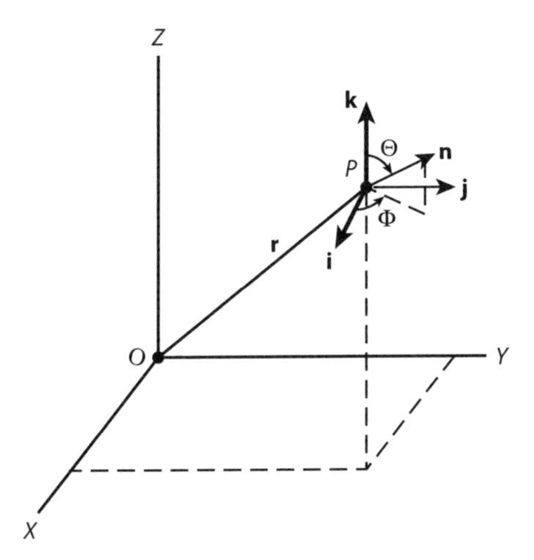

Figure 11.2 Geometry for transfer equation in Cartesian coordinates.

Or defining $\mu \equiv \cos\Theta$,

$$\frac{1}{c}\frac{\partial I_{\mathbf{n}\nu}}{\partial t} + (1-\mu^2)^{1/2}\cos\Phi\frac{\partial I_{\mathbf{n}\nu}}{\partial x} + (1-\mu^2)^{1/2}\sin\Phi\frac{\partial I_{\mathbf{n}\nu}}{\partial y} + \mu\frac{\partial I_{\mathbf{n}\nu}}{\partial z}$$
$$= \eta_{\mathbf{n}\nu} - \chi_\nu I_{\mathbf{n}\nu}. \qquad (11.31)$$

Equations (11.30) and (11.31) apply in the laboratory frame. In general they are four-dimensional partial differential equations in spacetime, with three independent parameters (Θ, Φ, ν). In stationary material, χ_ν is isotropic; $\eta_{\mathbf{n}\nu}$ may depend on $\mathbf{n}$ if it contains scattering terms. In moving material, both become anisotropic because of Doppler shifts and aberration.

Equations (11.30) and (11.31) can be applied in moving material if $\eta_{\mathbf{n}\nu}$ and χ_ν are transformed from the fluid frame to the laboratory frame; see Appendix B. When χ_ν or $\eta_{\mathbf{n}\nu}$ includes scattering terms, the transfer equation is an *integro-partial-differential equation* containing angle and frequency integrals of I. Radiative transfer problems are difficult to solve because of the large dimensionality and the integro-differential nature of the system.

Spherical Geometry

To a first approximation, stars are spherical. In the laboratory frame, both the direction $\mathbf{n}$ and the frequency ν of photons are constant. In a general spherical medium, a position is specified by the coordinate triplet (r, θ, ϕ), where r is the distance of P from the origin O; θ, $(0 \le \theta \le \pi)$, is the polar angle between its propagation vector $\mathbf{n}$ and the z axis; and ϕ, $(0 \le \phi \le 2\pi)$, is the azimuthal angle between the projection of $\mathbf{n}$ on the x-y plane, measured counterclockwise around the z axis from the x axis; see figure 11.3.

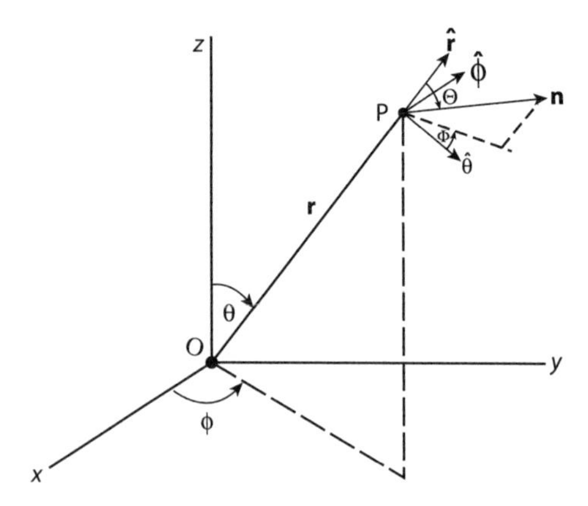

Figure 11.3 Geometry for transfer equation in spherical coordinates.

At position P, write $\mathbf{n}$ in terms of the angles (Θ, Φ), where $0 \leq \Theta \leq \pi$ is the polar angle between the local unit radial vector $\hat{\mathbf{r}}$ to $\mathbf{n}$, and $0 \leq \Phi \leq 2\pi$ is the azimuthal angle between the projection of $\mathbf{n}$ on the plane perpendicular to $\hat{\mathbf{r}}$, measured counterclockwise from $\hat{\boldsymbol{\theta}}$; i.e.,

$$\mathbf{n} = \cos\Theta\ \hat{\mathbf{r}} + \sin\Theta\cos\Phi\ \hat{\boldsymbol{\theta}} + \sin\Theta\sin\Phi\ \hat{\boldsymbol{\phi}}, \tag{11.32}$$

or

$$\begin{aligned}\mathbf{n} = {} & [\sin\theta\cos\phi\cos\Theta + (\cos\theta\cos\phi\cos\Phi - \sin\phi\sin\Phi)\sin\Theta]\,\mathbf{i} \\ & + [\cos\theta\sin\phi\cos\Theta + (\cos\theta\sin\phi\cos\Phi + \cos\phi\sin\Phi)\sin\Theta]\,\mathbf{j} \\ & + (\cos\theta\cos\Theta - \sin\theta\cos\Phi\sin\Theta)\,\mathbf{k}.\end{aligned} \tag{11.33}$$

For an infinitesimal displacement ds along $\mathbf{n}$ to a point P',

$$\mathbf{ds} = (\cos\Theta\,\hat{\mathbf{r}} + \sin\Theta\cos\Phi\,\hat{\boldsymbol{\theta}} + \sin\Theta\sin\Phi\ \hat{\boldsymbol{\phi}})\,ds. \tag{11.34}$$

Now $dr = \mathbf{ds}\cdot\hat{\mathbf{r}} = \cos\Theta\,ds$; $r\,d\theta = \mathbf{ds}\cdot\hat{\boldsymbol{\theta}} = \sin\Theta\cos\Phi\,ds$; and $r\sin\theta\,d\phi = \mathbf{ds}\cdot\hat{\boldsymbol{\phi}} = \sin\Theta\sin\Phi\,ds$. Therefore,

$$(\partial r/\partial s) = \cos\Theta;\ (\partial\theta/\partial s) = \sin\Theta\cos\Phi/r;\ (\partial\phi/\partial s) = \sin\Theta\sin\Phi/r\sin\theta. \tag{11.35}$$

At P', both Θ and Φ have changed because in addition to a change in coordinates (r, θ, ϕ), the local basis set $(\hat{\mathbf{r}}, \hat{\boldsymbol{\theta}}, \hat{\boldsymbol{\phi}})$ has rotated. *But the components of* $\mathbf{n}$ *expressed relative to the fixed basis* $(\mathbf{i}, \mathbf{j}, \mathbf{k})$ *must remain unchanged.* From (11.33), one can derive expressions for the changes (dn_x, dn_y, dn_z) resulting from hypothetical changes $(d\theta, d\phi, d\Theta, d\Phi)$. Setting them to zero gives three equations for $(d\theta, d\phi, d\Theta, d\Phi)$, which, using the expressions in (11.35) for $(\partial\theta/\partial s)$ and $(\partial\phi/\partial s)$, can be solved to obtain

$$(\partial\Theta/\partial s) = -\sin\Theta/r \quad \text{and} \quad (\partial\Phi/\partial s) = -\sin\Theta\sin\Phi\cot\theta/r. \tag{11.36}$$

Allowing for differences in notation (i.e., $\theta_P \equiv \Theta, \phi_P \equiv \Phi, \Theta_P \equiv \theta$, and $\Phi_P \equiv \phi$, where the variable subscripted P is the notation used by Pomraning), the formulae in (11.35) and (11.36) are the same as in [877, p. 24].

Inserting the results in (11.35) and (11.36) into (11.22), we obtain finally

$$\frac{1}{c}\frac{\partial I_{\mathbf{n}\nu}}{\partial t} + \cos\Theta \frac{\partial I_{\mathbf{n}\nu}}{\partial r} + \frac{\sin\Theta\cos\Phi}{r}\frac{\partial I_{\mathbf{n}\nu}}{\partial \theta} + \frac{\sin\Theta\sin\Phi}{r\sin\theta}\frac{\partial I_{\mathbf{n}\nu}}{\partial \phi} - \frac{\sin\Theta}{r}\frac{\partial I_{\mathbf{n}\nu}}{\partial \Theta} - \frac{\sin\Theta\sin\Phi\cot\theta}{r}\frac{\partial I_{\mathbf{n}\nu}}{\partial \Phi} = \eta_{\mathbf{n}\nu} - \chi_\nu I_{\mathbf{n}\nu}. \tag{11.37}$$

Equation (11.37) applies in the laboratory frame for a general spherical medium, with a time-dependent radiation field.

Symmetry

If the thickness of a star's atmosphere is very small compared to its radius, we may treat it as consisting of homogeneous planar layers or homogeneous spherical shells in which the material properties are functions of z or r only. In this case the radiation field in a planar atmosphere is a function of $(z, t; \mu, \nu)$. In particular it has *azimuthal symmetry*, and (11.31) reduces to

$$\frac{1}{c}\frac{\partial I_{\mu\nu}(z,t)}{\partial t} + \mu\frac{\partial I_{\mu\nu}(z,t)}{\partial z} = \eta_{\mu\nu}(z,t) - \chi_\nu I_{\mu\nu}(z,t), \tag{11.38}$$

which is a partial differential equation in two variables and two parameters.

Likewise, in a spherically symmetric atmosphere, I depends on the radial coordinate r only, does not depend on θ and ϕ, and by symmetry is independent of the azimuthal angle Φ. So $I(\mathbf{r}, t; \mathbf{n}, \nu)$ is azimuthally symmetric and reduces to $I(r, t; \mu, \nu)$, and (11.37) reduces to

$$\frac{1}{c}\frac{\partial I_{\mu\nu}(r,t)}{\partial t} + \mu\frac{\partial I_{\mu\nu}(r,t)}{\partial r} + \frac{(1-\mu^2)}{r}\frac{\partial I_{\mu\nu}(r,t)}{\partial \mu} = \eta_{\mu\nu}(r,t) - \chi_\nu(r,t)I_{\mu\nu}(r,t), \tag{11.39}$$

now a partial differential equation in three variables and two parameters.

Equations (11.38) and (11.39) apply in the laboratory frame for stationary material. They may also be used for moving material if the velocities are vertical (or radial) only, and the material properties are Lorentz transformed from the laboratory frame to the comoving frame.

⋆ Cylindrical Geometry

Cylindrical geometry is useful for broad, thin accretion disks and long thread-like objects such as relativistic jets from galactic nuclei. In a general cylindrical medium the position of a point P is specified by (r, z, ϕ), where r is its distance from the

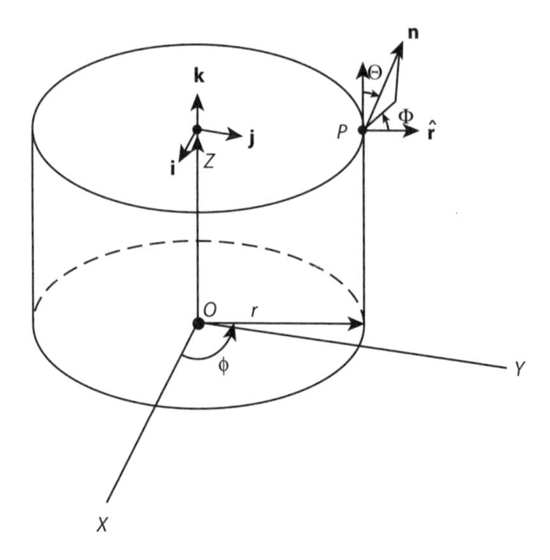

Figure 11.4 Geometry for transfer equation in cylindrical coordinates.

z axis; z is its distance above the x-y plane; and ϕ, $0 \le \phi \le 2\pi$, is its azimuthal angle, in the x-y plane, counterclockwise around the z axis from the x axis; see figure 11.4.

The orthonormal unit vectors along which (r, z, ϕ) are measured are

$$\hat{\mathbf{r}} = \cos\phi \; \mathbf{i} + \sin\phi \; \mathbf{j}, \quad \hat{\mathbf{z}} = \mathbf{k}, \quad \text{and} \quad \hat{\boldsymbol{\phi}} = -\sin\phi \; \mathbf{i} + \cos\phi \; \mathbf{j}. \tag{11.40}$$

In the laboratory frame, both the direction $\mathbf{n}$ and frequency ν of a ray of radiation are constant. At position P write $\mathbf{n}$ in terms of (Θ, Φ), where $0 \le \Theta \le \pi$ is the polar angle between the z axis and $\mathbf{n}$; and $0 \le \Phi \le 2\pi$ is the azimuthal angle measured counterclockwise from the projection of $\mathbf{n}$ onto the x-y plane to the radial unit vector $\hat{\mathbf{r}}$. Then

$$\mathbf{n} = \sin\Theta \cos\Phi \,\hat{\mathbf{r}} + \cos\Theta \,\mathbf{k} + \sin\Theta \sin\Phi \,\hat{\boldsymbol{\phi}}, \tag{11.41}$$

or

$$\mathbf{n} = (\cos\Phi \cos\phi - \sin\Phi \sin\phi) \sin\Theta \,\mathbf{i} + (\cos\Phi \sin\phi + \sin\Phi \cos\phi) \sin\Theta \,\mathbf{j} + \cos\Theta \,\mathbf{k}. \tag{11.42}$$

For an infinitesimal displacement ds along $\mathbf{n}$, to a point P',

$$\mathbf{ds} = (\sin\Theta \cos\Phi \,\hat{\mathbf{r}} + \cos\Theta \,\mathbf{k} + \sin\Theta \sin\Phi \,\hat{\boldsymbol{\phi}}) \, ds. \tag{11.43}$$

Now from (11.43) $dr = \mathbf{ds} \cdot \hat{\mathbf{r}} = \sin\Theta \cos\Phi \, ds$, $dz = \mathbf{ds} \cdot \mathbf{k} = \cos\Theta$, and $r \, d\phi = \mathbf{ds} \cdot \hat{\boldsymbol{\phi}} = \sin\Theta \sin\Phi \, ds$; therefore,

$$(\partial r/\partial s) = \sin\Theta \cos\Phi, \quad (\partial z/\partial s) = \cos\Theta, \quad \text{and} \quad (\partial\phi/\partial s) = \sin\Theta \sin\Phi / r. \tag{11.44}$$

Equation (11.44) shows that as a result of the displacement **ds**, the basis vectors $\hat{\mathbf{r}}$ and $\hat{\boldsymbol{\phi}}$ at P' will have rotated by an angle $d\phi$ around the z axis with respect to those at point P. This change in the basis vectors implies that, in principle, Φ may change to $(\Phi + d\Phi)$, and Θ to $(\Theta + d\Theta)$. But because $\mathbf{n}$ is fixed in space, its components (n_x, n_y, n_z) must remain unchanged relative to the fixed basis $(\mathbf{i}, \mathbf{j}, \mathbf{k})$. From (11.42), one can derive expressions for changes (dn_x, dn_y, dn_z) in (n_x, n_y, n_z) that result from hypothetical changes $(d\phi, d\Theta, d\Phi)$. Then demanding that these changes be zero gives us three equations relating $(d\phi, d\Theta, d\Phi)$, which, using (11.44) for $(\partial\phi/\partial s)$, can be solved to yield $(\partial\Theta/\partial s) \equiv 0$; $(\partial\Phi/\partial s) = -\sin\Theta\sin\Phi/r$; and, in the laboratory frame $(\partial\nu/\partial s) \equiv 0$. Using these results in (11.36) we obtain the laboratory-frame transport equation for general cylindrical geometry:

$$\frac{1}{c}\frac{\partial I_{\mathbf{n}\nu}}{\partial t} + \sin\Theta\cos\Phi\frac{\partial I_{\mathbf{n}\nu}}{\partial r} + \cos\Theta\frac{\partial I_{\mathbf{n}\nu}}{\partial z} + \frac{\sin\Theta\sin\Phi}{r}\frac{\partial I_{\mathbf{n}\nu}}{\partial\phi} - \frac{\sin\Theta\sin\Phi}{r}\frac{\partial I_{\mathbf{n}\nu}}{\partial\Phi} = \eta_{\mathbf{n}\nu} - \chi_\nu I_{\mathbf{n}\nu}, \tag{11.45}$$

which is a five-dimensional partial differential equation with two parameters (Θ and ν). Equation (11.45) applies in the laboratory frame, for both static and moving media. If the medium has *cylindrical symmetry* (i.e., its properties are independent of ϕ), one can omit $\partial I/\partial\phi$. If the cylinder is of infinite length, one can omit $\partial I/\partial z$. If the system is in steady state, one can omit $\partial I/\partial t$. In each case the dimensionality of the system decreases by one. But the radiation field is still a function of Θ and Φ, i.e., in general does *not* have azimuthal symmetry.

11.3 MOMENTS OF THE TRANSFER EQUATION

Laboratory-Frame Equations of Radiation Dynamics

Angular moments of the transport equation have both physical significance and mathematical utility. Suppressing reference to $(\mathbf{x}, t)$, we can write the first three angular moments of I in dyadic notation as

$$\oint \begin{bmatrix} 1 \\ \mathbf{n} \\ \mathbf{n}\,\mathbf{n} \end{bmatrix} I_{\mathbf{n}\nu}\, d\Omega = 4\pi \begin{bmatrix} J_\nu \\ \mathbf{H}_\nu \\ \mathsf{K}_\nu \end{bmatrix} = \begin{bmatrix} cE_\nu \\ \mathbf{F}_\nu \\ c\,\mathsf{P}_\nu \end{bmatrix}. \tag{11.46}$$

Planar Geometry

The first two angular moments of (11.38) are obtained by multiplying it by 1 and $\mathbf{n}/c$ and integrating over all solid angles; the extra factor of $(1/c)$ is included because the monochromatic radiation momentum density is $\mathbf{M}_\nu \equiv c^{-2}\mathbf{F}_\nu$. Thus for planar geometry we have

$$\frac{\partial E_\nu}{\partial t} + \nabla\cdot\mathbf{F}_\nu = \oint (\eta_{\mathbf{n}\nu} - \chi_\nu I_{\mathbf{n}\nu})\, d\Omega, \tag{11.47}$$

and

$$\frac{1}{c^2}\frac{\partial \mathbf{F}_\nu}{\partial t} + \nabla\cdot\mathsf{P}_\nu = \frac{1}{c}\oint (\eta_{\mathbf{n}\nu} - \chi_\nu I_{\mathbf{n}\nu})\,\mathbf{n}\,d\Omega. \tag{11.48}$$

Equation (11.47) is the *monochromatic radiation energy equation*. It states that the rate of change of radiation energy density at frequency ν equals the net rate radiant energy is gained from the material minus the net flow of radiant energy through the surface of the volume element (i.e., the divergence of the radiation energy flux). Likewise, (11.48) is the *monochromatic radiation momentum equation*. It states that the rate of change of the radiation momentum density at frequency ν equals the net radiant momentum gain from the material minus the net flow of radiant momentum through the surface of the volume element (i.e., the divergence of the radiation pressure tensor).

Integrated over frequency, (11.47) is the *total radiation energy equation*

$$\frac{\partial E}{\partial t} + \nabla\cdot\mathbf{F} = \int_0^\infty d\nu \oint (\eta_{\mathbf{n}\nu} - \chi_\nu I_{\mathbf{n}\nu})\,d\Omega \equiv -cG^0. \tag{11.49}$$

In (11.49), cG^0 [erg/cm^3/sec] is the *net rate of radiative energy deposition* into the material, i.e., the rate radiant energy is absorbed by the material minus the rate radiant energy is emitted, per unit volume. The frequency integral of (11.48) is the *total radiation momentum equation*

$$\frac{1}{c^2}\frac{\partial \mathbf{F}}{\partial t} + \nabla\cdot\mathsf{P} = \frac{1}{c}\int_0^\infty d\nu \oint (\eta_{\mathbf{n}\nu} - \chi_\nu I_{\mathbf{n}\nu})\,\mathbf{n}\,d\Omega \equiv -\mathbf{G}, \tag{11.50}$$

where $\mathbf{G}$ is the *net radiation force density* on the material, i.e., the rate of radiation momentum deposited in the material minus the rate of radiation momentum emitted by the material, per unit volume. All quantities in these equations are measured in the laboratory frame.[5]

In static material, χ_ν is isotropic, and we will see in § 11.6 that in planar and spherical symmetry, Thomson scattering $\eta_{\mathbf{n}\nu}$ is an *even* function of $\mathbf{n}$.[6] Defining $\langle\eta_\nu\rangle \equiv \frac{1}{2}\int_{-1}^{1} \eta_{\mu\nu}d\mu$, in static material (11.49) and (11.50) reduce to

$$\frac{\partial E}{\partial t} + \nabla\cdot\mathbf{F} = \int_0^\infty (\langle\eta_\nu\rangle - \chi_\nu J_\nu)\,d\nu \equiv -cG^0, \tag{11.51a}$$

and

$$\frac{1}{c^2}\frac{\partial \mathbf{F}}{\partial t} + \nabla\cdot\mathsf{P} = -\frac{1}{c}\int_0^\infty \chi_\nu \mathbf{F}_\nu\,d\nu \equiv -\mathbf{G} \tag{11.51b}$$

for both planar and spherically symmetric media. In physical terms, $\eta_{\mathbf{n}\nu}$ does not appear in (11.51b) because the momentum of an emitted photon is canceled by an identical photon emitted in the opposite direction.

[5] We use this notation for the energy and momentum exchange rates because we show in chapter 19 that $G^\alpha \equiv (G^0, \mathbf{G}) \equiv (G^0, G^1, G^2, G^3)$ is a *four-vector*, which can be transformed between frames by Lorentz transformation.

[6] This is not the case for Compton scattering.

Equations (11.49)–(11.51b) have the structure of kinetic equations:

$$\frac{\partial}{\partial t}(\text{density of quantity}) + (\text{divergence of its flux}) = \text{sources} - \text{sinks}.$$

They are *dynamical equations* for the radiation field and show that, in contrast to calculating the spectrum of a static atmosphere, where an "instantaneous" *snapshot* suffices, when computing the dynamics of a radiating medium, the time derivatives must be retained when radiation contributes significantly to the total energy and momentum density, or their transport, in the *radiating fluid*. This is the case for all high-temperature flows. For example, at the center of a star, radiation dominates the energy and momentum content of the fluid by orders of magnitude.

Spherical Symmetry

Consider (11.39) for a static spherically symmetric atmosphere. First rewrite it in *conservative form*:

$$\frac{1}{c}\frac{\partial I_{\mu\nu}}{\partial t} + \mu\frac{\partial I_{\mu\nu}}{\partial r} + \frac{(1-\mu^2)}{r}\frac{\partial I_{\mu\nu}}{\partial \mu} \longrightarrow$$

$$\frac{1}{c}\frac{\partial I_{\mu\nu}}{\partial t} + \mu\frac{\partial I_{\mu\nu}}{\partial r} + \frac{1}{r}\frac{\partial[(1-\mu^2)I_{\mu\nu}]}{\partial \mu} + \frac{2\mu}{r}I_{\mu\nu} \longrightarrow$$

$$\frac{1}{c}\frac{\partial I_{\mu\nu}}{\partial t} + \frac{\mu}{r^2}\frac{\partial(r^2 I_{\mu\nu})}{\partial r} + \frac{1}{r}\frac{\partial[(1-\mu^2)\,I_{\mu\nu}]}{\partial \mu} = \eta_{\mu\nu} - \chi_\nu I_{\mu\nu}. \qquad (11.52)$$

Integration of (11.52) over $d\Omega$ gives the spherically symmetric version of the radiation energy equation (11.47).

Now recast (11.38) multiplied by μ into conservative form:

$$\frac{\mu}{c}\frac{\partial I_{\mu\nu}}{\partial t} + \mu^2\frac{\partial I_{\mu\nu}}{\partial r} + \frac{\mu(1-\mu^2)}{r}\frac{\partial I_{\mu\nu}}{\partial \mu} \longrightarrow$$

$$\frac{\mu}{c}\frac{\partial I_{\mu\nu}}{\partial t} + \mu^2\frac{\partial I_{\mu\nu}}{\partial r} + \frac{1}{r}\frac{\partial[\mu(1-\mu^2)I_{\mu\nu}]}{\partial \mu} + \frac{(3\mu^2-1)}{r}I_{\mu\nu} \longrightarrow$$

$$\frac{\mu}{c}\frac{\partial I_{\mu\nu}}{\partial t} + \frac{\mu^2}{r^2}\frac{\partial(r^2 I_{\mu\nu})}{\partial r} + \frac{1}{r}\frac{\partial[\mu(1-\mu^2)I_{\mu\nu}]}{\partial \mu} + \frac{(\mu^2-1)}{r}I_{\mu\nu} = \mu(\eta_{\mu\nu} - \chi_\nu I_{\mu\nu}). \qquad (11.53)$$

Integration of (11.53) over $d\Omega$ gives the spherically symmetric version of the radiation momentum equation (11.48).

⋆ General Curvilinear Coordinates

Equations (11.51) are tensor equations in ordinary three-space; hence they apply in any system of coordinates if the correct expressions for the divergence of a vector or tensor are used. One method is to employ *covariant derivatives*, using standard techniques of tensor calculus [1068]. The divergence of a vector **V** in general curvilinear coordinates is

$$\nabla\cdot\mathbf{V} \equiv V^i_{;i} = \frac{\partial V^i}{\partial x^i} + \begin{Bmatrix} i \\ i \;\; j \end{Bmatrix} V^j, \tag{11.54}$$

and the divergence of a 2-index tensor **T** in curvilinear coordinates is

$$\nabla\cdot\mathbf{T} \equiv T^{ij}_{;j} = \frac{\partial T^{ij}}{\partial x^j} + \begin{Bmatrix} i \\ i \;\; k \end{Bmatrix} T^{kj} + \begin{Bmatrix} j \\ j \;\; k \end{Bmatrix} T^{ik}. \tag{11.55}$$

Summation on repeated indices is implied. Here $\begin{Bmatrix} i \\ j \;\; k \end{Bmatrix}$ is a *Christoffel symbol of the second kind*. For an orthogonal coordinate system with metric $ds^2 = h_1(dx^1)^2 + h_2(dx^2)^2 + h_3(dx^3)^2$,

$$\begin{aligned} \begin{Bmatrix} i \\ i \;\; i \end{Bmatrix} &= \frac{1}{2h_i}\frac{\partial h_i}{\partial x^i}, & \begin{Bmatrix} i \\ i \;\; j \end{Bmatrix} &= \frac{1}{2h_i}\frac{\partial h_i}{\partial x^j}, \\ \begin{Bmatrix} i \\ j \;\; j \end{Bmatrix} &= -\frac{1}{2h_i}\frac{\partial h_j}{\partial x^i}, & \begin{Bmatrix} i \\ j \;\; k \end{Bmatrix} &= 0. \end{aligned} \tag{11.56}$$

For example, for a general (non-symmetric) medium in spherical coordinates, $ds^2 = dr^2 + r^2\, d\theta^2 + r^2 \sin^2\theta\, d\phi^2$, and the nonzero Christoffel symbols are

$$\begin{aligned} \begin{Bmatrix} 1 \\ 2 \;\; 2 \end{Bmatrix} &= -r, & \begin{Bmatrix} 1 \\ 3 \;\; 3 \end{Bmatrix} &= -r\sin^2\theta, \\ \begin{Bmatrix} 2 \\ 2 \;\; 1 \end{Bmatrix} &= \frac{1}{r}, & \begin{Bmatrix} 2 \\ 3 \;\; 3 \end{Bmatrix} &= -\sin\theta\cos\theta, \\ \begin{Bmatrix} 3 \\ 3 \;\; 1 \end{Bmatrix} &= \frac{1}{r}, & \begin{Bmatrix} 3 \\ 3 \;\; 2 \end{Bmatrix} &= \cot\theta. \end{aligned} \tag{11.57}$$

Thus from (11.54),

$$F^i_{;i} = \frac{\partial F^1}{\partial r} + \frac{\partial F^2}{\partial \theta} + \frac{\partial F^3}{\partial \phi} + \frac{2F^1}{r} + \cot\theta\, F^2. \tag{11.58}$$

In (11.58) the vector components are abstract *contravariant components*; they are related to *physical components* by $F_i^{\text{phys}} \equiv \sqrt{h_i}\, F^i$. With this substitution, (11.58) becomes

$$\nabla\cdot\mathbf{F} = \frac{1}{r^2}\frac{\partial(r^2 F_r)}{\partial r} + \frac{1}{r\sin\theta}\frac{\partial(\sin\theta F_\theta)}{\partial\theta} + \frac{1}{r\sin\theta}\frac{\partial F_\phi}{\partial\phi}. \tag{11.59}$$

Similarly, using (11.55) and (11.57), and converting to physical components via the relation $P_{ij}^{\rm phys} = \sqrt{h_i h_j}\, P^{ij}$, we find

$$(\nabla\cdot\mathbf{P})_r = \frac{1}{r^2}\frac{\partial(r^2 P_{rr})}{\partial r} + \frac{1}{r\sin\theta}\frac{\partial(\sin\theta\, P_{r\theta})}{\partial\theta} + \frac{1}{r\sin\theta}\frac{\partial P_{r\phi}}{\partial\phi} + \frac{1}{r}(P_{rr} - E), \tag{11.60a}$$

$$(\nabla\cdot\mathbf{P})_\theta = \frac{1}{r^2}\frac{\partial(r^2 P_{r\theta})}{\partial r} + \frac{1}{r\sin\theta}\frac{\partial(\sin\theta\, P_{\theta\theta})}{\partial\theta} + \frac{1}{r\sin\theta}\frac{\partial P_{\theta\phi}}{\partial\phi} + \frac{1}{r}(P_{r\theta} - \cot\theta P_{\phi\phi}), \tag{11.60b}$$

$$(\nabla\cdot\mathbf{P})_\phi = \frac{1}{r^2}\frac{\partial(r^2 P_{r\phi})}{\partial r} + \frac{1}{r\sin\theta}\frac{\partial(\sin\theta\, P_{\phi\theta})}{\partial\theta} + \frac{1}{r\sin\theta}\frac{\partial P_{\phi\phi}}{\partial\phi} + \frac{1}{r}(P_{r\phi} + \cot\theta P_{\theta\phi}). \tag{11.60c}$$

The tensor methods illustrated above are powerful. They can be used for any set of nonorthogonal coordinates with a general metric $ds^2 = \sum_{i,j} g_{ij} dx^i dx^j$.

Closure

The moment equations (11.51) are insufficient to calculate $E_\nu, \mathbf{F}_\nu$, and $\mathbf{P}_\nu$. Even taking advantage of the symmetry of $\mathbf{P}_\nu$, we need (3, 6, 10) quantities in (1D, 2D, 3D) geometries. But (11.51) provides only (2, 3, 4) equations, leaving us short (1, 2, 6) equations. This *closure problem* is common to all systems of moment equations: the equation order n always contains a moment of order $n+1$. One way to close the system is to write it to some order and then make an ad hoc assumption relating the highest-order moment in the last equation to those of lower order.

A better way to close the radiation moment equations in one-dimensional geometries is to use current estimates of source-sink terms to calculate $I_{\mu\nu}, J_\nu$, and K_ν with a formal solution [cf. (11.100)] of the transfer equation and then use these results to (1) connect $K_\nu(\tau)$ and $J_\nu(\tau)$ with a non-dimensional *variable Eddington factor* (which depends on frequency), defined as

$$f_\nu^{\rm K}(\tau) \equiv K_\nu(\tau)/J_\nu(\tau), \tag{11.61}$$

as a function of frequency and depth, and (2) connect $H_\nu(0)$ to $J_\nu(0)$ with a *flux boundary ratio* defined as

$$g_\nu^{\rm K}(\tau = 0) \equiv H_\nu(\tau = 0)/J_\nu(\tau = 0) \tag{11.62}$$

at the surface. With these ratios the angle variable has been eliminated, and we can rewrite the system of two moment equations in terms of the mean intensity and Eddington flux, reducing the order of the system (and the cost of solving it) [54].

This process is iterated until the Eddington factors and the radiation variables are determined consistently with the desired accuracy. For more details, refer to § 12.5.

In the analysis of stellar spectra, one invariably deals with the moment equations because with the exception of the Sun, stars are observed as *unresolved* point sources; hence we obtain data only for the frequency-dependent emergent flux.

11.4 TIME-INDEPENDENT, STATIC, PLANAR ATMOSPHERES

The hydrodynamic adjustment time in a stellar atmosphere is short: minutes to a fraction of a day; any imbalance in its structure is quickly erased by sound waves (unless it is driven by an explosion). In contrast, in most stars thermonuclear energy sources last for tens of millions to billions of years, so the energy flux through their atmospheres can be considered to be essentially invariable. Moreover, a photon's *flight time* $t_\lambda \equiv \lambda_p/c$ over a mean free path is much smaller than other characteristic timescales in the atmosphere, e.g., its *pulsation time* $t_p \approx \ell/v_{\text{sound}}$. Here ℓ is a fraction of the stellar radius, so t_p is of the order hours to days. In this limit radiation field is *quasi-stationary*, i.e., adjusts almost instantaneously to changes in the atmosphere's structure; hence it can be computed with a time-independent snapshot. An exception is a distended envelope that is so large that photon flight times are of the same order as timescales for changes in its structure. Then one has to account for *time retardation* of radiation coming from remote locations. In this section we assume that the atmosphere's structure is *planar and static*, and its *radiation field is time-independent.*[7]

Transfer and Moment Equations

In a *one-dimensional planar atmosphere*, $(\partial I/\partial x)$ and $(\partial I/\partial y)$ are zero, and (11.31) becomes[8]

$$\frac{1}{c}\frac{\partial I_{\mu\nu}}{\partial t} + \mu\frac{\partial I_{\mu\nu}}{\partial z} = \eta_{\mu\nu} - \chi_\nu I_{\mu\nu}. \tag{11.63}$$

For time-independent transfer the time derivative in (11.63) is dropped:

$$\mu\frac{\partial I_{\mu\nu}}{\partial z} = \eta_{\mu\nu} - \chi_\nu I_{\mu\nu}. \tag{11.64}$$

For given $\eta_{\mu\nu}$ and χ_ν, (11.64) is an ordinary differential equation with parameters $-1 \le \mu \le 1$ and ν. If $\eta_{\mu\nu}$ contains scattering terms that are explicitly angle-dependent, it is an integro-differential equation. The first two moments of (11.64) are

$$\frac{\partial H_\nu}{\partial z} = \langle\eta_\nu\rangle - \chi_\nu J_\nu \tag{11.65a}$$

[7] Of course, it is essential to retain the time dependence of the radiation field when moments of the radiation field are used in the dynamical equations of § 11.3.

[8] In (11.63) z increases *upward* in the atmosphere.

and

$$\frac{\partial K_\nu}{\partial z} = -\chi_\nu H_\nu, \tag{11.65b}$$

and their frequency-integrated counterparts are

$$\frac{\partial H}{\partial z} = \int_0^\infty (\langle \eta_\nu \rangle - \chi_\nu J_\nu)\, d\nu \tag{11.66a}$$

and

$$\frac{\partial K}{\partial z} = -\int_0^\infty \chi_\nu H_\nu \, d\nu. \tag{11.66b}$$

If

$$\int_0^\infty (\langle \eta \rangle_\nu - \chi_\nu J_\nu)\, d\nu \equiv 0, \tag{11.67}$$

i.e., exactly as much radiant energy is thermally emitted and outscattered by the material as it absorbs, then it is in *radiative equilibrium*.

Optical Depth and Source Function

The radiation field at any point in an atmosphere depends on how many *photon mean free paths* it is away from the boundary surface, not how far away it is geometrically. If a coherently scattered photon starts N mean free paths from the surface, it will be scattered $O(N^2)$ times to reach a level from which it can escape in a single flight, at which point it is in the *transport regime*. On the other hand, if the number of mean free paths from where a photon is emitted and the boundary is extremely large, the cumulative probability that it will be destroyed and thermalized by an absorption process approaches unity. It is then in the *diffusion regime* (§ 11.9).[9]

For radiative transfer, the most meaningful measurement of position in a one-dimensional atmosphere is *optical depth*. At frequency ν, the optical depth of a point at geometric depth z is defined to be the integrated opacity of the material along a vertical (or radial) line of sight from the boundary to that point:

$$\tau_\nu(z) \equiv \int_z^{z_{\max}} \chi_\nu(z')\, dz'. \tag{11.68}$$

τ_ν is zero at the boundary, where $z = z_{\max}$. Note that $d\tau_\nu(z) \equiv -\chi_\nu(z)\, dz$. The negative sign indicates that optical depth increases *inward* into the atmosphere, whereas z increases *upward*.

Recalling that $1/\chi_\nu$ is the photon mean free path at frequency ν, we see that $\tau_\nu(z)$ is the *number of photon mean free paths* at frequency ν along a vertical line of sight

[9] As shown in chapter 14, this picture does not apply to scattering in spectral lines, where a photon can scatter many times in the opaque line core and move but little in physical space, get redistributed to the line's wing, where the opacity is much lower, and make a large jump in physical space.

from z_{max} down to z. If we observe (inward) along a ray that is inclined with angle cosine μ with respect to the vertical, then the optical depth along the ray is

$$\tau_{\mu\nu}(z) = \int_{z}^{z_{max}} \chi_\nu(z')\, dz'/\mu. \tag{11.69}$$

The projection factor μ^{-1} applies *only* in planar media. Equation (11.69) shows that one can see deeper into an atmosphere along the vertical line of sight ($\mu = 1$) than along a slant ray ($\mu < 1$). Therefore, as mentioned in § 3.3, if the disk of a star can be resolved, one can in principle infer some information about the depth variation of physical quantities.

Another basic concept for transfer problems is the *source function*, which is defined to be the ratio of the total emissivity to total opacity,

$$S_{\mu\nu}(z) = \eta_{\mu\nu}(z)/\chi_\nu(z). \tag{11.70}$$

To simplify notation, we suppress explicit reference to z. In terms of τ_ν and $S_{\mu\nu}$, (11.64) becomes

$$\mu \frac{\partial I_{\mu\nu}}{\partial \tau_\nu} = I_{\mu\nu} - S_{\mu\nu}. \tag{11.71}$$

If there is no explicit dependence of S upon angle (i.e., the absorption and emission terms are isotropic), then

$$\mu \frac{\partial I_{\mu\nu}}{\partial \tau_\nu} = I_{\mu\nu} - S_\nu. \tag{11.72}$$

Equation (11.72) is the "standard" transfer equation for plane-parallel model atmosphere calculations. If we were to assume LTE, then

$$\mu \frac{\partial I_{\mu\nu}}{\partial \tau_\nu} = I_{\mu\nu} - B_\nu, \tag{11.73}$$

where B_ν is the Planck function.

Physical Significance of the Source Function

The physical meaning of S_ν can be understood as follows. Suppose both the emission and absorption coefficients in the material are isotropic. The number of photons at frequency ν emitted into all directions from a volume with an elementary surface area dS and length ds is then $dN_\nu^{em} = (4\pi/h\nu)\langle\eta_\nu\rangle\, d\nu\, dt\, dS\, ds$. The factor 4π comes from integration over all solid angles, and $h\nu$ transforms energy to photon numbers. Using the definitions of τ_ν and S_ν, we can rewrite $\langle\eta_\nu\rangle\, ds$ as $(\langle\eta\rangle_\nu/\chi_\nu)\, \chi_\nu\, ds = S_\nu(\tau_\nu)\, d\tau_\nu$. Then

$$\frac{dN_\nu^{em}}{d\tau_\nu} = S_\nu(\tau_\nu) \left(\frac{4\pi}{h\nu}\, d\nu\, dt\, dS\right), \tag{11.74}$$

which shows that the source function is proportional to the *number of photons emitted per unit optical depth interval*. Likewise, the number of photons *absorbed* per unit optical depth interval (from all solid angles) is

$$\frac{dN_\nu^{\mathrm{abs}}}{d\tau_\nu} = J_\nu(\tau_\nu)\left(\frac{4\pi}{h\nu}\,d\nu\,dt\,dS\right). \tag{11.75}$$

These expressions are useful in making physical interpretations of solutions of the transfer equation.

Boundary Conditions

To solve the transfer equation one needs *boundary conditions*. Three problems of importance are a *finite planar slab*, a *spherical shell*, and a *semi-infinite atmosphere*, i.e., a medium with an open boundary on one side but so optically thick it effectively extends to infinity.

Finite Planar Slab

To calculate the radiation field in a finite slab that has total optical depth $\mathtt{T}_\nu$ ($\tau_\nu \equiv 0$ at the side nearest the observer) and total geometrical thickness Z ($z \equiv 0$ at the side farthest from the observer), we must specify the intensity incident on both of its faces. Taking μ to be positive toward the observer, we require two known functions $\mathtt{I}_\nu^-$ and $\mathtt{I}_\nu^+$ such that

$$I_\nu(\tau_\nu = 0, -\mu) \equiv \mathtt{I}_\nu^-(-\mu), \qquad (0 \le \mu \le 1), \tag{11.76}$$

at the upper boundary, and at the lower boundary

$$I_\nu(\tau_\nu = \mathtt{T}_\nu, \mu) \equiv \mathtt{I}_\nu^+(\mu), \qquad (0 \le \mu \le 1). \tag{11.77}$$

If the slab is symmetric about its midpoint and is illuminated symmetrically, then the boundary condition at its surfaces is

$$\mathtt{I}_\nu^-(-\mu) \equiv \mathtt{I}_\nu^+(\mu), \tag{11.78}$$

where $\mathtt{I}_\nu^-(-\mu)$ is given. The boundary condition at its midpoint is

$$I_\nu\left(\tfrac{1}{2}\mathtt{T}_\nu, \mu\right) = I_\nu\left(\tfrac{1}{2}\mathtt{T}_\nu, -\mu\right). \tag{11.79}$$

In this case, we can either integrate a ray all the way from one boundary to the other, or integrate from one boundary to the center and then back in the opposite direction along the same ray. Either way, we get the radiation field in both directions by applying the symmetry condition.

Spherical Shell

Consider a spherical shell with an outer radius R and inner radius r_i, surrounding a central source of radius r_s, which emits a specific intensity $\mathtt{I}_\nu^+(r_s, \mu_s)$ as shown in figure 11.5.

At the outer boundary, $r = R$ (11.76) still applies. At the inner boundary, for any ray that intersects the central source, e.g., ABC,

$$I_\nu(r = r_i, \mu) = \mathtt{I}_\nu^+(r = r_s, \mu_s), \qquad (0 \le \mu_s \le 1), \tag{11.80}$$

where

$$\mu^2 = 1 - (r_s/r_i)^2(1 - \mu_s^2). \tag{11.81}$$

For rays that miss the central source, e.g., $A''BC''$ or $A'BB'C'$, one sees that by symmetry

$$I_\nu(r = r_i, +\mu) \equiv I_\nu(r = r_i, -\mu) \tag{11.82a}$$

for

$$\mu \le \sqrt{1 - (r_s/r_i)^2}. \tag{11.82b}$$

In the limit that $r_s \to 0$, as for a planetary nebula containing a central point source (star) radiating with intensity I^+,

$$I_\nu(r = r_i, +\mu) = I_\nu(r = r_i, -\mu) + \mathtt{I}^+\delta(\mu - 1), \qquad (0 \le \mu \le 1). \tag{11.83}$$

Equation (11.83) is the *Milne boundary condition* [761]. Note that if there is no central source, $I_\nu(r = r_i, +\mu) \equiv I_\nu(r = r_i, -\mu)$, $(0 \le \mu \le 1)$.

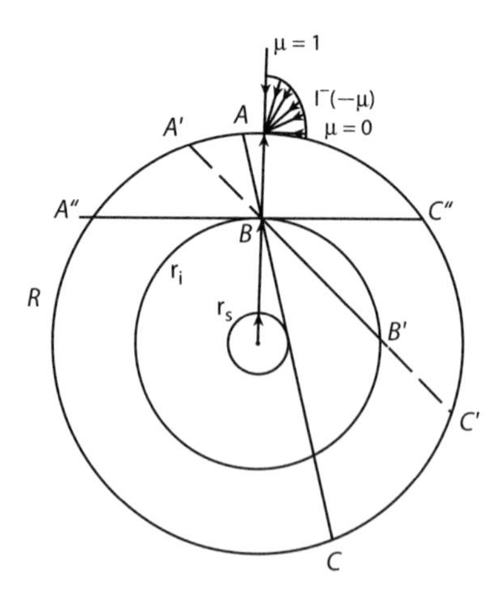

Figure 11.5 Radiation field in a spherical shell around a central source.

Semi-Infinite Atmosphere

In the semi-infinite stellar atmospheres problem (planar or spherical), it is usually assumed that the radiation incident upon the upper boundary is $I_\nu^- \equiv 0$ (obviously this would not be true in a binary system). At the lower boundary condition we invoke a *boundedness condition*, which is useful in analytical work when the limit $\tau_\nu \to \infty$ is taken. Specifically, we require

$$\lim_{\tau_\nu \to \infty} I_\nu(\tau_\nu, \mu) e^{-\tau_\nu/\mu} = 0. \tag{11.84}$$

The reasons for this particular choice will become clear in the discussion below. Alternatively, at great depth in the atmosphere we may write $I_\nu(\tau_\nu, \mu)$ in terms of the local value of S_ν and its gradient, or we may specify the net flux entering the atmosphere from below; these conditions follow naturally from physical considerations in the diffusion limit (§ 11.9).

Geometric Nature of the Boundary Conditions

The discussion above shows that to determine the radiation field in a medium we must solve a problem with separated boundary conditions. In a 1D problem, we must specify boundary conditions at *two points* along the axis perpendicular to level surfaces in the material. In 2D problems, boundary conditions are needed on as many as *four curves*; and in 3D problems, as many as *six surfaces* bounding the material are needed.

The separated nature of the boundary conditions leads to mathematical complications in solving transfer problems. For example, if the emissivity contains a scattering term, we cannot generate the solution by independent directional sweeps through the material, because information from *all* directions is required to compute the scattering integral in each sweep.

Simple Examples

To study methods of solving (11.71), consider some very simple archetype expressions for S_ν, assuming LTE. If there is no scattering term in the extinction coefficient, then $\chi_\nu \equiv \kappa_\nu$ and the source function reduces to

$$S_\nu = B_\nu. \tag{11.85}$$

If there is also a coherent and isotropic scattering term, then

$$\chi_\nu = \kappa_\nu + \sigma_\nu, \tag{11.86}$$

and

$$\eta_\nu = \kappa_\nu B_\nu + \sigma_\nu J_\nu, \tag{11.87}$$

so

$$S_\nu = (\kappa_\nu B_\nu + \sigma_\nu J_\nu)/(\kappa_\nu + \sigma_\nu). \tag{11.88}$$

In terms of the thermal coupling parameter we can rewrite (11.88) as

$$S_\nu = \epsilon_\nu B_\nu + (1 - \epsilon_\nu) J_\nu. \tag{11.89}$$

For a spectral line with an overlapping (thermal) background continuum,

$$\chi_\nu = \kappa_c + \ell\phi_\nu, \tag{11.90}$$

where κ_c and $\ell\phi_\nu$ are the continuum and line opacities. If a fraction ϵ of the line emission comes from thermal processes and the remainder from isotropic coherent scattering, then

$$\eta_\nu = \kappa_c B_\nu + \ell\phi_\nu[\epsilon B_\nu + (1 - \epsilon) J_\nu]. \tag{11.91}$$

Then defining $r \equiv \kappa_c/\ell$,

$$S_\nu = \frac{r + \epsilon\,\phi_\nu}{r + \phi_\nu} B_\nu + \frac{(1 - \epsilon)\,\phi_\nu}{r + \phi_\nu} J_\nu \equiv \xi_\nu B_\nu + (1 - \xi_\nu)\phi_\nu J_\nu. \tag{11.92}$$

The variation of the continuum over a line width is negligible.

If electrons in the upper level of a spectral line transition are strongly perturbed by collisions with neighboring atoms and randomly distributed over its profile before they emit a photon, then the frequencies of the incoming and outgoing photons are uncorrelated, and we have complete redistribution. If we make the further approximation that the incident radiation is nearly isotropic, we can use *angle-averaged complete redistribution*. In this case,

$$\eta_\nu = \kappa_c B_\nu + \ell\,\phi_\nu \left[\epsilon B_\nu + (1 - \epsilon) \int_0^\infty \phi_{\nu'} J_{\nu'}\, d\nu'\right]. \tag{11.93}$$

Combining (11.90) with (11.93), we have

$$S_\nu = \frac{r + \epsilon\phi_\nu}{r + \phi_\nu} B_\nu + \frac{(1 - \epsilon)\phi_\nu}{r + \phi_\nu} \int_0^\infty \phi_{\nu'} J_{\nu'}\, d\nu' \tag{11.94}$$

$$\equiv \xi_\nu B_\nu + (1 - \xi_\nu) \int_0^\infty \phi_{\nu'} J_{\nu'}\, d\nu'. \tag{11.95}$$

The line-scattering terms in the source functions in (11.92) and (11.94) are based on heuristic arguments; hence they are *only illustrative*. Rigorous expressions have to be derived from the equations of kinetic equilibrium, with physically realistic redistribution functions; cf. chapters 9, 10, and 15.

Consider now some simple 1D plane-parallel transfer problems. In all cases we assume $\mu > 0$, i.e., radiation moving toward an external observer.

No Absorption, No Emission

Suppose $\chi_\nu \equiv 0$, and $\eta_\nu \equiv 0$. The transfer equation reduces to $dI_\nu/dz = 0$; its solution is

$$I_\nu = \text{const}, \tag{11.96}$$

consistent with the proof in § 3.1 of the invariance of the specific intensity when no sources or sinks are present.

No Absorption, Only Emission

Suppose $\chi_\nu \equiv 0$, but $\eta_{\mu\nu} > 0$, as would apply to radiation from, say, forbidden lines in a planetary nebula, the solar corona, or optically thin lines in the solar transition region. Then the radiation emerging from an optically thin radiating slab is

$$I_\nu(z,\mu) = \mathrm{I}_\nu{}^+(0,\mu) + \int_0^z \eta_\nu(z')\,dz'/\mu. \tag{11.97}$$

No Emission, Only Absorption

Or suppose $\eta_{\mu\nu} \equiv 0$, but $\chi_\nu > 0$. Then $\mu(dI_\nu/d\tau_\nu) = I_\nu$, which in a slab of total optical depth T_ν has the solution

$$\begin{aligned} I_\nu(\tau_\nu,\mu) &= I_\nu^+(0,\mu)\,e^{-\chi_\nu(z_{\max}-z)/\mu} \\ &\equiv I_\nu^+(\mathrm{T}_\nu,\mu)\,e^{-(\mathrm{T}_\nu-\tau_\nu)/\mu}. \end{aligned} \tag{11.98}$$

This is the usual exponential attenuation law for an absorbing medium, e.g., a photometric filter. Typically the energy absorbed is converted to heat, which, if radiated, emerges in the far infrared.

Absorption and Emission

Now let both the absorption and emission coefficients be nonzero ($\chi_\nu > 0, \eta_{\mu\nu} > 0$) and be known functions of (z,ν). Such a solution is called a *formal solution* because in general both χ_ν and $\eta_{\mu\nu}$ may depend on the radiation field and cannot be given a priori without having already solved the transfer problem. But for now, regarding S_ν as given, (11.71) is a linear first-order differential equation with constant coefficients, which has the integrating factor $e^{-\tau_\nu/\mu}$. Thus

$$\mu\frac{\partial}{\partial\tau_\nu}\left[I(\tau_\nu,\mu)e^{-\tau_\nu/\mu}\right] = -e^{-\tau_\nu/\mu}S_\nu(\tau_\nu). \tag{11.99}$$

Integration of (11.99) gives

$$I_\nu(\tau_1,\mu) = I_\nu(\tau_2,\mu)e^{-(\tau_2-\tau_1)/\mu} + \int_{\tau_1}^{\tau_2} S_\nu(t_\nu)e^{-(t_\nu-\tau_1)/\mu}\,dt_\nu/\mu. \tag{11.100}$$

This expression, valid both for $\tau_1 \le \tau_2$, in which case $\mu > 0$, and for $\tau_2 \le \tau_1$, in which case $\mu < 0$, has a simple physical interpretation. The specific intensity at τ_1

is composed of two parts. The first term is the incoming intensity $I_\nu(\tau_2, \mu)$ at optical depth τ_2, reduced by attenuation between τ_1 and τ_2. The second term is the integrated contribution of photons created in all elementary intervals $[t_\nu/\mu, (t_\nu + dt_\nu)/\mu]$ between τ_1/μ and τ_2/μ [recall that $S_\nu(t_\nu)dt_\nu/\mu$ represents the number of photons created in the slant-length element dt_ν/μ], reduced by the intervening attenuation between τ_1 and t_ν.

The following results are special cases of (11.100) and have a similar physical interpretation. Note that the boundedness condition (11.84) assures that the emergent radiation field from a semi-infinite medium is unaffected by radiation imposed at the lower boundary.

Semi-Infinite Atmosphere

For the special case of radiation emergent ($\tau_1 = 0$) from a semi-infinite atmosphere ($\tau_2 = \infty$) the formal solution (11.100) reads

$$I_\nu(0, \mu) = \int_0^\infty S_\nu(t_\nu) e^{-t_\nu/\mu}\, dt_\nu/\mu. \tag{11.101}$$

Mathematically, (11.101) states that the specific intensity is the *Laplace transform* of the source function, a property that may be used to solve for S in certain classes of problems.

Semi-Infinite Atmosphere with a Linear Source Function

Another special case of the formal solution (11.100) is the emergent intensity from a semi-infinite atmosphere in which the source function is linear in optical depth, $S_\nu(\tau_\nu) = a + b\tau_\nu$. We find

$$I_\nu(0, \mu) = a + b\mu = S_\nu(\tau_\nu = \mu). \tag{11.102}$$

This expression is called the *Eddington–Barbier relation*. It shows that for a linear S_ν, $I_\nu(0, \mu)$ equals the value of S_ν at $\tau_\nu \approx \mu$ along that ray. Thus values of $I_\nu(0, \mu)$ for angle cosines $1 \geq \mu \geq 0$ map values of the S_ν at radial optical depths between $1 \gtrsim \tau_\nu \gtrsim 0$. In general, S_ν is not a linear function of τ_ν; yet a linear fit may be a fair approximation on the limited range ($0 \leq \tau_\nu \leq 1$), and the Eddington–Barbier relation can give a reasonable first estimate of the emergent intensity.

Finite Homogeneous Slab

The radiation emergent from a homogeneous finite slab, ($\mu = 1$), is

$$I_\nu(0, 1) = \left(1 - e^{-\mathrm{T}_\nu}\right) S_\nu. \tag{11.103}$$

For $\mathrm{T}_\nu \gg 1$, $I(0, 1) = S_\nu$, whereas for $\mathrm{T}_\nu \ll 1$, $I_\nu(0, 1) \to S_\nu \times \mathrm{T}_\nu$.

Both limits are easily understood by recalling that S_ν gives the number of photons emitted per unit optical depth. In the optically thin case ($\mathrm{T}_\nu \ll 1$), there is little attenuation, so essentially all emitted photons emerge from the medium. The total optical depth is T_ν, so I_ν is $S_\nu \times \mathrm{T}_\nu$. In the optically thick case, most photons created deeper than $\tau_\nu = 1$ are absorbed, so only photons emitted at $\tau_\nu \lesssim 1$ contribute to I_ν; therefore, I_ν saturates to S_ν regardless of the actual optical thickness of the slab.

11.5 SCHWARZSCHILD-MILNE EQUATIONS

Schwarzschild's Equation

The general formal solution of the radiative transfer equation for a semi-infinite plane-parallel atmosphere, with no incoming radiation at the surface ($\tau = 0$), is

$$I_{\mu\nu}(\tau_\nu) = \int_{\tau_\nu}^{\infty} S_\nu(t_\nu) e^{-(t_\nu - \tau_\nu)/\mu} \frac{dt_\nu}{\mu}, \quad \text{for} \quad \mu \geq 0, \tag{11.104a}$$

$$I_{\mu\nu}(\tau_\nu) = \int_{0}^{\tau_\nu} S_\nu(t_\nu) e^{-(\tau_\nu - t_\nu)/(-\mu)} \frac{dt_\nu}{-\mu}, \quad \text{for} \quad \mu < 0. \tag{11.104b}$$

Positive (negative) values of μ correspond to outward (inward) rays. The mean intensity is obtained by integrating the specific intensity over μ:

$$J_\nu(\tau_\nu) = \tfrac{1}{2}\left[\int_0^1 d\mu \int_{\tau_\nu}^{\infty} S_\nu(t_\nu) e^{-(t_\nu - \tau_\nu)/\mu} \frac{dt_\nu}{\mu} - \int_{-1}^{0} d\mu \int_0^{\tau_\nu} S_\nu(t_\nu) e^{(\tau_\nu - t_\nu)/\mu} \frac{dt_\nu}{\mu}\right]. \tag{11.105}$$

Interchanging the order of integration and substituting $\mu = \pm 1/w$ in the first and second integrals, respectively, we obtain

$$J_\nu(\tau_\nu) = \tfrac{1}{2}\left[\int_{\tau_\nu}^{\infty} S_\nu(t_\nu)\, dt_\nu \int_1^{\infty} e^{-w(t_\nu - \tau_\nu)} \frac{dw}{w} + \int_0^{\tau_\nu} S_\nu(t_\nu)\, dt_\nu \int_1^{\infty} e^{-w(\tau_\nu - t_\nu)} \frac{dw}{w}\right]. \tag{11.106}$$

The integral over w is a higher transcendental function known as the first *exponential integral* $E_1(x)$, which is the first member of the family

$$E_n(x) \equiv \int_1^{\infty} e^{-xt} t^{-n}\, dt \tag{11.107}$$

for integer $n > 0$. Thus

$$J_\nu(\tau_\nu) = \tfrac{1}{2} \int_0^{\infty} S_\nu(t_\nu) E_1(|\, t_\nu - \tau_\nu|)\, dt_\nu. \tag{11.108}$$

An advantage of (11.108) is that the angle integration is exact. This expression was first derived by K. Schwarzschild in one of the fundamental papers of radiative transfer theory. See [721, p. 35] and also [511, 653].

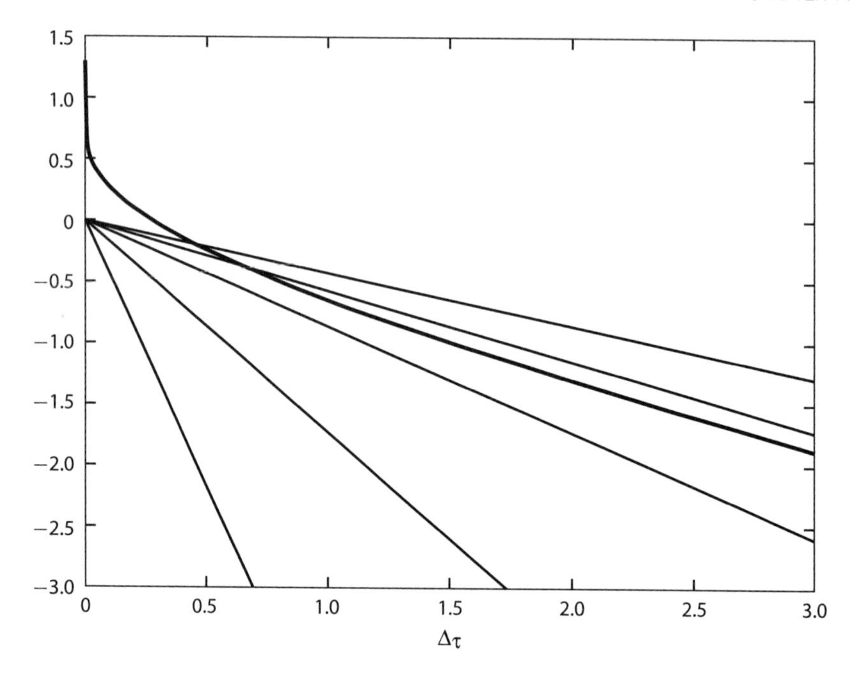

Figure 11.6 Kernel functions for mean intensity and specific intensity.

Lambda Operator

Equation (11.108) is usually written in terms of the Λ-*operator*:

$$J_\nu(\tau_\nu) = \Lambda_{\tau_\nu}[S_\nu(t_\nu)], \tag{11.109}$$

where

$$\Lambda_\tau[f(t)] \equiv \tfrac{1}{2}\int_0^\infty f(t)\, E_1(\,|t-\tau|\,)\, dt. \tag{11.110}$$

The exponentials $e^{-|\Delta\tau/\mu|}$ in (11.104) and the exponential integral $E_1|\Delta\tau|$ (11.108) are the *kernel functions* of $\Lambda_{\mu\nu}$ and Λ_ν. See figure 11.6. The ordinate is $\log_{10}$ of the kernel function; the abscissa is $|\Delta\tau|$. The heavy curve is for E_1, which has a logarithmic singularity at $|\Delta\tau| = 0$. The lighter (straight) lines are for $e^{-|\Delta\tau/\mu|}$, with $\mu = (1.0, 0.75, 0.5, 0.25, 0.10)$ from top to bottom.

The effective width $|\,t_\nu - \tau_\nu\,|$ of the kernel $\Lambda_{\mu\nu}$ at a given position in the atmosphere decreases quickly with decreasing μ, because the number of mean free paths a photon travels when moving along a ray with angle cosine μ relative to the normal is a factor of $1/\mu$ larger than the vertical optical depth measured to the same physical depth. On the other hand, the kernel for the specific intensity propagating in the normal direction, ($\mu = 1$), has a greater width than the kernel for the mean intensity, which contains contributions from all angles. Note that for $|\Delta\tau| \gtrsim 0.6$, E_1 lies below $e^{-|\Delta\tau/\mu|}$ for $\mu = 1$, as can be expected, because the mean intensity is an average over angle. Note also that as $\mu \to 0, e^{-|\Delta\tau/\mu|}$ becomes a delta function at $|\Delta\tau| = 0$, which says physically that the source function is constant along a ray that is perpendicular to the vertical.

Although historically the Λ-operator was introduced as an operator that gives the mean intensity, variants of the symbol Λ are now used for other operators as well. This change was motivated by the *Accelerated lambda iteration* method, described in chapter 13.

Specifically, we may write $\Lambda_{\mu\nu}$ for an operator that gives the specific intensity as

$$I_{\mu\nu} \equiv \Lambda_{\mu\nu}[S_\nu], \tag{11.111}$$

which is just a compact form of (11.104). The traditional Λ-operator, which gives the mean intensity,

$$J_\nu = \Lambda_\nu[S_\nu], \tag{11.112}$$

is the integral of the elementary $\Lambda_{\mu\nu}$-operator over μ:

$$\Lambda_\nu = \tfrac{1}{2}\int_{-1}^{1} \Lambda_{\mu\nu}\, d\mu. \tag{11.113}$$

In addition, we can write a *frequency-averaged* operator $\overline{\Lambda}$, which gives the frequency-averaged mean intensity in a spectral line:

$$\overline{J} \equiv \int_0^\infty \phi_\nu J_\nu\, d\nu = \overline{\Lambda}\,[S], \tag{11.114}$$

where

$$\overline{\Lambda} \equiv \int_0^\infty \phi_\nu \Lambda_\nu\, d\nu, \tag{11.115}$$

and ϕ_ν is the normalized line-absorption profile. In general, computation of $I_{\mu\nu}, J_\nu$, or $\overline{J}$ from (11.111), (11.112), and (11.114) requires the solution of an integral equation.

Finite Slabs

The above expressions apply for a semi-infinite atmosphere with no incident radiation. A different formulation of the Λ-operator is needed for finite slabs and/or nonzero incident radiation. Even for a semi-infinite atmosphere we can consider only a finite numerical domain, say, $0 \le \tau \le \tau_{\max}$. As τ approaches $\tau_{\max}$, the range of the Λ-operator necessarily runs beyond the edge of the domain on which S_ν is known. We must therefore apply an appropriate boundary condition at $\tau_{\max}$. In practice, one can use the diffusion approximation, see § 11.9, to extrapolate S_ν into the range $\tau \ge \tau_{\max}$.

For a finite slab with the total monochromatic optical depth T_ν at frequency ν, (11.104) should be replaced by the general formal solution, (11.100). Then for an outgoing ray, ($\mu > 0$),

$$I_{\mu\nu}(\tau_\nu) = \mathrm{I}^{+}_{\mu\nu} e^{-(\mathrm{T}_\nu - \tau_\nu)/\mu} + \int_{\tau_\nu}^{\mathrm{T}_\nu} S_\nu(t_\nu) e^{-(t_\nu - \tau_\nu)/\mu} \, dt_\nu/\mu, \tag{11.116}$$

and for an incoming intensity, ($\mu < 0$),

$$I_{\mu\nu}(\tau_\nu) = \mathrm{I}^{-}_{\mu\nu} e^{-\tau_\nu/(-\mu)} + \int_{0}^{\tau_\nu} S_\nu(t_\nu) e^{-(\tau_\nu - t_\nu)/(-\mu)} \, dt_\nu/(-\mu). \tag{11.117}$$

Comparison with (11.104) shows there is an extra term containing the incident intensity on the opposite side of the slab, attenuated by the optical depth to that boundary. This contribution does not depend on the source function. In this case (11.111) can be modified to read

$$I_{\mu\nu} = \Lambda_{\mu\nu}[S] + \mathrm{I}^{\mathrm{BC}}_{\mu\nu}, \tag{11.118}$$

where $\mathrm{I}^{\mathrm{BC}}_{\mu\nu}$ is the attenuated intensity imposed at the opposite boundary, as given by the first term on the right-hand side of (11.116) or (11.117).

Another possibility is to leave (11.111) unchanged and to absorb the incoming intensity into the definition of the source function instead. Noting that (e.g., for the outgoing intensity),

$$\int_{\tau_\nu}^{\mathrm{T}_\nu} \delta(t_\nu - \mathrm{T}_\nu) e^{-(t_\nu - \tau_\nu)/\mu} \, dt_\nu/\mu = e^{-(\mathrm{T}_\nu - \tau_\nu)/\mu}, \tag{11.119}$$

one sees that the source function can be modified as follows:

$$S_\nu(t_\nu) \rightarrow S_\nu(t_\nu) + \mathrm{I}_{\mu\nu}(\mathrm{T}_\nu)\delta(t_\nu - \mathrm{T}_\nu) + \mathrm{I}_{\mu\nu}(0)\delta(t_\nu). \tag{11.120}$$

Milne's Equations

By an analysis similar to that used to derive (11.108), Milne obtained expressions for H_ν and K_ν, [721, p. 77]:

$$H_\nu(\tau_\nu) = \tfrac{1}{2} \int_{\tau_\nu}^{\infty} S_\nu(t_\nu) E_2(t_\nu - \tau_\nu) \, dt_\nu - \tfrac{1}{2} \int_{0}^{\tau_\nu} S_\nu(t_\nu) E_2(\tau_\nu - t_\nu) \, dt_\nu, \tag{11.121}$$

and

$$K_\nu(\tau_\nu) = \tfrac{1}{2} \int_{0}^{\infty} S_\nu(t_\nu) E_3(|t_\nu - \tau_\nu|) \, dt_\nu. \tag{11.122}$$

We can also define the corresponding operators[10]

$$\Phi_\tau[f(t)] \equiv \tfrac{1}{2} \int_{\tau}^{\infty} f(t) E_2(t - \tau) \, dt - \tfrac{1}{2} \int_{0}^{\tau} f(t) E_2(\tau - t) \, dt, \tag{11.123}$$

[10] *Note that our definitions of the* Φ *and* X *operators are* $\frac{1}{4}$ *times the definitions of Kourganoff's* Φ *and* X *operators*, see [633, eq. 9.32–9.34]. We have made this change to make our notation consistent with Eddington's moments, i.e., so that $J_\nu = \Lambda[S_\nu]$, $H_\nu = \Phi[S_\nu]$, and $K_\nu = X[S_\nu]$.

and

$$X_\tau[f(t)] \equiv \tfrac{1}{2}\int_0^\infty f(t)E_3(|\, t-\tau\,|)\,dt. \tag{11.124}$$

⋆ Mathematical Properties of the Λ, Φ, and X Operators

The mathematical properties of the exponential integrals are discussed in detail in [5, pp. 228 – 231], [225, Appendix I], and [633, Chap. 2]. Only a few useful results are summarized here. By direct integration of (11.107), we find

$$E_n(0) = 1/(n-1), \qquad n \geq 1. \tag{11.125}$$

Differentiation of (11.107) with respect to x yields

$$E_n'(x) = -E_{n-1}(x), \qquad n > 1, \tag{11.126}$$

and integration of (11.107) by parts gives

$$E_n(x) = [e^{-x} - xE_{n-1}(x)]/(n-1), \qquad n > 1. \tag{11.127}$$

A uniformly convergent series for all x, useful mainly for small x, is

$$E_1(x) = -\gamma - \ln x + \sum_{k=1}^{\infty}(-1)^{k-1}\frac{x^k}{k\,k!}, \qquad x > 0. \tag{11.128}$$

Here $\gamma = 0.57721\ 56649\ \ldots$ is the *Euler–Mascheroni constant.* Several useful numerical approximation formulae for $E_1(x)$ are known; see [5, pp. 228–237]. An asymptotic expansion for $E_1(x)$ valid for $x \gg 1$ is

$$E_1(x) = \frac{e^{-x}}{x}\left[1 - \frac{1}{x} + \frac{2!}{x^2} - \frac{3!}{x^3} + \ldots\right], \tag{11.129}$$

showing that for large x, $E_1(x) \sim e^{-x}/x$. Equation (11.125) shows that $E_1(x)$ is singular at the origin, which implies that the derivative of $E_2(x)$ is singular at $x = 0$, even though $E_2(0) = 1$ is finite. The physical implications of these mathematical facts will become clear in later discussions.

We list here the Λ, Φ, and X transforms [633, § 14] of some elementary functions, which will be of use in later work.

$$\Lambda_\tau[1] = 1 - \tfrac{1}{2}E_2(\tau), \tag{11.130a}$$

$$\Lambda_\tau[t] = \tau + \tfrac{1}{2}E_3(\tau), \tag{11.130b}$$

$$\Lambda_\tau[t^n] = \tfrac{1}{2}n!\left[\sum_{m=0}^{n}\frac{\tau^m}{m!}\delta_\alpha + (-1)^{n+1}E_{n+2}(\tau)\right], \tag{11.130c}$$

$$\Lambda_\tau[e^{-a\tau}] = \frac{e^{-a\tau}}{2a}\left[\ln\frac{a+1}{|a-1|} - E_1(\tau - a\tau)\right] + \frac{1}{2a}E_1(\tau), \tag{11.130d}$$

where $a > 0, a \neq 1$, and $\delta_\alpha = 0$ if $\alpha = n+1-m$ is even, and $\delta_\alpha = 2/\alpha$ if α is odd. Similarly,

$$\Phi_\tau[1] = \tfrac{1}{2} E_3(\tau), \tag{11.131a}$$

$$\Phi_\tau[t] = \tfrac{1}{3} - \tfrac{1}{2} E_4(\tau), \tag{11.131b}$$

$$\Phi_\tau[t^n] = \tfrac{1}{2} n! \left[\sum_{m=0}^{n} \frac{\tau^m}{m!} \delta_\beta + (-1)^n E_{n+3}(\tau) \right], \tag{11.131c}$$

$$\begin{aligned}\Phi_\tau[e^{-a\tau}] = & \frac{e^{-a\tau}}{2a^2} \left[2a + \ln \frac{|a-1|}{a+1} + E_1(\tau - a\tau) \right] \\ & - \frac{1}{2a^2} \left[aE_2(\tau) + E_1(\tau) \right],\end{aligned} \tag{11.131d}$$

where $a > 0, a \neq 1, \delta_\beta = 0$ if $\beta = n+2-m$ is even, and $\delta_\beta = 2/\beta$ if β is odd. Finally,

$$X_\tau[1] = \tfrac{1}{3} - \tfrac{1}{2} E_4(\tau), \tag{11.132a}$$

$$X_\tau[t] = \tfrac{1}{3} \tau + \tfrac{1}{2} E_5(\tau), \tag{11.132b}$$

$$X_\tau[t^n] = \tfrac{1}{2} n! \left[\sum_{m=0}^{n} \frac{\tau^m}{m!} \delta_\gamma + (-1)^{n+1} E_{n+4}(\tau) \right], \tag{11.132c}$$

$$\begin{aligned}X_\tau[e^{-a\tau}] = & \frac{e^{-a\tau}}{2a^2} \left[-2a + \ln \frac{a+1}{|a-1|} - E_1(\tau - a\tau) \right] \\ & + \frac{1}{2a^2} \left[a^2 E_3(\tau) + aE_2(\tau) + E_1(\tau) \right],\end{aligned} \tag{11.132d}$$

where $a > 0, a \neq 1$, where $\delta_\gamma = 0$ if $\gamma = n+4-m$ is even, and $\delta_\gamma = 2/\gamma$ if γ is odd. By definition, $0! \equiv 1$. More properties of the Λ, Φ, and X operators are discussed in [633, Chap. 2].

For a linear source function $S(\tau) = a + bt$, (11.130)–(11.132) give

$$\Lambda_\tau[a + bt] = (a + b\tau) + \tfrac{1}{2} [bE_3(\tau) - aE_2(\tau)], \tag{11.133a}$$

$$\Phi_\tau[a + bt] = \tfrac{1}{3} b + \tfrac{1}{2} [aE_3(\tau) - bE_4(\tau)], \tag{11.133b}$$

$$X_\tau[a + bt] = \tfrac{1}{3} (a + b\tau) + \tfrac{1}{2} [bE_5(\tau) - aE_4(\tau)]. \tag{11.133c}$$

These expressions reveal some important facts. Because the exponential integrals decay asymptotically as e^{-x}/x, (11.133a) shows that $J(\tau) = \Lambda_\tau(S)$ closely approaches the local value of $S(\tau)$ for $\tau \gg 1$; i.e., *at great depth, the* Λ*-operator reproduces the local source function.* Equation (11.133b) shows that for $\tau \gg 1$, $H \to \frac{1}{3} b$; i.e., *at depth, the flux depends only on the gradient of the source function.* Likewise, at great depth, $K(\tau)/J(\tau) \to \frac{1}{3}$. These results are the same as those given by the asymptotic analysis in § 11.9.

At the surface, $E_2(0) = 1$ and $E_3(0) = \frac{1}{2}$; hence

$$J(0) = \Lambda_0(S) = \tfrac{1}{2}a + \tfrac{1}{4}b \equiv \tfrac{1}{2}S(\tau = \tfrac{1}{2}). \tag{11.134}$$

In particular, if there is no gradient in S, i.e., $b \equiv 0$, then $J(0) = \frac{1}{2}a = \frac{1}{2}S(0)$. This result makes sense physically because J at the surface is the average over a hemisphere of empty space containing no radiation and a hemisphere in which $I \equiv S$. In general, J departs most strongly from S at the surface.

Similarly, the flux at the surface is

$$H(0) = \tfrac{1}{4}a + \tfrac{1}{6}b \equiv \tfrac{1}{4}S(\tau = \tfrac{2}{3}), \tag{11.135}$$

which implies that the emergent flux $F(0) = 4H(0)$ is formed at $\tau \approx \frac{2}{3}$, a result consistent with the simplest model of a gray atmosphere in radiative equilibrium (see § 17.1). In fact, this is a good estimate because from a probabilistic viewpoint a photon emitted at $\tau = \frac{2}{3}$ has a chance of about $e^{-0.67} \approx 0.5$ to emerge from the surface. Note also that the effect of a gradient, relative to the case $b = 0$, is larger for $H(0)$ than for $J(0)$. Both (11.134) and (11.135) are consistent with the Eddington-Barbier relation; they provide useful estimates of I, J, and H at the surface.

11.6 SECOND-ORDER FORM OF THE TRANSFER EQUATION

In 1905, Schuster [979] derived a *second-order* form of the transfer equation, which uses *symmetric and antisymmetric averages of the specific intensity along a ray.* Lacking automated computational capability, he assumed isotropic radiation in the incoming and outgoing hemispheres (i.e., the *two-stream approximation*). His idea was rediscovered by P. Feautrier [323–325], who developed it into an effective algorithm well suited to digital computers. It treats the complete angle-frequency variation of the radiation field and can handle complicated scattering terms in the source function. It is a powerful numerical method for solving difficult NLTE transfer problems. We discuss its numerical implementation in § 12.2.[11]

Derivation

The symmetric and antisymmetric averages of the specific intensity at frequency ν propagating along a ray in the $\pm\mu$ direction are

$$j_{\mu\nu} \equiv \tfrac{1}{2}[I_\nu(\mu) + I_\nu(-\mu)] \quad \text{and} \quad h_{\mu\nu} \equiv \tfrac{1}{2}[I_\nu(\mu) - I_\nu(-\mu)], \tag{11.136}$$

where $(0 \le \mu \le 1)$. We have replaced Schuster's archaic notation with one suggestive of angle-dependent versions of Eddington's J_ν and H_ν. For the moment, we

[11] Although we consider only time-independent problems in this book, Feautrier's method can also be adapted to the calculation of radiation fields in dynamical media.

will assume that (1) the source function $S_\nu(\mu)$ contains only thermal absorption and emission terms so that it is isotropic, i.e., independent of μ, or (2) if it contains scattering terms, they are isotropic (which is unphysical) or have forward-backward symmetry so that $S_\nu(\mu) \equiv S_\nu(-\mu)$.

Then the transfer equations for two rays moving in directions $\pm\mu$ are

$$\mu \frac{\partial I_\nu(\mu)}{\partial \tau_\nu} = I_\nu(\mu) - S_\nu(\mu), \quad \text{and} \quad -\mu \frac{\partial I_\nu(-\mu)}{\partial \tau_\nu} = I_\nu(-\mu) - S_\nu(-\mu). \tag{11.137}$$

Adding equations (11.137), we have

$$\mu \frac{\partial h_{\mu\nu}}{\partial \tau_\nu} = j_{\mu\nu} - S_\nu, \qquad (0 \le \mu \le 1), \tag{11.138}$$

and by subtracting them, we find

$$\mu \frac{\partial j_{\mu\nu}}{\partial \tau_\nu} = h_{\mu\nu}, \qquad (0 \le \mu \le 1). \tag{11.139}$$

Then substituting the expression for $h_{\mu\nu}$ given by (11.139) into (11.138), we obtain *Schuster's second-order transfer equation*,

$$\mu^2 \frac{\partial^2 j_{\mu\nu}}{\partial \tau_\nu^2} = j_{\mu\nu} - S_\nu, \qquad (0 \le \mu \le 1), \tag{11.140}$$

which is to be solved on the domain $0 \le \tau_\nu \le \mathrm{T}_\nu$.

Anisotropic Thomson Scattering

We saw in § 6.1 that Thomson scattering is not isotropic, but has a phase function $g(\mathbf{n}', \mathbf{n}) = \frac{3}{4}[1 + (\mathbf{n}' \cdot \mathbf{n})^2]$, where $\mathbf{n}'$ and $\mathbf{n}$ are unit vectors along the directions of the incoming and outgoing photons. We must now examine whether this anisotropy might invalidate (11.139), which is crucial in combining the two first-order equations (11.138) and (11.139) into the second-order equation (11.140). Using a simplified notation, we write

$$\eta_s(\mathbf{n}, \nu) = \frac{n_e \sigma_\mathrm{T}}{4\pi} \oint I(\mathbf{n}', \nu) g(\mathbf{n}', \mathbf{n})\, d\Omega'. \tag{11.141}$$

Or in terms of μ and ϕ,

$$S_s(\mu, \phi; \nu) = \frac{1}{4\pi} \int_{-1}^{1} d\mu' \int_0^{2\pi} g(\mu', \phi'; \mu, \phi) I(\mu', \phi'; \nu)\, d\phi'. \tag{11.142}$$

Assuming azimuthal symmetry without any loss of generality, we may set ϕ' to an arbitrary value, say, $\phi' = 0$. Then

$$S_s(\mu, \nu) \equiv \frac{1}{2} \int_{-1}^{1} g(\mu', \mu) I(\mu', \nu)\, d\mu', \tag{11.143}$$

where

$$g(\mu', \mu) \equiv \frac{1}{2\pi} \int_0^{2\pi} g(\mu', \mu; \phi)\, d\phi. \tag{11.144}$$

For the Rayleigh phase function,

$$g(\mathbf{n}', \mathbf{n}) = \tfrac{3}{4}\big[1 + (\mathbf{n}' \cdot \mathbf{n})^2\big]. \tag{11.145}$$

Expressing the dot product of $\mathbf{n}'$ and $\mathbf{n}$ in terms of μ and trigonometric functions of ϕ, and performing the integration in (11.145), one finds [544]

$$g(\mu', \mu) = \tfrac{3}{8}\big[(3 - \mu^2) + (3\mu^2 - 1)\mu'^2\big]. \tag{11.146}$$

Then from (11.143),

$$S_s(\mu, \nu) = \tfrac{3}{8}\big[(3J_\nu - K_\nu) + \mu^2(3K_\nu - J_\nu)\big]. \tag{11.147}$$

At great depth where $K_\nu \rightarrow \frac{1}{3}J_\nu$, $S_s(\mu, \nu) \rightarrow J_\nu$.

The crucial point shown in (11.147) is that for Thomson scattering in planar or spherically symmetric media, $S_s(\mu, \nu)$ is an *even* function of μ, which validates the cancellation of the source function in (11.139), which in turn is needed to derive (11.140).

Boundary Conditions

The solution of a second-order differential equation demands that two boundary conditions be met. In the Schuster-Feautrier method these are enforced explicitly, which ensures the correct behavior of the solution at the upper and lower (or outer and inner) boundaries of the physical domain. Older methods, which applied one boundary condition and attempted to match the other by an equivalent of an "eigenvalue" or "shooting" method, are found to be exponentially unstable (see, e.g., [731, §6.2]).

Given explicit expressions for $\mathtt{I}^+$ and $\mathtt{I}^-$ as defined in (11.76)–(11.83), at the upper boundary we have

$$h_{\mu\nu}\,(\tau_\nu = 0) = j_{\mu\nu}\,(\tau_\nu = 0) - \mathtt{I}_\nu^-(0, -\mu), \qquad (0 \leq \mu \leq 1). \tag{11.148}$$

If there is no imposed exterior radiation field, then $\mathtt{I}_\nu^-(0, -\mu) \equiv 0$, so

$$h_{\mu\nu}\,(\tau_\nu = 0) \equiv j_{\mu\nu}\,(\tau_\nu = 0)\,, \tag{11.149}$$

and

$$\mu\left(\partial j_{\mu\nu}/\partial \tau_\nu\right)_0 = j_{\mu\nu}(\tau_\nu = 0). \tag{11.150}$$

In planar geometry, at the lower boundary $\tau_\nu \equiv \mathrm{T}_\nu$, we specify $\mathtt{I}_\nu^+(\mathrm{T}_\nu, +\mu) \equiv I^+(\mu, \nu)$. Then

$$h_{\mu\nu}\,(\tau_\nu = \mathrm{T}_\nu) = I^+(\mu, \nu) - j_{\mu\nu}\,(\tau_\nu = \mathrm{T}_\nu)\,, \qquad (0 \leq \mu \leq 1) \tag{11.151}$$

and

$$\mu\left(\partial j_{\mu\nu}/\partial\tau_\nu\right)_{\mathrm{T}_\nu} = \mathrm{I}_\nu^+(\mu) - j_{\mu\nu}\left(\tau_\nu = \mathrm{T}_\nu\right). \tag{11.152}$$

In laboratory problems there may be *reflecting boundaries* at which $h_{\mu\nu}(\tau_\nu = 0) \equiv 0$, and/or $h_{\mu\nu}(\tau_\nu = \mathrm{T}_\nu) \equiv 0$, or boundaries with an *imposed net flux*, $h_{\mu\nu}(\tau_\nu = \mathrm{T}_\nu) \equiv f_{\mu\nu}$, where $f_{\mu\nu}$ is a given function, and similarly at the upper boundary. *One of the great strengths of the Feautrier formulation is the ease with which precise boundary conditions can be set.* Using (11.140) with (11.148) and (11.151) (or their imposed-flux counterparts) at the boundaries, we have a system that contains only $j_{\mu\nu}$ and known incident intensities I_ν^+ and I_ν^-. This system is a well-posed two-point boundary-value problem for $j_{\mu\nu}(\tau_\nu)$, $0 \le \tau_\nu \le \mathrm{T}_\nu$.

Finally, once the solution for $j_{\mu\nu}$ and $h_{\mu\nu}$ is obtained, we can compute

$$\int_0^1 j_{\mu\nu}\, d\mu = J_\nu = cE_\nu/4\pi, \tag{11.153a}$$

$$\int_0^1 h_{\mu\nu}\, \mu\, d\mu = H_\nu = F_\nu/4\pi, \tag{11.153b}$$

$$\int_0^\infty d\nu \int_0^1 j_{\mu\nu}\, d\mu = J = cE/4\pi, \tag{11.153c}$$

and

$$\int_0^\infty d\nu \int_0^1 h_{\mu\nu}\, \mu\, d\mu = H = F/4\pi. \tag{11.153d}$$

11.7 DISCRETIZATION

The discussion in § 11.5 and § 11.6 describes the calculation of moments of the radiation field *for a given source function*. But in § 11.4 we found that the source function in a transfer equation for the specific intensity, e.g., (11.89) and (11.92), may contain the mean intensity (the integral of the specific intensity over angle), or even an integral of the mean intensity over frequency, i.e., a *double integral* of the specific intensity over angle and frequency, e.g., (11.94). Hence in general the transfer equation can be an integro-partial-differential equation with a double integral containing the dependent variable. Solving such equations numerically requires efficient techniques.

Angle Quadratures

In solving diffusion problems G. Wick [1161] introduced the idea of converting integro-differential equations to *a set of coupled ordinary differential equations* by replacing the integrals with discrete *quadrature sums*:

$$\int_a^b f(x)\, dx \to \sum_{i=1}^{I} w_i f(x_i). \tag{11.154}$$

The distribution of the sample points x_m and the values of the quadrature weights w_i are chosen in such a way as to optimize the approximation.

As mentioned in chapter 1, Chandrasekhar and his associates, see, e.g., [209, 218, 220, 225, 227], used quadratures to discretize the angle integral in the gray atmosphere problem with set of *quadrature points* $\{\mu_m\}$ in the range $(-1,\ 1)$ and a set of *quadrature weights* $\{w_m\}$ such that

$$\int_{-1}^{1} I(\mu)\,d\mu \approx \sum_{m=-M}^{M} w_m I(\mu_m). \tag{11.155}$$

In these formulae the points and weights are chosen symmetrically about $\mu=0$; that is, $\mu_{-m}=-\mu_m$, and $w_{-m}=w_m$. If we set $I(\mu)\equiv 1$, we see that the weights must be normalized so that

$$\sum_{m=1}^{M} w_m = 1. \tag{11.156}$$

The accuracy of the quadrature formula depends both on the number of points and on their distribution on $[-1,\ 1]$. Chandrasekhar chose Gaussian quadratures in which the $2M$ points are the roots of Legendre polynomials of order $2M$ on the *open* interval $(-1, 1)$. They give exact results if $I(\mu)$ is a polynomial of order up to $2(2M-1)$. But at $\tau=0$, the outgoing radiation field $I^+(\mu)$ is large for $(0\le\mu\le 1)$, and the incoming field is $I^-(\mu)\equiv 0$ for $(-1\le\mu\le 0)$. Therefore, $I(0,\mu)$ is *discontinuous* at $\mu=0$. Fitting a single polynomial passing through the discontinuity at $\mu\to 0^+$ and $\mu\to 0^-$ diminishes the quadrature's accuracy.

Sykes [1067] suggested using a *double-Gauss quadrature*, in which we take M points to be the roots of the Legendre polynomial of order M, scaled from $[-1,1]$ to $[0,1]$ in the outgoing direction, and another M points scaled to $[-1,0]$ in the incoming direction. On each half-range the quadrature is exact for polynomials of order only up to $2M-1$. But because it avoids the discontinuity at $\mu=0$, it has superior accuracy and converges more quickly to the exact result as M increases, near the upper boundary ($\tau\to 0$). In the original method, $\mu=\pm 1/\sqrt{3}$ for $M=1$, which captures the exact boundary condition $J(0)\equiv\sqrt{3}H$. The same choice is made for a double-Gauss quadrature with $M=1$ instead of $\mu=\pm\frac{1}{2}$. Extensive sets of points and weights for Gaussian quadratures are given in [5, pp. 916–922].

Newton-Cotes formulae have equally spaced points on the *closed* intervals $[-1,0]$ and $[0,1]$. They also can be chosen so that each half-range of μ is spanned by a separate formula. If M points are used in each range, the quadrature is exact for polynomials of order up to $M-1$ if M is even, and of order M if it is odd. Because the positions of the points in a Gaussian quadrature can be optimized, their order of accuracy is formally higher than a Newton-Cotes quadrature. But for some problems a Newton-Cotes formula gives better results.

Another method is to employ a lower-order formula, e.g., Simpson's rule, repeatedly to cover each half-range. Although its formal accuracy is low, in some cases it can give better results because it allows us to put points where they are most needed. For example, in an extended envelope, the disk of a star of radius R, seen at a distance r, subtends an angle $\sim \mathsf{R}/r \ll 1$ radians, so quadrature points must be put near $\mu \approx 1 - \frac{1}{2}(\mathsf{R}/r)^2 \approx 1$. Here we could "tailor" a quadrature having endpoints exactly at ± 1 by using a low-order formula repeatedly, so that as many points are near $\mu = 1$ as are needed. Likewise, for a strongly radiating shock of thickness d propagating vertically in an atmosphere in which the vertical length scale ℓ corresponding to unit optical depth is $\gg d$, there must be quadrature points near $\mu \approx d/\ell \ll 1$.

Before fast computers were available, one tried to get maximum accuracy with a minimum number of points. Today, it can be preferable to use a uniform (or tailored) point distribution that will give a reliable answer even when the angular distribution of the intensity has "unusual" properties.

Frequency Quadratures

An atmosphere's spectrum contains broad expanses of continuum and large numbers of spectral lines with narrow profiles. In a smooth continuum it is usually satisfactory to use a simple formula such as Simpson's rule repeatedly. One must design the quadrature to handle steep gradients or abrupt breaks in the continuum (such as at an atomic absorption edge). A spectral line typically will have a very narrow core, e.g., a Doppler core, and extended wings. The core is important in fixing the line's radiative excitation/de-excitation rates in the kinetic equilibrium equations, whereas the wings are important for spectroscopic diagnostics. The frequency quadrature in the core can often be a simple trapezoidal rule, which happens to be surprisingly accurate in integrating a Gaussian profile. The accuracy of the quadrature can be verified by using more closely spaced sample points. But from a practical viewpoint extreme accuracy is probably unnecessary because it must be kept in mind that numerous other approximations have been made in virtually every aspect of the physics and numerics entering the calculation of a model.

In constructing a model atmosphere, having evaluated the radiation field at all depths and all frequencies in the star's spectrum, we must integrate over frequency in order to (1) check that the radiation field satisfies the constraint of radiative equilibrium; (2) calculate the radiation force on the material in the equation of hydrostatic equilibrium; (3) evaluate both the radiative excitation/de-excitation and ionization/recombination rates in the equations of kinetic equilibrium; and (4) compute the emergent radiation field. In practice, we almost always deal with double integrals, over angle and frequency, of the radiation field times a weighting function (e.g., a line profile), at all depths. These quadratures become nested sums of the products of the angle and frequency weights times the specific intensity (or $j_{\mu\nu}$ and $h_{\mu\nu}$ for Schuster's equation). For example, in a 1D medium they are of the form

$$\mathsf{M}^L(\tau_d) \approx \sum_{n=1}^{N} \sum_{m=1}^{M} w_m \mu_m^L f(\tau_d, \mu_m, \nu_n) I(\tau_d, \mu_m, \nu_n). \tag{11.157}$$

Depth Integrals

To compute the emergent intensity using (11.104), or J, H, and K using the Λ, Φ, and X operators in § 11.5, we have to integrate a source function known on a discrete grid $\{\tau_d\}$ or $\{m_d\}$, over depth. In those formulae, the best way to represent $I(\tau_d, \mu_m, \nu_n)$ is with a *cubic spline* [892, § 3.8], which gives a piecewise continuous and doubly differentiable approximation for I as a function of τ. A spline fit to a function f on a grid $\{\tau_d\}$ gives us sets of the four coefficients in the formulae $f(x) = c_0 + c_1 x + c_2 x^2 + c_3 x^3$ that apply on each interval $[\tau_d, \tau_{d+1}]$ between two successive grid points; here $x \equiv (\tau - \tau_d)$. Then the integrations over the appropriate kernels can be carried out analytically. The procedure described above is computationally expensive because it is costly to calculate the kernels. Differential equation methods that are more efficient will be described in § 12.4.

11.8 PROBABILISTIC INTERPRETATION

The specific intensity represents the statistical behavior of a large *ensemble* of photons. Another approach is to describe consecutive absorptions, emissions, and scatterings in probabilistic terms for a *single* photon. Here one follows the path of a photon from its place of origin until it is destroyed by an absorption process or escapes from the medium. The two views are complementary. In some cases the probabilistic view helps one to understand more easily some of the results obtained using the intensity-based statistical view.

Along a definite ray in a medium, let Δx be an increment of length measured from an arbitrary origin $x = 0$ located at optical depth τ. The corresponding optical depth increment (suppressing mention of frequency and angle) along the ray is $\Delta\tau \equiv \chi|\Delta x|$. Let the intensity at $x = 0$ be $I(0)$. From (11.98), the specific intensity at $\tau \pm \Delta\tau$ is $I(\tau) = I(0)e^{-\Delta\tau}$. Hence the probability that a photon is *not* absorbed in $\Delta\tau$ is

$$p_{\rm na}(\Delta\tau) = e^{-\Delta\tau}. \tag{11.158}$$

And the probability that it *is* absorbed between τ and $\tau \pm \Delta\tau$ is

$$p_{\rm a}(\Delta\tau) = 1 - e^{-\Delta\tau}. \tag{11.159}$$

If $\Delta\tau \ll 1$, then in (11.158) and (11.159), $e^{-\Delta\tau} = 1 - \Delta\tau + \ldots$, so that

$$p_{\rm a}(\Delta\tau) = \Delta\tau, \qquad \Delta\tau \ll 1. \tag{11.160}$$

Therefore, the *joint probability* that a photon emitted at τ travels an optical distance $\Delta\tau$ and then is absorbed within the interval $(\tau, \tau \pm \Delta\tau)$ is the product of the probabilities in (11.158) and (11.160):

$$p(\Delta\tau)\Delta\tau = e^{-\Delta\tau}\Delta\tau. \tag{11.161}$$

Consider now a plane-parallel atmosphere, in which photons travel in all directions. Let τ be the optical depth measured along the normal to the atmospheric layers; the corresponding optical depth along a ray with direction cosine μ relative to the normal is τ/μ. Then the probability that a photon emitted at $\tau = 0$, traveling with an arbitrary angle cosine μ, is absorbed in the optical depth range $(\tau, \tau + d\tau)$ measured in the normal direction is $p(\tau)\, d\tau = e^{-\tau/\mu} d\tau/\mu$. Defining $x \equiv 1/\mu$, and integrating over rays at all angles, we find

$$\bar{p}(\tau)\, d\tau = \int_0^1 \frac{e^{-\tau/\mu}}{\mu}\, d\mu\, d\tau = \int_1^\infty \frac{e^{-\tau x}}{x}\, dx\, d\tau \equiv E_1(\tau)\, d\tau. \tag{11.162}$$

With these results we can understand (11.108) probabilistically. Rewrite it, dropping the frequency subscript, as

$$J(\tau)\, d\tau = \int_\tau^\infty \tfrac{1}{2}[S(t)\, dt]\, [E_1(t-\tau)\, d\tau] + \int_0^\tau \tfrac{1}{2}[S(t)\, dt]\, [E_1(\tau - t)\, d\tau]. \tag{11.163}$$

The source function $S(t)\, dt$ in (11.163) is the number of photons created in the interval $(t, t+dt)$; hence $\frac{1}{2}S(t)\, dt$ is the number of photons created in that interval and emitted in one hemisphere (either toward or away from the surface). And $E_1(|t - \tau|)\, d\tau$ is the probability that an average photon travels an optical distance between t and τ and is then absorbed in the interval $(\tau, \tau + d\tau)$. Therefore, the product of these two terms is the number of photons emitted in $(t, t+dt)$ *and* absorbed in $(\tau, \tau + d\tau)$. The integral of this number over dt gives the total number of photons absorbed in $(\tau, \tau + d\tau)$, which in turn is $J(\tau)\, d\tau$.

11.9 DIFFUSION LIMIT

At great optical depths both the properties of the radiation field and the form of the transfer equation are simple, so we can obtain an asymptotic solution that applies throughout the interior of a star and provides a lower boundary condition for a stellar atmosphere. At depths much larger than a photon mean free path, the radiation is trapped, is very nearly isotropic, and approaches thermal equilibrium, i.e., $S_\nu \to B_\nu$. Choose a reference depth τ_ν and expand S_ν in a Taylor series:

$$S_\nu(t_\nu) = \sum_{n=0}^\infty \frac{d^n B_\nu}{d\tau_\nu^n} \frac{(t_\nu - \tau_\nu)^n}{n!}. \tag{11.164}$$

Substitution of this expression into the formal solution (11.100) gives

$$I_\nu(t_\nu, \mu) = \sum_{n=0}^\infty \mu^n \frac{d^n B_\nu}{d\tau_\nu^n} = B_\nu(\tau_\nu) + \mu \frac{dB_\nu}{d\tau_\nu} + \mu^2 \frac{d^2 B_\nu}{d\tau_\nu^2} + \ldots \tag{11.165}$$

for outgoing radiation, $0 \le \mu \le 1$. The same expression, differing only by terms $O(e^{-\tau_\nu/\mu})$, is obtained for incoming radiation, $-1 \le \mu \le 0$. In the limit $\tau_\nu \gg 1$ the exponential terms can be dropped and (11.165) used for the full range $-1 \leq \mu \leq 1$. Substituting this expression into the definitions of the angular moments, we obtain

$$J_\nu(\tau_\nu) = \sum_{n=0}^{\infty} \frac{1}{2n+1} \frac{d^{2n} B_\nu}{d\tau_\nu^{2n}} = B_\nu(\tau_\nu) + \frac{1}{3}\left(\frac{d^2 B_\nu}{d\tau_\nu^2}\right) + \ldots \tag{11.166}$$

$$H_\nu(\tau_\nu) = \sum_{n=0}^{\infty} \frac{1}{2n+3} \frac{d^{(2n+1)} B_\nu}{d\tau_\nu^{(2n+1)}} = \frac{1}{3}\left(\frac{dB_\nu}{d\tau_\nu}\right) + \frac{1}{5}\left(\frac{d^3 B_\nu}{d\tau_\nu^3}\right) + \ldots \tag{11.167}$$

and

$$K_\nu(\tau_\nu) = \sum_{n=0}^{\infty} \frac{1}{2n+3} \frac{d^{2n} B_\nu}{d\tau_\nu^{2n}} = \frac{1}{3} B_\nu(\tau_\nu) + \frac{1}{5}\left(\frac{d^2 B_\nu}{d\tau_\nu^2}\right) + \ldots. \tag{11.168}$$

Note that only even-order terms survive in the even moments J_ν and K_ν and only odd-order terms in H_ν.

How rapidly do these series converge in a star? To get order of magnitude estimates, consider frequency-integrated quantities; approximate derivatives by finite differences, e.g., $|d^n B/d\tau^n| \sim B/\Delta\tau^n$; and set B equal to its value at the depth Δz, where essentially all the Sun's luminosity has been generated by thermonuclear reactions in its core. The ratio of successive terms in the series is of the order $O(1/\Delta\tau^2)$, where $\Delta\tau$ is the frequency-averaged optical depth measured inward from the surface to that radius. In terms of a frequency- and depth-averaged photon mean free path λ_p, the convergence factor is $O(\lambda_p^2/\Delta z^2)$, which is $\ll 1$. Hence a photon must travel a huge number of mean free paths before it trickles to the surface and can fly into interstellar space. Thus radiative transfer at great optical depth is accurately described as a *diffusion process*.

For example, in the Sun, take $\Delta z \sim 0.7 R_\odot \approx 5 \times 10^{11}$ cm; at this depth in the solar envelope, $T \sim 7 \times 10^6$ K. From an inspection of stellar opacity tables, we find that $\lambda_p/\Delta z \ll 1$, i.e., $\Delta\tau \gg 1$. A quantitative estimate of $\Delta\tau$ can be obtained from the following argument: if $n = \Delta\tau$ is the number of mean free paths a photon must random walk before it "percolates" outward to the surface, then it will experience $O(n^2)$ absorptions and emissions. In free flight a photon moves with speed c, so its *free-flight time* to the surface is $t_f = n\lambda_p/c$. But because the process is a random walk, the *diffusion time* for it to reach the surface is $t_d = n^2\lambda_p/c$, so its *effective diffusion speed* outward is $v_d \sim c/n$.

We can estimate v_d, and hence n, and hence $\Delta\tau$, and hence the convergence factor $O(1/\Delta\tau^2)$ from the facts that the solar luminosity is $L_\odot = 4 \times 10^{33}$ erg/s, the solar radius $R_\odot = 7 \times 10^{10}$ cm, so the energy flux at $r = 0.3\, R_\odot$ is $F = L/4\pi r^2 \sim 10^{12}$ erg/cm^2/s. The radiant energy density at this depth is $E = a_R T^4 \sim 7 \times 10^{-15} \times 2.5 \times 10^{27} \sim 2 \times 10^{13}$ erg/cm^3.

The effective diffusion speed needed to transport this energy density and give the Sun's energy flux at its surface is $v_d \sim E/F \sim 20$ cm/s, which is minuscule compared to the speed of light, $c = 3 \times 10^{10}$ cm/s, and consistent with the physical characterization of diffusion given above. Therefore, the number of absorption/emission processes required for a photon to emerge at the Sun's surface is

$n = c/v_d \sim 1.5 \times 10^9$. Hence $\Delta\tau \sim 10^9$, and the convergence factor of the series above is of the order 10^{-18}!

Thus in the deep interior of a star only the leading terms are needed:

$$I_\nu(\tau_\nu, \mu) \approx B_\nu(\tau_\nu) + \mu(dB_\nu/d\tau_\nu), \tag{11.169}$$

$$J_\nu(\tau_\nu) \approx B_\nu(\tau_\nu), \tag{11.170}$$

$$H_\nu(\tau_\nu) \approx \tfrac{1}{3}(dB_\nu/d\tau_\nu), \tag{11.171}$$

$$K_\nu(\tau_\nu) \approx \tfrac{1}{3} B_\nu(\tau_\nu). \tag{11.172}$$

From (11.170) and (11.172) we see that, at great depth, (1) the mean intensity thermalizes, i.e., approaches the Planck function, (2) the radiation field is almost perfectly isotropic, so the Eddington factor $f_\nu^{\mathrm{K}} \to \frac{1}{3}$, and (3) the monochromatic flux is proportional to the derivative of the monochromatic Planck function with respect to optical depth. In this regime we have a purely local specification of the moments of the radiation field.

The Planck function is a function of (ν, T) only, so the flux can be written in terms of the temperature gradient:

$$H_\nu = \frac{1}{3}\frac{dB_\nu}{d\tau_\nu} = -\frac{1}{3\chi_\nu}\frac{dB_\nu}{dz} = -\left(\frac{1}{3\chi_\nu}\frac{\partial B_\nu}{\partial T}\right)\frac{dT}{dz}. \tag{11.173}$$

Integrating over frequency, we find that the *total radiation flux* in the diffusion approximation is

$$H = -\frac{1}{3\chi_R}\frac{dB}{dT}\frac{dT}{dz}, \tag{11.174}$$

where the average opacity

$$\frac{1}{\chi_R} \equiv \int_0^\infty \frac{1}{\chi_\nu}\frac{dB_\nu}{dT}\,d\nu \bigg/ \frac{dB}{dT} \tag{11.175}$$

is the *Rosseland mean opacity* [909]. *This mean opacity gives the correct total radiation flux at great optical depths in a star.* For this reason it is universally used in the theory of stellar interiors. The Rosseland opacity is a *harmonic mean*, i.e., it measures the average of $1/\chi_\nu$; wavelengths having the lowest monochromatic opacities, and hence the largest monochromatic flux make the largest contribution to the integral. In essence, χ_R is a measure of the effective bandwidth in the continuum that is able to transmit flux.

Combining (11.173) and (11.174), we get

$$H_\nu = \left(\frac{\chi_R}{\chi_\nu}\right)\left(\frac{\partial B_\nu/\partial T}{dB/dT}\right)H, \tag{11.176}$$

which is a very useful lower boundary condition in a stellar atmosphere.

The diffusion approximation is sometimes called the *radiation conduction approximation* because the form of (11.174) is similar to a material conduction equation, i.e., flux $\propto$ ($-\nabla$ source). The coefficient $(1/3\chi_R)(dB/dT)$ is sometimes called the *radiative conductivity*, which is appropriate because $1/\chi_R = \lambda_p$ is an average photon mean free path.

From a different viewpoint, (11.170) and (11.171) show that the ratio of the anisotropic to isotropic parts of the radiation field, i.e., the size of the leak compared to the reservoir, is

$$\frac{\text{Anisotropic term}}{\text{Isotropic term}} \sim \frac{3H}{B} = \left(\frac{\sigma_R T_{\text{eff}}^4}{4\pi}\right) \bigg/ \left(\frac{\sigma_R T^4}{\pi}\right) \sim \left(\frac{T_{\text{eff}}^4}{T^4}\right). \tag{11.177}$$

Thus the rate of energy leakage is very small when $T \gg T_{\text{eff}}$. For example, in the Sun this ratio is $\sim (6000/10^7)^4 \sim 10^{-13}$. Alternatively, the radiant energy content per unit volume is $(4\pi J/c) \approx (4\pi B/c)$, and the rate of radiant energy flow per unit volume by photons moving with speed c is (F/c). Again, (energy flow/energy content) $= (F/4B) \sim (T_{\text{eff}}/T)^4$.

One of the most important advantages of the Schuster-Feautrier method is that it allows accurate calculation of the net flux $|H_\nu|$ for all τ_ν, even when $|H_\nu| \ll |J_\nu|$; hence it guarantees a smooth transition to the diffusion limit.

Chapter Twelve

Direct Solution of the Transfer Equation

The solution of transfer problems with scattering terms is central to computing the radiation field emergent from a star. In a hot stellar atmosphere, scattering can be a major contributor to the opacity, can strongly affect energy balance, and can change the profiles of spectral lines used as diagnostic tools to determine a star's effective temperature, surface gravity, and composition. The way scattering enters a problem may be obvious, as when a source function contains explicit scattering terms, or it can be more subtle, as when the non-equilibrium occupation numbers of the energy levels of atoms, ions, and molecules (and hence the opacity and emissivity of the material) are determined by kinetic equations that contain dominant radiative transition rates, or when the atmosphere's structure (temperature and density distribution) is determined by the requirement of energy and momentum balance, using equations of hydrostatic and radiative equilibrium if it is static and dynamical equations if it is moving (e.g., expanding).

If the source function at a given wavelength has no scattering term, the specific intensity of the radiation field as a function of position, angle, and frequency can be calculated by direct integration with a formal solution of the transfer equation using one of the methods described in §12.4. But scattering terms contain integrals of the specific intensity over angle and/or frequency. We are then confronted with an impasse because the source function cannot be computed until these angle-frequency integrals are evaluated; yet the intensity, and hence these integrals, cannot be computed without knowing the source function.

In § 12.1 we explain qualitatively how scattering can couple the properties of the radiation field to source/sink terms within a huge volume in an atmosphere and analyze a plausible, but ineffective, iterative scheme, Λ-*iteration* (also called *source iteration* in neutron transport theory), for solving transfer equations with scattering terms. We show that in optically thick media a prohibitively large number of iterations is needed to obtain a solution of the transfer equation when the thermal source term is smaller than several percent of the total source function.

In § 12.2 and § 12.3 we describe two very effective methods to obtain solutions of transfer equations having dominant scattering terms: (1) *Feautrier's method* [323–325], a powerful algorithm that can handle very general scattering terms and that underlies the methods used to construct NLTE atmospheres, and (2) *Rybicki's method* [917], based on a clever reorganization of the Feautrier difference equations, which is very effective for solving line-formation problems that have simple scattering terms and also for determining both the radiation field and the temperature-density structure in LTE atmospheres. Both of these methods are formulated in terms of differential equations and treat radiative transfer as a two-point boundary value problem.[1] This formulation is advantageous, because it allows other physical requirements (kinetic equilibrium and energy/momentum-balance equations) to be coupled directly into the transfer equation; see, e.g., [53, 55, 56].

Although these algorithms give accurate self-consistent solutions of problems in which the state of the material and the radiation field are tightly coupled, in their original form they are often computationally too demanding to use in making model atmospheres that have, say, thousands of spectral lines. In chapter 13, we show that this problem can be overcome using *approximate transport operators*, which are orders of magnitude faster than direct solutions and can be iterated to fully converged solutions. A key component of these iterative methods is efficient calculation of formal solutions of the transfer equation, the topic of § 12.4. Closure of the angle-integrated moments of the difference equations in terms of variable Eddington factors is described in § 12.5.

12.1 THE PROBLEM OF SCATTERING

When scattering is important in the continuum, the source function is of the form of equation (11.88). If scattering in a spectral line is coherent, the source function has the form of equation (11.92). If the scattering process in the line redistributes photons over the entire line profile, it has the more complicated form of equation (11.94), or a yet more complicated form when the scattering process is described by a partial redistribution function such as those in chapters 10 and 15.

The scattering problem is ubiquitous in the study of stellar atmospheres. For example, in very hot atmospheres, the material is very strongly ionized, electron scattering is a major contributor to the continuum opacity, and the thermal coupling parameter ϵ may be $\lesssim 10^{-4}$ into large optical depths, until the density becomes large enough for free-free absorption to dominate. In cooler (near-solar T_{eff}) metal-poor stars, thermal absorption from H^- is dominated by Rayleigh scattering by H and H_2, so $\epsilon \ll 1$ to great depth, until H becomes excited and ionized, H^- forms, and ϵ ultimately $\rightarrow 1$. In very strong spectral lines ϵ can be $\sim 10^{-8}$.

Transfer Equation

To take a concrete example, consider a source function with an LTE thermal emission component and a coherent scattering term:

$$S_\nu = (1 - \epsilon_\nu)J_\nu + \epsilon_\nu B_\nu, \tag{12.1}$$

where the *thermal coupling parameter* (destruction probability) is

$$\epsilon_\nu \equiv \kappa_\nu / \chi_\nu = \kappa_\nu / (\kappa_\nu + \sigma_\nu), \tag{12.2}$$

[1] Classical integral equation methods are described in [39, 40, 41, 100, 264, 606, 609].

and $(1-\epsilon_\nu)$ is the *single scattering albedo*. In terms of the Λ-operator (§ 11.5), the solution of the transfer equation for the mean intensity is

$$J_\nu(\tau_\nu) = \Lambda_{\tau_\nu}[S_\nu] = \Lambda_{\tau_\nu}[(1-\epsilon_\nu)J_\nu] + \Lambda_{\tau_\nu}[\epsilon_\nu B_\nu], \tag{12.3}$$

or, rewriting (12.3) as an equation for $J_\nu - B_\nu$,

$$J_\nu(\tau_\nu) - B_\nu(\tau_\nu) = \Lambda_{\tau_\nu}[(1-\epsilon_\nu)(J_\nu - B_\nu)] + \Lambda_{\tau_\nu}[B_\nu] - B_\nu(\tau_\nu). \tag{12.4}$$

If there were no scattering term in the source function ($\epsilon_\nu \equiv 1$), the first term on the right-hand side of (12.3) would be zero. Then $J_\nu = \Lambda_{\tau_\nu}[B_\nu]$ can be calculated by simple integration. But when $\epsilon < 1$, (12.4) is an *integral equation* for J_ν, which is mathematically more difficult to solve.

Interaction Sphere and Thermalization Length

Scattering makes transfer problems more complex because it leads to long-range interactions between the radiation field and matter, and the source term in the transfer problem becomes nonlocal. For ease of discussion, assume that all material properties are independent of position in the atmosphere; as photons emerge from each scattering event, they are distributed isotropically; and they then travel a mean free path before interacting with the material again. If the scattering process is coherent, the opacity χ_ν of the material "seen" by a photon is the same before and after each scattering event, so its mean free path $\lambda_\nu = 1/\chi_\nu$ remains the same, and it travels in a classical random walk. If the destruction probability per scattering of a photon is ϵ_ν, the photon will have a high probability, i.e., approaching unity, of being destroyed after $N \sim O(1/\epsilon_\nu)$ interactions. Thus if ϵ_ν is very small, scattering allows photons to flow between regions in an atmosphere that are remote from one another without coupling to local conditions; therefore, it *delocalizes gas-radiation equilibration*.

From standard statistical arguments, see, e.g., [927, § 1.7], the radius L_ν of a photon's *interaction sphere*, i.e., mean distance it can random-walk in steps equal to its mean free path before being destroyed after N coherent scattering events, is its *thermalization length*,

$$L_\nu = \sqrt{N}\lambda_\nu = \lambda_\nu/\sqrt{\epsilon_\nu} = 1/\sqrt{\kappa_\nu(\kappa_\nu + \sigma_\nu)}, \tag{12.5}$$

Note that for coherent scattering, L_ν is the geometric average of the mean free path $1/(\kappa_\nu + \sigma_\nu)$ set by the total opacity and the mean free path $1/\kappa_\nu$ set by the thermal opacity.[2] Further, the *thermalization depth*, i.e., the monochromatic optical depth at which photons with frequency ν are destroyed, is

$$\tau_{\text{th}} \approx L_\nu/\lambda_\nu = (\lambda_\nu/\sqrt{\epsilon_\nu})/\lambda_\nu = 1/\sqrt{\epsilon_\nu}. \tag{12.6}$$

[2] We emphasize that the scaling rule $L_\nu \sim \lambda_\nu/\sqrt{\epsilon_\nu}$ applies *only* to coherent scattering. As will be shown in chapters 14 and 15, when the scattering process is noncoherent, the scaling may be $L_\nu \sim \lambda_\nu/\epsilon_\nu$, or even $L_\nu \sim \lambda_\nu/\epsilon_\nu^2$. In such cases the interaction sphere can be immense, and the problem described here is much worse.

The analysis above reveals four basic features of radiative transfer:

1. If there is an abrupt change in the material at some depth, it is "felt," i.e., information about it is transported by photons, through the entire interaction sphere, up to a distance $\sim L_\nu$.
2. Thus the presence of a boundary of the medium is "felt," and the radiation field modified, into an optical depth $\tau_{\text{th}} \approx 1/\sqrt{\epsilon_\nu}$, which may be much larger than a photon mean free path. If radiation is important in determining the physical state of the material (we will show that this is virtually always the case in the observable parts of a stellar atmosphere), then the physical state of the material is influenced into an optical depth τ_{th} from the boundary. In particular, if the medium is a finite slab, and its entire optical thickness is smaller than τ_{th}, it is *effectively thin*, i.e., *all* the material in the slab will respond to the presence of the boundary.
3. Suppose that at some stage of an iterative procedure the computed temperature structure of an atmosphere is not correct. Because scattering couples the radiation field throughout the whole interaction sphere, any local error in the energy balance of the material, and hence in the radiation field, is communicated over a distance L_ν (averaged over frequency). Thus to correct the temperature structure, we need an algorithm that propagates information about the response of the radiation field to the temperature structure, and vice versa, *throughout an entire interaction sphere*.
4. If we use an algorithm that propagates this information over only one mean free path in each iteration, *at least* $O(1/\sqrt{\epsilon_\nu})$ iterations will be needed. Not only is such an iteration process extremely inefficient, but it most likely will not converge in an acceptable number of repetitions.

Lambda Iteration

To solve the integral equation (12.3), we could represent J_ν, B_ν, and ϵ_ν with, say, spline approximations on a discrete grid of optical depths τ_j, as described in § 11.7. Analytically integrating the splines times the first exponential integral in the Λ-operator, we would generate a matrix equation of the form

$$\sum_{j=1}^{ND} \Lambda_{ij} J_j = b_i, \quad (i = 1, ND), \tag{12.7}$$

which can easily be solved with a computer. But the calculation of the matrix elements of Λ_{ij} and its inversion are costly, so before fast computers were available, Strömgren proposed solving (12.3) or (12.4) iteratively using a process known today as Λ-iteration. We shall see below that this method does not converge when scattering is the dominant term in the source function.

It is very important to understand the physical reason Λ-iteration fails to converge in order to motivate the development of the *direct methods* for solving the transfer equation, discussed in § 12.2–§ 12.5 below.

Suppose we have an initial estimate of $J_\nu^{(0)}$; a reasonable choice would be the Planck function, which becomes accurate deep in the atmosphere. Using $J_\nu^{(0)}$ in (12.3), we compute a new estimate $J_\nu^{(1)}$:

$$J_\nu^{(1)} = \Lambda_{\tau_\nu}\left[(1-\epsilon_\nu)J_\nu^{(0)}\right] + \Lambda_{\tau_\nu}[\epsilon_\nu B_\nu]. \tag{12.8}$$

Because both of the integrands on the right-hand side are known, (12.8) is no longer an integral equation, and we can evaluate $J_\nu^{(1)}$ by direct integration.[3] Repeating this procedure again and again, we have in general

$$J_\nu^{(n+1)} = \Lambda_{\tau_\nu}\left[(1-\epsilon_\nu)J_\nu^{(n)}\right] + \Lambda_{\tau_\nu}[\epsilon_\nu B_\nu]. \tag{12.9}$$

Alternatively, we may use the iteration process to determine the source function. Then from (12.3) we have

$$S_\nu^{(n+1)} = (1-\epsilon_\nu)\Lambda_{\tau_\nu}\left[S_\nu^{(n)}\right] + \epsilon_\nu B_\nu, \tag{12.10}$$

or

$$(S_\nu - B_\nu)^{(n+1)} = (1-\epsilon_\nu)\left\{\Lambda_{\tau_\nu}\left[(S_\nu - B_\nu)^{(n)}\right] + \Lambda_{\tau_\nu}[B_\nu] - B_\nu\right\}. \tag{12.11}$$

Again we can take $S_\nu^{(0)} = B_\nu$.

The iteration procedure in (12.9)–(12.11) is continued until a prechosen "convergence criterion"

$$\left\| \left(S_\nu^{(n)} - S_\nu^{(n-1)}\right) / S_\nu^{(n-1)} \right\| < e_n \tag{12.12}$$

is satisfied for $e_n \ll 1$. But even if this criterion is satisfied, there is no guarantee the solution is *accurate* to the fractional error e_n. See table 12.1.

In principle, it is not necessary to use an iteration method to solve the problem posed by equations (12.9)–(12.11). Because the equation is linear in the source function (or mean intensity), it could also be treated with an operator formalism. Thus the solution of equation (12.10) can be written symbolically as

$$S_\nu = \left[1 - (1-\epsilon_\nu)\Lambda_{\tau_\nu}\right]^{-1}(\epsilon_\nu B_\nu). \tag{12.13}$$

[3] In practice, (12.8) would be evaluated using one of the fast formal solutions in § 12.4. Note that $\Lambda_{\tau_\nu}[\epsilon_\nu B_\nu]$ is computed only once.

Table 12.1 Iteration History

	$\epsilon = 10^{-2}$		$\epsilon = 10^{-4}$		$\epsilon = 10^{-8}$	
n	e_n	E_n	e_n	E_n	e_n	E_n
1	7×10^{-1}	4	7×10^{-1}	49	7×10^{-1}	5000
10	4×10^{-2}	1	5×10^{-2}	17	5×10^{-2}	1800
10^2	1×10^{-3}	7×10^{-2}	5×10^{-3}	5	5×10^{-3}	560
10^3	1×10^{-8}	1×10^{-6}	4×10^{-4}	1	5×10^{-4}	180
10^4			1×10^{-5}	7×10^{-2}	5×10^{-5}	55
10^5			1×10^{-10}	1×10^{-6}	5×10^{-6}	17
10^6					5×10^{-7}	5
10^7					4×10^{-8}	1
10^8					1×10^{-9}	7×10^{-2}
10^9					1×10^{-14}	1×10^{-6}

In this formalism Λ-iteration is the result of expanding the operator (dropping the subscripts ν and τ_ν for brevity) as

$$[1-(1-\epsilon)\,\Lambda]^{-1} \approx 1+(1-\epsilon)\,\Lambda+(1-\epsilon)^2\,\Lambda^2+\cdots+(1-\epsilon)^n\,\Lambda^n+\ldots$$
$$=\sum_{i=0}^{\infty}\,[(1-\epsilon)\,\Lambda]^i. \tag{12.14}$$

Then the nth iterate of the source function S is

$$S^{(n)}=\sum_{i=0}^{n}\,(1-\epsilon)^i\,\Lambda^i\,[\,\epsilon B\,], \tag{12.15}$$

which is equivalent to (12.9) or (12.10).

If $(1-\epsilon_\nu) \ll 1$, the iteration procedures in (12.9)–(12.15) will converge, because successive corrections $\|S_\nu^{(n)}-S_\nu^{(n-1)}\|$ are $O(1-\epsilon_\nu)^n$ relative to B_ν. *But if* $(1-\epsilon_\nu) \approx 1$, *i.e., the destruction probability* $\epsilon_\nu \ll 1$ *over a large depth in the medium, the iteration will converge only very slowly or fail to converge altogether.* The symptom of failure is that the solution *stabilizes*: the fractional difference e_n between successive iterations is small, but the changes are *monotonic* and are nearly equal iteration after iteration. So even though $e_n \ll 1$, we do not know whether, say, $1/e_n$ more iterations may be required to reach the real solution. See figures 12.1–12.3 and table 12.1.

Another way to understand the failure of Λ-iteration is to consider the meaning of (12.15) probabilistically. The individual terms for $i = 0, 1, 2, \ldots$ can be interpreted as contributions to the source function from photons that have been scattered $0, 1, 2, \ldots$ times after their creation. If $\epsilon \ll 1$, there will be a large number of uninterrupted consecutive scatterings, so a correspondingly large number of iterations are needed.

We wrote the solution above in terms of integral equations containing the Λ-operator, but equivalent iterative solutions using differential equations have the

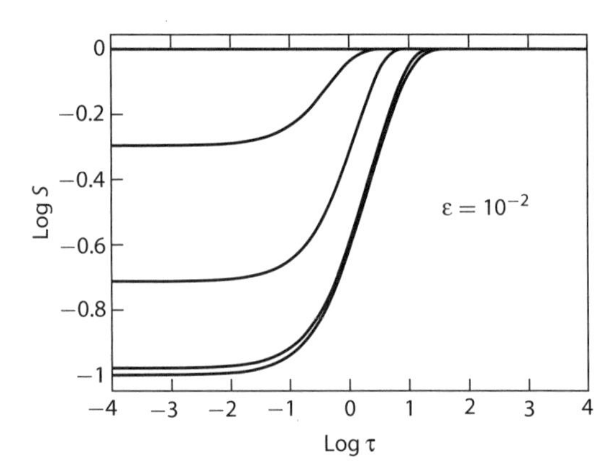

Figure 12.1 $S_n(\tau)$ for $\epsilon = 10^{-2}$ with $n = 1, 10, 100, 1000$ Λ-iterations.

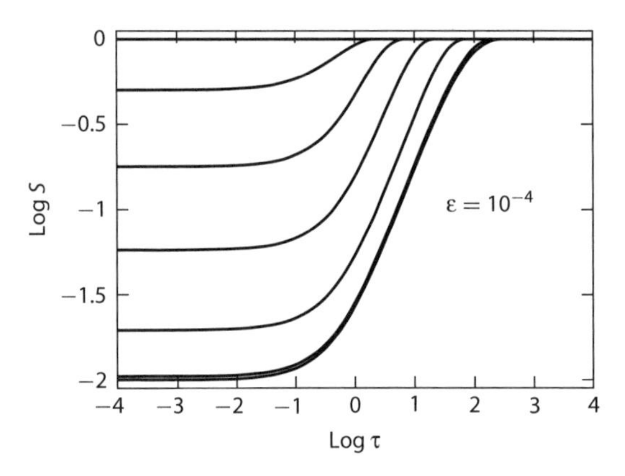

Figure 12.2 $S_n(\tau)$ for $\epsilon = 10^{-4}$ with $n = 1, 10, \ldots, 10^5$ Λ-iterations.

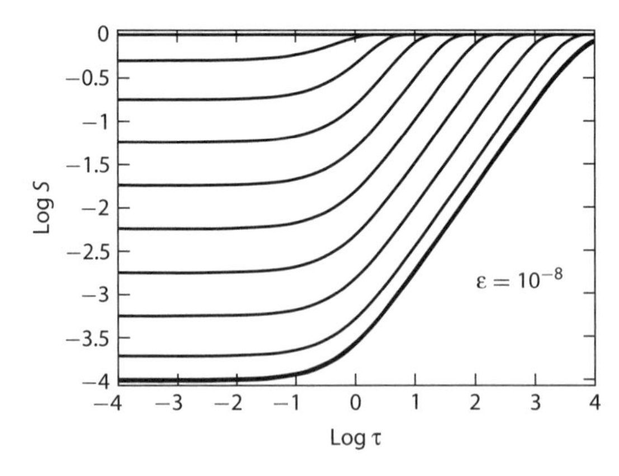

Figure 12.3 $S_n(\tau)$ for $\epsilon = 10^{-8}$ with $n = 1, 10, \ldots, 10^9$ Λ-iterations.

same problems. *We will call any iteration process that determines new values of the source function by using the previous iterate in a formal solution of the transfer equation a "Λ-iteration,"* even if the Λ-operator of equation (11.109) is not actually employed.

Coherent Scattering with a Linear Thermal Source

To demonstrate the failure of Λ-iteration quantitatively, let us apply it to a highly simplified scattering problem for which an analytical solution can be obtained to see how well that solution can be recovered using Λ-iteration. Assume isotropic coherent scattering, and for brevity drop the subscript ν; then $S = (1-\epsilon)J + \epsilon B$. Further, assume that ϵ is depth-independent.

The first two moments of the transfer equation, (11.65a) and (11.65b), are

$$\frac{\partial H}{\partial \tau} = J - S = \epsilon(J - B) \tag{12.16}$$

and

$$\frac{\partial K}{\partial \tau} = H. \tag{12.17}$$

Here is an example of the closure problem described in § 11.3: we have two equations containing three moments, J, H, and K. To find an additional relation among them we need to know the angular distribution of I.

One possibility is to use Schuster's *two-stream approximation* [979]. Suppose the specific intensity is essentially isotropic in the incoming and outgoing hemispheres, i.e., $I(\tau,\mu) = I^+(\tau)$, $0 \le \mu \le 1$ for outgoing radiation, and $I(\tau,\mu) = I^-(\tau)$, $-1 \le \mu \le 0$ for incoming radiation. Then

$$J = \tfrac{1}{2}\int_{-1}^{1} I(\mu)\,d\mu = \tfrac{1}{2}(I^+ + I^-), \tag{12.18}$$

$$H = \tfrac{1}{2}\int_{-1}^{1} I(\mu)\mu\,d\mu = \tfrac{1}{4}(I^+ - I^-), \tag{12.19}$$

$$K = \tfrac{1}{2}\int_{-1}^{1} I(\mu)\mu^2\,d\mu = \tfrac{1}{6}(I^+ + I^-) \equiv \tfrac{1}{3}J. \tag{12.20}$$

This relation between K and J is the same as that given by the diffusion approximation for $\tau \gg 1$, but now we apply it regardless of the values of I^+ and I^-, even at $\tau = 0$ where $I^- = 0$, so it is reasonable to set $K = \frac{1}{3}J$ at all depths. Then combining (12.17) and (12.16), we have

$$\frac{\partial^2 J}{\partial \tau^2} = 3\epsilon(J - B). \tag{12.21}$$

If we take a linear expansion $B(\tau) = a + b\,\tau$ for the Planck function, the second derivative of $B(\tau)$ is zero, so (12.21) can be rewritten as

$$\frac{\partial^2 (J - B)}{\partial \tau^2} = 3\epsilon(J - B), \tag{12.22}$$

which has the general solution

$$J(\tau) - B(\tau) = \alpha \exp\left(-\sqrt{3\epsilon}\tau\right) + \beta \exp\left(\sqrt{3\epsilon}\tau\right). \tag{12.23}$$

β must be $\equiv 0$ to avoid exponential divergence of $J(\tau)$ from $B(\tau)$ as $\tau \to \infty$.

To determine α, we need an independent relation between $J(0)$ and $H(0)$, which can be found by evaluating the integrals in (12.18) and (12.19). Assume $I^- \equiv 0$; then in Schuster's approximation $H(0) = \frac{1}{2}J(0)$. A more accurate boundary condition is obtained by replacing the angular integrals with quadrature formulae, i.e.,

$$\int_{-1}^{1} f(\mu)\, d\mu \longrightarrow \sum_{m=-M}^{M} w_m f(\mu_m).$$

The simplest choice is a one-point quadrature: $\mu_1 = \pm 1/\sqrt{3}$ and $w_1 = \frac{1}{2}$. Then

$$J(0) = \tfrac{1}{2}I^+ \quad \text{and} \quad H(0) = \tfrac{1}{2}\mu_1 I^+ \equiv J(0)/\sqrt{3}, \tag{12.24}$$

and hence

$$J(0)/\sqrt{3} = H(0) = (dK/d\tau)_0 = \tfrac{1}{3}(dJ/d\tau)_0. \tag{12.25}$$

From (12.23) and (12.25) we obtain

$$(a + \alpha)/\sqrt{3} = \tfrac{1}{3}(b - \sqrt{3\epsilon}\,\alpha). \tag{12.26}$$

Solving for α, and using it in (12.23), we find

$$J(\tau) = a + b\tau + \left[\frac{b - \sqrt{3}a}{\sqrt{3}(1 + \sqrt{\epsilon})}\right] \exp\left(-\sqrt{3\epsilon}\tau\right). \tag{12.27}$$

Equation (12.27) reveals the essential physics of the problem:

1. Suppose the atmosphere is isothermal. Then $b = 0$ and $B \equiv a$, so

$$J(0) = \sqrt{\epsilon}B/\left(1 + \sqrt{\epsilon}\right). \tag{12.28}$$

2. Thus if $\epsilon \ll 1$, as it is for many spectral lines, we find

$$J(0) = \sqrt{\epsilon}B. \tag{12.29}$$

   Therefore, *when $\epsilon \ll 1$, $J(0)$ is much smaller than B.*
3. In contrast, for a purely thermal source function (i.e., $\epsilon \equiv 1, B = \text{constant}$), $J(0) \equiv \frac{1}{2}B$ because a hemisphere with no incoming radiation is averaged with a hemisphere with isotropic outgoing radiation. In particular, the emergent intensity in a pure absorption line formed in an isothermal atmosphere would be the

same at all frequencies. To an observer outside that atmosphere, *the line would have disappeared.*

4. This large discrepancy is easily understood: *If the source function has a dominant scattering term, photons are scattered back into the atmosphere, reabsorbed, and destroyed, and the emergent intensity is depleted.*
5. The departure of J from B extends *very deep* into the atmosphere because of the slow decay of the exponential term. Thus (12.27) shows that $J(\tau) \to B(\tau)$, i.e., is thermalized, only at depths $\tau_{\text{th}} \gtrsim 1/\sqrt{3\epsilon}$. In real problems, ϵ may be very small, so τ_{th} can be very large.

The failure of Λ-iteration is shown by numerical examples in figures 12.1–12.3 for three small values of ϵ. The light curves in each panel are sequences of source functions $S_n = (1-\epsilon)J_n + \epsilon B$ obtained from Λ-iteration with a Feautrier formal solution (see § 12.4), for $n = 1, 10, 100, 1000, \ldots, 10^9$ iterations, starting with $S = B = 1$. The heavy curve at the bottom of these figures is the source function S_∞ obtained from a direct solution of the scattering problem using the Feautrier method (see § 12.2). We calculated S_∞ numerically instead of using the analytical formula (12.27) so that S_n and S_∞ have the same discretization errors.

Table 12.1 lists $e_n \equiv \|(S_n - S_{n-1})/S_{n-1}\|$, the maximum fractional *change* between successive Λ-iterations, and $E_n \equiv \|(S_n - S_\infty)/S_\infty\|$, the actual fractional *error* in S_n. The large values of E_n for small n when $\epsilon \ll 1$ reflect the fact that $S_\infty(0)$ is $O(\sqrt{\epsilon})$, whereas $S_1 \approx J_1$ is ~ 0.5. Note particularly that even though e_n may be very small, the real error E_n is always orders of magnitude larger.

Table 12.1 shows that in all cases $O(1/\epsilon)$ Λ-iterations are required to obtain a source function good even to several percent. To get a "fully converged" source function, an additional factor of 3 or so more iterations would be required. That amount of computation is prohibitive. The scaling as $1/\epsilon$ is easily understood. Information about the existence of a boundary must be propagated to the thermalization depth $\tau \approx 1/\sqrt{\epsilon}$. Because a Λ-iteration propagates information over $\Delta\tau \approx 1$, and the process is a strict random walk (coherent scattering), $n \approx \left(1/\sqrt{\epsilon}\right)^2 \sim O(1/\epsilon)$ iterations would be needed.

The basic conclusion is that any useful method must account for scattering terms in the source function from the outset and employ either a direct solution or an *effective* iterative solution when they are present.

12.2 FEAUTRIER'S METHOD

Transfer Equation

Feautrier's method is based on Schuster's second-order form of the transfer equation, § 11.6. We repeat its derivation here for convenience. For two oppositely directed rays,

$$\pm\mu \frac{\partial I(z, \pm\mu, \nu)}{\partial z} = \chi(z, \nu)[\,S(z, \mu, \nu) - I(z, \pm\mu, \nu)\,]. \tag{12.30}$$

On the half-range $0 \le \mu \le 1$, define the symmetric and antisymmetric averages

$$j(z,\mu,\nu) \equiv \tfrac{1}{2}[\, I(z,\mu,\nu) + I(z,-\mu,\nu)\,] \tag{12.31}$$

and

$$h(z,\mu,\nu) \equiv \tfrac{1}{2}[\, I(z,\mu,\nu) - I(z,-\mu,\nu)\,], \tag{12.32}$$

which have a mean-intensity-like and flux-like character. Adding (12.30) for $+\mu$ and $-\mu$, we find for the source function being an even function of μ, i.e., $S(z,\mu,\nu) = S(z,-\mu,\nu)$,

$$\mu \frac{\partial h(z,\mu,\nu)}{\partial z} = \chi(z,\nu)[\, S(z,\mu,\nu) - j(z,\mu,\nu)\,], \tag{12.33}$$

and subtracting them, we obtain

$$\mu \frac{\partial j(z,\mu,\nu)}{\partial z} = -\chi(z,\nu) h(z,\mu,\nu). \tag{12.34}$$

Substituting (12.34) for h into (12.33), we have

$$\frac{\mu^2}{\chi(z,\nu)} \frac{\partial}{\partial z}\left[\frac{1}{\chi(z,\nu)} \frac{\partial j(z,\mu,\nu)}{\partial z} \right] = j(z,\mu,\nu) - S(z,\mu,\nu). \tag{12.35}$$

Defining $d\tau(z,\nu) = -\chi(z,\nu)dz \equiv d\tau_\nu$ and abbreviating the notation, we rewrite (12.35) as

$$\mu^2 \frac{\partial^2 j_{\mu\nu}}{\partial \tau_\nu^2} = j_{\mu\nu} - S_{\mu\nu}, \tag{12.36}$$

which holds for scattering source functions of the form:

$$S_\nu = \alpha_\nu J_\nu + \beta_\nu, \tag{12.37a}$$

$$S_\nu = \alpha_\nu \int_0^\infty \phi_{\nu'} J_{\nu'}\, d\nu' + \beta_\nu, \tag{12.37b}$$

$$S_\nu = \alpha_\nu \int_0^\infty R(\nu', \nu) J_{\nu'}\, d\nu' + \beta_\nu, \tag{12.37c}$$

$$S_{\mu\nu} = \tfrac{3}{8}[\, (3J_\nu - K_\nu) + \mu^2 (3K_\nu - J_\nu)\,], \tag{12.37d}$$

i.e., isotropic, coherent, continuum scattering; complete redistribution in a spectral line; angle-averaged partial redistribution in a line; or anisotropic Thomson scattering. Other terms are required for angle-dependent redistribution, cf. chapter 15, or in moving material, cf. chapters 19 and 20. In (12.37), the α's are essentially scattering coefficients divided by the total opacity, and the β's are thermal terms. We stress that these equations are only illustrative. We will show later that those coefficients may depend on the radiation field in an entire transition array (for multi-level atoms) or even in the entire spectrum (when the constraint of radiative equilibrium must be satisfied).

Equations (12.33) and (12.34) give us two equations in the angle-dependent symmetric average $j_{\mu\nu}$ and antisymmetric average $h_{\mu\nu}$ of the radiation field. As described in § 11.4–§ 11.6, they allow us to handle the two-point boundary conditions of the problem, to make the transition to the diffusion approximation at great depth, and to calculate the flux at the lower boundary accurately; see § 11.9. When combined, they yield the single equation (12.35) in the angle-dependent symmetric average of the radiation field only. Its angle integral contains two moments: J_ν and K_ν. That moment equation is closed using variable Eddington factors, defined in § 11.3, which can be iterated to consistency.

Boundary Conditions

As discussed in § 11.4, boundary conditions are needed at $\tau = 0$ and at $\tau = \tau_{\max}$. The latter is either the optical thickness (or half-thickness) of a finite slab or a great depth in a semi-infinite atmosphere, where the diffusion approximation applies. For stellar atmospheres, one typically has $I(0, -\mu, \nu) = 0$ at $\tau = 0$, i.e., zero intensity incident from above,[4] which in the present notation implies that $h_{\mu\nu}(0) \equiv j_{\mu\nu}(0)$. Then (12.34) becomes

$$\mu \left.\frac{\partial j_{\mu\nu}}{\partial \tau_\nu}\right|_0 = j_{\mu\nu}(0). \tag{12.38}$$

At $\tau = \tau_{\max}$, we specify that $I(\tau_{\max}, +\mu, \nu) \equiv I^+(\mu, \nu)$, a given function of (μ, ν), and write $h_{\mu\nu}(\tau_{\max}) = I^+(\mu, \nu) - j_{\mu\nu}(\tau_{\max})$. Then

$$\mu \left.\frac{\partial j_{\mu\nu}}{\partial \tau_\nu}\right|_{\tau_{\max}} = I^+(\mu, \nu) - j_{\mu\nu}(\tau_{\max}). \tag{12.39}$$

In particular, if the diffusion approximation is valid at $\tau_{\max}$, then

$$I^+(\mu, \nu) = B_\nu(\tau_{\max}) + \mu \left.\frac{\partial B_\nu}{\partial \tau_\nu}\right|_{\tau_{\max}}. \tag{12.40}$$

Discretization

In astrophysics we encounter a wide variety of ordinary, partial, and integro-differential equations. In rare cases, an analytical solution is possible; usually it is not. These systems may be highly nonlinear, be coupled to other equations that make the whole system nonlinear, or contain higher transcendental functions whose properties cannot be expressed analytically. In this event, the problem must be solved numerically.

[4] Henceforth we shall assume $I(0, -\mu, \nu) = 0$. Generalization to nonzero incident intensity at the surface is straightforward; see § 17.7 and § 18.1.

The underlying method is to re-express *analytical operators* in terms of *algebraic operations*. To deal with nonlinearities in the resulting algebra, one needs a strongly convergent iteration procedure. Basic steps needed to make the transition from analysis to algebra are the following.

Geometric Grids

Digital computers are finite-state devices; they cannot deal with, or yield, continuous functions in the mathematical sense. Instead, a continuous function $f(x, y, z)$ is defined by its values $\{f(x_i, y_j, z_k)\}$ on a discrete grid of points $\{x_i\}$, $(i = 1, \ldots, I)$; $\{y_j\}$, $(j = 1, \ldots, J)$; $\{z_k\}$, $(k = 1, \ldots, K)$.

Interpolation

The value of a function $f(x, y, z)$ between these grid points must be determined by interpolation. The interpolating function must pass exactly through the known values on the grid points. Equally important, it must be smooth, i.e., not have spurious "bumps" and/or "wiggles" inside one of the domains $[(x_i, x_{i\pm1}), (y_j, y_{j\pm1}), (z_k, z_{k\pm1})]$, and it must be continuous and differentiable across the boundaries between domains.

Finite Differences

Derivatives are replaced by finite differences. For example, one might write $dy/dx = f(x, y)$ as $(y_i - y_{i-1})/(x_i - x_{i-1}) \approx g(x, y)$, where $g(x, y)$ is an appropriate approximation to $f(x, y)$ on the ranges $[\, x_{i-1},\, x_i \,]$ and $[\, y_{i-1},\, y_i \,]$. The discretization is not unique; e.g., we could take $g(x, y) \approx f(x_{i-1}, y_{i-1})$ or $g(x, y) \approx f(x_i, y_i)$, or use a multi-step method such as the *Runge–Kutta procedure*; see [5, p. 896].

Quadratures

As in chapter 11, definite integrals, over either angle or frequency, are represented by *quadrature sums*; then an integro-differential equation is decomposed into a coupled set of differential equations. An extensive set of quadrature formulae is given in [5, chapter 25].

Indexing

In stellar atmospheres, physical quantities depend on optical depth, angle, and frequency. To indicate this dependence, we use mnemonic subscripts d for the depth, m for the angle cosine μ, and n for the frequency ν. Thus for a variable v, we write $v(\tau_d, \mu_m, \nu_n)$ as v_{dmn}. For each choice of frequency ν_n and angle cosine μ_m, the transfer equation is differenced on a discrete set of optical depths $\{\tau_{dmn}\}$, $(d = 1, \ldots, ND)$, where ND is the number of depth points on the grid. τ is measured downward or inward into the atmosphere, so $\tau_{1n} < \tau_{2n} < \cdots < \tau_{ND,n}$.

The increments $\Delta\tau_{d+\frac{1}{2},n} \equiv \tau_{d+1,n} - \tau_{dn}$, $(d = 1, ND - 1)$ give the number of photon mean free paths between grid points. The source function is in general a double quadrature of the specific intensity over a set of angle points $\{\mu_m\}$, $(m = 1, \ldots, NA)$ and frequency points $\{\nu_n\}$, $(n = 1, \ldots, NF)$ for a line profile (spectral line formation) or the entire frequency spectrum (radiative equilibrium).

Depth Grids

As in § 11.7, let $\{z_d\}$, $(d = 1, \ldots, ND)$ be a set of discrete values of the geometric height measured vertically upward (or radially outward) in the atmosphere in a planar (or spherical) atmosphere, with $z_1 = z_{\max}$. The total *column mass*, starting with $m_1 = 0$, is

$$m_d = \sum_{i=1}^{d} \tfrac{1}{2}(\rho_{i-1} + \rho_i)(z_{i-1} - z_i), \quad (d = 2, \ldots, ND) \quad [\mathrm{gm/cm^2}] \tag{12.41}$$

measured downward (or inward) into the atmosphere. The *optical depth scale* $\{\tau_{dn}\}$ at frequency ν_n is calculated from $\{m_d\}$ as

$$\begin{aligned}\Delta\tau_{d+\frac{1}{2},n} &\equiv \tau_{d+1,n} - \tau_{dn} \\ &\equiv \tfrac{1}{2}\left[(\chi_{d+1,n}/\rho_{d+1}) + (\chi_{dn}/\rho_d)\right](m_{d+1} - m_d), \quad d = 1, \ldots, ND. \end{aligned} \tag{12.42}$$

The grid $\{m_d\}$ is always the fundamental depth coordinate used in stellar atmospheres computations because χ_{dn} can differ by orders of magnitude at different frequencies, e.g., in the core, wings, and adjacent continuum of a spectral line, and vary strongly as a function of depth at a given frequency. To deal with these variations, the depth scale is usually chosen to have equally spaced steps in $\{\log m_d\}$. The number of steps per decade, say, 5–10, should be large enough to assure adequate resolution of the $\{\tau_{dn}\}$ grids at all frequencies. For simple benchmark problems, e.g., to test the accuracy of a method, one might use twice as many depth points per decade.

In current model calculations, including line blanketing, where all frequencies are ultimately coupled by the constraint of radiative equilibrium, one may need to use $O(10^5)$ frequency points. Typically, the error made by using fewer depth points is much smaller than the error made by not having sufficient resolution in frequency. So one has to make trade-offs.

We will find in chapter 13 that for the fast iterative methods of solving the transfer equation, the rate of convergence is actually *inversely* proportional to the number of points per decade of optical depth. Of course, after a converged solution, satisfying simultaneously the physical constraints on the structure of the model (e.g., plane parallel in radiative and hydrostatic equilibrium or a spherically symmetric flow that satisfies the appropriate energy and momentum conservation equations) and the kinetic equations for all bound and free states in the material has been obtained, one can use a finer, interpolated depth grid to calculate the emergent radiation field at a given frequency with higher accuracy.

Algorithmic Reorganization

When differential equations are converted to difference equations and integrals to quadrature sums, it is legitimate to *interchange* the order of the operations equivalent to differentiation and integration. For example, in § 12.3 we show how the Feautrier algorithm can be "turned inside-out" to achieve greater computational efficiency for some problems.

Second-Order Feautrier Method

Internal Points

The differential equations (12.36) are converted to a set of difference equations by discretization of all variables. Thus: (1) choose a set of column masses $\{m_d\}$; (2) a set of angle cosines $\{\mu_m\}$; and (3) a set of frequencies $\{\nu_n\}$. (4) Group angles and frequencies into a single serial set of values, subscripted l $(l = 1, \ldots, NL \equiv NA \times NF)$, such that $(\mu_l, \nu_l) \equiv (\mu_m, \nu_n)$ for $l = m + (n-1)\,NA$. (5) For any variable f, write $f(m_d, \mu_m, \nu_n) = f_{dmn} \equiv f_{dl}$. (6) Write the derivatives in (12.33) as finite differences. (7) For any X, write

$$(dX/d\tau)_{d-\frac{1}{2}} \approx (\Delta X/\Delta\tau)_{d-\frac{1}{2}} \equiv (X_d - X_{d-1})/(\tau_d - \tau_{d-1}), \tag{12.43a}$$

$$(dX/d\tau)_{d+\frac{1}{2}} \approx (\Delta X/\Delta\tau)_{d+\frac{1}{2}} \equiv (X_{d+1} - X_d)(\tau_{d+1} - \tau_d), \tag{12.43b}$$

and

$$\left(\frac{d^2X}{d\tau^2}\right)_d \approx \frac{(dX/d\tau)_{d+\frac{1}{2}} - (dX/d\tau)_{d-\frac{1}{2}}}{\frac{1}{2}\left(\Delta\tau_{d+\frac{1}{2}} + \Delta\tau_{d-\frac{1}{2}}\right)}. \tag{12.43c}$$

Then defining

$$\Delta\tau_{dl} \equiv \tfrac{1}{2}\left(\Delta\tau_{d-\frac{1}{2},l} + \Delta\tau_{d+\frac{1}{2},l}\right), \tag{12.44}$$

(12.36) is converted to the difference equation:

$$\frac{\mu_l^2 j_{d-1,l}}{\Delta\tau_{d-\frac{1}{2},l}\Delta\tau_{dl}} - \frac{\mu_l^2 j_{dl}}{\Delta\tau_{d,l}}\left(\frac{1}{\Delta\tau_{d-\frac{1}{2},l}} + \frac{1}{\Delta\tau_{d+\frac{1}{2},l}}\right) + \frac{\mu_l^2 j_{d+1,l}}{\Delta\tau_{d+\frac{1}{2},l}\Delta\tau_{dl}} = j_{dl} - S_{dl},$$
$$(d = 2, \ldots, ND-1), \quad (l = 1, \ldots, NL). \tag{12.45}$$

Source Function

Replacing integrals in S by quadrature sums, we have

$$S_{dl} \equiv \alpha_{dl} \sum_{l'=1}^{NL} w_{dl'} \phi_{dl'} j_{dl'} + \beta_{dl}, \quad (l = l, \ldots, NL) \tag{12.46}$$

and

$$S_{dl} \equiv \alpha_{dl} \sum_{l'=1}^{NL} \mathcal{R}_{dl'l} j_{dl'} + \beta_{dl}, \quad (l = 1, \ldots, NL), \tag{12.47}$$

for (12.37b) and (12.37c), respectively. Here $\mathcal{R}_{dl'l}$ is a discrete representation of $R(\nu', \nu)$.

Notice that the source functions in (12.46) and (12.47) do not depend on angle; hence they contain redundant information, a fact that is exploited in Rybicki's method in § 12.3.

Boundary Conditions

For each $l = 1 \ldots NL$, the $(ND-2)$ equations (12.45) are supplemented by discretized versions of the boundary conditions (12.38) at the upper boundary and either (12.39) or (12.40) at the lower boundary. Thus at the upper boundary we might write

$$\mu_l (j_{2l} - j_{1l})/\Delta\tau_{\frac{3}{2},l} = j_{1l}. \tag{12.48}$$

But (12.48) is only first-order accurate because its right-hand side is evaluated at $d = 1$, and the left-hand side approximates $(dj/d\tau_{dl})$ at $d = \frac{3}{2}$. Auer [45] showed how the use of a Taylor's series expansion,

$$j_{2l} = j_{1l} + \Delta\tau_{\frac{3}{2},l} \left. \frac{dj_l}{d\tau_l} \right|_1 + \tfrac{1}{2} \left(\Delta\tau_{\frac{3}{2},l} \right)^2 \left. \frac{d^2 j_l}{d\tau_l^2} \right|_1, \tag{12.49}$$

removes this mismatch, and second-order accuracy is achieved. Then, using (12.36) for the second derivative in the last term, we obtain

$$\mu_l \left(j_{2l} - j_{1l} \right) / \Delta\tau_{\frac{3}{2},l} = j_{1l} + \Delta\tau_{\frac{3}{2},l} \left(j_{1l} - S_{1l} \right) / 2\mu_l. \tag{12.50}$$

The second-order terms in the boundary conditions improve the solution only if $\frac{1}{2}\Delta\tau/\mu \ll 1$. This requirement is easily met at the upper boundary where optically thin layers track the swift variation of j near the open boundary. At the lower boundary, because of the rapid convergence of $S_{dl} \to B_{dl}$ (see § 11.9), we can use first-order discretizations of (12.39) and (12.40) in a semi-infinite atmosphere, provided the optical depth at the lower boundary is large enough. For a finite slab, we can apply (12.39) at depth points $ND - 1$ and ND for small $\Delta\tau_{ND-\frac{1}{2}}$. For given values of the opacity and emissivity, or the coefficients and profile functions in source functions such as (12.37), equations (12.45) and its boundary conditions pose a *system of linear difference equations* for j_{dl}.

Two-Point Boundary Value Nature of the Problem

We wish to solve for I_{dmn} at all angles and frequencies, depth by depth. But there is an unresolvable problem: starting values for I_{mn} at the boundaries are required at *all* angle cosines and frequencies. As described above, the boundary conditions fall into two groups: $I^{-}_{\mu\nu}(0) = 0$ for incoming intensities in the range $-1 \le \mu \le 0$ at the upper boundary, and $I^{+}_{\mu\nu}(\tau_{\max}) = g(\mu, \nu)$ for outgoing rays in the range $0 \le \mu \le 1$ at the lower boundary; $g(\mu, \nu)$ is either a given function of (μ, ν) for a finite slab or set by the diffusion approximation in a semi-infinite atmosphere.

The problem is this: suppose we try to start the integration at $\tau = 0$ and proceed inward step by step. This cannot be done because the intensities $I^{+}_{\mu\nu}(0)$ on outgoing rays $0 \le \mu \le 1$, needed in order to evaluate the scattering term [the integrals on the right-hand sides of (12.37)] are not known. Similarly, at $\tau_{\max}$ the intensity on incoming rays, $I^{-}_{\mu\nu}(\tau_{\max})$, $-1 \le \mu \le 0$, are unknown.

We could attempt to overcome the problem by taking trial values for the missing intensities $I^{+}_{\mu\nu}(0)$ or $I^{-}_{\mu\nu}(\tau_{\max})$ and then integrating into the medium or toward its surface. If we had chosen trial values for the incoming intensity at $\tau_{\max}$, at the end of the integration we should find $I^{-}_{\mu\nu}(0)$, for $-1 \le \mu \le 0 \equiv 0$. *In general, this result will not be obtained.* One could try to adjust the trial values so as to satisfy the boundary condition at the end of the integration, but this *eigenvalue approach is exponentially unstable* [731, § 6.2] and can work only if the medium is nearly optically thin. The same problem arises if the integration were started at the surface: in general, the specific intensity field computed at the lower boundary would not match the diffusion limit. In short, we must face the two-point boundary-value nature of the problem from the outset.

Structure of the System

Define a set of column vectors $\mathbf{j}_d$ each of length *NL* that contain the angle-frequency components of the specific intensity at column-mass m_d. Then the discretized Feautrier's equations are a *block tridiagonal system* of the form

$$\mathbf{T}\cdot\mathbf{j} = \mathbf{R}, \tag{12.51}$$

or

$$-\mathbf{A}_d\,\mathbf{j}_{d-1} + \mathbf{B}_d\,\mathbf{j}_d - \mathbf{C}_d\,\mathbf{j}_{d+1} = \mathbf{R}_d \quad (d = 1, \ldots, ND). \tag{12.52}$$

$\mathbf{A}_d$, $\mathbf{B}_d$, and $\mathbf{C}_d$ are $(NL \times NL)$ matrices. $\mathbf{A}_d$ and $\mathbf{C}_d$ are *diagonal*, containing the finite-difference terms in (12.45) that couple the angle-frequency components of $\mathbf{j}_{ld}$ at depth point d to those at depth points $d-1$ and $d+1$, respectively.

$\mathbf{B}_d$ is an $(NL \times NL)$ *full* matrix. It has finite-difference terms coupling depth points $d-1$, d, and $d+1$ down its diagonal and off-diagonal terms from the quadrature sum coupling an angle-frequency component of the radiation field at depth point d to all other angle-frequencies in the discretized representation of the scattering integral in the source function at depth d.

$\mathbf{R}_d$ is a column vector of length *NL*, containing the thermal (i.e., non-scattering) source terms in (12.37).

The sign conventions in (12.52) and (12.53) are the result of multiplying both (12.45) and the discretized boundary conditions by -1, so that the elements of $\mathbf{A}$, $\mathbf{B}$, and $\mathbf{C}$ and the coefficients of S_{dl} on the right-hand side are all positive. Note that in this method, the scattering term can be of arbitrary complexity, e.g., as in equation (12.37c).

The grand matrix system is

$$\begin{pmatrix} \mathbf{B}_1 & -\mathbf{C}_1 & \mathbf{0} & \mathbf{0} & \cdots & \cdots & \mathbf{0} \\ -\mathbf{A}_2 & \mathbf{B}_2 & -\mathbf{C}_2 & \mathbf{0} & \cdots & \cdots & \mathbf{0} \\ \mathbf{0} & -\mathbf{A}_3 & \mathbf{B}_3 & -\mathbf{C}_3 & \cdots & \cdots & \mathbf{0} \\ \mathbf{0} & \mathbf{0} & \ddots & \ddots & \ddots & \cdots & \vdots \\ \vdots & \vdots & \vdots & \ddots & \ddots & \ddots & \mathbf{0} \\ \vdots & \vdots & \vdots & \cdots & -\mathbf{A}_{\mathrm{ND}-1} & \mathbf{B}_{\mathrm{ND}-1} & -\mathbf{C}_{\mathrm{ND}-1} \\ \mathbf{0} & \cdots & \cdots & \cdots & \mathbf{0} & -\mathbf{A}_{\mathrm{ND}} & \mathbf{B}_{\mathrm{ND}} \end{pmatrix} \begin{pmatrix} \mathbf{j}_1 \\ \mathbf{j}_2 \\ \mathbf{j}_3 \\ \vdots \\ \vdots \\ \mathbf{j}_{\mathrm{ND}-1} \\ \mathbf{j}_{\mathrm{ND}} \end{pmatrix} = \begin{pmatrix} \mathbf{R}_1 \\ \mathbf{R}_2 \\ \mathbf{R}_3 \\ \vdots \\ \vdots \\ \mathbf{R}_{\mathrm{ND}-1} \\ \mathbf{R}_{\mathrm{ND}} \end{pmatrix}. \tag{12.53}$$

At an interior point ($d = 2, \ldots, ND-1$), ($l = 1, \ldots, NL$), and ($k = 1, \ldots, NL$), and with the source function given by (12.46),

$$\mathbf{A}_{dlk} = \mu_l^2 / \left(\Delta\tau_{d-\frac{1}{2},l}\, \Delta\tau_{d,l}\right) \delta_{lk}, \tag{12.54}$$

$$\mathbf{C}_{dlk} = \mu_l^2 / \left(\Delta\tau_{d+\frac{1}{2},l}\, \Delta\tau_{d,l}\right) \delta_{lk}, \tag{12.55}$$

$$\mathbf{B}_{dlk} = \delta_{lk} + \mathbf{A}_{dlk} + \mathbf{C}_{dlk} - \alpha_{dl} w_{m_k} w_{n_k} \phi_{dk}. \tag{12.56}$$

Here δ_{lk} is the Kronecker symbol. n_k and m_k are the frequency and angle indices corresponding to the serial angle-frequency index k. We use this notation only for quadrature weights; we keep the global indices l and k for α and ϕ. The vector $\mathbf{R}_d = \beta_{dl}$ contains only thermal terms in the source function, which are independent of the radiation field.

At the upper boundary, $\mathbf{A}_{1lk} \equiv 0$,

$$\mathbf{B}_{1lk} = \left[1 + \left(2\mu_l / \Delta\tau_{\frac{3}{2},l}\right) + 2\left(\mu_l / \Delta\tau_{\frac{3}{2},l}\right)^2\right] \delta_{lk} - \alpha_{1l} w_{n_k} \phi_{1k} w_{m_k}, \tag{12.57}$$

$$\mathbf{C}_{1lk} = 2\left(\mu_l / \Delta\tau_{\frac{3}{2},l}\right)^2 \delta_{lk}, \tag{12.58}$$

and

$$\mathbf{R}_{1l} = \beta_{1l}. \tag{12.59}$$

At the lower boundary, $\mathbf{C}_{\mathrm{ND},lk} \equiv 0$,

$$\mathbf{B}_{\mathrm{ND},lk} = \left[1 + \left(2\mu_l / \Delta\tau_{\mathrm{ND}-\frac{1}{2},l}\right) + 2\left(\mu_l / \Delta\tau_{\mathrm{ND}-\frac{1}{2},l}\right)^2\right] \delta_{lk} - \alpha_{\mathrm{ND},k} w_{n_k} \phi_{\mathrm{ND},k} w_{m_k}, \tag{12.60}$$

$$\mathbf{A}_{\mathrm{ND},lk} = 2\left(\mu_l / \Delta\tau_{\mathrm{ND}-\frac{1}{2},l}\right)^2 \delta_{lk}, \tag{12.61}$$

and

$$\mathbf{R}_{\mathrm{ND},l} = \beta_{\mathrm{ND},l} + \left(2\mu_l / \Delta\tau_{\mathrm{ND}-\frac{1}{2},l}\right) I^+_{\mathrm{ND},l}. \tag{12.62}$$

Solution Algorithm

To solve the system, we use an efficient recursive forward-elimination and back-substitution procedure [323–325]. In this scheme, we express $\mathbf{j}_d$ in the dth line in terms of $\mathbf{j}_{d+1}$. Thus from the first line of (12.53) we have

$$\mathbf{j}_1 = \mathbf{B}_1^{-1}(\mathbf{C}_1\,\mathbf{j}_2 + \mathbf{R}_1) \equiv \mathbf{D}_1\,\mathbf{j}_2 + \mathbf{E}_1. \tag{12.63}$$

This expression is substituted in the second line of (12.53), to yield

$$\mathbf{j}_2 = (\mathbf{B}_2 - \mathbf{A}_2\mathbf{D}_1)^{-1}[\mathbf{C}_2\,\mathbf{j}_3 + (\mathbf{R}_2 + \mathbf{A}_2\mathbf{E}_1)] \equiv \mathbf{D}_2\,\mathbf{j}_3 + \mathbf{E}_2. \tag{12.64}$$

We proceed with this elimination scheme, line by line, obtaining in general

$$\mathbf{j}_d \equiv \mathbf{D}_d\,\mathbf{j}_{d+1} + \mathbf{E}_d, \quad d = 2, \ldots, ND - 1, \tag{12.65}$$

where

$$\mathbf{D}_d = (\mathbf{B}_d - \mathbf{A}_d\mathbf{D}_{d-1})^{-1}\mathbf{C}_d, \tag{12.66}$$

and

$$\mathbf{E}_d = (\mathbf{B}_d - \mathbf{A}_d\mathbf{D}_{d-1})^{-1}(\mathbf{R}_d + \mathbf{A}_d\mathbf{E}_{d-1}). \tag{12.67}$$

Here $\mathbf{D}_d$ and $\mathbf{E}_d$ are column vectors of length ND.

Starting at $d = 1$, we use (12.63) to compute $\mathbf{D}_1$ and $\mathbf{E}_1$. Then we use (12.66) and (12.67) to compute $\mathbf{D}_d$ and $\mathbf{E}_d$ for $(d = 2, \ldots, ND - 1)$. At $d = ND$, $\mathbf{C}_{ND} \equiv 0$, so $\mathbf{D}_{ND} \equiv 0$; hence $\mathbf{j}_{ND} = \mathbf{E}_{ND}$. Having found $\mathbf{j}_{ND}$, we make a reverse sweep with (12.65) to obtain $\mathbf{j}_d$, $(d = ND - 1, \ldots, 1)$.

The main difficulty in solving the system (12.53) is that the $(NL \times NL)$ matrix $(\mathbf{B}_d - \mathbf{A}_d\mathbf{D}_{d-1})$ must be inverted in order to evaluate (12.66) and (12.67). Because $\mathbf{A}_d$ is diagonal, only $O(NA^2 \cdot NF^2)$ operations are required to form $\mathbf{A}_d\mathbf{D}_{d-1}$ even though $\mathbf{D}_{d-1}$ is full. Hence most of the work goes into "inverting" the full matrix $\mathbf{B}_d$, which requires $O(NA^3 \cdot NF^3)$ operations. With a single-processor computer, it is a factor of 2 faster to compute the *LU decomposition* of $(\mathbf{B}_d - \mathbf{A}_d\mathbf{D}_{d-1})$ and use it to solve the set of equations with the column vectors of $\mathbf{C}_d$ as right-hand sides. With a massively parallel machine, the solutions for the set of right-hand sides can be carried out simultaneously.

$(\mathbf{R}_d + \mathbf{A}_d\mathbf{E}_{d-1})$ is a vector, so computing $\mathbf{E}_d$ from (12.67) also takes only $O(NA^2 \cdot NF^2)$ operations. Thus the time required by the Feautrier method scales as $t_F \propto c_F(ND \cdot NA^3 \cdot NF^3)$. This scaling is unfavorable in NA and NF, so the representation for these variables must be economized as much as possible. For the trivial case of a coherent and isotropic scattering problem, $NF = 1$ and NA need not be large, and the Feautrier method is nearly optimum. In other cases it is possible to reduce the number of angle points by using variable Eddington factors; see § 11.3.

Summary

In physical terms, the Feautrier method applies the upper boundary condition for any incoming intensity; then calculates the symmetric average of the intensity as a function of depth in terms of an as-yet-to-be-determined outgoing intensity; applies the lower boundary condition to fix the outgoing intensity at the deepest part of the medium; then works recursively backwards to obtain the average intensity at each depth point, using the now-known average intensity from the next deepest depth point.

Feautrier's method is stable and has desirable properties. At depth, the system becomes diagonal ($1/\Delta\tau^2 \to 0$), so $\mathbf{j}_d \to \mathbf{S}_d$. Indeed, $\mathbf{j}_d \to \mathbf{S}_d + \mu^2(d^2\mathbf{S}/d\tau^2)$, which recovers the diffusion limit. It is ideal for solving archetype problems that illustrate the physics of spectral line formation (chapters 14 and 15).

Additionally, it provides a basis for the computation of LTE and NLTE model atmospheres (chapters 17 and 18). The power of the method is far greater than shown above. However, in problems with very large numbers of frequencies, e.g., the computation of the complete NLTE spectrum of a star subject to the constraints of energy and momentum balance, even the direct solution of the transfer equation by the Feautrier method becomes prohibitively expensive, so we use the modern iterative methods presented in chapter 13 instead.

12.3 RYBICKI'S METHOD

The Feautrier method groups all angle-frequency information together at a given depth and then solves for the source function depth by depth. Using that method we can treat a fully *frequency-dependent* source function [e.g., equation (12.37c)] with a general redistribution function; but the computing time scales as the cube of the number of frequency points. Rybicki [917] pointed out that, in the commonly considered case of complete redistribution in spectral line formation, much of the frequency-dependent information is redundant, because in such cases we need only the single quantity $\overline{J} \equiv \int \phi_\nu J_\nu d\nu$ to specify the source function [equation (12.37b)]. He showed how the solution can be reorganized in this case to yield a system for $\overline{J}$ as a function of depth. For certain classes of problems this method has very favorable computing-time requirements.

Instead of describing the frequency variation of the specific intensity at a given depth, we can turn the overall structure of the problem "inside out" by reversing the grouping and working with a set of vectors that give the depth variation of the intensity at a given frequency-angle point. Thus, define a set of column vectors of length *ND*,

$$\mathbf{j}_l \equiv (j_{1l}, j_{2l}, \ldots, j_{\mathrm{ND},l})^{\mathrm{T}}, \tag{12.68}$$

where l denotes a particular angle-frequency point. Similarly, let

$$\overline{\mathbf{J}} \equiv (\overline{J}_1, \overline{J}_2, \ldots, \overline{J}_{\mathrm{ND}})^{\mathrm{T}}. \tag{12.69}$$

Then for angle-frequency point l, equations (12.52)–(12.62) are equivalent to the system

$$\mathbf{U}_l \cdot \mathbf{j}_l + \mathbf{V}_l \cdot \overline{\mathbf{J}} = \mathbf{E}_l, \quad (l = 1, \ldots, NL). \tag{12.70}$$

Here $\mathbf{U}_l$ is a $(ND \times ND)$ tridiagonal matrix representing the depth variation of the difference operator at angle-frequency l. $\mathbf{V}_l$ is a $(ND \times ND)$ diagonal matrix containing the depth variation of the effective scattering albedo $[\alpha_{dl}]$, and $\mathbf{E}_i$ is a column vector of length *ND* containing the depth variation of the thermal term $[\beta_{dl}]$ in (12.56)–(12.62).[5]

There is one set of equations (12.70) for each angle-frequency point. In addition, there are *ND* equations that define $\overline{J}_d$ as a function of depth, namely,

$$\sum_{l'=1}^{NL} w_{l'}\phi_{dl'}j_{dl'} - \overline{J}_d = 0, \quad (d = 1, \ldots, ND). \tag{12.71}$$

The overall structure of the grand matrix for all angle-frequencies and all depths is

$$\begin{pmatrix} \mathbf{U}_1 & \mathbf{0} & \mathbf{0} & \cdots & \mathbf{0} & \mathbf{0} & \mathbf{V}_1 \\ \mathbf{0} & \mathbf{U}_2 & \mathbf{0} & \cdots & \vdots & \vdots & \mathbf{V}_2 \\ \vdots & \mathbf{0} & \ddots & \cdots & \vdots & \vdots & \vdots \\ \vdots & \vdots & \mathbf{0} & \ddots & \mathbf{0} & \vdots & \vdots \\ \vdots & \vdots & \vdots & \cdots & \mathbf{U}_{NL-1} & \mathbf{0} & \mathbf{V}_{NL-1} \\ \mathbf{0} & \mathbf{0} & \mathbf{0} & \cdots & \mathbf{0} & \mathbf{U}_{NL} & \mathbf{V}_{NL} \\ \mathbf{X}_1 & \mathbf{X}_2 & \cdots & \cdots & \cdots & \mathbf{X}_{NL} & \mathbf{A} \end{pmatrix} \begin{pmatrix} \mathbf{j}_1 \\ \mathbf{j}_2 \\ \vdots \\ \vdots \\ \mathbf{j}_{NL-1} \\ \mathbf{j}_{NL} \\ \overline{\mathbf{J}} \end{pmatrix} = \begin{pmatrix} \mathbf{E}_1 \\ \mathbf{E}_2 \\ \vdots \\ \vdots \\ \mathbf{E}_{NL-1} \\ \mathbf{E}_{NL} \\ \mathbf{F} \end{pmatrix}. \tag{12.72}$$

Here $\mathbf{X}_l$ is an $(ND \times ND)$ diagonal matrix containing the depth variation of the quadrature weights and profiles at frequency ν_l in the definition of $\overline{\mathbf{J}}$ in (12.71). If we are solving the line-formation problem (12.72) for a single line, $\mathbf{A}$ would be $-\mathbf{I}$, the identity matrix, and $\mathbf{F}$ would be void. But as we will see in chapter 17, more general systems of the form of (12.72) arise in the computation of LTE model atmospheres. Comparison of (12.72) and (12.52) shows that *the inner and outer structure of the system of equations has been interchanged.*

The solution of system (12.72) is quite efficient. We reduce each line of the grand matrix by calculating

$$\mathbf{j}_l = (\mathbf{U}_l^{-1}\mathbf{E}_l) - (\mathbf{U}_l^{-1}\mathbf{V}_l)\,\overline{\mathbf{J}}, \quad (l = 1, \ldots, NL). \tag{12.73}$$

The inversion of $\mathbf{U}_l$ in (12.73) is done with *LU* decomposition and solutions of the linear equations having the column vectors $\mathbf{E}_l$ or $\mathbf{V}_l$ on the right-hand side.[6] Then, multiplying (12.73) by $\mathbf{X}_i$ and subtracting the result from the last line, we cancel

[5] Note that here the matrix elements of $\mathbf{U}_l$ are the same as in the second-order Feautrier scheme at angle-frequency point (μ_l, ν_l). But they could also be the matrix elements of the fourth-order scheme, along with the Rybicki-Hummer modification of the elimination scheme, described in § 12.4.

[6] This procedure is not time-consuming because in the Rybicki method the $\mathbf{U}$ matrices are only tridiagonal and the $\mathbf{V}$ matrices are diagonal.

the "elements" in its lth column to zero. Using this procedure for all values of l, the final system for $\bar{\mathbf{J}}$ is

$$\mathbf{P}\bar{\mathbf{J}} = \mathbf{L}, \tag{12.74}$$

where the full $(ND \times ND)$ matrix $\mathbf{P}$ is

$$\mathbf{P} = \mathbf{A} - \sum_{l=1}^{NL} \mathbf{X}_l(\mathbf{U}_l^{-1}\mathbf{V}_l), \tag{12.75}$$

and $\mathbf{L}$, a vector of length ND, is

$$\mathbf{L} = \mathbf{F} - \sum_{l=1}^{NL} \mathbf{X}_l(\mathbf{U}_l^{-1}\mathbf{E}_l). \tag{12.76}$$

We then obtain $\bar{\mathbf{J}}$ from (12.74) and can calculate $\mathbf{S}$, a vector of length ND for the depth variation of the source function. Then, if desired, the full angle-frequency variation of the specific intensity can be reconstructed using the already-available quantities $(\mathbf{U}_l^{-1}\mathbf{V}_l)$ and $(\mathbf{U}_l^{-1}\mathbf{E}_l)$ in (12.73).

Solving the NL tridiagonal systems in (12.73) requires $O(ND^2 \cdot NL)$ operations, and the final system (12.74) requires $O(ND^3)$ operations, so the total computing time scales as $t_R \propto c_R(ND^2 \cdot NL) + c_R'(ND^3)$. Unlike the Feautrier system, in which the computing time scales as the cube of the number of angle-frequency points (i.e., $ND \cdot NL^3$), *the Rybicki algorithm is linear in NL*. Thus for the calculation of the source function of a line formed with complete redistribution, the Rybicki algorithm is more economical than Feautrier's when a large number of angle-frequency points is required.

The Rybicki algorithm becomes more costly if more than a single quantity like $\bar{\mathbf{J}}$ is required to solve the problem. For example, if we solve for the *coupled* source functions in two spectral lines, then we need *two* quantities, $\bar{\mathbf{J}}_1$ and $\bar{\mathbf{J}}_2$. In this case the solution scales as $t_R = c'(ND^2 \cdot NL) + c''(2ND)^3$; i.e., eight times as much computation is needed. Nonetheless, as will be seen in § 17.3, the Rybicki algorithm is optimum for the construction of full-spectrum LTE model atmospheres. Even though we must find two depth-dependent quantities, namely, corrections to the current estimates of the temperature and density distributions in the model, NL is very large, so the Feautrier algorithm is much more costly.

The Rybicki algorithm is equivalent to the integral equation method in which one writes $\mathbf{j}_l = \boldsymbol{\Lambda}_l\bar{\mathbf{J}} + \mathbf{M}_l$, where the $\boldsymbol{\Lambda}_l$ matrix is generated by analytical integration of the kernel function against a set of basis functions representing the depth variation of $\bar{\mathbf{J}}$. In fact, $\mathbf{U}_l^{-1}$ from (12.73) *is* the $\boldsymbol{\Lambda}_l$ matrix. But by using the Rybicki algorithm to generate it, we can achieve higher accuracy (if fourth-order terms are included), the boundary conditions are more easily handled, and "inversion" of the tridiagonal matrix $\mathbf{U}_l$ is much less costly than any other way of generating $\boldsymbol{\Lambda}_l$.

12.4 FORMAL SOLUTION

In this section we discuss efficient methods for computing the full depth-angle-frequency dependence of the specific intensity, *given the total source function $S_{\mu\nu}$, including scattering terms*. Formal solutions are only a subset of the calculations treated in § 12.2 and § 12.3. Indeed, in work done before about 1986, they were almost an afterthought used to evaluate the emergent radiation field from a model atmosphere. But they now have an important role in the extremely fast iterative methods (see chapter 13) developed to solve the most general sets of equations used in the construction of NLTE model atmospheres including thousands of lines, subject to constraints such as radiative and hydrostatic equilibrium, or with velocity fields.

Use of equations (11.104), (11.108), (11.121), or (11.122) for a formal solution is inefficient because the kernel functions are relatively costly to compute. But current methods are much faster and more flexible in application, and we will see later that *speedy formal-solution algorithms are critical in the most complex modeling applications because they can reduce the computation time by orders of magnitude, making it possible to solve problems that would otherwise be intractable.*

We will discuss two types of formal solutions: (1) a numerical scheme using an improved version of the Feautrier method described in § 12.2, and (2) methods that solve the first-order form of the transfer equation, using (a) *integrals on a set of discrete ray segments* or (b) the *discontinuous finite-element method*. In astrophysics, methods of type (2a) are commonly called *short-characteristics methods*.

At the end of this section we give a comparison of some of the advantages and disadvantages of each method.

Scalar Feautrier Method

The Feautrier method described in § 12.2 allows great generality in the form of the source function; the price paid for it is considerable matrix algebra. But to calculate the specific intensity for a single angle and frequency with a *given* source function, the algorithm uses only *scalars* (i.e., simple numbers), so it is linear in the number of depths, angles, and frequencies. Hence the time required to calculate radiation field at *ND* depths, for *NA* angles, and for *NF* frequencies scales as $t_F \propto c'_F\ (ND \cdot NA \cdot NF)$, which is minimal for the amount of information to be found.

The second-order accurate method has been extended to have fourth-order accuracy, and the elimination scheme modified to overcome problems arising from the finite length of a computer word when there are very small optical depth intervals at frequencies where the opacity is very small.

Fourth-Order Accurate Algorithm

The difference formula (12.45) of the original Feautrier algorithm makes use of the analytical second derivative of j_l in (12.36) only at depth point d and is second-order accurate. Auer [48] showed that its accuracy can be increased to fourth order by using a Hermite formula constructed so as to include information about the second derivative of j_l at $d \pm 1$ in the difference scheme. We assume the source function at angle cosine μ and frequency ν, denoted with the serial number l as described in § 12.2, is given.

Define the new variable $\tilde{\tau} \equiv \tau/\mu$; then $j' \equiv (dj/d\tilde{\tau}) = \mu\,(dj/d\tau)$, and $j'' \equiv (d^2j/d\tilde{\tau}^2) = \mu^2(d^2j/d\tau^2)$. Then the transfer equation (12.36) at depth point d is

$$j_d'' = j_d - S_d. \tag{12.77}$$

To improve accuracy, replace the original finite difference scheme in (12.45) with the more general form

$$-A_d\, j_{d-1} + B_d\, j_d - C_d\, j_{d+1} + \alpha_d\, j_{d-1}'' + \beta_d\, j_d'' + \gamma_d\, j_{d+1}'' = S_d - j_d. \tag{12.78}$$

The notation and signs in (12.78) have been chosen to parallel those in (12.52). The coefficients $(A_d, \ldots, C_d, \alpha_d, \beta_d, \gamma_d)$ are found by expanding $j(\tilde{\tau})$ and $j''(\tilde{\tau})$ to fourth order in Taylor series on the interval $\tilde{\tau}_{d-1} < \tilde{\tau} < \tilde{\tau}_{d+1}$:

$$j_{d\pm1} = j_d \pm j_d'\,\Delta\tilde{\tau}_{d\pm\frac{1}{2}} + \tfrac{1}{2}j_d''\,\Delta\tilde{\tau}_{d\pm\frac{1}{2}}^2 \pm \tfrac{1}{6}j_d'''\,\Delta\tilde{\tau}_{d\pm\frac{1}{2}}^3 + \tfrac{1}{24}j_d^{iv}\,\Delta\tilde{\tau}_{d\pm\frac{1}{2}}^4 \cdots, \tag{12.79}$$

and

$$j_{d\pm1}'' = j_d'' \pm j_d'''\,\Delta\tilde{\tau}_{d\pm\frac{1}{2}} + \tfrac{1}{2}\,j_d^{iv}\,\Delta\tilde{\tau}_{d\pm\frac{1}{2}}^2 \cdots, \tag{12.80}$$

where $\Delta\tilde{\tau}_{d-\frac{1}{2}} \equiv (\tilde{\tau}_d - \tilde{\tau}_{d-1})$ and $\Delta\tilde{\tau}_{d+\frac{1}{2}} \equiv (\tilde{\tau}_{d+1} - \tilde{\tau}_d)$. Then substituting (12.79) and (12.80) into (12.78), collecting the terms multiplying the same derivative of j, and dropping the subscript d on $A, B, C, \alpha, \beta, \gamma$, and j, we have

$$\begin{aligned}
&j\,[-A + (B-1) - C] \\
&+ j'\left[A\,\Delta\tilde{\tau}_{d-\frac{1}{2}} - C\,\Delta\tilde{\tau}_{d+\frac{1}{2}}\right] \\
&+ j''\left[-\tfrac{1}{2}\left(A\,\Delta\tilde{\tau}_{d-\frac{1}{2}}^2 + C\,\Delta\tilde{\tau}_{d+\frac{1}{2}}^2\right) + \alpha + \beta + \gamma\right] \\
&+ j'''\left[\tfrac{1}{6}\left(A\,\Delta\tilde{\tau}_{d-\frac{1}{2}}^3 - C\,\Delta\tilde{\tau}_{d+\frac{1}{2}}^3\right) - \alpha\,\Delta\tilde{\tau}_{d-\frac{1}{2}} + \gamma\,\Delta\tilde{\tau}_{d+\frac{1}{2}}\right] \\
&+ j^{iv}\left[-\tfrac{1}{24}\left(A\,\Delta\tilde{\tau}_{d-\frac{1}{2}}^4 + C\,\Delta\tilde{\tau}_{d+\frac{1}{2}}^4\right) + \tfrac{1}{2}\left(\alpha\,\Delta\tilde{\tau}_{d-\frac{1}{2}}^2 + \gamma\,\Delta\tilde{\tau}_{d+\frac{1}{2}}^2\right)\right] = 0.
\end{aligned} \tag{12.81}$$

The coefficients in (12.81) are determined only to within an arbitrary multiplicative constant, so we can find the coefficients $A, B, C, \alpha, \beta, \gamma$ by setting each term in a square bracket in (12.81) identically to zero. Solving the resulting set of equations, we find

$$A_d \equiv \mu^2/\left[\tfrac{1}{2}(\Delta\tau_{d-\frac{1}{2}} + \Delta\tau_{d+\frac{1}{2}})\Delta\tau_{d-\frac{1}{2}}\right], \tag{12.82}$$

$$C_d \equiv \mu^2/\left[\tfrac{1}{2}(\Delta\tau_{d-\frac{1}{2}} + \Delta\tau_{d+\frac{1}{2}})\Delta\tau_{d+\frac{1}{2}}\right], \tag{12.83}$$

$$B_d \equiv 1 + A_d + C_d, \tag{12.84}$$

$$\alpha_d \equiv \tfrac{1}{6}\left(1 - \tfrac{1}{2}A_d\,\Delta\tau_{d+\frac{1}{2}}^2/\mu^2\right), \tag{12.85}$$

$$\gamma_d \equiv \tfrac{1}{6}\left(1 - \tfrac{1}{2}C_d\,\Delta\tau_{d-\frac{1}{2}}^2/\mu^2\right), \tag{12.86}$$

$$\beta_d \equiv 1 - \alpha_d - \gamma_d. \tag{12.87}$$

The resulting equation for the fourth-order scheme is

$$(-A_d + \alpha_d) j_{d-1} + (B_d + \beta_d) j_d + (-C_d + \gamma_d) j_{d+1}$$
$$= \alpha_d S_{d-1} + \beta_d S_d + \gamma_d S_{d+1}. \tag{12.88}$$

Taking account of the higher-order terms changes the coefficients of j in the difference formula and also introduces information about the source terms at $d-1$ and $d+1$.

Fourth-order boundary conditions spoil the tridiagonal form of the system, so we accept third order. The second-order upper boundary condition (12.50) is of the generic form $a j_1 + b j_2 + c j'_1 + d j''_1 = 0$. To extend it to third order we add information about the source term at $d = 2$. Applying the same technique used to obtain (12.82)–(12.87), we find

$$\mu (j_1 - j_2)/\Delta\tau_{\frac{3}{2}} + j_1 + \left(\tfrac{1}{3}\Delta\tau_{\frac{3}{2}}/\mu\right)(j_1 - S_1) + \left(\tfrac{1}{6}\Delta\tau_{\frac{3}{2}}/\mu\right)(j_2 - S_2) = 0. \tag{12.89}$$

For small $\Delta\tau_{\mathrm{ND}-\frac{1}{2}}$ the third-order lower boundary condition is

$$\mu (j_{\mathrm{ND}} - j_{\mathrm{ND}-1})/\Delta\tau_{\mathrm{ND}-\frac{1}{2}} + (j_{\mathrm{ND}} - I^+) + \left(\tfrac{1}{6}\Delta\tau_{\mathrm{ND}-\frac{1}{2}}/\mu\right)(j_{\mathrm{ND}-1} - S_{\mathrm{ND}-1})$$
$$+ \left(\tfrac{1}{3}\Delta\tau_{\mathrm{ND}-\frac{1}{2}}/\mu\right)(j_{\mathrm{D}} - S_{\mathrm{D}}) = 0. \tag{12.90}$$

For a semi-infinite atmosphere, one might use the diffusion approximation instead of (12.90).

In matrix notation,

$$\mathbf{T}_{\mathrm{H}} \cdot \mathbf{j} = \mathbf{H} \cdot \mathbf{S}. \tag{12.91}$$

$\mathbf{T}_{\mathrm{H}}$ and $\mathbf{H}$ are tridiagonal matrices (the subscript denotes “Hermite”) with elements $A_d, B_d, C_d, \alpha_d, \beta_d, \gamma_d$ defined in equations (12.82)–(12.90). The computer time required to generate the additional elements of $\mathbf{T}_{\mathrm{H}}$ and $\mathbf{H}$ and to solve the system with a tridiagonal right hand side is negligible. The solution proceeds in essentially the same way as described in § 12.2. The difference is that we are now dealing with ordinary numbers instead of matrices and vectors, so the time required to obtain a full formal solution at all depths, angles, and frequencies scales as only $t_H \propto c_H\,(ND \cdot NA \cdot NF)$. Auer [48] showed that the Hermitian system is more accurate and more stable than all other formal-solution methods and generalized it to the case where the source function contains a scattering term.

Improved Rybicki-Hummer Solution Algorithm

Rybicki and Hummer [925] developed a variant of the Feautrier solution that gives higher numerical accuracy when optical depth increments in some region (specifically near the surface) become very small. A problem arises because of a mismatch between the *precision* of (i.e., number of significant figures in) a computer's word (roughly 13 s.f. for a 64-bit word) and its *exponent range*. If $\Delta\tau$ is quite small, in (12.56) the term in B_d is proportional to $(\Delta\tau)^{-2}$ and can be computed without difficulty. But the result, if sufficiently large, will cause the "one" from the Kroneker delta to be lost off the end of the machine word. When this happens, the solution of the *difference equation* no longer has the same mathematical properties as the solution of the *differential equation*, and serious errors can occur.

The cure for this problem is to change the algorithm used for B_d and D_d in (12.65) by taking linear combinations of the original coefficients in such a way as to preserve the unit term on the diagonal. Thus define

$$H_d \equiv -A_d + B_d - C_d, \quad H_1 \equiv B_1 - C_1, \quad H_{\rm ND} \equiv -A_{\rm ND} + B_{\rm ND}. \tag{12.92}$$

The cancellations in H are to be made *analytically*. For the second-order Feautrier scheme in § 12.2, $H_d \equiv 1$ for $(2 \le d \le ND - 1)$; but we keep this notation so it can be applied more generally. In addition, introduce a new auxiliary variable

$$F_d \equiv D_d^{-1} - 1 \tag{12.93}$$

so that

$$D_d = (1 + F_d)^{-1}. \tag{12.94}$$

In terms of these new variables, the upper boundary condition is

$$F_1 \equiv C_1^{-1} H_1, \tag{12.95}$$

and

$$E_1 = (1 + F_1)^{-1} \cdot C_1^{-1} R_1. \tag{12.96}$$

For $2 \le d \le ND - 1$ the forward sweep is

$$F_d = C_d^{-1}\{H_d + A_d \cdot [1 - (1 + F_{d-1})^{-1}]\}, \tag{12.97}$$

$$E_d = (1 + F_d)^{-1} C_d^{-1} (R_d + A_d E_{d-1}), \tag{12.98}$$

and at the lower boundary,

$$E_{\rm ND} = \{H_{\rm ND} + A_{\rm ND} \cdot [1 - (1 + F_{\rm ND-1})^{-1}]\}^{-1} \cdot (R_{\rm ND} + A_{\rm ND} E_{\rm ND-1}). \tag{12.99}$$

In the reverse sweep $j_{\rm ND+1} \equiv 0$; hence $j_{\rm ND} = E_{\rm ND}$ and

$$j_d = (1 + F_d)^{-1} j_{d+1} + E_d, \quad d = ND - 1, \ldots, 1. \tag{12.100}$$

We have written (12.93)–(12.100) as if all quantities are matrices. But in the scalar case (i.e., a formal solution), equations (12.92)–(12.99) simplify to

$$F_1 = H_1/C_1, \quad E_1 = R_1/B_1, \tag{12.101}$$

and

$$F_d = \{H_d + [A_d F_{d-1}/(1 + F_{d-1})]\}/C_d, \qquad (12.102)$$

$$E_d = (R_d + A_d E_{d-1})/[C_d(1 + F_{d-1})]. \qquad (12.103)$$

for $(d = 1, ND - 1)$.

This form shows why this algorithm is preferable to the original: only positive signs appear in (12.101)–(12.103), indicating that the auxiliary quantities are all positive and hence cancellation problems cannot occur. This improved algorithm can be extended to the fourth-order form of the equations, it is efficient, is numerically stable, is very accurate, and guarantees recovery of the diffusion limit.

Short Characteristics

In this approach [816, 1187], one describes the variation of physical quantities analytically on finite intervals and develops an analytical expression for the intensity integral.[7] From equation (11.100), the formal solution for the radiation field at frequency ν (ignored here for notational simplicity) along an outgoing ($\mu > 0$) ray in a 1D finite slab of total optical thickness T is

$$I(\tau, \mu) = I(T, \mu)e^{-(T-\tau)/\mu} + \int_{\tau}^{T} S(t)\, e^{-(t-\tau)/\mu}\, dt/\mu, \quad \mu > 0. \qquad (12.104)$$

The equation for the incoming intensity ($\mu < 0$) is analogous. Divide the slab into a set of finite layers and use (12.104) for the formal solution across each layer. The formal solution between optical depths τ_{d+1} and τ_d (for outgoing radiation, $\mu > 0$) is

$$I(\tau_d, \mu) = I(\tau_{d+1}, \mu) \exp[-(\tau_{d+1} - \tau_d)/\mu] + \int_{\tau_d}^{\tau_{d+1}} S(t) \exp[-(t - \tau_d)/\mu]\, dt/\mu, \quad \mu > 0, \qquad (12.105)$$

and between τ_{d-1} and τ_d (for incoming radiation, $\mu < 0$), it is

$$I(\tau_d, \mu) = I(\tau_{d-1}, \mu) \exp[-(\tau_d - \tau_{d-1})/(-\mu)] + \int_{\tau_{d-1}}^{\tau_d} S(t) \exp[-(\tau_d - t)/(-\mu)]\, dt/(-\mu), \quad \mu < 0. \qquad (12.106)$$

This notation is unambiguous for 1D problems. But in multidimensional problems, two angle cosines are required to specify the direction of a ray. Generalizing

[7] The term "short characteristics" as used here is a misnomer: a formal solution of the 1D, time-independent transport equation in static material is simply an integration along a single ray at a fixed frequency. The concept of characteristics is correctly applied to partial differential equations in multiple space dimensions, or in space and time. See [345, chapter 17 and 18], [761, chapter 4 and 5], [904, § 9.19, § 10.9, and § 13.8], [1019, chapter 4] and the authoritative discussion of their mathematical properties in [257].

the notation to indicate these angles explicitly is cumbersome. Instead, it is more useful to work with the optical depth $d\mathbf{t}$ *along a given line of sight*. In 1D,

$$d\mathbf{t} \equiv d\tau/|\mu|. \tag{12.107}$$

But in other geometries (e.g., spherical), or in moving media, there is not such a simple connection [168, 627, 647, 650, 801]. We take $\mathbf{t}$ to *increase* along the direction of propagation for incoming radiation ($\mu < 0$) and *decrease* along the direction of propagation for outgoing radiation ($\mu > 0$). Using this convention, and omitting explicit indication of the angle dependence of the intensity, the 1D equations (12.105) and (12.106) become

$$I(\mathbf{t}_d) = I(\mathbf{t}_{d+1})\exp(-\Delta\mathbf{t}_{d+\frac{1}{2}}) + \int_{\mathbf{t}_d}^{\mathbf{t}_{d+1}} S(t)\exp[-(\mathbf{t}-\mathbf{t}_d)]\,d\mathbf{t}, \tag{12.108}$$

and

$$I(\mathbf{t}_d) = I(\mathbf{t}_{d-1})\exp(-\Delta\mathbf{t}_{d-\frac{1}{2}}) + \int_{\mathbf{t}_{d-1}}^{\mathbf{t}_d} S(t)\exp[-(\mathbf{t}_d-\mathbf{t})]\,d\mathbf{t}, \tag{12.109}$$

where

$$\Delta\mathbf{t}_{d\pm\frac{1}{2}} \equiv |\mathbf{t}_d - \mathbf{t}_{d\pm1}|. \tag{12.110}$$

We emphasize again that τ refers to the normal optical depth, whereas $\mathbf{t}$ refers to optical depth along a given ray.

In this method we approximate $S(\mathbf{t})$ by simple functions for which the integrals in (12.105) and (12.106) [or (12.108) and (12.109)] can be evaluated analytically. One choice is a piecewise polynomial:

$$S(\mathbf{t}) = a_d^0 + a_d^1\mathbf{t} + a_d^2\mathbf{t}^2 + \ldots, \quad \mathbf{t}_{d-1} < \mathbf{t} < \mathbf{t}_d, \tag{12.111}$$

where the coefficients $a_d^0,\ a_d^1,\ldots$ are different for different depth points d. But choosing the degree of the polynomial (or functions other than polynomials) is not straightforward. Whereas use of higher-order polynomials may yield a more accurate representation, they may also lead to "ringing," i.e., a functional representation that exhibits steep variations with depth, sometimes reaching unphysical negative values.

Currently, the three most common implementations in astrophysical applications are the following:

1. First-order (linear) short characteristics: $S(\mathbf{t})$ is assumed to vary linearly between d and $d \pm 1$. This representation is safe because it cannot lead to ringing, but in many cases of interest it may not be sufficiently accurate unless points d and $d \pm 1$ are quite closely spaced. Even when they are closely spaced, use of linear segments systematically *overestimates* the intensity when the source function has upward curvature everywhere and systematically *underestimates* it when the

source function has downward curvature everywhere. These situations arise in reality.

2. Second-order short characteristics: $S(\mathbf{t})$ is represented by a quadratic polynomial. The coefficients of the expansion depend on S_{d-1}, S_d, and S_{d+1}. This representation is in principle more accurate, but more time-consuming, and may lead to the above-mentioned ringing problem.
3. Modern techniques that retain the accuracy of higher-order schemes, while avoiding the ringing problem: For example, *Bezier interpolants* or *monotonized splines* [51]. These potentially powerful techniques have not yet been widely used for astrophysical radiative transfer.

Besides these three methods, a number of other formulations, e.g., integral-operator methods, have been used, but they require much more computational effort, and hence are mainly of historical interest and will not be discussed here; the interested reader is referred to [606, 608, 609, 962, 965].

First-Order Short Characteristics

The integrals in equations (12.105) and (12.106) [or (12.108) and (12.109)] are evaluated assuming that the source function $S(t)$ is a linear function of t between adjacent grid points on the ray:

$$S(\mathbf{t}) = S_d \frac{\mathbf{t} - \mathbf{t}_{d-1}}{\mathbf{t}_d - \mathbf{t}_{d-1}} + S_{d-1} \frac{\mathbf{t}_d - \mathbf{t}}{\mathbf{t}_d - \mathbf{t}_{d-1}} \qquad \mathbf{t}_{d-1} \le \mathbf{t} \le \mathbf{t}_d, \tag{12.112}$$

where $S_d \equiv S(\tau_d)$. Substituting (12.112) into (12.108) and (12.109), we obtain

$$I(\tau_d, \mu) = I(\tau_{d+1}, \mu)\exp(-\Delta\mathbf{t}_{d+\frac{1}{2}}) + \lambda^+_{d,d} S_d + \lambda^+_{d,d+1} S_{d+1} \tag{12.113}$$

for the outward-directed intensity, $\mu > 0$, and

$$I(\tau_d, \mu) = I(\tau_{d-1}, \mu)\exp(-\Delta\mathbf{t}_{d-\frac{1}{2}}) + \lambda^-_{d,d} S_d + \lambda^-_{d,d-1} S_{d-1} \tag{12.114}$$

for the inward-directed intensity, $\mu < 0$. Here

$$\lambda^+_{d,d} \equiv y_{d+\frac{1}{2}} / \Delta\mathbf{t}_{d+\frac{1}{2}}, \tag{12.115}$$

$$\lambda^+_{d,d+1} \equiv -y_{d+\frac{1}{2}} / \Delta\mathbf{t}_{d+\frac{1}{2}} + x_{d+\frac{1}{2}}, \tag{12.116}$$

$$\lambda^-_{d,d} = \lambda^+_{d-1,d-1} \equiv y_{d-\frac{1}{2}} / \Delta\mathbf{t}_{d-\frac{1}{2}}, \tag{12.117}$$

$$\lambda^-_{d,d-1} = \lambda^+_{d-1,d} \equiv -y_{d-\frac{1}{2}} / \Delta\mathbf{t}_{d-\frac{1}{2}} + x_{d-\frac{1}{2}}, \tag{12.118}$$

$$x_{d\pm\frac{1}{2}} \equiv 1 - \exp(-\Delta\mathbf{t}_{d\pm\frac{1}{2}}), \tag{12.119}$$

$$\begin{aligned} y_{d\pm\frac{1}{2}} &\equiv \Delta\mathbf{t}_{d\pm\frac{1}{2}} - x_{d\pm\frac{1}{2}} \\ &= \Delta\mathbf{t}_{d\pm\frac{1}{2}} - 1 + \exp(-\Delta\mathbf{t}_{d\pm\frac{1}{2}}). \end{aligned} \tag{12.120}$$

The solution is calculated on a discrete optical-depth grid, $\{\tau_d\}$, $d = 1, \ldots, D$, from $\tau_1 = 0$ to $\tau_D = \mathrm{T}$ for a finite grid or to τ_D greater than the thermalization depth for a semi-infinite atmosphere. The boundary conditions are

$$I(\tau_1, \mu) = I^-(0, \mu) \tag{12.121}$$

at the upper boundary (for $\mu < 0$), and

$$I(\tau_D, \mu) = I^+(\mathrm{T}, \mu) \quad \text{finite slab} \tag{12.122a}$$

$$I(\tau_D, \mu) = B(\tau_D) + \mu(dB/d\tau)_D \quad \text{diffusion approximation} \tag{12.122b}$$

$$I(\tau_D, \mu) = I(\tau_D, -\mu) \quad \text{symmetric slab} \tag{12.122c}$$

at the lower boundary (for $\mu > 0$). In (12.121) and (12.122a), I^- and I^+ are given functions.

Second-Order Short Characteristics

In this method, the source function is interpolated by a parabola going through S_{d-1}, S_d, S_{d+1} using the Lagrangian interpolation formula

$$S(\mathbf{t}) = \sum_{i=-1}^{1} \frac{\prod_{j \neq i}(\mathbf{t} - \mathbf{t}_{d+j})}{\prod_{j \neq i}(\mathbf{t}_{d+i} - \mathbf{t}_{d+j})} S_{d+i}. \tag{12.123}$$

Substituting this expression into equations (12.108) and (12.109) and performing necessary integrations, we obtain [816]

$$I(\tau_d, \mu) = I(\tau_{d+1}, \mu) \exp(-\Delta\mathbf{t}_{d+\frac{1}{2}}) + \lambda^+_{d,d-1} S(\mathbf{t}_{d-1}) + \lambda^+_{d,d} S(\mathbf{t}_d) + \lambda^+_{d,d+1} S(\mathbf{t}_{d+1}) \tag{12.124}$$

for the outward-directed intensity, $\mu > 0$, and

$$I(\tau_d, \mu) = I(\tau_{d-1}, \mu) \exp(-\Delta\mathbf{t}_{d-\frac{1}{2}}) + \lambda^-_{d,d-1} S(\mathbf{t}_{d-1}) + \lambda^-_{d,d} S(\tau_d) + \lambda^-_{d,d+1} S(\mathbf{t}_{d+1}) \tag{12.125}$$

for the inward-directed intensity, $\mu < 0$. Here

$$\lambda^+_{d,d} \equiv \frac{2y_{d+\frac{1}{2}} \Delta\mathbf{t}_d - z_{d+\frac{1}{2}}}{\Delta\mathbf{t}_{d-\frac{1}{2}} \Delta\mathbf{t}_{d+\frac{1}{2}}}, \tag{12.126}$$

$$\lambda^+_{d,d-1} \equiv \frac{z_{d+\frac{1}{2}} - y_{d+\frac{1}{2}} \Delta\mathbf{t}_{d+\frac{1}{2}}}{2\Delta\mathbf{t}_d \Delta\mathbf{t}_{d-\frac{1}{2}}}, \tag{12.127}$$

$$\lambda^+_{d,d+1} \equiv \frac{z_{d+\frac{1}{2}} - y_{d+\frac{1}{2}} (2\Delta\mathbf{t}_d + \Delta\mathbf{t}_{d+\frac{1}{2}})}{\Delta\mathbf{t}_d \Delta\mathbf{t}_{d+\frac{1}{2}}} + x_{d+\frac{1}{2}}, \tag{12.128}$$

$$\lambda^-_{d,d} \equiv \frac{2y_{d-\frac{1}{2}} \Delta\mathbf{t}_d - z_{d-\frac{1}{2}}}{\Delta\mathbf{t}_{d-\frac{1}{2}} \Delta\mathbf{t}_{d+\frac{1}{2}}}, \tag{12.129}$$

$$\lambda^-_{d,d+1} \equiv \frac{z_{d-\frac{1}{2}} - y_{d-\frac{1}{2}}\Delta \mathbf{t}_{d-\frac{1}{2}}}{2\Delta \mathbf{t}_d \Delta \mathbf{t}_{d+\frac{1}{2}}}, \tag{12.130}$$

$$\lambda^-_{d,d-1} \equiv \frac{z_{d-\frac{1}{2}} - y_{d-\frac{1}{2}}(2\Delta \mathbf{t}_d + \Delta \mathbf{t}_{d-\frac{1}{2}})}{2\Delta \mathbf{t}_d \Delta \mathbf{t}_{d-\frac{1}{2}}} + x_{d-\frac{1}{2}}, \tag{12.131}$$

where

$$\Delta \mathbf{t}_d \equiv \tfrac{1}{2}(\Delta \mathbf{t}_{d-\frac{1}{2}} + \Delta \mathbf{t}_{d+\frac{1}{2}}), \tag{12.132}$$

x and y are given by (12.118) and (12.120), and

$$z_{d\pm\frac{1}{2}} \equiv (\Delta \mathbf{t}_{d\pm\frac{1}{2}})^2 - 2y_{d\pm\frac{1}{2}}. \tag{12.133}$$

Discontinuous Finite Elements

The discontinuous finite element (DFE) method, first used for solving neutron transport problems, has been adapted to astrophysical radiative transfer [200]. This algorithm appears to have several advantages and is likely to be used increasingly in numerical radiative transfer work. The notation and analysis in the paper cited are used here.

Mathematically, the method is an application of the Galerkin method; see, e.g., [1191]. The domain of the solution (optical depths between $\tau = 0$ and $\tau = \tau_{max}$) is divided into a set of zones, bounded by the discretized depth points τ_d and τ_{d+1}, for $d = 1, \ldots, D-1$. In each zone, we choose a set of *basis functions*. The parameters in these functions determine the degrees of freedom in the representation of the specific intensity. The transfer problem is reduced to finding the parameters of the basis functions in each zone. The specific intensity moving in a specified direction in a given zone is written as

$$I(\tau) = \sum_{j=1}^{L} w_j(\tau)\, I_j, \tag{12.134}$$

where w_j are the basis functions, I_j are the parameters to be determined, and L is the number of the degrees of freedom (number of basis functions).

For 1D problems, $L = 2$, and the parameters I_1 and I_2 are associated with the values of the specific intensity at the boundaries of the given zone. *This method does not require that these parameters for adjacent zones be continuous across an interface. In fact, allowing discontinuities at the zone boundaries greatly enhances the power of the method.*

The basis functions are essentially arbitrary; they need only satisfy the condition

$$w_j(\tau_k) = \delta_{jk} \tag{12.135}$$

and the completeness relation

$$\sum_{j=1}^{L} w_j(\tau) = 1. \tag{12.136}$$

The simplest choice is linear basis functions. We change notation and write $I_1 \rightarrow I_d^+$ as $\tau_1 \rightarrow \tau_d$ from the right, and $I_2 \rightarrow I_{d+1}^-$ as $\tau_2 \rightarrow \tau_{d+1}$ from the left. Then, on the open interval (τ_d, τ_{d+1}) the basis functions are

$$w_1(\tau) = \frac{\tau_{d+1} - \tau}{\Delta\tau_{d+\frac{1}{2}}} \tag{12.137}$$

and

$$w_2(\tau) = 1 - w_1(\tau) = \frac{\tau - \tau_d}{\Delta\tau_{d+\frac{1}{2}}}, \tag{12.138}$$

which satisfy conditions (12.135) and (12.136). Notice that the basis functions are independent of the angle cosine μ because they contain ratios of optical depth differences.

At any grid point, say d, there are two *associated finite elements* ("specific intensities"), namely, I_d^-, coming from the interval (τ_{d-1}, τ_d) as $\tau \rightarrow \tau_d$ from the left, and I_d^+, coming from the interval (τ_d, τ_{d+1}) as $\tau \rightarrow \tau_d$ from the right. In general,

$$I_d^+ \neq I_d^-, \tag{12.139}$$

hence the name "discontinuous finite element." We emphasize that the notation I_d^+ and I_d^- used here does not refer to intensities in opposite directions at τ_d, as is usual in radiative transfer theory. Although these quantities are loosely referred to as "specific intensities," it must be remembered that they are *not* true specific intensities, but are only *parameters* from which the specific intensity can be evaluated. Because of the discontinuity at the grid points, we have freedom in defining the specific intensity at the grid point d as a function of I_d^+ and I_d^-. A judicious choice of a linear combination of these two quantities leads to an increased overall accuracy of the method.

The transfer problem is solved by determining the unknown parameters I_d^+ and I_{d+1}^- for all d; they are found by forming its *projections* onto the set of basis functions. Thus if the transfer equation is multiplied by w_j $(j = 1, 2)$ and integrated over the given zone, one obtains a set of two linear equations for the two unknowns I_d^+ and I_{d+1}^-. The crucial point is not to integrate only over the open interval (τ_d, τ_{d+1}), but instead over an interval in which the boundary on the *upwind side* of the zone (the direction from which the photons come into the zone) is extended infinitesimally into the upwind zone. For instance, for incoming radiation (for which the optical depth increases along the direction of propagation) we integrate over the interval $(\tau_d - \epsilon, \tau_{d+1})$. This *ansatz* is not a strict mathematical requirement of the method; it is physically motivated because it ensures that information is propagated mathematically in the direction of propagation of radiation along the chosen beam.

Assume now that there is a beam of incoming radiation with a given angle cosine μ; the derivation of the expressions for outgoing radiation is similar. Let

$$d\mathbf{t} = d\tau/(-\mu) \tag{12.140}$$

($\mu < 0$ for incoming radiation). The transfer equation is

$$\frac{dI}{d\mathbf{t}} = -I + S. \tag{12.141}$$

Combining (12.134)–(12.138) into a single expression for I, we have

$$I(\mathbf{t}) = \frac{\mathbf{t}_{d+1} - \mathbf{t}}{\Delta\mathbf{t}_{d+\frac{1}{2}}} I_d^+ + \frac{\mathbf{t} - \mathbf{t}_d}{\Delta\mathbf{t}_{d+\frac{1}{2}}} I_{d+1}^-, \tag{12.142}$$

so the basis functions are

$$w_1(\mathbf{t}) = \frac{\mathbf{t}_{d+1} - \mathbf{t}}{\Delta\mathbf{t}_{d+\frac{1}{2}}} \quad \text{and} \quad w_2(\mathbf{t}) = \frac{\mathbf{t} - \mathbf{t}_d}{\Delta\mathbf{t}_{d+\frac{1}{2}}}. \tag{12.143}$$

The projections are

$$\lim_{\epsilon\to 0} \int_{\mathbf{t}_d-\epsilon}^{\mathbf{t}_{d+1}} \left(\frac{dI}{d\mathbf{t}} + I - S\right) w_j(\mathbf{t})\, d\mathbf{t} = 0, \quad j = 1, 2. \tag{12.144}$$

The individual terms in (12.144) are evaluated as follows: the derivative of specific intensity contains a regular term

$$\frac{dI}{d\mathbf{t}} = \frac{I_{d+1}^- - I_d^+}{\Delta\mathbf{t}_{d+\frac{1}{2}}}, \quad \mathbf{t}_d < \mathbf{t} \le \mathbf{t}_{d+1} \tag{12.145}$$

and a singular term

$$\left(\frac{dI}{d\mathbf{t}}\right)_{\mathbf{t}_d} = \left(I_d^+ - I_d^-\right) \delta(\mathbf{t} - \mathbf{t}_d). \tag{12.146}$$

The integral of the first term in (12.144) is

$$\begin{aligned}
\lim_{\epsilon\to 0} \int_{\mathbf{t}_d-\epsilon}^{\mathbf{t}_{d+1}} \frac{dI}{d\mathbf{t}} w_1(\mathbf{t})\, d\mathbf{t} &= \frac{1}{\Delta\mathbf{t}_{d+\frac{1}{2}}} (I_{d+1}^- - I_d^+) \int_{\mathbf{t}_d}^{\mathbf{t}_{d+1}} w_1(\mathbf{t})\, d\mathbf{t} + (I_d^+ - I_d^-) \\
&= \tfrac{1}{2}(I_{d+1}^- + I_d^+) - I_d^-,
\end{aligned} \tag{12.147}$$

because $\int_{\mathbf{t}_d}^{\mathbf{t}_{d+1}} w_1(\mathbf{t})\, d\mathbf{t} = \int_{\mathbf{t}_d}^{\mathbf{t}_{d+1}} w_2(\mathbf{t})\, d\mathbf{t} = \frac{1}{2}\Delta\mathbf{t}_{d+\frac{1}{2}}$. Similarly, the integral of the second term in (12.144) is

$$\lim_{\epsilon\to 0} \int_{\mathbf{t}_d-\epsilon}^{\mathbf{t}_{d+1}} I(\mathbf{t}) w_1(\mathbf{t})\, d\mathbf{t} = \tfrac{1}{3}\Delta\mathbf{t}_{d+\frac{1}{2}} \left(I_d^+ + \tfrac{1}{2} I_{d+1}^-\right). \tag{12.148}$$

To evaluate the last contribution to the integral in (12.144) one must make an assumption about the depth dependence of the source function S. The most straightforward assumption is that it is piecewise linear, so that

$$S(\mathbf{t}) = S_d\, w_1(\mathbf{t}) + S_{d+1}\, w_2(\mathbf{t}), \tag{12.149}$$

where $S_d = S(\tau_d)$ and $S_{d+1} = S(\tau_{d+1})$. Then the source function is a continuous function of τ or $\mathbf{t}$, and the last integral is the same as (12.148):

$$\lim_{\epsilon \to 0} \int_{\mathbf{t}_d - \epsilon}^{\mathbf{t}_{d+1}} S(\mathbf{t}) w_1(\mathbf{t})\, d\mathbf{t} = \tfrac{1}{3} \Delta \mathbf{t}_{d+\frac{1}{2}} \left(S_d + \tfrac{1}{2} S_{d+1} \right). \tag{12.150}$$

Putting it all together, the projection on to w_1 is

$$\frac{I_{d+1}^- + I_d^+ - 2I_d^-}{\Delta \mathbf{t}_{d+\frac{1}{2}}} = \tfrac{2}{3} \left(S_d - I_d^+ \right) + \tfrac{1}{3} \left(S_{d+1} - I_{d+1}^- \right), \tag{12.151}$$

and the projection on to w_2 is

$$\frac{I_{d+1}^- - I_d^+}{\Delta \mathbf{t}_{d+\frac{1}{2}}} = \tfrac{1}{3} \left(S_d - I_d^+ \right) + \tfrac{2}{3} \left(S_{d+1} - I_{d+1}^- \right). \tag{12.152}$$

The latter equation does not contain a contribution from the upwind cell, I_d^-, because the singular term (12.146) vanishes when integrated against w_2. In the vocabulary of the Galerkin method [1191], the symmetric matrix

$$\begin{pmatrix} 2/3 & 1/3 \\ 1/3 & 2/3 \end{pmatrix}$$

multiplying the vector of terms containing $S - I$ is called the *mass matrix* of this problem. It has been found by test calculations that replacing this mass matrix with the unit matrix results in improved stability of the system. Such a technique is called *mass lumping*.

The final linear relations we obtain for the quantities I^+ and I^- are

$$\frac{I_{d+1}^- + I_d^+ - 2I_d^-}{\Delta \mathbf{t}_{d+\frac{1}{2}}} = S_d - I_d^+ \tag{12.153}$$

and

$$\frac{I_{d+1}^- - I_d^+}{\Delta \mathbf{t}_{d+\frac{1}{2}}} = S_{d+1} - I_{d+1}^-. \tag{12.154}$$

By eliminating I_d^+, we obtain a linear recurrence relation for I_d^-:

$$a_d I_{d+1}^- = 2I_d^- + \Delta \mathbf{t}_{d+\frac{1}{2}} S_d + b_d S_{d+1}, \tag{12.155}$$

and I_d^+ likewise follows from

$$a_d I_d^+ = 2(\Delta \mathbf{t}_{d+\frac{1}{2}} + 1)I_d^- + b_d S_d - \Delta \mathbf{t}_{d+\frac{1}{2}} S_{d+1}, \tag{12.156}$$

where

$$a_d \equiv \Delta \mathbf{t}^2_{d+\frac{1}{2}} + 2\Delta \mathbf{t}_{d+\frac{1}{2}} + 2, \tag{12.157}$$

and

$$b_d \equiv \Delta \mathbf{t}_{d+\frac{1}{2}} (\Delta \mathbf{t}_{d+\frac{1}{2}} + 1) \,. \tag{12.158}$$

Finally, the resulting specific intensity at τ_d is given by a linear combination of the "discontinuous" intensities I_d^- and I_d^+:

$$I_d = \frac{I_d^- \Delta \mathbf{t}_{d+\frac{1}{2}} + I_d^+ \Delta \mathbf{t}_{d-\frac{1}{2}}}{\Delta \mathbf{t}_{d+\frac{1}{2}} + \Delta \mathbf{t}_{d-\frac{1}{2}}} \,. \tag{12.159}$$

With this definition of the specific intensity at grid points, the discontinuous finite element method recovers the diffusion limit for large $\mathbf{t}$, i.e.,

$$I \approx S - \frac{dS}{d\mathbf{t}} + \frac{d^2 S}{d\mathbf{t}^2} - \dots \,. \tag{12.160}$$

To verify this important property, expand S_d in a Taylor series about $\mathbf{t}_{d+1}$ and let $x \equiv \Delta \mathbf{t}_{d+\frac{1}{2}}$ to simplify notation. Then

$$S_d \approx S_{d+1} - x S'_{d+1} + \tfrac{1}{2} x^2 S''_{d+1} + \mathrm{O}(\mathbf{t}^{-3}). \tag{12.161}$$

Substituting this expression into (12.155), and keeping only terms $\mathrm{O}(x)$ and $\mathrm{O}(x^2)$, we obtain

$$I^-_{d+1} \approx S_{d+1} - S'_{d+1} + \left(\tfrac{1}{2}x - 1\right) S''_{d+1} + \mathrm{O}(\mathbf{t}^{-3}). \tag{12.162}$$

To get an analogous expression for I_d^+ from (12.156), make a similar expansion of S_{d+1} around $\mathbf{t}_d$, and use (12.162) applied at depth d, because (12.156) contains the term $2(x+1)I_d^-$. Writing $\Delta \mathbf{t}_{d-\frac{1}{2}} \equiv \overline{x}$, we find

$$I_d^+ \approx S_d - S'_d + \left(1 + \frac{\overline{x}}{x} - \tfrac{1}{2}x\right) S''_d + \mathrm{O}(\mathbf{t}^{-3}). \tag{12.163}$$

Alternatively, rewriting (12.162) for depth d we have

$$I_d^- \approx S_d - S'_d + \left(\tfrac{1}{2}\overline{x} - 1\right) S''_d + \mathrm{O}(\mathbf{t}^{-3}). \tag{12.164}$$

Equations (12.163) and (12.164) show that both I_d^- and I_d^+ have an error of $\mathrm{O}(xS'')$, i.e., $\mathrm{O}(\mathbf{t}^{-1})$, but in the opposite sense. Therefore, we use them in a linear combination

$$I_d = (1-\alpha) I_d^- + \alpha I_d^+, \tag{12.165}$$

with the coefficient α chosen to minimize the error in the term proportional to S''. Adding (12.163) and (12.164), multiplied by α and $(1-\alpha)$, respectively, we find that the coefficient of the term in S'' reduces to

$$C = \tfrac{1}{2}(y - y\alpha - \alpha) + (2\alpha - 1) + ay, \tag{12.166}$$

where $y \equiv \bar{x}/x$. Because both a and y are O(1) when x is large, C is effectively minimized when the first term $y - y\alpha - \alpha = 0$, i.e., when $\alpha = y/(1+y)$. The suggestion for α in [200], written in the present notation, is

$$\alpha_{\rm CDK} = \bar{x}/(x+\bar{x}) = y/(1+y), \tag{12.167}$$

which is the optimum value found above. The basic condition for the validity of this conclusion is that $y \approx O(1)$, i.e., that the run of the optical depth increments is smooth. If the optical depth points are taken to be logarithmically equidistant with N points per decade, i.e., $\log(\tau_{d+1}/\log\tau_d) = 1/N$, then $y = 10^{-(1/N)}$, which is indeed O(1). The corresponding value of α,

$$\alpha = \frac{10^{-(1/N)}}{1+10^{-(1/N)}} = \frac{1}{1+10^{(1/N)}}, \tag{12.168}$$

can be used to evaluate the intensity I_d in (12.165) instead of (12.159). We find that *the method has second-order accuracy*, the same as the original Feautrier algorithm. This fact might seem counterintuitive because we have used *linear* segments. It is the result of the cancellation of the dominant error in the linear combinations (12.159) or (12.165).

In applying the discontinuous finite element method to plane-parallel atmospheres, one must remember that the expressions written above were derived assuming that optical depth *increases* along the ray of propagation of the radiation of interest. In this geometry, optical depth increases along a ray for incoming radiation, but *decreases* along a ray for outgoing radiation. Therefore, for plane-parallel atmospheres, we must solve the above recurrence relations separately for incoming ($\mu < 0$) and outgoing intensities ($\mu > 0$). For incoming intensity, the boundary condition at $d = 1$ is

$$I_1^- = I(\tau_1, \mu) \equiv I^-(0, \mu), \tag{12.169}$$

as in (12.118), and the recursion proceeds from $d = 1$ to $d = D$. For outgoing radiation, the recursion goes in the opposite way, from $d = D$ to $d = 1$. The lower boundary condition is

$$\begin{aligned} I_D^- &= I(\tau_D, \mu) = I^+(\mathrm{T}, \mu) && \text{finite slab} \\ &= B(\tau_D) + \mu \frac{dB}{d\tau} && \text{diffusion approximation.} \end{aligned} \tag{12.170}$$

If the lower boundary is the center of a symmetric slab, the outgoing intensity I_D^- is set equal to the incoming intensity I_D^+ from the previous sweep.

Comparison of Methods

We have discussed three basic methods for computing formal solutions of the transfer equation: Feautrier (F), short characteristics (SC), and discontinuous finite elements (DFE). The Feautrier and short characteristic methods have two variations, so there are five schemes to intercompare:

(a) second-order (original) Feautrier method (F2),
(b) fourth-order Feautrier (Hermite) method (F4),
(c) first-order short characteristics (SC1),
(d) second-order short characteristics (SC2), and
(e) the discontinuous finite element (DFE) method.

The most important criteria are accuracy and speed. Theoretical analysis can offer guidance, but the performance of these methods is best tested on actual problems.

Accuracy

We took a test source function of the form

$$S(t) = at^2 + bt + c + d\exp(-ft), \tag{12.171}$$

where t is the monochromatic optical depth along a given line of sight. In order to study a wide range of optical properties, let t be a combination of line and continuum opacity:

$$t \equiv \tau_{\mu\nu} = \tau_0(\phi_\nu + r)/\mu. \tag{12.172}$$

Here τ_0 is a standard optical depth scale, ϕ_ν is a Doppler profile with frequency displacement from line center measured in Doppler widths, and r is the ratio of continuum to line-center opacity. In the computations, the depths points are equidistant in $\log\tau_0$ between -2.5 and 4; the first depth point is at $\tau_0 = 0$. The upper boundary condition has no incoming radiation at $\tau_0 = 0$, and the lower boundary condition at $\tau_0 = \tau_{\max} = 10^4$ is the diffusion approximation.

The emergent specific intensity for the source function (12.171) and (12.172) can be computed analytically:

$$I_{\rm exac}(0,\mu,\nu) = 2a\left(\frac{\mu}{\phi_\nu + r}\right)^2 + b\frac{\mu}{\phi_\nu + r} + c + \frac{d}{1 + f(\phi_\nu + r)/\mu}. \tag{12.173}$$

The absolute angle- and frequency-dependent error $e(\mu,\nu)$ of each numerical method is

$$e(\mu,\nu) \equiv \left|I_{\rm num}(0,\mu,\nu) - I_{\rm exac}(0,\mu,\nu)\right| / I_{\rm exac}(0,\mu,\nu). \tag{12.174}$$

The accuracy of the five formal solvers was tested for three problems:

1. A linear source function plus an exponential term, which simulates a temperature rise at the surface

$$a = 0, \quad b = 0.1, \quad c = 1, \quad d = 10, \quad f = 0.1,$$

2. A quadratic source function plus an exponential term

$$a = 0.01, \quad b = 0.1, \quad c = 1, \quad d = 10, \quad f = 1,$$

3. A pure quadratic source function.

In figures 12.4–12.6 the maximum absolute error e for $\mu = 1$ at any frequency is plotted as a function of the number of depth points D. For problems **1** and **2** we expect that $e \propto D^{-2}$.

For problem **1**, F2 and DFE have about the same accuracy. SC1 is slightly more accurate. SC2 is more accurate than SC1; its error $e \propto D^{-3}$. For both problems, F4 is the most accurate; its error $e \propto D^{-4}$. For problem **2**, DFE is the least accurate, followed by SC1 and F2. Again, SC2 is more accurate than SC1; its error $e \propto D^{-3}$. For $D \sim 100$, SC2 is about an order of magnitude more accurate than F2, SC1, and DFE. At $D \approx 100$, F4 is about an order of magnitude more accurate than SC2 and about two orders of magnitude more accurate than the other methods. For problem **3**, SC2 is exact; its small error plotted in figure 12.6 is merely the round-off error of the computer.

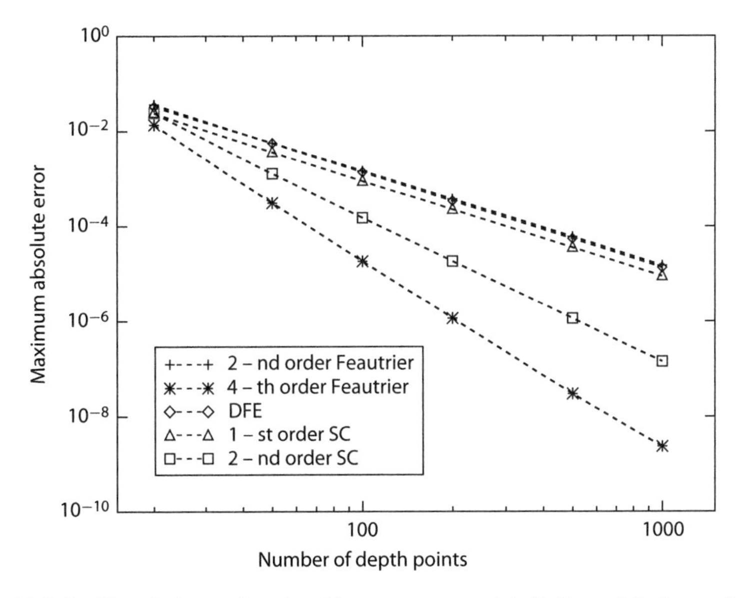

Figure 12.4 Problem **1.** *Source function*: Linear + exponential. *Ordinate*: Maximum absolute error over all frequencies for rays moving along $\mu = 1$. *Abscissa:* Number of depth points.

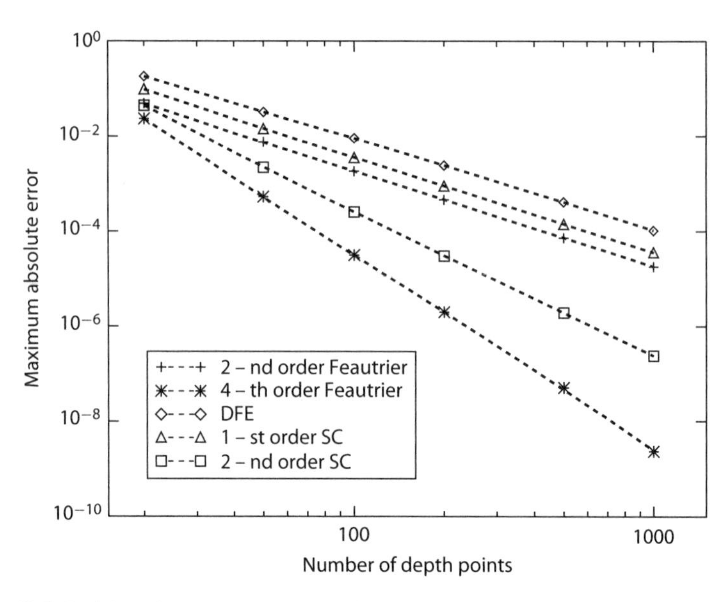

Figure 12.5 Problem **2.** *Source function*: Quadratic + exponential. *Ordinate*: Maximum absolute error over all frequencies for rays moving along $\mu = 1$. *Abscissa:* Number of depth points.

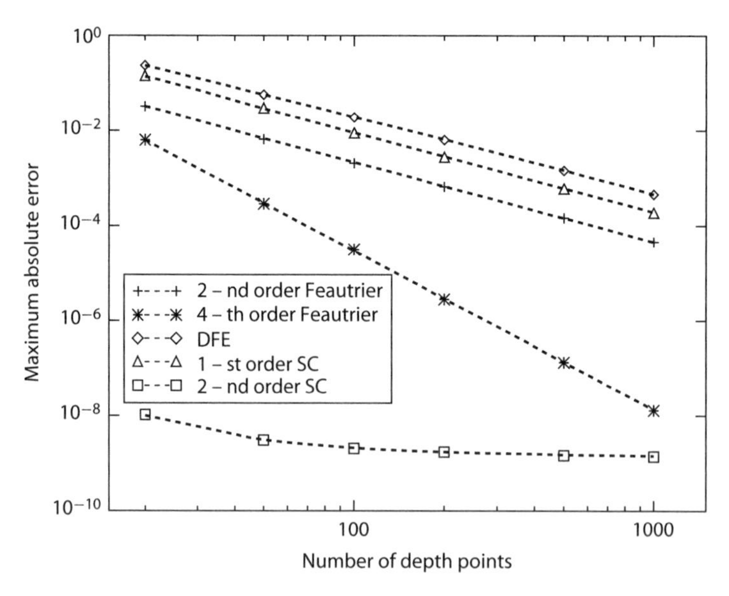

Figure 12.6 Problem **3.** *Source function:* Pure quadratic. *Ordinate*: Maximum absolute error over all frequencies for rays moving along $\mu = 1$. *Abscissa:* Number of depth points.

Speed

A crucial quantity is the computer time used by each method. We wrote FORTRAN subroutines in which each method was coded to work as efficiently as possible.

Table 12.2 Relative Time Used by Each Formal Solver on Different Platforms

Method	DEC	SUN	LINUX
F2	0.81	0.70	0.71
F4	1.00	1.00	1.00
DFE	0.67	0.96	1.10
SC1	0.90	0.71	1.73
SC2	1.49	1.41	2.31

Tests were run for 10^4 to 10^6 individual formal solutions. In all cases, the time is essentially linear in the number of depth points. Instead of expressing the execution time for a given method in, say, μsec, which is set by the internal clock of a machine,[8] to judge the relative efficiency of the individual methods we normalized the time used by each method to the time used by the F4 scheme. Because F2 and F4 solve for $j_{\mu\nu}$ and $h_{\mu\nu}$, from which both $I_{\mu\nu}$ and $I_{-\mu,\nu}$ can be calculated, we counted the computer time needed by the other three methods to solve the transfer equation for two angles, $\pm\mu$. We compared the relative performance on three platforms: a DEC Alpha workstation (XP-1000), a SUN Ultra workstation, and a LINUX machine with the g77 compiler. The results are summarized in table 12.2. The number given is the ratio of the time consumed by each method compared to that used by the fourth-order Feautrier scheme.

F2 takes about 70–80% of the time of F4 on all platforms. Its greater speed must be balanced against F4's superior accuracy. The efficiency of the other three methods depends sensitively on the compiler. In particular, we see that the free-distribution g77 compiler computes exponentials slowly. Hence under the LINUX system we used, the time for SC1 and SC2 is substantially longer than for F4. *Note that neither the Feautrier nor the DFE methods require exponentials.* The DFE method offers a noticeable reduction of computer time on DEC Alphas ($\approx$ 70% of F4), but its speed is lower under the SUN and LINUX systems. Keep in mind that if these tests were performed on modern computers with current compilers, the results could differ somewhat. Moreover, we have no indication what the relative speeds might be on massively parallel machines.

For problems having an angle-independent source function, the fourth-order Feautrier scheme is the method of choice. But inasmuch as F2 and F4 would take about twice as much time for a general angle-dependent source function, SC2 would be the most advantageous method if high accuracy is needed, or SC1 if lower accuracy is sufficient. *However, effective use of short characteristic methods requires a compiler that can compute exponentials quickly.*

[8] The absolute speeds of computers increase by orders of magnitude every few years.

12.5 VARIABLE EDDINGTON FACTORS

Method

In § 12.3 we mentioned that in common with all systems of moment equations, the system of angular moments of the first-order transfer equation does not close. The first moment equation contains the mean intensity $J_\nu \equiv \frac{1}{2}\int_{-1}^{1} I_{\mu\nu} d\mu$ and the Eddington flux $H_\nu \equiv \frac{1}{2}\int_{-1}^{1} I_{\mu\nu}\mu d\mu$; the second moment equation contains H_ν and $K_\nu \equiv \frac{1}{2}\int_{-1}^{1} I_{\mu\nu}\mu^2 d\mu$. Thus we have only two equations containing three unknowns. In contrast, with Schuster's second-order transfer equation (§ 11.6) we get a system of the two equations containing only the two angle-frequency dependent variables $j_{\mu\nu}$ and $h_{\mu\nu}$.

With Feautrier's method we can solve for the full angle-frequency-depth variation of the radiation field even when the source function contains a scattering term. The time needed scales as $t_F \propto c_F(ND \cdot NM^3 \cdot NF^3)$, where ND is the number of depth points, NM the number of angle cosines, and NF the number of frequencies. Clearly, the scaling in NM and NF is unfavorable, and many problems require a large number of frequency points. On the other hand, the angular information is redundant because only an angle-integrated quantity, J_ν, not $j_{\mu\nu}$ and $h_{\mu\nu}$, is needed.

By integrating Schuster's transfer equation over angle we eliminate the angular information [54] in terms of variable Eddington factors $f_\nu(\tau_\nu) \equiv K_\nu(\tau_\nu)/J_\nu(\tau_\nu)$ and $g_\nu \equiv H_\nu(0)/J_\nu(0)$ and can write

$$\frac{\partial^2(f_\nu J_\nu)}{\partial\tau_\nu^2} = J_\nu - S_\nu \qquad (12.175)$$

with boundary conditions

$$\left.\frac{\partial(f_\nu J_\nu)}{\partial\tau_\nu}\right|_0 = g_\nu J_\nu(0) \qquad (12.176)$$

and

$$\left.\frac{\partial(f_\nu J_\nu)}{\partial\tau_\nu}\right|_{\tau_{\max}} = \tfrac{1}{2}(B_\nu - J_\nu) + \frac{1}{3}\left.\left(\frac{\partial B_\nu}{\partial\tau_\nu}\right)\right|_{\tau_{\max}}, \qquad (12.177)$$

which can be differenced in the same way as the angle-dependent equations.[9] If f_ν is known for all frequencies at all depths, the time required to solve this system scales as $t' \propto c'(ND \cdot NF^3)$, which is a factor of NM^3 smaller than the solution of the original Feautrier equations. The procedure is the following:

1. From a given first estimate of S_ν (e.g., $S_\nu \equiv B_\nu$) we calculate $j_{d\mu\nu}$ one angle and frequency at a time using any of the formal solutions in § 12.4. These calculations require O($ND \cdot NM \cdot NF$) operations.

[9] S_ν is the angle integral of the source function, which might have an angle-dependent scattering term.

2. Given $j_{d\mu\nu}$, we compute the Eddington factors:

$$f_{dn} = \frac{\sum_m w_m \mu_m^2 j_{dmn}}{\sum_m w_m j_{dmn}}, \quad (d = 1, \ldots, ND), \ (n = 1, \ldots, NF), \quad (12.178)$$

$$g_{1n} = \frac{\sum_m w_m \mu_m j_{1mn}}{\sum_m w_m j_{1mn}}, \quad (n = 1, \ldots, NF), \quad (12.179)$$

and angle-integrated source functions

$$S_{dn} \equiv \sum_m w_m S_{dmn}. \quad (12.180)$$

3. With f_{dn} known, we solve (12.175)–(12.177) for J_ν, using explicit expressions for the source function such as those in (12.37) [or even their angle-dependent versions if the redistribution functions in (12.37c) are angle-dependent].
4. We then re-evaluate S_ν using the new values of J_ν. If S_ν found in step (3) differs significantly from that in step (1), we repeat steps (1)–(3) until S_ν is known to the desired accuracy.

Only a few iterations are needed because the radiation field departs from isotropy only in a layer a few photon mean free paths in from the open surface. Hence $t_{\text{vef}} \ll t_{\text{F}}$. This method has the advantage that if the radiation field is known only to modest precision, the Eddington factors are more precise. For example, they would be unchanged if j_{dmn} were wrong by the same scale factor at all depths. If I is the number of iterations required to achieve convergence, the computing time scales as $t_{\text{vef}} \propto c'(I \cdot ND \cdot NF^3) + c''(I \cdot ND \cdot NM \cdot NF)$.

At great depth in an atmosphere, $f_{dn} \to \frac{1}{3}$. At its surface, $f_{dn} > \frac{1}{3}$ because outgoing rays with $\mu \approx 1$ are more heavily weighted in (12.178). In a highly extended spherical atmosphere, $f_{dn} \to 1$ at radii much larger than the radius of its photosphere. It is often said that the Eddington factor lies in the range $\frac{1}{3} \leq f_{dn} \leq 1$. This statement is untrue: suppose an atmosphere has an embedded hydrodynamic shock. The shock will be transparent (one photon mean free path thick) in the vertical direction but extends to $\pm\infty$ horizontally. Because the shock is hotter than the ambient atmosphere, the radiation field along a line of sight in the shock layer is more intense than in any other direction; hence the angular distribution of the radiation field peaks near $\mu = 0$. In reality, one can say only that $0 \leq f_{dn} \leq 1$. For time-dependent and relativistic problems attempts have been made to compute $f \equiv K/J$ with a formula containing only J and H. Generally they fail because the second moment of the radiation field contains information not contained in the first two moments. They may even allow a radiation front to propagate faster than the speed of light! Although variable Eddington factors save computing time, the savings are small compared to those given by the iterative methods in chapter 13.

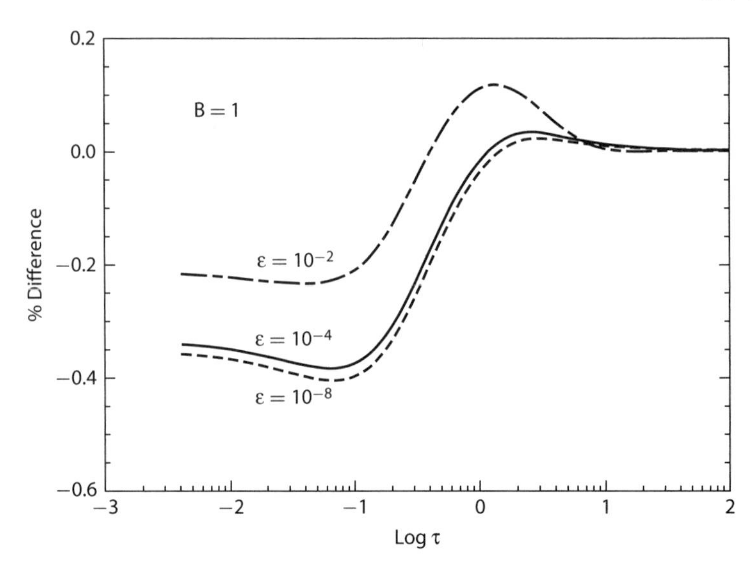

Figure 12.7 *Ordinate*: Percent difference between source function obtained assuming isotropic and anisotropic scattering. *Abscissa:* log τ.

Anisotropic Thomson Scattering

The differences between isotropic and anisotropic Thomson scattering for a planar atmosphere using (12.37d) as the source term with $B \equiv 1$, and $\epsilon = 10^{-8}, 10^{-4}$, and 10^{-2} are shown in figure 12.7. The ordinate gives the difference $(J_{\mathrm{anisotropic}} - J_{\mathrm{isotropic}})/J_{\mathrm{isotropic}}$ in percents. The effects are very small; indeed, the error in the mean intensity is larger if the order of the quadrature used to compute J and K is too low. But in an extended atmosphere, $K/J \to 1$ at large distance from the "photosphere," where the radial optical depth $\sim$ unity, so the difference may be significantly larger.

Chapter Thirteen

Iterative Solution of the Transfer Equation

We ... distinguish two classes of methods for numerical operations in linear algebra. In ... *direct methods* we normally perform a sequence of operations once only, and the results obtained are an approximation to the true results. ... The *indirect methods* attempt solution by a process of successive approximation. The same sequence of operations, usually shorter than that of a direct method, is repeated several times, the results either increasing steadily in accuracy (converging), or showing no definite trend to constancy (diverging). Here the numerical analyst is interested ... in securing the most rapid rate of convergence. Finally, we may combine the two types of method. For example, a direct method which yields a few accurate figures might be included as part of an iterative scheme, which may yield better accuracy with a minimum of effort.
L. Fox [346, p. 51]

13.1 ACCELERATED LAMBDA ITERATION: A HEURISTIC VIEW

In this chapter we discuss one of the most important advances in modern radiative transfer theory, the *accelerated lambda iteration*, or ALI, method.[1] In astrophysical quantitative spectroscopy it has allowed us to construct theoretical models of unprecedented realism and complexity. A number of review papers dealing with these methods can be found in [264, 272, 467, 536, 541].

ALI comes in many varieties, depending on the form of the source function and its dependence on the intensity of the radiation field. Here we illustrate the method using a radiative transfer equation that has a schematic source function containing a linear scattering term. In reality, the source function may be a complicated nonlinear function of the specific intensity and other structural quantities. Applications of ALI range from the solution of the radiative transfer equation with a scattering term in the source function for a given atmospheric structure to radiative transfer combined with NLTE kinetic equilibrium equations and a full set of structural equations, e.g., for a stellar atmosphere, accretion disk, or other medium. ALI has also been applied to some advanced problems,

[1] This name was coined by Hamann in [430] and by Werner and Husfeld in [1154]. The acronym has also been used to stand for *approximate lambda iteration*; *operator perturbation* [607]; and *approximate operator iteration* (AOI) [623] methods.

such as transfer of polarized radiation and transport in multidimensional media. We will discuss the application of ALI to specific astrophysical problems in later chapters.[2]

As discussed in chapter 12, the simplest scheme to deal with a scattering term is the (notorious) Λ-iteration, which computes a new estimate of the specific (or mean) intensity from a source function that uses their current values and then solves the transfer equation. Specifically, for the source function

$$S = (1 - \epsilon)J + \epsilon B, \tag{13.1}$$

a formal solution of the transfer equation yields $J = \Lambda[S]$, so (13.1) is equivalent to the integral equation

$$S = (1 - \epsilon)\Lambda\big[S\big] + \epsilon B, \tag{13.2}$$

which, when discretized, produces a linear algebraic equation with Λ represented by a matrix.

The Λ-iteration procedure can be written as

$$S^{(n+1)} = (1 - \epsilon)\Lambda\big[S^{(n)}\big] + \epsilon B. \tag{13.3}$$

This method is simple. It treats all coupling iteratively, so it avoids the need to invert the Λ-operator (matrix). But as we saw in §12.1, at best it converges extremely slowly and may just stabilize without converging.

More efficient methods to treat such problems iteratively date back to the work of Jacobi in the mid-19th century. The first application of those methods to astrophysical radiative transfer was made in two important papers by Cannon [176, 177], who introduced an iterative scheme he called *operator splitting*. His idea was to write

$$\Lambda = \Lambda^* + (\Lambda - \Lambda^*), \tag{13.4}$$

where Λ^* is a "judiciously chosen" *approximate* Λ-operator. Then the iteration scheme for solving (13.2) becomes

$$S^{(n+1)} = (1 - \epsilon)\Lambda^*\big[S^{(n+1)}\big] + (1 - \epsilon)(\Lambda - \Lambda^*)\big[S^{(n)}\big] + \epsilon B. \tag{13.5}$$

The action of the exact Λ-operator is thus split into two contributions: an approximate Λ^*-operator that acts on the *new* iterate of the source function, and the difference between the exact and approximate operators $(\Lambda - \Lambda^*)$, acting on the previous, i.e., *old*, and thus known, source function. As was shown in § 12.4, the latter contribution can readily be evaluated by a formal solution.

[2] Here we will normally suppress explicit indication that physical quantities such as the mean intensity J, source function S, thermalization parameter ϵ, and transport operators such as Λ are functions of depth. We may also omit frequency subscripts on J and S so they could mean either J_ν and S_ν for a purely thermal or coherent scattering problem, or $\overline{J}$ and $\overline{S}$ for a spectral line formed with complete redistribution of the scattered photons. Also, in keeping with the traditional practice in mathematics, in this chapter we generally do not make a typographical distinction between scalars, vectors, and matrices.

We emphasize that although it contains an *approximate* operator, (13.2) is *repeated until convergence is obtained*; hence it is *exact* at the converged limit. If we were to choose $\Lambda^* = 0$, we would recover ordinary Λ-iteration. On the other hand, the choice $\Lambda^* = \Lambda$ gives the exact solution without iteration, but only with inversion of the exact Λ-operator, which in general is costly. Thus if Λ^* is to give an essential improvement over both methods, it must incorporate all of the *essential* properties of the exact Λ-operator (so that the iteration process converges quickly), but at the same time it must be easy (and inexpensive) to invert. Thus the construction of the optimum Λ^* is a delicate matter.

In Cannon's original papers, a special variant of the Λ^*-operator was employed, namely, an angular and/or frequency quadrature of low order. The full significance of this particular ALI method was not recognized until about a decade later, when Scharmer [955, 957] reformulated this concept of ALI in a more physical way in terms of Rybicki's ideas of *core saturation* in spectral lines and *preconditioning* of transfer problems in general [918, 920]. His work motivated an intensive development of ALI-based approaches. The early period of development of the ALI method culminated in a seminal paper by Olson, Auer, and Buchler [815] who put the ALI method on a solid mathematical foundation.[3] We follow, and extend, their analysis in the next section.

One may write (13.3) in a somewhat different form. First, we introduce an "intermediate source function," obtained from the old source function by a formal solution

$$S^{\mathrm{FS}} \equiv (1-\epsilon)\Lambda\big[S^{(n)}\big] + \epsilon B; \tag{13.6}$$

here the superscript FS denotes "formal solution." Using this definition, (13.5) can be rewritten as

$$\delta S^{(n)} \equiv S^{(n+1)} - S^{(n)} = [I - (1-\epsilon)\Lambda^*]^{-1}\big[S^{\mathrm{FS}} - S^{(n)}\big], \tag{13.7}$$

where I is the unit (identity) operator. This expression is particularly instructive. To put it in better perspective, we can use (13.6) to rewrite (13.3), the traditional Λ-iteration, as

$$\delta S^{(n)} = S^{\mathrm{FS}} - S^{(n)}. \tag{13.8}$$

Then we see that the ALI iteration process is driven, as is ordinary Λ-iteration, by the difference between the old source function and the newer source function obtained from a formal solution.

But (13.7) shows that for ALI this difference is effectively *amplified*[4] by an *acceleration operator* $[1-(1-\epsilon)\Lambda^*]^{-1}$. For example, any diagonal (i.e., local) Λ^*-operator is constructed such that $\Lambda^*(\tau) \to 1$ for large τ (because $I_\nu \to S_\nu$ for large τ). In typical cases $\epsilon \ll 1$. Thus the acceleration operator does in fact act as a large amplification factor ϵ^{-1} in optically thick regions of the medium, where a photon requires $O(\tau^2)$ scatterings, and hence $O(\tau^2)$ iterations by classical Λ-iteration, to escape from the surface. The term *accelerated Λ-iteration* should

[3] We shall refer to this fundamental paper as "OAB."

[4] This interpretation was first recognized in by Hamann and by Werner and Husfeld in the papers cited above.

not be confused with the concept of *acceleration of convergence*, which is discussed in § 13.4.

We note that one can use an analogous formulation of ALI in which one iterates on the mean intensity J instead of the source function S. Using (13.1), one writes

$$J = \Lambda[S] = \Lambda[(1-\epsilon)J + \epsilon B], \tag{13.9}$$

and the ALI iteration scheme as

$$J^{(n+1)} = \Lambda^*[(1-\epsilon)J^{(n+1)}] + (\Lambda - \Lambda^*)[(1-\epsilon)J^{(n)}] + \Lambda[\epsilon B]. \tag{13.10}$$

Introducing $J^{\rm FS} \equiv \Lambda[S^{(n)}]$ and $\delta J^{(n)} = J^{(n+1)} - J^{(n)}$, one obtains

$$\delta J^{(n)} = [I - \Lambda^*(1-\epsilon)]^{-1}[J^{\rm FS} - J^{(n)}]. \tag{13.11}$$

Equation (13.7) can be also derived using slightly different reasoning. Suppose that we have a current estimate of the source function, S_0. Write the correct source function as $S = S_0 + \delta S$. Further, write the exact Λ-operator as $\Lambda = \Lambda^* + \delta\Lambda$, i.e., as an approximate operator Λ^* plus a "perturbation" $\delta\Lambda$. Then require that the source function satisfy (13.2). Substituting the above perturbation expansions for S and Λ, we find

$$S_0 + \delta S = (1-\epsilon)\Lambda^*[S_0] + (1-\epsilon)\Lambda^*[\delta S] + (1-\epsilon)\delta\Lambda[S_0] + \epsilon B, \tag{13.12}$$

where we dropped the second-order contribution $\delta\Lambda[\delta S]$. Rearranging terms and using (13.6), we obtain

$$[1 - (1-\epsilon)\Lambda^*][\delta S] = (1-\epsilon)\Lambda[S_0] + \epsilon B - S_0 \equiv S^{\rm FS} - S_0. \tag{13.13}$$

Equation (13.13) is equivalent to (13.7), which explains why this formulation is sometimes called the "operator perturbation method."

Hence the ALI iteration scheme proceeds as follows:

(a) For a given $S^{(n)}$ (with an initial estimate $S^{(0)} = B$, or some other suitable value), perform a formal solution, one frequency and angle at a time, to obtain new values of the specific intensity $I_{\mu\nu} \equiv \Lambda[S^{(n)}]$.
(b) Calculate the new source function $S^{\rm FS}$ from (13.6), using the new values of the specific intensity.
(c) Then use (13.7) to evaluate a new iterate of the source function, $S^{(n+1)}$.
(d) Because in general the source function found in step (c) will differ from that used in step (a), repeat steps (a) through (c) to convergence.

In applying this idea in astrophysical radiative transfer problems, the basic issue is to choose an optimum Λ^*-operator that leads to fast convergence, while requiring only a small amount of computer time to invert Λ^*. In early work the Λ^*-operator was constructed using physical considerations,[5] but starting with OAB it became

[5] For a review of the historical development, see [522].

clear that the optimum approximate operator must be constructed using purely mathematical analysis. We summarize the necessary and useful findings from linear algebra in the next section.

13.2 ITERATION METHODS AND CONVERGENCE PROPERTIES

Many problems of radiative transfer with scattering can be formulated in terms of systems of linear algebraic equations. Some examples have been shown in previous chapters, and more appear throughout this book. We must therefore acquire some basic knowledge about the numerical solution of (large) systems of linear algebraic equations.

We first define some terminology. Consider a general linear system

$$A x = b, \tag{13.14}$$

where the size and form of the matrix A and the right-hand side vector b depend on the problem in question. For instance, for the problem defined by (13.2), $A = I - (1-\epsilon)\Lambda$, $b = \epsilon B$, and $x = S$. Although the solution of (13.14) can be written as

$$x = A^{-1}\, b, \tag{13.15}$$

it is not useful in most applications because A is typically a large matrix and its inversion is too costly. Therefore, one seeks less expensive iterative schemes.

A general iterative method for solving (13.14) can be written symbolically as

$$x^{(n+1)} = G^{(n)}(x^{(n)}, x^{(n-1)}, \ldots, x^{(n-k+1)}, A,\, b). \tag{13.16}$$

In (13.16), k is the *order* of the method. First-order methods are the most common; they are also sometimes called *one-point iteration methods*. When $k = 1$, the next iterate $x^{(n+1)}$ depends only on the current iterate $x^{(n)}$.

Iterative schemes can also be classified according to the form of function $G^{(n)}$. If $G^{(n)}$ does not depend on n, the method is called *stationary*. If $G^{(n)}$ is a linear function, the iteration process is called *linear*. The most common and widely used iteration methods are linear, first-order, stationary methods.

There are several ways to describe iteration schemes. We shall use a formulation based on the concept of preconditioning [918], sometimes also called "matrix (operator) splitting." In this method, the matrix A is split into two parts,

$$A = P + (A - P), \tag{13.17}$$

and the iteration scheme is

$$P\, x^{(n+1)} + (A - P)x^{(n)} = b. \tag{13.18}$$

Thus the new iterate $x^{(n+1)}$ is obtained by inverting the *preconditioner* P, instead of the original matrix A. The approximate operator Λ^* used in § 13.1 is an example of a preconditioner; in that case $P = I - (1-\epsilon)\Lambda^*$.

For the scheme to be useful, P must be chosen such that inverting it is much less costly than inversion of the original matrix A; at the same time, the overall scheme must converge reasonably quickly. Therefore, the choice of a preconditioner is a crucial part of the problem.

The iteration scheme (13.18) can also be written as

$$x^{(n+1)} = P^{-1}(P - A)\,x^{(n)} + P^{-1}b \equiv Fx^{(n)} + b', \tag{13.19}$$

where

$$F \equiv P^{-1}(P - A) = I - P^{-1}A \tag{13.20}$$

is called the *amplification matrix* of the iteration scheme.

Another version of the iteration scheme (13.18) is obtained by writing it in terms of the *residual* in the nth iteration,

$$r^{(n)} \equiv b - A\,x^{(n)}, \tag{13.21}$$

and the *correction*

$$\delta x^{(n+1)} \equiv x^{(n+1)} - x^{(n)}. \tag{13.22}$$

Then (13.18) reduces to a simple and revealing form

$$P\,\delta x^{(n+1)} = r^{(n)}. \tag{13.23}$$

Thus in the nth step of the iteration process, we compute the residual vector $r^{(n)}$ from (13.21). The new correction $\delta x^{(n+1)}$ to the solution is obtained by solving (13.23), which is a linear system with A replaced by the preconditioner matrix P, and with the residual vector $r^{(n)}$ as the right-hand side.

As shown above, any iterative scheme can be written in the form

$$x^{(n+1)} = Fx^{(n)} + b'. \tag{13.24}$$

The converged solution is $A\,x^{(\infty)} = b$, or

$$x^{(\infty)} = Fx^{(\infty)} + b'. \tag{13.25}$$

The *true error* in the nth iteration of the solution is

$$e^{(n)} \equiv x^{(n)} - x^{(\infty)}. \tag{13.26}$$

Subtracting (13.25) from (13.24), we have

$$e^{(n+1)} = F\,e^{(n)}, \tag{13.27}$$

which shows that the iteration process (13.18) or (13.23) will converge quickly if the matrix F is in some sense “small.”

A more exact mathematical expression of this requirement is that

$$||e^{(n+1)}|| \leq ||F|| \cdot ||e^{(n)}|| \approx \sigma \, ||e^{(n)}||. \tag{13.28}$$

Here the symbol $||\ldots||$ denotes a norm of a vector or matrix, and σ is the *spectral radius* of F. There are several choices for the norm. For vectors we shall use the Euclidean norm

$$||x|| = \left(\sum_{i=1}^{N} x_i^2 \right)^{1/2}, \tag{13.29}$$

where N is the dimension of vector x. For a matrix, we use the spectral radius, which is its eigenvalue with the largest absolute value. We denote it as σ also. The iteration process will not converge at all if $\sigma > 1$, will converge rapidly if $\sigma \ll 1$, and will converge, but very slowly, if $\sigma \lesssim 1$. Thus *the spectral radius, or the maximum eigenvalue, of the amplification matrix determines the convergence properties of the iteration scheme*. This property provides an objective criterion by which to judge the quality of convergence and also guides a choice of the optimum preconditioner. In astrophysical radiation transport, this connection was first realized and clearly formulated in OAB.

Another way of seeing that the maximum eigenvalue of the amplification matrix is crucial is to expand the error vector in terms of eigenvectors of the amplification matrix F, i.e., $e^{(n)} = \sum c_i^{(n)} u_i$. The eigenvectors are defined by $F u_i = \lambda_i u_i$, where λ_i is the corresponding eigenvalue. Then (13.27) becomes

$$e^{(n+1)} = F \cdot e^{(n)} = \sum_i c_i \, F \cdot u_i = \sum_i \lambda_i \, c_i \, u_i. \tag{13.30}$$

After several iterations the eigenvector with the largest eigenvalue will dominate, so $e^{(n+1)} \to \lambda_{\rm max} e^{(n)}$. Therefore, the smaller $\lambda_{\rm max}$ is, the faster the convergence will be.

We can now understand the nature of *false convergence*. In practical applications, we do not know the exact solution $x^{(\infty)}$; thus we do not know the true error $e^{(n)}$. What we know, and what is often used as a convergence criterion, is the *relative error* $\delta x^{(n+1)} \equiv x^{(n+1)} - x^{(n)}$. The convergence criterion is taken to be $||\delta x^{(n+1)}|| < E ||x^{(n)}||$, where E is a small number, say, $E \approx 10^{-3}$. But this choice may be misleading.

To find the relation between the true error and the relative error, we apply (13.26) for n and $n+1$, to write the norm of the relative error as $||x^{(n+1)} - x^{(n)}|| \equiv ||e^{(n+1)} - e^{(n)}||$. Further, we can rewrite (13.27) as

$$e^{(n+1)} = F e^{(n)} = F e^{(n+1)} + F(e^{(n)} - e^{(n+1)}). \tag{13.31}$$

Therefore,

$$e^{(n+1)} = (I - F)^{-1} F (e^{(n)} - e^{(n+1)}), \tag{13.32}$$

and hence the norms satisfy the relation

$$||e^{(n+1)}|| \leq ||(I-F)^{-1}F|| \cdot ||e^{(n)} - e^{(n+1)}|| \approx \frac{\sigma}{1-\sigma} ||e^{(n)} - e^{(n+1)}||. \quad (13.33)$$

From (13.33) we see that if $\sigma \approx 1$, i.e., $1-\sigma \ll 1$, the true error may be much larger than the relative error. For instance, if $\sigma = 1-\epsilon$ and ϵ is small, say, $\epsilon = 10^{-4}$, then $||e^{(n+1)}|| \approx 10^4 \, ||e^{(n)} - e^{(n+1)}||$; i.e., the absolute error is 10^4 times larger than the relative error! Note also that (13.33) implies that

$$||e^{(n+1)}|| \approx \sigma ||e^{(n)}||, \quad (13.34)$$

so that the error is reduced by a factor σ in each iteration. Hence in order to reduce the initial error by a factor of 10^{-f} we need O(N) iterations, where

$$\sigma^N \approx 10^{-f}, \quad \text{or} \quad N \approx -f \ \ln 10/\ln\sigma. \quad (13.35)$$

From the previous example, in order to reduce the error by a factor of 10^{-3} for small ϵ, we would need $N \approx 7/\epsilon$ iterations. For $\epsilon = 10^{-4}$, this means ~70,000 iterations!

Linear, Stationary, First-Order Schemes

There are many possible choices for a preconditioner. Often the choice is specifically tailored to the form of the master matrix A, but there are also several general methods, which we briefly describe below.

Relaxation Method

This is the simplest possibility. Here we simply back-substitute the previous iterate into the original system to calculate a "correction" in order to obtain an "improved" result. That is, we take

$$x^{(n+1)} = x^{(n)} + (b - Ax^{(n)}). \quad (13.36)$$

Obviously, no matrix inversion is required.

In the language of preconditioners, we see from (13.19) that this scheme amounts to using

$$P = I, \quad (13.37)$$

or

$$x^{(n+1)} - x^{(n)} \equiv \delta x^{(n+1)} = r^{(n)}. \quad (13.38)$$

This process is nothing more than classical Λ-iteration. Thus we can draw two conclusions: (1) Λ-iteration is indeed a relaxation method, as stated earlier. (2) For a realistic transfer problems we have shown that Λ-iteration is at best extremely slow, and often useless. Hence the much more strongly convergent methods described below should always be used instead.

Jacobi Method

This is the simplest non-trivial, and one of the most popular, methods. In this case, the preconditioner is the diagonal part of A [346, chapter 8]. Here we write matrix A as

$$A = -L + D - U, \tag{13.39}$$

where $D = \mathrm{diag}(A)$ (i.e., $D_{ij} = A_{ij}\delta_{ij}$) is the diagonal part of A; L is the lower triangular part, $L_{ij} = -A_{ij}$ for $j < i$, and $L_{ij} = 0$ otherwise; and U is the upper triangular part $U_{ij} = -A_{ij}$ for $j > i$, and $U_{ij} = 0$ otherwise.

In this method, the preconditioner is

$$P = D, \tag{13.40}$$

and the iteration scheme is

$$D\,x^{(n+1)} - (L + U)\,x^{(n)} = b, \tag{13.41}$$

or

$$x^{(n+1)} = D^{-1}(L + U)\,x^{(n)} + D^{-1}b, \tag{13.42}$$

or

$$\delta x^{(n+1)} = D^{-1}r^{(n)}. \tag{13.43}$$

In this method inversion of matrices is completely avoided, because the inverse of a diagonal matrix is just a series of algebraic divisions; hence it is very fast.

To find the convergence properties of the method one would need to compute the largest eigenvalue of the amplification matrix $D^{-1}(L+U)$. But it is apparent that the method will converge quickly if matrix A is diagonally dominated. In practice, a useful bound on the size of the largest eigenvalue can sometimes be obtained using *Gershgorin's theorem* [346, pp. 276 and 288], [892, p. 486], [1166, p. 72 and pp. 638–646]. In the special case of real matrices it states that the eigenvalues λ_i of a matrix A lie within a certain distance from the corresponding diagonal element A_{ii}; this distance is given by the sum of the absolute values of the off-diagonal elements. Thus the maximum eigenvalue of matrix A is bounded by

$$\lambda_{\max} \leq \max\left|A_{ii} \pm \sum_{j\neq i} |A_{ij}|\right|. \tag{13.44}$$

In § 13.3 we will use (13.44) to estimate the maximum eigenvalue of the amplification matrix.

Gauss-Seidel Method

Here one takes the preconditioner as the diagonal plus the lower subdiagonal part of A,

$$P = D - L, \tag{13.45}$$

and the iteration scheme proceeds as

$$-Lx^{(n+1)} + Dx^{(n+1)} = Ux^{(n)} + b. \qquad (13.46)$$

Although at first sight this method looks more complicated than the Jacobi method, actually it is not. Thus write $x^{(n)} = x^{\text{old}}$ and $x^{(n+1)} = x^{\text{new}}$, and solve (13.46) recursively, starting with x_1, as follows:

$$x_1^{\text{new}} = D_{11}^{-1}\Big(b_1 - \sum_{j>1} U_{1j}\, x_j^{\text{old}}\Big) = A_{11}^{-1}\Big(b_1 - \sum_{j>1} A_{1j}\, x_j^{\text{old}}\Big). \qquad (13.47)$$

Then

$$x_2^{\text{new}} = D_{22}^{-1}\Big(b_2 - L_{21}\, x_1^{\text{new}} - \sum_{j>2} U_{2j}\, x_j^{\text{old}}\Big) = A_{22}^{-1}\Big(b_2 - \sum_{j\neq 2} A_{2j}\, x_{j,i}^{\text{old or new}}\Big), \qquad (13.48)$$

where in the second equality we use x_j^{new} for $j < 2$ and x_j^{old} for $j > 2$. The algorithm in general is

$$x_i^{\text{new}} = A_{ii}^{-1}\Big(b_i - \sum_{j\neq i} A_{ij}\, x_{j,i}^{\text{old or new}}\Big), \qquad (13.49)$$

where

$$x_{j,i}^{\text{old or new}} = \begin{cases} x_j^{\text{new}} \text{ for } j < i, \\ x_j^{\text{old}} \text{ for } j > i. \end{cases} \qquad (13.50)$$

Thus, one sweeps through the mesh from $i = 1$ to $i = N$, updating one element of the unknown vector x_i^{new} at a time, using the newly computed values x_j^{new} for previous cells $j < i$ and old values for $j > i$. The method can also be formulated with $D - U$ as preconditioner; in this case the sweep must be started on the other side of the mesh.

Successive Over-Relaxation (SOR)

This method can be viewed as a variant of either Jacobi or Gauss-Seidel iteration, although it is more often used with the latter.[6] In this case we take the preconditioner to be

$$P = (D/\omega) - L, \qquad (13.51)$$

where ω is a tunable (scalar) over-relaxation parameter, $1 < \omega < 2$, and write the iteration scheme as

$$[\,(D/\omega) - L\,]x^{(n+1)} = [\,U - (\omega - 1)(D/\omega)\,]x^{(n)} + b. \qquad (13.52)$$

The name "over-relaxation" comes from the fact that for $\omega > 1$ the corrections are somewhat larger than in the case of original Gauss-Seidel or Jacobi methods, thus accelerating the convergence.

[6] When used with the Jacobi method it is sometimes called the "extrapolated Jacobi method."

But one still needs a good choice for ω. As suggested in [1086], a nearly optimum value of ω is given by

$$\omega^{\text{opt}} = \frac{2}{1 + \sqrt{1 - \lambda_{\max}^2}}, \tag{13.53}$$

where $\lambda_{\max}$ is the largest eigenvalue of the Jacobi iteration matrix. If ω has this optimum value, then the largest eigenvalue of the SOR amplification matrix can be shown to be $\omega^{\text{opt}} - 1$. Using this value we may get a large speed-up in convergence.

All the previous methods were stationary schemes, except the SOR method, which may be stationary or nonstationary, depending on whether we choose the over-relaxation parameter ω to be constant or allow it to vary from iteration to iteration.

Double-Splitting and Multiple-Splitting Methods

In the double-splitting method, one applies preconditioning twice, and the calculation of the new solution $x^{(n+1)}$ is split into two steps. In the first step, we apply a preconditioner P_1 to obtain an improved solution $x^{(n+1/2)}$, and in the second step, we apply another preconditioner P_2 to advance the solution from $x^{(n+1/2)}$ to $x^{(n+1)}$. Specifically,

$$P_1 x^{(n+1/2)} = (P_1 - A)\, x^{(n)} + b, \tag{13.54}$$

and

$$P_2 x^{(n+1)} = (P_2 - A)\, x^{(n+1/2)} + b. \tag{13.55}$$

The amplification matrix in this case is given by

$$F = \left(I - P_2^{-1} A\right)\left(I - P_1^{-1} A\right). \tag{13.56}$$

The method can be extended to an arbitrary number (NS) of intermediate steps obtaining $x^{(n+1)}$ from $x^{(n)}$; in this case the amplification matrix can be written

$$F = \prod_{j=1}^{NS} \left(I - P_j^{-1} A\right). \tag{13.57}$$

Block-Matrix Methods

Any of the above methods can also be used in a "block" sense. Write

$$A \equiv \begin{pmatrix} A_{11} & A_{12} & A_{13} & \dots & A_{1N} \\ A_{21} & A_{22} & A_{23} & \dots & A_{2N} \\ A_{31} & A_{32} & A_{33} & \dots & A_{3N} \\ \vdots & \vdots & \vdots & \ddots & \vdots \\ A_{N1} & A_{N2} & A_{N3} & \dots & A_{NN} \end{pmatrix}, \tag{13.58}$$

where the individual elements A_{ij} are not numbers but matrices of some order lower than N. For the block-Jacobi method we then understand that the preconditioner is the matrix formed from the diagonal blocks A_{ii}. Here one has to invert the individual A_{ii} matrices.

An analogous block version can be constructed for the Gauss-Seidel, SOR, and double - (or multiple-) splitting methods. The block-Jacobi or block-Gauss-Seidel methods are advantageous if the inversions of the matrices A_{ii} can be done simply, e.g., when these matrices are sparse or have some simple structure, such as when they are tridiagonal matrices.

Nonlinear, Nonstationary Schemes

The goal of these methods is to find efficient methods to drive the true error $e^{(n)}$ of the current iterate $x^{(n)}$ to "zero" as efficiently as possible. They accelerate convergence to the correct solution and hence are referred to as *acceleration schemes*. We consider only first-order methods here.

Gradient (Steepest Descent) Methods

This class of methods attempts to minimize the errors (residuals) in the solution with respect to a set of *conjugate vectors* rather than minimize the residuals themselves. Then the iteration process has the general form

$$x^{(n+1)} = x^{(n)} + C^{(n)} p^{(n)}. \tag{13.59}$$

Possible choices for $C^{(n)}$ are a scalar, $\alpha^{(n)}$; a diagonal matrix; or a constant matrix. The vector $p^{(n)}$ is called a *search vector*. Possible choices for $p^{(n)}$ include the residual; the gradient of some error measure; the direction A-*orthogonal* to the gradient or the residual direction; or the previous vectors $p^{(n-i)}$. We shall outline only the most useful methods.

Recall that the scalar product of two general vectors x and y is defined as

$$(x, y) \equiv \sum_{i=1}^{N} x_i y_i. \tag{13.60}$$

They are orthogonal if $(x, y) = 0$, and A-orthogonal if $(x, Ay) = 0$. Suppose we take a scalar $\alpha^{(k)}$ for $C^{(k)}$, so that (13.59) now reads

$$x^{(n+1)} = x^{(n)} + \alpha^{(n)} p^{(n)}. \tag{13.61}$$

The search vector is chosen to be the direction of the gradient of $\phi^{(n)}$, which is a measure of the current error, and α is set to minimize the error. ϕ may be defined in several ways, for instance, $\phi^{(n)} = (r^{(n)}, r^{(n)}) = ||r^{(n)}||^2$, where $r^{(n)}$ is the current residual defined in (13.21). After some algebra, one finds

$$p^{(n)} = A^{\mathrm{T}} r^{(n)}, \quad \alpha^{(n)} = \frac{\left(r^{(n)}, Ap^{(n)}\right)}{\left(Ap^{(n)}, Ap^{(n)}\right)}, \tag{13.62}$$

where A^{T} is the transpose of A.

Conjugate-Gradient (CG) and Bi-Conjugate Gradient (BCG) Methods

In this powerful class of methods one generates a set of A-orthogonal vectors and then minimizes the error measure in the direction of each of them. The most popular choice is to take the search vector $p^{(n)}$ to be A-orthogonal to the residual $r^{(n)}$. The iteration process is given by (13.61). In the case of a symmetric, positive-definite matrix A, it proceeds as follows.

First, we choose a suitable starting estimate $x^{(0)}$ and set $p^{(0)} = r^{(0)} = b - Ax^{(0)}$. Then, for each $i = 0, 1, \ldots, N-1$ we compute

$$\alpha_i = \frac{\left(r^{(i)}, r^{(i)}\right)}{\left(Ap^{(i)}, p^{(i)}\right)}, \tag{13.63}$$

$$x^{(i+1)} = x^{(i)} + \alpha_i p^{(i)}, \tag{13.64}$$

$$r^{(i+1)} = r^{(i)} - \alpha_i Ap^{(i)}, \tag{13.65}$$

$$\beta_i = \frac{\left(r^{(i+1)}, r^{(i+1)}\right)}{\left(r^{(i)}, r^{(i)}\right)}, \tag{13.66}$$

and

$$p^{(i+1)} = r^{(i+1)} + \beta_i p^{(i)}. \tag{13.67}$$

For symmetric positive-definite matrices the vectors $p^{(n)}$ form the basis of the corresponding vector space; hence any vector, including the exact solution x, can be written as linear combination of N such vectors. This is what is done when applying (13.63)–(13.67). Without rounding errors, the exact solution would be obtained after $N-1$ steps.

A variant of this method is the *bi-conjugate gradient* (BCG) method. It consists of developing two subspaces, one based on A and one based on A^{T}. The basis vectors are mutually orthogonal instead of being orthogonal within each set. Equations (13.64), (13.65), and (13.67) are unchanged; equations (13.63) and (13.66) are modified to read

$$\alpha_i = \frac{\left(r^{(i)}, \overline{r}^{(i)}\right)}{\left(Ap^{(i)}, \overline{p}^{(i)}\right)} \tag{13.68}$$

$$\beta_i = \frac{\left(r^{(i+1)}, \overline{r}^{(i+1)}\right)}{\left(r^{(i)}, \overline{r}^{(i)}\right)}, \tag{13.69}$$

where

$$\overline{r}^{(i+1)} = \overline{r}^{(i)} - \alpha_i A^{\mathrm{T}} \overline{p}^{(i)}, \quad \text{and} \quad \overline{p}^{(i+1)} = \overline{r}^{(i+1)} + \beta_i \overline{p}^{(i)}. \tag{13.70}$$

with the initial values $\overline{r}^{(0)} = r^{(0)}$, $\overline{p}^{(0)} = \overline{r}^{(0)}$. For more details refer to [928]. In the astrophysical context, the method was used for radiative transfer in spherical media in [34].

Generalized Minimum Residual Method (GMRES)

This is a generalization of the previous method to the case of nonsymmetric matrices [315, 929, 930]. The simplest approach is to use the theorem of linear algebra that states that for any nonsingular matrix A, the matrix $A^{\mathrm{T}}A$ is symmetric and positive-definite. We replace the original system $Ax = b$ by an equivalent system $(A^{\mathrm{T}}A)\,x = A^{\mathrm{T}}b$ and then apply (13.63)–(13.67) using $(A^{\mathrm{T}}A)$ and $A^{\mathrm{T}}b$ instead of A and b.

A variant of GMRES is called the *generalized conjugate residual* (GCR) method, and a special form of that is called ORTHOMIN. The GCR iterations proceed as follows. We start with $x^{(0)}$ and set $p^{(0)} = r^{(0)} = b - Ax^{(0)}$. Then, for each $i = 0, 1, \ldots,$ we compute

$$\alpha_i = \frac{\left(Ap^{(i)}, r^{(i)}\right)}{\left(Ap^{(i)}, Ap^{(i)}\right)}, \tag{13.71}$$

$$x^{(i+1)} = x^{(i)} + \alpha_i p^{(i)}, \tag{13.72}$$

$$r^{(i+1)} = r^{(i)} - \alpha_i\, Ap^{(i)}, \tag{13.73}$$

$$\beta_{ij} \equiv -\frac{\left(Ap^{(j)}, Ar^{(i+1)}\right)}{\left(Ap^{(j)}, Ap^{(j)}\right)}, \qquad j = 0, 1, \ldots, i, \tag{13.74}$$

and

$$p^{(i+1)} = r^{(i+1)} + \sum_{j=0}^{i} \beta_{ij} p^{(j)}. \tag{13.75}$$

Thus the new search vector $p^{(i+1)}$ is made A-orthogonal to all previous search vectors $p^{(i)}, p^{(i-1)}$, etc. This procedure may be cumbersome or demand too much memory, so one may stop and restart the orthogonalization process or limit it to the k most recent search vectors. In this case, the summation in (13.75) is replaced by $\sum_{j=i'}^{i} \beta_{ij} p^{(j)}$, where $i' \equiv i - k + 1$. This variant is called ORTHOMIN(k). The first application of this method to astrophysical radiative transfer calculations was in [623].

Other Methods

Many other methods are possible. We mention here only the Chebyshev method [200, 296]. It is not clear whether it can be used as a universal method for radiative transfer problems because it requires some specific properties of the eigenvalues of matrix A, but it has been successfully applied to solving sets of kinetic equilibrium equations.

13.3 ACCELERATED LAMBDA ITERATION (ALI)

Construction of an Approximate Operator

The breakthrough for astrophysical work was made in the OAB paper [815] where, using Gershgorin's theorem, it was shown that the nearly optimum Λ^* is simply the *diagonal* part of the exact Λ matrix and that this choice always produces sufficiently low maximum eigenvalues of the amplification matrix. This scheme is essentially equivalent to the Jacobi method with a particular choice for the preconditioner. For the problem defined by (13.2), i.e., for $A = I - (1-\epsilon)\Lambda$, in which all the matrix elements are positive because of the physical meaning of the Λ matrix, the largest eigenvalue of the amplification matrix (13.20) is bounded by

$$|\lambda_{\max}| \le \max \left\{ [1-(1-\epsilon)\Lambda_{ii}]^{-1}(1-\epsilon)\Big[\sum_{j\neq i} \Lambda_{ij} - \Lambda_{ii}\Big] \right\}. \tag{13.76}$$

The term $L \equiv \sum_j \Lambda_{ij}$ can be thought of as being the result of a formal solution with the source function set everywhere to unity. Near the outer boundary, $\Lambda^*(\tau \approx 0) \approx 0$, and $L \approx \frac{1}{2}$ (which follows from the Eddington-Barbier relation), so that $\lambda_{\max} \lesssim (1-\epsilon)/2$. Thus there is no difficulty in obtaining a reasonable convergence rate at the boundary. At large depths, the Λ matrix becomes diagonally dominated; hence $\lambda_{\max}$ in these layers goes to zero.

In OAB a detailed study was made of the dependence of the maximum eigenvalue of the amplification matrix on various parameters. They found the interesting feature that $\lambda_{\max}$ depends sensitively on the coarseness of the spatial mesh in the sense that the maximum eigenvalue *decreases* with a *decreasing* number of depth points per decade of optical depth. Therefore, the convergence of the method is fastest for coarse grids (of course, the coarseness should not be exaggerated because the overall accuracy would then deteriorate).

The dependence of $\lambda_{\max}$ on the coarseness of the grid can be understood intuitively in terms of communication of information along the computational grid. In a pure Λ-iteration, communication of the solution propagates on the order of one optical depth per iteration. In an accelerated Λ-iteration, the Λ^*-operator subtracts out an analytic solution that mimics the true solution locally. And because Λ^* is moved to the left-hand side of the equation for the source function, it is implicit in the solution for the next iteration values. Therefore, information is no longer limited to propagating on the optical depth scale, but is instead limited to propagating on the discrete grid. In a finer grid, the information must go through more points in order to propagate the same distance in the optical depth space; therefore, the iteration process is slower.

For classical Λ-iteration, $\Lambda^* = 0$, thus $F = (1-\epsilon)\Lambda$. Another important result obtained by OAB is that in this case the maximum eigenvalue of F is

$$\lambda_{\max} \approx (1-\epsilon)(1-T^{-1}), \tag{13.77}$$

where $T > 1$ is the total optical thickness of the radiating slab. For $T \gg 1$, $\lambda_{\max}$ is very close to unity for $\epsilon \ll 1$. This is yet another demonstration of the failure of classical Λ-iteration for an optically thick slab with a small photon destruction probability ϵ.

It is also natural to choose the diagonal part of the exact Λ as an approximate operator from a physical point of view. Recall that the matrix element Λ_{ji} tells us what portion of photons created in an elementary interval around depth point i

[i.e., $S(\tau_i)$] are absorbed at depth point j [described by $J(\tau_j)$]. Most photons are absorbed close to their point of creation, so the diagonal term Λ_{ii} is much larger than the off-diagonal terms. In other words, approximating the exact Λ by a diagonal operator means replacing the kernel function for the mean intensity, (11.108), by a δ-function, which, as seen in figure 11.6, is reasonable. These considerations also show that the next simplest approximation for the Λ-operator would be its tridiagonal part; here the information on the source function at a given depth is communicated to its nearest neighbors.

Evaluation of Diagonal Operators

The remaining problem is the evaluation of the diagonal of the Λ matrix. This calculation can be done either approximately or exactly. In either case, one uses the observation that the elements of the Λ matrix can be evaluated by setting the source function to be the unit pulse function, $S(\tau_{d'}) = \delta_{d'd}$, and solving the transfer equation with this source function. In matrix notation,

$$J_d = \sum_{d'} \Lambda_{dd'} S_{d'}. \tag{13.78}$$

Taking the unit pulse function $S(\tau_{d'}) = \delta_{d'd}$ for S, we obtain

$$J_d = \Lambda_{dd} \, . \tag{13.79}$$

In other words, the diagonal element of the Λ matrix at depth point d is equal to the mean intensity (or the specific intensity, in the case of elementary, frequency- and angle-dependent $\Lambda_{\mu\nu}$) computed for the source function having the zero value everywhere but in the point d, i.e., $S_d = 1$, and $S_i = 0$, $(i \neq d)$; or, written explicitly as a function of τ,

$$\Lambda_{dd'} = \Lambda_{\tau_d}[\delta(\tau_{d'} - \tau)]. \tag{13.80}$$

There are several variants of constructing the approximate Λ^*-operators using the above prescription. The most widely used are the following two schemes.

(i) Olson and Kunasz [816] suggested using the method of short characteristics to evaluate the Λ^*-operator. As follows from (13.80), to evaluate the diagonal operator at τ_d we merely take the coefficients located at $S(\tau_d)$, which yields, for the first-order short characteristics method

$$\Lambda^*_{\mu\nu} = 1 - \frac{1 - e^{-\Delta\tau_{d+1/2}}}{\Delta\tau_{d+1/2}}, \quad \mu > 0 \tag{13.81}$$

and

$$\Lambda^*_{\mu\nu} = 1 - \frac{1 - e^{-\Delta\tau_{d-1/2}}}{\Delta\tau_{d-1/2}}, \quad \mu < 0. \tag{13.82}$$

Recall that $\Delta\tau_{d-1/2} = (\tau_d - \tau_{d-1})/|\mu|$. We have dropped the frequency subscript on the right-hand side but keep it on the left-hand side to emphasize that the Λ-operator is an elementary operator yielding the specific intensity $I_{\mu\nu}$. The approximate operator corresponding to the mean intensity of radiation, Λ^*_ν, is given by

$$\Lambda^*_\nu = 1 - \frac{1}{2}\int_0^1 \left(\frac{1 - e^{-\Delta\tau_{d-1/2}}}{\Delta\tau_{d-1/2}} + \frac{1 - e^{-\Delta\tau_{d+1/2}}}{\Delta\tau_{d+1/2}} \right) d\mu. \tag{13.83}$$

If one uses second-order short characteristic method the resulting formulae for Λ^* are somewhat more complicated [816]. They are given by appropriate integrals over angles of the quantities $\lambda^+_{d,d}$ and $\lambda^-_{d,d}$ given by equations (12.126) and (12.129).

(ii) Another approach is to use the Feautrier method. Here one solves the tridiagonal system $\mathbf{T}_{\mu\nu}\cdot\mathbf{j}_{\mu\nu} = \mathbf{S}_{\mu\nu}$ for the vector $\mathbf{j}_{\mu\nu}$, which contains the symmetrized intensity $j_{\mu\nu} \equiv \frac{1}{2}[I(\mu,\nu) + I(-\mu,\nu)]$, at all depth points, given the vector $\mathbf{S}_{\mu\nu}$, which contains the source function at all inner depth points, with a modified meaning at the boundary points; see § 13.2. Thus

$$\mathbf{j}_{\mu\nu} = \mathbf{T}^{-1}_{\mu\nu}\cdot\mathbf{S}_{\mu\nu}, \tag{13.84}$$

which shows that $\Lambda_{\mu\nu}$ is

$$\tfrac{1}{2}\left(\Lambda_{\mu\nu} + \Lambda_{-\mu\nu}\right) = \mathbf{T}^{-1}_{\mu\nu}. \tag{13.85}$$

The Λ_ν matrix corresponding to the mean intensity is

$$\Lambda_\nu = \int_0^1 \mathbf{T}^{-1}_{\mu\nu}\, d\mu. \tag{13.86}$$

This idea was first applied in OAB; the authors argued that evaluating the exact diagonal would be time-consuming and suggested evaluation of an *approximate* diagonal. But Rybicki and Hummer [925], in a very important paper, demonstrated that one can use an ingenious method to calculate the *exact* diagonal elements of the inverse of a tridiagonal matrix efficiently. In fact, the entire set of diagonal elements of the matrix $\mathbf{T}^{-1}$ can be found in order D operations. This feature makes it the method of choice, because it avoids computing costly exponentials, a problem inherent in both previous approaches. This method is particularly useful if one performs the formal solution of the transfer equation by the Feautrier method, because then the evaluation of the Λ^*-operator requires almost no additional computational effort.

The procedure, following [925], is as follows. Let $\mathbf{T}$ be an $N \times N$ tridiagonal matrix and let its inverse be $\boldsymbol{\lambda} \equiv \mathbf{T}^{-1}$. The equation for the inverse can be written as $\mathbf{T}\cdot\boldsymbol{\lambda} = 1$ or, in component form,

$$-A_i\lambda_{i-1,j} + B_i\lambda_{ij} - C_i\lambda_{i+1,j} = \delta_{ij}. \tag{13.87}$$

For any fixed value of j this equation can be solved by one of the forms of Gaussian elimination. In the usual implementation the elimination proceeds from $i=1$ to $i=N$, followed by back-substitution from $i=N$ to $i=1$,

$$D_i = (B_i - A_i D_{i-1})^{-1} C_i, \tag{13.88}$$

$$Z_{ij} = (B_i - A_i D_{i-1})^{-1} (\delta_{ij} + A_i Z_{i-1,j}), \tag{13.89}$$

and

$$\lambda_{ij} = D_i \lambda_{i+1,j} + Z_{ij}. \tag{13.90}$$

It is also possible to implement the method in reverse order,

$$E_i = (B_i - C_i E_{i+1})^{-1} A_i, \tag{13.91}$$

$$W_{ij} = (B_i - C_i E_{i+1})^{-1} (\delta_{ij} + C_i W_{i+1,j}), \tag{13.92}$$

and

$$\lambda_{ij} = E_i \lambda_{i-1,j} + W_{ij}. \tag{13.93}$$

The crucial idea of the method is to use parts of *both* of these implementations to find the diagonal elements λ_{ii}.

Inasmuch as $\delta_{ij}=0$ for $i \neq j$, it follows from (13.89) and (13.92) that $Z_{ij}=0$ for $i<j$, and $W_{ij}=0$ for $i>j$. Thus, from (13.89) and (13.90) we obtain, for special choices of i and j,

$$Z_{ii} = (B_i - A_i D_{i-1})^{-1}, \tag{13.94}$$

$$\lambda_{ii} = D_i \lambda_{i+1,i} + Z_{ii}, \tag{13.95}$$

$$\lambda_{i-1,i} = D_{i-1} \lambda_{ii}, \tag{13.96}$$

and, from (13.92) and (13.93),

$$W_{ii} = (B_i - C_i E_{i+1})^{-1}, \tag{13.97}$$

$$\lambda_{ii} = E_i \lambda_{i-1,i} + W_{ii}, \tag{13.98}$$

and

$$\lambda_{i+1,i} = E_{i+1} \lambda_{ii}. \tag{13.99}$$

Using (13.94), (13.95), and (13.99), we eliminate Z_{ii} and $\lambda_{i+1,i}$ to obtain

$$\lambda_{ii} = (1 - D_i E_{i+1})^{-1} (B_i - A_i D_{i-1})^{-1}. \tag{13.100}$$

The right-hand side now depends only on the single-indexed quantities A_i and B_i, which are given, and D_i and E_i, which can be found by two passes through the depth grid, using the recursion relations (13.88) and (13.91). Thus λ_{ii} can be found in order N operations.

Because the approximate operator is typically evaluated during the formal solution, there is little extra work involved in its determination. The quantities A_i, B_i, C_i, and D_i are common to both problems, and one needs only to include the recursion relation (13.91) as a part of the back-substitution to find the auxiliary quantities E_i.

Other Types of Operators

Tridiagonal Λ^*-Operator

Olson and Kunasz [816] suggested using the tridiagonal part of the exact Λ-operator as a better approximate Λ^*-operator. They give expressions for off-diagonal elements obtained from the short characteristic method; they are similar to, but more complicated than, those in (13.83). The off-diagonal elements can also be evaluated even more economically with the Rybicki-Hummer scheme, using (13.96) and (13.99).

From a physical point of view, a tridiagonal operator is clearly a better approximation to the exact operator, because it permits transfer of information between a given depth point and its immediate neighbors. Expressions for the individual elements of the matrix elements of the approximate operator are given in [816]. Explicit expressions are also given in chapter 12, equations (12.126)–(12.133), where, for instance, $\lambda_{d,d}$ is replaced by $\Lambda^*_{d,d}$, etc. The use of these quantities in formulating the approximate Λ-operator explains the notation employed in chapter 12.

Higher Multiband Operators

Similarly, one may use a higher multiband part of the exact Λ-operator. As can be expected, using pentadiagonal and higher-order multiband operators will increase a speed of convergence, but this gain is somewhat outweighed by increasing numerical work to evaluate the corresponding matrix elements. These methods are used only rarely in actual numerical work. A systematic study of the properties of the ALI iterations as a function of the number of bands is presented in [707] and [458].

Gauss-Seidel Approximate Operator

This approach was first used for astrophysical radiative transfer by Trujillo-Bueno and Fabiani-Bendicho [1086]. The essential problem is to avoid constructing the individual elements of the matrix $A = I - (1 - \epsilon)\Lambda$ explicitly. To address it, the authors developed an ingenious scheme based on the second-order short characteristics formal solver. In essence, they perform the formal solution in two sweeps: first a "downward" sweep for incoming radiation ($\mu < 0$), in which they use the "old" source function at all depth points. However, this result is inadequate because as shown in (13.49) and (13.50), the Gauss-Seidel scheme requires both the "old and new" source functions at the same time. Hence, in the upward sweep ($\mu > 0$), they correct the source function to its "new" value depth by depth, using the information from the previous depth points. A full description of the procedure is rather technical, so the reader is referred to the original paper for further details.

Successive Over-Relaxation

Successive over-relaxation was suggested in [1086] as an efficient modification of the Gauss-Seidel method. The authors also designed a simple procedure to evaluate an approximate optimum over-relaxation parameter ω; see (13.50). Essentially, they use the observation that after several iterations the error vector is dominated by the largest eigenvalue of the amplification matrix (see § 13.2), and thus the ratio of the maximum relative errors in the subsequent iterations is essentially the maximum eigenvalue, $\lambda_{\max} \approx R_c^{(i+1)}/R_c^{(i)}$, where $R_c^{(i)}$ is the maximum relative change in the ith iteration. This value of $\lambda_{\max}$ is used in (13.53) to evaluate the over-relaxation parameter ω.

Double-Splitting Methods

One application of the double-splitting method was suggested in [623] and is being used in the code ALTAIR [200, 296, 623], in the context of transfer of radiation in a gas of multi-level atoms.

13.4 ACCELERATION OF CONVERGENCE

Although all the iterative methods discussed above can be used as stand-alone methods, it is often advantageous to combine them. In particular, we can obtain *acceleration of convergence* by combining a specific preconditioner method with a specific nonlinear, nonstationary method such as shown in (13.59)–(13.75).

For example, recall that the original system $Ax = b$ when solved by iteration has a current residual given by $r^{(n)} = b - Ax^{(n)}$. If we use a preconditioner, P, the original system is effectively replaced by an equivalent system

$$(P^{-1}A)\,x = P^{-1}b, \quad \text{or} \quad \widetilde{A}\,x = \widetilde{b}, \tag{13.101}$$

which has the same solution x as the original system. The current "preconditioned residual" of this equation, $\widetilde{r}^{(n)}$, is given by [50]:

$$\begin{aligned}\widetilde{r}^{(n)} &= \widetilde{b} - \widetilde{A}x^{(n)} = P^{-1}(b - Ax^{(n)}) \\ &= P^{-1}r^{(n)} = \delta x^{(n+1)} \equiv x^{(n+1)} - x^{(n)},\end{aligned} \tag{13.102}$$

where the fourth equality follows from equation (13.23). Equation (13.102) shows that *the residual of the preconditioned equation is the correction to the current estimate.*

The nonlinear, nonstationary methods described above need only the residuals (current and previous), the matrix A, and the right-hand side vector b to construct the individual search vectors $p^{(j)}$. Thus we can combine a choice of preconditioner scheme with a nonlinear (acceleration) method by using the algorithms for the nonlinear scheme, while replacing

$$\begin{aligned} A &\longrightarrow \widetilde{A} = P^{-1}A, \\ b &\longrightarrow \widetilde{b} = P^{-1}b, \\ r^{(n)} &\longrightarrow \widetilde{r}^{(n)} = x^{(n+1)} - x^{(n)}. \end{aligned} \tag{13.103}$$

As was already stressed above, in order to apply any acceleration scheme to the radiative transfer problems, it has to be formulated in such a way that the matrix A is *not assembled explicitly*, nor is a *multiplication of matrix A with any vector performed explicitly*. This can be done by applying an ingenious approach suggested by Auer [50].

For example, consider the GMRES (ORTHOMIN) method, and examine how it is combined with an arbitrary preconditioning scheme. Using (13.102), the preconditioned residual $\widetilde{r}^{(n-1)}$ is given by

$$\widetilde{r}^{(n-1)} = \widetilde{x}^{(n)} - x^{(n-1)}, \tag{13.104}$$

and $\widetilde{A}\,\widetilde{r}^{(n-1)}$ can be expressed as

$$\begin{aligned}\widetilde{A}\,\widetilde{r}^{(n-1)} &= P^{-1}A\left(\widetilde{x}^{(n)} - x^{(n-1)}\right)\\ &= P^{-1}\left[\left(b - A\,x^{(n-1)}\right) - \left(b - A\,\widetilde{x}^{(n)}\right)\right]\\ &= \widetilde{r}^{(n-1)} - \left(\widetilde{x}^{(n+1)} - x^{(n)}\right),\end{aligned} \tag{13.105}$$

where the first equality follows from (13.104) and the last equality from the identity $P^{-1}\left(b - A\,x^{(n-1)}\right) \equiv \widetilde{r}^{(n-1)}$ and from equation (13.102). Here $\widetilde{x}^{(i)}$ is the ith iterate obtained by an application of an ordinary (unaccelerated) preconditioning approach, using a previously calculated $x^{(i-1)}$. This notation is chosen to stress that the final (accelerated) ith iterate, denoted as $x^{(i)}$, is modified by an application of the ORTHOMIN (or possibly another nonlinear) scheme.

Equations (13.104) and (13.105) show that $\widetilde{r}$ and $\widetilde{A}\,\widetilde{r}$ can indeed be calculated without a construction of, and explicit multiplication with, matrix $\widetilde{A}$, using just the iterates in three consecutive iteration steps.

The iterations proceed as follows. The first iteration is done as in the ordinary preconditioning scheme, yielding $x^{(1)} = \widetilde{x}^{(1)}$. One then performs the second iteration of the ordinary preconditioning scheme, yielding $\widetilde{x}^{(2)}$, and thus also $\widetilde{r}^{(1)}$ and $\widetilde{A}\,\widetilde{r}^{(1)}$. One sets $p^{(1)} = \widetilde{r}^{(1)}$ and $\widetilde{A}\,p^{(1)} = \widetilde{A}\,\widetilde{r}^{(1)}$. In the subsequent iterations, $i \geq 2$, one uses equations (13.104) and (13.105) for $\widetilde{r}^{(i-1)}$ and $\widetilde{A}\,\widetilde{r}^{(i-1)}$ and appropriately modified equations (13.71)–(13.75), namely,

$$\alpha_{i-1} = \frac{\left(\widetilde{A}p^{(i-1)}, \widetilde{r}^{(i-1)}\right)}{\left(\widetilde{A}p^{(i-1)}, \widetilde{A}p^{(i-1)}\right)}, \tag{13.106}$$

$$x^{(i)} = x^{(i-1)} + \alpha_{i-1}p^{(i-1)}, \tag{13.107}$$

$$\widetilde{r}^{(i)} = \widetilde{r}^{(i-1)} - \alpha_{i-1}\widetilde{A}p^{(i-1)}, \tag{13.108}$$

$$\beta_{ij} = -\frac{\left(\widetilde{A}p^{(j)}, \widetilde{A}r^{(i)}\right)}{\left(\widetilde{A}p^{(j)}, \widetilde{A}p^{(j)}\right)}, \qquad j = 1, \ldots, i-1, \tag{13.109}$$

$$p^{(i)} = r^{(i)} + \sum_{j=1}^{i-1} \beta_{ij}p^{(j)}, \tag{13.110}$$

and

$$\widetilde{A}p^{(i)} = \widetilde{A}r^{(i)} + \sum_{j=1}^{i-1} \beta_{ij}\widetilde{A}p^{(j)}, \tag{13.111}$$

which follows from (13.110). Hence we indeed need not perform a multiplication with the matrix $\widetilde{A}$.

A preconditioned variant of the bi-conjugate gradient method, using a different strategy for dealing with a multiplication with matrix A and avoiding computing the transposed matrix A^T was developed in [34], to which the reader is referred for computational details.

Residual Minimization–Ng Method

An alternative acceleration method is to minimize the residuals themselves. The underlying theory is described in [794, 1008, 1017]. In the astrophysical literature, perhaps the most widely used method in this category was originally developed by Ng [794] for plasma physics problems and introduced into astrophysical radiative transfer in [49, 815]. It is somewhat similar to ORTHOMIN. As discussed in detail in [49], it is based on minimizing a distance between two consecutive "extrapolated estimates" of the solution. The extrapolated estimate of the solution in the nth iteration is given by

$$\widetilde{x}^{(n)} = \left(1 - \sum_{m=1}^{M} \alpha_m\right) x^{(n)} + \sum_{m=1}^{M} \alpha_m x^{(n-m)}$$

or

$$\widetilde{r}^{(n)} = \left(1 - \sum_{m=1}^{M} \alpha_m\right) r^{(n)} + \sum_{m=1}^{M} \alpha_m r^{(n-m)}. \tag{13.112}$$

The coefficients α_m are determined by minimizing the norm $||\widetilde{x}^{(n)} - \widetilde{x}^{(n-1)}||^2$. In effect, we solve for α_m by least squares. It is advantageous to generalize the definition of the scalar product and the norm to $||x||^2 \equiv (x, x) = \sum_j w_j (x_j)^2$, where w_j are weights, the choice of which depend on the problem; see below. For the most common choice of $M = 2$, the coefficients α are given by

$$\alpha_1 = (\delta_{01}\delta_{22} - \delta_{02}\delta_{21}) \,/\, (\delta_{11}\delta_{22} - \delta_{12}\delta_{21}), \tag{13.113}$$

and

$$\alpha_2 = (\delta_{02}\delta_{11} - \delta_{01}\delta_{21}) \,/\, (\delta_{11}\delta_{22} - \delta_{12}\delta_{21}), \tag{13.114}$$

where

$$\delta_{ij} \equiv \left(\Delta x^{(n)} - \Delta x^{(n-i)}\right) \cdot \left(\Delta x^{(n)} - \Delta x^{(n-j)}\right), \tag{13.115}$$

for $i = 1, 2$, and $j = 1, 2$; and $\delta_{0j} \equiv \Delta x^{(n)} \cdot (\Delta x^{(n)} - \Delta x^{(n-j)})$; with

$$\Delta x^{(n)} \equiv x^{(n)} - x^{(n-1)}. \tag{13.116}$$

The method does not specify how the individual iterates $x^{(n)}$ are evaluated. Hence it can be used in conjunction with any method that provides an algorithm for computing $x^{(n)}$.

13.5 ASTROPHYSICAL IMPLEMENTATION

ALI has proved to be a very reliable tool for solving a wide variety of transfer equations. As will be seen in chapters 14, 15, 18, 19, and 20, it allows us to solve problems in which the state of the material and the radiation field are mutually strongly coupled, and both depart markedly from LTE. In addition, ALI can be combined with Newton-Raphson iteration to yield very powerful algorithms that permit accurate calculation of realistic and complex NLTE model atmospheres in planar and spherical geometry, which allow for departures from LTE and account for blanketing by thousands to millions of spectral lines.

Formal Solution of the Transfer Problem with a Coherent Scattering Term

To illustrate an actual implementation of ALI in the context of astrophysical radiative transfer, we consider the simplest nontrivial case of a source function composed of a known thermal term plus an isotropic scattering contribution, such as the electron (Thomson) scattering or Rayleigh scattering. In this case,

$$S_{\nu,\mathbf{n}} = \frac{\eta_{\nu,\mathbf{n}}}{\chi_{\nu,\mathbf{n}}} + \frac{\sigma_\nu}{\chi_{\nu,\mathbf{n}}} J_\nu, \tag{13.117}$$

where $S_{\nu,\mathbf{n}}$ is the total source function, generally frequency- and direction-dependent, $\eta_{\nu,\mathbf{n}}$ the thermal emission coefficient, $\chi_{\nu,\mathbf{n}}$ the total absorption coefficient (including the true absorption and scattering; $\chi_{\nu,\mathbf{n}} = \kappa_{\nu,\mathbf{n}} + \sigma_\nu$), and σ_ν the scattering coefficient. In the case of Thomson scattering, it is given by $\sigma_\nu = n_e \sigma_e$, where n_e is the electron density and σ_e the Thomson cross section, which is frequency-independent.

As was demonstrated in § 12.2, one can solve this problem directly (non-iteratively) using the second-order form of the transfer equation and employing the Feautrier or Rybicki (§ 12.3) algorithms. However, an application of the short characteristics or discontinuous finite element scheme requires the knowledge of the *total* source function; in this case the ALI scheme outlined below may be used to advantage. The solution algorithm as specified below can also be directly applied in a general three-dimensional case.

For a given frequency and given direction, the specific intensity given by a formal solution using the total source function is formally written as

$$I_{\nu,\mathbf{n}} = \Lambda_{\nu,\mathbf{n}}[S_{\nu,\mathbf{n}}]. \tag{13.118}$$

Although written in an operator form, it should be stressed that the Λ-operator does not have to be assembled explicitly; equation (13.118) should rather be understood as a *process* of obtaining the specific intensity from the known source function.

Using the basic idea of ALI, this expression is rewritten in an iterative form

$$I_{\nu,\mathbf{n}}^{\rm new} = \Lambda_{\nu,\mathbf{n}}^*[S_{\nu,\mathbf{n}}^{\rm new}] + (\Lambda_{\nu,\mathbf{n}} - \Lambda_{\nu,\mathbf{n}}^*)\big[S_{\nu,\mathbf{n}}^{\rm old}\big], \tag{13.119}$$

where Λ^* is a suitably chosen approximate operator. Using equation (13.119), the mean intensity can be written as

$$J_\nu^{\rm new} \equiv \frac{1}{4\pi}\oint d\Omega\, I_{\nu,\mathbf{n}}^{\rm new} = \frac{1}{4\pi}\oint d\Omega\, \Lambda_{\nu,\mathbf{n}}^*[S_{\nu,\mathbf{n}}^{\rm new}] + \frac{1}{4\pi}\oint d\Omega\, (\Lambda_{\nu,\mathbf{n}} - \Lambda_{\nu,\mathbf{n}}^*)\big[S_{\nu,\mathbf{n}}^{\rm old}\big]. \tag{13.120}$$

The "new" and "old" values of the source function are given by (13.117) with $J_\nu^{\rm new}$ and $J_\nu^{\rm old}$, respectively. Notice that the coefficients $\chi_{\nu,\mathbf{n}}$, $\eta_{\nu,\mathbf{n}}$, and σ_ν are fixed, so that they are the same for the "new" and "old" source function. Substituting (13.117) into (13.120), we obtain

$$\begin{aligned} J_\nu^{\rm new} = &\frac{1}{4\pi}\oint d\Omega\, \Lambda_{\nu,\mathbf{n}}^*\left[\frac{\eta_{\nu,\mathbf{n}}}{\chi_{\nu,\mathbf{n}}} + \frac{\sigma_\nu}{\chi_{\nu,\mathbf{n}}} J_\nu^{\rm new}\right] \\ &+ \frac{1}{4\pi}\oint d\Omega\, \Lambda_{\nu,\mathbf{n}}\left[S_{\nu,\mathbf{n}}^{\rm old}\right] \\ &- \frac{1}{4\pi}\oint d\Omega\, \Lambda_{\nu,\mathbf{n}}^*\left[\frac{\eta_{\nu,\mathbf{n}}}{\chi_{\nu,\mathbf{n}}} + \frac{\sigma_\nu}{\chi_{\nu,\mathbf{n}}} J_\nu^{\rm old}\right]. \end{aligned} \tag{13.121}$$

The second term can be written as

$$J_\nu^{\rm FS} \equiv \frac{1}{4\pi}\oint d\Omega\, \Lambda_{\nu,\mathbf{n}}\left[S_{\nu,\mathbf{n}}^{\rm old}\right], \tag{13.122}$$

i.e., as a "newer" mean intensity obtained by a formal solution using an old source function. Introducing a correction to the mean intensity,

$$\delta J_\nu \equiv J_\nu^{\rm new} - J_\nu^{\rm old}, \tag{13.123}$$

equation (13.121) can be written as

$$\delta J_\nu = \left(I - \frac{1}{4\pi}\oint d\Omega\, \Lambda_{\nu,\mathbf{n}}^* \frac{\sigma_\nu}{\chi_{\nu,\mathbf{n}}}\right)^{-1}\left[J_\nu^{\rm FS} - J_\nu^{\rm old}\right], \tag{13.124}$$

where I is a unit (identity) operator. The most practical choice for the approximate operator is a diagonal (local) operator, in which case the action of Λ^* is simply an algebraic multiplication, and $\Lambda_{\nu,\mathbf{n}}^*$ is understood as a real number. One can define an averaged Λ^*-operator as

$$\overline{\Lambda_\nu^*} \equiv \frac{1}{4\pi}\oint d\Omega \left(\frac{\sigma_\nu}{\chi_{\nu,\mathbf{n}}}\right)\Lambda_{\nu,\mathbf{n}}^* \tag{13.125}$$

so that the correction δJ_ν, using a diagonal operator, can simply be written as

$$\delta J_\nu = \frac{J_\nu^{\rm FS} - J_\nu^{\rm old}}{1 - \overline{\Lambda}_\nu^*}. \tag{13.126}$$

The iteration scheme proceeds as follows. For each frequency ν:

(a) For a given $S^{\rm old}$ (with an initial estimate $S^{\rm old} = B$, or some other suitable value), perform a formal solution, one frequency and direction at a time, to obtain new values of the specific intensity $I_{\nu,\mathbf{n}}$.
(b) By integrating over directions using equation (13.122), obtain the value of the formal-solution mean intensity $J_\nu^{\rm FS}$.
(c) Using (13.126), evaluate a new iterate of the mean intensity, $J_\nu^{\rm new} = J_\nu^{\rm old} + \delta J_\nu$.
(d) If the mean intensity found in step (c) differs from that used in step (a), update the source function from (13.117) using the newly found mean intensity and repeat steps (a) through (c) to convergence.

In fact, step (b) is not done independently after completing all formal solutions in step (a); instead, one gradually updates the integral over directions to obtain $J_\nu^{\rm FS}$ and $\overline{\Lambda}_\nu^*$ after the formal solution for each direction. Step (b) thus takes a negligible time compared to (a); similarly, step (c), which is just one division per depth point, also takes a negligible time. The whole procedure is thus completely dominated by the formal solutions.

In the following chapters of this book we will describe an application of ALI in more complex cases, such as radiative transfer in a gas of two-level atoms with and without background continuum (§ 14.2); general multi-level transfer with specified temperature and density (§ 14.5); two-level (§ 15.4) and multi-level (§ 15.5) transfer with partial frequency redistribution; calculation of NLTE model stellar atmospheres (§ 18.4); radiative transfer in moving atmospheres in the observer's frame (§ 19.2), comoving frame (§ 19.3), and mixed frame (§ 19.4); multi-level atom problem in moving atmospheres (§ 19.5); and the construction of global expanding model atmospheres (§ 20.4).

Comparison of Various Approximate Operators and Acceleration Techniques

The most important question to be answered now is "How well do ALI methods perform?" As an example, consider the archetype problem

$$S(\tau) = (1-\epsilon)\Lambda[\,S\,] + \epsilon B(\tau). \tag{13.127}$$

In figure 13.1 we show a comparison of convergence of the maximum relative error using classical Λ-iteration, a diagonal operator without acceleration, and a diagonal operator with four different acceleration schemes. Obviously, the convergence rate of classical Λ-iteration is glacially slow. The results in table 12.1 show that for $\epsilon = 10^{-4}$ one would need $O(10^4)$ classical Λ-iterations to get a

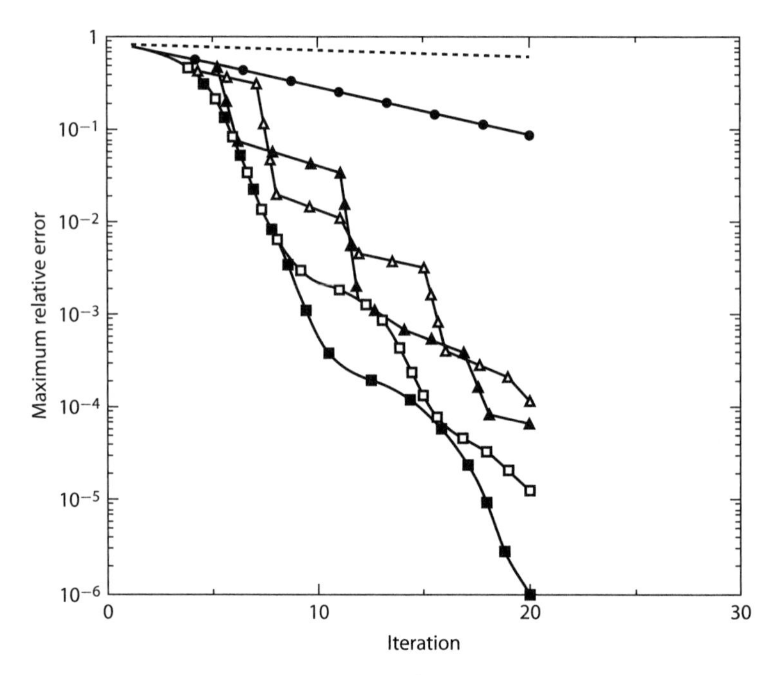

Figure 13.1 Archetype problem with $\epsilon = 10^{-6}$. Λ-iteration: ....; unaccelerated diagonal: •; Ng acceleration, M = 2: △; Ng acceleration, $M = 4$: ▲; ORTHOMIN, $M = 2$:□; ORTHOMIN, $M = 4$: ■. Adapted from [523].

result good to a few percent, and $\sim 3 \times 10^4$ to get one accurate to three digits. The rate using an unaccelerated diagonal operator is much better; the result would be good to approximately three digits in 60 iterations. That number is not prohibitive because each iteration requires only a formal solution, which, using the second-order Feautrier or the discrete finite element methods, is very fast. Much greater speeds are obtained with accelerated operators. The speed-up with Ng acceleration using either two or four previous solutions is impressive. After a few iterations, ORTHOMIN acceleration is about a factor of 10 faster than Ng acceleration.

The convergence rate can also be improved by increasing the bandwidth of the acceleration operators, i.e., to include adjacent grid points. In figure 13.2 we see again that classical Λ-iteration is useless. An unaccelerated tridiagonal operator converges faster than an unaccelerated diagonal operator; but a diagonal operator with Ng acceleration converges as quickly as, or even more quickly than, the unaccelerated tridiagonal operator. An accelerated tridiagonal operator reduces the error by about four orders of magnitude after the first acceleration. An increase in the bandwidth allows information to flow more quickly over the grid.

An extensive study of convergence rates of different algorithms and the time they consume on a variety of computers was reported in [458]. The timing data given there are no longer very useful because the speed of computers has increased enormously compared to those available then; nevertheless, the relative times of different algorithms still provide guidance. The accuracy of ALI for solving transfer

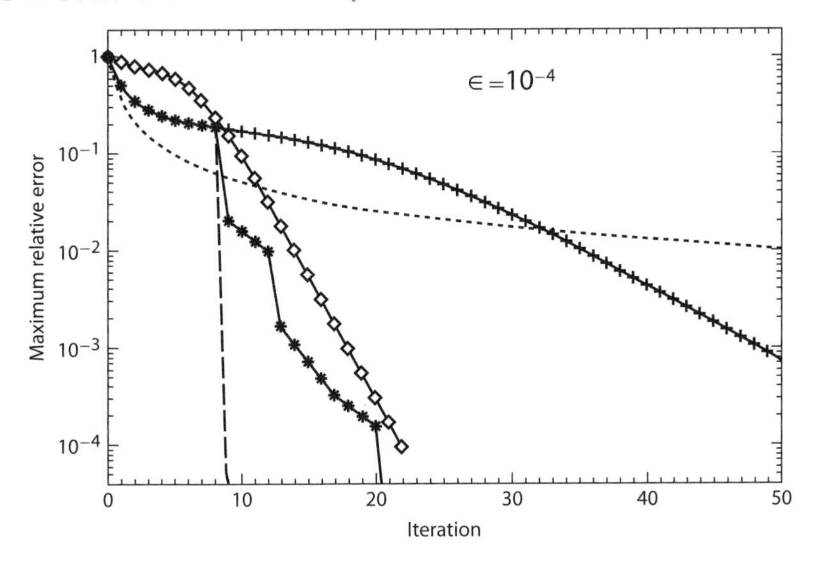

Figure 13.2 Archetype problem with $\epsilon = 10^{-4}$. Λ-iteration: ....; unaccelerated diagonal: $+$; unaccelerated tridiagonal: $\diamond$; diagonal with Ng: $*$; tridiagonal with Ng: $--$. Adapted from [523].

problems was questioned in [235]. But an ALI solution, *properly implemented*, will *always* converge to the "exact" solution (to within round-off errors) of the *discrete difference-equation representation* of the problem. Of course, if too coarse a grid is used, the quality of the solution can deteriorate.

Chapter Fourteen

NLTE Two-Level and Multi-Level Atoms

Great progress in understanding the physics of spectral line formation has resulted from the study of solutions of the coupled equations of transfer and kinetic equilibrium. In this chapter we consider the kinetic equilibrium problem for an "impurity" atom that has no effect on the structure of the atmosphere. We regard the model atmosphere as given, i.e., the temperature T, electron density n_e, mass density ρ, and the total number density $n_{\rm atom}$ of that species are specified functions of spatial position (i.e., depth in 1D models). The atom is assumed to have *NL* discrete energy levels. These levels may be distributed over several ionization stages. The kinetic equations describe all interactions among all levels.

14.1 FORMULATION

Basic Equations

We summarize here the basic equations, approximations, and underlying concepts needed to treat the problem of NLTE radiation transfer in a gas of multi-level atoms. We number the energy levels as $i = 1, 2, \ldots, NL$, which are ordered by increasing energy, $E_1 < E_2 < \ldots < E_{NL}$. These levels may encompass several ionization stages, in which case those with the lowest indices have the lowest ionization stage.

The kinetic equilibrium equation for level i is [cf. equation (9.18)]

$$n_i \sum_{j \neq i}^{NL} \left(R_{ij} + C_{ij}\right) = \sum_{j \neq i}^{NL} n_j \left(R_{ji} + C_{ji}\right). \tag{14.1}$$

The left-hand side gives the total number of transitions out of level i per unit time; the right-hand side gives the total number of transitions per unit time into level i from all other levels. R_{ij} and C_{ij} are the radiative and collisional rates in the transition $i \to j$, respectively. Assuming that only collisions with electrons are important (generally true), the collisional rates are known functions of temperature and electron density; and having taken the atmospheric structure as fixed, the collisional rates are fully specified.

The radiative rates for bound-bound transitions $i < j$ (level j is a level of the same ionization stage as i) are given by

$$R_{ij} = B_{ij} \int_0^\infty J_\nu \, \phi_{ij}(\nu) \, d\nu, \quad i < j, \tag{14.2}$$

where $\phi_{ij}(\nu)$ is the absorption profile for the transition $i \to j$ and B_{ij} is the Einstein absorption probability. The radiative rate for a downward transition $j \to i$ is given by

$$R_{ji} = A_{ji} + B_{ji} \int_0^\infty J_\nu \, \psi_{ji}(\nu) \, d\nu, \quad i < j, \tag{14.3}$$

where A_{ji} is the Einstein spontaneous emission probability, B_{ji} is the Einstein induced emission probability, $B_{ji} \int_0^\infty J_\nu \psi_{ji}(\nu) d\nu$ is the total rate of induced downward transitions, and $\psi_{ji}(\nu)$ is the emission profile for the transition $j \to i$. In this chapter, we will assume complete redistribution of the emitted photons, in which case $\psi_{ji} \equiv \phi_{ij}$ for all transitions between levels i and j. The more general case of departures from complete redistribution is treated in the next chapter.

If levels i and j belong to different ionization stages, then the process $i \to j$ is a photoionization from level i to level j of the next ion. Usually level j is the ground state of the next ion, but in some cases a photoionization transition from level i to the ground state of the next ion may be dipole forbidden; then level j is an excited state of the next ion. The radiative rates for $i < j$ are given by

$$R_{ij} = 4\pi \int_{\nu_{ij}}^\infty \frac{\sigma_{ij}(\nu) J_\nu}{h\nu} \, d\nu \tag{14.4}$$

and

$$R_{ji} = (n_i/n_j)^* \, 4\pi \int_{\nu_{ij}}^\infty \frac{\sigma_{ij}(\nu)}{h\nu} \left(\frac{2h\nu^3}{c^2} + J_\nu \right) e^{-h\nu/kT} \, d\nu. \tag{14.5}$$

With all the radiative and collisional rates specified, (14.1) is a *linear* equation for the level populations. If (14.1) were to be written for all levels of the atom, the resulting equations would be a linearly dependent system because the last equation is a linear combination of the remaining equations because all the transitions into and out of the last level were already accounted for in the equations for the previous levels.

Therefore, one of the equations has to be replaced by some other relation. Typically, one uses the expression for the total number density of the atom, sometimes called the *abundance definition equation*,

$$\sum_{j=1}^{NL} n_j = n_{\rm atom}. \tag{14.6}$$

Here $n_{\rm atom}$ is the total number of the given atomic species in all excitation and ionization states, which can also be written as $n_{\rm atom} = A_{\rm atom} n_H$, where A is the *chemical abundance* of that atom relative to hydrogen, and n_H the *total number density of hydrogen atoms and ions.* There is still freedom as to which equation is to be replaced by the abundance definition condition. Two possibilities are used in the literature. The common one is to replace the equation for the last level, $i = NL$. Another choice, suggested in [200], sometimes more advantageous in terms of numerical stability, is to replace the rate equation for the level with the largest population.

The set of equations (14.1) and (14.6) can be written in matrix notation as

$$\mathbf{A} \cdot \mathbf{n} = \mathbf{b}, \tag{14.7}$$

where $\mathbf{n}$ is a vector of level populations, $\mathbf{n} = (n_1, n_2, \ldots, n_{NL})^T$, and $\mathbf{b}$ is a right-hand-side vector, which, in the case of the abundance definition equation replacing the NLth equation, is $\mathbf{b} = (0, 0, \ldots, 0, n_{\rm atom})^T$. The matrix $\mathbf{A}$, often called the *rate matrix*, has elements

$$A_{ii} = \sum_{j \neq i}^{NL} (R_{ij} + C_{ij}), \tag{14.8a}$$

$$A_{ij} = -(R_{ji} + C_{ji}) \quad \text{for} \quad j \neq i \quad \text{and} \quad i \neq NL, \tag{14.8b}$$

$$A_{NL,j} = 1, \quad \text{for} \quad j = 1, \ldots, NL. \tag{14.8c}$$

For simplicity, we neglect here the upper sums, defined by (9.86). Although the set of rate equations (14.7) is formally linear in level populations, in reality it is not. If the radiation field were known, we could compute the radiative rates, and the kinetic equilibrium equations would be a *linear* set of equations for the level populations $n_i, i = 1, \ldots, NL$. Or, if the atomic level populations were known, we could compute the absorption and emission coefficients and solve the radiative transfer equation, which is a linear differential equation for the specific intensity. However, we do not know the level populations a priori (unless we assume LTE), and when solving the kinetic equilibrium equations, we cannot take the radiative rates as known because the radiation field depends on the atomic level populations we wish to determine. Hence we have to view the mean intensities of radiation as unknown quantities as well. In short, we have to solve a *coupled* system of kinetic equilibrium and radiative transfer equations, which makes the problem difficult.

In this chapter, to describe radiation we use the second-order form of the transfer equation with a variable Eddington factor:

$$\frac{\partial^2 (f_\nu J_\nu)}{\partial \tau_\nu^2} = J_\nu - \frac{\eta_\nu}{\chi_\nu}, \tag{14.9}$$

with boundary conditions

$$\left[\frac{\partial (f_\nu J_\nu)}{\partial \tau_\nu} \right]_0 = g_\nu J_\nu(0), \tag{14.10}$$

and

$$\left[\frac{\partial (f_\nu J_\nu)}{\partial \tau_\nu} \right]_{\tau_{\max}} = \tfrac{1}{2}(B_\nu - J_\nu) + \left[\frac{1}{3\chi_\nu} \left(\frac{\partial B_\nu}{\partial z} \right) \right]_{\tau_{\max}}, \tag{14.11}$$

where $g_\nu \equiv H_\nu(0)/J_\nu(0)$.

Using equations (11.2) and (11.5), the opacity and emissivity in a bound-bound transition $l \leftrightarrow u$ are given by, assuming complete redistribution,

$$\chi_{lu}(\nu) = (h\nu/4\pi)(n_l B_{lu} - n_u B_{ul})\,\phi_{lu}(\nu) \equiv \kappa_{lu}\phi_{lu}(\nu), \tag{14.12a}$$

$$\eta_{lu}(\nu) = (h\nu/4\pi) n_u A_{ul}\phi_{lu}(\nu). \tag{14.12b}$$

The quantity κ_{lu} has the meaning of frequency-averaged line absorption coefficient. Notice that the stimulated emission term in (14.12a) and the spontaneous emission in (14.12b) should generally contain the emission profile, $\psi_{ul}(\nu)$, but in the case of complete redistribution they are equal:

$$\psi_{ul}(\nu) \equiv \phi_{lu}(\nu). \tag{14.13}$$

The source function in the case of complete redistribution is found from (14.12) to be

$$S_{lu} = \frac{n_u A_{ul}}{n_l B_{lu} - n_u B_{ul}} = \frac{2h\nu_{lu}^3}{c^2}\frac{1}{(n_l g_u/n_u g_l) - 1} \tag{14.14}$$

and is *frequency independent*.

The total absorption and emission coefficients are

$$\chi_\nu = \chi_{lu}(\nu) + \chi_\nu^{\text{back}}, \quad \eta_\nu = \eta_{lu}(\nu) + \eta_\nu^{\text{back}}, \tag{14.15}$$

where χ_ν^{back} and η_ν^{back} are the background opacity and emissivity produced by sources other than the transition $l \leftrightarrow u$. We will often assume that χ_ν^{back} and η_ν^{back} vary negligibly with frequency over the line profile and take them to be frequency-independent. In this case, they are denoted as χ_c and η_c.

One sometimes uses a unified notation for bound-bound and bound-free transitions,

$$\chi_{lu}(\nu) = \alpha_{lu}(\nu)[n_l - G_{lu}(\nu) n_u], \tag{14.16a}$$

$$\eta_{lu}(\nu) = (2h\nu^3/c^2)\alpha_{lu}(\nu) G_{lu}(\nu) n_u, \tag{14.16b}$$

where

$$G_{lu}(\nu) \equiv \begin{cases} g_l/g_u, & \text{for bound-bound transitions,} \\ (n_l/n_u)^* e^{-h\nu/kT}, & \text{for bound-free transitions.} \end{cases} \tag{14.17}$$

The cross sections α are given by

$$\alpha_{lu}(\nu) \equiv \begin{cases} (h\nu_{lu}/4\pi) B_{lu}\,\phi_{lu}(\nu), & \text{for bound-bound transitions,} \\ \sigma_{lu}(\nu), & \text{for bound-free transitions.} \end{cases} \tag{14.18}$$

As noted previously, it is convenient to work with the dimensionless frequency variable x, measured from the line center in units of Doppler width:

$$x \equiv (\nu - \nu_0)/\Delta\nu_D, \tag{14.19}$$

where ν_0 is the line center's frequency, and $\Delta\nu_D$ is the Doppler width. In terms of this variable, the normalized Doppler and Voigt profiles are given by

$$\phi_D(x) = e^{-x^2}/\sqrt{\pi} \qquad \text{(Doppler)}, \tag{14.20a}$$

$$\phi_V(a,x) = \frac{a}{\pi^{\frac{3}{2}}}\int_{-\infty}^{\infty} \frac{e^{-y^2}\,dy}{(x-y)^2 + a^2} \qquad \text{(Voigt)}. \tag{14.20b}$$

Both are normalized such that

$$\int_{-\infty}^{\infty} \phi(x)dx = 1. \tag{14.21}$$

We will often use the mean optical depth τ associated with the averaged line opacity κ_{lu},

$$d\tau \equiv -\kappa_{lu}\,dz, \tag{14.22}$$

through which the monochromatic optical depth is expressed as $\tau_\nu = \tau\phi_{lu}(\nu)$ or $\tau_x = \tau\phi_{lu}(x)$.

Photon Diffusion, Destruction, Escape, and Thermalization

The equations above fully specify the problem of line and continuum formation for multi-level atoms in a gas having a prescribed temperature and density structure. We have already pointed out that the overall system is nonlinear. Yet another major complication is somewhat hidden: that the equations are highly *nonlocal*. That is, although the local level populations depend on the local radiation field, the local radiation intensity is determined by solving the radiative transfer equation and thus depends on the absorption and emission coefficients (and hence level populations) essentially *everywhere* in the medium or, in practice, on the properties of a large volume around the point of interest. This is because the mean free path of photons is typically much larger than the mean free path of material particles.

Much of the basic physics of line formation can be understood in terms of the characteristic lengths for photon diffusion, destruction, and thermalization, which are determined by a photon's escape and destruction probabilities. The following discussion is somewhat reminiscent of our previous discussion of coherent scattering in § 12.1. Although coherent scattering (e.g., in a continuum) and non-coherent line scattering have features in common, it is essential to understand the important differences between them.

The *mean free path* l of a line photon at a given frequency is the (geometric) distance over which it can, on the average, move in the atmosphere between successive interactions with the material (absorption or scattering). That is, the distance whose optical thickness at that frequency is of the order of unity—i.e., $l_\nu \approx 1/\chi_\nu = 1/(\kappa_{lu}\phi_\nu + \chi_c)$. In the case of complete redistribution, the probability of emission at frequency ν is ϕ_ν, so

$$l = \langle l_\nu \rangle = \int_0^\infty l_\nu \phi_\nu \, d\nu = \int_0^\infty \frac{\phi_\nu}{\kappa_{lu}\phi_\nu + \chi_c} \, d\nu. \tag{14.23}$$

Equation (14.23) shows that when $\kappa_{lu} \gg \chi_c$, photons emitted in the line core travel relatively small distances, while those in the wings can travel much larger distances, up to a distance corresponding to unit optical distance in continuum.

When a photon is "absorbed" in the line and excites an atom to its upper state, it is usually *scattered*, i.e., re-emitted by a radiative de-excitation, and travels another mean free path. This process may occur again and again before the photon is ultimately *destroyed* and its energy deposited into the thermal pool, either by a collisional de-excitation or by being absorbed in the overlapping continuum. Notice that unlike the case of coherent scattering, an "identical" incident photon may produce an emergent photon with a different frequency after each consecutive scattering in a line. Thus there exists a characteristic length L, the *destruction length*, over which a photon can travel before it is destroyed. The destruction length is more basic physically than the mean free path, for it measures the distance over which a photon created at a given point retains its identity, and hence can "communicate" information about conditions at that point to another. L sets the size of an *interaction region*: the volume containing those points that can influence one another via photon exchange.

The relative sizes of L and l depend upon a photon's *destruction probability*, P_d, which gives the average probability it is destroyed when it next interacts with the material. The probability that a photon is re-emitted (i.e., "survives" the scattering event) is $P_s = A_{ul}/P_u$, where A_{ul} is the spontaneous emission rate, and P_u is the total rate of transitions out of level u. In the simple case of a two-level atom we have $P_s = A_{ul}/(C_{ul} + A_{ul})$, where C_{ul} is the rate of collisional de-excitation; hence $P_d = 1 - P_s = C_{ul}/(C_{ul} + A_{ul})$.

Because $C_{ul} \propto n_e$, the destruction probability approaches unity in the deeper, denser regions of an atmosphere, whereas it can be quite small in the uppermost, less dense layers. If we assume that all photons absorbed in the continuum are destroyed thermally, the contribution of continuum processes to P_d is the average of $\chi_c/(\kappa_{lu}\phi_\nu + \chi_c)$ over the line profile. The continuum sets an upper bound on both l and L because a photon cannot travel more than a unit optical depth in the continuum before it is absorbed and destroyed instead of scattered.

At great depths in the atmosphere, P_d approaches unity because densities (and hence collisional rates) are large, and possibly also because of the strength of an overlapping continuum. Then $L \to l$, and the photon is almost surely destroyed when it is next "absorbed." The radiation field thus becomes strongly coupled to local conditions and *thermalizes* to its local thermodynamic equilibrium value (i.e., the Planck function at the local temperature). In contrast, when the destruction probability is very small, $L \gg l$, the interaction region may become enormous compared to the volume over which a photon can propagate in a single flight. In this case, the radiation field is dominated by *nonlocal* influences and is the result of physical conditions that may be quite inhomogeneous. For example, within the volume there may be large variations of kinetic temperature that imply strong variations of thermodynamic properties. The radiation field may then depart markedly from its local equilibrium value, and this departure will extend throughout the entire (possibly very large) interaction region.

The most important effects are produced by the presence of a *boundary* of the medium. If the interaction region at some point in the atmosphere contains the boundary, photons may escape from that point out into interstellar space, without depositing their energy into the thermal pool. Therefore, the radiation field at test points whose interaction regions extend beyond the atmosphere's surface must be depressed below the value that it would have had if there were no boundary.[1] That is not the end of the story, because while photon escape leads to a deficiency of radiation throughout an entire interaction volume, which extends to distance L from the boundary, the radiation field at *these* points influences that at points that lie yet another destruction length L deeper. Thus there is a "compounding" of the effect, which leads to a departure of the radiation field from its equilibrium value at positions quite far from the boundary.

The depth at which the radiation field (or source function) ultimately approaches closely its equilibrium value is called the *thermalization depth* τ_{th}. This concept, introduced by Jefferies [577] and independently by the Russian school [573], is very fruitful. To obtain a quantitative estimate of τ_{th}, we compare the frequency-averaged probability that photons in a line escape, P_{e}, with their destruction probability P_{d}. Deep within the medium where $P_{\text{e}}(\tau) \ll P_{\text{d}}$, photons are surely thermalized before they escape; hence $J_\nu \to B_\nu$. At the surface where $P_{\text{e}}(\tau) \gg P_{\text{d}}$, photons escape freely before thermalization; hence the radiation field (and the source function) depart from B_ν. It is therefore reasonable to choose the thermalization depth τ_{th} to be the point at which

$$P_{\text{e}}(\tau_{\text{th}}) \approx P_{\text{d}}. \tag{14.24}$$

Defined in this way, τ_{th} is the greatest depth from which photons have a significant chance to escape before being destroyed. The escape probability (and hence τ_{th}) depends sensitively upon (a) the nature of photon redistribution over the line profile when emitted and (b) the amount of background continuum absorption. The probability of photon "absorption" is highest at line center. If photons are scattered *coherently*, then those removed at line center will be re-emitted there and tend to be trapped by the large line-core optical depth out to the uppermost layers, where the line core finally becomes optically thin. Hence large departures of the line-core radiation field from its equilibrium value are inhibited everywhere but in regions very close to the atmosphere's surface. In contrast, if photons are *completely redistributed* over the line profile, then there is a significant chance that, after a number

[1] Here we have assumed that there is no radiation incident from an external source, e.g., from a companion star in a binary. If there *is* external illumination at the boundary from an exterior source, then the influence of that illumination is in turn "felt" in the atmosphere to the distance L from the boundary.

of scatterings, a photon absorbed at line center will be emitted in the line wing where the opacity is low, and the probability of escape is high. Thus photons that would have been trapped if emitted coherently now escape freely, depressing the intensity in the line core in much deeper layers in the atmosphere than would be the case if the scattering process were coherent. Hence the radiation field in the line as a whole responds to the fact that the boundary lies within a mean free path at some frequencies, even if not at others.

Thermalization Depth

Expressions for the thermalization depth can be derived using several different approaches: (a) analysis of the asymptotic behavior of an integral equation form of the transfer equation [66]; (b) calculation of the distribution of the distances photons can travel from their points of creation to their places of destruction [551,923]; and (c) calculation of the probability distribution for photon escape from a given point of origin in the atmosphere [332–334].

A simple and physically instructive derivation can be made using the criterion stated in (14.24). Consider the case of a two-level atom, having complete redistribution in the scattering process, and with no background opacity. The destruction probability per scattering event is

$$P_{\rm d} = \frac{C_{ul}}{A_{ul} + C_{ul}} \equiv \epsilon. \tag{14.25}$$

Further, assume a homogeneous atmosphere in which ϵ does not depend on depth.

The escape probability will be considered in detail in § 14.3. Here we use an approximate but powerful picture introduced by Osterbrock [820]. Take a line photon and express its frequency with a dimensionless value x. Further, express the depth in the medium with the mean optical depth τ and assume that $\tau > 1$. The monochromatic optical depth at frequency x is $\tau_x = \tau\phi(x)$. The frequencies in the line fall into two categories: "optically thick" frequencies for which $\tau_x > 1$, and "optically thin" frequencies for which $\tau_x < 1$. The *critical frequency* that divides these two regimes is given by

$$\tau\phi(x_{\rm c}) = 1, \tag{14.26}$$

i.e., it is a function of mean optical depth. One can approximate the escape probability by invoking a dichotomous model. Let us assume that all photons in the "optically thick" frequencies are re-absorbed, whereas those at "optically thin" frequencies escape. A photon has a probability $\frac{1}{2}\phi(x)dx$ of being emitted into the outward hemisphere in the frequency range $(x, x+dx)$, so the total (one-sided) escape probability in this picture is given by

$$P_{\rm e}(\tau) \approx \tfrac{1}{2}\int_{-\infty}^{-x_{\rm c}} \phi(x)\,dx + \tfrac{1}{2}\int_{x_{\rm c}}^{\infty} \phi(x)\,dx = \int_{x_{\rm c}}^{\infty} \phi(x)\,dx. \tag{14.27}$$

This integral is easy to evaluate for $|x_1| \gg 1$, which it will be when $\tau \gg 1$. Using equations (14.20) in (14.27), and requiring that $|x_c| \gg 1$, we find

$$P_e^D(\tau) = \tfrac{1}{2}\,\mathrm{erfc}(x_c)$$
$$\approx \exp(-x_c^2)/2\sqrt{\pi}x_c \quad \text{(Doppler)}, \tag{14.28a}$$
$$P_e^V(\tau) \approx a/\pi x_c \quad \text{(Voigt)}. \tag{14.28b}$$

Now applying the requirement that $\tau\phi(x_c) = 1$, we find

$$x_c = \sqrt{\ln(\tau/\pi^{\frac{1}{2}})} \quad \text{(Doppler)}, \tag{14.29a}$$
$$x_c = (a\tau/\pi)^{\frac{1}{2}} \quad \text{(Voigt)}. \tag{14.29b}$$

Substituting (14.29) into (14.28), we obtain

$$P_e^D(\tau) \approx \frac{1}{2\tau\sqrt{\ln(\tau/\pi^{\frac{1}{2}})}} \quad \text{(Doppler)}, \tag{14.30a}$$
$$P_e^V(\tau) \approx (a/\pi\tau)^{\frac{1}{2}} \quad \text{(Voigt)}. \tag{14.30b}$$

Finally, setting $P_e(\tau_{th}) = P_d = \epsilon$, and solving for τ_{th}, we find

$$\tau_{th}^D \sim C/\epsilon \quad \text{(Doppler)}, \tag{14.31a}$$
$$\tau_{th}^V \sim a/\epsilon^2 \quad \text{(Voigt)}, \tag{14.31b}$$

where C is a number of order unity that depends implicitly upon ϵ; other factors of order unity have been suppressed.

Equations (14.31) justify the basic conclusion drawn from the physical arguments earlier, namely, that the thermalization depth for a line with small ϵ is enormous. Recall that for coherent scattering the thermalization depth from (12.6) is $\tau_{th}^{coh} \sim 1/\sqrt{\epsilon}$. It is clear that noncoherence, with its attendant increase of photon diffusion in the line wings, greatly increases the depth in the atmosphere over which the radiation field and the source function can depart from the local Planck function. These results also show that the thermalization depth of a Voigt profile exceeds that of a Doppler profile when $a > \epsilon$ (usually the case), because the redistribution of photons from the line core to the stronger line wings is larger, thus increasing the photons' escape probability.

It is worth remembering that the results above depend upon the assumptions of (a) no background opacity and (b) complete redistribution. We will see later in this chapter that the effect of a background opacity can greatly reduce τ_{th}. In chapter 15 we will find that in the case of resonance lines, scattering in strong radiation-damping wings is nearly *coherent*, and τ_{th} is better approximated by equation (14.31a) than by (14.31b). Equations (14.31) show that Λ-iteration starting from LTE is futile for lines having small ϵ. Correct solutions can be obtained only by explicitly including the scattering term in the transfer equation and solving it with a direct solution or by accelerated lambda iteration (ALI).

With the qualitative arguments given above we have extracted much of the physical essence of the problem. But to obtain quantitative results, and to extend them to cases where heuristic discussion becomes ineffective, we must turn to a more detailed mathematical analysis.

14.2 TWO-LEVEL ATOM

In this section we describe line-formation problems that are simple enough to be solved readily, but that nevertheless provide a good description of some of the physically important processes and also yield considerable insight. It will be obvious that many of the assumptions made are oversimplifications and are not valid in actual astrophysical media, for which elaborate numerical calculations are generally required to yield accurate results. On the other hand, our real goal is to *understand* the answers, not merely obtain them. This can best be done with a clear grasp of the prototype problems discussed in this chapter, which provide a conceptual framework of great utility for the interpretation of results from computations with detailed model atmospheres and complicated model atoms.

We assume that the material is composed of a gas of atoms that have only two bound levels, plus free electrons. We assume that the structure of the medium is given, i.e., the temperature, density, and electron number density are known functions of depth. The most important difference between the LTE and NLTE treatments of line formation is the way in which radiation is coupled to the gas. In the LTE approach, it is *assumed* that the *local* values of two thermodynamic variables, T and N (or n_e), are sufficient to determine completely the excitation and ionization state of the gas, and hence its opacity, emissivity, and source function, independent of the state of gas at other points in the atmosphere.

But as has been emphasized in previous chapters, the excitation and ionization state of the gas is actually strongly influenced by the radiation field, which, in turn, is determined by the state of the gas in large volumes of the medium, via the transfer of radiation. Therefore, the two problems of radiative transfer and state of the material (which is in kinetic equilibrium with the radiation field) are inextricably coupled and must be solved *simultaneously*.

Source Function

Consider first the case of a two-level atom without a background continuum. The source function in the case of complete redistribution is given by (14.14),

$$S_{lu} = \frac{n_u A_{ul}}{n_l B_{lu} - n_u B_{ul}} = \frac{2h\nu_{lu}^3}{c^2} \frac{1}{(n_l g_u / n_u g_l) - 1} \equiv S_{\mathrm{L}}. \tag{14.32}$$

The source function S_{lu} is *frequency independent*. Notice that in the case of LTE, $n_l/n_u = (n_l/n_u)^* = (g_l/g_u)\exp(h\nu_{lu}/kT)$ and one recovers $S_L = B_{\nu_{lu}}$, as it should be.

Equation (14.32) is an *implicit* form of the source function because the level populations depend upon the radiation field. This dependence can be displayed explicitly by using the equations of kinetic equilibrium that determine n_l and n_u, namely,

$$n_l(B_{lu}\bar{J}_{lu} + C_{lu}) = n_u(A_{ul} + B_{ul}\bar{J}_{lu} + C_{ul}), \tag{14.33}$$

where the frequency-averaged mean intensity in the line $l \to u$ is given by $\bar{J} = \int_0^\infty J_\nu \phi_\nu d\nu$.

Equation (14.33) can be used to write an expression for (n_l/n_u). Upon substituting this expression into (14.32) and making use of the Einstein relations and the detailed balance result $C_{lu} = (n_u/n_l)^* C_{ul}$, we find

$$S_L = \frac{\int J_\nu \phi_\nu d\nu + \epsilon' B}{1 + \epsilon'} \equiv (1 - \epsilon)\bar{J} + \epsilon B, \tag{14.34}$$

where

$$\epsilon' \equiv \frac{C_{ul}}{A_{ul}}\left(1 - e^{-h\nu_{lu}/kT}\right), \tag{14.35}$$

and

$$\epsilon \equiv \frac{\epsilon'}{1 + \epsilon'} = \frac{C_{ul}}{C_{ul} + A_{ul}/(1 - e^{-h\nu_{lu}/kT})}. \tag{14.36}$$

Equation (14.34) is the basic equation of the problem. It is interesting that it was derived in the 1930s by Milne [721], but was not used in astrophysical work until the late 1950s and early 1960s.

The terms in (14.34) admit a straightforward physical interpretation. The source function contains a *non-coherent scattering term* $\bar{J}$ and a *thermal source* term $\epsilon' B$. The thermal source term represents photons that are *created* by collisional excitation, followed by radiative de-excitation. The term ϵ' in the denominator is a *sink term* that represents those photons that are destroyed by collisional de-excitation of the atom following a photoexcitation.

The second equality of (14.34) has a similar interpretation. The quantity ϵ defined by (14.36) is interpreted as the *photon destruction probability*. The term $1 - \exp(-h\nu/kT)$ represents a correction for stimulated emission; if stimulated emission were neglected, we would recover (14.25). These two terms describe completely the coupling of the radiation field to the local state of the gas. The scattering term is a *reservoir term* that represents the end result of the cumulative contributions of the source and sink terms over the entire interaction region.

If densities are made sufficiently large, then the collision rate C_{ul} may eventually exceed A_{ul}, so that $\epsilon' \gg 1$, and $\epsilon \to 1$; then $S_{lu} \to B_\nu(T)$, and LTE is recovered. However, in virtually all situations of astrophysical interest, $\epsilon \ll 1$ in regions of line formation, and, in general, the source function cannot be expected to have a value close to the Planck function. This fact was partially recognized in the classical theory by the division of lines into the categories of "absorption" and "scattering" lines. The division, however, was largely ad hoc, and the thermal coupling parameter had to be guessed from heuristic arguments. In the present analysis the coupling parameter follows directly and uniquely from the kinetic equilibrium equation. Further, in the classical treatments of "pure" scattering lines, it was sometimes incorrectly argued that the small thermal terms could be discarded. However, the important point to bear in mind is that although the thermal term ϵB may be small compared to the scattering term *locally*, when integrated over the entire interaction region it accumulates to a value of importance. Moreover, at depths larger than the thermalization depth the intensity must ultimately couple to the thermal pool. But if the thermal terms are discarded, the transfer equation becomes *homogeneous* in the radiation field and the scale of solution is unknown; this scale is, in fact, *fixed by the (small) thermal term* at the point of thermalization.

In the presence of background continuum the absorption coefficient is given by; see equation (14.15)

$$\chi_\nu = \chi_{lu}(\nu) + \chi_c(\nu) = \kappa_{lu}\phi_\nu + \chi_c, \tag{14.37}$$

where χ_c is the absorption coefficient in the continuum, which in most cases we can assume to be independent of frequency over the line width. The emission coefficient is given by

$$\eta_\nu = \eta_{lu}(\nu) + \eta_c(\nu) = \chi_{lu}(\nu)S_{lu} + \chi_c S_C, \tag{14.38}$$

where η_c is the emission coefficient in the continuum, and $S_C \equiv \eta_c/\chi_c$ is the continuum source function.

The total source function is thus

$$S_\nu = \frac{\chi_{lu}(\nu)S_{lu} + \chi_c S_C}{\chi_{lu}(\nu) + \chi_c} = \frac{\phi_\nu}{\phi_\nu + r}S_{lu} + \frac{r}{\phi_\nu + r}S_C, \tag{14.39}$$

where

$$r \equiv \chi_c/\kappa_{lu}. \tag{14.40}$$

The first term represents the fraction of photons created by the line processes, and the second term corresponds to photons created by continuum processes. The total source function now depends explicitly upon frequency, which complicates the mathematical analysis. But if we assume that ϵ, B, ϕ_ν, r, and S_C are given functions of position, the frequency dependence is simple because the total source function can be fully specified by a single frequency-independent quantity, S_L, the line source function. This is particularly important for applying the ALI scheme to solve this problem numerically. We return to this point below.

For the case of no background continuum, the source function (dropping the subscripts l and u indicating levels) is

$$S(\tau) = [1 - \epsilon(\tau)] \int_{-\infty}^{\infty} \phi(\tau,x) J(\tau,x)\, dx + \epsilon(\tau)B(\tau) \tag{14.41}$$

and the transfer equation is

$$\mu \frac{dI(\tau,x)}{d\tau} = \phi(\tau,x)[I(\tau,x) - S(\tau)]; \tag{14.42}$$

whereas with a background continuum, it becomes

$$\mu \frac{dI(\tau,x)}{d\tau} = [\phi(\tau,x) + r]I(\tau,x) - \phi(\tau,x)\, S(\tau) - rS_C(\tau). \tag{14.43}$$

Here, τ is the mean optical depth associated with the averaged line opacity κ_{lu}. The task now is to solve equations (14.42) and (14.41), or (14.43) and (14.39), simultaneously. This can be done with a number of methods; we describe the most important of these below.

A standard problem of radiative transfer, whose solution occupies a large literature, is the case of a two-level atom with no background continuum in a constant-property medium, i.e., with ϵ, B, and $\phi(\tau,x)$ independent of depth:

$$S(\tau) = (1-\epsilon)\bar{J}(\tau) + \epsilon B. \tag{14.44}$$

We turn now to methods for solving the radiative transfer problem with a source function given by (14.44).

Direct Solution of the Transport Problem

Because the two-level atom problem is linear, one can use direct methods, without the need to employ an iteration scheme. Two such methods are described below.

Integral Equation Method

The formal solution of the transfer equation (14.42) is

$$J(\tau,x) = \tfrac{1}{2}\int_0^\infty S(t)E_1\left[\left|\int_\tau^t \phi(t',x)\,dt'\right|\right]\phi(t,x)\,dt. \tag{14.45}$$

For a depth-dependent profile, the argument of the exponential integral E_1 depends upon both τ and t, which complicates the analysis. So here we will assume the profile is depth-independent, in which case only the *displacement* $(t-\tau)$ enters:

$$J(\tau,x) = \tfrac{1}{2}\phi(x)\int_0^\infty S(t)E_1[\,|t-\tau|\,\phi(x)]\,dt. \tag{14.46}$$

Substitution of this expression for J into equation (14.41) yields an integral equation for S,

$$S(\tau) = [1-\epsilon(\tau)]\int_0^\infty S(t)K_1(|t-\tau|)\,dt + \epsilon(\tau)B(\tau), \tag{14.47}$$

where the *kernel function* K_1 is defined as

$$K_1(s) \equiv \tfrac{1}{2}\int_{-\infty}^{\infty} E_1(\phi_x s)\phi_x{}^2\,dx = \int_0^{\infty} E_1(\phi_x s)\phi_x{}^2\,dx. \qquad (14.48)$$

One of the basic advantages of the integral equation formulation is that it displays explicitly the intimate dependence of the source function upon the mathematical behavior of the kernel function, and valuable insight can be gained from an analytical study of the kernel. In particular, the *asymptotic form* of $K_1(s)$, for $s \gg 1$, shows that a characteristic of line-formation problems with non-coherent scattering is an extremely *long-range interaction* of one part of the medium with another.[2] For *coherent scattering* the kernel function is $E_1(|t-\tau|)$ [cf. (11.108)]. This kernel decays relatively rapidly, falling off as $e^{-|t-\tau|}/|t-\tau|$, thereby limiting the range of depths that are directly coupled in the scattering process. In contrast, the asymptotic behavior of $K_1(s)$ is found in [66, Appendix I] to be

$$K_1(s) \approx \frac{1}{4s^2\sqrt{\ln(s/\pi^{\frac{1}{2}})}}, \qquad s \gg 1, \qquad (14.49)$$

for a Doppler profile and

$$K_1(s) \approx a^{\frac{1}{2}}/6s^{\frac{3}{2}}, \qquad s \gg 1, \qquad (14.50)$$

for a Voigt profile. The range of these kernel functions is very large compared to that for coherent scattering, which implies that the radiation fields at widely separated points in the medium become mutually interdependent.

Numerical methods for solving equations of the form of (14.47) are discussed in detail in [62] and [39, chapter 8]. In essence, the solution is obtained by using a functional representation (usually a polynomial approximation) of $S(\tau)$ on a discrete grid $\{\tau_d\}$. These functions are integrated *analytically* against the kernel to construct a final matrix system of the form

$$\mathbf{S} = \mathbf{K}\cdot\mathbf{S} + \mathbf{L}, \qquad (14.51)$$

where $\mathbf{S} = (S_1, S_2, \ldots, S_D)^T$ is a column vector of the values of the source function at all discretized depth points. One thus ends up with a system of *linear* equations for the components of $\mathbf{S}$, which can be solved by any standard method of linear algebra.

Differential Equation Method

Alternatively, one may use any of the differential equation methods described in chapter 12. The transfer equation (14.9) together with boundary conditions (14.10) and (14.11) are discretized in depth and solved numerically using either the Feautrier

[2] This result was discussed in physical terms in §14.1.

or Rybicki schemes described in § 12.2 and 12.3, respectively. The source function has the form of equation (12.37b), with

$$\alpha_\nu = (1 - \epsilon), \quad \beta_\nu = \epsilon B \tag{14.52}$$

in the case of no overlapping continuum, or

$$\begin{aligned} \alpha_\nu &= (1 - \epsilon)\,\phi_\nu/(\phi_\nu + r), \\ \beta_\nu &= (\phi_\nu \epsilon B + r S_c)/(\phi_\nu + r) \end{aligned} \tag{14.53}$$

in the case of overlapping continuum. One can follow the formalism in § 12.2 or § 12.3 to obtain the mean intensity of radiation and the source function. Notice also that all of the parameters (ϵ, B, and ϕ) may be depth-dependent without causing any difficulty in the calculation.

Iterative Solution of the Transport Problem

Although this problem can be solved directly, we now describe some iterative methods. This has a dual purpose. First, iterative methods, in particular an application of the ALI method, provide a fast and robust solution of the problem, and second, it has a significant pedagogical value because one can explain iterative methods for a simple case.

ALI Method with No Background Continuum

An application of the ALI scheme to solve the two-level atom problem without background continuum is straightforward. Indeed, this was the case we used to explain the nature of the method in § 13.1. We repeat the basic steps.

(a) For a given, current value of the source function $S^{(n)}$ [with an initial estimate $S^{(0)} = B$, or some other suitable value], we compute a formal solution of the transfer equation for one frequency and angle at a time. We can use any method described in § 12.4, e.g., the short characteristics method, the discontinuous finite element method, or the Feautrier method. We then have new values of the specific intensity $I_{\mu\nu}$, and also of the elementary approximate Λ-operator $\Lambda^*_{\mu\nu}$, whether diagonal or tridiagonal, as explained in § 13.3. For instance, if one uses the Feautrier method, the approximate operator is evaluated with the Feautrier elimination scheme using equations (13.87)–(13.100). Similarly, if one uses the short characteristic scheme, the elementary approximate operator is given by (13.81)–(13.82).

(b) Having obtained the specific intensities, one evaluates a new estimate of the source function, using equation (14.44),

$$S^{\rm FS}(\tau) = \tfrac{1}{2}(1 - \epsilon) \int_{-1}^{1} d\mu \int_0^\infty d\nu\, \phi_\nu I_{\mu\nu}(\tau) + \epsilon B(\tau), \tag{14.54}$$

and the corresponding integrated approximate Λ-operator,

$$\overline{\Lambda}^*(\tau) = \tfrac{1}{2}\int_{-1}^{1} d\mu \int_0^\infty d\nu\, \phi_\nu \Lambda^*_{\mu\nu}(\tau). \tag{14.55}$$

Recall that in the case of a diagonal operator $\Lambda^*_{\mu\nu}(\tau)$ is a *scalar* function of τ.

(c) The heart of the ALI scheme is the evaluation of the new iterate of the source function, using equation (13.7), i.e.,

$$S^{(n+1)} = S^{(n)} + [I - (1-\epsilon)\Lambda^*]^{-1}[S^{\rm FS} - S^{(n)}]. \tag{14.56}$$

Again, in the case of diagonal Λ^*, the inverse of $[1-(1-\epsilon)\Lambda^*]$ is a division by $[1-(1-\epsilon)\Lambda^*(\tau)]$. In the case of a tridiagonal operator [writing $\delta S_d \equiv S^{(n+1)}(\tau_d) - S^{(n)}(\tau_d)$], (14.56) becomes

$$-(1-\epsilon_d)\Lambda^*_{d,d-1}\delta S_{d-1} + [1-(1-\epsilon_d)\Lambda^*_{d,d}]\delta S_d - (1-\epsilon_d)\Lambda^*_{d,d+1}\delta S_{d+1} = S_d^{\rm FS} - S_d^{(n)}. \tag{14.57}$$

This tridiagonal system of equations for the unknown corrections δS is solved by the standard elimination scheme described in § 12.2.

(d) Steps (a) through (c) are iterated to convergence, possibly using the acceleration schemes described in § 13.4 to speed up the overall procedure.

Note that all the parameters of the problem, i.e., ϵ, B, and ϕ_ν, may be depth-dependent without causing any difficulty because the only place where nonlocal coupling of the various physical quantities enters is in the formal solution of monochromatic transfer equation along a single ray, which is a simple linear differential equation.

ALI Method with Background Continuum

The case of background continuum is somewhat more complicated, basically because then the total source function depends on frequency; see (14.39),

$$S_\nu^{\rm tot} = \frac{\phi_\nu}{\phi_\nu + r} S_{\rm L} + \frac{r}{\phi_\nu + r} S_{\rm C}. \tag{14.58}$$

Fortunately, the dependence of the total source function on frequency is very simple. The continuum source function $S_{\rm C}$ is a known function, so the only unknown function is the frequency-independent line source function $S_{\rm L}$.

To construct the appropriate approximate Λ-operator, we first write an exact expression. The specific intensity is given by the elementary Λ-operator acting on the total source function,

$$I_{\mu\nu} = \Lambda_{\mu\nu}\left[S_\nu^{\rm tot}\right] = \Lambda_{\mu\nu}\left[\frac{\phi_\nu}{\phi_\nu + r} S_{\rm L} + \frac{r}{\phi_\nu + r} S_{\rm C}\right]. \tag{14.59}$$

The line source function is given by equation (14.44). Substituting (14.44) into (14.59) and rearranging the terms, we obtain

$$S_{\rm L} = (1-\epsilon)\overline{\Lambda}[S_{\rm L}] + S_0, \tag{14.60}$$

where

$$\overline{\Lambda}[S_{\rm L}] = \tfrac{1}{2}\int_{-1}^{1} d\mu \int_0^\infty d\nu\, \phi_\nu\, \Lambda_{\mu\nu}\left[\frac{\phi_\nu}{\phi_\nu + r} S_{\rm L}\right], \tag{14.61}$$

and

$$S_0 = \tfrac{1}{2}(1-\epsilon)\int_{-1}^{1} d\mu \int_0^\infty d\nu\, \phi_\nu\, \Lambda_{\mu\nu}\left[\frac{r}{\phi_\nu + r} S_{\rm C}\right] + \epsilon B, \tag{14.62}$$

so S_0 is a known function.

Equation (14.60) is similar to the expression for the (line) source function without a background continuum. Hence we may use the same iterative scheme as described earlier, with the difference that the integrated approximate Λ^*-operator is given by

$$\overline{\Lambda}^*[S_{\rm L}] = \tfrac{1}{2}\int_{-1}^{1} d\mu \int_0^\infty d\nu\, \phi_\nu\, \Lambda^*_{\mu\nu}\left[\frac{\phi_\nu}{\phi_\nu + r} S_{\rm L}\right]. \tag{14.63}$$

Again, in the case of a diagonal approximate operator, $\overline{\Lambda}^*$ is a scalar function of τ, given by

$$\overline{\Lambda}^*(\tau) = \tfrac{1}{2}\int_{-1}^{1} d\mu \int_0^\infty d\nu\, \frac{\phi_\nu^2}{\phi_\nu + r}\, \Lambda^*_{\mu\nu}(\tau). \tag{14.64}$$

The solution is obtained with the same procedure as outlined above for the case of no continuum.

Behavior of the Solution

We now describe the basic features of the results obtained by any of the methods discussed above and explain the behavior of the source function and the predicted line profile.

Constant-Property Media with No Continuum

Consider first the standard problem of line formation in a homogeneous semi-infinite slab (i.e., with depth-independent ϵ, B, and ϕ_ν) with no background continuum. A solution of this problem was first obtained in [66].

Numerical solutions for the full depth variation of the source function are shown in figures 14.1 and 14.2. The cases in figure 14.1 are for constant Planck function ($B \equiv 1$), a Doppler profile, and various values of ϵ. Two features are seen immediately:

(i) In each case, the surface value of the source function is given by

$$S(0) = \sqrt{\epsilon} B. \tag{14.65}$$

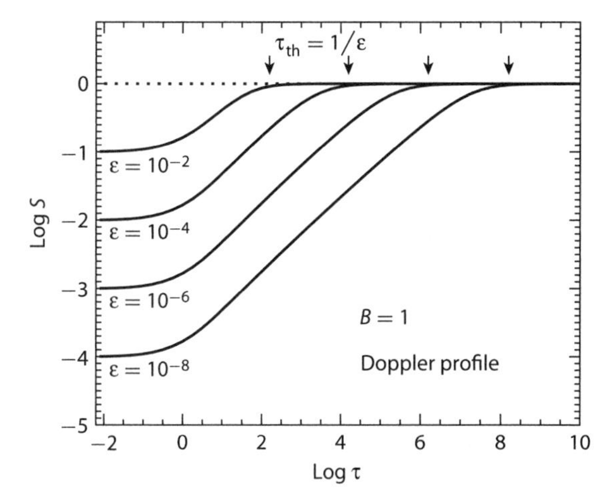

Figure 14.1 Line source function in a semi-infinite atmosphere with $B=1$ for a line with a pure Doppler profile ($a=0$) and with various values of ϵ. The arrows on the top indicate the position of the thermalization depth $\tau_{\rm th}$.

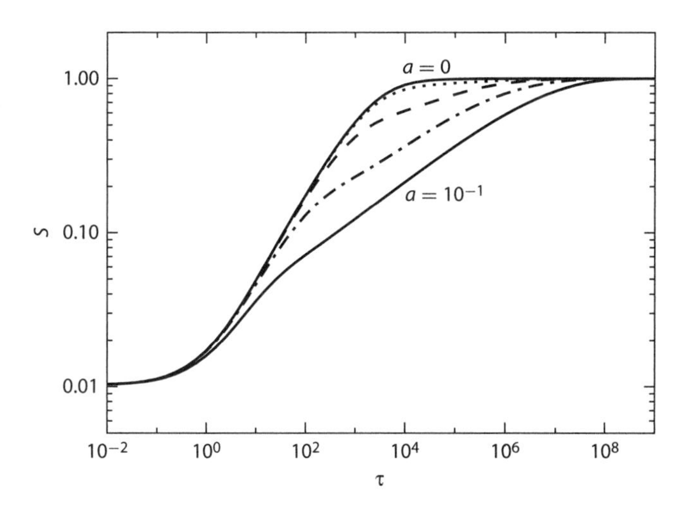

Figure 14.2 Line source function in a semi-infinite atmosphere with $B=1$ for lines with $\epsilon=10^{-4}$ and Voigt profile with $a=0$ (pure Doppler profile), $a=10^{-4}$ (dotted line), $a=10^{-3}$ (dashed line), $a=10^{-2}$ (dot-dashed line), and $a=10^{-1}$ (full line).

This is a robust exact result that earned a special name, the $\sqrt{\epsilon}$ law. The simplicity of this result is intriguing; we discuss it below.

(ii) In each case, the local source function approaches the Planck function, $S(\tau)\rightarrow B$, at $\tau\approx 1/\epsilon$. As discussed previously, this is the *thermalization depth* for the Doppler profile; see (14.31a).

The results shown in figure 14.2 are for lines with $\epsilon = 10^{-4}$ and for Voigt profiles ranging from a pure Doppler profile ($a = 0$) to $a = 10^{-1}$. The increase in thermalization depth from ϵ^{-1} to $a\epsilon^{-2}$ shows plainly.

Figure 14.5 below shows the emergent line profiles for some of the models displayed in figures 14.1 and 14.2. In all cases we obtain *absorption lines* with dark cores; in contrast, the LTE solution with $S^{\mathrm{L}} = B$ yields *no line* whatsoever. This difference is the result of the effects of scattering; the classical theory would also have predicted a line of the same central depth for the same ϵ. However, there remain two important differences:

(a) Both the upper- and lower-state occupation numbers differ from their LTE values (so that assuming LTE for them, as was done in many earlier studies, would be incorrect).

(b) The dark portion of the NLTE profiles, where *non-coherent scattering* is assumed, is wider. Recall that the classical theory assumes coherent scattering. The Eddington-Barbier relation implies that $I_\nu(0) \approx S(\tau = 1)$; hence the lines will be dark for $|x| \leq x_{\mathrm{th}}$, where $\tau_{\mathrm{th}}\phi(x_{\mathrm{th}}) = 1$. Furthermore, because the thermalization depth τ_{th} is much larger for non-coherent scattering ($\sim\epsilon^{-1}$ to ϵ^{-2} instead of $\epsilon^{-\frac{1}{2}}$), the corresponding values of x_{th} are also larger.

$\sqrt{\epsilon}$ *Law*

There are several mathematical proofs of the $\sqrt{\epsilon}$ law (for a review, see [573]); here we follow an elegant mathematical analysis presented in [350]. Differentiating equation (14.47) with respect to τ, multiplying by $S(\tau)$, and integrating from 0 to ∞, we find

$$\int_0^\infty d\tau\, S(\tau)\frac{\partial S(\tau)}{\partial \tau} = (1-\epsilon)\int_0^\infty d\tau\, S(\tau)\frac{\partial}{\partial \tau}\int_0^\infty d\tau' K_1(\tau-\tau')\, S(\tau'). \quad (14.66)$$

We now use the Frisch and Frisch lemma [350] with $\sigma = 0$,

$$\int_0^\infty d\tau\, S(\tau)\frac{\partial}{\partial \tau}\int_0^\infty d\tau'\, K_1(\tau-\tau')\, S(\tau') = \tfrac{1}{2}A\, S_\infty^2, \quad (14.67)$$

to obtain

$$\int_0^\infty d\tau\, S(\tau)\frac{\partial S(\tau)}{\partial \tau} = \tfrac{1}{2}(S_\infty^2 - S_0^2) = \tfrac{1}{2}(1-\epsilon)AS_\infty^2, \quad (14.68)$$

where A is the normalization of the kernel function; see equation (14.48). In a static, semi-infinite atmosphere, $A = 1$. We are then left with $S_0^2 = \epsilon S_\infty^2$, and as we have discussed before, $S_\infty = B$ (an exact mathematical proof is given in [66]), so we obtain the $\sqrt{\epsilon}$ law.

Note that the Frisch and Frisch lemma requires only that the kernel function K_1 is integrable; the $\sqrt{\epsilon}$ law is thus valid for *any kind of absorption profile*. The presence of the square root may seem to suggest that the $\sqrt{\epsilon}$ law is somehow connected to the $\sqrt{N}$ law for random walk of diffusion. However, this not the case. To see this

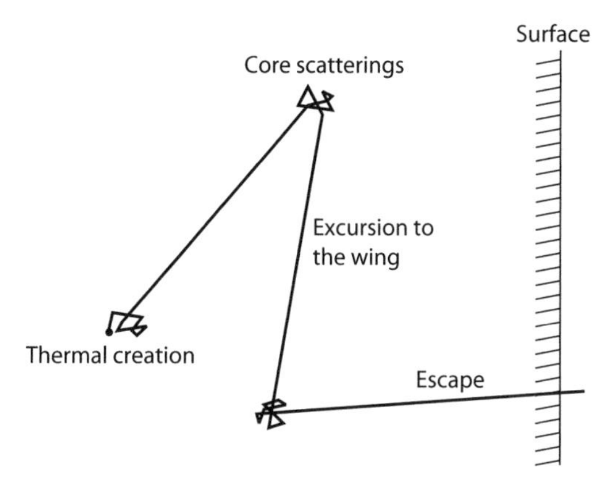

Figure 14.3 Schematic representation of the trajectory of a photon in a gas of two-level atoms.

mathematically, we use the asymptotic expression for the kernel function K_1, say, for the Doppler profile (14.49). The variance of the corresponding random walk is

$$\int_{-\infty}^{\infty} \tau^2 K_1(\tau)\, d\tau = \infty. \tag{14.69}$$

Inasmuch as the integral diverges, the $\sqrt{N}$ law is inapplicable. Physically, the process is indeed *not* a random walk with equal steps. As we have noted before, the process of line transfer is fundamentally different from a random walk; it is instead a process of many short displacements (each a transfer in the line core), followed by very infrequent excursions to the line wing, accompanied by a large displacement in the physical space. This is schematically illustrated in figure 14.3.

The $\sqrt{\epsilon}$ law is thus very interesting. It is valid for any type of line profile. If it is not a consequence of the random walk arguments, what is its physical interpretation? It turns out that the $\sqrt{\epsilon}$ law is a consequence of only two assumptions: the semi-infiniteness of the atmosphere and the translational symmetry of the problem, i.e., the fact that the corresponding kernel function K_1 is a function of only the optical depth difference, $|t - \tau|$, not of the values of the individual optical depths. An explanation of the $\sqrt{\epsilon}$ law in purely physical terms was given in [520]; we summarize it below.

In physical terms, the semi-infiniteness of the medium may be naturally expressed through the *principle of invariance*, first introduced by Ambartsumyan [27] and extended in [225], which states that the radiation emergent from a semi-infinite homogeneous atmosphere is invariant under the addition (or subtraction) of a layer of an arbitrary, finite thickness anywhere in the medium. Then the $\sqrt{\epsilon}$ law follows from placing an infinitesimal layer *on the surface* of the atmosphere. Such a layer has essentially two effects: (i) it absorbs radiation that would otherwise escape; and (ii) it creates new photons that can ultimately escape. The principle of invariance stipulates that these two contributions should exactly cancel. To express this statement mathematically, first recall the results from § 11.4 that a number of photons created in an elementary optical depth interval $(\tau, \tau + d\tau)$ across area dS in a time interval

dt is given by [cf. equation (11.74)] $\delta N_\nu^{\rm em} = S_\nu(\tau)\, d\tau (4\pi/h\nu)\, d\nu\, dt\, dS$. The total number of line photons is given by $\delta N^{\rm em} = \int \delta N_\nu^{\rm em} \phi_\nu\, d\nu = S(\tau)\, d\tau (4\pi/h\nu)\, dt\, dS$ (because the source function is frequency-independent). Similarly, as follows from (11.75), the total number of absorbed line photons in the same elementary volume is given by $\delta N^{\rm ab} = \bar{J}(\tau) d\tau (4\pi/h\nu)\, dt\, dS$.

The invariance of emergent radiation under addition of an infinitesimal layer at the top of the atmosphere is given by

$$S(0) d\tau\, [1 - p_T(0)] = \bar{J}(0) d\tau, \tag{14.70}$$

where $p_T(\tau)$ is the probability that a photon emitted at depth τ will ultimately be destroyed (thermalized) anywhere in the atmosphere, after an arbitrary number of intermediate scatterings. The quantity $1 - p_T(\tau)$ is thus the probability that photons created at depth τ will ultimately escape from the medium (again, after an arbitrary number of intermediate scatterings). Notice that this probability is *not* equal to the escape probability considered in § 14.3, which has the simpler meaning of the probability of escape within a single flight. The left-hand side then gives the number of photons created in the additional layer at the surface that *escape*, while the right-hand side gives the number of photons *absorbed* in this layer.

It remains to express the thermalization probability p_T. It can be shown [520] that for the given model (constant property media, no background continuum), this probability is given by

$$p_T(\tau) = \bar{J}(\tau)/B. \tag{14.71}$$

The proof goes as follows. First, one introduces a more elementary thermalization probability $\theta(\tau, \tau')$ such that $\theta(\tau, \tau') d\tau'$ is the probability that a photon emitted at depth τ will be thermalized in the optical depth interval $(\tau', \tau' + d\tau')$. The total thermalization probability is then

$$p_T(\tau) = \int_0^\infty \theta(\tau, \tau')\, d\tau'. \tag{14.72}$$

On the other hand, the total number of photons thermalized in the interval $(\tau, \tau + d\tau)$ is given by $\epsilon \bar{J}(\tau)\, d\tau dA$ [where $dA \equiv (4\pi/h\nu) dt dS$], because the number of photons absorbed in this range is $\bar{J}(\tau) d\tau dA$, and the fraction ϵ of those absorbed photons is destroyed (thermalized). However, photons that were destroyed at the range must have been previously created somewhere in the medium, so that the number of thermalized photons in the range $(\tau, \tau + d\tau)$ is also given by $\epsilon\, B \int_0^\infty \theta(\tau', \tau) d\tau' dA$ (ϵB is the number of created photons, and the integral over θ accounts for their ultimate thermalization). Equating these two expressions for the number of thermalized photons, we obtain

$$\bar{J}(\tau) = B \int_0^\infty \theta(\tau', \tau)\, d\tau'. \tag{14.73}$$

Comparing equations (14.72) and (14.73), we see that equation (14.71) is satisfied if

$$\theta(\tau, \tau') = \theta(\tau', \tau), \tag{14.74}$$

which may be called the *reciprocity relation*.

A proof of (14.74) is given in [520]; here we give only an intuitive verification. The thermalization probability $\theta(\tau, \tau')$ is the sum of all contributions from individual photon trajectories $\tau \to \tau', \tau \to \tau_1 \to \tau', \tau \to \tau_1 \to \tau_2 \to \ldots \to \tau'$, where the intermediate points $\tau_1, \tau_2, \ldots$, are the places where the photon is re-emitted. Now taking each individual trajectory in the reverse order, $\tau' \to \ldots \to \tau_1 \to \tau$, we find the same contribution as the original trajectory if (i) the probability that a photon is emitted at depth t' and is absorbed at the interval $(t, t+dt)$ depends only on the difference $|t - t'|$, not on the individual values of t and t'; (ii) the photon destruction probability ϵ does not depend on depth; and (iii) the re-emission probability of a photon does not depend on depth, nor on its previous history. Conditions (i) and (ii) are satisfied because of the homogeneity of the atmosphere, and condition (iii) holds because of the assumption of complete frequency redistribution and isotropic scattering.

Equation (14.70) then reads (writing all quantities at depth $\tau = 0$),

$$S(1 - \overline{J}/B) = \overline{J}. \tag{14.75}$$

The $\sqrt{\epsilon}$ law now follows trivially. By substituting $\overline{J} = (S - \epsilon B)/(1-\epsilon)$ into (14.75), we obtain

$$S^2 = \epsilon B^2, \tag{14.76}$$

i.e., the $\sqrt{\epsilon}$ law. We also note that from (14.75) follow the intriguing expressions

$$(\overline{J}/S) + (\overline{J}/B) = 1, \quad \text{or} \quad (1/S) + (1/B) = (1/\overline{J}), \quad \text{at} \quad \tau = 0, \tag{14.77}$$

which are equivalent to the $\sqrt{\epsilon}$ law, but do not contain ϵ explicitly. This is again a consequence of the fact that the $\sqrt{\epsilon}$ law is merely an expression of the semi-infiniteness of the medium, its homogeneity, and the reversibility of photon trajectories.

Finite Slabs

A *finite slab* atmosphere of total thickness T is a case of considerable astrophysical interest. It can be used to represent nebulae, or limited zones in an atmosphere to which a particular ion is confined owing to changes in the ionization balance (e.g., chromospheric lines of, say, He II are limited to layers bounded above by the corona and below by the photosphere). In finite atmospheres two physically distinct behaviors are found, depending on whether the atmosphere is *effectively thick* or *effectively thin*. If $T \gg \tau_{\rm th}$, then photons from the slab center will not escape before they thermalize; in this case $S(0)$ will attain its semi-infinite value for the corresponding value of ϵ and approaches B at depths $\tau \gtrsim \tau_{\rm th}$ from the surfaces. If, however, $T \ll \tau_{\rm th}$, then the solution never thermalizes, and $S(\tau)$ becomes proportional to the creation rate—i.e., $S(\tau) = \epsilon B f(\tau)$, where $f(\tau)$ is independent of ϵ for a given T.

An estimate of S at the slab center can be obtained as follows (see also [42]). The ratio of the *total* number of emissions along a column through the slab to those *thermally created* must be equal to the mean number of times, $\langle N \rangle$, a photon is scattered before it escapes or is destroyed, i.e.,

$$\langle N \rangle = \frac{\int_{-\infty}^{\infty} d\nu\, (4\pi h\nu\phi_\nu) \int_0^T d\tau\, S(\tau)}{\int_{-\infty}^{\infty} d\nu\, (4\pi h\nu\phi_\nu) \int_0^T d\tau\, \epsilon(\tau)B(\tau)} = \frac{\int_0^T S(\tau)\, d\tau}{\int_0^T \epsilon(\tau)B(\tau)\, d\tau}. \tag{14.78}$$

Here we have used the relationtion $\eta_\nu = \chi_\nu S_\nu$ and noted that $S(\tau)$ is frequency-independent. For a finite slab the dominant photon loss mechanism is escape; hence $\langle N \rangle \approx 1/P_e(T)$. At slab center $P_e \approx [2\, P_e(T/2)]_\infty$, where the subscript denotes the escape probability from the indicated depth in a semi-infinite slab, and the factor of 2 accounts for losses through both faces.

To obtain an order of magnitude estimate from (14.78), we replace $S(\tau)$ with $S_{\max} = S(T/2)$ and assume ϵB is constant so that $\langle N \rangle \approx S_{\max}/(\epsilon B)$. Then, using (14.28a) and (14.28b) to calculate $P_e(T)$, and again ignoring numerical factors of order unity, we find

$$S_{\max} \approx \epsilon\, T(\ln T)^{\frac{1}{2}} B \quad \text{(Doppler)}, \tag{14.79}$$

and

$$S_{\max} \approx \epsilon\, (T/a)^{\frac{1}{2}} B \quad \text{(Voigt)}. \tag{14.80}$$

The behavior described above is seen in the numerical results (first obtained in [66]) shown in figure 14.4, which displays $S(\tau)$ for lines with Doppler profiles ($a = 0$), having various values of ϵ, in an atmosphere with $T = 10^4$. The dashed curve gives

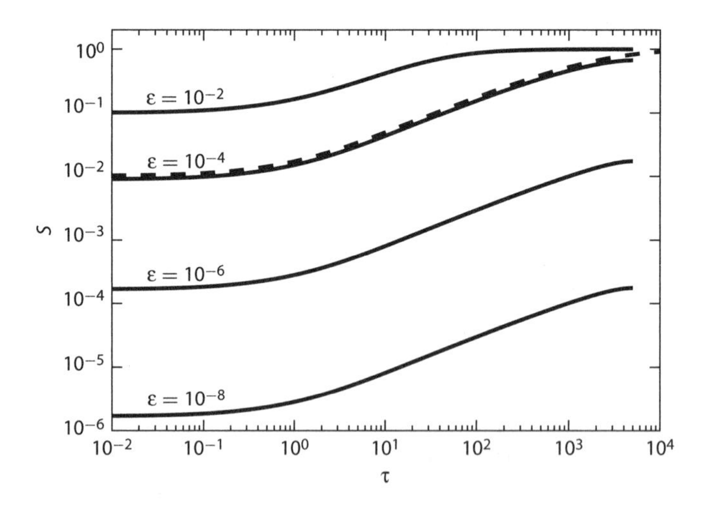

Figure 14.4 Line source function in finite atmospheres with total thickness $T = 10^4$ and $B = 1$, for a line with a pure Doppler profile ($a = 0$). Dashed curve corresponds to semi-infinite atmosphere with $\epsilon = 10^{-4}$. As $S(\tau)$ is symmetric about $T/2$, the region between $T/2$ and T is not shown owing to use of a logarithmic scale for abscissa.

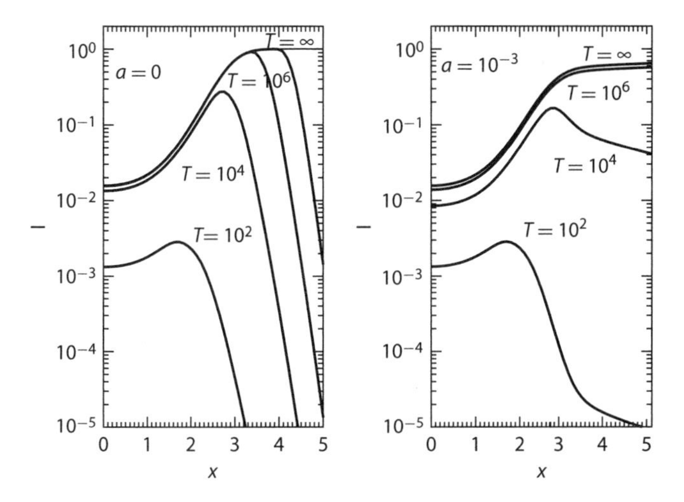

Figure 14.5 Emergent intensity for finite and semi-infinite atmosphere in Doppler and Voigt profiles with $\epsilon = 10^{-4}$.

the solution for a semi-infinite atmosphere with $\epsilon = 10^{-4}$. It can be seen that for $\epsilon \geq 10^{-4}$ the solutions closely resemble semi-infinite atmosphere solutions at the corresponding ϵ, whereas for $\epsilon < 10^{-4}$ the atmosphere becomes effectively thin and S_L falls below the corresponding semi-infinite curve and, in fact, scales linearly with ϵ. Emergent intensities are shown in figure 14.5 for lines with $\epsilon = 10^{-4}$ and for various values of a and T. For $T = \infty$, an absorption line is obtained in every case, with central intensity independent of a.

In finite atmospheres, emission lines are obtained, for the line wing becomes completely transparent for sufficiently large x, and the intensity must go to zero. At smaller x the intensity rises rapidly and, for effectively thick atmospheres, saturates to the semi-infinite atmosphere value. Finally, in the line core, scattering leads to a self-reversal. The line profiles shown in figure 14.5 strongly resemble those from laboratory emission sources with saturated lines and from hot chromospheric layers above a relatively cool photosphere.

14.3 APPROXIMATE SOLUTIONS

Throughout this book we will encounter many powerful numerical methods that are capable of providing quite sophisticated modeling techniques for stellar atmospheres and other astrophysical media. We should therefore ask whether inherently approximate and relatively crude methods, loosely described as *escape probability methods* or *probabilistic methods*, are still meaningful in view of the overall development of the field. The answer is positive, essentially for the two following reasons.

First, there is a methodological reason. Although the numerical simulations ultimately provide detailed quantities to be compared to observations, we need to understand the *physical meaning* of the results obtained. Escape probability methods provide the desired physical insight. As we will see below, they are, for instance, able to assign a clear physical meaning to many of the integral expressions of radiative transfer. Using probabilistic arguments, one is often able to design most suitable approximations to exact expressions.

Second, there is a practical reason. Numerical simulations are capable of providing exact solutions in certain cases, but they can be quite expensive in terms of computer resources. Moreover, they offer high accuracy for solutions of structural equations, but such a high accuracy may not be needed for astrophysical applications, because usually the equations being solved are only approximations to reality. The status of current astrophysical radiative transfer theory is that exact numerical methods are practical for one-dimensional static media, but are either extremely demanding or even completely out of the question for multi-dimensional coupled radiation (magneto)hydrodynamics. Even in one-dimensional simulations, it is worthwhile to have fast numerical methods that allow us to explore wide ranges of parameter space easily, which would otherwise be impossible with detailed numerical methods.

In those situations, it makes sense to use some approximate methods. The most popular and efficient among those are the escape probability methods. The topic has a long history. A comprehensive review of the topic appears in [919]; in this section we summarize its basic concepts. We concentrate on static, one-dimensional media; applications of escape probability ideas to other problems (e.g., moving media) will be discussed in later chapters of this book.

The essence of the escape probability approach is that it provides a simple approximate relation between the radiation intensity and the source function. Having such a relation, one can use it to simplify the problem of coupled radiative transfer equation and kinetic equilibrium equations. Further, it may also provide directly the emergent radiation from the medium. In some cases, the physical meaning of the escape probability methods may be hidden in the formalism, but one should bear in mind that the heart of all escape probability approaches is an approximate relation between intensity and the source function. We first summarize here some results obtained in § 11.8, define the escape probability and related quantities, and then derive some general relations.

Concept of the Net Radiative Bracket

The net rate $R_{ji}^{\rm net}$ for the transition $j \to i$ is defined by

$$n_j R_{ji}^{\rm net} \equiv n_j A_{ji} + n_j B_{ji} \overline{J}_{ij} - n_i B_{ij} \overline{J}_{ij}. \tag{14.81}$$

The frequency-averaged mean intensity is defined as $\overline{J}_{ij} = \int_0^\infty J_\nu \phi_{ij}(\nu)\, d\nu$; n_i and n_j are the atomic level populations; and A and B are the Einstein coefficients. The first term represents spontaneous emission; the second, stimulated emission; and the third, photoexcitation (absorption of a photon).

It is very useful to express the net rate of the transition between levels j and i as the spontaneous rate times a correction factor,

$$n_jA_{ji} + n_jB_{ji}\overline{J}_{ij} - n_iB_{ij}\overline{J}_{ij} \equiv n_jA_{ji}Z_{ji}, \tag{14.82}$$

where the correction factor Z_{ji} is known as the *net radiative bracket* [1077], the *escape coefficient* [39], or the *flux divergence coefficient* [175].

Noting that the line source function for the transition $i \leftrightarrow j$ is

$$S_{ij} = \frac{n_jA_{ji}}{n_iB_{ij} - n_jB_{ji}}, \tag{14.83}$$

we can rewrite the *net radiative bracket* as

$$Z_{ji} = 1 - (\overline{J}_{ij}/S_{ij}), \tag{14.84}$$

or express $\overline{J}_{ij}$ through Z as

$$\overline{J}_{ij} = (1 - Z_{ji})S_{ij}. \tag{14.85}$$

The net radiative bracket as such does not immediately help to solve a coupled radiative transfer problem because it depends on the mean intensity, so it can be evaluated only when the solution of the radiative transfer problem is already known. Nevertheless, the concept of the net radiative bracket has utility: suppose that we are able to estimate Z somehow, independently of the radiation field. In this case, we can solve the set of kinetic equilibrium equations for all the atomic level populations, and thus evaluate the line source functions, and finally, compute the radiation intensities by a formal solution of the transfer equation with a known source function. In other words, the difficulties with treating the coupling of radiation and matter (i.e., atomic level populations) would be avoided.

We show below that the escape probability approach is able to provide the desired approximate form of the net radiative bracket.

Concept of Escape Probability

We adopt here a conventional definition of the escape probability to be the probability that a photon *escapes the medium in a single direct flight*, without an intervening interaction with material particles (i.e., without undergoing a scattering process). We note that one may also define a *probability of quantum exit*, which is the probability that a photon will escape the medium directly or after a number of intermediate scatterings [573, 1030]. This kind of escape probability is quite powerful and is often used in analytical radiative transfer (it is essentially a Green function for the problem), but its determination is actually equivalent to a full solution of the transfer problem, so we will not consider this concept any further.

There are different kinds of escape probability, depending on the properties of the initial photon. The *elementary escape probability* is defined for a photon at a specified position in the medium, with a specified frequency, propagating in a specified direction. Let t_ν be the monochromatic optical depth along the ray from the given point to the boundary of the medium; then the escape probability is given by

$$p_\nu(t_\nu) = e^{-t_\nu}, \tag{14.86}$$

which is equivalent to (11.158).

Consider a plane-parallel, horizontally homogeneous slab. Any ray is specified by its direction cosine μ. In this case the elementary escape probability is both frequency- and angle-dependent:

$$p_{\mu\nu}(\tau_{\mu\nu}) = e^{-\tau_{\mu\nu}}; \tag{14.87}$$

or, writing $\tau_{\mu\nu} = \tau_\nu/\mu$, where τ_ν is the monochromatic optical depth measured inward,

$$p_{\mu\nu}(\tau_\nu) = e^{-\tau_\nu/\mu}, \quad \text{for } \mu > 0, \tag{14.88}$$

and

$$p_{\mu\nu}(\tau_\nu) = e^{-(T_\nu - \tau_\nu)/\mu}, \quad \text{for } \mu < 0, \tag{14.89}$$

because for the opposite direction, the optical distance toward the surface is $(T_\nu - \tau_\nu)/\mu$, where T_ν is the total optical thickness of the slab. Notice that in the case of semi-infinite atmosphere, $T_\nu = \infty$, the escape probability in any inward direction is 0, which is obvious from the basic meaning of "escape."

Averaging over all directions, we obtain the angle-averaged monochromatic escape probability,

$$p_\nu(\tau_\nu) = \tfrac{1}{2}\int_{-1}^{1} p_{\nu\mu} d\mu = \tfrac{1}{2}\int_{-1}^{0} e^{-(T_\nu - \tau_\nu)/\mu} d\mu + \tfrac{1}{2}\int_{0}^{1} e^{-\tau_\nu/\mu} d\mu. \tag{14.90}$$

This equation can be recast into a different form. Taking the second integral (the first one is similar), we write $\int_0^1 e^{-\tau/\mu} d\mu = \int_1^\infty e^{-\tau x}/x^2 dx$. The last integral is the second exponential integral, E_2. Thus we have

$$p_\nu(\tau_\nu) = \tfrac{1}{2}\left[E_2(T_\nu - \tau_\nu) + E_2(\tau_\nu)\right]. \tag{14.91}$$

In the case of a semi-infinite slab, the first term vanishes, and we are left with

$$p_\nu(\tau_\nu) = \tfrac{1}{2} E_2(\tau_\nu). \tag{14.92}$$

These expressions are consistent with those derived in § 11.8. There we introduced the probability, averaged over angles (in one hemisphere), that a photon emitted at $\tau = 0$ will be absorbed in the elementary optical depth range $(\tau, \tau + d\tau)$. This probability is given (see equation 11.162) by $\bar{p}(\tau) d\tau = E_1(\tau) d\tau$. Considering now the photons emitted at optical depth τ in all directions toward the surface at $\tau = 0$, the probability (averaged over angles in the corresponding hemisphere) that such a photon is absorbed between t and $t+dt$ after a direct flight between τ and t is given by $E_1(\tau - t)dt$, and thus the probability that a photon is absorbed anywhere between τ and 0 is $\int_0^\tau E_1(\tau - t)dt = 1 - E_2(\tau)$.

Consequently, the probability that the photon is *not* absorbed between τ and 0, i.e., it *escapes* from the medium, is given by $1-[1-E_2(\tau)] = E_2(\tau)$. The probability that the original photon is emitted in a direction toward the surface (as opposed to being emitted into the other hemisphere) is $\frac{1}{2}$, and the final escape probability (in the semi-infinite medium) is given by $\frac{1}{2}E_2(\tau)$, which agrees with (14.92).

It is sometimes convenient to use the *one-sided escape probability*, which represents the probability that a photon emitted isotropically into one hemisphere will escape through the corresponding boundary in a single flight. We denote this probability as $\mathcal{P}_\nu(\tau_\nu)$. In this case, we have

$$\mathcal{P}_\nu(\tau_\nu) = \tfrac{1}{2}E_2(\tau_\nu). \tag{14.93}$$

Finally, we define a *frequency- and angle-averaged escape probability* for an ensemble of photons emitted with probability $\phi(\nu)$:

$$P_e = \int_0^\infty p_\nu \phi(\nu)\, d\nu. \tag{14.94}$$

Consider now the average escape probability for an ensemble of photons in a given line. In this case, $\phi(\nu)$ is the emission profile coefficient. Assume, for simplicity, complete frequency redistribution, in which case the emission profile coefficient is equal to the absorption profile (the more general case where the two profiles may be different is considered in chapter 10). We note that the line absorption profile is normalized to unity, $\int_0^\infty \phi(\nu)d\nu = 1$, and that the monochromatic optical depth in a line is $\tau_\nu = \tau\phi(\nu)$. The one-sided averaged escape probability for line photons is thus given by

$$\mathcal{P}_e(\tau) = \tfrac{1}{2}\int_0^\infty E_2[\tau\phi(\nu)]\phi(\nu)\, d\nu. \tag{14.95}$$

The integral on the right-hand side is an important function of the radiative transfer theory; it is usually denoted as K_2, after [66].

The one-sided averaged line escape probability is thus

$$\mathcal{P}_e(\tau) = \tfrac{1}{2}K_2(\tau). \tag{14.96}$$

Thus the total escape probability for a finite slab is given by

$$P_e(\tau) = \mathcal{P}_e(\tau) + \mathcal{P}_e(T - \tau) = \tfrac{1}{2}K_2(\tau) + \tfrac{1}{2}K_2(T - \tau). \tag{14.97}$$

It is also possible to average the escape probability $P_e(\tau)$ over depth (for finite slabs; the depth average for the semi-infinite medium would be identically zero); this quantity is called the *mean escape probability* or the *escape factor*. These quantities are useful as rough approximations in cases where the medium is treated as a single zone. However, here we are interested in methods that are able to give spatial information, so we do not treat this case.

The Irons Theorem

The escape probability and the net radiative bracket may be expected to behave in a similar way. Indeed, at large depths ($\tau \gg 1$), we have $J_\nu \to S$, and thus $\overline{J} \to S$. Consequently, $Z \to 0$. The line is said to be in the *detailed radiative balance*. Physically, photons are not able to escape from the large depths; therefore, the total number of radiative transitions $j \to i$ is exactly balanced by the total number of radiative transitions $i \to j$. The escape probability $P_e(\tau)$ also goes to zero for $\tau \gg 1$. Close to the surface, both the net radiative bracket and the escape probability attain their largest values.

Are the escape probability and the net radiative bracket equal at all points in the medium? As we will see below, they are in fact *approximately equal*. However, an interesting *exact relation* also holds, namely, that they are equal in the average sense,

$$\langle Z \rangle = \langle P_e \rangle, \tag{14.98}$$

where the angle brackets denote an emission-weighted (or source-function-weighted) average over the whole volume, i.e.,

$$\langle f \rangle \equiv \frac{\int f(\tau) S(\tau)\, d\tau}{\int S(\tau)\, d\tau}. \tag{14.99}$$

The relation (14.98) is called the *Irons theorem*, because Irons [562] was the first to provide a mathematical proof of what had been a folk theorem for some time. The proof goes as follows. First, one derives a general expression that applies for a single frequency and angle, ν and μ. The emergent intensity along this ray is given by [see equation (11.101)]

$$I_{\mu\nu}(0) = \int_0^\infty S_\nu(\tau_{\mu\nu}) e^{-\tau_{\mu\nu}} d\tau_{\mu\nu} = \int_0^\infty S_\nu(\tau_{\mu\nu}) p_{\mu\nu}(\tau_{\mu\nu})\, d\tau_{\mu\nu}. \tag{14.100}$$

This expression has a simple physical interpretation. The term $S_\nu(\tau_{\mu\nu}) d\tau_{\mu\nu}$ represents the number of photons created on the optical depth range $(\tau_{\mu\nu}, \tau_{\mu\nu} + d\tau_{\mu\nu})$, per elementary intervals $d\nu$ and $d\mu$; see § 11.4. This number, multiplied by the escape probability, $p_{\mu\nu}(\tau_{\mu\nu})$, gives the number of emergent photons.

At the same time, the emergent intensity may be obtained by integrating the radiative transfer equation $(dI_{\mu\nu}/d\tau_{\mu\nu}) = S_\nu - I_{\mu\nu}$ without any integrating factor, i.e.,

$$I_{\mu\nu}(0) = \int_0^\infty (S_\nu - I_{\mu\nu})\, d\tau_{\mu\nu}. \tag{14.101}$$

Equating the right-hand sides of (14.100) and (14.101), we obtain

$$\int_0^\infty \left(1 - \frac{I_{\mu\nu}}{S_\nu}\right) S_\nu(\tau_{\mu\nu})\, d\tau_{\mu\nu} = \int_0^\infty p_{\mu\nu}(\tau_{\mu\nu}) S_\nu(\tau_{\mu\nu})\, d\tau_{\mu\nu}, \tag{14.102}$$

or, in the notation of (14.99),

$$\left\langle 1 - \frac{I_{\mu\nu}}{S} \right\rangle = \langle p_{\mu\nu} \rangle. \tag{14.103}$$

When equation (14.103) is averaged over frequencies and angles, we obtain the Irons theorem.

Physically, the Irons theorem expresses the energy balance of photons in the line. The left-hand side of equation (14.103) represents the excess of the number of emitted photons over the number of absorbed photons, integrated over the whole medium. This number equals the total number of escaping photons, as expressed by the right-hand side.

Escape Probability Treatments

The fact that the escape probability and the net radiative bracket are equal in an averaged sense does not mean that they should be equal locally, at every point in the radiating slab. However, one finds that although the detailed equality of the frequency-averaged mean intensity and the net radiative bracket does not hold *generally*, it is nevertheless a satisfactory *approximation*.

The formal solution for the mean intensity of radiation is

$$J_\nu(\tau_\nu) = \int_0^{\tau_\nu} S_\nu(t) E_1(\tau_\nu - t)\, dt + \int_{\tau_\nu}^{T_\nu} S_\nu(t) E_1(t - \tau_\nu)\, dt. \tag{14.104}$$

As discussed in § 11.5, the kernel $E_1(t)$ has a width of the order of one optical depth unit. In contrast, the scale of depth variation of the source function $S(t)$ may be much larger. If we assume that the source function is constant over the region where the kernel E_1 contributes significantly to the integral, then the source function can be taken out of the integral, setting

$$S_\nu(t) = S_\nu(\tau_\nu). \tag{14.105}$$

Equation (14.104) is then modified to read

$$\begin{aligned} J_\nu(\tau_\nu) &= \left[1 - \tfrac{1}{2}E_2(\tau_\nu) - \tfrac{1}{2}E_2(T_\nu - \tau_\nu)\right] S_\nu(\tau_\nu) \\ &= [1 - p_\nu(\tau_\nu)] S_\nu(\tau_\nu). \end{aligned} \tag{14.106}$$

Integrating equation (14.106) over frequencies with weighting factor $\phi(\nu)$, and assuming that the source function is independent of frequency (i.e., the case of a single line with a complete redistribution), we obtain

$$\bar{J}(\tau) = \left[1 - \tfrac{1}{2}K_2(T_\nu - \tau) - \tfrac{1}{2}K_2(\tau)\right] S(\tau) = [1 - P_{\rm e}(\tau)]\, S(\tau). \tag{14.107}$$

In this case, the net radiative radiative bracket is equal, at all points in the medium, to the escape probability,

$$Z(\tau) = P_{\rm e}(\tau), \tag{14.108}$$

which can be seen by comparing equations (14.85) and (14.107).

This approximation is called the *first-order escape probability* method. Its computational advantage is immediately clear: If we write the kinetic equilibrium equations in terms of net rates, we may replace Z by the escape probability P_e for all transitions. The rate equations no longer contain an unknown radiation field, so they can be solved easily. Nevertheless, they still must be solved by iteration because the escape probabilities depend on optical depths, which in turn depend on the level populations. But as these iterations are not related to consecutive photon scattering, the iteration process is quite different from the Λ-iteration scheme and is typically much faster.

The derivation above is a purely mathematical one. The only physical point there is the argument concerning the scale of variation of the source function and the kernel function. Therefore, it is useful to examine a more physical derivation, which could shed more light on the nature and limitations of the escape probability method.

Consider first a limiting case of very large optical depth in a medium with constant (or very slowly varying) properties. In this case, the variation of the source function with depth arises only because of the presence of a boundary or boundaries. As discussed above, deep in the medium the escape probability is essentially zero. On the microscopic level, every downward radiative transition is immediately balanced by the upward transition. The balancing transition does not necessarily occur at the same point in the medium, since a photon will travel a distance of the order of one unit of monochromatic optical depth. Nevertheless, because the properties of the medium do not vary over the mean free path of the photon, the resulting picture is the same as if *every emitted photon is immediately re-absorbed at the same point* in the medium. This is the reason why this approximation was historically called the *on-the-spot approximation*; it is also sometimes called *complete line saturation* or, perhaps most frequently, *detailed radiative balance*. In this case,

$$n_j A_{ji} + n_j B_{ji} \overline{J}_{ij} - n_i B_{ij} \overline{J}_{ij} = 0; \qquad (14.109)$$

thus

$$\overline{J}_{ij} = S_{ij}. \qquad (14.110)$$

Therefore, in this approximation,

$$Z_{ij} = 0, \qquad (14.111)$$

and inasmuch as $P_{\rm e} \approx 0$, here we again have the case where $Z \approx P_{\rm e}$.

A more general case is provided by the so-called *dichotomous model*. Instead of assuming that all emitted photons are re-absorbed on the spot, we divide them into two groups. The first group of photons is indeed re-absorbed on the spot, while in contrast the rest of photons escape the medium altogether. This is of course an approximation; in reality there is a continuous distribution of distances that a newly created photon can travel, ranging from zero all the way to the optical distance toward the boundary. The dichotomous model essentially approximates the real distribution by a bi-modal distribution. This procedure may seem rather crude, but, as we will see below, it reflects the basic physics of line transfer.

The fraction of photons that do escape is given by the escape probability. The net rate in the transition, i.e., a difference between the downward and upward transition rate and therefore a difference between the number of photons created and number of those destroyed, is given by the fraction of the spontaneous emission rate that produces the escaping photons, i.e.,

$$n_j A_{ji} + n_j B_{ji} \overline{J}_{ij} - n_i B_{ij} \overline{J}_{ij} = n_j A_{ji} P_{\mathrm{e}}. \tag{14.112}$$

Comparing this equation to (14.82), we see that

$$Z = P_{\mathrm{e}}, \tag{14.113}$$

i.e., the equality of the net radiative bracket and the escape probability is *exact* here. This model is also called the *normalized on-the-spot* approximation, after [1106].

We will see later (chapter 19) that the first-order escape probability (i.e., dichotomous model) is an excellent approximation for media having a large velocity gradient; the approximation is called there the *Sobolev approximation*. In static media, the situation is more complex. We return to this point later.

Finally, we give asymptotic expressions for the one-sided escape probability as $\tau \to \infty$ when the absorption profile coefficient $\phi(\nu)$ is given by Doppler and Voigt profiles; see [66]:

$$P^{\mathrm{D}}(\tau) \approx \frac{1}{4\tau\sqrt{\ln(\tau/\pi^{\frac{1}{2}})}}, \tag{14.114a}$$

$$P^{\mathrm{V}}(\tau) \approx a^{\frac{1}{2}}/3\tau^{\frac{1}{2}}. \tag{14.114b}$$

The approximate expressions derived using the Osterbrock picture (14.30) agree with the exact asymptotic expression (14.114) within a factor of 2 for the Doppler profile and a factor of $\sqrt{\pi}/3$ for the Voigt profile. Thus despite the approximate nature of the Osterbrock picture, its results are in surprisingly good agreement with exact asymptotic results. This shows that it provides a valuable insight into the nature of the escape probability and enables us to derive some basic expressions very easily.

Core-Saturation Method

We have seen above that the idea of a core-wing separation of line photons enables us to calculate an approximate escape probability very easily. In his *core-saturation* method, Rybicki [918] showed that this approach can be used not only to evaluate escape probabilities, but also to treat the entire radiative transfer process.

Consider a beam of radiation along a ray in a plane-parallel, horizontally homogeneous medium. Let τ_ν be the monochromatic optical depth along this ray measured inward from the boundary. The frequency dependence of the opacity and therefore τ_ν may be arbitrary; for the sake of simplicity we assume here radiative transfer in a single line, in which case $\tau_\nu = \phi_\nu \tau$, where τ is the frequency-averaged optical depth in the line, and ϕ_ν is its absorption profile coefficient.

At any depth in the medium, divide frequency space into two parts, the *core* and the *wing*. To this end, choose a parameter $\gamma \approx 1$ such that in the core region, $\tau_\nu \geq \gamma$. Then make the approximation that

$$I_\nu = S_\nu \quad \text{for} \quad \tau_\nu \geq \gamma. \tag{14.115}$$

The remaining part of frequency space, defined by $\tau_\nu < \gamma$, is the *wing* region. In this region, we do not impose any approximation on S_ν. The core-wing separation is dependent on the position in the medium. The division frequency between the core and the wing region, $x_{\rm d}$, is given by

$$\tau\phi(x_{\rm d}) = \gamma. \tag{14.116}$$

For instance, for a Doppler profile we have, in analogy to equation (14.29a),

$$x_{\rm d} = \sqrt{\ln(\tau/\gamma\pi^{\frac{1}{2}})}. \tag{14.117}$$

The next step is to write down the corresponding expression for the net radiative rate. First, we express the frequency-averaged mean intensity $\overline{J}$ as

$$\overline{J} = 2\int_0^\infty J_x\phi(x)\,dx = 2\int_0^{x_{\rm d}} J_x\phi(x)\,dx + 2\int_{x_{\rm d}}^\infty J_x\phi(x)\,dx \equiv \overline{J}_{\rm c} + \overline{J}_{\rm w}, \tag{14.118}$$

i.e., we split the frequency-averaged mean intensity into the core and wing contributions. Assuming a frequency-independent source function, $S_\nu = S$, and using (14.115), the core contribution is given by

$$\overline{J}_{\rm c} \equiv 2\int_0^{x_{\rm d}} J_x\phi(x)\,dx = 2S\int_0^{x_{\rm d}} \phi(x)\,dx \equiv SN_{\rm c}, \tag{14.119}$$

where $N_{\rm c}$ is called the *core normalization*. We use a *wing normalization*:

$$N_{\rm w} = 1 - N_{\rm c} = 2\int_{x_{\rm d}}^\infty \phi(x)\,dx. \tag{14.120}$$

Here we begin to see an intimate relation between the core-saturation method and the Osterbrock picture. Setting the parameter γ to 1, the escape probability in the Osterbrock picture is given by

$$P(\tau) = \tfrac{1}{2}N_{\rm w}. \tag{14.121}$$

Now, using the notion of wing normalization, we can express the net radiative rate as

$$n_jA_{ji} + n_jB_{ji}\overline{J}_{ij} - n_iB_{ij}\overline{J}_{ij} = n_jA_{ji} + (n_jB_{ji} - n_iB_{ij})[S_{ij}(1 - N_{\rm w}) + \overline{J}_{\rm w}]. \tag{14.122}$$

So far, the expression is exact. Assuming further that the total source function is given by the line source function (i.e., we neglect the contribution from a continuum opacity as well as an overlap of other lines), the source function is given by

$$S = S_{ij} = \frac{n_j A_{ji}}{n_i B_{ij} - n_j B_{ji}}. \tag{14.123}$$

Substituting (14.123) into (14.122), we are left with

$$n_j A_{ji} + n_j B_{ji} \overline{J} - n_i B_{ij} \overline{J} = n_j A_{ji} N_{\mathrm{w}} + (n_j B_{ji} - n_i B_{ij}) \overline{J}_{\mathrm{w}}. \tag{14.124}$$

This equation also has a profound significance in NLTE radiative transfer. The expression for the net radiative rate is very similar to the original one, the difference being that the spontaneous emission term is multiplied by the wing normalization N_{w}, and the net absorption term contains the wing part of the frequency-averaged mean intensity $\overline{J}_{\mathrm{w}}$ instead of $\overline{J}$. Physically, this result corresponds to the following picture: a photon created in a given line has a largest probability to be created with frequencies near the line core. Such a photon travels a short distance (since the opacity it "sees" is large), until it is absorbed. When it is re-emitted, it is emitted again most likely with a frequency close to the line center. Only in the rare event when the photon has a sufficiently large frequency separation from the core can it travel a large distance in physical space or escape altogether from the medium. This was already mentioned in § 14.2 and depicted schematically in figure 14.3. In other words, roughly speaking, the core frequencies are inefficient for line transfer, and the only frequency region that is mainly responsible for a transfer of line radiation is the line wing.

Another closely related view is that (14.124) introduces the concept of *preconditioning* of the rate equations. Typically, the net rate is given by a difference of two large terms that nearly cancel. For instance, deep in the medium, most absorptions are balanced by emissions, more or less at the same spot (or, if not, very close to the original spot; see the previous section). Such a situation is very unfavorable for any iterative numerical method, since a small error in either of the two rates may lead to disastrously large errors in the current value of the *net* rate. The idea of preconditioning is to remove analytically the large contributions that balance each other and leave only the active terms.

This is exactly what was accomplished by the core-saturation method. Then, instead of having a difference of two large terms, the total emission and absorption rate, we have a difference between two "effective" transition rates. Taking again deep layers, the wing normalization is very small, $N_{\mathrm{w}} \ll 1$, because most of the line profile is optically thick. The effective spontaneous emission rate, $n_j A_{ji} N_{\mathrm{w}}$, which is thus much smaller than the total spontaneous rate, $n_j A_{ji}$, reflects the number of transitions that are *not* immediately balanced by inverse transitions. The same applies for the absorption rate; we may therefore say that the rates are *preconditioned*.

Generally, the important point is that the core components, which we know are inefficient for line transfer, were completely eliminated from the problem. But, at the same time, the form of the rate equation is unchanged. However, one should bear in mind that although elegant and intuitively clear, the preconditioning based on the core-saturation method is only *approximate*. We consider a more exact, although conceptually similar, version of the preconditioning of the rate equations below in § 14.5.

The method outlined by Rybicki [918, 919] may be used as an approximate numerical method for solving a NLTE line transfer problem. Rybicki has also suggested an iterative extension of the method that would allow one to obtain an *exact* solution [918]. Nevertheless, in view of modern, fast, and accurate numerical schemes like the ALI method, this method is not used any longer in current computational work. It is the *concept* of core saturation that has significant value for understanding the basic physics of line transfer. As was pointed out in chapter 13, the core-saturation method was one of the basic inspirations of the early versions of the ALI scheme [955]; in fact, the early approximate Λ-operators were based on the core-saturation idea.

14.4 EQUIVALENT-TWO-LEVEL-ATOM APPROACH

Classical Scheme

As we have seen in § 14.2, a coupling the kinetic equilibrium equations and the radiative transfer equation is easy to handle for a two-level atom, because in that case the kinetic equilibrium equation can easily be eliminated. One obtains a single integral equation for the source function. A straightforward application of this idea to the case of multi-level atoms consists of selecting a single transition, say, $l \leftrightarrow u$, for which the coupling of transfer and kinetic equilibrium equations is treated explicitly, while for the remaining transitions one assumes that the level populations and the radiation field are known. In other words, only two selected levels, l and u, are directly coupled to the radiation field in the transition $l \leftrightarrow u$, so the formalism will resemble that of a two-level atom. The procedure is called the *equivalent-two-level-atom* (ETLA) method.

Before more powerful and robust numerical schemes for treating a general multi-level atom problem were developed, in particular, those based on the application of the ALI scheme, the ETLA method was popular. But one can immediately see the drawback of this approach: because the information about other transitions is lagged, and in fact the other transitions are treated essentially by a Λ-iteration, the overall process may converge very slowly or not converge at all.

Nevertheless, it is still being used in several popular computer codes such as PANDORA [68–70, 1118] and ALTAIR [200]. It is also successful for line transfer in expanding atmospheres [740], as discussed in § 19.6, and some ideas based on ETLA are useful in treating lines transfer in multi-level atoms with partial frequency redistribution (see chapter 15).

The kinetic equilibrium equations for levels l and u are written as

$$n_l(R_{lu} + a_1) = n_u(R_{ul} + C_{ul}) + a_2, \tag{14.125a}$$

$$n_u(R_{ul} + a_3) = n_l(R_{lu} + C_{lu}) + a_4, \tag{14.125b}$$

where

$$a_1 = \sum_{j \neq l,u} (R_{lj} + C_{lj}) + C_{lu}, \quad (14.126a)$$

$$a_2 = \sum_{j \neq l,u} n_j (R_{jl} + C_{ji}), \quad (14.126b)$$

$$a_3 = \sum_{j \neq l,u} (R_{uj} + C_{uj}) + C_{ul}, \quad (14.126c)$$

$$a_4 = \sum_{j \neq l,u} n_j (R_{ju} + C_{ju}). \quad (14.126d)$$

Here we have separated the rates for transitions between levels l and u from all other transition rates into and out of level l. The sums in equations (14.126) extend over all levels of the species, including possibly different ionization stages. No notational distinction between bound-bound and bound-free transitions is being made here.

Now we must solve equations (14.125) for the ratio n_l/n_u. We express the radiative rates in the transition $l \leftrightarrow u$ explicitly, $R_{lu} = B_{lu}\overline{J}_{lu}$, $R_{ul} = A_{ul} + B_{ul}\overline{J}_{lu}$, with $\overline{J}_{lu} = \int J_\nu \phi_{lu}(\nu) d\nu$, and substitute the result into the expression for the source function,

$$S_{lu} = \frac{2h\nu_{lu}^3}{c^2} \frac{1}{(n_l g_u / n_u g_l) - 1}. \quad (14.127)$$

We then make use of the Einstein relations among the transition probabilities. After a fair amount of algebra, we obtain

$$S_{lu} = \frac{\int_0^\infty J_\nu \phi_{lu}(\nu)\, d\nu + (\epsilon'_{lu} + \theta_{lu}) B_\nu(T)}{1 + \epsilon'_{lu} + \eta'_{lu}}, \quad (14.128)$$

where

$$\epsilon'_{lu} \equiv \frac{C_{ul}(1 - e^{-h\nu_{lu}/kT})}{A_{ul}}, \quad (14.129a)$$

$$\eta'_{lu} \equiv \frac{a_2 a_3 - (g_l/g_u) a_1 a_4}{A_{ul}(a_2 + a_4)}, \quad (14.129b)$$

$$\theta_{lu} \equiv \frac{n_l^* a_1 a_4 (1 - e^{-h\nu_{lu}/kT})}{n_u^* A_{ul}(a_2 + a_4)}. \quad (14.129c)$$

The terms a_1 and a_2 represent, respectively, the total rate of transitions out of level l to other all levels but u, and a total transition rate into level l from all other levels but u; a_3 and a_4 represent similar quantities for the upper level u. Notice that the populations of *all other levels* appear in a_2 and a_4, and therefore in η' and θ. The resemblance of (14.128) to the analogous equation for a two-level atom, (14.34), is obvious. Indeed, (14.129) is the same as (14.34) if $\eta' = \theta = 0$.

There is another form of the ETLA source function, which is not as intuitively appealing, but which is advantageous in numerical applications, and also for formulating the ETLA procedure for line transfer with partial redistribution; see § 15.5. The line source function is written as

$$S_{lu} = \frac{\overline{J}_{lu} + \eta_{lu}}{1 + \epsilon_{lu}}, \tag{14.130}$$

where

$$\epsilon_{lu} = [\alpha_{lu} - (g_l/g_u)\beta_{lu}]/A_{ul}, \tag{14.131a}$$

$$\eta_{lu} = \beta_{lu}/B_{lu}, \tag{14.131b}$$

and

$$\alpha_{lu} = (a_2 a_3 + a_4 C_{ul})/(a_2 + a_4), \tag{14.132a}$$

$$\beta_{lu} = (a_1 a_4 + a_2 C_{lu})/(a_2 + a_4). \tag{14.132b}$$

One can easily verify that this form is equivalent to the original formulation (14.128)–(14.129).

The iteration procedure required to obtain the solution of the problem, which consists of the determination of the occupation numbers of all the levels as functions of depth, is fairly straightforward in principle, though often quite complicated in practice. One starts with an estimate of all level populations (say, LTE values), so one can construct provisional optical depth scales to compute opacities and emissivities in all lines and continua of interest and to solve the radiative transfer equation to determine the radiation field in all transitions. One then uses these values to compute radiative rates in all transitions. The rate equations (14.1) are then re-solved at each depth to obtain an improved estimate of the occupation numbers n. The new radiative rates and occupation numbers are then used to compute the quantities a_1–a_4; and hence ϵ', η', and θ for all transitions, from which one obtains new estimates of $\overline{J}$ and S in every line (and, by similar expressions, for every continuum). The transfer equations using this source function are re-solved, and the process is iterated to convergence. Once a converged solution is obtained, the source functions in all transitions are known, and line profiles and radiation fields in all continua may be calculated for each line and continuum.

As mentioned above, the practical implementation of a successful iteration procedure using the ETLA formalism is often complicated because the rate of convergence (or lack thereof!) may be strongly affected by technical details, such as the way the radiative rates are computed and, in particular, the order in which one performs the series of ETLA calculations for the atomic transitions. A more fundamental difficulty is that this approach can *stabilize* on an *inconsistent* solution; see the discussion in [39, § 4.2] and [41, pp. 27–63]. This failure, which can be overcome by special procedures, is, nevertheless, not surprising; there are many physical situations in which the radiation fields in different lines are strongly interdependent, in contradiction to the basic assumption of the equivalent-two-level-atom approach. The inconsistency problem is wholly overcome by global schemes, such as the complete-linearization method and the ALI-based method discussed below.

Preconditioning

Another scheme that can to a certain extent overcome the problems mentioned above, while keeping the overall framework of the equivalent-two-level-atom scheme intact, is based on applying the idea of preconditioning, as was done by Avrett and coworkers in [63, 71].

We will first describe the idea in the context of the simplest case of two-level atom with no background continuum. One expresses the frequency-averaged mean intensity through the averaged Λ-operator [cf. (11.114)], written in a discretized form as

$$\overline{J}_i = \sum_j \overline{\Lambda}_{ij} S_j. \tag{14.133}$$

Unlike most of the methods considered in the book, the matrix elements of the Λ-operator are constructed *explicitly*, using a quadrature representation of the integral defining $\overline{\Lambda}$ [cf. (11.110) and (11.115)],

$$\overline{J}(\tau) = \tfrac{1}{2} \int_0^\infty dt\, S(t) \int_0^\infty d\nu\, E_1(|t-\tau|\phi_\nu)\, \phi_\nu. \tag{14.134}$$

Equation (14.133) can also be written as

$$\overline{J}_i = S_i + \sum_j L_{ij} S_j, \tag{14.135}$$

where $L_{ij} \equiv \overline{\Lambda}_{ij} - \delta_{ij}$, or, in an operator form, $\mathbf{L} \equiv \overline{\Lambda} - \mathbf{I}$, where $\mathbf{I}$ is the unit operator. Further, operator $\mathbf{L}$ can be split into the local and nonlocal parts, or matrix $\mathbf{L}$ into the diagonal and off-diagonal parts,

$$L_{ij} = d_i \delta_{ij} + c_{ij}, \tag{14.136}$$

where $d_i \equiv L_{ii}$ are the diagonal elements of $\mathbf{L}$, and $c_{ij} \equiv L_{ij}(1-\delta_{ij})$ are its off-diagonal elements.

The net radiative bracket, defined by (14.84), namely, $Z \equiv 1 - (\overline{J}/S)$, is then given by

$$Z = -(1/S_i) \sum_j L_{ij} S_j = -d_i - (1/S_i) \sum_j c_{ij} S_j. \tag{14.137}$$

This expression, which is still exact, expresses the net radiative bracket solely through the source function, but this expression is nonlinear. To proceed further, the source function in the denominator of (14.137) is expressed as

$$S_i \equiv q_i (n_u/n_l)_i, \tag{14.138}$$

where $(n_u/n_l)_i$ is the ratio of the populations of the upper and lower level at depth i, and

$$q_i \equiv (g_l/g_u)(2h\nu_{lu}^3/c^2)[1 - (n_u/n_l)_i(g_l/g_u)]^{-1}. \tag{14.139}$$

Notice that if the stimulated emission is neglected, $q_i = (g_l/g_u)(2h\nu_{lu}^3/c^2)$ is a constant. In a general case, q_i is only weakly dependent on the (unknown) population ratio, unless stimulated emission is very important. With these definitions, the net rate in transition $l \to u$ becomes

$$n_u A_{ul} - (n_l B_{lu} - n_u B_{ul})\bar{J} \equiv n_u A_{ul} Z = (-n_u d_{ul} - n_l e_{ul}) A_{ul}, \tag{14.140}$$

where now subscripts lu at d and e refer to the indices of the levels involved in the transition. All depth-dependent quantities are understood to be given at the same depth i, and $e_i \equiv (e_{ul})_i$ is given by

$$e_i \equiv (1/q_i) \sum_j c_{ij} S_j. \tag{14.141}$$

Equation (14.140) is in fact a preconditioned form of the net rate. Instead of dealing with a subtraction of two large quantities as expressed in the first equality in (14.140), one now has a difference of two much smaller quantities. Indeed, at large depths $\Lambda_{ii} \to 1$; hence $d_i \equiv 1 - \Lambda_i$ is small. Similarly, c is small as well. The final expression is similar in spirit to the form of the net rate derived by the core-saturation approach, equation (14.124), but does not involve any approximation. It forms the basis for the *preconditioned ETLA method.*

In the case of two-level atom, one writes the kinetic equilibrium equation as

$$n_u(A_{ul} Z_{ul} + C_{ul}) = n_l C_{lu}, \tag{14.142}$$

which becomes, using (14.140),

$$n_u(-A_{ul} d_{ul} + C_{ul}) = n_l(A_{ul} e_{ul} + C_{lu}), \quad \text{or} \quad \frac{n_u}{n_l} = \frac{C_{lu} + A_{ul} e_{ul}}{C_{ul} - A_{ul} d_{ul}}. \tag{14.143}$$

It is instructive to consider the case of negligible stimulated emission. We drop subscripts l, u of the levels and restore the depth subscripts i and j. The source function can be written, using (14.143), (14.138), and (14.141), as

$$S_i = \frac{e_i + \epsilon_i B_i}{\epsilon_i - d_i} = \frac{\sum_j c_{ij} S_j + \epsilon_i B_i}{\epsilon_i - d_i}, \tag{14.144}$$

where $\epsilon_i \equiv C_{ul}/A_{ul}$. Equations (14.144) form a linear set for the unknown source function at all depths, S_i, and thus can be solved directly. As suggested in [63], it can also be solved iteratively, using a simple, Λ-iteration-type procedure, namely,

$$S_i^{\text{new}} = \frac{\sum_j c_{ij} S_j^{\text{old}} + \epsilon_i B_i}{\epsilon_i - d_i}. \tag{14.145}$$

Although different in appearance, this method is equivalent to the traditional ALI scheme with a local (diagonal) approximate operator. To demonstrate this, let us rewrite equation (13.5) with an approximate operator given as a local operator with elements $\Lambda_i^*\delta_{ij}$,

$$S_i^{\rm new} = (1-\epsilon)\left[\Lambda_i^* S_i^{\rm new} + \sum_j (\Lambda_{ij} - \Lambda_i^*\delta_{ij})S_j\right] + \epsilon_i B_i. \tag{14.146}$$

Equation (14.146) is equivalent to (14.145) if $1-(1-\epsilon\Lambda_i^*) = \epsilon_i - d_i$, where $d_i \equiv \Lambda_{ii} - 1$, and therefore

$$\Lambda_i^* = \frac{\Lambda_{ii} - \epsilon_i}{1-\epsilon_i}. \tag{14.147}$$

The new scheme is thus equivalent to the traditional ALI with a diagonal operator, not given by the diagonal of the exact Λ as in the Jacobi method, known in astrophysics as the Olson-Auer-Buchler method, but by a slightly modified form.

An application to multi-level atoms within the ETLA scheme is straightforward. Kinetic equilibrium equations for general levels l and u are written, in analogy to (14.125), and using equation (14.140), as

$$n_l(A_{ul}e_{ul} + a_1) = n_u(-A_{ul}d_{ul} + C_{ul}) + a_2, \tag{14.148a}$$

$$n_u(-A_{ul}d_{ul} + a_3) = n_l(A_{ul}e_{ul} + C_{lu}) + a_4, \tag{14.148b}$$

where

$$a_1 = \sum_{j<l}(-A_{lj}d_{dj} + C_{lj}) + \sum_{j>l,\, j\neq u}(A_{jl}e_{jl} + C_{lj}) + C_{lu}, \tag{14.149a}$$

$$a_2 = \sum_{j<l} n_j C_{jl} + \sum_{j>l,\, j\neq u} n_j(-A_{jl}\, d_{jl} + C_{jl}), \tag{14.149b}$$

and analogously for a_3 and a_4. Solution of (14.148) is, similarly to equation (14.143),

$$\frac{n_u}{n_l} = \frac{\beta_{lu} + A_{ul}e_{ul}}{\alpha_{lu} - A_{ul}d_{ul}}, \tag{14.150}$$

where α and β are given by equations (14.132). The source function is given by (14.138). As in the case of two-level atom, one applies a Λ-iteration, using the old source function to evaluate e_{ul}. As discussed above, such a "preconditioned Λ-iteration" is in fact equivalent to the ALI scheme with a diagonal approximate operator and therefore converges significantly faster than a genuine Λ-iteration. The overall ETLA procedure is done analogously as in the case of traditional ETLA described above, i.e., by updating the source function (population ratio) for one transition at a time. This scheme forms the basis of more recent versions of the PANDORA program [70].

14.5 NUMERICAL SOLUTION OF THE MULTI-LEVEL ATOM PROBLEM

In this section we describe methods that are able to provide a global and consistent solution of the multi-level atom problem. They were developed in the last three decades, motivated by the ever-increasing quality of stellar observations, which necessitated improved analysis methods.

Complete Linearization

The complete linearization method was the first robust global method that was able to solve a multi-level atom problem consistently. In fact, it was originally used to treat a more complex problem, namely, computing NLTE model atmospheres [53] and only later was adapted to deal with the problems defined in this chapter [47]. After the advent of ALI schemes, the method ceased to be the most popular numerical method to treat the multi-level atom problem, mainly because of its large demands on computer time.[3] Nevertheless, we discuss the method in some detail because its ideas are still in use for computing NLTE model atmospheres and because the present case is simpler and easier to understand.

The complete linearization method is an application of the Newton-Raphson method of solving a nonlinear set of algebraic equations. The basic equations of the problem are the radiative transfer equations, (14.9), written for a set of discretized frequency points $\nu_i, i = 1, \ldots, NF$, and the rate equations, (14.1), for the atomic level populations. All integrals over frequency are replaced by quadrature sums, i.e.,

$$\int F(\nu)\, d\nu = \sum_{j=1}^{NF} w_j F(\nu_j). \tag{14.151}$$

The set of frequencies $\nu_i, i = 1, \ldots, NF$ must sample all lines and continua of the atom under study with sufficient numerical accuracy.

The transfer equation is a differential equation. To proceed, it first must be discretized in depth to transform it into a difference, and hence algebraic, equation. We choose a set of discretized depth points labeled by an index d, $d = 1, \ldots, ND$, where ND is the total number of discretized depth points.

The temperature, electron density, and total particle density are specified at each point in the atmosphere. We arrange the unknown quantities to be determined into a state vector,

$$\boldsymbol{\psi}_d \equiv (J_{d1}, \ldots, J_{d,NF}, n_{d1}, \ldots, n_{d,NL})^T, \tag{14.152}$$

at every depth point. Notice that because we use the radiative transfer equation in the variable Eddington factor form, the state parameters are the NF *mean intensities*,

[3] But thanks to the huge increases in the speed of even personal computers, this caveat is no longer severe.

not the $NF \times NA$ specific intensities (NA being the number of discretized angles) as it would be for the original transfer equation.

The discretized radiative transfer equation for the ith frequency point reads (cf. § 12.2) for the inner depth points,

$$\frac{f_{d-1,i}}{\Delta\tau_{d-\frac{1}{2},i}\Delta\tau_{di}} J_{d-1,i} - \frac{f_{di}}{\Delta\tau_{di}}\left(\frac{1}{\Delta\tau_{d-\frac{1}{2},i}} + \frac{1}{\Delta\tau_{d+\frac{1}{2},i}}\right) J_{di} + \frac{f_{d+1,i}}{\Delta\tau_{d+\frac{1}{2},i}\Delta\tau_{di}} J_{d+1,i} = J_{di} - \frac{\eta_{di}}{\chi_{di}}, \quad d = 2, \ldots, ND-1, \tag{14.153}$$

for the upper boundary condition ($d = 1$),

$$(f_{2i}J_{2i} - f_{1i}J_{1i})/\Delta\tau_{\frac{3}{2},i} = g_i J_{1i} + \tfrac{1}{2}\Delta\tau_{\frac{3}{2},i}(J_{1i} - \eta_{1i}/\chi_{1i}), \tag{14.154}$$

and for the lower boundary condition ($d = ND$),

$$\begin{aligned}(f_{ND,i}J_{ND,i} - f_{ND-1,i}J_{ND-1,i})/\Delta\tau_{ND-\frac{1}{2},i} \\ = (H_i^+ - \tfrac{1}{2}J_{ND,i}) - \tfrac{1}{2}\Delta\tau_{ND-\frac{1}{2},i}\left(J_{ND,i} - \eta_{ND,i}/\chi_{ND,i}\right).\end{aligned} \tag{14.155}$$

Here

$$\begin{aligned}\Delta\tau_{d\pm\frac{1}{2},i} &\equiv \tfrac{1}{2}(\chi_{d\pm1,i} + \chi_{d,i})|z_{d\pm1} - z_d| \\ &= \frac{1}{2}\left(\frac{\chi_{d\pm1,i}}{\rho_{d\pm1}} + \frac{\chi_{d,i}}{\rho_{d,i}}\right)|m_{d\pm1} - m_d|,\end{aligned} \tag{14.156}$$

and

$$\Delta\tau_{di} \equiv \tfrac{1}{2}\left(\Delta\tau_{d-\frac{1}{2},i} + \Delta\tau_{d+\frac{1}{2},i}\right), \tag{14.157}$$

and $H_i^+ = \int_0^1 I_i^+(\mu)\mu\, d\mu$ is the flux from the incoming radiation at the lower boundary. In the diffusion approximation, $H_i^+ = \frac{1}{2}B_i + \frac{1}{3}\partial B_i/\partial\tau_i$, where $B_i \equiv B(\nu_i)$ is the value of the Planck function at the lower boundary at frequency ν_i.

Suppose the required solution $\boldsymbol{\psi}_d$ can be written in terms of the current, but imperfect, solution $\boldsymbol{\psi}_d^0$ as $\boldsymbol{\psi}_d = \boldsymbol{\psi}_d^0 + \delta\boldsymbol{\psi}_d$. And let us write, formally, the entire set of structural equations acting on $\boldsymbol{\psi}_d$ as

$$\mathbf{P}_d(\boldsymbol{\psi}_d) = 0. \tag{14.158}$$

To obtain the solution, we demand that $\mathbf{P}_d(\boldsymbol{\psi}_d^0 + \delta\boldsymbol{\psi}_d) = 0$, and assuming that $\delta\boldsymbol{\psi}_d$ is "small" compared to $\boldsymbol{\psi}_d$, we use a Taylor expansion of $\mathbf{P}$,

$$\mathbf{P}_d(\psi_d^0) + \sum_j \frac{\partial\mathbf{P}_d}{\partial\boldsymbol{\psi}_{d,j}}\delta\boldsymbol{\psi}_{d,j} = 0, \tag{14.159}$$

to solve for $\delta\boldsymbol{\psi}_d$. Because only a first-order (i.e., linear) term of the expansion is taken into account, this approach is called a *linearization*. To obtain the corrections $\delta\boldsymbol{\psi}_d$, one has to form a matrix of partial derivatives of all the equations with respect to all the unknowns at all depths—the so-called *Jacobi matrix*, or *Jacobian*—and then solve it. The kinetic equilibrium equations are *local*, i.e., for depth point d they contain the unknown quantities n_{di} and J_{dj} only at depth d. The radiative transfer equations, however, couple depth point d to two neighboring depths $d-1$ and $d+1$; see equations (14.153)–(14.155). Then the system of linearized equations (14.159) can be written as

$$-\mathbf{A}_d\,\delta\boldsymbol{\psi}_{d-1} + \mathbf{B}_d\,\delta\boldsymbol{\psi}_d - \mathbf{C}_d\,\delta\boldsymbol{\psi}_{d+1} = \mathbf{L}_d, \tag{14.160}$$

where $\mathbf{A}, \mathbf{B}$, and $\mathbf{C}$ are $(NF+NL)\times(NF+NL)$ matrices, and $\mathbf{L}$ is a residual error vector, given by $\mathbf{L}_d = -\mathbf{P}_d(\boldsymbol{\psi}_d^0)$. At convergence, $\mathbf{L}\to 0$ and thus $\delta\boldsymbol{\psi}_d \to 0$. The system of equations (14.160) is block tridiagonal and is solved by the standard Gauss-Jordan elimination procedure, essentially the same as that used in the Feautrier elimination procedure for solving the radiative transfer equation. In the astrophysical literature, the whole procedure is sometimes inaccurately called "Feautrier elimination," although that name should be reserved for the method that puts the radiative transfer equation into its second-order form and solves its discretized version by algebraic elimination.

The first NF rows of matrices $\mathbf{A}$, $\mathbf{B}$, and $\mathbf{C}$ correspond to the radiative transfer equations at the frequency points $i=1,\ldots,NF$. For interior depth points, $d=2,\ldots,ND-1$, they are given by

$$(A_d)_{ij} = \frac{f_{d-1,i}}{\Delta\tau_{d-\frac{1}{2},i}\Delta\tau_{d,i}}\,\delta_{ij}, \quad i,j \le NF, \tag{14.161a}$$

$$(B_d)_{ij} = \left[\frac{f_{di}}{\Delta\tau_{di}}\left(\frac{1}{\Delta\tau_{d-\frac{1}{2},i}} + \frac{1}{\Delta\tau_{d+\frac{1}{2},i}}\right) + 1\right]\delta_{ij}, \quad i,j \le NF, \tag{14.161b}$$

$$(C_d)_{ij} = \frac{f_{d+1,i}}{\Delta\tau_{d+\frac{1}{2},i}\Delta\tau_{d,i}}\,\delta_{ij}, \quad i,j \le NF. \tag{14.161c}$$

Here δ_{ij} is the Kronecker delta, $\delta_{ij}=1$ for $i=j$, and $=0$ otherwise. Its presence reflects the fact that the transfer equation (14.9) is monochromatic, i.e., it has no coupling to different frequencies. An explicit coupling of frequencies occurs if scattering is non-coherent, e.g., in the case of partial frequency redistribution (cf. chapters 10 and 15), or for Compton scattering.

The remaining elements of the first NF rows of the matrices **A**, **B**, and **C** correspond to the derivatives of the transfer equations with respect to occupation numbers. For $d = 2, \ldots, ND - 1$, and $j = 1, \ldots, NL$, they are given by

$$(A_d)_{i,NF+j} = f_{d-1,i} J_{d-1,i} \frac{\partial}{\partial n_{d-1,j}} \left(\frac{1}{\Delta\tau_{d-\frac{1}{2},i} \Delta\tau_{di}} \right), \tag{14.162a}$$

$$(C_d)_{i,NF+j} = f_{d+1,i} J_{d+1,i} \frac{\partial}{\partial n_{d+1,j}} \left(\frac{1}{\Delta\tau_{d+\frac{1}{2},i} \Delta\tau_{di}} \right), \tag{14.162b}$$

$$(B_d)_{i,NF+j} = f_{di} J_{di} \frac{\partial}{\partial n_{dj}} \left(\frac{1}{\Delta\tau_{d-\frac{1}{2},i} \Delta\tau_{di}} + \frac{1}{\Delta\tau_{d+\frac{1}{2},i} \Delta\tau_{di}} \right) - \frac{1}{\chi_{d,i}} \frac{\partial \eta_{di}}{\partial n_{dj}} + \frac{\eta_{di}}{\chi_{di}^2} \frac{\partial \chi_{di}}{\partial n_{dj}}. \tag{14.162c}$$

For the linearized boundary conditions we have from (14.154) and (14.155), for $i, j \leq NF$,

$$(B_1)_{ij} = \left[f_{1i}/\Delta\tau_{\frac{3}{2},i} + g_i + \tfrac{1}{2}\Delta\tau_{\frac{3}{2},i} \right] \delta_{ij}, \tag{14.163a}$$

$$(C_1)_{ij} = (f_{2i}/\Delta\tau_{\frac{3}{2},i})\, \delta_{ij} \tag{14.163b}$$

at the upper boundary, and

$$(A_{ND})_{ij} = (f_{ND-1,i}/\Delta\tau_{ND-\frac{1}{2},i})\, \delta_{ij}, \tag{14.164a}$$

$$(B_{ND})_{ij} = \left[f_{ND,i}/\Delta\tau_{ND-\frac{1}{2},i} + \tfrac{1}{2} + \tfrac{1}{2}\Delta\tau_{ND-\frac{1}{2},j} \right] \delta_{ij} \tag{14.164b}$$

at the lower boundary. The expressions for the remaining columns that correspond to the derivatives of the boundary conditions of the transfer equation with respect to the occupation numbers are similar to (14.162).

The components of the right-hand side vector **L** for $i \leq NF$ are

$$(L_d)_i = \frac{f_{d-1,i} J_{d-1,i}}{\Delta\tau_{d-\frac{1}{2},i} \Delta\tau_{di}} - \frac{f_{di} J_{di}}{\Delta\tau_{di}} \left(\frac{1}{\Delta\tau_{d-\frac{1}{2},i}} + \frac{1}{\Delta\tau_{d+\frac{1}{2},i}} \right) + \frac{f_{d+1,i} J_{d+1,i}}{\Delta\tau_{d+\frac{1}{2},i} \Delta\tau_{di}} - J_{di} + \frac{\eta_{di}}{\chi_{di}}, \quad d = 2, \ldots, ND - 1, \tag{14.165a}$$

$$(L_1)_i = (f_{2i} J_{2i} - f_{1i} J_{1i})/\Delta\tau_{\frac{3}{2},i} - g_i J_{1i} - \tfrac{1}{2}\Delta\tau_{\frac{3}{2},i} (J_{1i} + \eta_{1i}/\chi_{1i}), \tag{14.165b}$$

$$(L_{ND})_i = (f_{ND-1,i} J_{ND-1,i} - f_{ND,i} J_{ND,i})/\Delta\tau_{ND-\frac{1}{2},i} + (H_i^+ - \tfrac{1}{2} J_{ND,i}) - \tfrac{1}{2}\Delta\tau_{ND-\frac{1}{2},i} \left(J_{ND,i} - \eta_{ND,i}/\chi_{ND,i} \right). \tag{14.165c}$$

The rows $i = NF+1, \ldots, NF+NL$ of the matrices **A**, **B**, and **C** correspond to the linearized kinetic equilibrium equations (14.1). Because, as mentioned above, these equations are local and do not couple neighboring depth points, we have

$$(A_d)_{ij} = (C_d)_{ij} = 0, \quad \text{for} \quad i \geq NF+1,\ j = 1, \ldots, NF+NL. \tag{14.166}$$

To express the elements of the matrix **B**, one can use two different forms. First, one linearizes the original rate equation, $\mathbf{An} = \mathbf{b}$, in which case (dropping, for simplicity, the depth index d on the right-hand sides of the following equations)

$$(B_d)_{NF+i,j} = \sum_{k=1}^{NL} \frac{\partial A_{i,k}}{\partial J_j} n_k \quad i = 1, \ldots, NL,\ j = 1, \ldots, NF, \tag{14.167}$$

$$(B_d)_{NF+i,NF+j} = A_{ij} \quad i,j = 1, \ldots, NL, \tag{14.168}$$

and

$$(L_d)_{NF+i} = b_i - \sum_{j=1}^{NL} A_{ij} n_j. \tag{14.169}$$

Another choice is to express the kinetic equilibrium equations as $\mathbf{n} = \mathbf{A}^{-1}\mathbf{b}$ and to linearize this equation. To this end, we express the derivatives of the inverse of the rate matrix as follows: $\mathbf{A}\mathbf{A}^{-1} = 1$, so $\partial(\mathbf{A}\mathbf{A}^{-1})/\partial x = (\partial\mathbf{A}/\partial x)\mathbf{A}^{-1} + \mathbf{A}(\partial\mathbf{A}^{-1}/\partial x) = 0$; therefore,

$$\partial\mathbf{A}^{-1}/\partial x = -\mathbf{A}^{-1}(\partial\mathbf{A}/\partial x)\mathbf{A}^{-1}, \tag{14.170}$$

where x is any quantity. The corresponding Jacobian is expressed as

$$\begin{aligned}\partial(\mathbf{n} - \mathbf{A}^{-1}\mathbf{b})/\partial x &= \partial\mathbf{n}/\partial x - \mathbf{A}^{-1}(\partial\mathbf{A}/\partial x)\mathbf{A}^{-1}\mathbf{b} \\ &= \partial\mathbf{n}/\partial x - \mathbf{A}^{-1}(\partial\mathbf{A}/\partial x)\mathbf{n}.\end{aligned} \tag{14.171}$$

Having assembled the matrices **A**, **B**, **C** and the vector **L**, the block tridiagonal system (14.160) is solved by a forward-backward sweep. As discussed in § 12.2, one first constructs auxiliary matrices $\mathbf{D}_d$ and vectors $\mathbf{E}_d$, proceeding from $d = 1$ to ND (the forward elimination step):

$$\mathbf{D}_d = (\mathbf{B}_d - \mathbf{A}_d\mathbf{D}_{d-1})^{-1}\mathbf{C}_d, \tag{14.172}$$

and

$$\mathbf{E}_d = (\mathbf{B}_d - \mathbf{A}_d\mathbf{D}_{d-1})^{-1}(\mathbf{L}_d + \mathbf{A}_d\mathbf{E}_{d-1}), \tag{14.173}$$

where we use the convention that $\mathbf{D}_0 = \mathbf{E}_0 = 0$. The solution vectors are obtained with a backward sweep:

$$\delta\boldsymbol{\psi}_d \equiv \mathbf{D}_d\,\delta\boldsymbol{\psi}_{d+1} + \mathbf{E}_d, \quad d = ND-1, \ldots, 1, \tag{14.174}$$

with $\delta\boldsymbol{\psi}_{ND} = \mathbf{E}_{ND}$. The main difficulty in solving the system (14.160) is that the $(NF + NL) \times (NF + NL)$ matrix $(\mathbf{B}_d - \mathbf{A}_d\mathbf{D}_{d-1})$ must be inverted at every depth in order to evaluate (14.172) and (14.173). Because $\mathbf{A}_d$ is partly diagonal and partly empty, only $O(NF^2)$ operations are required to form $\mathbf{A}_d\mathbf{D}_{d-1}$ even though $\mathbf{D}_{d-1}$ is full. Hence most of the work goes into inverting the full matrix $(\mathbf{B}_d - \mathbf{A}_d\mathbf{D}_{d-1})$, which requires $O[(NF + NL)^3]$ operations.

Variants of Complete Linearization

When the complete linearization method was first introduced to construct NLTE model stellar atmospheres [53], it was realized that its application to multi-level atoms can be made more efficient by eliminating level populations from the state vector [47]. One can write

$$\delta \mathbf{n} = \sum_{j=1}^{NF} (\partial \mathbf{n}/\partial J_j)\delta J_j, \tag{14.175}$$

where from (14.171) it follows that (substituting $x \to J_j$)

$$(\partial \mathbf{n}/\partial J_j) = \mathbf{A}^{-1}(\partial \mathbf{A}/\partial J_j)\,\mathbf{n}. \tag{14.176}$$

This equation expresses the linearized corrections of the level populations directly in terms of corrections to the mean intensities.

The linearized transfer equation (omitting the index i that indicates frequency) can be written

$$-a_d\delta J_{d-1} + b_d\delta J_d - c_d\delta J_{d+1} - \boldsymbol{\alpha}_d\boldsymbol{\delta}\mathbf{n}_{d-1} + \boldsymbol{\beta}_d\boldsymbol{\delta}\mathbf{n}_d - \boldsymbol{\gamma}_d\boldsymbol{\delta}\mathbf{n}_{d+1} = l_d, \tag{14.177}$$

where a_d, b_d, and c_d are given by (14.161) without the δ_{ij} term. For instance, $a_d = (A_d)_{ij}/\delta_{ij}$, and there are similar expressions for b_d and c_d. The vector $\boldsymbol{\alpha}_d$ has components given by (14.162), i.e., $\alpha_{d,j} = (A_d)_{i,NF+j}$, etc. Substituting (14.175) and (14.176) into (14.177) eliminates the corrections $\boldsymbol{\delta}\mathbf{n}$ completely.

The resulting system of linear algebraic equations is still written in the form of equation (14.160), where now $\boldsymbol{\psi}_d = (J_1, \ldots, J_{NF})^T$. The matrices **A**, **B**, and **C** are now $(NF \times NF)$ and are full. The number of operations required to obtain a solution of this system scales as $O(NF^3)$, in contrast to the scaling $O[(NF + NL)^3]$ for the original approach. If the number of frequencies and the number of levels are comparable, the present scheme leads to a significant reduction of the number of operations required. However, if the number of frequencies is much larger than the number of populations, which in fact is a typical situation, this method does not offer any substantial improvement.

To cope with this problem, another variant of complete linearization was suggested [52] that employs the idea of Rybicki-type elimination. It assumes that at frequency ν_j there is only one transition, denoted t, connecting levels l and u, that contributes to the opacity and emissivity. At this particular frequency, the linearized transfer equation can be written schematically as

$$\mathbf{T}_j\delta\mathbf{J}_j - \mathbf{L}_j\boldsymbol{\delta}\mathbf{n}_l - \mathbf{U}_j\boldsymbol{\delta}\mathbf{n}_u = \mathbf{R}_j, \tag{14.178}$$

where the $\boldsymbol{\delta}$-vectors contain the *depth* variation of a quantity, e.g.,

$$\boldsymbol{\delta}\mathbf{J}_j = (\delta J_{1j}, \delta J_{2j} \ldots, \delta J_{ND,j})^T, \tag{14.179}$$

and the matrices $\mathbf{T}_j$, $\mathbf{L}_j$, and $\mathbf{U}_j$ are of dimension ($ND \times ND$) and are tridiagonal. Equation (14.178) can be solved to obtain an equation of the form

$$\boldsymbol{\delta}\mathbf{J}_j = \mathbf{T}_j^{-1}\mathbf{L}_j\boldsymbol{\delta}\mathbf{n}_l + \mathbf{T}_j^{-1}\mathbf{U}_j\boldsymbol{\delta}\mathbf{n}_u + \mathbf{T}_j^{-1}\mathbf{R}_j, \tag{14.180}$$

where $\mathbf{T}_j^{-1}\mathbf{L}_j$ and $\mathbf{T}_j^{-1}\mathbf{U}_j$ are now full matrices.

Now the basic radiation-dependent quantities entering the rate equations are radiative rates integrated over the transitions in question. We therefore introduce variations in the net rates defined as

$$\begin{aligned}(\boldsymbol{\delta}\mathbf{Z}_t)_d &= n_{d,l}\delta R_{d,lu} - n_{d,u}\delta R_{d,ul} \\ &= \sum_j [4\pi w_j \alpha_{lu}(\nu_j)/h\nu_j][n_{d,l} - G_{lu}(\nu_j) n_{d,u}]\, \delta J_{dj},\end{aligned} \tag{14.181}$$

where the sum extends over only those frequencies contained within transition t. If we substitute equations of the form of (14.180) into (14.181) and perform the indicated summations, we obtain

$$\boldsymbol{\delta}\mathbf{Z}_t + \mathbf{A}_t\boldsymbol{\delta}\mathbf{n}_l + \mathbf{B}_t\boldsymbol{\delta}\mathbf{n}_u = \mathbf{C}_t, \tag{14.182}$$

where $\mathbf{A}_t$ and $\mathbf{B}_t$ are full matrices. From (14.2)–(14.5) and (14.179), it follows that one can write

$$\boldsymbol{\delta}\mathbf{n}_m = \sum_t \mathbf{D}_{mt}\boldsymbol{\delta}\mathbf{Z}_t, \tag{14.183}$$

where $\mathbf{D}_{mt}$ is a diagonal matrix with elements

$$(\mathbf{D}_{mt})_d = (\partial n_m/\partial Z_t)_d = (\mathbf{A}_d)^{-1}_{mj} - (\mathbf{A}_d)^{-1}_{mi}. \tag{14.184}$$

Here $\mathbf{A}_d$ is the unperturbed rate matrix at depth point d, and i and j are the lower and upper states in transition t. Using (14.183) in (14.182), we obtain the system

$$\mathbf{E}_t\boldsymbol{\delta}\mathbf{Z}_t \equiv (\mathbf{I} + \mathbf{A}_t\mathbf{D}_{lt} + \mathbf{B}_t\mathbf{D}_{ut})\boldsymbol{\delta}\mathbf{Z}_t = \mathbf{C}_t - \mathbf{A}_t\sum_{t'\neq t}\mathbf{D}_{lt'}\boldsymbol{\delta}\mathbf{Z}_{t'} - \mathbf{B}_t\sum_{t'\neq t}\mathbf{D}_{ut'}\boldsymbol{\delta}\mathbf{Z}_{t'}, \tag{14.185}$$

with one such equation for each transition t.

The size of the system (14.185), which contains the full transition-to-transition coupling (interlocking) at each depth point, is ($ND \cdot NT \times ND \cdot NT$), where NT is the total number of transitions. A direct solution of these equations would require a time that scales as $T_D = c\, ND^3 NT^3$, which, despite the huge increases in speed of ordinary workstations, may quickly become impractically large even for modest values, say, $NT \approx 20$ and $ND \approx 50$.

The system is therefore often solved by iteration, using a *successive over-relaxation* (SOR) *method*; see, e.g., [892]. In this approach there are *two* basic iteration cycles: (i) the SOR iteration, to obtain a definite set of $\delta\mathbf{Z}_t$'s ($t = 1, \ldots, NT$) within a given stage of linearization, and (ii) the overall linearization procedure, where successive sets of $\delta\mathbf{Z}_t$'s are used to update rates, and the full kinetic equilibrium equations are then re-solved. The SOR procedure is started by computing solutions for the systems $\mathbf{E}_t\delta\mathbf{Z}_t^{(0)} = \mathbf{C}_t$ ($t = 1, \ldots, NT$); this initial solution requires $c\,ND^3NT$ operations, and the resolved systems (equivalent to $\mathbf{E}_t^{-1}$) are saved. Then with any current set of estimates of the $\delta\mathbf{Z}$'s, the right-hand side of equation (14.185) can be evaluated for each transition in turn (note that only vector multiplications are involved, so the procedure is very fast). The result is a single vector of known value on the right-hand side, and, using the previously resolved $\mathbf{E}_t$, a new value of $\delta\mathbf{Z}_t$ is obtained. Each cycle of the SOR procedure requires $c\,ND^2NT^2$ operations, so if NI iterations are necessary, the overall computing time scales as $T_{\rm SOR} = c\,ND^3NT + c'NI\,ND^2NT^2$, which is favorable compared to T_D if $NI < ND\,NT$.

Actual tests of the SOR method [52] showed that this method works well, even though the SOR iterations resemble the equivalent-two-level-atom approach, because only one transition at a time is treated. The reason is that this part of the calculation is required only to determine the $\delta\mathbf{Z}_t$'s, which are merely one step of the overall linearization procedure, in itself designed to handle the interlocking problem self-consistently. Inasmuch as further steps in the linearization are presumed, the $\delta\mathbf{Z}_t$'s at any given stage need not be known perfectly, but only with sufficient accuracy that the *error* in the current estimate of $\delta\mathbf{Z}_t$ is smaller than the full size of the $\delta\mathbf{Z}_t$'s of the next linearization step. In other words, this step calculates only *corrections* to the net rates, not the net rates themselves, and therefore one can afford a simplified and less accurate treatment of them.

ALI Methods

In chapter 13 we explained the accelerated lambda iteration, or ALI, method in detail. The method was presented essentially as a fast and robust scheme for solving large linear systems. A typical application of the method was considered in § 14.2 for the case of a two-level atom. The coupled problem of solving the radiative transfer equation together with kinetic equilibrium equation was recast to the equation for the line source function, which is a linear integral equation. Upon discretization, it forms a linear algebraic equation, and therefore the standard version of ALI is straightforward to apply.

A general multi-level atom problem is, however, intrinsically nonlinear. Therefore, the basic ALI scheme needs to be modified to solve it. Let us first recast the kinetic equilibrium equation (14.1) into a more convenient form:

$$\sum_{l<i}\left[n_iA_{il} - (n_lB_{li} - n_iB_{il})\overline{J}_{il}\right] - \sum_{j>i}\left[n_jA_{ji} - (n_iB_{ij} - n_jB_{ji})\overline{J}_{ij}\right] + \sum_{j\neq i}(n_iC_{ij} - n_jC_{ji}) = 0. \tag{14.186}$$

This equation can also be written as

$$\sum_{j<i} n_i R_{ij}^{\rm net} - \sum_{j>i} n_j R_{ji}^{\rm net} + \sum_{j\neq i} (n_i C_{ij} - n_j C_{ji}) = 0, \tag{14.187}$$

where the net radiative rate in the transition $j \to i$ is

$$n_j R_{ji}^{\rm net} \equiv n_j A_{ji} - (n_i B_{ij} - n_j B_{ji}) \overline{J}_{ij}. \tag{14.188}$$

The basic idea is to apply ALI to eliminate the radiation intensities. To this end, one uses an expression that lies in the heart of the ALI formalism, namely, the formal solution of the radiative transfer equation expressed as

$$J_\nu = \Lambda_\nu^*[S_\nu^{\rm new}] + (\Lambda_\nu - \Lambda_\nu^*)[S_\nu^{\rm old}]. \tag{14.189}$$

In other words, the current (or "new") mean intensity of radiation is composed of two terms: an approximate Λ^*-operator acting on a "new" source function, plus a correction term given by the difference between the exact and approximate Λ-operator acting on the "old," and therefore known, source function. The frequency-averaged mean intensity of radiation, to which the induced radiative transition rates are proportional, is thus schematically given by

$$\overline{J}_{ij} = \overline{\Lambda}_{ij}^*[S^{\rm new}] + (\overline{\Lambda}_{ij} - \overline{\Lambda}_{ij}^*)[S^{\rm old}] \equiv \overline{\Lambda}_{ij}^*[S^{\rm new}] + \overline{J}_{ij}^{\rm eff}, \tag{14.190}$$

where

$$\overline{\Lambda}_{ij} = \int_0^\infty \Lambda_\nu \phi_{ij}(\nu)\, d\nu, \tag{14.191}$$

and a similar form for $\overline{\Lambda}^*$. The "new" averaged mean intensities of radiation are thus given by an action of a presumably simple Λ^*-operator on the new source function, which itself is a function of the level populations, plus a term known from the previous iteration. This effectively eliminates the mean intensities of radiation from the state vector, and the only quantities that need to be determined are the atomic level populations.

But there is a complication here, namely that the source function $S^{\rm new}$ is a nonlinear function of the "new" populations. By applying the ALI idea, one succeeds in eliminating the radiation intensity from the rate equations, but at the expense of ending with a set of nonlinear equations for the populations. One can cope with this problem by one of two possible ways.

(i) *Linearization*: The usual way of solving the set of nonlinear equations is by applying the Newton-Raphson method. Each iteration requires one to set up and to invert the Jacobi matrix of the system. This linearization should not be confused with "complete linearization" described above because here we linearize only the system of equations for level populations.

(ii) *Preconditioning*: This is an ingenious way to remove inactive (scattering) parts of radiative rates analytically from the rate equations and thus to achieve a linearity of the ALI form of the rate equations. It was shown [1031] that these approaches, linearization and preconditioning, are essentially equivalent. We will not describe the linearization scheme here, but rather concentrate on preconditioning, which is intuitively clearer and easier to implement in computer codes.

Preconditioning for Non-Overlapping Lines, No Background Continuum

The idea of preconditioning is easy to explain for a simple situation, namely, the case of non-overlapping lines and no background continuum, and with the total source function is given by line source function $S_{ij} = n_j A_{ji}/(n_i B_{ij} - n_j B_{ji})$. Let us further assume that we have a local (diagonal) approximate Λ^*-operator; in that case the action of the Λ^*-operator is equivalent to a multiplication by a real number, denoted here as Λ^* also. The net rate (14.188) can then be written as

$$n_j R_{ji}^{\rm net} = n_j A_{ji} - (n_i B_{ij} - n_j B_{ji}) \left[\overline{\Lambda}_{ij}^* \frac{n_j A_{ji}}{n_i B_{ij} - n_j B_{ji}} + \overline{J}_{ij}^{\rm eff} \right]$$
$$= n_j A_{ji}(1 - \overline{\Lambda}_{ij}^*) - (n_i B_{ij} - n_j B_{ji}) \overline{J}_{ij}^{\rm eff}. \tag{14.192}$$

Note that the offending nonlinear term $(n_i B_{ij} - n_j B_{ji})$ in the denominator of the expression for the source function exactly cancels the same term coming from the expression for the net rate, so the result is indeed *linear* in the level populations.

This is a very interesting expression. Notice first that the original net rate, (14.188), is represented by a subtraction of two large contributions, *all* emissions (i.e., radiative transitions $j \to i$) minus *all* absorptions (transitions $i \to j$), while the result, the net rate, is rather small. Physically, this follows from the fact that most radiative transitions $j \to i$ are those that immediately follow a previous absorption of a photon, transitions $i \to j$, i.e., they are the part of a *scattering* process and thus do not contribute to the *net* rate of transitions $j \to i$. In order to improve the numerical conditioning of the system of rate equations, one has to eliminate the scattering contributions somehow, i.e., to "precondition" the rates.

In the ALI form of the net rate, (14.192), we see that deep in the atmosphere, $\Lambda^* \to 1$, so that the first term is indeed very small. Similarly, the second term is also small because $\overline{J}_{ij}^{\rm eff}$ is small. In other words, the radiative rates are indeed preconditioned. In the context of the ALI approach, this idea was first used in [1154]; a systematic study can be found in [925] and [926].

Substituting (14.192) into the kinetic equilibrium equation (14.187), we obtain

$$\sum_{j<i} \left[n_i A_{ij}(1 - \overline{\Lambda}_{ij}^*) - (n_j B_{ji} - n_i B_{ij}) \overline{J}_{ij}^{\rm eff} \right]$$
$$- \sum_{j>i} \left[n_j A_{ji}(1 - \overline{\Lambda}_{ij}^*) - (n_i B_{ij} - n_j B_{ji}) \overline{J}_{ij}^{\rm eff} \right]$$
$$+ \sum_{j \neq i} (n_i C_{ij} - n_j C_{ji}) = 0. \tag{14.193}$$

The result is to leave the form of rate equations the same as before, except that the Einstein coefficient A has been multiplied by the factor $(1 - \overline{\Lambda}_{ij}^*)$, and the frequency-averaged mean intensity $\overline{J}_{ij}$ has been replaced by $\overline{J}_{ij}^{\text{eff}}$. These are the same types of modifications that appeared in the core-saturation method, equation (14.124), but now Λ^*_{ij} has replaced the core normalization N_c. The conditioning of the system of the rate equations is now considerably improved because much of the transfer in the "core" of the line (described by the local part of the Λ-operator) has cancelled out analytically. Further, these preconditioned equations are clearly linear in the level populations, where the linearity was achieved automatically by an exact cancelation of the $(n_i B_{ij} - n_j B_{ji})$ terms.

Preconditioning with a Background Continuum

Here we follow the analysis and formalism of [925] for the simplest non-trivial case, namely, using a local (diagonal) approximate operator for a multi-level problem with non-overlapping lines, but allowing for a background continuum.

The total opacity and emissivity in a line $l \rightarrow u$ are given by

$$\chi_\nu = \chi_{lu}(\nu) + \chi_c(\nu), \quad \text{and} \quad \eta_\nu = \eta_{lu}(\nu) + \eta_c(\nu), \tag{14.194}$$

where χ_{lu} and η_{lu} are, respectively, the absorption and emission coefficients for the line $l \rightarrow u$, and χ_c and η_c are the absorption and emission coefficients for the background continuum. The source function at frequency ν is given by

$$S(\nu) = \rho_{lu}(\nu) S_{lu} + [\,1 - \rho_{lu}(\nu)\,] S_c(\nu), \tag{14.195}$$

where S_{lu} and $S_c(\nu)$ are the source functions corresponding to a pure line and the background continuum, respectively,

$$S_{lu} = \frac{n_u A_{ul}}{n_l B_{lu} - n_u B_{ul}}, \tag{14.196a}$$

$$S_c(\nu) = \frac{\eta_c(\nu)}{\chi_c(\nu)}, \tag{14.196b}$$

$$\rho_{lu}(\nu) = \frac{\chi_{lu}(\nu)}{\chi_{lu}(\nu) + \chi_c(\nu)}. \tag{14.196c}$$

Rybicki and Hummer [925,926] denoted $\rho_{lu}(\nu)$ as $r_{lu}(\nu)$; we use the notation $\rho_{lu}(\nu)$ to avoid confusion with the traditional continuum-strength parameter r defined by (14.40), which is related to ρ as $\rho_{lu}(\nu) = \phi_{lu}(\nu)/[(\phi_{lu}(\nu) + r]$. Notice that equations (14.195)–(14.196) clearly show that the remarkable cancellation of factors that led to linear equations for the case of no background continuum does not occur here because the denominator is no longer proportional to $(n_i B_{ij} - n_j B_{ji})$. However, as suggested in [925], one way to overcome this difficulty is to replace equation (14.195) with

$$S(\nu) = \rho_{lu}^{\text{old}}(\nu) S_{lu} + [\,1 - \rho_{lu}^{\text{old}}(\nu)\,] S_c(\nu), \tag{14.197}$$

i.e., one evaluates ρ_{lu} using the "old" populations. The procedure is now analogous to that employed in the case of no background continuum.

The expression for the "new" mean intensity using the approximate operator reads

$$J_\nu = \Lambda^*_\nu S^{\rm new}_\nu + (\Lambda_\nu - \Lambda^*_\nu)[\, S^{\rm old}_\nu \,] = \Lambda^*_\nu \rho^{\rm old}_{lu}(\nu) S_{lu} + J^{\rm eff}_\nu, \tag{14.198}$$

where

$$\begin{aligned} J^{\rm eff}_\nu &= (\Lambda_\nu - \Lambda^*_\nu) S^{\rm old}_\nu + \Lambda^*_\nu [\, 1 - \rho^{\rm old}_{lu}(\nu)\,] S_c(\nu) \\ &= \Lambda_\nu [\, S^{\rm old}_\nu \,] - \Lambda^*_\nu [\, S^{\rm old}_\nu - S_c(\nu) + \rho^{\rm old}_{lu}(\nu) S_c(\nu)\,] \\ &= \Lambda_\nu [\, S^{\rm old}_\nu \,] - \Lambda^*_\nu [\, \rho^{\rm old}_{lu}(\nu)\, S^{\rm old}_{lu} \,] \\ &\equiv J^{\rm FS}_\nu - \Lambda^*_\nu [\, \rho^{\rm old}_{lu}(\nu) S^{\rm old}_{lu} \,]. \end{aligned} \tag{14.199}$$

The quantity $J^{\rm FS}_\nu = \Lambda_\nu[\, S^{\rm old}_\nu \,]$ is the mean intensity obtained by a formal solution with the old populations. Notice that the Λ^*-operator, which is local, acts as an ordinary multiplication in its action of the source function. Hence the brackets [...] are omitted on Λ^*.

The frequency-averaged mean intensity for the transition $l \rightarrow u$ is given by

$$\overline{J}_{lu} = \overline{\Lambda}^*_{lu} S_{lu} + \overline{J}^{\rm eff}_{lu}, \tag{14.200}$$

where

$$\overline{\Lambda}^*_{lu} = \int_0^\infty \Lambda^*_\nu \rho^{\rm old}_{lu}(\nu)\, \phi_{lu}(\nu)\, d\nu, \tag{14.201a}$$

$$\overline{J}^{\rm eff}_{lu} = \int_0^\infty J^{\rm eff}_\nu \,\phi_{lu}(\nu)\, d\nu = \overline{J}^{\rm FS}_{lu} - \overline{\Lambda}^*_{lu} S^{\rm old}_{lu}, \tag{14.201b}$$

$$\overline{J}^{\rm FS}_{lu} = \int_0^\infty J^{\rm FS}_\nu \,\phi_{lu}(\nu)\, d\nu. \tag{14.201c}$$

When substituting (14.198) into (14.188), the net rate in the transition $l \rightarrow u$ is written

$$\begin{aligned} n_u R^{\rm net}_{ul} &= n_u A_{ul} - (n_l B_{lu} - n_u B_{ul}) \overline{J}_{lu} \\ &= n_u A_{ul}(1 - \overline{\Lambda}^*_{lu}) - (n_l B_{lu} - n_u B_{ul}) \overline{J}^{\rm eff}_{lu}, \end{aligned} \tag{14.202}$$

and the complete set of the kinetic equilibrium equations can be written as

$$\begin{aligned} &\sum_{l<i} \left[n_i A_{il}(1 - \overline{\Lambda}^*_{il}) - (n_l B_{li} - n_i B_{il}) \overline{J}^{\rm eff}_{li} \right] \\ &- \sum_{j>i} \left[n_j A_{ji}(1 - \overline{\Lambda}^*_{ji}) - (n_i B_{ij} - n_j B_{ji}) \overline{J}^{\rm eff}_{ij} \right] + \sum_{j\neq i} (n_i C_{ij} - n_j C_{ji}) = 0. \end{aligned} \tag{14.203}$$

This equation is *linear* in the new populations and therefore can be solved easily. Notice that it is the replacement $\rho_{lu} \rightarrow \rho^{\rm old}_{lu}$ that plays the key role in achieving the linearity of (14.203).

Preconditioning with a Nonlocal Operator

Next, consider the case of a nonlocal approximate operator. In this case one has to modify equations (14.198) and (14.199) to read

$$J_\nu = \Lambda_\nu^*[\rho_{lu}^{\text{old}}(\nu)S_{lu}] + J_\nu^{\text{eff}}, \tag{14.204}$$

(where we have again replaced ρ_{lu} with ρ_{lu}^{old}), and

$$J_\nu^{\text{eff}} = \Lambda_\nu[S_\nu^{\text{old}}] - \Lambda_\nu^*[\rho_{lu}^{\text{old}}(\nu)S_{lu}^{\text{old}}] = J_\nu^{\text{FS}} - \Lambda_\nu^*[\rho_{lu}^{\text{old}}(\nu)S_{lu}^{\text{old}}]. \tag{14.205}$$

The frequency-averaged mean intensity is now

$$\overline{J}_{lu} = \widetilde{\Lambda}_{lu}^*[S_{lu}] + \overline{J}_{lu}^{\text{eff}}, \tag{14.206}$$

where

$$\widetilde{\Lambda}_{lu}^*[\ldots] = \int_0^\infty \Lambda_\nu^*[\rho_{lu}^{\text{old}}(\nu)\ldots]\,\phi_{lu}(\nu)\,d\nu, \tag{14.207a}$$

$$\overline{J}_{lu}^{\text{eff}} = \int_0^\infty J_\nu^{\text{eff}}\,\phi_{lu}(\nu)\,d\nu = \overline{J}_{lu}^{\text{FS}} - \widetilde{\Lambda}_{lu}^*[S_{lu}^{\text{old}}]. \tag{14.207b}$$

The net rate is given by

$$\begin{aligned} n_u R_{ul}^{\text{net}} = & \, n_u A_{ul} - (n_l B_{lu} - n_u B_{ul})\widetilde{\Lambda}_{lu}^*[S_{lu}] \\ & - (n_l B_{lu} - n_u B_{ul})\overline{J}_{lu}^{\text{eff}}. \end{aligned} \tag{14.208}$$

The term $\widetilde{\Lambda}_{lu}^*[S_{lu}]$ can be written

$$\begin{aligned} \widetilde{\Lambda}_{lu}^*[S_{lu}] = & \int_0^\infty \Lambda_\nu^*\left[\frac{\chi_{lu}^{\text{old}}}{\chi_{lu}^{\text{old}} + \chi_c^{\text{old}}}\left(\frac{n_u A_{ul}}{n_l B_{lu} - n_u B_{ul}}\right)\right]\phi_{lu}(\nu)\,d\nu \\ \approx & \, A_{ul}\int_0^\infty \frac{h\nu}{4\pi}\Lambda_\nu^*\left[\frac{\phi_{lu}(\nu)}{\chi_{lu}^{\text{old}} + \chi_c^{\text{old}}}\,n_u\right]\phi_{lu}(\nu)\,d\nu, \end{aligned} \tag{14.209}$$

where we have again replaced $n_l B_{lu} - n_u B_{ul}$ by $n_l^{\text{old}} B_{lu} - n_u^{\text{old}} B_{ul}$ in the denominator of the expression for the line source function. The term $\widetilde{\Lambda}_{lu}^*[S_{lu}]$ is now linear in the new population n_u, but the overall linearity of (14.208) is spoiled by the presence of $(n_l B_{lu} - n_u B_{ul})$ in the second term of the right-hand side.

As suggested in [925], in order to achieve linearity, one evaluates this term using the old populations: i.e., as $(n_l^{\text{old}} B_{lu} - n_u^{\text{old}} B_{ul})\widetilde{\Lambda}_{lu}^*[S_{lu}]$. To obtain a more convenient form of this expression, it is advantageous to define the operator $\widehat{\Lambda}_{lu}^*$ (acting on level populations) as

$$A_{ul}\widehat{\Lambda}_{lu}^*[n_u] \equiv (n_l^{\text{old}} B_{lu} - n_u^{\text{old}} B_{ul})\widetilde{\Lambda}_{lu}^*[S_{lu}], \tag{14.210}$$

which can be also expressed as

$$\widehat{\Lambda}^*_{lu}[\ldots] = \kappa_{lu}^{\rm old} \int_0^\infty \Lambda^*_\nu \left[\frac{\rho_{lu}^{\rm old}(\nu)}{\kappa_{lu}^{\rm old}} \ldots \right] \phi_{lu}(\nu)\, d\nu. \tag{14.211}$$

Substituting these expressions into the kinetic equilibrium equations (14.187), we obtain

$$\sum_{l<i} \left[A_{il}(\mathbf{I} - \widehat{\Lambda}^*_{il})[n_i] - (n_l B_{li} - n_i B_{il}) \overline{J}_{li}^{\rm eff} \right]$$
$$- \sum_{j>i} \left[A_{ji}(\mathbf{I} - \widehat{\Lambda}^*_{ji})[n_j] - (n_i B_{ij} - n_j B_{ji}) \overline{J}_{ij}^{\rm eff} \right] + \sum_{j \neq i} (n_i C_{ij} - n_j C_{ji}) = 0, \tag{14.212}$$

where $\mathbf{I}$ is the identity operator for which $\mathbf{I}[n_i] = n_i$. For a tridiagonal Λ^*-operator, one obtains a block tridiagonal set of equations for the population numbers at all depths, which can be solved by Gaussian elimination, as is done in the Feautrier method.

Full Preconditioning

The final stage is to allow for overlaps of the line and continua of the same species. Here we follow closely the formalism of [926], which introduces a formulation of the radiative rates and opacities/emissivities without an explicit separation of the transitions into lines and continua. Define, for a line transition $l \leftrightarrow u$ (l and u belong to the same ion),

$$U_{ul}(\nu) \equiv (h\nu/4\pi) A_{ul} \phi_{lu}(\nu), \quad u > l, \tag{14.213a}$$

$$U_{lu}(\nu) \equiv 0, \quad u > l, \tag{14.213b}$$

$$V_{lu}(\nu) \equiv (h\nu/4\pi) B_{lu} \phi_{lu}(\nu), \tag{14.213c}$$

and similarly for the continuum (bound-free) transitions,

$$U_{ul}(\nu) \equiv n_{\rm e} \Phi_{ul}(T)(2h\nu^3/c^2) \exp(-h\nu/kT) \sigma_{lu}(\nu), \quad u > l, \tag{14.214a}$$

$$U_{lu}(\nu) \equiv 0, \quad u > l, \tag{14.214b}$$

$$V_{ul}(\nu) \equiv n_{\rm e} \Phi_{ul}(T) \exp(-h\nu/kT) \sigma_{lu}(\nu), \quad u > l, \tag{14.214c}$$

$$V_{lu}(\nu) \equiv \sigma_{lu}(\nu), \quad u > l, \tag{14.214d}$$

where $\sigma_{lu}(\nu)$ is the photoionization cross section, and $\Phi_{ul}(T)$ is the Saba-Boltzmann factor, defined by

$$\Phi_{ul}(T) = \frac{g_l}{2g_u} \left(\frac{h^2}{2\pi m_{\rm e} kT} \right)^{\frac{3}{2}} \exp[(E_u - E_l)/kT]. \tag{14.215}$$

The quantities U correspond to spontaneous radiative processes, while the quantities V correspond to induced processes. With these definitions, the opacity and emissivity in the transition $i \leftrightarrow j$ are expressed as

$$\chi_{lu}(\nu) = n_l V_{lu} - n_u V_{ul}, \quad u > l, \tag{14.216a}$$

$$\eta_{ul}(\nu) = n_u U_{ul}, \tag{14.216b}$$

and the total opacity and emissivity can be written as

$$\chi_\nu = \sum_{u>l} (n_l V_{lu} - n_u V_{ul}) + \chi_c, \tag{14.217a}$$

$$\eta_\nu = \sum_{u,l} n_u U_{ul} + \eta_c. \tag{14.217b}$$

It is possible to express the total emissivity using an unrestricted summation over u and l because the terms with $u < l$ vanish because of the definitions in (14.213) and (14.214).

In terms of U and V, the radiative rate for the transition $u \to l$ is given by

$$R_{ul} = \int_0^\infty \frac{4\pi}{h\nu} [U_{ul}(\nu) + V_{ul}(\nu) J_\nu] \, d\nu, \tag{14.218}$$

which applies for both lines and continua and for $u > l$ and $u < l$. The kinetic equilibrium equation for level u can now be written in the form

$$\begin{aligned} \sum_l n_l C_{lu} &+ \sum_l \int_0^\infty \frac{4\pi}{h\nu} (n_l U_{lu} + n_l V_{lu} J_\nu) \, d\nu \\ &= \sum_l n_u C_{ul} + \sum_l \int_0^\infty \frac{4\pi}{h\nu} (n_u U_{ul} + n_u V_{ul} J_\nu) \, d\nu. \end{aligned} \tag{14.219}$$

This expression clearly shows the role of the radiation intensity in the kinetic equilibrium equation. There is no explicit separation of the integration into "line" and "continuum" frequencies. This formulation automatically takes into account any overlaps of lines and continua, as well as the overlapping of lines.

Now applying the basic idea of ALI, we can write the mean intensity of radiation as

$$J_\nu = \Lambda_\nu^*[S_\nu] + (\Lambda_\nu - \Lambda_\nu^*)[S_\nu^{\text{old}}]. \tag{14.220}$$

Using (14.217), the source function can be written as

$$S_\nu = \frac{\sum_{u,l} n_u U_{ul} + \eta_c}{\sum_{u>l} (n_l V_{lu} - n_u V_{ul}) + \chi_c}. \tag{14.221}$$

The expression for the "old" source function is similar, replacing n_l and n_u by $n_l^{\rm old}$ and $n_u^{\rm old}$, respectively.

As suggested in [926], we evaluate the action of the approximate Λ^*-operator on the source function using the *old* absorption coefficient:

$$\Lambda_\nu^*[S_\nu] = \Lambda_\nu^*[\eta_\nu/\chi_\nu^{\rm old}] = \Lambda_\nu^*\Big[(\chi_\nu^{\rm old})^{-1}\Big(\sum_{u,l} n_u U_{ul} + \eta_c\Big)\Big]. \tag{14.222}$$

Rybicki and Hummer [926] used the notation $\Lambda_\nu[(\chi_\nu^{\rm old})^{-1}\ldots] \equiv \Psi_\nu[\ldots]$, but we keep the original notation here. Then, using equations (14.220) and (14.222), the mean intensity can be written as

$$J_\nu = \Lambda_\nu[\eta_\nu^{\rm old}/\chi_\nu^{\rm old}] - \sum_{ij} \Lambda_\nu^*[n_i^{\rm old} U_{ij}/\chi_\nu^{\rm old}] + \sum_{ij} \Lambda_\nu^*[n_i U_{ij}/\chi_\nu^{\rm old}]. \tag{14.223}$$

Here we used the fact that the background emissivity η_c is the same when evaluating both the new and old emissivity. Using this expression for the mean intensity in the kinetic equilibrium equation (14.219), we obtain

$$\sum_l n_l C_{lu} + \sum_l \int_0^\infty \frac{4\pi}{h\nu}\Big(n_l U_{lu} + n_l V_{lu}\Lambda_\nu[\eta_\nu^{\rm old}/\chi_\nu^{\rm old}]$$

$$- \sum_{ij} n_l V_{lu}\Lambda_\nu^*[n_i^{\rm old} U_{ij}/\chi_\nu^{\rm old}] + \sum_{ij} n_l V_{lu}\Lambda_\nu^*[n_i U_{ij}/\chi_\nu^{\rm old}]\Big)\, d\nu$$

$$= \sum_l n_u C_{ul} + \sum_l \int_0^\infty \frac{4\pi}{h\nu}\Big(n_u U_{ul} + n_u V_{ul}\Lambda_\nu[\eta_\nu^{\rm old}/\chi_\nu^{\rm old}]$$

$$- \sum_{ij} n_u V_{ul}\Lambda_\nu^*[n_i^{\rm old} U_{ij}/\chi_\nu^{\rm old}] + \sum_{ij} n_u V_{ul}\Lambda_\nu^*[n_i U_{ij}/\chi_\nu^{\rm old}]\Big)\, d\nu. \tag{14.224}$$

This equation is not linear in the new populations because the last summations on each side (called the *critical summations* in [926]) contain products of n_l and n_i on the left and the product of n_u and n_i on the right. As was done in simpler cases of preconditioning, this equation may be made linear by replacing some level populations by the "old" ones. That is, for the critical summation of the left, either n_l or n_i should be replaced by its old value, but not both. As suggested in [926], the best choice is to replace n_l by its old value, while adopting the new population for all the n_i values. The reason is that were we to use old values for n_i, the last two summations on the both left and right side would cancel exactly, and one would then be using classical Λ-iteration.

With the choice described above, one obtains the *fully preconditioned* form of the kinetic equilibrium equation, which is now linear in the new populations:

$$\sum_l n_l C_{lu} + \sum_l \int_0^\infty \frac{4\pi}{h\nu} \Big(n_l U_{lu} + n_l V_{lu} \Lambda_\nu [\eta_\nu^{\text{old}} / \chi_\nu^{\text{old}}]$$

$$- \sum_{ij} n_l V_{lu} \Lambda_\nu^* [n_i^{\text{old}} U_{ij} / \chi_\nu^{\text{old}}] + \sum_{ij} n_l^{\text{old}} V_{lu} \Lambda_\nu^* [n_i U_{ij} / \chi_\nu^{\text{old}}] \Big) \, d\nu$$

$$= \sum_l n_u C_{ul} + \sum_l \int_0^\infty \frac{4\pi}{h\nu} \Big(n_u U_{ul} + n_u V_{ul} \Lambda_\nu [\eta_\nu^{\text{old}} / \chi_\nu^{\text{old}}]$$

$$- \sum_{ij} n_u V_{ul} \Lambda_\nu^* [n_i^{\text{old}} U_{ij} / \chi_\nu^{\text{old}}] + \sum_{ij} n_u^{\text{old}} V_{ul} \Lambda_\nu^* [n_i U_{ij} / \chi_\nu^{\text{old}}] \Big) \, d\nu. \quad (14.225)$$

As shown in [926], this full preconditioning strategy is unnecessarily complex for many problems. For example, in the critical summations in equation (14.225), the product $V_{lu} U_{ij} = 0$ for all frequencies when lu and ij refer to non-overlapping transitions. Even when some frequency overlap occurs, the mathematical overlap of the functions V_{lu} and U_{ij} is clearly much larger for one transition with itself than between two different transitions.

These observations suggest that a useful simplification of the full preconditioning strategy is to *precondition within the same transition only*. This is accomplished by using the new populations n_i in the critical summations only for the two terms for which ij is the same as ul and lu. In the other terms the use of n_i^{old} will cancel an identical term in the preceding summation. Consequently, all summations in equation (14.225) reduce to just two terms corresponding to the transitions ul and lu:

$$\sum_l n_l C_{lu} + \sum_l \int_0^\infty \frac{4\pi}{h\nu} \Big(n_l U_{lu} + n_l V_{lu} \Lambda_\nu [\eta_\nu^{\text{old}} / \chi_\nu^{\text{old}}]$$

$$- n_l V_{lu} \Lambda_\nu^* [n_u^{\text{old}} U_{ul} / \chi_\nu^{\text{old}}] - n_l V_{lu} \Lambda_\nu^* [n_l^{\text{old}} U_{lu} / \chi_\nu^{\text{old}}]$$

$$+ n_l^{\text{old}} V_{lu} \Lambda_\nu^* [n_u U_{ul} / \chi_\nu^{\text{old}}] + n_l^{\text{old}} V_{lu} \Lambda_\nu^* [n_l U_{lu} / \chi_\nu^{\text{old}}] \Big) \, d\nu$$

$$= \sum_l n_u C_{ul} + \sum_l \int_0^\infty \frac{4\pi}{h\nu} \Big(n_u U_{ul} + n_u V_{ul} \Lambda_\nu [\eta_\nu^{\text{old}} / \chi_\nu^{\text{old}}]$$

$$- n_u V_{ul} \Lambda_\nu^* [n_l^{\text{old}} U_{lu} / \chi_\nu^{\text{old}}] - n_u V_{ul} \Lambda_\nu^* [n_u^{\text{old}} U_{ul} / \chi_\nu^{\text{old}}]$$

$$+ n_u^{\text{old}} V_{ul} \Lambda_\nu^* [n_l U_{lu} / \chi_\nu^{\text{old}}] + n_u^{\text{old}} V_{ul} \Lambda_\nu^* [n_u U_{ul} / \chi_\nu^{\text{old}}] \Big) \, d\nu. \quad (14.226)$$

This set of linear equations gives us an explicit formulation of preconditioning within the same transition, which, as stated above, is the most advantageous preconditioning strategy for the multi-level atom problem.

14.6 PHYSICAL INTERPRETATION

Using any global method described above, one can obtain a numerical solution of the multi-level atom problem. However, as we have stressed several times throughout this book, our aim is not only to *obtain* results, but also to *understand* them. As we have seen in this chapter, the problem of NLTE line formation in a gas of multi-level atoms is a nonlinear, highly coupled, and highly nonlocal problem. Moreover, the solution sensitively depends on a large number of atomic parameters. Therefore, it is essentially impossible to make any general statements about the nature of the NLTE multi-level atom solutions. However, one can view the complex solution of the problem as a superposition of several competing elementary processes, which are described below.

Basic NLTE Line Effects

It is advantageous to define the NLTE *departure coefficient*, also called the b-factor, as

$$b_i \equiv n_i/n_i^*, \tag{14.227}$$

where n_i^* is the LTE population of level i. We stress that the LTE population is defined with respect to the ground state of the next higher ion,

$$n_i^* = n_1^+ n_e \Phi_i(T), \tag{14.228}$$

where n_1^+ is the actual population of the ground state of the next ion, n_e is the electron density, and $\Phi_i(T)$ is the Saha-Boltzmann factor defined by equation (9.5), or by (14.215) with $l=i$ and $u=1^+$ (the ground state of the next ion).

Two-Level Atom

We illustrate the utility of departure coefficients in a simple case of two-level atom without continuum, studied in detail in § 14.2. Using b-factors, equation (14.14) for the source function is written

$$S_{lu} = \frac{2h\nu^3}{c^2} \frac{1}{(b_l/b_u)\exp(h\nu_{lu}/kT) - 1}, \tag{14.229}$$

which follows from (14.227) and the Boltzmann relation for LTE populations, $n_l^*/n_u^* = (g_l/g_u)\exp(E_{lu}/kT)$. Again, the factor ν^3 varies negligibly over the line profile, so we may put $\nu \approx \nu_{lu}$. The source function becomes

$$S_{lu} = \frac{b_u}{b_l} B_{\nu_{lu}} \frac{1 - \exp(-h\nu_{lu}/kT)}{1 - (b_u/b_l)\exp(-h\nu_{lu}/kT)}. \tag{14.230}$$

In the case of negligible stimulated emission, $\exp(-h\nu_{lu}/kT) \ll 1$, the expression for the source function is particularly simple,

$$S_{lu} \approx (b_u/b_l)\, B_{\nu_{lu}}. \tag{14.231}$$

We know from the results of § 14.2 that $S < B$ for optical depths smaller than the thermalization depth, which means that $b_u/b_l < 1$. Close to the surface, and for $\epsilon \ll 1$, we have $b_u/b_l \approx \sqrt{\epsilon} \ll 1$. Line transfer in a strong line (its opacity being much larger than the background opacity) thus causes the upper level of the transition being *underpopulated* with respect to LTE, while the lower level becomes *overpopulated*. The degree of overpopulation, however, depends on additional parameters, such as actual values of ϵ and $h\nu/kT$. For instance, if $n_u \ll n_l$ in a two-level atom, then $n_l \approx N_{\rm total}$, and the NLTE effects do not change the population of the ground state in any appreciable way, so that $b_l \approx 1$ and $b_u \approx S_{lu}/B$.

This can also be understood in physical terms. In LTE, the line is in detailed radiative balance; i.e., the number of radiative transitions upward is exactly balanced by the number of radiative transitions downward. The number of transitions upward is proportional to the number of available photons, while the number of transitions downward is independent of the number of photons (assuming, for simplicity, that spontaneous emissions dominate, which is usually the case). In the atmospheric regions where the line photons start to "feel" the boundary, and hence start to escape (possibly after many intermediate scattering events), there is a lack of them, and more so when going closer to the boundary. The number of upward and downward transitions cannot be balanced any longer; there are fewer transitions in the upward direction, and hence the population of the upper level decreases below its LTE value.

If $\epsilon \lesssim 1$, the source function at depths as well as at the surface is close to the Planck function, so $b_l \approx b_u \approx 1$. One essentially recovers LTE, as it should be.

Three-Level Atom

The levels are denoted as 1, 2, and 3. There are two different cases.

(i) Levels 2 and 3 have very close energies. This means that the collisional rates C_{23} and C_{32} dominate over all the other transition rates into and out of level 3, so that the kinetic equilibrium equation for level 3 is simply $n_2C_{23} \approx n_3C_{32}$. Writing $n_i = b_i n_i^*$, one obtains

$$n_2^* b_2 C_{23} \approx n_3^* b_3 C_{32}, \tag{14.232}$$

and, because of the detailed balance relation $n_2^*C_{23} = n_3^*C_{32}$, one finds

$$b_2 \approx b_3. \tag{14.233}$$

Hence collisions between two close-lying levels force the levels to share the same b-factor. This finding forms the basis of the common practice of treating the entire multiplet as one "level" when solving the multi-level atom kinetic equilibrium equations. If the energies of levels 2 and 3 are very close, one may approximate $n_2/n_3 = (g_2/g_3)\exp(E_{32}/kT) \approx (g_2/g_3)$. One then introduces level J, composed of levels 2 and 3, with $n_J \equiv n_2 + n_3$,

$g_J \equiv g_2 + g_3$. The transition rates from this combined level J to level 1 are given, e.g., $n_J A_{J1} \equiv n_2 A_{21} + n_3 A_{31} = n_J[(n_2/n_J)A_{21} + (n_3/n_J)A_{21}]$; hence $A_{J1} = (g_2 A_{21} + g_3 A_{31})/(g_2 + g_3)$, and similarly for other transition rates. The resulting atom behaves exactly as a two-level atom. If $\epsilon \approx C_{J1}/(A_{J1} + C_{J1}) \ll 1$, then level J is significantly underpopulated at the upper atmospheric layers; hence both original levels 2 and 3 are underpopulated, moreover, with the same departure coefficient b_J.

The same idea is behind the concept of *superlevels*, discussed in detail in § 18.5. A superlevel is composed of a large number of genuine levels with close energies, which are supposed to have the same b-factor. This concept is very useful for reducing the number of levels for a numerical solution of kinetic equilibrium equations for complex atoms and ions.

(ii) Energy differences between levels 1, 2, and 3 are similar, $E_{32} \sim E_{21}$. If these energy differences are *small* compared to the thermal energy, then the collisional rates tend to dominate, or at least are comparable to the radiative rates; hence the NLTE effects will be weak. A more interesting case is when the energy differences are *large* compared to the thermal energy. In this case $n_1 \gg n_j$, $j = 2, 3$, and thus the optical depth in the resonance lines τ_{12} and τ_{13} is much larger than that in the subordinate line τ_{23}. The following scenario then develops: Deep down in the atmosphere all lines are opaque, and all the populations attain their LTE values. Going upward, one first reaches the region where the optical depth τ_{23} decreases below the thermalization depth for the line $2 \rightarrow 3$, while the resonance lines are opaque so that they are in detailed radiative balance. Due to a similar argument as above, line $2 \rightarrow 3$ behaves as a line in a two-level atom in such layers, so that the level 2 becomes *overpopulated* and level 3 becomes *underpopulated* with respect to LTE, $b_2 > 1$, $b_3 < 1$. However, at still higher layers, optical depths τ_{12} and τ_{13} become smaller than the respective thermalization depths; therefore, the basic NLTE line effect takes place, which will lead to an ever greater underpopulation of level 3, and diminishing an overpopulation of level 2 and driving it eventually to an *underpopulation*, if the corresponding ϵ_{12} is low enough.

Multi-Level Atom

The behavior of b-factors of the individual levels in a complex multi-level atom can be thought of as a competition of two elementary effects:

(i) Radiative transfer in a line with a small destruction probability (i.e., with the collisional rates much smaller than the radiative rates) causes the upper level to be underpopulated and the lower level to be overpopulated.

(ii) Collisional transitions between any two levels tend to drive them closer to a Boltzmann equilibrium within each other, i.e., forcing them to have the same NLTE departure coefficient. This effect is stronger for levels with close energies or for high densities (because the collisional rates are proportional to density).

In realistic atmospheres, there are always gradients of basic structural parameters, such as temperature and density. Although these gradients influence a detailed behavior of the b-factors, they do not change the basic physical picture outlined above, at least for monotonic gradients (e.g., for the temperature and density monotonically decreasing outward). If, for instance, the gradient in temperature changes sign, e.g., if the temperature first decreases outward and then starts to increase (as in stellar chromospheres), then the source functions in the lines respond to the temperature rise, and if ϵ is not very small, the source function may rise as well, thus changing the values of the departure coefficients. We do not go into detail on the topic of line formation in stellar chromospheres; the interested reader is referred to [39, 1118, 1119].

Basic NLTE Continuum Effects

NLTE formation of bound-free transitions (continua) depends sensitively on whether the continuum is *active*, i.e., it is the dominant opacity source in its frequency region and therefore it builds its own radiation field, or is a *passive* continuum, which is formed in the frequency region of another, stronger continuum. Examples of the former type in hot stars are the hydrogen and ionized helium Lyman continua, and possibly the neutral helium ground-state continuum.

Active Continua

Since they dominate the opacity, and since the cross section usually decreases rapidly with increasing frequency (e.g., a $\sigma \sim \nu^{-3}$ decrease for hydrogenic atoms), they can in fact be viewed as broad lines, with a profile decreasing from the values at the threshold to both sides. The drop of the cross section on the low-frequency side may be taken as either an abrupt decrease to zero or a more gradual one if the effects of level dissolution—cf. § 9.4—are taken into account.

One can take a "two-level" atom with level 1 being the ground state of an atom/ion and level 2 the ground state of the next higher ion, and view the transition $1 \rightarrow 2$ as a broad "line." Sine the behavior of the source function, and hence the departure coefficients b, does not depend on the type of profile but only on ϵ, one recovers the same behavior as for a genuine two-level atom, i.e., $S \sim \sqrt{\epsilon} B$ at the surface and $b_2/b_1 \sim \sqrt{\epsilon} < 1$. The higher ion thus becomes underpopulated and the lower ion overpopulated. This can again be explained in terms of missing photons for the induced transition upward (photoionization), leading to an excess of recombinations with respect to photoionizations. As for lines in a two-level atom, this effect is not very sensitive to temperature and density gradients, again, as long as they are smooth and monotonic.

In realistic atoms, there are many transitions out and into the levels involved in an active continuum (typically the respective ground states) so that the above picture is modified by these intervening processes. However, since the cross sections for bound-free transitions are much smaller than the cross section for bound-bound transitions, the continua tend to be formed deeper than lines. Therefore, even the ground-state continuum (say, the hydrogen Lyman continuum in hot stars) may be less opaque than not only Lyman lines but also Balmer lines. The thermalization depth of the Lyman continuum may then be located deeper than any other thermalization depths, which would lead to the hydrogen ground state being overpopulated in very deep layers. An analogous situation applies for the Balmer continuum in A-stars, which now behaves like a ground-state continuum because the $L\alpha$ line is extremely opaque and thus in detailed radiative balance, which leads to an overpopulation of the $n = 2$ level in deep layers.

Passive Continua

It should be realized that most continua are passive ones. They do not contribute much to the opacity and radiation field in their frequency regions, but they are very important for the overall kinetic equilibrium of the atoms and ions to which they belong. Since the continua are typically formed deeper than lines, the continua tend to determine directly the NLTE level populations in deep layers, and indirectly everywhere because they influence the thermalization depths for lines (which are also typically located in deep layers) and thus the formation of lines all the way to the surface.

The most important effect of passive continua was mentioned in § 9.5. Assuming that the photoionization/recombination rates dominate the other rates, and assuming one-level ions, the ground-state population of the lower ion is approximated by equation (9.90),

$$(n_1/n_1^*) = \left[\int_{\nu_0}^{\infty} (\alpha_\nu B_\nu / h\nu)\, d\nu + C_{1k}\right] \Big/ \left[\int_{\nu_0}^{\infty} (\alpha_\nu J_\nu / h\nu)\, d\nu + C_{1k}\right], \tag{14.234}$$

where α_ν is the photoionization cross section. Stimulated emission is neglected. Expressing $J_\nu = W B_\nu(T_R)$, with W being the *dilution factor* and T_R the *radiation temperature*, neglecting the collisional rates, and assuming that the cross section is a sharply peaked function of frequency, (14.234) can be written as

$$b_1 = (n_1/n_1^*) \approx B_{\nu_0}(T)/J_{\nu_0} \approx B_{\nu_0}(T) \big/ [W B_{\nu_0}(T_R)]. \tag{14.235}$$

In the upper layers of an atmosphere one can use the Eddington-Barbier relation (see § 11.4), namely, $J_{\nu_0}(\tau_0 \to 0) \approx \frac{1}{2} B_{\nu_0}(\tau_0 = \frac{1}{2})$, where τ_0 is the optical depth at frequency ν_0. In other words, $W = \frac{1}{2}$ and $T_R = T(\tau_0 = \frac{1}{2})$ in this case. If ν_0 is larger than the frequency corresponding to the maximum of the Planck function at the local temperature, $B_\nu(T)$ decreases exponentially with T (the Wien tail). If the temperature in the atmosphere decreases outward (which is usually the case; see chapters 17 and 18), then the decrease of the Planck function from $\tau_0 \sim \frac{1}{2}$ to $\tau_0 \sim 0$ may vastly dominate over the dilution factor $W \sim \frac{1}{2}$ coming from the Eddington-Barbier relation, so that $B_{\nu_0}(\tau_0 \sim 0) \ll \frac{1}{2} B_{\nu_0}(\tau_0 = \frac{1}{2})$, and one obtains $b_1 \ll 1$, i.e., an *underpopulation* of level 1. If level 1 contains most of the population of the given ion, the underpopulation of this level propagates to other levels of the ion due to their coupling with the ground state by collisional processes, as well as radiative processes involving weak lines. Therefore, essentially *all* levels of the lower ion become underpopulated. One can then describe the above effect as NLTE *overionization*; i.e., the lower ion being underpopulated, while the upper ion is overpopulated.

In physical terms, the overionization occurs because the ionization is produced by an intense radiation coming from deep, hot layers, while the recombination is produced by electrons whose velocity corresponds to the temperature of the local, cooler layers.

In realistic atoms, a similar imbalance of photoionizations and recombinations may occur not only for the ground-state continuum, but also for the continua arising from low-lying excited levels, which enhances the overionization effect described above. In fact, this turns out to be one of the most important NLTE effects in multi-level atoms.

The basic NLTE effect for passive continua is thus directly linked to the temperature gradient, and in the usual case of the temperature decreasing outward, the effect is opposite to the basic effect for active continua. For intermediate-strength continua there is a competition between these two, plus other effects due to an interplay with other lines, so that the result is impossible to predict without detailed numerical calculations. But it can always be verified, after the fact, that the results are physically reasonable, using the reasoning outlined above.

Chapter Fifteen

Radiative Transfer with Partial Redistribution

15.1 FORMULATION

We first review the basic expressions for the emission coefficient and the source function, derived in chapter 10, for the case of departures from complete redistribution. The absorption coefficient for the line transition $i \leftrightarrow j$ is given either by

$$\chi_{ij}(\nu) = \frac{h\nu_{ij}}{4\pi}\,[n_i B_{ij}\phi_{ij}(\nu) - n_j B_{ji}\psi_{ji}(\nu)], \tag{15.1}$$

where $\phi(\nu)$ and $\psi(\nu)$ are the normalized profiles of absorption and emission, respectively, or by

$$\chi_{ij}(\nu) = \frac{h\nu_{ij}}{4\pi} n_i B_{ij}\phi_{ij}(\nu), \tag{15.2}$$

in which case it does not include the stimulated emission contribution as a negative absorption, but instead includes it in the emission coefficient; see (15.9).

Unlike the case of complete redistribution, the form of (15.1) does not offer any practical advantage because the stimulated emission term is nonlinear in the radiation intensity because of the explicit dependence of the emission profile coefficient ψ on the specific intensity; see (15.5). Nevertheless, this form is often used in iterative numerical methods with the emission profile coefficient given by its current value; see § 15.5.

Another possibility is to consider stimulated emission with an absorption profile coefficient ϕ, which is not physically correct, but permissible if the stimulated emission is almost, but not completely, negligible. In that case one can use the usual expression

$$\chi_{ij}(\nu) = \frac{h\nu_{ij}}{4\pi}[n_i B_{ij} - n_j B_{ji}]\phi_{ij}(\nu). \tag{15.3}$$

The spontaneous emission coefficient is now written as

$$\eta_{ji}(\nu) = \frac{h\nu_{ij}}{4\pi} n_j A_{ji}\psi_{ji}(\nu). \tag{15.4}$$

In the case of complete frequency redistribution, both profiles are identical, $\psi_{ji}(\nu) \equiv \phi_{ij}(\nu)$. As discussed in chapter 10, departures from complete redistribution, *partial frequency redistribution* (PFR), is characterized by the inequality

$\psi_{ji}(\nu) \neq \phi_{ij}(\nu)$. As derived in § 10.5, the emission profile coefficient is given by [cf. equation (10.174)]

$$\psi_{ji}(\nu, \mathbf{n}) = \phi_{ij}(\nu) + \frac{n_i B_{ij} \overline{J}_{ij}}{n_j P_j} \left[p_i^* \mathcal{R}_{iji}(\nu, \mathbf{n}) - \phi_{ij}(\nu) \right]$$
$$+ \sum_{\ell < j, \ell \neq i} \frac{n_\ell B_{\ell j} \overline{J}_{\ell j}}{n_j P_j} \left[p_\ell^* \mathcal{R}_{\ell j i}(\nu, \mathbf{n}) - \phi_{ij}(\nu) \right], \tag{15.5}$$

and the source function for this transition (neglecting stimulated emission) is [cf. equation (10.175)]

$$S_{ji}(\nu, \mathbf{n}) = \frac{n_j A_{ji} \psi_{ji}(\nu, \mathbf{n})}{n_i B_{ij} \phi_{ij}(\nu)}$$
$$= \frac{n_j A_{ji}}{n_i B_{ij}} + \frac{A_{ji}}{P_j} \overline{J}^R_{iji}(\nu, \mathbf{n}) + \sum_{\ell < j, \ell \neq i} \frac{n_\ell B_{\ell j}}{n_i B_{ij}} \frac{A_{ji}}{P_j} \overline{J}^R_{\ell j i}(\nu, \mathbf{n}). \tag{15.6}$$

One recognizes the first term as S^{CFR}, the source function in the case of complete redistribution. Here, P_j is the total rate of transitions out of level j, $\mathcal{R}_{iji}(\nu, \mathbf{n})$ and $\mathcal{R}_{\ell ji}(\nu, \mathbf{n})$ are the normalized scattering integrals for the resonance scattering process $i \to j \to i$ and the resonant Raman scattering process $\ell \to j \to i$, respectively, which are given by equations (10.154) and (10.155),

$$\mathcal{R}_{ief}(\nu, \mathbf{n}) = \int_0^\infty \oint R_{ief}(\nu', \mathbf{n}'; \nu, \mathbf{n}) \, I(\nu', \mathbf{n}') \, (d\Omega'/4\pi) \, d\nu' / \overline{J}_{ie}, \tag{15.7}$$

where R is the appropriate redistribution function, and $\overline{J}_{ie}$ is the frequency-averaged mean intensity, $\overline{J}_{ie} = \int I(\nu', \mathbf{n}') \phi_{ie}(\nu') \, d\nu' \, d\Omega'/4\pi$. Finally, the redistribution integral $\overline{J}^R$ is given by [cf. (10.176)]

$$\overline{J}^R_{ief}(\nu, \mathbf{n}) = \int_0^\infty \oint \left[p_i^* \frac{R_{ief}(\nu', \mathbf{n}'; \nu, \mathbf{n})}{\phi_{ef}(\nu)} - \phi_{ie}(\nu') \right] I(\nu', \mathbf{n}') \, (d\Omega'/4\pi) \, d\nu'. \tag{15.8}$$

There are several levels of complication that result from departures from complete frequency redistribution:

(i) The most difficult problem would have occurred had we postulated that the rate of stimulated emission is given by the emission profile, $R_{ji} \propto A_{ji} + B_{ji} \int \psi_{ji}(\nu, \mathbf{n}) I(\nu, \mathbf{n}) d\nu (d\Omega/4\pi)$. But within the scope of the semiclassical approach that forms the basis of the astrophysical treatment of redistribution, this is not necessary and in fact not entirely consistent, as explained in chapter 10. The total rate for the transition $j \to i$ is then $R_{ji} \propto A_{ji} + B_{ji} \overline{J}_{ij}$, where $\overline{J}_{ij} = \int \phi_{ij}(\nu, \mathbf{n}) I(\nu, \mathbf{n}) d\nu (d\Omega/4\pi)$, exactly as in the case of CFR. The transition rates, and therefore the overall set of kinetic equilibrium equations, thus remain the same as in the case of CFR. Notice that before this point was clarified, some treatments, e.g., [750], considered the emission profile in the stimulated emission rate, which led to considerable complications in solving the rate equations.

(ii) In case the stimulated emission is important, either it is treated as negative absorption, as in (15.1), or the total emission coefficient may be given by

$$\eta_{ji}(\nu, \mathbf{n}) = \frac{h\nu_{ij}}{4\pi} n_j [A_{ji} + B_{ji} I(\nu, \mathbf{n})] \, \psi_{ji}(\nu, \mathbf{n}). \tag{15.9}$$

Because ψ_{ji} explicitly depends on the radiation intensity through the scattering integral, the emission coefficient would be *nonlinear* in the radiation intensity. This is a complication, but not a significant one because as we saw repeatedly, the transfer equation is implicitly nonlinear in the radiation intensity anyway as a result of the implicit dependence of atomic level populations, and hence the absorption and emission coefficients, on the radiation intensity. Nevertheless, as will be demonstrated later, partial redistribution effects are significant only for strong resonance lines in the UV region where the stimulated emission contribution is negligible. It is therefore not unreasonable to neglect stimulated emission when dealing with transitions for which the partial redistribution effects are taken into account; this is done for convenience and a simplification of expressions, not for conceptual necessity. But we stress again that modern numerical methods are capable of handling stimulated emission exactly, albeit iteratively, using the traditional way of treating it as a negative absorption.

(iii) Even when stimulated emission is neglected, the emission coefficient and therefore the line source function are more complicated because they explicitly depend on radiation field and thus on frequency. Recall that in the case of CFR, the line source function is frequency-independent.

Now consider the example of a two-level atom. The source function, neglecting stimulated emission [cf. equation (10.166)], is written as

$$\begin{aligned} S_{21}(\nu, \mathbf{n}) &= S^{\mathrm{CFR}} + \frac{A_{21}}{P_2} \overline{J}^R_{121}(\nu, \mathbf{n}) = (1-\epsilon)\overline{J}_{12} + \epsilon B + (1-\epsilon)\overline{J}^R_{121}(\nu, \mathbf{n}) \\ &= (1-\epsilon) \int_0^\infty \oint R_{121}(\nu', \mathbf{n}'; \nu, \mathbf{n}) \, I(\nu', \mathbf{n}') \, (d\Omega'/4\pi) \, d\nu' \Big/ \phi_{12}(\nu) + \epsilon B, \end{aligned} \tag{15.10}$$

because in the absence of stimulated emission $A_{21}/P_2 = 1-\epsilon$. The last equality follows from (15.8).

Using the angle-averaged approximation, the source function can be written as

$$S_{21}(\nu) = (1-\epsilon) \int_0^\infty R_{121}(\nu', \nu) J(\nu') \, d\nu' \Big/ \phi_{12}(\nu) + \epsilon B. \tag{15.11}$$

The redistribution function is generally given as a superposition of two terms corresponding to coherent scattering and to complete redistribution, as derived in chapter 10 using both a quantum and a semiclassical derivation. In the case of the two-level atom, the redistribution function is given by [cf. (10.151)]

$$\begin{aligned} R_{121}(\nu',\nu) &= \gamma R_{\rm II}(\nu',\nu) + (1-\gamma) R_{\rm III}(\nu',\nu) \\ &\approx \gamma R_{\rm II}(\nu',\nu) + (1-\gamma)\phi(\nu')\phi(\nu), \end{aligned} \tag{15.12}$$

where the branching ratio (coherence fraction) γ is given by [cf. (10.152)]

$$\gamma = \frac{P_2}{P_2 + Q_2} = \frac{A_{21} + C_{21}}{A_{21} + C_{21} + Q_2}, \tag{15.13}$$

where Q_2 is the elastic collision rate. The source function is then given by

$$S_{21}(\nu) = (1-\epsilon)\left[\gamma \int_0^\infty R_{\rm II}(\nu',\nu) J(\nu')\, d\nu' \big/ \phi_{12}(\nu) + (1-\gamma)\bar{J}_{12}\right] + \epsilon B. \tag{15.14}$$

Equation (15.14) can also be written by dropping the subscript 12, which is not necessary for a two-level atom, and introducing the subscript L, which signifies the source function for a *line*, as

$$S_{\rm L}(\nu) = (1-\epsilon)\left\{\bar{J} + \gamma\left[\left[\int R_{\rm II}(\nu',\nu) J(\nu')\, d\nu' \big/ \phi(\nu) - \bar{J}\right]\right]\right\} + \epsilon B. \tag{15.15}$$

This equation shows that if the elastic collision rate dominates over the spontaneous emission rate, the branching ratio is small, $\gamma \ll 1$, and one recovers the complete frequency redistribution form of the source function, $S \approx (1-\epsilon)\bar{J} + \epsilon B$.

For the purposes of later discussion, we will also consider the case of a subordinate line. In this case the redistribution function is given by (see § 10.4)

$$\begin{aligned} R(\nu',\nu) &= \gamma R_{\rm V}(\nu',\nu) + (1-\gamma) R_{\rm III}(\nu',\nu) \\ &\approx \gamma\big[\alpha R_{\rm II}(\nu',\nu) + (1-\alpha)\phi(\nu')\phi(\nu)\big] + (1-\gamma)\phi(\nu')\phi(\nu) \\ &\approx \alpha\gamma R_{\rm II}(\nu',\nu) + (1-\alpha\gamma)\phi(\nu')\phi(\nu). \end{aligned} \tag{15.16}$$

The second line follows from the usual replacement of the $R_{\rm III}$ redistribution function by complete redistribution, (10.130), and from equation (10.131). The parameter α is given by

$$\alpha = a_u/(a_u + a_l) \approx \sum_{j<u} A_{uj} \Big/ \Big(\sum_{j<u} A_{uj} + \sum_{j<l} A_{lj}\Big), \tag{15.17}$$

where a_l and a_u are the total damping parameters from the lower and upper level; their ratio is approximately given by the ratio of appropriate sums of the Einstein coefficients for spontaneous emission.

15.2 SIMPLE HEURISTIC MODEL

Before outlining several numerical methods of solving the problem of line transfer with partial frequency redistribution, we first develop a simple heuristic model of line transfer with partial frequency redistribution that allows us to understand physics of the problem.

Doppler Diffusion

First, recall the basic picture of a photon transport in the case of CFR. Because the absorption profile (and thus the emission profile as well) has a sharp peak at the line center as a function of frequency, after being absorbed in a line a photon is most likely re-emitted with a frequency close to the line center. After its next absorption, it is again re-emitted close to the line center. The photon thus experiences a large number of scatterings in the line-core region. Because the opacity in the core is large, the photon does not move very far in physical space. Spatial transfer of photons is made possible by their infrequent excursions to the line wing, where the opacity is much lower; this means the photon may travel a large distance in physical space. If a photon does not escape during its excursion to the wing, it will again very likely be re-emitted in the core during its next absorption + re-emission event, and the process of multiple absorptions + emissions is repeated until the photon finally escapes or is destroyed by a collisional process.

In the case of partial frequency redistribution, the sequence of repeated scatterings in the line *core* is essentially the same as in the case of complete redistribution because the emission profile coefficient in the core is very similar to the absorption profile coefficient—recall that the redistribution function in the core behaves as $R(x', x) \approx \phi(x')\phi(x)$. However, significant differences arise when the photon is re-emitted in the *wing*. Instead of being immediately absorbed and re-emitted in the core as in the case of CFR, the photon is re-emitted essentially coherently, and hence is trapped in the wing and experiences a random walk in physical space.

To a zero-order approximation, one may treat the partial frequency redistribution in the line wing as coherent scattering. There are in fact approximate approaches that do exactly that. They are called the *partial coherent scattering* (PCS) approximation, and will be described in more detail in § 15.3. However, this is not the whole story. As explained in § 10.3, line scattering that is coherent in the atom's frame is not completely coherent in the laboratory frame. Because of Doppler motions of the absorbing atoms, the redistribution function $R_{\rm II}(x', x)$ has a narrow peak that has a width of about one Doppler width (i.e., 1 in units of x) for large x' ($x' \gg 1$). Moreover, this peak is not symmetric about x', in the sense that there is a slightly larger probability that a photon is re-emitted closer to the line center than it was absorbed. If a photon experiences many repeated scatterings, it gradually moves toward the line center. This process is called *Doppler diffusion*.

Specifically, the mean shift per scattering is given by

$$\overline{\Delta x}(x) = \int_{-\infty}^{\infty} \frac{R_{\rm II}(x', x)}{\phi(x')} (x - x')dx', \tag{15.18}$$

because $R_{\rm II}(x', x)/\phi(x')$ represents the conditional probability that a photon with frequency x will be emitted after previous absorption of a photon with frequency x'. Substituting the approximate form of $R_{\rm II}$ given by (10.128), one obtains

$$\overline{\Delta x}(x) \approx -1/x. \tag{15.19}$$

This result was first obtained in [820].

Neglecting for a moment a gradual decrease of the step length, the total distance traveled by a photon before it arrives to the line center is

$$\tau \approx \bar{l}\, n^{1/2} \propto \bar{l}\, x \propto (\pi x^2/a)\, x \propto \pi x^3/a, \tag{15.20}$$

where $\bar{l}$ is the average mean free path of the photon, given by the mean free path of the original photon, $\bar{l} = \phi(x)^{-1} \approx \pi x^2/a$; n is the total number of steps, which is given by x—the distance between the original frequency and the line core—divided by the mean shift in frequency, $1/x$; thus $n \approx x^2$. Here we used the line wing approximation for the Voigt function (cf. equation 8.24), $\phi(x) \approx a/(\pi x^2)$. If we take into account a gradual decrease of the photon mean free path and the fact that the mean frequency shift per scattering increases with decreasing frequency, we obtain $\tau \approx \pi x^3/4a$. A rigorous analysis was performed in [517], where a more accurate expression was derived. This is, however, unnecessary for the present purposes. We call this optical depth the *frequency-thermalization depth* corresponding to frequency x,

$$\tau_{\rm FT}(x) = \pi x^3/4a, \tag{15.21}$$

which has the meaning that for a given frequency x, the specific intensity for smaller optical depths behaves similarly to the case of coherent scattering. For those depths, a photon has a small chance to drift into the core, and hence can be viewed as being trapped in the wing, and its transfer is essentially a random walk in physical space, in sharp contrast to the case of CFR. This diffusive behavior does influence the overall escape of photons in a line, and hence the large-scale behavior of the source function. Incidentally, these considerations provide insight into certain approximate numerical methods, such as the modified partial coherent scattering approximation, discussed in more detail in § 15.3.

An inverse quantity, mentioned in § 10.3, is a "characteristic frequency,"

$$x_D(\tau) = (4a/\pi)^{1/3} \tau^{1/3}, \tag{15.22}$$

which has the meaning that at a given depth τ, photons created with smaller frequencies will very likely be frequency-thermalized, while photons created with frequency larger than $x_D(\tau)$ will not be frequency-thermalized and thus can roughly be considered to behave as photons that experience a chain of coherent scatterings.

So far, we have neglected inelastic collisions that destroy the photon and thus interrupt the chain of correlated scattering events. Let us now assume that a photon has a probability ϵ of being destroyed by a collision, $\epsilon \approx C_{21}/(A_{21} + C_{21})$ for a two-level atom. The probability that a photon is not destroyed after n scattering events is $(1-\epsilon)^n$. The probability that a photon originally created with frequency x will be able to experience so many scattering events that it eventually drifts to the line center is given by

$$p_\epsilon(x) \approx (1-\epsilon)^{x^2} \approx 1 - x^2\epsilon \tag{15.23}$$

for $\epsilon \ll 1$. We may then introduce a characteristic frequency, x_ϵ, for which $1 - x_\epsilon^2\epsilon \approx 1/2$, i.e.,

$$x_\epsilon \approx \epsilon^{-1/2} \tag{15.24}$$

(here again we disregard constants of the order of unity). Another way to derive this expression is to realize that if the global photon destruction probability is ϵ (regardless of frequency), a line photon may experience on the order of $1/\epsilon$ scattering events before being destroyed. In order to be able to drift from frequency x to the line center, it needs about x^2 consecutive scattering events, as explained above; hence $x^2 = 1/\epsilon$ defines a characteristic frequency. The meaning of x_ϵ is that a photon created with frequency x_ϵ has a probability of about 1/2 to be frequency-thermalized, i.e., to drift to the line center. Photons created with larger frequencies will not arrive at line center, and the transfer solution for those frequencies will approach that of coherent scattering.

The above analysis applies to a two-level atom. However, it can easily be extended to a multi-level atom. According to (15.16), the part of the redistribution integral that contains $R_{\rm II}$, and that therefore describes a photon correlation during the scattering process, is multiplied by $\alpha\gamma$. This quantity may be viewed as the probability that the photon correlation is not destroyed during the scattering process, and $(1-\alpha\gamma)$ as the *probability of destruction of photon correlation*.

The probability that a photon survives the scattering event, as well as that its correlation with the previously absorbed photon is not destroyed, is thus given by $(1-\epsilon)\gamma'$, where $\gamma' = \alpha\gamma$. The probability of frequency thermalization is then given by $p_\epsilon \approx [(1-\epsilon)\gamma']^{x^2}$, which represents a generalization of equation (15.23). Hence the above formalism applies if we replace ϵ by ϵ', where

$$1 - \epsilon' \equiv (1-\epsilon)\gamma'; \tag{15.25}$$

hence

$$\epsilon' = 1 - \gamma' + \epsilon\gamma'. \tag{15.26}$$

This result shows that when $\gamma' < (1-\epsilon)$, the total destruction parameter ϵ' is dominated by γ', i.e., by the destruction of photon correlation. For subordinate lines, γ' is not very close to unity; in fact, its typical value is $\gamma' \approx 1/2$ (if the total spontaneous rates out of the upper and the lower level are similar). Hence the characteristic frequency x_ϵ is of the order of unity; consequently, partial redistribution effects cannot fully develop. For a line connected to the ground state, with the upper state not being the lowest excited non-metastable state (i.e., a state from which there exist other allowed transition(s) to excited states; an example being the hydrogen $L\beta$ line),

$\alpha = 1$, but the γ parameter is significantly smaller than unity. Take the example of a three-level atom; then $\gamma_{31} \approx A_{31}/(A_{31} + A_{32})$, assuming for simplicity that the other transition rates out of level 3 are small, and the elastic collision rates are also smaller that the Einstein A-coefficients. Consequently, γ is again of the order of 1/2, and partial redistribution effects do not fully develop.

Next, consider the case of an overlapping continuum. In contrast to the case of collisional destruction of a photon where the destruction probability ϵ is independent of frequency, an overlapping continuum introduces a frequency dependence to the destruction probability. Specifically, the probability that a photon with frequency x survives one scattering event is given by

$$p_x^1 = \frac{\phi_x}{\phi_x + r}(1 - \epsilon) \approx \left(1 + \beta x^2\right)^{-1} (1 - \epsilon), \tag{15.27}$$

where $r = \kappa_C/\kappa_L$, with κ_C being the continuum opacity, κ_L the frequency-averaged line opacity, and $\beta = \pi r/a$. Here we again used the usual line wing approximation of the Voigt function, $\phi_x \approx a/(\pi x^2)$. In order to isolate the effect of overlapping continuum, we disregard the collisional destruction of a photon and set $\epsilon = 0$. The probability that a photon survives n scattering events is then (cf. [517])

$$p_x^n = \left(1 + \beta \sum_i^n x_i^2\right)^{-1} \approx [1 + \beta n x^2 (1 - n x^{-2})]^{-1}, \tag{15.28}$$

which applies for $x \ll \beta^{-1/2}$ (i.e., $\beta x^2 \ll 1$, which is the only interesting frequency region because otherwise the overlapping continuum dominates the process). As $n \approx x^2$, the probability that a photon with frequency x will be frequency-thermalized is $p_c(x) \approx (1 + \beta x^4)^{-1}$. The characteristic frequency for the destruction of frequency correlation resulting from the continuum process is then defined by $p_c(x_c) \approx \frac{1}{2}$; hence

$$x_c \approx \beta^{-1/4} \approx a^{1/4} r^{-1/4}, \tag{15.29}$$

and the corresponding depth, $\tau_c \equiv \tau_{\rm FT}(x_c)$,

$$\tau_c \approx a^{-1/4} r^{-3/4}. \tag{15.30}$$

These two equations only apply if $\beta x^2 \ll 1$ for all frequencies of interest. For frequencies larger than x_c, the photon is likely to be destroyed before it is frequency-thermalized, and so radiation in frequencies larger than x_c behaves more like it experiences CFR. We are thus left with the condition $\beta x_c^2 \ll 1$, or $r \ll a\pi^{-2}$. Indeed, for $r \gtrsim a\pi^{-2}$, the characteristic frequency lies in the line core, and the whole frequency region where PFR effects are expected to be significant is dominated by the continuum.

When Are Departures from Complete Redistribution Important?

The discussion above helps us to draw some general conclusions about the practical importance of PFR effects, namely, the effects of partial frequency redistribution in a line are most significant for the following:

- **Low-density media:** Where $C_{ul} \ll A_{ul}$, and hence $\epsilon \ll 1$, which is the same criterion as for the importance of NLTE effects.
- **A strong line:** That is, $r < a\pi^{-2}$. A typical value of the damping parameter is $a \approx 10^{-3}$, so $r < 10^{-4}$ is a necessary (but not sufficient) condition for a line to exhibit significant PFR effects.
- **A resonance line:** A line connecting the ground state with the first excited state, or with a state that is not connected by a dipole-allowed transition to another excited state.

Typical examples of lines that satisfy these conditions are the hydrogen $L\alpha$ line, the Mg II h and k lines, and the Ca II H and K lines in low-density media, such as atmospheres of giants and supergiants, and the outer atmospheres (chromospheres) of main-sequence stars that are not so hot or so cool that neutral hydrogen and once-ionized magnesium and calcium still exist in significant concentrations. Consequently, most practical applications of line transfer with partial redistribution were devoted to these lines. Of course, there are other lines for which PFR effects are important, such as the C II resonance lines and the resonance lines of alkali ions. Even if some of the above three conditions are not satisfied, partial redistribution effects may still be important. For example, the hydrogen $L\beta$ line in the solar atmosphere exhibits only modest PFR effects if only the effects of photon correlation in a resonance scattering $1s \rightarrow 3p \rightarrow 1s$ are taken into account. However, when a resonance Raman scattering $2s \rightarrow 3p \rightarrow 1s$ is taken into account, the PFR effects are significant [535]. We return to this point in § 15.6.

15.3 APPROXIMATE SOLUTIONS

In this section we outline several methods that yield approximate solutions for line transfer with partial frequency redistribution, with emphasis on a two-level atom. As we will see in § 15.4, one can relatively easily obtain an exact numerical solution of this problem. But as was explained in § 15.1, there are two complications when compared to the standard complete redistribution problem. First, one has to evaluate the redistribution function itself, which may be time-consuming, and, second, the solution of the transfer equation with a frequency-dependent (and possibly also angle-dependent) source function is significantly more time-consuming than the complete redistribution problem. Therefore, several methods were suggested that avoid dealing with the redistribution functions completely and that make the partial redistribution problem no more computationally extensive than the complete redistribution one. On the other hand, with the advent of ALI-based methods, the original approximate methods lost most of their former appeal, so we describe them only briefly because they are still being used in some studies.

In this section we consider resonance scattering in a single line with a sharp lower level that is not interlocked with another line; i.e., essentially a two-level atom situation. In this case, the appropriate redistribution function is a linear combination of $R_{\rm II}(x', x)$ and of complete redistribution, $\phi(x')\phi(x)$.

Partial Coherent Scattering Approximation

As already mentioned in § 10.3, the partial coherent scattering (PCS) approximation is based on the observation that if a photon is absorbed in the line wing, $x \gg 1$, it will be most likely re-emitted with a similar frequency. In other words, the redistribution function $R_{\rm II}(x', x)$ exhibits a sharp peak for $x = x'$ with a width of about one Doppler width. As first suggested in [582], it is reasonable to replace the re-emission probability $R_{\rm II}/\phi$ by a δ-function, $\delta(x' - x)$, for large x'. It was subsequently realized [624] that for the redistribution function to be symmetric and normalized, it has to be written as (cf. 10.121)

$$R_{\rm II}(x', x) \approx [1 - a(x', x)]\,\phi(x')\,\phi(x) + a(x', x)\phi(x')\,\delta(x' - x), \tag{15.31}$$

where $a(x', x)$ is an empirical function. The original form of this function, suggested in [582], sometimes called the Jefferies-White approximation, was $a(x', x) = \overline{a}(x)$, where $\overline{a}(x) = 1$ for $x > 3$, and $\overline{a}(x) = 0$ for $x \le 3$. An improved form of the function, sometimes called the Kneer approximation, was suggested in [624]. It is given by

$$a(x', x) = \widetilde{a}[\max(|x'|, |x|)], \quad \text{for} \quad x \neq x', \tag{15.32a}$$

$$a(x', x') = \int_{-\infty}^{\infty} a(x', x)\,\phi(x)\,dx, \tag{15.32b}$$

where $\widetilde{a}$ is a somewhat arbitrary function that satisfies $\widetilde{a}(x) \to 0$ for $x \to 0$, and $\widetilde{a}(x) \to 1$ for $x \to \infty$. A detailed analysis in [624] suggested the following form:

$$\widetilde{a}(x) = \begin{cases} 1 - \exp\left(-[(x - x_0)/\Delta]^2\right), & \text{for} \quad x \ge x_0, \\ 0, & \text{for} \quad x < x_0, \end{cases} \tag{15.33}$$

with $x_0 = 2$ and $\Delta = 2$.

In the following, we assume the line is symmetric about the center, $J(-x) = J(x)$, so we consider only positive frequencies, $x \ge 0$. Further, we will set the exponential term in (15.33) to 0, and let the division frequency have an arbitrary value x_0. The auxiliary function $\widetilde{a}(x)$ then becomes a unit step function, $\widetilde{a}(x) = 1$ for $x < x_0$, and $\widetilde{a}(x) = 1$ for $x \ge x_0$. Notice that this form is not identical to the original Jefferies-White form, because the coefficients of the linear combination of the complete redistribution and the coherent scattering terms are still given by (15.32), so that proper normalization of the redistribution function is preserved.

In this approximation, the redistribution integral is given by

$$\int_0^\infty \frac{R_{\rm II}(x',x)}{\phi(x)} J(x')\,dx' = \begin{cases} \overline{J} - \int_{x_0}^\infty \phi(x') J(x')\,dx' + J(x)\Phi_0, & x \le x_0, \\ J(x), & x > x_0, \end{cases} \tag{15.34}$$

where $\Phi_0 \equiv \int_{x_0}^\infty \phi(x')\,dx'$; this term originates due to the Kneer's normalization, equation (15.32). The source function (15.15) becomes

$$S(x) = \begin{cases} (1-\epsilon)\gamma \left[\overline{J} - \int_{x_0}^\infty \phi(x') J(x')\,dx' + J(x)\Phi_0\right] + \epsilon B, & x \le x_0, \\ (1-\epsilon)\left[(1-\gamma)\overline{J} + \gamma J(x)\right] + \epsilon B, & x > x_0. \end{cases} \tag{15.35}$$

The transfer problem with a source function given by (15.35) can be solved by any standard method because a dependence of the line source function on frequency is contained only in the mean intensity. In other words, the line source function $S(x)$ contains only two different frequency-integrated terms (for $x \le x_0$), plus the mean intensity $J(x)$ at the same frequency. Thus solution of the transfer equations does not pose any additional problems; it is numerically equivalent to the solution for a line with complete redistribution and background electron scattering.

Modified Partial Coherent Scattering Approximation

A modification of the standard partial coherent scattering approximation was suggested in [517]. This approach still uses equation (15.35) for the line source function; the only difference is that the division frequency x_0 is allowed to be depth-dependent. As noted in § 15.2, it is natural to set the division frequency to be equal to the characteristic frequency for Doppler diffusion given by equation (15.22), namely,

$$x_0 \equiv x_D(\tau) = (4a/\pi)^{1/3} \tau^{1/3}, \tag{15.36}$$

where τ is the optical depth associated with the frequency-averaged line opacity, and $d\tau = -\chi_0 dz = -(h\nu_0/4\pi) n_i B_{ij} dz$.

This method retains the ease of implementation of the standard partial coherent scattering approximation, while the accuracy of the results is significantly improved. The reason, as explained in § 15.2, is that the traditional PCS approximation completely neglects Doppler diffusion, while the present modified scheme takes it into account, at least approximately. We see these facts demonstrated in figure 15.1 for transfer in a line of a two-level atom in a constant-property medium with $\epsilon = 10^{-6}$, $B = 1$, and $r = 0$, i.e., no background continuum. In this case the effects of partial redistribution are largest.

We see that in the case of a constant-property medium the modified partial coherent scattering approach provides an excellent approximation, while the traditional PCS approximation, in both the original Jefferies-White form and the Kneer modification, although providing a reasonable approximation in the line core, fails in the wing, precisely because of their neglect of Doppler diffusion.

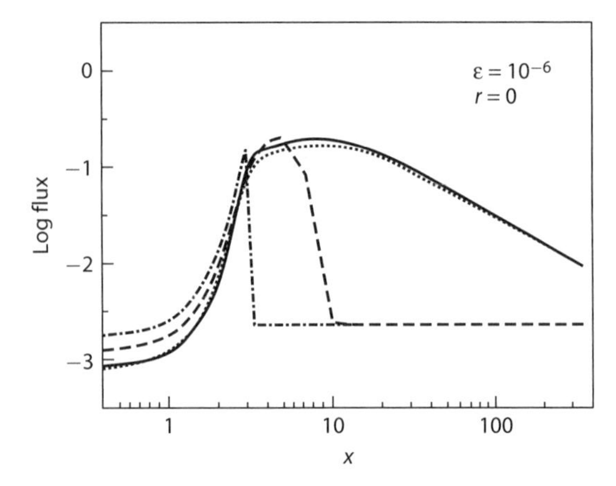

Figure 15.1 Comparison of the emergent flux from a semi-infinite atmosphere composed of two-level atoms, without a background continuum, for various numerical treatments of partial frequency redistribution. The Voigt parameter is $a = 10^{-3}$, the destruction probability is $\epsilon = 10^{-6}$, and the Planck function is set to unity, $B = 1$. The profiles are computed with the exact $R_{\rm II}$ (dotted line); with the standard PCS approximation, taking $x_0 = 3$ and $\Delta = 0$ (dot-dashed line); with Kneer's variant of the PCS approximation, taking $x_0 = 2, \Delta = 2$ (dashed line); and with the modified partial coherent scattering approximation with a depth-dependent $x_0(\tau)$ given by (15.36) (full line). The profiles are symmetric about the line center, so only one half is shown. From [517].

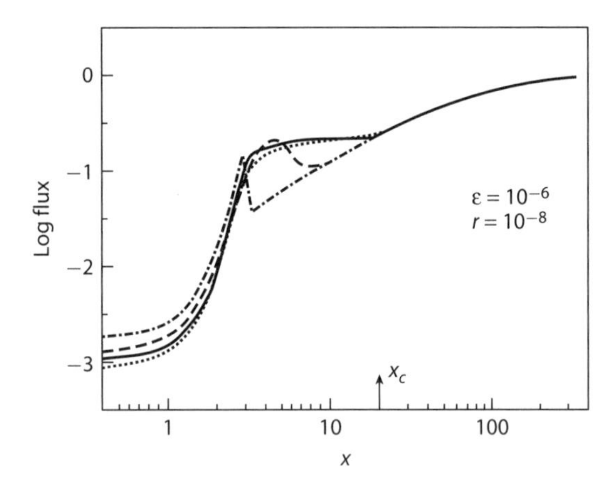

Figure 15.2 The same as in figure 15.1, but with $r = 10^{-8}$. The arrow at the bottom indicates the position of the characteristic frequency for destruction of photon correlation by continuum processes, x_c, defined by equation (15.29). From [517].

The effects of partial redistribution are less pronounced, and the depth-independent PCS approximation works better, when an overlapping continuum is taken into account. This is shown in figure 15.2, which displays a model similar to that in figure 15.1, the only difference being a (relatively weak) continuum with

$r = 10^{-8}$. There one sees that, as discussed in § 15.2, the results of all the methods essentially agree for $x > x_c$.

Fokker-Planck Approach

This method, first applied in the context of line transfer with partial redistribution in [442], is conceptually very similar to a treatment of Compton scattering in the Kompaneets approximation (e.g., [927]). The basic idea is as follows. When evaluating the redistribution integral, one has to take into account two basic features. (i) The redistribution function $R_{\rm II}(x', x)$ is a sharply peaked function of frequency, but (ii) $J(x')$ is a relatively smooth function of frequency beyond the line core. These properties suggest that one may expand the mean intensity in a Taylor expansion and perform the necessary integrations over frequency analytically, using an approximate form of $R_{\rm II}$. Specifically, $J(x')$ is expanded around its value at x, i.e.,

$$J(x') = J(x) + \frac{\partial J(x)}{\partial x}(x' - x) + \frac{1}{2}\frac{\partial^2 J(x)}{\partial x^2}(x' - x)^2 + \cdots . \qquad (15.37)$$

The redistribution function $R_{\rm II}$ is expressed by the Unno approximation [cf. (10.128)],

$$R_{\rm II}(x', x) \approx \frac{1}{2}\operatorname{erfc}(|r| + |s|) + \frac{a}{\pi}\frac{1}{s^2}\operatorname{ierfc}(|r|) + \cdots , \qquad (15.38)$$

where $r = (x' - x)/2$, and $s = (x' + x)/2$. In the line wing, the first term is negligible, and the Voigt function can be approximated by $\phi(x) \approx a/\pi x^2$. Hence

$$\frac{R_{\rm II}(x', x)}{\phi(x)} \approx \frac{x^2}{s^2}\operatorname{ierfc}(|r|) \approx \left(1 + \frac{2r}{x}\right)\operatorname{ierfc}(|r|), \qquad (15.39)$$

where the last approximation follows from neglecting terms of the order $(r/s)^2$ when evaluating the expression x^2/s^2.

Upon integration, after some algebra one obtains

$$\int_{-\infty}^{\infty} \frac{R_{\rm II}(x', x)}{\phi(x)} J(x')\, dx' = J(x) - \frac{1}{x}\frac{\partial J(x)}{\partial x} + \frac{1}{2}\frac{\partial^2 J(x)}{\partial x^2} + \cdots . \qquad (15.40)$$

The first term of equation (15.40) corresponds to coherent scattering, while the next two terms represent a change of photon frequency, described as a diffusion in the frequency space. Notice also that the coefficient of the second term, $-1/x$, represents a mean frequency shift per scattering event; see also equation (15.19). This approximation thus accounts for the two basic features of redistribution in resonance lines, namely, coherent scattering plus diffusion in the frequency space.

One may employ this formalism in several different ways. The original approach developed in [442] substituted equation (15.40) into the second-order form of the transfer equation written in the Eddington approximation. Introducing a suitable transformation of variables, the resulting equation assumes a form of the Poisson equation,

$$\frac{\partial^2 J}{\partial \tau^2} + \frac{\partial^2 J}{\partial \sigma^2} = \epsilon(J - B)\sqrt{6}\,\delta(\sigma), \tag{15.41}$$

where σ is a transformed frequency variable defined as $dx/d\sigma = (3/2)^{1/2}\phi(x)$. In some special cases (e.g., constant ϵ and B), (15.41) can be solved analytically; see [442].

Another possibility would be to represent derivatives by differences and to solve the resulting discretized form of the transfer equation numerically. The computational effort would be essentially the same as for an exact solution (see § 15.4); the only advantage would be avoiding numerical evaluation of the redistribution function and the associated scattering integral.

15.4 EXACT SOLUTIONS

Numerical Representation of the Scattering Integral

A direct numerical solution of the partial redistribution problem can be obtained by a variety of methods. The first task is to provide an accurate numerical scheme for evaluating the scattering integral, in particular of the type

$$\int_0^\infty R_{\rm II}(x', x) J(x')\, dx'. \tag{15.42}$$

Simple quadrature methods exist for representing the scattering integral in the case of complete redistribution. However, such methods are no longer satisfactory when dealing with integrals of the type in (15.42) because the function $R_{\rm II}(x', x)$ has a discontinuous first derivative at $x' = x$ and changes very rapidly on the scale of one Doppler width. Inasmuch as several discretized frequency points per one Doppler width are needed to represent the redistribution function accurately, and one has to cover the range of several tens to hundreds of Doppler widths, the total number of frequency points quickly becomes prohibitively large. However, this problem is somewhat artificial because the radiation field itself does not vary nearly as rapidly with frequency as the redistribution function.

The problem was successfully tackled in an important paper [7]. It is the first application of the idea of *natural cubic splines* in astrophysical radiative transfer theory. In the following, we follow the formalism of this paper closely. The numerical solution of the transfer equation requires that the value of integral (15.42) at discrete frequencies x_i be evaluated as a quadrature sum of weighted values of J at the same frequency points:

$$\int_0^\infty R_{\rm II}(x', x_i) J(x')\, dx' \approx \int_{x_1}^{x_N} R_{\rm II}(x', x_i) J(x')\, dx' \approx \sum_{j=1}^{N} R_{ij} J(x_j). \tag{15.43}$$

This is usually accomplished with an application of a quadrature rule that is equivalent to assuming that the integrand can be approximated by some simple function, such as a polynomial.

This method works well when the integrand varies smoothly over the relevant range of x' so that only a moderate number of discrete frequency points are needed. However, this is not the case for the integral in equation (15.43). Although $J(x')$ indeed varies slowly with x' (outside the line-core region), $R_{\rm II}$ is sharply peaked so that the primary contribution to the integral always comes from a narrow range of x' near x_i, of the order of one Doppler width. As mentioned above, the required frequency bandwidth to describe the line can be fairly large, which would lead to an impractically large number of discrete frequencies required to cover that bandwidth with a resolution smaller than a Doppler width. Moreover, the discontinuity of the slope of $R_{\rm II}$ at $x' = x$ rules out simple functional representations such as polynomials.

To overcome this problem, [7] suggested use of a more general type of quadrature scheme in which $J(x')$ is assumed to be slowly varying and well represented by a simple function, while $R_{\rm II}$ is treated essentially exactly. It was suggested to represent $J(x')$ with *natural cubic spline functions*. A cubic spline function is defined by the following properties [7, 11]: (i) it passes through the given functional values $J(x_j)$ at the discrete points x_j; (ii) the function and its first two derivatives are continuous; (iii) it is a cubic polynomial in all the intervals between the successive discrete points; and (iv) the second derivatives at the first and the last points vanish. It can be shown that this function is uniquely determined by these properties.

Natural cubic splines are advantageous because the more usual high-order polynomial approximations tend to be unstable in the sense that the coefficients R_{ij} may possess large positive and negative values, which leads to severe cancellation in the summation (15.43). Natural cubic splines are much more stable. Moreover, as we have seen in § 15.2 and 15.3, the behavior of the transfer problem is describable as a diffusion in frequency, which depends on the second derivative of J with respect to frequency. This suggests that a suitable representation of J should posses a continuous second derivative at all points. Natural cubic spline functions are the simplest functions that satisfy this requirement while avoiding instability problems.

The natural cubic spline functions are conveniently constructed by first introducing *cardinal* natural cubic splines $\psi_j(x), j = 1, \dots, N$, on the discrete frequency grid $x_k, k = 1, \dots, N$, by the requirements

$$\psi_j(x_k) = \delta_{jk} \tag{15.44}$$

and

$$\sum_{j=1}^{N} \psi_j(x') = 1 \tag{15.45}$$

for all x'. The natural cubic spline approximation for $J(x')$ is then

$$\widetilde{J}(x') = \sum_{j=1}^{N} J(x_j)\psi_j(x'). \tag{15.46}$$

Using $\widetilde{J}$ instead of J in (15.43), we obtain an explicit formula for the coefficients R_{ij},

$$R_{ij} = \int_{x_1}^{x_N} R_{\rm II}(x', x_i)\psi_j(x')\, dx'. \tag{15.47}$$

It remains to evaluate the actual spline function. As shown in [11, p. 10], if $y(x)$ is a natural cubic spline on the discrete grid $x_k, k = 1, \ldots, N$, then on the interval (x_{j-1}, x_j) one has

$$\begin{aligned} y(x) = {} & M_{j-1}\frac{(x_j - x)^3}{6h_j} + M_j\frac{(x - x_{j-1})^3}{6h_j} \\ & + \left(y_{j-1} - \frac{M_{j-1}h_j^2}{6}\right)\frac{x_j - x}{h_j} + \left(y_j - \frac{M_j h_j^2}{6}\right)\frac{x - x_{j-1}}{h_j}, \end{aligned} \tag{15.48}$$

where $y_j \equiv y(x_j)$, $M_j \equiv y''(x_j)$, and $h_j \equiv x_j - x_{j-1}$. The representation is determined once the values of $M_j, j = 1, \ldots, N$ are given. These are found from the following tridiagonal system:

$$M_1 = M_N = 0, \tag{15.49a}$$

$$\alpha_j M_{j-1} + 2M_j + \gamma_j M_{j+1} = d_j, \quad j = 2, \ldots, N-1, \tag{15.49b}$$

where

$$\begin{aligned} \alpha_j &= \frac{h_j}{h_j + h_{j+1}}, \quad \gamma_j = 1 - \alpha_j, \\ d_j &= \frac{6}{h_j + h_{j+1}}\left(\frac{y_{j+1} - y_j}{h_{j+1}} - \frac{y_j - y_{j-1}}{h_j}\right). \end{aligned} \tag{15.50}$$

The system is solved by a standard elimination procedure. Considering i to be fixed and taking $y(x) = \psi_i(x)$, equations (15.44) and (15.50) determine the quantities d_j. Solving the tridiagonal system (15.49), one determines the values of M_k, and the values of $\psi_i(x)$ are computed from (15.48).

The coefficients R_{ij} are determined using the methods to generate the redistribution function $R_{\rm II}$ mentioned in § 10.3 (for more details, refer to [7]), together with an evaluation of the cardinal splines $\psi_j(x)$ as described above. The integral in (15.43) is computed separately in each interval (x_{k-1}, x_k) in order to avoid integrating across discontinuities in slope and because the natural cubic splines are simple cubic polynomials in each interval. Essentially any simple quadrature formula can be used.

Finally, the coefficients R_{ij} must be properly normalized before they can be used. The normalization condition

$$\int_{-\infty}^{\infty} R_{\rm II}(x', x_i)\, dx' = \phi(x_i) \tag{15.51}$$

must be satisfied. This means that the coefficients R_{ij} have to satisfy

$$\sum_{j=1}^{N} R_{ij} = \phi(x_i). \tag{15.52}$$

If the coefficients R_{ij} were calculated exactly, they would satisfy this condition automatically. However, in practice, it is not satisfied because of the approximate nature of numerical integrations. Therefore, the coefficients R_{ij} have to be replaced by $\widetilde{R}_{ij} = c_i R_{ij}, j = 1, \ldots, N$, where

$$c_i = \phi(x_i) \Big/ \sum_{j=1}^{N} R_{ij}, \quad i = 1, \ldots, N. \tag{15.53}$$

This procedure slightly distorts the redistribution function, but that is much less dangerous than normalization errors, which would have an effect of introducing spurious sources and sinks of photons into the transfer calculations.

Direct Methods

Having obtained an accurate and efficient numerical representation of the scattering integral, the numerical solution of the transfer equation is straightforward. To illustrate the numerical implementation, let us consider a realistic case of line transfer with the source function given by

$$S(x) = \frac{\phi(x)}{\phi(x) + r} S^{\mathrm{L}}(x) + \frac{r}{\phi(x) + r} S^{\mathrm{C}}, \tag{15.54}$$

where S^{L} and S^{C} are the source function in the line and the continuum, respectively. The line source function is given by (15.14) or (15.15); i.e., it describes redistribution in a line with a sharp lower level (resonance line). The discretized form of the source function is given by

$$S_i = \frac{\phi_i}{\phi_i + r} \left\{ (1-\epsilon) \left[\gamma \sum_{j=1}^{N} \frac{R_{ij}}{\phi_i} J_j + (1-\gamma) \sum_{j=1}^{N} \phi_j w_j J_j \right] + \epsilon B_i \right\} + \frac{r}{\phi_i + r} S_i^{\mathrm{C}}, \tag{15.55}$$

where we used a shorthand notation $S_i \equiv S(x_i)$, $\phi_i \equiv \phi(x_i)$, etc. The quantities w_i are the quadrature weights for the integration over frequency, and R_{ij} are the normalized coefficients representing the redistribution integral, determined as described above. By rearranging the summations, the source function can be written as

$$S_i = \sum_{j=1}^{N} \alpha_{ij} J_j + \beta_i, \tag{15.56}$$

where

$$\alpha_{ij} = \frac{\phi_i}{\phi_i + r}\left\{(1-\epsilon)\left[\gamma\frac{R_{ij}}{\phi_i} + (1-\gamma)\phi_i w_i \delta_{ij}\right]\right\}, \tag{15.57}$$

and

$$\beta_i = \frac{\phi_i}{\phi_i + r}\,\epsilon B_i + \frac{r}{\phi_i + r} S_i^{\mathrm{C}}. \tag{15.58}$$

The most straightforward direct method is an application of the Feautrier scheme together with the variable Eddington factor technique. The method represents a generalization of the formalism outlined in § 12.2. In the present case, one introduces a set of vectors $\mathbf{J}_d = (J_{1,d}, \ldots, J_{N,d})^T$ at all discretized depth points d, $d = 1, \ldots, ND$. The discretized transfer equation is still written as a block tridiagonal system for the vectors $\mathbf{J}_d$, namely,

$$-\mathbf{A}_d \mathbf{J}_{d-1} + \mathbf{B}_d J_d - \mathbf{C}_d \mathbf{J}_{d+1} = \mathbf{L}_d, \tag{15.59}$$

where $\mathbf{A}_d$, $\mathbf{B}_d$, and $\mathbf{C}_d$ are $N \times N$ matrices. Because the redistribution integral is *local*, the coupling of the individual frequency points is described by the matrices $\mathbf{B}_d$, while the coupling of the depth points is described through the matrices $\mathbf{A}_d$ and $\mathbf{C}_d$. Consequently, the matrices $\mathbf{B}_d$ are full, while the matrices $\mathbf{A}_d$ and $\mathbf{C}_d$ are diagonal; moreover, their diagonal elements are given by the same expressions as in the case of complete redistribution. Specifically, the matrix elements are given by (considering, for simplicity, only the internal depth points, $d = 2, \ldots, ND - 1$; the expressions for $d = 1$ and $d = ND$ are similar)

$$(A_d)_{ij} = \frac{f_{d-1,i}}{\Delta\tau_{d-\frac{1}{2},i}\,\Delta\tau_{d,i}}\,\delta_{ij}, \tag{15.60a}$$

$$(C_d)_{ij} = \frac{f_{d+1,i}}{\Delta\tau_{i,d+\frac{1}{2},i}\,\Delta\tau_{d,i}}\,\delta_{ij}, \tag{15.60b}$$

$$(B_d)_{ij} = \alpha_{ij} + \left[\frac{f_{di}}{\Delta\tau_{di}}\left(\frac{1}{\Delta\tau_{d-\frac{1}{2},i}} + \frac{1}{\Delta\tau_{d+\frac{1}{2},i}}\right) + 1\right]\delta_{ij}, \tag{15.60c}$$

and the right-hand side vector is given by

$$(L_d)_i = \beta_i = \frac{\phi_i}{\phi_i + r}\,\epsilon B_i + \frac{r}{\phi_i + r}\, S_i^{\mathrm{C}}. \tag{15.61}$$

Application of Accelerated Lambda Iteration

Early applications of the ALI-based schemes for partial redistribution problem were made in [178], and later in [956]. A systematic development of the ALI approach for solving the two-level atom problem with partial frequency redistribution was done in an important paper [835]. We follow their analysis closely, using, however, a notation close to that used in chapter 13.

The total source function is given by

$$S_x = \rho_x S_x^{\mathrm{L}} + (1 - \rho_x) S^{\mathrm{C}}, \tag{15.62}$$

where $\rho_x = \phi(x)/[\phi(x) + r]$. In terms of a notation that expresses the dependence on frequency by a subscript, the line source function is written as

$$S_x^{\mathrm{L}} = (1 - \epsilon) \int_{-\infty}^{\infty} g(x', x) J_{x'}\, dx' + \epsilon B, \tag{15.63}$$

where we introduced a shorthand notation $g(x', x) \equiv R(x', x)/\phi(x)$. Here, we use a general redistribution function, without writing down explicitly its separation into the R_{II} and R_{III} parts.

The application of ALI is done in a similar manner as in the case of complete redistribution. We write the mean intensity of radiation, quite generally, as the action of the Λ-operator on the total source function:

$$J_x = \Lambda_x [S_x]. \tag{15.64}$$

The iteration scheme is set up as follows. Suppose that the current estimate of the source function, $S_x^{(n)}$, and the mean intensity, $J_x^{(n)}$, are known. The next iterate of the mean intensity is written as

$$J_x^{(n+1)} = \Lambda_x^* [S_x^{(n+1)}] + (\Lambda_x - \Lambda_x^*)[S_x^{(n)}], \tag{15.65}$$

i.e., as an action of the approximate operator Λ^* on the new iterate of the source function, plus a "correction," $\Lambda - \Lambda^*$, acting on the previous, known iterate of the source function, exactly as in the case of complete redistribution. The new iterate of the line source function is then written as

$$S_x^{\mathrm{L}(n+1)} = (1 - \epsilon) \int_{-\infty}^{\infty} g(x', x) \left\{ \Lambda_{x'}^* [S_{x'}^{(n+1)}] + (\Lambda_{x'} - \Lambda_{x'}^*)[S_{x'}^{(n)}] \right\} dx' + \epsilon B. \tag{15.66}$$

We write

$$S_x^{\mathrm{SF}} \equiv (1 - \epsilon) \int_{-\infty}^{\infty} g(x', x) \Lambda_{x'} [S_{x'}^{(n)}]\, dx' + \epsilon B, \tag{15.67}$$

which denotes a newer source function that is obtained from the previous source function by a formal solution of the transfer equation. Equation (15.66) can be rewritten as

$$S_x^{\mathrm{L}(n+1)} = (1 - \epsilon) \int_{-\infty}^{\infty} g(x', x) \Lambda_{x'}^* \left[\rho_{x'} S_{x'}^{\mathrm{L}(n+1)} - \rho_{x'} S_{x'}^{\mathrm{L}(n)} \right] dx' + S_x^{\mathrm{FS}}, \tag{15.68}$$

because the term $(1 - \rho_x) S^{\mathrm{C}}$ is fixed and does not change between iterations. Introducing

$$\delta S_x^{\mathrm{L}(n)} = S_x^{\mathrm{L}(n+1)} - S_x^{\mathrm{L}(n)}, \tag{15.69}$$

we obtain an equation for changes of the *line* source function,

$$\delta S_x^{\mathrm{L}(n)} - (1-\epsilon)\int_{-\infty}^{\infty} g(x',x)\Lambda^*_{x'}[\,\rho_{x'}\delta S_{x'}^{\mathrm{L}(n)}\,]\,dx' = S_x^{\mathrm{FS}} - S_x^{\mathrm{L}(n)}. \qquad (15.70)$$

This equation is written for a general Λ^*-operator. However, the problem is greatly simplified by using an approximate *local* operator; see chapter 13. In this case, the action of the approximate Λ^*-operator is simply a multiplication by a real number, namely, $\Lambda^*_x[S_x] = \Lambda^*_x S_x$, where Λ^*_x is understood as a scalar variable. Introducing a set of discretized frequency points $x_i, i = 1,\ldots,N$, (15.70) is discretized to yield a set of linear algebraic equations for the corrections $\delta S_{x_i}^{\mathrm{L}(n)}$,

$$\delta S_{x_i}^{\mathrm{L}(n)} - (1-\epsilon)\sum_j g_{ij}\rho_j\Lambda^*_j\delta S_{x_j}^{\mathrm{L}(n)} = S_i^{\mathrm{FS}} - S_i^{\mathrm{L}(n)}. \qquad (15.71)$$

In a concise matrix notation, (15.71) is written as

$$\mathbf{A}\cdot\delta\mathbf{S}^{\mathrm{L}} = \mathbf{r}, \qquad (15.72)$$

where $\delta\mathbf{S}^{\mathrm{L}}$ is a column vector composed of $\delta S_{x_i}^{\mathrm{L}(n)}$ at all frequency points, $\mathbf{r}$ is a vector of the right-hand sides, $r_i = S_i^{\mathrm{FS}} - S_i^{\mathrm{L}(n)}$, and the elements of the system matrix $\mathbf{A}$ are

$$A_{ij} = \delta_{ij} - (1-\epsilon)\,g_{ij}\,\Lambda^*_j\,\rho_j. \qquad (15.73)$$

The iterations proceed as follows:

(a) given an initial estimate of the source function, for instance, $S^{(0)} = B$,
(b) compute S^{FS} from (15.67),
(c) at each depth, solve the linear set (15.71) for the corrections $\delta S^{\mathrm{L}(n)}$,
(d) update S^{L} from (15.69) and S from (15.62),
(e) test for convergence and return to (b) if not converged.

However, solving the set (15.71) requires the inversion of an $(N \times N)$ matrix at each depth, which is still computationally expensive. Recall that one needs of the order of 50–100 frequency points to represent the redistribution function in a line with sufficient accuracy. But this computational cost may be significantly reduced. The point is that the matrix $\mathbf{A}$ is fixed, while only the residuum vector $\mathbf{r}$ changes from iteration to iteration. One can perform an inversion of matrix $\mathbf{A}$ at each depth in the first iteration and use the inverted matrices in subsequent iterations. An even more efficient procedure, suggested in [835], is to solve the system (15.72) by *LU* decomposition and to store the decomposed matrices in memory. In the first iteration, one computes matrix $\mathbf{A}$ and its *LU* decomposition at each depth, and the matrices are stored in the decomposed form. In the subsequent iterations, one uses the decomposed matrices to solve system (15.72) for the new right-hand side vector $\mathbf{r}$. The computer time needed to perform these iterations is thus negligible compared to the computer time needed for the first iteration.

Nevertheless, it is still possible to speed up calculations. The important issue here is that equation (15.71) is used to calculate the *corrections* of the source function, not the source function itself. Therefore, one can afford approximations. One such approximation was suggested in the first modern ALI-type approach to solving the partial redistribution problem in [956]. The approximation is based on replacing $g(x', x)$ in (15.70) by $\phi_{x'}$, i.e., essentially assuming complete redistribution for evaluations of the *corrections* to the source function. Equation (15.70) now reads

$$\delta S_x^{\mathrm{L}(n)} - (1-\epsilon) \int_{-\infty}^{\infty} \phi_{x'} \Lambda_{x'}^* [\, \rho_{x'} \delta S_{x'}^{\mathrm{L}(n)} \,]\, dx' = r_x^{(n)}. \tag{15.74}$$

The integral on the left-hand side is independent of frequency x. After [956], it is denoted as $\Delta T^{(n)}$,

$$\Delta T^{(n)} \equiv (1-\epsilon) \int_{-\infty}^{\infty} \phi_{x'} \Lambda_{x'}^* [\, \rho_{x'} \delta S_{x'}^{\mathrm{L}(n)} \,]\, dx'. \tag{15.75}$$

The equation for the corrections of the line source function then reads

$$\delta S_x^{\mathrm{L}(n)} - \Delta T^{(n)} = r_x^{(n)}. \tag{15.76}$$

Substituting (15.76) into (15.75), one obtains an equation for the frequency-independent correction ΔT,

$$\Delta T^{(n)} - (1-\epsilon) \int_{-\infty}^{\infty} \phi_{x'} \Lambda_{x'}^* [\rho_{x'} \Delta T^{(n)}]\, dx' = \bar{r}^{(n)}, \tag{15.77}$$

where

$$\bar{r}^{(n)} = (1-\epsilon) \int_{-\infty}^{\infty} \phi_{x'} \Lambda_{x'}^* [\rho_{x'} r_{x'}^{(n)}]\, dx'. \tag{15.78}$$

If the approximate operator is local, then (15.77) has a particularly simple form,

$$\left(1 - (1-\epsilon) \int_{-\infty}^{\infty} \phi_{x'} \Lambda_{x'}^* \rho_{x'} dx' \right) \Delta T^{(n)} = \bar{r}^{(n)}. \tag{15.79}$$

Both equations (15.74) and (15.79) are analogous to the corresponding equations for the ALI solution of the complete redistribution problem and may be solved in the same way.

The method outlined above is fast and simple, but in some cases it may not converge well. The reason is obvious—for a strong line with $\epsilon \ll 1$, one has a long chain of uninterrupted scatterings, and the transport of line wing photons is much better described by (almost) coherent scattering than by complete redistribution, as was explained in § 15.2.

This suggests that a better scheme would take into account a coherent nature of line scattering in the wings. The simplest such scheme is the partial coherent scattering approximation, described in § 15.3. This possibility had been suggested in the original ALI paper [956] and was explicitly formulated and tested in [835]. Briefly, the integral in (15.63) is replaced by

$$\int_{-\infty}^{\infty} g(x',x) J_{x'}\, dx' = (1-\alpha_x) \int_{\rm core} \phi_{x'} J_{x'}\, dx' + \alpha_x \int_{\rm wing} \delta(x-x') J_{x'}\, dx', \qquad (15.80)$$

where the exact extent of the "core" and the "wing" region, as well as the exact value of the splitting coefficient α, is left unspecified. Numerical experience showed that the core defined as $|x| < x_c = 3.5$, and $\alpha_x = 0$ in the core, and $\alpha_x = g(x,x)$ in the wing yield the best results, i.e., the fastest convergence.

Substituting (15.80) into equation (15.70), and using a local operator Λ^*, one obtains

$$\delta S_x^{{\rm L}(n)} - (1-\alpha_x)(1-\epsilon) \int_{\rm core} \phi_{x'} \rho_{x'} \Lambda^*_{x'} \delta S_{x'}^{{\rm L}(n)}\, dx'$$
$$- \alpha_x (1-\epsilon) \int_{\rm wing} \delta(x-x')\, \rho_{x'} \Lambda^*_{x'} \delta S_{x'}^{{\rm L}(n)}\, dx' = r_x. \qquad (15.81)$$

The crucial feature of this approximation is the simplicity of the coupling between the core and the wing frequencies. For the core, $|x| < x_c$, the wing frequencies drop out, and in the wing, the core frequencies only appear through a single integral. This makes it possible to solve for all the $\delta S_x^{{\rm L}(n)}$ by a single scalar equation. Specifically, for the core frequencies we can define a modified ΔT,

$$\Delta T_{\rm core}^{(n)} = (1-\epsilon) \int_{\rm core} \phi_{x'} \Lambda^*_{x'} \rho_{x'} \delta S_{x'}^{{\rm L}(n)}\, dx', \qquad (15.82)$$

which implies $\delta S_x^{{\rm L}(n)} - \Delta T_{\rm core}^{(n)} = r_x$ and, upon integrating over the core,

$$\left(1 - (1-\epsilon) \int_{\rm core} \phi_{x'} \Lambda^*_{x'} \rho_{x'} dx' \right) \Delta T_{\rm core}^{(n)} = \overline{r}_{\rm core}^{(n)}, \qquad (15.83)$$

where $\overline{r}_{\rm core}^{(n)}$ is given by an equation similar to (15.78), where the integral extends only over the core frequencies. Once $\Delta T_{\rm core}^{(n)}$ is determined, the correction to the wing source functions is obtained from a simple scalar expression,

$$\delta S_x^{{\rm L}(n)} = \frac{r_x + (1-\alpha_x) \Delta T_{\rm core}^{(n)}}{1 - \alpha_x (1-\epsilon) \Lambda^*_x \rho_x}. \qquad (15.84)$$

In the original paper [835] the method is called CRDCS (complete redistribution over the core and coherent scattering in the wings). The method converges even in most difficult cases where the previous approximate ALI scheme based on complete redistribution, equation (15.79), fails completely.

15.5 MULTI-LEVEL ATOMS

The majority of studies of line transfer with partial frequency redistribution were done in the context of a two-level atom. However, for any realistic calculation one has to take into account many levels of an atom/ion.

As discussed above, the kinetic equilibrium equation in the presence of partial frequency redistribution remains formally the same as in the case of complete redistribution. Compared to the case of complete redistribution, the effects of partial redistribution come through the modified emission profile coefficients. In the case of CFR, these are known functions of frequency and the state parameters: temperature and density. However, in the general case of partial frequency redistribution, they are no longer known functions but depend on atomic level populations and the radiation field, as explained in chapter 10. This fact introduces an additional nonlinearity into the multi-level atom problem. Moreover, the modified emission profile coefficients influence the radiation field, and hence the radiative rates, and consequently the atomic level populations. Because the rate equations couple all the levels, the presence of partial redistribution in one transition may in principle influence the populations of all levels. The problem seems quite complicated, but may be significantly simplified by the following observations.

First, although the number of emissions in line $j \rightarrow i$ in the frequency range $(\nu, \nu + d\nu)$ is $n_j A_{ji} \psi_{ji}(\nu)$, where ψ is the emission profile coefficient (generally different from the absorption profile ϕ_{ij}), the total number of spontaneous emissions is given by $n_j A_{ji} \int \psi_{ji}(\nu)\, d\nu = n_j A_{ji}$. Obviously, partial redistribution effects influence the frequency redistribution of re-emitted radiation in a line scattering but do not change the total number of emissions. Therefore, formally, the spontaneous emission rate is the same as in the case of complete redistribution.

The only formal difference from CFR would arise if the stimulated emission rate were written as $B_{ji} \int \psi_{ji}(\nu) J_\nu d\nu$. However, as was explained in § 15.1, this is not necessary. The atomic level populations depend on the radiation intensity through the radiative rates. Radiative rates in the continua are not directly influenced by partial redistribution effects. Radiative rates in the lines are dominated by the radiation intensity in the line core. Recall that the rate $R_{ij} \propto \int \phi_{ij}(\nu) J(\nu)\, d\nu$, where $\phi(\nu)$ is a sharply peaked function of frequency with a maximum at the line center. On the other hand, departures from complete frequency redistribution are typically seen in the line wings, while they are quite modest in the line core. Therefore, the effects of partial frequency redistribution on the radiative rates in the lines are also small.

This suggests that the interplay between partial redistribution and kinetic equilibrium may be treated iteratively. Moreover, since the presence of partial frequency redistribution does not influence significantly other transitions (an exception may be a resonance Raman scattering; see below), one may employ the idea of the equivalent-two-level-atom (ETLA) scheme to treat partial redistribution in all transitions separately. To avoid confusion, we stress that the ETLA scheme is *not* applied here to solve for a NLTE multi-level atom problem, but only to treat *departures from complete frequency redistribution* for one transition at a time.

Equivalent-Two-Level-Atom Approach

The departures from complete redistribution are naturally described by the ratio of the emission to the absorption profile coefficients, written here in a general, frequency- and angle-dependent form,

$$\rho_{ij}(\nu, \mathbf{n}) \equiv \psi_{ji}(\nu, \mathbf{n})/\phi_{ij}(\nu). \tag{15.85}$$

As follows from (15.5) and (15.6), it is given by

$$\rho_{ij}(\nu, \mathbf{n}) = 1 + \sum_{\ell<j} \frac{n_\ell B_{\ell j}}{n_j P_j} \overline{J}^R_{\ell ji}(\nu, \mathbf{n}), \tag{15.86}$$

where P_j is the total rate of transitions out of level j and $\overline{J}^R_{\ell ji}$ the redistribution integral defined by (10.176) or (15.8),

$$\overline{J}^R_{\ell ji}(\nu, \mathbf{n}) = \int_0^\infty \oint \left[p^*_\ell \frac{R_{\ell ji}(\nu', \mathbf{n}'; \nu, \mathbf{n})}{\phi_{ij}(\nu)} - \phi_{\ell j}(\nu') \right] I(\nu', \mathbf{n}')\,(d\Omega'/4\pi)\,d\nu', \tag{15.87}$$

where the terms with $\ell = i$ correspond to the usual resonance scattering within the line $i \to j$, while the terms with $\ell \neq i$ correspond to resonance Raman scattering (or "cross-redistribution") $\ell \to j \to i$.

The line source function for the transition $i \leftrightarrow j$ is given by

$$S^{\rm L}_{ij} = \frac{n_j A_{ji}}{n_i B_{ij}} \frac{1 + (n_i B_{ij}/n_j P_j)[\overline{J}^R_{iji} + \overline{P}_{ji}]}{1 - (n_j B_{ji}/n_i B_{ij})\rho_{ij}}, \tag{15.88}$$

where $\overline{P}_{ji}$ represents the contribution of all cross-redistribution (resonance Raman scattering) transitions,

$$\overline{P}_{ji} = \sum_{\ell<j, \ell\neq i} \frac{n_\ell B_{\ell j}}{n_i B_{ij}} \overline{J}^R_{\ell ji}. \tag{15.89}$$

The basis of the equivalent two-level atom approach is to eliminate the unknown populations n_i and n_j from the expression for the line source function using the corresponding kinetic equilibrium equations. The procedure is similar to that described in § 14.4. The resulting expression for the line source function is [518, 535]

$$S^{\rm L}_{ij}(\nu, \mathbf{n}) = \frac{\overline{J}_{ij} + \eta_{ij} + \zeta_{ij}[\overline{J}^R_{iji}(\nu, \mathbf{n}) + \overline{P}_{ji}(\nu, \mathbf{n})]}{1 + \epsilon_{ij}(\nu, \mathbf{n})}, \tag{15.90}$$

where

$$\eta_{ij} = \beta_{ij}/B_{ij}, \tag{15.91a}$$

$$\zeta_{ij} = (A_{ji} + B_{ji}\overline{J}_{ji} + \alpha_{ij})/A_{ji}, \tag{15.91b}$$

$$\epsilon_{ij}(\nu, \mathbf{n}) = \{\alpha_{ij} + B_{ji}\overline{J}_{ij}[1 - \rho_{ij}(\nu, \mathbf{n})] - \eta_{ij} B_{ji} \rho_{ij}(\nu, \mathbf{n})\}/A_{ji}, \tag{15.91c}$$

where α and β are given by equations (14.132). There are two important points to realize. First, unlike the case of complete redistribution, the line source function is now a nonlinear function of the radiation intensity. There are two sources of nonlinearity here: an occurrence of $\overline{J}$ in the expression for ζ and the presence of $\overline{J}$ and ρ in the expression for ϵ. Neither of these terms contributes in the case of complete redistribution, in which case the line source function is given by

$$S_{ij}^{L(\mathrm{CFR})} = \frac{\overline{J}_{ij} + \eta_{ij}}{1 + \epsilon_{ij}}. \tag{15.92}$$

Both nonlinear terms in equation (15.90) result from stimulated emission. However, as explained in chapter 10, it is sound to treat these offending nonlinear terms iteratively, by using current values of $\overline{J}$ and ρ.

Second, another important difference from the case of complete redistribution is the presence of the cross-redistribution term $\overline{P}$ in the expression for the line source function. This term is formally linear in the radiation intensity, but involves the radiation field in the transitions $\ell \to j$. As discussed in [535], in order to maintain the spirit of the standard equivalent-two-level-atom approach, this term is also treated iteratively, using current values of the radiation intensity in the transitions $\ell \to j$.

A full solution of the multi-level atom problem with partial redistribution proceeds by the following steps.

(0) Compute a full multi-level atom solution assuming complete frequency redistribution (CFR) in all transitions. Actually, if one does not intend to compare results with CFR and partial frequency redistribution (PFR) explicitly, the starting CFR solution does not have to be fully converged, since one is going to iterate anyway.
(1) Using current values of populations and radiation intensities, set up the ETLA quantities η, ζ, and ϵ for a selected transition $i \to j$.
(2) Solve a coupled system of radiative transfer equations for the frequency points within the selected transition using the source function (15.90). As explained above, this source function is linear in the radiation intensity, because the potential nonlinear terms are being treated with current, known values of the intensity and level populations. Any numerical method for solving the two-level problem with partial redistribution, described in § 15.4, can be used here.
(3) Once the solution of the transfer equation is obtained, the quantity ρ_{ij} can be calculated from (15.86).
(4) After completing similar ETLA procedures for all the transitions to be treated with partial redistribution, proceed with one more iteration of the global complete-linearization multi-level scheme, where ρ_{ij} is now viewed as a *known quantity* and is fixed. Then steps (1)–(4) are iterated to convergence.

In this scheme, the frequency coupling resulting from partial frequency redistribution is *separated* from the global multi-level coupling; therefore, any method capable of solving the multi-level problem can be employed here.

An implementation of the present scheme into a complete-redistribution multi-level code proceeds as follows [535, 1091]. One writes the total absorption and emission coefficient as

$$\chi(\nu,\mathbf{n}) = \chi_c + \sum_{ij} V_{ij}(\nu,\mathbf{n})[n_i - n_j G_{ij}(\nu,\mathbf{n})], \tag{15.93a}$$

$$\eta(\nu,\mathbf{n}) = \eta_c + \sum_{ij} (2h\nu^3/c^2)\, n_j G_{ij}(\nu,\mathbf{n}) V_{ij}(\nu,\mathbf{n}), \tag{15.93b}$$

where V_{ij} is defined by equation (14.213) and (14.214), i.e., $V_{ij} = (h\nu/4\pi)B_{ij}\phi_{ij}$ for bound-bound transitions, and $V_{ij} = \sigma_{ij}$ for bound-free transitions; G is given by

$$G_{ij}(\nu,\mathbf{n}) = \begin{cases} (g_i/g_j)\rho_{ij}(\nu,\mathbf{n}), & \text{(bound-bound)}, \\ n_e \Phi_{ij}(T)\exp(-h\nu/kT), & \text{(bound-free)}. \end{cases} \tag{15.94}$$

The latter expression is the same as in the case of complete redistribution. The radiative rates can be written as $(i<j)$

$$R_{ij} = \oint d\Omega \int_0^\infty \frac{d\nu}{h\nu} V_{ij}(\nu,\mathbf{n}) I(\nu,\mathbf{n}), \tag{15.95a}$$

$$R_{ji} = \oint d\Omega \int_0^\infty \frac{d\nu}{h\nu} \left[(2h\nu^3/c^2) + I(\nu,\mathbf{n})\right] G_{ij}(\nu,\mathbf{n}) V_{ij}(\nu,\mathbf{n}), \tag{15.95b}$$

which are valid for both bound-bound and bound-free transitions.

Equations (15.93)–(15.95) are identical to those applied in the case of complete linearization; the only difference is in the definition of G_{ij}, which now contains the factor ρ_{ij}. Therefore, the only change required in a complete redistribution multi-level code is to modify G_{ij} for the lines that are selected to be treated with partial redistribution from $G_{ij} = g_i/g_j$ to $G_{ij} = (g_i/g_j)\rho_{ij}$, which is a rather trivial modification.

For instance, the present scheme was implemented [535] using the global complete redistribution, multi-level code MULTI [183] and applied for studying hydrogen line formation in the solar atmosphere; see § 15.6.

Another scheme based on the ETLA method for treating partial redistribution in multi-level atoms was developed in [71] as a generalization of the preconditioned ETLA [63] (cf. § 14.4), employing the parameter ρ. It was also applied for line formation in the solar atmosphere [71].

ALI Scheme with Preconditioning

A generalization of the ALI method with full preconditioning, described in § 14.5, was developed in [1091]. The formalism follows that of (14.213)–(14.226); here we describe only modifications to this formalism that arise as a result of the partial redistribution treatment. The U and V quantities for lines in (14.213) are generalized to read

$$U_{ul}(\nu, \mathbf{n}) \equiv (h\nu/4\pi) A_{ul} \psi_{ul}(\nu, \mathbf{n}), \quad u > l, \tag{15.96a}$$

$$U_{lu}(\nu, \mathbf{n}) \equiv 0, \quad u > l, \tag{15.96b}$$

$$V_{lu}(\nu, \mathbf{n}) \equiv (h\nu/4\pi) B_{lu} \phi_{lu}(\nu), \quad u > l, \tag{15.96c}$$

$$V_{ul}(\nu, \mathbf{n}) \equiv (h\nu/4\pi) B_{ul} \psi_{ul}(\nu, \mathbf{n}), \quad u > l. \tag{15.96d}$$

The U and V quantities for bound-free transitions (14.214) remain unchanged, and so do the formal expressions for the absorption and emission coefficients, (14.217); more precisely, they are written in a slightly more general form that allows for a dependence of U and V on direction:

$$\chi(\nu, \mathbf{n}) = \sum_{u>l} \left[n_l V_{lu}(\nu, \mathbf{n}) - n_u V_{ul}(\nu, \mathbf{n}) \right] + \chi_c(\nu), \tag{15.97a}$$

$$\eta(\nu, \mathbf{n}) = \sum_{u,l} n_u U_{ul}(\nu, \mathbf{n}) + \eta_c(\nu). \tag{15.97b}$$

The expression for the bound-bound radiative rates (14.218) also remains the same, again generalized to allow for a dependence of U and V on direction,

$$R_{ul} = \oint \int_0^\infty \frac{4\pi}{h\nu} \left[U_{ul}(\nu, \mathbf{n}) + V_{ul}(\nu, \mathbf{n}) I(\nu, \mathbf{n}) \right] d\nu (d\Omega/4\pi). \tag{15.98}$$

The application of ALI and preconditioning is similar to that described in § 14.5; we will use a more general form that allows for an explicit dependence of all the radiation-dependent quantities on direction. Also, we use the original notation introduced in [926] and used by [1091], which expresses the radiation intensity through the Ψ-operator instead of the Λ-operator,

$$I_{\nu,\mathbf{n}} = \Lambda_{\nu,\mathbf{n}}[\eta_{\nu,\mathbf{n}}/\chi_{\nu,\mathbf{n}}] \equiv \Psi_{\nu,\mathbf{n}}[\eta_{\nu,\mathbf{n}}], \tag{15.99}$$

where we express a dependence of ν and $\mathbf{n}$ with subscripts. Also, instead of the superscript "old" to denote the quantities from the previous iteration, we will use a dagger, $\dagger$, as was done in [926]. The basic ALI expression is written as

$$I_{\nu,\mathbf{n}} = \Psi^*_{\nu,\mathbf{n}}[\eta_{\nu,\mathbf{n}}] + (\Psi_{\nu,\mathbf{n}} - \Psi^*_{\nu,\mathbf{n}})[\eta^\dagger_{\nu,\mathbf{n}}]. \tag{15.100}$$

Substituting (15.97) into (15.100), one obtains a generalization of (14.223):

$$I(\nu, \mathbf{n}) = \Psi_{\nu,\mathbf{n}}[\eta^\dagger_{\nu,\mathbf{n}}] - \sum_{ij} \Psi^*_{\nu,\mathbf{n}}[n^\dagger_j U^\dagger_{ji}] + \sum_{ij} \Psi^*_{\nu,\mathbf{n}}[n_i U_{ij}]. \tag{15.101}$$

In the present case, one has to distinguish between U_{ji} and $U^\dagger_{ji}$, because unlike the case of complete linearization, U changes from iteration to iteration because it depends on the radiation field through the emission profile coefficient ψ_{ji}.

Substituting (15.101) into the expression for the radiative rates, (15.98), the kinetic equilibrium equation for level u corresponding to (14.224) now reads

$$\sum_l n_l C_{lu} + \sum_l \oint d\Omega \int_0^\infty \frac{d\nu}{h\nu}\left(n_l U_{lu} + n_l V_{lu} I^{\rm eff}_{\nu,\mathbf{n}} + \sum_{ij} n_l V_{lu} \Psi^*_{\nu,\mathbf{n}}\big[n_i U_{ij}\big]\right)$$
$$= \sum_l n_u C_{ul} + \sum_l \oint d\Omega \int_0^\infty \frac{d\nu}{h\nu}\left(n_u U_{ul} + n_u V_{ul} I^{\rm eff}_{\nu,\mathbf{n}} + \sum_{ij} n_u V_{ul} \Psi^*_{\nu,\mathbf{n}}\big[n_i U_{ij}\big]\right), \tag{15.102}$$

where

$$I^{\rm eff}_{\nu,\mathbf{n}} = I^{\rm FS}_{\nu,\mathbf{n}} - \sum_{ij} \Psi^*_{\nu,\mathbf{n}}\big[n_j^\dagger U_{ji}^\dagger\big], \tag{15.103}$$

where $I^{\rm FS}_{\nu,\mathbf{n}} \equiv \Psi_{\nu,\mathbf{n}}\big[\eta^\dagger_{\nu,\mathbf{n}}\big]$ is the "newer" specific intensity obtained by a formal solution with the old emissivity.

In the case of complete redistribution, only the last terms on both sides of (15.102) are nonlinear in the populations. However, if one or more transitions are treated with partial redistribution, there are additional nonlinearities from the terms with U and V because they depend on the radiation field through the emission profile coefficient ψ. As suggested in [1091], these terms are replaced by the "old" values, $U^\dagger$ and $V^\dagger$, in equation (15.102). The nonlinearities in the last terms on both sides in the products $n_l n_i$ and $n_u n_i$ are removed in the same way as in the case of complete redistribution, by replacing them with $n_l^\dagger n_i$ and $n_u^\dagger n_i$. Moreover, as suggested in [926] and [1091], it is sufficient to take for the approximate Λ^* or Ψ^* its diagonal (local) part, in which case the action of these operators is simply a multiplication by a scalar quantity.

The final system of linear, preconditioned kinetic equilibrium equations then reads

$$\sum_l n_l C_{lu} + \sum_l \oint d\Omega \int_0^\infty \frac{d\nu}{h\nu}\left(n_l U^\dagger_{lu} + n_l V^\dagger_{lu} I^{\rm eff}_{\nu,\mathbf{n}} + n_l^\dagger V^\dagger_{lu} \sum_{ij} n_i \Psi^*_{\nu,\mathbf{n}} U^\dagger_{ij}\right)$$
$$= \sum_l n_u C_{ul} + \sum_l \oint d\Omega \int_0^\infty \frac{d\nu}{h\nu}\left(n_u U^\dagger_{ul} + n_u V^\dagger_{ul} I^{\rm eff}_{\nu,\mathbf{n}} + n_u^\dagger V^\dagger_{ul} \sum_{ij} n_i \Psi^*_{\nu,\mathbf{n}} U^\dagger_{ij}\right). \tag{15.104}$$

An implementation in a complete redistribution multi-level code is done again by employing the quantity ρ, the ratio of the emission and absorption profile coefficients defined by equation (15.85), which is computed using (15.86) at the end of a given iteration step for the transitions that are selected for a treatment with partial redistribution, and held fixed in the subsequent iteration of the ALI cycle.

15.6 APPLICATIONS

We briefly consider here two types of applications: idealized models that aim to demonstrate the basic effects of partial redistribution in the simplest situations and applications on real stellar atmospheres that aim to produce useful spectroscopic diagnostics.

Idealized Models

A great deal of insight into the nature of partial redistribution effects can be obtained from highly idealized studies that simplify the problem as much as possible. A typical example of such simplification is to consider a two-level atom in a constant-property medium and with the stimulated emission treated with the absorption profile coefficient ϕ_ν instead of the emission profile coefficient ψ_ν, both in the emission coefficient, as well as in the stimulated emission rate. In this case the line source function is given by [cf. (15.11)]

$$S_l(\nu) = (1-\epsilon)\,\phi_\nu^{-1}\int_0^\infty R(\nu',\nu)J_{\nu'}\,d\nu' + \epsilon B. \tag{15.105}$$

This form of the source function allows a direct solution of the transfer problem, using the methods described in § 15.4. This approach has been used to assess the differences between the frequency-dependent source function $S_l(\nu)$, obtained when partial redistribution effects are taken into account, and the frequency-independent source function $S_l^{\rm CFR} = (1-\epsilon)\bar{J} + \epsilon B$ obtained from complete redistribution.

The results with $R(\nu',\nu) = R_{\rm I}(\nu',\nu)$ and $R_{\rm II}(\nu',\nu)$ were first obtained in [547], those with $R(\nu',\nu) = R_{\rm III}(\nu',\nu)$ in [331], and those with $R(\nu',\nu) = R_{\rm V}(\nu',\nu)$ in [530]. Besides numerical studies, there were also rigorous mathematical analyses of the transport with various basic redistribution functions, aimed at deducing the large scale and asymptotic behavior of the source function. In particular, [349] summarized previous work and performed a rigorous mathematical analysis of the solutions with all the basic redistribution functions.

- Results with $R_{\rm I}$: A number of solutions were obtained in [547] for both finite and semi-infinite isothermal atmospheres, assuming zero continuum opacity, for $\epsilon = 10^{-4}$ and 10^{-6}. Although there are some differences in the behavior of $S_l(\nu)$ and $S_l^{\rm CFR}$, the resulting emergent line profiles are virtually identical.
- Results with $R_{\rm III}$: The qualitative conclusions reached in [331] are similar to those for $R_{\rm I}$, namely that while there are some differences in the behavior of $S_l(\nu)$ and $S_l^{\rm CFR}$, the resulting emergent line profiles were virtually identical.
- Results with $R_{\rm II}$: As was discussed extensively throughout this chapter, this is the case that should show significant differences between the results from partial and complete redistribution. This was indeed demonstrated in the pioneering study [547]. Results for an isothermal atmosphere of total thickness $T = 10^6$ for a line with $a = 10^{-3}$ and $\epsilon = 10^{-4}$ are shown in figure 15.3.

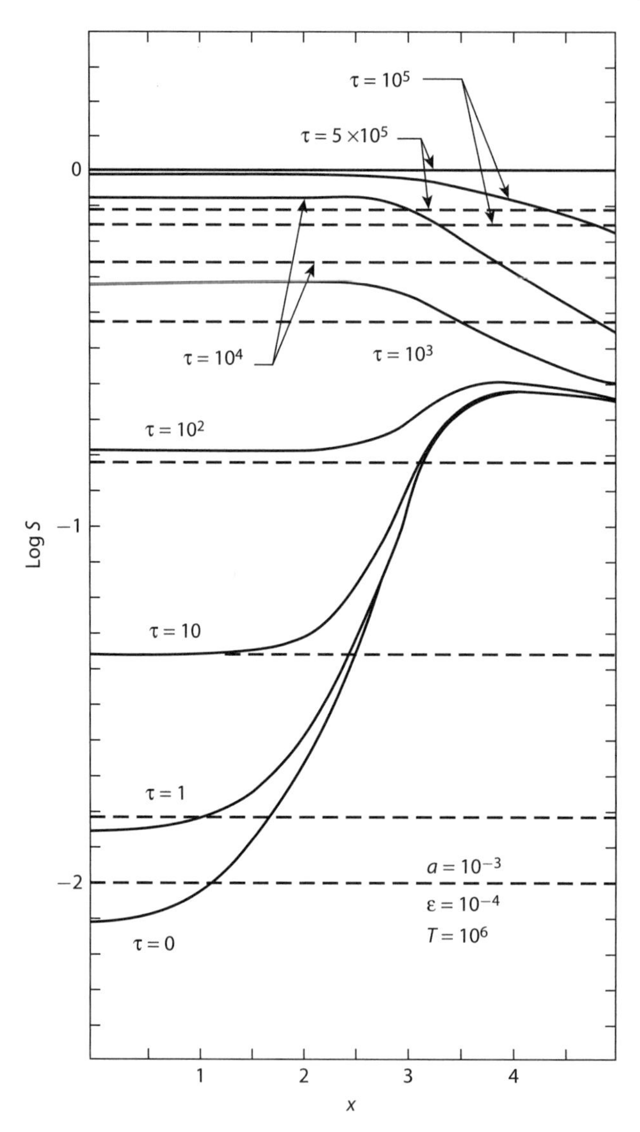

Figure 15.3 Source functions in an isothermal atmosphere with $T = 10^6$ for a line with $a = 10^{-3}$ and $\epsilon = 10^{-4}$. Dashed curves: frequency-independent source functions obtained assuming complete redistribution. Solid curves: frequency-dependent source functions obtained using the redistribution function $R_{\rm II}$. From [547].

At the line center, $S_l(\nu)$ is near $S_l^{\rm CFR}$ for small optical depths, but rises above $S_l^{\rm CFR}$ at greater depths, and approaches the Planck function sooner than the complete redistribution source function. This is because the essentially coherent nature of the scattering process in the wing *inhibits* photon escape because it traps photons in a wing whenever they have been re-emitted there during the scattering

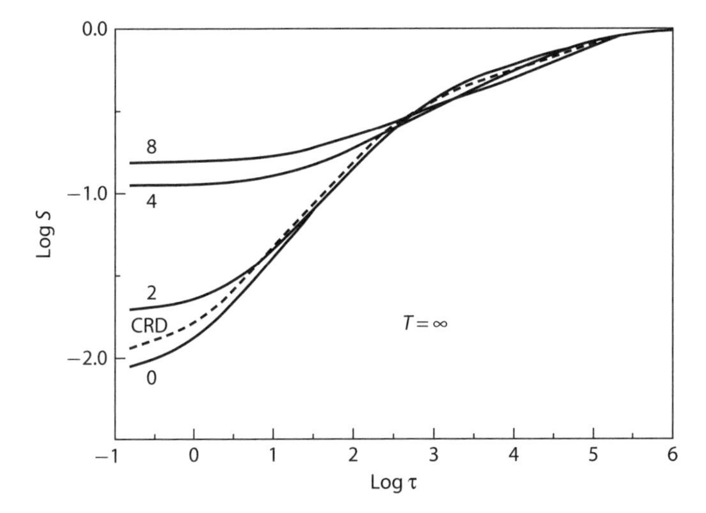

Figure 15.4 Source functions obtained for an isothermal semi-infinite atmosphere with $\epsilon = 10^{-4}$, using redistribution function R_V. From [530].

process. Because the photon escape probability is reduced, thermalization to the Planck function occurs more rapidly than in the case of complete redistribution. In contrast, in the line wings, the frequency-dependent source function lies significantly *below* $S_l^{\rm CFR}$, because photons are no longer being fed into the wing from the core as efficiently.

- Results with R_V: The numerical results obtained in [530] showed that the behavior of the source function and the emergent intensity for R_V is intermediate between those obtained for complete redistribution and for $R_{\rm II}$, and typically closer to results with CFR. This can be expected on physical grounds, as the redistribution function R_V behaves as a linear combination of $R_{\rm II}$ and complete redistribution, and the significant probability of photon destruction γ' inhibits Doppler diffusion and forces the source function to behave similarly to the CFR source function. This is illustrated in figure 15.4.

Applications to Solar and Stellar Lines

Besides studies of the partial redistribution effects in the idealized models, described above, most actual applications to real astronomical objects were done for resonance lines in the solar atmosphere, more specifically, in the solar chromosphere. There were many studies; here we mention only the most important ones.

One of the first examples of the importance of partial redistribution effects in resonance-line formation arose from attempts to fit the observed solar chromospheric $L\alpha$ profile [1118]. It was found that when the line profile is calculated under the assumption of complete redistribution, using models that otherwise provide good fits to the continuum data formed at the same layers as the wings of $L\alpha$, the intensity in the predicted wing profile is *much* larger (by factor of 5 to 6) than observed. The study suggested that the effects of partial redistribution are responsible for the disagreement and found empirically that a good fit to the observed profile is obtained if the scattering is assumed to be about 93% coherent and 7% completely redistributed. In the present terminology, treatment of collisional redistribution was used with the redistribution function given by (10.151), with the branching ratio between coherent scattering and complete redistribution set *empirically*. Moreover, the function R_{II} was approximated using the partial coherent scattering approximation.

Subsequent work [749, 750] showed that one can obtain a good fit to observations when the redistribution function is taken as (10.151) with *realistic* values of the branching ratio γ, and when the redistribution function R_{II} and the radiative transfer are treated exactly. However, the theoretical prediction still failed to fit the solar $L\beta$ line wings.

The first study of cross-redistribution was aimed at explaining the observed center-to-limb variation of the solar Ca II H and K lines. Previous studies using complete redistribution were able to fit the line profiles, but not the center-to-limb variations. Calculations using a five-level model atom were made in [997]. Similar calculations were made later [1091]. These studies showed that allowing for the effects of partial redistribution gave an excellent fit of the observed line profiles with predictions.

A more recent study [535] used an essentially exact non-impact redistribution function for resonance scattering in $L\alpha$ and $L\beta$, and also a proper generalized redistribution function for resonance Raman scattering $2s \rightarrow 3p \rightarrow 1s$ (i.e., scattering from $H\alpha$ to $L\beta$) derived in [251] (described in § 10.4), and used the ETLA-based method to solve the multi-level atom problem as described in § 15.5. Some results are shown in figures 15.5–15.7.

Figure 15.5 displays the basic parameter ρ as a function of distance from the line center, $\Delta\lambda$, and of the line center optical depth, τ_0 for both $L\alpha$ and $L\beta$. The basic features are $\rho \approx 1$ for large depths and for wavelengths close to the line center, which is easily understood from the discussion in this chapter. For $L\alpha$, ρ first exhibits a slight increase for all frequencies at optical depths of the order of 10^6 to 10^8, followed by a sharp drop for smaller optical depths. This behavior is explained by the decoupling of the core and wing photons. Unlike the CFR case, photons are no longer efficiently redistributed from the core to the wings; therefore, the emissivity, and hence ρ, drops—see also figure 15.6. The peaks at the near-wing frequencies ($\Delta\lambda \approx 0.05$ nm) are the results of an interplay between the chromospheric temperature rise and transfer effects.

The behavior of $L\beta$ is different. The most conspicuous feature is a sharp rise of ρ at optical depths of the order of 10^6 for wing frequencies, which is the combined result of the chromospheric temperature rise and the redistribution of photons from the $H\alpha$ line. The decrease of ρ at small optical depths for wing frequencies is much less pronounced than in $L\alpha$ because the coherence fraction does not approach unity as for $L\alpha$, but rather $\gamma_{31} \approx A_{3p1s}/(A_{3p1s} + A_{3p2s}) \approx 0.88$. Hence, while a photon in the wing of $L\alpha$ can survive 10^3 to 10^4 consecutive scatterings before the coherence is destroyed, the coherence effects in $L\beta$ are destroyed after roughly $1/(1 - 0.88) \approx 8$ scatterings. Therefore, the coherence effects cannot develop fully for $L\beta$.

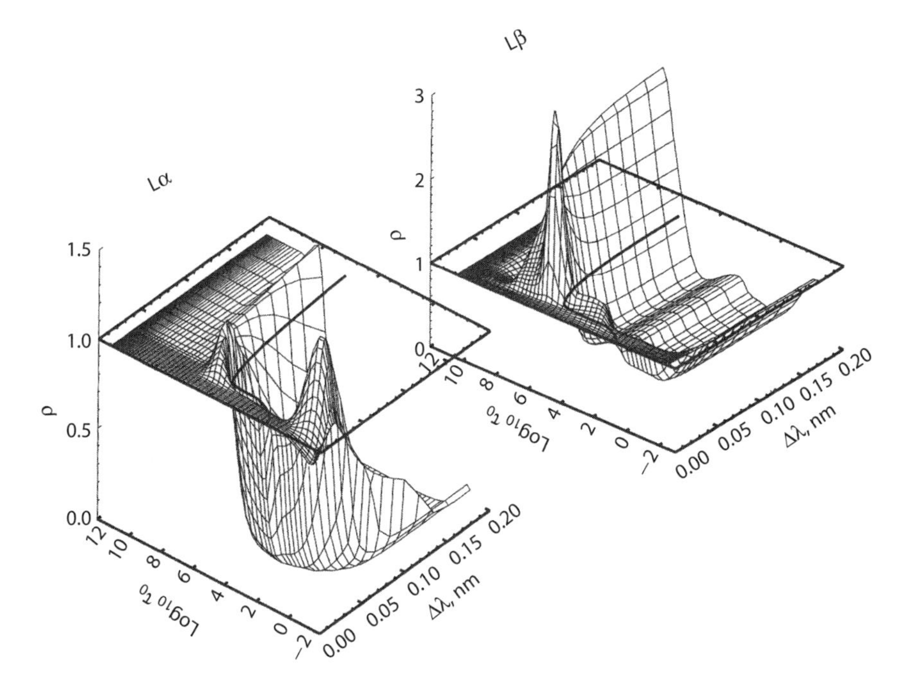

Figure 15.5 Parameter $\rho(\nu)$ as a function of distance from the line center, $\Delta\lambda$, and optical depth at line center τ_0, for the $L\alpha$ (left panel) and $L\beta$ (right panel) lines, for the solar atmosphere model VAL-C from [1119]. The heavy curves in the planes corresponding to $\rho = 1$ (i.e., to complete redistribution) indicate the location of $\tau(\Delta\lambda) = 1$. From [535].

Figure 15.6 displays monochromatic source functions for four different frequencies corresponding to the line center, the emission peak, the near wing, and the far wing. The behavior of the $L\alpha$ source function exhibits several well-known features. First, the source function is nearly identical for all frequencies at large depths. Then, progressively, frequencies closer and closer to the line center began to decouple from the core frequencies. Due to the above-mentioned lack of redistribution of photons from the core, the source function decreases below the core source function.

The source function for $L\beta$ exhibits an increase in the line wing frequencies, which results from an indirect effect of $L\alpha$ on the populations of the $n=2$ and $n=3$ levels (discussed in more detail in [535]), and also the result of the feeding of photons from $H\alpha$. The usual approximation of complete non-correlation of the photons absorbed in $H\alpha$ and re-emitted in $L\beta$ means that

$$\int_{H\alpha} P_{\rm II}(\nu'_{\rm H\alpha}, \nu_{\rm L\beta}) J(\nu'_{\rm H\alpha})\, d\nu'_{\rm H\alpha} \Big/ \phi(\nu'_{\rm H\alpha}) - \overline{J}_{\rm H\alpha} = 0. \tag{15.106}$$

However, when cross-redistribution is taken into account, the left-hand side of (15.106) is allowed to depart from zero. The function $P_{\rm II}(\nu'_{\rm H\alpha}, \nu_{\rm L\beta})$ exhibits a large degree of "coherence," in the sense that the function $P_{\rm II}/\phi$ has a sharp peak at $x'_{\rm H\alpha} = x_{\rm L\beta}$, where x is the frequency difference from the line center measured in

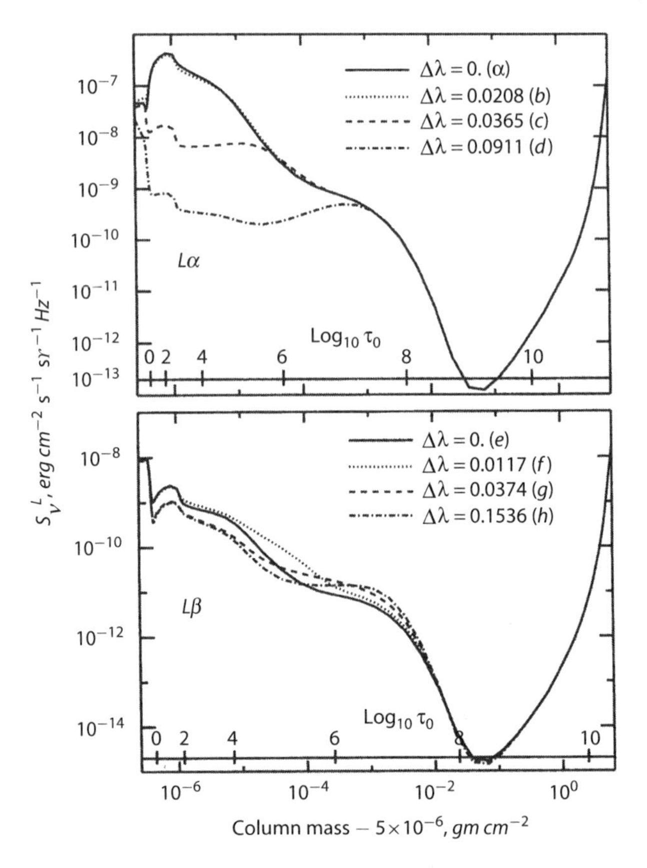

Figure 15.6 Line source functions are displayed as functions of column mass in the VAL-C model from [1119], for $L\alpha$ (upper panel) and $L\beta$ (lower panel), and for various distances $\Delta\lambda$ from the line center, as indicated. The letters *a–h* correspond to the positions in the line profiles indicated in figure 15.7. From [535].

units of the respective Doppler widths. However, the Doppler width for $H\alpha$ is 6.4 times smaller than that for $L\beta$. This means that a photon re-emitted at, say, 3 Doppler widths from the center of $L\beta$ was likely absorbed 19.2 Doppler widths from the center of $H\alpha$, which is in the continuum where the radiation intensity is larger than within the line because $H\alpha$ is an absorption line. Consequently, the left-hand side of (15.106) is positive; hence there is an excess of photons in the near wings of $L\beta$ when cross-redistribution is accounted for, as compared to the cases when it is neglected.

All these considerations are illustrated in figure 15.7, where the emergent flux in $L\alpha$ and $L\beta$ is shown. The largest effect is a significant decrease of flux in the wing of $L\alpha$ starting at about 0.03 nm in the PFR case, as found in [749, 750]. The decrease of the flux in the PFR case is a direct consequence of the behavior of the source function shown in figure 15.6. On the other hand, there is little difference between three PFR models that differ in the treatment of $L\beta$, which shows that the most

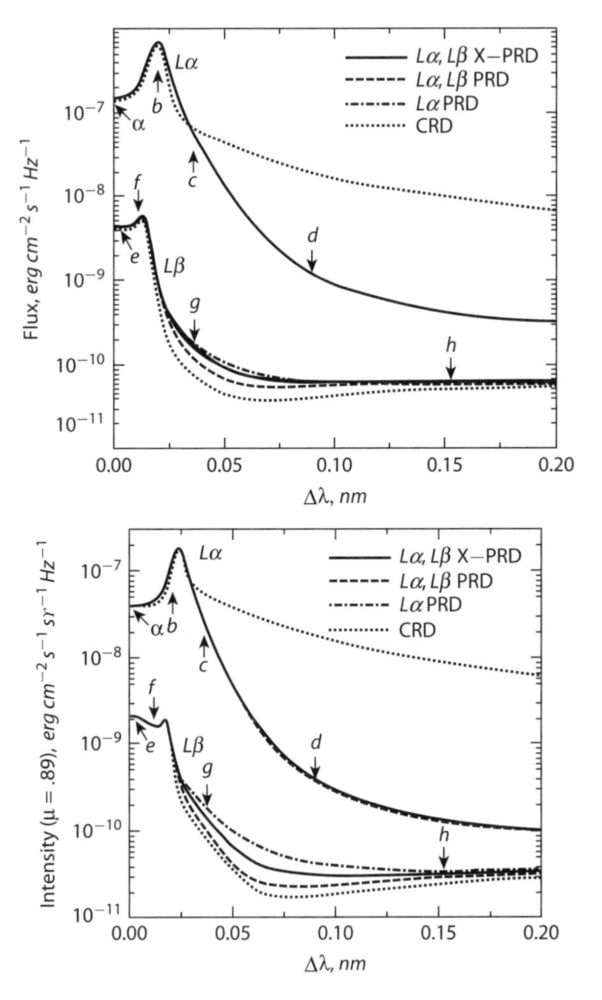

Figure 15.7 Emergent flux as a function of distance from the liner center $\Delta\lambda$ for the $L\alpha$ and $L\beta$ lines. Curves are shown for various assumption regarding the scattering. Full lines: flux computed with partial redistribution in both lines, and with cross-redistribution (resonance Raman scattering) from $H\alpha$ to $L\beta$. Dashed lines: flux computed with partial redistribution in both lines, but without cross-redistribution. Dash-dotted lines: only $L\alpha$ taken with partial redistribution, while $L\beta$ is taken with complete redistribution. Dotted lines: complete redistribution in all lines. From [535].

important mechanism for the formation of the $L\alpha$ line is the resonance scattering within the line itself.

In contrast, formation of the $L\beta$ line is very sensitive to model assumptions. When PFR is taken into account, the flux in the wings of $L\beta$ decreases as a result of the same mechanism for $L\alpha$. However, when the cross-redistribution $H\alpha \rightarrow L\beta$ is taken into account, the flux in the near wings increases, because of the feeding mechanism explained above.

Chapter Sixteen

Structural Equations

In the preceding chapters we described the properties of radiation, its interactions with material, and its transport. We turn now to the equations that determine a star's structure. In decreasing generality these are (1) 3D time-dependent Newtonian hydrodynamics, (2) 1D time-dependent flows (e.g., stellar pulsation), (3) 1D steady flows (e.g., stellar winds), (4) their static limit, in radiative and hydrostatic equilibrium (the "standard stellar atmospheres problem"), (5) a phenomenological treatment (mixing-length theory) of energy transport by convection, which competes with radiative transport in early-type stars and dominates it in late-type stars, and (6) a brief summary of the equations used to study stellar structure and evolution.

The hydrodynamic equations are *nonlinear* and admit a wide variety of flows, most of them unstable on some length and/or time scale. In their greatest generality one cannot solve them even with the fastest computers now available; hence we limit attention to simple examples. There are many good books on hydrodynamics, e.g., [36,674,1001,1007,1124], so only the barest outline of derivations is given here. We restrict the discussion to the Newtonian limit, i.e., the fluid speed v is $\ll c$, the speed of light, in which $O(v/c)$ radiative terms can be neglected. Flows with relativistic radiative terms are discussed in chapters 19 and 20 and in [198,746,877].

16.1 EQUATIONS OF HYDRODYNAMICS

Kinematics

Eulerian and Lagrangian Time Derivative

The physical quantities in a moving fluid depend on both time t [s] and a set of spatial coordinates $\mathbf{x} = (x^1, x^2, x^3)$ [cm]. The time variation of some physical quantity f at *a fixed point in space* is given by its *Eulerian time derivative* $(\partial f/\partial t)|_{\mathbf{x}}$. The time variation of some physical quantity f over a time interval Δt seen by an observer moving with a *material element* (i.e., composed of a definite set of particles) is

$$\Delta f = f(\mathbf{x} + \mathbf{\Delta x}, t + \Delta t) - f(\mathbf{x}, t) \approx \left.\frac{\partial f}{\partial t}\right|_{\mathbf{x}} \Delta t + \left.\frac{\partial f}{\partial x^i}\right|_t \Delta x^i, \tag{16.1}$$

where $x^i = x^i(t)$ are the coordinates of the material element's path, and where we adopt the usual convention that a summation is performed over repeated indices. Passing to the limit $\Delta t \to 0$, we obtain the *Lagrangian time derivative,*

$$\frac{Df}{Dt} \equiv \lim_{\Delta t \to 0} \frac{\Delta f}{\Delta t} = \frac{\partial f}{\partial t} + v^i \frac{\partial f}{\partial x^i} = \frac{\partial f}{\partial t} + \mathbf{v} \cdot \nabla f. \tag{16.2}$$

Both (16.1) and (16.2) hold for scalars and vector or tensor components. In those equations we assumed Cartesian coordinates; they can be generalized to other metric coordinate systems by replacing partial derivatives in space by covariant derivatives (cf. § 11.3); for more details refer to [746, Appendix A.3.10],

$$\frac{Df}{Dt} = \frac{\partial f}{\partial t} + v^i f_{;i}. \tag{16.3}$$

The Eulerian and Lagrangian descriptions are physically equivalent, but in some cases, one is more intuitively clear than the other. For example, by definition, the *velocity* [cm/s] of a fluid element is

$$\mathbf{v} \equiv \frac{D\mathbf{x}}{Dt}; \tag{16.4}$$

hence its *acceleration* [cm/s^2] is

$$\mathbf{a} \equiv \frac{D\mathbf{v}}{Dt}. \tag{16.5}$$

Likewise, in describing a fluid's dynamical behavior, it is sometimes conceptually easier to consider a *volume V bounded by a surface S* in the fixed laboratory frame ("Eulerian picture"). In other cases it is better to use a *volume $\mathcal{V}$ with a bounding surface $\mathcal{S}$* in the moving fluid frame ("Lagrangian picture"). To connect the two descriptions, we need two additional kinematic results.

Euler Expansion Formula

At $t = 0$ suppose an element of fluid is located at a position $\boldsymbol{\xi}$ with a laboratory-frame volume dV_0 and that it follows a path $\mathbf{x} = \mathbf{x}(\boldsymbol{\xi}, t)$, while at the same time the volume element evolves to dV. Under this mapping,

$$dV = dx^1 dx^2 dx^3 = J dV_0 \equiv J \left(\frac{x^1, x^2, x^3}{\xi^1, \xi^2, \xi^3} \right) d\xi^1 d\xi^2 d\xi^3, \tag{16.6}$$

where J is the *Jacobian* of the transformation. Taking the Lagrangian time derivative of the Jacobian, one finds the *Euler expansion formula* (see, e.g., [36, § 4.21]):

$$\frac{DJ}{Dt} = (\nabla \cdot \mathbf{v}) J. \tag{16.7}$$

Reynolds Transport Theorem

Let $F(\mathbf{x},t)$ be a scalar or vector or tensor component and $\mathcal{V}(\mathbf{x},t)$ a material fluid element. The time variation of the integral of F in $\mathcal{V}$ is

$$\mathsf{F}(t) \equiv \int_{\mathcal{V}} F(\mathbf{x},t)d\mathcal{V}. \tag{16.8}$$

Using (16.7) in (16.8), we obtain the *Reynolds transport theorem* [36, § 4.22],

$$\frac{D\mathsf{F}}{Dt} = \frac{D}{Dt}\int_{\mathcal{V}_0} F(\mathbf{x},t)J\, d\mathcal{V}_0 = \int_{\mathcal{V}}\left(\frac{DF}{Dt} + F\,\nabla\cdot\mathbf{v}\right)d\mathcal{V}. \tag{16.9}$$

Applying the divergence theorem to the second term on the right, we have

$$\frac{D\mathsf{F}}{Dt} = \int_{\mathcal{V}} \frac{\partial F}{\partial t}\, d\mathcal{V} + \int_{\mathcal{S}} F\,\mathbf{v}\cdot d\boldsymbol{\mathcal{S}}, \tag{16.10}$$

where $d\boldsymbol{\mathcal{S}}$ is an oriented element of the surface $\mathcal{S}$.

Mass Conservation

To express mass conservation it is easiest use the Eulerian picture for an arbitrary laboratory-frame volume V with a bounding surface S, which has an outward directed normal $\mathbf{n}$. The rate of increase of a quantity f (scalar, vector, or tensor component) in V equals the net inward flux of f, i.e., $-(f\mathbf{v})\cdot\mathbf{n}$, integrated over S:

$$\int_V \frac{\partial f}{\partial t}\, dV = -\int_S (f\mathbf{v})\cdot\mathbf{n}\, dS. \tag{16.11}$$

Using the divergence theorem in the right-hand side of (16.11), we find

$$\frac{\partial f}{\partial t} + \nabla\cdot(f\,\mathbf{v}) = 0. \tag{16.12}$$

In particular, if f is the mass density ρ [gm/cm^3], we have the *equation of continuity*,

$$\frac{\partial \rho}{\partial t} + \nabla\cdot(\rho\,\mathbf{v}) = 0, \tag{16.13}$$

or, equivalently, from (16.2),

$$\frac{D\rho}{Dt} + \rho\,\nabla\cdot\mathbf{v} = 0. \tag{16.14}$$

Momentum Conservation

The momentum conservation equation (*Cauchy's equation of motion*) is

$$\rho \frac{D\mathbf{v}}{Dt} = \mathbf{f} + \nabla \cdot \mathbf{T}. \tag{16.15}$$

The density of a definite fluid element times its acceleration following its motion equals the *total external force density* $\mathbf{f}$ [dynes/cm^3] from gravity and radiation, plus *surface forces* from the divergence of the *material's stress tensor* $\mathbf{T}$ [dynes/cm^2]. The stress tensor $\mathbf{T}$ is the sum of an isotropic, compressive (fluids at rest do not support tension), *hydrostatic pressure*[1] $-p\,\mathbf{I}$ and a *viscous stress* $\boldsymbol{\sigma}$ from internal frictional forces when the fluid is in motion:

$$\mathbf{T} = -p\,\mathbf{I} + \boldsymbol{\sigma}, \tag{16.16}$$

or $T^{ij} = -p\,\delta^{ij} + \sigma^{ij}$, where δ^{ij} is the Kronecker symbol. Then

$$\rho \frac{D\mathbf{v}}{Dt} = \mathbf{f} - \nabla p + \nabla \cdot \boldsymbol{\sigma}. \tag{16.17}$$

If there are to be no internal frictional forces when the fluid is in rigid rotation, one can show that $\boldsymbol{\sigma}$ must be symmetric [36, §5.13]. If we also require that viscous terms depend only linearly on the spatial gradients of the velocity (a *Newtonian fluid*), the most general tensor of second rank that satisfies these requirements is

$$\sigma_{ij} = \mu\left[\left(\frac{\partial v_i}{\partial x^j}\right) + \left(\frac{\partial v_j}{\partial x^i}\right)\right] + \lambda\left(\frac{\partial v_k}{\partial x^k}\right)\delta_{ij} = 2\mu E_{ij} + \lambda(\nabla \cdot \mathbf{v})\delta_{ij}, \tag{16.18}$$

where E_{ij} is the *rate of strain tensor*, defined as

$$E_{ij} \equiv \frac{1}{2}\left(\frac{\partial v_i}{\partial x^j} + \frac{\partial v_j}{\partial x^i}\right). \tag{16.19}$$

In (16.18) μ is the coefficient of *shear viscosity* or *dynamical viscosity* and λ is the coefficient of *dilatational viscosity* [gm/cm/s].

It is useful to rewrite (16.18) in a *trace-free* form:

$$\sigma_{ij} = \mu\left[\left(\frac{\partial v_i}{\partial x^j}\right) + \left(\frac{\partial v_j}{\partial x^i}\right) - \frac{2}{3}\left(\frac{\partial v_k}{\partial x^k}\right)\delta_{ij}\right] + \zeta\left(\frac{\partial v_k}{\partial x^k}\right)\delta_{ij}, \tag{16.20}$$

where $\zeta \equiv \lambda + \frac{2}{3}\mu$ is the coefficient of *bulk viscosity*. When the fluid dilates symmetrically, i.e., such that $(\partial v_1/\partial x^1) = (\partial v_2/\partial x^2) = (\partial v_3/\partial x^3)$, and

[1] Throughout this chapter p without a subscript denotes *gas* pressure only.

$(\partial v_i/\partial x^j)=0, i\neq j$, the term in square brackets is identically zero. Kinetic theory predicts ζ is zero for an ideal gas. (It is nonzero for a gas that is temporarily out of equilibrium, e.g., has just passed through a shock.) Thus for an ideal gas $\sigma_{ij}\equiv 0$ for symmetric dilation, as is consistent with intuition [314, p. 19] because in that case there is no slip of one part of the fluid relative to another. With the specific choice of (16.20) for $\boldsymbol{\sigma}$, the momentum-conservation equation (16.17) is the *Navier-Stokes equation.*

Material Momentum Equation

By adding $\mathbf{v}$ times the continuity equation (16.13) to (16.17), we obtain the *material momentum equation*,

$$\frac{\partial(\rho\,\mathbf{v})}{\partial t}+\nabla\cdot\left[\rho\,(\mathbf{v}\otimes\mathbf{v})+p\,\mathsf{I}-\boldsymbol{\sigma}\right]=\mathbf{f}, \tag{16.21}$$

which is in conservation form. Here $\otimes$ denotes the *outer product* of two vectors, $(\mathbf{a}\otimes\mathbf{b})_{ij}\equiv a_i b_j$.

Radiating Fluid Momentum Equation

For a non-radiating fluid, the external force density $\mathbf{f}$ is the gravitational force density $\rho\,\mathbf{g}$. For a radiating fluid, it is $\mathbf{f}=\rho\mathbf{g}+\mathbf{G}$, where $\mathbf{G}$ is the radiative force density defined in equation (11.50),

$$c\mathbf{G}\equiv\int d\nu\oint d\Omega\,\mathbf{n}\,[\chi(\mathbf{n},\nu)I(\mathbf{n},\nu)-\eta(\mathbf{n},\nu)]. \tag{16.22}$$

One can use either the expression on the right-hand side of (11.50) for $\mathbf{G}$ or the left-hand side to write the *radiating-fluid momentum equation*,

$$\frac{\partial}{\partial t}\left(\rho\,\mathbf{v}+\frac{\mathbf{F}}{c^2}\right)+\nabla\cdot\left[\rho\,(\mathbf{v}\otimes\mathbf{v})+(p\,\mathsf{I}-\boldsymbol{\sigma})+\mathsf{P}\right]=\rho\,\mathbf{g}. \tag{16.23}$$

In general, radiation stress must be represented by a full tensor P. The radiation field becomes essentially isotropic at great optical depth inside the material; there P can be replaced by the diagonal tensor $P_r\mathsf{I}$.

Energy Conservation

There are several forms of the Newtonian energy conservation equation.

Total Energy

Conservation of the *total energy* of a viscous fluid in a material volume element $\mathcal{V}$ requires that

$$\frac{D}{Dt}\int_{\mathcal{V}} \rho\left(e+\tfrac{1}{2}v^2\right)\, d\mathcal{V}$$
$$= \int_{\mathcal{V}} \mathbf{f}\cdot\mathbf{v}\, d\mathcal{V} + \int_{\mathcal{S}} \mathbf{t}\cdot\mathbf{v}\, d\mathcal{S} - \int_{\mathcal{S}} \mathbf{q}\cdot d\boldsymbol{\mathcal{S}} + \int_{\mathcal{V}} (\epsilon_N + cG^0) d\mathcal{V}. \qquad (16.24)$$

The left-hand side of (16.24) is the rate of change of internal plus kinetic energy in $\mathcal{V}$. The terms on the right are the rate of (1) work done by external forces, (2) work done by frictional forces, where $\mathbf{t} = -p\mathbf{n}$ is the surface force, (3) energy loss by thermal conduction $\mathbf{q}$ [erg/cm^2/sec], and (4) deposition of thermonuclear energy ϵ_N and radiative energy cG^0 [erg/cm^3/sec], given by equation (11.49),

$$cG_0 \equiv \int d\nu \oint d\Omega\, [\chi(\mathbf{n},\nu)I(\mathbf{n},\nu) - \eta(\mathbf{n},\nu)]. \qquad (16.25)$$

Material Total Energy Equation

Using (16.2), and applying the divergence theorem to the surface terms in (16.24), we get the *material total energy equation*:

$$\frac{\partial}{\partial t}\left[\rho\,(e+\tfrac{1}{2}v^2)\right] + \nabla\cdot\left[\rho\,(e+\tfrac{1}{2}v^2)\mathbf{v} + (p\mathsf{I}-\boldsymbol{\sigma})\cdot\mathbf{v} + \mathbf{q}\right]$$
$$= \mathbf{v}\cdot\mathbf{f} + \epsilon_N + cG^0. \qquad (16.26)$$

Radiating Fluid Total Energy Equation

Or, substituting the left-hand side of equation (11.49) for cG^0, we obtain the *radiating fluid total energy equation*:

$$\frac{\partial}{\partial t}\left[\rho\,(e+\tfrac{1}{2}v^2) + E\right]$$
$$+ \nabla\cdot\left[\rho\,(e+\tfrac{1}{2}v^2)\mathbf{v} + (p\mathsf{I}-\boldsymbol{\sigma})\cdot\mathbf{v} + \mathbf{q} + \mathbf{F}\right] = \mathbf{v}\cdot\mathbf{f} + \epsilon_N. \qquad (16.27)$$

Simpler equations are obtained if viscosity and thermal conduction are ignored. Equation (16.27) shows explicitly the fundamental fact that *all the internal, kinetic, and luminous energy in a star results, ultimately, from the release of thermonuclear energy or from work by gravitational and radiative forces* ($\mathbf{v}\cdot\mathbf{f}$).

Mechanical Energy Equation

By forming the dot product of the fluid velocity $\mathbf{v}$ with Cauchy's equation of motion (16.15), we get the *mechanical energy equation*,

$$\rho v_i \frac{Dv^i}{Dt} \equiv \tfrac{1}{2}\rho \frac{Dv^2}{Dt} = \mathfrak{f}_i v^i + v_i \frac{\partial T^{ij}}{\partial x^j}. \tag{16.28}$$

Gas Energy Equation

Subtracting the mechanical energy equation (16.28) from the material total energy equation (16.26), we obtain the *gas energy equation*,

$$\rho \frac{De}{Dt} = \frac{\partial v_i}{\partial x^j} T^{ij} - \frac{\partial q^j}{\partial x^j} + cG^0 + \epsilon_N. \tag{16.29}$$

Entropy Generation

From (16.16) and (16.20) for a Newtonian fluid we have

$$\frac{\partial v_i}{\partial x^j} T^{ij} = -p\,\nabla \cdot \mathbf{v} + \Phi, \tag{16.30}$$

where Φ is the *dissipation function*,

$$\Phi = 2\mu E_{ij}E^{ij} + \left(\zeta - \tfrac{2}{3}\mu\right)(\nabla \cdot \mathbf{v})^2. \tag{16.31}$$

Φ accounts for viscous energy dissipation by internal frictional forces in the gas and can be shown to be *always non-negative*.

Using the equation of continuity and (16.30), (16.29) becomes

$$\rho \left[\frac{De}{Dt} + p\frac{D}{Dt}\left(\frac{1}{\rho}\right)\right] = \Phi - \nabla \cdot \mathbf{q} + \epsilon_N + cG^0. \tag{16.32}$$

Thus the gas energy equation is equivalent to the *first law of thermodynamics*,

$$\frac{De}{Dt} + p\frac{D}{Dt}\left(\frac{1}{\rho}\right) = T\frac{Ds}{Dt}, \tag{16.33}$$

with an *entropy generation* term of the form

$$T\frac{Ds}{Dt} = \rho^{-1}\left(\Phi - \nabla \cdot \mathbf{q} + \epsilon_N + cG^0\right). \tag{16.34}$$

If $\mathbf{q}$ is given by *Fourier's law* $\mathbf{q} = -K_q \nabla T$, where K_q [erg/cm/sec/K] is the *coefficient of thermal conduction*, then the second term on the right-hand side of (16.34) can also be shown to be non-negative, and ϵ_N is positive. Hence in the absence of radiation (cG^0 can be either positive or negative), $(Ds/Dt) \geq 0$, consistent with the second law of thermodynamics. For an ideal gas without radiative or thermonuclear sources $(Ds/Dt) \equiv 0$.

Auxiliary Conditions

For the moment, ignore the presence of radiation. Then the dynamical equations have four independent variables (space coordinates and time) and six primary dependent variables: three velocity components, density, pressure, and internal energy. But there are only five equations: conservation of mass, three momentum components, and energy. To solve for all the primary dependent variables, additional information is needed to close the system.

Constitutive Relations

If the fluid is assumed to be in LTE it can be closed with *constitutive relations*. By introducing yet another dependent variable, the temperature, and using thermodynamics or statistical mechanics, one can derive a *mechanical equation of state*,

$$p = p\,(\rho, T, \text{chemical composition}), \tag{16.35}$$

for the gas pressure, and a *caloric equation of state*,

$$e = e\,(\rho, T, \text{chemical composition}), \tag{16.36}$$

for the internal energy. Partial ionization needs to be taken into account in the calculation of both p and e. These additional relations permit the (now seven) dependent variables to be found. With all the primary dependent variables known, secondary quantities such as viscosity, conductivity, nuclear energy generation rate, opacity, and emissivity can be computed.

Initial and Boundary Conditions

To obtain a unique solution of the time-dependent equations given above by means of a step-by-step integration in time, both *initial conditions*, i.e., values of all dependent variables over the entire spatial domain at a given instant, and *boundary conditions*, i.e., the values of, or relations among, dependent variables at the boundaries of the domain at all times, are required. The boundary conditions must be specified at two distinct points for 1D flows, at most four distinct curves for 2D flows, and at most six distinct surfaces for 3D flows. In this book, we consider only steady flows (stellar winds), static atmospheres, and quasi-static stellar interiors. In these cases initial conditions in the sense described above are irrelevant. Only an approximate initial solution is required, which is refined iteratively until the constraints of energy and momentum balance are satisfied accurately.

Radiation

When radiation is taken into account, the situation becomes much more complex. In the fluid equations, the radiative energy and momentum exchange terms G^0 and $\mathbf{G}$ between the radiation field and the material [cf. equations (11.49) and (11.50)] imply a delocalization of the solution. And to evaluate G^0 and $\mathbf{G}$ we must solve angle-frequency-dependent radiative transfer equations. If spectral lines are included, this must be done at a huge number of frequencies because Doppler shifts can desaturate the opacity where there are strong lines and shift that opacity into other spectral regions. Further, in the outer layers of a stellar atmosphere, LTE does not apply. Then the radiation field can determine the state of the material, but the material's opacity and emissivity determine the radiation field. In this case, the difficulty is not just that we have to solve large numbers of transfer equations, but we also must deal with the non-equilibrium coupling between the radiation field and the material.

16.2 1D FLOW

Mass Conservation

For a 1D flow the hydrodynamic equations are much simpler. In this case we assume planar or spherically symmetric geometry. In planar geometry, (16.13) reduces to

$$\frac{\partial \rho}{\partial t} + \frac{\partial(\rho v_z)}{\partial z} = 0. \tag{16.37}$$

In spherically symmetric geometry it becomes

$$\frac{\partial \rho}{\partial t} + \frac{1}{r^2}\frac{\partial(r^2 \rho v_r)}{\partial r} = 0. \tag{16.38}$$

Momentum Conservation

In a planar atmosphere, the momentum equation (16.23) for a radiating gas with zero bulk viscosity reduces to

$$\frac{\partial(\rho v_z)}{\partial t} + \frac{\partial(\rho v_z^2)}{\partial z} + \frac{\partial p}{\partial z} - \frac{\partial}{\partial z}\left(\tfrac{4}{3}\mu\frac{\partial v_z}{\partial z}\right) = f_z - \rho g. \tag{16.39}$$

Here g is the (downward) acceleration of gravity in cm/s^2; it always tends to decelerate an outward flow. f_z [dynes/cm^3] is the z component of the radiation force density on the material. It is positive in the optically thin regions near the surface of a star and in some cases can accelerate an outward flow (a stellar wind).

In spherically symmetric geometry the momentum equation is

$$\frac{\partial(\rho v_r)}{\partial t} + \frac{1}{r^2}\frac{\partial(r^2\rho v_r^2)}{\partial r} + \frac{\partial p}{\partial r} - \frac{1}{r^2}\frac{\partial}{\partial r}\left[\tfrac{4}{3}\mu r^3\frac{\partial}{\partial r}\left(\frac{v_r}{r}\right)\right] = f_r - \rho\frac{G\mathsf{M}}{r^2}. \tag{16.40}$$

Here G is the Newtonian gravitational constant, and M is the mass of the star.

Energy Conservation

In planar geometry, the material total energy equation (16.26) for a radiating gas with zero bulk viscosity reduces to

$$\frac{\partial}{\partial t}\left[\rho\left(e+\tfrac{1}{2}v_z^2\right)\right]+\frac{\partial}{\partial z}\left[\rho v_z\left(h+\tfrac{1}{2}v_z^2-\tfrac{4}{3}\nu\frac{\partial v_z}{\partial z}\right)-K_q\frac{\partial T}{\partial z}\right]$$
$$=-\rho v_z g+\epsilon_N+cG^0. \tag{16.41}$$

Here $h=e+p/\rho$ is the *enthalpy* of the material, and $\nu\equiv\mu/\rho$ is its *kinematic viscosity*. In spherically symmetric geometry it becomes

$$\frac{\partial}{\partial t}\left[\rho(e+\tfrac{1}{2}v_r^2)\right]+\frac{1}{r^2}\frac{\partial}{\partial r}\left\{r^2\rho v_r\left[h+\tfrac{1}{2}v_r^2-\tfrac{4}{3}\nu r\frac{\partial}{\partial r}\left(\frac{v_r}{r}\right)\right]-r^2K_q\frac{\partial T}{\partial r}\right\}$$
$$=-\rho v_r\frac{\mathrm{GM}}{r^2}+\epsilon_N+cG^0. \tag{16.42}$$

Alternatively, the radiating-fluid total energy equation is

$$\frac{\partial}{\partial t}\left[\rho\left(e+\tfrac{1}{2}v_r^2\right)+E\right]$$
$$+\frac{1}{r^2}\frac{\partial}{\partial r}\left\{r^2\rho v_r\left[h+\tfrac{1}{2}v_r^2-\tfrac{4}{3}\nu r\frac{\partial}{\partial r}\left(\frac{v_r}{r}\right)\right]-r^2\left(K_q\frac{\partial T}{\partial r}-F\right)\right\}$$
$$=-\rho v_r\frac{\mathrm{GM}}{r^2}+\epsilon_N. \tag{16.43}$$

Auxiliary Conditions

To close the system, we again need (16.35) and (16.36) for p and e as well as initial and boundary conditions.

16.3 1D STEADY FLOW

Models of stellar winds usually assume steady flow in spherical geometry. The time derivatives in the equations of § 16.2 are dropped. In some cases analytical integrals of the conservation laws can be found. For steady flow the Eulerian time derivative $(\partial f/\partial t)$ of a quantity f is $\equiv 0$, but its Lagrangian time derivative need not be zero because of advection terms $v_\ell(\partial f/\partial \ell)$ along a fluid element's path ℓ.

Mass Conservation

For steady flow, mass conservation requires $\nabla\cdot(\rho\,\mathbf{v})=0$. In planar geometry the outward *mass flux* [gm/cm^2/s] is

$$\rho v_z\equiv\dot{m}=\text{constant}, \tag{16.44}$$

where

$$m = \int_z^\infty \rho(z')\,dz' \tag{16.45}$$

is the *column mass* [gm/cm^2] above height z. In spherical geometry the *mass loss rate* (usually measured in units of $\mathsf{M}_\odot$/year) is

$$4\pi\rho r^2 v_r \equiv \dot{\mathsf{M}} = \text{constant.} \tag{16.46}$$

Momentum Conservation

In view of (16.44) and (16.46), in planar geometry the momentum conservation equation for a radiating steady flow reduces to

$$\rho v_z \frac{dv_z}{dz} + \frac{d}{dz}\left(p - \tfrac{4}{3}\mu\frac{dv_z}{dz}\right) = f_z - \rho g. \tag{16.47}$$

If we assume that as $z \to \infty, \rho \to 0$, $p \to 0$, and $v \to v_\infty$, then for an ideal gas (16.47) can be integrated to give

$$\dot{m}\,[v_z(\infty) - v_z(z)] - p(z) = \int_z^\infty f_z(z')\,dz' - gm. \tag{16.48}$$

For steady flow in spherically symmetric geometry the momentum conservation equation reduces to

$$\rho v_r \frac{dv_r}{dr} + \frac{dp}{dr} - \frac{1}{r^2}\frac{d}{dr}\left[\tfrac{4}{3}\mu r^3 \frac{d}{dr}\left(\frac{v_r}{r}\right)\right] = f_r - \rho\frac{G\mathsf{M}}{r^2}. \tag{16.49}$$

There is no analytical integral of this equation.

Energy Conservation

Again calling on (16.44) and (16.45), the material total energy equation for a radiating steady flow in planar geometry reduces to

$$\dot{m}\frac{d}{dz}\left(h + \tfrac{1}{2}v_z^2 - \tfrac{4}{3}\nu\frac{dv_z}{dz}\right) - \frac{d}{dz}\left(K_q\frac{dT}{dz}\right) = -\dot{m}g + \epsilon_N + cG^0. \tag{16.50}$$

In spherical geometry it is

$$\rho v_r \frac{d}{dr}\left[h + \tfrac{1}{2}v_r^2 - \tfrac{4}{3}\nu r\frac{d}{dr}\left(\frac{v_r}{r}\right)\right] - \frac{1}{r^2}\frac{d}{dr}\left(r^2 K_q \frac{dT}{dr}\right)$$
$$= -\rho v_r \frac{G\mathsf{M}}{r^2} + \epsilon_N + cG^0. \tag{16.51}$$

In the absence of radiative energy exchange and thermonuclear energy release, (16.50) and (16.51) can be integrated to obtain

$$\dot{\mathsf{m}}\left[h + \tfrac{1}{2}v_z^2 + gz - \tfrac{4}{3}\nu\left(\frac{dv_z}{dz}\right)\right] - K_q\left(\frac{dT}{dz}\right) = \text{constant}, \tag{16.52}$$

and

$$\dot{\mathsf{M}}\left[h + \tfrac{1}{2}v_r^2 - \frac{G\mathsf{M}}{r} - \tfrac{4}{3}\nu r\frac{d}{dr}\left(\frac{v_r}{r}\right)\right] - 4\pi r^2 K_q \frac{dT}{dr} = \text{constant}. \tag{16.53}$$

16.4 STATIC ATMOSPHERES

The goal of most work on stellar atmospheres is an accurate computation of stellar spectra. In the "standard stellar atmospheres problem," intrinsically variable stars are excluded, and the atmosphere is taken to be *static*, i.e., all velocities are zero. Normally it is assumed to have 1D geometry (planar or spherical), to be in hydrostatic equilibrium, and to be in radiative (+ convective if appropriate) equilibrium, with $\epsilon_N \equiv 0$. Even so, the problem of modeling a stellar atmosphere is more difficult than a problem in pure hydrodynamics because in addition to finding fluid variables, one must compute the radiation field for a wide range of frequencies to (a) obtain energy balance, (b) determine non-equilibrium occupation numbers of bound and free states of the material, and (c) to compare with stellar spectra.

Hydrostatic Equilibrium

Pressure Balance

In planar geometry, (16.47) with $\dot{\mathsf{m}} = 0$ becomes the *equation of hydrostatic equilibrium*,

$$\frac{dp}{dz} = f_z - \rho g, \tag{16.54}$$

or, from the definition of **G** in equation (11.51b),

$$\frac{dp}{dm} = g - \frac{1}{c\rho}\int_0^\infty \chi_\nu F_\nu \, d\nu. \tag{16.55}$$

When the radiative force can be neglected (a cool atmosphere),

$$p = g\,m + \text{constant}, \tag{16.56}$$

where m is the column mass defined in (16.45). That is, the local gas pressure is just sufficient to support the weight of the overlying layers.

Radiative Equilibrium

The static limit of the material energy equation (16.26) with $\epsilon_N \equiv 0$ is

$$cG^0 + \nabla \cdot \left(K_q \nabla T\right) = 0. \tag{16.57}$$

When there is a net deposition of radiant energy ($G^0 > 0$), a temperature gradient results, which allows heat to flow away, leading to equilibration. Alternatively, from (11.49),

$$\nabla \cdot \left[\mathbf{F} + \left(K_q \nabla T\right)\right] = 0. \tag{16.58}$$

In normal stellar atmospheres, conduction terms can be ignored.[2] Then (16.58) reduces to $G^0 \equiv 0$, or from equation (11.51),

$$\int_0^\infty (\eta_\nu - \chi_\nu J_\nu)\, d\nu = 0, \tag{16.59}$$

so the material is in radiative equilibrium, i.e., it emits exactly as much energy as it absorbs. Alternatively, from (16.58),

$$\nabla \cdot \mathbf{F} = 0; \tag{16.60}$$

so at all depths $F = \sigma_R T_{\text{eff}}^4 =$ constant in planar geometry, and $L = 4\pi r^2 F =$ constant in spherically symmetric geometry.

Integration Procedure

For simplicity, suppose we wish to construct an LTE atmosphere. Given an initial model, including a preliminary estimate of the run of the temperature with depth, we make a step-by-step integration of (16.54) or (16.55) for the prescribed surface gravity g; this gives us the the gas pressure $p_{\text{gas}}(T)$ and the material density $\rho(T)$ each depth point. We can then compute opacities and emissivities and solve frequency-dependent transfer equations. In general, the resulting model yields a radiation field that does not satisfy the constraint of radiative equilibrium. Several efficient procedures to find the temperature distribution that *does* satisfy this constraint are described in detail in chapters 17 and 18.

16.5 CONVECTION

Thus far we have assumed that energy transport in the star is by radiation. As mentioned in chapter 1, in his study of the solar photosphere K. Schwarzschild [980] derived the criterion for radiative transfer to be stable against *convection* and

[2] In stellar coronae, temperatures $\sim 10^6$ K and thermal conductivity becomes important.

found that the layers of the photosphere he studied were essentially in radiative equilibrium.[3]

Mixing Length Theory

L. Prandtl [884, 885] proposed that the chaotic motions in a convection zone be viewed as a turbulent boundary layer. L. Biermann [119] developed Prandtl's ideas into a mathematical formalism, which was generalized and extended by E. Vitense [130, 134, 1125]. The physical picture is that upward (downward) moving fluid elements have an excess (deficiency) of thermal energy relative to their surroundings. At the end of a characteristic *mixing length* ℓ, they "dissolve" smoothly into the surroundings, delivering any excess energy they possess or absorbing any deficiency. The resulting temperature gradient is smaller than if radiation were the only transport mechanism. Clearly, convection is a time-dependent 3D flow; so how can we expect to describe it in quasi-static, 1D geometry? As we shall see below, in the Sun the velocities of these elements are small. Hence there is time for radiative exchange between them. In this sense the convection pattern may be considered to be instantaneously "frozen." And if the fluid elements are small compared to the mixing length, say, a scale height, there are many fluid elements at a given height in the atmosphere, so an average over a horizontal layer (or spherical shell) should yield a reasonable estimate of the net rate of energy transport upward.

Schwarzschild Stability Criterion

Consider a small element of gas in a gravitationally stratified atmosphere. Assume it is initially hotter than, and in pressure equilibrium with, its surroundings. Because of its higher temperature, it has a lower density than the background and will tend to rise. Assume it travels upward a distance Δr, that its motion is so slow that it remains in pressure equilibrium with its surroundings, and that it exchanges no thermal energy with them. The background pressure drops as the element rises; hence it expands, and its density decreases by $(d\rho/dr)_{\rm el}\,\Delta r$. The density in the background decreases by $(d\rho/dr)_{\rm bg}\,\Delta r$. At the new position, the density in the fluid element has changed by an amount

$$\Delta\rho = \left[(d\rho/dr)_{\rm el} - (d\rho/dr)_{\rm bg}\right]\Delta r \qquad (16.61)$$

relative to the background. If $\Delta\rho > 0$, the density in the fluid element at its new position is greater than the ambient density, so the gravitational force acting on it

[3] When examined at higher resolution than Schwarzschild could attain, we see a fine pattern of *granules*, which are the tops of the rising convective cells that have become optically thin and radiate away their excess energy efficiently. They are separated by narrow darker lanes in which the cooled material flows back down.

will tend to draw it back toward its original position. Thus *the atmosphere is stable against convection if*

$$(d\rho/dr)_{\rm el} > (d\rho/dr)_{\rm bg} \quad \text{or} \quad -(d\rho/dr)_{\rm el} < -(d\rho/dr)_{\rm bg}. \tag{16.62}$$

This is the most basic statement of the *Schwarzschild stability criterion.*[4]

As we shall see in chapters 17 and 18, density is usually not taken as one of the primary dependent variables, but only as an implicit function of pressure and temperature through the equation of state. It is more useful to rewrite (16.62) in terms of gradients of the primary variables. Assume that the material is an ideal gas, that the background is in radiative equilibrium, and that the fluid element moves adiabatically (i.e., does not exchange energy with its surroundings).

In the radiative background, $\ln \rho_{\rm bg} = \ln p_{\rm bg} - \ln T_{\rm bg} + C'$. In the adiabatic fluid element, $\ln \rho_{\rm el} = (1/\gamma)\ln p_{\rm el} + C$. Using these expressions in (16.62), we find *the material is stable against convection when*

$$-(1/\gamma)(d\ln p/dr)_{\rm el} < -(d\ln p/dr)_{\rm bg} + (d\ln T/dr)_{\rm bg}, \tag{16.63a}$$

or, with pressure equilibrium between the fluid elements and the background,

$$-(d\ln T/dr)_{\rm bg} < -[(\gamma-1)/\gamma](d\ln p/dr)_{\rm el}. \tag{16.63b}$$

or

$$\nabla_{\rm bg} \equiv (d\ln T/d\ln p)_{\rm bg} < [(\gamma-1)/\gamma] = (d\ln T/d\ln p)_{\rm ad} \equiv \nabla_{\rm ad}. \tag{16.63c}$$

Convective and Adiabatic Gradients

We can recast (16.63a) into its customary form in terms of scale heights.

- The temperature near the surface of an atmosphere in radiative equilibrium varies only slowly. Then for an ideal gas in a layer where T and μ are nearly constant and radiation forces can be neglected, (16.55) becomes

$$\frac{d\ln \rho}{dz} = -\frac{\mu m_0 g}{kT} \equiv -\frac{1}{H_\rho}. \tag{16.64}$$

  Here $H_\rho \equiv kT/\mu m_0 g$ is the *density scale height*, μ is the effective mean molecular weight [see equation (17.82)], and m_0 is the atomic mass unit. Integration of (16.64) yields $\rho = \rho_0 \exp(-z/H_\rho)$, showing that density decreases exponentially outward in an isothermal layer.
- In parallel with (16.64) define the *local pressure scale height* as

$$H_P \equiv -(dr/d\ln p) = -p(dr/dp) = p/(\rho g). \tag{16.65}$$

[4] The annoying minus signs in (16.62) and (16.63a) appear because the density, pressure, and temperature gradients are all < 0 as a function of increasing radius.

- Multiplying (16.63a) through by (16.65) and referring back to (11.174), we find that in the limit of the diffusion approximation the *stability against convection* requires that

$$\nabla_{\rm rad} \equiv \left(\frac{d\ln T}{d\ln p}\right)_{\rm rad} = \frac{3\chi_R}{16}\left(\frac{T_{\rm eff}}{T}\right)^4 \frac{p}{\rho g} < \nabla_{\rm ad} \equiv \left(\frac{d\ln T}{d\ln p}\right)_{\rm ad} = \frac{\Gamma_2 - 1}{\Gamma_2}. \tag{16.66}$$

  That is, the material in a star is stable against convection only when the radiative gradient $\nabla_{\rm rad}$ *is less than the adiabatic gradient* $\nabla_{\rm ad}$. Both $\nabla_{\rm rad}$ and $\nabla_{\rm ad}$ are known because they are functions of only the dependent variables obtained at each step of the integration of the system (16.54)–(16.59).

Three factors contribute to violation of this criterion:

- To generalize (16.63ab) to a nonideal gas, we replaced the ideal gas ratio $(\gamma-1)/\gamma$ with $(\Gamma_2-1)/\Gamma_2 \equiv (\partial \ln T/\partial \ln p)_s$, where Γ_2 is the adiabatic exponent for an ionizing gas plus radiation; see, e.g., [262, § 9.12–§ 9.18], [746, § 71], [1102, § 56], [1113]. Both ionization and radiation reduce Γ_2, and hence lower the adiabatic gradient. For example, for a monatomic perfect gas, $\gamma = \frac{5}{3}$, and $\nabla_{\rm ad} = \frac{2}{3}/\frac{5}{3} = 0.4$; for pure radiation, $\gamma = \frac{4}{3}$, and $\nabla_{\rm ad} = \frac{1}{3}/\frac{4}{3} = 0.25$; when hydrogen ionizes, $\Gamma_2 \approx 1.1$ and $\nabla_{\rm ad} \sim 0.1$. Unsöld [1099] pointed out that ionization of hydrogen must produce an extensive subphotospheric convection zone in the Sun.
- Instability in ionizing material is enhanced not only when $\Gamma_2 \to 1$, but also when $\chi_{\rm ion}/kT$ becomes ~ 1 for some element, and its excited states become populated; see (4.20). If this element is a major constituent of the material, then χ_R increases by a large factor. From (11.174) we see that for a given temperature gradient, radiation flow is severely inhibited when $\chi_R \gg 1$, so energy will be transported more efficiently by convection.
- The basic point is that *convection is a very efficient transport mechanism.* When either (a) the material ionizes, and energy that would have been transported by radiation gets "soaked up" by the material, or (b) the opacity is large, so that the efficiency of radiative transport is strongly inhibited, convection then takes over.

Convective Flux

To describe the process, we need the following gradients: $\nabla_{\rm ad}$, the adiabatic gradient; $\nabla_{\rm el}$, the gradient in the convective elements; ∇, the actual gradient in the final state of the material when radiation plus convection transport the total flux; and $\nabla_{\rm rad}$, the radiative gradient that would hold in the absence of convection. In general,

$$\nabla_{\rm ad} \le \nabla_{\rm el} \le \nabla \le \nabla_{\rm rad}. \tag{16.67}$$

When convection occurs, the temperature gradient is reduced below $\nabla_{\rm rad}$. It continues to decrease until the radiative and convective fluxes added together reach the value necessary to satisfy the requirement of energy conservation in (16.82). The goal is to determine the final gradient ∇.

Consider a rising element of material. If ΔT is the difference between its temperature and the background's, the excess energy it delivers per unit volume when it dissolves into the background is $\rho c_p \Delta T$. Elements moving upward a distance Δr with an average speed $\langle v \rangle$ transport (assuming pressure equilibrium) a net flux

$$F_{\text{conv}} = \rho c_p \langle v \rangle \Delta T = \rho c_p \langle v \rangle \left[\left(-\frac{dT}{dr} \right) - \left(-\frac{dT}{dr} \right)_{\text{el}} \right] \Delta r. \tag{16.68}$$

The fluid elements passing through a given level in the material are randomly distributed over the paths they travel; take $\Delta r = \ell/2$ as the average over all elements. Using (16.65) to rewrite (16.68) in terms of the gradients defined in (16.67), we get

$$F_{\text{conv}} = \tfrac{1}{2} \rho c_p \langle v \rangle T \left(\nabla - \nabla_{\text{el}} \right) \left(\ell / H_P \right). \tag{16.69}$$

To estimate $\langle v \rangle$, equate the work done by buoyancy forces on a fluid element to its kinetic energy plus the work lost to "friction" in pushing aside other turbulent elements. If $\Delta\rho$ is the density difference between the element and its surroundings, the buoyant force is $f_b = -g\,\Delta\rho$. To account for ionization effects and radiation, allow μ in (16.64) to be variable and write $d \ln \rho = d \ln p - Q d \ln T$, where $Q \equiv 1 - (\partial \ln \mu / \partial \ln T)_p$. Then in pressure equilibrium (i.e., $\Delta p \equiv 0$), $\Delta\rho = -Q\rho \Delta T / T$ and

$$f_b = \left(\frac{\rho g Q}{T} \right) \Delta T = \left(\frac{\rho g Q}{T} \right) \left[\left(-\frac{dT}{dr} \right) - \left(-\frac{dT}{dr} \right)_{\text{el}} \right] \Delta r. \tag{16.70}$$

The buoyancy force is linear in the displacement Δr. Integrating over a total displacement Δ and averaging the result over all elements by setting $\Delta = \frac{1}{2}\ell$, the average work done on all elements passing a given point is

$$\langle w \rangle = \int_0^{\Delta} f_b(\Delta r)\, d(\Delta r) = \tfrac{1}{8} \left(\rho g Q H_P \right) \left(\nabla - \nabla_{\text{el}} \right) \left(\ell / H_P \right)^2. \tag{16.71}$$

If half this work is converted to kinetic energy, i.e., $\frac{1}{2}\langle w \rangle \approx \frac{1}{2} \rho \langle v \rangle^2$, we find

$$\langle v^2 \rangle = \tfrac{1}{8} \left(g Q H_P \right) \left(\nabla - \nabla_{\text{el}} \right) \left(\ell / H_P \right)^2. \tag{16.72}$$

The left-hand side of (16.72) is quadratic in $\langle v \rangle$, and the right-hand side is quadratic in the displacement; hence it applies to both upward- and downward-moving fluid elements. Using (16.72) in (16.69), we find

$$F_{\text{conv}} = \left(g Q H_P / 32 \right)^{1/2} \rho c_p T \left(\nabla - \nabla_{\text{el}} \right)^{3/2} \left(\ell / H_P \right)^2. \tag{16.73}$$

An uncertainty is the value to choose for ℓ, which is essentially a parameter. Usually ℓ is specified in terms of H_P, with $\ell/H_P \sim O(1)$. In favorable cases, from a comparison of the properties of a series of convective models with data from real stars, it is possible to choose a reasonable value for ℓ.

Efficiency of Convection

It is illuminating to make some order of magnitude estimates of the efficiency of convection deep within a stellar interior. Suppose we adopt values for the physical variables at a radius $r = \frac{1}{2}R_\odot$ in a typical solar model [18, §76]. From (16.68) we find that the temperature gradient needed to transport the entire solar flux at this depth is $\Delta\nabla T \equiv [(dT/dr) - (dT/dr)_{\rm ad}] \sim 8.5 \times 10^2/\ell^{4/3}$ K/cm, where ℓ is the mixing length. If we take $\ell \approx 2H_P \sim 10^9$ cm, we find $\Delta\nabla T \sim 8.5 \times 10^{-10}$, which is minuscule compared to the average temperature gradient $T_c/R_\odot \approx 2 \times 10^{-4}$ in the Sun.

Over an entire mixing length, the temperature difference that would transport the total solar flux by convection amounts to $\ell \times \Delta\nabla T \sim 1$ K! Clearly, convection in the deep interior is very efficient, so when (16.66) is violated, one simply switches from the radiative gradient $\nabla_{\rm rad}$ to the adiabatic gradient $\nabla_{\rm ad}$ and continues the integration of the system (16.66)–(16.69). If (16.66) is satisfied once again, one switches back to the radiative gradient.

With the same solar model, (16.72) yields $\langle v \rangle \sim .05$ km/sec, which is much smaller than typical thermal velocities (several km/sec) in the solar interior. The turnover time for a typical convective element is $\ell/\langle v \rangle \sim 1.4 \times 10^6$ sec, or about 12 days. This time interval is many orders of magnitude smaller than a nuclear timescale, which implies that a convective zone is well mixed. If thermonuclear reactions alter the composition anywhere in the zone, the entire zone will equilibrate rapidly to the new composition.

Closure of the System

Unfortunately, switching from the radiative gradient to the adiabatic gradient and back fails to apply in regions where there are very steep density and/or temperature gradients, e.g., near the surface of a star. In such cases we must determine the actual gradient ∇ reached when both radiation and convection transport the flux. Observe that (16.73) provides only one equation in two unknowns: ∇ and $\nabla_{\rm el}$. Another relation is needed in order to solve for these gradients in terms of $\nabla_{\rm ad}$ and $\nabla_{\rm rad}$.

Following Unsöld, we can obtain it by considering the efficiency of the convective transport. The temperature in a rising fluid element exceeds that in the ambient atmosphere (the reason for a net energy transport). The element can radiate into its surroundings, thus diminishing its excess energy and decreasing the amount it yields when it dissolves into its surroundings at the end of a mixing length. We describe this loss with an "efficiency ratio,"

$$\varepsilon \equiv \frac{\text{excess energy at time of dissolution}}{\text{energy lost by radiation during lifetime of element}}. \tag{16.74}$$

The excess energy in a fluid element is proportional to $(\nabla - \nabla_{\text{el}})$. Had it moved adiabatically, its energy content would be proportional to $(\nabla - \nabla_{\text{ad}})$. Its radiation loss is proportional to $(\nabla - \nabla_{\text{ad}}) - (\nabla - \nabla_{\text{el}}) = (\nabla_{\text{el}} - \nabla_{\text{ad}})$, so

$$\varepsilon = (\nabla - \nabla_{\text{el}})/(\nabla_{\text{el}} - \nabla_{\text{ad}}). \tag{16.75}$$

ε can also be computed in terms of local variables. The excess energy content of a fluid element with volume V and excess temperature ΔT is $\rho c_p V \Delta T$. The radiative loss depends on whether it is optically thick or thin. In the optically thin limit, where $\tau_{\text{el}} \equiv \chi_R \ell \ll 1$, the rate of energy loss from a volume V is $de/dt = [4\pi \chi_R \Delta B(T)]V$. Using equation (4.69) for $B(T)$, an excess temperature $\approx \frac{1}{2}\Delta T$ for the average over all elements, and taking $(\ell/\langle v\rangle)$ for the element's lifetime, we have

$$\varepsilon_{\text{thin}} = \frac{\rho c_p V \Delta T}{(4\pi \chi_R)(4\sigma_R T^3/\pi)(\Delta T/2)V(\ell/\langle v\rangle)} = \frac{\rho c_p \langle v\rangle}{8\sigma_R T^3 \tau_{\text{el}}}. \tag{16.76}$$

At the opposite extreme, $\tau_{\text{el}} \gg 1$, we use the diffusion approximation to estimate the flux lost by an element of size ℓ and surface area A, which has a temperature excess ΔT. Take $(-dT/dr) \approx (\Delta T)/\ell$, and a lifetime $(\ell/\langle v\rangle)$. Then

$$\varepsilon_{\text{thick}} = \frac{\rho c_p V \Delta T}{[(16\sigma_R T^3/3\chi_R)(\Delta T/\ell)A(\ell/\langle v\rangle)]} = \frac{3(V/A)\chi_R(\rho c_p \langle v\rangle)}{16\sigma_R T^3}. \tag{16.77}$$

The ratio (V/A) depends on the element's geometry; if we assume it is spherical $(V/A) = \ell/3$, then

$$\varepsilon_{\text{thick}} = \left(\tfrac{1}{2}\tau_{\text{el}}\rho c_p \langle v\rangle\right)/(8\sigma_R T^3). \tag{16.78}$$

A reasonable interpolation between the two extreme cases is

$$\varepsilon = (\rho c_p \langle v\rangle/8\sigma_R T^3) \times \left[\left(1 + \tfrac{1}{2}\tau_{\text{el}}^2\right)/\tau_{\text{el}}\right]. \tag{16.79}$$

Combining (16.75) and (16.79), and substituting (16.72) for $\langle v\rangle$, we obtain

$$\frac{\nabla_{\text{el}} - \nabla_{\text{ad}}}{(\nabla - \nabla_{\text{el}})^{1/2}} = \frac{16\sqrt{2}\,\sigma_R T^3}{\rho c_p (g Q H_P)^{1/2}(\ell/H_P)} \times \frac{\tau_{\text{el}}}{\left(1 + \frac{1}{2}\tau_{\text{el}}^2\right)} \equiv \mathcal{B}. \tag{16.80}$$

Using this equation, one can determine $(\nabla - \nabla_{\text{el}})$ needed for evaluating the convective velocity and flux. Noting that $(\nabla_{\text{el}} - \nabla_{\text{ad}}) = (\nabla_{\text{el}} - \nabla) + (\nabla - \nabla_{\text{ad}})$, (16.80) can be written as $x^2 + \mathcal{B}x - (\nabla - \nabla_{\text{ad}}) = 0$, where $x \equiv (\nabla - \nabla_{\text{el}})^{1/2}$. Solving this equation for x, one obtains

$$\nabla - \nabla_{\text{el}} = \tfrac{1}{2}\mathcal{B}^2 + (\nabla - \nabla_{\text{ad}}) \mp \mathcal{B}\left[\tfrac{1}{4}\mathcal{B}^2 + (\nabla - \nabla_{\text{ad}})\right]^{1/2}. \tag{16.81}$$

Finally, the convective and radiative fluxes must sum to the correct total flux:

$$F_{\text{rad}} + F_{\text{conv}} = \sigma_R T_{\text{eff}}^4. \tag{16.82}$$

Convective transport is very efficient in a star's interior; small changes in the temperature gradient can transport a star's entire energy flux. But in its outer layers, the convective elements become optically thin, and this phenomenological approach may yield an inaccurate description of its atmospheric structure and an imprecise value for the convective flux.

16.6 STELLAR INTERIORS

The nuclear-burning timescale of a star is huge (10^6–10^{11} years) compared to its dynamical timescales (days to $\sim$ a year). For most of its life, a star's structure can be represented by a sequence of *quasi-static* models obtained by solving a set of nonlinear ordinary differential equations. But even in that limit, changes in its composition resulting from thermonuclear reactions can proceed on a much shorter timescale and must be taken into account. A good discussion of stellar interiors is given in [619].

Basic Equations

Mass Conservation

If ρ_r is the density at radius r, the mass increment dM_r contained in a shell $(r, r+dr)$ is

$$\left(dM_r/dr\right) = 4\pi r^2 \rho_r. \tag{16.83}$$

Hydrostatic Equilibrium

In hydrostatic equilibrium,

$$\left(dP_r/dr\right) = -GM_r\rho_r/r^2, \tag{16.84}$$

where $P_r \equiv p_{\text{gas}}(r) + p_{\text{rad}}(r)$ is the total pressure (gas + radiation) at r.

Energy Generation

A star's luminosity is normally set by the energy released from thermonuclear reactions. But when the fuel for a particular reaction is exhausted, the main source of energy release is a (generally brief) gravitational contraction of the star as a whole. The increment in luminosity dL_r resulting from an energy release rate ϵ_r [erg/s/cm^3] by all sources is

$$\left(dL_r/dr\right) = 4\pi r^2 \epsilon_r. \tag{16.85}$$

Energy Transport

In regions of a stellar interior that are in radiative equilibrium, a photon mean free path is generally short, so we can use the diffusion-approximation solution of the transport equation to determine the temperature gradient. From (11.174),

$$\left(dT_r/dr\right) = -3\chi_R H_r/\left(dB_r/dT_r\right). \tag{16.86}$$

By definition $H_r = F_r/4\pi = L_r/16\pi^2 r^2$, and $B_r = (a_R c/4\pi)T_r^4$; hence the temperature gradient is

$$\left(dT_r/dr\right) = -3\chi_R L_r/(16\pi a_R c r^2 T_r^3). \tag{16.87}$$

Here we used the Rosseland mean opacity χ_R [cm^{-1}] defined in (11.75) because $1/\chi_R$ is a photon mean free path. In stellar interiors work, it is customary to use *opacity per unit mass*: $\kappa_R' \equiv \chi_R/\rho$ [cm^2/gm].

Bear in mind that the diffusion approximation is valid only at great optical depth. If the material is opaque, i.e., $1/\chi_\nu \lesssim 1$, the radius given by (16.87) differs negligibly from the radius of the star, i.e., where the photosphere is formed ($\tau \approx 1$). But in stars with distended envelopes, (16.87) has to be replaced by a more accurate boundary condition.

Independent Variable

The independent variable in equations (16.83)–(16.87) can be either r or M_r. It is advantageous to use M_r, because then a definite total mass for the model can be specified at the outset. The transformation from r to M_r is made by inverting (16.83) and dividing (16.84)–(16.87) by $4\pi r^2 \rho_r$. The system of equations is then the following:

$$\left(dr \,/\, dM_r\right) = 1/4\pi r^2 \rho_r, \tag{16.88}$$

$$\left(dP_r/dM_r\right) = -GM_r/4\pi r^4, \tag{16.89}$$

$$\left(dL_r/dM_r\right) = \epsilon/\rho_r, \tag{16.90}$$

$$\left(dT_r/dM_r\right) = -3\chi_R L_r/64\pi^2 a_R c r^4 \rho_r T_r^3. \tag{16.91}$$

Calculation of Stellar Structure

The Mathematical Problem

Mathematically, the calculation of stellar structure poses a two-point boundary value problem for a system of four first-order differential equations. We have already encountered two-point boundary conditions in solving the radiative transfer equation in § 11.4 and § 11.6. At a star's center, the mass $M_r \equiv 0$ and luminosity $L_r \equiv 0$ because the central point has zero volume. But the temperature and pressure $T(0)$ and $P(0)$ at $r = 0$ are unknown eigenvalues, which have to be determined consistently with the boundary conditions at the surface. At the surface we know that $M(r) \equiv \mathsf{M}$, the prechosen mass, and we can estimate $P(\mathsf{M})$ and $T(\mathsf{M})$ from approximate model atmospheres, but $\mathsf{L}(\mathsf{M})$ and $\mathsf{R}(\mathsf{M})$ are unknown eigenvalues.

The computational scheme used today is the Henyey method [480,481]. In it, the differential equations (16.88)–(16.91) are replaced by nonlinear difference equation representations on a mass-zone grid, which allow for the two-point boundary conditions. A detailed description of this procedure is given in [437,619]. The differential equations are first order. Thus the system of difference equations has a *bidiagonal* structure of (4×4) matrices for the four dependent variables $r(M_r)$, $P(M_r)$, $T(M_r)$, and $L(M_r)$.

The equations are nonlinear. To solve them, we linearize them by replacing a dependent variable x, in the difference equations with $x_0 + \Delta x$, as described in §14.5. The resulting algebraic equations are expanded to first order in the Δx and higher-order terms are discarded. The zero-order terms (i.e., in x_0 only) are put on the right-hand side as source terms. The result is a *linear system of algebraic equations*, which can be solved for the Δx's by Gaussian elimination. If the initial estimate of x_0 is close to the true value x, the method converges quadratically (a fractional error ϵ in the first iteration becomes ϵ^2, ϵ^4, etc. in successive iterations.) In chapters 17 and 18 the same method in a more complicated form is used to compute LTE and NLTE model atmospheres. Henyey's method is powerful, reliable, and able to handle stellar structures of great complexity.

Initialization

As mentioned above, the Henyey method requires an initial model of a star's structure, and the more accurate that model, the faster the convergence. In this connection, obtaining good estimates for the boundary conditions is nontrivial. In modeling a whole star, it could be argued that its atmosphere can be viewed as merely a thin outer skin (recall the remark by Salpeter quoted in chapter 1). We know that pressure and temperature at the boundary are extremely small compared to their values at the center. One might guess that to first order the state of the atmosphere results from conditions set in the interior of the star, which contains essentially all the star's mass and is where energy generation takes place. So in early work, "zero boundary conditions," i.e., $T(\mathsf{M}) = 0$ and $P(\mathsf{M}) = 0$, were often used. Schwarzschild [982, § 11] showed that if the outer envelope is in radiative equilibrium, even these crude "zero" boundary conditions actually do give envelope solutions that converge rapidly to the same interior solution (see [619, figure 10.2]). But it should be noted that the structure in the atmospheric layers themselves is markedly different; see [982, figure 11.1]. If the outer layers are in radiative equilibrium, better estimates of $T(\mathsf{M})$ and $P(\mathsf{M})$ can be obtained by using approximate model atmospheres; see [619, § 10] for details. However, if there is a strong ionization/convection zone in the envelope, then, as Schwarzschild put it, "this convenient convergence does not hold," and the structure of deep interior parts of the star depends sensitively on the surface boundary conditions. A similar problem arises at the center: a test calculation with trial values for $P(0)$ and $T(0)$ most likely will not reach the correct surface values, because outward integrations that differ only a little at the center diverge strongly as they approach the surface.

The historical method [982] for dealing with the situation was to nondimensionalize all the equations and make a set of inward integrations for choices of parameters that represent $T(\mathsf{M})$ and $P(\mathsf{M})$. The values and slopes of these solutions are matched to a different set of interior solutions integrated outward from the center for values of parameters that represent choices of $T(0)$ and $P(0)$. This approach works for stars near the main sequence, which are nearly homogeneous chemically, but breaks down as nuclear reactions change the composition of the innermost layers of the envelope and the core. Then we apply the more powerful Henyey method. To start it, we can use approximate model atmospheres at the surface and make analytical expansions for the behavior of the dependent variables near the core; see [619, § 10] for details.

Observations show that great changes in the structure of a star as a whole and its atmosphere occur as it ages. As massive stars age, they move quickly away from the main sequence and become much more luminous, their outer envelope expands by a huge factor, and they become red supergiants. Less massive stars linger near the main sequence for a longer time, but again increase in luminosity and radius to become red giants. These distended envelopes are all strongly unstable to convection, for which we do not yet have a fully satisfactory theory. Realistic model atmospheres, in some cases including stellar winds, are now used for the outer boundary condition in many studies of stellar interior models; see, e.g., [233, 265, 642, 643, 700, 725, 950, 969, 1051, 1080].

Auxiliary Conditions

In addition to partial ionization in the equation of state, one must account for *configurational terms* that correct for multi-particle interactions in the plasma [1079, 1254–1256] and degeneracy of the electron gas in the high-density interior [619, chapters 13–16]. In the interior of a star radiation is very nearly isotropic and in thermal equilibrium, so radiative forces in the equation of hydrostatic equilibrium are given by the gradient of radiation pressure $p_{\text{rad}} = \frac{1}{3}a_R T^4$. In calculations of stellar interiors, tables are usually used for the equation of state and the Rosseland mean opacity, and the emissivity is computed from the Kirchhoff-Planck relation. An expression for energy release by thermonuclear reactions is also needed:

$$\epsilon_N = \epsilon_N(\rho, T, \text{chemical composition}). \tag{16.92}$$

The thermonuclear source term is actually a reaction network. As a star's composition changes when, say, hydrogen is transmuted to helium or helium to carbon, etc., different reactions come into play.

Chapter Seventeen

LTE Model Atmospheres

In this chapter we use the information presented above about the interaction of radiation with, and its transport through, material to construct mathematical models of stellar atmospheres. We consider planar models having an imposed radiation flux at "infinite" optical depth, which are in hydrostatic, radiative, and local thermodynamic equilibrium, and touch briefly on those having a convective zone.

By comparing a star's observed continuum energy distribution and the profiles and equivalent widths of spectral lines of different chemical elements to those computed with these models, one can estimate its effective temperature, surface gravity, and chemical composition. Classical descriptions of this procedure are given in [20–22, 777, 782]. *But we caution again that LTE models do not accurately represent the physical state of a star's outermost layers, so the inferred values of* T_{eff}, $\log g$, *and element abundances can have substantial errors.* Nevertheless, they (1) illustrate the physical requirements to be met in constructing model atmospheres, (2) introduce some of the mathematical methods needed, and (3) motivate the development of the more physically realistic models described in chapter 18.

17.1 GRAY ATMOSPHERE

A *gray atmosphere* is assumed to be composed of a hypothetical material having an opacity that does not vary with frequency. Of course, no such substance exists, so the "gray atmosphere problem" should be viewed more as a study of the methodology of radiative transfer than as providing a model for a real stellar atmosphere. It merits study because its exact solution has been found; thus it provides a standard against which the results of numerical methods can be compared. With the additional assumption of LTE, it gives an estimate of an atmosphere's temperature distribution that can be used as a starting solution in constructing non-gray atmospheres. That temperature structure is not accurate near a star's surface because the opacity of stellar material actually varies rapidly with frequency, and the state of the material departs from LTE. But the solution is valid at great depth on a Rosseland mean optical depth scale. In addition, the gray problem is important for neutron transport in heavy-water nuclear reactors; indeed, its exact solution first appeared in the neutron transport literature; see [711, 872, 873].

Transfer and Moment Equations

In a gray planar medium, the transfer equation (11.71) is

$$\mu \frac{\partial I_{\mu\nu}}{\partial \tau} = I_{\mu\nu} - S_{\nu}. \tag{17.1}$$

Integrating (17.1) over frequency, defining

$$\int_0^\infty I_{\mu\nu}\,d\nu \equiv I_\mu \quad \text{and} \quad \int_0^\infty S_\nu\,d\nu \equiv S, \tag{17.2}$$

and noting that τ does not depend on ν, we have

$$\mu \frac{\partial I_\mu}{\partial \tau} = I_\mu - S. \tag{17.3}$$

The constraint of radiative equilibrium, equation (11.67), requires that

$$\int_0^\infty \chi_\nu S_\nu\,d\nu \equiv \int_0^\infty \chi_\nu J_\nu\,d\nu. \tag{17.4}$$

For gray material, $\chi_\nu = \text{constant}$; hence $S = J \equiv \int_0^\infty J_\nu\,d\nu$. Thus

$$\mu \frac{\partial I_\mu}{\partial \tau} \equiv I_\mu - J. \tag{17.5}$$

This is a linear, homogeneous, integro-partial-differential equation for I_μ. At first sight it appears to imply that the gray atmosphere problem is a pure scattering problem. Actually it is not; only because the material is gray and in radiative equilibrium can we set $S \equiv J$. Note that because it is homogeneous in the radiation field, (17.5) does not set the scale of I_μ.

The zeroth angular moment of (17.5) is $dH/d\tau = 0$, which shows that the Eddington flux $H \equiv F/4\pi$ is constant as a function of depth. So at large optical depth we take F, the outward energy flux generated by thermonuclear reactions in the deep interior of the star, as given, which provides a lower boundary condition on the solution. With the additional assumption of LTE, we can write

$$J(\tau) = S(\tau) = B[T(\tau)] = \sigma_R T(\tau)^4/\pi, \tag{17.6}$$

which allows us to associate a temperature with the radiative-equilibrium radiation field. Using Schwarzschild's equation (11.109), we can also write

$$J(\tau) \equiv \Lambda_\tau[S(t)] = \Lambda_\tau[J(t)]. \tag{17.7}$$

This linear integral equation is known as *Milne's equation*; the gray atmosphere problem is also called the *Milne problem*. The solution of (17.7) satisfies *both* the transfer equation and the constraint of radiative equilibrium.

The first angular moment of (17.5) is $dK/d\tau = H$; because H is constant, we get the exact integral

$$K(\tau) = H(\tau + C), \tag{17.8}$$

which in physical terms shows that radiation pressure in a planar gray medium increases linearly with optical depth. It also helps us to choose an appropriate functional form for $S(\tau) \equiv J(\tau)$ because we know that in the diffusion limit $J(\tau) \to 3K(\tau) \to 3H\tau$ to very high accuracy as $\tau \to \infty$.[1]

[1] The same result is found even at the surface in Schuster's two-stream approximation; see § 11.6.

Thus we take $J(\tau)$ to have the general form

$$J(\tau) = 3H[\tau + q(\tau)] = \frac{3}{4\pi}\sigma_R T_{\text{eff}}^4[\tau + q(\tau)], \tag{17.9}$$

where the *Hopf function* $q(\tau)$ [507–511] remains to be determined. Combining (17.8) and (17.9), we see that

$$\lim_{\tau\to\infty}[\tfrac{1}{3}J(\tau) - K(\tau)] = H \lim_{\tau\to\infty}[\tau + q(\tau) - \tau - C] \equiv 0; \tag{17.10}$$

hence $C \equiv q(\infty)$. We thus obtain the exact expression

$$K(\tau) = H[\tau + q(\infty)]. \tag{17.11}$$

The solution of the gray atmosphere problem revolves entirely around the determination of $q(\tau)$.

Approximate Solutions

Eddington Approximation

In the absence of knowledge of the mathematical form of the Hopf function, Eddington [302, p. 320] proposed that as a trial solution we assume it is constant as a function of depth. Then, at all depths,

$$J_{\text{E}}(\tau) \equiv 3H(\tau + q_{\text{E}}). \tag{17.12}$$

To evaluate q_{E}, we demand that the emergent flux $H(0)$ computed using this expression for $S(\tau)$ in (11.133b) equal the nominal value H:

$$3H\Phi_0(0 + q_{\text{E}}) = H\left[1 - \tfrac{3}{2}E_4(0) + \tfrac{3}{2}E_3(0)\,q_{\text{E}}\right] = \left(\tfrac{1}{2} + \tfrac{3}{4}q_{\text{E}}\right)H \equiv H, \tag{17.13}$$

which implies that $q_{\text{E}} = \frac{2}{3}$, so in the Eddington approximation,

$$J_{\text{E}}(\tau) = 3H(\tau + \tfrac{2}{3}). \tag{17.14}$$

- *Accuracy*: The value $q_{\text{E}}(\tau) \equiv 0.666\ldots$ is near the average of the exact values $q(0) = 1/\sqrt{3} \approx 0.577$ and $q(\infty) \approx 0.710$. Compared to the exact solution, the fractional error in q_{E} is $+15.5\%$ at $\tau = 0$ and $-6.2\,\%$ at $\tau = \infty$. A table of the errors at several values of τ is given in [633, p. 89].
- *Emergent intensity*: Using (17.14) in (11.101), the emergent intensity is

$$I_{\text{E}}(0,\mu) = 3H\int_0^\infty \left(\tau + \tfrac{2}{3}\right) e^{-\tau/\mu}\, d\tau/\mu = 3H\left(\mu + \tfrac{2}{3}\right), \tag{17.15}$$

which is consistent with the Eddington-Barbier relation (11.102).

- Further, note that

$$I_E(0,0) = J_E(0), \tag{17.16}$$

a result that also holds for the exact solution. Indeed, in a planar atmosphere the relation $I(\tau, \mu=0) = S(\tau) \equiv J(\tau)$ must hold for *all* τ, because a ray at $\mu=0$ always samples the same depth in the atmosphere.

- *Limb darkening*: As seen by an external observer, the center of a star's disk corresponds to $\Theta = 0^\circ$ or $\mu = 1$, and the edge (or *limb*) of the disk to $\Theta = 90^\circ$ or $\mu = 0$. The ratio $\phi(\mu) \equiv I(0,\mu)/I(0,1)$ is the *limb-darkening function*. In the Eddington approximation,

$$\phi_E(\mu) = I_E(0,\mu)/I_E(0,1) = \tfrac{3}{5}\left(\mu + \tfrac{2}{3}\right), \tag{17.17}$$

which predicts that the intensity at the limb is about 40% of the disk-center value, in fair agreement with observations of the Sun.

- *Eddington factor*: In the Eddington approximation, the Eddington factor defined in equation (11.61) is $f_\nu^K(\tau) \equiv K_\nu(\tau)/J_\nu(\tau) \equiv \frac{1}{3}$, and the flux boundary ratio defined in equation (11.62) is $g_\nu^K(0) \equiv H_\nu(0)/J_\nu(0) = \frac{1}{2}$; these can be used as starting values in iterative solutions of the transfer equation.

Two-stream Approximation

Following Schuster's analysis of the second-order transfer equation (see § 12.2), one could take as a trial solution $I(\tau,\mu) \equiv I^+(\tau)$ for $0 \le \mu \le 1$, and $I(\tau,\mu) \equiv I^-(\tau)$ for $-1 \le \mu \le 0$ at all depths. Then

$$J(\tau) \equiv \tfrac{1}{2}\int_{-1}^{1} I(\tau,\mu)\,d\mu = \tfrac{1}{2}[I^+(\tau) + I^-(\tau)], \tag{17.18a}$$

$$H(\tau) \equiv \tfrac{1}{2}\int_{-1}^{1} I(\tau,\mu)\mu\,d\mu = \tfrac{1}{4}[I^+(\tau) - I^-(\tau)], \tag{17.18b}$$

$$K(\tau) \equiv \tfrac{1}{2}\int_{-1}^{1} I(\tau,\mu)\mu^2\,d\mu = \tfrac{1}{6}[I^+(\tau) + I^-(\tau)] \equiv \tfrac{1}{3}J(\tau). \tag{17.18c}$$

From (17.18a) and (17.18c) we see that if the radiation field is separately isotropic in the incoming and outgoing hemispheres, $K(\tau) \equiv \frac{1}{3}J(\tau)$ and we recover the Eddington approximation. At the boundary ($\tau = 0$) there is no incoming radiation, i.e., $I^- \equiv 0$, so from (17.18a) and (17.18b) we have $J(0) = 2H(0)$, and from (17.11) we find $q_E \equiv \frac{2}{3}$; hence the two-stream approximation recovers the Eddington approximation $J_E(\tau) = 3H(\tau + \frac{2}{3})$.

As a variant two-stream approximation, suppose the radiation field is represented by two beams moving along $\pm\mu_0$: $I(\tau,\mu) \equiv I^+(\tau)\delta(\mu - \mu_0)$ for $0 \le \mu \le 1$, and

$I(\tau,\mu) \equiv I^-(\tau)\delta(\mu+\mu_0)$ for $-1 \le \mu \le 0$. Then

$$J(\tau) \equiv \tfrac{1}{2}\left[I^+(\tau)\int_0^1 \delta(\mu-\mu_0)\,d\mu + I^-(\tau)\int_{-1}^0 \delta(\mu+\mu_0)\,d\mu\right]$$
$$= \tfrac{1}{2}\,[I^+(\tau)+I^-(\tau)], \tag{17.19a}$$

$$H(\tau) \equiv \tfrac{1}{2}\left[I^+(\tau)\int_0^1 \delta(\mu-\mu_0)\mu\,d\mu + I^-(\tau)\int_{-1}^0 \delta(\mu+\mu_0)\mu\,d\mu\right]$$
$$= \tfrac{1}{2}\,[I^+(\tau)-I^-(\tau)]\mu_0, \tag{17.19b}$$

$$K(\tau) \equiv \tfrac{1}{2}\left[I^+(\tau)\int_0^1 \delta(\mu-\mu_0)\mu^2\,d\mu + I^-(\tau)\int_{-1}^0 \delta(\mu+\mu_0)\mu^2\,d\mu\right]$$
$$= \tfrac{1}{2}\,[I^+(\tau)+I^-(\tau)]\mu_0^2. \tag{17.19c}$$

If we choose $\mu_0 = 1/\sqrt{3}$, we recover the Eddington approximation $J \equiv 3K$. Further, $I^- \equiv 0$ at $\tau = 0$, so $J(0) = \sqrt{3}H(0)$, and from (17.12) we infer that $q(0) \equiv 1/\sqrt{3}$, which we shall see below is the *exact* value of $q(0)$.

Power-Series Expansion

At the surface, the Eddington-Barbier relation tells us that I^+ cannot be isotropic because the source function $S(\tau) \equiv J(\tau)$ has a gradient. Suppose that $I(\tau,\mu)$ for $\tau \lesssim 1$ can be expanded in odd powers of μ:

$$I(\tau,\mu) \approx I_0(\tau) + \sum_{k=1}^{\infty} I_{2k-1}(\tau)\mu^{2k-1}. \tag{17.20}$$

When multiplied by even powers of μ, all terms in the sum integrate to zero on $[-1,1]$. Thus the ratio $K(0)/J(0) = \frac{1}{3}$ is preserved. The dominant nonzero term in the expansion, $I_1(\tau)\mu$, accounts for the outward flux. But the slope $q'(\tau)$ of the exact solution is infinite at $\tau = 0$ [see (17.32)]; hence this expansion breaks down.

- *Plane wave*: For a plane wave, e.g., when an observer is so far from a stellar surface that it subtends a minuscule solid angle,

$$I(\mu) \to I_0\delta(\mu-1). \tag{17.21}$$

Then $J = H = K = I_0$, and $K/J = 1$.

Iterated Eddington Approximation

As shown in (17.7), for the gray problem $J(\tau) \equiv \Lambda_\tau[J(t)]$. Thus we might expect to obtain an improved estimate of $J(\tau)$ by applying the Λ-operator to J_E. Inserting (17.14) into equation (11.133a), we find

$$J_E^1(\tau) = \Lambda_\tau\left[3H\left(\tau+\tfrac{2}{3}\right)\right] = 3H\left[\tau+\tfrac{2}{3}-\tfrac{1}{3}E_2(\tau)+\tfrac{1}{2}E_3(\tau)\right]. \tag{17.22}$$

- *Accuracy*: $J_E^1(\tau) \to J_E(\tau)$ for $\tau \gg 1$ because the exponential integrals in (17.22) $\to 0$ as $\tau \to \infty$. The largest difference between $J_E^1(\tau)$ and $J_E(\tau)$ is at $\tau = 0$, where $J_E^1(0)/J_E(0) = \frac{7}{8}$, and $q_E^1(0) = \frac{7}{12} \approx 0.583$, in better agreement with the exact value $q(0) = 1/\sqrt{3} \approx 0.577$. However, q_E^1 is significantly better than q_E only for $\tau < 1$.
- *Derivative*: An important property of q_E^1 is that its derivative is infinite at $\tau = 0$, a property it shares with the exact solution; see (17.32).
- *Second moment*: Applying the X-operator to $J_E(\tau)$, we find

$$K_E^1(\tau) = X_\tau \left[3H\left(t + \tfrac{2}{3}\right)\right] = H\left[\tau + \tfrac{2}{3} - E_4(\tau) + \tfrac{3}{2}E_5(\tau)\right]. \tag{17.23}$$

- *Variable Eddington factor*: In this approximation, the variable Eddington factor is

$$f_E^1(\tau) = \frac{K_E^1(\tau)}{J_E^1(\tau)} = \frac{\left[\tau + \frac{2}{3} - E_4(\tau) + \frac{3}{2}E_5(\tau)\right]}{3\left[\tau + \frac{2}{3} - \frac{1}{3}E_2(\tau) + \frac{1}{2}E_3(\tau)\right]}. \tag{17.24}$$

  As $\tau \to \infty, f_E^1 \to \frac{1}{3}$. At $\tau = 0$, $f_E^1 = \frac{17}{42} \approx 0.405$, which agrees well with the exact value $f(0) = q(\infty)/3q(0) \approx 0.410$ and shows that the angular distribution of the radiation field at the surface is peaked in the outward direction. A comparison of $f_E^1(\tau)$ with $f_{\text{exact}}(\tau)$ in table 17.1 shows that even *one Λ-iteration of the source function produces quite accurate Eddington factors*. Indeed, this result motivated the variable Eddington factor method described in § 12.5. At $\tau = 0$, (17.23) gives $K_E^1(0)/H = \frac{17}{24} \approx 0.708$, in good agreement with the exact value $q(\infty) = 0.710$.
- *Limb darkening*: The emergent intensity in this approximation is found by using (17.22) in (11.101) and re-ordering the integrations:

$$\begin{aligned} I_E^1(0,\mu) &= 3H \int_0^\infty \left[\tau + \tfrac{2}{3} - \tfrac{1}{3}E_2(\tau) + \tfrac{1}{2}E_3(\tau)\right] e^{-\tau/\mu}\, d\tau/\mu \\ &= 3H\{\tfrac{7}{12} + \tfrac{1}{2}\mu + (\tfrac{1}{3}\mu + \tfrac{1}{2}\mu^2)\ln[(1+\mu)/\mu]\}. \end{aligned} \tag{17.25}$$

  Now $\lim_{\mu \to 0} (\mu \ln \mu) = 0$; thus $I_E^1(0,0) = 1.75H$, close to the exact value $\sqrt{3}H$.

Table 17.1 Comparison of $f_E^1(\tau)$ and $f_{\text{exact}}(\tau)$

τ	$f_E^1(\tau)$	$f_{\text{exact}}(\tau)$	τ	$f_E^1(\tau)$	$f_{\text{exact}}(\tau)$
0.00	0.4048	0.4102	0.60	0.3399	0.3397
0.01	0.3956	0.4015	0.80	0.3375	0.3372
0.10	0.3667	0.3711	1.00	0.3362	0.3357
0.20	0.3546	0.3572	2.00	0.3339	0.3336
0.40	0.3443	0.3449	5.00	0.3333	0.3333

Table 17.2 Comparison of $\phi_E(\mu)$, $\phi_E^1(\mu)$, and $\phi_{exact}(\mu)$

μ	$\phi_E(\mu)$	$\phi_E^1(\mu)$	$\phi_{exact}(\mu)$	μ	$\phi_E(\mu)$	$\phi_E^1(\mu)$	$\phi_{exact}(\mu)$
0.0	0.4000	0.3512	0.3439	0.6	0.7600	0.7562	0.7546
0.2	0.5200	0.5049	0.4988	0.8	0.8800	0.8784	0.8779
0.4	0.6400	0.6326	0.6291	1.0	1.0000	1.0000	1.0000

The improved limb-darkening function $\phi_E^1(\mu) \equiv I_E^1(0, \mu)/I_E^1(0, 1)$ is compared with $\phi_E(\mu)$ and $\phi_{exact}(\mu)$ in table 17.2. At $\mu = 0.3$, which is at 95% of the projected disk of a spherical star, $\phi_E^1(\mu)$ and $\phi_{exact}(\mu)$ differ by about 1%. For the Sun, it is difficult to make measurements beyond that point because (1) the solar surface brightness is inhomogeneous at the limb, and (2) ground-based observations are blurred by seeing effects. Even at $\mu = 0.0$, they differ by only 2%, which is close to the limit of observational accuracy.

- *Failure of Λ-iteration*: As follows from (17.6), the gray problem is formally equivalent to the scattering problem with $\epsilon = 0$. Despite the fact that one can prove [1178, p. 31] that $\lim_{n\to\infty} \Lambda_\tau^n[1] = 0$, so in principle an initial error at any depth can be removed by Λ-iteration, the analysis in § 12.1 shows that for small thermalization parameters ϵ, the procedure is too slow to converge in a useful number of steps. For the gray problem $\epsilon = 0$; hence Λ-iteration cannot converge to the exact solution even in an infinite number of steps.

Exact Solution

To fill the gaps in the physically oriented discussion presented above and obtain rigorous results, advanced analysis using Laplace and Fourier transforms, Wiener-Hopf methods, and the principle of invariance is required; see [27, chapter 33], [225, chapters 4 and 5], [1178, chapter 3]. Here we merely quote results; they can be obtained more easily by means of the discrete ordinates method discussed below.

Milne's Equation

In terms of the Hopf function,

$$\tau + q(\tau) = \Lambda_\tau[t + q(t)], \tag{17.26}$$

or

$$q(\tau) = \Lambda_\tau[q(t)] + \tfrac{1}{2}E_3(\tau). \tag{17.27}$$

Basic Properties of the Hopf Function

Assuming a trial solution of the form $J(\tau) \approx \tau + c$, Hopf was able to prove rigorously that $q(\tau)$ is *unique* [508], *bounded* $\frac{1}{2} < q(\tau) < 1$ [507], and *monotonically increasing* [510] in the range $0 \le \tau \le \infty$; see [1178, pp. 31–34]. In principle, the integral equation

$$q(\tau) = \Lambda_\tau[q(t)] + f(\tau) \tag{17.28}$$

can be solved by iteration in a *Neumann series*

$$q(\tau) = \sum_{n=0}^{\infty} \Lambda_\tau^n[f(t)], \tag{17.29}$$

where $\Lambda^0[f] \equiv f$. Therefore, the exact solution for $q(\tau)$ is

$$q(\tau) = \tfrac{1}{2}\left\{E_3(\tau) + \Lambda_\tau[E_3(t)] + \Lambda_\tau^2[E_3(t)] + \cdots + \Lambda_\tau^n[E_3(t)]\ldots\right\}. \tag{17.30}$$

However, this series has no practical value because it contains unknown functions and barely converges. Nevertheless, starting from (17.27) and using a series of the form (17.30) at $\tau = 0$, both Bronstein [151] and Hopf [509] found exact expressions for the sum and evaluated the exact value of $q(0)$. They showed that (see also [1178, pp. 35–37])

$$q(0) = \tfrac{4}{3} - q(0)[4q(0) - 1], \tag{17.31}$$

or $q(0) \equiv 1/\sqrt{3}$, which is known as the *Hopf-Bronstein relation*.[2]

Thus the exact solution for a gray planar scattering medium has the boundary ratio $q_{\text{exact}}(0) = 1/\sqrt{3}$, which is a bit larger than the Eddington approximation value because of forward peaking of the radiation field produced by the gradient in the source function. This *Hopf boundary ratio* is also known as the *Mark boundary ratio* in neutronics.

Differentiation of (17.27) gives

$$q'(\tau) = \Lambda_\tau[q'(t)] + \tfrac{1}{2}q(0)E_1(\tau) - \tfrac{1}{2}E_2(\tau), \tag{17.32}$$

which shows that $q'(0)$ is logarithmically infinite because $E_1(0)$ is. Physically, this singularity occurs because once τ is even slightly > 0, nonzero contributions to $J(\tau)$ are immediately made by incoming rays $I^-(\tau, \mu)$ with μ slightly < 0 that penetrate into the material near the boundary of a planar medium. This singularity is absent in spherical geometry.

[2] In an addendum to the paper by Bronstein [152], Milne remarks, "... Dr. Bronstein's paper is dated 1929 Nov. 4; Dr. Hopf's paper is dated 1930 Jan. 1.... We have a clear case of the independent and almost simultaneous discovery of an important result by two different investigators, though the actual priority of publication rests with Dr. Bronstein."

Exact Limb-Darkening Function

The exact expression for the limb-darkening function of a planar gray medium provides a path to finding the exact functional form of $q(\tau)$. Define the "theoretical" limb-darkening function as

$$\mathcal{H}(\mu) \equiv I(0,\mu)/I(0,0). \tag{17.33}$$

The reference point here is taken to be the limb, not disk center, as used earlier, because knowing the Hopf-Bronstein relation, we can use the exact identity $I(0,0) \equiv J(0) = \sqrt{3}H$.

Using (17.9) in equation (11.101), we find

$$\mathcal{H}(\mu) = \sqrt{3} \int_0^\infty [t + q(t)]\, e^{-t/\mu}\, dt/\mu, \tag{17.34}$$

which shows that $\mathcal{H}(\mu)$ is the Laplace transform of the source function. The exact expression for $\mathcal{H}(\mu)$ [511, p. 105] was derived by Hopf using Wiener-Hopf techniques [1162]. Hopf's derivation is complicated; simpler methods are discussed in [633, § 29], [872, 873], and [1178, pp. 37–45]. Different choices of variables yield equivalent expressions for $\mathcal{H}(\mu)$; a form useful for numerical computation is

$$\mathcal{H}(\mu) = \frac{1}{(1+\mu)^{1/2}} \exp\left[\frac{1}{\pi}\int_0^{\pi/2} \frac{\theta \tan^{-1}(\mu \tan\theta)}{1 - \theta\cot\theta}\, d\theta\right]. \tag{17.35}$$

Numerical values for $\mathcal{H}(\mu)$ can also be obtained from table 17.2 by renormalizing entries in the last column such that $\mathcal{H}(0) = 1$.

Exact $q(\tau)$

Given an exact expression for $\mathcal{H}(\mu)$, the exact $q(\tau)$ can be obtained by inverse Laplace transformation. First, using (17.35), we can obtain the exact value for $q(\infty)$ from the identity $q(\infty) = K(0)/H$, or

$$q(\infty) = \int_0^1 \mathcal{H}(\mu)\, \mu^2\, d\mu \bigg/ \int_0^1 \mathcal{H}(\mu)\, \mu\, d\mu, \tag{17.36}$$

which yields [873]

$$q(\infty) = \frac{6}{\pi^2} + \frac{1}{\pi}\int_0^{\pi/2} \left(\frac{3}{\theta^2} - \frac{1}{1 - \theta\cot\theta}\right) d\theta. \tag{17.37}$$

Numerical integration of (17.37) gives $q(\infty) = 0.71044\,60896$.[3]

[3] See [1123], where $q(\infty)$ is given to 59 significant figures!

Table 17.3 Exact Hopf Function

τ	$q(\tau)$	τ	$q(\tau)$
0.00	0.57735 02692	0.20	0.64955 03109
0.01	0.58823 54752	0.40	0.67309 12554
0.02	0.59539 08024	0.60	0.68580 13583
0.04	0.60628 62793	1.00	0.69853 93182
0.06	0.61578 87669	1.50	0.70513 01524
0.10	0.62791 87380	2.00	0.70791 66189

An exact analytical expression for $q(\tau)$ in the range $0 \leq \tau \leq \infty$ was found by. Mark [711]:

$$q(\tau) = q(\infty) - \frac{1}{2\sqrt{3}} \int_0^1 \frac{e^{-\tau/u}\, du}{\mathcal{H}(u)\left[(1 - u \tanh^{-1} u)^2 + \frac{1}{4}\pi^2 u^2\right]}; \tag{17.38}$$

see also [633, § 29.7] and [1178, pp. 45–47]. Numerical values of $q(\tau)$ can be obtained by numerical integration of (17.38). Some values of $q(\tau)$ from [1123] are listed in table 17.3.

Discrete Ordinate Method

This method limits to the exact solution of the gray problem and introduces a technique used in most modern methods for solving transfer problems with scattering terms. Recalling (17.5), the problem to be solved is

$$\mu \frac{\partial I(\tau, \mu)}{\partial \tau} \equiv I(\tau, \mu) - \tfrac{1}{2} \int_{-1}^{1} I(\tau, \mu')\, d\mu'. \tag{17.39}$$

As described in § 11.7, Chandrasekhar replaced the angle integral with a Gaussian quadrature having a set of quadrature points $\{\mu_j\}$ in the range $[-1, 1]$ and a set of quadrature weights $\{a_j\}$ such that

$$\int_{-1}^{1} I(\tau, \mu)\, d\mu \approx \sum_{j=-n}^{n} a_j I(\tau, \mu_j). \tag{17.40}$$

With this transformation, (17.39) becomes a coupled set of $2n$ linear ordinary differential equations. In these quadrature formulae the points are chosen symmetrically about $\mu = 0$; i.e., $\mu_{-j} = -\mu_j$, and $w_{-j} = a_j$; hence

$$\mu_i \frac{\partial I(\tau, \mu_i)}{\partial \tau} = I(\tau, \mu_i) - \tfrac{1}{2} \sum_{j=-n}^{n} a_j I(\tau, \mu_j), \quad (i = \pm 1, \ldots, \pm n). \tag{17.41}$$

The crux of the method is that (17.41) can be solved exactly for a set of continuous *analytical* functions of τ (exponentials), which give the value of $I(\mu)$ in *pencils* of radiation within discrete angle intervals. The method is a hybrid in the sense that the angular variation of $I(\tau, \mu)$ is discretized, whereas its depth dependence is described by continuous algebraic functions. On physical grounds alone we can expect this representation to become exact as $n \to \infty$.

Eigenvalue Problem

The first-order system (17.41) is linear, so take a trial solution of the form $I_i = g_i \exp(-k\tau)$, where the (depth-independent) values of g_i and k are to be determined. Then

$$g_i(1 + k\mu_i) = \tfrac{1}{2}\sum_{j=-n}^{n} a_j g_j = \text{constant} \equiv G, \tag{17.42}$$

or $g_i \equiv G/(1 + k\mu_i)$. Using this expression for g_i in (17.42), we find

$$\tfrac{1}{2}\sum_{j=-n}^{n} \frac{a_j}{1 + k\mu_j} = 1. \tag{17.43}$$

This *characteristic equation* is satisfied only by certain *eigenvalues* of k. Recalling that $\mu_{-j} = -\mu_j$, and $a_{-j} = a_j$, we add together the terms in $\pm j$, obtaining

$$\sum_{j=1}^{n} \frac{a_j}{1 - k^2\mu_j^2} = 1. \tag{17.44}$$

To solve (17.44), define the *characteristic function* $T(k^2)$ as

$$T(k^2) \equiv 1 - \sum_{j=1}^{n} \frac{a_j}{1 - k^2\mu_j^2}, \tag{17.45}$$

which has its roots, i.e., $T(k^2) = 0$, at the eigenvalues.

If $k = 0$ in (17.44), $T(k^2 = 0) \equiv 0$, so $k = 0$ is an eigenvalue. Note that $k^2 = \mu_j^{-2}$ is a *pole* of T. For $k^2 = \mu_j^{-2} - \epsilon$, $T(k^2) < 0$; and as $\epsilon \to 0$, $T(k^2) \to -\infty$. For $k^2 = \mu_j^{-2} + \epsilon$, $T(k^2) > 0$; and as $\epsilon \to 0$, $T(k^2) \to \infty$. Therefore, T passes through zero on the interval between successive poles. Thus there are $(n-1)$ nonzero roots $\mu_1^{-2} < k_1^2 < \mu_2^{-2} < k_2^2 < \ldots < k_{n-1}^2 < \mu_n^{-2}$, ordered such that $\mu_i > \mu_{i+1}$. There are a total of $2n-2$ nonzero values of k, in pairs of the form $\pm k_i$, $(i = 1, \ldots, n-1)$. The eigenvalues lie on bounded intervals, and hence can be found by standard numerical methods.

General and Particular Solutions

The *general solution* of (17.41) has the form

$$I(\tau, \mu_i) = b\left[\sum_{\alpha=1}^{n-1} \frac{L_\alpha e^{-k_\alpha \tau}}{1 + k_\alpha \mu_i} + \sum_{\alpha=1}^{n-1} \frac{L_{-\alpha} e^{+k_\alpha \tau}}{1 - k_\alpha \mu_i}\right]. \tag{17.46}$$

The scale factor b, yet to be determined, is needed because the original equation is homogeneous. For a given quadrature formula, and hence a specific set of eigenvalues k_α, $I(\tau, \mu_i)$ is an analytical *eigenfunction* of (17.41).

To get the *complete solution*, we also need a *particular solution* corresponding to the root $k^2 = 0$. The Eddington solution shows that J is linear in τ for $\tau \gg 1$. If we use $I_i(\tau) = b(\tau + q_i)$ as a trial solution in (17.41), we find

$$q_i = \mu_i + \tfrac{1}{2} \sum_{j=-n}^{n} a_j q_j. \tag{17.47}$$

A solution of (17.47) is $q_i = Q + \mu_i$, where Q is a constant, because the sum $\sum_{j=-n}^{n} a_j \mu_j$ represents $\int_{-1}^{1} \mu \, d\mu$, which is identically zero. Thus the particular solution is $I(\tau, \mu_i) = b(\tau + Q + \mu_i)$, and the complete solution is

$$I(\tau, \mu_i) = b\left[\tau + Q + \mu_i + \sum_{\alpha=1}^{n-1} \frac{L_\alpha e^{-k_\alpha \tau}}{1 + k_\alpha \mu_i} + \sum_{\alpha=1}^{n-1} \frac{L_{-\alpha} e^{+k_\alpha \tau}}{1 - k_\alpha \mu_i}\right]. \tag{17.48}$$

The unknown coefficients b, Q, and $L_{\pm\alpha}$ are found from boundary conditions.

- *Finite slab*: For a slab of finite optical thickness T, the imposed external radiation fields $I(0, -\mu_i)$ and $I(\mathrm{T}, +\mu_i)$ are both given functions of μ. Applying these boundary conditions, we have

$$I(0, -\mu_i) \equiv b\left[Q - \mu_i + \sum_{\alpha=1}^{n-1} \frac{L_\alpha}{1 - k_\alpha \mu_i} + \sum_{\alpha=1}^{n-1} \frac{M_\alpha e^{-k_\alpha \mathrm{T}}}{1 + k_\alpha \mu_i}\right], \tag{17.49}$$

  and

$$I(\mathrm{T}, +\mu_i) \equiv b\left[\mathrm{T} + Q + \mu_i + \sum_{\alpha=1}^{n-1} \frac{L_\alpha e^{-k_\alpha \mathrm{T}}}{1 + k_\alpha \mu_i} + \sum_{\alpha=1}^{n-1} \frac{M_\alpha}{1 - k_\alpha \mu_i}\right]. \tag{17.50}$$

  Here, to improve the numerical condition of the system, we defined $M_\alpha \equiv L_{-\alpha} e^{k_\alpha \mathrm{T}}$. Thus we have $2n$ linear equations in $2n$ unknowns: b, Q, L_α, and M_α, $(\alpha = 1, \ldots, n-1)$, which can be solved numerically.
- *Semi-infinite atmosphere*: For a *semi–infinite* atmosphere we demand that I remain bounded as $\tau \to \infty$, which implies that in (17.48) all the coefficients $L_{-\alpha} \equiv 0$. And recalling that Hopf proved that $q(\tau)$ is a monotonically increasing function

of τ, we see that the L_α's in (17.48) must be negative. Further, in modeling a stellar atmosphere we can assume there is no incoming radiation at the upper boundary, so $I(0, -\mu_i) \equiv 0$; hence at $\tau = 0$,

$$I(0, -\mu_i) = b\left[Q - \mu_i + \sum_{\alpha=1}^{n-1} \frac{L_\alpha}{1 - k_\alpha \mu_i}\right] \equiv bS(\mu_i) = 0, \quad (i = 1, \ldots n). \tag{17.51}$$

We thus have n linear equations for the unknowns L_α, $(\alpha = 1, \ldots, n-1)$ and Q. To set the scale of the solution, demand that at all depth points the net flux computed from (17.48) equal the nominal flux:

$$\tfrac{1}{2}\int_{-1}^{1} I(\mu)\,\mu\,d\mu = \tfrac{1}{2}\sum_{j=-n}^{n} a_j \mu_j I_j \equiv H. \tag{17.52}$$

Substituting (17.48) (with $L_{-\alpha} \equiv 0$) into (17.52), we find

$$H = \tfrac{1}{2}b\left[(\tau + Q)\sum_{j=-n}^{n} a_j\mu_j + \sum_{j=-n}^{n} a_j\mu_j^2 + \sum_{\alpha=1}^{n-1} L_\alpha e^{-k_\alpha\tau}\sum_{j=-n}^{n}\frac{a_j\mu_j}{1 + k_\alpha\mu_j}\right]. \tag{17.53}$$

The first sum in (17.53) is zero and the second is $\frac{2}{3}$. In the third term, set $\tau = 0$ and use the characteristic function as written in (17.45) to rewrite it as $\sum_{\alpha=1}^{n-1} L_\alpha T(k_\alpha^2)$, which is also zero because k_α is an eigenvalue. Thus $b = 3H$, where H is the constant Eddington flux; cf. (17.9).

In summary, the discrete ordinate representations of $I(\tau, \mu)$, $J(\tau)$, and $q(\tau)$ in a semi-infinite atmosphere are

$$I(\tau, \mu_i) = 3H\left[\tau + Q + \mu_i + \sum_{\alpha=1}^{n-1}\frac{L_\alpha e^{-k_\alpha\tau}}{(1 + k_\alpha\mu_i)}\right], \quad (i = 1 \ldots n), \tag{17.54}$$

$$J(\tau) = 3H\left(\tau + Q + \sum_{\alpha=1}^{n-1} L_\alpha e^{-k_\alpha\tau}\right), \tag{17.55}$$

and

$$q(\tau) = Q + \sum_{\alpha=1}^{n-1} L_\alpha e^{-k_\alpha\tau}, \tag{17.56}$$

with $H \equiv$ constant. Equation (17.56), coupled with the fact that the L_α's are negative, verifies the statement made in § 17.1 that q is a monotonically increasing function, approaching the limiting value $Q \equiv q(\infty)$.

Product Rule for Quadrature Points and Eigenvalues

A number of additional important results can be obtained from the characteristic equation. Define $x \equiv 1/k$ and $X \equiv 1/k^2$. Then from (17.44),

$$T_n(X) = 1 - X\sum_{j=1}^{n}\frac{a_j}{X-\mu_j^2} = \sum_{j=1}^{n} a_j - X\sum_{j=1}^{n}\frac{a_j}{X-\mu_j^2} = \sum_{j=1}^{n}\frac{a_j\mu_j^2}{\mu_j^2 - X}. \quad (17.57)$$

To clear $T_n(X)$ of fractions, multiply through by $\prod_{j=1}^{n}(\mu_j^2 - X)$, which gives

$$P(X) \equiv \prod_{j=1}^{n}(\mu_j^2 - X)T_n(X) = \sum_{i=1}^{n} w_i\mu_i^2 \prod_{j\neq i}^{n}(\mu_j^2 - X). \quad (17.58)$$

The right-hand side is a polynomial of order $(n-1)$ in X. On the other hand, $T_n(X)$ has the $(n-1)$ roots $X_1 = 1/k_1^2, \ldots, X_{n-1} = 1/k_{n-1}^2$, so $P(X)$ must also be of the form $C(X - X_1)\ldots(X - X_{n-1})$. C is the coefficient of the term in X^{n-1}, which from (17.58) is $(-1)^{n-1}\sum_{i=1}^{n} w_i\mu_i^2 = (-1)^{n-1}\frac{1}{3}$. Therefore, $P(X) = \frac{1}{3}$ $(X_1 - X)\ldots(X_{n-1} - X)$, and

$$T_n(X) = \prod_{j=1}^{n-1}(X_j - X)\Big/3\prod_{j=1}^{n}\left(\mu_j^2 - X\right). \quad (17.59)$$

The first form of $T_n(X)$ in (17.57) shows $T_n(X = 0) = 1$. Setting $X = 0$ in (17.59), we find that *independent of the order of approximation n*,

$$\mu_1\mu_2\ldots\mu_n k_1 k_2 \ldots k_{n-1} = 1/\sqrt{3}. \quad (17.60)$$

Emergent Intensity

Now consider the emergent intensity. The surface boundary conditions in (17.51) can be written as

$$I_n(0, -\mu_i) = \tfrac{3}{4}FS(\mu_i). \quad (17.61)$$

Strictly, (17.61) holds only at $\mu = \mu_1, \mu_2, \ldots, \mu_n$. But in the limit $n \to \infty$, it becomes exact.

Hopf-Bronstein Relation

Using the discrete ordinate method we can easily recover the Hopf-Bronstein relation $q(0) \equiv 1/\sqrt{3}$. Clearing fractions from (17.61), we have

$$S_n(\mu)\prod_{\alpha=1}^{n-1}(1 - k_\alpha\mu) = (Q - \mu)\prod_{\alpha=1}^{n-1}(1 - k_\alpha\mu) + \sum_{\alpha=1}^{n-1} L_\alpha \prod_{j\neq\alpha}^{n-1}(1 - k_j\mu). \quad (17.62)$$

The right-hand side of (17.62) is clearly a polynomial of order n in μ. But we showed above that $S_n(\mu)$ has the n roots $\mu = \mu_1, \mu_2, \ldots, \mu_n$. Therefore, the polynomial must be of the form $C(\mu - \mu_1) \ldots (\mu - \mu_n)$. The coefficient of μ^n on the right-hand side of (17.58) is $C = (-1)^n k_1 \ldots k_{n-1}$. Thus

$$S_n(\mu) = \frac{\prod_{j=1}^{n-1} k_j \prod_{i=j}^{n} (\mu_i - \mu)}{\prod_{j=1}^{n-1} (1 - k_j \mu)} = \frac{\prod_{j=1}^{n} (\mu_j - \mu)}{\prod_{j=1}^{n-1} (x_j - \mu)}. \tag{17.63}$$

From (17.63), $S_n(0) = \mu_1 \mu_2 \ldots \mu_n k_1 k_2 \ldots k_{n-1}$, so from (17.60), $S(0) \equiv 1/\sqrt{3}$, independent of the order n, which shows that the result $q(0) = 1/\sqrt{3}$ is exact.

Accuracy

Numerical results for $q_n(\tau)$ were obtained by Chandrasekhar [218] using a single-Gauss quadrature with ($n = 1, \ldots, 4$), and by Sykes [1067] using a double-Gauss quadrature with ($n = 1, \ldots, 3$). In every case, one finds the exact value $q(0) = 1/\sqrt{3}$. The maximum percentage absolute error in $J(\tau)$ for the single-Gauss solutions is (9.0, 4.1, 2.7, 2.0) for $n =$ 1, 2, 3, 4, and for double-Gauss it is (9.0, 1.8, 0.9) for $n =$ 1, 2, 3, respectively. The double-Gauss values for $q(\infty)$, i.e., Q, are 0.71132 and 0.71057 for $n = 2$ and 3; the latter compares well with the exact value 0.710446.

The double-Gauss solution with $n = 3$ gives the emergent intensity $I(0, \mu)/F$ with a root-mean-square error of 0.1% and is accurate to 0.02% for $\mu \geq 0.3$. The main importance of these results is that the discrete ordinate method yields the exact solution in the limit as $n \to \infty$ and also affords a powerful technique for more realistic problems.

Further results of double-Gauss discrete-ordinate solutions are shown in table 17.4 for several values of n (the number of angle points per hemisphere) using the quadrature points and weights in [5, pp. 916–922]. The computations were carried out with 31 significant-figure arithmetic; the eigenvalues were iterated until their fractional error was 10^{-32}. For all n, $q(0) = 1/\sqrt{3}$ to 16 digits. In the table, $\delta q(\infty) \equiv q_{\rm DO}(\infty) - q_{\rm exact}(\infty)$, not a fractional error.

Note that we get very accurate values for $q(\infty)$ even for modest n. $k_{\max}$ and $k_{\min}$ are a few digits of the maximum and minimum eigenvalues. As $k_{\min} \to 1$, the corresponding value of $\mu \to 1$; i.e., the more closely that ray approaches vertical, the more deeply its effect is felt in the solution. The larger $k_{\max}$ becomes, the

Table 17.4 Properties of Discrete-Ordinate Solutions for $q(\tau)$

n	$\delta q(\infty)$	$k_{\max}$	$k_{\min}$	n	$\delta q(\infty)$	$k_{\max}$	$k_{\min}$
3	1.2×10^{-4}	7.6	1.5203	32	1×10^{-10}	730	1.0021
4	2.5×10^{-5}	13.1	1.2272	48	1×10^{-11}	1600	1.0009
5	7.3×10^{-6}	20.0	1.1279	64	2×10^{-12}	2900	1.0005
8	5.1×10^{-7}	49.1	1.0421	80	6×10^{-13}	4500	1.0004
16	9.2×10^{-9}	187	1.0091	96	2×10^{-13}	6400	1.0002

corresponding value of $\mu \to 0$, i.e., more closely that ray becomes parallel to the surface of the atmosphere. The results in table 17.4 agree with those in [1067] to the number of digits in that work.

As n increases, we get finer and finer angular resolution. Despite the fact that $q(0)$ is exact, the maximum error in q_{DO} always occurs near the surface (recall that $q'(0)$ is infinite), where $I(\mu)$ varies most swiftly with μ. To achieve higher accuracy at small τ one would need a quadrature with points even closer to $\mu = 0$. This is precluded even with a higher-order Gaussian quadrature, because it covers only the *open* interval $(-1, +1)$. The smallest eigenvalue limits to $k_{max} = 1$, corresponding to a ray penetrating vertically into the medium. We could obtain this eigenvalue precisely by using a quadrature on the *closed* interval $[-1, +1]$ with a point at $\mu = 1$. In the limit $n \to \infty$, the discrete-ordinates method gives the exact solution; see [633, chapter 6] or [729, pp. 56–62].

Emergent Flux from a Gray Atmosphere

The premise of all the analysis above is that the opacity of the material is gray. From it we deduced that

$$J(\tau) \equiv S(\tau) \equiv 3H[\tau + q(\tau)], \tag{17.64}$$

where

$$H = F/4\pi = \sigma_R T_{\text{eff}}^4/4\pi. \tag{17.65}$$

F is the energy flux [erg/cm^2/sec] imposed at the lower boundary. If we make the additional assumption that the material is in LTE, then from (17.9),

$$S(\tau) \equiv B[T(\tau)] = \sigma_R T^4(\tau)/\pi \tag{17.66}$$

or

$$T^4(\tau) = \tfrac{3}{4} T_{\text{eff}}^4 [\tau + q(\tau)], \tag{17.67}$$

which gives the depth dependence of the temperature in terms of the effective temperature. With the Eddington approximation we get a simpler formula:

$$T^4(\tau) = \tfrac{3}{4} T_{\text{eff}}^4 (\tau + \tfrac{2}{3}), \tag{17.68}$$

which gives $T(\tau) = T_{\text{eff}}$ at $\tau = \frac{2}{3}$, suggesting that to a first approximation the continuum is formed at $\tau \approx \frac{2}{3}$.

But note that even though the *material* is gray, the temperature is a function of depth; hence the Planck function $B_\nu[T(\tau)]$ is a function of both depth and frequency, as is the radiation flux:

$$F_\nu(\tau) = 2\pi \left\{ \int_\tau^\infty B_\nu[T(\tau)] E_2(t - \tau)\, dt - \int_0^\tau B_\nu[T(\tau)] E_2(\tau - t)\, dt \right\}. \tag{17.69}$$

Temperature enters the Planck function only through the ratio $(h\nu/kT)$, and (17.67) shows that T/T_{eff} is a unique function of τ, say, $p(\tau)$. Thus we can simplify (17.69)

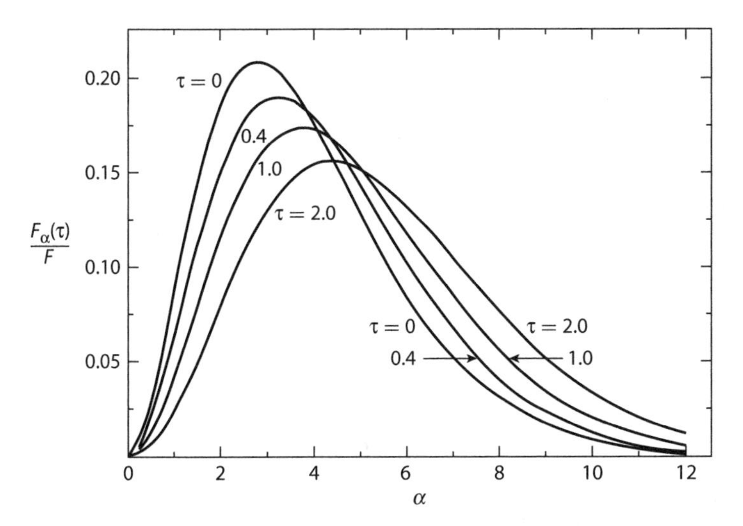

Figure 17.1 Frequency distribution of dimensionless radiation flux $F_\alpha(\tau)/F$ as a function of depth in a gray atmosphere. From [225].

by introducing the parameter $\alpha \equiv h\nu/kT_{\rm eff}$ and use the corresponding flux $F_\alpha(\tau) \equiv F_\nu(\tau)d\nu/d\alpha$ to write

$$\frac{F_\alpha(\tau)}{F} = \left(\frac{4\pi k^4}{c^2h^3\sigma_R}\right)\alpha^3 \left\{ \int_\tau^\infty \frac{E_2(t-\tau)\,dt}{\exp\left[\alpha/p(t)\right]-1} - \int_0^\tau \frac{E_2(\tau-t)\,dt}{\exp\left[\alpha/p(t)\right]-1} \right\}. \tag{17.70}$$

The terms in the brackets are functions of α and τ only and may be calculated once and for all. A plot of $F_\alpha(\tau)/F$ is shown in figure 17.1. It shows the degradation of photon energies as they transfer from depth to the open boundary at the surface. In particular, the most common photon energy at $\tau = 0$ is only about 75% of that at $\tau = 1$, reflecting the increase in temperature as a function of depth.

Physical Interpretation

An important point to realize is that the gray temperature structure follows solely from the radiative transfer equation and the radiative equilibrium equation. The hydrostatic equilibrium equation does not enter this derivation. In other words, the temperature in a gray atmosphere, as a function of optical depth, does not depend on the surface gravity. However, the hydrostatic equation determines the relation between the optical depth and the geometrical coordinate (m or z).

We see that the temperature is a monotonically increasing function of optical depth. It is easy to understand this behavior in physical terms. The condition of radiative equilibrium stipulates that the *total* radiation flux is constant with depth in the atmosphere. However, the radiation flux measures the *anisotropy* of the radiation field (for instance, the flux would be zero for perfectly isotropic radiation). From the radiative transfer equation, and in particular from the diffusion approximation, it follows that the anisotropy of radiation decreases with increasing depth in the atmosphere. The only way to maintain the constant flux in spite of decreasing anisotropy of radiation is to increase the total energy density of radiation (proportional to J), and hence the temperature (recall that $J = S = B = \sigma_R T^4/\pi$).

The fact the integrated J is equal to integrated B at all depths τ does not necessarily mean that the frequency-dependent J_ν has to be equal to B_ν for all frequencies. In fact, we should expect that there should be a frequency range for which $J_\nu > B_\nu$, i.e., $J_\nu - B_\nu > 0$; these regions may be called "heating" regions; while at the rest of frequencies $J_\nu < B_\nu$, i.e., $J_\nu - B_\nu < 0$; these regions may be called "cooling" regions. As is shown in chapter 11, J_ν is proportional to the number of photons absorbed per unit optical depth, while $S_\nu = B_\nu$ is proportional to the number of photons emitted per unit optical depth. Thus, $J_\nu > B_\nu$ means that more photons are absorbed than emitted at frequency ν; the energy of extra absorbed photons must then increase the internal energy, i.e., the temperature, of the medium.

Which frequency regions are the heating ones, and which are the cooling ones? In the case of an LTE-gray atmosphere, the answer follows from examining the behavior of the Planck function and from the Eddington-Barbier relation, written as $J(0) = \frac{1}{2} S(\tau = \frac{1}{2})$. Recall that in the high-frequency limit, $(h\nu/kT) \gg 1$, we obtain the Wien form of the Planck function,

$$B(\nu, T) \approx \frac{2h\nu^3}{c^2} \exp(-h\nu/kT), \tag{17.71}$$

while in the low-frequency limit, $(h\nu/kT) \ll 1$, the Rayleigh-Jeans tail, we have

$$B(\nu, T) \approx \frac{2k\nu^2}{c^2} T. \tag{17.72}$$

Let us consider surface layers of a gray atmosphere. If the frequency ν is "large," i.e., in the Wien regime, then a decrease of the local temperature between $\tau = \frac{1}{2}$ and the surface ($\tau = 0$) translates into a large decrease of $B_\nu(T(\tau))$, because for large frequencies the Planck function is very sensitive to T; see (17.71). In other words, B_ν at the surface may be significantly (even orders of magnitude) lower than B_ν at $\tau = \frac{1}{2}$. Since the mean intensity at the surface is about one half of $B_\nu(\tau = \frac{1}{2})$, it is clear that $J_\nu(0) > B_\nu(0)$ for these frequencies. The large frequencies are therefore the "heating" frequencies.

In contrast, for low frequencies (the Rayleigh-Jeans tail), B_ν is linearly proportional to T. The $T(\tau)$ relation for a gray atmosphere establishes that $T(0) \approx 0.83\, T(\tau = \frac{1}{2})$. The factor $\frac{1}{2}$ from the Eddington-Barbier relation will now dominate, so we get $J_\nu(0) = \frac{1}{2} B_\nu(\tau = \frac{1}{2}) < B_\nu(0)$. Consequently, the low frequencies are the "cooling" frequencies. One can make these considerations more quantitative, but this is not necessary; the only important point to remember is that the high-frequency part of the spectrum is responsible for heating, while the low-frequency part is responsible for cooling.

Two-Step Gray Model

The above considerations are interesting, but not particularly useful for a purely gray atmosphere. They are, however, very helpful if we consider an atmosphere with some simple departures from the grayness. Let us consider a *two-step gray* model, i.e., with the opacity given as a step function, $\chi_\nu = \chi$ (the original gray opacity) for $\nu < \nu_0$, and $\chi_\nu = a\chi$ for $\nu \geq \nu_0$, with $a \gg 1$, i.e., with a large opacity for high frequencies (one may visualize this as a schematic representation of a strong continuum jump, for instance, the Lyman discontinuity). We denote the original optical depth as $\tau^{\rm old}$ and the new one (for $\nu \geq \nu_0$) as $\tau_\nu^{\rm new}$. Let us further assume that the frequency ν_0 is high enough to be in the range of "heating" frequencies.

What are the changes of the temperature structure with respect to the original gray temperature distribution implied by the opacity jump? We will consider separately the surface layers $\tau \approx 0$ and the deep layers.

- *Surface layers*: Since the opacity for $\nu \geq \nu_0$ is much larger than the original opacity, we may neglect the contribution of the latter to the radiative equilibrium integral, so the modified radiative equilibrium equation becomes

$$\int_{\nu_0}^{\infty} J_\nu \, d\nu = \int_{\nu_0}^{\infty} B_\nu \, d\nu, \tag{17.73}$$

which, together with the Eddington-Barbier relation $J_\nu(0) = \frac{1}{2} B_\nu(\tau_\nu^{\rm new} = \frac{1}{2})$, yields for the new surface temperature, T_0, the expression

$$\frac{1}{2} \int_{\nu_0}^{\infty} B_\nu \left(T(\tau_\nu^{\rm new} = \tfrac{1}{2})\right) d\nu = \int_{\nu_0}^{\infty} B_\nu(T_0) \, d\nu, \tag{17.74}$$

from which follows that $T_0 < T(\tau_\nu^{\rm new} = \frac{1}{2})$. Since the temperature at $\tau_\nu^{\rm new} = \frac{1}{2}$ must be close to the original temperature at the surface (since $\tau_\nu^{\rm new} \gg \tau^{\rm old}$), the new temperature at the surface is *lower* than the original surface temperature, which gives rise to the term *surface cooling* effect.

  In physical terms, by *adding opacity in the heating portion* of the spectrum, we effectively suppress this heating. Therefore, we obtain a cooling. These considerations also suggest that by adding an additional opacity in the cooling, i.e., the low-frequency part of the spectrum, we may actually get a *surface heating* of the atmosphere.

- *Deep layers*: It is intuitively clear that the atmospheric layers that are optically thick in all frequencies will be little influenced by the additional opacity jump. However, an interesting region is the one that is opaque for large frequencies ($\nu \geq \nu_0$) (i.e., $\tau_\nu^{\rm new} \gg 1$ for these frequencies), while still transparent for the original opacity, $\tau^{\rm old} < 1$. Since the optical depth is large for $\nu \geq \nu_0$, $J_\nu \approx B_\nu$ for these frequencies; therefore, the monochromatic flux is close to zero. The condition of radiative equilibrium at those depths may be written as $J' = B'$, where the primed quantities are defined as partial integrals, e.g., $J' = \int_0^{\nu_0} J_\nu \, d\nu$, and analogously for B. From the radiative transfer equation and the Eddington approximation, we have $dJ'/d\tau = 3H$ (not H'; or, better speaking, $H' = H$, because there is no flux for $\nu \geq \nu_0$). We may formally write $J' = \sigma' T^4$, and by repeating the same procedure as in deriving the original gray temperature structure, we obtain (in the Eddington approximation)

$$T^4 \approx \tfrac{3}{4}(\sigma/\sigma')\, T_{\rm eff}^4\, (\tau + \tfrac{2}{3}). \tag{17.75}$$

We have $\sigma' < \sigma$, because $J' < J$. This is because the energy density of radiation for $\nu < \nu_0$ is smaller than the total energy density. Therefore, the new temperature is larger than the original one. Consequently, the phenomenon is called the *backwarming effect.*

In physical terms one concludes that by adding opacity, the flux in the high-opacity part *drops*. Therefore, the flux in the rest must *increase* in order to keep the total flux constant. However, the only way to accomplish it in LTE is to increase the temperature gradient, and therefore the temperature itself in the previously flat $T(\tau)$ region.

It should be stressed that the above-discussed phenomena of surface cooling and backwarming are quite general and are not at all limited to a gray approximation. In any model, including sophisticated NLTE models (see chapter 18), there are always frequencies that cause heating and those that cause cooling. Any process that changes opacity/emissivity in those regions changes the overall balance and therefore influences the temperature structure. In the NLTE models, there are typically several intervening or competing mechanisms, but the fundamental physics behind the temperature structure is basically the same as in the case of the gray model. Similarly, the mechanism of backwarming is quite general. The usefulness of the gray model is that one can describe all these phenomena by a simple analytical model.

17.2 EQUATION OF STATE

The main task in calculating the structure of an LTE atmosphere with a non-gray opacity is to derive the depth distribution of temperature that yields radiative equilibrium. With that information, all its other properties—thermodynamic variables such as density and pressure, its opacity χ_ν and emissivity η_ν, and its radiation field—can be computed by straightforward integration. Then the transfer equation can be solved for the depth variation of the frequency—dependent radiation field. However, it is *not* known a priori.

The construction of an LTE model atmosphere can be carried out using a general, stable, and effective computational method [53,54]. In this approach, one linearizes the system of all equations (i.e., the equation of state, the expressions for opacity and emissivity, the radiative transfer equation, and the constraints of hydrostatic equilibrium and radiative equilibrium) in terms of a *current solution*, which does not satisfy the constraints as accurately as desired, and a set of corrections to the fundamental variables (T, N) at each point in the current solution, which will give us new temperature and density distributions that more nearly satisfy the constraints. We thus allow for the corrections to *all* variables in the structural and transfer equations and for the coupling of the corrections at one point in the atmosphere to all other points. The procedure is essentially a Newton-Raphson iteration, so errors in the solution, ideally, decrease quadratically. Before doing that in general (cf. § 17.3), we describe a solution of the equation of state for given fundamental variables T and N.

Gas Pressure

The material in stellar atmospheres is a mixture of atoms, ions, and molecules of many elements, plus free electrons. The excitation and ionization potentials of the atoms and successive ions of a given element usually are significantly different. Thus the transition, with increasing temperature, from a neutral atom to its first ionization stage, or from one ionization stage to the next, is typically abrupt. An element often exists mainly in two successive ionization stages. Consequently, the ratios of line strengths of successive ionic spectra (say, He I and He II, or Ca I and Ca II) may vary rapidly as a function of temperature, which gives us a tool for inferring temperatures in the atmosphere.

In an LTE atmosphere, the total particle number density N and temperature T at a given depth are sufficient to determine all the physical and optical properties of the gas. According to equation (4.48), the gas pressure $p_{\rm gas}$ in a partially ionized plasma (ignoring molecule formation) is

$$p_{\rm gas} = NkT = (N_{\rm atoms} + N_{\rm ions} + n_{\rm e})kT \equiv (N_{\rm N} + n_{\rm e})kT. \tag{17.76}$$

Here N is the number density of *all* types of particles in the plasma; $N_{\rm N}$ is the number density of "nuclei" (i.e., all atoms plus ions); and $n_{\rm e}$ is the number density of *free* electrons. To specify the *chemical composition* of the gas, let α_k be the fractional number abundance of the element with atomic number k: $\left(\sum_k \alpha_k = 1\right)$. For *particle conservation*, the number density of atoms and ions of chemical element k is

$$N_k = \alpha_k N_{\rm N} = \alpha_k (N - n_{\rm e}). \tag{17.77}$$

In addition, the number of free electrons must equal the total ionic charge for the atmosphere's plasma to be electrically neutral. In the notation of chapter 4, *charge conservation* requires that

$$n_{\rm e} = \sum_k \sum_{j=1}^{Z_k} j N_{jk} = \sum_k N_k \sum_{j=1}^{Z_k} j f_{jk}(n_{\rm e}, T) = N_{\rm N} \sum_k \alpha_k \sum_{j=1}^{Z_k} j f_{jk}(n_{\rm e}, T). \tag{17.78}$$

Hence an alternative expression for the gas pressure is

$$p_{\rm gas} = N_{\rm N} \left[1 + \sum_k \alpha_k \sum_{j=1}^{Z_k} j f_{jk}(n_{\rm e}, T) \right] kT. \tag{17.79}$$

Let A_k be the *atomic weight*[4] of element k and m_0 be the *atomic mass unit* [amu]. Then the mass density is

$$\rho = N_{\rm N} m_0 \sum_k \alpha_k A_k \equiv (N - n_{\rm e})\langle m \rangle, \tag{17.80}$$

where $\langle m \rangle$ is the *average mass per nucleus* (atoms + ions).

Another commonly used expression for the gas pressure is

$$p_{\rm gas} = \rho kT / \mu m_0, \tag{17.81}$$

where μ is the *effective mean molecular weight* of the gas:

$$\mu \equiv \sum_k \alpha_k A_k \bigg/ \left(1 + \sum_k \alpha_k \sum_{j=1}^{Z_k} j f_{jk}\right). \tag{17.82}$$

Here μ is the number of amu per free particle. Its value depends on the material's composition and degree of ionization. For example, (a) in a pure H plasma $\mu = 1.00$ when the gas is neutral, and $\mu = 0.5$ when it is fully ionized. (b) In a mix of 90% H and 10% He by number, $\mu = 1.3$ when the gas is neutral; $\mu = 0.68$ when H is fully ionized but He is neutral; $\mu = 0.65$ when H is fully ionized and He is singly ionized; and limits to $\mu = 0.62$ when both H and He are fully ionized. (c) In a pure helium plasma, $\mu = 4$ when He is neutral; $\mu = 2$ when it is all He^+; and limits to $\mu = 1.3$ when it is all He^{2+}. If a heavier element, e.g., C, N, O is the main constituent in a plasma, $\mu = A$ when it is neutral, and limits to $\mu = A/(Z+1)$ when fully ionized.

Electron Density

Initialization

To evaluate (17.76)–(17.82), we need to know the electron density $n_{\rm e}$. Suppose the gas consists of hydrogen, which has an ionization potential $\chi_{\rm H} = 13.6$ eV, and a single "metal," which has a single ionization stage with $\chi_{\rm M} \ll \chi_{\rm H}$, say, 4 or 5 eV. For the metal, take $\alpha_{\rm M} \ll 1$, so $\alpha_{\rm H} = 1 - \alpha_{\rm M} \approx 1$. From (17.76) and (17.78),

$$n_{\rm e}/N = (\alpha_{\rm H} f_{\rm H} + \alpha_{\rm M} f_{\rm M}) \big/ [\alpha_{\rm H}(1 + f_{\rm H}) + \alpha_{\rm M}(1 + f_{\rm M})]. \tag{17.83}$$

When the temperature is high enough for hydrogen to be even slightly ionized, it contributes most of the electrons; at lower temperatures, hydrogen is essentially neutral, and $n_{\rm e}$ is determined by $f_{\rm M}$. Thus if $\alpha_{\rm M} \ll 1$, then at high temperatures, when $f_{\rm H} \to 1$, $(n_{\rm e}/N) \to \frac{1}{2}$. At intermediate temperatures, when $f_{\rm M} \approx 1$, but $\alpha_{\rm M} \ll f_{\rm H} \ll 1$, $(n_{\rm e}/N) \approx f_{\rm H}$. At low temperatures, where $f_{\rm H} \to 0$, and $f_{\rm M}/f_{\rm H} \gg 1$, $(n_{\rm e}/N) \to \alpha_{\rm M} f_{\rm M}$. Therefore, at high temperatures, the metal abundance

[4] The atomic weight of an element includes the mass of all its electrons.

is essentially irrelevant in determining the electron number density n_e, but it is decisive at low temperatures. Recalling that absorption by H^- is the major opacity source in cooler stars, and $n(H^-)/n(H) \propto n_e$, we see that the metal abundance can also fix the opacity for some stars.

We saw in chapter 4 that if we know (n_e, T), we can calculate f_{jk} for all ions of all elements from Saha's equation (4.36). But if we are given (N, T) or (p_g, T), we must find n_e from the nonlinear equation (17.78). This can be done with a strongly convergent linearization procedure, described below, which fits into the formalism used for constructing both LTE and NLTE model atmospheres.[5] To start, we need an approximate first estimate of n_e.

For example, given N and T, (4.36), (17.77), and (17.78) can be solved analytically for the electron density in a pure hydrogen gas:

$$n_e(H) = \left\{ [N\Phi_H(T) + 1]^{1/2} - 1 \right\} \Big/ \Phi_H(T). \qquad (17.84)$$

If only the "metal" in the gas is ionized (i.e., $f_H \ll \alpha_M$), then

$$n_e(M) = \left\{ \left[\alpha_M N \Phi_M(T) + \tfrac{1}{4}(1+\alpha_M)^2 \right]^{1/2} - \tfrac{1}{2}(1+\alpha_M) \right\} \Big/ \Phi_M(T). \qquad (17.85)$$

If $\alpha_M \ll 1$ and $\chi_M < \chi_H$,

$$n_e^0 \approx n_e(H) + n_e(M) \qquad (17.86)$$

is a fairly good first estimate of n_e. For multi-element mixes, an analytical solution is not feasible, so we use the procedure described below.

Linearization Procedure

In the integration of an LTE model atmosphere, N and T at a given depth are set by constraints of hydrostatic and radiative equilibrium. To determine the physical and optical properties of the material and continue the integration, we must find $n_e(N, T)$ as a function of depth. This is most effectively done by linearizing the charge conservation equation and solving it iteratively. This procedure is a simple example of the technique used to solve the more complicated problems faced when transfer equations are coupled to radiative and hydrostatic equilibrium constraints (this chapter) or to NLTE kinetic equilibrium equations (chapter 18).

Using a first estimate of n_e^0 to evaluate the right-hand side of (17.78), we obtain a new value n_e^1, which generally $\neq n_e^0$. Then we replace n_e^0 with $n_e^0 + \delta n_e$, where δn_e is meant to satisfy (17.78) more closely. Because (17.78) is nonlinear, δn_e cannot be determined exactly. So we expand all terms to first order in δn_e to obtain a linear

[5] In addition, unlike table interpolation, it permits one to allow for changes in the chemical composition of the material at different values of (p_g, T), e.g., caused by thermonuclear reactions in a stellar interior.

equation for δn_e and use this to update our estimate of n_e. If our initial estimate is fairly accurate, i.e., $\delta n_e/n_e \ll 1$, this iteration converges quickly.

For brevity, from equation (17.78) define

$$\bar{\Sigma}(n_e, T) \equiv \sum_k \alpha_k \sum_{j=1}^{Z_k} j f_{jk}(n_e, T) = \sum_k [\alpha_k / S_k(n_e, T)] \sum_{j=1}^{Z_k} j P_{jk}(n_e, T). \quad (17.87)$$

Then the linearized version of (17.78) is

$$n_e^0 + \delta n_e \approx (N - n_e^0 - \delta n_e)\bar{\Sigma} + (N - n_e^0)(\partial \bar{\Sigma}/\partial n_e)_T \, \delta n_e, \quad (17.88)$$

which implies that

$$\delta n_e \approx \left[(N - n_e^0)\bar{\Sigma} - n_e^0\right] \Big/ \left[1 + \bar{\Sigma} - (N - n_e^0)(\partial \bar{\Sigma}/\partial n_e)_T\right]. \quad (17.89)$$

This value of δn_e is not exact, so we use $n_e^0(\text{new}) = n_e^0(\text{old}) + \delta n_e$ as a new estimate in all terms on the right-hand side of (17.89) and re-evaluate δn_e. If $\delta n_e/n_e < 1$ is small, then this process converges quadratically, and we obtain an accurate result quickly.

From the definitions of P_{jk} and S_k in equation (4.36), one can calculate analytical expressions for the derivatives $(\partial P_{jk}/\partial n_e)_T$ and $(\partial S_{jk}/\partial n_e)_T$, which are then assembled into $(\partial \bar{\Sigma}/\partial n_e)_T$:

$$\left(\frac{\partial \bar{\Sigma}}{\partial n_e}\right)_T = \sum_k \alpha_k \left[S_k^{-1} \sum_j j \frac{\partial P_{jk}}{\partial n_e} - S_k^{-2} \frac{\partial S_k}{\partial n_e} \sum_j j P_{jk}\right]. \quad (17.90)$$

Finally, using equations (4.33)–(4.35) with a converged value for n_e, we can evaluate

$$n_{ijk}^* \equiv n_e N_{j+1,k}^* \Phi_{ijk}(T) = \alpha_k n_e (N - n_e) f_{j+1,k}(n_e, T) \Phi_{ijk}(T). \quad (17.91)$$

The solution for $n_e(N, T)$ discussed above assumes N and T are given. But those quantities are determined by the requirement of hydrostatic and radiative equilibrium, which implies that their values at a given depth are coupled to other depths by differential (or finite-difference) equations. We address this problem in the next section. For now, we observe that N and T are both known only approximately at any stage of the calculation of a model atmosphere and that other physical quantities (e.g., opacities) depend on N, T, and n_e. Therefore, we also need to calculate the response of occupation numbers to perturbations in both N and T.

From (17.78) we find

$$n_e + \delta n_e \approx (N + \delta N - n_e - \delta n_e)\bar{\Sigma} + (N - n_e)\left[\left(\frac{\partial \bar{\Sigma}}{\partial n_e}\right)_T \delta n_e + \left(\frac{\partial \bar{\Sigma}}{\partial T}\right)_{n_e} \delta T\right], \quad (17.92)$$

or, if n_e is the solution of (17.89) for the current values of N and T,

$$\delta n_e = \left[\bar{\Sigma}\, \delta N + (N - n_e) \left(\frac{\partial \bar{\Sigma}}{\partial T} \right)_{n_e} \delta T \right] \Big/ \left[1 + \bar{\Sigma} - (N - n_e) \left(\frac{\partial \bar{\Sigma}}{\partial n_e} \right)_T \right]$$

$$\equiv \left(\frac{\partial n_e}{\partial N} \right)_T \delta N + \left(\frac{\partial n_e}{\partial T} \right)_N \delta T. \tag{17.93}$$

Again, $(\partial \bar{\Sigma}/\partial T)_{n_e}$ and $(\partial \bar{\Sigma}/\partial n_e)_T$ can be calculated analytically. Finally, from (17.91) one can develop an expression for δn^*_{ijk} of the form $\delta n^*_{ijk} = c_1\, \delta N + c_2\, \delta T + c_3\, \delta n_e$, which by use of (17.93) can be collapsed to an expression of the form

$$\delta n^*_{ijk} = \left(\frac{\partial n^*_{ijk}}{\partial N} \right)_T \delta N + \left(\frac{\partial n^*_{ijk}}{\partial T} \right)_N \delta T. \tag{17.94}$$

The two equations above provide the information needed in § 17.3 to find the response of the opacity and emissivity $(\delta\chi, \delta\eta)$ to changes $(\delta N, \delta T)$ in a model's structure.

17.3 NON-GRAY LTE RADIATIVE-EQUILIBRIUM MODELS

Here we outline two types of methods for constructing an LTE model atmosphere. Modern methods are based on linearization. The traditional methods, based on moment equations and the so-called temperature correction procedure, seem to be obsolete. However, they provide a valuable insight into the physical properties of atmospheric models, which would otherwise be lost if using a purely numerical linearization procedure. These schemes may even become useful, after appropriate generalization, in the context of modern, ALI-based methods, for computing NLTE model atmospheres (cf. chapters 18 and 20). Finally, they are used to provide a starting solution for the linearization method. We describe these methods first. In this section, we consider convectively stable atmospheres; model with convection will be described in the next section.

Moment Equations and Mean Opacities

The frequency-dependent moment equations of the transfer equation can be written as

$$(dH_\nu/dz) = -\kappa_\nu (J_\nu - B_\nu), \tag{17.95}$$

$$(dK_\nu/dz) = -\chi_\nu H_\nu, \tag{17.96}$$

and the frequency-integrated moment equations as

$$(dH/dz) = -(\kappa_J J - \kappa_B B), \tag{17.97}$$

$$(dK/dz) = -\chi_H H, \tag{17.98}$$

where $J \equiv \int_0^\infty J_\nu \, d\nu$ and similarly for H, K, and

$$\kappa_J \equiv \int_0^\infty \kappa_\nu J_\nu \, d\nu / J, \tag{17.99}$$

$$\kappa_B \equiv \int_0^\infty \kappa_\nu B_\nu \, d\nu / B, \tag{17.100}$$

$$\chi_H \equiv \int_0^\infty \chi_\nu H_\nu \, d\nu / H, \tag{17.101}$$

are the *absorption mean*, *Planck mean*, and *flux-mean* opacities, respectively. Note that the scattering coefficient is included in χ_H, but not in other mean opacities. This is because the scattering contributions cancel on the right-hand side of equation (17.95).

The above definition of mean opacities uses the monochromatic absorption coefficient *per unit length*. In fact, many studies use mean opacities that are defined using the monochromatic opacities *per unit mass*. They are defined as

$$\bar{\chi}_H \equiv \int_0^\infty (\chi_\nu / \rho) H_\nu \, d\nu / H, \tag{17.102}$$

and analogously for $\bar{\kappa}_J$ and $\bar{\kappa}_B$; we use the notation with a bar to distinguish these two definitions.

Moment equations (17.97) and (17.98) are *exact* (within LTE), but we see that one needs *three* generally different mean opacities. Moreover, the absorption mean and flux-mean opacities are not known a priori because they depend on unknown moments J_ν and H_ν. For frequency-independent opacity (true gray atmosphere), one obtains $\kappa_J = \kappa_B$, but they are not generally equal to χ_H unless the scattering is negligible. In the diffusion limit, one has $J_\nu \approx B_\nu$, and $H_\nu = \frac{1}{3}(dB_\nu/d\tau_\nu) = -\frac{1}{3}(dB_\nu/dT)(dT/dz)\chi_\nu^{-1}$ (cf. § 11.9). Therefore, $\kappa_J = \kappa_B$, and

$$\chi_H^{\rm diff} = \int_0^\infty (dB_\nu/dT) \, d\nu \Big/ \int_0^\infty \chi_\nu^{-1} (dB_\nu/dT) \, d\nu \equiv \chi_R, \tag{17.103}$$

where χ_R is the *Rosseland mean opacity*, defined by equation (11.175). As pointed out in § 11.9, the Rosseland mean opacity gives the correct value of the total radiation flux at great depths. Therefore, the temperature structure for a non-gray model given by

$$T^4(\tau_R) = \tfrac{3}{4} T_{\rm eff}^4 [\tau_R + Q(\tau_R)] \tag{17.104}$$

is asymptotically correct at great depth in an atmosphere once the diffusion approximation is valid. $Q(\tau_R)$ is a slowly varying bounded function analogous to the gray Hopf function. A reasonable approximation is to use the gray Hopf function, $Q(\tau_R) \approx q(\tau_R)$.

Unsöld-Lucy Temperature Correction Scheme

Before we describe the linearization method, we first outline one of the most popular temperature correction schemes, the Unsöld-Lucy procedure, which is directly related to the moment equations.

Using the optical depth scale corresponding to the Planck mean opacity,

$$d\tau \equiv -\kappa_B \, dz, \tag{17.105}$$

the integrated moment equations (17.97) and (17.98) become

$$(dH/d\tau) = (\kappa_J/\kappa_B)\, J - B, \tag{17.106}$$

$$(dK/d\tau) = (\chi_H/\kappa_B)\, H. \tag{17.107}$$

At a given stage of an iterative method, equations (17.106) and (17.107) hold individually for current (imperfect) values J, H, and K, but the total flux does not necessarily satisfy radiative equilibrium $H^{\mathrm{ex}} = (\sigma_R/4\pi)T_{\mathrm{eff}}^4$. One then introduces a correction $\Delta H(\tau)$ such that $H(\tau) + \Delta H(\tau) = H^{\mathrm{ex}}$, and corresponding corrections to J and K. Equations (17.106) and (17.107) are then written as

$$[d(H_0 + \Delta H)/d\tau] = (\kappa_J/\kappa_B)\,(J_0 + \Delta J) - (B_0 + \Delta B), \tag{17.108}$$

$$[d(K_0 + \Delta K)/d\tau] = (\chi_H/\kappa_B)\,(H_0 + \Delta H). \tag{17.109}$$

Since equations (17.106) and (17.107) are satisfied for the current values J_0, H_0, and K_0, one can drop them in equations (17.108) and (17.109) and write these equations as perturbation equations for corrections ΔJ, ΔH, and ΔK. Notice that the mean opacities κ_J, κ_B, and χ_H are left unperturbed, which is a reasonable approximation as long as they are not very sensitive to temperature. Equations (17.108) and (17.109) may be combined into one using the Eddington approximation to express $\Delta J = 3\,\Delta K$, and solving equation (17.109),

$$\Delta J(\tau) = \Delta J(0) + 3\int_0^\tau (\chi_H/\kappa_B)\Delta H(\tau')\, d\tau'. \tag{17.110}$$

Using the surface Eddington approximation $\Delta H(0) = \frac{1}{2}\Delta J(0)$, equation (17.108) gives the perturbation equation for ΔB:

$$\Delta B(\tau) = -\frac{d\Delta H(\tau)}{d\tau} + \frac{\kappa_J}{\kappa_B}\left[3\int_0^\tau \frac{\chi_H}{\kappa_B}\Delta H(\tau')\, d\tau' + 2\,\Delta H(0)\right]. \tag{17.111}$$

Assuming the corrections to $B(\tau)$ are small, one can use a linear approximation $\Delta B = (4\sigma_R T^3/\pi)\Delta T$, so that equation (17.111) can be viewed as an equation for corrections to temperature that give the correct total flux. It shows that a *local* change of temperature is driven by the differences of the current and exact Eddington flux ΔH in the whole atmospheric region between the surface and the current depth. Experience has shown that the Unsöld-Lucy procedure is quite effective in constructing LTE radiative-equilibrium models. The scheme can be made more accurate by dropping the Eddington approximation and introducing the averaged Eddington factors through

$$K \equiv \int_0^\infty K_\nu \, d\nu = \int_0^\infty J_\nu f_\nu \, d\nu \equiv fJ, \tag{17.112}$$

and similarly the surface Eddington factor

$$H(0) \equiv \int_0^\infty H_\nu(0) \, d\nu = \int_0^\infty J_\nu(0) \, g_\nu \, d\nu \equiv \bar{g} J(0), \tag{17.113}$$

so that equation (17.111) is generalized to

$$\frac{4\sigma_R T^3}{\pi} \Delta T = -\frac{d\Delta H}{d\tau} + \frac{\kappa_J}{\kappa_B} \frac{1}{f} \left[\int_0^\tau \frac{\chi_H}{\kappa_B} \Delta H(\tau') \, d\tau' + \frac{f(0) \Delta H(0)}{\bar{g}} \right]. \tag{17.114}$$

As follows from (17.108) and (17.106), $-d\Delta H/d\tau = dH_0/d\tau = (\kappa_J/\kappa_B)J_0 - B_0$, so that equation (17.114) can also be written as (dropping subscript 0)

$$\frac{4\sigma_R T^3}{\pi} \Delta T = \frac{\kappa_J}{\kappa_B} J - B + \frac{\kappa_J}{\kappa_B} \frac{1}{f} \left[\int_0^\tau \frac{\chi_H}{\kappa_B} \Delta H(\tau') \, d\tau' + \frac{f(0) \Delta H(0)}{\bar{g}} \right]. \tag{17.115}$$

The method can also be generalized to the NLTE case (cf. § 18.4), and even to moving atmospheres (cf. § 20.4).

Starting Procedure for Linearization

Before we can apply the iteration scheme described in § 17.2, we need a starting solution that is approximately correct. The basic parameters that define a model are its effective temperature, surface gravity, and chemical composition.

Temperature Distribution

As was shown above, a temperature distribution of the form

$$T^4(\tau_R) = \tfrac{3}{4} T_{\rm eff}^4 [\tau_R + Q(\tau_R)] \tag{17.116}$$

is asymptotically correct at great depth in an atmosphere and can be used as a reasonable first guess everywhere. A reasonable first guess to get started would be to use the gray Hopf function for Q in (17.116). Or if we know the temperature distribution $T_0(\tau_R)$ in a converged model having about the same surface gravity and effective temperature, we might adopt the scaled distribution

$$T^4(\tau_R) = T_{\rm eff}^4 [T^4(\tau_R)/T_{\rm eff}^4]_0 \equiv \tfrac{3}{4} T_{\rm eff}^4 [\tau_R + Q(\tau_R)] \tag{17.117}$$

to give us a first estimate for $T(\tau_R)$ or, equivalently, $Q(\tau_R)$. The distribution (17.117) will also be asymptotically correct at great optical depth.

Pressure Distribution

On a fixed column mass grid $\{m_d\}$, $(d = 1, \ldots, ND)$ one integrates the hydrostatic equilibrium equation (16.55) in the approximate form

$$(dp_g/dm) = g - \frac{4\pi}{c\rho}\int_0^\infty \chi_\nu H_\nu \, d\nu \approx g - (\sigma_R T_{\mathrm{eff}}^4 \chi_R/c\rho) \tag{17.118}$$

simultaneously with the definition of the optical depth scale

$$d\tau_R = (\chi_R/\rho)dm. \tag{17.119}$$

Boundary Values

To obtain starting values for the fundamental variables (T, N) at m_1, we assume that "at infinity" outside the atmosphere the mass coordinate, temperature, and density are zero. For the first grid point at $d = 1$, we choose m_1 so small (say $m_1 \sim 10^{-6}$) that τ_R will also be very small. In a star with a high effective temperature, the material at m_1 will be essentially completely ionized. For complete ionization, (17.78) and (17.80) give

$$(n_e/\rho)_1 = \sum_k \alpha_k Z_k \Big/ m_0 \sum_k \alpha_k A_k. \tag{17.120}$$

In completely ionized (rarefied) material, the opacity is dominated by electron scattering, which is frequency independent. Hence

$$(\chi_R/\rho)_1 = \sigma_e \sum_k \alpha_k Z_k \Big/ m_0 \sum_k \alpha_k A_k. \tag{17.121}$$

For example, if the material is composed of hydrogen and helium only, and $Y \equiv \alpha_{He}/\alpha_H$, then

$$(\chi_R/\rho)_1 = (\sigma_e/m_0)(1 + 2Y)/(1 + 4Y) = 0.4(1 + 2Y)/(1 + 4Y). \tag{17.122}$$

For $Y = 0.1$, $(\chi_R/\rho)_1 \approx 0.34\,\mathrm{cm}^2/\mathrm{gm}$, so for $m_1 \sim 10^{-6}$, $\tau_{R,1} \sim 3 \times 10^{-7}$. At this very small optical depth, a reasonable first guess for the boundary temperature is $T_1 = \left[\frac{3}{4}Q(0)\right]^{1/4} T_{\mathrm{eff}}$, and for the boundary pressure,

$$p_{g,1} = \left[g - (\sigma_R T_{\mathrm{eff}}^4/c)\,(\chi_R/\rho)_1\right] m_1 \equiv N_1 k T_1, \tag{17.123}$$

which gives us an estimate of N_1.

Continuing On

Given initial values for T_1 and N_1, the integration can proceed step by step inward with a "predictor-corrector" method.[6] To simplify notation, we omit the subscript R

[6] For this discussion we assume only one pass through the procedure.

on opacities and optical depths. Indicating a provisional value with a superscript 0, the "predictor" step is

$$\tau_d^0 = \tau_{d-1} + \left(\frac{d\tau}{dm}\right)^0_{d-\frac{1}{2}} (m_d - m_{d-1}) \approx \tau_{d-1} + \left(\frac{\chi}{\rho}\right)_{d-1} (m_d - m_{d-1}), \quad (17.124a)$$

$$T_d^0 \approx T_{\text{eff}} \left\{ \tfrac{3}{4} \left[\tau_d^0 + Q(\tau_d^0) \right] \right\}^{1/4}, \quad (17.124b)$$

$$\begin{aligned} p_{g,d}^0 &= N_d^0 k T_d^0 = N_{d-1} k T_{d-1} + \left(\frac{dp}{dm}\right)^0_{d-\frac{1}{2}} (m_d - m_{d-1}) \\ &\approx N_{d-1} k T_{d-1} + \left[g - \left(\frac{\sigma_R T_{\text{eff}}^4}{c}\right) \left(\frac{\chi}{\rho}\right)_{d-1} \right] (m_d - m_{d-1}). \quad (17.124c) \end{aligned}$$

Using these estimates of N_d^0 and T_d^0 in the linearization procedure of § 17.2, we can calculate an improved electron density $n_{e,d}^0$ and opacity $(\chi^0/\rho^0)_d$. Then the "corrector" step is

$$\tau_d = \tau_{d-1} + \frac{1}{2} \left(\frac{\chi_{d-1}}{\rho_{d-1}} + \frac{\chi_d^0}{\rho_d^0} \right) (m_d - m_{d-1}), \quad (17.125a)$$

$$T_d = T_{\text{eff}} \left\{ \tfrac{3}{4} \left[\tau_d + Q(\tau_d) \right] \right\}^{1/4}, \quad (17.125b)$$

$$p_{g,d} = N_{d-1} k T_{d-1} + \left[g - \left(\frac{\sigma_R T_{\text{eff}}^4}{2c}\right) \left(\frac{\chi_{d-1}}{\rho} + \frac{\chi_d^0}{\rho_d^0} \right) \right] (m_d - m_{d-1}). \quad (17.125c)$$

One could repeat this procedure to get more accurate integration from m_{d-1} to m_d. But doing that has little value inasmuch as we used a quasi-gray expression in (17.116) to obtain temperatures and, worse, ignored the frequency dependence of the opacity and emissivity.

Given T_d and N_d on the grid $\{m_d\}, (d = 1, \ldots, ND)$, we solve the equation of state for the electron density $n_e^*(N_d, T_d)$ and compute $n_{ijk}^*(N_d, T_d)$ for all atomic and ionic levels *ijk* at all points on the mass grid. Next we calculate χ_{nd}^* and η_{nd}^* on a frequency grid $\{\nu_n\} = \nu_1, \ldots, \nu_{NF}$ plus (isotropic) electron scattering and evaluate mean intensities J_{dn} and Eddington factors f_{dn} from a formal solution of the transfer equation with source functions S_{dn}; see chapter 12.

Finally, having an estimate of the frequency-dependent radiation field, we can evaluate a better estimate of the radiation force terms in (17.118) and (17.123) and re-integrate the hydrostatic equation to get better values of $\{N_d\}, d = 1 \ldots ND$,[7]

$$N_1 k T_1 = m_1 \left[g - \frac{4\pi}{c} \sum_{n=1}^{NF} w_n (\chi_{1n}/\rho_1) g_n J_{1,n} \right], \quad (17.126a)$$

[7] Note that both (17.126a) and (17.126b) are valid for both LTE and NLTE atmospheres, provided χ, f, and J have been evaluated consistently.

and

$$
\begin{aligned}
&N_d k T_d - N_{d-1} k T_{d-1} \\
&\quad = g(m_d - m_{d-1}) - \frac{4\pi}{c} \sum_{n=1}^{NF} w_n (f_{dn} J_{dn} - f_{d-1,n} J_{d-1,n}). \qquad (17.126b)
\end{aligned}
$$

Computational Method

We use this starting model to calculate the radiation field from discretized forms of (12.175)–(12.177), with $\kappa^{\mathrm{sc}} \equiv \chi - \kappa$ representing the scattering part of the total absorption coefficient, and the diffusion limit lower boundary condition (11.176).[8] Thus at $d = 1$,

$$
(f_{2n} J_{2n} - f_{1n} J_{1n}) / \Delta\tau_{\frac{3}{2},n} = g_n J_{1n}, \qquad (17.127a)
$$

at $d = 2, \ldots, ND - 1$,

$$
\begin{aligned}
&\frac{f_{d-1,n} J_{d-1,n}}{\Delta\tau_{d-\frac{1}{2},n} \Delta\tau_{dn}} - \frac{f_{dn}}{\Delta\tau_{dn}} \left(\frac{1}{\Delta\tau_{d-\frac{1}{2},n}} + \frac{1}{\Delta\tau_{d+\frac{1}{2},n}} \right) J_{dn} + \frac{f_{d+1,n} J_{d+1,n}}{\Delta\tau_{d+\frac{1}{2},n} \Delta\tau_{dn}} \\
&\quad = \left(1 - \frac{\kappa_{dn}^{\mathrm{sc}}}{\chi_{dn}} \right) J_{dn} - \frac{\eta_{dn}}{\chi_{dn}}, \qquad (17.127b)
\end{aligned}
$$

and at $d = ND$,

$$
\frac{(f_{ND,n} J_{ND,n} - f_{ND-1,n} J_{ND-1,n})}{\Delta\tau_{ND-\frac{1}{2},n}} = \left(\frac{H}{\chi_{\nu_n}} \frac{\partial B_{\nu_n}}{\partial T} \Big/ \sum_{n'} \frac{w_{n'}}{\chi_{\nu_{n'}}} \frac{\partial B_{\nu_{n'}}}{\partial T} \right)_{ND}, \qquad (17.127c)
$$

where

$$
\Delta\tau_{d\pm\frac{1}{2},n} \equiv \tfrac{1}{2} \left[(\chi_{d\pm1,n}/\rho_{d\pm1}) + (\chi_{dn}/\rho_d) \right] |m_{d\pm1} - m_d|, \qquad (17.127d)
$$

and

$$
\Delta\tau_{dn} \equiv \tfrac{1}{2} (\Delta\tau_{d-\frac{1}{2},n} + \Delta\tau_{d+\frac{1}{2},n}). \qquad (17.127e)
$$

We would find the resulting radiation field does not satisfy the discretized constraint of radiative equilibrium

$$
\sum_{n=1}^{NF} (\eta_{nd} - \kappa_{nd} J_{nd}) \equiv 0, \quad (d = 1, \ldots, ND). \qquad (17.128)
$$

[8] Here we assume that both χ and η are isotropic and use first-order boundary conditions for simplicity. It is straightforward to include second-order terms.

Therefore, we must revise the structure of the model, i.e., $T(m)$ and $N(m)$, such that the transfer, radiative-equilibrium, and hydrostatic-equilibrium equations are all satisfied simultaneously. There are two difficulties: (a) the problem is *nonlinear*, and (b) the coupling is *global*. For example, a change δT_d at *any* depth point d implies a change δN_d (through hydrostatic equilibrium); and hence changes δn_{dijk} in all occupation numbers, and thus changes $\delta\chi_{dn}$ and $\delta\eta_{dn}$ in the opacity and emissivity at all frequencies ν_n at that depth. And therefore it changes $\delta J_{d'n'}$ in the radiation field at all frequencies $\nu_{n'}$ and *all* depths d' throughout the entire atmosphere, which in turn drive changes $\delta T'_d$ and $\delta N'_d$ at all other depth points d'.

To deal with these problems we linearize the transfer and structural equations by replacing each variable x with $x_0 + \delta x$, expanding all expressions to first order in the δx's, and solving the resulting linear system for them. The power of this method is that (a) it can be applied for a wide variety of constraints, and (b) it accounts for the effects of a change in any variable at a given point in the atmosphere on all variables at all other points. In particular it describes, to first order, the global effects of changes in any local variable. The result is a Newton-Raphson procedure, which can yield convergence after only a few iterations.

In linearizing the transfer equations we assume that Eddington factors remain unchanged and will be updated later using updated values of the material properties. Then at each frequency ν_n, we have

$$\begin{aligned}
&\frac{f_{d-1,n}\delta J_{d-1,n}}{\Delta\tau_{d-\frac{1}{2},n}\Delta\tau_{dn}} - \left[\frac{f_{dn}}{\Delta\tau_{dn}}\left(\frac{1}{\Delta\tau_{d-\frac{1}{2},n}} + \frac{1}{\Delta\tau_{d+\frac{1}{2},n}}\right) + \left(1 - \frac{\kappa^{\text{sc}}_{dn}}{\chi_{dn}}\right)\right]\delta J_{dn} \\
&\qquad + \frac{f_{d+1,n}\delta J_{d+1,n}}{\Delta\tau_{d+\frac{1}{2},n}\Delta\tau_{dn}} + a_{dn}\delta\omega_{d-1,n} + b_{dn}\delta\omega_{dn} + c_{dn}\delta\omega_{d+1,n} \\
&\qquad - (\eta_{dn} + \kappa^{\text{sc}}_{dn}J_{dn})\frac{\delta\chi_{dn}}{\chi^2_{dn}} + \frac{\delta\eta_{dn}}{\chi_{dn}} + \frac{\delta\kappa^{\text{sc}}_{dn}}{\chi_{dn}}J_{dn} \\
&\qquad = \beta_{dn} + J_{dn} - (\kappa^{\text{sc}}_{dn}J_{dn} + \eta_{dn})/\chi_{dn},
\end{aligned} \tag{17.129}$$

where

$$\alpha_{dn} \equiv (f_{dn}J_{dn} - f_{d-1,n}J_{d-1,n})/(\Delta\tau_{d-\frac{1}{2},n}\Delta\tau_{dn}), \tag{17.130a}$$

$$\gamma_{dn} \equiv (f_{dn}J_{dn} - f_{d+1,n}J_{d+1,n})/(\Delta\tau_{d+\frac{1}{2},n}\Delta\tau_{dn}), \tag{17.130b}$$

$$\beta_{dn} = \alpha_{dn} + \gamma_{dn}, \tag{17.130c}$$

$$a_{dn} \equiv \left[\alpha_{dn} + \tfrac{1}{2}\beta_{dn}(\Delta\tau_{d-\frac{1}{2},n}/\Delta\tau_{dn})\right]/(\omega_{dn} + \omega_{d-1,n}), \tag{17.130d}$$

$$c_{dn} \equiv \left[\gamma_{dn} + \tfrac{1}{2}\beta_{dn}(\Delta\tau_{d+\frac{1}{2},n}/\Delta\tau_{dn})\right]/(\omega_{dn} + \omega_{d+1,n}), \tag{17.130e}$$

$$b_{dn} \equiv a_{dn} + c_{dn}, \tag{17.130f}$$

$$\omega_{dn} \equiv \chi_{dn}/\rho_d. \tag{17.130g}$$

Similar expressions can be derived for the boundary conditions. They are considered in detail in § 18.2.

Assuming LTE, the corrections to all material properties in equations (17.129)–(17.130g) can be expressed in terms of δN and δT only. Thus from (17.80),

$$\delta\rho_d = (\delta N_d - \delta n_{e,d})\langle m\rangle, \tag{17.131}$$

which, using (17.93), results in an expression of the form

$$\delta\rho_d = \left(\frac{\partial\rho}{\partial N}\right)_d \delta N_d + \left(\frac{\partial\rho}{\partial T}\right)_d \delta T_d. \tag{17.132}$$

From (11.11) and (11.12) we have

$$\delta\chi_{dn} = \left(\frac{\partial\chi_n^*}{\partial T}\right)_d \delta T_d + \left(\frac{\partial\chi_n^*}{\partial n_e}\right)_d \delta n_{e,d} + \sum_{ijk}\left(\frac{\partial\chi_n^*}{\partial n_{ijk}^*}\right)_d \delta n_{ijk,d}^*, \tag{17.133}$$

and similarly for $\delta\kappa_{dn}$ and $\delta\eta_{dn}$. In (17.133) and its counterpart for $\delta\eta$, $(\partial/\partial T)$ is applied only to *explicit* appearances of T in $\exp(-h\nu/kT), \alpha_{kk}(\nu, T)$, etc., and similarly for $(\partial/\partial n_e)$. Finally, using (17.93) and (17.94), (17.133) can be collapsed to an expression of the form

$$\delta\chi_{dn} = \left(\frac{\partial\chi_{dn}^*}{\partial N}\right)_d \delta N_d + \left(\frac{\partial\chi_{dn}^*}{\partial T}\right)_d \delta T_d, \tag{17.134}$$

with a similar expression for $\delta\kappa_{dn}$ and $\delta\eta_{dn}$.

The linearized constraint of radiative equilibrium is

$$\sum_n w_n\left[\kappa_{dn}\delta J_{dn} + \delta\kappa_{dn}J_{dn} - \delta\eta_{dn}\right] = \sum_n w_n[\eta_{dn} - \kappa_{dn}J_{dn}], \tag{17.135}$$

and the linearized constraint of hydrostatic equilibrium is

$$\begin{aligned}\frac{4\pi}{c}\sum_n & w_n(f_{dn}\delta J_{dn} - f_{d-1,n}\delta J_{d-1,n})\\ &+ k\,(T_d\delta N_d + N_d\delta T_d - T_{d-1}\delta N_{d-1} - N_{d-1}\delta T_{d-1})\\ &= g\,(m_d - m_{d-1}) - N_d kT_d + N_{d-1}kT_{d-1}\\ &- \frac{4\pi}{c}\sum_n w_n(f_{dn}J_{dn} - f_{d-1,n}J_{d-1,n}).\end{aligned} \tag{17.136}$$

The upper boundary condition is given in § 18.2.

The whole system for all depths and frequencies can be organized into a form suitable for a Rybicki-method solution. Let

$$\boldsymbol{\delta}\mathbf{J}_n \equiv (\delta J_{1n}, \delta J_{2n}, \ldots, \delta J_{ND,n})^{\mathrm{T}}, \quad (n = 1, \ldots, NF) \tag{17.137}$$

$$\boldsymbol{\delta}\mathbf{T} \equiv (\delta T_1, \delta T_2, \ldots, \delta T_{ND})^{\mathrm{T}}, \tag{17.138}$$

$$\boldsymbol{\delta}\mathbf{N} \equiv (\delta N_1, \delta N_2, \ldots, \delta N_{ND})^{\mathrm{T}}, \tag{17.139}$$

$$\mathbf{E}_n \equiv (\mathbf{E}_1, \mathbf{E}_2, \ldots, \mathbf{E}_{ND})^{\mathrm{T}}. \tag{17.140}$$

Here $\delta\mathbf{J}_n$, $\delta\mathbf{N}$, $\delta\mathbf{T}$, and $\mathbf{E}_n$ are column vectors of length *ND* containing, respectively, the depth variation of coefficients of the corrections δJ_{dn} to the radiation field at frequency ν_n; the corrections δN_d and δT_d to the total particle number density and temperature; and E_{dn} is the right-hand side of (17.129) for frequency ν_n.

At each frequency ν_n, the linearized transfer equation (17.129) reduces to a set of linear equations of the general form

$$\sum_{d'=d-1}^{d+1} \mathbf{U}_{dd',n}\delta\mathbf{J}_{d'n} + \sum_{d'=d-1}^{d+1} \mathbf{V}_{dd',n}\delta\mathbf{N}_{d'} + \sum_{d'=d-1}^{d+1} \mathbf{W}_{dd',n}\delta\mathbf{T}_{d'} = \mathbf{E}_{dn} \tag{17.141}$$

for $(n = 1, NF)$. $\mathbf{U}_n$, $\mathbf{V}_n$, and $\mathbf{W}_n$ are $ND \times ND$ tridiagonal matrices that account for the coupling of the corrections to the radiation field at frequency ν_n, and material properties as functions of T and N, at the three adjacent grid points $(d-1, d, d+1)$.

The contribution to the linearized radiative equilibrium equation from each frequency n at depth point d is of the form

$$\mathbf{X}_n\delta\mathbf{J}_n + \mathbf{A}_n\delta\mathbf{N} + \mathbf{B}_n\delta\mathbf{T} = \mathbf{F}_n. \tag{17.142}$$

Here, because there is no coupling between depth points, $\mathbf{X}$, $\mathbf{A}$, and $\mathbf{B}$ are $ND \times ND$ diagonal matrices, and $\mathbf{F}$ is a column vector of length *ND* representing the right-hand side of (17.135).

The contribution to the linearized hydrostatic equilibrium equation from each frequency ν_n at depth point d is of the form

$$\mathbf{Y}_n\delta\mathbf{J}_n + \mathbf{C}_n\delta\mathbf{N} + \mathbf{D}_n\delta\mathbf{T} = \mathbf{G}_n. \tag{17.143}$$

Here $\mathbf{Y}$, $\mathbf{C}$, and $\mathbf{D}$ are $ND \times ND$ bidiagonal matrices that account for the coupling of the corrections to the radiation field and material properties at adjacent grid points $(d, d+1)$, and $\mathbf{G}$ is a column vector of length *ND* representing the right-hand side of (17.136).

Equations (17.141)–(17.143) yield the system

$$\begin{pmatrix} \mathbf{U}_1 & 0 & \cdots & \cdots & \cdots & 0 & \mathbf{V}_1 & \mathbf{W}_1 \\ 0 & \mathbf{U}_2 & 0 & \cdots & \cdots & \vdots & \mathbf{V}_2 & \mathbf{W}_2 \\ \vdots & 0 & \mathbf{U}_3 & \cdots & \cdots & \vdots & \mathbf{V}_3 & \mathbf{W}_3 \\ \vdots & \vdots & \vdots & \cdots & \cdots & \vdots & \vdots & \vdots \\ 0 & 0 & 0 & \cdots & \cdots & \mathbf{U}_{NF} & \mathbf{V}_{NF} & \mathbf{W}_{NF} \\ \mathbf{X}_1 & \mathbf{X}_2 & \cdots & \cdots & \cdots & \mathbf{X}_{NF} & \mathbf{A} & \mathbf{B} \\ \mathbf{Y}_1 & \mathbf{Y}_2 & \cdots & \cdots & \cdots & \mathbf{Y}_{NF} & \mathbf{C} & \mathbf{D} \end{pmatrix} \begin{pmatrix} \delta\mathbf{J}_1 \\ \delta\mathbf{J}_2 \\ \delta\mathbf{J}_3 \\ \vdots \\ \delta\mathbf{J}_{NF} \\ \delta\mathbf{N} \\ \delta\mathbf{T} \end{pmatrix} = \begin{pmatrix} \mathbf{E}_1 \\ \mathbf{E}_2 \\ \mathbf{E}_3 \\ \vdots \\ \mathbf{E}_{NF} \\ \mathbf{F} \\ \mathbf{G} \end{pmatrix}. \tag{17.144}$$

To solve (17.144), express $\delta\mathbf{J}_n$ in the nth "row" of the grand matrix in terms of $\delta\mathbf{N}$ and $\delta\mathbf{T}$ as

$$\delta\mathbf{J}_n = (\mathbf{U}_n^{-1}\mathbf{E}_n) - (\mathbf{U}_n^{-1}\mathbf{V}_n)\delta\mathbf{N} - (\mathbf{U}_n^{-1}\mathbf{W}_n)\delta\mathbf{T}. \tag{17.145}$$

The $ND \times ND$ matrices $\mathbf{U}_n^{-1}$ are in general *full*. Then we eliminate each $\boldsymbol{\delta}\mathbf{J}_n$ from the last two "rows" of the grand matrix one frequency at a time to generate a final system of the form

$$\begin{pmatrix} \mathbf{P} & \mathbf{Q} \\ \mathbf{R} & \mathbf{S} \end{pmatrix} \begin{pmatrix} \boldsymbol{\delta}\mathbf{N} \\ \boldsymbol{\delta}\mathbf{T} \end{pmatrix} = \begin{pmatrix} \mathbf{L} \\ \mathbf{M} \end{pmatrix}, \tag{17.146}$$

where

$$\mathbf{P} = \mathbf{A} - \sum_{n=1}^{NF} \mathbf{X}_n(\mathbf{U}_n^{-1}\mathbf{V}_n) \qquad \mathbf{Q} = \mathbf{B} - \sum_{n=1}^{NF} \mathbf{X}_n(\mathbf{U}_n^{-1}\mathbf{W}_n) \tag{17.147a}$$

$$\mathbf{R} = \mathbf{C} - \sum_{n=1}^{NF} \mathbf{Y}_n(\mathbf{U}_n^{-1}\mathbf{V}_n) \qquad \mathbf{S} = \mathbf{D} - \sum_{n=1}^{NF} \mathbf{Y}_n(\mathbf{U}_n^{-1}\mathbf{W}_n) \tag{17.147b}$$

$$\mathbf{L} = \mathbf{F} - \sum_{n=1}^{NF} \mathbf{X}_n(\mathbf{U}_n^{-1}\mathbf{E}_n) \qquad \mathbf{M} = \mathbf{G} - \sum_{n=1}^{NF} \mathbf{Y}_n(\mathbf{U}_n^{-1}\mathbf{E}_n). \tag{17.147c}$$

$\mathbf{P}, \mathbf{Q}, \mathbf{R}$, and $\mathbf{S}$ are $(ND \times ND)$ full matrices. Then the final system (17.146) is solved for $\boldsymbol{\delta}\mathbf{N}$ and $\boldsymbol{\delta}\mathbf{T}$.

Using $\boldsymbol{\delta}\mathbf{N}$ and $\boldsymbol{\delta}\mathbf{T}$, we revise the particle number density and temperature at each grid point m_d; solve the equation of state for the electron density $n_{ed}^*(N_d, T_d)$; compute $n_{ijk}^*(N_d, T_d)$ for all atomic and ionic levels *ijk*; calculate $\chi_{nd}^*(N_d, T_d)$ and $\eta_{nd}^*(N_d, T_d)$ on the frequency grid $\{\nu_n\}$ using LTE versions of (11.11) and (11.12); and evaluate new values for the mean intensities J_{dn} and Eddington factors f_{dn} from a formal solution of the transfer equation. We use these revised values to reconstruct (17.144) and iterate. As the solution improves, $\mathbf{E}, \mathbf{F}$, and $\mathbf{G}$ all $\rightarrow 0$; hence $\boldsymbol{\delta}\mathbf{N}$ and $\boldsymbol{\delta}\mathbf{T} \rightarrow 0$.

It must be remembered that the changes $\boldsymbol{\delta}\mathbf{N}$ and $\boldsymbol{\delta}\mathbf{T}$ given by this procedure are *not exact* because they do not account for nonlinearities in the equations. In the first few iterations there is the possibility that the "corrections" obtained in this way are inaccurate enough that if we applied the full changes, the iteration process might diverge. So if at any depth a fractional change $|\delta N_d/N_d|$ or $|\delta T_d/T_d|$ exceeds some pre-specified maximum, say 10%, then both the $\boldsymbol{\delta}\mathbf{N}$ and $\boldsymbol{\delta}\mathbf{T}$ vectors are scaled by the same constant factor so the largest fractional change made in any value of either variable is less than this maximum. The value of the allowed maximum change has to be determined by experience. If the problem has been programmed correctly, after a few iterations the corrections decrease rapidly, and their full values can be applied safely.

In this scheme the computing time per iteration scales as $t_c = c\,(2NF)ND^2 + c'(2ND)^3$. Because t_c is *linear* in the number of frequencies NF, many frequencies may be used, e.g., to model line blanketing. But there is a penalty because we must solve for *two* depth-dependent quantities. Experience has shown [426] that when the radiation force term in the hydrostatic equation (17.118) is small relative to the gravitational force, as it is for stars with low T_{eff}, the pressure stratification is essentially determined by gravity, so p_g is a known linear function of m_d. In this case we set $\delta p_g \equiv 0$, which implies $\delta N = -(N/T)\delta T$, and use this relation to combine terms containing the elements of $\boldsymbol{\delta}\mathbf{N}$ with terms containing the elements of $\boldsymbol{\delta}\mathbf{T}$ in (17.141), (17.142), and (17.143). Then $\boldsymbol{\delta}\mathbf{N}$ has been eliminated from the calculation, and the computing time per iteration now scales as $t_c = c\,(NF)ND^2 + c'(ND)^3$, a substantial saving.

17.4 MODELS WITH CONVECTION

The atmosphere is convectively unstable if the Schwarzschild criterion for convective instability is satisfied,

$$\nabla_{\rm rad} > \nabla_{\rm ad}, \tag{17.148}$$

where $\nabla_{\rm rad} = (d \ln T/d \ln P)_{\rm rad}$ is the logarithmic temperature gradient in radiative equilibrium, and $\nabla_{\rm ad}$ is the adiabatic gradient. The latter is viewed as a function of temperature and pressure, $\nabla_{\rm ad} = \nabla_{\rm ad}(T, P)$. As in chapter 16, the density ρ is considered to be a function of T and P through the equation of state.

If the condition (17.148) is satisfied, the radiative equilibrium equation is modified to become

$$\int_0^\infty H_\nu\, d\nu + \frac{F_{\rm conv}}{4\pi} = \int_0^\infty \frac{d(f_\nu J_\nu)}{d\tau_\nu}\, d\nu + \frac{F_{\rm conv}}{4\pi} = \frac{\sigma_R}{4\pi}\, T_{\rm eff}^4, \tag{17.149}$$

where $F_{\rm conv}$ is the convective flux, given by [cf. equation (16.73)]

$$F_{\rm conv} = (gQH_P/32)^{1/2}(\rho c_P T)(\nabla - \nabla_{\rm el})^{3/2}(\ell/H_P)^2, \tag{17.150}$$

and where

$$H_P \equiv -(d \ln P/dz)^{-1} = P/(\rho g) \tag{17.151}$$

is the pressure scale height, $Q \equiv -(d \ln \rho/d \ln T)_P$, c_P is the specific heat at constant pressure, ℓ/H_P is the ratio of the convective mixing length to the pressure scale height, taken as a free parameter of the problem, ∇ is the actual logarithmic temperature gradient, and $\nabla_{\rm el}$ is the gradient of the convective elements. The latter is determined by considering the efficiency of the convective transport. As derived in § 16.5, the gradient $\nabla_{\rm el}$ is obtained by solving [see equation (16.80)]

$$\nabla_{\rm el} - \nabla_{\rm ad} = \mathcal{B}\sqrt{\nabla - \nabla_{\rm el}}, \tag{17.152}$$

where

$$\mathcal{B} \equiv \frac{16\sqrt{2}\sigma_R T^3}{\rho c_p (gQH_P)^{1/2}(\ell/H_P)}\, \frac{\tau_{\rm el}}{1 + \frac{1}{2}\tau_{\rm el}^2}, \tag{17.153}$$

and $\tau_{\rm el} = \chi_R \ell$ is the optical thickness of the characteristic element size ℓ.

Starting Model

To obtain a starting, gray model with convection, we proceed as follows. We first construct the starting model in the absence of convection, as described in § 17.3. Then at each point we can calculate $\nabla_{\rm rad} = \nabla_{\rm rad}(T, \rho, P)$ and $\nabla_{\rm ad} = \nabla_{\rm ad}(T, \rho, P)$. If at some point we find that the criterion for stability against convection is violated, we must determine the true gradient ∇, where $(\nabla_{\rm ad} \le \nabla \le \nabla_{\rm rad})$, that satisfies (17.149). If the instability occurs deep enough for the diffusion approximation to be valid, then $(F_{\rm rad}/F) = (\nabla/\nabla_{\rm rad})$, and (17.149) and (17.150) reduce to

$$\mathcal{A}\big(\nabla - \nabla_{\rm el}\big)^{3/2} = \nabla_{\rm rad} - \nabla, \tag{17.154}$$

where

$$\mathcal{A} = (\nabla_{\rm rad}/\sigma_{\rm R} T_{\rm eff}^4)(gQH_P/32)^{1/2}(\rho c_P T)(\ell/H_P)^2. \tag{17.155}$$

We see that $\mathcal{A}$ depends only on local variables. Adding $\big(\nabla - \nabla_{\rm el}\big) + \big(\nabla_{\rm el} - \nabla_{\rm ad}\big)$ to both sides of (17.154), and using (17.152) to eliminate $\big(\nabla_{\rm el} - \nabla_{\rm ad}\big)$, we obtain a cubic equation for $x \equiv \big(\nabla - \nabla_{\rm el}\big)^{1/2}$, namely,

$$\mathcal{A}\big(\nabla - \nabla_{\rm el}\big)^{3/2} + \big(\nabla - \nabla_{\rm el}\big) + \mathcal{B}\big(\nabla - \nabla_{\rm el}\big)^{1/2} = \big(\nabla_{\rm rad} - \nabla_{\rm ad}\big) \tag{17.156a}$$

or

$$\mathcal{A}x^3 + x^2 + \mathcal{B}x = \big(\nabla_{\rm rad} - \nabla_{\rm ad}\big), \tag{17.156b}$$

which can be solved numerically for the root x_0. We thus obtain the true gradient $\nabla = \nabla_{\rm ad} + \mathcal{B}x_0 + x_0^2$ and can proceed with the integration, now regarding T as a function of P.

Linarization

It is convenient to write the integral form of the radiative + convective equilibrium equation, obtained by differentiating (17.149) with respect to m,

$$\int_0^\infty (\kappa_\nu J_\nu - \eta_\nu)\, d\nu + \frac{\rho}{4\pi}\frac{dF_{\rm conv}}{dm} = 0. \tag{17.157}$$

If convection is taken into account, one must first evaluate the radiative and adiabatic temperature gradients and test for the Schwarzschild stability criterion (17.148). The temperature gradient is evaluated as

$$\nabla_{d-\frac{1}{2}} = \frac{T_d - T_{d-1}}{T_d + T_{d-1}} \cdot \frac{P_d + P_{d-1}}{P_d - P_{d-1}}. \tag{17.158}$$

The corresponding adiabatic gradient also has to be evaluated at the midpoint $d - \frac{1}{2}$, $\nabla_{\rm ad} = \nabla_{\rm ad}(T_{d-\frac{1}{2}}, P_{d-\frac{1}{2}})$, with $T_{d-\frac{1}{2}} = \frac{1}{2}(T_d + T_{d-1})$, and a similar expression for

$P_{d-\frac{1}{2}}$. The convective flux is a quantity that also corresponds to the grid midpoints; hence it should be written as $F_{\rm conv} \equiv F_{{\rm conv},d-\frac{1}{2}}$.

At the depth points where the Schwarzschild criterion is satisfied, the energy balance equation (17.157) is discretized to read

$$\sum_i w_i(\kappa_i J_i - \eta_i) + (\rho_d/4\pi)(F_{{\rm conv},d+\frac{1}{2}} - F_{{\rm conv},d-\frac{1}{2}})/\Delta m_d = 0, \qquad (17.159)$$

where $\Delta m_d = \frac{1}{2}(m_{d+\frac{1}{2}} - m_{d-\frac{1}{2}})$. The first, radiative-equilibrium, term of (17.159) is discretized as in equation (17.135). The convective term is linearized using a numerical differentiation. Introduce $\widetilde{H}_{{\rm conv},d\pm\frac{1}{2}} \equiv (\rho_d/4\pi)F_{{\rm conv},d\pm\frac{1}{2}}$. Then

$$\delta\widetilde{H}_{{\rm conv},d\pm\frac{1}{2}} = \left(\frac{\partial\widetilde{H}_{{\rm conv},d\pm\frac{1}{2}}}{\partial N_d}\right)\delta N_d + \left(\frac{\partial\widetilde{H}_{{\rm conv},d\pm\frac{1}{2}}}{\partial T_d}\right)\delta T_d + \left(\frac{\partial\widetilde{H}_{{\rm conv},d\pm\frac{1}{2}}}{\partial N_{d\pm1}}\right)\delta N_{d\pm1} + \left(\frac{\partial\widetilde{H}_{{\rm conv},d\pm\frac{1}{2}}}{\partial T_{d\pm1}}\right)\delta T_{d\pm1}, \qquad (17.160)$$

where the partial derivatives $\partial\widetilde{H}_{\rm conv}/\partial T$ and $\partial\widetilde{H}_{\rm conv}/\partial N$ are evaluated numerically, for instance, $\partial\widetilde{H}_{{\rm conv},d\pm1}/\partial T_d = [\widetilde{H}_{{\rm conv},d\pm1}(T_d + \Delta T_d) - \widetilde{H}_{{\rm conv},d\pm1}(T_d)]/\Delta T_d$, where $\Delta T_d = \alpha T_d$ with α being a small number, e.g., $\alpha = 0.01$. A different, but essentially analogous, procedure for linearizing the convective term will be considered in chapter 18.

17.5 LTE SPECTRAL LINE FORMATION

A spectral line is more opaque in its core than in its wings, so radiation in its observed profile comes from a wide range of atmospheric depths: from high layers seen in its core, to deep layers where the continuum is formed, seen in its wings. The strength of a spectral line depends on the number of absorbing atoms along the line of sight, and hence on that element's abundance in the atmosphere. From measurements of line strengths we can make quantitative analyses of stellar compositions. Further, spectral lines have narrow frequency widths, so their profiles are sensitive to velocity fields in a star's atmosphere.

In this section we describe some classical approaches to the analysis of stellar line strengths. Fuller discussions of these methods and their application can be found in [20, 21, 246, 379, 391, 731, 859, 1006, 1102, 1178]. But we emphasize from the outset that *because they assume LTE, these older methods are only heuristic, having made assumptions that do not have a solid physical basis.* Nevertheless, they can provide some basic orientation, and as mentioned in chapter 2, important discoveries were made using them.

Transfer Equation

Much of the classical work on spectral line formation was based on an extremely simplified version of the transfer equation. Write the opacity in a line transition $(i \to j)$ as $\chi_\ell \phi_\nu$, where $\chi_l \equiv (\pi e^2/mc) f_{ij}[n_i - (g_i/g_j)n_j]$, and ϕ_ν is the line's normalized profile. Assume a fraction ϵ of the line photons are absorbed, destroyed, and their energy fed into the thermal pool. This energy loss from the radiation field is balanced by an equal amount emitted thermally, contributing

$$\eta_l^{\mathrm{t}}(\nu) = \epsilon \chi_l \phi_\nu B_\nu(T) \tag{17.161}$$

to the total emission coefficient. The remaining photons are assumed to be scattered coherently and isotropically, so that

$$\eta_l^{\mathrm{s}}(\nu) = (1-\epsilon)\chi_l \phi_\nu J_\nu. \tag{17.162}$$

Write the continuum opacity as $\chi_{\mathrm{c}} = \kappa_{\mathrm{c}} + \sigma_{\mathrm{c}}$, where κ_{c} is the thermal absorption coefficient and σ_{c} the (isotropic, coherent) electron scattering coefficient. In terms of these sources and sinks, the transfer equation is

$$\begin{aligned} \mu \frac{dI_\nu}{dz} =& -(\kappa_{\mathrm{c}} + \sigma_{\mathrm{c}} + \chi_l\, \phi_\nu)\, I_\nu \\ &+ (\kappa_{\mathrm{c}} + \epsilon \chi_l \phi_\nu) B_\nu + [\sigma_{\mathrm{c}} + (1-\epsilon)\chi_l \phi_\nu] J_\nu. \end{aligned} \tag{17.163}$$

Most spectral lines are so narrow that the frequency dependence of κ_{c} and σ_{c} is negligible compared to the swift variation of ϕ_ν over the line profile. Let $d\tau_\nu \equiv -(\kappa_{\mathrm{c}} + \sigma_{\mathrm{c}} + \chi_\ell\, \phi_\nu)\, dz$, $\beta_\nu \equiv \chi_\ell\, \phi_\nu/(\kappa_{\mathrm{c}} + \sigma_{\mathrm{c}})$, and $r_{\mathrm{c}} \equiv \sigma_{\mathrm{c}}/(\kappa_{\mathrm{c}} + \sigma_{\mathrm{c}})$. Then we can rewrite (17.163) as

$$\mu \frac{dI_\nu}{d\tau_\nu} = I_\nu - \frac{\epsilon\, \beta_\nu + (1 - r_{\mathrm{c}})}{1+\beta_\nu} B_\nu - \frac{(1-\epsilon)\, \beta_\nu + r_{\mathrm{c}}}{1+\beta_\nu} J_\nu. \tag{17.164}$$

Or, further defining

$$\xi_\nu \equiv [\epsilon\, \beta_\nu + (1 - r_{\mathrm{c}})]/(1+\beta_\nu), \tag{17.165}$$

we have

$$\mu \frac{dI_\nu}{d\tau_\nu} = I_\nu - \xi_\nu B_\nu - (1-\xi_\nu)\, J_\nu. \tag{17.166}$$

Equation (17.166) is known as the *Milne-Eddington equation*; see [302], [758, pp. 159–187], [908], [1061].

From a physical point of view, (17.166) is a severe idealization of line formation and can be criticized on several counts:

(1) Scattering in a spectral line is *not* coherent. This inadequacy may partially be overcome by replacing (17.166) with

$$\mu \frac{dI_\nu}{d\tau_\nu} = I_\nu - \xi_\nu B_\nu - \frac{(1-\epsilon)\beta_\nu}{(1+\beta_\nu)\phi_\nu} \int_0^\infty R(\nu', \nu)\, J_{\nu'}\, d\nu' - \frac{r_{\mathrm{c}}}{1+\beta_\nu} J_\nu, \tag{17.167}$$

where $R(\nu', \nu)$ is an appropriate (angle-averaged) redistribution function (cf. chapters 10 and 15).

(2) To solve (17.166) or (17.167), both the occupation numbers n_i and n_j and the thermalization parameter ϵ must be known. In classical treatments of line formation one assumes the occupation numbers are given by their LTE values: $n_i = n_i^*$, and $n_j = n_j^*$. But as we saw in chapter 14, *there is no justification for this assumption, and in general it may yield results that are seriously in error.* It could possibly be valid if $\epsilon \approx 1$; but even then the occupation numbers n_l and n_u can be driven away from their LTE values by other processes (e.g., photoionization). When the line scatters radiation, it is certainly inconsistent to suppose the occupation numbers have their LTE values, because they depend on the line's radiation field via the kinetic equilibrium equations.

(3) An equation of the form (17.166) was derived by Milne for coherent scattering by a strict two-level atom (i.e., no continuum); his analysis yielded the correct value for ϵ in that case [758, pp. 172–178]. However, analysis of multi-level atomic models shows (cf. chapter 14) that *other kinds* of terms may appear in the source function, which depend upon the radiation fields in transitions (both continua and lines) *other than the one under consideration*, so, in principle, *all lines and continua in the spectrum are coupled together in a collective photon pool.* Clearly, both (17.166) and (17.167) are incomplete physically.

Line Profiles

An exact solution of (17.166) in a semi-infinite atmosphere can be found [228] when r_c, β_ν, and ϵ, and hence ξ_ν, are constant with depth, and the Planck function B_ν is a linear function of the continuum optical depth at frequency ν:

$$B_\nu = a_\nu + b_\nu \tau_c = a_\nu + b_\nu \tau_\nu/(1 + \beta_\nu) \equiv a_\nu + c_\nu \tau_\nu. \tag{17.168}$$

It differs little from the approximate solution below, which shows the main features of the problem. The zero-order moment of (17.166) is

$$\frac{dH_\nu}{d\tau_\nu} = J_\nu - (1 - \xi_\nu) J_\nu - \xi_\nu B_\nu = \xi_\nu (J_\nu - B_\nu). \tag{17.169}$$

In the Eddington approximation its solution for the mean intensity is

$$J_\nu(\tau_\nu) = a_\nu + c_\nu \tau_\nu + \left[\frac{c_\nu - \sqrt{3}\, a_\nu}{\sqrt{3}\,(1 + \sqrt{\xi_\nu})} \right] \exp(-\sqrt{3\, \xi_\nu}\, \tau_\nu), \tag{17.170}$$

which is the same as equation (12.27) with ϵ generalized to ξ_ν as defined in (17.165). The emergent flux in the line is

$$H_\nu(0) = J_\nu(0)/\sqrt{3} = (\sqrt{3\, \xi_\nu}\, a_\nu + c_\nu)/3(1 + \sqrt{\xi_\nu}). \tag{17.171}$$

Equation (17.170) shows that thermalization ($J_\nu \to B_\nu$) occurs only at optical depths $\tau > 1/\sqrt{\xi_\nu}$. From (17.165), this depth is $\sim 1/\sqrt{1-r_c}$ in the continuum ($\beta_\nu = 0$) and $\sim 1/\sqrt{\epsilon}$ in a strong line ($\beta_\nu \to \infty$). In both cases thermalization occurs at a depth $1/\sqrt{p_d}$, where p_d is the probability that a photon is destroyed and converted to thermal energy as it interacts with the material. These results are compatible with the random-walk arguments in § 12.1. But keep in mind that all the results in the remainder of this section apply only for coherent scattering and for the highly idealized assumptions listed above (17.168).

The frequency profile of the flux in a line is given by (17.171). In the continuum, $\beta_\nu \equiv 0$, so $c_\nu \equiv b_\nu$ and $\xi_\nu \equiv (1 - r_c)$; hence the emergent continuum flux is

$$H_c(0) = \left[\sqrt{3(1-r_c)}a_\nu + b_\nu\right] \Big/ \left[3\left(1+\sqrt{1-r_c}\right)\right]. \tag{17.172}$$

Combining (17.171) and (17.172), we find that the residual flux in the line is

$$R_\nu(0) \equiv \frac{H_\nu(0)}{H_c(0)} = \left[\frac{\sqrt{3\xi_\nu}\, a_\nu + c_\nu}{1+\sqrt{\xi_\nu}}\right]\left[\frac{1+\sqrt{1-r_c}}{a_\nu\sqrt{3(1-r_c)}+b_\nu}\right]. \tag{17.173}$$

In the early work on line formation this classical theory led to the following categorization.

- **Scattering Lines**
  Set $r_c = 0$ (no continuum scattering) and $\epsilon = 0$, i.e., pure scattering in the line. Then $\xi_\nu = 1/(1+\beta_\nu)$, $H_c = \frac{1}{6}(b_\nu + \sqrt{3}a_\nu)$, and (17.173) becomes

$$R_\nu^{\text{scat}}(0) = 2\left[\sqrt{3}a_\nu + \frac{b_\nu}{\sqrt{1+\beta_\nu}}\right] \Big/ \left[\left(1+\sqrt{1+\beta_\nu}\right)\left(\sqrt{3}a_\nu + b_\nu\right)\right]. \tag{17.174}$$

  As $\beta_\nu \to \infty$ in the core, $R_\nu^{\text{scat}}(0) \to 0$; i.e., *the core of a strong scattering line is dark.*

- **Absorption Lines**
  In LTE, $r_c = 0$ and $\epsilon = 1$; then $\xi_\nu \equiv 1$, and

$$R_\nu^{\text{abs}}(0) = \left[\sqrt{3}\, a_\nu + b_\nu/(1+\beta_\nu)\right] \Big/ \left(\sqrt{3}\, a_\nu + b_\nu\right). \tag{17.175}$$

  As $\beta_\nu \to \infty$, *the residual flux in the core of an absorption line remains finite*:

$$R_\nu^{\text{abs}}(\beta_\nu \to \infty) = 1 \Big/ \left[1 + (b_\nu/\sqrt{3}\, a_\nu)\right]. \tag{17.176}$$

  As $\beta_\nu \to \infty$, we receive radiation from only the surface layers of the atmosphere, and the emergent flux is determined by the surface value of the Planck function, which is nonzero. If there were no temperature gradient, $b_\nu = 0$, so $R_\nu^{\text{abs}}(0) \equiv 1$, i.e., the line disappears. In contrast, in a scattering line, photons are continually diverted out of a pencil of radiation, and in the limit $\beta_\nu \to \infty$, none survive to emerge at the surface.

Solar Lines

To test the predictions of these formulae, one can compare them with observed solar line profiles. Assume a strong line is formed in LTE and use (17.176) to calculate its profile. Rewrite (17.168) as

$$B_\nu(\tau_\nu) = a_\nu + b_\nu \tau_c \equiv B_\nu(T_0) + \left(\frac{\partial B_\nu}{\partial T}\right)_0 \left(\frac{\partial T}{\partial \tau_R}\right)_0 \left(\frac{\partial \tau_R}{\partial \tau_c}\right) \tau_c; \tag{17.177}$$

then $a_\nu = B_\nu(T_0)$. To calculate b_ν we use the Eddington-approximation gray atmosphere temperature distribution (17.68): $T^4(\tau_R) = T_0^4\left(1 + \frac{3}{2}\tau_R\right)$. Then we have $(\partial B_\nu/\partial T)_0 = (u_0/T_0)B_\nu(T_0)/(1 - e^{-u_0})$, $(\partial T/\partial \tau_R)_0 = \frac{3}{8}T_0$, $(\partial \tau_R/\partial \tau_c) = \kappa_R/\kappa_c$, so $b_\nu = \frac{3}{8}B_\nu(T_0)X_0(\kappa_R/\kappa_c)$, where $u_0 \equiv h\nu/kT_0$, and $X_0 \equiv u_0/(1 - e^{-u_0})$. Hence

$$R_0^{\rm abs} = 1 \Big/ \left\{1 + \left[\tfrac{1}{8}\sqrt{3}X_0(\kappa_R/\kappa_c)\right]\right\}. \tag{17.178}$$

For the Sun, the gray temperature distribution gives $T_0 \approx 4800$ K. If we choose $\lambda = 5000$ Å, then $u_0 \approx 6$, $X_0 \approx 6$, and $\kappa_c \simeq \kappa_R$. Then (17.178) gives

$$R_0^{\rm abs} = 1 \Big/ \left(1 + \tfrac{3}{4}\sqrt{3}\right) = 0.44. \tag{17.179}$$

This value is in fair agreement with the depths of some of the stronger lines observed in that wavelength region of the solar spectrum. Some lines, however, are deeper, specifically resonance lines such as the sodium D-lines. In the classical picture, these facts led to the concept that resonance lines are "scattering lines," and subordinate lines (e.g., $H\alpha$) are "absorption lines."

The classification of lines into these two types was thought to be reasonable because in a resonance line, the most probable exit mode of the excited electron from the upper state is direct radiative decay to the lower state, whereas in subordinate lines, a number of decay modes may exist, and photons may effectively be removed from the line and destroyed. It was thought that the central intensity of subordinate lines gives information about the surface temperature of the atmosphere. But the agreement is fortuitous. The characterization of resonance lines as "scattering lines" and subordinate lines as "absorption lines" is at best schematic, and as shown in chapter 14 is incompatible with the actual physics of line formation. For example, the boundary value of the source function of the $H\alpha$ line (a subordinate line) has little to do with the temperature of the outer solar atmospheric layers.

Center-to-Limb Variation

Using S_ν from (17.166), the emergent specific intensity at frequency ν and angle $\theta = \cos^{-1}\mu$ with respect to the normal is

$$I_\nu(0,\mu) = \int_0^\infty S_\nu(\tau_\nu)\, e^{-\tau_\nu/\mu} d\tau_\nu/\mu$$
$$= \int_0^\infty [B_\nu + (1-\xi_\nu)(J_\nu - B_\nu)]\, e^{-\tau_\nu/\mu}\, d\tau_\nu/\mu. \quad (17.180)$$

Using (17.170) for J_ν we find

$$I_\nu(0,\mu) = (a_\nu + c_\nu\,\mu) + \left[\frac{(1-\xi_\nu)(c_\nu - \sqrt{3}\,a_\nu)}{\sqrt{3}\,(1+\sqrt{\xi_\nu})\,(1+\sqrt{3\xi_\nu}\,\mu)}\right]. \quad (17.181)$$

In the continuum, $\beta_\nu = 0$, and $r_c \equiv 0$; then $\xi_\nu \equiv 1$ and $c_\nu \equiv b_\nu$. Hence

$$I_c(0,\mu) = a_\nu + b_\nu\mu. \quad (17.182)$$

From these expressions we find that the residual intensity is $r_\nu(\mu) \equiv I_\nu(0,\mu)/I_c(0,\mu)$ and the absorption depth is $a_\nu(\mu) \equiv 1 - r_\nu(\mu)$.

Consider first an "absorption" line ($\epsilon = 1$). Then

$$r_\nu(\mu) = \big[a_\nu + b_\nu\mu/(1+\beta_\nu)\big]/(a_\nu + b_\nu\mu). \quad (17.183)$$

In this picture, as the limb is approached, i.e., $\mu \to 0, I_\nu \to I_c$ and $r_\nu(\mu) \to 1$ because then only the surface layer, and hence the same value of the Planck function, is seen throughout the profile, and the contrast between the line and continuum vanishes. On the other hand, for $\epsilon = 0$ (a "pure scattering line"), both $\xi_\nu \to 0$ and $c_\nu \to 0$ as $\beta_\nu \to \infty$, and (17.183) gives $I_\nu(0,\mu) \equiv 0$ for all μ. Thus the cores of "pure scattering lines" remain dark even at the limb, so there is a distinction between the center-to-limb variation of "absorption" and "scattering" lines.

In the solar spectrum, some lines do weaken toward the limb, whereas others do not. This behavior supported the notion that lines can be classified as "absorption" and "scattering" lines, even though that classification might conflict with that based on their central intensities. Actually, neither categorization is adequate because LTE level populations were assumed even for "scattering" lines, and noncoherent scattering has been ignored. In short, the approach described above is at best heuristic and ignores many fundamental physical phenomena.

Theoretical Curve of Growth

For faint stars, it may be difficult to obtain line profiles with high-dispersion spectrographs, so a procedure to analyze equivalent-width data is needed. In such work the goal is to get estimates of the relative abundances of the elements in the stellar material and to infer information about a star's effective temperature and surface gravity. One assumes that spectral lines are formed in a layer having a known average temperature and electron density.[9] Line profiles are calculated for a range of assumed abundances, and their equivalent widths are obtained by integrating over frequency. A *curve of growth* is a plot of a line's equivalent width versus the number of absorbing atoms along the line of sight.

One assumes LTE to compute the occupation number $n_i^*(N,T)$, the continuum opacity χ_c, and the line opacity χ_{ij} for a transition $(i \to j)$ as

$$\chi_{ij} = \frac{\pi e^2}{mc} f_{ij}\, n_i^*\, (1 - e^{-h\nu/kT}). \tag{17.184}$$

For the line profile we use a normalized Voigt function, so $\chi_{ij}(\nu) = \chi_0\, H(a, v)$, where $\chi_0 \equiv \chi_{ij}/\sqrt{\pi}\,\Delta\nu_D$, and

$$H(a, v) = \frac{a}{\pi} \int_{-\infty}^{\infty} \frac{e^{-y^2}\, dy}{(v-y)^2 + a^2}. \tag{17.185}$$

Here $v \equiv \Delta\nu/\Delta\nu_D$, $a \equiv \Gamma/(4\pi\,\Delta\nu_D)$, and $\Delta\nu_D \equiv \nu\,\xi_0/c$, where ξ_0 is the most probable velocity of the absorbing atoms along the line of sight.

In the Milne-Eddington model, $\Delta\nu_D$, the parameter a, and the ratio $\beta_\nu \equiv \chi_{ij}(\nu)/\kappa_c$ are assumed to be constant with depth in the region of line formation. The approximation of a depth-independent β_ν is actually fairly good for some spectral lines, e.g., Mg II λ4481 and Si II $\lambda\lambda$ 4128, 4151 in B-stars. With these assumptions, the emergent flux at frequency ν in a line is

$$F_\nu = 2 \int_0^\infty B_\nu[T(\tau_c)]\, E_2\left[\int_0^{\tau_c} (1+\beta_\nu)\, dt\right] (1+\beta_\nu)\, d\tau_c. \tag{17.186}$$

As before, take $B_\nu[T(\tau_c)] = B_0 + B_1\tau_c$, where τ_c is the optical depth in the continuum adjacent to the line. Then for β_ν constant with depth,

$$F_\nu = 2 \int_0^\infty (B_0 + B_1\tau_c)\, E_2\,[(1+\beta_\nu)\tau_c]\,(1+\beta_\nu)\, d\tau_c \tag{17.187a}$$

$$= 2B_0 \int_0^\infty E_2(x)\, dx + \frac{2B_1}{1+\beta_\nu} \int_0^\infty E_2(x)\, x\, dx \tag{17.187b}$$

$$= B_0 + \tfrac{2}{3}\, B_1/(1+\beta_\nu). \tag{17.187c}$$

From (17.187c) the continuum flux is $F_c = B_0 + \frac{2}{3}B_1$, so the line depth in the flux profile is

$$A_\nu = [\beta_\nu/(1+\beta_\nu)]/\left[1 + \tfrac{3}{2}(B_0/B_1)\right]. \tag{17.188}$$

[9] Clearly, it is an oversimplification to represent a stellar atmosphere as a uniform layer with a unique temperature and density because both have significant gradients in regions where spectral lines are formed. This assumption may be appropriate for analysis of laboratory data obtained from a homogeneous vapor in a back-lighted absorption tube.

Taking the limit of (17.188) as $\beta_\nu \to \infty$, we find the central depth of an infinitely opaque line is $A_0 = 1/[1 + \frac{3}{2}(B_0/B_1)]$; hence

$$A_\nu = A_0 \beta_\nu / (1 + \beta_\nu). \tag{17.189}$$

Assuming the line to be symmetric about its center, its equivalent width is

$$W = \int_0^\infty A_\nu \, d\nu = 2A_0 \, \Delta\nu_D \int_0^\infty \frac{\beta(v)\, dv}{1 + \beta(v)}. \tag{17.190}$$

In LTE, $\beta_\nu \equiv \chi_\ell \, \phi_\nu / \kappa_c$; so write $\beta(v) = (\chi_0/\kappa_c)\, H(a, v) \equiv \beta_0 H(a, v)$ and define the *reduced equivalent width* as $W_0 \equiv W_\nu/(2A_0 \Delta\nu_D)$; then

$$W_0(a, \beta_0) = \int_0^\infty \frac{\beta_0 H(a, v)\, dv}{1 + \beta_0 H(a, v)}. \tag{17.191}$$

Qualitative Expectations

Consider how a line develops as more atoms absorb radiation along the line of sight. (1) At first, the absorbers present each remove photons from the radiation field, so the line strength is proportional to the number of absorbing atoms. In this limit, the Doppler core may be opaque, so its opacity sets the equivalent width. The line wings are still essentially transparent and do not reduce the observed flux. (2) As more absorbing atoms are added, the core becomes opaque and reaches a limiting depth. It is then *saturated*; in this regime, the addition of more absorbers increases the equivalent width only slowly. (3) Finally, when a large enough number of absorbers are present, the opacity in the line wings is sufficient to increase the equivalent width.

Linear Part

As a schematic representation of the Voigt profile, write $H(a, v) \sim \exp(-v^2) + a/\sqrt{\pi}\, v^2$. In this formula we use the first term in the core ($v \le v^*$) and the second in the wings ($v \ge v^*$), where v^* is chosen as the transition point where the two terms are equal.

Consider now the contribution from the core only; take $\beta_0 \ll 1$, and write $\beta(v) \approx \beta_0 \, e^{-v^2}$. Then (17.191) becomes

$$W_0 = \beta_0 \int_0^\infty \frac{e^{-v^2} dv}{1 + \beta_0 \, e^{-v^2}} = \beta_0 \int_0^\infty e^{-v^2} (1 - \beta_0 e^{-v^2} + \ldots)\, dv, \tag{17.192}$$

or

$$W_0 = \tfrac{1}{2}\sqrt{\pi}\beta_0 \left[1 - (\beta_0/\sqrt{2}) + (\beta_0^2/\sqrt{3}) - \ldots\right]. \tag{17.193}$$

For small values of β_0 (weak lines) the linear term dominates. This limit is called the *linear part* of the curve of growth. *On the linear part of the curve, the equivalent width is directly proportional to the number of absorbers.* Note that β_0 varies as $1/\Delta\nu_D$, as does W_0; therefore, on the linear part, the equivalent width W is independent of $\Delta\nu_D$.

Saturation Part

On the *flat* or *saturation part* of the curve of growth, $\beta_0 \gg 1$ is large enough that the line core has reached its limiting depth, but is not so large that the line wings contribute significantly to the equivalent width. Take $\beta(v) \approx \beta_0\, e^{-v^2}$, and let $u = v^2$ so that $dv = du/2\sqrt{u}$; then (17.191) becomes

$$W_0 = \frac{1}{2}\int_0^\infty \frac{\beta_0\, e^{-u}\, du}{\sqrt{u}(1+\beta_0\, e^{-u})} = \frac{1}{2}\int_0^\infty \frac{du}{\sqrt{u}(1+e^{u-\alpha})}, \tag{17.194}$$

where $\beta_0 \equiv e^\alpha$. This integral may be rewritten [225, p. 390] as

$$2W_0 = \int_0^\alpha \frac{du}{\sqrt{u}} + \alpha \int_0^\infty \frac{\sqrt{\alpha(1+t)}}{1+e^{\alpha t}}\, dt - \alpha \int_0^\alpha \frac{\sqrt{\alpha(1-t)}}{1+e^{\alpha t}}\, dt. \tag{17.195}$$

Following Sommerfeld, replace the upper limit in the third integral by ∞ (because $\alpha \gg 1$) and expand $\sqrt{\alpha(1 \pm t)}$ as a power series in t around $t = 0$. Sommerfeld showed that the result can be written in closed form using the Riemann zeta function, and ultimately we obtain the asymptotic expression

$$W_0 = \sqrt{\ln \beta_0}\left[1 - \frac{\pi^2}{24(\ln \beta_0)^2} - \frac{7\pi^4}{384(\ln \beta_0)^4} - \ldots\right]. \tag{17.196}$$

This expansion is semi-convergent; it is useful for $\beta_0 \gtrsim 55$. On the saturation (or "flat") part of the curve of growth, the equivalent width grows very slowly with increasing number of absorbers: $W_0 \propto \sqrt{\ln \beta_0}$. The weak dependence of W_0 on β_0 implies that W on this part of the curve is proportional to $\Delta\nu_D$ because the depth of the line profile is fixed at A_0; hence its equivalent width is proportional to the line width (cf. figure 17.2). Similarly, W depends on β_0 because the optical depth in the line must exceed unity before the continuum optical depth does in order to produce a depression in the continuum. This condition is met when $v \le v_0$, where $\beta_0 \exp(-v_0^2) \approx 1$. Thus v_0, which sets the width of the dark core (and hence determines W_0), varies as $\sqrt{\ln \beta_0}$.

Damping Part

For very large numbers of absorbers, the line wings finally become opaque enough to give the dominant contribution to the equivalent width. This regime is called the *damping* or *square-root part* of the curve of growth. Taking $H(a, v) \propto a/\sqrt{\pi} v^2$ and defining $K \equiv a\beta_0/\sqrt{\pi}$, (17.191) gives

$$W_0 = \int_0^\infty \frac{dv}{1+v^2/K} = \tfrac{1}{2}\pi\sqrt{K} = \tfrac{1}{2}\sqrt{\pi a \beta_0}. \tag{17.197}$$

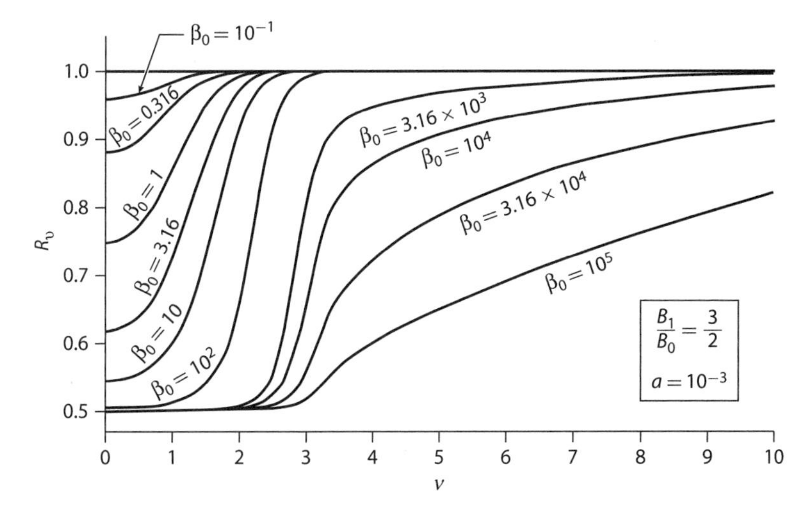

Figure 17.2 Development of an LTE line profile with increasing numbers of atoms along the line of sight. R_v is residual intensity.

Hence the reduced equivalent width $W_0 \propto \sqrt{\beta_0}$. And recalling that both a and β_0 vary as $1/\Delta\nu_D$, one finds that W is independent of $\Delta\nu_D$ on this part of the curve. The entire curve of growth is shown in figure 17.3. Notice that the larger the value of the damping parameter a, the sooner the wings dominate W, and hence the sooner the damping part of the curve rises away from the flat part. Useful sets of curves

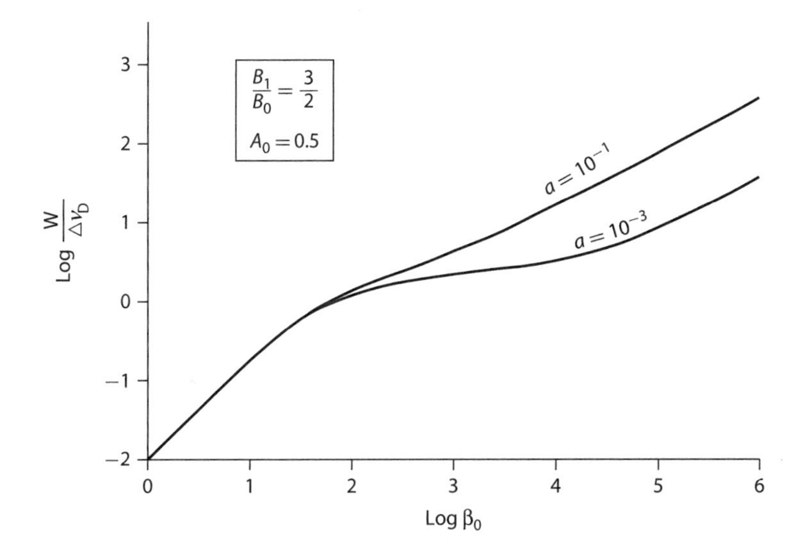

Figure 17.3 Milne-Eddington curves of growth for two values of $a = \Gamma/4\pi\,\Delta\nu_D$.

of growth are given in [1180–1182] for a range of the parameters B_0 and B_1 and different assumptions about the atmospheric model and the transfer problem.

Empirical Curve of Growth

Curves of growth provide estimates of some important properties of an atmosphere and require only equivalent width data, so they can be used for faint stars.

Procedure

In the theoretical curve of growth one plots $\log(W_\nu/\Delta\nu_D) = \log(W_\lambda/\Delta\lambda_D) = \log(W_\lambda c/\lambda\xi_0)$ as a function of $\log\beta_0$. Here ξ_0 is the total random velocity of the atoms forming the line, and

$$\beta_0 = \frac{\chi_\ell}{\kappa_c} = \left(\frac{\sqrt{\pi}e^2 f_{ij}}{mc\Delta\nu_D}\right)\frac{n^*_{ijk}}{\kappa_c} = \left(\frac{\sqrt{\pi}e^2}{mc}\right)(f_{ij}\lambda)\left(\frac{n^*_{ijk}}{\xi_0\kappa_c}\right). \tag{17.198}$$

The stimulated emission correction factor $(1 - e^{-h\nu/kT})$ that appears in both the line and continuum opacities has canceled out.

In a stellar spectrum one typically observes lines from several multiplets of an ion, in one or more ionization stages of an element, for several elements. Consider first lines from a single ion of one element. Because of the factor $\exp(-\chi_{ijk}/kT)$ in n^*_{ijk}, each multiplet has its own curve of growth. In view of equation (4.20), (17.198) can be rewritten as

$$\log\beta_0 = \log(g_{ijk}f_{ijk}\lambda) - \theta\chi_{ijk} + \log C_{jk}, \tag{17.199}$$

where $\theta \equiv 5040/T$, χ is in eV, and

$$C_{jk} \equiv [N_{jk}/U_{jk}(T)][(\sqrt{\pi}e^2/mc)/(\xi_0\kappa_c)]. \tag{17.200}$$

To construct an empirical curve of growth, one plots the value of $\log(W_\lambda/\lambda)$ versus $\log(gf\lambda)$ for each line of ion j of element k. Then assuming there is a unique relation between β_0 and W_λ, θ in (17.199) is adjusted to minimize the scatter around a mean curve; the corresponding value of T is called the *excitation temperature* $T_{\rm exc}$, which is regarded as the characteristic temperature in the line-forming region. A good example is shown in figure 17.4.

Results

The empirical curve corrected for excitation effects is then compared with a theoretical curve. To superimpose them, a relative shift in both the abscissa and the ordinate is normally needed. This procedure yields two results.

(1) The ordinate of the empirical curve is $\log(W_\lambda/\lambda)$; for the theoretical curve it is $\log(W_\lambda/\Delta\lambda_D) = \log(W_\lambda/\lambda) - \log(\xi_0/c)$. Their difference yields ξ_0, which

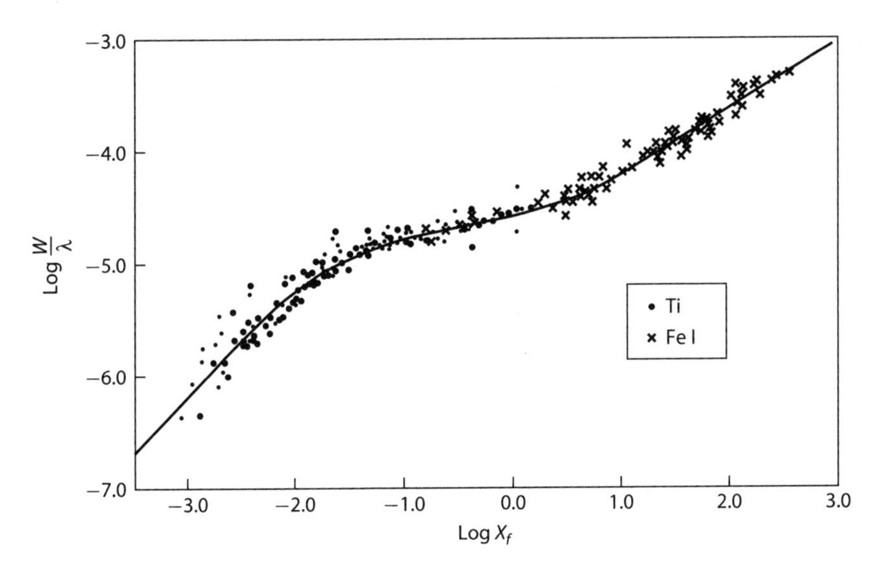

Figure 17.4 An empirical curve of growth. X_f in this figure $\equiv \beta_0$ defined above. From [1179].

often exceeds the most probable thermal velocity $\xi_{\rm th} = (2kT_{\rm exc}/Am_{\rm H})^{\frac{1}{2}}$ at the excitation temperature $T_{\rm exc} = 5040/\theta_{\rm exc}$. The discrepancy is attributed to non-thermal motions, called *microturbulence* [1064], on scales comparable to a photon mean path. If they have a Gaussian distribution around an average speed $\xi_{\rm turb}$, then

$$\xi_0 = (\xi_{\rm th}^2 + \xi_{\rm turb}^2)^{1/2}. \tag{17.201}$$

As noted in chapter 2, these motions are not turbulence in the hydrodynamic sense, but likely the result of the overshoot of motions in a deeper unstable layer into a star's photosphere [1048].

(2) The difference between the abscissae of the theoretical curve of growth and the average of empirical curves of growth yields

$$\log C_{jk} = \log \beta_0 - [\log(gf\lambda) - \theta_{\rm exc}\chi_{ijk}] \tag{17.202}$$

and thus an estimate of the abundance of ion j of element k.

LTE Spectrum Synthesis

The curve of growth method has many approximations that limit the accuracy of the results. Nevertheless, curves of growth revealed the huge difference between the "metals" content of Population I and Population II stars and showed the progressive enrichment of the primeval hydrogen/helium plasma with higher-Z material ejected from supernovae. Today, having reliable atomic data and fast computers, we can do much better. For bright stars (e.g., the Sun [659]) we have high-quality spectroscopic data, so we can make a point-by-point synthesis of the spectrum.

Choice of the Model

First, we select a model that closely resembles the stellar atmosphere to be analyzed by comparing observed and computed parameters in the star's spectrum, e.g., (1) the continuum energy distribution (UV, visible, and IR); (2) measures of continuum features such as the Balmer jump D_B; (3) broad- and intermediate-band colors; (4) density-sensitive line profiles; and (5) line-strength ratios of two or more ionization stages of a given element. For example, for B-stars, continuum parameters can be used to estimate $T_{\rm eff}$; hydrogen-line profiles are sensitive to $\log g$; and the ratio of the equivalent widths of, say, Si III $\lambda 4552$ to Si II $\lambda\lambda\, 4128, 31$ is a function of both $T_{\rm eff}$ and $\log g$ but is not too sensitive to the abundance of Si.

A useful technique is to make a plot in the ($\log g$ versus $T_{\rm eff}$) plane of the loci of ($T_{\rm eff}, \log g$) values where the computed value of a chosen parameter [say, $W_\lambda(H\gamma)$] equals the observed value. The loci of different criteria have different dependencies on temperature and gravity, and hence can intersect, ideally in a single point but in practice typically in a small area, thus defining optimum values for ($T_{\rm eff}, \log g$) with an estimate of their uncertainty.

Line Profiles

Using the temperature/density distribution in the atmosphere, we compute the excitation-ionization balance for levels of an element that produce observed spectral lines and the continuum opacity at the line's wavelength.

The absorption and emission coefficients are given by equations (11.11) and (11.12), with all the level populations given by their LTE values, $n_{ijk} = n^*_{ijk}$. In particular, for a line $l \to u$ of the species k and ionization state j, the absorption coefficient is

$$
\begin{aligned}
\kappa_{lu}(\nu) &= \frac{h\nu}{4\pi}(n_l^* B_{lu} - n_u^* B_{ul})\,\phi_{lu}(\nu) = \frac{\pi e^2}{m_e c} f_{lu} n_l^* \left(1 - \frac{n_u^* B_{ul}}{n_l^* B_{lu}}\right)\phi_{lu}(\nu) \\
&= \frac{\pi e^2}{m_e c} f_{lu} \frac{g_l}{U_{jk}} N_{jk} e^{-E_l/kT}\left(1 - e^{-h\nu/kT}\right)\phi_{lu}(\nu) \\
&= C\, g_l f_{lu} \frac{N_{jk}}{U_{jk}}\, e^{-E_l/kT} \frac{1}{\Delta\nu_D} H(a, v)\left(1 - e^{-h\nu/kT}\right),
\end{aligned}
\qquad (17.203)
$$

where $C \equiv \pi^{\frac{1}{2}} e^2/m_e c$ is a constant, $C = 0.01497$ in cgs units; U_{jk} and N_{jk} are the partition function and the total number density of ion j of element k; g_l and E_i are the statistical weight and the energy (with respect to the ground state), respectively, of level l; $\Delta\nu_D$ is the Doppler width of line $l \to u$; and $H(a, v)$ is the Voigt function. Here we dropped subscripts j and k on g_l, E_l, f_{lu}, and $\Delta\nu_D$. The atomic properties of the line $l \to u$ are contained only in the product $g_l f_{lu}$, which explains why the atomic line lists tabulate the gf-values, the energy E_l, and the wavelength λ_{lu}; the latter enters the Doppler width and the Voigt function parameter $v \equiv (\nu - \nu_{lu})/\Delta\nu_D$.

The total absorption and emission coefficients at a given frequency ν are given by

$$\chi(\nu) = \kappa_c(\nu) + \sum_{lu} \kappa_{lu}(\nu) + \kappa^{\text{sc}}(\nu), \tag{17.204}$$

$$\eta(\nu) = \left[\kappa_c(\nu) + \sum_{lu} \kappa_{lu}(\nu)\right] B_\nu + \kappa^{\text{sc}}(\nu) J_\nu, \tag{17.205}$$

where $\kappa_c(\nu)$ is the absorption coefficient in the continuum (all bound-free and free-free processes contributing at frequency ν) and $\kappa^{\text{sc}}(\nu)$ the scattering coefficient [e.g., $\kappa^{\text{sc}}(\nu) = n_{\text{e}}\sigma_{\text{e}}$ for electron (Thomson) scattering]. The form of equation (17.205) assumes that the scattering is coherent and isotropic. If the scattering contribution is small, such as in the case of cool stars, one can approximate $\kappa^{\text{sc}}(\nu) J_\nu \approx \kappa^{\text{sc}}(\nu) B_\nu$; hence $\eta(\nu) = \chi(\nu) B_\nu$.

The transfer equation is solved frequency by frequency. Essentially any method described in chapter 12 can be used. Due to the presence of J_ν in the emission coefficient, the most straightforward way is to use the Feautrier scheme. One can use the short-characteristics scheme or the discontinuous finite element (DFE) scheme, but in this case the scattering term has to be treated iteratively, for instance, by the ALI approach.

If the scattering contribution is small, one can also use an integral-equation method, which computes the monochromatic optical depth as

$$\tau_\nu(z) = \int_z^{z_{\max}} \chi(z')\, dz', \tag{17.206}$$

and emergent flux

$$F_\nu = 2 \int_0^\infty B_\nu\, [T(\tau_\nu)]\, E_2(\tau_\nu)\, d\tau_\nu \tag{17.207}$$

is calculated by numerical integration. The advantage of this approach is that the integration over angles is performed *analytically* (through the exponential integral function E_2).

Using detailed spectrum synthesis to determine chemical abundances has obvious advantages over the curve-of-growth method. The variations of temperature, density, excitation-ionization equilibrium, Doppler width, and damping parameters, as well as blending of two or more lines, are fully taken into account.

Having computed a synthetic spectrum, one can proceed in two different ways. (i) One can compute a theoretical equivalent width of each line of a chosen element for a set of assumed values of $\alpha_k \equiv N_k/N_\odot$ and interpolate to find the α_k that best fits its observed equivalent width. These individual abundance estimates are averaged over all lines to get a final abundance. (ii) One can do a fit of the entire line profiles. This approach is more general, for it provides good results even in cases when the former scheme based on equivalent widths may not, such as in spectral regions crowded with lines where the placement of the observed "continuum" is uncertain.

When the latter scheme is used, one has to account for all additional mechanisms that cause observed broadening of spectral lines. There are essentially two types of such broadening: (a) broadening due to different velocities (projected on the line of sight toward a distant observer) of different positions on the stellar surface (physical broadening), and (b) a purely instrumental broadening due to a finite resolution of the spectrograph. Typical examples of the physical broadening are the *rotational broadening* and the *macroturbulent broadening* [1064], presumed to result from motions on scales that are a significant fraction of the size of the disk of the star. Detailed description of these types of broadening, their treatment in the spectrum synthesis, and modern Fourier-transform techniques for measuring them are given in [391, chapters 17 and 18].

17.6 LINE BLANKETING

The basic parameters determining the continuum energy distribution and line spectrum of a star are its *effective temperature, surface gravity, and chemical composition.* In early work (before fast computers), the gray-atmosphere temperature distribution, scaled to a specified effective temperature, was sometimes used to compute "non-gray models." Such models are inadequate to describe actual stellar atmospheres because the opacity of stellar material has large variations at photoionization edges and in spectral lines and depends on its (non-equilibrium) excitation and ionization state.

For example, at visible wavelengths major contributors to the bound-free opacity in the O-stars are He II and light-element ions (e.g., of C, N, O, Ne, Si); in the B-stars, He I becomes important; for A-stars, neutral H produces the Balmer jump; for solar-type stars, the dominant source is from H^- and in cooler stars, a number of molecules and negative ions. Free-free absorption by He^+, He, and H is important in the O-stars; by H in the A-stars; by H^- in the Sun; and by H_2 in M-stars. Electron scattering is a major opacity source in O-stars and Rayleigh scattering by H and H_2 in spectral types G and K. In addition there are *thousands to millions* of spectral lines from ions, atoms, and molecular bands. In LTE, for a given chemical composition, the state of the material is determined by its temperature and total particle number density.

The inward increase of temperature in a stellar atmosphere means that the layers from which we receive radiation in opaque spectral lines and continua are cooler, and hence emit less energy, than the layers where the material is more transparent. Thus the flux measured in a photometric band where there are numerous spectral lines band is diminished, which is called the *blocking effect*. The total flux integrated over all wavelengths must be conserved, so the blocked energy is redistributed to other wavelengths and emerges in more transparent regions of the spectrum. Also, because the bandwidth of the spectrum in which energy transport is efficient is restricted by spectral lines (or an opaque continuum), a steeper temperature gradient is needed to drive the flux through a given amount of material, leading to what is called the *backwarming effect*. The collective effects of spectral lines in a star's opacity are referred to as *line blanketing*.

Picket Fence Model

The most obvious way to treat spectral lines in a stellar atmosphere would be a *direct method* in which we use enough frequency points in the calculation to resolve the profiles of all the lines included in the opacity. Thanks to the immense increases in computer speed in the recent past, such a calculation is essentially possible. However, in practice one is usually forced to employ a *statistical method* to simulate line blanketing; see below and in § 18.5.

To gain insight, it is worthwhile to study even highly idealized models such as the *picket fence model* developed by Chandrasekhar [216, 225] and Münch [775–780]. In this model, one assumes that (a) the continuum is in LTE and has a frequency-independent opacity, i.e., $\kappa_\nu \equiv \kappa$; (b) the lines do not overlap and have square profiles of constant width and opacity ratio $\beta \equiv l/\kappa$ relative to the continuum; and (c) the lines are distributed uniformly throughout the spectrum such that within a given frequency band, a fraction w_1 has only pure continuum opacity, and a fraction $w_2 = 1 - w_1$ has both the background continuum and the line opacity. The pictorial representation of the problem in figure 17.5 shows why it is called the "picket fence model."

For frequencies in the continuum

$$\mu \frac{dI_\nu^{(1)}}{d\tau} = I_\nu^{(1)} - B_\nu. \tag{17.208a}$$

In a line where a fraction ϵ of the photons are emitted thermally,

$$\mu \frac{dI_\nu^{(2)}}{d\tau} = (1+\beta)I_\nu^{(2)} - (1-\epsilon)\beta J_\nu^{(2)} - (1+\epsilon\beta)B_\nu. \tag{17.208b}$$

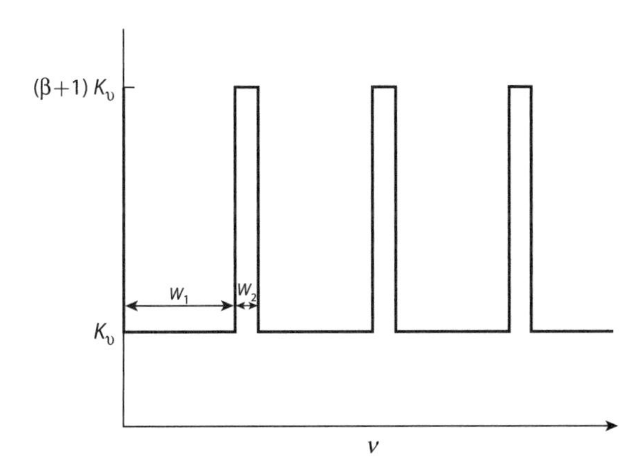

Figure 17.5 Schematic "picket fence" opacity.

Integrating over frequency, letting radiation-field quantities without a subscript ν denote integrated quantities, and accounting for the relative probabilities that a band is covered by line or continuum, we find

$$\mu \frac{dI^{(1)}}{d\tau} = I^{(1)} - w_1 B, \tag{17.209a}$$

and

$$\mu \frac{dI^{(2)}}{d\tau} = (1+\beta)I^{(2)} - (1-\epsilon)\beta J^{(2)} - (1+\epsilon\beta)w_2 B. \tag{17.209b}$$

These equations are to be solved simultaneously with a constraint of radiative equilibrium, obtained by integrating over angle and frequency and demanding that $F^{(1)} + F^{(2)} = \text{constant}$, which is true if

$$J^{(1)} + (1+\epsilon\beta)J^{(2)} = [w_1 + w_2(1+\epsilon\beta)]B. \tag{17.210}$$

Absorption Lines

When $\epsilon \equiv 1$, both the continuum and the lines are formed in LTE. Then equations (17.209) become

$$\mu \frac{dI^{(1)}}{d\tau} = (I^{(1)} - w_1 B), \tag{17.211a}$$

$$\mu \frac{dI^{(2)}}{d\tau} = (1+\beta)(I^{(2)} - w_2 B), \tag{17.211b}$$

where, from (17.210),

$$B = \frac{J^{(1)} + (1+\beta)J^{(2)}}{w_1 + w_2(1+\beta)}, \tag{17.212}$$

or, defining $\gamma_1 \equiv 1$ and $\gamma_2 \equiv 1+\beta$,

$$B = \sum_{b=1}^{2} \gamma_b J^{(b)} \Bigg/ \sum_{b=1}^{2} w_b \gamma_b. \tag{17.213}$$

To solve this system we choose a set of discrete ordinates $\{\mu_i\}$, $(i = \pm 1, \ldots, \pm n)$, with $\mu_{-i} = -\mu_i$, such that

$$J^{(b)} = \tfrac{1}{2} \sum_{j=-n}^{n} a_j I_j^{(b)}. \tag{17.214}$$

Substituting (17.213) and (17.214) into (17.211), we have

$$\frac{\mu_i}{\gamma_b}\frac{dI_i^{(b)}}{d\tau} = I_i^{(b)} - \left(\frac{w_b}{2\sum_m w_m\gamma_m}\right)\sum_{m=1}^{2}\gamma_m\sum_{j=-n}^{n} a_j I_j^{(m)},$$

$$(b = 1, 2),\ (i = \pm 1, \ldots, \pm n). \tag{17.215}$$

Given a hint from (17.42), we take a trial solution of the form

$$I_j^{(b)} = C\, w_b\, e^{-k\tau} \big/ (1 + k\mu_j/\gamma_b) \tag{17.216}$$

and find that k must satisfy the characteristic equation

$$\sum_{b=1}^{2} w_b\gamma_b = \sum_{b=1}^{2} w_b\gamma_b \sum_{j=1}^{n} a_j \Big/ (1 - k^2\mu_j^2/\gamma_b^2). \tag{17.217}$$

Applying a procedure analogous to that described in § 17.1, one obtains after some algebra

$$[B(0)/F] = (\sqrt{3}/4)\Big[\Big(\sum_b w_b\gamma_b\Big)\Big(\sum_b w_b\gamma_b^{-1}\Big)\Big]^{-1/2}. \tag{17.218}$$

For this problem the Planck mean opacity is

$$\overline{\kappa}_P = \kappa[w_1 + w_2(1+\beta)], \tag{17.219}$$

whereas the Rosseland mean opacity is

$$\overline{\kappa}_R = \kappa/[w_1 + w_2/(1+\beta)]. \tag{17.220}$$

Thus (17.218) reduces to

$$[B(0)/F] = (\sqrt{3}/4)(\overline{\kappa}_R/\overline{\kappa}_P)^{1/2} \tag{17.221a}$$

or

$$T_0/T_{\text{eff}} = \left(\sqrt{3}/4\right)^{1/4}(\overline{\kappa}_R/\overline{\kappa}_P)^{1/8}. \tag{17.221b}$$

In the limit of very opaque lines, $\overline{\kappa}_R$ saturates to $\overline{\kappa}_R \to \kappa/w_1$, whereas the Planck increases to $\overline{\kappa}_P \to \kappa w_2\beta$. Thus the effect of opaque LTE lines is to lower the boundary temperature, and the decreased bandwidth for transport of the radiation leads to backwarming in lower layers.

Examples are shown in figure 17.6, where $B(\tau)$ is plotted for the gray case and a solution with $\epsilon = 1, w_1 = 0.8, w_2 = 0.2$, and $\gamma_2 = 10$; see [776]. $B(0)/F$ decreases from the gray value 0.433 to 0.286, so $T(0)/T_{\text{eff}}$ drops from 0.811 to 0.721. Note the similar backwarming effects in both models. The large surface temperature drop in LTE contrasts with the near absence of a surface effect in the line-scattering case.

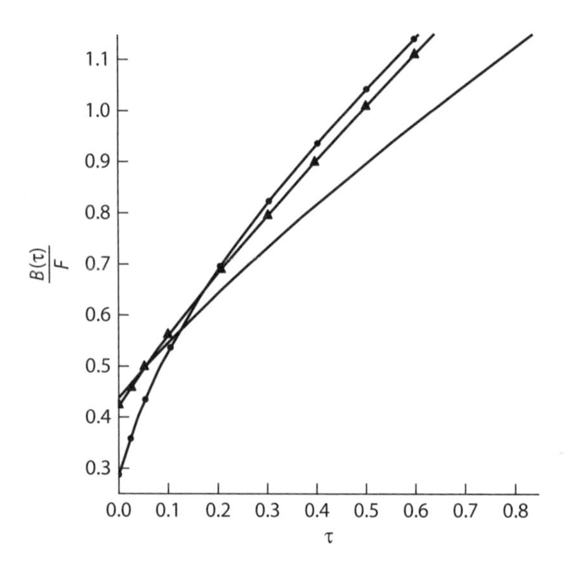

Figure 17.6 Depth variation of the integrated Planck function in picket fence models. *Plain curve*: gray solution, $\beta = 1$. *Dots*: $\epsilon = 1$ (LTE lines), $\gamma_2 = 10$, $w_1 = 0.8$, $w_2 = 0.2$. *Triangles*: $\epsilon = 0$ (scattering lines), $\gamma_2 = 10$, $w_1 = 0.8$, $w_2 = 0.2$.

Scattering Lines

We can treat scattering in spectral lines using the above framework by setting $\epsilon < 1$. Define $\gamma \equiv (1+\beta)$, $\lambda \equiv (1+\epsilon\beta)$ and $\sigma \equiv (w_1 + \lambda w_2)^{-1}$. Then (17.212) becomes

$$B = \sigma(J^{(1)} + \lambda J^{(2)}), \tag{17.222}$$

and equations (17.209) become

$$\mu \frac{dI^{(1)}}{d\tau} = I^{(1)} - w_1 \sigma (J^{(1)} + \lambda J^{(2)}), \tag{17.223a}$$

and

$$\mu \frac{dI^{(2)}}{d\tau} = \gamma I^{(2)} - [\gamma - \lambda(1 - w_2\sigma\lambda)]J^{(2)} - w_2\sigma\lambda J^{(1)}. \tag{17.223b}$$

Applying the discrete-ordinate method, one finds that in the limit $\epsilon \to 0$, the boundary temperature is only slightly below the gray value, with $B(0)/F = 0.4308$ compared with the gray value 0.4330 and the LTE value 0.286. This is not surprising from the physical standpoint because scattering lines do not contribute to the energy balance (an energy absorbed in a scattering process is immediately re-emitted, so the net gain of energy is essentially zero). In any case, *the effect of lines in the outermost layers of a star depends sensitively on the mechanism of line formation.*

Direct Simulation

Although the picket fence model has given us some basic insight into the effects of line blanketing, it is obviously not an adequate representation of a real stellar spectrum in which lines are created by different kinds of atoms and molecules. For the case of very hot stars, the largest effects of line blanketing result from strong lines in the ultraviolet part of the spectrum. Early work on this problem for O-, B-, and A-stars accounted for the effects of the most abundant ions of elements such as H, He, C, N, O, Si, S, Ar, and Fe; see, e.g., [8, 141, 238, 728, 734, 747, 865, 1057, 1058]. This work showed that because some of the emergent flux is blocked in the ultraviolet and forced to emerge in the visible part of the spectrum, from ground-based observations alone one would infer too high an effective temperature for a star using an unblanketed model. For example, the visible flux from a blanketed model with $T_{\rm eff} = 21,900$ K closely resembles that from an unblanketed model with $T_{\rm eff} = 24,000$ K.

Statistical Methods

For cooler stars, there are millions of overlapping lines from atoms, and even more in molecular bands. A direct simulation is then impractical; hence one turns to statistical representations of line blanketing.

Opacity Distribution Functions

In this approach we attempt to replace the complicated frequency variation of the line opacity within a given frequency band, such as sketched in figure 17.7, with a small number of parameters. One measures the fraction of the band covered by a line opacity $\geq$ some prechosen value $\chi_i(\nu)$, as shown by the heavier lines in the figure, and plots a graph of this fraction against frequency. The original wavelength band must be narrow enough to assure that the exact position of a spectral line does

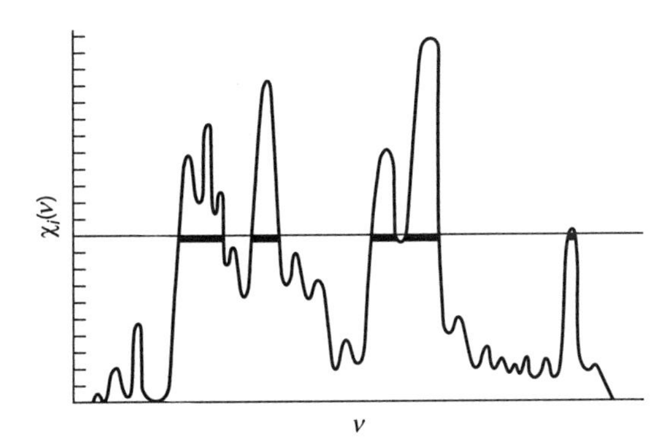

Figure 17.7 Schematic absorption coefficient of overlapping spectral lines. A large number of frequency points would be required to describe the detailed frequency variation of the opacity.

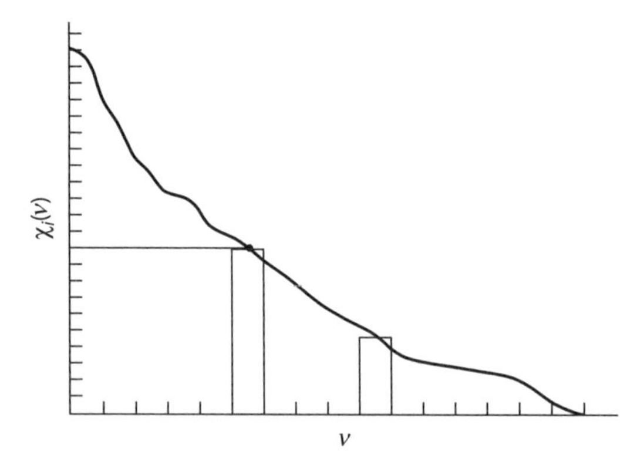

Figure 17.8 Schematic opacity distribution function (ODF) of the spectrum shown in figure 17.7. A relatively small number of representative opacities suffice to describe this smooth distribution.

not matter, i.e., that other properties such as the underlying continuum opacity and the Planck function do not vary much in that band.

The result is a smoother curve, such as that shown in figure 17.8, that can be well approximated by a small number of subintervals, possibly of differing widths, containing constant opacities. Detailed descriptions of this technique are given in [654–657, 663, 865, 1059]. A critical study of this approach [179] showed it can yield results that reproduce with satisfactory accuracy both the emergent fluxes and physical atmospheric structure given by more detailed direct calculations.

The main limitation of the opacity distribution function method is that it implicitly assumes that the wavelength positions of the lines do not change substantially as a function of optical depth, at least over a photon mean free path at a given wavelength. Put another way, it is important to the transfer problem whether or not a line in one layer of the atmosphere coincides in frequency with a line or a continuum band in an overlying layer, because photons could escape relatively freely in the latter case, but not in the former.

But marked depth variations in the line spectrum as a function of depth in the atmosphere that would invalidate the ODF approach do in fact occur. For example, (a) if molecular bands of two different species overlap, then at a given wavelength, the opacity of one species might show a rapid increase or decrease with depth relative to the other. Then even though the average opacity of the two bands together might not change markedly, the positions of the two sets of lines could be different enough to change the photon mean free paths at different frequencies significantly as a function of depth. (b) A strong shock in the atmosphere could produce an abrupt change in the excitation-ionization state of the gas over a short distance. Then the line spectra could change radically through the shock front. (c) Velocity shifts in an expanding atmosphere systematically moves lines away from their rest frequencies (see § 19.2 and § 20.2). In such cases one might need to employ either the direct method or a generalization of the statistical approach that allows for changes in the frequency positions of the lines such as that described below.

Opacity Sampling

An alternative approach is the *opacity sampling* method [146, 409, 410, 457, 477, 595, 660, 662, 866, 1023]. This technique is based on a random sampling approach that is free of the assumptions that obviate the opacity distribution method. Care must be used in this method to use enough sample points to assure that frequencies in both high-opacity and low-opacity regions of the spectrum are adequately represented. Advantages of this method documented in the references cited above are the that opacity at each frequency and depth is computed monochromatically for the given chemical composition, without any averaging. Therefore, it is possible to represent accurately the variations with depth and wavelength of complex spectra, e.g., for molecules. In principle one can also allow for large-scale velocity fields, e.g., expansion or pulsation, in the atmosphere. Although this method was initially considered to be computationally "expensive," this is no longer the case because present-day processor speeds have increased by several orders of magnitude, and computers having massively parallel architectures with multiple processors operating simultaneously now exist.

Finally, note that *both the opacity distribution function and opacity sampling methods described above assume LTE.* As shown in chapters 14 and 18, that assumption gives a poor description of the physics of both continuum and spectral line formation and can lead to serious systematic errors. Nevertheless, one can easily use the approach of opacity sampling, and to generalize the concept of ODF, to NLTE situations; cf. § 18.5.

17.7 MODELS WITH EXTERNAL IRRADIATION

As stressed above, a vast majority of model atmospheres assume no external irradiation. However, there are cases where an external irradiation is important, such as close binary systems in which a cool star is irradiated by a nearby hot star. An extreme case is represented by extrasolar giant planets located in very small distances from their parent stars.

Therefore, we briefly outline modifications of the modeling procedures and properties of the resulting models when external irradiation is taken into account.

Linearization Scheme

It is important to realize that when one performs an *exact* numerical solution, including external irradiation is very easy: the only modification is to allow for a nonzero incoming intensity $I_{-\mu,\nu}(0) \equiv I_{\mu\nu}^{\text{ext}} \neq 0$, $(\mu > 0)$, at the upper boundary. In terms of the Feautrier variables j and h, the upper boundary condition becomes [cf. (11.148)]

$$\mu\left(\partial j_{\mu\nu}/\partial\tau_\nu\right)_0 = j_{\mu\nu}(0) - I_{\mu\nu}^{\text{ext}}. \tag{17.224}$$

By multiplying both sides by μ and integrating over angles from 0 to 1, one obtains

$$(\partial K_\nu/\partial\tau_\nu)_0 = \int_0^1 j_{\mu\nu}(0)\mu\, d\mu - H_\nu^{\text{ext}}, \tag{17.225}$$

where

$$H_\nu^{\rm ext} \equiv \int_0^1 I_{\mu\nu}^{\rm ext} \mu \, d\mu. \tag{17.226}$$

Employing the Eddington factors, one obtains a closed equation for J_ν,

$$\partial(f_\nu J_\nu)/\partial\tau_\nu\big|_0 = g_\nu J_\nu(0) - H_\nu^{\rm ext}, \tag{17.227}$$

which is discretized as

$$(f_{2n}J_{2n} - f_{1n}J_{1n})/\Delta\tau_{\frac{3}{2},n} = g_n J_{1n} - H_n^{\rm ext}, \tag{17.228}$$

which represents a generalization of equation (17.127a).

In fact, it is more accurate to use the second-order boundary condition [cf. equation (12.49)], which follows from expanding j_2 by a Taylor series (dropping, for simplicity, an indication of the frequency and angle dependence),

$$j_2 = j_1 + \frac{dj}{d\tau}\bigg|_1 \Delta\tau_{\frac{3}{2}} + \frac{1}{2}\frac{d^2 j}{d\tau^2}\bigg|_1 (\Delta\tau_{\frac{3}{2}})^2. \tag{17.229}$$

Upon multiplying both sides by μ^2 and integrating over angles, one obtains (reintroducing the frequency subscript n)

$$(f_{2n}J_{2n} - f_{1n}J_{1n})/\Delta\tau_{\frac{3}{2},n} = g_n J_{1n} - H_n^{\rm ext} + (\Delta\tau_{\frac{3}{2},n}/2)(J_{1n} - S_{1n}), \tag{17.230}$$

where we used the Feautrier expression $d^2 j/d\tau^2 = j - S$. The source function is generally given by $S_{1n} = (\eta_{1n} + \kappa_{1n}^{\rm sc} J_{1n})/\chi_{1n}$. The rest of the discretized structural equations and their linearized forms (17.127b)–(17.136) remain unchanged, and so does the numerical scheme for solving them.

Pseudo-Gray Model

Although the procedure outlined above provides an *exact* (within the approximation of LTE) model and thus is sufficient for applications, it is worthwhile to develop an approximate model based on moment equations of the transfer equation in the spirit of classical gray model. Such a model serves a dual purpose: (i) it can be used as a starting model for the linearization procedure if irradiation is important and the classical gray model does not provide a satisfactory initial solution, and, in particular, (ii) it provides a physical insight into the effects of irradiation.

The frequency-integrated moment equations are [cf. (17.97) and (17.98)]

$$(dH/dz) = -(\kappa_J J - \kappa_B B), \tag{17.231}$$

$$(dK/dz) = -\chi_H H, \tag{17.232}$$

where κ_J and κ_B are the absorption mean and Planck mean opacities, respectively. Radiative equilibrium stipulates that $dH/dz = 0$ [with $H \equiv (\sigma_R/4\pi)T_{\rm eff}^4$], and thus

$$B = (\kappa_J/\kappa_B)J. \tag{17.233}$$

This equation determines the temperature through $B = (\sigma_R/\pi)T^4$. To determine J, one solves (17.232) to obtain $K(\tau_H) = H\tau_H + K(0)$, where $\tau_H \equiv -\chi_H dz$ is the optical depth associated with the flux-mean opacity. Using the Eddington factor $f \equiv K/J$, one can write the integrated mean intensity as

$$J(\tau_H) = [(\sigma_R/4\pi)T_{\rm eff}^4\,\tau_H + K(0)]/f. \tag{17.234}$$

To estimate $K(0)$ and the Eddington factor f, one invokes the two-stream approximation, in which the radiation field is represented by two beams moving along $\pm\mu_0$; $I(\tau,\mu) \equiv I^+(\tau)\delta(\mu-\mu_0)$ for $0 \le \mu \le 1$, and $I(\tau,\mu) \equiv I^-(\tau)\delta(\mu+\mu_0)$ for $-1 \le \mu \le 0$. The moments are then given by [cf. equations (17.19a–c)]

$$J(\tau) = \tfrac{1}{2}\,[I^+(\tau) + I^-(\tau)], \tag{17.235a}$$

$$H(\tau) = \tfrac{1}{2}\,[I^+(\tau) - I^-(\tau)]\,\mu_0, \tag{17.235b}$$

$$K(\tau) = \tfrac{1}{2}\,[I^+(\tau) + I^-(\tau)]\,\mu_0^2. \tag{17.235c}$$

From (17.235b) we have

$$I^+(0) = I^-(0) + 2H(0)/\mu_0 \equiv I^{\rm ext} + 2H/\mu_0, \tag{17.236}$$

and, therefore,

$$K(0) = (I^{\rm ext} + H/\mu_0)\,\mu_0^2. \tag{17.237}$$

It is instructive to express

$$I^{\rm ext} \equiv WB(T_{\rm irr}) = W(\sigma_R/\pi)T_{\rm irr}^4, \tag{17.238}$$

i.e., as the Planck function corresponding to the "irradiation temperature" $T_{\rm irr}$ times a dilution factor W. As follows from (17.235a) and (17.235c), $f = \mu_0^2$. In the Eddington approximation, $f = \frac{1}{3}$, and thus $\mu_0 = 1/\sqrt{3}$, and $g \equiv H(0)/J(0) = 1/\sqrt{3}$. The temperature structure can then be written as

$$T^4(\tau_H) = \frac{\kappa_J}{\kappa_B}\left[\frac{3}{4}T_{\rm eff}^4\left(\frac{1}{3f}\tau_H + \frac{1}{3g}\right) + WT_{\rm irr}^4\right], \tag{17.239}$$

where we used f and g in order to allow formally for departures from the Eddington approximation. This expression is *exact* within LTE, but is only formal because κ_J, κ_B, f, g, and τ_H are not a priori known. If $\kappa_J = \kappa_B$, and no irradiation, $W = 0$ (or $T_{\rm irr} = 0$), one recovers the classical gray temperature distribution, if one expresses $\tau + q(\tau) \equiv (1/3f)\tau + (1/3g)$.

In the Eddington approximation, one has

$$T^4(\tau_H) = \frac{\kappa_J}{\kappa_B}\frac{3}{4}T_{\rm eff}^4\left[\tau_H + \frac{1}{\sqrt{3}} + \frac{4W}{3}\left(\frac{T_{\rm irr}}{T_{\rm eff}}\right)^4\right]. \tag{17.240}$$

If $W^{1/4}T_{\rm irr} \ll T_{\rm eff}$, the irradiation causes only a small correction to $q(0)$ (or to $1/\sqrt{3}$ in the Eddington approximation). A much more interesting case is the opposite one when $W^{1/4}T_{\rm irr} \gg T_{\rm eff}$. This situation occurs, for instance, in close binaries when a cool star is irradiated by a very hot star (e.g., a pre-cataclysmic variable system with an M-star and a companion hot subdwarf or white dwarf), or, in an even more extreme case, a gaseous planet irradiated by a solar-type star. For instance, for the planet OGLE-TR-56b, $W \approx 2.2 \times 10^{-2}$, $T_{\rm itr} \approx 6000$ K, and $T_{\rm eff} \approx 100$ K; cf. [527]. In such cases, the upper layers are completely dominated by irradiation. The surface temperature is

$$T(0) \approx \gamma W^{1/4} T_{\rm irr} \equiv \gamma T_0, \tag{17.241}$$

where $\gamma \equiv (\kappa_J/\kappa_B)^{1/4}$. [We disregard an order-of-unity factor $(4/3)^{1/4}$.] Temperature T_0 is traditionally called the "effective temperature" in the planetary science literature. This may be confusing; the reader has to be aware which notation is being used, particularly in recent papers on extrasolar giant planets because the researchers in this field come from both the planetary science and the stellar atmosphere communities and use their respective traditional terminologies.

To gain a better insight, one can define a *penetration depth* [527],

$$\tau_{\rm pen} \equiv W\,(T_{\rm irr}/T_{\rm eff})^4\,, \tag{17.242}$$

as the flux-mean optical depth where the thermal contribution ($\propto T_{\rm eff}^4$) and the irradiation contribution ($\propto WT_{\rm irr}^4$) are nearly equal. For $\tau_H < \tau_{\rm pen}$, the irradiation dominates, and the local temperature is given by $T \approx \gamma T_0$ and is constant if $\gamma \approx 1$. For $\tau_H > \tau_{\rm pen}$, one essentially recovers the classical gray temperature distribution $T \propto \tau^{1/4} T_{\rm eff}$. In some cases, the penetration depth may be very large; for instance, in the case of OGLE-TR-56b, $\tau_{\rm pen} \approx 9 \times 10^4$!

In addition to the above-described obvious effects of irradiation on the temperature structure, there are two less obvious but very important effects of the strong irradiation, namely, (i) the flux-mean opacity may differ significantly from the Rosseland mean opacity; and (ii) the absorption mean and Planck mean opacities may be very different. In particular, the latter effect may even lead to the existence of two solutions for the temperature structure—a bifurcation—as suggested in [527]. The reason is as follows: the Planck mean opacity weights the opacity by the Planck function corresponding to the *local temperature*. On the other hand, the absorption mean opacity weights the opacity by the local mean intensity, which, at the surface layers of a strongly irradiated object, is roughly given as a Planck function corresponding to the *irradiation temperature*. Notice that in this case

$$\kappa_J \approx \frac{W\int \kappa_\nu(T) B_\nu(T_{\rm irr})\,d\nu}{W\int B_\nu(T_{\rm irr})\,d\nu} = \frac{\int \kappa_\nu(T) B_\nu(T_{\rm irr})\,d\nu}{\int B_\nu(T_{\rm irr})\,d\nu}, \tag{17.243}$$

i.e., the dilution factor cancels out, and only the spectral distribution of the irradiation intensity matters. In the above-mentioned case of a cool planet irradiated by a solar-type star, the Planck mean opacity is determined predominantly by the infrared opacity, while the absorption mean is determined by the opacity in the visible range. If there is a source of large opacity present in the visible range but missing in the infrared, or vice versa, the Planck mean and the absorption mean may be quite different. Such a source of opacity is provided by the TiO (and to a lesser extent VO) molecule, which has many strong lines in the optical spectrum but not in the infrared. Moreover, the TiO opacity sensitively depends on the local temperature (and, to a lesser extent, on density), in the sense that around 1500 K it condensates and rains away; hence for lower temperatures its opacity is essentially zero.

Because of the sensitivity of κ_J on the irradiation temperature as well as on the local temperature, the quantity γ may exhibit a strongly non-monotonic behavior as a function of local temperature; see, e.g., figure 1 of [527]. The actual surface temperature is approximately determined by solving the equation

$$T/T_0 = \gamma(T), \tag{17.244}$$

which, for γ exhibiting a non-monotonic behavior, may lead to two or even more solutions for the atmospheric structure at the upper layers [527].

From the physical point of view one may interpret the absorption mean opacity κ_J as the global absorption efficiency, and the Planck mean opacity as the global emission efficiency, of the medium. The integrated mean intensity J represents a total photon pool, i.e., a pool of potential source of energy for matter if photons are absorbed, and the integrated Planck function B as the total thermal pool. The radiative equilibrium stipulates that the total energy absorbed at a given spot is equal to the total energy emitted on the spot and therefore acts as a thermostat: radiative equilibrium sets the local temperature in such a way that $\kappa_J J = \kappa_B B$. In the case of strong irradiation, κ_J is determined by the spectral distribution of the irradiation spectrum, weighted by the local monochromatic opacity. If the latter is very sensitive to temperature (for instance, it is low for low T and high for high T), then the "radiative equilibrium thermostat" may find two solutions, either a high T with high κ_J or a low T with low κ_J; in both cases the radiative equilibrium $\kappa_J J = \kappa_B B$ is satisfied. For a comprehensive discussion and application to actual model atmospheres of extrasolar giant planets, see [527].

We stress that the bifurcation described above represents a *mathematical property* of the solution of a very simplified set of structural equations, based on LTE and also on chemical equilibrium in the case of cool objects (assuming that the formation of molecules can be described by a local equilibrium chemistry), so that it depends solely on the local thermodynamic parameters—temperature and density. The theory outlined above has to be viewed as a rough approximation to reality; nevertheless, it shows that the external irradiation may lead to a number of very interesting phenomena.

17.8 AVAILABLE MODELING CODES AND GRIDS

The most popular codes are Kurucz's ATLAS [654, 660] and the code by the Scandinavian group called MARCS [421, 422]. It should be noted that most of the NLTE codes (cf. § 18.6) can be used to calculate LTE models as well. The NLTE codes PHOENIX and TLUSTY were actually used for generating grids of LTE models.

The most extensive grid of LTE plane-parallel line-blanketed models is that of Kurucz [657] widely used by the astronomical community. Updated models are distributed via Kurucz's CD-ROMs and online.[10] The grid covers effective temperatures between 3500 K and 50,000 K, $\log g$ between -1 and 5, and several metallicities. The term "metallicity" traditionally means that all the chemical species heavier than helium share a common ratio of their abundance to the solar abundance; this ratio is called the metallicity. Numerically, the metallicity is often taken as a logarithm of the metal abundance ratio.

Using the MARCS code, [421] generated their original grid of models for cool stars, with $T_{\rm eff}$ between 3750 and 6000 K, $\log g$ between 0.75 and 3.0, and metallicities $-3.0 \leq [M/H] \leq 0$. Recently, [422] made public a new, very extensive grid of MARCS model atmospheres, with $T_{\rm eff}$ between 2500 and 8000 K, $\log g$ between -1 and 5, and metallicities $-5 \leq [M/H] \leq 1$. They also include "CN cycled" models with C/N = 4.07 (solar), 1.5, and 0.5 and C/O from 0.09 to 5, which represents stars of spectral types R, S, and N.

Hauschildt et al. [452, 453] used their code PHOENIX to generate several grids: a grid for M-subdwarfs [14] and an extended grid of LTE spherical models for cool stars called NEXTGEN [452], with $T_{\rm eff}$ between 3000 and 10000 K, with step 200 K; $\log g$ between 3.5 and 5.5, with step 0.5; and metallicities $-4.0 \leq [M/H] \leq 0$. Another grid [453] is for pre-main-sequence cool stars with $T_{\rm eff}$ between 2000 and 6800 K, $\log g$ between 2 and 3.5 with step 0.5, stellar mass $M = 0.1 M_\odot$, and metallicities $-4.0 \leq [M/H] \leq 0$. The models are available online.[11] A detailed comparison between the ATLAS and NEXTGEN models was done in [112].

[10] Kurucz CD-ROM No. 13, Smithsonian Astrophysical Observatory; http://kurucz.harvard.edu/grids.html.

[11] http://www.hs.uni-hamburg.de/EN/For/ThA/phoenix/index.html.

Chapter Eighteen

Non-LTE Model Atmospheres

In this chapter, we discuss the problem of classical non-LTE stellar atmospheres. By a classical atmosphere we mean a plane-parallel, horizontally homogeneous atmosphere in hydrostatic and radiative (or radiative + convective) equilibrium. The term non-LTE, or NLTE, refers to any description allowing for some kind of departure from LTE. In practice, one usually means that number densities of some selected energy levels of some selected chemical species are allowed to depart from their LTE values.

We show that thanks to enormous progress in the development of fast and efficient numerical methods and ever-increasing computer speeds, one can now construct model atmospheres with an unprecedented degree of realism, in which literally tens to hundreds of millions spectral lines are taken into account, out of LTE, to determine the atmospheric structure and the emergent radiation field.

18.1 OVERVIEW OF BASIC EQUATIONS

The relevant equations were derived in chapters 11 and 16. Here we give a brief summary of the structural equations for plane-parallel, horizontally homogenous atmospheres and discuss the most appropriate forms of these equations to be used for modeling purposes.

Structural Equations

Radiative Transfer Equation

As stated in earlier chapters, the most advantageous form of the transfer equation for use in model atmosphere construction is the second-order form:

$$\frac{\partial^2 (f_\nu J_\nu)}{\partial \tau_\nu^2} = J_\nu - S_\nu, \tag{18.1}$$

where τ_ν is the monochromatic optical depth and f_ν the variable Eddington factor. The upper boundary condition is written generally as

$$\left[\frac{\partial (f_\nu J_\nu)}{\partial \tau_\nu}\right]_0 = g_\nu J_\nu(0) - H_\nu^{\mathrm{ext}}, \tag{18.2}$$

where g_ν is the surface Eddington factor defined, in terms of the Feautrier variable j defined by (12.31), by

$$g_\nu \equiv \int_0^1 j_\nu(\mu,0)\mu\, d\mu / J_\nu(0), \tag{18.3}$$

and

$$H_\nu^{\rm ext} \equiv \int_0^1 I_\nu^{\rm ext}(\mu)\mu\, d\mu, \tag{18.4}$$

where $I_\nu^{\rm ext}(\mu)$ is an external incoming intensity at the top of the atmosphere. In most cases one assumes no incoming radiation, $I_\nu^{\rm ext}(\mu) = 0$, but in some cases it may be important, for instance, for close binary systems.

The lower boundary condition is written in a similar way

$$\left[\frac{\partial(f_\nu J_\nu)}{\partial\tau_\nu}\right]_{\tau_{\rm max}} = H_\nu^+ - \tfrac{1}{2}J_\nu, \tag{18.5}$$

where $H_\nu^+ = \int_0^1 I_\nu^+(\mu,\tau_{\rm max})\mu\, d\mu$. The last term on the right-hand side contains the factor $\frac{1}{2}$; one could have introduced an Eddington factor at the lower boundary, but because the radiation field is essentially isotropic there, it would be equal to $\frac{1}{2}$ anyway. One typically assumes the diffusion approximation at the lower boundary, in which case $I_\nu^+(\mu) = B_\nu + \mu(dB_\nu/d\tau_\nu)$, hence equation (18.5) is written as

$$\left[\frac{\partial(f_\nu J_\nu)}{\partial\tau_\nu}\right]_{\tau_{\rm max}} = \left[\tfrac{1}{2}(B_\nu - J_\nu) + \frac{1}{3}\frac{\partial B_\nu}{\partial\tau_\nu}\right]_{\tau_{\rm max}}. \tag{18.6}$$

Equations (18.1), (18.2), and (18.6) contain only the mean intensity of radiation, J_ν, a function of frequency and depth, but not the specific intensity, $I_{\mu\nu}$, which is also a function of the angle cosine μ. It is the mean intensity of radiation that enters other structural equations; therefore, mean intensities, not specific intensities, are the appropriate variables to take as atmospheric state parameters.

An important numerical advantage is that instead of dealing with $NF \times NA$ quantities describing the radiation field per depth point (NF and NA being the number of discretized frequency and angle points, respectively), we have only NF parameters. However, the Eddington factors f_ν and g_ν still have to be computed by a separate set of formal solutions for the specific intensities, one frequency at a time, and gradually updated using current values of the optical depth and the source function.

Hydrostatic Equilibrium Equation

This equation reads

$$\frac{dP}{dm} = g, \tag{18.7}$$

where P is the total (gas plus radiation) pressure, and m the Lagrangian mass, or column mass,

$$dm = -\rho\, dz, \tag{18.8}$$

where ρ is the mass density and z the geometrical distance measured along the normal to the surface from the bottom of the atmosphere to the top. g is the surface gravity, which is assumed constant throughout the atmosphere and given by $g = G\mathsf{M}/\mathsf{R}^2$, where M and R are the stellar mass and radius, respectively; G is the gravitational constant. The simplicity of equation (18.7) is one of the reasons why the Lagrangian mass m is chosen as the geometrical coordinate, because then the total pressure can be written as $P(m) = gm + P_0$.

The total pressure is generally composed of three parts: the gas pressure, $P_{\rm gas}$, the radiation pressure, $P_{\rm rad}$, and a "turbulent pressure," $P_{\rm turb}$. The gas pressure is given, assuming an ideal gas equation of state, by

$$P_{\rm gas} = NkT, \tag{18.9}$$

where N is the total particle number density, T the electron temperature, and k the Boltzmann constant. We assume that all the particles have the same kinetic temperature. The so-called "turbulent pressure" is not a well-defined quantity. It is introduced to mimic a pressure associated with random motion of "turbulent eddies" as $P_{\rm turb} \propto \rho v_{\rm turb}^2$, where $v_{\rm turb}$ is the microturbulent velocity. Although it is sometimes taken into account in model atmosphere construction, we will ignore it here. The hydrostatic equilibrium equation can then be written as

$$\frac{dP_{\rm gas}}{dm} = g - \frac{4\pi}{c} \int_0^\infty \frac{dK_\nu}{dm}\, d\nu = g - \frac{4\pi}{c} \int_0^\infty \frac{\chi_\nu}{\rho} H_\nu\, d\nu, \tag{18.10}$$

where H_ν and K_ν are the first and second angular moments of the specific intensity. The right-hand side of this equation is usually called the *effective gravitational acceleration*, resulting from the action of a true gravitational acceleration (acting downward) minus a radiative acceleration (acting outward).

Radiative Equilibrium Equation

The condition of radiative equilibrium can be written as an expression of the conservation of the total radiation flux—the *differential form*,

$$\int_0^\infty H_\nu\, d\nu = \int_0^\infty \frac{d(f_\nu J_\nu)}{d\tau_\nu}\, d\nu = \frac{\sigma_R}{4\pi} T_{\rm eff}^4, \tag{18.11}$$

where $T_{\rm eff}$ is the effective temperature and σ_R the Stefan-Boltzmann constant. The radiative equilibrium equation can also be written as an equality of the total absorbed and emitted energy—the *integral form*,

$$\int_0^\infty (\kappa_\nu J_\nu - \eta_\nu)\, d\nu = 0, \tag{18.12}$$

where κ_ν is the *thermal* absorption coefficient and η_ν the thermal emission coefficient (the scattering contributions cancel out when one assumes coherent scattering, which is usually the case in stellar atmospheres). Both equations are mathematically equivalent but have different numerical properties. The differential form (18.11) is more accurate at large depths, while the integral form (18.12) behaves better numerically at small depths. To improve a numerical accuracy of the solution, either one considers the integral form at the upper layers of the atmosphere and the differential form at deep layers [414, 521] or one takes a linear combination of both forms [533, 1151]. In the latter case, the radiative equilibrium equation becomes

$$\alpha\left[\int_0^\infty (\kappa_\nu J_\nu - \eta_\nu)\, d\nu\right] + \beta\left[\int_0^\infty \frac{d(f_\nu J_\nu)}{d\tau_\nu}\, d\nu - \frac{\sigma_R}{4\pi}\, T_{\rm eff}^4\right] = 0, \tag{18.13}$$

where α and β are empirical coefficients that satisfy $\alpha \to 1$ in the upper layers, and $\alpha \to 0$ for deep layers, while the opposite applies for β. The division between the "surface" and "deep" layers is a free parameter and is determined empirically, the transition being chosen close to the depth where the Rosseland mean optical depth is about unity.

Although equation (18.13) is advantageous for numerical calculations (e.g., [533]), one should bear in mind that, mathematically speaking, satisfying equation (18.13) does not guarantee satisfying the constraint of radiative equilibrium (18.11) or (18.12). For instance, equation (18.13) is satisfied if $\int(\kappa_\nu J_\nu - \eta_\nu)d\nu = \epsilon$ and $\int[d(f_\nu J_\nu)/d\tau_\nu]\, d\nu - (\sigma/4\pi)\, T_{\rm eff}^4 = -\epsilon$. Therefore, when using equation (18.13) for modeling, one should check whether both forms of the radiative equilibrium equation (18.11) and (18.12) are satisfied individually as well.

An alternative expression of radiative equilibrium is the condition of thermal balance of electrons, stating that total energy added to the electron thermal pool by radiative and collisional processes is equal to the energy removed from the thermal pool. This condition is usually used in modeling photoionized media [821]; it was used for constructing model stellar atmospheres by [641]. It can also be employed to check a numerical accuracy of the radiative + kinetic equilibrium solutions, e.g., [493].

Kinetic Equilibrium Equations

In any NLTE model stellar atmosphere study, one first selects the chemical species that are to be considered out of LTE, i.e., for which one writes down the set of kinetic equilibrium equations (also called statistical equilibrium or rate equations). The kinetic equilibrium equations are assumed to be separate for each individual atomic species. The latter assumption is valid as long as the collisional rates depend only on temperature and electron density (i.e., only collisions with free electrons are important), but not on level populations of other species. Even if this is not the case, for instance, when taking into account charge transfer reactions with neutral hydrogen, one can view the corresponding perturber number density to be given, for instance, from the previous iteration.

For each atomic species, I, the set of kinetic equilibrium equations may be written as (see chapters 9 and 14)

$$\mathcal{A}_I \cdot \mathbf{n}_I = \mathbf{b}_I. \tag{18.14}$$

In the following, we will drop the subscript I unless absolutely necessary.

The elements the right-hand side vector **b** are given below by (18.22), and the elements of the rate matrix $\mathcal{A}$ are given by (cf. § 9.5 and § 14.1)

$$\mathcal{A}_{ii} = \sum_{j \neq i} (R_{ij} + C_{ij}), \tag{18.15a}$$

$$\mathcal{A}_{ij} = -(R_{ji} + C_{ji}), \ \text{for } j \neq i \text{ and } i \neq k, \tag{18.15b}$$

$$\mathcal{A}_{kj} = 1 + S_j, \tag{18.15c}$$

where k is the index of the *characteristic level*, i.e., the level for which the rate equation is replaced by the particle conservation (abundance definition) equation, and S_j is the upper sum defined by (9.86). Assuming $i < j$, the radiative rates are given by (14.4) and (14.5):

$$R_{ij} = \frac{4\pi}{h} \int_0^\infty \frac{\sigma_{ij}(\nu)}{\nu} J_\nu d\nu, \tag{18.16a}$$

$$R_{ji} = \frac{4\pi}{h} \int_0^\infty \frac{\sigma_{ij}(\nu)}{\nu} G_{ij}(\nu) \left(\frac{2h\nu^3}{c^2} + J_\nu \right) d\nu, \tag{18.16b}$$

where $G_{ij}(\nu)$ is defined by equation (14.17), i.e.,

$$G_{ij}(\nu) \equiv \begin{cases} g_i/g_j, & \text{for bound-bound,} \\ n_e \Phi_i(T) \exp(-h\nu/kT), & \text{for bound-free.} \end{cases} \tag{18.17}$$

and $\Phi_i(T)$ is the Saha-Boltzmann factor,

$$\Phi_i(T) = \frac{g_i}{2g_1^+} \left(\frac{h^2}{2\pi m_e kT} \right)^{3/2} e^{(E_I - E_i)/kT}. \tag{18.18}$$

E_I is the ionization potential of the ion to which level i belongs, E_i is the excitation energy of level i, and g_1^+ is the statistical weight of the ground state of the next ion.

If one adopts an occupation probability formalism that describes bound level dissolution resulting from perturbations with neighboring particles, the above equation remains the same, replacing $\sigma_{ij}(\nu) \to \sigma_{ij}(\nu) w_j$, and

$$G_{ij}(\nu) = \begin{cases} (g_i w_i)/(g_j w_j), & \text{for bound-bound,} \\ (w_i/w_j) n_e \Phi_i(T) \exp(-h\nu/kT), & \text{for bound-free.} \end{cases} \tag{18.19}$$

The collisional rates, assuming that only collisions with free electrons are important (again, $i < j$), are given by

$$C_{ij} = n_e \Omega_{ij},$$
$$C_{ji} = (n_i^*/n_j^*) C_{ij}. \tag{18.20}$$

The set of rate equations for all levels of an atom would form a linearly dependent system. Therefore, one equation of the set has to be replaced by the *number conservation*, also called the *abundance definition equation*:

$$\sum_{i=1}^{NL_I} n_i(1 + S_i) = N_I = (N - n_e)\alpha_I, \tag{18.21}$$

where NL_I is the total number of levels of species I (in all ionization stages), and $\alpha_I = A_I / \sum_J A_J$ is the *fractional abundance* of the chemical element I, with $A_I \equiv N_I/N_H$ being the *abundance* of the species I, defined here as a ratio of the total number of atoms I, in all degrees of ionization, to the total number of hydrogen atoms, per unit volume. The summation extends over all species, including hydrogen (for which, by definition, $A_H = 1$). As defined in § 9.5, the upper sum S_i, which is nonzero only for i being a ground state of an ion, accounts for the total population of upper levels that are not taken into account explicitly.

As mentioned above and in § 9.5, the level for which the rate equation is replaced by the abundance definition equation, the so-called characteristic level, is arbitrary. Usually, one chooses either the last level ($i = NL_I$) or a level with the highest population. Let k be the characteristic level. The elements of the right-hand-side vector **b** are given by

$$b_i = (N - n_e)\alpha_I\, \delta_{ki}, \tag{18.22}$$

i.e., the only nonzero element of **b** is the term corresponding to level k.

The complete set of kinetic equilibrium equations is written as

$$\boldsymbol{\mathcal{A}} \cdot \mathbf{n} = \mathbf{b}, \tag{18.23}$$

where the full rate matrix $\boldsymbol{\mathcal{A}}$ is a block-diagonal matrix composed of all the individual matrices $\boldsymbol{\mathcal{A}}_I$ and in a similar manner for the vector of populations **n** and the right-hand side vector **b**.

Charge Conservation Equation

This equation expresses the global electrical neutrality of the medium:

$$\sum_i n_i Z_i - n_e = 0, \tag{18.24}$$

where Z_i is the charge associated with level i; i.e., $Z_i = 0$ for levels of neutral atoms, $Z_i = 1$ for levels for once ionized ions, etc., and n_e is the electron density. The summation extends over all levels of all ions of all species.

Mass Density and Fictitious Massive Particle Density Equations

The mass density is expressed in terms of atomic level populations as

$$\rho = \sum_i m_i n_i, \tag{18.25}$$

where m_i is the mass of the atom to which level i belongs. It can also be expressed as

$$\rho = (N - n_e)\,\mu m_H, \tag{18.26}$$

where $m_H = 1.67333 \times 10^{-24}$ g is the mass of the hydrogen atom and μ the *mean molecular weight*, defined by

$$\mu = \sum_I \alpha_I (m_I/m_H) = \sum_I A_I (m_I/m_H) \Big/ \sum_I A_I, \tag{18.27}$$

where m_I is the mass of an atom of species I.

One sometimes introduces a *fictitious massive particle density*, defined as

$$n_m \equiv (N - n_e)\mu, \tag{18.28}$$

so that the mass density can be written as

$$\rho = n_m m_H. \tag{18.29}$$

Absorption and Emission Coefficients

The above set of structural equations has to be complemented by equations defining the absorption and emission coefficients.

The absorption coefficient (or opacity) is given by (see chapter 5 for details)

$$\chi_\nu = \kappa_\nu + \kappa_\nu^{\rm sc}, \tag{18.30}$$

where κ_ν is the coefficient of true absorption (or extinction coefficient), and $\kappa_\nu^{\rm sc}$ is the scattering coefficient. The extinction coefficient is given by

$$\begin{aligned}\kappa_\nu = &\sum_i \sum_{j>i} \left[n_i - n_j G_{ij}(\nu)\right] \sigma_{ij}(\nu) \\ &+ \sum_\kappa n_e n_\kappa \sigma_{\kappa\kappa}(\nu, T) \left(1 - e^{-h\nu/kT}\right) + \kappa_\nu^{\rm add},\end{aligned} \tag{18.31}$$

where the first term represents the contribution of the bound-bound and bound-free transitions, and the second term the free-free transitions. The summations extend over all level of all species. The term $\kappa_\nu^{\rm add}$ represents any additional opacity. It is used to account for opacity sources that are not written in terms of detailed bound-bound or bound-free transitions and may represent, for instance, the bulk of opacities of less important species lumped together, or possibly the total opacity of molecular species. For the purposes of model atmosphere construction, it is assumed to be specified as a function of temperature and density, or possibly taken from a previously constructed opacity table.

The scattering part of the absorption coefficient is given by

$$\kappa_\nu^{\rm sc} = n_e \sigma_e + \sum_i n_{{\rm Ray},i} \sigma_{{\rm Ray},i}, \tag{18.32}$$

where σ_e is the Thomson cross section, $\sigma_{Ray,i}$ is the Rayleigh scattering cross section of species i, and n_i is the number density of species i. The summation extends over all species for which Rayleigh scattering gives a non-negligible contribution to the total scattering opacity. Hence it depends on the atmospheric conditions. For hot stars (type A and hotter), where hydrogen is fully or at least partially ionized, Rayleigh scattering is negligible. For cooler stars (F, G, and K types), Rayleigh scattering by neutral hydrogen is important. In this case, neutral hydrogen is predominantly in its ground state, so $n_{Ray,i} = n_1^H$. For even cooler objects, Rayleigh scattering by the hydrogen molecule, and/or other molecular species, becomes important.

The total emission coefficient is also given as a sum of thermal and scattering contributions. The latter refers only to *continuum scattering*; scattering in spectral lines, discussed in chapters 10 and 15, is described in the "thermal part" with the emission profile coefficient ψ_ν. The continuum scattering part is usually treated separately from the thermal part, and the "thermal emission coefficient" is usually called the "emission coefficient." Specifically,

$$\eta_\nu^{tot} = \eta_\nu + \eta_\nu^{sc}, \tag{18.33}$$

where

$$\eta_\nu = (2h\nu^3/c^2)\left[\sum_i \sum_{j>i} n_j G_{ij}(\nu)\sigma_{ij}(\nu) + \sum_\kappa n_e n_\kappa \sigma_{\kappa\kappa}(\nu, T) e^{-h\nu/kT}\right] + \eta_\nu^{add}. \tag{18.34}$$

The additional emissivity, if included, is usually given by $\eta_\nu^{add} = \kappa_\nu^{add} B_\nu$. The form of the scattering part of the emission coefficient depends on additional assumptions. In the simple case that we assume an isotropic phase function and electron scattering is treated as coherent Thomson scattering, then

$$\eta_\nu^{sc} = \kappa_\nu^{sc} J_\nu. \tag{18.35}$$

Convection

Although our emphasis is on radiative equilibrium model atmospheres for hot stars (type A and earlier), we give below a brief outline of the simplest treatment of convection, namely, the mixing-length theory. Such a description can easily be incorporated into the basic structural equations, thus extending the range of applicability of the numerical schemes aimed at constructing model stellar atmospheres.

As discussed in chapters 16 and 17, the atmosphere is convectively unstable if the Schwarzschild criterion for convective instability is satisfied,

$$\nabla_{rad} > \nabla_{ad}, \tag{18.36}$$

where $\nabla_{rad} = (d \ln T/d \ln P)_{rad}$ is the logarithmic temperature gradient in radiative equilibrium, and ∇_{ad} is the adiabatic gradient. The latter is viewed as a function of temperature and pressure, $\nabla_{\rm ad} = \nabla_{\rm ad}(T, P)$. As in chapter 16, the density ρ is considered to be a function of T and P through the equation of state.

If the condition (18.36) is satisfied, the radiative equilibrium equation in its differential form, (18.11), is modified to become

$$\int_0^\infty \frac{d(f_\nu J_\nu)}{d\tau_\nu}\, d\nu + \frac{F_{\rm conv}}{4\pi} = \frac{\sigma_R}{4\pi} T_{\rm eff}^4, \tag{18.37}$$

where $F_{\rm conv}$ is the convective flux, given by [cf. equation (16.73)]

$$F_{\rm conv} = (gQH_P/32)^{1/2}(\rho c_P T)(\nabla - \nabla_{\rm el})^{3/2}(\ell/H_P)^2, \tag{18.38}$$

where $H_P \equiv -(d \ln P/dz)^{-1} = P/(\rho g)$ is the pressure scale height, c_P is the specific heat at constant pressure, $Q \equiv -(d \ln \rho/d \ln T)_P$, ℓ/H_P is the ratio of the convective mixing length to the pressure scale height, taken as a free parameter of the problem, ∇ is the actual logarithmic temperature gradient, and $\nabla_{\rm el}$ is the gradient of the convective elements. The latter is determined by considering the efficiency of the convective transport. As derived in § 16.5, the gradient $\nabla_{\rm el}$ is obtained by solving [see equation (16.80)]

$$\nabla_{\rm el} - \nabla_{\rm ad} = \mathcal{B}\sqrt{\nabla - \nabla_{\rm el}}, \tag{18.39}$$

where

$$\mathcal{B} \equiv \frac{16\sqrt{2}\sigma_R T^3}{\rho c_p (gQH_P)^{1/2}(\ell/H_P)} \frac{\tau_{\rm el}}{1 + \frac{1}{2}\tau_{\rm el}^2}, \tag{18.40}$$

and $\tau_{\rm el} = \chi_R \ell$ is the optical thickness of the characteristic element size ℓ. Equation (18.39) is solved expressing $(\nabla_{\rm el} - \nabla_{\rm ad}) = (\nabla_{\rm el} - \nabla) + (\nabla - \nabla_{\rm ad})$ to yield

$$\nabla - \nabla_{\rm el} = (\nabla - \nabla_{\rm ad}) + \tfrac{1}{2}\mathcal{B}^2 - \mathcal{B}\sqrt{\tfrac{1}{2}\mathcal{B}^2 - (\nabla - \nabla_{\rm ad})}. \tag{18.41}$$

The gradient of the convective elements is then a function of temperature, pressure, the actual gradient, $\nabla_{\rm el} = \nabla_{\rm el}(T, P, \nabla)$, and the convective flux can also be regarded as a function of T, P, and ∇.

To write the integral form of the radiative + convective equilibrium equation, one differentiates (18.37) with respect to m to obtain

$$\int_0^\infty (\kappa_\nu J_\nu - \eta_\nu)\, d\nu + \frac{\rho}{4\pi}\frac{dF_{\rm conv}}{dm} = 0. \tag{18.42}$$

Discretization

Regardless of which method is chosen for solving the set of basic structural equations, they must be discretized in all continuous variables, namely, the vertical geometrical coordinate (depth), frequency, and angle. We are essentially free to choose all these discretizations, but they must satisfy the following conditions.

The discretization in depth must guarantee that the derivative with respect to the depth coordinate is sufficiently accurate. Because the spatial derivatives in the radiative transfer and the radiative equilibrium equations are written as derivatives with respect to the monochromatic optical depth τ_ν, the chosen depth discretization must guarantee that the implied discretizations of τ_ν for all frequencies ν is of sufficient resolution and covers a sufficient range of optical depths. Moreover, it should not have unnecessarily dense spacing in some regions and inadequate spacing in others. As was suggested in [53], and used in NLTE stellar atmosphere modeling ever since, the most advantageous depth coordinate is the column mass (Lagrangian mass) m, defined by equation (18.8). The discretized depth points are denoted as $m_d, d = 1, \ldots, ND$, where ND is the total number of depth points. The points are chosen to be approximately logarithmically equidistant. The reason is that the spacing in τ_ν, defined by

$$d\tau_\nu = -\chi_\nu dz = (\chi_\nu/\rho)\, dm, \tag{18.43}$$

is also roughly logarithmically equidistant because χ_ν/ρ is only weakly dependent on depth, in contrast to χ_ν itself, which may vary by many orders of magnitude through the atmosphere.

The uppermost depth point m_1 has a small, but nonzero, value because setting $m_1 = 0$ would result in $\rho_1 = 0$, and hence all number densities would be zero, which would render the kinetic equilibrium equation and charge and particle conservation equation meaningless. Hence m_1 should be set so that ideally $\tau_{\nu,1} \ll 1$ for all frequencies. The deepest point, m_{ND}, is also arbitrary; in view of the lower boundary condition for the transfer equation written in the diffusion approximation, (18.6), m_{ND} should be set so that $\tau_{\nu,ND} \gg 1$ for all frequencies.

The discretization in frequency should be chosen so that (i) it covers the "whole" frequency range, i.e., the intensity of radiation in the frequency range beyond the lowest and highest frequency contributes negligibly to frequency integrals from zero to infinity at all depths in the atmosphere, and (ii) it provides a sufficiently accurate representation of all integrals over the finite frequency range, as for instance over line profiles. The resulting set of discretized frequency points is denoted as $\nu_i, i = 1, \ldots, NF$, where NF is the total number of frequencies.

A discretization in angle enters only indirectly into the procedure that determines the variable Eddington factors. Again, it must be chosen to provide a sufficiently accurate representation of the angular integrations. In most stellar atmosphere work, a Gaussian quadrature with six directions (three for incoming and three for outgoing directions) is satisfactory.

Below we summarize all the discretized structural equations. We use a subscript d to indicate the depth index and a subscript i or j to indicate the frequency index. To simplify the notation, we denote discretized frequency- and depth-dependent quantities as $f_{di} \equiv f(m_d, \nu_i)$.

Radiative Transfer Equation

Discretization of the transfer equation was described in detail in chapters 11, 12, and 14, see, e.g., equations (14.153)–(14.157).

For inner depth points, $d = 2, \ldots, ND - 1$, one has

$$\frac{f_{d-1,i}}{\Delta\tau_{d-\frac{1}{2},i}\Delta\tau_{di}} J_{d-1,i} - \frac{f_{di}}{\Delta\tau_{di}} \left(\frac{1}{\Delta\tau_{d-\frac{1}{2},i}} + \frac{1}{\Delta\tau_{d+\frac{1}{2},i}} \right) J_{di} + \frac{f_{d+1,i}}{\Delta\tau_{d+\frac{1}{2},i}\Delta\tau_{di}} J_{d+1,i} = J_{di} - \frac{\eta_{di}^{\text{tot}}}{\chi_{di}}. \tag{18.44}$$

The right-hand side of equation (18.44) can be written as

$$J_{di} - \frac{\eta_{di}^{\text{tot}}}{\chi_{di}} = \epsilon_{di} J_{di} - \frac{\eta_{di}}{\chi_{di}}, \tag{18.45}$$

with

$$\epsilon_{di} \equiv 1 - \kappa_{di}^{\text{sc}} / \chi_{di} = \kappa_{di} / \chi_{di}. \tag{18.46}$$

The upper boundary condition ($d = 1$) in the second-order form is [cf. equation (17.230)]

$$\frac{f_{2i}J_{2i} - f_{1i}J_{1i}}{\Delta\tau_{\frac{3}{2},i}} = g_i J_{1,i} - H_i^{\text{ext}} + \frac{\Delta\tau_{\frac{3}{2},i}}{2} \left(\epsilon_{1i} J_{1i} - \frac{\eta_{1i}}{\chi_{1i}} \right). \tag{18.47}$$

The lower boundary condition ($d = ND$), also in the second-order form, is

$$\frac{f_{ND,i}J_{ND,i} - f_{ND-1,i}J_{ND-1,i}}{\Delta\tau_{ND-\frac{1}{2},i}} = \tfrac{1}{2}(B_{ND,i} - J_{ND,i}) + \frac{1}{3} \frac{B_{ND,i} - B_{ND-1,i}}{\Delta\tau_{ND-\frac{1}{2},i}} - \frac{\Delta\tau_{ND-\frac{1}{2},i}}{2} \left(\epsilon_{ND,i} J_{ND,i} - \frac{\eta_{ND,i}}{\chi_{ND,i}} \right), \tag{18.48}$$

where

$$\begin{aligned} \Delta\tau_{d\pm\frac{1}{2},i} &\equiv \tfrac{1}{2}(\chi_{d\pm1,i} + \chi_{d,i})|z_{d\pm1} - z_d| \\ &= \tfrac{1}{2}\left(\frac{\chi_{d\pm1,i}}{\rho_{d\pm1}} + \frac{\chi_{d,i}}{\rho_{d,i}} \right) |m_{d\pm1} - m_d| \\ &\equiv \tfrac{1}{2}(\omega_{d\pm1,i} + \omega_{d\pm1,i})|m_{d\pm1} - m_d|, \end{aligned} \tag{18.49}$$

and

$$\Delta\tau_{d,i} \equiv \tfrac{1}{2}(\Delta\tau_{d-\frac{1}{2},i} + \Delta\tau_{d+\frac{1}{2},i}). \tag{18.50}$$

Two remarks are in order. First, even in the case of no external irradiation, the fact that the first depth point, m_1, is not at the very top of the atmosphere means that there is radiation that originates in the region between $m=0$ and $m=m_1$, which comes from outside to the surface at m_1; hence $I^{\rm ext}$ and $H^{\rm ext}$ are larger than zero. As suggested in [960], such incoming specific intensity can be approximated as $I_i(\tau_{i1},\mu)=S_i(\tau_{i1})[1-\exp(-\tau_{i1}/\mu)]$, which follows from the formal solution of the transfer equation with a constant source function for $m<m_1$, i.e., $S_i(\tau<\tau_1)=S_i(\tau_1)$. Here, τ_1 is the monochromatic optical depth at $m=m_1$, given approximately by $\tau_{1,i}\approx\frac{1}{2}\chi_{1i}/\rho_1$. The external flux $H^{\rm ext}$ is easily obtained from equation (18.4). However, with a judicious choice of m_1 this term is small and is usually neglected in actual models.

Second, as suggested in [4], one can introduce a *wind albedo*, in which case $I_i(\tau_{i1},\mu)=aI_i(\tau_{i1},-\mu)$, where a is an empirical albedo parameter, $0<a<1$. Such an approach may be useful for approximating the influence of an external wind on an inner static part of the atmosphere. This was used in the past [4], but modern approaches, described in chapters 19 and 20, have rendered it obsolete.

Hydrostatic Equilibrium Equation

The discretized form of (18.10) reads

$$N_d kT_d - N_{d-1}kT_{d-1} + \frac{4\pi}{c}\sum_{i=1}^{NF} w_i(f_{di}J_{di} - f_{d-1,i}J_{d-1,i}) = g(m_d - m_{d-1}). \quad (18.51)$$

The upper boundary condition is derived from equation (18.10), assuming that the radiation force remains constant from the boundary surface upward:

$$N_1 kT_1 = m_1\left[g - (4\pi/c)\sum_{i=1}^{NF} w_i(\chi_{1i}/\rho_1)g_i J_{1i}\right]. \quad (18.52)$$

Radiative Equilibrium Equation

By analogy, discretizing equation (18.13), one obtains

$$\alpha_d \sum_{i=1}^{NF} w_i(\kappa_{di}J_{di} - \eta_{di}) + \beta_d\left[\sum_{i=1}^{NF} w_i(f_{di}J_{di} - f_{d-1,i}J_{d-1,i})/\Delta\tau_{d-\frac{1}{2},i} - (\sigma_R/4\pi)T_{\rm eff}^4\right] = 0. \quad (18.53)$$

Other Structural Equations

The kinetic equilibrium equations and the charge conservation equations are *local* equations; hence no depth discretization is needed. The radiative rates are written as quadrature sums,

$$R_{lu} = (4\pi/h) \sum_{i=1}^{NF} w_i \sigma_{lu}(\nu_i) J_{di}/\nu_i, \tag{18.54a}$$

$$R_{ul} = (4\pi/h) \sum_{i=1}^{NF} w_i \sigma_{lu}(\nu_i) G_{lu}(\nu_i) \left[(2h\nu_i^3/c^2) + J_{di}\right]/\nu_i, \tag{18.54b}$$

where the summations formally extend over all frequency points. It should be kept in mind that the appropriate cross sections differ from zero only in limited ranges of frequencies.

Convection

If convection is taken into account, one must first evaluate the radiative and adiabatic temperature gradients and test for the Schwarzschild stability criterion (18.36). The temperature gradient is evaluated as

$$\nabla_{d-\frac{1}{2}} = (\ln T_d - \ln T_{d-1})/(\ln P_d - \ln P_{d-1}). \tag{18.55}$$

The gradient may also be evaluated as

$$\nabla_{d-\frac{1}{2}} = \frac{T_d - T_{d-1}}{T_d + T_{d-1}} \cdot \frac{P_d + P_{d-1}}{P_d - P_{d-1}}. \tag{18.56}$$

In the following, we use equation (18.56) to represent the logarithmic temperature gradient. The corresponding adiabatic gradient also has to be evaluated at the midpoint $d - \frac{1}{2}$, $\nabla_{\rm ad} = \nabla_{\rm ad}(T_{d-\frac{1}{2}}, P_{d-\frac{1}{2}})$, with $T_{d-\frac{1}{2}} = (T_d + T_{d-1})/2$, and a similar equation for $P_{d-\frac{1}{2}}$. The midpoint values may also be evaluated as geometric means, i.e. $T_{d-\frac{1}{2}} = (T_d T_{d-1})^{1/2}$, but in the following we use the arithmetic mean. The convective flux is a quantity that also corresponds to the grid mid-points, hence it should be written as $F_{\rm conv} = F_{{\rm conv},d-\frac{1}{2}}$.

At the depth points where the Schwarzschild criterion is satisfied, the radiative equilibrium equation, written in the integral form, has to be modified to read

$$\sum_i w_i(\kappa_i J_i - \eta_i) + (\rho_d/4\pi)(F_{{\rm conv},d+\frac{1}{2}} - F_{{\rm conv},d-\frac{1}{2}})/\Delta m_d, \tag{18.57}$$

where $\Delta m_d \equiv (m_{d+\frac{1}{2}} - m_{d-\frac{1}{2}}) = (m_{d+1} - m_{d-1})/2$.

18.2 COMPLETE LINEARIZATION

Formulation

The discretized structural equations (18.44), (18.47), (18.48), (18.51), (18.52), (18.53), (18.23), and (18.24), together with a number of auxiliary relations, form a set of highly coupled, nonlinear algebraic equations. The fundamental problem of stellar atmosphere modeling is to find a robust and efficient method for a numerical solution of these equations. The decisive breakthrough, and in fact the beginning of the modern era of stellar atmosphere models, was the development of the *complete linearization* (CL) method by Auer and Mihalas [53]. This was the first scheme to treat all equations on the same footing, thus solving all structural equations simultaneously. Before that, the equations were typically solved one at a time, iterating between them. In many cases, those iterations were slow, or the scheme failed to converge at all.

The complete linearization method was discussed in chapter 14, where it was used to solve a simpler problem of the simultaneous solution of the radiative transfer and kinetic equilibrium equations. Here we briefly outline the extension of the complete linearization method to solve the full stellar atmosphere problem.

The physical state of an atmosphere is fully described by the set of vectors ψ_d for every depth point d, ($d = 1, \ldots, ND$). The state vector ψ_d is given by

$$\boldsymbol{\psi}_d = \{J_1, \ldots, J_{NF}, N, T, n_\mathrm{e}, n_1, \ldots, n_{NL}\}, \tag{18.58}$$

where J_i is the mean intensity of radiation in the ith frequency point; we have omitted the depth subscript d. The dimension of the vector ψ_d is NN, $NN = NF + NL + NC$, where NF is the number of frequency points, NL the number of atomic energy levels for which the rate equations are solved, and NC the number of constraint equations ($NC = 3$ in the present case); we leave NC as a general number since in some cases it may be larger than 3. For instance, some studies [53, 292] added the fictitious massive particle density n_m in the set of state parameters. As we show below, it is advantageous to add the temperature gradient ∇ when one allows for convection.

The linearization procedure is analogous to that described in § 14.5. Suppose the required solution $\boldsymbol{\psi}_d$ can be written in terms of the current, but imperfect, solution $\boldsymbol{\psi}_d^0$ as $\boldsymbol{\psi}_d = \boldsymbol{\psi}_d^0 + \delta\boldsymbol{\psi}_d$. The entire set of structural equations can be formally written as an operator $\mathbf{P}$ acting on the state vector $\boldsymbol{\psi}_d$ as

$$\mathbf{P}_d(\boldsymbol{\psi}_d) = 0. \tag{18.59}$$

To obtain the solution, we demand that $\mathbf{P}_d(\boldsymbol{\psi}_d^0 + \delta\boldsymbol{\psi}_d) = 0$, and assuming that $\delta\boldsymbol{\psi}_d$ is "small" compared to $\boldsymbol{\psi}_d$, we use a Taylor expansion of $\mathbf{P}$,

$$\mathbf{P}_d(\psi_d^0) + \sum_j \frac{\partial \mathbf{P}_d}{\partial \boldsymbol{\psi}_{d\,j}} \delta\boldsymbol{\psi}_{d\,j} = 0, \tag{18.60}$$

to solve for $\delta\boldsymbol{\psi}_d$. Because only a first-order (i.e., linear) term of the expansion is taken into account, this approach is called a *linearization*. Mathematically, it is the *Newton-Raphson method* for solving a set of nonlinear algebraic equations. To obtain the corrections $\delta\boldsymbol{\psi}_d$, one has to form a matrix of partial derivatives of all the equations with respect to all the unknowns at all depths—the *Jacobi matrix*, or *Jacobian*—and solve equation (18.60). The kinetic equilibrium and charge conservation equations are *local*, i.e., for depth point d they contain the unknown quantities $n_{d,i}$ and $J_{d,j}$ only at depth d. The radiative equilibrium equation (in the differential form) and the hydrostatic equilibrium equation couple two neighboring depth points $d-1$ and d. The radiative transfer equations couple depth point d to two neighboring depths $d-1$ and $d+1$; see equations (18.1)–(18.5). Consequently, the system of linearized equations can be written as

$$-\mathbf{A}_d\,\delta\boldsymbol{\psi}_{d-1} + \mathbf{B}_d\,\delta\boldsymbol{\psi}_d - \mathbf{C}_d\,\delta\boldsymbol{\psi}_{d+1} = \mathbf{L}_d, \tag{18.61}$$

where **A**, **B**, and **C** are $NN \times NN$ matrices, and **L** is a residual error vector, given by

$$\mathbf{L}_d = -\mathbf{P}_d(\boldsymbol{\psi}_d^0). \tag{18.62}$$

At the convergence limit, $\mathbf{L} \to 0$ and thus $\delta\boldsymbol{\psi}_d \to 0$.

Evaluation of the individual elements of matrices **A**, **B**, and **C** can be tedious. Here we give only a brief outline. First, a comment about notation. The order of the individual quantities in the state vector $\boldsymbol{\psi}$ and the order of equations within the global operator **P** are arbitrary. Here we choose, in keeping with the original approach introduced in [53], the state vector in the form (18.58) and the corresponding order of equations: *NF* transfer equations, hydrostatic equilibrium equation, radiative equilibrium equation, charge conservation equation, and *NL* kinetic equilibrium equations supplemented by the corresponding particle conservation (abundance definition) equations. To simplify the notation, the index corresponding to hydrostatic equilibrium (or total particle density N) is denoted as *NH*, the index corresponding to radiative equilibrium (or temperature T) as *NR*, and that corresponding to charge conservation (and the electron density n_e) as *NP*. In the present convention,

$$NH = NF + 1, \quad NR = NF + 2, \quad NP = NF + 3. \tag{18.63}$$

Linearized Transfer Equation

The elements corresponding to the transfer equation (the first *NF* rows) and to derivatives with respect to the mean intensities (the first *NF* columns) were written in § 14.5, equations (14.161)–(14.165).

Let $i, i = 1, \ldots, NF$, be a row corresponding to the transfer equation, and let $P_i(\boldsymbol{\psi}) = 0$ be a formal expression of the transfer equation. Then the individual matrix elements are as follows.

- For the upper boundary condition, $d = 1$, and $j = 1, \ldots, NF$,

$$(B_1)_{ij} \equiv (\partial P_{1,i}/\partial J_{1,j}) = \left[f_{1,i}/\Delta\tau_{\frac{3}{2},i} + g_i + (\Delta\tau_{\frac{3}{2},i}/2)\epsilon_{1,i}\right]\delta_{ij}, \tag{18.64a}$$

$$(C_1)_{ij} \equiv -(\partial P_{1,i}/\partial J_{2,j}) = (f_{2,i}/\Delta\tau_{\frac{3}{2},i})\,\delta_{ij}. \tag{18.64b}$$

The other columns corresponding to the components $\psi_{d,k}$, for $k > NF$, i.e., corresponding to the temperatures and the number densities (total, electron, and level populations) are given by

$$(B_1)_{ik} \equiv \frac{\partial P_{1,i}}{\partial \psi_{1,k}} = \left[-\frac{f_{1,i}J_{1,i} - f_{2,i}J_{2,i}}{\Delta\tau^2_{\frac{3}{2},i}} + \frac{1}{2}\left(\epsilon_{1,i}J_{1,i} - \frac{\eta_{1,i}}{\chi_{1,i}}\right)\right] \frac{\partial \Delta\tau_{\frac{3}{2},i}}{\partial \psi_{1,k}}$$
$$+ \frac{\Delta\tau_{\frac{3}{2},i}}{2}\left[\frac{\partial \epsilon_{1,i}}{\partial \psi_{1,k}} J_{1,i} - \frac{\eta_{1,i}}{\chi_{1,i}}\left(\frac{1}{\eta_{1,i}}\frac{\partial \eta_{1,i}}{\partial \psi_{1,k}} - \frac{1}{\chi_{1,i}}\frac{\partial \chi_{1,i}}{\partial \psi_{1,k}}\right)\right], \quad (18.65a)$$

$$(C_1)_{ik} \equiv -\frac{\partial P_{1,i}}{\partial \psi_{2,k}} = \left(\frac{f_{1,i}J_{1,i} - f_{2,i}J_{2,i}}{\Delta\tau^2_{\frac{3}{2},i}}\right) \frac{\partial \Delta\tau_{\frac{3}{2},i}}{\partial \psi_{2,k}}, \quad (18.65b)$$

where, generally,

$$\frac{\partial \Delta\tau_{d\pm\frac{1}{2},i}}{\partial \psi_{d,k}} = \frac{\Delta\tau_{d\pm\frac{1}{2},i}}{\omega_d + \omega_{d\pm1}} \frac{\partial \omega_{d,i}}{\partial \psi_{d,k}}, \quad \frac{\partial \Delta\tau_{d\pm\frac{1}{2},i}}{\partial \psi_{d\pm1,k}} = \frac{\Delta\tau_{d\pm\frac{1}{2},i}}{\omega_d + \omega_{d\pm1}} \frac{\partial \omega_{d\pm1,i}}{\partial \psi_{d\pm1,k}}, \quad (18.66)$$

and $\omega_{di} = \chi_{di}/\rho_d$. The components of the right-hand side vector $\mathbf{L}_1$ are given by

$$L_{1i} \equiv -P_{1,i} = -\frac{f_{1,i}J_{1,i} - f_{2,i}J_{2,i}}{\Delta\tau_{\frac{3}{2},i}} - g_i J_{1,i} + H_i^{\rm ext} - \frac{\Delta\tau_{\frac{3}{2},i}}{2}\left(\epsilon_{1,i}J_{1,i} - \frac{\eta_{1,i}}{\chi_{1,i}}\right). \quad (18.67)$$

- For the inner points, $d = 2, \ldots, ND-1$, we obtain for $j = 1, \ldots, NF$,

$$(A_d)_{ij} = \frac{f_{d-1,i}}{\Delta\tau_{d-\frac{1}{2},i}\Delta\tau_{d,i}} \delta_{ij}, \quad (18.68a)$$

$$(B_d)_{ij} = \left[\frac{f_{d,i}}{\Delta\tau_{d,i}}\left(\frac{1}{\Delta\tau_{d-\frac{1}{2},i}} + \frac{1}{\Delta\tau_{d+\frac{1}{2},i}}\right) + \epsilon_{d,i}\right] \delta_{ij}, \quad (18.68b)$$

$$(C_d)_{ij} = \frac{f_{d+1,i}}{\Delta\tau_{d+\frac{1}{2},i}\Delta\tau_{d,i}} \delta_{ij}, \quad (18.68c)$$

and, after some algebra, for $k > NF$,

$$(A_d)_{ik} = a_{di}(\partial\omega_{d-i,i}/\partial\psi_{d-1,k}), \quad (18.69a)$$

$$(C_d)_{ik} = c_{di}(\partial\omega_{d+1,i}/\partial\psi_{d+1,k}), \quad (18.69b)$$

$$(B_d)_{ik} = -(a_{di} + c_{di})\frac{\partial \omega_{d,i}}{\partial \psi_{d,k}} + \frac{\partial \epsilon_{d,i}}{\partial \psi_{d,k}} J_{di}$$
$$-\frac{\eta_{d,i}}{\chi_{d,i}}\left(\frac{1}{\eta_{d,i}}\frac{\partial \eta_{d,i}}{\partial \psi_{d,k}} - \frac{1}{\chi_{d,i}}\frac{\partial \chi_{d,i}}{\partial \psi_{d,k}}\right), \quad (18.69c)$$

where

$$\alpha_{di} = (f_{di}J_{di} - f_{d-1,i}J_{d-1,i})/(\Delta\tau_{d-\frac{1}{2},i}\Delta\tau_{di}), \tag{18.70a}$$

$$\gamma_{di} = (f_{di}J_{di} - f_{d+1,i}J_{d+1,i})/(\Delta\tau_{d+\frac{1}{2},i}\Delta\tau_{di}), \tag{18.70b}$$

$$\beta_{di} = \alpha_{di} + \gamma_{di}, \tag{18.70c}$$

$$a_{di} = \left[\alpha_{di} + \tfrac{1}{2}\beta_{di}(\Delta\tau_{d-\frac{1}{2},i}/\Delta\tau_{di})\right]/\left(\omega_{d-1,i} + \omega_{di}\right), \tag{18.70d}$$

$$c_{di} = \left[\gamma_{di} + \tfrac{1}{2}\beta_{di}(\Delta\tau_{d+\frac{1}{2},i}/\Delta\tau_{di})\right]/\left(\omega_{d+1,i} + \omega_{di}\right), \tag{18.70e}$$

and the right-hand side vector

$$L_{di} = -\beta_{di} - \epsilon_{di}J_{di} + \eta_{di}/\chi_{di}. \tag{18.71}$$

- For the lower boundary condition, $d = ND$, we have, for $j \leq NF$, $k > NF$,

$$(B_d)_{ij} = \left[f_{di}/\Delta\tau_{d-\frac{1}{2},i} + \tfrac{1}{2} + (\Delta\tau_{d-\frac{1}{2},i}/2)\,\epsilon_{di}\right]\delta_{ij}, \tag{18.72a}$$

$$(A_d)_{ij} = (f_{d-1,i}/\Delta\tau_{d-\frac{1}{2},i})\,\delta_{ij}, \tag{18.72b}$$

$$\begin{aligned}(B_d)_{ik} = &\left[-\frac{f_{di}J_{di} - f_{d-1,i}J_{d-1,i}}{\Delta\tau^2_{d-\frac{1}{2},i}} + b_i + \frac{1}{2}\left(\epsilon_{di}J_{di} - \frac{\eta_{di}}{\chi_{di}}\right)\right]\frac{\partial\Delta\tau_{d-\frac{1}{2},i}}{\partial\psi_{d,k}} \\ &+\frac{\Delta\tau_{d-\frac{1}{2},i}}{2}\left[\frac{\partial\epsilon_{di}}{\partial\psi_{d,k}}J_{di} - \frac{\eta_{di}}{\chi_{di}}\left(\frac{1}{\eta_{di}}\frac{\partial\eta_{di}}{\partial\psi_{d,k}} - \frac{1}{\chi_{di}}\frac{\partial\chi_{di}}{\partial\psi_{d,k}}\right)\right] \\ &-\left(\frac{1}{2} + \frac{1}{3\Delta\tau_{d-\frac{1}{2},i}}\right)\left(\frac{dB_i}{dT}\right)_d \delta_{k,NR},\end{aligned} \tag{18.72c}$$

$$\begin{aligned}(A_d)_{ik} = &\left(\frac{f_{di}J_{di} - f_{d-1,i}J_{d-1,i}}{\Delta\tau^2_{d-\frac{1}{2},i}} - b_i\right)\frac{\partial\Delta\tau_{d-\frac{1}{2},i}}{\partial\psi_{d-1,k}} \\ &-\frac{1}{3\Delta\tau_{d-\frac{1}{2},i}}\left(\frac{dB_i}{dT}\right)_{d-1}\delta_{k,NR},\end{aligned} \tag{18.72d}$$

where

$$b_i \equiv \frac{1}{3}\frac{B_{di} - B_{d-1,i}}{\Delta\tau^2_{d-\frac{1}{2},i}}. \tag{18.73}$$

The last terms in (18.72c,d) which only apply for $\psi_k = T$, i.e., $k = NR$, arise from derivatives of the Planck function with respect to temperature. Finally,

$$\begin{aligned}L_{di} = &-\frac{f_{di}J_{di} - f_{d-1,i}J_{d-1,i}}{\Delta\tau_{d-\frac{1}{2},i}} - \tfrac{1}{2}(J_{di} - B_{di}) + \frac{1}{3}\frac{B_{di} - B_{d-1,i}}{\Delta\tau_{d-\frac{1}{2},i}} \\ &-\frac{\Delta\tau_{d-\frac{1}{2},i}}{2}\left(\epsilon_{di}J_{di} - \frac{\eta_{di}}{\chi_{di}}\right).\end{aligned} \tag{18.74}$$

Linearized Hydrostatic Equilibrium Equation

The components of matrices **A**, **B**, and vector **L** corresponding to the hydrostatic equilibrium equation, the row $NH = NF + 1$, are given by

$$(B_1)_{NH,i} = (4\pi/c)\, w_i \omega_{1,i} g_i, \quad i \leq NF, \tag{18.75a}$$

$$(B_1)_{NH,NH} = kT_1, \tag{18.75b}$$

$$(B_1)_{NH,NR} = kN_1 + (4\pi/c) \sum_{j=1}^{NF} w_j g_j J_{1,j} (\partial \omega_{1,j} / \partial T_1), \tag{18.75c}$$

$$(B_1)_{NH,n} = (4\pi/c) \sum_{j=1}^{NF} w_j g_j J_{1,j} (\partial \omega_{1,j} / \partial \psi_{1,n}),\ n > NR, \tag{18.75d}$$

$$(L_1)_{NH} = g m_1 - N_1 k T_1 - (4\pi/c) \sum_{j=1}^{NF} w_j \omega_{1j} g_j J_{1j}, \tag{18.75e}$$

and, for $d > 1$,

$$(A_d)_{NH,i} = (4\pi/c) w_i f_{d-1,i}, \quad i \leq NF, \tag{18.76a}$$

$$(B_d)_{NH,i} = (4\pi/c) w_i f_{di}, \quad i \leq NF, \tag{18.76b}$$

$$(A_d)_{NH,NH} = kT_{d-1}, \tag{18.76c}$$

$$(B_d)_{NH,NH} = kT_d, \tag{18.76d}$$

$$(A_d)_{NH,NR} = kN_{d-1}, \tag{18.76e}$$

$$(B_d)_{NH,NR} = kN_d, \tag{18.76f}$$

$$(L_d)_{NH} = g(m_d - m_{d-1}) - N_d k T_d + N_{d-1} k T_{d-1} - (4\pi/c) \sum_{j=1}^{NF} w_j (f_{dj} J_{dj} - f_{d-1,j} J_{d-1,j}). \tag{18.77}$$

Linearized Radiative Equilibrium Equation

In order to simplify the notation, we consider the case of the integral and the differential forms separately. When considering a linear combination of both forms, equation (18.53), the matrix elements are given as a corresponding linear combination.

First, take the integral form. Equation (18.12) is local, and hence the only nonzero contributions to matrix **B**, for $d = 1, \ldots, ND$, are (with $k > NF$)

$$(B_d)_{NR,i} = w_i \kappa_{di}, \quad i \leq NF, \tag{18.78a}$$

$$(B_d)_{NR,k} = w_i [(\partial \kappa_{di} / \partial \psi_{dk}) J_{di} - (\partial \eta_{di} / \partial \psi_{dk})], \tag{18.78b}$$

and the right-hand side

$$(L_d)_{NR} = -\sum_{j=1}^{NF} w_j(\kappa_{dj}J_{dj} - \eta_{dj}). \tag{18.78c}$$

The matrix elements for the differential form, for $d = 2, \ldots, ND$, are given by

$$(A_d)_{NR,i} = w_i f_{d-1,i}/\Delta\tau_{d-\frac{1}{2},i}, \quad i \leq NF, \tag{18.79a}$$

$$(B_d)_{NR,i} = w_i f_{di}/\Delta\tau_{d-\frac{1}{2},i}, \quad i \leq NF, \tag{18.79b}$$

$$(A_d)_{NR,k} = \sum_{j=1}^{NF} w_j \frac{f_{dj}J_{dj} - f_{d-1,j}J_{d-1,j}}{\Delta\tau^2_{d-\frac{1}{2}j}} \frac{\partial\Delta\tau_{d-\frac{1}{2}j}}{\partial\psi_{d-1,k}}, \tag{18.79c}$$

$$(B_d)_{NR,k} = -\sum_{j=1}^{NF} w_j \frac{f_{dj}J_{dj} - f_{d-1,j}J_{d-1,j}}{\Delta\tau^2_{d-\frac{1}{2}j}} \frac{\partial\Delta\tau_{d-\frac{1}{2}j}}{\partial\psi_{d,k}}, \tag{18.79d}$$

where the last two equations apply for $k \geq NR$, and

$$(L_d)_{NR} = \frac{\sigma_R}{4\pi} T^4_{\text{eff}} - \sum_{j=1}^{NF} w_j \frac{f_{dj}J_{dj} - f_{d-1,j}J_{d-1,j}}{\Delta\tau_{d-\frac{1}{2}j}}. \tag{18.80}$$

and for the upper boundary condition,

$$(B_1)_{NR,i} = w_i g_i, \tag{18.81a}$$

$$(L_1)_{NR} = \frac{\sigma_R}{4\pi} T^4_{\text{eff}} - \sum_{j=1}^{NF} w_j(g_j J_{1j} - H^{\text{ext}}_j). \tag{18.81b}$$

Linearized Charge Conservation Equations

The equation is local and simple, so the linearization is straightforward:

$$(B_d)_{NP,NP} = -1, \tag{18.82a}$$

$$(B_d)_{NP,i} = Z_i, \quad i > NP, \tag{18.82b}$$

where Z_i is the charge of the ion to which level i belongs, and

$$(L_d)_{NP} = n_e - \sum_{i=1}^{NL} n_i Z_i. \tag{18.83}$$

All other elements of the NPth row of matrix **B** and all elements of the NPth row of matrices **A** and **C** are zero.

Linearized Kinetic Equilibrium Equations

The matrix elements corresponding to the kinetic equilibrium equations were considered in detail in § 14.5 and do not need to be repeated here. The only modification of the formalism presented there is to set up matrix elements $B_{m,NR}$ and $B_{m,NP}$ ($m > NP$) that correspond to the derivatives of the transition rates with respect to temperature and electron density, respectively. The transition rates do not explicitly depend on N; hence the matrix elements $B_{i,NH} = 0$. The kinetic equilibrium equations are local; therefore, $A_{ij} = C_{ij} = 0$ for $i > NP$ and all j. Because the form of matrix elements is the same for all depth points, we drop the depth index d.

The matrix elements are as follows:

$$B_{m,i} = \sum_{j=1}^{NL} (\partial \mathcal{A}_{mj} / \partial J_i)\, n_j, \quad i \le NF, \tag{18.84a}$$

$$B_{m,NR} = \sum_{j=1}^{NL} (\partial \mathcal{A}_{mj} / \partial T)\, n_j, \tag{18.84b}$$

$$B_{m,NP} = \sum_{j=1}^{NL} (\partial \mathcal{A}_{mj} / \partial n_e)\, n_j, \tag{18.84c}$$

$$B_{mj} = \mathcal{A}_{mj}, \quad j > NP, \tag{18.84d}$$

$$L_m = b_m - \sum_{j=1}^{NL} \mathcal{A}_{mj} n_j. \tag{18.84e}$$

For each atomic species I, the rate equation for a characteristic level, say, k, is replaced by the particle conservation equation, (18.21). The corresponding matrix elements are

$$B_{ki} = 1 + S_i, \tag{18.85a}$$

$$B_{k,NH} = -\alpha_I, \tag{18.85b}$$

$$B_{k,NP} = \alpha_I + \sum_{j=1}^{NL} n_j (\partial S_j / \partial n_e) \tag{18.85c}$$

$$B_{k,NR} = \sum_{j=1}^{NL} n_j (\partial S_j / \partial T) \tag{18.85d}$$

where i labels all levels considered for species I, and $\alpha_I = A_I / \sum_J A_J$ is the fractional abundance of species I.

Linearized Equation for Convection

When convection is taken into account, it is advantageous to consider the logarithmic gradient of temperature ∇ as one of the state parameters and include it in the state vector $\boldsymbol{\psi}$,

$$\boldsymbol{\psi}_d = \{J_{d1}, \dots, J_{d,NF}, N, T, n_{e,d}, n_{d1}, \dots, n_{d,NL}, \nabla_d\}, \tag{18.86}$$

where we adopt the convention that the gradient ∇_d is the one corresponding to depth $d - \frac{1}{2}$, i.e.,

$$\nabla_d \equiv \nabla_{d-\frac{1}{2}} = \frac{T_d - T_{d-1}}{T_d + T_{d-1}} \cdot \frac{P_d + P_{d-1}}{P_d - P_{d-1}}. \tag{18.87}$$

There are four modifications of matrices **A**, **B**, **C**, and vector **L** when convection is taken into account:

- a modification of the row $NR = NF + 2$, corresponding to radiative equilibrium, now being modified to radiative + convective equilibrium;
- an addition of a column in all matrices corresponding to ∇; using the convention of equation (18.86), it is the column $NN = NF + NL + 4$;
- an addition of a row NN, corresponding to ∇;
- a modification of vector **L**, namely, changing the NR element and adding the NN element.

Because the convective flux depends only on T, P, and ∇, and writing $P = NkT$, the only new elements of the row NR of matrices **A**, **B**, **C** are those corresponding to N, T, and ∇. In the radiative zone, $\nabla < \nabla_{\rm ad}$, all new elements are zero. In the convection zone, $\nabla \geq \nabla_{\rm ad}$, there are the following additions to the matrix elements, denoted by superscript "conv":

In the differential equation form, that is linearizing equation (18.37), where we introduce $H_{\rm conv} \equiv F_{\rm conv}/4\pi$,

$$A^{\rm conv}_{NR,NH} = -\tfrac{1}{2} \left(\frac{\partial H_{\rm conv}}{\partial P}\right)_{d-\frac{1}{2}} kT_{d-1}, \tag{18.88a}$$

$$A^{\rm conv}_{NR,NR} = -\tfrac{1}{2} \left(\frac{\partial H_{\rm conv}}{\partial T}\right)_{d-\frac{1}{2}} - \tfrac{1}{2} \left(\frac{\partial H_{\rm conv}}{\partial P}\right)_{d-\frac{1}{2}} kN_{d-1}, \tag{18.88b}$$

$$B^{\rm conv}_{NR,NH} = \tfrac{1}{2} \left(\frac{\partial H_{\rm conv}}{\partial P}\right)_{d-\frac{1}{2}} kT_d, \tag{18.88c}$$

$$B^{\rm conv}_{NR,NR} = \tfrac{1}{2} \left(\frac{\partial H_{\rm conv}}{\partial T}\right)_{d-\frac{1}{2}} + \tfrac{1}{2} \left(\frac{\partial H_{\rm conv}}{\partial P}\right)_{d-\frac{1}{2}} kN_d, \tag{18.88d}$$

$$B^{\rm conv}_{NR,NN} = \frac{\frac{3}{2} H_{{\rm conv},d-\frac{1}{2}}}{\nabla_d - \nabla_{{\rm el},d-\frac{1}{2}}}, \tag{18.88e}$$

$$L^{\rm conv}_{NR} = -H_{{\rm conv},d-\frac{1}{2}} \tag{18.88f}$$

where we used $\partial T_{d-\frac{1}{2}}/\partial T_d = 1/2$. The partial derivatives of the convective flux with respect to temperature and pressure are obtained numerically,

$$\frac{\partial H_{\rm conv}}{\partial T} = \frac{H_{\rm conv}(T + \delta T, P, \nabla) - H_{\rm conv}(T, P, \nabla)}{\delta T}, \tag{18.89}$$

and analogously for the derivative with respect to pressure. Here δT is an arbitrary small quantity; it is typically chosen $\delta T = 0.01\,T$. The derivative with respect to ∇ is computed analytically.

In the integral form, linearization of equation (18.57) leads to the following additions to the matrix elements:

$$A^{\rm conv}_{NR,NH} = \left(\frac{\partial H_{\rm conv}}{\partial P}\right)_{d-\frac{1}{2}} \frac{\rho_d}{2\Delta m_d} kT_{d-1}, \tag{18.90a}$$

$$A^{\rm conv}_{NR,NR} = \left[\left(\frac{\partial H_{\rm conv}}{\partial T}\right)_{d-\frac{1}{2}} + \left(\frac{\partial H_{\rm conv}}{\partial P}\right)_{d-\frac{1}{2}} kN_{d-1}\right] \frac{\rho_d}{2\Delta m_d}, \tag{18.90b}$$

$$A^{\rm conv}_{NR,NN} = \frac{H_{{\rm conv},d-\frac{1}{2}}}{(\nabla_d - \nabla_{{\rm el},d-\frac{1}{2}})} \frac{3\rho_d}{4\Delta m_d}, \tag{18.90c}$$

$$B^{\rm conv}_{NR,NH} = \left[\left(\frac{\partial H_{\rm conv}}{\partial P}\right)_{d+\frac{1}{2}} - \left(\frac{\partial H_{\rm conv}}{\partial P}\right)_{d-\frac{1}{2}}\right] \frac{\rho_d}{2\Delta m_d} kT_d + \frac{\rho_d}{(N_d - n_{{\rm e},d})} \frac{H_{{\rm conv},d+\frac{1}{2}} - H_{{\rm conv},d-\frac{1}{2}}}{\Delta m_d}, \tag{18.90d}$$

$$B^{\rm conv}_{NR,NR} = \left[\left(\frac{\partial H_{\rm conv}}{\partial T}\right)_{d+\frac{1}{2}} + \left(\frac{\partial H_{\rm conv}}{\partial P}\right)_{d+\frac{1}{2}} kN_d - \left(\frac{\partial H_{\rm conv}}{\partial T}\right)_{d-\frac{1}{2}} - \left(\frac{\partial H_{\rm conv}}{\partial P}\right)_{d-\frac{1}{2}} kN_d\right] \frac{\rho_d}{2\Delta m_d}, \tag{18.90e}$$

$$B^{\rm conv}_{NR,NP} = -\frac{\rho_d}{(N_d - n_{{\rm e},d})} \frac{H_{{\rm conv},d+\frac{1}{2}} - H_{{\rm conv},d-\frac{1}{2}}}{\Delta m_d}, \tag{18.90f}$$

$$B^{\rm conv}_{NR,NN} = \left[\frac{H_{{\rm conv},d+\frac{1}{2}}}{\nabla_{d+1} - \nabla_{{\rm el},d+\frac{1}{2}}} - \frac{H_{{\rm conv},d-\frac{1}{2}}}{\nabla_d - \nabla_{{\rm el},d-\frac{1}{2}}}\right] \frac{3\rho_d}{2\Delta m_d}, \tag{18.90g}$$

$$C^{\rm conv}_{NR,NH} = -\left(\frac{\partial H_{\rm conv}}{\partial P}\right)_{d+\frac{1}{2}} \frac{\rho_d}{2\Delta m_d} kT_{d+1}, \tag{18.90h}$$

$$C^{\rm conv}_{NR,NH} = -\left[\left(\frac{\partial H_{\rm conv}}{\partial T}\right)_{d+\frac{1}{2}} + \left(\frac{\partial H_{\rm conv}}{\partial P}\right)_{d+\frac{1}{2}} kN_{d-1}\right] \frac{\rho_d}{2\Delta m_d}, \tag{18.90i}$$

$$C^{\rm conv}_{NR,NN} = -\frac{H_{{\rm conv},d+\frac{1}{2}}}{(\nabla_d - \nabla_{{\rm el},d+\frac{1}{2}})} \frac{3\rho_d}{2\Delta m_d}, \tag{18.90j}$$

$$L^{\rm conv}_{NR} = -\frac{\rho_d}{\Delta m_d}(H_{{\rm conv},d+\frac{1}{2}} - H_{{\rm conv},d-\frac{1}{2}}). \tag{18.90k}$$

The additional row corresponding to ∇, i.e., a linearized equation (18.56), is simple:

$$A^{\rm conv}_{NN,NH} = -\frac{2P_d\nabla_d}{P_d^2 - P_{d-1}^2}\, kT_{d-1}, \tag{18.91a}$$

$$A^{\rm conv}_{NN,NR} = \frac{2T_d\nabla_d}{T_d^2 - T_{d-1}^2} - \frac{2P_d\nabla_d}{P_d^2 - P_{d-1}^2}\, kN_{d-1}, \tag{18.91b}$$

$$B^{\rm conv}_{NN,NH} = -\frac{2P_{d-1}\nabla_d}{P_d^2 - P_{d-1}^2}\, kT_d, \tag{18.91c}$$

$$B^{\rm conv}_{NN,NR} = \frac{2T_{d-1}\nabla_d}{T_d^2 - T_{d-1}^2} - \frac{2P_{d-1}\nabla_d}{P_d^2 - P_{d-1}^2}\, kN_d, \tag{18.91d}$$

$$B^{\rm conv}_{NN,NN} = -1, \tag{18.91e}$$

and

$$L_{NN} = \nabla_d - \frac{T_d - T_{d-1}}{T_d + T_{d-1}} \cdot \frac{P_d + P_{d-1}}{P_d - P_{d-1}}. \tag{18.92}$$

An Application to LTE Models

It should be mentioned that the complete linearization method is capable of computing an LTE model without significant changes of the overall computational scheme. Such a method is not an optimal one when computing only LTE models because more efficient schemes specifically designed to compute LTE models exist (such as the Rybicki-type or the temperature-correction schemes; see chapter 17). Yet the complete linearization is still useful for computing an LTE model for the purpose of providing an input model for subsequent NLTE model calculation, or for a comparison between LTE and NLTE results, because it uses the same input atomic physics and the same internal numerical approximations and representations (such as a choice of discretization in depth, frequency, and angle and a numerical representation of the transfer equation).

When computing an LTE model atmosphere, the state vector $\boldsymbol{\psi}$ is the same as in the case of NLTE—equation (18.58). The structural equations, i.e., radiative transfer, hydrostatic equilibrium, radiative equilibrium, charge, and particle conservation equations remain unchanged; the only change is replacing the kinetic equilibrium equations with the expression for LTE level populations,

$$n_i = n_i^* = n_e n_1^+ \Phi_i(T), \tag{18.93}$$

where $\Phi_i(T)$ is the Saha-Boltzmann factor, defined by equation (18.18). The set of equations for level populations can still be formally represented by the rate equation

$\mathcal{A} \cdot \mathbf{n} = \mathbf{b}$, where the only nonzero elements of the rate matrix are given by

$$\mathcal{A}_{ii} = 1, \tag{18.94a}$$

$$\mathcal{A}_{i,p} = -n_e \Phi_i(T), \tag{18.94b}$$

$$\mathcal{A}_{k,j} = 1 + S_j. \tag{18.94c}$$

Here p denotes the ground state of the next ionization stage to level i, and k denotes the characteristic level, i.e., the one that represents the abundance definition equation (18.21). The expression for the elements of the right-hand side vector $\mathbf{b}$ remains in the form of equation (18.22). The elements of the matrix $\mathbf{B}$ are given by (18.84), which for the rate matrix given by (18.94) become (dropping again the depth index d)

$$B_{m,i} = 0, \quad (i \leq NF), \tag{18.95a}$$

$$B_{m,NR} = -n_e n_p (d\Phi_i/dT), \quad B_{m,NP} = -n_p \Phi_i(T), \tag{18.95b}$$

$$B_{mm} = 1, \quad B_{mp} = -n_e \Phi_i(T), \tag{18.95c}$$

and the matrix elements for the characteristic level k remain in the form of (18.85).

Solution of the Linearized Equations

The system of equations (18.61) is block tridiagonal and is solved by the standard Gauss-Jordan elimination procedure, essentially the same as that used in the Feautrier elimination procedure for solving the radiative transfer equation. Specifically, the block tridiagonal system (18.61) is solved by a forward-backward sweep. As discussed in § 12.2, one first constructs auxiliary matrices $\mathbf{D}_d$ and vectors $\mathbf{E}_d$, proceeding from $d = 1$ to ND (the forward elimination step),

$$\mathbf{D}_d = (\mathbf{B}_d - \mathbf{A}_d \mathbf{D}_{d-1})^{-1} \mathbf{C}_d, \tag{18.96}$$

and

$$\mathbf{E}_d = (\mathbf{B}_d - \mathbf{A}_d \mathbf{D}_{d-1})^{-1} (\mathbf{L}_d + \mathbf{A}_d \mathbf{E}_{d-1}), \tag{18.97}$$

where we use the convention that $\mathbf{D}_0 = \mathbf{E}_0 = 0$. The solution vectors are obtained with a backward sweep,

$$\delta \boldsymbol{\psi}_d \equiv \mathbf{D}_d \, \delta \boldsymbol{\psi}_{d+1} + \mathbf{E}_d, \quad d = ND - 1, \ldots, 1, \tag{18.98}$$

with $\delta \boldsymbol{\psi}_{ND} = \mathbf{E}_{ND}$.

The main difficulty in solving the system (18.61) is that the $NN \times NN$ matrices $(\mathbf{B}_d - \mathbf{A}_d \mathbf{D}_{d-1})$ for $d = 1, \ldots ND$ must be inverted in order to evaluate (18.96) and (18.97). Therefore, the total computer time for ordinary complete linearization scales roughly as

$$t \propto (NF + NL + NC)^3 \times ND \times N_{\text{iter}}, \tag{18.99}$$

where N_{iter} is the number of iterations need to obtain a converged model.

It is immediately clear that the original complete linearization, despite its inherent power and robustness, cannot be used as a general numerical scheme because in realistic calculations one must use a very large number of frequency points *NF* to describe the radiation field sufficiently accurately—of the order of 10^5 to 10^6 points. Inverting matrices of this dimension is generally out of question. (This may not be the case if one has access to a high-speed massively parallel computer where each node has a large memory.) Generally one must seek less global, but much faster, schemes. We shall return to this point in the next section.

Computational Procedure

The basic computational procedure was established in early applications of the complete linearization method [53, 55, 56, 734, 737] and is still used even if the iteration scheme was significantly modified and upgraded, as will be described in the next section. The procedure consists of the following basic steps.

Selecting Explicit Atoms, Ions, Levels, Transitions

Depending on the problem at hand, which is defined by the value of the effective temperature, surface gravity, and chemical composition, one selects chemical species to be treated *explicitly*. The explicit species, or *explicit atoms*, are those that are expected to contribute to the total opacity and emissivity or to the total number of particles, or are important electron donors.

For each explicit species, one has to select a set of ionization stages that are to be treated explicitly. In most applications, it is not practical to select all ionization stages because they may be too numerous (for instance, 27 ionization stages for iron); moreover, only a few of them make an appreciable contribution. However, the selection of ionization stages should be done judiciously in order not to neglect any important stages that may contribute, if not everywhere, then at some depths in the atmosphere. A resulting set of ionization stages constitutes a set of *explicit ions*.

For each explicit ion, one has to select levels that are to be treated in detail, i.e., for which the kinetic equilibrium equation is solved—called *explicit levels*. These levels can be either true eigenstates of an atom/ion or a group of energy levels lumped together, called *superlevels*; see § 18.5.

Finally, one has to select transitions between explicit levels that are to be treated in detail, i.e., their transition rates are calculated explicitly from equations (18.16)–(18.20). To this end, one has to choose a set of frequency points that can give an accurate numerical representation of the appropriate integrals over frequency. By setting frequency points for all selected transitions, one arrives at the global set of frequencies. This set should be inspected for possible overlaps of lines and unnecessarily dense spacing of frequencies in such regions, and some frequency points may be removed. Also, care should be taken to select the lowest and the highest frequency of the global set so that frequency integrals from 0 to infinity are represented accurately enough.

For all explicit levels and transitions one has to collect all the necessary atomic data. For levels, these are the level energies and statistical weights, which are typically easy to find. For explicit transitions, the situation is usually more difficult. For bound-free transitions, one needs photoionization cross sections as functions of frequency and collisional ionization rates. For bound-bound transitions, one needs oscillator strengths and collisional excitation rates. The latter quantities are often not well known. Generally, the accuracy of atomic data sets the limits of internal accuracy of the resulting model atmospheres.

Starting Model

Complete linearization is an iteration scheme; therefore, it needs a starting solution. In fact, the quality of the starting solution, i.e., some measure of how close it is to the final solution, in many cases determines the convergence properties of the iteration scheme or, indeed, whether the method converges at all. Therefore, finding a good starting model is a very important, perhaps crucial, part of the overall strategy.

As a starting model, one can use a previously computed NLTE model atmosphere with similar basic parameters. If no such model exists, one has to construct the starting model from scratch. Such a procedure often involves several steps, each usually requiring an iteration procedure of its own, with increasing physical complexity. The first step is to compute an LTE-gray model atmosphere, as described in § 17.1. and § 17.3.

Any stellar atmosphere exhibits more or less pronounced departures from LTE in its upper part, but it eventually reaches LTE when optical depths in all frequencies become sufficiently large. Going even deeper, one reaches layers where the diffusion approximation starts to hold. It is, therefore, reasonable to employ this LTE-gray model as a starting solution for the subsequent iteration scheme, because it is guaranteed to provide a correct solution at the deepest layers of the atmosphere.

The important point is that for the purposes of providing the starting solution, one does not use just any LTE-gray model. This should be a model in which the Rosseland mean opacity is computed by a frequency quadrature using the same opacity sources, and the same numerical treatment of frequency quadratures, as for the subsequently calculated models.

Computational Strategy

From the initial LTE-gray model, one can start the iteration procedure to obtain a desired NLTE model, or, more often, one proceeds through a series of intermediate models with increasing physical complexity and degree of realism. For instance, the original applications of the complete linearization [53,55,56,734,737] used the following sequence of models:

$$\text{LTE-gray} \rightarrow \text{LTE} \rightarrow \text{NLTE-C} \rightarrow \text{NLTE-L}.$$

Starting from the LTE-gray model, one first converges a full LTE model. Taking it as a starting model, one converges a NLTE model in which all lines are assumed to be in the detailed radiative balance, called the NLTE-continua-only (NLTE-C) model. The final stage is to converge a full NLTE model, including all lines: the

NLTE-lines, or NLTE-L model. This procedure has an obvious advantage in that it consecutively improves the model from deep layers toward the surface.

Such a sequence has been used more or less ever since. However, there are variations. In some cases one can go directly from LTE-gray to NLTE-L. In contrast, in some cases, typically when calculating very complex models, one has to proceed through several stages of NLTE-L models, with increasing number of lines considered explicitly.

Iteration Process

One iteration of the complete-linearization scheme consists of the following steps:

- computing the elements of the matrices $\mathbf{A}_d$, $\mathbf{B}_d$, $\mathbf{C}_d$, and vectors $\mathbf{L}_d$;
- solving the linearized system by a forward-backward sweep [(18.96)–(18.98)], and computing new state vectors $\boldsymbol{\psi}_d$,
- before entering the next iteration step, performing a "formal solution."

The last step is a very important, sometimes even crucial, ingredient of the iteration process. By a formal solution we mean a solution of *one equation at a time*, using the current values of all other state parameters. Specifically, the corrections $\delta\boldsymbol{\psi}_d$ may be applied to obtain new temperature, electron density, and mean intensities, but not the new populations. They are obtained by using new T, n_e, and J_ν to calculate new radiative and collisional rates and to solve the kinetic equilibrium equations. This procedure has the advantage that new populations are consistent with the new values of the radiation field.

One then solves the transfer equation for all frequencies and angles in order to calculate new variable Eddington factors.

One can even employ an internal iteration process, in which T and n_e are kept fixed, and one updates *both* the population numbers and mean intensities of radiation. Here one can use classical Λ-iteration; i.e., iterating between the kinetic equilibrium and the radiative transfer equations (using current values of level populations to compute opacities and emissivities), to solve the transfer equation, use the calculated mean intensities to compute new radiative rates, and recompute level populations from the kinetic equilibrium equation. It is reasonable to use Λ-iteration here despite its drawbacks because one does not aim to compute the full solution, but only to provide more consistent values of level populations and radiation intensities to improve the convergence of the global iteration scheme. This approach proved to be helpful in the early studies [53,55,56,521,734,737].

With the advent of ALI-based methods for solving the multi-level atom problem, it became clear that instead of performing several Λ-iterations, one can perform several iterations of an ALI scheme with preconditioning (see § 14.5), essentially without any increase of computer time, because in both cases the total computational load is dominated by the formal solutions of the transfer equation. This is the approach used, for instance, in [533,682,683].

When convection is taken into account, one has to calculate a new gradient ∇_d and compare it to the adiabatic gradient $\nabla_{\rm ad}$. Because during the iteration process one deals with imperfect values of T and P, at certain depth d one may obtain $\nabla_d < \nabla_{\rm ad}$, while in the surrounding points $\nabla \geq \nabla_{\rm ad}$. Had one entered the next linearization step with such values of T and P, the atmosphere at depth point d would be considered as stable against convection, which would lead to potentially disastrous changes of temperature in the next linearization step because the numerical scheme would try to enforce radiative equilibrium there. One has to modify the temperature at such points to have $\nabla \geq \nabla_{\rm ad}$.

18.3 OVERVIEW OF POSSIBLE ITERATIVE METHODS

The complete linearization method described above provides a global solution of the classical stellar atmospheres problem. Moreover, thanks to the mathematical nature of the Newton-Raphson scheme, the iterations converge quadratically (if they converge at all). However, the main drawback of the scheme is the need to invert the individual blocks of the Jacobi matrix.

Consider first the size of matrices to be inverted, or, equivalently, the number of necessary state parameters to describe a realistic a model atmosphere. The basic goal is to construct as realistic a model atmosphere as possible. Within the context of "classical models," i.e., for plane-parallel, horizontally homogeneous, hydrostatic models, this means to take into account "all" opacity sources, continua and lines, that contribute to the total opacity. Because a majority of these opacity sources are provided by metal lines, the corresponding models are often called *metal line-blanketed* model atmospheres. They will be discussed in detail in § 18.5.

The number of frequency points depends sensitively on the degree of sophistication of the model and on the actual numerical scheme to treat the line-blanketing opacity. Early applications considered very simplified models with about 100 frequency points; modern line-blanketed models require at least a few times 10^4 to several times 10^5 frequency points.

In LTE line-blanketed models, one can treat opacity in a statistical way (see § 17.6), and, in any case, the opacity, albeit a complicated function of frequency, is a function of temperature and density only. In contrast, in NLTE, the opacity in each single line depends on (generally non-equilibrium) populations of the lower and upper level of the corresponding transition. Therefore, one has to solve a kinetic equilibrium equation for all levels that are in transitions that contribute to the opacity, i.e., essentially all levels of all species, including, of course, the iron-peak elements (or at least iron alone), which provide a huge number, literally millions, of lines. Since each ion of iron has of the order of 10^4 levels, the total number of levels to be treated in NLTE may well be of the order of 10^5.

With these numbers, an application of the standard complete linearization technique for realistic NLTE atmospheres is hopeless and completely out of the question. Therefore, one has to seek less global, less costly, and more efficient approaches.

Any method should solve all the structural equations, although not necessarily simultaneously, as the complete linearization method does. The art is to find a procedure that allows one not to solve all the equations simultaneously (i.e., some information is lagged behind), while remaining sufficiently efficient and robust.

Taking equation (18.99) as a guide, we list below a number of options that can be used to reduce total computer time. We first outline all the options and then discuss some of the most important possibilities in detail.

- Modifications within the framework of standard linearization:
  - reducing *NF*, number of linearized frequencies;
  - reducing *NL*, number of linearized population numbers;
  - reducing *NC*, number of "constraint" equations;
  - reducing *ND*, number of depth points;
  - reducing N_{iter}, number of iterations needed for convergence.
- Avoiding repeated inversions of the Jacobian.
- Approximate Newton-Raphson schemes.

We consider all these possibilities in more detail below.

Improvements of Standard Linearization

Here we discuss in more detail the first category of options, namely, those in which the global framework of complete linearization remains intact, while efficiency is gained by reducing the size of the state vector, i.e., by reducing the size of matrices to be inverted.

Reducing NF

There are several ways to achieve this goal.

(i) The obvious possibility is to consider only a small number of frequencies. This was the approach that had to be adopted in early applications of the original complete linearization method. However, as we aim at constructing a method that is able to provide an "exact" solution for the stellar atmosphere problem, we do not consider this possibility further.

(ii) The next simplest approach is to linearize mean intensities only for those frequency points that are "essential," while keeping the mean intensities for other frequencies fixed during linearization and updating them during a subsequent formal solution. Such an idea was implemented, for instance, in the original complete-linearization computer program [737], which actually used more sophisticated equivalent-two-level-atom (ETLA) procedures, and in early complete-linearization-based model atmosphere codes, e.g., [521]. The fixed-rates approach has proved to be useful in some cases, but if too many transitions are treated in this mode, one basically recovers a Λ-iteration type of behavior, which is a serious drawback. In particular, the convergence rate is slow, and the solution tends to stabilize rather than truly converge. Using an ETLA procedure partly overcomes this problem, but it is much more time-consuming.

(iii) Another way to reduce the size of matrices to be inverted is to adopt the multi-frequency/multi-gray method [29–31]. Here, one substitutes

$$\{J_1, \ldots, J_{NF}\} \rightarrow \{\widetilde{J}_1, \ldots, \widetilde{J}_{NB}\}, \tag{18.100}$$

where $\widetilde{J}_i$ represents a mean intensity characteristic of ith frequency block, and NB is the number of blocks. In order to achieve a substantial reduction of the time, one needs to have $NB \ll NF$. Each block groups together all frequencies for which the radiation is formed in a similar way. It is not necessary that the block be composed of a continuous frequency interval (for instance, one block may represent all wings of weak lines; see [29]). The essence of the method consists of selecting appropriate frequency bands and the individual frequency points belonging to them. However, this is also a drawback of the method, because the bands have to be set essentially by hand.

(iv) Finally, the accelerated lambda iteration (ALI) method reduces the number of unknowns even more, because it is able to eliminate *all* or most frequency points completely. This is the critical improvement. It will be considered in detail in § 18.4.

Reducing NL

While the ALI-based methods are able to reduce the value of the dominant component of the state vector (*NF*), the second largest component, the total number of atomic energy levels for which the kinetic equilibrium equations are solved (*NL*), may still be prohibitively large. For instance, if several iron-peak elements, in several degrees of ionization, are treated in NLTE, the total number of levels may become comparable to the total number of frequency points. Again, there are several ways to cope with this problem.

(i) By analogy with the case of *NF*, an obvious possibility is to use simplified model atoms, in which only several of the most important (usually low-lying) levels are considered explicitly. The remaining levels are either neglected or treated in LTE. Generally, such a strategy may work for light elements but cannot be used for iron-peak elements because of the large number of levels in the whole energy range.

(ii) A much better approach is to consider the so-called *superlevels* [31,292,533]. The idea consists of grouping several (many) individual energy levels together to form a "superlevel." This is the second crucial ingredient of modern powerful numerical methods. It will be described in detail in § 18.5.

(iii) Another idea, which may significantly reduce the number of level populations to be linearized, is the idea of *level grouping*. A level group is a set of several levels whose populations are assumed to vary in a coordinated way in the linearization. More precisely, instead of linearizing individual level populations, one linearizes the total population of the group, assuming that the ratios of the individual level populations within the group to the total population of the group are unchanged during the linearization. In the formal solution step, one solves for all the individual level populations. The concept of level groups should not be confused with the concept of superlevels; in the latter case, the level groups are only a numerical trick to make the complete linearization matrices smaller, while the level populations are determined exactly; the former case, superlevels, approximates the individual populations of the components of the superlevel by assuming that they are in Boltzmann equilibrium with respect to each other. In fact, one may group the individual superlevels into level groups as well.

Reducing NC

One possibility is to solve the hydrostatic equilibrium equation separately [414, 1148, 1151]. Yet another possibility is to solve the radiative equilibrium equation separately by a *temperature-correction procedure*. In the context of modified complete linearization, these possibilities usually do not offer a substantial advantage. In some cases they may lead to a somewhat more stable solution, but typically the iteration process converges more slowly or fails to converge at all. The reasons are similar to those for the failure of classical Λ-iteration, as were explained in §§ 12.1 and 13.2. However, these methods may be helpful to speed up calculations when using an ALI-based method of solution; cf. § 18.4.

Reducing ND

Again, a trivial possibility is using a coarse depth mesh. This has the additional benefit that the ALI method converges faster when using a coarser depth resolution, as pointed out in chapter 13. However, many problems (e.g., models in which one encounters multiple ionization fronts; convective models) may require a large number of depths. But as the computation time is linear in number of depths, increasing *ND* is usually not a crucial problem. Nevertheless, convergence of ALI-based methods may deteriorate. In such cases one could use *multi-grid schemes* (e.g., [1086]). However, such an approach has not yet been used in the context of the full stellar atmosphere problem.

Another potentially promising possibility is to use *adaptive mesh refinement* (AMR) techniques (e.g., [287]), which, so far, have been used with great success in radiation-hydrodynamics calculations. Again, these have not yet been used in the context of stellar atmosphere models. It may be one of the most important remaining improvements of classical 1D stellar atmosphere modeling techniques.

Reducing N_{iter}

Such a reduction can certainly be achieved by an appropriate modification of the global iteration scheme; for instance, the hybrid CL/ALI method (see § 18.4) method decreases the number of iterations significantly with respect to the ALI method by itself. What is meant here, however, is a reduction of the number of iterations for a given global scheme. There are different ways to achieve that.

(i) The first possibility is to improve the "formal solution," i.e., all calculations that are done between two subsequent iterations of the linearization scheme. Typically, linearization provides new values of the components of the state vector; one may then keep some parameters fixed (e.g., temperature and density) and compute more consistent values of others (i.e., mean intensities and level populations, and possibly electron density), by a simultaneous solution of the transfer and the rate equations. Typically, most modeling programs offer some kind of Λ-iteration treatment to improve level populations. Some approaches (e.g., [533]) use another ALI approach with preconditioning (see § 14.5) to update the atomic level populations. The basic idea behind all such approaches is to determine values of all state parameters as consistently as possible before entering a new iteration of complete linearization, with the expectation that this helps the overall iteration process. In many cases, it does indeed help significantly.

(ii) One can use mathematical *acceleration of convergence* procedures. Because of their importance, they will be discussed in more detail below.

(iii) Another potential possibility is to use the *successive over-relaxation* (SOR) method. It consists of multiplying the corrections $\delta\psi$ by a certain coefficient α. This coefficient can be set by an educated guess, or one can use a procedure suggested in [1086] (see also § 13.3), namely, to express α in terms of the spectral radius of the appropriate iteration operator, which in turn may be approximated by a ratio of maximum relative changes of the components of the state vector in two subsequent previous iterations.

Acceleration Methods

The most popular acceleration method is the Ng acceleration [49, 50, 794], used in the context of accelerating a complete-linearization-based scheme to calculate model stellar atmospheres in [532]. Here we give only a brief outline.

The general idea is to construct a new iterate of the state vector based on the information not only from the previous iteration step, as in a standard version of an iterative linearization procedure, but also from still earlier steps. Essentially all astrophysical applications, including the one developed in [532], use the three-point version, in which the "accelerated" iterate is written as a linear combination of the three previous iterates,

$$\mathbf{x}^* = (1 - a - b)\mathbf{x}^{(n-1)} + ax^{(n-2)} + b\mathbf{x}^{(n-3)}, \tag{18.101}$$

where a and b are coefficients to be determined. Here we write $\mathbf{x}$ as the general state vector. We also write a general iterative process as

$$\mathbf{x}^{(n+1)} = F(\mathbf{x}^n). \tag{18.102}$$

In the case of Newton-Raphson iterations, the iteration operator F is given by

$$\mathbf{F}(\mathbf{x}) = \mathbf{x} - \mathbf{J}(\mathbf{x})^{-1}\mathbf{P}(\mathbf{x}), \tag{18.103}$$

where $\mathbf{J}$ is the Jacobian of the system. Substituting equation (18.102) into (18.101), one obtains

$$\mathbf{F}(\mathbf{x}^*) = (1 - a - b)\mathbf{x}^{(n)} + a\mathbf{x}^{(n-1)} + b\mathbf{x}^{(n-2)}. \tag{18.104}$$

The coefficients a and b are found by minimizing the quantity

$$\Omega = \sum_i W_i [x_i^* - F(x_i)^*]^2, \tag{18.105}$$

where W_i is an arbitrary weighting factor. As recommended in [815], it is taken as $W_i = 1/x_i^{(n)}$. After some algebra, one obtains the coefficients a and b ([532, 815]); see also § 13.4,

$$a = (\delta_{01}\delta_{22} - \delta_{02}\delta_{21}) \,/\, (\delta_{11}\delta_{22} - \delta_{12}\delta_{21}), \tag{18.106a}$$

$$b = (\delta_{02}\delta_{11} - \delta_{01}\delta_{21}) \,/\, (\delta_{11}\delta_{22} - \delta_{12}\delta_{21}), \tag{18.106b}$$

where

$$\delta_{ij} \equiv \left(\mathbf{\Delta x}^{(n)} - \mathbf{\Delta x}^{(n-i)}\right) \cdot \left(\mathbf{\Delta x}^{(n)} - \mathbf{\Delta x}^{(n-j)}\right), \tag{18.107}$$

for $i, j = 1, 2$; and $\delta_{0j} \equiv \Delta x^{(n)} \cdot (\Delta x^{(n)} - \Delta x^{(n-j)})$; with

$$\mathbf{\Delta x}^{(n)} \equiv \mathbf{x}^{(n)} - \mathbf{x}^{(n-1)}. \tag{18.108}$$

The method does not specify how the individual iterates $\mathbf{x}^{(n)}$ are evaluated. Hence it can be used in conjunction with any method that provides an algorithm for computing $\mathbf{x}^{(n)}$.

In the context of complete linearization, one can choose a particular flavor of minimization and the appropriate weighting factors. As demonstrated in [532] for actual numerical examples, the most advantageous choice is to use a global minimization, in which case one minimizes the quantity

$$\Omega = \sum_{d=1}^{ND} \sum_{i=1}^{NN} [\psi_{di}^* - F(\psi_{di}^*)]^2 / \psi_{di}^{(n)}. \tag{18.109}$$

The reason for such a choice is that the elements of the state vector $\boldsymbol{\psi}$ have very different numerical values both for the individual elements $\boldsymbol{\psi}_d$ at one depth point and for one of the elements ψ_{di} at different depth points. Using equation (18.109) avoids putting unreasonably large weight in the summation for numerically large quantities.

The experience showed that in a vast majority of cases Ng acceleration improves convergence significantly; the acceleration is usually performed for the first time at or around the seventh iteration of the complete-linearization scheme and is done

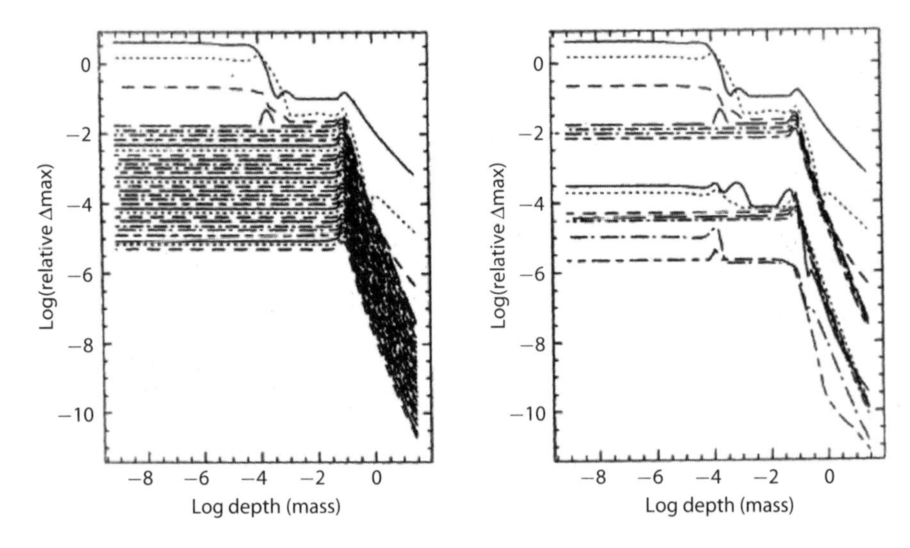

Figure 18.1 Example of convergence behavior of models calculated without (left panel) and with Ng (right panel) acceleration. The plots show the maximum relative changes of all elements of vector ψ_d as a function of depth, represented by the column mass m in gm cm^{-2}. The line pattern is repeated every eighth iteration, the upper full line corresponds to the first iteration, the dotted line to the second iteration, etc. The Ng acceleration was performed first in the seventh iteration, and then after each four iterations. The models were computed using the Kantorovich acceleration as well, switched on after the third iteration. From [532].

typically every four iterations afterwards. This is illustrated in figure 18.1. In some cases, like in models with convection or in specific models with sharp ionization fronts, Ng acceleration does not help and may even lead to numerical problems and divergence. Therefore, one should apply Ng acceleration judiciously.

Linearization without Inverting the Jacobian

The next class of improvements, which can be used in conjunction with any of the above methods, is to avoid somehow inverting the Jacobian of the system, which clearly leads to substantial computer-time savings. In these methods, one has to perform a few first iterations using the original Newton-Raphson scheme where the Jacobian is inverted; only when the current estimate of solution is "close" to the true solution may one use a simplified treatment. There are essentially two possibilities.

Broyden Method

The first is an application of the Broyden method, also called the "least change secant method," first used in the context of stellar atmospheres in [628]. Instead of using a Jacobi matrix (which is an analog of a tangent in one dimension), one uses a

Broyden matrix, whose analog in 1D is a secant. The inverse Broyden matrix in the subsequent iteration can be calculated directly from the previous inverse Broyden matrix. This means that inverting $NN \times NN$ matrices is avoided. The method and its actual application to stellar atmosphere models are described § 18.4 and in more detail in [1151].

Kantorovich Method

An even simpler method is called the Kantorovich method, which keeps the Jacobian fixed after a certain iteration, so the subsequent iterations of complete linearization use the same Jacobian (more accurately, the inverse of Jacobian is kept fixed for future use); only the right-hand side vectors $\mathbf{L}$ are re-evaluated after each iteration. In a one-dimensional analog, the Newton method computes a new iterate using the current slope of the tangent, while the Kantorovich variant keeps the tangent fixed. The method was used in [492]; the properties of the method as applied to the complete-linearization scheme were studied in detail in [532], who also coined the term "Kantorovich method," because Kantorovich [610] first proved rigorously the convergence of the method. Experience with the method showed that it is surprisingly robust. Usually, one needs to perform two to four iterations of the full linearization scheme, depending on the problem at hand and the quality of the initial estimate. Also, it is sometimes very advantageous to "refresh" the Jacobian (i.e., set it using the current solution and invert it) after a certain number of Kantorovich iterations.

18.4 APPLICATION OF ALI AND RELATED METHODS

From the point of view of pure radiative transfer, ALI is to be understood as an efficient numerical scheme to solve large sparse linear systems. However, from the point of view of modeling stellar atmospheres, application of the accelerated lambda iteration (ALI) method is an approach that reduces the number of frequency points at which the radiative transfer equation is linearized or may eliminate such points altogether. We shall describe the three most widely used variants of such a scheme below.

Standard ALI Scheme

In analogy to the problem of simultaneous solution of the transfer equation together with a set of kinetic equilibrium equations, referred to as the multi-level transfer problem, the essence of applying ALI is to express the mean intensity of radiation as

$$J_\nu^{(n)} = \Lambda_\nu^* S_\nu^{(n)} + \left(\Lambda_\nu - \Lambda_\nu^*\right) S_\nu^{(n-1)} \equiv \Lambda_\nu^* S_\nu^{(n)} + \Delta J_\nu, \tag{18.110}$$

where Λ_ν and Λ_ν^* are the exact and the approximate Λ-operators and S_ν the source function, all at frequency ν. The superscript n indicates the iteration number. The mean intensity of radiation is thus represented by two terms. The second one—the "correction" term ΔJ_ν—is known from the previous iteration, and the first one represents the action of an approximate (and, therefore, simple) operator, Λ^*, on the source function, which is expressed as a function of temperature, density, and atomic level populations. The radiative transfer equations are thus eliminated from the coupled system of structural equations. Similarly, as in the case of the hybrid CL/ALI method described below, the source function employed in (18.110) is defined as the *thermal* source function, $S_\nu \equiv \eta_\nu/\kappa_\nu$, without the scattering terms. The relation between the Λ^*-operator acting on the thermal source function and that defined as acting on the *total* source function that includes continuum scattering is described by equations (18.132)–(18.134).

As in the case of a multi-level atom problem, eliminating mean intensities from the structural equations leads to nonlinear equations for the reduced state vector,

$$\widetilde{\boldsymbol{\psi}}_d = \{N, T, n_e, n_1, \ldots, n_{NL}\}. \tag{18.111}$$

However, the equations are linearized anyway, so this does not lead to any additional problems.

In the context of NLTE model stellar atmospheres, such an ALI-based scheme was first applied in [1148] using a diagonal Λ^*-operator; later [1149] it was extended to a tridiagonal operator. We outline below an implementation of the method as used in [1148]. Using equation (18.110), one can write the radiative rates entering the kinetic equilibrium equation as, for instance, for an upward rate,

$$R_{ij} = 4\pi \int_0^\infty \frac{\sigma_{ij}(\nu)}{h\nu} (\Lambda_\nu^* S_\nu + \Delta J_\nu)\, d\nu, \tag{18.112}$$

and in a similar way for the downward rates. Because the source function is a generally a nonlinear function of the population numbers, temperature, and electron density, the system of kinetic equilibrium equations becomes nonlinear in these state parameters, and hence also needs to be linearized. The system of kinetic equilibrium equations, written formally as

$$\boldsymbol{\mathcal{A}} \cdot \mathbf{n} = \mathbf{b}, \tag{18.113}$$

is linearized by writing

$$\delta\mathbf{n} = \frac{\partial \mathbf{n}}{\partial T}\delta T + \frac{\partial \mathbf{n}}{\partial n_e}\delta n_e + \sum_{k=1}^{NF} \frac{\partial \mathbf{n}}{\partial S_k}\delta S_k, \tag{18.114}$$

where the derivatives of the population numbers are given by

$$\frac{\partial \mathbf{n}}{\partial x} = \boldsymbol{\mathcal{A}}^{-1}\left(\frac{\partial \mathbf{b}}{\partial x} - \frac{\partial \boldsymbol{\mathcal{A}}}{\partial x} \cdot \mathbf{n}\right), \tag{18.115}$$

for any variable x, i.e., $x = T, n_e, S_k$. The correction to the source function can be written in terms of the corrections to the temperature, electron density, and population numbers,

$$\delta S_k = \frac{\partial S_k}{\partial T}\delta T + \frac{\partial S_k}{\partial n_{\rm e}}\delta n_{\rm e} + \sum_{i=1}^{NL} \frac{\partial S_k}{\partial n_i}\delta n_i. \tag{18.116}$$

This equation is used to eliminate δS_k to end up with a linear system for the corrections $\delta \mathbf{n}$ in terms of δT and $\delta n_{\rm e}$:

$$\begin{aligned}\delta\mathbf{n} = {} & \delta T\left[\frac{\partial \mathbf{n}}{\partial T} + \sum_{k=1}^{NF} \frac{1}{\kappa_k}\frac{\partial \mathbf{n}}{\partial S_k}\left(\frac{\partial \eta_k}{\partial T} - S_k\frac{\partial \kappa_k}{\partial T}\right)\right] \\ & + \delta n_{\rm e}\left[\frac{\partial \mathbf{n}}{\partial n_{\rm e}} + \sum_{k=1}^{NF} \frac{1}{\kappa_k}\frac{\partial \mathbf{n}}{\partial S_k}\left(\frac{\partial \eta_k}{\partial n_{\rm e}} - S_k\frac{\partial \kappa_k}{\partial n_{\rm e}}\right)\right] \\ & + \sum_{i=1}^{NL}\delta n_i\left[\frac{\partial \mathbf{n}}{\partial n_i} + \sum_{k=1}^{NF} \frac{1}{\kappa_k}\frac{\partial \mathbf{n}}{\partial S_k}\left(\frac{\partial \eta_k}{\partial n_i} - S_k\frac{\partial \kappa_k}{\partial n_i}\right)\right].\end{aligned} \tag{18.117}$$

Linearization of the radiative equilibrium equation in the ALI formalism is similar. For simplicity, let us assume, as in [1148], that one employs the integral form (18.12). Substituting for J_ν from (18.110), one obtains

$$\int_0^\infty \kappa_\nu[(1 - \Lambda^*_\nu)S_\nu - \Delta J_\nu]\, d\nu = 0. \tag{18.118}$$

This equation may be viewed as a "preconditioned" form of the radiative equilibrium equation, in analogy to the preconditioned form of the radiative rates and the kinetic equilibrium equations described in § 14.5. Equation (18.118) is discretized and linearized as before. After some algebra, one obtains

$$\begin{aligned}\sum_{k=1}^{NF} w_k\kappa_k(l^*_k S_k - \Delta J_k) = {} & \sum_{k=1}^{NF} w_k\left[\delta T\left(\frac{\partial \kappa_k}{\partial T}\Delta J_k - \frac{\partial \eta_k}{\partial T}l^*_k\right)\right. \\ & \left. +\delta n_{\rm e}\left(\frac{\partial \kappa_k}{\partial n_{\rm e}}\Delta J_k - \frac{\partial \eta_k}{\partial n_{\rm e}}l^*_k\right) + \sum_{i=1}^{NL}\delta n_i\left(\frac{\partial \kappa_k}{\partial n_i}\Delta J_k - \frac{\partial \eta_k}{\partial n_i}l^*_k\right)\right] = 0,\end{aligned} \tag{18.119}$$

where

$$l^*_k \equiv 1 - \Lambda^*_k. \tag{18.120}$$

Linearization of the charge conservation equation is the same as in the traditional complete linearization because it does not involve radiation field,

$$\delta n_{\rm e} - \sum_{i=1}^{NL} Z_i\delta n_i = \sum_{i=1}^{NL} Z_i n_i - n_{\rm e}. \tag{18.121}$$

The hydrostatic equilibrium equation can be linearized in an analogous way, replacing J_ν by its ALI expression (18.110). Whereas this is possible, it would lead to the complication that the resulting set of equations is nonlocal. Therefore, the early studies, such as [1148], chose to leave the hydrostatic equilibrium equation out of the set of linearized equations. Instead, it was solved iteratively after each completed ALI iteration, using radiation intensities and Eddington factors obtained from the formal solution. This procedure does not cause any numerical instabilities or a significant slowdown of iterations, as long as a model is not too close to the Eddington limit. In this case, the reduced state vector does not contain N,

$$\widetilde{\boldsymbol{\psi}}_d = \{T, n_e, n_1, \ldots, n_{NL}\}. \tag{18.122}$$

The system of linearized equations (18.117), (18.119), and (18.121), with Λ^* being a diagonal operator, so that its action is a simple multiplication by a scalar quantity Λ^*, can be written as

$$\mathbf{M}_d \cdot \delta\widetilde{\boldsymbol{\psi}}_d = \mathbf{L}_d \tag{18.123}$$

for each depth point d. This equation is local, so the corrections $\delta\boldsymbol{\psi}_d$ can be obtained independently for each depth point. If one considers a tridiagonal Λ^*-operator [1149], the resulting set of equations for the corrections is block tridiagonal, as in the original complete linearization.

Broyden Scheme

As mentioned in § 18.3, the Broyden method is designed to avoid repeated matrix inversions of the Jacobian in the Newton-Raphson method. In the context of equation (18.123), which works in terms of a block diagonal Jacobian composed of matrices $\mathbf{M}_d$, the solution is simply

$$\delta\widetilde{\boldsymbol{\psi}}_d^k = (\mathbf{M}_d^k)^{-1}\mathbf{L}_d^k, \tag{18.124}$$

independently for each depth, where superscript k indicates the kth iteration. In the original application of the method, an inversion of $\mathbf{M}_d^k$ has to be done in each iteration. The Broyden method gives an update formula for the *inverse* Jacobian without actually inverting it,

$$\left(\mathbf{M}_d^{k+1}\right)^{-1} = \left(\mathbf{M}_d^k\right)^{-1} + \frac{\left[\delta\widetilde{\boldsymbol{\psi}}_d^k - \left(\mathbf{M}_d^k\right)^{-1}\mathbf{y}_d^k\right] \otimes \left[\left(\delta\widetilde{\boldsymbol{\psi}}_d^k\right)^{\mathrm{T}}\left(\mathbf{M}_d^k\right)^{-1}\right]}{\left(\delta\widetilde{\boldsymbol{\psi}}_d^k\right)^{\mathrm{T}}\left(\mathbf{M}_d^k\right)^{-1}\mathbf{y}_d^k}, \tag{18.125}$$

where $\mathbf{y}_d^k \equiv \mathbf{L}_d^{k+1} - \mathbf{L}_d^k$, and $\otimes$ denotes the diadic (outer) product of two vectors, $(\mathbf{a} \otimes \mathbf{b})_{ij} \equiv a_i b_j$. The scheme leads to a significant speed-up of the iteration procedure, but requires much larger memory because one has to store the inverse matrices $(\mathbf{M}_d^k)^{-1}$ from the previous iteration at all depths.

If a tridiagonal Λ^*-operator is used, an application of the Broyden scheme is more complicated. The grand matrix $\mathbf{P}$ is block tridiagonal, and its inverse cannot be updated because it is never constructed explicitly (because one uses the forward-backward sweep, sometime referred to, inaccurately, as Feautrier elimination). To avoid this problem, one uses the Schubert formula [977] to update the grand matrix, preserving its block tridiagonal structure, as

$$\mathbf{P}^{k+1} = \mathbf{P}^k + \frac{\left(\mathbf{y}^k - \mathbf{P}^k \delta\widetilde{\boldsymbol{\psi}}^k\right) \otimes \left(\mathbf{x}^k\right)^{\mathrm{T}}}{\left(\mathbf{x}^k\right)^{\mathrm{T}} \cdot \mathbf{x}^k}, \tag{18.126}$$

where $\mathbf{x}^k \equiv \mathbf{Z}\,\delta\widetilde{\boldsymbol{\psi}}^k$; the *structure matrix* $\mathbf{Z}$ is defined by $Z_{ij} = 1$ if $P_{ij} \neq 0$, and $Z_{ij} = 0$ otherwise. Using (18.126) one obtains updated submatrices $\mathbf{A}_d$, $\mathbf{B}_d$, and $\mathbf{C}_d$ of the grand matrix $\mathbf{P}$ (without computing necessary derivatives) and then uses the original Broyden formula (18.125) to update the inverse of the auxiliary matrix $(\mathbf{B}_d - \mathbf{A}_d \mathbf{D}_{d-1})^{-1}$ that is created in the forward sweep to solve the block tridiagonal system. For more details refer to [1151].

Temperature Correction

An inspection of equations (18.122) and (18.123) reveals that one has to invert an $(NL+2) \times (NL+2)$ matrix per depth. However, NL counts all levels of *all* chemical species, while the kinetic equilibrium equations are independent for the individual species. In many cases of interest (particularly for hot stars where the material is almost fully ionized), the charge conservation equation may be solved separately from other constraint equations. This leaves the radiative equilibrium equation as the only structural equation that couples *all* atomic level populations. Had one succeeded in finding an efficient scheme to solve the radiative equilibrium separately, then one would be faced with inverting a set of matrices $NL_I \times NL_I$, where NL_I is the number of level of species I, which would take much less computer time than the original inversion of an $(NL+2) \times (NL+2)$ matrix per depth.

Such a scheme, suggested in [291, 292], is based on a generalization of the LTE Unsöld-Lucy temperature correction scheme, described in § 17.3. In LTE, the correction to the current temperature is given by [cf. equation (17.115)]

$$\frac{4\sigma_R T^3}{\pi} \Delta T = \frac{\kappa_J}{\kappa_B} J - B + \frac{\kappa_J}{\kappa_B} \frac{1}{f} \left[\int_0^\tau \frac{\chi_H}{\kappa_B} \Delta H \, d\tau' + \frac{f(0) \Delta H(0)}{\bar{g}} \right], \tag{18.127}$$

where κ_J, κ_B, and χ_H are the absorption mean, Planck mean, and flux-mean opacities, respectively [cf. equations (17.99)–(17.101)], and f and $\bar{g}$ are the averaged Eddington factors defined by equations (17.112) and (17.113). ΔH is the correction to the Eddington flux needed to obtain radiative equilibrium, $\Delta H = (\sigma_R/4\pi) T_{\mathrm{eff}}^4 - H$, with H (and similarly J and K) being the current values of the moments. Equation (18.127) applies if the right-hand side of the moment equation, $-dH_\nu/dz = \chi_\nu (J_\nu - S_\nu)$, is given by $\chi_\nu (J_\nu - S_\nu) = \kappa_\nu (J_\nu - B_\nu)$. If LTE does not apply, one can still write

$$\chi_\nu S_\nu = \kappa_\nu^B B_\nu + \gamma_\nu J_\nu, \tag{18.128}$$

where $\kappa_\nu^B \neq \kappa_\nu$ and γ_ν are quantities to be specified. Changing the definition of the absorption and Planck means to $\tilde{\kappa}_J = \int_0^\infty (\chi_\nu - \gamma_\nu) J_\nu \, d\nu / J$, and $\tilde{\kappa}_B = \int_0^\infty \kappa_\nu^B B_\nu \, d\nu / B$, the correction to temperature is given by equation (18.127) even in the case of NLTE, replacing κ_J with $\tilde{\kappa}_J$ and κ_B with $\tilde{\kappa}_B$.

It remains to determine κ_ν^B and γ_ν. It should be stressed that since these parameters are used to compute *corrections* to the temperature, not the temperature itself, they may be approximate. The simplest approximation is to use the standard definition, but *skipping the strong lines*. The rationale for that approach is that for a strong line, one can write the source function in the two-level atom approximation as $S = (1-\epsilon)\bar{J} + \epsilon B$; hence

$$\int_{\text{line}} \chi_\nu (J_\nu - S_\nu) \, d\nu \approx \epsilon (\bar{J} - B_{\nu_0}), \tag{18.129}$$

which is small if $\epsilon \ll 1$. In [291] it was suggested that contributions of all lines that are optically thick at a given depth be omitted in evaluating κ_ν^B. Experience showed that better convergence properties are achieved if one further empirically modifies (18.127) to

$$\Delta T = \frac{\pi}{4\sigma_R T^3} \left[d_1 \left(\frac{\tilde{\kappa}_J}{\tilde{\kappa}_B} J - B \right) + \frac{\tilde{\kappa}_J}{\tilde{\kappa}_B} \left(\frac{d_2}{f} \int_0^\tau \frac{\chi_H}{\tilde{\kappa}_B} \Delta H \, d\tau' + d_3 \frac{f(0) \Delta H(0)}{f \bar{g}} \right) \right], \tag{18.130}$$

where $d_1 = c_1 e^{-\tau_0/\tau}$, $d_{2,3} = c_{2,3}\left(1 - e^{-\tau_0/\tau}\right)$, where c_i are constants of the order of unity, and τ_0 is an empirical ad hoc damping parameter, also of the order of unity. This parametrization damps the first, local term at deep layers ($\tau \gg 1$) and leaves the terms with ΔH intact, while the opposite applies at the surface layers, which is exactly what is needed to obtain better numerical conditioning of the procedure. The scheme works well in some cases, but causes numerical problems (slow convergence or even divergence) in others, so it should be used with caution.

Hybrid Complete-Linearization/Accelerated Lambda Iteration Method

The method developed in [533] is a variant of the ALI method that offers significant numerical advantages. The ALI-based method discussed above starts with the multi-level formulation of ALI and linearizes the resulting set of the transfer plus the kinetic equilibrium equations together with other structural equations. In contrast, the hybrid CL/ALI scheme starts with the linearized structural equations and then eliminates δJ_ν from the linearized set using the idea of ALI. Although the approaches are very similar, the hybrid scheme offers a significant benefit, as pointed out in [533]: Whereas the ALI treatment is used for most frequency points, the radiation intensity in several selected frequency points may still be linearized. Hence the method offers a wide spectrum of options, ranging from full CL to the full ALI method.

We shall briefly outline the formulation of the scheme. The mean intensity at some selected frequency points, called *"explicit frequencies,"* is linearized, exactly as in the standard complete linearization, using the transfer equation (18.1) together with boundary conditions (18.2) and (18.5). The remaining frequency points, called *"ALI frequencies,"* are treated with ALI; i.e., the mean intensity is expressed by equation (18.110). In other words, the radiative transfer equation for these frequency points is replaced by

$$J_{dj} = \Lambda^*_{d,j}(\eta_{dj}/\kappa_{dj}) + \Delta J_{dj}. \tag{18.131}$$

Here we consider a diagonal Λ^* for simplicity. An extension to a tridiagonal Λ^*-operator [534] is straightforward, but involves cumbersome algebraic expressions. Numerical experience showed that although an application of the tridiagonal operator speeds up the convergence of the iterations, a similar or even faster convergence is obtained with a diagonal operator applied together with the Ng acceleration.

As before, Λ^* is defined as an operator acting on the *thermal* source function, $S_\nu = \eta_\nu/\kappa_\nu$, not on the source function defined as $S_\nu = \eta_\nu^{\rm tot}/\chi_\nu$. This is a somewhat subtle point that needs an explanation. Let Λ^0_ν be the operator acting on the *total* source function, so that

$$\begin{aligned} J_\nu = \Lambda^0_\nu[S_\nu] &= \Lambda^0_\nu\left[\frac{\eta_\nu}{\kappa_\nu + \kappa_\nu^{\rm sc}}\right] + \Lambda^0_\nu\left[\frac{\kappa_\nu^{\rm sc} J_\nu}{\kappa_\nu + \kappa_\nu^{\rm sc}}\right] \\ &\equiv \Lambda^0_\nu\left[\epsilon_\nu \frac{\eta_\nu}{\kappa_\nu}\right] + \Lambda^0_\nu\left[(1-\epsilon_\nu)J_\nu\right], \end{aligned} \tag{18.132}$$

where $\chi_\nu = \kappa_\nu + \kappa_\nu^{\rm sc}$, and $\epsilon_\nu \equiv \kappa_\nu/(\kappa_\nu + \kappa_\nu^{\rm sc})$, with $\kappa_\nu^{\rm sc}$ being the scattering coefficient (typically given by $\kappa_\nu^{\rm sc} = n_{\rm e}\sigma_{\rm e}$). From (18.132) follows

$$J_\nu = (\mathbf{I} - \Lambda^0_\nu[1-\epsilon_\nu])^{-1}\Lambda^0_\nu[\epsilon_\nu(\eta_\nu/\kappa_\nu)], \tag{18.133}$$

where $\mathbf{I}$ is a unit operator. Therefore, operator Λ^*_ν used in equation (18.131) is related to the original operator Λ^{0*}_ν as

$$\Lambda^*_\nu = (\mathbf{I} - \Lambda^{0*}_\nu[1-\epsilon_\nu])^{-1}\Lambda^{0*}_\nu[\epsilon_\nu]. \tag{18.134}$$

Equation (18.131) is linearized as follows

$$\delta J_{dj} = \Lambda^*_{dj}\delta S_{dj} = D^T_{dj}\delta T_d + D^{n_{\rm e}}_{dj}\delta n_{\rm e} + \sum_{i=1}^{NL} D^i_{dj}\delta n_{di}, \tag{18.135}$$

where

$$D^x_{dj} = \Lambda^*_{dj}\frac{\eta_{dj}}{\kappa_{dj}}\left(\frac{1}{\eta_{dj}}\frac{\partial\eta_{dj}}{\partial x} - \frac{1}{\kappa_{dj}}\frac{\partial\kappa_{dj}}{\partial x}\right) \tag{18.136}$$

represents a derivative of the mean intensity in the "ALI" frequency points with respect to x, where $x = T, n_{\rm e}, n_i$.

As an illustration of the implementation of the hybrid scheme, we consider the radiative equilibrium equation in the integral form, (18.12). It is discretized as follows:

$$\sum_i^{NFEXP} w_i(\kappa_{di}J_{di} - \eta_{di}) + \sum_i^{NFALI} w_i(\kappa_{di}J_{di} - \eta_{di}) = 0, \tag{18.137}$$

where *NFXEP* is the total number of "explicit" frequencies and *NFALI* the total number of "ALI" frequencies, $NFEXP + NFALI = NF$. Using equation (18.131), equation (18.137) can be written as

$$\sum_{i}^{NFEXP} w_i(\kappa_{di}J_{di} - \eta_{di}) + \sum_{i}^{NFALI} w_i[\kappa_{di}\Delta J_{di} + (\Lambda^*_{di} - 1)\eta_{di}] = 0. \quad (18.138)$$

Linearization of equation (18.138) is straightforward. Denoting the left-hand side of equation (18.138) as P_d, the linearized form reads

$$\sum_{j}^{NFEXP} \frac{\partial P_d}{\partial J_{dj}}\delta J_{dj} + \frac{\partial P_d}{\partial T_d}\delta T_d + \frac{\partial P_d}{\partial n_{e,d}}\delta n_{e,d} + \sum_{i=1}^{NL} \frac{\partial P_d}{\partial n_{di}}\delta n_{di} = -P_d. \quad (18.139)$$

The matrix elements of the Jacobian are the partial derivatives standing at the corresponding corrections δx. This equation is local; hence only matrix elements of $\mathbf{B}_d$ have nonzero values. They are given by

$$(B_d)_{NR,i} = w_i \kappa_{di}, \quad i \le NFEXP, \quad (18.140a)$$

$$(B_d)_{NR,k} = \sum_{i}^{NFEXP} w_i \left[\frac{\partial \kappa_{di}}{\partial \psi_{dk}} J_{di} - \frac{\partial \eta_{di}}{\partial \psi_{dk}} \right] + \sum_{j}^{NFALI} w_i \left[\frac{\partial \kappa_{dj}}{\partial \psi_{dk}} \Delta J_{dj} + \frac{\partial \eta_{dj}}{\partial \psi_{dk}} (\Lambda^*_{dj} - 1) \right], \quad (18.140b)$$

where $\psi_{dk} = T$ (for $k = NR$), $\psi_d = n_e$ (for $k = NP$), or $\psi_d = n_i$ (for $i = 1, \ldots NL$, i.e., $k > NP$).

Equation (18.140a) and the first term of (18.140b) are similar to those for standard complete linearization, equations (18.78), whereas the second term of (18.140b) arises from the elimination of the mean intensities from the state vectors using the ALI approach and is specific to the hybrid CL/ALI treatment. One employs an analogous procedure for the differential form of the radiative equilibrium equation and for the hydrostatic equilibrium equation. The kinetic equilibrium equations are treated similarly. Because they exhibit the largest formal difference from the standard application of ALI described above, we give a brief outline.

The rate equations are written in the form

$$\mathbf{n} - \mathcal{A}^{-1} \cdot \mathbf{b} = 0, \quad (18.141)$$

from which one obtains

$$(\partial \mathbf{n}/\partial x) = -\mathcal{A}^{-1} \cdot \mathbf{V}_x. \quad (18.142)$$

This quantity represents the column of the Jacobi matrix corresponding to quantity x; x stands for any quantity of vector $\boldsymbol{\psi}$ defined by equation (18.58). Here

$$\mathbf{V}_x = (\partial \mathcal{A}/\partial x) \cdot \mathbf{n} - (\partial \mathbf{b}/\partial x). \quad (18.143)$$

In the case of standard complete linearization, the corresponding elements of vector $\mathbf{V}_x$ are evaluated as outlined in § 18.2. Let the transition $l \leftrightarrow u$ be represented by an arbitrary combination of the "explicit" and "ALI" frequency points; either subset is allowed to be empty. Generally, the contribution from this transition comes only to the two following components of vector $\mathbf{V}_x$, namely,

$$(\mathbf{V}_x)_l = \frac{\partial(R_{lu}+C_{lu})}{\partial x} n_l - \frac{\partial(R_{ul}+C_{ul})}{\partial x} n_u,$$
$$(\mathbf{V}_x)_u = -(\mathbf{V}_x)_l. \tag{18.144}$$

The radiative rate is written in a discretized form as

$$R_{lu} = (4\pi/h)\left[\sum_i^{NFEXP} w_i \sigma_{lu}(\nu_i) J_i/\nu_i + \sum_j^{NFALI} w_j \sigma_{lu}(\nu_j) J_j/\nu_j,\right], \tag{18.145}$$

and similarly for the downward rate. The contribution to $\mathbf{V}_x$ from the collisional rates and from "explicit" frequency points is the same as in the standard CL, whereas the ALI contribution is given by

$$(V_x)_l^{\mathrm{ALI}} = [n_l - n_u G_{lu}(\nu)](4\pi/h) \sum_j^{NFALI} w_j \sigma_{lu}(\nu_j) D_j^x/\nu_j, \tag{18.146}$$

where D_j^x is given by equation (18.136). The important difference from the standard complete linearization is that because the derivatives D_j^x are generally nonzero for x being the individual level populations, the Jacobian contains contributions from the populations. This, of course, expresses the fact stated already that within the ALI formalism the kinetic equilibrium equations are nonlinear in the level populations.

It was shown that by selecting a few (typically 10–30) frequency points judiciously (typically, at the head of the most opaque continua, like the hydrogen and He II Lyman continua and in the centers of strongest lines), the computer time per iteration is essentially the same as in the case of full ALI, while the number of iterations is essentially the same as in the case of full CL, i.e., much lower than in the case of true ALI. The method thus combines two major advantages of its two constituents, namely, the convergence rate being virtually as high as that for the standard CL method, and the computer time per iteration is almost as low as that for the standard ALI method.

Approximate Newton-Raphson Method

The method was first suggested in [478] and [967] in the context of line formation with velocity fields and was subsequently elaborated and extended in [492] and [493] to treat the full spherically expanding model atmosphere problem.

The method is conceptually very close to the ALI-type methods; it differs in implementation and in some subtle points. The idea is as follows: we start with the traditional CL method. Instead of linearizing all the components of the state vector, we use the transfer equations to eliminate the corrections δJ_i from the solution vector. From the linearized transfer equation, we have (for $i = 1, \ldots, NF$ and $d = 1, \ldots, ND$)

$$\delta J_{id} = \sum_{d'=1}^{ND} \sum_{j=1}^{NL+NC} \frac{\partial J_{id}}{\partial x_{jd'}} \delta x_{jd'} , \tag{18.147}$$

where x_{jd} are the components of the state vector other than radiation intensities (i.e., level populations, T, N, and n_e) and the $\partial J_{id} / \partial x_{jd'}$ terms can be derived from linearized transfer equations. The important point is that all components of this matrix are, in general, nonzero. This is because a discretized transfer equation (18.44) can be written in matrix form as

$$\mathrm{T}_i \, \mathbf{J}_i = \mathbf{S}_i, \tag{18.148}$$

where $\mathbf{J}_i = (J_1, \ldots, J_{ND})^T$ is the vector of mean intensities at frequency i, $\mathbf{S}_i$ is the source vector, and T_i is a tridiagonal matrix. Consequently, $\mathbf{J}_i = \mathrm{T}_i^{-1} \, \mathbf{S}_i$, where T_i^{-1} is a full matrix.

Equation (18.147) by itself does not offer any advantage, because by using it the global tridiagonality of the block system would be destroyed. However, the trick is to consider an approximate form of equation (18.147), namely,

$$\delta J_{id} = \sum_{d'=d-a}^{d+a} \sum_{j=1}^{NL+NC} \frac{\partial J_{id}}{\partial x_{jd'}} \delta x_{jd'}, \tag{18.149}$$

where a is set to $a = 0$ (diagonal form); or $a = 1$ (tridiagonal form); or possibly even $a = 2$ (pentadiagonal form). Using equation (18.149), δJ_i can be eliminated from linearized equations, so one is left with linearizing only $NC + NL$ quantities. Because the method uses an approximate expression for δJ_i, it earned the name "approximate Newton-Raphson" (ANR) method.

The method is very similar to the hybrid CL/ALI method described earlier (if the latter uses a full ALI form). The only difference is that in ALI methods one linearizes $\delta J_i = \Lambda^* \, \delta S_i$, where δS_i is expressed through corrections of other state parameters ($\delta n_j, \delta T, \delta n_e$), but with Λ^* computed on a fixed optical depth scale. In other words, ANR automatically takes into account a response of the radiation field to changes in source function as well as to the optical depth, whereas the standard variant of the CL/ALI method takes into account only the response to changes in the source function.

More details about the implementation of the method, in the context of NLTE models of expanding atmospheres, is presented in § 20.4.

18.5 NLTE METAL LINE BLANKETING

By the term line-blanketed model atmospheres we understand models that take into account effects of "all" lines of all important species. There are literally millions of lines that contribute to the opacity; their number is even a few orders of magnitude higher when considering molecular lines.

To be able to treat metal line blanketing numerically, a useful method must be able to work efficiently with a large number of frequency points and a large number of energy levels (populations). The problem of a large number of frequencies is dealt with by the application of the ALI method or by the approximate Newton-Raphson method. The problem of a large number of energy levels is effectively solved by using the concept of superlevels, possibly together with level grouping if one works within the framework of a linearization method. These methods are described below.

Superlevels and Superlines

This idea consists of grouping several, possibly many, individual energy levels together, forming a *superlevel*. The basic physical assumption is that all genuine levels j within a superlevel J are in Boltzmann equilibrium with respect to each other,

$$n_j/n_{j'} = g_j/g_{j'} \exp[-(E_j - E_{j'})/kT]. \quad (18.150)$$

There is a certain flexibility in choosing a partitioning of levels into superlevels. However, in order to provide a realistic description, the levels forming a superlevel have to possess close energies and have similar properties, for instance, belonging to the same multiplet, belonging to the same spin system, or having the same parity. The requirement of close energies is needed because collisional rates between levels with a small energy difference tend to be large, and hence dominate over the radiative rates. With dominant collisional rates, one indeed recovers LTE.

The idea of superlevels can be used for any type of atom or ion. For light elements, which do not have very complex energy level structure, one can treat most or all levels individually or select a certain number of lower levels to be treated individually, and treat the rest by introducing several superlevels. For the iron-peak elements, there are so many levels that typically there are no levels to be treated separately; instead, all levels are lumped into superlevels. In early applications [31, 292], all levels of a particular ion of iron were partitioned into a relatively small number of superlevels, typically seven to nine.

Choosing a small number of superlevels has the obvious advantage that one deals with a small number of levels for which the kinetic equilibrium equations are solved, but it also has several drawbacks. Since the energy widths of the superlevels are relatively large, the corresponding transitions between them span wide frequency intervals. Moreover, if one superlevel mixes radiatively decaying and metastable states, the assumption of their mutual Boltzmann equilibrium, equation (18.150), becomes questionable. In setting up the system of superlevels, one has to make a compromise between setting very narrow superlevels, which is physically more realistic but computationally more demanding because one deals with a large number of superlevels, and setting a small number of wide superlevels, which yields less accurate results. A good strategy [533, 1151] is to inspect a histogram of level energies as a function of level excitation energy. The gaps and peaks in this distribution are used to define the limiting energies for the superlevels. This is illustrated in figure 18.2.

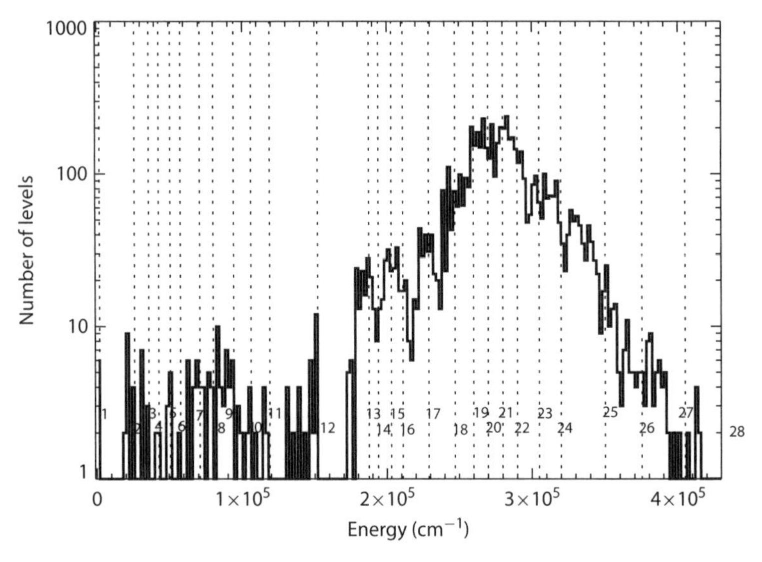

Figure 18.2 A histogram of the actual level energies for Fe III. The ordinate shows a number of levels per 2000 cm^{-1} energy bins. Only even-parity levels are shown. Dotted lines indicate the limiting energies for the individual superlevels.

In the following, we denote by lowercase letters i and j the genuine energy levels, while the superlevels are denoted by uppercase letters I and J, with the convention that a superlevel I is formed of several levels i. The population of the superlevel I is given by

$$N_I \equiv \sum_i n_i, \tag{18.151}$$

and the statistical weight as

$$g_I \equiv \sum_i g_i. \tag{18.152}$$

Some studies [31, 292, 1151] use a somewhat different definition, in which the statistical weights are depth-dependent. The present definition is adopted from [533] and is chosen this way because it allows for general occupation probabilities of the involved levels. The depth dependence, which arises because of differences in the individual energies of levels forming a superlevel, is absorbed here in the generalized occupation probability,

$$w_I \equiv \frac{\exp(E_I/kT)}{g_I} \sum_i g_i w_i \exp(-E_i/kT), \tag{18.153}$$

where the averaged energy is defined by

$$E_I \equiv \frac{\sum_i g_i w_i E_i \exp(-E_i/kT)}{\sum_i g_i w_i \exp(-E_i/kT)}. \tag{18.154}$$

The quantity w_i is the occupation probability of level i. In the standard formalism, $w_i = 1$ (the level dissolution is neglected), then w_I for a superlevel is still generally $w_I \neq 1$, so that the "occupation probability" of a superlevel I does not have a meaning of a true occupation probability, but rather the (depth-dependent) correction to statistical weights and level energies arising because of a finite span of the individual level energies and the assumption that the populations of levels within a superlevel is given by the Boltzmann distribution.

In practical applications, although it is possible to work in terms of depth-dependent superlevel energies, one usually considers them as constant, corresponding to a temperature for which the given ion is the dominant ionization stage within the species [533, 1151].

Superlines

Transitions between superlevels are called *superlines*. The absorption coefficient for a transition $I \to J$, not corrected for stimulated emission, is given by

$$\kappa_{IJ}(\nu) = \sum_i \sum_j n_i w_j \sigma_{ij}(\nu), \tag{18.155}$$

where $\sigma_{ij}(\nu) = (\pi e^2/m_e c) f_{ij} \phi_{ij}(\nu)$, is the cross section for the transition $i \to j$, f_{ij} is the oscillator strength, and $\phi_{ij}(\nu)$ is the (normalized) absorption profile coefficient.

Within the superlevel formalism, the absorption coefficient for transition $I \to J$ has to be given by

$$\kappa_{IJ}(\nu) = n_I w_J \sigma_{IJ}(\nu); \tag{18.156}$$

therefore, the cross section has to be given, using equation (18.153), (18.155), and (18.156), by

$$\sigma_{IJ}(\nu) = \frac{g_J \exp(-E_J/kT) \sum_i \sum_j g_i w_i w_j \sigma_{ij}(\nu) \exp(-E_i/kT)}{\left[\sum_i g_i w_i \exp(-E_i/kT)\right]\left[\sum_j g_j w_j \exp(-E_j/kT)\right]}. \tag{18.157}$$

Similarly, from the definition of the emission coefficient (assuming complete frequency redistribution in the lines),

$$(c^2/2h\nu^3)\eta_{ji}(\nu) = \sum_i \sum_j n_j w_i (g_i/g_j) \sigma_{ij}(\nu) \equiv n_J w_I (g_I/g_J) \sigma_{JI}(\nu), \tag{18.158}$$

one obtains for the emission cross section

$$\sigma_{JI}(\nu) = \frac{g_J \exp(-E_I/kT) \sum_i \sum_j g_i w_i w_j \sigma_{ij}(\nu) \exp(-E_j/kT)}{\left[\sum_i g_i w_i \exp(-E_i/kT)\right]\left[\sum_j g_j w_j \exp(-E_j/kT)\right]}. \tag{18.159}$$

The absorption and emission cross sections are generally different, even with the assumption of complete redistribution. They would be equal if $E_j - E_i = E_J - E_I$ for all i, j, i.e., if the frequencies of all transitions forming a superlevel are the same.

Since the superlevels are constructed to be composed of levels of nearly equal frequency, the approximation $\sigma_{IJ}(\nu) \approx \sigma_{JI}(\nu)$ is usually well justified.

For completeness, the photoionization cross section from a superlevel is given by

$$\sigma_{I\kappa}(\nu) = \frac{\sum_i g_i w_i \sigma_{i\kappa}(\nu) \exp(-E_i/kT)}{\sum_i g_i w_i \exp(-E_i/kT)}. \tag{18.160}$$

The collisional rates are given by expressions analogous to equations (18.157), (18.159), and (18.160), replacing σ with C.

Opacity Sampling and Opacity Distribution Functions

The absorption cross section as defined by equation (18.157) is a complicated and highly non-monotonic function of frequency. This is illustrated in figure 18.3, where a typical cross section is plotted. The upper panel shows the actual cross section, which was calculated using some 15,500 frequency points; the lower panel will be discussed later.

However, from the point of view of constructing a model stellar atmosphere, a detailed form of cross section does not matter. The only quantities that do matter are the corresponding integrals over the frequency range covered by a superline; the integrals may represent, for instance, the radiative equilibrium integrand, $(\kappa_\nu J_\nu - \eta_\nu)$,

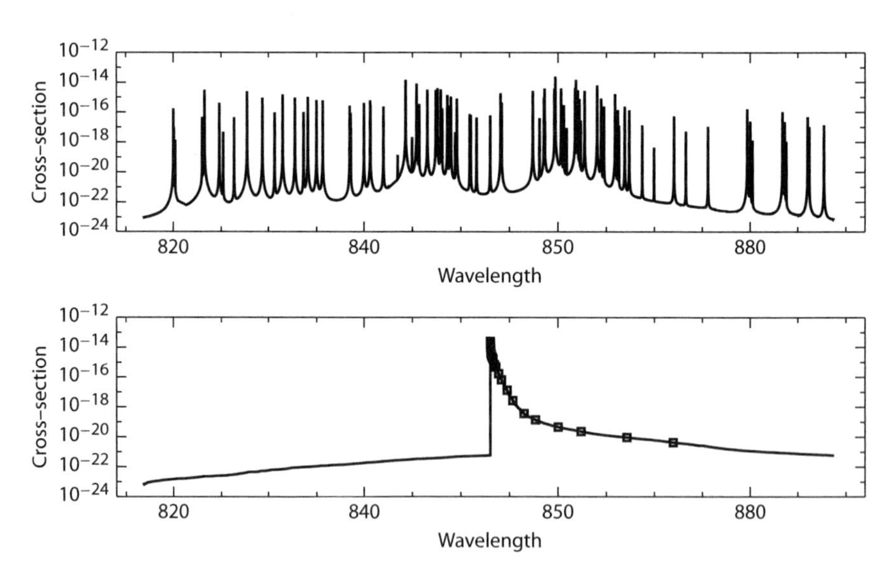

Figure 18.3 A typical superlevel cross section (transition 1–13 of Fe III from [533]). The upper panel shows the actual cross section calculated for $T = 19,221$ K, $n_e = 10^{14}$ cm^{-3}; a total of 15,485 frequency points were used. The lower panel shows the corresponding opacity distribution function (ODF), constructed as described in the text. Small squares, with the total number of 24, indicate the frequency points used to represent this ODF in model atmosphere calculations. From [533].

or integrals occurring in evaluating radiative rates, with integrands (schematically) $\sigma(\nu)J_\nu$. We write the integrals to be dealt with schematically as

$$\int_{\nu_0}^{\nu_1} f(\nu)\sigma(\nu)J_\nu \, d\nu, \tag{18.161}$$

where ν_0 and ν_1 are the minimum and maximum frequency within a superline, and $f(\nu)$ is a function of frequency, which is typically a smooth function of frequency or a constant. Numerical evaluation of integral (18.161) has to be done carefully, because in addition to a wildly varying function of frequency, $\sigma(\nu)$, there is another rapidly varying function, J_ν. More precisely, J_ν is a smooth function of frequency for large optical depths where $J_\nu \to B_\nu$, but it obviously reflects the variations of $\sigma(\nu)$ for low optical depths. There are two possibilities for evaluating the integrals (18.161), which we describe in turn below.

Opacity Sampling

An obvious possibility is to use a direct approach, and to evaluate the integral straightforwardly by choosing a sufficient number of frequency points to represent all frequency variations of function $\sigma(\nu)$ and J_ν. In fact, one can choose a random sampling of frequencies, provided it is reasonably dense. The individual lines within a superline are distributed randomly; hence, statistically speaking, the errors in the integral (18.161) that arise because of missing frequency regions with a high opacity (line centers) or a low opacity (regions between lines) essentially cancel. Because lines are distributed randomly, a reasonable frequency sampling is achieved by using equidistant (or logarithmically equidistant) spacing of frequency points. Obviously, when using a spacing that is around, or lower than, a characteristic Doppler width of lines, one basically recovers an "exact" quadrature scheme.

To estimate the number of frequency points ideally needed, let us assume that metal lines are practically everywhere, so one should cover all the frequency range by frequency points that are spaced proportionally to the value of a fiducial Doppler width; $\Delta\nu_D^* = \nu\, v^*/c$, where ν is the frequency, v^* a characteristic velocity (given by a thermal velocity of a characteristic species, e.g., iron) or a characteristic "turbulent velocity," whichever is larger, and c the light speed. One requires a frequency resolution of $a\,\Delta\nu_D^*$, where a is an adjustable parameter (ideally, a value below 1).

The total number of frequency points would then be

$$NF \approx \int_{\nu_{\min}}^{\nu_{\max}} \frac{1}{a\,\Delta\nu_D^*}\, d\nu = \frac{c}{a\, v^*}\, \ln(\nu_{\max}/\nu_{\min}). \tag{18.162}$$

Assuming, for example, that we need to cover about three decades in frequency, and taking $a = 0.75$, one would need about 2×10^5 to 4×10^5 frequency points for hot models ($v^* \approx 10$–20 km/s), while one needs about 10^6 points for cool models ($v^* \approx 3$ km/s), or even more for cooler and low-microturbulence models. Such numbers were prohibitive in the past, but thanks to modern efficient methods, such as an application of ALI or approximate Newton-Raphson scheme, these large numbers of frequency points can be accommodated into model atmosphere programs and resulting models (e.g., [292,524,533,682,683,1150,1151,1156]).

Opacity Distribution Functions

In the context of NLTE models, this approach is now more or less outdated thanks to fast numerical methods that allow treating large numbers of frequency points, but it was very attractive in the early stages of development of NLTE blanketed models in the 1980s and 1990s. Because it is still occasionally used, we briefly describe it here.

The approach consists of resampling the cross section, in a given *individual superline*, to yield a monotonic or step-wise monotonic function of frequency, called the *opacity distribution function* (ODF), and to represent this function by a relatively small number of frequency points. This procedure avoids all problems of missing low and high opacities but suffers from the serious problem of treating an overlap of two distribution functions (see below). The idea of ODF is used routinely in the context of LTE models (cf. § 17.6); the basic difference from the present case is that the LTE ODFs consider all lines of all species to form an ODF. An LTE ODF is then defined only by its frequency limits, and the opacity itself is a function of temperature and electron density only—see § 17.6.

In the context of NLTE, the idea of ODF was first used in [31] and later in [292] and [533]. In the approach of [533], the peak of the ODF is placed at the position of the mean frequency of the individual superline components, weighted by the cross section, i.e.,

$$\bar{\nu}_{ij} = \int_{\nu_0}^{\nu_1} \nu\sigma_{ij}(\nu)d\nu \bigg/ \int_{\nu_0}^{\nu_1} \sigma_{ij}(\nu)d\nu. \qquad (18.163)$$

The ODF is then taken as a monotonically decreasing function in the direction toward the strongest individual line of the superline.

The construction of the ODF is illustrated in the lower panel of figure 18.3. In this particular case, the strongest individual line happens to be in the direction of decreasing frequency (increasing wavelength) from the mean frequency, so the ODF is chosen to decrease monotonically from the peak in this direction. Once the frequency limit of the superline is reached (corresponding to $\lambda = 888$ Å in figure 18.3), the rest of the original resampled function is continued on the opposite side of the peak.

The basic advantage of the opacity distribution function approach is that the individual ODFs may be represented by a relatively small number of discretized frequency points (for example, 24 points represent the particular ODF displayed in figure 18.3). Consequently, the total number of frequencies needed to cover all ODFs is of the order of few times 10^4; i.e., about an order of magnitude lower than for the opacity sampling approach. This was the main reason for the popularity of the ODF approach in the past.

There are two basic types of problems connected with an application of ODFs. The first type concerns the inherent limitations of the statistical representation of the superline cross section, and the second one concerns the treatment of line overlaps. Roughly speaking, while the first type of problem is usually minor or can be dealt with easily, the problem of superlevel overlaps is basically a flaw that cannot be simply remedied.

The inherent reliability of a distribution function representation rests on several assumptions. The most important one, discussed in detail in [181] and [533], is that the individual points that represent a discretized ODF (in other words, a certain frequency bin of an ODF represented by a step-wise function) are always identifiable with definite frequency subintervals of the original cross section, *at all depths in the atmosphere*. As a simple example, the peak of an ODF corresponds to the center of the strongest line within a superline. The strongest line must always be the same line, at all depths. This may actually be a problem for LTE ODFs that combine lines of all species and all ionization stages, so that the strongest line formed deep in the atmosphere with a high temperature may easily be different from the strongest line formed near the surface (cf. § 17.6). In the context of NLTE, this problem does not usually arise because all lines within a superline do belong to the same ion; moreover, they originate from levels with very similar energies.

There are essentially two types of overlaps that may cause problems: (i) an overlap of a superlevel ODF with an ordinary line, and (ii) an overlap of two superlines. In the first case, if a superline accidentally overlaps a strong line of a light element, this results in a modified radiation field, and hence the radiative rates for this superline, which may be spurious if in reality the strong ordinary line happens not to overlap any of the individual metal lines that form the superline. And, vice versa, if a peak of the ODF of a strong superline accidentally overlaps an ordinary line of a light element, the radiation field and radiative rates in such an ordinary line may be significantly modified, which may lead to spurious effects in the NLTE level populations of this particular species. One can in principle check for such situations and remove them by hand, for instance, by shifting an offending superline ODF to prevent spurious overlaps, but the process may become quite cumbersome.

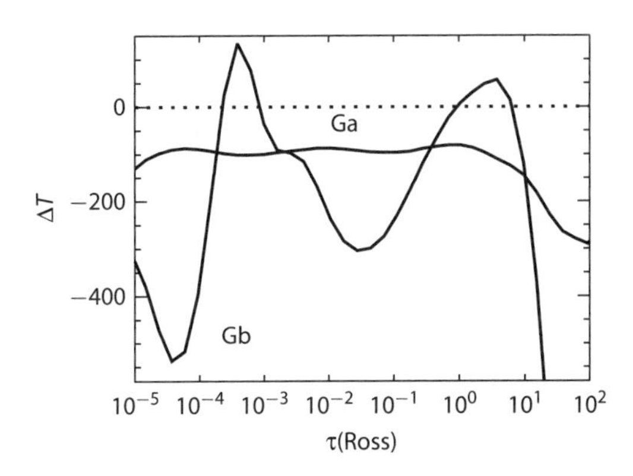

Figure 18.4 Differences in the atmospheric temperature structure computed using the opacity sampling and the ODF approach. All models are for $T_{\rm eff} = 35,000$ K, and $\log g = 4$. The reference model is computed with opacity sampling with frequency step 0.75 fiducial Doppler widths. Model Ga is similar, but with the step 30 Doppler widths. Model Gb is computed with ODFs. From [681].

Overlaps of two superlines are essentially impossible to deal with. For instance, one obtains spurious effects if the peaks of two ODFs accidentally overlap even if the strongest individual lines do not. Or one may have the opposite effect, in which one neglects a real overlap of two strong lines belonging to two different superlines because the corresponding ODFs overlap only weakly. Hence, when applying an ODF approach, one has to assume that the spurious effects that arise because of overlaps of two superlines cancel in the statistical sense. Finally, the concept of representing superlines by ODFs breaks down completely when computing model atmospheres with velocity fields; see chapter 19.

Nevertheless, the ODF approach, if applied carefully (e.g., removing spurious overlaps with ordinary lines), provides results similar to those obtained with opacity sampling; for details, see [681]. This study compares the ODF results with various sampling steps. Some representative results are shown in figure 18.4.

18.6 APPLICATIONS: MODELING CODES AND GRIDS

Finally, we briefly describe some applications of the theory outlined in this chapter. We do not deal with results of spectroscopic analyses of individual objects, which is the topic of a vast amount of literature. Instead, we summarize the currently available NLTE model atmosphere computer programs and current NLTE model atmosphere grids, which serve as a basis of spectroscopic analysis of individual objects.

Available NLTE Model Atmosphere Codes

The first publicly available NLTE hydrostatic model atmosphere code was the "NCAR code" [737]. It was based on the standard complete linearization and written for computers available in the 1970s; hence its application was limited to simple NLTE models, with a maximum number of frequency points equal to 120 and a maximum of NLTE levels of about 15. Several authors later extended this code to be able to treat a slightly larger numbers of frequencies and NLTE levels, but the fundamental limitation of the standard CL scheme (cf. § 18.2) prevented a wider application of the code to compute more realistic models.

More recent, powerful, and widely used codes for computing NLTE metal line-blanketed model atmospheres are TMAP (Tübingen Model Atmosphere Package; originally named PRO2) [292, 1150, 1151], based on the ALI scheme developed in [1148, 1149], and TLUSTY [521, 533], based on the hybrid CL/ALI method. They are described in more detail below. Static models are also being constructed by codes originally designed for expanding atmospheres (by setting the expansion velocity to a very low value, specifically, to a small fraction of the sound speed), such as CMFGEN [493] or PHOENIX [456, 458]. The two latter programs are described in more detail in § 20.4.

TMAP

This program solves the basic structural equations: radiative transfer, hydrostatic equilibrium, radiative equilibrium, kinetic equilibrium, and charge and particle conservation equations. It is fully data-oriented as far as the choice of atomic species, ions, energy levels, transitions, and opacity sources to be taken into account. There are no default opacities built in, but the program offers a wide range of default expressions for various cross sections, in which the user supplies values of appropriate free parameters. Handling of atomic data for TMAP is described in [894]. Both options ODF and opacity sampling are offered for a treatment of metal line blanketing.

The program uses the ALI scheme to eliminate the radiative transfer equation, and the remaining state equations are solved by linearization, as described in § 18.4. A diagonal or tridiagonal Λ^*-operator is used. To speed up calculations, the Broyden inversion methods, as well as Kantorovich acceleration, are offered.

The program is written in FORTRAN 77. It was developed by Werner, Dreizler, and collaborators. The numerical scheme is described in [1148] (standard ALI scheme with a diagional operator), [1149] (an extension to tridiagonal operator), [292] (NLTE line blanketing using superlevels and superlines), and [1151] (an overview and the Broyden scheme).

TLUSTY

Similarly to TMAP, this program solves the basic structural equations: radiative transfer, hydrostatic equilibrium, radiative equilibrium, kinetic equilibrium, and charge and particle conservation equations. However, not all the structural equations have to be solved, and some state parameters may be held fixed. For instance, temperature may be fixed and the radiative equilibrium equation is not solved. This procedure sometimes helps to converge otherwise difficult models.

Again, the program is fully data-oriented as far as the choice of atomic species, ions, energy levels, transitions, and opacity sources to be taken into account. There are no default opacities built in, but the program offers a wide range of default expressions for various cross sections, in which the user supplies values of appropriate free parameters (for instance, the parameter $\bar{g}$ in the Van Regemorter formula for collisional excitation rates [cf. equation (9.59)]. Handling of atomic data is described in [521, 681] and in the user's guide.[1]

As in TMAP, both ODF and opacity sampling are offered for a treatment of metal line blanketing in TLUSTY, although a detailed setup is somewhat different. With TLUSTY, first NLTE metal line-blanketed models used the ODF approach [529, 679, 680, 684]; however, recent models (including the OSTAR2002 and BSTAR2006 grids—see below) use the opacity sampling method with a sampling of 0.75 fiducial Doppler widths., which requires several times 10^5 frequency points.

The basic difference with respect to TMAP code is the numerical method used for solving the structural equations. TLUSTY uses the hybrid CL/ALI scheme. There are several formal solvers of the transfer equation available: Feautrier (second or fourth order) or the discontinuous finite element scheme (DFE). The Λ^*-operator can be either diagonal or tridiagonal, computed as the corresponding part of the exact Λ. To reduce the overall number of iterations as well as computer time per iteration,

[1] http://nova.astro.umd.edu/.

the Ng acceleration, the Kantorovich method, and the successive overrelaxation method are implemented (cf. § 18.3), again with all setups being driven by input data.

The program is written in FORTRAN 77. In the original version, it was developed by Hubeny [521], using the traditional complete linearization, and subsequently modified and upgraded by Hubeny and Lanz in [532] (Kantorovich and Ng acceleration), [531] (occupation probabilities and pseudocontinuum opacities), and [533] (an implementation of the hybrid CL/ALI method, together with the superlevel and superline approach to treat NLTE metal line blanketing).

Although originally developed for model atmospheres of hot stars (types O, B, A), it was later extended both to higher and lower temperatures. For convectively unstable atmospheres, the convection is treated as described in this chapter. Molecular opacities are not computed on the fly; instead, one interpolates in a precalculated opacity table. This table contains opacities of all sources that are treated in LTE (molecules and less important atomic species); therefore, the opacities are functions of only temperature, density, and frequency. There is a separate variant called CoolTlusty that is designed to model atmospheres of substellar mass objects, such as brown dwarfs and giant planets [527], where all the opacity sources (both molecular and condensates) are supplied from precalculated tables.

While it is not the topic of this book, we mention that TLUSTY can also be used to compute models of vertical structure of accretion disks because the structural equations describing the vertical structure are similar to structural equations describing a stellar atmosphere. Calculations of the disk is set up by a simple switch; the advantage of this approach is that any upgrades in numerical methods as well as atomic physics automatically apply for both atmospheres and disks. For instance, a treatment of Compton scattering, as described in [525], was developed for modeling hot accretion disks around black holes but can also be used for stellar atmospheres, if needed.

There is a accompanying program SYNSPEC[2] that performs detailed spectrum synthesis for the model atmospheres provided by TLUSTY, as well as from other sources, for instance, the Kurucz models. It solves, frequency by frequency, the radiative transfer equation taking the atmospheric structure, i.e., temperature, mass density, electron density, and (for NLTE models) atomic level populations as given functions of depth. Necessary parameters for the continuum opacities are taken from the input data for TLUSTY, while the parameters for lines are taken from an independent line list that contains gf values, level energies and statistical weights, and line broadening parameters. For the species that were treated in NLTE by TLUSTY, an association between levels referred to in the line list and those considered as explicit levels in TLUSTY is done automatically by SYNSPEC based on the level energy and statistical weight. The lines of other species are treated in LTE.

[2] http://nova.astro.umd.edu/Synspec49/synspec.html

NLTE Multi-Level Codes

It should be noted that besides the codes that solve for the complete NLTE model atmosphere structure, there are several codes that take an atmospheric structure (temperature, density) given and fixed, and solve for kinetic equilibrium + radiative transfer for a selected chemical element. Although recently these codes have lost much of their former appeal because the modern codes are capable of solving the structure plus detailed NLTE rate equations for many species without significant problems, they are still being used for analyzing stellar spectra. The first modern NLTE multi-level code was the "Kitt Peak code" [47]; other popular codes of this sort are DETAIL/SURFACE [172] and the still widely used codes PANDORA [68–70] and MULTI [183].

Available NLTE Model Atmosphere Grids

The first publicly available grid of NLTE model atmospheres of hot (O and B) stars was presented in [730]. In these models, departures from LTE were taken into account for the five lowest levels of hydrogen, the first two levels of He I and He II, and an "average light ion" element (which represents C, N, and O) consisting of five stages of ionization, each with a ground state only. These models allowed for six hydrogen line transitions: $L\alpha$, $L\beta$, $L\gamma$, $H\alpha$, $H\beta$, and $P\alpha$ for O-stars and $H\alpha$, $H\beta$, $H\gamma$, $P\alpha$, $P\beta$, and $B\alpha$ for B-stars (for which the Lyman lines were set in detailed balance). The grid was widely used for many spectroscopic analyses of hot stars until it was surpassed by NLTE metal line-blanketed models that started to appear in the mid- to late 1990s. To account for NLTE effects in metals, the following strategy was usually used: the atmospheric structure (T, N, n_e) was taken from the H-He model grid and held fixed for subsequent NLTE multi-level atom line-formation solution for a selected chemical element, using one of the NLTE line-formation codes mentioned above.

Once the methodology for computing NLTE metal line-blanketed model atmospheres became available in the 1990s, the individual studies usually considered only a limited number of such models because they were computationally demanding. Production of more extended model grids only started in the 2000s. There are several partial grids of NLTE models for various stellar types, mostly hot stars. Models constructed by the TMAP code for very hot white dwarfs, subdwarfs, and pre-white dwarfs (also known as the PG 1159 stars) are available online.[3] A different strategy is described in [897], who introduce a Virtual Observatory service *TheoSSA* (Theoretical Simple Spectra Access[4]) within the framework of the *German Astrophysical Virtual Observatory*, a web-based interface that enables a user to either extract already computed models or generate a specific model using TMAP, for very hot objects (hottest white dwarfs; super-soft X-ray sources, etc.).

The effort of the developers of TLUSTY culminated in the construction of a grid of NLTE fully blanketed model atmospheres for O-stars (OSTAR2002 [682]) and

[3] http://astro-uni-tuebingen.de/rauch/TMAP/TMAP.html

[4] http://vo.ari.uni-heidelberg.de/ssatr-0.01/TrSpectra.jsp

early B-stars (BSTAR2006 [683]). The intent was to provide more or less definitive grids of models in the context of 1D plane-parallel geometry, with hydrostatic and radiative equilibrium and without any unnecessary numerical approximations. The basic limitation of these models is the quality and availability of atomic data, which, despite recent efforts (such as the *Opacity Project*, *OPAL*, or *IRON Project*) are still incomplete (for instance, the lack of available collisional excitation cross sections for dipole-forbidden transitions in the iron-peak elements).

The OSTAR2002 grid contains model atmospheres for $T_{\rm eff}$ between 27,500 and 55,000 K, log g between 4.75 and a value that corresponds to an approximate location of the Eddington limit, and for 10 metallicities: 2, 1, 1/2, 1/5, 1/10, 1/30, 1/50, 1/100, 1/1000, and 0 times the solar metal composition, so that the grid is useful for studies of typical environments of massive stars: the Galactic center, the Magellanic Clouds, blue compact dwarf galaxies like I Zw-18, and galaxies at high redshifts. Departures from LTE are allowed for the following species: H, He, C, N, O, Ne, Si, P, S, Fe, and Ni, in all important stages of ionization. There are altogether over 1,000 (super)levels to be treated in NLTE, about 10^7 lines, and about 250,000 frequency points to describe the spectrum.

The BSTAR2006 grid is similar. It contains models for $T_{\rm eff}$ between 15,000 and 30,000 K, log g values are set similarly to the OSTAR2002 grid, and for six metallicities: 2, 1, 1/2, 1/5, 1/10, and 0 times solar. The species treated in NLTE are the same as in OSTAR2002, adding Mg and Al but removing Ni, which is less important for B-stars. There are altogether about 1,450 (super)levels treated in NLTE, about 10^7 lines, and about 400,000 frequency points. The models for both grids are available online.[5]

Several representative results from the grids are shown below. Figure 18.5 displays the temperature structure for three representative effective temperatures of the OSTAR2002 grid. The temperature distribution nicely illustrates the basic features of line blanketing, namely, the backwarming (a heating of moderately deep atmospheric layers between Rosseland optical depths 0.01 and 1), and a surface cooling; cf. chapter 17. The zero-metallicity and low-metallicities models exhibit a temperature rise at the surface, a typical NLTE effect discovered by Auer and Mihalas [53] and explained as an indirect heating effect of the hydrogen Lyman and Balmer lines. This effect competes with surface cooling caused by metal lines, and these effects nearly cancel at metallicities 1/50 (for hotter models) to 1/10 (cooler models). Interestingly, the temperature curves for all metallicities cross in a very narrow range of optical depths.

A sensitivity of the predicted spectra to the effective temperature is depicted in figure 18.6, which shows (from top to bottom) emergent spectra for 50,000, 45,000, 40,000, 35,000, and 30,000 K. Notice a diminishing Lyman jump at 911 Å when going to higher temperatures; this is a consequence of increased ionization of hydrogen. A similar sensitivity to metallicity is shown in figure 18.7.

[5] http://nova.astro.umd.edu

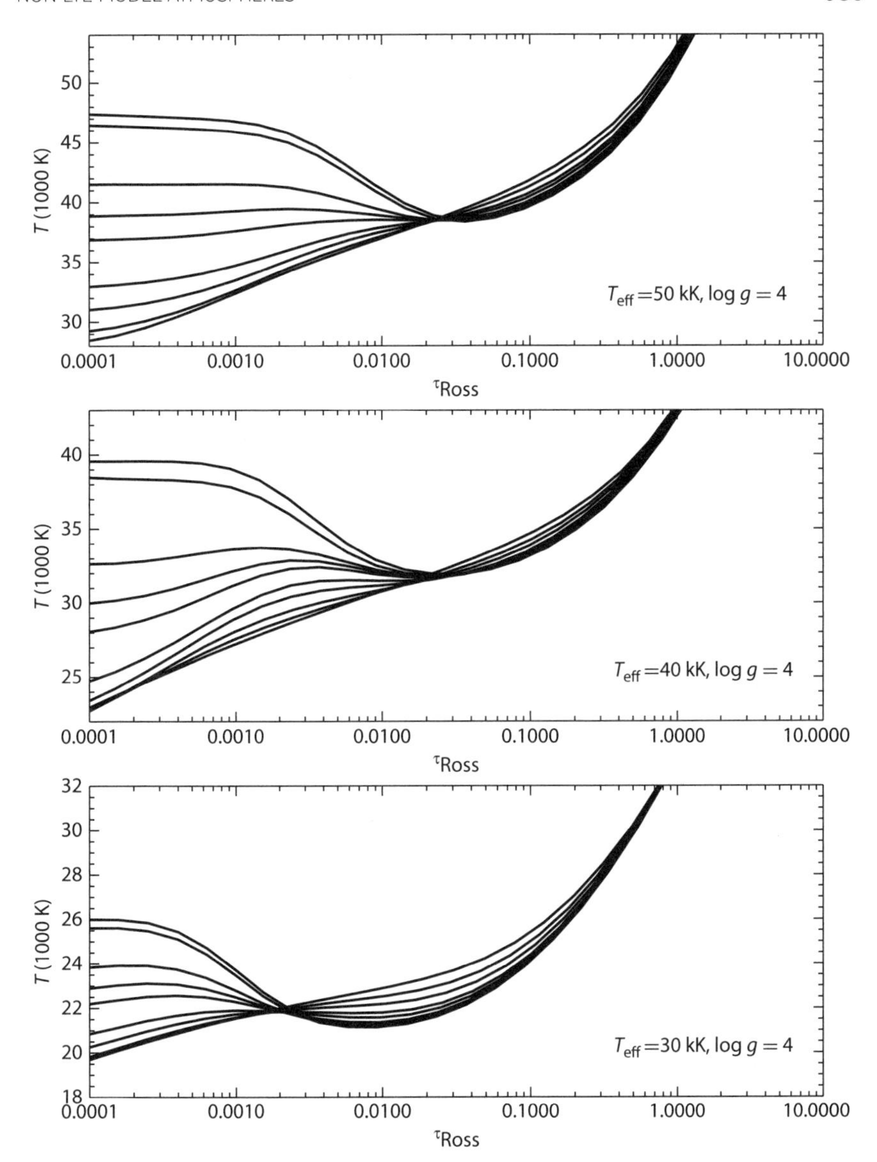

Figure 18.5 Temperature as a function of Rosseland optical depth for OSTAR2002 model atmospheres with $T_{\rm eff} = 50{,}000$ K (top), 40,000 K (middle), and 30,000 K (bottom); $\log g = 4.0$, and various metallicities. At low optical depths ($\tau_{\rm Ross} < 10^{-3}$), the top curves are for a pure H-He model, and temperature is progressively lower when increasing the metallicity, while the reverse applies in deep layers, $\tau_{\rm Ross} < 10^{-3}$. From [682].

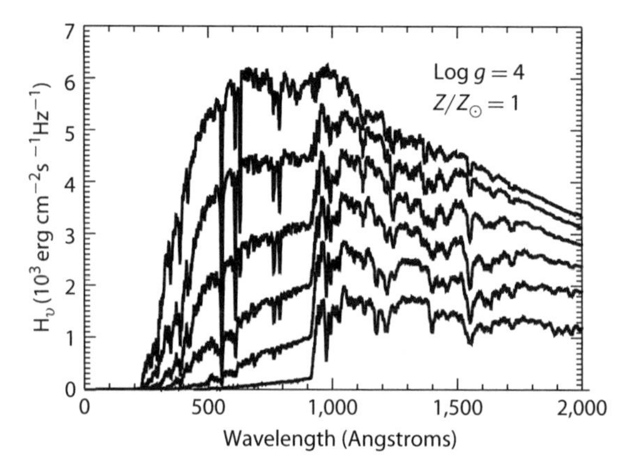

Figure 18.6 Predicted flux for six solar composition OSTAR2002 model atmospheres with $T_{\rm eff}$ between 55,000 and 30,000 K with a step of 5,000 K, and for $\log g = 4$. From [682].

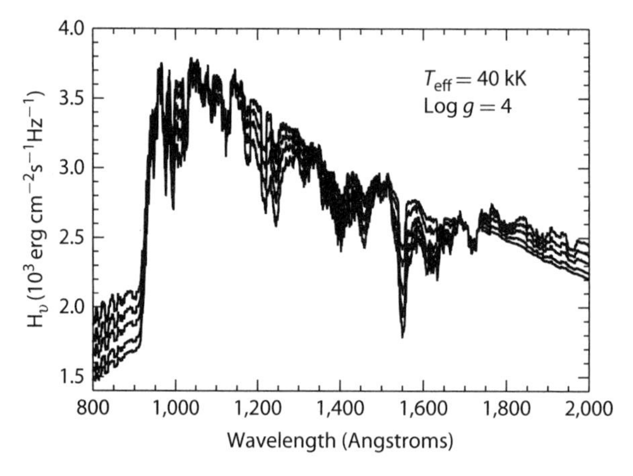

Figure 18.7 Predicted flux for five OSTAR2002 model atmospheres with $T_{\rm eff} = 40{,}000$ K, $\log g = 4$, for five different metallicities. The models with higher metallicities exhibit deeper lines but higher continuum flux. From [682].

Chapter Nineteen

Extended and Expanding Atmospheres

19.1 EXTENDED ATMOSPHERES

In the preceding chapters we have assumed that stellar atmospheres are stratified in plane-parallel layers. This is a good approximation for a star on the main sequence because the density scale-height in its atmosphere is very small compared to its radius. Hence the density of material in the atmosphere falls sharply with increasing radius, and the outward-directed escape probability for photons increases to 1 in a relatively shallow layer. But some stars, e.g., a giant or a supergiant, have an *extended atmosphere* whose thickness is a significant fraction of (or even exceeds) the radius of its dense stellar interior.

Atmospheric extension has important observational and physical implications. To a good approximation these atmospheres can be taken to be *spherically symmetric.* In an extended atmosphere, optical depth unity at different frequencies occurs at significantly different radii in the star's atmosphere. Hence the observed stellar "disk" at frequencies at which the material is relatively transparent is confined to a smaller solid angle around the radial direction than at frequencies that are more opaque. As a result, stars with extended envelopes show continuum energy distributions that have an anomalously low radiation temperature compared to the excitation temperature inferred from spectral lines. That is, their continuum energy distributions are "flatter" as a function of frequency than those of main-sequence stars of the same spectral type, in the sense that they show "excess" emission in the infrared and a "deficiency" in the ultraviolet. Thus there are ambiguities in the meaning of the "radius" and the "effective temperature" of such stars; a penetrating discussion of these matters can be found in [93]. In many cases there are also indications of rapid atmospheric expansion in stars with extended envelopes, so dynamical models that include the effects of outward flow are necessary, e.g., for Wolf-Rayet or Of stars.

These facts imply that the temperature structure of an extended atmosphere is significantly different from that in a planar model and that there are mathematical complications because of sharp angular peaking of the radiation field in the radial direction at small optical depths.

Transfer Equation

In spherical geometry the equation of transfer is

$$\mu \frac{\partial I_{\mu\nu}}{\partial r} + \frac{(1-\mu^2)}{r} \frac{\partial I_{\mu\nu}}{\partial \mu} = \eta_\nu - \chi_\nu I_\nu, \tag{19.1}$$

which has angular moments

$$\frac{1}{r^2}\left[\frac{\partial(r^2 H_\nu)}{\partial r}\right] = \eta_\nu - \chi_\nu J_\nu \tag{19.2}$$

and

$$\frac{\partial K_\nu}{\partial r} + \frac{1}{r}(3K_\nu - J_\nu) = -\chi_\nu H_\nu, \tag{19.3}$$

or, using the Eddington factor,

$$\frac{\partial(f_\nu J_\nu)}{\partial r} + \frac{1}{r}(3f_\nu - 1)J_\nu = -\chi_\nu H_\nu. \tag{19.4}$$

These equations are clearly more complicated than their planar counterparts. Note that (19.1) is a partial differential equation in two independent variables. And that if one used (19.3) or (19.4) to eliminate H_ν from (19.2), the resulting moment equations do not have a simple form. Even though the variable Eddington factor $f_\nu \to \frac{1}{3}$ at depth, where the radiation field approaches isotropy, $f_\nu \to 1$ near the surface, where the radiation flows radially outward, so the Eddington approximation, which can give satisfactory results for the planar case, becomes unacceptably inaccurate.

Spherical Gray LTE Atmospheres

McCrea [718], Kosirev [632], and Chandrasekhar [214, 225] showed that one can construct a consistent approximation for a tenuous atmosphere enveloping a parent star of radius R by taking averages of the outward radiation field in the range $\mu_* \leq 1$, where $\mu \leq \mu_* \equiv (1 - \mathsf{R}^2/r^2)^{1/2}$, separately from those computed for the inward radiation field in the range $-1 \leq \mu \leq \mu_*$. But this approach works only if r_* can be chosen unambiguously. For extended photospheres having appreciable density and optical depth, it breaks down. So it is not surprising that effective methods for treating transfer problems in spherical geometry were relatively slow to develop.

If we assume that the atmospheric material is *gray*, i.e., $\chi_\nu \equiv \chi$, the problem becomes more tractable. Integrating (19.2) and (19.4) over frequency, omitting the subscript ν for frequency-integrated quantities, we find

$$\frac{1}{r^2}\left[\frac{\partial(r^2 H)}{\partial r}\right] = 0, \tag{19.5}$$

and

$$\frac{\partial(fJ)}{\partial r} + \frac{1}{r}(3f - 1)J = -\chi H. \tag{19.6}$$

To obtain (19.5) we demanded the atmosphere be in radiative equilibrium:

$$\int_0^\infty \chi_\nu B_\nu d\nu \equiv \chi B = \int_0^\infty \chi_\nu J_\nu d\nu \equiv \chi J. \tag{19.7}$$

Further, (19.5) yields the integral

$$r^2 H \equiv H_0 = \mathsf{L}/16\pi^2, \tag{19.8}$$

where L is the luminosity of the star.

The optical depth into the atmosphere, measured radially from an arbitrarily large outer radius R, is

$$\tau(r) = \int_r^R \chi(r')\, dr'. \tag{19.9}$$

Deep within the atmosphere, i.e., $r \ll R$ and $\tau \gg 1$, the radiation field becomes isotropic, so $f \to \frac{1}{3}$. In this limit, (19.6) becomes

$$\frac{\partial J}{\partial r} = -3\chi H = -3\chi H_0/r^2, \tag{19.10}$$

which yields the integral

$$J(\tau) = H_0 \left(3 \int_0^\tau r^{-2}\, d\tau' + C \right). \tag{19.11}$$

If one applies the Eddington boundary condition at $\tau = 0$ and $R = R_0$, then $J(0) = 2H(0) = 2H_0/R^2$ so that

$$J(\tau) = R^{-2} H_0 \left[3 \int_0^\tau (R^2/r^2)\, d\tau' + 2 \right], \tag{19.12}$$

a result first obtained by Chandrasekhar [214].

Further, if one assumes LTE, then $J(\tau) = B(\tau) = \sigma T^4/\pi$, which would give a temperature profile in the atmosphere. But (19.10) and (19.11) are valid only at great depth. Near the surface, the radiation field is in the free-flow regime, and $f \to 1$, so (19.6) becomes

$$\frac{\partial (r^2 J)}{\partial r} = -\chi r^2 H = -\chi H_0, \tag{19.13}$$

which yields

$$J(\tau) = r^{-2} H_0 (\tau + 1), \tag{19.14}$$

a result valid when $\tau \ll 1$, and $r \approx \mathsf{R}$.

Further progress can be made if one adopts a *power-law opacity*, i.e., $\chi = C_n r^{-n}$, as was done in [214] and [632]. As noted by Kosirev, there can be physical motivation for this choice in an expanding atmosphere, because the equation of continuity demands that $\rho v r^2 =$ constant, where ρ is the density and v is its expansion velocity. In the limit of very rapid expansion, where $v > v_{\text{escape}}$ (actually observed), the material moves at nearly constant velocity so that $\rho \sim r^{-2}$.

Then the opacity of the material can be expected to vary as a power of ρ (e.g., linearly for electron scattering or quadratically for free-free absorption), and hence as some power of $1/r$. Using a power-law expression for χ in (19.9) and adopting, for simplicity, $R = \infty$, we would have

$$\tau(r) = C_n r^{-(n-1)}/(n-1). \tag{19.15}$$

Combining (19.15) with (19.14), the limiting form for $J(\tau)$ as $r \to 0$ and $\tau \gg 1$ is

$$J \to [3(n-1)/(n+1)]H_0 r^{-2}\tau, \tag{19.16}$$

while (19.12), valid for $\tau < 1$, remains unchanged. In this idealized situation one could interpolate between these two extremes with an expression of the form [685]

$$J(\tau) = \left(\frac{3H_0}{r^2}\right)\left(\frac{n-1}{n+1}\right)\left[\tau + \frac{1}{3}\left(\frac{n+1}{n-1}\right)\right]. \tag{19.17}$$

Comparison with precise calculations show that (19.17) would be a "reasonable" first estimate if the material in atmosphere were in fact gray and in LTE. Then using the identity of J and B, and writing T_1 for the temperature at $\tau = 1$, (19.17) can be rewritten as

$$T(\tau) = T_1 \tau^{1/2(n-1)}\{[\tau + (n+1)/3(n-1)]/[1 + (n+1)/3(n-1)]\}^{1/4}. \tag{19.18}$$

Equation (19.18) shows an important difference between planar and spherical geometry, namely, $T \to 0$ as $\tau \to 0$ in an extended atmosphere rather than approaching a finite value. Thus the outer cool layers, which occupy a large volume, enhance the flux observed at longer wavelengths and lead to the distinctively "flatter" energy distribution mentioned above.

To calculate the flux seen by an observer at a distance D from the center of the star we use the (p, z) coordinate system shown in figure 19.1. The *impact parameter* p is the perpendicular distance of a ray from the parallel ray passing through the center of the star, and z is the distance along the ray measured from the plane through the center of the star perpendicular to the central ray. We take z to be positive toward the observer and place the observer at $z = \infty$ for the purpose of calculating integrals.

The (p, z) coordinates are related to polar coordinates (r, θ) by $p = r\sin\theta$ and $z = r\cos\theta$, and $r = (p^2 + z^2)^{1/2}$. For a given value of p the transfer equation along the ray[1] in the direction of increasing z is

$$[\partial I_\nu(p,z)/\partial z] = \eta_\nu(p,z) - \chi(p,z)I_\nu(p,z). \tag{19.19}$$

The formal solution of (19.19) for the intensity emerging at $z = \infty$ along the ray with impact parameter p is

$$I_\nu(p,\infty) = \int_{-\infty}^{\infty} B_\nu[T(p,z)]\exp[-\tau(p,z)]\chi(p,z)\,dz, \tag{19.20}$$

[1] Which is a *characteristic ray* of the differential operator.

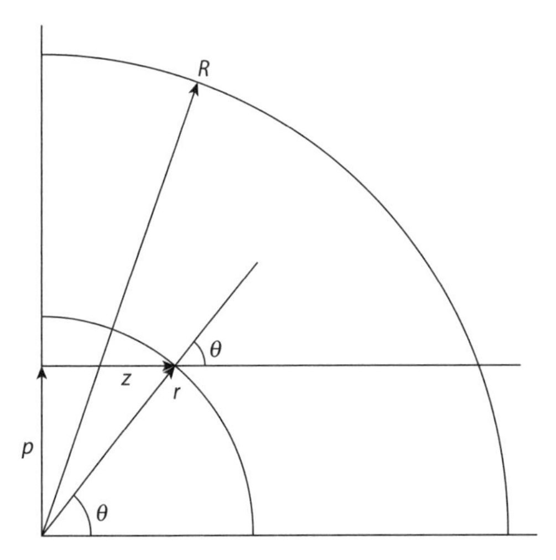

Figure 19.1 Coordinate system for solution of transfer equation in spherical symmetry.

where $\tau(p, z)$ is the optical depth measured from $z = \infty$ inward along the ray. The total flux received by an observer per unit receiver area is

$$F_\nu = 2\pi D^{-2} \int_0^\infty I_\nu(p, \infty) p \, dp. \tag{19.21}$$

Following Kosirev, one changes variables in (19.20) to $\theta = \cos^{-1}(z/r)$ and writes $\tau(p, z) = \tau(p, \theta)$ which, using a power-law opacity, becomes $\tau(p, \theta) = C_n p^{-(n-1)} \psi_n(\theta)$, where

$$\psi_n(\theta) \equiv \int_0^\theta \sin^{n-2} \theta' d\theta'. \tag{19.22}$$

Then

$$I_\nu(p, \infty) = C_n p^{-(n-1)} \int_0^\pi B_\nu[T(p, \theta)] \exp[-\tau(p, \theta)] \sin^{n-2} \theta \, d\theta. \tag{19.23}$$

Substituting (19.23) into (19.21) and making successive transformations from $p = r \sin\theta$ to $\tau(r)$, the radial optical depth given by (19.15), we obtain

$$F_\nu = \pi (R_1/D)^2 \int_0^\infty B_\nu[T(\tau)] \tau^{[-2/(n-1)]} \Phi_n(\tau) \, d\tau, \tag{19.24}$$

where R_1 is the radius at which $\tau = 1$, and

$$\Phi_n(\tau) \equiv 2 \int_0^\pi \exp[-(n-1)\tau \csc^{n-1} \theta \ \psi_n(\theta)] \sin\theta \, d\theta. \tag{19.25}$$

Using (19.18) for the temperature in (19.24), we obtain the flux emerging from a gray spherical atmosphere for a choice of T_1 as the characteristic temperature in the atmosphere and the index n, which determines its degree of extension ($n \to \infty$ for a planar model, and n is small for a large extent).

Observationally, the *color temperature* T_c at a wavelength λ_c is measured using a filter system or the spectrophotometric gradient; both measure the slope of the continuum. Van Blerkom [1105] calculated T_c at $\lambda = 5000$ Å for models with $T_1 = 50,000$ K. He found that for $n = (\infty, 10, 5, 3, 2)$, $T_c/10^4 = (5, 4.3, 3.5, 2.2, 1.2)$. Thus increasing atmospheric extent produces effects that simulate lower atmospheric temperatures. For example, a model with $T_1 = 50,000$ K and $n = 3$ has a flux distribution nearly identical to one with $T_1 = 30,000$ K and $n = 5$. These results show why supergiants and WR stars have lower color temperatures than main-sequence stars of the same spectral type. Thus the effects of atmospheric extension introduce ambiguities in the choice of a structural model for a star because of the trade-off between temperature and envelope size in fitting the data.

In figure 19.2 we see the frequency variation of the flux from a spherical gray atmosphere compared to that from a Planck function at temperature T_c, the color

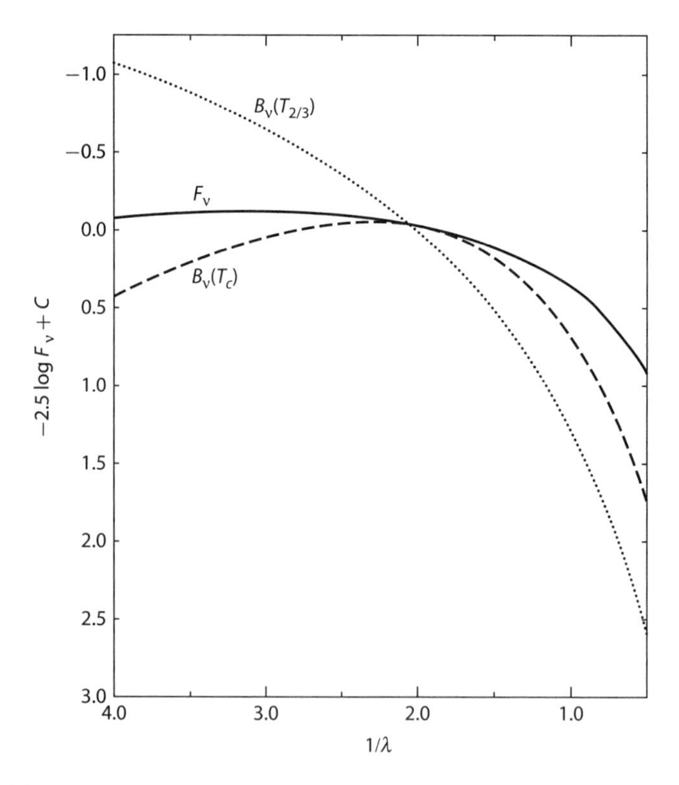

Figure 19.2 *Solid curve*: emergent flux F_ν from a spherical gray atmosphere with $T_c = 5 \times 10^4$ K and $n = 2$. *Dashed curve*: black-body curve with color temperature T_c that matches the slope of F_ν at λ 5000 Å. *Dotted curve*: black-body curve at $T = T(\tau = \frac{2}{3})$, a representative value for a planar atmosphere. *Abscissa*: $1/\lambda$, where λ is in microns. From [1105].

temperature of the flux at $\lambda_c = 5000$ Å, and a Planck function with $T = (\tau = \frac{2}{3})$, roughly the distribution that would emerge from a planar gray atmosphere of that temperature.

One sees that the flux from an extended atmosphere has both an *ultraviolet excess* and an *infrared excess* compared to a Planck function having the same color temperature T_c. In contrast, it has an *ultraviolet deficiency* and an *infrared excess* compared to that from a planar atmosphere having $T(\tau = \frac{2}{3})$ equal to T_c. Both discrepancies show that the temperature structure in an extended atmosphere differs greatly from that in a planar atmosphere.

Solution of the Transfer Equation

In the past, many methods were developed to solve the transfer equation in spherical geometry. Most of them are now of historical interest and have been surpassed by more efficient approaches. A review of an older work can be found, e.g., in [860]. We describe here the differential-equation technique, which is stable, general, and efficient. It uses a Feautrier solution along individual impact parameters, tangent to discrete shells, to treat the angular dependence of the radiation field required to evaluate the Eddington factors. Once the Eddington factors are determined, the scheme employs a Feautrier scheme to solve a combined moment equation obtained by using an elegant transformation introduced by Auer [46]. Alternatively, a direct solution can be obtained using a Rybicki-type method, if the scattering integral in the source function is independent of frequency [738]. An equivalent integral equation method has also been developed [962].

Introducing the radial optical depth scale $d\tau_\nu = -\chi_\nu \, dr$, the moment equations are written

$$\frac{\partial(r^2 H_\nu)}{\partial \tau_\nu} = r^2 (J_\nu - S_\nu), \tag{19.26}$$

and

$$\frac{\partial(f_\nu J_\nu)}{\partial \tau_\nu} - (3f_\nu - 1)\frac{J_\nu}{r\chi_\nu} = H_\nu. \tag{19.27}$$

There are two problems with solving equations (19.26) and (19.27): (i) a direct elimination of H does not yield a simple equation, but rather a complicated one involving both first and second derivatives, and (ii) the factor $(\chi_\nu r)^{-1}$ in the second term in equation (19.27) tends to diverge strongly near the surface because χ_ν, which is the opacity per unit *volume*, decreases there due to decreasing particle density over several orders of magnitude. This term destabilizes the system. Both these difficulties are eliminated by introducing [46] a *sphericality factor* q_ν, defined by

$$\ln(r^2 q_\nu) \equiv \int_{r_c}^{r} [(3f_\nu - 1)/(r' f_\nu)] \, dr' + \ln r_c^2, \tag{19.28}$$

where r_c is the "core radius" corresponding to the deepest point in the atmosphere considered in the solution. Clearly, q_ν is known only if the Eddington factor f_ν is known. Using q_ν, equation (19.27) can be rewritten as

$$\partial(f_\nu q_\nu r^2 J_\nu)/\partial\tau_\nu = q_\nu r^2 H_\nu, \tag{19.29}$$

which, when substituted into equation (19.26), yields a *combined moment equation*

$$\frac{\partial}{\partial\tau_\nu}\left[\frac{1}{q_\nu}\frac{\partial(f_\nu q_\nu r^2 J_\nu)}{\partial\tau_\nu}\right] = r^2(J_\nu - S_\nu), \tag{19.30}$$

or, introducing a new depth variable $dX_\nu \equiv q_\nu d\tau_\nu$,

$$\frac{\partial^2(f_\nu q_\nu r^2 J_\nu)}{\partial X_\nu^2} = \frac{r^2}{q_\nu}(J_\nu - S_\nu). \tag{19.31}$$

Equation (19.31) has to be supplemented by two boundary conditions on the outer and inner sides. To obtain an outer (upper) boundary condition one defines

$$h_\nu \equiv \int_0^1 I(r,\mu,\nu)\mu\, d\mu \Big/ \int_0^1 I(r,\mu,\nu)\, d\mu; \tag{19.32}$$

hence, from (19.29) one obtains

$$[\partial(f_\nu q_\nu r^2 J_\nu)/\partial X_\nu]_{r=R} = h_\nu (r^2 J_\nu)_{r=R}, \tag{19.33}$$

which is the desired outer boundary condition. At the lower (inner) boundary, one applies a planar diffusion approximation to express H as

$$H_\nu(r_c) = \tfrac{1}{3}\left(\chi_\nu^{-1}|\partial B_\nu/\partial\tau_\nu|\right)_{r_c} \tag{19.34}$$

and sets the gradient of the Planck function by demanding that the integral of $H_\nu(r_c)$ over all frequencies yield the correct integrated flux $H_c = \mathsf{L}/(16\pi^2 r_c^2)$. Then

$$\left[\frac{\partial(f_\nu q_\nu r^2 J_\nu)}{\partial X_\nu}\right]_{r=r_c} = r_c^2 H_c \left[\frac{\chi_\nu^{-1}(\partial B_\nu/\partial T)}{\int_0^\infty \chi_\nu^{-1}(\partial B_\nu/\partial T)\, d\nu}\right]_{r=r_c}. \tag{19.35}$$

As discussed in § 11.9, the diffusion approximation, and hence equations (19.34) and (19.35), will be valid if the photon mean free path $\chi_\nu^{-1} \ll R$; this criterion can be met by choosing r_c sufficiently deep in the atmosphere. For other physical situations, for instance, in a nebula, alternative inner boundary conditions must be posed [648].

To solve equations (19.31), (19.33), and (19.35), one employs the *tangent-ray approach*. One introduces a discrete radial mesh $\{r_d\}, d = 1, \ldots, D$, with $R = r_1 > r_2 > \cdots > r_D = r_c$, and a frequency mesh $\{\nu_u\}, n = 1, \ldots, N$. The equations are discretized by replacing derivatives by differences, using a standard approach (cf. §12.2), or using splines [648, 738], or a Hermite formula [48], and the frequency integrals, if any, in the source function with a quadrature formula. Equations (19.31), (19.33), and (19.35) will then have a standard tridiagonal form and are solved with the usual Feautrier-type elimination scheme. The computing time scales as $t \sim cDN^3$, which is modest if the source function is purely thermal or has a coherent scattering term ($N = 1$).

The system of equations (19.31), (19.33), and (19.35) can be solved only if the Eddington factors f and h are known. As in the case of plane-parallel atmospheres, they are determined by performing, at each frequency, a formal solution of the transfer equation with the current estimate of the source function. Specifically, one performs a ray-by-ray solution along a grid of impact parameters $\{p_i\}$, chosen to be *tangent* to each discrete radial shell, supplemented by an additional set of C impact parameters that are chosen to intersect the core and that include the central ray. The geometry of the situation is shown in figure 19.3.

The impact parameters are labeled with an index $i (i = 1, \ldots, NI)$, where $NI = D + C$. It is customary to denote p_1 as the central ray, p_C the last ray *inside* the core (i.e., $p_C < r_c$), $p_{C+1} = r_c$, and $p_{NI} = \mathrm{R}$. Each ray intersects all shells with $r_d \geq p_i$; these intersections define a mesh of z-points along a tangent ray, $\{z_{di}\}$ $(d = 1, \ldots D_i)$, where $D_i = D + C + 1 - i$ for $i > C$ and

$$z_{di} = (r_d^2 - p_i^2)^{\frac{1}{2}}. \tag{19.36}$$

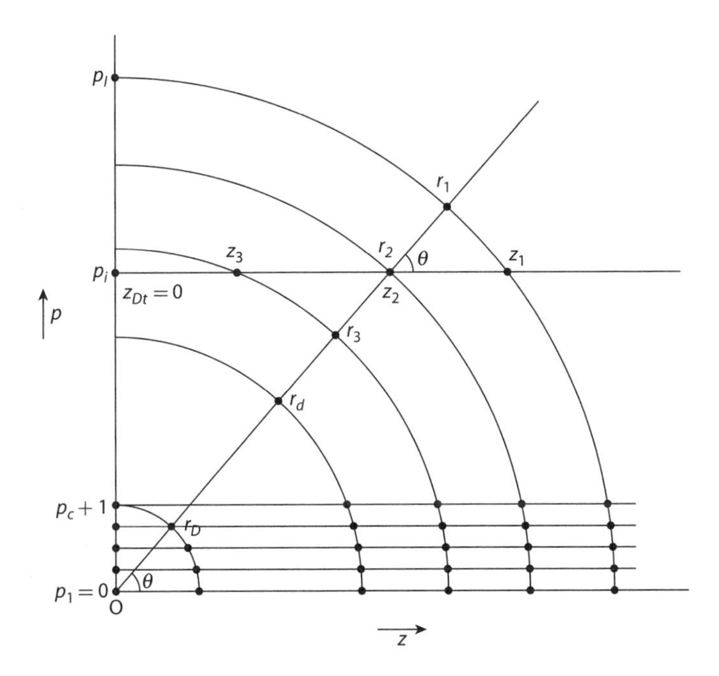

Figure 19.3 Discrete (p, z) mesh used in solution of spherical transfer equation. The impact parameters $\{p_i\}$ are chosen to be parallel to the central ray and tangent to spherical shells chosen to describe the depth variation of physical properties of the envelope. The intersections of the rays with the radial shells define a z-mesh along each ray.

As seen in figure 19.3, the ray p_i intersects the radial shell r_d at an angle whose cosine is

$$\mu_{di} \equiv \mu(r_d, p_i) = (r_d^2 - p_i^2)^{\frac{1}{2}}/r_d = z_{di}/r_d. \tag{19.37}$$

Hence, if one first constructs the solution along *all* rays $\{p_i\}$ and chooses a *particular value* r_d, then knowledge of the variation of $I_\nu(z_{di}, p_i)$ $(i = 1, \ldots, NI_d)$ is *equivalent to knowledge of the* μ*-variation of* $I_d(r_d, \mu)$ *on the mesh* $\{\mu_{di}\}$ $(i = 1, \ldots, NI_d)$, which spans the interval $1 \geq \mu \geq 0$. Here $NI_d \equiv NI + 1 - d$. Therefore, the ray-by-ray solution described above replaces an integration over angles μ by an integration over impact parameters and thus allows determination of the Eddington factors. The geometrical trick employed here to synthesize the angular information from ray solutions is actually nontrivial and makes direct use of the *symmetry* of the problem, which allows one to treat all points on a given shell as equivalent. Without strict spherical symmetry, the problem is more complex.

Consider now the ray specified by p_i. The transfer equation along the ray is

$$\pm[\partial I^{\pm}(z, p_i, \nu)/\partial z] = \eta(r, \nu) - \chi(r, \nu) I^{\pm}(z, p_i, \nu), \tag{19.38}$$

where the + and − signs refer, respectively, to radiation moving to or from the external observer. The radial variable r is understood to be a function of p_i and z, $r = (p_i^2 + z^2)^{\frac{1}{2}}$. One defines the optical depth along the ray, $d\tau(z, p_i, \nu) \equiv -\chi(r, \nu)\, dz$, and the source function $S(r, \nu) \equiv \eta(r, \nu)/\chi(r, \nu)$, which is assumed to be known. Introducing the mean-intensity-like and flux-like variables

$$u(z, p_i, \nu) \equiv \tfrac{1}{2}[I^+(z, p_i, \nu) + I^-(z, p_i, \nu)], \tag{19.39}$$

and

$$v(z, p_i, \nu) \equiv \tfrac{1}{2}[I^+(z, p_i, \nu) - I^-(z, p_i, \nu)], \tag{19.40}$$

one obtains the second-order system

$$\partial^2 u(z, p_i, \nu)/\partial \tau^2(z, p_i, \nu)] = u(z, p_i, \nu) - S[r(z, p_i), \nu], \tag{19.41}$$

with an upper (outer) boundary condition

$$\partial u(z, p_i, \nu)/\partial \tau(z, p_i, \nu)|_{z_{\max}} = u(z_{\max}, p_i, \nu), \tag{19.42}$$

where $z_{\max} = (R^2 - p_i^2)^{\frac{1}{2}}$. The inner boundary condition depends upon whether the ray intersects the core or not. In the former case, one applies the diffusion approximation as was done to obtain equation (19.35); in the latter, one has $v(0, p_i, z) \equiv 0$ because of the symmetry of the problem; hence

$$\partial u(z, p_i, \nu)/\partial \tau(z, I, \nu)|_{z=0} = 0. \tag{19.43}$$

Equations (19.41)–(19.43) are discretized to yield a single tridiagonal system of the standard Feautrier form, which is solved by the usual algorithm. The computing time for N frequencies, C core rays, and D radial points scales as $t \sim cN(D{\cdot}C + \sum D_i) \approx c'ND^2$ for $D \gg C$. Having calculated the complete solution $u_{din} \equiv u(z_d, p_i, \nu_n)$, one can construct the moments

$$J_{dn} = \sum_{i=1}^{NI_d} w_{di}^{(0)} u_{din}, \tag{19.44}$$

$$K_{dn} = \sum_{i=1}^{NI_d} w_{di}^{(2)} u_{din}, \tag{19.45}$$

and thus the Eddington factor $f_{dn} \equiv K_{dn}/J_{dn}$. Here w's are appropriate quadrature weights, obtained analytically by integration of moments of a piecewise polynomial representation of $u(r_d, \mu)$ on the mesh $\{\mu_i\}$ generated by the intersections of the rays $\{p_i\}$ with the radial shell r_d [550]. Using the new Eddington factors, the moment equations are re-solved, and the whole process is iterated to convergence. In most cases, convergence of the process is very fast.

The above scheme can be employed to solve general problems where the source function contains a frequency-dependent scattering term of the form $S_\nu = \alpha_\nu \int R(\nu', \nu) J_{\nu'} d\nu' + \beta_\nu$ that occurs in the case of partial redistribution (cf. chapter 15), or even for angle-dependent source function, as long it is a symmetric function of μ, $S_\nu(-\mu) = S_\nu(\mu)$.

In the case that the source function contains a single frequency-independent scattering integral, as for instance for a two-level or equivalent two-level atom where the source function is given through $\bar{J}$, one may avoid the iterative solution between the ray equations and moment equations and develop a *direct* solution using a Rybicki-type scheme [738]. Along each ray $p_i (i = 1, \ldots, NI)$ and for each frequency $\nu_n (n = 1, \ldots, N)$, one has a tridiagonal system of the form

$$\mathbf{T}_{in}\mathbf{u}_{in} = \mathbf{U}_{in}\bar{\mathbf{J}} + \mathbf{W}_{in}, \tag{19.46}$$

where the vector $\bar{\mathbf{J}} \equiv (\bar{J}_1, \ldots, \bar{J}_D)^T$ describes the depth variation of $\bar{J}$ and $\mathbf{u}$ the variation of $u(z, p_i, \nu_n)$ along the ray. $\mathbf{T}$ is a tridiagonal matrix, and $\mathbf{W}$ a vector of the right-hand side that contains the thermal part of the source function. Equation (19.46) is solved for $\mathbf{u}$ as

$$\mathbf{u}_{in} = \mathbf{A}_{in}\bar{\mathbf{J}} + \mathbf{B}_{in}, \tag{19.47}$$

where $\mathbf{A}_{in} = \mathbf{T}_{in}^{-1} \cdot \mathbf{U}_{in}$, and $\mathbf{B}_{in} = \mathbf{T}_{in}^{-1} \cdot \mathbf{W}_{in}$. This solution is substituted into the definition of $\bar{J}$,

$$\bar{J}_d = \sum_{n=1}^{N} w_n \phi_{dn} \sum_{i=1}^{NI_d} w_{di}^{(0)} u_{din}, \tag{19.48}$$

(w_n being the frequency quadrature weights) to yield the final system for $\bar{\mathbf{J}}$, namely,

$$\mathbf{C}\bar{\mathbf{J}} = \mathbf{D}. \tag{19.49}$$

The computing time for this solution scales as $t \sim cND^3$ for $D \gg C$. Details of this procedure, which is general, stable, and economical, can be found in the reference cited.

Static Extended Model Atmospheres

Static LTE spherically symmetric model atmospheres were computed in [188, 189, 196]. The first NLTE model atmospheres of static spherically symmetric atmospheres were constructed in [738], using a generalization of the complete linearization for static atmospheres [53] (cf. § 18.2). The same method was employed in [411] to develop an independent code for static spherical NLTE atmospheres. A generalization of the multi-frequency/multi-gray method [29, 31] was reported in [906], and an application of the idea of the accelerated lambda iteration method was developed in [640].

The extension effects are caused by a near-cancellation of gravity by radiation forces on the material. However, there is strong evidence that atmospheric extension is essentially always associated with large-scale expansion; hence static models can be expected to yield, at best, only qualitative information. We shall briefly discuss these models here, while we will consider *dynamical* models, which are more difficult to construct, in chapter 20.

Of all the structural equations, only the radiative transfer equation and the hydrostatic equilibrium equations are non-local, and hence their form depends on the choice of coordinate system. The remaining equations—radiative equilibrium, kinetic equilibrium equations (in NLTE), and charge conservation equation—are *local* equations, and hence are written in the same form as in the case of static plane-parallel atmospheres.

The hydrostatic equilibrium equation, which has to account for a variation of the gravitational force with radius, $g(r) = G\mathsf{M}/r^2$, with M being the stellar mass, becomes

$$dp_g/dr = -\rho G\mathsf{M}/r^2 + (4\pi/c) \int_0^\infty \chi_\nu H_\nu \, d\nu. \tag{19.50}$$

A more convenient form of this equation is

$$dp_g/d(1/r) = \rho \left[G\mathsf{M} - (4\pi r^2/c) \int_0^\infty (\chi_\nu/\rho) H_\nu \, d\nu \right]. \tag{19.51}$$

As suggested in [738], a numerically stable form is obtained by replacing χ_ν with $\chi_R(\chi_\nu/\chi_R)$, where χ_R is the Rosseland mean opacity. The quantity in parentheses is essentially a "profile function" for the frequency dependence of the opacity and should be insensitive to small changes in state parameters during an iteration process. Similarly, one replaces H_ν with $H(H_\nu/H^0)$, where H is the nominal total flux, $H \equiv \mathsf{L}/(16\pi^2 r^2)$, L being the stellar luminosity, and H^0 is the current integrated flux,

$H^0 = \int_0^\infty H_\nu \, d\nu$. With these replacements, the hydrostatic equilibrium equation can be written in the useful form

$$\frac{dp_g}{d(1/r)} = \rho \left(GM - \frac{L \chi_R \gamma}{4\pi c \rho} \right), \tag{19.52}$$

where

$$\gamma \equiv \int_0^\infty (\chi_\nu H_\nu)/(\chi_R H^0) \, d\nu. \tag{19.53}$$

Quantity γ is an example of a *form factor* (another example being the Eddington factor), which is a ratio of two similar quantities computed for the current values of the state parameters and held fixed during the next step of an iteration process.

An upper boundary condition poses a problem, which reflects an intrinsic physical inconsistency of the spherical static configuration. For an isothermal sphere the mass integrated to infinity is infinite, the pressure and density at infinity do not vanish, and the optical depth to infinity diverges [188]. One has to adopt some kind of cutoff procedure that yields a reasonable boundary pressure. The approach suggested in [738] is based on the assumption that all material properties except pressure and density are constant for $r > r_1$ (r_1 being the uppermost radial layer considered in the atmosphere), while the gas pressure and density are forced to decay with a constant scale height h, so that $\rho = \rho_1 \exp[-(r - r_1)/h]$, and analogously for p_g. One chooses an arbitrary value of τ_1 that represents a radial optical depth from r_1 to infinity and calculates the corresponding r_1 assuming that the material is completely ionized and that all the opacity comes from electron scattering. This allows one to express the optical depth τ_1 through the density, scale height, and parameter s_e as $\tau_1 = s_e \rho_1 h$, where $s_e = (\sigma_e/m_{\rm H})\alpha$, σ_e is the Thomson cross section, $m_{\rm H}$ is the mass of the hydrogen atom, and α is a correction for electrons provided by species other that hydrogen; in the case of H-He atmosphere with helium abundance Y, one has $\alpha = (1 + 2Y)/(1 + 4Y)$. This expression is easily generalized to the case of an arbitrary chemical composition and for incomplete ionization. The desired upper boundary condition is then written as

$$p_1 = (\tau_1 / s_e r_1^2)[GM - (\gamma L s_e)/(4\pi c)]. \tag{19.54}$$

In the application of standard complete linearization [738], the state vector at a given depth is analogous to that for plane-parallel atmospheres, equation (18.58), namely,

$$\boldsymbol{\psi}_d = \{J_1, \ldots, J_{NF}, N, T, n_{\rm e}, n_1, \ldots, n_{NL}\}, \tag{19.55}$$

where J_i is the mean intensity of radiation in the ith frequency point; we have omitted the depth subscript d. The dimension of the vector ψ_d is NN, $NN = NF + NL + NC$, where NF is the number of frequency points, NL the number of atomic energy levels for which the rate equations are solved, and NC the number of constraint equations. One discretizes and linearizes the radiative transfer equation, (19.31), together with boundary conditions (19.33) and (19.35); the hydrostatic equilibrium equation (19.52) with boundary condition (19.54); the radiative equilibrium equation (18.12), kinetic equilibrium equations (18.14), and the charge conservation equation (18.24); and ends up with the usual block tridiagonal system for the corrections to the state vector (for details and actual expressions, refer to [738]):

$$-\mathbf{A}_d\,\delta\boldsymbol{\psi}_{d-1} + \mathbf{B}_d\,\delta\boldsymbol{\psi}_d - \mathbf{C}_d\,\delta\boldsymbol{\psi}_{d+1} = \mathbf{L}_d, \tag{19.56}$$

which is solved by the standard elimination, exactly as in the case of plane-parallel atmospheres.

An application of the ALI method to construction of spherical static NLTE model atmospheres was developed in [640]. The approach represents a generalization of the ALI scheme employed for static atmospheres by [1148]. The mean intensities of radiation are eliminated from the state vector by expressing them through the approximate Λ^*-operator as [cf. equation (18.110)]

$$J_\nu^{(n)} = \Lambda_\nu^*[S_\nu^{(n)}] + \left(\Lambda_\nu - \Lambda_\nu^*\right)[S_\nu^{(n-1)}] \equiv \Lambda_\nu^*[S_\nu^{(n)}] + \Delta J_\nu, \tag{19.57}$$

and expressing the corrections as

$$\delta J_\nu^{(n)} = \Lambda_\nu^* \delta[S_\nu^{(n)}] + \Delta J_\nu, \tag{19.58}$$

where the corrections to the source function are given by [cf. equation (18.116)]

$$\delta S_k = \frac{\partial S_k}{\partial T}\delta T + \frac{\partial S_k}{\partial n_{\rm e}}\delta n_{\rm e} + \sum_{i=1}^{NL} \frac{\partial S_k}{\partial n_i}\delta n_i. \tag{19.59}$$

For details, refer to [640]. In fact, the approach developed in this reference reduces the state vector even further by eliminating level populations from the state vector by using implicit linearization

$$\delta n_i = \sum_m (\partial n_i/\partial \psi_m)\delta\psi_m, \tag{19.60}$$

where

$$(\partial n_i/\partial \psi_m) = \sum_j (\mathcal{A}^{-1})_{ij}[(\partial b_i/\partial \psi_m) - (\partial \mathcal{A}_{im}/\partial \psi_m) n_m], \tag{19.61}$$

where ψ_m are the individual components of the state vector, which is now reduced to a two-component vector $\{T, n_{\rm e}\}$ (plus possibly a third component, r; depending on details of the linearization strategy), and $\mathcal{A}_{ij}$ and b_j are the elements of the rate matrix and the right-hand side vector of the set of kinetic equilibrium equations (cf. § 18.1). A drawback of this approach is that it neglects a nonlinearity of the rate matrix in the atomic level populations when it is formulated within the ALI formalism (cf. § 18.4).

In principle, any method used for plane-parallel atmospheres can be applied, such as the direct application of the Werner's [1148] ALI-based scheme with explicit linearization of the level populations, the hybrid CL/ALI method [533], or the approximate Newton-Raphson scheme [492, 493]. These methods were not used to build specific programs designed for computing static spherical models because, as was mentioned earlier, atmospheric extension is essentially always accompanied by mass outflows (velocity fields); hence constructing static extended model atmospheres has only a limited meaning. In fact, a usual strategy to compute static extended models became to employ computer programs designed for extended moving atmospheres, setting the velocity to a very low value, typically a fraction of the thermal velocity.

19.2 MOVING ATMOSPHERES: OBSERVER'S-FRAME FORMULATION

The most straightforward approach to solve the problem of radiative transfer in a moving atmosphere is to formulate the transfer equation in the *inertial*, or *observer's, frame*. When the material in the atmosphere moves with velocity $\mathbf{v}(\mathbf{r})$ relative to an external observer at rest, there is a *Doppler shift* between frequencies measured in the observer's frame and the local frame in which a radiating atom is at rest. A photon with frequency ν traveling in direction $\mathbf{n}$, as measured in the observer's frame, has in the atom's frame the frequency

$$\nu' = \nu - \nu_0(\mathbf{n} \cdot \mathbf{v}/c). \tag{19.62}$$

Because of this intrinsic dependence on direction, the opacity and emissivity of the material, as seen by external observer, become *angle-dependent*.

Planar Geometry

The transfer equation for a time-independent moving medium in planar geometry is then

$$\mu[\partial I(z, \mu, \nu)/\partial z] = \eta(z, \mu, \nu) - \chi(z, \mu, \nu) I(z, \mu, \nu). \tag{19.63}$$

When describing a spectral line, it is convenient to measure frequency displacements from the line center in units of a Doppler width, $\Delta\nu_D \equiv \nu_0 v_{\rm th}/c$, and to measure velocities in the same units, $V = v/v_{\rm th}$. Here, $v_{\rm th}$ is the thermal velocity. Equation (19.62) then becomes

$$x' = x - \mu V, \tag{19.64}$$

where $x \equiv (\nu - \nu_0)/\Delta\nu_D$, and x' is defined similarly. We will assume the material velocity to be such that the effects of Doppler shifts on continuum are negligible, because the opacity and the emissivity in the continua do not vary much over the frequency range implied by velocity shifts unless the velocities become extremely large. Hence we account only for changes of frequency in line terms and write

$$\chi(z,\mu,x) = \chi_c(z) + \chi_l(z)\phi(z,\mu,x), \tag{19.65}$$

$$\eta(z,\mu,x) = \eta_c(z) + \eta_l(z)\phi(z,\mu,x), \tag{19.66}$$

where the normalized line profile coefficient is defined by

$$\phi(z,\mu,x) \equiv \phi(z; x - \mu V), \tag{19.67}$$

for example, for a Doppler profile,

$$\phi(z,\mu,x) = \pi^{-\frac{1}{2}} \exp\{-[x - \mu V(z)]^2\}. \tag{19.68}$$

We define line and continuum source functions

$$S_l(z) = \eta_l(z)/\chi_l(z), \tag{19.69}$$

$$S_c(z) = \eta_c(z)/\chi_c(z). \tag{19.70}$$

The total source function is then written

$$S(z,\mu,x) = \frac{\phi(z,\mu,x)S_l(z) + r(z)S_c(z)}{\phi(z,\mu,x) + r(z)}, \tag{19.71}$$

where $r(z) \equiv \chi_c(z)/\chi_l(z)$. Finally, we define the optical depth scale measured along a ray specified by μ,

$$\tau(z,\mu,x) \equiv \int_z^{z_{\max}} \chi(z',\mu,x)\, dz'/\mu, \tag{19.72}$$

where $z_{\max}$ denotes the upper surface of the atmosphere. Then the transfer equation becomes

$$[\partial I(z,\mu,x)/\partial\tau(z,\mu,x)] = I(z,\mu,x) - S(z,\mu,x). \tag{19.73}$$

Formal solution of this equation is given by

$$I(z_{\max},\mu,x) = I(0,\mu,x)e^{-\tau(0,\mu,x)} + \int_0^{\tau(z,\mu,x)} S(z,\mu,x)e^{-\tau(z,\mu,x)}\, d\tau(z,\mu,x). \tag{19.74}$$

Equation (19.74) is written for a finite slab with a given incident intensity at $z = 0$; for a semi-infinite atmosphere we set $\tau(0,\mu,x) = \infty$ and omit the first term with $I(0,\mu,x)$. Equations (19.72) to (19.73) are completely analogous to the corresponding equations for static planar atmosphere; the only difference is an explicit dependence of the source function on μ. The formal solution (19.74) allows a direct evaluation of the effects of velocity fields on the emergent intensity by accounting for the frequency shifts induced by velocities in the opacity and the emissivity of the material.

Equation (19.74) represents a *formal solution*, i.e., a solution of the transfer problem when the source function is fully specified. However, the line source function will, in general, contain a scattering term and therefore depend on the radiation field. Hence the source function can be strongly affected by the material motions. For example, an expansion of an atmosphere can displace a line away from a dark absorption feature into the bright continuum nearby, thus raising the mean intensity, and hence the line source function, dramatically. In other words, the photon escape probability increases enormously in moving atmospheres, because a photon that would experience a large number of scatterings in the line core in a static atmosphere can have its frequency shifted away from the core, and hence escape after a few scatterings. This observation is a basis of the *Sobolev approximation*, which will be described in detail in § 19.5.

To understand difficulties connected with the observer's-frame formulation of the transfer problem, consider a simple case of a two-level atom. The line source function is given by

$$S_l(z) = \tfrac{1}{2}(1-\epsilon)\int_{-\infty}^{\infty} dx \int_{-1}^{1} d\mu\, I(z,\mu,\nu)\phi(z,\mu,\nu) + \epsilon B, \tag{19.75}$$

where ϵ is the usual thermalization parameter. Note that in the scattering term one can no longer replace I with J because the line profile coefficient ϕ is angle-dependent. Note also that the intensity can no longer be assumed to be symmetric around the line center; hence the *full* profile has to be considered. Accurate evaluation of the scattering integral in equation (19.75) with a quadrature sum poses a numerical difficulty for two reasons: (1) The line profile coefficient $\phi(x-\mu V)$ is shifted by an amount $2V$ in frequency as μ varies from -1 to 1. Thus, in the frequency quadrature, an amount equal to twice the maximum macroscopic velocity must be added to the line bandwidth required to cover the static line profile. This requirement is not severe in studies involving macroscopic velocities up to few times the thermal velocity but may become prohibitive for atmospheres in supersonic motion where $v/c \sim 0.01$ or $2(\nu_0 v/c)/\Delta\nu_D \sim 200$. (2) The angle quadrature must employ a large number of angles. Because the argument of the line-profile coefficient is $(x-\mu V)$, there is an *inextricable coupling between the angular and frequency variations of the intensity*. Hence, if some maximum frequency increment $\Delta x_{\max}$ is required to obtain a reasonable accurate frequency quadrature (typically, $\Delta x_{\max} \sim \frac{1}{2}$), the corresponding maximum increment in the angle quadrature will be $\Delta\mu_{\max} = \Delta x_{\max}/V$, which can be quite stringent. These difficulties are largely avoided by transforming to the comoving frame (cf. § 19.3).

However, the above-mentioned numerical difficulties were truly prohibitive in the past when the computer speed and memory were relatively low. When computer capabilities increase, the difficulties with the observer's-frame formulation of the transfer problem are no longer stringent; in fact, the conceptual simplicity of the observer's-frame approach becomes advantageous. For instance, a generalization of the problem to more spatial dimensions is much easier in the observer's frame than in the comoving frame, which becomes very cumbersome in two or three dimensions. Also, in one-dimensional atmospheric models, due to the fact that modern NLTE line-blanketed models have to consider essentially the full frequency range with a sufficient resolution (cf. § 18.5), both difficulties associated with the observer's-frame formulation do not bring any significant additional problems.

As in the static case, it is convenient to cast equation (19.73) into second-order form. If the line profile is symmetric about the line center, then $\phi(-x+\mu V)=\phi(x-\mu V)$, which suggests that we should group together $I(z,\mu,x)$ and $I(z,-\mu,-x)$, because $d\tau(z,\mu,x)=d\tau(z,-\mu,-x)$ and $S(z,\mu,x)=S(z,-\mu,-x)$. Hence we introduce

$$u(z,\mu,x)\equiv\tfrac{1}{2}[I(z,\mu,x)+I(z,-\mu,-x)], \tag{19.76}$$

$$v(z,\mu,x)\equiv\tfrac{1}{2}[I(z,\mu,x)-I(z,-\mu,-x)], \tag{19.77}$$

so that

$$[\partial^2 u(z,\mu,x)/\partial\tau^2(z,\mu,x)] = u(z,\mu,x)-S(z,\mu,x). \tag{19.78}$$

The upper boundary condition, in the case of no incident radiation, becomes

$$[\partial u(z,\mu,x)/\partial\tau(z,\mu,x)]_{z_{\max}} = u(z_{\max},\mu,x). \tag{19.79}$$

At the lower boundary we assume either that the incident radiation is *specified*, in which case

$$[\partial u(z,\mu,x)/\partial\tau(z,\mu,x)]_{z=0} = I(0,\mu,x)-u(0,\mu,x), \tag{19.80}$$

or that, in a semi-infinite atmosphere, the lower boundary is chosen to be so deep that the *diffusion approximation* is valid, which demands that the velocity gradient is small enough that $\chi^{-1}(dV/dz)\ll 1$; i.e., there is a negligible change of velocity over a photon mean free path. Then

$$\left.\frac{\partial u(z,\mu,x)}{\partial\tau(z,\mu,x)}\right|_{z=0} = \left[\frac{\mu}{\chi(z,\mu,x)}\left(\frac{\partial B_\nu}{\partial T}\right)\left|\frac{dT}{dz}\right|\right]_{z=0}. \tag{19.81}$$

As in the static case, we introduce a discrete depth-mesh, $\{z_d\}$ $(d=1,\ldots,D)$, angle-mesh, $\{\mu_m\}$ $(m=1,\ldots,M)$, and frequency-mesh, $\{x_n\}$ $(n=1,\ldots,N)$. The angle points are distributed on the interval $[0,1]$, and frequencies must span a range $[x_{\min},x_{\max}]$, with $x_{\min}<0$ and $x_{\max}>0$, large enough to contain the whole line profile and to allow for Doppler shifts $\pm 2V_{\max}$. As usual, the frequency and angle quadratures are combined together; the corresponding quadrature set is denoted as $\{\mu_l,x_l\}\equiv\{\mu_m,\nu_n\}$ where $l=m+(n-1)M$; $(l=1,\ldots,L)$ with $L=MN$. We then replace equations (19.78)–(19.81) with difference equations and write

$$S_{dl}=S(z_d,\mu_l,\nu_l)=\alpha_{dl}\bar{J}_d+\beta_{dl}, \tag{19.82}$$

where α and β are the appropriate functions of r_d, ϕ_{dl}, and ϵ_d, and

$$\overline{J}_d \equiv \sum_{l=1}^{L} w_l \phi_{dl} u_{dl}, \qquad (19.83)$$

where $\phi_{dl} \equiv \phi(z_d, x_l - \mu_l V_d)$. The resulting system is then of the *standard Rybicki form* [cf. equation (12.72)] and is solved for $\overline{\mathbf{J}}$ as described in §12.3. An analogous integral-equation approach can also be devised [408, p. 120], but in applications the differential-equation method described here is easier to use. The procedure is stable, general, and quite efficient. The computing time scales as $t \approx cLD^2 + c'D^3$, so it is only *linear* in the number of angle-frequency points. The depth-mesh should be chosen sufficiently dense to ensure that only modest changes in $V(z_d)$, say $\lesssim \frac{1}{2}$, occur between successive depth points; otherwise, ϕ_{dl} may exhibit unacceptably large changes between depth points, which would lead to significant inaccuracies in the optical depth increments. Except for highly supersonic winds, this is not a very stringent requirement. Note also that the same method can be used to construct the *formal solution*, when S is given, by solving a single tridiagonal system of the form $\mathbf{T}_l \mathbf{u}_l = \mathbf{S}_l$ at each angle-frequency point required. The computing time required is only $t \approx cLD$, which is minimal.

Spherical Geometry

In spherical geometry the equation of transfer in the observer's (inertial) frame has the same form as the transfer equation for static atmospheres; the only difference being that the absorption and emission coefficients depend on frequency as well as angle,

$$\mu \frac{\partial I(r,\mu,\nu)}{\partial r} + \frac{(1-\mu^2)}{r} \frac{\partial I(r,\mu,\nu)}{\partial \mu} = \eta(r,\mu,\nu) - \chi(r,\mu,\nu) I(r,\mu,\nu). \qquad (19.84)$$

The method of solution is analogous to to that described in § 19.1 for static extended atmospheres. It is based on a ray-by-ray solution in the same (p,z) coordinate system. The transfer equation along the ray is

$$\pm[\partial I^{\pm}(z,p,x)/\partial z] = \eta(z,p,x) - \chi(z,p,x) I^{\pm}(z,p,x), \qquad (19.85)$$

where $\chi(z,p,x) = \chi_c(r) + \chi_l(r)\phi(z,p,x)$, and a similar expression defines $\eta(z,p,x)$; we use the relations $r(z,p) = (z^2+p^2)^{\frac{1}{2}}$ and $\mu(z,p) = z/(z^2+p^2)^{\frac{1}{2}}$. The profile coefficient is defined as $\phi(z,p,x) \equiv \phi[r(z,p), x - \mu(z,p)V(r)]$. In the case of a *radially expanding atmosphere*, $V(r)$ is positive in the direction of increasing r. Introducing the optical depth along the ray as

$$\tau(z,p,x) \equiv \int_z^{z_{\max}} \chi(z',p,x)\, dz', \qquad (19.86)$$

and defining

$$u(z,p,x) \equiv \tfrac{1}{2}[I^+(z,p,x) + I^-(z,p,-x)], \tag{19.87}$$

$$v(z,p,x) \equiv \tfrac{1}{2}[I^+(z,p,x) - I^-(z,p,-x)], \tag{19.88}$$

equation (19.85) can be rewritten in the second-order form

$$[\partial^2 u(z,p,x)/\partial\tau^2(z,p,x)] = u(z,p,x) - S(z,p,x), \tag{19.89}$$

where $S(z,p,x) = \eta(z,p,x)/\chi(z,p,x)$.

In formulating the boundary conditions a difficulty arises. On the axis $z = 0$ one can no longer write $v(0,p,x) = 0$ because *two* frequencies, $(\pm x)$, of radiation are involved. The way to circumvent the problem is to follow the ray for the entire length—i.e., to consider the whole interval $[-z_{\max}, z_{\max}]$. The lower and upper boundary conditions for rays that do not intersect the core then become

$$[\partial u(z,p,x)/\partial\tau(z,p,x)]_{\pm z_{\max}} = \pm u(z,p,x)_{\pm z_{\max}}. \tag{19.90}$$

For rays that intersect the core, $p \le r_c$, we either apply a *diffusion approximation* at an opaque core (stellar surface), which yields $u(z_{\min}, \mu, x)$ directly, or for a *hollow core* (nebular case), apply equation (19.89) at $z_{\min}$ (forcing the points at $\pm z_{\min}$ to be identical) and equation (19.90) at the ends of the ray.

To solve the system one introduces the same discrete meshes $\{r_d\}$ and $\{p_i\}$ used in § 19.1 to solve the static problem. The frequency mesh now includes the *whole* profile $\{x_n\}$, $n = \pm 1, \ldots, \pm N$, with $x_{-n} = -x_n$; one shall, however, be able to eliminate half of these (see below). One again obtains an equation in the form of (19.46),

$$\mathbf{T}_{in}\mathbf{u}_{in} = \mathbf{U}_{in}\bar{\mathbf{J}} + \mathbf{W}_{in}, \tag{19.91}$$

and hence can apply the Rybicki method to obtain $\bar{J}$. Because $\bar{J}(r_d)$ need be defined only for $\{r_d\}$, $1 \le d \le D$, while $u_{din} = u(z_d, p_i, x_n)$ is defined on a mesh $\{z_{d_i}\}$, $d_i = 1, \ldots, D_i$, which runs the whole length of the ray, it now turns out that, while the tridiagonal **T**-matrix is square, the **U**-matrix is *rectangular* and is a *chevron* matrix. Solution of these systems for each choice of (i, n) yields an expression of the form

$$\mathbf{u}_{in} = \mathbf{A}_{in}\bar{\mathbf{J}} + \mathbf{B}_{in}. \tag{19.92}$$

Equation (19.48) defining $\bar{\mathbf{J}}$ can be written in the discrete form

$$\bar{J}(r_d) = \sum_{n=-N}^{N} w_n \sum_{i=1}^{NI_d} a_{di}\phi[r_d, x_n - \mu(r_d, p_i)V(r_d)]\, u_{din}. \tag{19.93}$$

But, from the spherical symmetry of the problem, $I^\pm(z,p,x) = I^\pm(-z,p,x)$, and thus $u(z,p,-x) = u(-z,p,x)$, and $v(z,p,-x) = -v(z,p,-x)$. These relations allow elimination of the values of u at negative x and positive z, in equation (19.93), in terms of u at positive x and negative z. Thus

$$\overline{J} = \sum_{n=1}^{N} w_n \sum_{i=1}^{NI_d} a_{di} \left[\phi(r_d, x_n - \mu_{di} V_d)\, u_{din} + \phi(r_d, x_n + \mu_{di} V_d)\, u_{d'in}\right], \quad (19.94)$$

where $d' = D_i + 1 - d$. Equation (19.94), when used in the Rybicki method, yields **V**-matrices that are rectangular chevron matrices. Using equations (19.92) for all values of i and n in equation (19.94), one obtains a final system for $\overline{\mathbf{J}}$, which is then solved. The computing time required for the solution scales as $t \approx cND^3 + c'D^3$; this is less favorable than the scaling for the planar case, because there are now about as many angles (i.e., impact parameters) as there are depths.

The method is stable and easy to use and very efficient for small velocities (up to a few times thermal). For larger velocities, the number of depth points needed to resolve the velocity field becomes large; in this case one may use a comoving-frame method (§ 19.3) or use an iterative (ALI) scheme described below. A significant advantage of the observer's-frame method is that it can be used for arbitrary variations of the velocity field, such as non-monotonic flows, which is not true for comoving-frame methods as currently formulated.

Iterative Methods

Although one can obtain a direct solution of the problem as explained above, it is still advantageous to employ iterative schemes that can be significantly faster than a direct solution. The most efficient methods are based, as in the case of no velocity field, on the ideas of the accelerated lambda iteration. Its application to the radiative transfer problem with velocity fields in the observer's-frame formulation is a direct generalization of the approaches described in § 13.3 and will be only briefly outlined here.

We assume for simplicity a two-level atom source function without continuum, $S = (1-\epsilon)\overline{J} + \epsilon B$. A generalization is straightforward and is done analogously as was explained in chapter 14. A formal solution for the specific intensity at frequency ν and angle μ can be written as

$$I_{\mu\nu} = \Lambda_{\mu\nu}[S], \quad (19.95)$$

and for the averaged mean intensity as

$$\overline{J} = \Lambda[S], \quad (19.96)$$

where $\overline{J} = \frac{1}{2}\int_0^\infty \int_{-1}^1 \phi(\mu,\nu) I(\mu,\nu)\, d\mu\, d\nu$, and

$$\Lambda = \frac{1}{2}\int_0^\infty \int_{-1}^1 \phi(\mu,\nu)\Lambda_{\mu\nu}\, d\mu\, d\nu. \quad (19.97)$$

The exact Λ-operator is split into two parts,

$$\Lambda = \Lambda^* + (\Lambda - \Lambda^*), \tag{19.98}$$

with Λ^* being a suitably chosen approximate operator. In an iteration process, $\overline{J}$ is written as $\overline{J}^{\rm new} = \Lambda^*[S^{\rm new}] + (\Lambda - \Lambda^*)[S^{\rm old}]$, and the iterative procedure to evaluate the source function is set up as [cf. equation (13.5)]

$$S^{\rm new} = (1-\epsilon)\Lambda^*[S^{\rm new}] + (1-\epsilon)(\Lambda - \Lambda^*)[S^{\rm old}] + \epsilon B, \tag{19.99}$$

or

$$[I - (1-\epsilon)\Lambda^*][S^{\rm new} - S^{\rm old}] = S^{\rm FS} - S^{\rm old}, \tag{19.100}$$

where $S^{\rm FS}$ is a "newer" source function obtained from the old source function by a formal solution,

$$S^{\rm FS} = (1-\epsilon)\Lambda[S^{\rm old}] + \epsilon B. \tag{19.101}$$

The main ingredient of the method is a construction of the approximate Λ^*-operator (matrix), which is obtained by an integration of the elementary $\Lambda^*_{\mu\nu}$-operators over all frequencies and angles. In the discretized form, an integration over frequencies is treated as a quadrature sums, and in the tangent-ray approach, an integration over angles is replaced with an appropriate summation over the impact parameters, as explained earlier. In analogy with equation (19.93) one has

$$\Lambda^* = \sum_n w_n \sum_i a_i \phi_{in} \Lambda^*_{in}, \tag{19.102}$$

where Λ^*_{in} is an approximate operator corresponding to impact parameter i and frequency point n, In the following, we drop the subscript i indicating the given tangent ray.

As was discussed at length in chapter 13, there are many different types of an approximate operator. The simplest possibility is to use a *local* operator, whose optimum form is a diagonal part of the exact Λ-operator. This is a fundamental result obtained in [815]. The diagonal elements are generally given by

$$\Lambda^*_{dd} = (\Lambda[\mathbf{e}_d])_d = \sum_j \Lambda_{dj}\delta_{dj}, \tag{19.103}$$

where $\mathbf{e}_d$ is a vector having all components zero except at d; $(\mathbf{e}_d)_{d'} = \delta_{dd'}$, where δ is the Kronecker's δ-symbol. The diagonal values of the Λ-operator can thus be obtained by solving the transfer equation with the source function replaced with the unit-pulse function. As shown in § 13.3, the actual procedure depends on the adopted scheme of the formal solution (a solution of the transfer equation with known source function). We described earlier a formal solution based on a Feautrier-type approach; it should be stressed that a formal solution along the ray can be performed by any method described in § 12.4. If the short characteristics or the discontinuous finite element schemes are applied, one simply takes coefficients standing at the local value of the source function, exactly as in the case of no velocities.

If the Feautrier scheme is applied, the transfer equation is written as (dropping the frequency subscript n) $\mathbf{Tu} = \mathbf{S}$, where $\mathbf{S}$ is a vector of the source functions along the ray, and $\mathbf{T}$ is a tridiagonal matrix. The system has a solution $\mathbf{u} = \mathbf{T}^{-1}\mathbf{S}$, or in the component form $u_d = \sum_j (T^{-1})_{dj} S_j$. Substituting $\mathbf{S} = \mathbf{e}_d$, one obtains

$$\Lambda^*_{dd} \equiv u^*_d = \sum_j (T^{-1})_{dj}\delta_{dj} = (T^{-1})_{dd}. \tag{19.104}$$

The diagonal matrix elements of $\mathbf{T}^{-1}$ are evaluated using an efficient algorithm proposed by [925], described in § 13.3, which proceeds along with the formal solution with essentially no additional computational cost.

The ALI iteration scheme proceeds as follows.

For each impact parameter i:

(i) For an appropriate initial estimate of the source function $S^{(0)}$ perform a formal solution, for one frequency at a time, to obtain new values of the specific intensity (or the Feautrier intensity u_n), as well as the elements of the elements of the approximate Λ^*_{in}-operator.
(ii) Calculate the new value of the source function $S^{\rm FS}$ from (19.101), using the new computed values of the intensity.
(iii) Once this is done for all rays, construct the total Λ^* using equation (19.102) and evaluate a new iterate of the source function $S^{\rm new}$ from (19.100).
(iv) Repeat steps (i) to (iii) to convergence.

Since the computing time needed for step (iii) is negligible, the computation is dominated by the set of formal solutions for all rays and all frequencies. The total computing time per iteration scales as $t \approx c'ND^2$, so it is much more favorable than that for a direct solution if the number of iterations required is small. As discussed in § 13.5, this is indeed the case, in particular when one employs one of the acceleration schemes discussed in § 13.4.

19.3 MOVING ATMOSPHERES: COMOVING-FRAME FORMULATION

The formulation of the transfer equation discussed previously is based in the stationary frame of the observer, who views the material as moving. The complication in this approach is that the opacity and emissivity of the material become angle-dependent, owing to the effects of Doppler shifts and aberration of light. These effects lead to an inextricable coupling between angle and frequency that may present difficulties in the calculation of scattering terms using discrete quadrature schemes, in particular when the material velocities are large compared to the thermal velocity. It then may become attractive to treat the problem in a frame comoving with the fluid.

There are two motivations for working in the comoving, Lagrangian frame. (i) Both opacity and emissivity are isotropic in this frame. Further, in problems involving partial redistribution effects one may use standard static redistribution functions. Moreover, in calculation of scattering integrals one needs consider only a frequency bandwidth wide enough to contain the (static) line profile; this bandwidth is independent of the fluid velocity. Finally, the angle quadrature may be chosen on the basis of angular variation of the radiation alone. (ii) Hydrodynamical calculations of spherical flows often use the Lagrangian coordinate system, i.e., the comoving frame. It is thus desirable to treat the radiation field in a closely parallel way. Moreover, the quantities that describe an interaction of radiation and matter are related directly to proper variables that specify the thermodynamic state of the material. On the other hand, a disadvantage of the comoving-frame formulation is that present methods of solution work for relatively simple velocity fields (i.e., velocities monotonically increasing or decreasing in the radial direction); otherwise, it would be too difficult and cumbersome to pose boundary conditions on the problem.

The inertial-frame transfer equation for a spherically symmetric medium is

$$\frac{1}{c}\frac{\partial I(\mu,\nu)}{\partial t}+\mu\frac{\partial I(\mu,\nu)}{\partial r}+\frac{(1-\mu^2)}{r}\frac{\partial I(\mu,\nu)}{\partial \mu}=\eta(\mu,\nu)-\chi(\mu,\nu)I(\mu,\nu). \tag{19.105}$$

The goal of this approach is to rewrite equation (19.105) with all *material and radiation field quantities measured in the comoving frame*. To obtain expressions describing how the physical variables change between the inertial and the comoving frames, Lorentz transformations are applied. However, one encounters here a problem: strictly speaking, a Lorentz transformation applies only when the velocity v of one frame relative to the other is *uniform* and *constant*. But in stellar atmospheres the velocity fields generally depend on position and time, $\mathbf{v}=\mathbf{v}(\mathbf{r},t)$, and hence the fluid frame is *not an inertial frame*. Photon trajectories in the comoving frame are therefore not straight lines, but are *geodesics* whose shapes are determined by the metric of the curved spacetime through which the photons move. Furthermore, photon frequencies are not constant in this spacetime. Hence one has to account, in effect, for changes in the Lorentz transformation from one point of the medium to another. Corresponding terms can be derived by application of the differential operator $(c^{-1}\partial/\partial t+\mathbf{n}\cdot\nabla)$ in the transfer equation to the transformation coefficient of the specific intensity.

Lorentz Transformation of the Transfer Equation

We consider here transformations between the system at rest, specified by four coordinates $(x^0,x^1,x^2,x^3)=(ct,x,y,z)$, and the fluid system, specified by (ct_0,x_0,y_0,z_0), moving relative to the rest system with a constant velocity v in the z direction. This choice for v simplifies the formalism; generalizations for an arbitrary orientation of $\mathbf{v}$ are given in [1074]. Changes from one system to another are described by means of a *Lorentz transformation*, which corresponds to a proper rotation in four-dimensional spacetime. Physically, the Lorentz transformation is chosen in such a way that the equation for the wavefront of a light wave is of the same form (i.e., covariant) in both systems. In other words, the velocity of light is equal to c in both reference frames. Mathematically, the transformation is expressed in the form $x_0{}^\alpha = L_\beta{}^\alpha x^\beta$ $(\alpha = 0, \ldots, 3)$, where the Einstein convention of summing over repeated indices is employed. The transformation can also be expressed by means of the matrix

$$\mathbf{L} = \begin{pmatrix} \gamma & 0 & 0 & -\beta\gamma \\ 0 & 1 & 0 & 0 \\ 0 & 0 & 1 & 0 \\ -\beta\gamma & 0 & 0 & \gamma \end{pmatrix}, \tag{19.106}$$

where

$$\beta \equiv v/c, \tag{19.107}$$

$$\gamma \equiv (1 - \beta^2)^{-\frac{1}{2}}. \tag{19.108}$$

The inverse transformation from the comoving frame to the laboratory frame is $\mathbf{x} = \mathbf{L}^{-1}\mathbf{x}_0$, where

$$\mathbf{L}^{-1} = \begin{pmatrix} \gamma & 0 & 0 & \beta\gamma \\ 0 & 1 & 0 & 0 \\ 0 & 0 & 1 & 0 \\ \beta\gamma & 0 & 0 & \gamma \end{pmatrix}, \tag{19.109}$$

This result is to be expected on physical grounds (cf. Appendix A).

The Lorentz transformation can be applied to arbitrary four-vectors and four-tensors of rank two. The transformation rules of tensor analysis ensure that these quantities are covariant under the Lorentz transformation because it is a proper rotation in the four-space. Hence physical laws written in terms of four-vectors and four-tensors are *automatically covariant.*

The photon four-momentum is given by

$$P^\alpha = (h\nu/c)\, n^\alpha \equiv (h\nu/c)(1, n_x, n_y, n_z) \equiv (h\nu/c)(1, \mathbf{n}), \tag{19.110}$$

where $\mathbf{n}$ is the direction of photon propagation. Applying a Lorentz transformation to equation (19.110), one finds

$$[\nu_0, \nu_0 n_x{}^0, \nu_0 n_y{}^0, \nu_0 n_z{}^0] = [\nu\gamma(1 - n_z\beta), \nu n_x, \nu n_y, \nu\gamma(n_z - \beta)], \tag{19.111}$$

from which it follows that

$$\begin{aligned} &[\nu_0; \phi_0; (1 - \mu_0^2)^{\frac{1}{2}}; \mu_0] \\ &\quad = [\nu\gamma(1 - \mu\beta); \phi; \gamma^{-1}(1 - \mu^2)^{\frac{1}{2}}/(1 - \mu\beta); (\mu - \beta)/(1 - \mu\beta)]. \end{aligned} \tag{19.112}$$

Notice that (19.112) is not an equality of two four-vectors; it is instead a synoptic expression of four separate equalities. The inverse transformation gives

$$[\nu, \nu n_x, \nu n_y, \nu n_z] = [\nu_0\gamma(1 + n_z{}^0\beta), \nu_0 n_x{}^0, \nu_0 n_y{}^0, \nu_0\gamma(n_z{}^0 + \beta)], \tag{19.113}$$

From these equations, one easily obtains relationships between inertial- and comoving-frame frequencies and angles;

$$\nu = \nu_0 \gamma (1 + \beta \mu_0), \tag{19.114}$$

$$\mu = (\mu_0 + \beta)/(1 + \beta \mu_0), \tag{19.115}$$

$$(1 - \mu^2)^{\frac{1}{2}} = (1 - \mu_0^2)^{\frac{1}{2}} / [\gamma (1 + \beta \mu_0)], \tag{19.116}$$

and

$$\nu_0 = \nu \gamma (1 - \beta \mu_0), \tag{19.117}$$

$$\mu_0 = (\mu - \beta)/(1 - \beta \mu), \tag{19.118}$$

$$(1 - \mu_0^2)^{\frac{1}{2}} = (1 - \mu^2)^{\frac{1}{2}} / [\gamma (1 - \beta \mu)]. \tag{19.119}$$

Equations (19.117)–(19.119) express general results for Doppler shift and aberration; the classical expressions are obtained by keeping only terms to $O(v/c)$; i.e., setting $\gamma \equiv 1$.

From equations (19.111) and (19.113) it is seen that $d\nu_0 = (\nu_0/\nu)\, d\nu$, and $d\mu_0 = (\nu/\nu_0)^2\, d\mu$; hence using $d\Omega = \sin\theta\, d\theta d\phi = d\mu d\phi$, one finds

$$\nu\, d\nu\, d\Omega = \nu_0\, d\nu_0\, d\Omega_0. \tag{19.120}$$

Transformations for the specific intensity, opacity, and emissivity are derived in Appendix B (based on the original work of Thomas [1074]), namely,

$$I(\mu, \nu) = (\nu/\nu_0)^3 I^0(\mu_0, \nu_0), \tag{19.121}$$

$$\chi(\mu, \nu) = (\nu_0/\nu) \chi^0(\nu_0), \tag{19.122}$$

$$\eta(\mu, \nu) = (\nu/\nu_0)^2 \eta^0(\nu_0), \tag{19.123}$$

where all quantities with the suffix 0 are measured in the comoving frame.

Between the two frames moving *uniformly* with respect to one another, the differential operator in the transfer equation transforms as

$$c^{-1}(\partial/\partial t) + (\mathbf{n} \cdot \nabla) = (\nu_0/\nu)[c^{-1}(\partial/\partial t_0) + (\mathbf{n}^0 \cdot \nabla^0). \tag{19.124}$$

From equations (19.121)–(19.123) and (19.124) it is seen that the transfer equation,

$$c^{-1}(\partial I_\nu/\partial t) + (\mathbf{n} \cdot \nabla) I_\nu = \eta_\nu - \chi_\nu I_\nu, \tag{19.125}$$

transforms to

$$\begin{aligned}(\nu_0/\nu)[c^{-1}(\partial/\partial t_0) + (\mathbf{n}^0 \cdot \nabla^0)]\,[(\nu/\nu_0)^3 I^0(\mu_0, \nu_0)]\\ = (\nu/\nu_0)^2[\eta^0(\nu_0) - \chi^0(\nu_0) I^0(\mu_0, \nu_0)].\end{aligned} \tag{19.126}$$

If (ν/ν_0) is *constant*—as it will be if and only if the two frames are in *uniform* motion with respect to one another—equation (19.126) can be written as

$$[c^{-1}(\partial/\partial t_0) + (\mathbf{n}^0 \cdot \nabla^0)]I^0(\mu_0, \nu_0) = \eta^0(\nu_0) - \chi^0(\nu_0)I^0(\mu_0, \nu_0)]. \quad (19.127)$$

In this case, equations (19.125) and (19.127) are of the same *form*—the transfer equation is covariant. Two important points must be stressed. (i) Despite the similar form of the two equations, equation (19.127) is actually much simpler because of the *isotropy* of $\eta^0(\nu_0)$ and $\chi^0(\nu_0)$. (ii) The reduction of equation (19.126) to (19.127) is *not valid if the two frames do not move uniformly with respect to each other*; i.e., this equation does not apply in, for instance, an expanding or pulsating atmosphere. A general case will be considered below; we shall first discuss a transformation of moments of the radiation intensity.

Transformation of Moments

The total energy density, energy flux, and radiation stress tensor are specified by the frequency-integrated quantities [cf. equations (3.19), (3.71), (3.74)]

$$E(\mathbf{r}, t) = c^{-1} \int_0^\infty d\nu \oint d\Omega\, I(\mathbf{r}, \mathbf{n}, \nu, t), \quad (19.128)$$

$$\mathbf{F}(\mathbf{r}, t) = \int_0^\infty d\nu \oint d\Omega\, I(\mathbf{r}, \mathbf{n}, \nu, t)\, \mathbf{n}, \quad (19.129)$$

$$\mathsf{P}(\mathbf{r}, t) = c^{-1} \int_0^\infty d\nu \oint d\Omega\, I(\mathbf{r}, \mathbf{n}, \nu, t)\, \mathbf{nn}. \quad (19.130)$$

These are related by the frequency-integrated moment equations [cf. equations (11.49) and (11.50)]

$$c^{-2}(\partial \mathbf{F}/\partial t) + \nabla \cdot \mathsf{P} = c^{-1} \int_0^\infty d\nu \oint d\Omega\, [\eta(\mathbf{r}, \nu, t) - \chi(\mathbf{r}, \nu, t) I(\mathbf{r}, \nu, t)]\, \mathbf{n}, \quad (19.131)$$

and

$$(\partial E/\partial t) + \nabla \cdot \mathbf{F} = \int_0^\infty d\nu \oint d\Omega\, [\eta(\mathbf{r}, \nu, t) - \chi(\mathbf{r}, \nu, t) I(\mathbf{r}, \nu, t)]. \quad (19.132)$$

As written, equations (19.131) and (19.132) are already covariant. To see that, recall that the momentum-energy conservation laws of electrodynamics can be written in a covariant form,

$$R^{\alpha\beta}_{;\beta} = -G^\alpha, \quad (19.133)$$

where $R^{\alpha\beta}$ is the radiation stress-energy tensor and G^α the radiation four-force density. As mentioned in § 3.5, the tensor R is given by

$$\mathsf{R} = \begin{pmatrix} E & c^{-1}\mathbf{F} \\ c^{-1}\mathbf{F} & \mathsf{P} \end{pmatrix}, \quad (19.134)$$

or, in the component form,

$$R^{\alpha\beta} = c^{-1}\int_0^\infty d\nu \oint d\Omega\, I(\mathbf{n},\nu)n^\alpha n^\beta. \tag{19.135}$$

The radiation four-force density is given by

$$G^\alpha = c^{-1}\int_0^\infty d\nu \oint d\Omega\, [\chi(\mathbf{n},\nu)I(\mathbf{n},\nu) - \eta(\mathbf{n},\nu)]\, n^\alpha. \tag{19.136}$$

One can verify that R is a four-tensor by noting that it is formed from the outer product of the four-vector $(\nu, \nu\mathbf{n})$ [recall equation (19.110)] with itself, times the invariant $(I_\nu d\nu d\Omega/\nu^2)$ [see equations (19.121) and (19.120)] integrated over all angles and frequencies. Similarly, G^α is a four-vector, because the four-divergence of a four-tensor is automatically a four-vector.

For many purposes of astrophysical radiation hydrodynamics, it is useful to have transformations accurate to $O(v/c)$ of E, $\mathbf{F}$, and P, or, equivalently, J, H, and K, between the observer's and comoving frames. These are most easily obtained by using equations (19.111), (19.121), and (19.120), expanded to first order in (v/c). Thus, setting $\gamma = 1$, one finds that

$$I_\nu^0 d\nu_0 d\Omega_0 = (\nu_0/\nu)^2 I_\nu\, d\nu d\Omega \approx (1 - 2\beta\mu) I_\nu\, d\nu d\Omega, \tag{19.137}$$

from which it follows that $J^0 = J - 2\beta H$. Using analogous expressions for $I_\nu^0 \mu_0 d\nu_0 d\Omega_0$ and $I_\nu^0 \mu_0^2 d\nu_0 d\Omega_0$, one obtains for the first three moments

$$[J^0, H^0, K^0] = [J - 2\beta H, H - \beta(J + K), K - 2\beta H]. \tag{19.138}$$

The inverse transformation yields

$$[J, H, K] = [J^0 + 2\beta H^0, H^0 + \beta(J^0 + K^0), K^0 + 2\beta H^0]. \tag{19.139}$$

Notice that these transformations are valid only for the frequency-integrated moments.

Transfer Equation in the Comoving Frame

The full transformation of the transfer equation for a non-uniform velocity field can be done rigorously using covariant differentiation [195, 746]; however, this approach uses an involved tensor calculus. We shall use here a simple first-order expansion method that yields results correct to $O(v/c)$. The equations of transfer that we consider are, in planar and spherical geometry, respectively,

$$\begin{aligned}&[c^{-1}(\partial/\partial t) + \mu(\partial/\partial r)]\,[(\nu/\nu_0)^3 I^0(r_0, \mu_0, \nu_0, t_0)]\\ &\quad = (\nu/\nu_0)^2[\eta^0(\nu_0) - \chi^0(\nu_0) I^0(r_0, \mu_0, \nu_0, t_0)],\end{aligned} \tag{19.140}$$

and

$$[c^{-1}(\partial/\partial t) + \mu(\partial/\partial r) + r^{-1}(1-\mu^2)(\partial/\partial\mu)]\,[(\nu/\nu_0)^3 I^0(r_0,\mu_0,\nu_0,t_0)]$$
$$= (\nu/\nu_0)^2[\eta^0(\nu_0) - \chi^0(\nu_0)I^0(r_0,\mu_0,\nu_0,t_0)]. \tag{19.141}$$

We consider one-dimensional flows and apply a *local* Lorentz transformation to a frame that *instantaneously* coincides with the moving fluid. We shall neglect terms of $O(v^2/c^2)$ and set $\gamma \equiv 1$. We then have

$$r_0 = r, \tag{19.142}$$

and

$$ct_0(r,t) = ct - c^{-1}\int_0^r v(r',t)\,dr'. \tag{19.143}$$

To evaluate the derivatives in equations (19.140) and (19.141), the chain rule is applied:

$$\left(\frac{\partial}{\partial r}\right) \equiv \left(\frac{\partial}{\partial r}\right)_{\mu\nu t} = \left(\frac{\partial r_0}{\partial r}\right)_{\mu\nu t}\frac{\partial}{\partial r_0} + \left(\frac{\partial \mu_0}{\partial r}\right)_{\mu\nu t}\frac{\partial}{\partial \mu_0}$$
$$+ \left(\frac{\partial \nu_0}{\partial r}\right)_{\mu\nu t}\frac{\partial}{\partial \nu_0} + \left(\frac{\partial t_0}{\partial r}\right)_{\mu\nu t}\frac{\partial}{\partial t_0}, \tag{19.144}$$

$$\left(\frac{\partial}{\partial \mu}\right) \equiv \left(\frac{\partial}{\partial \mu}\right)_{r\nu t} = \left(\frac{\partial r_0}{\partial \mu}\right)_{r\nu t}\frac{\partial}{\partial r_0} + \left(\frac{\partial \mu_0}{\partial \mu}\right)_{r\nu t}\frac{\partial}{\partial \mu_0}$$
$$+ \left(\frac{\partial \nu_0}{\partial \mu}\right)_{r\nu t}\frac{\partial}{\partial \nu_0} + \left(\frac{\partial t_0}{\partial \mu}\right)_{r\nu t}\frac{\partial}{\partial t_0}, \tag{19.145}$$

$$\left(\frac{\partial}{\partial t}\right) \equiv \left(\frac{\partial}{\partial t}\right)_{r\mu\nu} = \left(\frac{\partial r_0}{\partial t}\right)_{r\mu\nu}\frac{\partial}{\partial r_0} + \left(\frac{\partial \mu_0}{\partial t}\right)_{r\mu\nu}\frac{\partial}{\partial \mu_0}$$
$$+ \left(\frac{\partial \nu_0}{\partial t}\right)_{r\mu\nu}\frac{\partial}{\partial \nu_0} + \left(\frac{\partial t_0}{\partial t}\right)_{r\mu\nu}\frac{\partial}{\partial t_0}. \tag{19.146}$$

One uses the first-order expressions

$$(\nu/\nu_0) = 1 + \beta\mu_0, \tag{19.147}$$
$$(\nu_0/\nu) = 1 - \beta\mu_0, \tag{19.148}$$
$$\mu_0 = (\mu - \beta)(1 - \beta\mu)^{-1}, \tag{19.149}$$
$$\mu = (\mu_0 + \beta)(1 + \beta\mu_0)^{-1} \tag{19.150}$$

and makes the additional approximation that the fluid accelerations (which are identically zero for steady flow) are so small that the change in any velocity during the time of a photon flight over a mean free path is negligible compared to the velocity itself. We then neglect $(\partial v/\partial t)$ and derivatives of the form $(\partial x_0/\partial t)$ for $x_0 = r_0, \mu_0$ or ν_0, and retain only $(\partial t_0/\partial t) \equiv 1$. The remaining coefficients in equations (19.144)–(19.146), to $O(v/c)$, are easily derived from equations (19.142)–(19.143) and the first-order derivatives written above. One finds

$$\left(\frac{\partial}{\partial r}\right)_{\mu\nu t}[r_0, \mu_0, \nu_0, t_0] = [1, c^{-1}(\mu_0^2 - 1)(\partial v/\partial r_0), -c^{-1}\mu_0\nu_0(\partial v/\partial r_0), -\beta/c] \tag{19.151}$$

and

$$\left(\frac{\partial}{\partial \mu}\right)_{r\nu t}[r_0, \mu_0, \nu_0, t_0] = [0, (1 + 2\mu_0\beta), -\nu_0\beta, 0]. \tag{19.152}$$

For planar geometry, equation (19.140) is rewritten, using the approximations mentioned above, as

$$\begin{aligned}&\{(\nu/\nu_0)[c^{-1}(\partial/\partial t) + \mu(\partial/\partial r)] + 3\mu[\partial(\nu/\nu_0)/\partial r]\}I^0(r_0, \mu_0, \nu_0, t_0)\\ &= \eta^0(\nu_0) - \chi^0(\nu_0)I^0(r_0, \mu_0, \nu_0, t_0).\end{aligned} \tag{19.153}$$

Substituting from equations (19.144)–(19.152) and retaining only terms of first order in (v/c), one obtains

$$\begin{aligned}&\left[\frac{1}{c}\frac{\partial}{\partial t_0} + \left(\mu_0 + \frac{v}{c}\right)\frac{\partial}{\partial r_0} + \frac{\mu_0(\mu_0^2 - 1)}{c}\left(\frac{\partial v}{\partial r_0}\right)\frac{\partial}{\partial \mu_0}\right.\\ &\quad\left.- \frac{\nu_0\mu_0^2}{c}\left(\frac{\partial v}{\partial r_0}\right)\frac{\partial}{\partial \nu_0} + \frac{3\mu_0^2}{c}\left(\frac{\partial v}{\partial r_0}\right)\right] I^0(r_0, \mu_0, \nu_0, t_0)\\ &= \eta^0(\nu_0) - \chi^0(\nu_0)I^0(r_0, \mu_0, \nu_0, t_0).\end{aligned} \tag{19.154}$$

For spherical geometry, one has the additional term

$$\begin{aligned}&(\nu/\nu_0)^2 r^{-1}(1 - \mu^2)(\partial/\partial\mu)[(\nu/\nu_0)^3 I^0]\\ &= (\nu/\nu_0)r^{-1}(1 - \mu^2)(\partial I^0/\partial\mu) + 3r^{-1}(1 - \mu^2)[\partial(\nu/\nu_0)/\partial\mu]I^0.\end{aligned}$$

Again expanding to first order in (v/c) and using the result from equation (19.113) that $\nu^2(1 - \mu^2) = \nu_0^2(1 - \mu_0^2)$, and also $[\partial(\nu/\nu_0)/\partial\mu] = (v/c)$, one obtains for this additional term

$$r_0^{-1}(1 - \mu_0^2)\{[(1 + \beta\mu_0)(\partial/\partial\mu_0) - \beta\nu_0(\partial/\partial\nu_0)] + 3\beta\}I^0.$$

Thus the comoving-frame transfer equation to order $O(v/c)$ in spherical geometry is

$$\left\{\frac{1}{c}\frac{\partial}{\partial t_0} + \left(\mu_0 + \frac{v}{c}\right)\frac{\partial}{\partial r_0} + \frac{(1-\mu_0^2)}{r_0}\left[1 + \frac{\mu_0 v}{c}\left(1 - \frac{d\ln v}{d\ln r_0}\right)\right]\frac{\partial}{\partial \mu_0} - \left(\frac{\nu_0 v}{c r_0}\right)\left[1 - \mu_0^2\left(1 - \frac{d\ln v}{d\ln r_0}\right)\right]\frac{\partial}{\partial \nu_0} + \left(\frac{3v}{c r_0}\right)\left[1 - \mu_0^2\left(1 - \frac{d\ln v}{d\ln r_0}\right)\right]\right\} I^0(r_0, \mu_0, \nu_0, t_0)$$
$$= \eta^0(\nu_0) - \chi^0(\nu_0) I^0(r_0, \mu_0, \nu_0, t_0). \tag{19.155}$$

Equations (19.154) and (19.155) are the comoving-frame equations including *all terms* of $O(v/c)$ and were first derived consistently by Castor [195]. The time derivative written in these equations is, in essence, still in the inertial frame (although it allows for retardation); the *Lagrangian* time derivative, which follows from a motion of a fluid element (cf. § 16.1), consists of the two terms $(D/Dt) = (\partial/\partial t) + (v/c)(\partial/\partial r)$, the second term being the *advection* term. For steady flows, the term in $(\partial/\partial t) \equiv 0$. It turns out (as verified by detailed calculations in [744]) that one should retain all the terms in the moment equations, while for a solution of the transfer equation itself, it is sufficient to retain *only the frequency-derivative terms*. In this limit, equation (19.154) for *steady* flow reduces to

$$\mu_0 \frac{\partial I^0(z, \mu_0, \nu_0)}{\partial z} - \frac{\nu_0 \mu_0^2}{c}\left(\frac{\partial v}{\partial z}\right)\frac{\partial I^0(z, \mu_0, \nu_0)}{\partial \nu_0}$$
$$= \eta^0(z, \nu_0) - \chi^0(z, \nu_0) I^0(z, \mu_0, \nu_0) \tag{19.156}$$

and equation (19.155) to

$$\mu_0 \frac{\partial I^0(r, \mu_0, \nu_0)}{\partial r} + \frac{(1-\mu_0^2)}{r}\frac{\partial I^0(r, \mu_0, \nu_0)}{\partial \mu} - \left(\frac{\nu_0 v}{c r}\right)\left[(1-\mu_0^2) + \mu_0^2\left(\frac{\partial \ln v}{\partial \ln r}\right)\right]\frac{\partial I^0(r, \mu_0, \nu_0)}{\partial \nu_0}$$
$$= \eta^0(r, \nu_0) - \chi^0(r, \nu_0) I^0(r, \mu_0, \nu_0). \tag{19.157}$$

A numerical method for solving equation (19.157) will be described below.

Equation (19.157) can be derived from the observer's-frame equation (19.84) by noticing that the differential operator $\mu(\partial/\partial r)$ is evaluated in the observer's frame for constant frequency ν. When moving a distance Δr while holding the observer's-frame frequency ν constant, the comoving-frame frequency $\nu_0 = \nu_0(\nu, \mu, r)$ changes because v changes. Thus $(\partial/\partial r)_\nu = (\partial/\partial r)_{\nu_0} + (\partial \nu_0/\partial r)_\nu (\partial/\partial \nu_0)_{r_0}$ and $(\partial \nu_0/\partial r)_\nu = -(\nu_0 \mu/c)(\partial v/\partial r)$. Analogously, $(\partial/\partial \mu)_\nu = (\partial/\partial \mu)_{\nu_0} + (\partial \nu_0/\partial \mu)_\nu (\partial/\partial \nu_0)_{r_0}$ and $(\partial \nu_0/\partial \mu)_\nu = -(\nu_0 v/c)$. Substituting these expressions into (19.84), one arrives at equation (19.157).

For some purposes (for instance, for studies of line transfer with partial redistribution; see [745]), it is useful to derive frequency-dependent moment equations from equations (19.154) and (19.155). In spherical geometry, one obtains for the zero-order moment

$$\frac{1}{c}\frac{\partial J_\nu^0}{\partial t}+\frac{v}{c}\frac{\partial J_\nu^0}{\partial r}+\frac{1}{r^2}\frac{\partial(r^2H_\nu^0)}{\partial r}+\left(\frac{v}{cr}\right)(3J_\nu^0-K_\nu^0)+\frac{1}{c}(J_\nu^0+K_\nu^0)\frac{\partial v}{\partial r}$$
$$+\left(\frac{v}{cr}\right)\frac{\partial}{\partial \nu_0}[\nu_0(3K_\nu^0-J_\nu^0)]-\frac{1}{c}\left(\frac{2v}{r}+\frac{\partial v}{\partial r}\right)\frac{\partial(\nu_0 K_\nu^0)}{\partial \nu_0}$$
$$=\eta^0(\nu_0)-\chi^0(\nu_0)J_\nu^0, \tag{19.158}$$

and for the first-order moment

$$\frac{1}{c}\frac{\partial H_\nu^0}{\partial t}+\frac{v}{c}\frac{\partial H_\nu^0}{\partial r}+\frac{\partial K_\nu^0}{\partial r}+\frac{1}{r}(3K_\nu^0-J_\nu^0)+\frac{2}{c}\left(\frac{v}{r}+\frac{\partial v}{\partial r}\right)H_\nu^0$$
$$+\left(\frac{v}{cr}\right)\frac{\partial}{\partial \nu_0}[\nu_0(3N_\nu^0-H_\nu^0)]-\frac{1}{c}\left(\frac{2v}{r}+\frac{\partial v}{\partial r}\right)\frac{\partial(\nu_0 N_\nu^0)}{\partial \nu_0}$$
$$=-\chi^0(\nu_0)H_\nu^0, \tag{19.159}$$

where

$$N_\nu^0\equiv\tfrac{1}{2}\int_{-1}^{1} I^0(r,\mu_0,\nu_0,t)\,\mu_0^3\,d\mu_0, \tag{19.160}$$

and, for brevity, we have written $J_\nu^0\equiv J^0(r,\nu_0,t)$, etc., and suppressed the suffix 0 on r and t. Again, for practical transfer calculations with steady flow, it is sufficient to replace equations (19.158) and (19.159) with

$$r^{-2}[\partial(r^2H_\nu^0)/\partial r]-a[\partial(J_\nu^0-K_\nu^0)/\partial\nu^0+b(\partial K_\nu^0/\partial\nu_0)]=\eta^0(\nu_0)-\chi^0(\nu_0)J_\nu^0, \tag{19.161}$$

and

$$(\partial K_\nu^0/\partial r)+r^{-1}(3K_\nu^0-J_\nu^0)-a[\partial(H_\nu^0-N_\nu^0)/\partial\nu^0+b(\partial N_\nu^0/\partial\nu_0)]=-\chi^0(\nu_0)H_\nu^0, \tag{19.162}$$

where $a\equiv(\nu_0 v/cr)$ and $b\equiv(d\ln v/d\ln r)$; see [745].

For problems of radiation hydrodynamics we require the frequency-integrated moment equations, which follow immediately from (19.158) and (19.159):

$$(\partial E_R^0/\partial t)+v(\partial E_R^0/\partial r)+r^{-2}[\partial(r^2F_R^0)/\partial r]$$
$$+(v/r)(3E_R-p_R^0)+(\partial v/\partial r)(E_R^0+p_R^0)$$
$$=4\pi\int_0^\infty[\eta^0(\nu_0)-\chi^0(\nu_0)J_\nu^0]\,d\nu_0, \tag{19.163}$$

and

$$c^{-2}(\partial F_R^0/\partial t) + (v/c^2)(\partial F_R^0/\partial r) + (\partial p_R^0/\partial r) + (3p_R^0 - E^0)/r$$
$$+(2v/c^2 r)[1 + (d\ln v/d\ln r)]F_R^0 = -c^{-1}\int_0^\infty \chi^0(\nu_0)H_\nu^0\,d\nu_0, \quad (19.164)$$

where

$$E_R^0 = (4\pi/c)\int_0^\infty J^0(\nu_0)\,d\nu_0 \equiv (4\pi/c)J^0, \quad (19.165)$$

$$F_R^0 \equiv 4\pi H^0, \quad (19.166)$$

$$p_R^0 \equiv (4\pi/c)K^0. \quad (19.167)$$

In planar geometry, equations completely analogous to (19.158)–(19.164) can also be written down. Equations (19.163) and (19.164) contain additional velocity-dependent terms on the left-hand side compared to their counterparts, equations (19.131) and (19.132). But this is compensated by the tremendous simplification of the right-hand side, where the isotropy of the opacity and the emissivity in the comoving frame allow the integrals to be expressed in terms of moments themselves, instead of double integrals over the specific intensity.

Relativistic Flows

Proceeding in the same manner as above while retaining the terms with γ, one can obtain expressions that are valid to all orders of (v/c). We will not describe the derivation here but mention only the final results; the interested reader is referred to [732]. For notational simplicity, we suppress the suffix 0 that indicates the quantities in the comoving frame. The transfer equation reads

$$\frac{\gamma}{c}(1+\beta\mu)\frac{\partial I(\mu,\nu)}{\partial t} + \gamma(\mu+\beta)\frac{\partial I(\mu,\nu)}{\partial r}$$
$$+\gamma(1-\mu^2)\left[\frac{(1+\beta\mu)}{r} - \frac{\gamma^2}{c}(1+\beta\mu)\frac{\partial\beta}{\partial t} - \gamma^2(\mu+\beta)\frac{\partial\beta}{\partial r}\right]\frac{\partial I(\mu,\nu)}{\partial\mu}$$
$$-\gamma\left[\frac{\beta(1-\mu^2)}{r} + \frac{\gamma^2\mu}{c}(1+\beta\mu)\frac{\partial\beta}{\partial t} + \gamma^2\mu(\mu+\beta)\frac{\partial\beta}{\partial r}\right]\nu\frac{\partial I(\mu,\nu)}{\partial\nu}$$
$$+3\gamma\left[\frac{\beta(1-\mu^2)}{r} + \frac{\gamma^2\mu}{c}(1+\beta\mu)\frac{\partial\beta}{\partial t} + \gamma^2\mu(\mu+\beta)\frac{\partial\beta}{\partial r}\right]I(\mu,\nu)$$
$$=\eta(\mu,\nu) - \chi(\mu,\nu)I(\mu,\nu), \quad (19.168)$$

and the moment equations

$$
\begin{aligned}
&\frac{\gamma}{c}\left[\frac{\partial J(\nu)}{\partial t}+\beta\frac{\partial H(\nu)}{\partial t}\right]+\gamma\left[\frac{\partial H(\nu)}{\partial r}+\beta\frac{\partial J(\nu)}{\partial r}\right]\\
&\quad-\gamma\nu\left\{\frac{\beta}{r}\left[\frac{\partial J(\nu)}{\partial \nu}-\frac{\partial K(\nu)}{\partial \nu}\right]+\frac{\gamma^2}{c}\frac{\partial\beta}{\partial t}\left[\frac{\partial H(\nu)}{\partial \nu}+\beta\frac{\partial K(\nu)}{\partial \nu}\right]\right.\\
&\quad\left.+\gamma^2\frac{\partial\beta}{\partial r}\left[\frac{\partial K(\nu)}{\partial \nu}+\beta\frac{\partial H(\nu)}{\partial \nu}\right]\right\}\\
&\quad+\gamma\left\{\frac{2}{r}[H(\nu)+\beta J(\nu)]+\frac{\gamma^2}{c}\frac{\partial\beta}{\partial t}[H(\nu)+\beta J(\nu)]+\gamma^2\frac{\partial\beta}{\partial r}[J(\nu)+\beta H(\nu)]\right\}\\
&\quad=\eta(\nu)-\chi(\nu)J(\nu),
\end{aligned}
\tag{19.169}
$$

and

$$
\begin{aligned}
&\frac{\gamma}{c}\left[\frac{\partial H(\nu)}{\partial t}+\beta\frac{\partial K(\nu)}{\partial t}\right]+\gamma\left[\frac{\partial K(\nu)}{\partial r}+\beta\frac{\partial H(\nu)}{\partial r}\right]\\
&\quad-\gamma\nu\left\{\frac{\beta}{r}\left[\frac{\partial H(\nu)}{\partial \nu}-\frac{\partial N(\nu)}{\partial \nu}\right]+\frac{\gamma^2}{c}\frac{\partial\beta}{\partial t}\left[\frac{\partial K(\nu)}{\partial \nu}+\beta\frac{\partial N(\nu)}{\partial \nu}\right]\right.\\
&\quad\left.+\gamma^2\frac{\partial\beta}{\partial r}\left[\frac{\partial N(\nu)}{\partial \nu}+\beta\frac{\partial K(\nu)}{\partial \nu}\right]\right\}\\
&\quad+\gamma\left\{\frac{1}{r}[3K(\nu)-J(\nu)+\beta H(\nu)+\beta N(\nu)]\right.\\
&\quad+\frac{\gamma^2}{c}\frac{\partial\beta}{\partial t}[J(\nu)+2\beta H(\nu)-\beta N(\nu)]\\
&\quad\left.+\gamma^2\frac{\partial\beta}{\partial r}[2H(\nu)-N(\nu)+\beta J(\nu)]\right\}=-\chi(\nu)H(\nu).
\end{aligned}
\tag{19.170}
$$

Equation (19.168) is a complicated partial differential equation. The main difficulty presented by this equation is that the differential operator involves three independent variables and has a rather complex structure. Direct differencing may lead to a system of low accuracy and high dimensionality. However, one may reduce the complexity of this equation because it is linear (the coefficients standing at the derivatives do not depend on the solution), so for a given velocity field one may construct characteristic ray trajectories along which the spatial operator is a perfect differential. If one chooses a ray with impact parameter p (defined here as a distance of its closest approach to the origin), and let s be the path length along a ray from its point of the closest approach, then the trajectory defined by $[r(p,s),\mu(p,s)]$ is determined such that

$$
\frac{dI}{ds}=\frac{dr}{ds}\frac{\partial I}{\partial r}+\frac{d\mu}{ds}\frac{\partial I}{\partial \mu}.
\tag{19.171}
$$

From equations (19.168) and (19.171), it follows that the characteristic rays are described by the equations

$$\frac{dr}{ds} = \gamma(\mu + \beta), \tag{19.172}$$

$$\frac{d\mu}{ds} = \gamma(1 - \mu^2)\left[\frac{(1 + \beta\mu)}{r} - \gamma^2(\mu + \beta)\frac{\partial \beta}{\partial r}\right]. \tag{19.173}$$

For static media ($\beta \equiv 0$) the characteristics are the parallel straight lines $y \equiv (1 - \mu^2)^{\frac{1}{2}} r = \text{constant} = p$, which are exactly the tangent rays that were used in § 19.1 and § 19.2. Here the characteristic rays are no longer straight lines, and their shape has to be determined by solving equations (19.172) and (19.173) numerically. The appropriate boundary conditions are given by

$$\left(\frac{dr}{ds}\right)_{s=0} = 0, \tag{19.174}$$

because the origin $s = 0$ of the path-length variable s was chosen at the point of tangency to a radial shell, and

$$\left(\frac{d\mu}{ds}\right)_{s=0} = \frac{(1 - \beta^2)^{\frac{3}{2}}}{p}, \tag{19.175}$$

which follows from (19.173), realizing that $\mu(p, s = 0) = -\beta(p)$ as follows from equation (19.172).

Direct Solution for Spherically Symmetric Flows

We will now consider methods for solving the steady-state transfer equation in the comoving frame, retaining only the terms of $O(v/c)$ and, moreover, taking the limit that only Doppler shifts are taken into account. In other words, the aberration of light is neglected. The transfer equation becomes [cf. equation (19.157)]

$$\begin{aligned} \mu\frac{\partial I(r, \mu, \nu)}{\partial r} &+ \frac{(1 - \mu^2)}{r}\frac{\partial I(r, \mu, \nu)}{\partial \mu} \\ &- \left(\frac{\nu v}{cr}\right)\left[(1 - \mu^2) + \mu^2\left(\frac{\partial \ln v}{\partial \ln r}\right)\right]\frac{\partial I(r, \mu, \nu)}{\partial \nu} \\ &= \eta(r, \nu) - \chi(r, \nu) I(r, \mu, \nu). \end{aligned} \tag{19.176}$$

This notation will be used for the reminder of this section, keeping in mind that *all quantities are evaluated in the comoving frame*.

We will describe here an efficient method developed in [742] based on the Feautrier-type solution. It is applicable if the source function in the comoving frame is independent of angle. For treating angle-dependent partial redistribution one can use a scheme based on moment equations [e.g., equations (19.161) and (19.162)], thereby eliminating the angle variable, in which case a Feautrier-type solution again becomes practical [745].

We again adopt the (p, z) coordinate system introduced in § 19.1. Then, along a ray specified by constant p, equation (19.176) becomes

$$\pm\frac{\partial I^{\pm}(z,p,\nu)}{\partial z} - \widetilde{\gamma}(z,p)\frac{\partial I^{\pm}(z,p,\nu)}{\partial \nu} = \eta(r,\nu) - \chi(r,\nu)I^{\pm}(z,p,\nu), \quad (19.177)$$

where

$$\widetilde{\gamma}(z,p) \equiv [\nu v(r)/cr][1 - \mu^2(d\ln v/d\ln r)], \quad (19.178)$$

and $r \equiv (p^2 + z^2)^{\frac{1}{2}}$, $\mu \equiv (z/r)$. Introducing the optical depth along the ray, $d\tau(z,p,\nu) = -\chi(z,p,\nu)\,dz$, and the usual mean-intensity-like and flux-like variables

$$u(z,p,\nu) = \tfrac{1}{2}[I^{+}(z,p,\nu) + I^{-}(z,p,\nu)], \quad (19.179)$$

$$v(z,p,\nu) = \tfrac{1}{2}[I^{+}(z,p,\nu) - I^{-}(z,p,\nu)], \quad (19.180)$$

we obtain from equation (19.177) the system

$$\frac{\partial u(z,p,\nu)}{\partial \tau(z,p,\nu)} + \gamma(z,p,\nu)\frac{\partial v(z,p,\nu)}{\partial \nu} = v(z,p,\nu), \quad (19.181)$$

and

$$\frac{\partial v(z,p,\nu)}{\partial \tau(z,p,\nu)} + \gamma(z,p,\nu)\frac{\partial u(z,p,\nu)}{\partial \nu} = u(z,p,\nu) - S(z,p,\nu), \quad (19.182)$$

where $\gamma(z,p,\nu) \equiv \widetilde{\gamma}(z,p,\nu)/\chi(z,p,\nu)$, and the source function is assumed to have an equivalent two-level atom form $S(z,p,\nu) = S[r(z,p),\nu] = \alpha(r,\nu)\overline{J}(r) + \beta(r)$. The coefficients α and β contain the thermalization parameter ϵ, the opacity ratio χ_c/χ_l, and the profile coefficient. The frequency-averaged mean intensity $\overline{J}$ is given by

$$\overline{J}(r) = \int_{\nu_{\min}}^{\nu_{\max}} d\nu\,\phi(\nu)\int_0^1 d\mu\,u[z(r,\mu), p(r,\mu), \nu]. \quad (19.183)$$

Here, $\nu_{\min}$ and $\nu_{\max}$ are chosen to contain the whole line profile as seen in the comoving frame. Note particularly in equations (19.181) and (19.182) that *because we are working in the comoving frame*, we can now average I^+ and I^- at a *given value of* ν, in contrast to the situation in an observer's-frame formulation [cf. equations (19.76) and (19.77)].

Spatial boundary conditions are now required. At the outer radius, $r = R$, one has for no incident radiation $I^- = 0$ and therefore $u = v$; hence

$$[\partial u(z, p, \nu)/\partial\tau(z, p, \nu)]_{z_{\max}} + \gamma(z_{\max}, p, \nu)[\partial u(z_{\max}, p, \nu)/\partial\nu] = u(z_{\max}, p, \nu). \tag{19.184}$$

At the plane of symmetry, $z = 0$, one has $v(z, p, \nu) = 0$; hence for rays that do not intersect the core,

$$[\partial u(z, p, \nu)/\partial\tau(z, p, \nu)]_{z=0} = 0. \tag{19.185}$$

For rays that intersect the core (i.e., $p \leq r$), one either applies the diffusion approximation for an *opaque core* (stellar surface), which specifies v, or sets $v = 0$ (by symmetry) for a *hollow core* (nebular case).

In addition, an *initial condition in frequency* is required. For an *expanding atmosphere* [i.e., one in which $v > 0$ and $(dv/dr) > 0$], it is clear that the high-frequency edge of the line profile (in the *comoving* frame) cannot intercept *line* photons from any point in the atmosphere, because they will all be systematically red-shifted; any photon incident at the high-frequency edge must be a *continuum* photon. To specify the required initial condition one may therefore either (a) solve equations (19.181) and (19.182) in the continuum, omitting the frequency-derivative terms, which yields the standard second-order system to obtain $u(z, p, \nu_{\max}) = u_{\text{continuum}}$, or (b) fix the derivative $(\partial u/\partial\nu)_{\nu_{\max}}$ to any prescribed value given by the slope of the continuum. In particular, the choice $(\partial u/\partial\nu) = 0$ leads to equations identical to option (a) just mentioned.

The system is discretized using the same grids $\{r_d\}$, $\{p_i\}$, $\{z_{d_i}\}$ as were employed in § 19.1 and § 19.2. One now chooses the frequency grid $\{\nu_n\}$, $n = 1, \ldots, N$ in order of *decreasing values* $\nu_1 > \nu_2 > \cdots > \nu_N$, because the initial condition is posed at the highest frequency. Equation (19.183) is replaced with a quadrature sum

$$\bar{J}(r_d) = \sum_{n=1}^{N} w_n \sum_{i=1}^{NI_d} a_{di}\, \phi(r_d, \nu_n)\, u[z(r_d, p_i), p_i, \nu_n], \tag{19.186}$$

and equations (19.181) and (19.182) are replaced with difference approximations

$$(u_{d+1,in} - u_{din})/\Delta\tau_{d+\frac{1}{2},in} = v_{d+\frac{1}{2},in} + \delta_{d+\frac{1}{2},i,n-\frac{1}{2}}(v_{d+\frac{1}{2},in} - v_{d+\frac{1}{2},i,n-1}) \tag{19.187}$$

and

$$(v_{d+\frac{1}{2},in} - v_{d-\frac{1}{2},in})/\Delta\tau_{din} = u_{din} - S_{din} + \delta_{di,n-\frac{1}{2}}(u_{din} - u_{di,n-1}), \tag{19.188}$$

where u is presumed to be defined on the mesh-points $z_d = z(r_d, p_i)$, and $u_{din} \equiv u(z_d, p_i, \nu_n)$, while v is presumed to be defined on the interstices $z_{d\pm\frac{1}{2}} \equiv \frac{1}{2}(z_d + z_{d\pm1})$, and $v_{d\pm\frac{1}{2},in} \equiv v(z_{d\pm\frac{1}{2}}, p_i, \nu_n)$. One further defines

$$\chi_{d\pm\frac{1}{2},in} \equiv \tfrac{1}{2}(\chi_{d\pm1,in} + \chi_{din}), \tag{19.189}$$

$$\Delta\tau_{d\pm\frac{1}{2},in} \equiv \chi_{d\pm\frac{1}{2},in}|z_d - z_{d\pm1}|, \tag{19.190}$$

$$\Delta\tau_{din} \equiv \tfrac{1}{2}(\Delta\tau_{d+\frac{1}{2},in} + \Delta\tau_{d-\frac{1}{2},in}), \tag{19.191}$$

$$\delta_{di,n-\frac{1}{2}} \equiv \gamma_{din}/(\nu_{n-1} - \nu_n). \tag{19.192}$$

Similar difference equations can be written to represent the boundary conditions [742]. In equations (19.187) and (19.188), an *implicit* frequency differencing is used to assure stability [742, 904].

Equation (19.187) can be solved analytically for $v_{d+\frac{1}{2},in}$ to yield

$$v_{d+\frac{1}{2},in} = \frac{[(u_{d+1,in} - u_{din})/\Delta\tau_{d+\frac{1}{2},in}] + \delta_{d+\frac{1}{2},i,n-\frac{1}{2}} v_{d+\frac{1}{2},i,n-1}}{1 + \delta_{d+\frac{1}{2},i,n-\frac{1}{2}}}. \tag{19.193}$$

Organizing the solution into vectors that specify the depth-variation along a particular ray at a given frequency, i.e.,

$$\mathbf{u}_{in} \equiv (u_{1in}, u_{2in}, \ldots, u_{Din})^{\mathrm{T}}, \tag{19.194}$$

$$\mathbf{v}_{in} \equiv (v_{\frac{1}{2},in}, v_{\frac{3}{2},in}, \ldots, v_{D_i-\frac{1}{2},in})^{\mathrm{T}}, \tag{19.195}$$

equation (19.193) can be written in the form

$$\mathbf{v}_{in} = \mathbf{G}_{in}\mathbf{u}_{in} + \mathbf{H}_{in}\mathbf{v}_{i,n-1}, \tag{19.196}$$

where $\mathbf{G}$ is bidiagonal and $\mathbf{H}$ is diagonal. Equation (19.193) is used to eliminate $v_{d\pm\frac{1}{2},in}$ from (19.188); one then obtains a set of second-order equations for u_{din}, namely,

$$\begin{aligned}
&(u_{d+1,in} - u_{din})/[\Delta\tau_{d+\frac{1}{2},in}(1 + \delta_{d+\frac{1}{2},i,n-\frac{1}{2}})] \\
&\quad - (u_{din} - u_{d-1,in})/[\Delta\tau_{d-\frac{1}{2},in}(1 + \delta_{d-\frac{1}{2},i,n-\frac{1}{2}})]/\Delta\tau_{din} \\
&\quad = (1 + \delta_{di,n-\frac{1}{2}})u_{din} - S_{din} - \delta_{di,n-\frac{1}{2}} u_{di,n-1} \\
&\quad + [\delta_{d-\frac{1}{2},i,n-\frac{1}{2}}(1 + \delta_{d-\frac{1}{2},i,n-\frac{1}{2}})^{-1} v_{d-\frac{1}{2},i,n-1} \\
&\quad - \delta_{d+\frac{1}{2},i,n-\frac{1}{2}}(1 + \delta_{d+\frac{1}{2},i,n-\frac{1}{2}})^{-1} v_{d+\frac{1}{2},i,n-1}]/\Delta\tau_{din}.
\end{aligned} \tag{19.197}$$

Adding the boundary conditions to equation (19.197), one obtains the system

$$\mathbf{T}_{in}\mathbf{u}_{in} + \mathbf{U}_{in}\mathbf{u}_{i,n-1} + \mathbf{V}_{in}\mathbf{v}_{i,n-1} + \mathbf{W}_{in}\bar{\mathbf{J}} = \mathbf{X}_{in}, \tag{19.198}$$

where $\mathbf{T}_{in}$ is tridiagonal, $\mathbf{U}_{in}$ and $\mathbf{W}_{in}$ are diagonal, $\mathbf{V}_{in}$ is bidiagonal, and $\mathbf{X}_{in}$ is a vector.

To solve the complete system, one chooses a definite ray, specified by a given p_i, and carries out a frequency-by-frequency integration procedure, with n ranging from 1 to N. This is effected by noting that the initial condition in frequency implies that $\mathbf{U}_{i1}$, $\mathbf{V}_{i1}$, and $\mathbf{H}_{i1}$ are all exactly zero; thus one can obtain expressions of the form $\mathbf{u}_{i1} = \mathbf{A}_{i1} - \mathbf{B}_{i1}\overline{\mathbf{J}}$, and $\mathbf{v}_{i1} = \mathbf{C}_{i1} - \mathbf{D}_{i1}\overline{\mathbf{J}}$, where $\mathbf{A}_{i1} = \mathbf{T}_{i1}^{-1}\mathbf{X}_{i1}$ is a vector, $\mathbf{B}_{i1} = \mathbf{T}_{i1}^{-1}\mathbf{W}_{i1}$ is a matrix, $\mathbf{C}_{i1} = \mathbf{G}_{i1}\mathbf{A}_{i1}$, and $\mathbf{D}_{i1} = \mathbf{G}_{i1}\mathbf{B}_{i1}$. Similar substitutions are carried out for successive values of n to yield

$$\mathbf{u}_{in} = \mathbf{A}_{in} - \mathbf{B}_{in}\overline{\mathbf{J}}, \tag{19.199}$$

$$\mathbf{v}_{in} = \mathbf{C}_{in} - \mathbf{D}_{in}\overline{\mathbf{J}}, \tag{19.200}$$

where

$$\mathbf{A}_{in} = \mathbf{T}_{in}^{-1}(\mathbf{X}_{in} - \mathbf{U}_{in}\mathbf{A}_{i,n-1} - \mathbf{V}_{in}\mathbf{C}_{i,n-1}), \tag{19.201}$$

$$\mathbf{B}_{in} = \mathbf{T}_{in}^{-1}(\mathbf{W}_{in} - \mathbf{U}_{in}\mathbf{B}_{i,n-1} - \mathbf{V}_{in}\mathbf{D}_{i,n-1}), \tag{19.202}$$

$$\mathbf{C}_{in} = \mathbf{G}_{in}\mathbf{A}_{in} + \mathbf{H}_{in}\mathbf{C}_{i,n-1}, \tag{19.203}$$

$$\mathbf{D}_{in} = \mathbf{G}_{in}\mathbf{B}_{in} + \mathbf{H}_{in}\mathbf{D}_{i,n-1}. \tag{19.204}$$

Each result of the form of equation (19.199) for every frequency ν_n, along every ray p_i, is substituted into equation (19.186) to obtain the final system for $\overline{\mathbf{J}}$ of the form

$$\left(\mathbf{I} + \sum_{i,n} \mathbf{F}_{in}\mathbf{B}_{in}\right)\overline{\mathbf{J}} = \sum_{i,n} \mathbf{F}_{in}\mathbf{A}_{in}, \tag{19.205}$$

where the vectors $\mathbf{F}$ contain the quadrature weights. The solution of this final system yields $\overline{\mathbf{J}}$, and hence $S(r,\nu)$, and $u(z,p,\nu)$ and $v(z,p,\nu)$ from equations (19.199) and (19.200). Knowledge of $u(z,p,\nu)$ implies knowledge of $u(r,\mu,\nu)$, so one can calculate $J^0(r,\nu)$ and $K^0(r,\nu)$ in the comoving frame; similarly, one can calculate the flux $H^0(r,\nu)$ from $v(r,\mu,\nu)$. Thus one can obtain a complete solution for the radiation field and its moments in the comoving frame.

The number of operations required to obtain $\mathbf{A}_{in}$ and $\mathbf{B}_{in}$ in equation (19.199) is proportional to D_i^2, so by summing over all frequencies on all rays, one obtains an estimate of the computing time required, $t \approx cND^3 + c'D^3$, which is *linear* in the number of frequencies, but is *cubic* in the number of rays. In planar geometry one disposes with rays and uses M fixed angles; the computer time required scales as $t \approx cNMD^2 + c'D^3$.

Iterative Methods

Analogously to the case of static atmospheres, direct methods for solving the transfer equation in expanding atmospheres may put heavy demands on computer resources.

Therefore, one seeks more efficient, iterative methods. Among them, approaches based on an application of the idea of accelerated lambda iteration (ALI) became the method of choice in astrophysical work. The method was described in detail, in the context of static models, in chapter 13. Here we shall briefly outline several variants of the scheme.

ALI Scheme Based on a Feautrier-Type Solver

The first application of an ALI-based scheme for moving atmospheres was suggested by Hamann [431]. The method was based on the core-saturation scheme (cf. §14.4) and thus contained a free, adjustable parameter that corresponded to a core-wing separation frequency. A parameter-free formulation was first developed by Puls [889]. It is based on a direct solution of the comoving-frame transfer equation described above. The iteration scheme is analogous to the ALI scheme used in the observer's-frame approach, summarized by equations (19.95)–(19.104). In particular, one seeks to evaluate a diagonal (local) approximate Λ^* using equation (19.103),

$$\Lambda^*_{dd} \equiv u^*_d = (\Lambda[\mathbf{e}_d])_d, \tag{19.206}$$

where $\mathbf{e}_d$ is a unit pulse function corresponding to depth d, i.e., $e_{dj} \equiv \delta_{dj}$. This method of evaluation follows from [815], who showed that a nearly optimum local approximate operator is that given by the diagonal part of the exact Λ-operator. In the context of the Feautrier method of solution of the transfer equation in static atmospheres, or in moving atmospheres in the observer's-frame formulation, equation (19.206) reduces to an evaluation of diagonal elements of the inverse of a tridiagonal matrix, which can be done very efficiently using an ingenious procedure suggested in [925].

An application of ALI in the case of comoving-frame formulation is more complicated due to the presence of derivatives of the radiation intensity with respect to frequency. The first application of the ALI technique by [889] was done using the Feautrier-type scheme of the formal solution, equations (19.198) and (19.196). To simplify the notation, the index i that specifies the impact parameter is dropped here. The Feautrier variables satisfy

$$\mathbf{u}_n = \mathbf{T}_n^{-1}[\mathbf{S}_n - \mathbf{U}_n\mathbf{u}_{n-1} - \mathbf{V}_n\mathbf{v}_{n-1}], \tag{19.207}$$

$$\mathbf{v}_n = \mathbf{G}_n\mathbf{u}_n + \mathbf{H}_n\mathbf{v}_{n-1}. \tag{19.208}$$

For simplicity, we assume here a two-level atom without continuum, which allows us to write $\mathbf{X}_n = \mathbf{S}_n$, with $\mathbf{S}$ being the vector of source functions at all depth points along the ray; a more general formulation can be found in [889].

The intensities at the first frequency, u_1 and v_1, are specified by the initial condition at the blue wing of a line and are known. For the purposes of evaluating Λ^*, they may be thought of as being absorbed into the definition of $\mathbf{S}$ and formally set to 0. Hence, $\mathbf{u}_2 = \mathbf{T}_2^{-1}S_2$, exactly as in the case of the observer's-frame formulation. But at the third frequency, $\mathbf{u}_3 = \mathbf{T}_3^{-1}S_3 - \mathbf{T}_3^{-1}(\mathbf{U}_3\mathbf{u}_2 + \mathbf{V}_3\mathbf{v}_2)$, so when substituting $\mathbf{S}_3 = \mathbf{e}_d$, one obtains $(u_3^*)_d = (T_3^{-1})_{dd} - \sum_j (T_3^{-1})_{dj}(U_3)_{jj}(u_2^*)_j$ (plus a term with $\mathbf{v}_2$).

In the second term, $(u_2^*)_j = \sum_k (T_2^{-1})_{jk}\delta_{jd} = (T_2^{-1})_{dk}$. The inverse of a tridiagonal matrix is full; consequently all the elements $(u_2^*)_j$ are non-zero, and one obtains a nonlocal contribution to $(u_3^*)_d$ from *all* depth points. Therefore, a construction of a true diagonal element $\Lambda^*_{dd} = u^*_d$ would require an actual numerical solution of the system (19.207) and (19.208) with $\mathbf{S} = \mathbf{e}_d$ for each depth point d, which would make the whole idea of the ALI scheme meaningless. Therefore, one has to construct an *approximate diagonal* of the Λ-operator.

An efficient algorithm for evaluating elements of Λ^* was suggested in [889]. The first step is to write down a local representation of $u_{n,d}$ as

$$u^\dagger_{n,d} = \mathrm{diag}(T_n^{-1}) \left[S_{n,d} - U_n u^\dagger_{n-1,d} - VA_n v^\dagger_{n-1,d-\frac{1}{2}} + VB_n v^\dagger_{n-1,d+\frac{1}{2}} \right], \tag{19.209}$$

where one writes the elements of the bidiagonal matrix $\mathbf{V}_n$ as $(\mathbf{V}_n)_{ij} = VA_n\delta_{ij} - VB_n\delta_{i+1,j}$. Equation (19.208) is used to determine $v^\dagger_{n-1,d\pm\frac{1}{2}}$,

$$\begin{aligned} v^\dagger_{n-1,d-\frac{1}{2}} &= G_{n-1,d-\frac{1}{2}}(u^\dagger_{n-1,d} - u_{n-1,d-1}) + H_{n-1,d-\frac{1}{2}} v^\dagger_{n-2,d-\frac{1}{2}}, \\ v^\dagger_{n-1,d+\frac{1}{2}} &= G_{n-1,d+\frac{1}{2}}(u_{n-1,d+1} - u^\dagger_{n-1,d}) + H_{n-1,d+\frac{1}{2}} v^\dagger_{n-2,d+\frac{1}{2}}, \end{aligned} \tag{19.210}$$

which shows that one would still get additional nonlocal terms in v's. To avoid such nonlocal terms, one neglects the contribution of the off-diagonal elements $u_{n-1,d\pm1}$ to $v_{n-1,d\pm\frac{1}{2}}$ and writes

$$\begin{aligned} v^*_{n-1,d-\frac{1}{2}} &= G_{n-1,d-\frac{1}{2}} u^*_{n-1,d} + H_{n-1,d-\frac{1}{2}} v^*_{n-2,d-\frac{1}{2}}, \\ v^*_{n-1,d+\frac{1}{2}} &= -G_{n-1,d+\frac{1}{2}} u^*_{n-1,d} + H_{n-1,d+\frac{1}{2}} v^*_{n-2,d+\frac{1}{2}}, \end{aligned} \tag{19.211}$$

with $u^*_{n,d}$ given by equation (19.209) with $\dagger$ replaced with $*$. Using this algorithm, one can evaluate the elements of $u^*_{n,d}$, and hence $\Lambda^*_{n,dd}$, along with the formal solution of the transfer equation, consecutively for all frequencies n, $(n = 1, \ldots, N)$ and for each impact parameter p. For details and a discussion of the (favorable) mathematical properties of the resulting Λ^*-operator, refer to [889].

The iteration scheme proceeds as follows.

For each impact parameter:

(i) Start with $u^*_{1,d} = v^*_{1,d\pm\frac{1}{2}} \equiv 0$.

(ii) For each frequency point $n (n = 2, \ldots, N)$ compute the term in square brackets in equation (19.209) using previous solutions $u^*_{n-1,d}$ and $v^*_{n,d\pm\frac{1}{2}}$.

(iii) Compute the diagonal of $\mathbf{T}_n^{-1}$ using the Rybicki-Hummer procedure [925] and compute $u^*_{n,d}$ from (19.209), and subsequently $v_{n,d\pm\frac{1}{2}}$ from (19.211).

(iv) The vectors $u^*_{n-1,d}$ and $v^*_{n-1,d\pm\frac{1}{2}}$ are updated by the actual values computed by the exact formal solution, and the latter quantities are also being added to the total Λ^* with appropriate quadrature weights for spatial and frequency integration.

(v) When the loop over all frequencies and impact parameters is completed, one has the desired values for Λ^* and the new formal solution $\overline{J}_{\rm FS}$; one then updates the source function using (19.100).

(vi) Repeats steps (i) to (v) to convergence.

For more details and further discussion refer to [889].

ALI Scheme Based on a Short-Characteristics Solver

A different parameter-free formulation was developed by Hauschildt [449]. The advantage of the method is that it can be used to treat a fully relativistic case described by equation (19.168) without any significant complications as compared to the simplified case described by equation (19.176). In fact, the method was originally developed [449] to treat the fully relativistic case.

One first rewrites the transfer equation in a convenient form, assuming a stationary case ($\partial/\partial t \equiv 0$), and again dropping subscript 0, with the understanding that all quantities are measured in the comoving frame. Equation (19.168) becomes (cf. [732])

$$a_r \frac{\partial I}{\partial r} + a_\mu \frac{\partial I}{\partial \mu} - a_\nu \frac{\partial(\nu I)}{\partial \nu} + 4a_\nu I = \eta - \chi I, \tag{19.212}$$

where, in the fully relativistic case,

$$a_r = \gamma(\mu + \beta) \tag{19.213}$$

$$a_\mu = \gamma(1-\mu^2)\left[\frac{(1+\beta\mu)}{r} - \gamma^2(\mu+\beta)\frac{\partial \beta}{\partial r}\right], \tag{19.214}$$

$$a_\nu = \gamma\left[\frac{\beta(1-\mu^2)}{r} + \gamma^2\mu(\mu+\beta)\frac{\partial \beta}{\partial r}\right], \tag{19.215}$$

or, in the simplified case for $\gamma \to 1$, $\beta \ll \mu$,

$$a_r = \mu, \tag{19.216}$$

$$a_\mu = \frac{(1-\mu^2)}{r}\left[1 + \frac{\mu v}{c}\left(1 - \frac{d\ln v}{d\ln r}\right)\right], \tag{19.217}$$

$$a_\nu = \left(\frac{v}{cr}\right)\left[1 - \mu^2\left(1 - \frac{d\ln v}{d\ln r}\right)\right]. \tag{19.218}$$

Along the characteristic rays, equation (19.212) has the form

$$\frac{\partial I}{\partial s} - a_\nu \frac{\partial(\nu I)}{\partial \nu} = \eta - (\chi + 4a_\nu)I, \tag{19.219}$$

where s is the geometrical path along the ray. The characteristics are given by the coupled set of ordinary differential equations (19.172) and (19.173) with boundary conditions (19.174) and (19.175). The next step is to replace the frequency derivative in equation (19.219) by a difference. Again, to assure stability, one considers a *fully implicit* discretization, namely,

$$[\partial(\nu I)/\partial\nu]_{\nu_n} = [\nu_n I(\nu_n) - \nu_{n-1} I(\nu_{n-1})]/(\nu_n - \nu_{n-1}). \qquad (19.220)$$

Equation (19.219) then becomes

$$\frac{dI_n}{ds} - a_n \frac{\nu_n I_n - \nu_{n-1} I_{n-1}}{\nu_n - \nu_{n-1}} = \eta_n - (\chi_n + 4a_n) I_n, \qquad (19.221)$$

where one writes $I_n \equiv I(\nu_n)$, $a_n \equiv a_{\nu_n}$, etc. Defining an optical depth along the characteristics as

$$d\tau_n \equiv \left[\chi_n + a_n \left(4 - \frac{\nu_n}{\nu_n - \nu_{n-1}}\right)\right] ds \equiv \widetilde{\chi}_n \, ds, \qquad (19.222)$$

the transfer equation along the ray is written as

$$\frac{dI_n}{d\tau_n} = I_n - \widetilde{S}_n, \qquad (19.223)$$

where the modified source function is given by

$$\widetilde{S}_n = \frac{\eta_n}{\widetilde{\chi}_n} - \frac{a_n}{\widetilde{\chi}_n} \frac{\nu_{n-1}}{\nu_n - \nu_{n-1}} I_{n-1}. \qquad (19.224)$$

Equation (19.223) has the same form as the transfer equation along a ray in a static medium, and its solution can be done by applying methods described in § 12.4. However, the Feautrier scheme is not directly applicable because now the modified source function depends explicitly on μ due to the dependence of a_ν on μ. One can use either a short-characteristics scheme or the discontinuous finite element (DFE) method; most studies and computer programs (e.g., PHOENIX [456]) use the short characteristics method. In this scheme, solution of equation (19.223) is written as

$$I(\tau_d) = I(\tau_{d-1})\, e^{-(\tau_d - \tau_{d-1})} + \int_{\tau_{d-1}}^{\tau_d} \widetilde{S}(t) e^{-(t-\tau_{d-1})}\, dt, \qquad (19.225)$$

and the integral on the right-hand side of equation (19.225) is evaluated analytically by employing a simple piecewise polynomial approximation for $\widetilde{S}(t)$. Widely used approaches are to approximate $\widetilde{S}$ by a *linear* segment between the mesh points τ_{d-1} and τ_d (first-order scheme), or a parabola defined by the values of $\widetilde{S}$ at the mesh points τ_{d-1} and τ_d, and τ_{d+1} (second-order scheme). In either case, equation (19.225) is written as (cf. §12.4)

$$I_d = I_{d-1} \exp(-\Delta\tau_{d-\frac{1}{2}}) + a_d \widetilde{S}_{d-1} + b_d \widetilde{S}_d + c_d \widetilde{S}_{d+1}, \qquad (19.226)$$

where $\widetilde{S}_d \equiv \widetilde{S}(\tau_d)$, $\Delta\tau_{d-\frac{1}{2}} \equiv \tau_d - \tau_{d-1}$, and the coefficients a_d, b_d, and c_d are simple functions of $\Delta\tau_{d-\frac{1}{2}}$ and $\Delta\tau_{d+\frac{1}{2}}$; the explicit expressions are given in §12.4; for convenience, they are summarized here:

$$
\begin{aligned}
a_d &= x_{d-\frac{1}{2}} - y_{d-\frac{1}{2}}/\Delta\tau_{d-\frac{1}{2}}, \\
b_d &= y_{d-\frac{1}{2}}/\Delta\tau_{d-\frac{1}{2}}, \\
c_d &= 0
\end{aligned}
\tag{19.227}
$$

for the first-order scheme and

$$
\begin{aligned}
a_d &= [z_{d-\frac{1}{2}} - y_{d-\frac{1}{2}}(2\Delta\tau_d + \Delta\tau_{d-\frac{1}{2}})]/(2\Delta\tau_d\Delta\tau_{d-\frac{1}{2}}) + x_{d-\frac{1}{2}}, \\
b_d &= (2y_{d-\frac{1}{2}}\Delta\tau_d - z_{d-\frac{1}{2}})/(\Delta\tau_{d-\frac{1}{2}}\Delta\tau_{d+\frac{1}{2}}), \\
c_d &= (z_{d-\frac{1}{2}} - y_{d-\frac{1}{2}}\Delta\tau_{d-\frac{1}{2}})/(2\Delta\tau_d\Delta\tau_{d+\frac{1}{2}})
\end{aligned}
\tag{19.228}
$$

for the second-order scheme, where

$$
\begin{aligned}
\Delta\tau_d &= \tfrac{1}{2}(\Delta\tau_{d-\frac{1}{2}} + \Delta\tau_{d+\frac{1}{2}}), \\
x_{d-\frac{1}{2}} &= 1 - \exp(-\Delta\tau_{d-\frac{1}{2}}), \\
y_{d-\frac{1}{2}} &= \Delta\tau_{d-\frac{1}{2}} - x_{d-\frac{1}{2}}, \\
z_{d-\frac{1}{2}} &= (\Delta\tau_{d-\frac{1}{2}})^2 - 2y_{d-\frac{1}{2}}.
\end{aligned}
\tag{19.229}
$$

The formal solution—i.e., a solution of the transfer equation with the source function fully specified—proceeds in a similar way to the case of static atmospheres. For the rays that do not intersect the core, the formal solution starts at the leftmost point, where the outer boundary condition is used for $d = 1$ and proceeds all the way along the ray. For the core-intersecting rays, the formal solution is split into two parts: (i) integration from the leftmost point, where I_1 is given by the outer boundary condition, to the core, and (ii) integration from the surface of the core, where the intensity is given by the inner boundary condition, all the way to the outer boundary.

A solution in the general case, where the source function depends on the specific intensity in all frequencies and positions, is done by applying the idea of the accelerated lambda iteration method, as in the previously discussed cases of the observer's-frame formulation, or the comoving-frame formulation using a Feautrier-type formal solver. In the present case, there is a complication that arises due to the presence of the derivatives $\partial I/\partial\nu$ in the modified source function $\widetilde{S}$. To avoid these problems, the simplest possibility is to neglect the frequency derivatives altogether [449] and set $\widetilde{S} = S$ for the evaluation of Λ^*; i.e., the diagonal elements of the elementary (frequency- and ray-dependent) Λ^*_{dd} will be equal to the coefficients b_d given by equation (19.227) or (19.228). It is shown in the above-mentioned reference that such an approach results in a convergent scheme, albeit with a somewhat slower convergence rate than in the static case.

Having obtained the elementary Λ^*_{dd} at all points d along the ray, for all characteristic rays, and for all frequencies, the total Λ^* at a given radial position is evaluated by integrating over all frequencies and angles. As before, an integration over angles is replaced by the appropriate integration over the impact parameters. Proper care should be taken about dividing the tangent rays into those that do intersect the core and those that do not; for details and explicit expressions, refer to [449].

A more efficient possibility to treat the frequency derivatives, also suggested in [449], is to include the $\partial I/\partial\nu$ terms in an approximate way by adding an angle-averaged term involving the Λ^* matrix for the previous frequency point as

$$\Lambda^{*\prime} = \sum_n w_n \phi(\nu_n) \left(\Lambda^*_n - \overline{a}_{\nu_n} \frac{\nu_{n-1}}{\nu_n - \nu_{n-1}} \Lambda^*_{n-1} \right), \tag{19.230}$$

where Λ^*_n is calculated by the original procedure with neglected frequency derivative integrated over angles (impact parameters), w_n is the frequency quadrature weight, and $\overline{a}_{\nu_n}$ is a suitable angle average of a_{ν_n}, taken for simplicity as $\overline{a}_{\nu_n} = a_{\nu_n}(\mu = 1)$ [449].

Approximate Newton-Raphson Scheme

This approach is analogous to the approximate Newton-Raphson scheme for static atmospheres that was described in § 18.4. In the case of moving atmospheres, it was first applied in [478,967] and further developed and extended by Hillier [492]. His formalism is based on expressions for the first two moments of the specific intensity (J and H moments) that follow directly from equations (19.198) and (19.196) for the Feautrier variables u and v, integrated over impact parameters (i.e., angles), namely (cf. [492]),

$$\mathbf{T}_n \mathbf{J}_n = \mathbf{X}_n + \mathbf{U}_n \mathbf{J}_{n-1} + \mathbf{V}_n \mathbf{H}_{n-1}, \tag{19.231}$$

and

$$\mathbf{H}_n = \mathbf{A}_n \mathbf{J}_n + \mathbf{B}_n \mathbf{H}_{n-1}, \tag{19.232}$$

where $\mathbf{J}$ is the vector of mean intensity, $\{J_1, \ldots, J_D\}$, where D is the number of radial points, $\mathbf{H}$ is an analogous vector of the flux, $\mathbf{T}$ is a tridiagonal matrix, $\mathbf{U}$ is a diagonal matrix, $\mathbf{V}$ is a lower bidiagonal matrix, $\mathbf{A}$ is an upper bidiagonal matrix, $\mathbf{B}$ is a diagonal matrix, and $\mathbf{X}$ is a vector. Equations (19.231) and (19.232) are linearized to yield

$$\begin{aligned} \mathbf{T}_n \delta \mathbf{J}_n = {} & \mathbf{U}_n \delta \mathbf{J}_{n-1} + \mathbf{V}_n \delta \mathbf{H}_{n-1} \\ & + \delta \mathbf{X}_n - \delta \mathbf{T}_n \mathbf{J}_n + \delta \mathbf{U}_n \mathbf{J}_{n-1} + \delta \mathbf{V}_n \mathbf{H}_{n-1}, \end{aligned} \tag{19.233}$$

and

$$\delta \mathbf{H}_n = \mathbf{A}_n \delta \mathbf{J}_n + \mathbf{B}_n \delta \mathbf{H}_{n-1} + \delta \mathbf{A}_n \mathbf{J}_n + \delta \mathbf{B}_n \mathbf{H}_{n-1}. \tag{19.234}$$

The form of linearized equations (19.233) and (19.234) is identical to that of equations (19.231) and (19.232), so that the linearized equations can be solved in the same manner as the original equations. The corrections to the mean intensities are expressed through the corrections to the other structural parameters—atomic level populations, and possibly also the temperature and electron density (depending whether one assumes temperature and density to be fixed or to be determined together with the radiation field), namely [cf. equation (18.147)],

$$\delta J_{nd} = \sum_{d'=1}^{D} \sum_{j=1}^{NC} \frac{\partial J_{nd}}{\partial x_{jd'}} \, \delta x_{jd'}, \tag{19.235}$$

where $x_{jd}, j = 1, \ldots, NC$ is the set of the structural parameters at depth d; their number is NC, the number of "constraint" equations. As mentioned in § 18.4, the gist of the method is to replace the summation over the depth points in equation (19.235), which extends over all depths (radii) by a restricted summation taking just a several depths around the point d,

$$\delta J_{nd} = \sum_{d'=d-a}^{d+a} \sum_{j=1}^{NC} \frac{\partial J_{nd}}{\partial x_{jd'}} \, \delta x_{jd'}, \tag{19.236}$$

where a is set to a small number, typically $a = 1$, which corresponds to a tridiagonal approximate Newton-Raphson operator. In other words, in evaluating corrections to the mean intensities at certain radial point, one takes into account corrections of the state parameters at this point plus its immediate neighbors. For more details of the implementation refer to [492]. The method forms the basis of an efficient and popular code CMFGEN for computing NLTE spherically symmetric expanding model atmospheres [492, 493].

19.4 MOVING ATMOSPHERES: MIXED-FRAME FORMULATION

In the previous two sections we have discussed two different formulations of the radiative transfer in moving media, namely, those formulated in the inertial observer's-frame and those formulated in the Lagrangian comoving-frame. Both these approaches have specific advantages and disadvantages. The advantage of the observer's-frame formulation is that the differential operator in the transfer equation—the *streaming term*—contains only derivatives with respect to the geometrical coordinates (r and μ in spherically symmetric media), and hence is as simple as in the static case. However, the right-hand side—an *interaction* term—is cumbersome. The opposite applies for the comoving-frame formalism. Here, the streaming term is very complicated, while the interaction term is as simple as in the static case. Besides that, the most significant disadvantage of the comoving-frame approach is that its practical applications are restricted to monotonic velocity fields.

It is therefore tempting to formulate an approach that would combine the advantages of both standard formulations. This is the main idea behind the *mixed-frame* approach, in which the radiation quantities are defined in the inertial (observer's) frame, while the interaction between radiation and matter is described in the comoving frame, using a Taylor expansion about the inertial-frame values. Therefore, such a formulation of the interaction term is only approximate. This approach was suggested by Mihalas and Klein [739], but it has largely been neglected in radiation transport and stellar atmospheres work because it fails to treat spectral lines accurately, particularly for supersonic velocity fields.

However, many radiation-hydrodynamic problems do not require an elaborate treatment of spectral line transport, but rather a good treatment of continuum transport. In this situation, the mixed-frame formulation is ideally suited, thanks to several advantages. (1) Instead of having a number of velocity-dependent terms on the left-hand (differential) side of the transfer equation, many of which involve spatial velocity derivatives, there are no such terms on the left-hand side in the mixed-frame approach. (2) There are no terms with derivatives of the velocity. (3) The characteristics of the associated transport equation are straight lines. (4) There is no need for the monotonicity in the velocity field required by the comoving-frame formalism. (5) The mixed-frame approach is easily generalized to two and three dimensions, and the associated solvers are straightforward, albeit more expensive, extensions of those employed in one dimension.

The approach was generalized in [526] to include an anisotropic (and possibly non-coherent) continuum scattering term. It turned out that it is very well suited to neutrino transport, in which all the relevant neutrino-matter interaction cross sections are smooth functions of neutrino energy. We stress that although neutrino transport is not specifically considered in this book, the formalism is exactly the same as for photon transport; hence the same equations and same solution algorithms can be used in both fields.

Formulation of the Transfer Equation

The general transfer equation is written as

$$\left(\frac{1}{c}\frac{\partial}{\partial t} + \mathbf{n}\cdot\nabla\right) I(\nu,\mathbf{n}) = \eta_{\rm th}(\nu,\mathbf{n}) + \eta_{\rm sc}(\nu,\mathbf{n}) - [\kappa(\nu,\mathbf{n}) + \sigma(\nu,\mathbf{n})]\, I(\nu,\mathbf{n}), \tag{19.237}$$

where κ is the true absorption coefficient, σ is the scattering coefficient, $\eta_{\rm th}$ is the thermal emission coefficient, and $\eta_{\rm sc}$ the scattering part of the emission coefficient. The total absorption coefficient is $\chi(\nu,\mathbf{n}) = \kappa(\nu,\mathbf{n}) + \sigma(\nu,\mathbf{n})$.

Denoting by subscript 0 quantities in the comoving frame, the Lorentz transformations of the frequency and direction are (in a general case)

$$\nu_0 = \nu\gamma(1 - \mathbf{n}\boldsymbol{\beta}), \tag{19.238}$$

and

$$\mathbf{n}_0 = (\nu/\nu_0)\left\{\mathbf{n} - \boldsymbol{\beta}\left[\gamma - \mathbf{n}\boldsymbol{\beta}\gamma^2/(\gamma+1)\right]\right\}, \tag{19.239}$$

where $\boldsymbol{\beta} \equiv \mathbf{v}/c$. To $O(v/c)$, one has the usual expressions

$$\nu_0 = \nu\,(1 - \mathbf{n}\boldsymbol{\beta}), \tag{19.240}$$

$$\mathbf{n}_0 = (\nu/\nu_0)(\mathbf{n} - \boldsymbol{\beta}) = \mathbf{n}\,(1 + \mathbf{n}\boldsymbol{\beta}) - \boldsymbol{\beta}. \tag{19.241}$$

The absorption coefficient transforms as [cf. equation (19.122)]

$$\kappa(\nu, \mathbf{n}) = (\nu_0/\nu)\kappa^0(\nu_0), \tag{19.242}$$

and similarly for the scattering coefficient σ. The emission coefficient in equation (19.237) transforms as [cf. equation (19.123)]

$$\eta_{\text{th}}(\nu, \mathbf{n}) = (\nu/\nu_0)^2 \eta^0_{\text{th}}(\nu_0, \mathbf{n}_0). \tag{19.243}$$

The essential point is to express the quantities on the right-hand side of the transfer equations through the corresponding comoving-frame quantities and their derivatives, using a linear approximation of the Taylor expansion [739]. The absorption coefficient is expanded as

$$\kappa^0(\nu_0) = \kappa^0(\nu) + (\nu_0 - \nu)(\partial\kappa^0/\partial\nu); \tag{19.244}$$

hence we obtain from equations (19.238) and (19.242)

$$\kappa(\nu, \mathbf{n}) = \kappa^0(\nu) - \mathbf{n}\boldsymbol{\beta}\left[\kappa^0(\nu) + \nu(\partial\kappa^0/\partial\nu)\right]. \tag{19.245}$$

The transformation of σ is analogous. The thermal emission coefficient is written as [739]

$$\eta_{\text{th}}(\nu, \mathbf{n}) = \eta^0_{\text{th}}(\nu) + \mathbf{n}\boldsymbol{\beta}\left[2\eta^0_{\text{th}}(\nu) - \nu(\partial\eta^0_{\text{th}}/\partial\nu)\right], \tag{19.246}$$

where the thermal emission coefficient is assumed to be isotropic in the comoving frame.

The scattering emission term in the comoving frame is given by

$$\eta^0_{\text{sc}}(\nu_0, \mathbf{n}_0) = \sigma^0(\nu_0) \oint (d\Omega'_0/4\pi)\, I^0(\nu_0, \mathbf{n}'_0)\, g^0(\mathbf{n}'_0 \cdot \mathbf{n}_0), \tag{19.247}$$

where g^0 is the phase function for the scattering; for simplicity it is set here $g^0 \equiv 1$ (isotropic scattering). A generalization for anisotropic scattering is presented in [526]. The primed quantities refer to the properties of the absorbed photon. The scattering emission coefficient transforms according to equations (19.243) and (19.247) as

$$\eta_{\text{sc}}(\nu, \mathbf{n}) = (\nu/\nu_0)^2 \sigma^0(\nu_0) \oint (d\Omega'_0/4\pi)\, I^0(\nu_0, \mathbf{n}'_0). \tag{19.248}$$

The specific intensity transforms as

$$I^0(\nu_0, \mathbf{n}'_0) = (\nu_0/\nu)^3 I(\nu, \mathbf{n}'), \tag{19.249}$$

and the element of the solid angle as

$$d\Omega'_0 = (\nu/\nu_0)^2 d\Omega', \tag{19.250}$$

and analogously for $\sigma^0(\nu_0)$,

$$\sigma^0(\nu_0) = \sigma^0(\nu) - \mathbf{n}\boldsymbol{\beta}\, \nu\, (\partial\sigma^0/\partial\nu). \tag{19.251}$$

After some algebra, one obtains for the transfer equation in the mixed-frame formalism

$$\begin{aligned}[(1/c)(\partial/\partial t) + \mathbf{n}\cdot\nabla] I(\nu, \mathbf{n}) &= \eta^0_{\rm th} - \chi^0 I(\nu, \mathbf{n}) + \sigma^0 J(\nu) \\ &+ \mathbf{n}\boldsymbol{\beta}[\widetilde{\eta}^0_{\rm th} + \widetilde{\chi}^0 I(\nu, \mathbf{n}) + \sigma_J J(\nu)] - \boldsymbol{\beta}\boldsymbol{\sigma}_H \mathbf{H}(\nu),\end{aligned} \tag{19.252}$$

where

$$\widetilde{\kappa}^0 \equiv \kappa^0[1 + (\partial \ln \kappa^0/\partial \ln \nu)], \tag{19.253}$$

$$\widetilde{\sigma}^0 \equiv \sigma^0[1 + (\partial \ln \sigma^0/\partial \ln \nu)], \tag{19.254}$$

$$\widetilde{\chi}^0 \equiv \widetilde{\kappa}^0 + \widetilde{\sigma}^0, \tag{19.255}$$

$$\widetilde{\eta}^0 \equiv \eta^0_{\rm th}[2 - (\partial \ln \eta^0_{\rm th}/\partial \ln \nu)], \tag{19.256}$$

$$\sigma_J \equiv \sigma^0 \left[2 - (\partial \ln \sigma^0/\partial \ln \nu) - (\partial \ln J/\partial \ln \nu)\right], \tag{19.257}$$

$$\boldsymbol{\sigma}_H \equiv \sigma^0\, [1 - (\partial \ln \mathbf{H}/\partial \ln \nu)]\,, \tag{19.258}$$

and J and $\mathbf{H}$ are the usual moments of the specific intensity.

Spherically-Symmetric Atmospheres

In the spherically symmetric situation, the transfer equation (19.252) is written, in *conservative form*, as

$$\begin{aligned}&\frac{1}{c}\frac{\partial I}{\partial t} + \frac{\mu}{r^2}\frac{\partial(r^2 I)}{\partial r} + \frac{1}{r}\frac{\partial[(1-\mu^2)I]}{\partial \mu} \\ &= \left(\eta^0 + \mu\beta\widetilde{\eta}^0\right) - (\chi^0 - \mu\beta\widetilde{\chi}^0) I + (\sigma^0 + \mu\beta\sigma_J) J - \beta\sigma_H H,\end{aligned} \tag{19.259}$$

where the only nonzero component of $\mathbf{H}$ is the radial component, denoted by H, and analogously for $\boldsymbol{\beta}$ and $\boldsymbol{\sigma}_H$. The moment equations read, expressing the second moment K through the Eddington factor f, defined by $K \equiv fJ$,

$$\frac{1}{c}\frac{\partial J}{\partial t} + \frac{1}{r^2}\frac{\partial(r^2 H)}{\partial r} = \eta^0 - \kappa^0 J + \Xi H\,, \tag{19.260}$$

and

$$\frac{1}{c}\frac{\partial H}{\partial t}+\frac{1}{r^2}\frac{\partial(r^2fJ)}{\partial r}-\frac{1-f}{r}J=-\chi^0H+\tfrac{1}{3}\beta\widetilde{\eta}^0+\xi J, \tag{19.261}$$

where

$$\Xi=\beta\left[\widetilde{\kappa}^0+\sigma^0\left(\frac{\partial\ln\sigma^0}{\partial\ln\nu}+\frac{\partial\ln H}{\partial\ln\nu}\right)\right] \tag{19.262}$$

and

$$\xi=\beta(\tfrac{1}{3}\sigma_J+f\widetilde{\chi}^0). \tag{19.263}$$

Assuming a stationary situation ($\partial/\partial t\equiv 0$), one can rewrite equation (19.259) in non-conservative form,

$$\mu\frac{\partial I}{\partial r}+\frac{1-\mu^2}{r}\frac{\partial I}{\partial\mu}=(\eta^0+\mu\beta\widetilde{\eta}^0)-(\chi^0-\mu\beta\widetilde{\chi}^0)I$$
$$+(\sigma^0+\mu\beta\sigma_J)J-\beta\sigma_H H. \tag{19.264}$$

As before, it is advantageous to employ the tangent-ray approach. One considers transfer along the ray specified by a constant impact parameter p. The coordinate along p is z, where $r=(p^2+z^2)^{\frac{1}{2}}$ and $\mu=z/r$. As $\mu(\partial/\partial r)+r^{-1}(1-\mu^2)\partial/\partial\mu\equiv d/dz$, the characteristics of equation (19.264)—the tangent rays along which one solves the transfer equation—are straight lines.

The intensity propagating in the direction of increasing z is denoted by I^+ or $I(\mu)$, and that for decreasing z by I^- or $I(-\mu)$. The transfer equation along the ray is written in a compact form:

$$dI^{\pm}(p,z)/d\tau^{\pm}(p,z)=I^{\pm}(p,z)-S^{\pm}(p,z), \tag{19.265}$$

where $d\tau^{\pm}(p,z)\equiv\mp\chi^{\pm}\,dz$, and

$$S^{\pm}(p,z)=S_0^{\pm}(p,z)+a^{\pm}(p,z)J(r)+b^{\pm}(p,z)H(r), \tag{19.266}$$

where

$$\chi^{\pm}(p,z)=\chi^0(r)\mp\mu(p,z)\beta(r)\widetilde{\chi}^0(r)=\chi^0(r)\mp(z/r)\beta(r)\widetilde{\chi}^0(r), \tag{19.267}$$

with

$$S_0^{\pm}(p,z)=\frac{\eta^0\pm\mu\beta\widetilde{\eta}^0}{\chi^{\pm}(p,z)} \tag{19.268}$$

$$a^{\pm}(p,z)=\frac{\sigma^0\pm\beta\mu\sigma_J}{\chi^{\pm}(p,z)} \tag{19.269}$$

$$b^{\pm}(p,z)=-\frac{\beta\sigma_H}{\chi^{\pm}(p,z)}. \tag{19.270}$$

Equation (19.265) is discretized as described in § 19.1 and § 19.2. The radius grid is defined in terms of depth index $d = 1, \dots, D$, which increases inward from the surface: $r_1 = R > r_2 > \dots > R_c$, where R_c is the core radius. The impact parameters are labeled by the same index as the radii; i.e., the impact parameter for the jth ray is $p_j = r_j$. In addition, one introduces C core rays with $0 \le p_{D+j} \le R_c, j = 1, \dots, C$. The total number of rays (impact parameters) is, thus, $I = D + C$.

The moments J, H, and K, which are integrals over angles, are expressed as quadratures over the impact parameters, viz.

$$J(r_d) = \sum_{j=d}^{I} w_{jd} \left[I^+(p_j, z_d) + I^-(p_j, z_d) \right], \tag{19.271}$$

$$H(r_d) = \sum_{j=d}^{I} w_{jd}\, \mu_{jd} \left[I^+(p_j, z_d) - I^-(p_j, z_d) \right], \tag{19.272}$$

$$K(r_d) = \sum_{j=d}^{I} w_{jd}\, \mu_{jd}^2 \left[I^+(p_j, z_d) + I^-(p_j, z_d) \right]. \tag{19.273}$$

In the source function (equation 19.266) the parameters $a^{\pm}$, $b^{\pm}$, and $S_0^{\pm}$ are known functions of r, while the only unknowns are the moments J and H, which have to be solved for self-consistently with the transfer equation. Using the Feautrier scheme, one can in principle obtain a direct (non-iterative) solution, as in the case of the standard comoving-frame transfer equation in spherical geometry described in § 19.3. However, it is usually more efficient to use an iterative scheme, which typically offers a significant reduction of the required computing time. Such a scheme is outlined below.

ALI Iteration Scheme

The transfer equation (19.265), with the source function given by equation (19.266), is solved by an application of an accelerated lambda iteration (ALI) scheme. The solution of equation (19.265) can be written formally as

$$I_{j,d}^{\pm} = \sum_{d'=1}^{j} \Lambda_{j,dd'}^{\pm} S_{j,d'}^{\pm}, \tag{19.274}$$

where $I_{j,d}^{\pm} \equiv I^{\pm}(p_j, z_d)$ and $S_{j,d}^{\pm} \equiv S^{\pm}(p_j, z_d)$. The iteration scheme adopted here is based on approximate Λ-operator given by the diagonal (local) part of the exact Λ. (A generalization to a tridiagonal operator, which is also employed in [526], is straightforward.) Dropping the superscripts $\pm$, one writes

$$I_{j,d}^{\rm new} = \Lambda_{j,dd} S_{j,d}^{\rm new} + \sum_{d'=1,\, d' \neq d}^{j} \Lambda_{j,dd'} S_{j,d'}^{\rm old}. \tag{19.275}$$

Using this expression, one can express the moments as

$$J_d^{\text{new}} = \frac{1}{2} \sum_{j=d}^{I} w_{jd} \Lambda_{j,dd}^{\pm} [S_{jd}^0 + a_{jd}^{\pm} J_d^{\text{new}} + b_{jd}^{\pm} H_d^{\text{new}}] + \text{``old'' terms},$$

$$H_d^{\text{new}} = \frac{1}{2} \sum_{j=d}^{I} w_{jd} \mu_{jd} \Lambda_{j,dd}^{\pm} [S_{jd}^0 + a_{jd}^{\pm} J_d^{\text{new}} + b_{jd}^{\pm} H_d^{\text{new}}] + \text{``old'' terms}.$$

(19.276)

Here, the superscript $\pm$ indicates that the summation is carried over both the $+$ and $-$ terms. After some algebra, one obtains a set of two coupled equations for the new values of the moments J and H at the given radius r_d,

$$\begin{pmatrix} 1 - \Lambda_d^{JJ} & -\Lambda_d^{JH} \\ -\Lambda_d^{HJ} & 1 - \Lambda_d^{HH} \end{pmatrix} \cdot \begin{pmatrix} J_d^{\text{new}} - J_d^{\text{old}} \\ H_d^{\text{new}} - H_d^{\text{old}} \end{pmatrix} = \begin{pmatrix} J_d^{\text{FS}} - J_d^{\text{old}} \\ H_d^{\text{FS}} - H_d^{\text{old}} \end{pmatrix}, \tag{19.277}$$

where J^{FS} is given by equation (19.271) with the specific intensity given by the "old" intensity I_{jd}^{old} (and analogously for H), and the matrix elements are given by

$$\Lambda_d^{JJ} = \sum_{j=d}^{I} w_{jd} \, (a_{j,d}^{+} \Lambda_{j,dd}^{+} + a_{j,d}^{-} \Lambda_{j,dd}^{-}), \tag{19.278}$$

$$\Lambda_d^{JH} = \sum_{j=d}^{I} w_{jd} \, (b_{j,d}^{+} \Lambda_{j,dd}^{+} + b_{j,d}^{-} \Lambda_{j,dd}^{-}), \tag{19.279}$$

$$\Lambda_d^{HJ} = \sum_{j=d}^{I} w_{jd} \, \mu_{j,d} \, (a_{j,d}^{+} \Lambda_{j,dd}^{+} - a_{j,d}^{-} \Lambda_{j,dd}^{-}), \tag{19.280}$$

$$\Lambda_d^{HH} = \sum_{j=d}^{I} w_{jd} \, \mu_{j,d} \, (b_{j,d}^{+} \Lambda_{j,dd}^{+} - b_{j,d}^{-} \Lambda_{j,dd}^{-}). \tag{19.281}$$

To evaluate new values of the moments, one has to invert one simple 2×2 matrix per depth point. The individual matrix elements $\Lambda_{j,dd}$ are evaluated during the formal solution step. It was found [526] that one can use, without any deterioration of the convergence properties the ALI iteration procedure, a simplified approximate Λ-matrix that is obtained by dropping the off-diagonal terms of the global 2×2 Λ-matrix used in equation (19.277) by setting

$$\Lambda_d^{JH} = \Lambda_d^{HJ} = 0. \tag{19.282}$$

This procedure reduces inverting the 2×2 Λ-matrix to a simple scalar division.

It should be pointed out that if anisotropic scattering is taken into account using the phase function $g(\mathbf{n}', \mathbf{n}) = a + b\,\mathbf{n}' \cdot \mathbf{n}$, the source function (19.266) will contain in addition the second moment K; consequently, equation (19.277) should be formulated through a 3×3 matrix; see [526].

The iteration scheme proceeds as follows.

(i) For given moments $J^{(n)}, H^{(n)}$ (and with a suitable initial estimate of $J^{(0)}, H^{(0)}$), perform a set of formal solutions for all impact parameters p, so obtain new specific intensities, which are denoted $I^{\pm}_{\rm FS}(p, z)$.
(ii) Compute new values for the moments $J^{\rm FS}, H^{\rm FS}$ using equations (19.271)–(19.272), with the specific intensity $I_{\rm FS}$.
(iii) Solve equation (19.277), radius by radius, to obtain new values for the three moments.
(iv) As long as the new moments differ from the old moments, iterate steps (i) through (iii) to convergence.

Again, since the crucial part, the update of the source function (step iii), is very simple, the problem is effectively reduced to a set of formal solutions along the tangent rays. As before, there are three possibilities for solving the transfer equation along the ray, using the Feautrier method, the short characteristics scheme, or the discontinuous finite element (DFE) scheme.

19.5 SOBOLEV APPROXIMATION

In the previous three sections we have considered various approaches that are capable of providing an exact solution of the transfer problem in moving atmospheres (§ 19.2 and § 19.3) or approximate schemes applicable in the case of smooth variation of opacity with frequency (§ 19.4). All approaches developed there require rather elaborate and time-consuming numerical calculations.

It was realized by Sobolev [1027, 1029] long ago, and later elaborated and extended by Castor [194, 197] and in [553, 890, 924], that the presence of a large velocity gradient in an expanding medium actually simplifies line-transfer problems, for it dominates the photon escape and thermalization process and implies a *geometric localization of the source function* not present in static problems. In Sobolev's theory, the solution of the transfer problem is, in effect, replaced by the calculation of *escape probabilities*. Unlike the case of static atmospheres where the escape probability approach is always approximate and thus inadequate for modern theoretical spectroscopy, using the escape probability, or Sobolev, approach for rapidly expanding atmospheres is reasonably accurate, and hence still widely used.

Surfaces of Constant Radial Velocity

Most of the emission observed at frequency ν arises from regions where the Doppler-shifted line center frequency is at the observed value. The observed flow velocities in early-type atmospheres (Of and WR), which are up to 3000 km s^{-1}, vastly exceed the thermal velocity ($\sim 30\,\text{km s}^{-1}$). Therefore, the geometrical region from which the emission at any one frequency arises must be a very thin region, centered on a *surface of constant radial velocity*, also called *resonance surface*, such that $(\nu - \nu_0)/\nu_0 = v_r\mu/c$. Here, v_r is the velocity in the radial direction. Measuring frequency displacement from the line center in units of Doppler width, $x \equiv (\nu - \nu_0)/\Delta\nu_D = (\nu - \nu_0)c/(\nu_0 v_{\text{th}})$, and expressing velocities in terms of the thermal velocity, $V_r \equiv v_r/v_{\text{th}}$, the shape of the resonance surface is determined simply by $\mu V_r = x$.

The shape of resonance surfaces depends on the nature of the velocity field, which, ideally, should be given by hydrodynamic calculations. However, one can gain considerable insight by consideration of some simple velocity laws of an analytic form. Figure 19.4 (a and b) shows the constant-radial-velocity surface for two such cases.

(a) $V = V_\infty(1 - r_c/r)^{\frac{1}{2}}$, which mimics a transsonic wind, and which corresponds to the situation where the gas leaves the star with velocities greater than escape velocity and accelerates everywhere for $r \geq r_c$ (see chapter 20 for details).
(b) $V = r^{-\frac{1}{2}}$, which corresponds to the situation where the material is ejected with just the escape velocity and is decelerated by gravity. Here, r is expressed in units of the photospheric radius, $r_0 = r_c$.

Inspecting the results displayed in figure 19.4, some important conclusions can be drawn. The fact that the surfaces of constant radial velocity extend over large regions (even infinite if the flow is not decelerating) implies a complete *breakdown of the Eddington-Barbier relation* for expanding atmospheres. One can no longer associate a given frequency in the line profile with a specific position in the envelope, but only with a wide range of positions. From the viewpoint of an outside observer, geometric localization occurs only if variations in total particle density and ionization-excitation equilibria confine the region of high emissivity. What is worse, this conclusion is *independent of the intrinsic line strength* [548]. So, in principle, one can no longer obtain a depth analysis of atmospheric structure by examining weak and strong lines. These considerations imply significant modeling and diagnostic problems for expanding atmospheres.

Another type of problem, also seen in figure 19.4b, is that in the case of decelerating flows a particular line of sight may intersect a surface of constant radial velocity at two distinct points; hence two regions, which may have vastly different physical properties, contribute to the information received by the observer. Furthermore, these two distinct regions can also interact radiatively, so the Sobolev method as described below has to be reformulated [924].

Escape Probability in Expanding Atmospheres

Consider a line in an optically thick expanding envelope surrounding an opaque core of radius r_c and assume that the effects of a background continuum can be ignored. Consider a coordinate system that is at rest with respect to a particular fluid element. Along a ray that passes through the test point there will be a differential Doppler shift at each successive sample point along the ray relative to the test point. Eventually

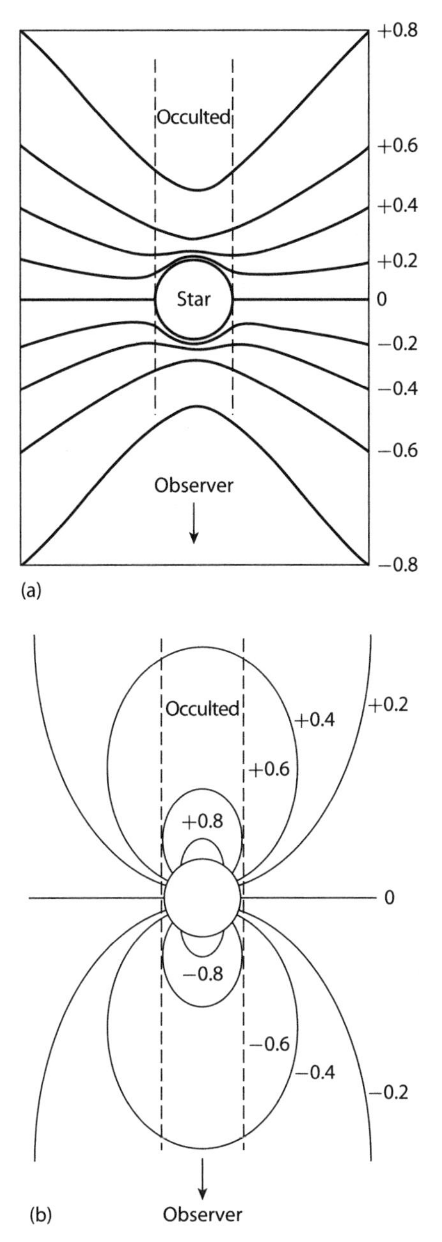

Figure 19.4 Surfaces of constant radial velocity, $v_z = \text{constant}$. (a) $V(r) = V_\infty(1 - r_c/r)^{\frac{1}{2}}$. Curves are labeled with V_z/V_∞. (b) $V(r) = r^{-\frac{1}{2}}$. From [639].

this shift becomes so large that no line photon emitted within the effective bounds of the line profile (assumed to be limited by some $\pm\Delta\nu_{\max}$, with $\Delta\nu_{\max} \sim \Delta\nu_D$, the Doppler width) can interact with the line profile at the test point. The velocity field has effectively introduced an *intrinsic escape mechanism* for photons, or a definite *interaction region*, which is a region within which photons emitted, or scattered, can have any effect upon the intensity in the line at the test point. Beyond the interaction region, as measured from the point of emission, photons can no longer be absorbed in a line by the material, even if it is of infinite extent, but escape freely to infinity. Moreover, in the limit of *large velocity gradients* the interaction region will be small, and hence may be assumed to be *homogeneous* in its physical properties (temperature, density, ionization state, etc.). The theory can then be formulated in terms of *local quantities* and through a parameter β that gives the *probability of photon escape* summed over all directions and line frequencies.

In analogy with the escape probability approximation in static media, one can write

$$\bar{J}(r) = (1-\beta)S(r) + \beta_c I_c. \tag{19.283}$$

The first term is derived from the value that $\bar{J}$ would have in the limit of no escapes, namely, $\bar{J} = S$, corrected for velocity-induced escapes. The parameter β_c measures the probability of penetration (summed over frequency and angle) of the specific intensity I_c emitted from the core to the test point. We must now calculate β and β_c.

As was done earlier, we express velocities in units of a thermal velocity, $V(r) \equiv v(r)/v_{\text{th}}$, and frequency displacements from the line center in Doppler units, $x \equiv (\nu - \nu_0)/\Delta\nu_{\text{D}}$, where $\Delta\nu_{\text{D}} = \nu_0 v_{\text{th}}/c$. The optical depth along a ray to an observer at infinity can be written

$$\tau(z,p,x) = \int_z^\infty \chi(z',p,x)\,dz' = \int_z^\infty \chi_l(r')\phi(x')\,dz', \tag{19.284}$$

where

$$\chi_l(r') = (\pi e^2/mc) f_{ij}[n_i(r') - (g_i/g_j)n_j(r')]/\Delta\nu_{\text{D}} \equiv \kappa_0(r')/\Delta\nu_{\text{D}}, \tag{19.285}$$

and where $r' \equiv (z'^2 + p^2)^{\frac{1}{2}}$, $\mu' \equiv z'/r'$, and

$$x' = x'(z',p,x) \equiv x - V_z(z') = x - \mu' V(r'). \tag{19.286}$$

The main contribution to the integral in equation (19.284) comes from the region close to the point where $x' \sim 0$, i.e., $z' = z_0(p,x)$, where z_0 is chosen such that $z_0 r_0^{-1} V(r_0) = x$; here $r_0 \equiv (z_0^2 + p^2)^{\frac{1}{2}}$. The surface of $z_0(p,x)$ is obviously just a surface of constant radial velocity, and the point $z = z_0$ or $r = r_0$ is usually referred to as the *resonance point*.

The essence of the Sobolev approximation is to assume that the material properties vary *slowly* within the region close to the resonance point. Specifically, for a given observer's-frame frequency ν, and along a given line of sight, the resonance point is defined by the condition $(\nu - \nu_0)/\nu_0 = v_l/c$, where v_l is the velocity projected to the line of sight and ν_0 the line-center frequency. Differentiating this expression, one obtains

$$\Delta l = \frac{c}{\nu_0}\frac{\Delta\nu}{|dv_l/dl|}. \tag{19.287}$$

If one takes for $\Delta\nu$ the width of the absorption profile of a line, then Δl represents a size of the resonance region. As the width of the absorption profile is roughly given by the Doppler width, $\Delta\nu \sim \Delta\nu_D$, the characteristic size of the resonance region is

$$\Delta l_S = \frac{c}{\nu_0}\frac{\Delta\nu_D}{|dv_l/dl|}, \tag{19.288}$$

which is called the *Sobolev length.* Furthermore, approximating $|dv_l/dl| \sim v_0/l_0$, where v_0 is a typical velocity and l_0 a typical length scale in the medium, and expressing the Doppler width as $\Delta\nu_D = (\nu_0/c)v_{\rm th}$, one obtains

$$\Delta l/l_0 \approx v_{\rm th}/v_0. \tag{19.289}$$

Therefore, the resonance region will be relatively sharp when the macroscopic velocities are much larger than thermal; for this reason the Sobolev approximation is sometimes called the *supersonic approximation.*

Under this approximation, one can replace $\chi_l(r')$ with $\chi_l(r_0)$ and remove this factor from the integral in (19.283). One can change the variable of integration from z' to x'. According to equation (19.286), the transformation is

$$\begin{aligned} -(\partial x'/\partial z)_p &= (\partial V_z/\partial z)_p = (\partial\{\mu(z,p)V[r(z,p)]\}/\partial z)_p \\ &= \mu^2(\partial V/\partial r) + (1-\mu^2)(V/r) \equiv Q(r,\mu), \end{aligned} \tag{19.290}$$

where μ and r are again understood to be functions of z and p. Thc quantity Q is exactly what was called dv_l/dl in equation (19.287). If the interaction region is small, the transformation coefficient written above may be assumed to be essentially constant and may be evaluated at the resonance point $z = z_0(p,x)$. One then defines

$$\Phi(x) \equiv \int_{-\infty}^{x} \phi(x')\,dx', \tag{19.291}$$

where $\Phi(-\infty) = 0$ and $\Phi(\infty) = 1$. Equation (19.284) can be rewritten as

$$\tau(z,p,x) = \tau(-\infty,p,x)\,\Phi[x'(z,p,x)], \tag{19.292}$$

where $x'(z,p,x)$ is defined by equation (19.286) and

$$\begin{aligned} \tau(-\infty,p,x) &= \chi_l(r_0)/Q(r_0,\mu_0) \\ &\equiv \tau_0(r_0)/\{1+\mu^2[(d\ln V/d\ln r)-1]\}_0, \end{aligned} \tag{19.293}$$

where

$$\tau_0(r_0) \equiv \chi_l(r_0)/(V/r)_0. \tag{19.294}$$

It should be stressed that in equations (19.293)–(19.294), r_0 and μ_0 are viewed as functions of p and x, i.e., $r_0 = r_0[z_0(p,x),p]$ and $\mu_0 = \mu_0[z_0(p,x),p]$. Using the expression $\chi_l = \kappa_0/\Delta\nu_D = \kappa_0 c/(\nu_0 v_{\rm th})$, equation (19.294) can also be written as

$$\tau_0 = \kappa_0 r_0 c/(\nu_0 v_0). \tag{19.295}$$

Sometimes one defines the Sobolev optical depth as a frequency-averaged line optical depth corresponding to the Sobolev length,

$$\tau_S \equiv \chi_0 \Delta l_S = (\kappa_0/\Delta\nu_D)(c/\nu_0)(\Delta\nu_D/|dv_l/dl|) = (\kappa_0 c/\nu_0)/|dv_l/dl|. \quad (19.296)$$

One often uses the Sobolev optical depth in the radial direction

$$\tau_S \equiv (\kappa_0 c/\nu_0)/|dv_l/dr| = \chi_l v_{\rm th}/|dv/dr|. \quad (19.297)$$

The escape probability $\beta(r)$ is calculated for a given value of r; an integration over μ can be effected by using the above results for various values of p. The escape probability along the ray is simply $\exp(-\Delta\tau_\infty)$ [cf. equation (11.158)], where $\Delta\tau_\infty$ is the optical path length from the test point to infinity. Integrating over angle and frequency, one obtains

$$\beta(r) = \tfrac{1}{2}\int_{-1}^{1} d\mu \int_{-\infty}^{\infty} dx\, \phi[x'(z,p,x)] \exp\{-\tau[z(r,\mu),p(r,\nu,x)]\}. \quad (19.298)$$

Here it is assumed that photons that hit the opaque core are absorbed, and hence lost. Equation (19.298) is evaluated using equations (19.291) through (19.293) and assuming that the material in the interaction region is sufficiently homogeneous that the distinction between r_0 and r may be ignored. Then

$$\begin{aligned}\beta(r) &= \tfrac{1}{2}\int_{-1}^{1} d\mu \int_0^1 d\Phi\, \exp[-\chi_l(r)\Phi/Q(r,\mu)] \\ &= [1/\chi_l(r)] \int_0^1 \{1-\exp[-\chi_l(r)/Q(r,\mu)]\}Q(r,\mu)\, d\mu. \quad (19.299)\end{aligned}$$

For the special case that $V = kr$, $Q(r,\mu) = k$, and equation (19.299) reduces considerably to

$$\beta(r) = \{1-\exp[-\tau_0(r)]\}/\tau_0(r), \quad (19.300)$$

where now $\tau_0(r) = k^{-1}\chi_l(r)$. The same result is obtained if the angle-dependent terms in equation (19.290) are merely *ignored.*

To calculate β_c, assume that the test point is relatively far from the core (i.e., that the surface of the core is at $-\infty$). Then, from its physical meaning, β_c can be written

$$\begin{aligned}\beta_c(r) &= \tfrac{1}{2}\int_{-1}^{-\mu_c} d\mu \int_0^1 d\Phi\, \exp[-\chi_l(r)\Phi/Q(r,\mu)] \\ &= [1/2\chi_l(r)] \int_{\mu_c}^1 \{1-\exp[-\chi_l(r)/Q(r,\mu)]\}Q(r,\mu)\, d\mu, \quad (19.301)\end{aligned}$$

where $\mu_c \equiv [1 - (r_c/r)^2]^{\frac{1}{2}}$. Again, for the special case of a linear velocity law one obtains a considerable simplification, namely, $\beta_c(r) = W\beta(r)$, where W is the usual dilution factor defined by

$$W = \tfrac{1}{2}\{1 - [1 - (r_c/r)^2]^{\frac{1}{2}}\}. \tag{19.302}$$

The result just quoted is what would be expected physically, because W is the fraction of the full sphere contained in the solid angle subtended by the disk, while β measures the probability of penetration from the disk to the test point.

Note that both β and β_c are defined in terms of *local* quantities: the total opacity and the velocity gradient. Given these values, one can compute $\overline{J}$ from equation (19.283) *without actually solving the transfer equation*. This is the main advantage of the Sobolev approximation, which offers an enormous simplification of the problem. For the particular case of a two-level atom, where the source function (assuming complete redistribution) is given by $S = (1 - \epsilon)\overline{J} + \epsilon B$, one may use equation (19.283) to write

$$S = [(1 - \epsilon)\beta_c I_c + \epsilon B]/[(1 - \epsilon)\beta + \epsilon], \tag{19.303}$$

which shows that knowledge of β and β_c is sufficient to determine S.

It is instructive to consider a *uniformly expanding plane-parallel atmosphere*, for then one can obtain expressions that show the effects of velocity gradient on the thermalization of the source function in a particularly transparent way ([408, p. 87] and [709]). Let τ denote integrated line optical depth defined for a medium at rest and assume that the velocity gradient $\gamma = \partial V/\partial\tau$ is constant. The specific intensity at a test point τ in direction μ is

$$I(\tau, \mu, x) = \int_\tau^\infty S(\tau') \exp\left[-\int_0^{\tau'-\tau} \phi(x + \gamma\mu t)\, dt/\mu\right] \phi[x + \gamma\mu(\tau' - \tau)]\, d\tau'/\mu. \tag{19.304}$$

Hence the source function for a two-level atom is given by the integral equation

$$S(\tau) = (1 - \epsilon)\overline{J}(\tau) + \epsilon B(\tau) = (1 - \epsilon)\int_{-\infty}^\infty K_\beta(|\tau' - \tau|) S(\tau')\, d\tau' + \epsilon B(\tau), \tag{19.305}$$

where the kernel function

$$K_\beta(s) = \tfrac{1}{2}\int_{-\infty}^\infty dx \int_0^\beta d\mu\, \mu^{-1}\phi(x)\phi(x + \gamma\mu s) \exp\left[-\int_0^s \phi(x + \gamma\mu t)\, dt/\mu\right]. \tag{19.306}$$

It is easy to show that, unlike the static case where the kernel is normalized to unity, in the present case the effects of escapes lead to

$$\int_{-\infty}^\infty K_\beta(|\tau|)\, d\tau = 1 - \beta, \tag{19.307}$$

where β is the plane-parallel escape probability that follows from equations (19.299) and (19.290) in the limit $1/r \to 0$, namely,

$$\beta = |\gamma| \int_0^1 \{1 - \exp[-1/(|\gamma|\,\mu^2)]\}\mu^2\, d\mu. \tag{19.308}$$

Equation (19.305) may be recast into the standard form for a two-level atom by renormalizing the kernel to $K^*(\tau) = K_\beta(\tau)/(1 - \beta)$, and defining $(1-\epsilon^*) = (1-\beta)(1-\epsilon)$ and $B^*(\tau) \equiv \epsilon B(\tau)/\epsilon^*$. Then

$$S(\tau) = (1-\epsilon^*) \int_{-\infty}^{\infty} K^*(|\tau' - \tau|)\, S(\tau')\, d\tau' + \epsilon^* B^*(\tau). \tag{19.309}$$

When thermalization is achieved, S varies slowly and may be removed from under the integral to yield $S(\tau) = B^*(\tau) = \epsilon B(\tau)/(\epsilon + \beta - \epsilon\beta)$. For $\epsilon \gg \beta$, $S(\tau) \to B(\tau)$, as expected. But for $\beta \gg \epsilon$, escape dominates and $S(\tau) \to \epsilon B(\tau)/\beta$, showing that S decreases to the local creation rate ϵB as $\beta \to 1$, which is reasonable on physical grounds. If the medium has a boundary surface and B is constant, then (cf. [709])

$$S(0) = (\epsilon^*)^{\frac{1}{2}} B^* = \epsilon B/(\epsilon^*)^{\frac{1}{2}}, \tag{19.310}$$

which represents a generalization of the $\sqrt{\epsilon}$ law (§14.2). Thus when $\epsilon \gg \beta$, one recovers the usual static result $S(0) = \epsilon^{\frac{1}{2}} B$, while for $\beta \gg \epsilon$, we find $S(0) = \epsilon B/\beta^{\frac{1}{2}} = \beta^{\frac{1}{2}} S_\infty$, where S_∞ denotes the asymptotic value for S at depth.

Line Profiles

We now derive expressions for the line profiles seen by an external observer. The flux emergent at frequency x is proportional to

$$\begin{aligned} F_x &= 2\pi \int_0^\infty I(\infty, p, x)\, p\, dp \\ &= 2\pi \int_{r_c}^\infty S(r_0)\{1 - \exp[-\tau(-\infty, p, x)]\}\, p\, dp \\ &\quad + 2\pi \int_0^{r_c} S(r_0)\{1 - \exp[-\tau(-\infty, p, x)\Phi(x_c)]\}\, p\, dp \\ &\quad + 2\pi I_c \int_0^{r_c} \{1 - \exp[-\tau(-\infty, p, x)\Phi(x_c)]\}\, p\, dp, \end{aligned} \tag{19.311}$$

where, as above, r_0 denotes the value of r at the surface of constant radial velocity specified by x, and x_c is the value of x' given by equation (19.286) at $r' = r_c$ and $\mu' = [1 - (p/r_c)^2]^{\frac{1}{2}}$. The first term gives the emission from the part of the envelope seen outside the disk (i.e., $p > r_c$). The second term gives the emission from the part of the envelope superposed on the core, the factor $\Phi(x_c)$ correcting for occultation of material by the core. Note that in an expanding atmosphere, $\Phi(x_c)$ equals zero for $x < 0$ and will be essentially unity for $x > 0$, showing immediately the effect of core occultation on the red wing of the profile. The last term gives the continuum contribution from the core; in view of the properties of $\Phi(x_c)$ just mentioned, one sees that it is unattenuated in the red wing and more or less heavily attenuated in the blue wing of the line.

The flux at the continuum outside the line is proportional to

$$F_c = 2\pi I_c \int_0^{r_c} p\,dp = \pi r_c^2 I_c. \tag{19.312}$$

Transforming the variable of integration from p to r on surfaces $(z/r)V(r) = x$, equations (19.311) and (19.312) can be combined to yield an expression for the line profile $R_x \equiv (F_x - F_c)/F_c$, namely,

$$\begin{aligned} R_x = &\frac{2}{r_c^2 I_c} \int_{r_{\min}(x)}^{\infty} S(r) \frac{\tau_0(r)}{\tau(-\infty,p,x)} \{1 - \exp[-\tau(-\infty,p,x)]\}\, r\,dr \\ &- \frac{2}{r_c^2 I_c} \int_0^{r_c} S(r_c)\{\exp[-\tau(-\infty,p,x)\Phi(x_c)] - \exp[-\tau(-\infty,p,x)]\}\, p\,dp \\ &- \frac{2}{r_c^2} \int_0^{r_c} \{1 - \exp[-\tau(-\infty,p,x)\Phi(x_c)]\}\, p\,dp, \end{aligned} \tag{19.313}$$

where $r_{\min}(x)$ is the radius at which $V(r) = x$, and p is regarded as $p(r,x)$. The sign convention used in equation (19.313) is such that positive numbers denote an *emission* in the line. Each term in equation (19.313) can be interpreted in parallel with terms in equation (19.311).

There are several interesting results to be derived from equation (19.313). The last two terms, which represent the contribution form the core, will be ignored here. Considering a two-level atom for which the envelope is so thick that the source function achieves its asymptotic value $S = \epsilon B/\beta$, and replacing $\tau(-\infty,p,z,)$ with τ_0 and using equation (19.300), one obtains

$$R_x = \frac{\langle \epsilon\beta\tau_0 \rangle}{I_c} \int_{r_c}^{\infty} \frac{\epsilon\beta\tau_0}{\langle \epsilon\beta\tau_0 \rangle} \frac{2r\,dr}{r_c^2} = \frac{A\langle \epsilon\beta\tau_0 \rangle}{I_c}, \tag{19.314}$$

where $\langle \epsilon\beta\tau_0 \rangle$ is a typical value of $\epsilon\beta\tau_0$ and A the *effective emitting area* expressed in core units. For sufficiently large effective emitting areas, the line can become fairly bright relative to the continuum. In fact, most strong emission lines result largely from this geometrical effect.

Another interesting result is obtained from equation (19.311); see, e.g., [27, chapter 28]. Consider an envelope with constant velocity of expansion V; then $Q(r,\mu) = (V/r)\sin^2\theta$, and the surfaces of constant radial velocity are given by

$\cos\theta = (x/V) = \text{constant}$. The transformation from p to r is $p = r\sin\theta$, and in the limit of negligible contribution from the core, one has

$$F_x = 2\pi \sin^2\theta \int_0^\infty S(r)\{1 - \exp[-\chi_l(r)r/(V\sin^2\theta)]\}\, r\, dr. \tag{19.315}$$

If the envelope is *opaque*, the exponential term vanishes, and the integral becomes a constant, so that $F_x = C\sin^2\theta = C[1 - (x/V)^2]$; the line profile in this case is rounded, namely, *parabolic*. This conclusion is of importance because it shows that rounded profiles occur naturally, as a results of optical depth effects, even if the velocity is constant. In contrast, an interpretation based upon an analysis that assumes that the lines are optically thin would necessarily have invoked an accelerating or decelerating velocity fields, which would have quite different dynamical implications. This demonstrates that the spectroscopic diagnostics must be carried out with care, and to a high degree of consistency, if physically meaningful results are to be obtained.

Detailed calculations of line profiles, using equation (19.313), have been made [194] using the two-level atom source function given by (19.303). The velocity law was taken to be of the form $v(r) = v_\infty(1 - r_c/r)^{\frac{1}{2}}$. Adopted distributions of $\tau_0(r)$ are all characterized by a maximum of the range $1.1. \leq (r/r_c) \leq 4$. A wide variety of profiles can be produced by suitable choice of parameters. Three characteristic types of profiles, similar to those observed in WR stars, are shown in figure 19.5: (a) rounded emission with blue absorption, such as observed in the C III λ4650 and N III λ6438 lines, (b) flat-topped emission with blue absorption, such as seen in He I lines, (c) intense rounded emission with no absorption, such as observed in He II lines. In each case the intensity of the emission is proportional to $A\langle \epsilon B\tau_0\rangle / I_c$, as expected from equation (19.314); note that the flat-topped profile results from an optically thin line.

Effects of the Background Continuum

The formalism presented so far neglects the influence of the background continuum on the source function and the specific intensity in the frequency range of a line under study. Note that although the continuum enters the expression for the line profile, (19.313), it is taken as passive; i.e., the source function S used in equation (19.311) for the flux in the line does not contain a contribution of the continuum opacity and emissivity, and equation (19.312) for the continuum flux merely expresses it as an attenuated photospheric flux.

The effects of the background continuum on the source function in the line were studied in [553] and [890]. We shall not go into any detail here, but mention only a final result. The interested reader is referred to the above references. Equation (19.283) is generalized to read

$$\bar{J}(r) = (1 - \beta)S^{\mathrm{L}}(r) + \beta_c I_c + [S^{\mathrm{C}}(r) - S^{\mathrm{L}}(r)]\widetilde{U}(r, \zeta), \tag{19.316}$$

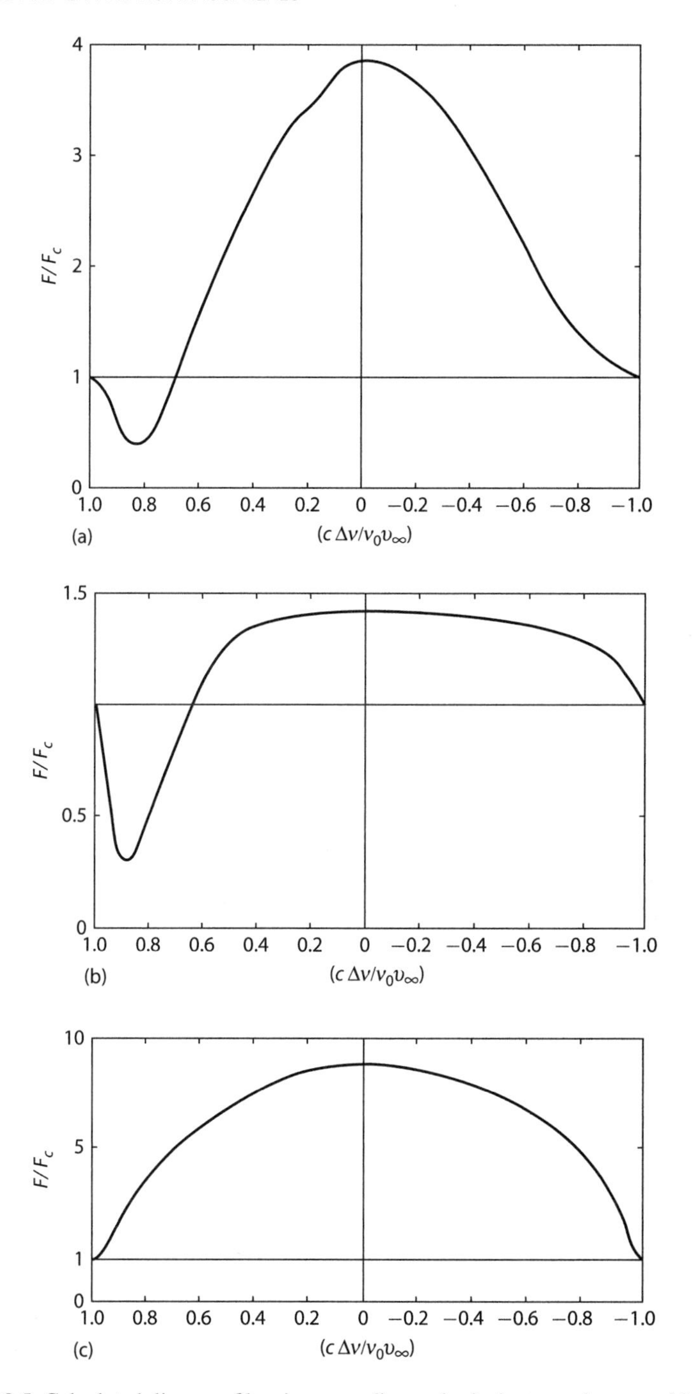

Figure 19.5 Calculated line profiles in expanding spherical atmospheres, with a two-level atom source function. (a) $\epsilon = 9.2 \times 10^{-3}$ and $\tau_0(\max) \approx 15$. (b) $\epsilon = 2 \times 10^{-3}$ and $\tau_0(\max) \approx 0.5$. (c) $\epsilon = 2.1 \times 10^{-2}$ and $\tau_0(\max) \approx 2$. From [194].

where $S^{\mathrm{L}}(r)$ and $S^{\mathrm{C}}(r)$ are the line and the continuum source functions, respectively, $\zeta \equiv \kappa^{\mathrm{C}}/\kappa^{\mathrm{L}}$, and $\widetilde{U}$ is defined by

$$\widetilde{U}(r,\zeta) = \tfrac{1}{2}\int_{-1}^{1} d\mu\, U[\widetilde{\tau}(r,\mu),\zeta], \tag{19.317}$$

where

$$U(\tilde{\tau},\zeta) \equiv \tilde{\tau}\int_{-\infty}^{\infty} dx\,\phi(x)\int_{x}^{\infty} dy\,\phi(y)\exp\left(-\tilde{\tau}\int_{x}^{y} ds\,\phi(s)\right)\{1-\exp[-\zeta\tilde{\tau}(y-x)]\}, \tag{19.318}$$

where $\widetilde{\tau}(r,\mu) \equiv \tau[-\infty, p(r,\mu), 0]$. Extensive tables of the function $\widetilde{U}(\tau,\zeta)$ are presented in [890].

The line source function for a two-level atom can now be written as

$$S^{\mathrm{L}} = \frac{(1-\epsilon)(\beta_c I_c + S^{\mathrm{C}}\widetilde{U}) + \epsilon B}{(1-\epsilon)(\beta + \widetilde{U}) + \epsilon}, \tag{19.319}$$

which represents a generalization of equation (19.303).

Sobolev with Exact Integration (SEI) Method

This method is based on the following observation [429]. When computing line profiles with the Sobolev theory, the basic approximation that all the quantities vary slowly on the scale of the Sobolev length is employed twice: once for evaluating the line source function using the escape probability, and once for expressing the formal solution of the transfer equation. If it is used only once, to obtain the line source function, while the formal solution is done by an exact integration, one obtains a quite accurate solution. One can also easily treat the doublet and multiplet lines.

The idea was incorporated [670] in the SEI code, which is often used to obtain line profiles in a fast and relatively accurate way.

19.6 NLTE LINE FORMATION

When dealing with expanding atmospheres, we have so far been concerned only with solution of the transfer equation. We have also seen that departures from LTE are absolutely crucial in this situation; in fact, the algorithms described in § 19.2–§ 19.4 did specifically consider the line source function in the (NLTE) two-level atom form, $S = (1-\epsilon)\overline{J} + \epsilon B$. Here, we shall describe several approaches to solve a NLTE multi-level atom problem in spherical atmospheres with velocity fields, with the structural parameters (temperature, density, velocity) assumed to be given. A more general problem, where all the structural parameters are determined simultaneously with the radiation field and atomic level populations, will be treated in § 20.4.

Frequency-Dependent Moment Equations

A direct solution of the comoving-frame transfer equation in terms of Feautrier variables u and v was presented in § 19.3. For some purposes, such as for line transfer with frequency-dependent source function, or a direct solution of the NLTE multi-level atom problem using the equivalent two-level-atom (ETLA) approach, it is advantageous to formulate a solution for the frequency-dependent moments of the specific intensity. As will be shown below, one can eliminate the H-moment to end up with an equation for the J-moment, similarly to the case of static atmospheres.

The formalism was developed in [745]; we will follow their notation closely. Using the Eddington and the spericality factors, the first two frequency-dependent moment equations are written [cf. equations (19.161) and (19.162)]

$$\frac{\partial(f_\nu q_\nu r^2 J_\nu)}{\partial X_\nu} + \gamma_\nu \left[\frac{\partial(1-g_\nu)r^2 H_\nu}{\partial \nu} + \beta \frac{\partial(g_\nu r^2 H_\nu)}{\partial \nu} \right] = r^2 H_\nu, \tag{19.320}$$

and

$$q_\nu \frac{\partial(r^2 H_\nu)}{\partial X_\nu} + \gamma_\nu \left[\frac{\partial(1-f_\nu)r^2 J_\nu}{\partial \nu} + \beta \frac{\partial(f_\nu r^2 J_\nu)}{\partial \nu} \right] = r^2 (J_\nu - S_\nu), \tag{19.321}$$

where

$$dX_\nu \equiv -\chi_\nu q_\nu dr \tag{19.322}$$

is a modified optical depth coordinate, where χ_ν is the absorption coefficient and q_ν the sphericality factor defined by [cf. equation (19.28)]

$$\frac{\partial \ln(r^2 q_\nu)}{\partial r} = \frac{3f_\nu - 1}{r f_\nu}, \quad \text{or} \quad \ln(r^2 q_\nu) \equiv \int_R^r [(3f_\nu - 1)/(r' f_\nu)]\, dr' + \ln R^2. \tag{19.323}$$

The Eddington factors are defined by $f_\nu \equiv K_\nu / J_\nu$ and $g_\nu \equiv N_\nu / H_\nu$, and the auxiliary quantities are defined by

$$\alpha_\nu \equiv \nu v(r)/cr, \quad \beta \equiv d \ln v(r) / d \ln r, \quad \gamma_\nu \equiv \alpha_\nu / \chi_\nu. \tag{19.324}$$

Equations (19.320) and (19.321) are discretized in frequency ν, with $\nu_n, (n = 1, \dots, NF)$, and radius r, with $r_d, (d = 1, \dots, D)$. As before, the notation $F_{n,d} \equiv F(\nu_n, r_d)$ is used. The discretized equation (19.320) reads

$$\begin{aligned} &\frac{f_{n,d+1} q_{n,d+1} r_{d+1}^2 J_{n,d+1} - f_{n,d} q_{n,d} r_d^2 J_{n,d}}{\Delta X_{n,d+\frac{1}{2}}} \\ &\quad + \frac{\gamma_{n-\frac{1}{2},d+\frac{1}{2}} r_{d+\frac{1}{2}}^2}{\Delta \nu_{n-\frac{1}{2}}} \left[\left(1 - g_{n-1,d+\frac{1}{2}} + \beta_{d+\frac{1}{2}} g_{n-1,d+\frac{1}{2}}\right) H_{n-1,d+\frac{1}{2}} \right. \\ &\quad \left. - \left(1 - g_{n,d+\frac{1}{2}} + \beta_{d+\frac{1}{2}} g_{n,d+\frac{1}{2}}\right) H_{n,d+\frac{1}{2}} \right] = r_{d+\frac{1}{2}}^2 H_{n,d+\frac{1}{2}}, \end{aligned} \tag{19.325}$$

where

$$\Delta X_{n,d+\frac{1}{2}} \equiv \tfrac{1}{2}(q_{n,d+1}\chi_{n,d+1} + q_{n,d}\chi_{n,d})(r_d - r_{d+1}), \tag{19.326}$$

$$\Delta \nu_{n-\frac{1}{2}} \equiv \nu_{n-1} - \nu_n, \tag{19.327}$$

$$r_{d+\frac{1}{2}} \equiv \tfrac{1}{2}(r_d + r_{d+1}), \tag{19.328}$$

and

$$\chi_{n-\frac{1}{2},d+\frac{1}{2}} \equiv \tfrac{1}{4}(\chi_{n,d+1} + \chi_{n,d} + \chi_{n-1,d+1} + \chi_{n-1,d}). \tag{19.329}$$

Solving equation (19.325) for $H_{n,d+\frac{1}{2}}$ in terms of $J_{n',d+1}$ and $J_{n',d}$ for all $n' \leq n$, one obtains

$$r^2_{d+\frac{1}{2}} H_{n,d+\frac{1}{2}} = \frac{f_{n,d+1}q_{n,d+1}r^2_{d+1}J_{n,d+1} - f_{n,d}q_{n,d}r^2_d J_{n,d}}{(1+\omega_{n,n,d+\frac{1}{2}})\Delta X_{n,d+\frac{1}{2}}} + \frac{\omega_{n,n-1,d*\frac{1}{2}}}{1+\omega_{n,n,d+\frac{1}{2}}} r^2_{d+\frac{1}{2}} H_{n-1,d+\frac{1}{2}}, \tag{19.330}$$

where

$$\omega_{n,n',d+\frac{1}{2}} \equiv \frac{\gamma_{n-\frac{1}{2},d+\frac{1}{2}}}{\Delta \nu_{n-\frac{1}{2}}} \left(1 - g_{n',d+\frac{1}{2}} + \beta_{d+\frac{1}{2}} g_{n',d+\frac{1}{2}}\right). \tag{19.331}$$

Equation (19.331) can be solved recursively in n by noting that at the highest frequency one considers only continuum radiation that is independent of frequency, $\partial I/\partial \nu = 0$; hence $\gamma_{\frac{1}{2},d+\frac{1}{2}} = 0$ and $\omega_{1,n,d+\frac{1}{2}} = 0$,

$$r^2_{d+\frac{1}{2}} H_{n,d+\frac{1}{2}} = \sum_{n'=1}^{n} \psi_{n,n',d+\frac{1}{2}} \left(\frac{f_{n',d+1}q_{n',d+1}r^2_{d+1}J_{n',d+1} - f_{n',d}q_{n',d}r^2_d J_{n',d}}{\Delta X_{n',d+\frac{1}{2}}} \right), \tag{19.332}$$

where

$$\psi_{n,n',d+\frac{1}{2}} \equiv \frac{1}{1+\omega_{n,n,d+\frac{1}{2}}} \prod_{m=n'+1}^{n} \frac{\omega_{m,m-1,d+\frac{1}{2}}}{1+\omega_{m,m,d+\frac{1}{2}}}, \tag{19.333}$$

where the product is taken to be unity for $n' = n$.

Equation (19.321) is discretized as

$$\frac{q_{n,d}}{\Delta X_{n,d}} \left(r^2_{d+\frac{1}{2}} H_{n,d+\frac{1}{2}} - r^2_{d-\frac{1}{2}} H_{n,d-\frac{1}{2}} \right) + \frac{\gamma_{n-\frac{1}{2},d}}{\Delta \nu_{n-\frac{1}{2}}} \left[(1 - f_{n-1,d} + \beta_d f_{n-1,d}) r^2_d J_{n-1,d} - (1 - f_{n,d} + \beta_d f_{n,d}) r^2_d J_{n,d} \right] = r^2_d (J_{n,d} - S_{n,d}). \tag{19.334}$$

Finally, using equation (19.332) to eliminate $H_{n,d\pm\frac{1}{2}}$ from equation (19.334), one obtains

$$\frac{q_{n,d}}{\Delta X_{n,d}}\left[\sum_{n'=1}^{n}\frac{\psi_{n,n',d+\frac{1}{2}}}{\Delta X_{n',d+\frac{1}{2}}}\left(f_{n',d+1}q_{n',d+1}r_{d+1}^2 J_{n',d+1}-f_{n',d}q_{n',d}r_d^2 J_{n',d}\right)\right.$$
$$\left.-\sum_{n'=1}^{n}\frac{\psi_{n,n',d-\frac{1}{2}}}{\Delta X_{n',d-\frac{1}{2}}}\left(f_{n',d}q_{n',d}r_d^2 J_{n',d}-f_{n',d-1}q_{n',d-1}r_{d-1}^2 J_{n',d-1}\right)\right]$$
$$=r_d^2[(1+\delta_{n,n,d})J_{n,d}-S_{n,d}-\delta_{n,n-1,d}J_{n-1.d}], \tag{19.335}$$

where

$$\delta_{n,m,d}\equiv\frac{\gamma_{n-\frac{1}{2},d}}{\Delta\nu_{n-\frac{1}{2}}}(1-f_{m,d}+\beta_d f_{m,d}). \tag{19.336}$$

Again, the initial condition at the highest frequency implies that $\delta_{1,m,d}=0$.

Equation (19.335) has to be supplemented by the boundary conditions. At the outer boundary, one assumes no incident radiation, $I(\nu_n,\mu)=0$ for $\mu<0$, and introduces two additional Eddington factors h_n and n_n, which are defined in terms of the outgoing intensity as

$$h_n\equiv H_{n,1}/J_{n,1}=\int_0^1 I(\nu_n,\mu)\mu\,d\mu\bigg/\int_0^1 I(\nu_n,\mu)\,d\mu, \tag{19.337}$$

and

$$n_n\equiv N_{n,1}/J_{n,1}=\int_0^1 I(\nu_n,\mu)\mu^3\,d\mu\bigg/\int_0^1 I(\nu_n,\mu)\,d\mu. \tag{19.338}$$

Using these quantities, equation (19.325) for $d=1$ reads

$$\frac{f_{n,2}q_{n,2}r_2^2 J_{n,2}-f_{n,1}q_{n,1}r_1^2 J_{n,1}}{\Delta X_{n,\frac{3}{2}}}=r_1^2[(h_n+\theta_{n,n})J_{n,1}-\theta_{n,n-1}J_{n-1,1}], \tag{19.339}$$

where

$$\theta_{n,m}\equiv\frac{\gamma_{n-\frac{1}{2},1}}{\Delta\nu_{n-\frac{1}{2}}}(h_m-n_m+\beta_1 n_m). \tag{19.340}$$

The initial condition at the highest frequency stipulates that $\theta_{1,m}=0$.

The inner boundary condition is applied sufficiently deep in the atmosphere where the velocity is essentially zero (i.e., much smaller than the thermal velocity) and where the diffusion approximation is valid for all frequencies, namely,

$$H_\nu=\frac{1}{3\chi_\nu}\left|\frac{\partial B_\nu}{\partial r}\right|, \tag{19.341}$$

which can be written in a discretized form as

$$\frac{f_{n,D}q_{n,D}r_D^2 J_{n,D} - f_{n,D-1}q_{n,D-1}r_{D-1}^2 J_{n,D-1}}{\Delta X_{n,D-\frac{1}{2}}} = \frac{r_D^2}{3\chi_{n,D}} \left| \frac{\partial B_{\nu_n}}{\partial r} \right|_D. \tag{19.342}$$

Introducing column vectors $\mathbf{J}_d \equiv (J_{1,d}, \ldots, J_{NF,d})^{\mathrm{T}}$ for $d = 1, \ldots, D$, equations (19.335), (19.339), and (19.342) can be written in the usual matrix form

$$-\mathbf{A}_d \mathbf{J}_{d-1} + \mathbf{B}_d \mathbf{J}_d - \mathbf{C}_d \mathbf{J}_{d+1} = \mathbf{L}_d, \tag{19.343}$$

for $d = 1, \ldots, D$. Here, $\mathbf{B}_d$ is a full matrix, and $\mathbf{A}_d$ and $\mathbf{C}_d$ are lower triangular matrices, all of dimension $(NF \times NF)$. When the Eddington factors $f_{n,d}$, $g_{n,d}$, h_n, and n_n are specified, equation (19.343) is readily solved by the standard elimination.

Equivalent-Two-Level-Atom Approach

This approach was formulated in [740]. One uses the discretized form of the transfer equation (19.335) with boundary conditions (19.339) and (19.342) and considers *one transition*, line or continuum, at a time. The source function in a given transition is expressed in the form of the equivalent two-level-atom (ETLA) approach, described in § 14.4. For a line transition $l \to u$, the source function is given by [cf. equations (14.130)–(14.132)]

$$S_{lu} = \frac{\int_0^\infty J_\nu \phi_{lu}(\nu)\, d\nu + \eta_{lu}}{1 + \epsilon_{lu}}, \tag{19.344}$$

where

$$\epsilon_{lu} = [\alpha_{lu} - (g_l/g_u)\beta_{lu}]/A_{ul},$$
$$\eta_{lu} = \beta_{lu}/B_{lu}, \tag{19.345}$$

and

$$\alpha_{lu} = (a_2 a_3 + a_4 C_{ul})/(a_2 + a_4),$$
$$\beta_{lu} = (a_1 a_4 + a_2 C_{lu})/(a_2 + a_4). \tag{19.346}$$

The terms a_1 and a_2 represent, respectively, the total rate of transitions out of level l to other all levels but u, and the total transition rate into level l from all other levels but u; a_3 and a_4 represent similar quantities for the upper level u. Explicit expressions are given by equations (14.126). Notice that the populations of *all other levels* appear in a_2 and a_4 and therefore in ϵ_{lu} and η_{lu}. For continuum transitions, the expressions are similar; for details refer to [740].

The integral in equation (19.344) is replaced by a quadrature sum. The discretized transfer equation (19.335) with boundary conditions (19.339) and (19.342), and with the source function given by (19.344), can be written in a matrix form as

$$-\mathbf{A}_d \mathbf{J}_{d-1} + \mathbf{B}_d \mathbf{J}_d - \mathbf{C}_d \mathbf{J}_{d+1} = \mathbf{L}_d, \tag{19.347}$$

where $\mathbf{J}_d$ now denotes a column vector of the mean intensities $\mathbf{J}_d = \{J_{n,d}\}^{\mathrm{T}}$ for all frequency points n that belong to the given transition; their number is denoted K. All matrices have dimension $(K \times K)$. Matrix $\mathbf{B}_d$ is a full matrix; matrices $\mathbf{A}_d$ and $\mathbf{C}_d$ are lower triangular for a line transition and diagonal for a continuum transition (because the $\partial/\partial\nu$ term in the transfer equation is neglected for them). The system (19.347) is solved by a standard forward-backward elimination scheme.

As pointed out in chapter 14, the basic drawback of the ETLA approach, when applied to static atmospheres, is that it treats only one transition at a time, so it may fail when strong interlocking of several transitions is present. However, as noted in [740], the situation is much more favorable in the case of moving atmospheres. Here the escape of photons in a line itself through the low-frequency wing usually dominates all other leakage mechanisms (which is in fact the basis for the Sobolev approximation); hence the interlocking effects are usually much less pronounced. In this sense, it is actually easier to solve the multi-level line-formation problem in an expanding medium than in a static atmosphere!

In [740] there also was suggested a generalization of the complete linearization method that was outlined in § 14.5. We will not describe the method here because the ALI-based approaches, outlined below, provide a more robust and efficient procedure. Its derivation is straightforward but tedious; the interested reader is referred to the original reference [740].

Solution Using the Sobolev Approximation

Before we discuss the most powerful and efficient methods for solving the multi-level transfer problem, namely, those based on the idea of ALI, we will discuss an application of the Sobolev approximation. As demonstrated in [889], these two approaches are closely related, in the sense that the Sobolev escape probabilities can be used as a special case of an approximate Λ^*-operator.

A scheme for solving the multi-level atom problem using Sobolev approximation is straightforward [622]. The net number of radiative transitions $j \rightarrow i$ $(j > i)$ is given by (assuming complete frequency redistribution)

$$Z_{ji} \equiv n_j R_{ji} - n_i R_{ij} = n_j A_{ji} - (n_i B_{ij} - n_j B_{ji}) \overline{J}_{ij}, \tag{19.348}$$

where the frequency-averaged mean intensity $\overline{J}_{ij}$ is expressed in a form that depends on whether the background continuum is taken into account or not.

Negligible Background Continuum

This is the original formalism developed in [622]. $\overline{J}_{ij}$ is written in the traditional Sobolev approximation as [cf. equation (19.283)]

$$\overline{J}_{ij} = (1 - \beta_{ij})S_{ij} + \beta_{c,ij}I_{c,ij}, \tag{19.349}$$

where S_{ij} is the line source function, β_{ij} is the Sobolev escape probability for the transition $i \to j$ defined by (19.299), and β_c is defined by (19.301). For simplicity, the total source function is taken as the line source function, which is given by

$$S_{ij} = n_j A_{ji}/(n_i B_{ij} - n_j B_{ji}). \tag{19.350}$$

Substituting equations (19.349) and (19.350) into (19.348), one obtains

$$Z_{ji} = n_j A_{ji}\beta_{ij} - (n_i B_{ij} - n_j B_{ji})\beta_{c,ij}I_{c,ij}. \tag{19.351}$$

Therefore, the kinetic equilibrium equations can be written in the usual form, with the radiative bound-bound rates modified to

$$\begin{aligned} R_{ij} &\longrightarrow B_{ij}\beta_{c,ij}I_{c,ij}, \\ R_{ji} &\longrightarrow A_{ji}\beta_{ij} + B_{ji}\beta_{c,ij}I_{c,ij}. \end{aligned} \tag{19.352}$$

The induced radiative rates no longer contain unknown averaged mean intensity $\overline{J}_{ij}$ but known quantity $\beta_{c,ij}I_{c,ij}$, and the spontaneous rate is modified by multiplying it by β_{ij}.

For continua, the radiative rates are written in the usual form,

$$\begin{aligned} R_{ij} &= \frac{4\pi}{h}\int_{\nu_0}^{\infty}\frac{\sigma_\nu}{\nu}J_\nu\, d\nu, \\ R_{ji} &= (n_i/n_j)^*\frac{4\pi}{h}\int_{\nu_0}^{\infty}\frac{\sigma_\nu}{\nu}\left(\frac{2h\nu^3}{c^2} + J_\nu\right)e^{-h\nu/kT}\, d\nu. \end{aligned} \tag{19.353}$$

One assumes that the continua are optically thin; hence the mean intensity is approximated as

$$J_\nu = 4WH_\nu^*, \tag{19.354}$$

where W is the dilution factor defined by equation (19.302), and H_ν^* is the Eddington flux at the top of the photosphere, provided by an independent model atmosphere. (If the photospheric model provides directly J_ν^*, then $J_\nu = 2WJ_\nu^*$.) Equation (19.354) expresses the Eddington approximation in the photosphere (since $W \to \frac{1}{2}$, $J_\nu \approx 2H_\nu^*$) and provides an accurate approximation at large distances [since $W \to (1/4)(R/r)^2$, $J_\nu \approx (R/r)^2 H_\nu^* \approx H_\nu$, where H_ν is the local Eddington flux].

If a particular continuum is optically thick, as for instance the He^+ ground-state continuum in Of stars, it is considered to be in detailed radiative balance, in which case the radiative rates are set to zero [622], $R_{ij} = R_{ji} = 0$.

With these approximations for the radiative rates, the kinetic equilibrium equations form a simple linear system that can be solved easily with any standard method.

Including Background Continuum

In this case, (19.349) is modified to (cf. equation 19.316)

$$\overline{J}_{ij} = (1 - \beta_{ij} - \widetilde{U}_{ij})S_{ij} + \beta_{c,ij}I_{c,ij} + \widetilde{U}_{ij}S^{\mathrm{C}}, \tag{19.355}$$

and, consequently, the replacement (19.352) becomes

$$\begin{aligned} R_{ij} &\longrightarrow B_{ij}(\beta_{c,ij}I_{c,ij} + \widetilde{U}_{ij}S^{\mathrm{C}}), \\ R_{ji} &\longrightarrow A_{ji}(\beta_{ij} + \widetilde{U}_{ij}) + B_{ji}(\beta_{c,ij}I_{c,ij} + \widetilde{U}_{ij}S^{\mathrm{C}}). \end{aligned} \tag{19.356}$$

An improved treatment of bound-free rates was developed in [942], to which the interested reader is referred for details.

Application of ALI

The formalism employed here is a relatively straightforward generalization of the approach described in § 14.5. It should be noted that the kinetic equilibrium equations are *local*. Therefore, once the radiation intensity is expressed through the approximate Λ-operator, all the geometry and global dynamics are contained in the form of Λ, and the formalism is essentially the same as in the case of static plane-parallel atmospheres. One way or another, i.e., using linearization or preconditioning, one ends up with a *linear* system for the corrections to the atomic level populations, as in the static case.

As before, there are approaches that use the Λ^*-operator explicitly [431, 450, 842, 889, 942] and related approaches that use the idea of the approximate Newton-Raphson approach [478, 492, 493, 967]. The latter approach used in the context of moving atmospheres and for solving all structural equations, not just kinetic + transfer equations, will be described in § 20.4, so it will not be considered here.

There are several flavors of the ALI-based methods. We will describe here the one developed in [889] and later elaborated in [942] because it offers an insight into an interesting relation between the ALI methods and the Sobolev approximation and even allows mixing them arbitrarily in the system of kinetic equilibrium equations.

In order to separate the line source function from the background contribution, the following procedure is used. The total source function in the frequency region of the transition $l \rightarrow u$ can be written [cf. (14.195)–(14.196)]

$$S(\nu) = \rho_{lu}(\nu)S_{lu} + [1 - \rho_{lu}(\nu)]S_c(\nu), \tag{19.357}$$

where

$$\rho_{lu}(\nu) = \frac{\chi_{lu}(\nu)}{\chi_{lu}(\nu) + \chi_c(\nu)}, \qquad S_{lu} = \frac{\eta_{lu}(\nu)}{\chi_{lu}(\nu)}, \tag{19.358}$$

where χ_{lu} and η_{lu} are, respectively, the absorption and emission coefficients for the line $l \to u$, and χ_c and η_c are the absorption and emission coefficients for all the opacity sources that overlap line $l \to u$. Traditionally, these sources are called the *background continuum*, although they may be free-free, bound-free, or bound-bound transitions. The line source function and the background source function are given by, respectively,

$$S_{lu} = \frac{n_u A_{ul}}{n_l B_{lu} - n_u B_{ul}}, \quad S_c(\nu) = \frac{\eta_c(\nu)}{\chi_c(\nu)}. \tag{19.359}$$

The ALI iteration process is set up by [cf. equations (14.198) and (14.199)]

$$J_\nu = \Lambda^*_\nu[S^{\rm new}_\nu] + (\Lambda_\nu - \Lambda^*_\nu)[S^{\rm old}_\nu] = \Lambda^*_\nu[\rho^{\rm old}_{lu}(\nu) S_{lu}] + J^{\rm eff}_\nu, \tag{19.360}$$

where the replacement $\rho_{lu} \to \rho^{\rm old}_{lu}$ is crucial to end up with a linear representation of the resulting kinetic equilibrium equations, as explained in § 14.5. The term $J^{\rm eff}_\nu$ is given by

$$\begin{aligned} J^{\rm eff}_\nu &= (\Lambda_\nu - \Lambda^*_\nu)[S^{\rm old}_\nu] + \Lambda^*_\nu[1 - \rho^{\rm old}_{lu}(\nu)] S_c(\nu) \\ &= \Lambda_\nu[S^{\rm old}_\nu] - \Lambda^*_\nu[\rho^{\rm old}_{lu}(\nu) S^{\rm old}_{lu}] = J^{\rm FS}_\nu - \Lambda^*_\nu[\rho^{\rm old}_{lu}(\nu) S^{\rm old}_{lu}]. \end{aligned} \tag{19.361}$$

The quantity $J^{\rm FS}_\nu = \Lambda[S^{\rm old}_\nu]$ is the mean intensity obtained by a formal solution with the old populations. Frequency-averaged mean intensity is given by

$$\overline{J}_{lu} = \overline{\Lambda}^*_{lu}[S_{lu}] + \overline{J}^{\rm eff}_{lu}, \tag{19.362}$$

where

$$\overline{\Lambda}^*_{lu} = \int \Lambda^*_{\mu\nu} \rho^{\rm old}_{lu}(\nu)\, \phi_{lu}(\nu)\, d\nu, \tag{19.363}$$

and

$$\overline{J}^{\rm eff}_{lu} = \int J^{\rm eff}_\nu\, \phi_{lu}(\nu)\, d\nu = \overline{J}^{\rm FS}_{lu} - \overline{\Lambda}^*_{lu} S^{\rm old}_{lu}, \tag{19.364}$$

where

$$\overline{J}^{\rm FS}_{lu} = \int J^{\rm FS}_\nu\, \phi_{lu}(\nu)\, d\nu. \tag{19.365}$$

As pointed out in [889], the representation (19.362) that relates $\overline{J}_{lu}$ to S_{lu} is of *affine type*, i.e., a linear operator ($\overline{\Lambda}^*_{lu}$) with an additional translation (displacement), $\overline{J}^{\rm eff}_{lu}$.

The above expressions are valid for any form of the approximate operator Λ^*. In the context of moving atmospheres, it is most advantageous to use a *local* operator, in which case its action is simply given by a scalar multiplication. As explained in § 13.2 and § 19.3, a nearly optimum local operator is a diagonal part of the exact Λ-operator. Its construction poses a non-trivial problem when using the comoving-frame formulation of the radiative transfer equation in moving atmospheres. A practical approach developed in [889] to obtain the local operator was described in § 19.3.

The net number of transitions $u \rightarrow l$ is then given by

$$Z_{ul} = n_u A_{ul} \overline{\Lambda}^*_{lu} - (B_{lu} n_l - B_{ul} n_u) \overline{J}^{\rm eff}_{lu} . \tag{19.366}$$

One can write the kinetic equilibrium equations in the usual form, with the radiative rates modified to

$$\begin{aligned} R_{lu} &\rightarrow B_{lu} J^{\rm eff}_{lu} , \\ R_{ul} &\rightarrow A_{ul} \overline{\Lambda}^*_{lu} + B_{ul} J^{\rm eff}_{lu} . \end{aligned} \tag{19.367}$$

These replacements are analogous to equations (19.352), which apply in the Sobolev approximation, using $\overline{\Lambda}^*_{lu}$ instead of the Sobolev escape probability β_{lu} and $J^{\rm eff}_{lu}$ instead of $\beta_{c,lu} I_{c,lu}$. As suggested in [889], one may therefore arbitrarily mix these two approaches in the system of kinetic equilibrium equations. One can view the Sobolev escape probability β_{lu} as an approximation for the local operator Λ^*_{lu}, which may offer a significant reduction of the computer time because an evaluation of β_{lu} is extremely fast.

Chapter Twenty

Stellar Winds

The outermost atmospheric layers of most stars are in a state of continuous rapid expansion; the material lost in such a flow is called a *stellar wind*. These winds have a wide range of properties. At one extreme are very massive flows, with a mass loss rate up to $\sim 10^{-5} M_\odot$/year (or more in some cases), which are optically thick in spectral lines, and even in some continua, and produce emission lines and P-Cygni profiles (see § 20.1). At the other extreme are tenuous flows such as that of the Sun, which is optically thin and inconsequential in terms of mass loss ($10^{-14} M_\odot$/year), but still of importance to the solar angular momentum balance.

There are several types of stellar winds, distinguished by their primary mechanisms.

- *Coronal winds* (or *thermally driven winds*) occur in stars with hydrogen convection zones, such as the G- and K-stars, that give rise to a mechanically heated *corona* of very high temperatures. Here, it is found that the corona cannot establish a static pressure balance with the interstellar medium but must inevitably expand supersonically, driving the flow by tapping the thermal energy of the gas.
- *Radiation-driven* or *line-driven winds* occur in early-type highly luminous stars (giants and supergiants of types O, B, A) where the radiation field is so intense that momentum imparted to the gas by photons drives the material to a transonic flow.
- *Porosity-regulated, continuum-driven winds* are believed to occur in stars that nominally exceed the Eddington limit, such as Luminous Blue Variables, and extremely massive, metal-free first-generation stars. The wind is driven by strong radiation in the continuum; the inhomogeneities (porosity) prevent the star from becoming gravitationally unbounded.

These three types of winds are the most common and will be described in some detail in this chapter. There are also several other types of mass outflows, which will be only mentioned here. More extended discussion may be found, for instance, in [668] and in a monograph [669].

- *Dust-driven winds* occur in cold supergiants (type M and AGB stars). The driving force is provided by an absorption of radiation by dust that is formed at the outer layers of their atmospheres where the temperature drops to values around or below $\sim 10^3$ K.
- *Pulsation-driven winds* occur in pulsating cool stars, such as Mira-type and AGB stars. The energy of the wind is provided by the kinetic energy of pulsation.

This mechanism can be enhanced when combined with dust formation and the resulting radiation pressure on dust.
- *Alfvén wave-driven winds* occur in stars with magnetic fields, where the mass loss is powered by the wave pressure in Alfvén waves.

Stellar winds have important implications for many astrophysical problems. In some cases, the mass loss rate is so large as to produce significant change in the star's mass on a thermonuclear-evolution time scale, and hence directly to affect the star's evolution track. In other cases, the occurrence of a sustained mass loss during its entire lifetime prevents the star from becoming a supernova and permits it to evolve to a white-dwarf configuration. Further, stellar winds represent important sources of mass and energy deposited into the interstellar medium.

In this book, we shall consider only winds from single, isolated stars. Stellar winds and mass outflows also occur in binaries, where they may induce a rapid mass exchange that radically alters the course of evolution of the individual components. The material presented here is a basis for study of the more complex cases mentioned above and provides a background for an approach to the literature.

There is a vast literature on spectroscopic diagnostics of stellar winds and comparisons of observed and theoretically predicted spectra (e.g., [645,669] and references therein). Here, we will not be concerned with these issues but instead concentrate on basic physical mechanisms that drive the wind and on model stellar atmospheres that encompass the whole atmospheric extent from the static photosphere to the supersonic wind. The main emphasis will be given to radiation-driven winds from hot stars.

20.1 QUALITATIVE PICTURE

Consider a spherically symmetric, radially expanding atmosphere that surrounds a star with a well-defined photospheric surface, as sketched in figure 20.1. For the subsequent discussion, we assume that the envelope is essentially transparent, so that every photon emitted toward an external observer can be received. The material behind the stellar disk is in an *occulted region* and is not seen by an external observer. The matter projected on the stellar disk can either (i) simply emit radiation without significant reabsorption or (ii) absorb the incident photospheric radiation and scatter it out of line of sight. From this material, in case (i) one would obtain a blue-shifted emission feature, while in (ii) one would obtain an absorption trough. From the matter in the emission lobes to the sides of the disk, one receives photons either emitted thermally or scattered from both the stellar and diffuse (from the envelope itself) radiation fields. The velocities along the line of sight in the emission-lobe material range from positive through negative values and produce a symmetric emission feature centered about the rest wavelength. Because some material is occulted by the star, the maximum redshifts that could be produced will not be observed. The result is a characteristic line profile, known as a P-Cygni profile after the star for which its significance as a signature of a mass outflow was first recognized [97].

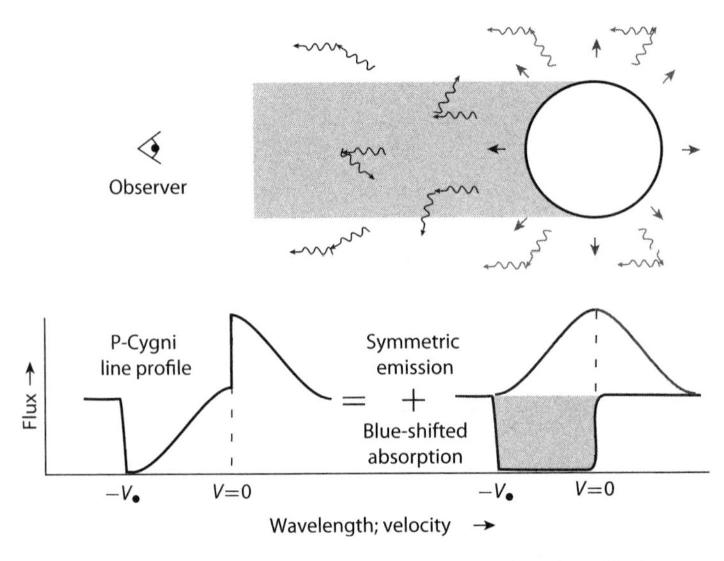

Figure 20.1 Schematic illustration of the formation of a P-Cygni profile in an expanding atmosphere. Figure courtesy of Stan Owocki.

The above picture corresponds to the case where an emission of a line photon results from a previous absorption of a photon in the same line; i.e., as a result of *line scattering*. Hence the mechanism of generating the P-Cygni profiles operates for resonance lines or for lines that behave like resonance lines (e.g., a subordinate line with a lower level being in a detailed balance with the ground state; for instance, the hydrogen $H\alpha$ line in A-stars, where the $L\alpha$ line is extremely optically thick). In the case where an emission of a line photon is a result of a process different from the resonance scattering, for instance, a recombination or collisional excitation, the occupation numbers of involved levels are close to their LTE values; hence the line source function is close to the Planck function. The resulting line profile is an essentially symmetric emission line (or an absorption line with a central emission reversal, or a pure absorption line, but weaker than formed by purely photospheric process). This mechanism was explained by means of a simple model in § 19.5. An example of such a line is the hydrogen $H\alpha$ line in O- and B-stars. Notice, however, that the $H\alpha$ line changes its character when going to A supergiants, where it starts to exhibit a P-Cygni profile (e.g., [645]).

20.2 THERMALLY DRIVEN WINDS

The thermally driven winds, which provide the basic physical picture of coronal winds in the Sun and solar-type stars, are not a primary topic of this book. We present here a brief outline of the basic physics because, thanks to its relative simplicity, the theory of thermally driven winds serves as a background for understanding other types of stellar winds.

Consider a steady, one-dimensional flow of ideal gas in spherical symmetry. The only external force acting on the gas is assumed to be the gravity. The structural equations to be solved are the following (cf. §16.3):

– the continuity equation

$$\frac{1}{r^2}\frac{d}{dr}(r^2\rho v) = 0, \tag{20.1}$$

– the momentum conservation equation

$$\rho v\frac{dv}{dr} = -\frac{dp}{dr} - \frac{\mathsf{G}\mathsf{M}\rho}{r^2}, \tag{20.2}$$

– and the energy equation

$$\frac{1}{r^2}\frac{d}{dr}[r^2\rho v(\tfrac{1}{2}v^2 + h)] = -\rho v\left(\frac{\mathsf{G}\mathsf{M}}{r^2}\right) + \frac{1}{r^2}\frac{d}{dr}\left(r^2 K_q\frac{dT}{dr}\right), \tag{20.3}$$

where $h = e + (p/\rho)$ is the specific enthalpy of the gas and K_q the coefficient of thermal conduction.

The continuity equation has the integral

$$4\pi r^2\rho v = \dot{\mathsf{M}}, \tag{20.4}$$

where $\dot{\mathsf{M}}$ is the *mass loss rate*. The energy equation has the integral

$$\dot{\mathsf{M}}[\tfrac{1}{2}v^2 + h - (G\mathsf{M}/r)] - 4\pi r^2 K_q(dT/dr) = \mathsf{E}, \tag{20.5}$$

where E is the total energy flux. The structure of a thermal wind is determined by solving two ordinary differential equations (20.2) and (20.5). When the system is integrated, two more integration constants appear; hence a total of four conditions, which can be boundary conditions, specifications of the behavior of the solution, or values of free parameters, are to be imposed to obtain a unique solution.

Isothermal Winds

Although physically unrealistic, it is very instructive to consider the case of an isothermal wind. In physical terms one invokes, in effect, some hypothetical heating mechanism to maintain a constant temperature to counteract the tendency of a gas to cool adiabatically. The continuity equation can be rewritten as

$$v^{-1}(dv/dr) = -\rho^{-1}(d\rho/dr) - (2/r). \tag{20.6}$$

The pressure can be expressed through the equation of state for ideal gas as

$$p = a^2\rho, \tag{20.7}$$

where a is the isothermal sound speed, assumed to be a function of r. From the equation of state and the continuity equation, one has

$$\rho^{-1}(dp/dr) = (da^2/dr) - (2a^2/r) - (a^2/v)(dv/dr); \qquad (20.8)$$

hence one can rewrite equation (20.2) as

$$[1 - (a^2/v^2)]\, v(dv/dr) = (2a^2/r) - (da^2/dr) - \mathsf{G}\mathsf{M}/r^2. \qquad (20.9)$$

For isothermal ideal gas one has $da^2/dr = 0$; hence equation (20.9) can be written as

$$\left(1 - \frac{a^2}{v^2}\right) v\frac{dv}{dr} = \frac{2a^2}{r} - \frac{\mathsf{G}\mathsf{M}}{r^2}. \qquad (20.10)$$

The right-hand side of equation (20.10) vanishes at the *critical radius*, $r = r_c \equiv \mathsf{G}\mathsf{M}/2a^2$. Therefore, at $r = r_c$ the left-hand side of (20.10) has to vanish as well. There are two possibilities: either

$$(dv/dr)_{r_c} = 0, \qquad (20.11)$$

or $(dv/dr)_{r_c}$ is nonzero, but v equals the *critical velocity*, which in this case is equal to the isothermal sound speed,

$$v(r_c) = v_c = a. \qquad (20.12)$$

Generally, the *sonic point* is defined as the point where $v(r) = a$. For an isothermal wind, the critical point thus coincides with the sonic point; this is not generally the case for other wind types, as we shall see in § 20.3. Note also that $v_c = \frac{1}{2}v_{\text{esc}}$, where $v_{\text{esc}}(r_c) = (2\mathsf{G}\mathsf{M}/r_c^2)^{\frac{1}{2}}$ is the escape velocity at the critical point.

If one demands that both v and (dv/dr) be single valued and continuous, one finds four types of solutions, as sketched in figure 20.2. First, supposing that $(dv/dr)_{r_c} = 0$, one can construct solutions in which $(v^2 - a^2)$ has the same sign for all r. If one chooses $v(r_c) < v_c$, then $v(r)$ has an absolute maximum at r_c, and v is everywhere subsonic (type 1). If $v(r_c) > v_c$, then $v(r)$ has an absolute minimum at r_c, and v is everywhere supersonic (type 2). If one assumes that $v(r_c) = v_c$, then one obtains two unique *critical solutions* that pass through the critical point (r_c, v_c) with finite slope. Both solutions are *transonic*, either with v increasing monotonically from subsonic ($v < v_c$) for $r < r_c$ to supersonic ($v > v_c$) for $r > r_c$ (type 3), or with v decreasing monotonically from supersonic to subsonic (type 4). If one drops the requirements that v be single valued, one finds two additional families of solutions, types 5 and 6 in figure 20.2.

In the astrophysical context, the theory outlined above has been applied to describe the solar wind. Since the wind originates in the solar corona, it is often called the *coronal wind*. There is a large volume of literature on this topic, starting with the pioneering studies of Parker [839, 840]; more recent reviews and monographs are [554, 841]. We will not describe the coronal wind here in detail; instead, we use

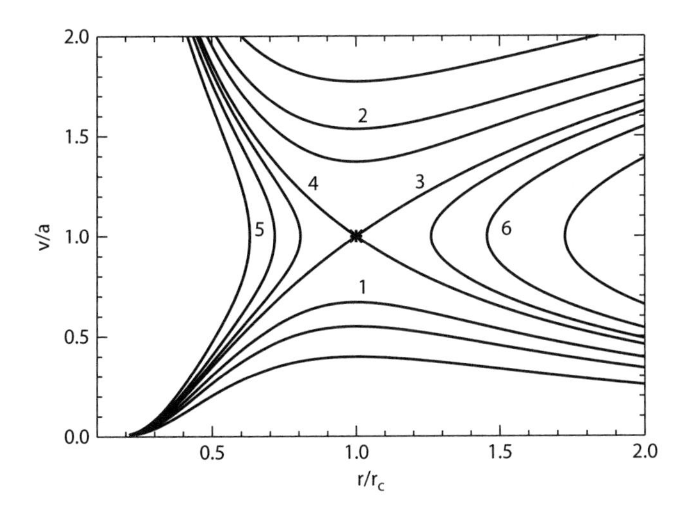

Figure 20.2 Schematic variation of velocity in units of the critical velocity $v_c = a$, as a function of radial distance in units of the critical radius r_c, for stellar winds and breezes. Solutions of type 1 are the subsonic breezes. Solutions of type 2 are everywhere supersonic. Solutions 3 and 4 are transonic critical solutions that pass through the critical point (asterisk) continuously. Solutions of types 5 and 6 are double-valued, but are important for fitting shock transitions.

the thermal wind model as a tool to gain physical understanding of a wind in more complex situations.

To choose which type of solution applies to real solar wind, one has to appeal to observations. Adopting for the temperature the value of $T \approx 1.5 \times 10^6$ K, the critical velocity is $v_c \approx 170\,\mathrm{km\,s^{-1}}$, which is much larger than observed flow velocities at the base of solar corona; hence solutions of type 2 and 4 are excluded. To choose between solutions of types 1 and 3 one has to analyze the asymptotic form of the solution [839, 840]. It turns out that the solutions of type 1 lead to a finite $p_\infty \equiv p(r \rightarrow \infty)$, which is by many orders of magnitude larger than the interstellar pressure, while the solution of type 3, the critical solution, can match a zero-pressure boundary condition at infinity. This fact led Parker to conclude that the solar wind is accelerating transonic flow, which later was verified by observations from space-based instruments.

Structure of the Critical Solution

General analytic solution of equation (20.10) is

$$\frac{v^2}{a^2} - \ln\frac{v^2}{a^2} - 4\ln\frac{r}{r_c} - \frac{4r_c}{r} = C, \tag{20.13}$$

where C is an integration constant. The critical solution is obtained for $C = -3$, for which indeed $v = a$ at $r = r_c$. Hence the velocity law for the isothermal transonic wind driven by gas pressure is

$$v \exp(-v^2/2a^2) = a(r_c/r)^2 \exp(-2r_c/r + 3/2), \tag{20.14}$$

which can also be written in terms of the critical velocity as

$$(v/v_c) \exp[-\tfrac{1}{2}(v/v_c)^2] = (r_c/r)^2 \exp(-2r_c/r + 3/2). \tag{20.15}$$

At the bottom of the subsonic region, specified by r_0, the velocity $v_0 \equiv v(r_0)$ is given by equation (20.14) or (20.15). As $v_0 \ll a$, it is given by

$$v_0 \approx a(r_c/r_0)^2 \exp(-2r_c/r_0 + 3/2). \tag{20.16}$$

Using the expression for the critical radius, $r_c = GM/2a^2$, equation (20.14) can be written as

$$(v/v_0) \exp(-v^2/2a^2) = (r_0/r)^2 \exp[(GM/a^2)(1/r_0 - 1/r)]. \tag{20.17}$$

At large distances, $r \gg r_c$, the velocity approaches

$$v(r) \approx 2a\,[\ln(r/r_0)]^{\frac{1}{2}}, \tag{20.18}$$

and thus $v(r) \to \infty$ as $r \to \infty$, which is obviously unphysical. This is an artifact of insisting the flow be isothermal, for this assumption implies a continuous deposition of thermal energy into the gas, leading to an infinite reservoir of energy that can accelerate the flow without limit. Therefore, one has to abandon the assumption of isothermal wind, at least beyond some distance from the star [839].

Mass Loss Rate

The mass loss rate is given by the equation of continuity (20.4), specified for either the lower boundary or the critical point

$$\dot{M} = 4\pi r_0^2 \rho_0 v_0 = 4\pi r_c^2 \rho_c a. \tag{20.19}$$

Using equation (20.16) and the expression for the escape velocity, the mass loss rate can also be written as

$$\dot{M} = 4\pi r_0^2 \rho_0 a [v_{\rm esc}(r_0)/2a]^2 \exp[-v_{\rm esc}^2(r_0)/2a^2 + \tfrac{3}{2}]. \tag{20.20}$$

This expression shows a very important conclusion, namely, that for an isothermal wind driven by gas pressure only, with specified conditions at the lower boundary (ρ_0, T_0, and r_0), *there is only one value of the mass loss rate for which the wind can exhibit the critical solution*, i.e., can reach supersonic velocities.

Effects of an Additional Outward Force

It is very instructive to consider a simple generalization of the previously developed model by adding an extra outward-directed force $f(r)$. The momentum equation becomes

$$v\frac{dv}{dr} = -\frac{1}{\rho}\frac{dp}{dr} - \frac{GM}{r^2} + f(r), \tag{20.21}$$

or, combined with the continuity equation,

$$(v^2 - a^2)\frac{1}{v}\frac{dv}{dr} = \frac{2a^2}{r} - \frac{GM}{r^2} + f(r). \tag{20.22}$$

In the supersonic region, the left-hand side of (20.22) is positive; therefore, an extra positive force will *increase the velocity gradient.* In contrast, in the subsonic region, the extra force will lead to *decreasing* the velocity gradient. This can be understood by noting that the extra force acts in the subsonic region to diminish the force of gravity and thus to increase the pressure scale height. The condition for the critical point is again $v_c = a$, and

$$r_c = \frac{GM}{2a^2} - \frac{f(r_c)r_c^2}{2a^2}. \tag{20.23}$$

The critical point is now closer to the star than in the case $f = 0$. Since the velocity gradient in the subsonic region is smaller, and the velocity at $r = r_c$ is the same, the velocity itself below the critical point must be higher than in the case $f = 0$. This applies at the lower boundary of the isothermal region as well; hence

$$\dot{M}_f = 4\pi r_0^2 \rho_0 v_0 > \dot{M}_0 \tag{20.24}$$

for given ρ_0; i.e., *the mass loss rate is higher.* This can also be understood in terms of density structure [668, 669]. An extra force in the subsonic region causes density to decrease more slowly compared to the extra force-free case, while at the same time the critical point is closer; hence the density at the critical point must be higher. Since the velocity at the critical point is the same ($v = a$), a higher density at the critical point results in a higher mass loss rate.

An important specific case of the outward force is the case where $f \sim r^{-2}$. It is instructive to express the force as $f(r) = GM/r^2\Gamma(r)$, where the factor Γ is considered here as a simple step function, $\Gamma(r) = 0$ for $r < r_d$, and $\Gamma(r) = \Gamma$ for $r > r_d$, where Γ is an empirical parameter. This mimics the situation where an extra $f \sim r^{-2}$ force is switched on at a certain distance from the star.

The location of the critical point is given by

$$r_c/[1 - \Gamma(r_c)] = GM/2a^2. \tag{20.25}$$

The location of the critical point for $\Gamma = 0$ is denoted as r_c^0. The behavior of the solution depends on the relation between r_c^0 and r_d.

If $r_d > r_c^0$, neither the structure of the subsonic region nor the location of the critical point is influenced by the extra force. Hence the mass loss rate remains the same as in the case of no extra force. This shows that an *outward directed force applied above the critical point does not affect the mass loss rate*. The velocity at the supersonic region will be larger because the right-hand side of equation (20.22) is larger and the term $(v^2 - a^2)$ is positive. So increasing Γ above the critical point will result in steeper velocity law and consequently larger velocities for $r > r_d$.

If $r_d < r_c(\Gamma)$, where $r_c(\Gamma) \equiv (1 - \Gamma)GM/2a^2 = (1 - \Gamma)r_c^0$, the critical point is at $r_c(\Gamma)$. If $r_c(\Gamma) < r_d < r_c^0$, then the location of the critical point depends on the actual value of Γ. For instance, for $\Gamma \geq 1$, the critical point is located at $r_c = r_d$. The velocity and density gradients are changed analogously to the case of constant force f, namely, the extra force will lead to a lower density gradient, and hence to larger density at the critical point, and thus to a higher mass loss rate.

20.3 RADIATION-DRIVEN WINDS

Basic Physics of Radiation Driving

Radiative momentum input to the gas results when photons are absorbed from an anisotropic radiation field and then are either destroyed or reemitted isotropically. Such a situation occurs in an extended stellar atmosphere, where at larger distances from the star the stellar radiation exhibits an almost purely radial streaming. The absorbed photons deposit all their outward-directed momentum into the material. Because the re-emission is isotropic, it produces no net change in the momentum of the gas. Therefore, the material experiences a net gain of momentum in the outward direction. The ions that are involved in this process are thus accelerated radially, and they drag along the rest of the gas through momentum exchange between the individual particles.

The outward acceleration of the gas due to radiation,

$$g_R = \frac{1}{\rho c} \int_0^\infty \chi_\nu F_\nu \, d\nu, \tag{20.26}$$

is to be compared with the inward acceleration of gravity, $g = GM/r^2$; if $g > g_R$ everywhere, then the atmosphere remains in hydrostatic equilibrium and does not expand. For convenience, define the *force ratio*

$$\Gamma \equiv g_R/g. \tag{20.27}$$

In hot stars (types O and B) the continuous opacity is dominated by electron scattering in the frequency regions where most of the flux emerges. Relevant energies of electrons and photons are still low enough that electron scattering can be treated as isotropic (Thomson) scattering, in which case the resulting Γ can be written as

$$\Gamma = \Gamma_e \equiv \kappa_e L/(4\pi c GM), \tag{20.28}$$

where $\kappa_e \equiv n_e \sigma_e/\rho$ is the electron scattering opacity per gram and L the stellar luminosity.

Consider now a spherically symmetric steady flow from a star. Using (20.27), the momentum equation (20.21) is written as

$$\rho v(dv/dr) = -(dp/dr) - G\mathsf{M}(1-\Gamma)\rho/r. \tag{20.29}$$

Using the same procedure as that to derive equation (20.9), one can rewrite equation (20.29) as

$$[1-(a^2/v^2)]\, v(dv/dr) = (2a^2/r) - (da^2/dr) - G\mathsf{M}(1-\Gamma)/r^2. \tag{20.30}$$

This equation allows one to gain insight into conditions under which one can have a continuous transonic flow with the simultaneous action of gravity and radiation. For simplicity, consider an isothermal atmosphere and thus drop the term (da^2/dr). It is then evident that to obtain a smooth transition from subsonic flow at small r to supersonic flow at large r, the right-hand side of (20.30) must satisfy the following three conditions: it (i) must be negative for $r < r_s$; (ii) must vanish at r_s, and (iii) must be positive for $r > r_s$, where r_s is the sonic radius where $v(r_s) \equiv v_s = a$. The condition for $r < r_s$ can be met only if $\Gamma < 1$ in that region; i.e., in the subsonic flow region the radiation force must be smaller than gravity. In contrast, in the supersonic flow region ($r > r_s$), Γ may become arbitrarily large. Indeed, the larger it is, the greater the momentum input to the gas, and the larger (dv/dr), and hence v_∞, will be. The situation here is similar to the case of thermal wind with an empirical extra force described in § 20.2; however, here the parameter Γ has a well-defined physical meaning.

If $\Gamma > 1$ everywhere in a stellar atmosphere, steady transonic flow is impossible; one has either an initially subsonic flow that decelerates outward or an initially supersonic flow that accelerates outward. Or, most likely, one will obtain a time-dependent flow in this situation.

Continuum Driving

A lower limit for Γ is $\Gamma_{\rm e}$. The luminosity corresponding to $\Gamma_{\rm e} = 1$ is called the *Eddington luminosity* , and is given by

$$L_{\rm E} = 4\pi c G\mathsf{M}/\kappa_{\rm e}. \tag{20.31}$$

The acceleration due to radiation may be written as $g_{\rm R} = \int_0^\infty \chi_\nu F_\nu \, d\nu/(\rho c) = \chi_{\rm F} F/c \approx \chi_{\rm R}\mathsf{L}/(4\pi r^2 c)$, where $\chi_{\rm F}$ and $\chi_{\rm R}$ are the flux-mean and Rosseland mean opacities, respectively, and F is the integrated flux. Therefore, with $\chi_{\rm R}$ being essentially constant, $g_{\rm R} \sim r^{-2}$, so that both the radiation force and force of gravity scale as r^{-2}. Consequently, as realized already by Eddington, if $\Gamma_{\rm e}$ rises to unity at some point of the (optically thick) envelope, one can expect $\Gamma \gtrsim 1$ throughout the reminder of the stellar interior, and the star as a whole becomes unstable.

Numerically, (20.28) gives $\Gamma_{\rm e} \approx 2.5 \times 10^{-5}\,(L/L_\odot)(\mathsf{M}_\odot/\mathsf{M})$. For an O-star, typical values are $\mathsf{L} \approx 10^6\,\mathsf{L}_\odot$ and $\mathsf{M} = 60\,\mathsf{M}_\odot$; hence $\Gamma_{\rm e} \approx 0.4$. This implies that normal O-stars are stable against radiative disruption and that the continuum radiation force alone cannot drive a transonic wind by itself. One has to look to spectral lines to provide the required force.

Effects of Inhomogeneities

Throughout this book we have been dealing only with a simple one-dimensional representation of a stellar atmosphere, assuming either a plane-parallel, horizontally homogeneous stratification (for geometrically thin atmospheres of main-sequence and dwarf stars) or a spherically symmetric geometry (giants and supergiants). We stress that we consider here objects that are indeed macroscopically either spherical (extended atmospheres) or plane-parallel (main-sequence atmospheres), so that the three-dimensional features refer to small-scale inhomogeneities in such global structures (as opposed to structures that do not posses such a simple geometrical framework, such as diffuse nebulae and complex circumstellar features).

The usefulness of the one-dimensional approximation for spectroscopic diagnostics is based on an implicit assumption that when solving the (possibly time-dependent) structural equations in three spatial dimensions and then appropriately averaging the resulting structure over the horizontal planes (in a planar atmosphere) or spherical shells (in a spherical atmosphere), and possibly over time, this procedure gives similar results as when solving the structural equations that are being averaged over planes or shells. In other words, it is assumed that inhomogeneities do not fundamentally alter the global physics of the problem.

Recent studies (e.g., [830]) demonstrated that in certain cases the atmospheric inhomogeneities may radically change the basic physics of radiation-driven stellar winds. The main point is that the radiation force may be quite different when employing these two types of averaging,

$$\frac{1}{c}\left\langle \int_0^\infty \kappa_\nu \mathbf{F}_\nu \, d\nu \right\rangle \neq \frac{1}{c}\int_0^\infty \left\langle \kappa_\nu \right\rangle \left\langle \mathbf{F}_\nu \right\rangle d\nu, \tag{20.32}$$

where $\kappa_\nu \equiv \chi_\nu/\rho$, and $\langle \cdots \rangle$ denotes an appropriate averaging over spherical surfaces (or horizontal planes). The reason for the inequality is that the radiation avoids patches with high opacity, resulting in a complex pattern of the radiation flux over the spherical shells. The high-opacity patches, or blobs, in which the radiation flux is significantly diminished as compared to the averaged case, tend to have higher densities and thus contain a larger fraction of the total mass. The net radiation force per unit mass (the acceleration) may then be significantly reduced compared to the averaged case.

This reasoning was used to describe mass outflows from objects that nominally exceed the Eddington limit and for which the outflows are driven by the continuum radiation. Since this case is very important and represents a new type of a stellar wind, moreover that of extreme strength, it will be briefly outlined in the last part of this section. Until then, we will continue with the traditional one-dimensional approach applicable to homogeneous or nearly homogeneous winds.

Line Driving

To estimate the maximum force that can result from a single line, assume that the line intercepts *unattenuated* continuum radiation; i.e., $F_\nu = F_c \approx B_\nu(T_{\rm eff})$. Then the upper limit to the acceleration of the material by a single line of an atom of chemical species k, ionization state j that originates between a lower excitation state i and an upper state u is

$$g_R^0 = (\pi e^2/mc) f_{iu} B_\nu(T_{\text{eff}})(n_{ijk}/N_k)(A_k X/m_H), \tag{20.33}$$

where n_{ijk} is the population of the particular level, N_k is the total number of atoms of species k, A_k is the abundance of species k relative to hydrogen, and X is the mass fraction of the stellar material that is hydrogen. In a pioneering study, Lucy and Solomon [705] considered the C IV line at $\lambda 1548$ Å and adopted $f_{iu} = 0.2$, $T_{\text{eff}} = 25,000$ K (to maximize B_ν), $A_C = 3 \times 10^{-4}$, $X = 1$, and $n_{ijC}/N_C = 1$ (i.e., all carbon being concentrated in the ground state of C^{3+}; again, to maximize the effect of this line); they found

$$\log(g_R^0)_{\lambda 1548} = 5.47 + \log(N_{ijC}/N_C). \tag{20.34}$$

For a typical O-supergiant, $\log g \approx 3$; hence the upper limit on the force, obtained when $(N_{ijC}/N_C) = 1$, from even this one line exceeds the force of gravity by about 2.5 dex, i.e., by a factor of 300!

Of course, the estimate just derived is (purposely) an upper limit, because the carbon atoms at the photosphere of the star produce a dark absorption line in which $F_\nu < F_c$. To account for this, [705] solved the transfer equation approximately and found that above a certain level in the atmosphere the radiation force computed using equation (20.26) still exceeds gravity. (Similar results were also found in realistic, NLTE metal line-blanketed plane-parallel model atmosphere calculations; e.g., [682]. For the purposes of constructing static model photospheres, one has to artificially suppress the resulting line radiation force.) In summary, one finds that for O-stars, the forces obtained when the atmosphere is assumed to be static are incompatible with that assumption; hence hydrostatic equilibrium in the outermost layers is not possible, and an outflow of stellar matter must occur.

The essential physics of the line driving is the following. Once the uppermost layer begins to move, the lines will be Doppler-shifted away from their rest positions, and will begin to intercept a fresh flux from the adjacent continuum; this significantly enhances the momentum input to the material, and hence increases the acceleration. The underlying layers must expand to fill the rarefaction left by acceleration of upper layers. Furthermore, the lines in these lower layers become desaturated due to the implied Doppler shift; hence these underlying layers also begin to experience a radiative force that exceeds the gravity. In this manner, the flow can be initiated; it remains to be shown that (i) the *amount* of mass loss produced is significant, and (ii) the variation of the radiation force with depth must be consistent with transonic flow.

The mass loss rate caused by a single strong line can be roughly estimated as follows. Assume, as before, that a line is formed at a wavelength close to the maximum of the Planck function corresponding to the stellar effective temperature, λ_{max}. Because of the deshadowing effect outlined above, the line can absorb, within the extent of the wind, stellar radiation between λ_{max} and $\lambda_{\text{max}}(1 - v_\infty/c)$. The total radiation momentum absorbed by this line in the wind is then $L_{\lambda_{\text{max}}} \lambda_{\text{max}} v_\infty / c^2$. The total rate of momentum loss in the wind is $\dot{M} v_\infty$; hence the total mass loss rate due to one line is $L_{\lambda_{\rm max}}\lambda_{\rm max}/c^2 \approx \mathsf{L}/c^2$, where L is the total (integrated) stellar luminosity. The latter approximate equality follows from the properties of the Planck function. For instance, for a typical Of star ζ Puppis (O4 If), which has $\mathsf{L} = 8 \times 10^5\, \mathsf{L}_\odot$, $v_\infty = 2200$ km s^{-1}, one obtains [668] $\dot{\mathsf{M}}$(one strong line) $\approx 6 \times 10^{-8}\, \mathsf{M}_\odot$. Taking the measured mass loss rate from the star, $\dot{\mathsf{M}} \approx 5 \times 10^{-6} \mathsf{M}_\odot$ yr^{-1}, one concludes that the wind of the star can be driven by a mere 80 strong lines located near the maximum of the flux. In reality, as will be shown below, the wind is driven by a large number of both strong and weak lines.

Radiation Force Due to a Single Line

In order to calculate the force exerted by the radiation on the material, it is essential to account for the saturation of the lines, so that the transition between the optically thick and thin lines is handled correctly. The problem was treated in detail by Castor [197]; here we will present simple heuristic arguments that contain the essence of the physics that led to his result. Suppose that the absorbing lines are confined to a discrete layer above the photosphere; the lines are assumed to have a rectangular box profile of width $\Delta\nu_{\rm D}$ and (constant) opacity χ_l. The momentum absorbed (per unit mass) from the unattenuated continuum flux F_c is given by

$$g_{R,l}(0) = F_c \chi_l \Delta\nu_{\rm D}/(c\rho). \tag{20.35}$$

As only the net momentum input is needed, reemissions, which are presumed to be isotropic, can be ignored. In this case the incident flux decays as $e^{-\tau_l}$, where τ_l is the line optical depth computed taking into account Doppler shifts. Thus the average rate of momentum input into the layer is

$$\tau_l \langle g_{R,l} \rangle = g_{R,l}(0) \int_0^{\tau_l} e^{-\tau'} d\tau', \tag{20.36}$$

or

$$\langle g_{R,l} \rangle = (F_c \chi_l \Delta\nu_{\rm D})/(c\rho)(1 - e^{-\tau_l})/\tau_l. \tag{20.37}$$

The term $(1 - e^{-\tau_l})/\tau_l$ can be replaced [197] by $\min(1, \tau_l^{-1})$, which provides an adequate approximation. To evaluate τ_l, one employs the Sobolev approximation (cf. § 19.5), in which case the effective optical thickness is equal to radial Sobolev optical thickness given by (19.297),

$$\tau_l \approx \chi_l v_{\rm th} |dv/dr|^{-1}, \tag{20.38}$$

where $v_{\rm th}$ is the thermal velocity. Note that the effective optical thickness given by (20.38) is significantly different from the static case where it is given by $\tau_l = \int_R^\infty \chi_l\, dr$, where R is the photospheric radius.

The force exerted by line l per unit mass of the gas can be written, using (20.37), as

$$f_l \equiv \langle g_{{\rm R},l} \rangle = (\Delta\nu_{\rm D} F_\nu/c)(\chi_l/\rho\tau_l)(1 - e^{-\tau_l}). \tag{20.39}$$

As pointed out in [1], the three factors in (20.39) have the following physical interpretation: (i) $(\Delta\nu_{\rm D}F_\nu/c)$ is the rate at which the radiative momentum is supplied by the photosphere in a frequency interval $\Delta\nu_{\rm D}$, (ii) $(\rho\tau_l/\chi_l)=\rho v_{\rm th}/|dv/dr|$ is the column mass that can absorb this momentum, and (iii) $(1-e^{-\tau_l})$ is the probability that such an absorption actually occurs.

It is convenient to introduce a depth scale which is independent of line strength, $t\equiv\tau_l/\beta_l$, where $\beta_l\equiv\chi_l/n_{\rm e}\sigma_{\rm e}$. The equivalent electron optical depth scale t, for expanding atmospheres, is given as

$$t = n_{\rm e}\sigma_{\rm e}v_{\rm th}(dv/dr)^{-1}. \tag{20.40}$$

With these variables, the radiative acceleration due to one line can be written as

$$\begin{aligned}\langle g_{{\rm R},l}\rangle &= (\Delta\nu_{\rm D}F_\nu/c\rho)(n_{\rm e}\sigma_{\rm e}/t)(1-e^{-\beta_l t})\\ &= (\Delta\nu_{\rm D}F_\nu/c)(\kappa_{\rm e}/t)\begin{cases}1, & \text{for} \quad \tau_l\gg 1,\\ \beta_l t, & \text{for} \quad \tau_l \ll 1,\end{cases}\end{aligned} \tag{20.41}$$

where $\kappa_{\rm e}\equiv n_{\rm e}\sigma_{\rm e}/\rho$. This equation shows that the force contribution of an optically thick line is independent of its strength and depends mainly on the velocity gradient, while the contribution of an optically thin line is proportional to its strength but is independent of the velocity gradient, i.e., on the kinematics of the expansion.

There is a subtle point here. Several quantities are proportional to the thermal velocity and thus to temperature. However, this dependence is in fact spurious. The dependence of the line force of the thermal velocity arises due to the following quantities: $\Delta\nu_{\rm D}\propto v_{\rm th}$, $t\propto v_{\rm th}$, and $\beta_l\propto v_{\rm th}^{-1}$. In the expression for the line force, (20.41), the terms with the thermal velocity exactly cancel, so the line acceleration from a single line is independent of the thermal velocity. When evaluating the line force, [2] argued that one can use an arbitrary $v_{\rm th}$ and choose the formula $v_{\rm th}=(2kT_{\rm eff}/m_H)^{\frac{1}{2}}$, where $T_{\rm eff}$ is the effective temperature.

The spurious dependence of g_R on $v_{\rm th}$ arises due to historical reasons because in early days of the theory the line was often viewed as having a box profile. This is also used here in writing equation (20.35), where the dependence on $v_{\rm th}$ would also immediately cancel out because the quantity $\chi_l\propto v_{\rm th}^{-1}$. Equation (20.35) may be written more exactly as

$$g_{R,l}(0) = \frac{1}{c}\int_0^\infty \frac{\chi_l(\nu)}{\rho}F_\nu\,d\nu = F_c\kappa_{ij}/(c\rho), \tag{20.42}$$

because the absorption coefficient (per unit length) for a line $l\equiv i\rightarrow j$ is given by

$$\chi_l(\nu) = \frac{h\nu_0}{4\pi}(n_iB_{ij}-n_jB_{ji})\phi_{ij}(\nu)\equiv\kappa_{ij}\phi_{ij}(\nu), \tag{20.43}$$

where n_j and n_j are the level populations, B's are the Einstein coefficients, and $\phi_{ij}(\nu)$ is the *normalized* absorption profile for the line $i\rightarrow j$, which is given, for instance, for the Doppler profile by $\phi(\nu)=\exp\{[(\nu-\nu_0)/\Delta\nu_{\rm D}]^2\}/(\pi^{\frac{1}{2}}\Delta\nu_{\rm D})$. The Doppler width $\Delta\nu_{\rm D}$ is absorbed into the expression for the absorption profile, as it should be in order for ϕ be normalized to unity, $\int_0^\infty \phi_{ij}(\nu)d\nu = 1$. The quantity κ_{ij}, which has the meaning of the frequency-averaged line opacity (per unit length), does not contain any dependence on the thermal velocity.

Radiation Force due to an Ensemble of Lines

Original Castor, Abbott, Klein Parametrization

The total line force is obtained by summing equation (20.41) over all lines and can be written

$$g_R = (F_c n_{\rm e}\sigma_{\rm e}/c\rho)M(t) = (\kappa_{\rm e}\mathsf{L}/4\pi c r^2)\,M(t), \tag{20.44}$$

where

$$M(t) \equiv F^{-1}\sum_l F_c(\nu_l)\Delta\nu_{{\rm D},l}\min(\beta_l, t^{-1}) \tag{20.45}$$

is the *radiation-force multiplier*. The calculation of the radiation force due to all lines is thus reduced to the evaluation of $M(t)$, which is a function of only one parameter t.

In the fundamental paper, Castor, Abbott, and Klein (CAK) [199] used a simplified approach to evaluate $M(t)$ and made a crucial observation that the force multiplier $M(t)$ can be well fitted by the formula

$$M(t) = kt^{-\alpha}, \tag{20.46}$$

where $k \approx 1/30$ and $\alpha \approx 0.7$. Their calculations were based on the assumption that the line spectrum is the same as that of C III, for which an extensive set of f-values was available, and assigning a total abundance of C^{++} relative to hydrogen of 10^{-3} (which is the total abundance of C, N, and O taken together). The occupation numbers were computed using LTE. Although these assumptions are somewhat rough, the results are qualitatively correct. Subsequent studies [2, 843] performed more realistic calculations of the force multiplier using extensive line lists for all species H-Zn. It turned out that a more accurate fitting formula is given by

$$M(t) = kt^{-\alpha}s^{\delta}, \tag{20.47}$$

where s is defined by

$$s \equiv 10^{-11}(n_{\rm e}/W) \propto 10^{-11}\rho/(m_H W), \tag{20.48}$$

and W is the dilution factor given by

$$W = \tfrac{1}{2}\{1 - [1 - (\mathsf{R}/r)^2]^{\frac{1}{2}}\}. \tag{20.49}$$

The values derived in [2] are $k = 0.28$, $\alpha = 0.56$, and $\delta = 0.09$.

The dependence of the force multiplier on density arises from the dependence of the ionization balance on electron density. The higher the density, the lower the overall state of ionization. Lower ionization stages of important species that contribute to the line force (C, N, O, Fe-group elements) have typically more lines, which, moreover, are usually nearer to the maximum to the photospheric flux; hence the acceleration generally increases with increasing density. Equation (9.95) explains why the number of lines, and hence the force multiplier, scales with (n_e/W).

In view of equation (20.41) it is clear that in the limit of only optically thin lines contributing to the force $\alpha \to 0$, while in the opposite limit of optically thick lines $\alpha \to 1$. Therefore, α generally satisfies $0 \le \alpha \le 1$, and its numerical value can be thought of as a measure of a relative contribution of strong lines to the radiation force. Similarly, k can be interpreted as a number of effective strong lines. In the following, we absorb the density dependence into the normalization parameter k, $k \to ks^\delta$.

Using equations (20.44)–(20.45), one obtains

$$g_R = \left(\frac{\kappa_e \mathsf{L} k}{4\pi c r^2}\right)\left(\frac{1}{n_e \sigma_e v_{th}}\frac{dv}{dr}\right)^\alpha = \frac{C}{r^2}\left(r^2 v \frac{dv}{dr}\right)^\alpha. \tag{20.50}$$

The second equality follows from writing $n_e \sigma_e = \kappa_e \rho$ and expressing ρ through the equation of continuity. The constant is given by

$$C = \frac{\kappa_e \mathsf{L} k}{4\pi c}\left(\frac{4\pi}{\kappa_e v_{th} \dot{\mathsf{M}}}\right)^\alpha. \tag{20.51}$$

Equation (20.51) may be recast to another form that becomes useful later, by using equation (20.28) to express $\kappa_e/4\pi = G\mathsf{M}\Gamma_e(\mathsf{L}/c)^{-1}$, namely,

$$C = k(v_{th}/c)^{-\alpha}(G\mathsf{M}\Gamma_e)^{1-\alpha}\left(\frac{\mathsf{L}}{\dot{\mathsf{M}} c^2}\right)^\alpha. \tag{20.52}$$

Gayley Parameterization

The form of parameterization of the line force expressed by (20.50)–(20.52) is a crucial ingredient of the CAK theory that enables one to obtain an analytical solution of the momentum equation and is widely used in astrophysical work. However, the normalization parameter k suffers from several conceptual drawbacks: (i) it is derived as a fitting parameter without having a well-defined physical meaning of its own; (ii) it depends on the thermal velocity; and (iii) it contains hidden dependences on the degree of shadowing of the lines by themselves and by neighboring lines. Therefore, Gayley [368] suggested an improved parameterization that eliminates these drawbacks and introduces a parameterization that offers a clear physical interpretation, yet at the end is equivalent to the original CAK parameterization.

We present here a simplified derivation of the parameterization that follows that of [368], to which the reader is referred for additional details and an illuminating discussion. The ratio of the radiation force on a bound electron to that on a free electron or, in other words, the ratio of the force due to a single line to that of electron scattering can be written as

$$\frac{g_{\rm line}}{g_{\rm el}} = \frac{\int_0^\infty n_i \sigma_i(\nu) F_\nu d\nu}{\int_0^\infty n_{\rm e} \sigma_{\rm e} F_\nu d\nu} = \frac{n_i \sigma_i F_i}{n_{\rm e} \sigma_{\rm e} F}, \tag{20.53}$$

where n_i is the population of the lower level of the line, $n_{\rm e}$ the electron density, $\sigma_i(\nu)$ the line cross section, $\sigma_i(\nu) = \sigma_i \phi(\nu)$, where $\phi(\nu)$ is the normalized profile; $\sigma_{\rm e}$ is the Thomson cross section. $F_i \equiv \int_0^\infty F(\nu)\phi(\nu)d\nu \approx F(\nu_i)$ where ν_i is the line center frequency. The force ratio can be written as

$$\frac{g_{\rm line}}{g_{\rm el}} = W_i\, q_i, \tag{20.54}$$

where $W_i \equiv \nu_i F_i / F$ is a flux-weighting coefficient (of the order of unity), and $q_i \equiv n_i \sigma_i / (n_{\rm e} \sigma_{\rm e} \nu_i)$. The quantity q_i has an interesting physical interpretation, as shown in [368]. Writing $\sigma_i = (\pi e^2/mc) f_i$, where f_i is the oscillator strength (or a quantum correction to the classical cross section), and $\sigma_{\rm e} = 8\pi r_{\rm e}^2/3$, where $r_{\rm e} = e^2/mc^2$ is the classical electron radius, q_i can be expressed as

$$q_i = \pi^2 f_i \, \frac{\nu_i}{\Gamma_{\rm rad}} \, \frac{n_i}{n_{\rm e}}, \tag{20.55}$$

where $\Gamma_{\rm rad} = 8\pi^2 \nu^2 r_{\rm e}/3c$ is the classical damping rate. Quantity $\nu_i/\Gamma_{\rm rad}$ is a product of the frequency and the lifetime of a line, i.e., a number of coherent oscillation cycles, which is sometimes called the Q-value of the resonator. Since $\pi^2 f_i$ is of the order of unity, q_i has the meaning of the Q-value times the bound-state population ratio $n_i/n_{\rm e}$. For typical values $\nu_i \sim 10^{15}\ {\rm s}^{-1}$ and $\Gamma_{\rm rad} \sim 10^8\ {\rm s}^{-1}$, one obtains $q_i \sim 10^7 (n_i/n_{\rm e})$. For a single strong line, $n_i/n_{\rm e} \sim 10^{-5}$, and hence $q_i \sim 10^2$, which again demonstrates the already stated fact that a force due to one strong line may exceed that of electron scattering by large factors. In order to take into account the self-shadowing of a line, the force ratio has to be multiplied by the penetration probability, given by [cf. equation (20.37)] $p_i = (1 - e^{-\tau_i})/\tau_i$, where τ_i is the Sobolev optical depth of the line.

The relative strength of the total line force is obtained by summing the contributions for all the lines.

$$\frac{F_{\rm lines}}{F_{\rm el}} = \sum_i W_i q_i p_i \equiv \overline{Q}\,\overline{P}, \tag{20.56}$$

where

$$\overline{Q} \equiv \sum_i W_i q_i, \tag{20.57}$$

$$\overline{P} \equiv \sum_i W_i q_i p_i / \overline{Q}, \tag{20.58}$$

which separates the total line strength, $\overline{Q}$, and the averaged self-shadowing correction $\overline{P}$. The value of $\overline{Q}$ can be estimated as $\overline{Q} \sim A_v \nu / \Gamma_{\rm rad}$, where A_v is the relative abundance of valence electrons in chemical species important for line driving, its typical value being $A_v \sim 10^{-4}$; ν is the typical frequency of the driving lines, $\nu \sim 10^{15}$; and $\Gamma_{\rm rad} \sim 10^8$ is a typical radiative damping. Therefore, $\overline{Q} \sim 10^3$. This again shows an enormous potential of the lines for driving the wind.

To make a connection to the original CAK formalism, i.e., to relate $\overline{Q}$ and $\overline{P}$ to k and α, one proceeds as follows (for details, and a more general derivation, refer to [368]). One replaces a discrete line list with a continuous one, which leads to replacing the sum $\sum_i W_i$ by an integral of the form $\int_0^\infty dq(dN/dq)$, where $N(q)$ is the total number of lines with strength greater than q. One then has

$$\overline{Q}\,\overline{P} = \int_0^\infty dq\,|dN/dq|\,q\,(1 - e^{-\tau})/\tau, \tag{20.59}$$

where $\tau = tqc/v_{\rm th}$. Assuming a power-law distribution of the line strengths with an exponential cutoff, $|dN/dq| \propto (q/Q_0)^{\alpha-2}\,e^{-q/Q_0}$, and using equations (20.45) and (20.46), one finds

$$M(t) = kt^{-\alpha} = \overline{Q}\,\overline{P} = \frac{1}{1-\alpha}\left(\frac{v_{\rm th}}{c}\right)^{\alpha} \overline{Q} Q_0^{-\alpha} t^{-\alpha}. \tag{20.60}$$

The final assumption concerns the value of the empirical cutoff parameter Q_0. It is argued in [368] that a reasonable *ansatz* is $Q_0 = \overline{Q}$. This leads to the desired relation between $\overline{Q}$ and the CAK parameter k,

$$k = \left(\frac{v_{\rm th}}{c}\right)^{\alpha} \frac{\overline{Q}^{1-\alpha}}{1-\alpha}. \tag{20.61}$$

Using $\overline{Q}$, (20.52) can be written as

$$C = \frac{(\overline{Q} G M \Gamma_{\rm e})^{1-\alpha}}{1-\alpha}\left(\frac{L}{\dot{M}c^2}\right)^{\alpha}. \tag{20.62}$$

Equation (20.62) removes the artificial dependence of the line driving force, or C, on the thermal velocity. It formally corresponds to setting $v_{\rm th} = c$ in all expressions that contain $v_{\rm th}$. The line normalization $\overline{Q}$ offers a significant advantage of being a *dimensionless* measure of line opacity that is independent on assumed thermal velocity. As shown in [368], its numerical values for hot, O- and B-star atmospheres is $\overline{Q} \approx 2000$.

The use of the $\overline{Q}$ normalization offers significant conceptual advantages, as was stressed in [368, 825, 826, 829]. For instance, from its definition it is clear that $\overline{Q}$ is roughly proportional to the metallicity, $\overline{Q} \propto Z$. Since the solar metallicity is $Z_\odot \approx 0.02$, one can write a simple formula [368]

$$\overline{Q} \approx 10^5 Z. \tag{20.63}$$

On the other hand, the original CAK formalism that uses the normalization k became a standard approach that is used in many studies. In order to not to depart too much from the standard approach, yet to provide a more modern view, both normalizations, using k and $\overline{Q}$, are going to be used in most important expressions.

Castor, Abbott, Klein (CAK) Theory

Using (20.50) for the line radiation force, the equation of motion (20.30) can be rewritten as

$$\left(1-\frac{a^2}{v^2}\right)v\frac{dv}{dr}=\frac{2a^2}{r}-\frac{da^2}{dr}-\frac{\mathsf{GM}(1-\Gamma_\mathrm{e})}{r^2}+\frac{C}{r^2}\left(r^2v\frac{dv}{dr}\right)^{\alpha}. \qquad (20.64)$$

Unlike equation (20.10) or (20.22) for thermal winds, (20.64) is *nonlinear* in (dv/dr) and thus has quite different mathematical properties. In particular, notice that the sonic point ($v=a$) is not the critical point of (20.64), because as the left-hand side vanishes, the right-hand side can be made to vanish as well with a suitable choice of (dv/dr). This difference from thermal wind theory is a direct consequence of a different form of the force that exhibits an *explicit* dependence on the velocity gradient (dv/dr). One can envisage an iterative procedure in which one uses an estimate of g_R (obtained, for instance, from a numerical line transfer calculation), which depends explicitly on r but only *implicitly* on (dv/dr); in this case one concludes that the critical point r_c is at the sonic point. The wind solution would then proceed as in thermal wind theory, but there is no guarantee that such a procedure will converge well. Also, one would lose important physical insight into the physics of line-driven wind.

Equation (20.64) is rewritten using substitutions $w\equiv\frac{1}{2}v^2$, $u\equiv-1/r$, and $w'\equiv(dw/du)$, as

$$F(u,w,w')\equiv[1-\tfrac{1}{2}(a^2/w)]w'-h(u)-C(w')^{\alpha}=0, \qquad (20.65)$$

where

$$h(u)\equiv-\mathsf{GM}(1-\Gamma_e)-2(a^2/u)-(da^2/du). \qquad (20.66)$$

The differential equation (20.65) has a *singular point* at which solutions terminate, have cusps, or show discontinuities. The locus of singular points is defined by the condition

$$\partial F(u,w,w')/\partial w'=1-\tfrac{1}{2}(a^2/w)-\alpha C(w')^{\alpha-1}=0. \qquad (20.67)$$

Not every points on this locus yield an acceptable solution; one has to impose the *regularity condition* that guarantees that the solution passes smoothly through the singular point; i.e., that w' is continuous there. This requirement can be met only if the solution is tangent to the singular locus at its point of contact; the regularity condition thus reads

$$(dF/du)_c=[(\partial F/\partial u)+w'(\partial F/\partial w)]_c=0. \qquad (20.68)$$

Equations (20.65)–(20.68) uniquely determine the critical point u_c (or r_c) for a given C or, conversely, C for a given r_c.

One can gain physical insight by reformulating the momentum equation by introducing

$$w \equiv v^2/v_{\rm esc}^2, \quad x \equiv 1 - \mathsf{R}/r, \quad \text{and} \quad w' \equiv dw/dx, \tag{20.69}$$

with $v_{\rm esc}$ being the escape velocity at the stellar surface, $v_{\rm esc}^2 = 2G\mathsf{M}(1-\Gamma_{\rm e})/\mathsf{R}$. The quantity w now represents the ratio of wind kinetic energy to the effective gravitational binding energy at the stellar surface. With these definitions, and neglecting for simplicity the gas-pressure contribution (i.e., assuming a zero sound-speed limit), equation (20.64) can be written in a particularly simple and revealing dimensionless form,

$$1 + w' = C'(w')^\alpha, \tag{20.70}$$

where the constant C' is given by

$$C' = C(v_{\rm esc}^2\mathsf{R}/2)^{\alpha-1} = C[G\mathsf{M}(1-\Gamma_{\rm e})]^{\alpha-1}, \tag{20.71}$$

or, using equations (20.51) and (20.52),

$$C' = k(v_{\rm th}/c)^{-\alpha}\left(\frac{\Gamma_{\rm e}}{1-\Gamma_{\rm e}}\right)^{1-\alpha}\left(\frac{\mathsf{L}}{\dot{\mathsf{M}}c^2}\right)^{\alpha} = \frac{1}{1-\alpha}\left(\frac{\overline{Q}\Gamma_{\rm e}}{1-\Gamma_{\rm e}}\right)^{1-\alpha}\left(\frac{\mathsf{L}}{\dot{\mathsf{M}}c^2}\right)^{\alpha}. \tag{20.72}$$

Figure 20.3 illustrates the solution of equation (20.70) for various values of the constant C' in graphical form. For fixed values of the stellar parameters ($\Gamma_{\rm e}$ and L) and the line opacity distribution parameters (α and k or $\overline{Q}$), the constant $C' \propto 1/\dot{\mathsf{M}}^\alpha$. For high $\dot{\mathsf{M}}$, hence low C', there is no solution, while for low $\dot{\mathsf{M}}$ (high C') there are two solutions. For the critical case, where the curve $C'(w')^\alpha$ is *tangent* to the line $1+w'$, there is one solution that represents the critical case of the *maximum* mass loss rate. The condition of tangency requires that $\alpha C'_c(w'_c)^{\alpha-1} = 1$, which together with the original equation (20.70) gives the critical solution

$$w'_c = \alpha/(1-\alpha), \quad \text{and} \quad C'_c = 1/[\alpha^\alpha(1-\alpha)^{1-\alpha}]. \tag{20.73}$$

The critical value of C' given by (20.73) specifies the maximum value of the mass loss rate in the CAK theory,

$$\begin{aligned}\dot{\mathsf{M}} &= \frac{\mathsf{L}}{c^2}\left(\frac{\Gamma_{\rm e}}{1-\Gamma_{\rm e}}\right)^{(1-\alpha)/\alpha}\frac{\alpha}{1-\alpha}[k(1-\alpha)]^{1/\alpha}\left(\frac{v_{\rm th}}{c}\right)^{-1}\\ &= \frac{\mathsf{L}}{c^2}\left(\frac{\overline{Q}\Gamma_{\rm e}}{1-\Gamma_{\rm e}}\right)^{(1-\alpha)/\alpha}\frac{\alpha}{1-\alpha}.\end{aligned} \tag{20.74}$$

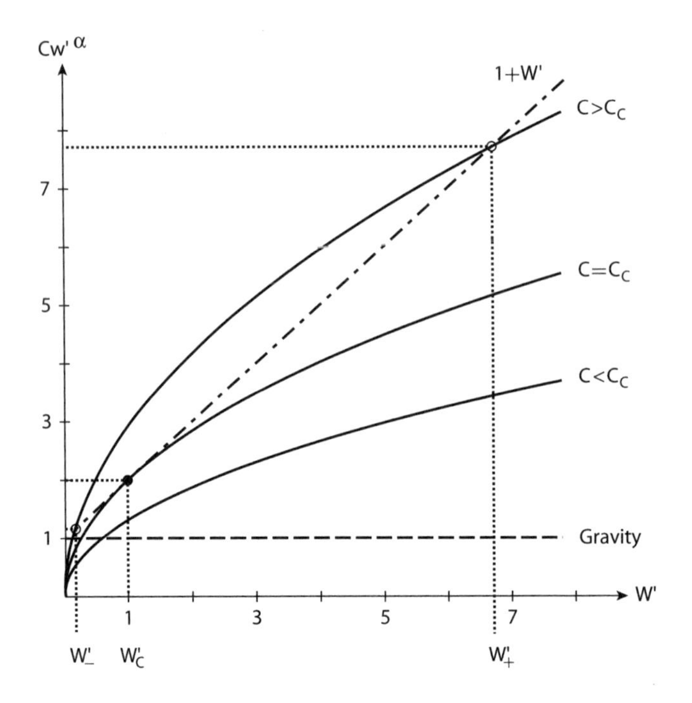

Figure 20.3 Graphical solution of the dimensionless equation of motion for different values of $C' \propto 1/\dot{\mathsf{M}}$. If $\dot{\mathsf{M}}$ is large, there are no solutions; if $\dot{\mathsf{M}}$ is small, there are two solutions. A maximum mass loss rate in the CAK wind model corresponds to a single, "critical" solution. From [823].

Equation (20.70) does not contain any explicit dependence on the radius; hence the critical conditions (20.73) apply to all radii. Upon integrating over all radii, one obtains

$$\int_0^1 w'_c \, dx = \frac{\alpha}{1-\alpha} = w_c(1) - w_c(0), \tag{20.75}$$

from which one obtains for the integration constant $v_\infty \equiv v(x=1)$

$$v_\infty^2 = v_{\rm esc}^2 \frac{\alpha}{1-\alpha}. \tag{20.76}$$

Similarly, by evaluating the integral in (20.75) from x to 1, one obtains the velocity structure

$$v^2(r) = v_\infty^2 \left(1 - \frac{\mathsf{R}}{r}\right). \tag{20.77}$$

To describe the velocity field in practical applications, one often uses an empirical β-type velocity law,

$$v(r) = v_\infty [1 - (\mathsf{R}/r)]^\beta. \tag{20.78}$$

The CAK velocity law corresponds to $\beta = \frac{1}{2}$. Moreover, for $0.5 \lesssim \alpha \lesssim 0.7$, CAK theory predicts $1 \lesssim v_\infty/v_{\rm esc} \lesssim 1.5$.

In the original CAK study, [199] obtained explicit analytical expressions for the mass loss rate, the velocity law, and the critical radius, without assuming a negligible gas pressure (sound speed). The resulting expressions are quite complicated; in the limit $v \geq a$, they become

$$\dot{\mathsf{M}} = \left(\frac{4\pi G\mathsf{M}}{\kappa_e v_{\rm th}}\right) \alpha \left(\frac{1-\alpha}{1-\Gamma_e}\right)^{(1-\alpha)/\alpha} (k\Gamma_e)^{1/\alpha}, \tag{20.79}$$

which can be shown to be equivalent to (20.74). The velocity law is

$$v^2 = \frac{2G\mathsf{M}(1-\Gamma_e)\alpha}{1-\alpha}\left(\frac{1}{r_s} - \frac{1}{r}\right), \tag{20.80}$$

where r_s is the sonic radius, essentially equal to the photospheric radius R. Again, equation (20.80) is equivalent to (20.75)–(20.77). The ratio of the critical radius to the sonic radius is

$$r_c/r_s = 1 + \left\{-\tfrac{1}{2}n + \left[\tfrac{1}{4}n^2 + 4 - 2n(n+1)\right]^{\frac{1}{2}}\right\}^{-1}, \tag{20.81}$$

which is based on the assumption that $a^2 \propto T \propto r^{-n}$. Likely values of n lie between 0 (isothermal) and $\frac{1}{2}$ (radiative equilibrium), which implies that $1.5 \lesssim (r_c/r_s) \lesssim 1.74$; the critical point thus lies well outside the sonic point.

Momentum Transport in the Wind

A stellar wind deposits its momentum, which was originally photon momentum, to the interstellar medium at a rate $\dot{\mathsf{M}}v_\infty$. If one assumes that every photon emitted by the star is scattered once in the wind, then the upper bound on the mass loss rate is

$$\dot{\mathsf{M}} \leq \mathsf{L}/v_\infty c \approx 7 \times 10^{-12}(\mathsf{L}/\mathsf{L}_\odot)(3000/v_\infty), \tag{20.82}$$

where $\dot{\mathsf{M}}$ is measured in $\mathsf{M}_\odot\ \mathrm{yr}^{-1}$ and v_∞ in km s^{-1}. For a typical O-star, $\mathsf{L} \approx 10^6\mathsf{L}_\odot$ and $v_\infty = 3000\ \mathrm{km\ s}^{-1}$; hence $\dot{\mathsf{M}} \approx 7 \times 10^{-6}\mathsf{M}_\odot\ \mathrm{yr}^{-1}$, which is in fact close to the typical observed value. The parameter

$$\epsilon \equiv c v_\infty \dot{\mathsf{M}}/\mathsf{L} \tag{20.83}$$

represents a measure of the efficiency with which matter is radiatively ejected in a wind and is called the *wind efficiency parameter*. For single scattering of all photons in the wind, ϵ cannot exceed unity.

A more complete picture of the momentum distribution in a wind emerges from integrating the momentum equation (20.29) over the whole region of the wind. For a general force law $f(r, v, dv/dr)$ one obtains

$$\int_0^{v_\infty} 4\pi r^2 \rho v\, dv + \int_{\mathsf{R}}^{\infty} 4\pi r^2 \rho \left[\frac{\mathsf{GM}(1-\Gamma_e)}{r^2} + \frac{1}{\rho}\frac{dp}{dr}\right] dr = \int_{\mathsf{R}}^{\infty} 4\pi r^2 \rho f\, dr. \tag{20.84}$$

The first integral in (20.84) is simply $\dot{\mathsf{M}}v_\infty$. The second integral is divided into the integrals over the subsonic and the supersonic regions. In the subsonic region, the matter is very nearly in hydrostatic equilibrium, in which case the integrand vanishes. In the supersonic region, the gas pressure gradient is negligible compared to gravity because the line force dominates. Using (20.28) one can approximate the second integral as

$$\frac{\mathsf{L}(1-\Gamma_e)}{c\Gamma_e}\int_{\tau_e}^{\infty} n_{\mathrm{e}}\sigma_{\mathrm{e}}\, dr = \frac{\mathsf{L}(1-\Gamma_e)\tau_e}{c\Gamma_e}, \tag{20.85}$$

where τ_e is the electron scattering optical depth of the whole supersonic region. For the CAK model, one can evaluate τ_e analytically [1], which gives

$$\tau_e = \frac{(1-\alpha)\Gamma_e}{\alpha(1-\Gamma_e)}\frac{\dot{\mathsf{M}}v_\infty c}{\mathsf{L}}. \tag{20.86}$$

Finally, using (20.40) and (20.44), one can write the third integral in (20.85) as $\beta\mathsf{L}/c$, where

$$\beta = v_{\mathrm{th}}^{-1}\int_0^{v_\infty} M(t)t\, dv \tag{20.87}$$

is essentially the equivalent number of strong lines a photon encounters when it traverses the wind. For a single-scattering model, $\beta \leq 1$.

Momentum conservation in the wind implies that

$$\dot{\mathsf{M}}v_\infty + [\tau_e(1-\Gamma_e)/\Gamma_e](\mathsf{L}/c) = \beta\mathsf{L}/c, \tag{20.88}$$

which shows that the momentum transferred from photons to the gas goes partly into the momentum lost in the wind and partly into supporting the extended envelope against gravity. One sees that the parameter ϵ defined by (20.83) underestimates the total momentum spent in driving the wind of given $\dot{\mathsf{M}}v_\infty$ because it omits the momentum transfer rate required to support the envelope. This was first pointed out in [1].

Correction for the Finite Size of the Star

The original CAK theory computed the line force in the wind assuming that the star is a point source; i.e., the stellar radiation has only the radial direction. This is a good approximation far from the star, but is inaccurate close to the star. In particular, around the critical point the stellar radiation exhibits a significant amount of non-radial momentum, and hence the total absorbed momentum and consequently the mass loss rate are *overestimated* by assuming a purely radial streaming on incoming photons.

A generalization of the CAK theory taking into account the finite size of the star [348, 843] employed the force multiplier given by

$$M(t) = kT^{-\alpha} s^{\delta} D_f, \tag{20.89}$$

where D_f is the *finite disk correction factor*, given by [199]

$$D_f = \frac{(1+\sigma)^{1+\alpha} - (1+\sigma\mu_c^2)^{1+\alpha}}{(1-\mu_c^2)(1+\alpha)\sigma(1+\sigma)^{\alpha}}, \tag{20.90}$$

$\sigma \equiv (d \ln v / d \ln r) - 1$, and $\mu_c \equiv [1 - (\mathsf{R}/r)^2]^{\frac{1}{2}}$. The momentum equation (20.64) now reads

$$\left(1 - \frac{a^2}{v^2}\right) v\frac{dv}{dr} = \frac{2a^2}{r} - \frac{da^2}{dr} - \frac{\mathsf{GM}(1-\Gamma_e)}{r^2} + \frac{C}{r^2} D_f s^{\delta} \left(r^2 v \frac{dv}{dr}\right)^{\alpha}. \tag{20.91}$$

Unlike equation (20.64), equation (20.91), which contains factors s and D_f, which are complicated functions of r, does not have an analytic solution; hence it has to be solved numerically. A simple program for calculating mass loss rate and terminal velocity following from equation (20.91) was developed in [644]. A comparison of the wind models with and without the finite disk correction reveals the following.

(i) The predicted mass loss rate with the finite disk correction is smaller by a factor of about 0.5. This is because the radiative acceleration close to the star is diminished. As demonstrated in § 20.2, a reduction of the acceleration in the subcritical wind region leads to a decrease of the density scale height there, and hence to a lower density at critical point, and hence to a lower mass loss rate.

(ii) The predicted terminal velocity of the wind is larger by a factor of about 2. This is explained by two combined effects. First, as the mass loss rate is smaller, the wind density is smaller; hence the Sobolev optical depth of the lines is smaller. Because the radiative acceleration $g_{\mathrm{R}} \sim t^{-\alpha}$, it becomes larger, which results in a higher terminal velocity. Second, the correction factor $D > 1$ in the supersonic region, which leads to an additional increase of v_∞.

(iii) The velocity as a function of r rises more gradually; i.e., an approximate β is now $\beta \approx 0.8$ rather that $\beta \approx 0.5$ as in the original CAK theory.

Wind Momentum-Luminosity Relation

Observationally, the parameters that can be deduced directly from observations are v_∞ (determined from the position of the blue edge of the P-Cygni profiles) and

$\dot{M}$ (determined by fitting the overall line profiles with theoretical predictions). It should be realized that the latter is still model-dependent.

The CAK theory, with a modification for finite size of the star, gives for the terminal velocity and the mass loss rate the following predictions [644]:

$$v_\infty \approx 2.25 \frac{\alpha}{1-\alpha} v_{\rm esc}, \tag{20.92}$$

and

$$\dot{M} \propto k^{1/\alpha'} L^{1/\alpha'} [M(1-\Gamma_e)]^{1-1/\alpha'} \propto \overline{Q}^{1/\alpha'-1} L^{1/\alpha'} [M(1-\Gamma_e)]^{1-1/\alpha'}, \tag{20.93}$$

where $\alpha' = \alpha - \delta$. The mass loss rate depends primarily on the stellar luminosity L and the effective number of strong lines, k or $\overline{Q}$ (which is obvious, because we deal with a line-driven wind), and also on the "effective mass" $M(1-\Gamma_e)$. While one can compare the "observed" $\dot{M}$ and v_∞ to theoretical predictions separately, one uses instead a *modified stellar wind momentum* defined by [891]

$$D \equiv \dot{M} v_\infty R^{\frac{1}{2}}, \tag{20.94}$$

which scales as

$$D \propto \overline{Q}^{1/\alpha'-1} L^{1/\alpha'} [M(1-\Gamma_e)]^{\frac{3}{2}-\frac{1}{\alpha'}}, \tag{20.95}$$

which is called a *wind momentum-luminosity relation*. For a typical value of the exponent $\alpha' \approx 2/3$, the dependence on $M(1-\Gamma_e)$ essentially disappears; hence the scatter of the observed $\dot{M}$ versus luminosity relation should be significantly reduced. This was indeed verified by observational studies (for a review, see [645, 669]).

Effects of Rotation

The theory presented so far has neglected the influence of stellar rotation on the wind structure. The simplest approach to the problem is to modify the gravity acceleration by the impact of centrifugal acceleration. Since the latter is a function of radius r as well as of co-latitude θ, the problem is no longer spherically symmetric. The pioneering studies of line-driven wind in the presence of rotation [348, 843] considered only the equatorial plane. The rotational speed is given by $v_\phi(r) = v_{\rm rot}(R, \theta = \pi/2)/r$. The effective gravity is then modified to

$$g_{\rm eff} = \frac{GM(1-\Gamma_e)}{r^2}\left(1 - \frac{\Omega^2}{r}\right), \tag{20.96}$$

where $\Omega \equiv v_{\rm rot}(R)/(v_{\rm esc}/\sqrt{2})$. Hence the only difference with respect to non-rotating winds is a modification of the effective stellar mass roughly by a factor of $(1-\Omega^2)$, so that the CAK mass loss rate and the terminal velocity are modified to

$$\dot{M}(\Omega) = \dot{M}(0)(1-\Omega^2)^{1-1/\alpha}, \tag{20.97}$$

and

$$v_\infty(\Omega) = v_\infty(0)(1 - \Omega^2)^{\frac{1}{2}}. \tag{20.98}$$

These expressions are valid only for the equatorial plane, so that they provide an upper limit for the influence of rotation on the wind properties. A general case was studied in [121], who introduced the concept of *wind-compressed disk*. Assuming a purely radial force exerted by radiation on the wind matter, specific angular moment is conserved for all particles, so that they remain in their own orbital plane. It can be shown [121] that in this case one can describe the wind with the latitude-dependent mass loss rate and terminal velocities that are obtained from (20.97) and (20.98) by replacing Ω with $\Omega \sin\theta$, i.e.,

$$\dot{\mathsf{M}}(\Omega, \theta) = \dot{\mathsf{M}}(0)(1 - \Omega^2 \sin^2\theta)^{1-1/\alpha}, \tag{20.99}$$

and

$$v_\infty(\Omega, \theta) = v_\infty(0)(1 - \Omega^2 \sin^2\theta)^{\frac{1}{2}}. \tag{20.100}$$

Therefore, $\dot{\mathsf{M}}$ increases and v_∞ decreases toward the equator.

During the motion of an individual particle, its azimuthal angle in the orbital plane increases, so it is deflected toward the equator. If the ratio $v_{\rm rot}/v_\infty$ is sufficiently large, the particle may actually cross the equator, in which case it may collide with particles coming from the opposite hemisphere, and a wind-compressed disk is formed. However, this picture relies on the assumption of a radial radiation force. Once a general direction of radiation force is taken into account [828], its polar component will create a pole-ward acceleration that will in fact inhibit a formation of the disk. An analysis of this effect is quite involved, and its discussion would take us afield.

Effects of Multiple Line Scattering

The picture presented so far is one in which a photon, after depositing its momentum to the ion by which it was scattered and leaving the resonance zone, is no longer scattered and escapes the atmosphere. Consequently, it deposits its momentum only once. This sets an upper limit of the mass loss rate that can be achieved. This can be estimated as follows. The line force has to overcome the wind inertia,

$$v(dv/dr) = g_l. \tag{20.101}$$

For a single optically thick line located near the maximum of the stellar flux, the acceleration due to the line is

$$g_l = \frac{F_c \chi_l \Delta\nu_{\rm D}}{c\rho}\frac{1}{\tau_l} = \frac{F_c \Delta\nu_{\rm D}}{c\rho v_{\rm th}}\frac{dv}{dr} = \frac{\nu L_\nu}{c^2 \rho\, 4\pi r^2}\frac{dv}{dr} \approx \frac{\mathsf{L}}{\dot{\mathsf{M}} c^2} v \frac{dv}{dr}, \tag{20.102}$$

where the first equality follows from (20.37) in the limit $\tau_l \gg 1$ and the second equality from (20.38). The last approximate equality follows from $F_\nu \sim B_\nu$, together with the property of the Planck function that $\nu B_\nu \approx B$ if ν is close to the peak of $B(\nu)$, and from the equation of continuity $\dot{\mathsf{M}} = 4\pi r^2 \rho v$. When combining (20.101) and (20.102), the accelerations (dv/dr) cancel, and one obtains $\dot{\mathsf{M}} \approx \mathsf{L}/c^2$ for the mass loss driven by one optically thick line. The wind is driven by many lines; assuming that the lines are independent of each other and non-overlapping, the total mass loss rate is $\dot{\mathsf{M}} \approx N_{\rm thick}\mathsf{L}/c^2$. Each line sweeps a fraction v_∞/c of the spectrum; therefore, there can be at most about $N_{\rm thick} \approx c/v_\infty$ non-overlapping optically thick lines evenly spread throughout the spectrum to drive the wind. Consequently, the *single-scattering limit* for the mass loss rate is

$$\dot{\mathsf{M}} v_\infty < \mathsf{L}/c. \tag{20.103}$$

One can introduce a *wind efficiency factor*, $\eta \equiv \dot{\mathsf{M}} v_\infty/(\mathsf{L}/c)$, so that the single-scattering limit corresponds to $\eta < 1$.

Observed values indicate that the winds of O- and B-stars are usually within this limit, $\eta < 1$. However, winds of Wolf-Rayet (WR) stars can exceed this limit by significant factors, $\eta \sim 10$ or more. This excess is sometimes referred to as the *Wolf-Rayet wind momentum problem*. Some studies even suggested that the WR winds cannot be radiatively driven. However, the large efficiency factors η merely mean that the WR winds cannot be treated by a standard single-scattering description that assumes that the thick lines do not overlap within the wind. In fact, $\eta > 1$ can be achieved by multiple scattering between *overlapping* optically thick lines with frequency separation, in velocity units, $\Delta v < v_\infty$.

As was shown in [370], if the optically thick lines are spread evenly with the separation Δv, then the wind efficiency parameter would be $\eta \approx v_\infty/\Delta v$, which can indeed lead to η of the order of a few tens required to explain the observed WR wind parameters. However, in reality the lines are not spread evenly; instead, they are distributed in bunches, with significant gaps between them. In order to describe these winds properly, one has to perform detailed calculations using realistic line lists and realistic radiative transfer calculations (note that although the Sobolev approximation may be extended to the case of a few overlapping lines, it is not feasible to treat general overlaps of many lines). Such calculation were not yet performed in full generality.

Stability of Line-Driven Winds

The theory of stellar winds described so far in this chapter assumes a steady wind. Despite the success of the line-driven wind theory in explaining the observed properties of winds in hot stars, the steady model fails to explain the presence of very high-temperature regions in the wind that are implied by the detected presence of high-temperature ions, such as O VI (e.g., [671]) or observed soft X-ray flux (for a review, see [669]), which requires temperatures of the order of $T \sim 10^6$ K. One of the mechanisms proposed to explain these observations is based on conclusions from several studies (e.g., [182, 705, 706]) that the radiation-driven wind is intrinsically unstable. As a result of this instability, small perturbations grow to large amplitudes and eventually steepen into shocks that heat the gas. However, [1] analyzed the stability of the CAK wind model and came to the conclusion that it is marginally stable, in the sense that small perturbations propagate as modified sound waves, now known as *Abbott waves*, without being amplified. This controversial situation was clarified by Owocki and Rybicki [832, 833] who showed that both previous results represent limiting cases of the assumed perturbation wavelength: if it is smaller than the Sobolev length, one gets an unstable situation, while when it is larger than the Sobolev length, one obtains a stable situation. Moreover, [832] formulated an explicit "bridging law" that connects these two limits.

Here we shall briefly outline the basic physical reasoning put forward in [832]. Let quantities with subscript 0 indicate the values for the original, unperturbed flow: the velocity $v_0(r)$ and the velocity gradient $v_0' \equiv dv_0/dr$. The Sobolev length is $L \equiv v_{\rm th}/v_0'$. Consider a single line with mean opacity χ_0. The Sobolev optical depth $\tau_0 = \chi_0 L$ represents the optical thickness of the resonant absorption layer whose geometrical thickness is L. Assume that the Sobolev approximation applies; i.e., L is small compared to any scale length for changes in mean flow properties. The mean force per unit mass exerted by the line is given by equation (20.37), which can be written as

$$g_0 \approx g_{\rm thin} \min(1, v_0'/\chi_0 v_{\rm th}), \tag{20.104}$$

where $g_{\rm thin} = F_c \Delta\nu_D \chi_l/\rho c$.

Apply now a sinusoidal perturbation

$$\delta v = \delta v_0 \, e^{i(kr-\omega t)}, \tag{20.105}$$

with wavenumber k and circular frequency ω. The wavelength of the perturbation is $\lambda \equiv k^{-1}$. To account for the possibility that the perturbation amplitude may grow or decay, the frequency ω may be complex. Disturbances in a wind are transmitted to other parts of the flow by the propagation of waves. These disturbances propagate under the combined influence of radiative and pressure forces, called *radiative-acoustic*, or *Abbott* waves [1]. The propagation and modification of these waves are described by linearized equations of continuity and momentum, which may be written to the first order as [706, 832]

$$-i\omega(\delta\rho/\rho_0) + ik\delta v = 0, \tag{20.106}$$

and

$$-i\omega\delta v = -a^2 ik(\delta\rho/\rho_0) + \delta g, \tag{20.107}$$

where a is the isothermal sound speed. For simplicity, the perturbations in the temperature are ignored, so one does not have to consider the linearized energy conservation equation. Combining equations (20.106) and (20.107), one obtains the *dispersion relation*

$$\omega^2 - i\omega(\delta g/\delta v) - a^2k^2 = 0. \tag{20.108}$$

Furthermore, ignoring pressure (density) perturbations, one arrives at a simple relation

$$\omega \approx i(\delta g/\delta v), \tag{20.109}$$

which shows that an unstable growth, which requires $\mathrm{Im}\,\omega > 0$, happens when $\mathrm{Re}(\delta g/\delta v) > 0$. In contrast, when $\mathrm{Im}\,\omega \leq 0$, and hence $\mathrm{Re}(\delta g/\delta v) < 0$, one obtains a stable wave. The response of the driving force to the velocity perturbations is thus critical to determining the stability of the wave propagation.

Assume first, as in [1], that both the mean flow and the perturbed flow satisfy the Sobolev approximation. In other words, the perturbation wavelength is assumed to be much larger than the Sobolev length, $\lambda \gg L$. The optically thick line force resulting from the perturbation (20.105) becomes

$$\delta g(r) = (\omega_0/\chi_0)\delta v(r) = (\omega_0/\chi_0) ik\, \delta v(r), \tag{20.110}$$

where $\omega_0 = g_{\mathrm{thin}} v_{\mathrm{th}}$. In this case, $(\delta g/\delta v) \propto ik$, giving ω that is purely real; hence a stable wave results that propagate *inward* at phase speed

$$(\omega/k) = -(\delta g/\delta v') = -U, \tag{20.111}$$

which became known as the *Abbott speed.*

The amplitude of the Abbott wave does not grow. The lack of amplification results from the dependence of the perturbed force on the *derivative* of the perturbed velocity, which introduces a $\pi/2$ phase shift between the velocity and force perturbations (since $\delta g \propto i\,\delta v$). Hence the net work done over a cycle by the line force is zero; consequently, the perturbation amplitude cannot grow and the flow is stable.

To consider the opposite case, one has to reexamine the perturbed line force that assumes that the perturbation neither is optically thin nor satisfies the Sobolev approximation. In other words, the perturbation wavelength is not larger than the Sobolev length. In this case, one has to employ a detailed expression for the driving force. Assuming a single line, one has

$$g(r) = g_{\mathrm{thin}} \int_{-\infty}^{\infty} dx\, \phi[x - v(r)/v_{\mathrm{th}} + v_0(r)/v_{\mathrm{th}}]\, e^{-t(x,r)}, \tag{20.112}$$

where

$$t(x, r) \equiv \int_0^r dr' \chi(r') \phi[x - v(r')/v_{\mathrm{th}} + v_0(r)/v_{\mathrm{th}}]. \tag{20.113}$$

The unperturbed optical depth can be approximated by

$$t_0(x, r) = \tau_0 \int_x^{\infty} \phi(x')\, dx' \equiv \tau_0 \Phi(x), \tag{20.114}$$

where

$$\tau_0 \equiv \chi_l v_{\mathrm{th}}/v_0' = \chi_0 L. \tag{20.115}$$

To first order in perturbed quantities, the perturbed line force δg that results from perturbations in both the optical depth, $t = t_0 + \delta t$, and the Doppler shift in the line profile, $\phi(x - \delta v/v_{\rm th}) \approx \phi(x) - \phi'(x)\delta v/v_{\rm th}$, is, from (20.112) and (20.113), given by

$$\delta g(r) = -g_{\rm thin} \int_{-\infty}^{\infty} dx\, [\phi(x)\delta v(r)/v_{\rm th} + \phi(x)\delta t(x, r)] e^{-\tau_0 \Phi(x)}, \qquad (20.116)$$

where it is assumed that the line opacity per unit mass, χ/ρ, is not affected by the perturbations; hence $\delta g_{\rm thin} = 0$ and

$$\delta t(x, r) = \tau_0 \int_x^{\infty} dx' \left\{ \phi(x') \frac{\delta\chi[r + L(x - x')]}{\chi_0} - \phi(x') \frac{\delta v[r + L(x - x')]}{v_{\rm th}} \right\}. \qquad (20.117)$$

Here one used the assumption of constant χ/ρ, which gives $\delta\chi/\chi_0 = \delta\rho/\rho_0$. By integrating equations (20.116) and (20.117) by parts, one obtains

$$\delta g(r) = -g_{\rm thin} \int_{-\infty}^{\infty} dx\, \phi(x) e^{-\tau_0 \Phi(x)} \int_x^{\infty} dx' \phi(x') \delta\tau[r + L(x - x')], \qquad (20.118)$$

where the perturbed Sobolev optical depth is given by

$$\delta\tau/\tau_0 = \delta\chi/\chi_0 - \delta v'/v_0' \approx -\delta v'/v_0'. \qquad (20.119)$$

Applying now a sinusoidal perturbation (20.105), the perturbed force (20.118) becomes

$$(\delta g/\delta v) = iK\omega_0\tau_o \int_{-\infty}^{\infty} dx\, \phi(x) e^{-\tau_0 \Phi(x)} \int_x^{\infty} dx' \phi(x') e^{-iK(x' - x)}, \qquad (20.120)$$

where $K = kL$ is the wavenumber expressed in units of the Sobolev length. The value of the perturbed force thus depends on both the perturbation wavelength and the line optical depth. Unlike the mean force g_0 [cf. equation (20.104)], it also depends on the actual form of the line profile coefficient $\phi(x)$. As was shown in [832], the perturbed force given by (20.120) can be approximated (for $\tau_0 \gg 1$) by a remarkably simple analytic expression

$$\frac{\delta g}{\delta v} \approx \omega_b \frac{ik}{\chi_b + ik}, \qquad (20.121)$$

where $\chi_b \equiv \chi_0 \phi(x_b)$, $\omega_b \equiv \omega_0 \phi(x_b)$, and x_b is the frequency of the blue edge of the local absorption profile, defined by

$$\Phi(x_b) \equiv 1/\tau_0. \qquad (20.122)$$

Equation (20.121) recovers both opposite limits of the force considered in the literature. In the short-wavelength limit, where $k^{-1} \equiv \bar{\lambda} \ll \bar{\lambda}_b \equiv \chi_b^{-1}$, one obtains $\delta g \propto \delta v$ (hence $\mathrm{Im}\,\omega > 0$), and one recovers the optically thin perturbation result of [706] and [182] that shows that the flow is unstable against small perturbations. In the opposite limit $\bar{\lambda} \gg \bar{\lambda}_b$, one obtains $\delta g \propto ik\,\delta v \propto \delta v'$, so one recovers the Sobolev-theory limit of [1], which implies that the flow is stable. Equation (20.121) thus bridges the two limits, and the characteristic length $\bar{\lambda}_b$ is called the *bridging length*. In the case of a Doppler profile and a strong line, $\tau_0 \gg 1$, one has $1/\tau_0 = \Phi(x_b) \approx \phi_{\mathrm{D}}(x_b)/2x_b$, and the blue-edge frequency is $x_b \approx [\ln(\tau_0/2\pi^{\frac{1}{2}} x_b)]^{\frac{1}{2}}$. The blue-edge frequency depends on τ_0 but is of the order of unity so that the bridging length is given by $\bar{\lambda}_b = [\chi_0 \phi_{\mathrm{D}}(x_b)]^{-1} = \tau_0/(2x_b\chi_0) = L/(2x_b)$.

The instability for $\bar{\lambda} \ll \bar{\lambda}_b$ is illustrated in figure 20.4. It arises because of deshadowing of the line by the extra Doppler shift from the velocity perturbation. For this reason, it is sometimes called the *line-deshadowing instability* (LDI).

A schematic representation of the stable situation ($\lambda \gg L$), Abbott waves, and the unstable situation ($\lambda \ll L$), line-deshadowing instability, is illustrated in figure 20.5. The left panel displays the case with $\lambda \gg L$ and explains the basic cause of stable inward propagation of Abbott waves in a line-driven stellar wind. The mean linear increase in velocity is superposed with a small-amplitude sinusoidal perturbation (wavy solid curve). The parts of the perturbation with an increased or decreased velocity gradient have an associated increase or decrease in the line force, as denoted by the up/down vertical arrows. The net effect is to accelerate or decelerate the various parts of the perturbation in such a way that after a short time the overall wave pattern (dashed curve) is shifted inward, against the direction of the stellar radiation (wavy arrows at left).

The right panel explains an unstable growth of perturbations for $\lambda \ll L$. Here the perturbed force is proportional to the velocity perturbation, which leads to an increase of the perturbations, as displayed by the arrows.

For an ensemble of lines, the perturbed force is given by an analogous expression [832],

$$\frac{\delta G}{\delta v} \approx \Omega \frac{ik}{\bar{\chi} + ik} = \frac{\Omega}{\bar{\chi}} \frac{ik\Lambda}{1 + ik\Lambda}, \tag{20.123}$$

where G denotes the *total* force by an ensemble of lines, and Ω and $\bar{\chi}$ are the appropriate averages of quantities ω_b and χ_b; explicit expressions are given in [832]. Quantity Λ is a "bridging length," being again of the order of the Sobolev length, $\Lambda \sim L$.

Finally, the previous analysis was limited to the case of purely absorbing lines. However, many strong lines that drive the wind are resonance or metastable lines that exhibit a large portion of scattering, which may change the dynamical properties of the perturbed force. The effects of scattering were studied in [704] and in more detail in [833], who have concluded that the scattered radiation exerts a drag force on velocity perturbations, which can reduce the contribution of scattering lines to the instability of the flow. This is illustrated in the lower panel of figure 20.4.

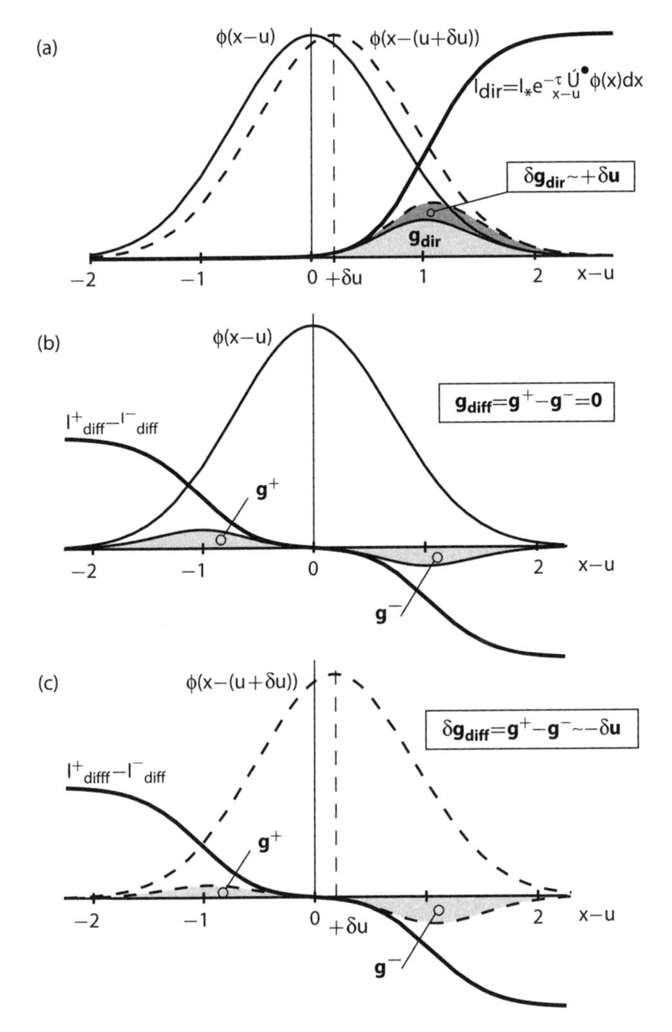

Figure 20.4 An illustration of the line-deshadowing instability. (a) The line profile ϕ and direct specific intensity I are displayed as functions of the comoving-frame frequency $x-u = x-v/v_{\rm th}$. The dashed profile shows the effect of the Doppler shift due to a perturbed velocity δv, which results in an extra area in the overlap with the blue-edge intensity giving a perturbed line force δg that scales in proportion to this perturbed velocity. (b) The comoving-frame frequency variation of the forward (+) and backward (−) streaming parts of the scattered (diffuse) radiation. This illustrates a cancellation of the overlap with the line profile that causes the net diffuse force to essentially vanish in a smooth supersonic flow. (c) However, because of the Doppler shift from the perturbed velocity, the dashed profile has a stronger interaction with the backward-streaming diffuse radiation, resulting in a diffuse line-drag force that scales with $-\delta u$, and hence tends to counter the instability of the direct line force. From [824].

Wind Clumping

The analysis of wind instability outlined above was based on a *linear* approximation, assuming that the density and velocity perturbations are small compared to the

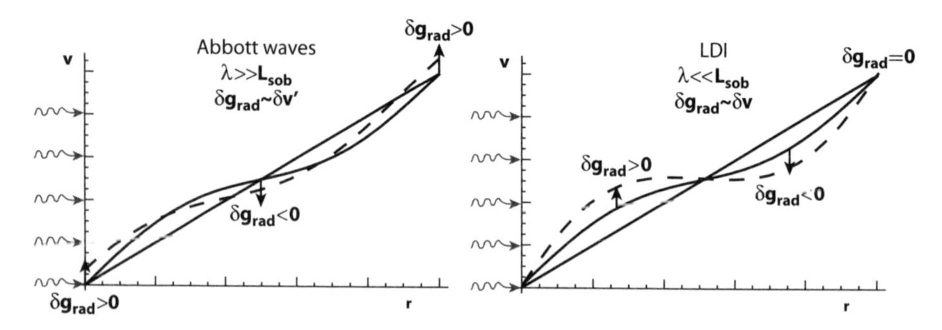

Figure 20.5 A schematic representation of the Abbott waves and the line-deshadowing instability (LDI). See the text for an explanation. Figure courtesy of Stan Owocki.

values for the undisturbed wind. However, an analytical model of [832, 833], as well as previous analyses by [706] demonstrated that if the wind is unstable, the perturbations grow very rapidly. Hydrodynamical simulations were done in [827, 831], which demonstrated that the growing instability leads to extensive structure in density and velocity. In the outer wind, the velocity variations become highly nonlinear and non-monotonic, with amplitudes approaching 1000 km s^{-1}, leading to a formation of strong shocks. However, these high-velocity regions have a low density, and hence represent only very little material. Therefore, for most of the wind mass, the dominant overall effect of the instability is that the material become concentrated into dense clumps.

An exact treatment of these instability-generated inhomogeneities will require sophisticated radiation hydrodynamic simulations, which are computationally very demanding. One the other hand, one needs at least a rudimentary way to treat clumping for a routine analysis of stellar observations. As a zero-order estimate of the effect of the clumping on the radiation field produced by the wind, one often uses the following procedure. One assumes that the mean density and velocity structure are given by the standard theory, for instance, CAK, or by an empirical velocity law (the β-law). One further assumes that the material is not distributed smoothly, but is composed of discrete blobs, or clumps, while the inter-blob space is empty. The "covering factor" of blobs is denoted f; its meaning is that the density of blobs is $\rho_b = \rho/f$. The absorption and emission coefficients, per unit length, are modified to

$$\chi_\nu(n_e, \mathbf{n}) \to \chi_\nu(n_e/f, \mathbf{n}/f)f, \tag{20.124}$$

and analogously for η. Here $\mathbf{n}$ is the vector of all level populations. The original opacity is a function of (smooth) electron density and level populations. The modified opacity is evaluated for number densities increased by the factor $1/f$ that accounts for an increased density in the blobs; to account for the fact that not the whole space is filled by the high-density material, but only a fraction f of it, such opacity should be multiplied by f. If the opacity is a linear function of populations (as, for instance, for resonance lines of the dominant ionization stage of a species), the total opacity and emissivity are not modified by the clumping. However, if the opacity exhibits a quadratic dependence on density, for instance, for recombination lines, the effects are clearly visible. This approach is often used in astrophysical work.

"Photon Tiring" Limit

It should be realized that the classical CAK theory breaks down as a star approaches the Eddington limit, $\Gamma_e \to 1$. As follows from equation (20.74), the CAK mass loss rate would diverge as $\dot{M} \sim 1/(1 - \Gamma_e)^{(1-\alpha)/\alpha}$, while the terminal velocity approaches zero as $v_\infty \sim (1 - \Gamma_e)^{\frac{1}{2}}$. However, as pointed out in [829], such a divergence would require an unlimited supply of driving radiation energy. Therefore, the finite energy available in the stellar luminosity L sets up an upper limit on the mass loss rate,

$$\tfrac{1}{2}\dot{M}v_{esc}^2 = L, \tag{20.125}$$

in which all the available radiation luminosity is used to lift mass out of the star's gravitational potential at the surface. Here v_{esc} is the escape velocity at the stellar surface. After [829], this limit is called the "photon tiring" limit,

$$\dot{M}_{tiring} = \frac{L}{v_{esc}^2/2} = 0.032\,(M_\odot/\text{yr})\, L_6 (R/R_\odot)(M/M_\odot)^{-1}, \tag{20.126}$$

where $L_6 \equiv L/10^6 L_\odot$. A comparison with the CAK mass loss rate given by equation (20.74) reveals that photon tiring would limit the CAK rate whenever

$$(1 - \Gamma_e) < \overline{Q}\Gamma_e \left(\frac{\alpha}{1-\alpha} \frac{2v_{esc}^2}{c^2} \right)^{\alpha/(1-\alpha)}. \tag{20.127}$$

This condition is extremely sensitive to α. For typical values for O-star winds, $\overline{Q} \sim 2000$ and $2v_{esc}^2/c^2 \sim 10^{-5}$, one obtains $(1 - \Gamma_e) < 8 \times 10^{-7}$ for $\alpha = 2/3$, so photon tiring is unimportant. However, with $\alpha = 1/2$, $(1 - \Gamma_e) < 2 \times 10^{-2}$, and the photon tiring limit becomes important, already within a few percent below the Eddington limit. These considerations are important when considering winds driven by continuum radiation (see below).

Continuum-Driven, Super-Eddington Winds

Finally, we will briefly mention a recent rapidly developing topic of extremely strong winds occurring in Luminous Blue Variables (LBV) and possibly in very massive, metal-poor or metal-free stars (the "first" stars).

The theory of line-driven winds described above was very successful in explaining winds in O- and B-stars, with mass loss rates ranging up to about $10^{-5} M_\odot$/yr, which is roughly the maximum mass loss rate that follows from the CAK theory with its recent extensions. However, observed mass loss rates in some objects, most notably LBVs, may be several orders of magnitude higher than this value. An extreme example is the giant eruption of massive LBV star η Carinae, which is estimated to have lost mass at a rate 0.1–0.5 $M_\odot$/yr, which is about four orders of magnitude greater than can be explained by the line-driven wind paradigm.

We follow here the theory developed in [830]. The crucial ingredient of the theory is the observation that while a spherically symmetric star cannot exceed the Eddington limit (because the force parameter $\Gamma = g_{\rm rad}/g_{\rm grav}$ is independent of r as both accelerations scale as r^{-2}, so $\Gamma > 1$ at any point would imply $\Gamma > 1$ everywhere, hence the whole star would become gravitationally unbounded), the situation is different in spatially inhomogeneous atmospheres.

In an inhomogeneous atmosphere, radiation transport will selectively avoid regions of high density in favor of low density, "porous" channels between them. The effective opacity is thus reduced, which may provide a mechanism for a transition from effectively sub-Eddington to super-Eddington conditions.

An important ingredient of the overall picture is the observation [604] that when approaching the Eddington limit, the stellar subatmospheric layers become convectively unstable. Convection thus may carry a substantial fraction of the total energy flux, which reduces the radiation flux. When approaching the surface, convection becomes inefficient, and a consequently increased radiation force may exceed gravity. A convection-induced outflow may be initiated, but as was shown in [830], it cannot be sustained because the resulting mass loss rate would exceed the photon tiring limit defined by equation (20.126). The stellar material that was lifted by the convective energy would likely fall back, creating a complex three-dimensional, time-dependent pattern. This can lead to a "porosity" of the atmosphere, i.e., a formation of higher-density clumps floating in a lower-density background. Deeper in the atmosphere, clumps tend to have a high density and be optically thick, which leads to a reduction of the effective radiation force. In the upper atmosphere, clumps become less dense and may no longer be optically thick, so they cease to provide an efficient radiation force-reducing mechanism, and the upper atmosphere may thus become super-Eddington.

The radiative transfer in such structures would be very complex and is beyond the scope of this book. However, one can envisage a simple one-dimensional description reminiscent of the "wind-clumping" approximation mentioned above, but being more general. One assumes that a medium is composed of an ensemble of identical blobs or clumps, with a characteristic dimension l_b and mass m_b; their density is thus $\rho_b \approx m_b/l_b^3$. In a medium with characteristic opacity κ (per gram, assumed frequency-independent), the characteristic blob optical depth is $\tau_b \approx \kappa\rho_b l_b \approx \kappa m_b/l_b^2$. The effective opacity of a blob (per gram) is given by $\kappa_{\rm eff} \approx \sigma_{\rm eff}/m_b$, where $\sigma_{\rm eff} \approx l_b^2$ for an optically thick blob and is given by $\sigma_{\rm eff} \approx l_b^2[1 - \exp(\tau_b)]$ in a general case. Therefore,

$$\kappa_{\rm eff} = (l_b^2/m_b)[1 - \exp(\tau_b)] \;\propto\; \kappa[1 - \exp(\tau_b)]/\tau_b. \tag{20.128}$$

If the mean distance between blobs is L_b, then the mean density of the medium is $\rho \approx m_b/L_b^3 = (l_b/L_b)^3\rho_b$, and the blob effective optical thickness can be written as $\tau_b \approx \rho\kappa L_b^3/l_b^2$. It is advantageous to write this expression through a "critical density" ρ_c as a mean medium density at which the blobs become optically thin, as

$$\tau_b = (\rho/\rho_c), \quad \rho_c \approx l_b^2/(\kappa L_b^3) \equiv 1/(\kappa h), \tag{20.129}$$

where the newly defined quantity $h \equiv L_b^3/l_b^2$ is called the "porosity length." It represents a photon mean free path in the high-density limit, $\rho \gg \rho_c$.

These considerations are applied to develop a simple heuristic model of porosity-regulated mass loss of a highly super-Eddington atmosphere with the Eddington parameter, derived from microscopic continuum opacity, $\Gamma \gg 1$. (The case with mildly super-Eddington atmosphere is similar, but algebraically more complicated.) In view of equations (20.128) and (20.129), the effective Eddington parameter $\Gamma_{\rm eff}$ is given by

$$\Gamma_{\rm eff} \approx \Gamma(\rho_c/\rho)\left(1 - e^{-\rho/\rho_c}\right). \tag{20.130}$$

At large densities at the base of an atmosphere $\Gamma_{\rm eff} < 1$. A wind can be initiated at the point where $\Gamma_{\rm eff} = 1$; at the upper layers with low densities $\Gamma_{\rm eff} \approx \Gamma \gg 1$. The mean density ρ_0 at the point where the wind is initiated can be obtained by solving (20.130) for $\Gamma_{\rm eff} = 1$, taking into account the fact that for $\Gamma \gg 1$ its reduction to 1 has to occur for large values of ρ/ρ_c,

$$\rho_0 \approx \Gamma \rho_c = \frac{\mathsf{L}}{4\pi G \mathsf{M} c h}. \tag{20.131}$$

Moreover, since the wind is assumed to be transonic, the expansion velocity at this point has to be equal to the sound speed local a_0. Therefore, the porosity-regulated mass loss rate is given by

$$\dot{\mathsf{M}}_{\rm por} = 4\pi a_0 \mathsf{R}^2 \rho_0 = \frac{a_0 \mathsf{L}}{gch} = \frac{\mathsf{L}}{a_o c}\frac{H}{h}, \tag{20.132}$$

where $g = G\mathsf{M}/\mathsf{R}^2$ is the usual surface gravity and $H = a_0^2/g$ the pressure scale height for the atmosphere in the absence of radiative forces.

The final step is to estimate the scaling of the porosity length h. By considering arguments similar to those used to derive the mixing-length theory for convection, [830] argued that a reasonable scaling is taking h proportional to the pressure scale height,

$$(h/H) \approx 1/(\Gamma - 1), \tag{20.133}$$

which yields

$$\dot{\mathsf{M}}_{\rm por} \approx \frac{\mathsf{L}}{a_0 c}(\Gamma - 1) \approx 5.6 \times 10^{-4}\,(\mathsf{M}_\odot/{\rm yr})\, L_6 T_5^{-\frac{1}{2}}(\Gamma - 1), \tag{20.134}$$

where $L_6 \equiv \mathsf{L}/10^6\mathsf{L}_\odot$ and $T_5 \equiv T/10^5$ K. Recall that the CAK mass loss rate for line-driven winds, as given by equation (20.74), yields for typical values $\overline{Q} \approx 2000$, $\alpha = 2/3$, and $\Gamma_e \approx 0.5$

$$\dot{\mathsf{M}}_{\mathrm{CAK}} \approx 7.2 \times 10^{-6}\,(\mathsf{M}_\odot/\mathrm{yr})\,L_6. \tag{20.135}$$

Comparing (20.134) to (20.135) shows that the porosity-regulated, continuum-driven mass loss rate can be *much larger* than that for the line-driven wind.

20.4 GLOBAL MODEL ATMOSPHERES

Model stellar atmospheres considered in chapters 17 and 18 were based on the assumption of hydrostatic equilibrium; hence they describe static stellar *photospheres*. In chapters 19 and 20 we have specifically considered a part of the atmosphere that is expanding or, generally, has a nonzero bulk velocity with respect to the center of the star. Describing these two regions separately brings about an artificial separation between the stellar photosphere and the wind. What is needed is a self-consistent description that avoids such an artificial separation and considers a stellar atmosphere all the way from the hydrostatic photosphere to the wind. Such a model was called a "unified model" because up to the late 1980s the prevalent scheme of treating hot star atmospheres was to split the static photosphere and the wind into two independently modeled regions. The conditions at the top of the photosphere were taken as boundary conditions for the wind, while the influence of the wind on the photosphere was largely neglected (with rare exceptions, for instance, the "wind-blanketing" scheme suggested in [4]). The first models that encompass the whole atmospheric extent in a self-consistent, albeit approximate, manner were constructed in [355]; to stress the difference from previous models the authors coined the term *unified model atmospheres*. With the advent of ALI-based numerical methods, constructing the global—photosphere + wind—models became more or less routine; hence the term "unified models" may be misleading, and the term "global models" seems to be more appropriate.

Overview of the Basic Assumptions and Structural Equations

Basic Stellar Parameters

Standard global models assume a spherically symmetric atmosphere in a steady state. The basic parameters of the problem are the stellar mass M, luminosity L, radius R, and the chemical composition. While the first two parameters are unambiguous, the radius is not well defined because of the finite extent of the atmosphere. One usually sets the "stellar (or photospheric) radius" to a value of the radius corresponding to a certain radial optical depth. One may use some averaged opacity (e.g., the Rosseland mean opacity), or the electron scattering opacity, to define the reference optical depth, τ_{ref}. Then the photospheric radius is defined as

$$\mathsf{R} = r(\tau_{\mathrm{ref}} = \tau_*), \tag{20.136}$$

where τ_* is chosen to be sufficiently large to ensure that the monochromatic radial optical depth in all frequencies is sufficiently large (at least of the order of few tens) so that the diffusion approximation applies. This procedure is analogous to setting the lower boundary for plane-parallel model atmospheres. To pursue this analogy even further, one defines the reference effective temperature and gravity acceleration by

$$T^*_{\rm eff} \equiv T_{\rm eff}(\tau_{\rm ref} = \tau_*) = \left[\mathsf{L}/(4\pi \mathsf{R}^2 \sigma_R)\right]^{1/4},$$
$$g_* \equiv g(\tau_{\rm ref} = \tau_*) = G\mathsf{M}/\mathsf{R}^2. \tag{20.137}$$

With these definitions, it is assured that the structure of the global model atmosphere approaches that of a plane-parallel model atmosphere with the same basic parameters $T^*_{\rm eff}$ and g_*.

Velocity and Density Structure

Ideally, one would determine a complete structure of the atmosphere by solving the appropriate hydrodynamical equations; i.e., the continuity equation, momentum equation, energy balance equation, and kinetic equilibrium equation, together with the radiative transfer equation. In most cases, however, one assumes the velocity field to be known, given either from previous hydrodynamical calculations or by an empirical β-law, written here in a somewhat modified form,

$$v(r) = v_\infty (1 - b\,\mathsf{R}/r)^\beta, \tag{20.138}$$

where the terminal velocity v_∞ and the slope β are the free parameters of the problem. The parameter b is taken in some studies as unity, but some authors use the definition

$$b = (1 - v_0/v_\infty)^{\frac{1}{\beta}}, \tag{20.139}$$

where v_0 is velocity at certain radius r_0 (see below). The density ρ is then given through the equation of continuity

$$4\pi r^2 v(r)\rho(r) = \dot{\mathsf{M}}, \tag{20.140}$$

where $\dot{\mathsf{M}}$ is the mass loss rate, which is another free parameter of the problem.

To assure a smooth transition from the essentially hydrostatic photosphere to the wind, one uses the following strategy. Below r_0, one solves the hydrostatic equilibrium equation to determine the pressure, and hence density. The velocity is determined from the equation of continuity (20.140). The hydrostatic equilibrium equation is written as [cf. equation (19.50)]

$$dp_g/dr = -\rho G\mathsf{M}/r^2 + (4\pi/c)\int_0^\infty \chi_\nu H_\nu\, d\nu, \tag{20.141}$$

and the gas pressure p_g is expressed, through the ideal-gas equation of state, as

$$p_g(r) = a^2(r)\rho(r), \tag{20.142}$$

where a is the isothermal sound speed, given by

$$a^2(r) = (k/\mu_I m_{\rm H})T(r), \tag{20.143}$$

k is the Boltzmann constant, μ_I the mean atomic weight that includes effects of ionization (see below), $m_{\rm H}$ the mass of the hydrogen atom, and T the temperature. This relation may also be written in the same form as that used for plane-parallel atmospheres, namely, $p_g = NkT$ and $\rho = (N - n_{\rm e})\mu m_{\rm H}$, where μ is the mean molecular weight defined by (18.27). It is related to the mean atomic weight used in equation (20.143) as $\mu_I = \mu(1 - n_{\rm e}/N)$; here N is the total particle number density and $n_{\rm e}$ the electron density. The transition point between the photosphere and the wind is specified by setting v_0 to a fraction of the isothermal sound speed, for instance, $v_0 = v(r_0) = a(r_0)/10$ (cf. [942]).

Energy Balance

In a spherically symmetric steady atmosphere the energy conservation equation is written as [cf. equation (16.51)

$$\rho v \frac{d}{dr}\left[h + \tfrac{1}{2}v^2 - \tfrac{4}{3}\nu r \frac{d}{dr}\left(\frac{v}{r}\right)\right] - \frac{1}{r^2}\frac{d}{dr}\left(r^2 K_q \frac{dT}{dr}\right) = -\rho v \frac{G\mathsf{M}}{r^2} + cG^0, \tag{20.144}$$

where $h = e + p/\rho$ is the enthalpy of the material, with e being the internal energy, ν the kinematic viscosity, K_q the coefficient of thermal conduction, and cG^0 the net rate of radiative energy input per unit volume, given (in the inertial frame) by

$$cG^0 = \int_0^\infty d\nu \oint d\Omega[\chi(\mathbf{n}, \nu)I(\mathbf{n}, \nu) - \eta(\mathbf{n}, \nu)]. \tag{20.145}$$

Under typical conditions occurring in stellar atmospheres, the viscosity and the thermal conduction terms are small and can be dropped, so the inertial-frame energy conservation equation reads

$$\rho v \frac{d}{dr}\left(h + \tfrac{1}{2}v^2\right) = -\rho v \frac{G\mathsf{M}}{r^2} + cG^0. \tag{20.146}$$

As discussed in chapter 19, the radiative deposition term cG^0 is complicated because one has to perform a double integration over frequencies and directions. It is therefore advantageous to write down the energy conservation equation in the comoving, Lagrangian frame. Under typical conditions in stellar atmospheres, the advection term (d/dr) is small compared to the radiative deposition term and can be dropped;

the energy conservation equation is then reduced to the usual radiative equilibrium equation,

$$4\pi \int_0^\infty (\chi_\nu J_\nu - \eta_\nu)\, d\nu = 0. \tag{20.147}$$

Here one has to bear in mind that all quantities are measured in the comoving frame.

Kinetic Equilibrium

The steady-state kinetic equilibrium equation in a spherically symmetric atmosphere is written as [cf. equation (9.15)]

$$\frac{1}{r^2}\frac{d(r^2 v n_i)}{dr} + n_i \sum_{j\neq i} P_{ij} - \sum_{j\neq i} n_j P_{ji} = 0, \tag{20.148}$$

where n_i is the population (occupation number) of level i, and P_{ij} is the total rate (both radiative and collisional) of the transition from state i to j. Most studies neglect the advection term (d/dr), in which case one deals with the standard kinetic equilibrium equation, discussed at length in chapters 9 and 14. In the inertial frame, the radiative rates are given by (for a radiative excitation)

$$R_{ij} = \int_0^\infty d\nu \oint d\Omega\, \frac{\alpha_{ij}(\mathbf{n}, \nu)}{h\nu} I(\mathbf{n}, \nu), \tag{20.149}$$

while in the comoving frame it is simplified to

$$R_{ij} = 4\pi \int_0^\infty \frac{\alpha_{ij}(\nu)}{h\nu} J_\nu\, d\nu. \tag{20.150}$$

Analogous expressions apply for the bound-free and downward rates.

Radiative Transfer

This topic was covered extensively in chapter 19; here we present only a brief outline. The transfer equation in the steady-state spherically symmetric atmosphere is written in the inertial frame as [cf. equation (19.84)]

$$\mu \frac{\partial I(\mu, \nu)}{\partial r} + \frac{(1-\mu^2)}{r}\frac{\partial I(\mu, \nu)}{\partial \mu} = \eta(\mu, \nu) - \chi(\mu, \nu) I(\mu, \nu), \tag{20.151}$$

and in the comoving frame as [cf. equation (19.175)]

$$\begin{aligned} &\mu \frac{\partial I(r, \mu, \nu)}{\partial r} + \frac{(1-\mu^2)}{r}\frac{\partial I(r, \mu, \nu)}{\partial \mu} \\ &\quad - \left(\frac{\nu v}{cr}\right)\left[(1-\mu^2) + \mu^2\left(\frac{d\ln v}{d\ln r}\right)\right]\frac{\partial I(r, \mu, \nu)}{\partial \nu} \\ &\quad = \eta(r, \nu) - \chi(r, \nu) I(r, \mu, \nu). \end{aligned} \tag{20.152}$$

In the inertial—observer's—frame, one has to solve for the frequency- and direction-dependent specific intensity of radiation because of the explicit dependence of the opacity and emissivity of frequency and direction. In contrast, in the comoving frame, one can solve the transfer equation for the *moments*, which significantly simplifies the solution. The moment equations in the comoving frame are given by equations (19.160) and (19.161) (we drop an explicit indication of the dependence of the moments on ν):

$$\frac{1}{r^2}\frac{\partial(r^2 H)}{\partial r} - \left(\frac{\nu v}{cr}\right)\left[\frac{\partial(J-K)}{\partial \nu} + \left(\frac{d\ln v}{d\ln r}\right)\frac{\partial K}{\partial \nu}\right] = \eta - \chi J, \tag{20.153}$$

and

$$\frac{\partial K}{\partial r} + \frac{(3K-J)}{r} - \left(\frac{\nu v}{cr}\right)\left[\frac{\partial(H-N)}{\partial \nu} + \left(\frac{d\ln v}{d\ln r}\right)\frac{\partial N}{\partial \nu}\right] = -\chi H, \tag{20.154}$$

where the moments are given by

$$[J,H,K,N] \equiv \tfrac{1}{2}\int_{-1}^{1} I(\mu,r)[1,\mu,\mu^2,\mu^3]\,d\mu. \tag{20.155}$$

Employing the Eddington and the sphericality factors,

$$f \equiv K/J, \qquad g \equiv N/H, \tag{20.156}$$

and

$$\ln(r^2 q) = \int_R^r (3f-1)/(r'f)\,dr' + \ln R^2, \tag{20.157}$$

the moment equations become [cf. equations (19.320) and (19.321)]

$$q\frac{\partial(r^2 H)}{\partial X} + \left(\frac{\nu v}{\chi rc}\right)\frac{\partial[(1+\sigma f)r^2 J]}{\partial \nu} = r^2\left(J - \frac{\eta}{\chi}\right), \tag{20.158}$$

and

$$\frac{\partial(fqr^2 J)}{\partial X} + \left(\frac{\nu v}{\chi rc}\right)\frac{\partial[(1+\sigma g)r^2 H]}{\partial \nu} = r^2 H, \tag{20.159}$$

where $\sigma \equiv (d\ln v/d\ln r) - 1$, and a new depth variable $dX \equiv -\chi q\,dr$. Unlike the case of plane-parallel atmospheres, one now has to solve for two moments, J and H, instead of just J; analogously, two Eddington factors, f and g, are needed. The second one, g, poses a potential problem. Unlike f, which is well-behaved, with its value typically between $\frac{1}{3}$ and 1, g may become very large or undefined when the flux-like moment $H \to 0$. Therefore, it was suggested (e.g., [493]) to introduce a new Eddington factor $g' \equiv N/J$ and to express N in the moment equations (20.158) and (20.159) as $N = (1-\epsilon)gH + \epsilon g'J$, with ϵ being a free parameter with a value between 0 and 1. Numerical experience showed [493] that setting $\epsilon = 1$ either for all frequencies and depths or for those frequencies and depths where the original g was poorly defined provides a stable numerical scheme.

As in the case of plane-parallel atmospheres, or static spherical atmospheres, the basic strategy is to keep the values of the Eddington and sphericality factors fixed during a current iteration of the global solution procedure and update them by a subsequent solution of the full angle- and frequency-dependent transfer equation for the specific intensity (20.152) once a given iteration step is completed. In this way, the unknown radiation parameters that have to be determined simultaneously with other structural parameters are the frequency-dependent moments J and H, but not frequency- and angle-dependent specific intensities I.

As in the case of plane-parallel atmospheres, the set of basic structural equations is supplemented by *charge conservation*, (18.24), and the definitions of the absorption and emission coefficients. One also has to formulate the appropriate boundary and initial conditions for the radiative transfer equation, as described in § 19.3. In summary, the initial boundary condition in frequency for expanding atmospheres is specified for the highest frequency, where only continuum opacities and emissivities are considered. The moment equations (20.158) and (20.159) are solved, neglecting the frequency derivatives. The inner boundary condition is formulated as the diffusion approximation for the deepest point considered, and an outer boundary condition is specified by setting the incident specific intensity to a prescribed value, typically 0 (for single stars). Care should be taken to select the outermost radial point to be large enough that the incoming intensity I^- is indeed zero; i.e., that the radial optical depth in all transitions is small. In some cases, one needs to employ specific techniques that correct for possible errors arising due to a truncation of the outer atmosphere (see, e.g., [493]).

Solution of the Structural Equations

There are two types of solution techniques employed to solve the structural equations outlined above. In the first type, one aims at an *exact*, or almost exact, solution of the problem. Although the exact solutions are now achievable, they require a considerable computational effort. Consequently, the number of such models that one can construct in a reasonable time is limited. Therefore, there have been developed a number of approaches that are inherently approximate, but require significantly less computational effort, and thus allow a wide range of the parameter space to be explored with modest computational resources and in a reasonable amount or time. We shall describe these two types of approaches in turn.

Exact Solution by Linearization

The exact numerical solution of the global expanding model atmosphere problem using the observer's-frame formulation of the transfer equation has not yet been developed, although it would be possible to do so. Its disadvantage is the need to use the specific intensity as a state parameter, which increases the number of structural parameters significantly. On the other hand, it would offer an important advantage for some applications, namely, that one can easily treat non-monotonic velocity fields that occur, for instance, in atmospheres of pulsating stars. As mentioned in § 19.4, one could in principle use the mixed-frame formalism, but it is applicable only when the opacity and emissivity are slowly varying functions of frequency; hence it is not applicable to model atmospheres where lines play an important role, such as in hot stars.

Therefore, essentially all modeling techniques applied in the context of expanding atmospheres of stars use the comoving-frame formalism. The comoving-frame moment equations (20.158) and (20.159) are discretized in frequency and space to yield [cf. equations (19.230) and (19.231)]

$$\mathbf{T}_n\mathbf{J}_n = \mathbf{X}_n + \mathbf{U}_n\mathbf{J}_{n-1} + \mathbf{V}_n\mathbf{H}_{n-1}, \tag{20.160}$$

and

$$\mathbf{H}_n = \mathbf{A}_n\mathbf{J}_n + \mathbf{B}_n\mathbf{H}_{n-1}, \tag{20.161}$$

where n is the frequency index, $\mathbf{J}_n$ is the column vector of mean intensities, $\{J_{1n}, \ldots, J_{Dn}\}^{\mathrm{T}}$, where D is the number of radial points, $\mathbf{H}_n$ is an analogous vector of the flux, $\mathbf{T}_n$ is a tridiagonal matrix, $\mathbf{U}_n$ is a diagonal matrix, $\mathbf{V}_n$ is a lower bidiagonal matrix, $\mathbf{A}_n$ is an upper bidiagonal matrix, $\mathbf{B}_n$ is a diagonal matrix, and $\mathbf{X}_n$ is a vector.

Equations (20.160) and (20.161) are linearized to yield

$$\begin{aligned}\mathbf{T}_n\delta\mathbf{J}_n = {}& \mathbf{U}_n\delta\mathbf{J}_{n-1} + \mathbf{V}_n\delta\mathbf{H}_{n-1}\\ &+ \delta\mathbf{X}_n - \delta\mathbf{T}_n\mathbf{J}_n + \delta\mathbf{U}_n\mathbf{J}_{n-1} + \delta\mathbf{V}_n\mathbf{H}_{n-1},\end{aligned} \tag{20.162}$$

and

$$\delta\mathbf{H}_n = \mathbf{A}_n\delta\mathbf{J}_n + \mathbf{B}_n\delta\mathbf{H}_{n-1} + \delta\mathbf{A}_n\mathbf{J}_n + \delta\mathbf{B}_n\mathbf{H}_{n-1}. \tag{20.163}$$

A discretization of other structural equations is straightforward, as it involves merely replacing the integrals over frequency by quadrature sums, and the derivatives with respect to the radial coordinate are replaced by differences. The derivative $(\partial/\partial r)$ occurs only in the hydrostatic equilibrium equation for the photospheric (subsonic) part of the atmosphere and/or if the advection terms in the energy equation or the kinetic equilibrium equations are being considered.

All discretized equations are then linearized. One could then form a state vector that contains all the structural parameters and solve the global linearized system for all of them, which would lead to a variant of the complete-linearization technique described in § 18.2. As explained there, its application involves repeated inversions of matrices whose number of rows and columns equals the total number of unknowns per depth. Hence it can be applied in practical applications only for a relatively low number of discretized frequencies, and a low number of selected atomic energy levels for which the populations are determined. Therefore, it is inapplicable to computing realistic models where tens or hundreds of thousands of frequencies and thousands of atomic energy levels are required.

The best option for using a linearization scheme efficiently is to eliminate some quantities from the set of unknown state parameters from the outset. As discussed at length in §18.3 and §18.4, this is achieved either by applying the idea of the accelerated lambda iteration (ALI) method or by using a (somewhat related) idea of the approximate Newton-Raphson scheme. Since one of the most powerful and widely used computer programs for calculating expanding model atmospheres—the program CMFGEN of Hillier and Miller [493]—is based on the latter scheme, it will be described first. A brief outline of the scheme in the context of static models was presented in § 18.4 and in the context of the formal solution of the comoving-frame transfer equation in § 19.3.

For the purposes of setting up the scheme, all the structural equations are divided into a set of discretized transfer equations and the rest, which are called *constraint equations*. The set of discretized constraint equations may be formally written as

$$P(T_d, n_{e,d}, n_{1d}, n_{2d}, \ldots, n_{NL,d}, J_{1d}, \ldots, J_{NF,d}) = 0, \tag{20.164}$$

where n_{id} is the population of level i at depth point d, NL is the total number of levels, J_{nd} is the mean intensity at the frequency point n and depth d, with the total number of frequency points being NF. It is advantageous to write the set of constraint equations as

$$P(T_d, n_{e,d}, n_{1d}, n_{2d}, \ldots, n_{NL,d}, J_{1d}, \ldots, J_{NF,d}, Z_{1,d}, \ldots, Z_{NT,d}) = 0, \tag{20.165}$$

where Z_{kd} is the net radiative rate for transition k; the total number of transitions is NT. Although the net rates can be written as functions of J_n and other state parameters, considering them as being part of the vector of unknowns simplifies the linearization procedure. The corrections to the mean intensities are expressed through the corrections to the other structural parameters—atomic level populations and temperature and electron density, namely [cf. equation (18.147)],

$$\delta J_{nd} = \sum_{d'=1}^{D} \sum_{j=1}^{NC} \frac{\partial J_{nd}}{\partial x_{jd'}} \, \delta x_{jd'}, \tag{20.166}$$

where $x_{jd}, j = 1, \ldots, NC$ is the set of the structural parameters at depth d; their number is NC, the number of constraint equations. In the present case $\mathbf{x}_d \equiv \{T, n_e, n_{1,d}, \ldots, n_{NL,d}\}$. In principle, the vector $\mathbf{x}$ may also contain N, the total particle number density (to allow for explicit linearization of the hydrostatic equilibrium equation in the subsonic region), but this was not done in [493], so for simplicity it will not be considered here.

As mentioned in § 18.4 and § 19.3, the basis of the method is to replace the summation over the depth points in equation (20.166), which extends over all depths (radii), by a restricted summation taking just several depths around the point d,

$$\delta J_{nd} = \sum_{d'=d-a}^{d+a} \sum_{j=1}^{NC} \frac{\partial J_{nd}}{\partial x_{jd'}} \, \delta x_{jd'}, \tag{20.167}$$

where a is set to a small number, typically $a = 1$, which corresponds to a tridiagonal approximate Newton-Raphson operator. In other words, in evaluating corrections to the mean intensities at a certain radial point, one takes into account corrections of the state parameters at this point plus its immediate neighbors.

So far, the crucial quantities $\partial J_{nd}/\partial x_{jd}$ are written formally. For practical applications, one has to devise a convenient procedure to evaluate them. As suggested in [493], it is done efficiently using the following algorithm. The mean intensities are considered to be functions of several independent variables, ψ_q, $q = 1, \ldots, Q$, which are directly related to the components of the vector $\mathbf{x}$ (for simplicity, we drop here the depth subscript d). Using these variables, one writes (for $d' = d-1, d, d+1$)

$$\frac{\partial J_{nd}}{\partial x_{jd'}} = \sum_{q=1}^{Q} \frac{\partial J_{nd}}{\partial \psi_{qd'}} \frac{\partial \psi_{qd'}}{\partial x_{jd'}}. \tag{20.168}$$

With a suitable choice of the "intermediate variables" ψ_q, the partial derivatives $\partial J_{nd}/\partial \psi_{qd'}$ as well as $\partial \psi_{qd'}/\partial x_{jd'}$ are readily evaluated. Hence the selection of the intermediate variables is important.

In static atmospheres, such intermediate variables are the absorption and emission coefficient; i.e., the mean intensity J_n and the flux-like moment H_n at frequency ν_n can be considered functions of two independent variables χ_n and η_n. (Strictly speaking, this applies only when one assumes coherent electron scattering, which is usually the case.) However, in the case of expanding atmospheres, J_n and H_n, measured in the comoving frame, are also directly coupled to the radiation field at higher frequencies, which is a consequence of the presence of the frequency derivative in the transfer equation. One can then write

$$\mathbf{J}_n = f(\eta_n, \chi_n, \mathbf{J}_{n-1}, \mathbf{H}_{n-1}). \tag{20.169}$$

For each of the Q quantities ψ_q, one can linearize equations (20.160) and (20.161) to yield

$$\begin{aligned} \mathbf{T}_n \boldsymbol{\delta} \mathbf{J}_{nq} = {} & \mathbf{U}_n \boldsymbol{\delta} \mathbf{J}_{n-1,q} + \mathbf{V}_n \boldsymbol{\delta} \mathbf{H}_{n-1,q} \\ & + \boldsymbol{\delta} \mathbf{X}_{nq} - \boldsymbol{\delta} \mathbf{T}_{nq} \mathbf{J}_n + \boldsymbol{\delta} \mathbf{U}_{nq} \mathbf{J}_{n-1} + \boldsymbol{\delta} \mathbf{V}_{nq} \mathbf{H}_{n-1}, \end{aligned} \tag{20.170}$$

and

$$\boldsymbol{\delta} \mathbf{H}_{nq} = \mathbf{A}_n \boldsymbol{\delta} \mathbf{J}_{nq} + \mathbf{B}_n \boldsymbol{\delta} \mathbf{H}_{n-1,q} + \boldsymbol{\delta} \mathbf{A}_{nq} \mathbf{J}_n + \boldsymbol{\delta} \mathbf{B}_{nq} \mathbf{H}_{n-1}, \tag{20.171}$$

where for each q, $\boldsymbol{\delta} \mathbf{J}_{nq}$ is a $D \times D$ matrix whose elements are the $\partial J_{nd}/\partial \psi_{qd'}$ factors of equation (20.168), and analogously for other $\boldsymbol{\delta}$-matrices and vectors.

A crucial point is that the form of equations (20.170) and (20.171) is identical to that of equations (20.160) and (20.161), or equations (19.230) and (19.231), and can be solved in the same way. Each equation (20.170) has D right-hand sides rather than a single right-hand side of equation (20.160). For more details and a discussion of several subtle points, see the original reference [493].

Exact Solution using Temperature Correction

The idea of this class of methods is to split the set of structural equations into two groups. The first one describes the basic structure of the atmosphere, essentially density, pressure, temperature, and velocity, while the second one describes a "restricted NLTE problem," i.e., a coupled system of the radiative transfer equation and the kinetic equilibrium equations for a given atmospheric structure. The density, pressure, and velocity are given by equations (20.138)–(20.143), so the problem is effectively reduced to finding the temperature, hence the name "temperature correction" procedure. An analogous procedure is often used to calculate LTE static model atmospheres (cf. § 17.3), and its variant was developed for NLTE static model atmospheres [292]; cf. § 18.4. Its application for expanding atmospheres was developed in [448] and forms a basis of another efficient program for calculating global model atmospheres, PHOENIX [456,458]. Since the program was originally designed for novae and supernovae (but later used for many other objects), it uses a fully relativistic formulation, which will be employed here as well.

The procedure is a generalization of the Unsöld–Lucy procedure, described in § 17.3 and § 18.4, for expanding atmospheres. One assumes radiative equilibrium in the comoving frame,

$$\int_0^\infty (\eta_\nu - \chi_\nu J_\nu)\, d\nu = 0. \tag{20.172}$$

To formulate the Unsöld–Lucy scheme, one needs to employ the equivalent form of radiative equilibrium, which follows from the frequency-integrated first-order moment of the transfer equation. It is given by [cf. (19.169) integrated over frequency]

$$\frac{\partial(r^2 H)}{\partial r} + \beta \frac{\partial(r^2 J)}{\partial r} + \frac{\beta}{r} r^2 (J - K) + \gamma^2 \frac{\partial \beta}{\partial r} r^2 (J + K + 2\beta H) = 0. \tag{20.173}$$

Here all quantities are measured in the comoving frame. The total integrated moment H expressed in the *observer's* (inertial) frame is given through the stellar luminosity, which is a basic stellar parameter, as

$$H_0^{\rm obs} = \mathsf{L}/(16\pi^2 r^2), \tag{20.174}$$

where the superscript "obs" denotes quantities in the observer's frame. The transformation of H from the comoving to the observer's frame is given by [746], which represents a generalization of equation (19.138),

$$H = \gamma^2 \left[(1 + \beta^2) H^{\rm obs} - \beta (J^{\rm obs} + K^{\rm obs}) \right]. \tag{20.175}$$

The basis of the scheme is to assume that the ratios $f^{\rm o} \equiv J^{\rm obs}/H^{\rm obs}$ and $g^{\rm o} \equiv K^{\rm obs}/H^{\rm obs}$ are only weakly dependent on temperature, so they may be held fixed during a given iteration step. One can then rewrite (20.175) as

$$H_1 = \gamma^2 \left[(1 + \beta^2) H_0^{\rm obs} - \beta H_0 (f^{\rm o} + g^{\rm o}) \right], \tag{20.176}$$

which represents an expression for improved Eddington flux H in the comoving frame. This equation is used as a boundary condition at the outer boundary, $\tau = 0$, while for other depths the improved Eddington flux is determined by equation (20.173) modified to read

$$\frac{\partial(r^2 H)}{\partial r} + \beta \frac{\partial(r^2 f^{\rm o} H)}{\partial r} + \frac{\beta}{r} r^2 H(f^{\rm o} - g^{\rm o}) + \gamma^2 \frac{\partial \beta}{\partial r} r^2 H(f^{\rm o} + g^{\rm o} + 2\beta) = 0, \quad (20.177)$$

where $f^{\rm o}$ and $g^{\rm o}$ are taken as known functions of r. Equation (20.177) can be solved as an ordinary differential equation for H. In a converged model, $H_1(r) - H(r) \equiv 0$. If the model is not converged, the difference $H_1(r) - H(r)$ is related to the errors in the state parameters. The temperature correction scheme assigns all this difference to the error in the temperature. In other words, it relates $\Delta H(r) \equiv H_1(r) - H(r) \propto \Delta T(r)$, where ΔT is a correction to the temperature that would give the correct total flux. Numerical experience showed, similarly to the case of static atmospheres, that obtaining the temperature correction from a difference in the total flux is numerically well conditioned in the inner parts of the atmosphere, while the condition of radiative equilibrium, equation (20.172), is more stable numerically in the outer parts. To make (20.172) directly useful for the temperature correction procedure, one writes $\eta_\nu = \kappa_\nu B_\nu + \sigma_\nu J_\nu$. In LTE this expression is exact, with κ_ν being the true absorption coefficient, and $\sigma_\nu = n_{\rm e}\sigma_{\rm e}$ (if the only scattering process taken into account is the coherent electron scattering). In NLTE, this is not generally correct, but one still may use the same expression, with a somewhat modified meaning of κ and σ, which may, for instance, take into account a contribution of scattering in the spectral lines [1151]; cf. § 18.4.

One introduces, after [448], a "correction function,"

$$C(r) = \begin{cases} \int_0^\infty \kappa_\nu [B_\nu(r) - J_\nu(r)]\, d\nu, & \text{for} \quad \tau < \tau_*, \\ H_1(r) - \int_0^\infty H_\nu(r)\, d\nu, & \text{for} \quad \tau \geq \tau_*, \end{cases} \quad (20.178)$$

where τ_* is an empirical "division depth" between the two forms of the flux correction function. Upon discretizing, the function C forms a column vector $\mathbf{C} \equiv \{C(r_1), \ldots, C(r_D)\}^{\rm T}$, with D being the total number of radial points. Analogously, one forms column vectors of current (imperfect) temperature $\mathbf{T}_0$ and a correction to the temperature $\mathbf{\Delta T}$ needed to satisfy the radiative equilibrium. The system $\mathbf{C}(\mathbf{T}_0 + \mathbf{\Delta T}) = 0$ is solved for the column vector of corrections $\mathbf{\Delta T}$ using the Newton-Raphson scheme,

$$\mathbf{\Delta T} = -[\partial \mathbf{C}/\partial \mathbf{T}]^{-1} \mathbf{C}(\mathbf{T}_0), \quad (20.179)$$

where $[\partial \mathbf{C}/\partial \mathbf{T}]$ is the Jacobi matrix (Jacobian) of the system, whose matrix elements are given by

$$[\partial \mathbf{C}/\partial \mathbf{T}]_{dd'} = \begin{cases} \int_0^\infty \kappa_\nu(r_d) \left[\frac{\partial B_\nu(T_d)}{\partial T_{d'}} \delta_{dd'} - \frac{\partial J_\nu(r_d)}{\partial T_{d'}} \right] d\nu, & \text{for} \quad \tau_d < \tau_*, \\ \frac{\partial H_1(r_d)}{\partial T_{d'}} - \int_0^\infty \frac{\partial H_\nu(r_d)}{\partial T_{d'}}\, d\nu, & \text{for} \quad \tau_d \geq \tau_*, \end{cases} \quad (20.180)$$

where the derivative $(\partial\kappa/\partial T)$ was set to zero for simplicity [448], but it is straightforward to include it if needed.

Finally, one has to express the derivatives $\partial J/\partial T$ and $\partial H/\partial T$. This is done by applying the idea of the accelerated lambda iteration (ALI) method. One writes $J = \Lambda[S] = \Psi[\eta]$, where Λ is the Λ-operator, $S = \eta/\chi$ is the source function, and Ψ is the Ψ-operator introduced in [925], as $\Psi[f] = \Lambda[f(1/\chi)]$. The total emission coefficient is $\eta = \eta_{\rm th} + \sigma J$, so that $J = \Psi[\eta_{\rm th}] + \Psi[\sigma J]$, and the mean intensity can be written as $J = (1 - \Psi\sigma)^{-1}[\eta_{\rm th}]$. Consequently, one can write $\partial J/\partial T = (1 - \Psi\sigma)^{-1}\Psi[\partial\eta_{\rm th}/\partial T]$. Since the Λ- or Ψ-operators are used in an iterative scheme to obtain merely *corrections* to the temperature, they may be replaced by corresponding approximate operators Λ^* or Ψ^*, exactly as in the usual ALI scheme, discussed at length in chapters 13 and 14. Therefore, one writes the derivative as (dropping the frequency subscript ν)

$$\frac{\partial J(r_d)}{\partial T_{d'}} = (1 - \Psi^*\sigma)^{-1}\Psi^*\left[\frac{\partial\eta_{{\rm th},d}}{\partial T'_d}\right]. \tag{20.181}$$

One uses an analogous operator Φ for H, defined by $H = \Phi[S]$, or a corresponding operator Υ as $H = \Upsilon[\eta]$, to express the matrix of derivatives $\partial H/\partial T$ as (again dropping the frequency subscript)

$$\frac{\partial H(r_d)}{\partial T_{d'}} = \Upsilon^*\left[\frac{\partial\eta_{{\rm th},d}}{\partial T_d}\delta_{dd'} + \frac{\partial J(r_d)}{\partial T_{d'}}\right], \tag{20.182}$$

where Υ^* is an approximate Υ-operator. The Ψ^*- and Υ^*-operators are typically chosen to be diagonal or tridiagonal. The elements of Λ^* or Ψ^* are evaluated in the formal solution of the transfer equation, as described in § 19.3, and the elements of Φ^* or Υ^* by an analogous procedure, with essentially no additional computational effort. Experience showed [448] that using tridiagonal operators is preferable because they lead to faster and more stable convergence of the temperature correction scheme.

Inherently Approximate Solutions

There are several types of approximate solutions. Either the constraint equations are solved more or less exactly, while the radiative transfer equation is treated in an approximate way, typically by the Sobolev approximation, or the transfer equation is treated exactly, but the constraint equations are approximated. From the point of view of providing reliable synthetic spectra to be compared to observations, the latter approach is usually preferable.

The latter approach is used in the popular computer program FASTWIND [942]. The authors introduce several interesting concepts. First, they introduce the so-called NLTE Hopf function. For a stellar photosphere, it is defined by

$$q_{\rm N}(\tau_{\rm R}) \equiv \frac{4}{3}\left(\frac{T(\tau_{\rm R})}{T_{\rm eff}}\right)^4 - \tau_{\rm R}, \tag{20.183}$$

where $\tau_{\rm R}$ is the Rosseland mean opacity. It has the meaning that for any computed NLTE model atmosphere, one can construct $q_{\rm N}$, with which the NLTE temperature structure can be expressed by the expression for an LTE-gray model atmosphere, replacing the true Hopf function q with $q_{\rm N}$,

$$T(\tau_{\rm R}) = \tfrac{3}{4}T_{\rm eff}^4[\tau_{\rm R} + q_{\rm N}(\tau_{\rm R})]. \tag{20.184}$$

For any database of computed NLTE model atmospheres one can construct a database of corresponding NLTE Hopf functions. To use this concept in the context of spherical expanding model atmospheres, [942] suggested to generalize the plane-parallel static NLTE Hopf function $q_{\rm N}$ to

$$q'_{\rm N}(\tau'_{\rm R}) \equiv [\tau'_{\rm R}/\tau_{\rm R}]q_{\rm N}(\tau_{\rm R}), \tag{20.185}$$

where the spherical generalization of the Rosseland mean optical depth is defined by

$$d\tau'_{\rm R} \equiv \chi_{\rm R}(\mathsf{R}/r)^2\, dr. \tag{20.186}$$

Equations (20.184)–(20.186) thus specify the temperature structure of the whole atmosphere without solving the energy conservation (radiative equilibrium) equation. However, it remains to relate the Rosseland mean optical depth scale to the r-coordinate or its equivalents (e.g., the column mass m; see below).

The hydrostatic equilibrium equation for the photospheric region is written as

$$\frac{dp}{dm} = g_* \left(\frac{\mathsf{R}}{r}\right)^2 - \frac{\sigma_R}{c}T_{\rm eff}^4 \left(\frac{\mathsf{R}}{r}\right)^2 \frac{\chi_{\rm R}}{\rho(r)}, \tag{20.187}$$

where $g_* \equiv (G\mathsf{M}/R^2)$ is the photospheric gravity acceleration, and

$$\frac{dr}{dm} = -\frac{1}{\rho(r)} = -\frac{c_1 T(r)}{p(r)}, \tag{20.188}$$

where $c_1 = k/(\mu_I m_{\rm H})$ [cf. equation (20.143)]. In equation (20.187), the radiation acceleration term $\int \chi_\nu H_\nu d\nu$ is expressed as $\chi_H H$, with $H = \mathsf{L}/(16\pi^2 r^2) \propto \sigma_R T_{\rm eff}^4(\mathsf{R}/r)^2$, and the flux-mean opacity χ_H was approximated by the Rosseland mean opacity, which is further approximated by a fitted Kramers-like formula

$$\chi_{\rm R}(r) = s_{\rm e}(r)\rho(r)\left[1 + \kappa\rho(r)T(r)^{-\xi}\right], \tag{20.189}$$

where $s_{\rm e}$ is the electron scattering opacity per unit mass, and κ and ξ are fitting parameters obtained by least-square fitting of calculated Rosseland opacities. These are evaluated iteratively: for the current atmospheric structure, one evaluates the fitting parameters, recomputes a new structures, updates the Rosseland mean opacity, and then re-determines the fitting parameters. In equations (20.187) and (20.188) one expresses ρ through T and p via equations (20.142) and (20.143). Finally, the temperature as a function of m is determined by the following differential equation:

$$\frac{dT}{dm} = \frac{3}{16}\left(\frac{\mathsf{R}}{r}\right)^2 \left(\frac{T_{\rm eff}}{T}\right)^3 T_{\rm eff}\, s_{\rm e}(r) \left(1 + \frac{\kappa}{c_1} p T^{-\xi-1}\right). \tag{20.190}$$

Equations (20.184)–(20.190) thus determine the photospheric structure, i.e., $T(m)$, $p(m)$, and $r(m)$ [and therefore also $\rho(m)$ and $v(m)$] for $m > m_0$, where m_0 is the column mass corresponding to the photosphere-wind transition point r_0 used in equation (20.139). For $m < m_0$ (the wind region), the structure is determined through the "wind equation" (20.138) together with (20.184)–(20.186) for the temperature. The whole process is iterated by updating the fitting parameters for the Rosseland mean opacity.

Having determined the basic structural parameters $T(m)$, $p(m)$, and $r(m)$, which go smoothly from the photospheric region to the wind, one performs a separate step of solving simultaneously the radiative transfer equation together with a set of kinetic equilibrium equations. This can be done either approximately, using the Sobolev approximation for determining the radiative rates, or exactly, using the ALI approach [889]. These approaches were described in detail in § 19.6. We stress that the only essential approximation compared to the "exact" approaches described earlier is a replacement of the radiative equilibrium equation with an approximate $T(r)$ relation specified by (20.190). The rest is exact.

Appendix A

Relativistic Particles

A.1 KINEMATICS AND DYNAMICS OF POINT PARTICLES

Coordinates

Frames

In the dynamics of rapidly moving material, e.g., in a stellar wind where $v/c \sim 0.1$ or larger, one must account for relativistic effects. Consider observers in two frames, the laboratory frame **F**, an inertial frame taken to be "at rest," and the *proper frame*[1] **F**$'$ moving at a constant speed v relative to **F**. For an observer in **F**$'$, the laws of thermodynamics, kinetic theory, and statistical mechanics will apply to an ensemble of particles (say, an ideal gas), and in **F**$'$ we can treat the interaction of radiation and the material using quantum mechanics. But as $v \to c$, an observer in **F** will measure different values for the properties of the material, the outcome of interactions of material particles with one other, and between the material and radiation.

Galilean Transformation

In Newtonian mechanics, time is assumed to be universal in all frames. The coordinate transformation between **F** and **F**$'$, which moves with uniform speed v along the z axis in **F**, is the *Galilean transformation*

$$z' = z - vt; \quad t' = t, \tag{A.1}$$

and

$$z = z' + vt; \quad t = t'. \tag{A.2}$$

Under this transformation the Newtonian equations of hydrodynamics have the same form in **F** and **F**$'$; i.e., those equations are *Galilean invariant*.

1D Lorentz Transformation

By penetrating gedanken experiments [314] Einstein deduced that in a vacuum, a light pulse moves isotropically at speed c in *all* frames, regardless of their relative motion. Therefore, if frame **F**$'$ with coordinates (x', y', z', t'), is in uniform motion

[1] Or *comoving frame.*

at speed v along the z axis relative to a stationary frame F with coordinates (x, y, z, t), the pulse in the two frames will be located at

$$x^2 + y^2 + z^2 - c^2t^2 \equiv x'^2 + y'^2 + z'^2 - c^2t'^2. \tag{A.3}$$

This relationship cannot be accommodated by the Galilean transformation (A.1) and (A.2), but requires the *Lorentz transformation* law, which Lorentz found in his study of the equations of electromagnetism.

In special relativity, the picture of time and a three-dimensional space as different physical entities is replaced by a unified four-dimensional *spacetime* by defining the *four-vector*

$$\mathbf{X} \equiv (x^0, x^1, x^2, x^3) = (ct, x, y, z). \tag{A.4}$$

Then using (A.3) alone, one can show [692, §1-§3], [767, §2.3] that for motion in one dimension along the z axis,

$$\mathbf{X}' = \mathsf{L}\mathbf{X}, \tag{A.5}$$

where the Lorentz transformation matrix L is

$$\mathsf{L} = \begin{pmatrix} \gamma & 0 & 0 & -\beta\gamma \\ 0 & 1 & 0 & 0 \\ 0 & 0 & 1 & 0 \\ -\beta\gamma & 0 & 0 & \gamma \end{pmatrix}. \tag{A.6}$$

Here

$$\beta \equiv v/c, \tag{A.7}$$

and

$$\gamma \equiv 1/\sqrt{1 - (v/c)^2} = 1/\sqrt{1 - \beta^2} \tag{A.8}$$

is the *Lorentz factor*. Thus

$$ct' = \gamma(ct - \beta z), \tag{A.9}$$

and

$$z' = \gamma(z - vt). \tag{A.10}$$

The inverse transformation from the comoving frame to the laboratory frame is

$$\mathbf{X} = \mathsf{L}^{-1}\mathbf{X}', \tag{A.11}$$

where

$$\mathsf{L}^{-1} = \begin{pmatrix} \gamma & 0 & 0 & \beta\gamma \\ 0 & 1 & 0 & 0 \\ 0 & 0 & 1 & 0 \\ \beta\gamma & 0 & 0 & \gamma \end{pmatrix}. \tag{A.12}$$

It is trivial to verify that (A.12) is the inverse of (A.6). Physically this must be so because as seen from **F**$'$, **F** moves with velocity $-\mathbf{v}$. Thus

$$ct = \gamma(ct' + \beta z'), \tag{A.13}$$

and

$$z = \gamma(z' + vt'). \tag{A.14}$$

3D Lorentz Transformation

By symmetry considerations (A.6) can be generalized to the case where **F**$'$ has a three-dimensional velocity $\mathbf{v}$ relative to **F** [767, §2.4]. One finds

$$\begin{pmatrix} ct' \\ \\ \mathbf{x}' \end{pmatrix} = \begin{pmatrix} \gamma & -\gamma\boldsymbol{\beta} \\ \\ -\gamma\boldsymbol{\beta} & \mathsf{I} + \dfrac{(\gamma - 1)\boldsymbol{\beta}\,\boldsymbol{\beta}}{\beta^2} \end{pmatrix} \begin{pmatrix} ct \\ \\ \mathbf{x} \end{pmatrix}. \tag{A.15}$$

Here $\mathbf{x}$ is the three-vector (x, y, z), and $c\boldsymbol{\beta} \equiv (v_x, v_y, v_z)$. Or, equivalently,

$$ct' = \gamma(ct - \boldsymbol{\beta}\cdot\mathbf{x}), \tag{A.16}$$

and

$$\mathbf{x}' = \mathbf{x} + \left[\frac{(\gamma - 1)\mathbf{v}\cdot\mathbf{x}}{v^2} - \gamma t\right]\mathbf{v}. \tag{A.17}$$

Even for a 3D relative velocity between frames, (A.16) and (A.17) reduce to (A.9) and (A.10) if one of the coordinate axes is chosen to lie along $\mathbf{v}$.

Lorentz Contraction and Time Dilation

Lorentz transformation has a number of consequences. Suppose one has a rod of length Δz as measured in the laboratory frame **F**, and an observer in a frame **F**$'$ moving with velocity $\mathbf{v}$ relative to **F** measures its length as $\Delta z'$ at a given instant t'. From (A.14), $\Delta z = \gamma \Delta z'$. Hence the rod is seen to be *shorter* by a factor of $1/\gamma = (1 - \beta^2)^{1/2}$ in the moving frame **F**$'$. This phenomenon is called *Lorentz contraction*. The length ℓ_0 of the rod measured in the frame in which it is seen at rest is called its *proper length*.

Or suppose that the time interval between two events at a fixed position z' in frame **F**$'$ moving with velocity $\mathbf{v}$ with respect to the laboratory frame **F** is measured to be $\Delta t'$ by a clock in the moving frame. Then from (A.9) the time interval between these two events measured by an observer in the fixed laboratory frame **F** is $\Delta t = \gamma \Delta t'$, i.e., a factor of γ longer in the laboratory frame than in the comoving frame. This phenomenon is called *time dilation*; it has been verified with great precision experimentally by comparing the decay time of an unstable particle (say, a meson) moving at very high speeds in an accelerator to its decay time when it is at rest in the laboratory frame.

Both of these effects are symmetric between frames: If a rod has length Δz in **F**, it will be observed to have length $\Delta z/\gamma$ in **F**$'$; and if a time interval is measured as Δt in **F**, it will be measured to be $\Delta t' = \gamma\Delta t$ in **F**$'$. This *must* be the case because the Lorentz transformation and its inverse depend only on the *relative* speed of one frame with respect to the other.

In summary, neither "space" nor "time" is absolute in the Newtonian sense. They are frame-dependent. Hence we abandon the Newtonian "3D space + time" picture and generalize to four-dimensional spacetime.

Four-Vectors, Four-Tensors, and World Scalars

In 4D spacetime we write *four-vectors* and *four-tensors* as

$$X^\alpha, \quad \alpha = (0, \ldots, 3) \quad \text{and} \quad T^{\alpha\beta}, \quad \alpha, \beta = (0, \ldots, 3). \tag{A.18}$$

An example of a four-vector is the coordinate system $\mathbf{X}$. An example of a four-tensor is the *Lorentz metric*, which in Cartesian coordinates is

$$\boldsymbol{\eta} \equiv \eta_{\alpha\beta} = \begin{pmatrix} -1 & 0 & 0 & 0 \\ 0 & 1 & 0 & 0 \\ 0 & 0 & 1 & 0 \\ 0 & 0 & 0 & 1 \end{pmatrix}. \tag{A.19}$$

Four-vectors and tensors have both "contravariant" (indices on top) and "covariant" (indices on bottom) forms. They represent the same physical quantity but have different geometric interpretations. A *world scalar*, formed by taking the inner product of the covariant and contravariant forms of a four-vector, $S = A_\alpha A^\alpha$, has the same value in all frames.[2] Multiplication of a covariant four-vector by the contravariant Lorentz metric converts it from covariant to contravariant form. Thus coordinates transform as $X_\alpha = \eta_{\alpha\beta} X^\beta$, and a contravariant four-tensor is converted to covariant form by two applications of the Lorentz metric: $T_{\alpha\beta} = \eta_{\alpha\mu}\eta_{\beta\nu}T^{\mu\nu}$. The net effect in Cartesian coordinates is to change the sign of the timelike component.

Spacetime Interval and Proper Time

An important example of a world scalar is the *spacetime interval ds*:

$$ds^2 \equiv dx^2 + dy^2 + dz^2 - c^2dt^2 = \eta_{\alpha\beta}\, dx^\alpha dx^\beta, \tag{A.20}$$

which is *invariant* (the same) in all frames. The spacetime interval *must be* invariant under Lorentz transformation because that transformation was *constructed* by invoking this invariance; see (A.1). Spacetime intervals fall uniquely into three groups—*spacelike*: $ds^2 > 0$; *timelike*: $ds^2 < 0$; and *null*: $ds^2 \equiv 0$. All physical events are either timelike or null. Null intervals are of special interest because they are photon paths.

[2] The repeated indices indicate summation over $\alpha = (0, \ldots, 3)$.

A relative of the spacetime interval is an interval of *proper time*, which is measured in the frame in which an event occurs at the same position:

$$c^2 d\tau^2 \equiv c^2 dt^2 - dx^2 - dy^2 - dz^2. \tag{A.21}$$

Proper time τ is the time measured in the frame of a moving particle. It is related to laboratory-frame time by

$$d\tau = \{1 - [(dx/dt)^2 - (dy/dt)^2 - (dz/dt)^2]/c^2\}^{1/2} dt = (dt/\gamma). \tag{A.22}$$

Spacetime Volume Element

The *4D spacetime volume element* is defined as

$$dV\, dt = dx\, dy\, dz\, dt. \tag{A.23}$$

In view of the discussion of Lorentz contraction and time dilation given above, (A.23) can be rewritten as

$$dV\, dt = dx\, dy\, dz\, dt = dx'\, dy' \gamma\, dz' \frac{dt'}{\gamma} \equiv dx'\, dy'\, dz'\, dt' = dV'\, dt'. \tag{A.24}$$

That is, *the spacetime volume element is an invariant.* We will use this result repeatedly.

Kinematics

To obtain kinematic and dynamical equations that are relativistically correct, a successful procedure is to re-express Newtonian equations in terms of world scalars, four-vectors, and four-tensors. Such a formulation is called *covariant* because the resulting expression is independent of the choice of coordinate system.[3]

Four-Velocity

The Newtonian definition of velocity is $\mathbf{v} = [(dx/dt), (dy/dt), (dz/dt)]$. But this expression is not satisfactory because dt is not a relativistic invariant. Therefore, use the invariant proper time (A.22) to define the *four-velocity* of a particle (or a fluid element) as

$$V^\alpha \equiv \frac{dx^\alpha}{d\tau} = \gamma(c, \mathbf{v}). \tag{A.25}$$

Applying the Lorentz metric, we find that the covariant components of V^α are

$$V_\alpha \equiv \eta_{\alpha\beta} V^\beta = \gamma(-c, \mathbf{v}). \tag{A.26}$$

[3] Note: "Covariant" as used here is unfortunately the same word used above for a certain geometric representation of the components of a four-vector, but the two meanings are distinct and should not be confused.

Then

$$V^\alpha V_\alpha = (-\gamma c)(\gamma c) + \gamma^2 v^2 = (v^2 - c^2)/(1 - v^2/c^2) = -c^2, \tag{A.27}$$

so the four-velocity is *timelike*.

Four-Acceleration

The relativistic expression for acceleration is

$$A^\alpha \equiv \frac{dV^\alpha}{d\tau}. \tag{A.28}$$

Because $V_\alpha V^\alpha = -c^2$,

$$V_\alpha \left(\frac{dV^\alpha}{d\tau} \right) = V_\alpha A^\alpha \equiv 0. \tag{A.29}$$

Hence four-acceleration is *spacelike* and orthogonal to the four-velocity.

Four-Momentum

In Newtonian mechanics the momentum of a point particle is

$$\mathbf{p} = m\mathbf{v}. \tag{A.30}$$

Here $\mathbf{v}$ is the particle's three-velocity, and m is its "mass." The relativistic generalization of (A.30) is the *four-momentum*

$$P^\alpha \equiv m_0 V^\alpha = m_0 \gamma (c, \mathbf{v}). \tag{A.31}$$

Because V^α is a four-vector, if P^α is to be a four-vector, then m_0 must be a world scalar, called the particle's *proper mass, i.e., mass measured in its own frame*. The concept of proper mass is important because it implies that a "particle" can be resolved into fundamental subatomic constituents (e.g., leptons, . . ., baryons, . . ., quarks, each of which has an intrinsic mass), plus binding energy. Note that the inner product of the four-momentum with itself is

$$P_\alpha P^\alpha \equiv m_0^2 \gamma^2 (-c^2 + v^2) = -m_0^2 c^2, \tag{A.32}$$

which is indeed a world scalar, as expected.

We can recover the Newtonian form (A.30) if we define $\mathfrak{m}$, the *effective mass* of a particle[4] to be

$$\mathfrak{m} \equiv \gamma m_0. \tag{A.33}$$

[4] Also called "relativistic mass" or "dynamical mass." The first name is misleading inasmuch as it implies that a particle's intrinsic "mass" changes with its speed. The other name is more satisfactory because it implies that the "real" mass of a particle is its proper mass, but in kinematical and/or dynamical interactions (momentum exchange) observed, say, when it emerges at very high energy from an accelerator and collides with a fixed target, the particle acts *as if* it has a mass $\mathfrak{m}$.

Then

$$P^\alpha = \gamma m_0(c, \mathbf{v}) \equiv (\mathfrak{m}c, \mathfrak{p}), \tag{A.34}$$

where now

$$\mathfrak{p} \equiv \mathfrak{m}\mathbf{v}. \tag{A.35}$$

Four-Force and Energy

In view of (A.35), the Newtonian formula for a force **f** acting on a point particle might be written as

$$\mathbf{f} \equiv \frac{d\mathfrak{p}}{dt} \longrightarrow \frac{d(\mathfrak{m}\mathbf{v})}{dt} = \frac{d(\gamma m_0 \mathbf{v})}{dt}. \tag{A.36}$$

But this formula is not covariant because dt is not a world invariant, so instead we adopt

$$F^\alpha \equiv \frac{dP^\alpha}{d\tau} = m_0 A^\alpha, \tag{A.37}$$

Both four-acceleration and four-force are orthogonal to four-velocity and are *spacelike*. Using (A.33) and (A.35) in (A.37), we have

$$\begin{aligned} F^\alpha &\equiv \frac{d(m_0 V^\alpha)}{d\tau} = \frac{d}{d\tau}[m_0\gamma(c, \mathbf{v})] \\ &= \frac{d}{d\tau}[\mathfrak{m}(c, \mathbf{v})] = \frac{dt}{d\tau}\frac{d}{dt}(\mathfrak{m}c, \mathfrak{p}) = \gamma(\dot{\mathfrak{m}}c, \mathbf{f}). \end{aligned} \tag{A.38}$$

Here "$\dot{\mathfrak{m}}$" means time derivative of $\mathfrak{m}$ with respect to coordinate time in the laboratory frame (not proper time). Then from (A.37) we get

$$V_\alpha A^\alpha \equiv 0 \Rightarrow \mathbf{f}\cdot\mathbf{v} = \dot{\mathfrak{m}}c^2 \tag{A.39}$$

and hence

$$F^\alpha = \gamma\left(\frac{\mathbf{f}\cdot\mathbf{v}}{c}, \mathbf{f}\right). \tag{A.40}$$

Classically, the work done by a force **f** on a particle equals the rate of change of its kinetic energy $\tilde{e}$, i.e., $d\tilde{e}/dt = \mathbf{f}\cdot\mathbf{v}$, so from (A.40),

$$F^\alpha = \gamma\left(\frac{1}{c}\frac{d\tilde{e}}{dt}, \mathbf{f}\right) = \gamma\left(\frac{\dot{\mathfrak{m}}c^2}{c}, \mathbf{f}\right). \tag{A.41}$$

Einstein took the $\tilde{e}$ to be the *total energy* of a particle, i.e., its kinetic energy plus *rest energy*, by writing $\tilde{e} \equiv \mathfrak{m}c^2 + \text{constant}$. The constant is indeterminable by experiment, so it can be set to zero. Then to first order in v^2/c^2,

$$\tilde{e} = \mathfrak{m}c^2 = \gamma m_0 c^2 = m_0 c^2/\sqrt{1 - (v^2/c^2)} \to m_0 c^2 + \tfrac{1}{2} m_0 v^2 + \dots. \tag{A.42}$$

The first term on the right is the particle's rest energy, and the second is its kinetic energy, which is now seen not to be fundamental, but only a *post-Newtonian* term.

Finally, from

$$P^\alpha \equiv (\mathfrak{m}c, \mathfrak{p}) = (\tilde{e}/c, \mathfrak{p}) \tag{A.43}$$

and (A.32), we can write

$$P_\alpha P^\alpha = -m_0^2 c^2 = \mathfrak{p}^2 - \tilde{e}^2/c^2, \tag{A.44}$$

which yields the very important relation

$$\tilde{e}^2 = m_0^2 c^4 + \mathfrak{p}^2 c^2. \tag{A.45}$$

Thus the total energy (including rest energy) of a free particle is

$$\tilde{e} = \mathfrak{m}c^2 = \gamma m_0 c^2 = \sqrt{m_0^2 c^4 + \mathfrak{p}^2 c^2}. \tag{A.46}$$

A.2 RELATIVISTIC KINETIC THEORY

In chapter 4 we discussed distribution functions for distinguishable particles, indistinguishable particles, and radiation, in the limit of strict thermodynamic equilibrium. All the results there refer to a fluid in its comoving frame. When the fluid moves, particularly at relativistic speeds, we must account for the frame dependence of its properties with kinetic theory.

Particle Distribution Function

In developing the theory, we must keep track of three reference frames.

(1) The laboratory frame: An inertial frame at rest. Variables in this frame are denoted with unembellished symbols.
(2) The comoving frame: A *set* of frames, each of which moves with a velocity $\mathbf{v}$ that instantaneously coincides with the frame of a fluid element at every point in the flow. Variables in this frame are subscripted with 0.
(3) An atom's frame: A set of frames, each of which moves with an individual material particle at a velocity $\mathbf{u}$ relative to the comoving frame of its fluid element. Here variables are superscripted with $'$. Each particle is at rest in its own frame. The fluid is locally at rest in the comoving frame. Neither is at rest in the laboratory frame.

Definition

Let $f(\mathbf{x}, \mathfrak{p}, t)$ be the *particle distribution function* in an element of fluid, *measured in the laboratory frame*. The number of particles at position $\mathbf{x}$, with momentum $\mathfrak{p}$, in a spatial volume element d^3x and a momentum-space volume element $d^3\mathfrak{p}$, at time t, is $N = f(\mathbf{x}, \mathfrak{p}, t)\, d^3x\, d^3\mathfrak{p}$. Write the laboratory-frame velocity of a particle in that fluid element as $\mathbf{u} = \mathbf{v} + \mathbf{U}$, where $\mathbf{v}$ is the *mean* or *flow velocity* of the fluid element, and $\mathbf{U}$ is the particle's random velocity relative to the average flow. Because $\langle \mathbf{U} \rangle \equiv 0$,

$$\mathbf{v}(\mathbf{x}, t) = \frac{\int \mathbf{u} f(\mathbf{x}, \mathfrak{p}, t)\, d^3\mathfrak{p}}{\int f(\mathbf{x}, \mathfrak{p}, t)\, d^3\mathfrak{p}}. \tag{A.47}$$

Mapping of Spatial Volume

Let $f_0(\mathbf{x}, \mathfrak{p}_0, t)$ be the particle distribution function at laboratory-frame position $\mathbf{x}$ and time t, measured in the comoving frame of an element of fluid. At a given time t, observers in both the laboratory and comoving frame will count the same *number* of particles in a definite fluid element. That is,

$$N = f(\mathbf{x}, \mathfrak{p}, t) d^3x\, d^3\mathfrak{p} \equiv N_0 = f_0(\mathbf{x}, \mathfrak{p}_0, t) d^3x_0\, d^3\mathfrak{p}_0. \tag{A.48}$$

Choose a group of particles that occupy a volume element d^3x' as seen in an atom's frame. Because of Lorentz contraction, that volume as seen in the comoving frame is

$$d^3x_0 = \sqrt{1 - (U^2/c^2)}\, d^3x', \tag{A.49}$$

where $\mathbf{U}$ is the random velocity of the atom relative to the comoving frame. And in the laboratory frame the same volume element is seen to be

$$d^3x = \sqrt{[1 - (u^2/c^2)}\, d^3x', \tag{A.50}$$

where now $\mathbf{u}$ is the total (fluid + random) velocity of the atom in the laboratory frame. Thus

$$d^3x_0 = \left[\frac{1 - (U^2/c^2)}{1 - (u^2/c^2)}\right]^{1/2} d^3x. \tag{A.51}$$

Is this conclusion correct? We know that $\mathbf{u}$ and $\mathbf{U}$ do not, in general, point in the same direction. So how can we make such simple transformations between the three frames? We can because we are making use of the *invariance of the spacetime volume element $dx\, dy\, dz\, dt$*. The transformation of t between the atom's frame and the comoving frame is $dt_0 = dt'/\sqrt{1 - (U/c)^2}$, and between the atom's frame and the laboratory frame $dt = dt'/\sqrt{1 - (u/c)^2}$. These transformations depend only on the *speed* of the atom's frame relative to the comoving frame or the rest frame, *regardless of the direction of the relative velocity between these frames*. Therefore, for material particles,

$$dx_0\, dy_0\, dz_0\, dt_0 \equiv dx\, dy\, dz\, dt, \tag{A.52a}$$

which implies

$$d^3x_0\,dt'/\sqrt{1-(U^2/c^2)} \equiv d^3x\,dt'/\sqrt{1-(u^2/c^2)}, \tag{A.52b}$$

which is identical to (A.51).

Mapping of Energy

The energy of a particle with rest mass m_0 as measured in the comoving frame is

$$\tilde{e}_0 = \gamma_0 m_0 c^2 = \frac{m_0 c^2}{[1-(U^2/c^2)]^{1/2}}, \tag{A.53}$$

and its energy as measured in the laboratory frame is

$$\tilde{e} = \gamma m_0 c^2 = \frac{m_0 c^2}{[1-(u^2/c^2)]^{1/2}}. \tag{A.54}$$

Combining (A.51), (A.53), and (A.54), we find that

$$\tilde{e}\,d^3x = \tilde{e}_0\,d^3x_0 \tag{A.55}$$

is a Lorentz invariant.

Mapping of Momentum

Now examine the relation between $d^3\mathbf{p}$ and $d^3\mathbf{p}_0$. Boost a particle's four-momentum from the comoving frame to the laboratory frame using the 3D Lorentz transformation. We find

$$\tilde{e} = \gamma(\tilde{e}_0 + \mathbf{v}\cdot\mathbf{p}_0), \tag{A.56}$$

and

$$\mathbf{p} = \mathbf{p}_0 + \left[\frac{\gamma\tilde{e}_0}{c^2} + (\gamma-1)\frac{\mathbf{v}\cdot\mathbf{p}_0}{v^2}\right]\mathbf{v}. \tag{A.57}$$

In general,

$$d^3\mathbf{p} = J\left(\frac{\mathbf{p}_1, \mathbf{p}_2, \mathbf{p}_3}{\mathbf{p}_{01}, \mathbf{p}_{02}, \mathbf{p}_{03}}\right) d^3\mathbf{p}_0, \tag{A.58}$$

where J is the Jacobian of the transformation from the comoving frame to the laboratory frame.

To simplify the calculation, do not use (A.57) directly. Instead, rotate the coordinate axes so that the comoving frame moves with speed v along the z axis in the new coordinate system. Then from (A.56) and (A.57) we have

$$(\tilde{e}, \mathfrak{p}_1, \mathfrak{p}_2, \mathfrak{p}_3) = \left[\gamma\left(\frac{\tilde{e}_0}{c} + \beta\,\mathfrak{p}_{03}\right), \mathfrak{p}_{01}, \mathfrak{p}_{02}, \gamma\left(\mathfrak{p}_{03} + \beta\,\frac{\tilde{e}_0}{c}\right)\right], \tag{A.59}$$

and the Jacobian in (A.58) is

$$J = \begin{vmatrix} 1 & 0 & 0 \\ 0 & 1 & 0 \\ 0 & 0 & \gamma\left(1 + \dfrac{\beta}{c}\dfrac{\partial \tilde{e}_0}{\partial \mathfrak{p}_{03}}\right) \end{vmatrix} = \gamma\left(1 + \frac{\beta}{c}\frac{\partial \tilde{e}_0}{\partial \mathfrak{p}_{03}}\right). \tag{A.60}$$

From (A.45), $\tilde{e}_0^2 = m_0^2 c^4 + \mathfrak{p}_0^2 c^2$, so

$$\frac{\partial \tilde{e}_0}{\partial \mathfrak{p}_{03}} = \frac{\mathfrak{p}_{03} c^2}{\tilde{e}_0}. \tag{A.61}$$

Hence

$$J = \gamma\left(1 + \frac{v\mathfrak{p}_{03}}{\tilde{e}_0}\right) = \frac{\tilde{e}}{\tilde{e}_0}, \tag{A.62}$$

where the second equality follows from (A.56). This expression contains no reference to the orientation of the coordinate axes; hence it holds in general. Using (A.62) in (A.58), we obtain

$$d^3\mathfrak{p}/\tilde{e} = \mathfrak{p}^2 d\mathfrak{p}\, d\Omega/\tilde{e} = d^3\mathfrak{p}_0/\tilde{e}_0, \tag{A.63}$$

so it is Lorentz invariant; here $d\Omega$ is an element of solid angle.

Invariance

Combining (A.63) with (A.55), we find

$$d^3x\, d^3\mathfrak{p} = d^3x_0\, d^3\mathfrak{p}_0, \tag{A.64}$$

which shows that for material particles, a *phase-space volume element is Lorentz invariant*. Finally, from (A.63) and (A.48) we obtain

$$f(\mathbf{x}, \mathfrak{p}, t) \equiv f_0(\mathbf{x}, \mathfrak{p}_0, t), \tag{A.65}$$

so the *particle distribution function is Lorentz invariant*, as stated earlier.

Relativistic Thermal Velocity Distribution

According to (A.46) the particle velocity distribution function in thermal equilibrium is proportional to

$$f(p) = e^{-\tilde{e}/kT} = e^{-(\gamma m_0 c^2/kT)}, \tag{A.66}$$

where $f(p)$ is normalized in momentum space such that

$$4\pi \int_0^\infty f(p)\, p^2 dp = A^{-1} \int_0^\infty e^{-(\gamma m_0 c^2/kT)} p^2 dp = 1. \tag{A.67}$$

A is a normalization constant to be determined.

Because the exponential contains γ, it is easiest to evaluate the integral in (A.67) by transforming to γ as the variable of integration. Now $p = m_0 \gamma v$. Therefore,

$$p^2 dp = m_0^3 \gamma^2 v^2 d(\gamma v). \tag{A.68}$$

From (A.8) we find

$$d(\gamma v) = \gamma dv + v\, d\gamma = \gamma^3 dv; \tag{A.69a}$$

therefore,

$$p^2 dp = m_0^3 \gamma^5 v^2 dv. \tag{A.69b}$$

Further, from (A.7),

$$v\, dv = c^2 \gamma^{-3} d\gamma, \tag{A.69c}$$

and

$$v^2 dv = c^3 \gamma^{-4} (\gamma^2 - 1)^{1/2} d\gamma, \tag{A.69d}$$

thus

$$p^2 dp = m_0^3 c^3 (\gamma^2 - 1)^{1/2} \gamma\, d\gamma. \tag{A.69e}$$

Using (A.69e), and defining $\alpha \equiv m_0 c^2/kT$, we can rewrite (A.67) as

$$\begin{aligned} A = 4\pi \int_0^\infty f(p)\, p^2 dp &= m_0^3 c^3 \int_1^\infty (\gamma^2 - 1)^{1/2} e^{-\alpha\gamma} \gamma\, d\gamma \\ &\equiv \frac{4\pi\, m_0^3 c^3}{\alpha^3} \int_\alpha^\infty (x^2 - \alpha^2)^{1/2} e^{-x} x\, dx. \end{aligned} \tag{A.70}$$

The integral in (A.70) equals $\alpha^2 K_2(\alpha)$ [386, eq. 3.389 (4)], where K_2 is the modified Bessel function of the second kind of order two [5, eq. 9.6.23]. Thus

$$A = 4\pi\, m_0^2\, ckT K_2(m_0 c^2/kT), \tag{A.71}$$

and

$$f(p) = \frac{e^{-(m_0 c^2 \gamma/kT)}}{4\pi\, m_0^2\, ckT K_2(m_0 c^2/kT)}. \tag{A.72}$$

In chapter 6 we needed $f(v)$. From (A.69b) and (A.72), we have the relativistic Maxwellian or Jüttner [607] distribution

$$f(v) = f(p)\,\frac{p^2\,dv}{v^2\,dv} = \frac{m_0\gamma^5 e^{-(\gamma m_0 c^2/kT)}}{4\pi\,ckTK_2(m_0\,c^2/kT)}, \tag{A.73}$$

which is normalized such that $\int_0^c f(v)\,v^2 dv = 1$.

Except in very high energy astrophysical sources, $z = m_0c^2/kT \gg 1$, and one can usually use the asymptotic expression [386, eq. 8.451 (6)]:

$$K_2(z) \approx \sqrt{\frac{\pi}{2z}}e^{-z}\left(1 + \frac{1}{2z} + \ldots\right). \tag{A.74}$$

Appendix B

Photons

B.1 LORENTZ TRANSFORMATION OF THE PHOTON FOUR-MOMENTUM

A photon's energy is $\tilde{e} = h\nu$ and the magnitude of its momentum is $h\nu/c$, so (A.45) implies that

$$(h\nu)^2 = (h\nu)^2 + m_0^2 c^4, \tag{B.1}$$

which shows that for a photon $m_0 \equiv 0$, i.e., *photons have no rest mass*. Therefore, $m_0 V^\alpha$ is not useful as an expression for a photon's four-momentum. Indeed, V^α for a photon is undefined because $d\tau \equiv 0$ for a photon. Instead, we can use the expression $P^\alpha = (\tilde{e}/c, \mathfrak{p})$, which gives

$$P^\alpha = \frac{h\nu}{c}\,(1, \mathbf{n}). \tag{B.2}$$

Photon momentum is a null vector: $P^\alpha P_\alpha \equiv 0$.

The components of P^α in two frames are related by the general Lorentz transformation (A.15). The transformation from the laboratory frame to the comoving frame is

$$\nu_0 = \gamma(1 - \mathbf{n}\cdot\boldsymbol{\beta})\,\nu, \tag{B.3}$$

and

$$\nu_0\,\mathbf{n}_0 = \nu\left[\mathbf{n} - \gamma\boldsymbol{\beta}\left(1 - \frac{\gamma\mathbf{n}\cdot\boldsymbol{\beta}}{\gamma+1}\right)\right]. \tag{B.4}$$

Taking the O(v/c) limits of (B.3) and (B.4), we get the standard formulae for Doppler shift

$$\nu_0 = (1 - \mathbf{n}\cdot\mathbf{v}/c)\,\nu \tag{B.5}$$

and aberration of light

$$\mathbf{n}_0 = (\mathbf{n} - \mathbf{v}/c)/(1 - \mathbf{n}\cdot\mathbf{v}/c). \tag{B.6}$$

The inverse transformations from the comoving frame to the laboratory frame are

$$\nu = \gamma(1 + \mathbf{n}_0\cdot\boldsymbol{\beta})\,\nu_0 \tag{B.7}$$

and

$$\nu\,\mathbf{n} = \nu_0 \left[\mathbf{n}_0 + \gamma \boldsymbol{\beta} \left(1 + \frac{\gamma \mathbf{n}_0 \cdot \boldsymbol{\beta}}{\gamma + 1} \right) \right]. \tag{B.8}$$

From (B.8) one can show that

$$(1 - \mathbf{n}_0 \cdot \mathbf{n}_0') = \frac{(1 - \mathbf{n} \cdot \mathbf{n}')}{\gamma^2 (1 + \boldsymbol{\beta} \cdot \mathbf{n})(1 + \boldsymbol{\beta} \cdot \mathbf{n}')}. \tag{B.9}$$

B.2 PHOTON DISTRIBUTION FUNCTION

In chapter 3 we defined the photon distribution function as

$$f_{\mathrm{R}}(\mathbf{x}, t; \mathbf{n}, \nu) = \frac{c^2}{h^4 \nu^3} I(\mathbf{x}, t; \mathbf{n}, \nu). \tag{B.10}$$

We must now show that it is relativistically invariant, as claimed in § 3.5.

Momentum-Space Invariant

To simplify the analysis, take the relative velocity between the rest frame and the comoving frame to be along the z axis. Then from (B.3) and (B.4) we have

$$\begin{aligned}&(\nu_0;\ \nu_0\, n_{x0};\ \nu_0\, n_{y0};\ \nu_0\, n_{z0})\\ &\quad = [\gamma \nu (1 - \beta n_z);\ \nu\, n_x;\ \nu\, n_y;\ \gamma \nu (n_z - \beta)]\end{aligned} \tag{B.11}$$

or

$$\begin{aligned}&[\nu_0;\ \mu_0;\ (1 - \mu_0^2)^{1/2};\ \Phi_0]\\ &\quad = [\gamma \nu (1 - \beta \mu);\ (\mu - \beta)/(1 - \beta \mu);\ (1 - \mu^2)^{1/2}/\gamma (1 - \beta \mu);\ \Phi].\end{aligned} \tag{B.12}$$

From (B.12) we obtain

$$d\nu_0 = (\nu_0/\nu) d\nu, \tag{B.13a}$$

$$d\mu_0 = (\nu/\nu_0)^2 d\mu, \tag{B.13b}$$

and

$$d\Phi_0 = d\Phi. \tag{B.13c}$$

Recall that $d\Omega = d\mu\, d\Phi$. Combining this identity with the results in (B.13), we find the important Lorentz invariant

$$\nu_0\, d\nu_0\, d\Omega_0 = \nu\, d\nu\, d\Omega. \tag{B.14}$$

Equation (B.14) also follows from (A.65), which applies for any kind of particle, not just material particles. For photons $p = h\nu/c$ and $\tilde{e} = h\nu$; therefore, (A.65) becomes

$p_0^2\, dp_0\, d\Omega_0/\tilde{e}_0 = p^2 dp\, d\Omega/\tilde{e}$, implying that $h^3 \nu_0^2 d\nu_0\, d\Omega_0/h\nu_0 = h^3 \nu^2 d\nu\, d\Omega/h\nu$, which is identical to (B.14).

To complete the proof that the distribution function f_R is invariant in spacetime, we also need to show invariance of an element of coordinate space. But photons travel with speed c in all frames, so (A.53) has no meaning. This difficulty was overcome in a penetrating analysis by L. H. Thomas [1075].

B.3 THOMAS TRANSFORMATIONS

Specific Intensity

An observer in the fixed laboratory frame would count

$$N = \frac{I(\mu, \nu)}{h\nu}\,(d\nu\, d\Omega)(\mu\, dS\, dt). \tag{B.15}$$

photons in a beam of light with specific intensity $I(\mu, \nu)$ passing through a surface area dS oriented at an angle $\Theta = \cos^{-1}\mu$ relative to the z axis into a solid angle $d\Omega$, in a frequency interval $d\nu$ and a time interval dt. To an observer moving with speed $-v$ along the z axis, dS appears to be moving with speed v in the positive z direction, into the radiation field; see figure B.1.

This observer would count

$$N_0 = \frac{I_0(\mu_0, \nu_0)}{h\nu_0}(d\nu_0\, d\Omega_0)\left[\mu_0\, dS\, dt_0 + \frac{v}{c} dS\, dt_0\right] \tag{B.16}$$

photons passing through dS. The first term on the right-hand side is the number of photons that would have been counted if dS had been stationary in the observer's frame, and the second term is the comoving-frame photon number density $\varphi_0 = I_0/ch\nu_0$ times the volume element $dS\, v dt_0$ swept out by the apparent motion of dS into the incoming beam.

But N must be identical to N_0 because the *same* photons were counted by both observers. From the Lorentz transformations between the two frames we know that

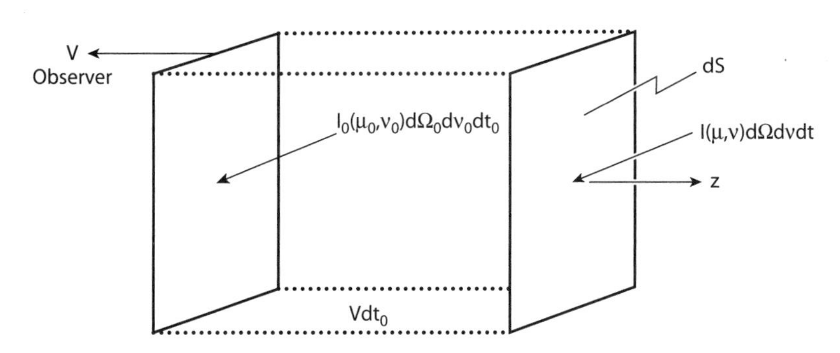

Figure B.1 Invariant photon counting experiment.

$dt_0 = \gamma dt$; $\nu = \gamma \nu_0(1 + \beta \mu_0)$; $\nu d\nu d\Omega = \nu_0 d\nu_0 d\Omega_0$; and $\mu = (\mu_0 + \beta)/(1 + \beta \mu_0)$. Using these relations in (B.16) and equating (B.16) to (B.15), we get the fundamental result that

$$I(\mathbf{x}, t; \mu, \nu)/\nu^3 \equiv I_0(\mathbf{x}, t; \mu_0, \nu_0)/\nu_0^3. \tag{B.17}$$

Hence from (B.10),

$$f_R(\mathbf{x}, t; \mu, \nu) \equiv f_{R0}(\mathbf{x}, t; \mu_0, \nu_0). \tag{B.18}$$

If one were willing to assert that *any* particle distribution function in phase space must be a relativistic invariant, then Thomas's derivation would be redundant. But it is not legitimate to apply (A.53) to photons, so his analysis fills an important logical gap.

Emissivity

Have an observer in the laboratory frame count the *number* of photons emitted from a definite material element in dV into $d\Omega d\nu$ in a time dt:

$$N = \eta(\mathbf{x}, t; \mu, \nu)\, d\nu\, d\Omega\, dV\, dt/h\nu. \tag{B.19}$$

Assuming isotropy of emission in the comoving frame, the same *number* of photons emitted is counted by an observer in that frame:

$$N_0 = \eta_0(\mathbf{x}, t; \nu_0)\, d\nu_0\, d\Omega_0\, dV_0\, dt_0/h\nu_0. \tag{B.20}$$

Thus requiring that $N \equiv N_0$, and recalling that $dVdt$ and $\nu d\nu d\Omega$ are Lorentz invariants, we have

$$\eta(\mathbf{x}, t; \mu, \nu)/\nu^2 \equiv \eta_0(\mathbf{x}, t; \mu_0, \nu_0)/\nu_0^2. \tag{B.21}$$

Opacity

Let the number of photons removed from a beam of light in a definite element of material as seen by an observer in the laboratory frame be

$$N = [\chi(\mathbf{x}, t; \mu, \nu) I(\mathbf{x}, t; \mu, \nu)/h\nu]\, d\nu\, d\Omega\, dV\, dt. \tag{B.22}$$

Assuming isotropy of χ_0 in the comoving frame, an observer in that frame will count

$$N_0 = [\chi_0(\mathbf{x}, t; \mu_0, \nu_0) I_0(\mathbf{x}, t; \mu_0, \nu_0)/h\nu_0]\, d\nu_0\, d\Omega_0\, dV_0\, dt_0 \tag{B.23}$$

photons removed from the beam. Again, demand $N \equiv N_0$ and apply the invariants discussed above. We find

$$\nu\, \chi(\mathbf{x}, t; \mu, \nu) \equiv \nu_0 \chi_0(\mathbf{x}, t; \mu_0, \nu_0). \tag{B.24}$$

Thomas's derivations of the transformation laws for emissivity and opacity are unique. His equations (B.17), (B.21), and (B.24) are *central* to the theory of radiation transport in moving material.

Glossary of Symbols

Physical and mathematical symbols used in the text are listed below, along with a brief description of their meaning and the page number on which each first appears. Standard mathematical symbols, dummy variables and indices, and notations used only in one location are not included.

a	semi-major axis	47
a	damping (Voigt) parameter (in Doppler widths)	232
a	isothermal sound speed	767
$\mathbf{a}$	vector of acceleration	547
a_0	Bohr radius	94
a_{1-4}	parameters (summed transition rates) in the ETLA scheme	483
a_j	angular quadrature weights in the discrete ordinate method	578
$a_k(t)$	coefficients of the expansion of the wave function	128
$a_{r,\mu,\nu}$	coefficients of the transfer equation along characteristic rays	732
$\hat{\boldsymbol{a}}_{\mathbf{k}}^{\dagger}$	creation (raising) operator	82
$\hat{\boldsymbol{a}}_{\mathbf{k}}$	destruction (lowering) operator	82
$a_{\mathbf{k}\alpha}$	coefficients of expanding vector potential into plane waves	126
a_R	radiation constant	105
$a(x)$	branching ratio in the partial coherent scattering approximation	318
a_ν	coefficient of linear expansion of the Planck function	608
A	matrix of a general linear system	425
$\mathbf{A}$	vector potential	71
$\mathbf{A}$	matrix of the set of kinetic equilibrium equations, rate matrix	450
$\widetilde{A}$	preconditioned matrix of a liner system	440
$\mathcal{A}$	auxiliary quantity for evaluating the starring model atmosphere with convection	605
$\mathcal{A}$	full rate matrix	638
$\mathbf{A}_d$	matrices of the block tridiagonal Feautrier system	394
$\mathbf{A}_d$	matrices of the block tridiagonal system in the linearization method	490
A_I	abundance of species I relative to hydrogen	638
$\mathcal{A}_I$	rate matrix for species I	636
A_{ji}	Einstein coefficient for a spontaneous radiative transition	119
A_k	atomic weight of element k	590
$A(t)$	matrix element of the time-evolution operator	302
A^T	transpose matrix of A	432
A_V	absolute visual magnitude	40
A_ν	line depth	612
b	right-hand side of a general linear system	425

$\mathbf{b}$	vector of right-hand sides of the kinetic equilibrium equation	450
$\tilde{b}$	preconditioned right-hand side of a linear system	440
b_i	NLTE departure coefficient	505
b_ν	coefficient of linear expansion of the Planck function	608
B	rotation constant of a molecule	216
$\mathbf{B}$	magnetic induction vector	68
$\mathcal{B}$	parameter for evaluating the temperature gradient of convective elements	565
$B(T)$	frequency-integrated Planck function	104
$\mathbf{B}_d$	matrices of the block tridiagonal Feautrier system	394
$\mathbf{B}_d$	matrices of the block tridiagonal system in the linearization method	490
B_{ij}	Einstein coefficient for an induced radiative transition	118
$B_\nu(T)$	Planck function	104
$B.C.$	bolometric correction	48
$B-V$	color (difference of B and V magnitudes)	37
c	speed of light in vacuum	63
c_p	specific heat at constant pressure	562
C	number of discretized core rays in the (p, z) coordinate system	698
C	a constant in the expression for radiation force in the Castor, Abbott, Klein theory	779
$\mathbf{C}_d$	matrices of the block tridiagonal Feautrier system	394
$\mathbf{C}_d$	matrices of the block tridiagonal system in the linearization method	490
C_{ij}	color index	36
C_{ij}	collisional rate	266
C_p	interaction constant (impact broadening)	234
$C(r)$	correction function in the temperature-correction scheme for moving atmospheres	810
C_I	a constant in the Saha equation	96
C_λ	calibration factor	30
$\hat{d}$	electric dipole operator	130
d_i	auxiliary parameter in the preconditioned ETLA method	485
D	stellar distance	36
D	Debye length	245
D	"doubled" dipole operator	255
D	diagonal part of a matrix	429
D	number of discretized radial points in the (p, z) coordinate system	698
D	modified stellar wind momentum	788
$\mathbf{D}$	electric displacement vector	68
D_0	reduced dipole moment of a molecule	220
D_1, D_2	modified matrix element of the dipole operator	303
$\mathbf{D}_d$	auxiliary matrix for the solution of a block tridiagonal system	394
D_f	finite disk correction factor for the radiation force multiplier	787
D_i	number of mesh points on a given tangent ray in the (p, z) coordinate system	699

$\mathbf{D}(R,\theta,\phi)$	permanent dipole moment of a molecule	220
e	elementary charge	124
e	internal energy	551
$\mathbf{e}_d$	unit-pulse vector, $(\mathbf{e}_d)_{d'} = \delta_{dd'}$	712
$\mathbf{e}_{\mathbf{k}\alpha}$	polarization vector	126
$e^{(n)}$	true error in the nth iteration of the solution of a linear system	426
E	total radiation energy density	68
$\mathbf{E}$	electric field vector	68
$\mathcal{E}$	total energy of a thermodynamic ensemble	89
$\mathcal{E}$	total energy of a molecular system	209
$\mathcal{E}$	radiation energy	63
E	total energy flux	767
$\mathbf{E}_d$	auxiliary vector for the solution of a block tridiagonal system	394
E_{ij}	rate of strain tensor	549
E_k	energy eigenvalue of state k	128
$E_n(x)$	exponential integrals	361
E_F	Fermi energy	111
$E_{\rm Morse}(R)$	Morse potential	210
$E_{\rm rot}, E_{\rm vib}$	rotational and vibrational energy eigenvalue	99
E_ν	monochromatic radiation energy density	68
$E_{\rm V}$	correction term in evaluating $R_{\rm V}$	313
$E(B-V)$	color excess	36
$E(U-B)$	color excess	36
f	integrated Eddington factor	596
f_c	continuum oscillator strength	189
$f_i(t,\mathbf{r},\mathbf{p})$	particle distribution function	264
f_{ij}	oscillator strength	135
f_{jk}	ionization fraction of ion j of element k	94
$f_{\rm R}(\mathbf{x},t;\mathbf{p})$	photon distribution function	65
f_λ	star's absolute flux distribution	33
$f_\nu, f_\nu^{\rm K}$	Eddington factor	351
F	radiation flux	22
F	amplification matrix of an iteration scheme	426
$\mathbf{F}$	total radiation flux vector	74
$\mathbf{F}$	Lorentz force	69
F_0	normal field strength	243
F_c	continuum flux	27
$F_{\rm conv}$	convective flux	562
$F(v)$	spontaneous radiative recombination probability	121
F_x, F_y, F_z	components of the radiation flux vector	74
$\mathbf{F}_\nu$	monochromatic radiation flux vector	74
$F(\alpha,\beta,\gamma,x)$	hypergeometric function	185
$\mathcal{F}(\omega)$	energy spectrum	228
g	surface gravity	22
$\bar{g}$	integrated surface Eddington factor	596

g_{bb}	bound-bound Gaunt factor	186
g_{bf}	bound-free Gaunt factor	187
$g_e(v)$	statistical weight of the free electron	95
g_{ff}	free-free Gaunt factor	189
g_i	statistical weight	89
$g(\mathbf{n}', \mathbf{n})$	phase function	150
g_{trans}	translational statistical weight	94
g_{R}	radiative acceleration	772
g_ν, g_ν^{K}	surface Eddington factor (flux boundary ratio)	351
g_ν	Eddington-like factor, $g_\nu = N_\nu / H_\nu$, in the spherical radiative transfer	755
G	gravitational constant	7
G	boundary condition parameter in the discrete ordinate method	579
$\mathbf{G}$	net radiation force density	348
G^0	$1/c$ times the net rate of radiative energy deposition	348
$\mathbf{G}_{in}$	bidiagonal matrix in the discretized comoving frame transfer equation	728
$G_{lu}(\nu)$	universal factor for the number of downward transitions	451
$G(v)$	induced radiative recombination probability	121
h	Planck constant	63
h	enthalpy	555
h	porosity length in the continuum-driven wind theory	799
$\hbar$	reduced Planck constant, $h/2\pi$	82
$h_{\mu\nu}$	Schuster-Feautrier antisymmetric average of the specific intensity	367
h_ν	surface Eddington factor in spherical radiative transfer	698
$\mathbf{H}$	magnetic field vector	68
$\mathcal{H}$	relativistic atomic Hamiltonian	123
$\hat{\boldsymbol{H}}_{1,2}$	two parts of the quantized interaction Hamiltonian	132
$H(a, v)$	Voigt function	232
H_{conv}	convective flux divided by 4π	562
$\hat{\boldsymbol{H}}_{\mathbf{k}}$	Hamiltonian for a one-dimensional harmonic oscillator	82
$\mathbf{H}_{in}$	diagonal matrix in the discretized comoving-frame transfer equation	728
$\hat{\boldsymbol{H}}_A$	atomic Hamiltonian	124
$\hat{\boldsymbol{H}}_{\mathrm{BP}}$	Breit-Pauli Hamiltonian	201
$\hat{\boldsymbol{H}}_I$	interaction Hamiltonian	124
H_P	pressure scale height	560
$\hat{\boldsymbol{H}}_R$	Hamiltonian of free radiation field	133
$\mathcal{H}(\mu)$	theoretical limb-darkening function	577
H_ν	Eddington flux (1st moment of the specific intensity)	76
i	orbital inclination	50
I	identity operator (matrix)	423
$I(\mathbf{x}, t; \mathbf{n}, \nu)$	specific intensity of radiation	63
$\mathtt{I}_\nu^\pm$	specific intensity at the boundaries	355

$\hat{j}$	total angular momentum operator	180
$\mathbf{j}$	electric current density	69
$j_{\mu\nu}$	Schuster-Feautrier symmetric average of the specific intensity	367
$\mathbf{j}_d$	vector of Feautrier symmetric intensities	394
J	total angular momentum quantum number	195
J	Jacobian of a transformation	341
$\overline{J}$	frequency-averaged mean intensity in a line	295
$\overline{\mathbf{J}}$	vector of averaged mean intensities in the Rybicki method	398
$\overline{J}^{R}_{ief}$	redistribution integral	330
$\overline{J}^{\rm eft}_{ij}$	correction factor in the ALI scheme for multi-level atoms	496
J_ν	mean intensity of radiation	67
k	Boltzmann constant	90
k	fitting parameter for the radiation-force multiplier in the Castor, Abbott, Klein theory	778
$\mathbf{k}$	wave propagation vector	70
k_α	eigenvalues of the discrete-ordinate characteristic equation	579
$K_1(s)$	kernel function for the two-level atom source function	461
$K_2(\tau)$	frequency-averaged kernel function K_1	475
$K_2(x)$	second-order Bessel function	163
K_q	coefficient of thermal conduction	552
K_β	kernel function for the source function in the Sobolev approximation	749
K_ν	second moment of the radiation intensity	78
l	angular momentum quantum number	172
$\hat{l}$	angular momentum operator	175
$\bar{l}$	averaged mean free path of a photon	516
ℓ	mixing length	562
l_b	characteristic size of blobs in the continuum-driven wind theory	798
L	total orbital angular momentum quantum number	195
L	Liouville operator	267
L	lower triangular part of a matrix	429
L	total number of angle-frequency points in discretized transfer equation in planar moving atmospheres	708
L	Sobolev length (in the stellar wind stability analysis)	791
$\mathbf{L}$	auxiliary operator in the preconditioned ETLA method	485
$\mathbf{L}$	total angular momentum (of multi-electron atoms)	194
$\mathbf{L}$	matrix of the Lorentz transformation	715
$\mathcal{L}$	line shape operator	255
L	stellar luminosity	6
$\mathsf{L}_\odot$	solar luminosity	51
L_b	mean distance between blobs in the continuum-driven wind theory	798
$L^a_b(x)$	Legendre polynomial	174
$\mathbf{L}_d$	right-hand side vector of the linearization method	490
$L(x,\gamma)$	Lorentz profile	294
$L_{\rm E}$	Eddington luminosity	773

L^{α}_{β}	individual components of the Lorentz transformation matrix	715
L_{ν}	thermalization length	380
$L(\lambda)$	flux distribution of the laboratory source	32
$L(\omega)$	line shape (line profile) function	249
m	stellar magnitude	34
m	magnetic quantum number	172
m	reduced mass	99
$\mathfrak{m}$	effective mass (in relativistic kinematics)	820
m_0	rest mass of a particle	123
m_0	atomic mass unit	560
m_{e}	electron mass	95
m_H	mass of the hydrogen atom	8
M	number of angles in a numerical angular quadrature	371
M	stellar mass	5
M	total mass of the system	171
M	multiplicity (of multi-electron atoms)	195
$\dot{\mathsf{M}}$	mass loss rate	767
$\mathsf{M}_{\odot}$	solar mass	6
$\mathbf{M}_d$	local matrix of linearized structural equations in the ALI scheme	670
$\mathbf{M}_{\mathrm{photon}}(\nu)$	photon momentum flux	75
M_r	stellar mass within radius r	565
$M(t)$	radiation-force multiplier in the Castor, Abbott, Klein theory	778
M_V	absolute visual magnitude	37
n	principal quantum number	173
n	exponent of the power law for opacity in the spherical gray model	693
$\mathbf{n}$	propagation vector	41
$\mathbf{n}$	vector of level populations	450
n_{atom}	total number density of an atomic species	449
n_{e}	electron density	96
n_i	atomic level population (occupation number)	88
n_i^*	LTE level population	96
$n_{\alpha}(\mathbf{n}, \nu)$	photon occupation number	65
$\mathcal{N}$	number of particles in a thermodynamic ensemble	89
N_I	total number density of an ion	262
$n_{\mathbf{k}}$	radiation eigenfunction	83
$\hat{\boldsymbol{n}}_{\mathbf{k}}$	photon number operator	85
N	total particle number density	100
N	number of frequencies in a numerical frequency quadrature	372
N_{c}	core normalization in the core-saturation method	480
N_i	total number of ions i	94
N_{w}	wing normalization in the core-saturation method	480
N_N	number density of nuclei (atoms plus ions)	589
N_{ν}	third-order moment of the specific intensity	722

NA	number of angles in an angular quadrature	391
NC	number of constraint equations	646
ND	number of depth points	390
NF	number of frequency points in a frequency quadrature	371
NI	number of impact parameters in the (p, z) coordinate system	699
NI_d	number of impact parameters within radius r_d in the (p, z) coordinate system	700
NH	index of N in the state vector	647
NL	number of angle-frequency points	392
NL	number of NLTE levels	448
NM	number of angles in an angular quadrature	418
NN	number of components of the state vector	646
NR	index of T in the state vector	647
NP	index of n_e in the state vector	647
K_ν	second-order moment of the radiation intensity	722
$\hat{\boldsymbol{O}}_\mathbf{k}$	a part of the interaction Hamiltonian	133
p	impact parameter of a tangent ray in a (p, z) coordinate system	694
$p_{\rm gas}$	gas pressure	5
$\mathbf{p}$	photon momentum	64
$\mathfrak{p}$	four-momentum	123
$p_{\rm I-V}(\xi', \xi)$	basic atom's-frame generalized redistribution functions	309
$p(\rightarrow i^*)$	probability of natural population of level i	328
p_i^*	probability of natural population of level i	332
p_F	Fermi momentum	111
$\hat{\boldsymbol{p}}_\mathbf{k}$	momentum operator	82
$p^{(n)}$	search vector (iterative solution of a linear system)	426
$p_T(\tau)$	total probability of photon thermalization	468
$p_{\mu\nu}$	elementary escape probability	474
p_ν	photoionization probability	121
$p(\tau)$	photon escape probability	374
P	orbital period	48
P	gas pressure	100
P	preconditioner of a general linear system	425
P	total radiation stress tensor	78
$P^{\rm coh}$	coherence fraction	301
$P_{\rm d}$	probability of destruction of a line photon	454
$\mathbf{P}_d$	operator representing the full set of structural equation for depth d	646
$P_{\rm e}$	total escape probability for a line photon	454
$\mathcal{P}_{\rm e}(\tau)$	one-sided escape probability	475
P_i	total rate of transitions out of level i	329
$P_{\rm gas}$	gas pressure	635
P_ν^{ij}	components of the radiation pressure tensor	77
P_{ij}	total transition rate	266

P_{lm}	associated Legendre function	173
P_{lu}	probability of transition $l \to u$	131
P_{II}	observer's-frame generalized redistribution function	324
P_{M}^{ij}	components of the Maxwell stress tensor	79
$\hat{\boldsymbol{P}}_{rs}$	permutation operator	106
p^{α}	components of the photon four-momentum	716
P_{ν}	monochromatic radiation stress tensor	78
q	particle charge	69
$\mathbf{q}$	general coordinate system	324
$\mathbf{q}$	thermal conduction	551
q_i	auxiliary parameter in the preconditioned ETLA method	485
$q_{ij}(T)$	collisional rate per perturber	275
$\hat{\mathbf{q}}_{\mathbf{k}}$	position operator	82
q_{ν}	sphericality factor	697
$q(\tau)$	Hopf function	571
$q_{\mathrm{N}}(\tau_{\mathrm{R}})$	NLTE Hopf function	812
Q	thermodynamic parameter, $Q = 1 - (\partial\mu/\partial T)_p$	562
$\overline{Q}$	parameter in the Gayley parameterization of the line force	781
Q_i	elastic collision rate at level i	301
$Q(r, \mu)$	general velocity gradient in the Sobolev theory	747
$Q(\tau)$	non-gray analog to the Hopf function	594
r	radius, radial coordinate	6
r	ratio of the continuum and the averaged line opacity	459
r_0	classical electron radius	149
$r_{\mathrm{bf}}(\mathbf{n}, \nu)$	frequency- and angle-dependent bound-free radiative rate	121
r_c	portion of continuum scattering in the Milne-Eddington equation	607
r_c	core radius	698
r_c	critical radius (for a stellar wind)	768
$r_{lu}(\mathbf{n}, \nu)$	frequency- and angle-dependent line radiative rate	118
$r^{(n)}$	residual of a general linear system in nth iteration	426
$\widetilde{r}^{(n)}$	preconditioned residual of a general linear system in nth iteration	426
r_s	sonic radius (for a stellar wind)	785
$r(\xi', \mathbf{n}'; \xi, \mathbf{n})$	atom's-frame redistribution function	298
$r_{\mathrm{I-V}}(\xi', \xi)$	basic atom's-frame redistribution functions	309
$R(r)$, $R_{nl}(r)$	radial wave function	172
R	radiation stress-energy tensor	79
R	stellar radius	5
$\mathcal{R}$	Rydberg constant	174
$\mathrm{R}_{\odot}$	solar radius	21
$R_{\mathrm{I}} - R_{\mathrm{V}}$	basic observer's-frame redistribution functions	312
$\mathbf{R}_d$	right-hand side of the block tridiagonal Feautrier system	394
R_{ij}	radiative rate	266
$\overline{R}_{ief}$	scattering integral	327

$\mathcal{R}_{ief}$	normalized scattering integral	327
R_λ	residual flux	27
R_λ	spectrometer response	30
$R(\nu', \mathbf{n}'; \nu, \mathbf{n})$	observer's (laboratory)-frame redistribution function	311
s	spin quantum number	179
s	path length along a ray	340
s	entropy	552
s	scaled electron density for the parameterization of the radiation-force multiplier	778
$\hat{s}$	spin operator	179
$S, S_\nu, S_{\mu\nu}$	source function	354
S	entropy	90
S	total spin quantum number	195
$\mathbf{S}$	total spin (of multi-electron atoms)	194
$\mathbf{S}$	Poynting vector	71
$\mathbf{S}$	oriented surface area	63
$\widetilde{S}$	modified source function in the comoving-frame transfer equation	733
S_i	coefficient in the particle conservation equation	283
S_{ji}	source function for a transition $j \leftrightarrow i$	330
$S(l, u)$	line strength	135
S_C	continuum (background) source function	459
S^FS	source function (in iterative methods) obtained by the formal solution	423
S_L	line source function	457
$S(\alpha)$	Stark profile function	259
$S(\lambda)$	sensitivity function of a detector	34
t	time	63
t, t_ν	optical depths (in integrals)	360
t	equivalent electron optical depth in the radiation-driven wind theory	777
$\mathbf{t}, \mathbf{t}_d$	slanted optical depth (along a given line of sight)	405
T	temperature	6
T	total thermal energy (of a star)	7
$\mathbf{T}$	stress tensor	549
T_e	electron temperature	114
T_eff	effective temperature	22
$\mathbf{T}_{in}$	tridiagonal matrix of the Rybicki-type scheme in the (p, z) coordinate system	701
T_ion	kinetic temperature of ions	114
$T(k^2)$	characteristic function in the discrete-ordinate method	579
T_rad T_R	radiation temperature	37
$T(t)$	time-development operator	250
$T(\lambda)$	filter transmission function	34
$\mathbf{T}_{\mu\nu}$	tridiagonal matrix of the scalar Feautrier solver	437
$\mathtt{T}_\nu$	total optical thickness of a slab	355

u	thermal velocity	265
u	modified radial coordinate in the Castot, Abbott, Klein theory	782
$\mathbf{u}_{in}$	vector of discretized u in (p, z) coordinates	710
$u(z, p, \nu)$	symmetric average of the intensity in (p, z) coordinates	700
U	Abbott speed (phase speed of Abbott waves in a perturbed stellar wind)	792
$U, U(T)$	partition function	95
U	upper triangular part of a matrix	429
$\mathcal{U}^*$	partition function	92
$\mathbf{U}_d$	tridiagonal submatrices of the global Rybicki system	398
$\mathbf{U}_{in}$	diagonal matrix of the Rybicki-type scheme in the (p, z) coordinate system	701
$\underline{U}_{lu}(\nu)$	spontaneous radiative transition parameter	501
$\widetilde{U}(r, \zeta)$	auxiliary function in the generalized Sobolev theory with continuum	754
$U_{\rm rot}, U_{\rm vib}$	rotational and vibrational partition function	99
$U(t)$	time-evolution operator in the interaction representation	253
$U_{\rm trans}$	translational partition function	95
U_t	time-development operator	129
$U_{\rm v}$	vibrational temperature	99
$U - B$	color (difference of the U and B magnitudes)	37
v	velocity	94
v	vibrational quantum number	216
v	frequency difference in units of Doppler width	232
$\mathbf{v}$	velocity vector	41
$\overline{v}$	most probable speed	94
v_∞	terminal velocity	784
$v_{\rm esc}$	escape velocity	768
$\mathbf{v}_{in}$	vector of discretized v in the comoving-frame transfer equation	728
$v_{\rm rot}$	rotational velocity	788
$v_{\rm th}$	thermal velocity	311
$v(z, p, \nu)$	antisymmetric average of the intensity in (p, z) coordinates	700
V	volume	67
V	velocity in units of thermal velocity	705
V	total potential energy (of a star)	6
$\mathcal{V}$	volume of a thermodynamic ensemble	89
V_{cl}	classical interaction potential (in the quantum impact theory)	251
$\mathbf{V}_d$	diagonal submatrices of the global Rybicki system	398
$V_{lu}(\nu)$	induced radiative transition parameter	501
$\hat{\mathbf{V}}(t)$	operator of perturbing potential	128
w	Doppler width	311
w	modified velocity in the Castor, Abbott, Klein theory	782
$w_{1,2}(\mathbf{t})$	basis functions for the discontinuous finite element scheme	410
w_i	occupation probability of state i	278
w_i	quadrature weights for a numerical integration	370
W	thermodynamic probability	89

W	dilution factor	285
W	line equivalent width	613
W_0	reduced equivalent width	613
$\mathbf{W}_{in}$	right-hand side of the Rybicki-type scheme in the (p, z) coordinate system	701
$W(r)$	probability of a distance from the nearest neighbor	242
W_λ	equivalent width	28
$W_{\rm vib}$	vibrational part of the molecular wave function	216
x	frequency displacement from the line center in units of Doppler widths	95
x	solution of a general linear system	425
$\mathbf{x}$	position vector	63
x_c	critical frequency (in the Osterbrock picture of the escape probability)	455
$x_D(\tau)$	characteristic frequency for Doppler diffusion	516
$x^{(n)}$	nth iterate of the solution of a linear system	425
x_ϵ	characteristic frequency for collisional destruction of photon correlation	517
$\mathbf{X}_d$	diagonal submatrices of the global Rybicki system	398
$X(\lambda, z)$	fraction of transmitted light	32
X_ν	modified optical depth coordinate in spherical radiative transfer	698
X_τ	(Milne's) Ξ-operator for the K-moment	365
Y	helium abundance (by number)	597
$Y_{lm}(\theta, \phi)$	spherical harmonics function	173
z	vertical coordinate	342
z	distance along a tangent ray in a (p, z) coordinate system	694
Z	atomic number	4
Z	ion charge	94
Z_{ij}	net radiative bracket	473
Z_{ij}	net number of radiative transitions	759
α	fine-structure constant	177
α	wavelength difference divided by the normal field strength	247
α	empirical parameter in combined radiative equilibrium equation	636
α	fitting parameter for the radiation-force multiplier in the Castor, Abbott, Klein theory	778
α_*	angular diameter of a star	76
$\alpha_{\rm bf}(\nu)$	bound-free cross section	187
$\alpha_{\rm ff}(\nu)$	bound-free cross section	189
α_{ij}	general absorption cross section	271
α_k	fractional abundance of element k	589
α_{lu}	auxiliary source function parameter in the ETLA method	484
$\alpha(t)$	atomic wave function	251
β	velocity in units of speed of light	72
β	electric field strength in units of normal field strength	243

β	empirical parameter in the combined radiative equilibrium equation	636
β	Sobolev escape probability	746
β	parameter in the empirical β-law for wind velocity	801
$\boldsymbol{\beta}$	velocity vector in units of speed of light	72
β_c	probability of penetration of the continuum radiation in the Sobolev theory	746
β_l	line opacity scaled to the electron scattering opacity	777
β_{lu}	auxiliary source function parameter in the ETLA method	484
β_ν	line-strength parameter in the Milne-Eddington equation	607
γ	polytropic index	101
γ	damping parameter (damping constant)	147
γ	coherence fraction	327
γ	Euler-Mascheroni constant	365
γ	core-wind separation parameter in the core-saturation method	480
γ	adiabatic exponent	560
γ	Lorentz contraction factor	715
$\gamma(z, p, \nu)$	velocity-dependent factor in the left-hand side of the comoving-frame transfer equation	726
Γ	damping parameter (damping constant)	229
Γ	force ratio ($= g_R/g$)	772
Γ_2	adiabatic exponent for ionizing gas plus radiation	561
Γ_e	force ratio for pure electron scattering	772
Γ_e	correction factor in the Van Regemorter formula for collisional rates	276
$\Gamma(p)$	Gamma function	235
δ	fitting parameter for the radiation-force multiplier in the Castor, Abbott, Klein theory	778
δg	radiation force perturbation (in the stellar wind stability analysis)	791
δ_{ij}	Kronecker delta symbol	77
δ_{ij}	coefficients to obtain an extrapolated iterate in the Ng acceleration	442
$\delta J^{(n)}$	correction to the mean intensity in the nth iteration	423
$\delta \mathbf{n}$	correction to the vector of level populations	493
$\delta S^{(n)}$	correction to the source function at the nth iteration	423
δv	velocity perturbation (in the stellar wind stability analysis)	791
$\delta(x)$	Dirac delta function	151
$\delta x^{(n)}$	correction to the solution of a general linear system in nth iteration	426
$\delta\rho$	density perturbation (in the stellar wind stability analysis)	791
Δ	designation of the molecular electronic state with $\Lambda = 2$	217
Δl_S	Sobolev length	746
$\Delta \mathbf{t}$	increment of the slanted optical depth	405
ΔB	correction to the total Planck function in the Unsöld-Lucy method	595

$\Delta T^{(n)}$	auxiliary quantity in the ALI scheme for partial redistribution	531
$\Delta\lambda$	wavelength difference; wavelength band	35
$\Delta\lambda_{\rm D}$	Doppler width	95
$\Delta\nu_D$	Doppler width (in frequency units)	232
$\Delta\tau$	optical depth difference	362
ϵ, ϵ_ν	thermal coupling parameter	357
ϵ	wind efficiency parameter	785
ϵ_i	energy of a particle (in a thermodynamic ensemble)	89
ϵ_I	ionization potential	95
ϵ_N	thermonuclear energy generation rate	551
ε	dielectric constant	69
ε	efficiency of convection	564
ζ	coefficient of bulk viscosity	549
η	emission coefficient (emissivity)	101
η	impact phase shift	235
η	wind efficiency factor	790
$\boldsymbol{\eta}$	Lorentz metric	818
η_{lu}	emission coefficient for the transition $l \rightarrow u$	451
η_{lu}	source function parameter in the ETLA method	483
$\eta_\nu^{\rm back}$	background emission coefficient	451
θ	one of the spherical coordinates	146
θ_{lu}	source function parameter in the ETLA method	483
$\theta_T(\tau, \tau')$	elementary probability of photon thermalization	468
Θ	polar angle	67
Θ	scattering angle (for line scattering)	311
$\Theta_{lm}(\theta, \phi)$	polar wave function	172
$\Theta_{\rm rot}$	rotational temperature	99
κ	absorption coefficient (opacity)	101
$\kappa_{\rm e}$	electron scattering opacity per gram	772
κ_B	Planck mean opacity	594
κ_J	absorption mean opacity	594
λ	wavelength	27
λ	coefficient of dilatational viscosity	549
$\lambda^{\pm}_{d,d\pm1}$	coefficients for the short characteristics scheme	406
λ_i	eigenvalue of a matrix	427
$\lambda_{\rm C}$	Compton wavelength of the electron	160
$\lambda_{\rm D}$	Debye length	142
λ_ν	photon mean free path	380
Λ	total orbital quantum number of a molecule	217
Λ, Λ_τ	(Schwarzschild's) Λ-operator	362
Λ	bridging length in the Owocki-Rybicki theory of the stability of stellar winds	794
$\overline{\Lambda}$	frequency-averaged Λ-operator	363
Λ^*	approximate Λ-operator	422
$\overline{\Lambda}^*$	frequency-averaged approximate Λ-operator	463
μ	effective mean molecular weight	6

μ	cosine of the polar angle	67
μ	magnetic permeability	69
μ	chemical potential	107
μ	reduced mass of a molecule	214
μ	coefficient of sheer (dynamical) viscosity	549
$\boldsymbol{\mu}$	total magnetic moment (of multi-electron atoms)	197
μ_0	Bohr magneton	178
μ_0	characteristic angle in the two-stream approximation	573
$\hat{\boldsymbol{\mu}}_e$	intrinsic magnetic moment of electron	175
μ_I	mean atomic weight	802
ν	frequency	61
ν	kinematic viscosity	555
ξ	frequency in the atom's frame	290
ξ	fitting parameter for the Kramers-like formula for the Rosseland mean opacity	812
$\xi_{\rm turb}$	microturbulent velocity	617
ξ_ν	auxiliary parameter in the Milne-Eddington equation	607
π	parallax	40
$\pi(t)$	perturber wave function	251
Π	designation of the molecular electronic state with $\Lambda = 1$	217
ρ	mass density	5
ρ	impact parameter of a collision	234
ρ	density matrix	249
ρ_0	effective impact parameter (in impact broadening)	234
ρ_b	characteristic density of blobs in the continuum-driven wind theory	798
$\rho_{ij}(\nu, \mathbf{n})$	ratio of the emission and absorption profiles	534
$\rho_{lu}(\nu)$	ratio of the line to the total opacity	498
$\tilde{\rho}(E)$	density of atomic energy states	133
ϱ	electric charge density	68
σ	spectral radius of a matrix	427
$\boldsymbol{\sigma}$	viscous stress	549
$\sigma_{\rm bf}$	bound-free (photoionization) cross section	137
$\sigma_{\rm cl}$	classical (scattering) cross section	148
σ_{ij}	cross section for the transition $i \to j$	135
$\sigma(n', l'; n, l)$	auxiliary function (integral of two Legendre functions)	182
$\sigma_{\rm KN}$	Klein-Nishina cross section	161
σ_R	Stefan-Boltzmann constant	22
$\sigma_{\rm Ray}$	Rayleigh scattering cross section	156
$\sigma_{\rm T}$	Thomson cross section	149
Σ	designation of the molecular electronic state with $\Lambda = 0$	217
$\bar{\Sigma}$	number of electrons donated per heavy particle	592
Σ_I	upper sum for ion I	282
τ	optical depth	353
τ	proper time (in relativistic kinematics)	819
τ_b	characteristic optical depth of blobs in the continuum-driven wind theory	798

$\tau_{\rm pen}$	penetration depth	630
$\tau_{\rm th}$	thermalization depth	380
$\tau_{\rm FT}$	frequency-thermalization depth	516
$\tau_{\rm KH}$	Kelvin-Helmholtz timescale	7
τ_R	Rosseland optical depth	597
τ_S	Sobolev optical depth	748
τ_ν	monochromatic optical depth	353
Υ	an analogous operator to Λ, yielding the flux	811
ϕ	scalar potential (of the electromagnetic field)	71
ϕ	one of the spherical coordinates	145
$\phi_{lu}(\nu)$	absorption profile coefficient	118
$\phi_{\rm D}$	Doppler profile	94
$\phi(\omega)$	power spectrum	229
Φ	azimuthal angle	67
Φ	dissipation function	552
$\Phi_i(T)$	Saha-Boltzmann factor	96
$\Phi_m(\phi)$	azimuthal wave function	172
$\Phi(s)$	autocorrelation function	229
$\Phi(\phi)$	azimuthal wave function	172
Φ_τ	(Milne's) Φ-operator for the H-moment	364
$\varphi(\mathbf{n}, \nu)$	photon number density	64
$\varphi(\mathbf{x}, t)$	total photon number density	64
$\varphi(\xi)$	absorption profile coefficient in the atom's frame	290
χ	total absorption (extinction) coefficient	340
χ_i	relative energy of a sublevel i	293
χ_{lu}	absorption coefficient for the transition $l \to u$	451
χ_H	flux mean opacity	594
χ_R	Rosseland mean opacity	376
$\chi_\nu^{\rm back}$	background absorption coefficient	451
ψ	degeneracy parameter (in the Fermi-Dirac statistics)	109
$\boldsymbol{\psi}_d$	state vector in the linearization method	488
$\psi_j(x)$	natural cardinal cubic spline	519
$\psi_{lu}(\nu)$	emission profile coefficient	119
Ψ_i	wave function	109
$\Psi_{\nu,\mathbf{n}}$	operator analogous to Λ, acting on the emission coefficient	537
$\Psi^*_{\nu,\mathbf{n}}$	approximate Ψ-operator	537
ω	circular frequency	70
ω	over-relaxation parameter	430
ω_d	opacity per gram (at depth d)	600
Ω	solid angle	63
Ω	scaled rotational velocity in the stellar wind theory	788
∇	logarithmic temperature gradient	561
$\nabla_{\rm ad}$	adiabatic gradient	561
∇_d	discretized logarithmic temperature gradient	605
$\nabla_{\rm el}$	temperature gradient of convective elements	561
$\nabla_{\rm rad}$	radiative temperature gradient	561

Bibliography

[1] D. Abbott. The theory of radiatively driven stellar winds. I. A physical interpretation. *Astrophys. J.*, 242, 1183, 1980.

[2] D. Abbott. The theory of radiatively driven stellar winds. II. The line acceleration. *Astrophys. J.*, 259, 282, 1982.

[3] D. Abbott and P. Conti. Wolf-Rayet stars. *Ann. Rev. Astr. Astrophys.*, 25, 113, 1987.

[4] D. Abbott and D. Hummer. Photospheres of hot stars. I. Wind blanketed model atmospheres. *Astrophys. J.*, 294, 286, 1985.

[5] M. Abramowitz and I. Stegun. *Handbook of Mathematical Functions.* (Washington, DC: U.S. Government Printing Office), 1972.

[6] H. Abt, A. Meinel, W. Morgan, and J. Tapscott. *An Atlas of Low-Dispersion Grating Stellar Spectra.* Kitt Peak National Observatory, Steward Observatory, and Yerkes Observatory, 1969.

[7] T. Adams, D. Hummer, and G. Rybicki. Numerical evaluation of the redistribution function $R_{IIA}(x, x')$ and the associated scattering integral. *J. Quantit. Spectrosc. Radiat. Transf.*, 11, 1365, 1971.

[8] T. Adams and D. Morton. A model atmosphere for a B4 V star with line blanketing. *Astrophys. J.*, 152, 195, 1968.

[9] S. Adelman, F. Kupka, and W. Weiss, editors. *Model Atmospheres and Spectrum Synthesis.* (San Francisco: Astronomical Society of the Pacific), 1996.

[10] S. Adelman and W. Wiese, editors. *Astrophysical Applications of Powerful New Databases.* (San Francisco: Astronomical Society of the Pacific), 1995.

[11] J. Ahlberg, E. Nilson, and J. Walsh. *The Theory of Splines and Their Applications.* (New York: Academic Press), 1967.

[12] M. Ajmera and K. Chung. Photodetachment of negative hydrogen ions. *Phys. Rev. A*, 12, 475, 1975.

[13] B. Alder, S. Fernbach, and M. Rotenberg, editors. *Methods in Computational Physics.* Volume 10. Atomic and Molecular Scattering. (New York: Academic Press), 1971.

[14] F. Allard and P. Hauschildt. Model atmospheres for M subdwarf stars. I. The base model grid. *Astrophys. J.*, 445, 433, 1995.

[15] F. Allard, P. Hauschildt, and A. Schweitzer. Spherically Symmetric Model Atmospheres for Low-Mass Pre-Main-Sequence Stars with Effective Temperatures between 2000 and 6800 K. *Astrophys. J.*, 539, 366, 2000.

[16] N. Allard and D. Koester. Theoretical profiles of Lyα satellites and application to synthetic spectra of DA white dwarfs. *Astr. Astrophys.*, 258, 464, 1992.

[17] N. Allard, D. Koester, N. Feautrier, and A. Spielfiedel. Free-free quasi-molecular absorption and satellites in Lyα due to collisions with H and H^+. *Astr. Astrophys. Suppl.*, 108, 417, 1994.

[18] C. Allen. *Astrophysical Quantities*. (London: Athlone Press), 3rd edition, 1973.

[19] C. Allende Prieto, D. Lambert, I. Hubeny, and T. Lanz. Non-LTE model atmospheres for late-type stars. I. A collection of data for light neutral and singly ionized atoms. *Astrophys. J. Suppl.*, 147, 363, 2003.

[20] L. Aller. Quantitative analysis of normal stellar spectra. In Greenstein [397], chapter 4, page 156.

[21] L. Aller. Interpretation of normal stellar spectra. In Greenstein [397], chapter 5, page 252.

[22] L. Aller. *The Atmospheres of the Sun and Stars*. (New York: Ronald Press Company), 2nd edition, 1963.

[23] L. Aller. *Physics of Thermal Gaseous Nebulae*. (Dordrecht: Reidel), 1984.

[24] A. Alonso, S. Arribas, and C. Martinez-Roger. Broad band *JHK* infrared photometry of an extended sample of late-type dwarfs and subdwarfs. *Astr. Astrophys. Suppl.*, 107, 365, 1994.

[25] A. Alonso, S. Arribas, and C. Martinez-Roger. Determination of bolometric fluxes for F, G, and K subdwarfs. *Astr. Astrophys.*, 297, 197, 1995.

[26] A. Alonso, S. Arribas, and C. Martinez-Roger. Determination of effective temperatures for an extended sample of dwarfs and subdwarfs (F0-K5). *Astr. Astrophys. Suppl.*, 117, 227, 1996.

[27] V. Ambartsumyan, editor. *Theoretical Astrophysics*. (London: Pergamon Press), 1958.

[28] J. Andersen. Accurate masses and radii of normal stars. *Astr. Astrophys. Rev.*, 3, 91, 1991.

[29] L. Anderson. Line blanketing without local thermodynamic equilibrium. I. A hydrostatic stellar atmosphere with H, He, and C lines. *Astrophys. J.*, 298, 848, 1985.

[30] L. Anderson. A code for line blanketing without local thermodynamic equilibrium. In Beckman and Crivellari [100], page 225.

[31] L. Anderson. Line blanketing without local thermodynamic equilibrium. II. A solar-type model in radiative equilibrium. *Astrophys. J.*, 339, 558, 1989.

[32] L. Anderson. Non-LTE line blanketing with elements 1-28. In Garmany [362], page 77.

[33] W. Anderson. Die Beziehung zwischen dem Gasdruck und der translatorischen Energie der Gasmoleküle. *Z. für Phys.*, 54, 433, 1929.

[34] L. Anusha, K. Nagendra, F. Paletou, and L. Léger. Preconditioned bi-conjugate gradient method for radiative transfer in spherical media. *Astrophys. J.*, 704, 661, 2009.

[35] A. Arharov, V. Novopashenny, and E. Terez. Absolute calibration of the energy distribution in the spectrum of Vega in the near-infrared region. *Astr. Cirk. No.* 1046, 1979.

[36] R. Aris. *Vectors, Tensors, and the Basic Equations of Fluid Mechanics*. (New York: Dover Publications), 1989.

[37] H. Arp. $(U - B)$ and $(B - V)$ colors of black bodies. *Astrophys. J.*, 133, 874, 1961.

[38] G. Arutyunyan and A. Nikogosyan. The redistribution function for scattering by relativistic electrons. *Sov. Phys. Doklady*, 25, 918, 1980.

[39] R. Athay. *Radiation Transport in Spectral Lines*. (Dordrecht: Reidel), 1972.

[40] R. Athay, L. House, and G. Newkirk, editors. *Line Formation in the Presence of Magnetic Fields*. (Boulder: National Center for Atmospheric Research), 1972.

[41] R. Athay, J. Mathis, and A. Skumanich, editors. *Resonance Lines in Astrophysics*. (Boulder: National Center for Atmospheric Research), 1968.

[42] R. Athay and A. Skumanich. Thermalization lengths and mean number of scatterings for line photons. *Astrophys. J.*, 170, 605, 1971.

[43] R. Atkinson. Atomic synthesis and stellar energy. I. *Astrophys. J.*, 73, 250, 1931.

[44] R. Atkinson. Atomic synthesis and stellar energy. II. *Astrophys. J.*, 73, 308, 1931.

[45] L. Auer. Improved boundary conditions for the Feautrier method. *Astrophys. J.*, 150, L53, 1967.

[46] L. Auer. The stellar atmospheres problem. In Hunt et al. [555], page 573.

[47] L. Auer. Application of the complete-linearization method to the problem of non-LTE line formation. *Astrophys. J.*, 180, 469, 1973.

[48] L. Auer. An Hermitian method for the solution of radiative transfer problems. *J. Quantit. Spectrosc. Radiat. Transf.*, 16, 931, 1976.

[49] L. Auer. Acceleration of convergence. In Kalkofen [609], page 101.

[50] L. Auer. Acceleration of convergence. In Crivellari et al. [264], page 9.

[51] L. Auer. Formal solution: Explicit answers. In Hubeny et al. [536], page 3.

[52] L. Auer and J. Heasley. An alternative formulation of the complete linearization method for the solution of non-LTE transfer problems. *Astrophys. J.*, 205, 165, 1976.

[53] L. Auer and D. Mihalas. Non-LTE model atmospheres. III. A complete-linearization method. *Astrophys. J.*, 158, 641, 1969.

[54] L. Auer and D. Mihalas. On the use of variable Eddington factors in non-LTE stellar atmospheres computations. *Mon. Not. Roy. Astr. Soc.*, 149, 65, 1970.

[55] L. Auer and D. Mihalas. Non-LTE model atmospheres. IV. Results for multi-line computations. *Astrophys. J.*, 160, 233, 1970.

[56] L. Auer and D. Mihalas. Non-LTE model atmospheres. VII. The H and He spectra of the O stars. *Astrophys. J. Suppl.*, 24, 193, 1972.

[57] L. Auer and D. Mihalas. Analyses of light-ion spectra in stellar atmospheres. V. Ne I in B stars. *Astrophys. J.*, 184, 151, 1973.

[58] L. Auer and F. Paletou. Two-dimensional radiative transfer with partial frequency redistribution. I. General method. *Astr. Astrophys.*, 284, 675, 1994.

[59] J. Aufdenberg. Line-blanketed spherically extended model atmospheres of hot luminous stars with and without winds. *Pub. Astr. Soc. Pacific*, 113, 119, 2001.

[60] J. Aufdenberg, P. Hauschildt, and E. Baron. A non-local thermodynamic equilibrium spherical line-blanketed stellar atmosphere model of the early B giant β C Ma. *Mon. Not. Roy. Astr. Soc.*, 302, 599, 1999.

[61] J. Aufdenberg, P. Hauschildt, S. Shore, and E. Baron. A spherical non-LTE line-blanketed stellar atmosphere model of the early B giant ϵ Canis Majoris. *Astrophys. J.*, 498, 837, 1998.

[62] E. Avrett. Solutions of the two-level line transfer problem with complete redistribution. In Avrett et al. [65], page 101.

[63] E. Avrett. Rapidly-converging methods for solving multilevel transfer problems. In *Numerical Methods for Multidimensional Radiative Transfer*, E. Meinköhn, G. Kanschat, R. Rannacher, and R. Wehrse, editors. (Heidelberg, Springer), page 217, 2007.

[64] E. Avrett, O. Gingerich, and C. Whitney. Proceedings of the First Harvard-Smithsonian Conference on Stellar Atmospheres. Technical Report 167, Smithsonian Astrophysical Observatory, Cambridge, Massachusetts, 1964.

[65] E. Avrett, O. Gingerich, and C. Whitney. The Formation of Spectrum Lines: Proceedings of the Second Harvard-Smithsonian Conference on Stellar Atmospheres. Technical Report 174, Smithsonian Astrophysical Observatory, Cambridge, Massachusetts, 1965.

[66] E. Avrett and D. Hummer. Non-coherent scattering. II. Line formation with a frequency independent source function. *Mon. Not. Roy. Astr. Soc.*, 130, 295, 1965.

[67] E. Avrett and M. Krook. The temperature distribution in a stellar atmosphere. *Astrophys. J.*, 137, 874, 1963.

[68] E. Avrett and R. Loeser. The PANDORA atmosphere program. In Cool Stars, Stellar Systems, and the Sun, Proceedings of the 7th Cambridge Workshop, ASP Conference Series, San Francisco: ASP), 26, 489, 1982.

[69] E. Avrett and R. Loeser. Line transfer in static and expanding spherical atmospheres. In Kalkofen [606], page 341.

[70] E. Avrett and R. Loeser. Solar and stellar atmospheric modeling using the PANDORA computer program. In Piskunov et al. [881], page A21.

[71] E. Avrett and R. Loeser. Models of the solar chromosphere and transition region from SUMER and HRTS observations: Formation of the extreme-ultraviolet spectrum of hydrogen, carbon, and oxygen. *Astrophys. J. Suppl.*, 175, 229, 2008.

[72] T. Ayres. A physically realistic approximate form for the redistribution function R_{IIA}. *Astrophys. J.*, 294, 153, 1985.

[73] W. Baade. The resolution of Messier 32, NGC 205, and the central region of the Andromeda nebula. *Astrophys. J.*, 100, 137, 1944.

[74] W. Baade. *Evolution of Stars and Galaxies*. (Cambridge: Harvard University Press), 1965.

[75] W. Baade and F. Zwicky. On super-novae. *Pub. Nat. Acad. Sci.*, 20, 254, 1934.

[76] J. Babb. Effective oscillator strengths and transition energies for the hydrogen molecular ion. *Molec. Phys.*, 81, 17, 1994.

[77] J. Bahcall and R. Wolf. Fine-structure transitions. *Astrophys. J.*, 152, 701, 1968.

[78] L. Balona. Effective temperature, bolometric correction, and mass calibration of O-F stars. *Mon. Not. Roy. Astr. Soc.*, 268, 119, 1994.

[79] K. Bappu and J. Sahade, editors. *Wolf-Rayet and High-Temperature Stars*. IAU Symposium No. 49. (Dordrecht: Reidel), 1973.

[80] M. Baranger. Simplified quantum-mechanical theory of pressure broadening. *Phys. Rev.*, 111, 481, 1958.

[81] M. Baranger. Problem of overlapping lines in the theory of pressure broadening. *Phys. Rev.*, 111, 494, 1958.

[82] M. Baranger. General impact theory of pressure broadening. *Phys. Rev.*, 112, 855, 1958.

[83] M. Baranger. Spectral line broadening in plasmas. In Bates [95], page 493.

[84] M. Baranger and B. Mozer. Electric field distributions in an ionized gas. *Phys. Rev.*, 115, 521, 1959.

[85] B. Barbuy, J. Meléndez, M. Spite, F. Spite, E. Depagne, V. Hill, R. Cayrel, P. Bonifacio, A. Damineli, and C. Torres. Oxygen abundance in the template halo giant HD 122563. *Astrophys. J.*, 588, 1072, 2003.

[86] P. Barklem, A. Belyaev, M. Guitou, N. Feautrier, F. Gadéa, and A. Spielfiedel. On inelastic hydrogen atom collisions in stellar atmospheres. *Astr. Astrophys.*, 530, 94, 2011.

[87] T. Barman, P. Hauschildt, A. Schweitzer, P. Stancil, E. Baron, and F. Allard. Non-LTE effects of Na I in the atmosphere of HD 209458b. *Astrophys. J.*, 569, 51, 2002.

[88] A. Barnard and J. Cooper. Computed profiles of He I 5016Å at high electron densities. *J. Quantit. Spectrosc. Radiat. Transf.*, 10, 695, 1970.

[89] A. Barnard, J. Cooper, and L. Shamey. Calculated profiles of He I 4471 and 4922Å and their forbidden components. *Astr. Astrophys.*, 1, 28, 1969.

[90] A. Barnard, J. Cooper, and E. Smith. The broadening of He I lines including ion dynamic corrections with application to λ 4471Å. *J. Quantit. Spectrosc. Radiat. Transf.*, 14, 1025, 1974.

[91] A. Barnard, J. Cooper, and E. Smith. Stark broadening tables for He I λ 4922Å. *J. Quantit. Spectrosc. Radiat. Transf.*, 15, 429, 1975.

[92] B. Baschek and J. Oke. Effective temperatures and gravities of Ap, Am, and normal A-type stars. *Astrophys. J.*, 141, 1404, 1965.

[93] B. Baschek, M. Scholz, and R. Wehrse. The parameters R and $T_{\rm eff}$ in stellar models and observations. *Astr. Astrophys.*, 246, 374, 1991.

[94] D. Bates. Absorption of radiation by an atmosphere of H, H^+, and H_2^+, semi-classical treatment. *Mon. Not. Roy. Astr. Soc.*, 112, 40, 1952.

[95] D. Bates, editor. *Atomic and Molecular Processes*. (New York, Academic Press), 1962.

[96] W. Baum, W. Hiltner, H. Johnson, and A. Sandage. The main sequence of the globular cluster M 13. *Astrophys. J.*, 130, 749, 1959.

[97] C. Beals. The Wolf-Rayet stars. *Pub. Domin. Astrophys. Obs.*, 4, 271, 1930.

[98] C. Beals. The contours of emission bands in novae and Wolf-Rayet stars. *Mon. Not. Roy. Astr. Soc.*, 91, 966, 1931.

[99] A. Beck and P. Havas, editors. *Collected Papers Of Albert Einstein, Volume 2. The Swiss Years: 1900–1909*. (Princeton: Princeton University Press), 1989.

[100] J. Beckman and L. Crivellari, editors. *Progress in Stellar Spectral Line Formation Theory. NATO Advanced Science Institute*. (Dordrecht: Reidel), 1985.

[101] A. Beer, editor. *Vistas in Astronomy*. Volume 1. (Oxford: Pergamon Press), 1955.

[102] K. Bell. The free-free absorption coefficient of the negative ion of molecular hydrogen. *J. Phys. B*, 13, 1859, 1980.

[103] K. Bell and K. Berrington. Free-free absorption coefficient of the negative hydrogen ion. *J. Phys. B*, 20, 801, 1987.

[104] K. Bell, A. Kingston, and W. McIlveen. The total free-free absorption coefficient of the negative ion of molecular hydrogen. *J. Phys. B*, 8, 659, 1975.

[105] R. Bell. Theoretical colors for F and G dwarf stars. *Mon. Not. Roy. Astr. Soc.*, 154, 343, 1971.

[106] R. Bell, K. Eriksson, B. Gustafsson, and Å. Nordlund. A grid of model atmospheres for metal-deficient giant stars. II. *Astr. Astrophys. Suppl.*, 23, 37, 1976.

[107] R. Bell and B. Gustafsson. The colors of G and K type giant stars. II. *Astr. Astrophys. Suppl.*, 34, 229, 1978.

[108] R. Bell and S. Parsons. Theoretical colors for F and G supergiants. *Mon. Not. Roy. Astr. Soc.*, 169, 71, 1974.

[109] V. Berestetskii, E. Lifshitz, and L. Pitaevskii. *Quantum Electrodynamics*. (Oxford: Butterworth-Heinemann), 2nd edition, 1982.

[110] J. Berger. Absorption coefficients for free-free transitions in a hydrogen plasma. *Astrophys. J.*, 124, 550, 1956.

[111] K. Berrington, P. Burke, M. Le Dourneuf, W. Robb, K. Taylor, and V. Lan. A new version of the general program to calculate atomic continuum processes using the *R*-matrix method. *Comp. Phys. Commun.*, 14, 367, 1978.

[112] E. Bertone, A. Buzzoni, M. Chavez, and L. Rodriguez-Merino. ATLAS vs. NextGen model atmospheres: A combined analysis of synthetic spectral energy distributions. *Astr. J.*, 128, 289, 2004.

[113] T. Le Bertre, A. Lebre, and C. Waelkens, editors. *Asymptotic Giant Branch Stars*. (San Francisco: Astronomical Society of the Pacific), 1999.

[114] M. Bessell, F. Castelli, and B. Plez. Model atmospheres, broad-band colors, bolometric corrections and temperature calibrations for O-M stars. *Astr. Astrophys.*, 333, 231, 1998.

[115] H. Bethe. Energy production in stars. *Phys. Rev.*, 55, 434, 1939.

[116] H. Bethe and R. Jackiw. *Intermediate Quantum Mechanics*. (New York: W. A. Benjamin, Inc.), 2nd edition, 1968.

[117] H. Bethe and E. Salpeter. *Quantum Mechanics of One- and Two- Electron Atoms*. (Berlin: Springer-Verlag), 1957.

[118] P. Bhatnagar, M. Krook, D. Menzel, and R. Thomas. Turbulence, kinetic temperature, and electron temperature in stellar atmospheres. In Beer [101], page 296.

[119] L. Biermann. Untersuchungen über den inneren Aufbau der Sterne. IV. Konvektionszonen im Innern der Sterne. *Z. für Astrophys.*, 5, 117, 1932.

[120] J. Binney and M. Merrifield. *Galactic Astronomy*. (Princeton: Princeton University Press), 1998.

[121] J. Bjorkman and J. Cassinelli. Equatorial disk formation around rotating stars due to Ram pressure confinement by the stellar wind. *Astrophys. J.*, 409, 429, 1993.

[122] M. Blaha. Effective Gaunt factors g_{eff} for excitation of positive ions by electron collisions in a simplified Coulomb-Born approximation. *Astrophys. J.*, 157, 473, 1969.

[123] B. Bohannan and N. Walborn. The Ofpe/WN 9 class in the Large Magellanic Cloud. *Pub. Astr. Soc. Pacific*, 101, 520, 1989.

[124] R. Bohlin. Comparison of white dwarf models with STIS spectrophotometry. *Astr. J.*, 120, 437, 2000.

[125] R. Bohlin, M. Dickinson, and D. Calzetti. Spectrophotometric standards from the far-ultraviolet to the near-infrared: STIS and NICMOS fluxes. *Astr. J.*, 122, 2118, 2001.

[126] R. Bohlin and R. Gilliland. HST absolute spectrophotometry of Vega from the far UV to the IR. *Astr. J.*, 127, 3508, 2004.

[127] D. Bohm and L. Aller. The electron velocity distribution in gaseous nebulae and stellar envelopes. *Astrophys. J.*, 105, 1, 1947.

[128] K.-H. Böhm. Zur Deutung der Mitte Rand Variation der Fraunhofer Linien. *Z. für Astrophys.*, 35, 179, 1954.

[129] K.-H. Böhm. Basic theory of line formation. In Greenstein [397], chapter 3, page 88.

[130] E. Böhm-Vitense. Über die Wasserstoffkonvektionszone in Sternen verschiedener Effektivtemperaturen und Leuchtkräfte. *Z. für Astrophys.*, 46, 108, 1958.

[131] E. Böhm-Vitense. The *UBVr* colors of extreme Population II giants. *Astr. Astrophys.*, 24, 447, 1973.

[132] E. Böhm-Vitense. *Introduction to Stellar Astrophysics*. Volume 1: Basic Stellar Observations and Data. (Cambridge: Cambridge University Press), 1989.

[133] E. Böhm-Vitense. *Introduction to Stellar Astrophysics*. Volume 2: Stellar Atmospheres. (Cambridge: Cambridge University Press), 1989.

[134] E. Böhm-Vitense and G. Nelson. About the proper choice of the characteristic length in the convection theory. *Astrophys. J.*, 210, 741, 1976.

[135] H. Bohn and B. Wolf. Dissociation equilibrium and thermodynamics of diatomic molecules in astrophysical applications. *Astr. Astrophys.*, 130, 202, 1984.

[136] M. Born and J. R. Oppenheimer. Zur Quantentheorie del Molekeln. *Ann. Physik*, 84, 457, 1927.

[137] A. Borysow, U. Jorgensen, and C. Zheng. Model atmospheres of cool, low-metallicity stars: The importance of collision-induced absorption. *Astr. Astrophys.*, 324, 185, 1997.

[138] S. Bose. Planck's Gesetz und Lichtquantenhypothese. *Z. für Phys.*, 26, 178, 1924.

[139] R. Bowers and T. Deeming. *Astrophysics*. Volume 1: Stars. (Boston: Jones and Bartlett Publishers), 1984.

[140] R. Bracewell. *The Fourier Transform and Its Applications*. (New York: Mc Graw-Hill Book Company), 3rd edition, 1999.

[141] P. Bradley and D. Morton. Model atmospheres for O-type stars with ultraviolet line blanketing. *Astrophys. J.*, 156, 687, 1969.

[142] S. Bréchot and H. van Regemorter. L'éllargissement des raies spectrales par chocs. *Ann. d'Astrophys.*, 27, 432, 1964.

[143] S. Bréchot and H. van Regemorter. L'éllargissement des raies spectrales par chocs. II. Théorie générale de l'éllargissement par chocs éllectroniques. *Ann. d'Astrophys.*, 27, 739, 1964.

[144] R. Breene. *The Shift and Shape of Spectral Lines*. (Oxford: Pergamon Press), 1961.

[145] G. Breit. Dirac's equation and the spin-spin interactions of two electrons. *Phys. Rev.*, 39, 616, 1932.

[146] J. Brett. Opacity sampling model photospheres for M dwarfs. I. Computations, sensitivities and comparisons. *Astr. Astrophys.*, 324, 185, 1997.

[147] O. Brill and B. Goodham. Causality in the Coulomb gauge. *Amer. J. Phys.*, 35, 282, 1967.

[148] D. Brink and G. Satchler. *Angular Momentum*. (Oxford: Oxford University Press), 1962.

[149] A. Brissaud and U. Frisch. Theory of Stark broadening - II. Exact line profile with model microfield. *J. Quantit. Spectrosc. Radiat. Transf.*, 11, 1767, 1971.

[150] M. Bronstein. Zur strahlungsgleichgewichtes Problem von Milne. *Z. für Phys.*, 58, 696, 1929.

[151] M. Bronstein. Über das Verhältnis der effektiven Temperatur der Sterne zur Temperatur ihrer Oberfläche. *Z. für Phys.*, 59, 144, 1929.

[152] M. Bronstein. Note on the temperature distribution in the deep layers of stellar atmospheres. *Mon. Not. Roy. Astr. Soc.*, 91, 133, 1930.

[153] R. Brown. Measurement of stellar diameters. *Ann. Rev. Astr. Astrophys.*, 6, 13, 1968.

[154] R. Brown and R. Twiss. Interferometry of the intensity fluctuations in light. IV. Test of an intensity interferometer on Sirius A. *Proc. Roy. Soc. London A*, 248, 222, 1958.

[155] D. Brüggemann and M. Bollig. An efficient algorithm for frequency integration of Voigt profiles. *J. Quantit. Spectrosc. Radiat. Transf.*, 48, 111, 1992.

[156] J. de Bruijne, R. Hoogerwerf, and P. de Zeeuw. A HIPPARCOS study of the Hyades open cluster. Improved color-absolute magnitude and Hertzsprung-Russell diagrams. *Astr. Astrophys.*, 367, 111, 2001.

[157] T. ten Brummelaar, M. Creech-Eakman, and J. Monnier. Probing stars with optical and near-IR interferometry. *Physics Today*, 62, 28, 2009.

[158] R. Buckingham, S. Reid, and R. Spence. Continuous absorption by the H molecular ion. *Mon. Not. Roy. Astr. Soc.*, 112, 382, 1952.

[159] D. Buckley and A. Longmore. The distance to M 13 via a subdwarf fit in the optical-infrared color-magnitude plane. *Mon. Not. Roy. Astr. Soc.*, 257, 731, 1992.

[160] G. Burbidge, E. Burbidge, W. Fowler, and F. Hoyle. Synthesis of the elements in stars. *Rev. Mod. Phys.*, 29, 547, 1957.

[161] A. Burgess. A general formula for the estimation of dielectronic recombination coefficients in low-density plasmas. *Astrophys. J.*, 141, 1588, 1965.

[162] A. Burgess and M. Seaton. A general formula for the calculation of atomic photoionization cross-sections. *Mon. Not. Roy. Astr. Soc.*, 120, 121, 1960.

[163] A. Burgess and H. Summers. Radiative Gaunt factors. *Mon. Not. Roy. Astr. Soc.*, 226, 257, 1987.

[164] P. Burke, A. Hibbert, and W. Robb. Electron scattering by complex atoms. *J. Phys. B*, 4, 135, 1971.

[165] P. Burke and W. Robb. R-matrix theory of atomic processes. *Adv. Atm. Mol. Phys.*, 11, 143, 1975.

[166] P. Burke and M. Seaton. Numerical solutions of the integro-differential equations of electron-atom collision theory. In Alder et al. [13], page 1.

[167] P. Burke and T. Webb. Electron scattering by atomic hydrogen using a pseudostate expansion. III. Excitation of $2s$ and $2p$ states at intermediate energies. *J. Phys. B*, 3, L131, 1970.

[168] J. Busche and D. Hillier. An efficient short characteristic solution for the transfer equation in axisymmetric geometries using a spherical coordinate system. *Astrophys. J.*, 53, 1071, 2000.

[169] R. Buser and R. Kurucz. A systematic investigation of multicolor photometric systems. III. Theoretical *UBV* colors and the temperature scale for early-type stars. *Astr. Astrophys.*, 70, 555, 1978.

[170] R. Buser and R. Kurucz. A library of theoretical stellar flux spectra. I. Synthetic *UBVRI* photometry and the metallicity scale for F- to K-type stars. *Astr. Astrophys.*, 264, 557, 1992.

[171] S. Butler and A. Dalgarno. Charge transfer of multiply charged ions with hydrogen and helium Landau-Zener calculations. *Astrophys. J.*, 241, 838, 1980.

[172] K. Butler and J. Giddings. In Newsletter on Analysis of Astronomical Spectra. No. 9 (Daresbury, England: Daresbury Laboratory).

[173] C. Cacciari. UV fluxes of Population II stars. *Astr. Astrophys. Suppl.*, 61, 339, 1985.

[174] C. Cacciari, M. Malagnini, C. Morossi, and L. Rossi. Physical parameters for Population II stars. *Astr. Astrophys.*, 183, 314, 1987.

[175] R. Canfield and R. Puetter. Theoretical quasar emission line ratios. I. Transfer and escape of radiation. *Astrophys. J.*, 243, 381, 1981.

[176] C. Cannon. Angular quadrature perturbations in radiative transfer theory. *J. Quantit. Spectrosc. Radiat. Transf.*, 13, 627, 1973.

[177] C. Cannon. Frequency quadrature perturbations in radiative transfer theory. *Astrophys. J.*, 185, 621, 1973.

[178] C. Cannon, P. Lopert, and C. Magnan. Redistribution perturbations in radiative transfer theory. *Astr. Astrophys.*, 42, 347, 1975.

[179] D. Carbon. A comparison of the straight-mean, harmonic-mean, and multiple-picket approximations for the line opacities in cool model atmospheres. *Astrophys. J.*, 187, 135, 1974.

[180] D. Carbon. Model atmospheres for intermediate- and late-type stars. *Ann. Rev. Astr. Astrophys.*, 17, 513, 1979.

[181] D. Carbon. Line blanketing. In Kalkofen [606], page 395.

[182] R. Carlberg. The instability of radiation-driven stellar winds. *Astrophys. J.*, 241, 1131, 1980.

[183] M. Carlsson. A computer program for solving multi-level non-LTE radiative transfer problems in moving or static atmospheres. Technical Report 33, Uppsala Astronomical Observatory, Uppsala, Sweden, 1986.

[184] B. Carney. Subdwarf ultraviolet excesses and metal abundances. *Astrophys. J.*, 233, 211, 1979.

[185] B. Carney and M. Aaronson. Subdwarf bolometric corrections. *Astr. J.*, 84, 867, 1979.

[186] B. Carroll and D. Ostlie. *An Introduction to Modern Astrophysics*. (Reading: Addison-Wesley Publishing Company), 1996.

[187] T. Carson. Coulomb free-free Gaunt factors. *Astr. Astrophys.*, 189, 319, 1988.

[188] J. Cassinelli. Extended model atmospheres for the central stars of planetary nebulae. *Astrophys. J.*, 165, 265, 1971.

[189] J. Cassinelli. The continuous energy distribution from stars with hot extended atmospheres. *Astrophys. Let.*, 8, 105, 1971.

[190] J. Cassinelli, J. Castor, and J. Lamers. Expanding envelopes of early-type stars – Current status. *Pub. Astr. Soc. Pacific*, 90, 496, 1978.

[191] F. Castelli. Synthetic photometry from ATLAS 9 models in the *UBV* Johnson system. *Astr. Astrophys.*, 346, 564, 1999.

[192] F. Castelli and C. Cacciari. Stellar parameters of Population II A-type stars from IUE spectra and new ODF ATLAS 9 model atmospheres. *Astr. Astrophys.*, 380, 630, 2001.

[193] F. Castelli, H. Lamers, F. de Andres, and E. Müller. A comparison between the observed and predicted UV line blocking for blanketed model atmospheres of early-type stars. *Astr. Astrophys.*, 91, 32, 1980.

[194] J. Castor. Spectral line formation in Wolf-Rayet envelopes. *Mon. Not. Roy. Astr. Soc.*, 149, 111, 1970.

[195] J. Castor. Radiative transfer in spherically symmetric flows. *Astrophys. J.*, 178, 779, 1972.

[196] J. Castor. The effect of sphericity on stellar continuous energy distributions. *Astrophys. J.*, 189, 273, 1974.

[197] J. Castor. On the force associated with absorption of spectral line radiation. *Mon. Not. Roy. Astr. Soc.*, 169, 279, 1974.

[198] J. Castor. *Radiation Hydrodynamics*. (Cambridge: Cambridge University Press), 2004.

[199] J. Castor, D. Abbott, and R. Klein. Radiation-driven winds in Of stars. *Astrophys. J.*, 195, 157, 1975.

[200] J. Castor, P. Dykema, and R. Klein. A new scheme for multidimensional line transfer. II. ETLA method in one dimension with application to iron $K\alpha$ lines. *Astrophys. J.*, 387, 561, 1992.

[201] G. Cayrel de Strobel. Comparaison entre les gradients absolus stellaires de Göttingen et de Paris. *Ann. d'Astrophys.*, 20, 55, 1957.

[202] G. Cayrel de Strobel. How precise are spectroscopic abundance determinations today? In Hayes et al. [464], page 137.

[203] G. Cayrel de Strobel, N. Knowles, G. Hernandez, and C. Benrolila. In search of real solar twins. *Astr. Astrophys.*, 94, 1, 1981.

[204] G. Cayrel de Strobel, Y. Lebreton, C. Soubrian, and E. Friel. Old, low-mass, metal-rich (SMR) stars. *Astrophys. Space Sci.*, 265, 345, 1999.

[205] R. Cayrel. The location of a few subdwarfs in the theoretical H-R diagram and the He content of Population II. *Astrophys. J.*, 151, 997, 1968.

[206] R. Cayrel. The first generation of stars. *Astr. Astrophys. Rev.*, 7, 217, 1996.

[207] R. Cayrel, Y. Lebreton, M.-N. Perrin, and C. Turon. The HR diagram in the plane $\log(T_{\rm eff}), M_{\rm bol}$ of Population II stars with HIPPARCOS parallaxes. In *Proceedings of the ESA Symposium 'HIPPARCOS'*, ESA SP-402, page 219, 1997.

[208] R. Cayrel and G. Traving. Zur Frage der Druckverbreiterung der solaren Balmerlinien. *Z. für Astrophys.*, 50, 239, 1960.

[209] C. Cesco, S. Chandrasekhar, and J. Sahade. On the radiative equilibrium of a stellar atmosphere. IV. *Astrophys. J.*, 100, 355, 1944.

[210] B. Chaboyer, P. Demarque, P. Kernan, and L. Krauss. The age of globular clusters in light of HIPPARCOS: Resolving the age problem? *Astrophys. J.*, 494, 96, 1998.

[211] J. Chamberlain and L. Aller. The atmospheres of A-type subdwarfs and 95 Leonis. *Astrophys. J.*, 114, 52, 1951.

[212] S. Chandrasekhar. The maximum mass of ideal white dwarfs. *Astrophys. J.*, 74, 81, 1931.

[213] S. Chandrasekhar. The dissociation formula according to the relativistic statistics. *Mon. Not. Roy. Astr. Soc.*, 91, 446, 1931.

[214] S. Chandrasekhar. On the hypothesis of the radial ejection of high speed atoms for the Wolf-Rayet stars and the novae. *Mon. Not. Roy. Astr. Soc.*, 94, 522, 1934.

[215] S. Chandrasekhar. The highly collapsed configurations of a stellar mass. (Second paper). *Mon. Not. Roy. Astr. Soc.*, 95, 207, 1935.

[216] S. Chandrasekhar. The radiative equilibrium of the outer layers of a star with special reference to the blanketing effect of the reversing layer, *Mon. Not. Roy. Astr. Soc.*, 96, 21, 1935.

[217] S. Chandrasekhar. Stochastic problems in physics and astronomy. *Rev. Mod. Phys.*, 15, 1, 1943.

[218] S. Chandrasekhar. On the radiative equilibrium of a stellar atmosphere. II. *Astrophys. J.*, 100, 76, 1944.

[219] S. Chandrasekhar. The negative ions of hydrogen and oxygen in stellar atmospheres. *Rev. Mod. Phys.*, 16, 101, 1944.

[220] S. Chandrasekhar. On the radiative equilibrium of a stellar atmosphere. V. *Astrophys. J.*, 101, 95, 1945.

[221] S. Chandrasekhar. On the continuous absorption coefficient of the negative hydrogen ion. *Astrophys. J.*, 102, 223, 1945.

[222] S. Chandrasekhar. On the continuous absorption coefficient of the negative hydrogen ion. II. *Astrophys. J.*, 102, 395, 1945.

[223] S. Chandrasekhar. *An Introduction to the Study of Stellar Structure*. (Chicago: University of Chicago Press), 1939. Reprinted in 1957 (New York: Dover Publications).

[224] S. Chandrasekhar. On the continuous absorption coefficient of the negative hydrogen ion. IV. *Astrophys. J.*, 128, 114, 1958.

[225] S. Chandrasekhar. *Radiative Transfer*. (Oxford: Clarendon Press), 1950. Reprinted in 1960 (New York: Dover Publications).

[226] S. Chandrasekhar and F. Breen. On the continuous absorption coefficient of the negative hydrogen ion. III. *Astrophys. J.*, 104, 430, 1946.

[227] S. Chandrasekhar and F. Breen. On the radiative equilibrium of a stellar atmosphere. XVIII. *Astrophys. J.*, 105, 461, 1947.

[228] S. Chandrasekhar and F. Breen. On the radiative equilibrium of a stellar atmosphere. XX. *Astrophys. J.*, 106, 145, 1947.

[229] S. Chandrasekhar and M. Krogdahl. On the negative hydrogen ion and its absorption coefficient. *Astrophys. J.*, 98, 205, 1943.

[230] R. Chapman. Radiative transfer in extended stellar atmospheres. *Astrophys. J.*, 143, 61, 1966.

[231] R. Chapman. Rational approximations to Gaunt factors. II. *J. Quantit. Spectrosc. Radiat. Transf.*, 10, 1151, 1970.

[232] S. Chapman and T. Cowling. *The Mathematical Theory of Non-Uniform Gases*. (Cambridge, Cambridge University Press), 3rd edition, 1970.

[233] C. Charbonnel, G. Meynet, A. Maeder, and D. Schaerer. Grids of stellar models. VI. Horizontal branch and early asymptotic giant branch for low-mass stars ($Z = 0.020, 0.001$). *Astr. Astrophys. Suppl.*, 115, 339, 1996.

[234] C. Charbonnel, G. Meynet, A. Maeder, G. Schaller, and D. Schaerer. Grids of stellar models. III. From 0.8 to 120 $M_\odot$ at $Z = 0.004$. *Astr. Astrophys. Suppl.*, 101, 415, 1993.

[235] L. Chevallier, F. Paletou, and B. Rutily. On the accuracy of the ALI method for solving the radiative transfer equation. *Astr. Astrophys.*, 411, 221, 2003.

[236] C. Christensen. Absolute spectral energy distributions and [Fe/H] values of metal-poor stars and globular clusters. *Astr. J.*, 83, 244, 1978.

[237] W. Christie and O. Wilson. ζ Aurigae, The structure of a stellar atmosphere. *Astrophys. J.*, 81, 426, 1935.

[238] G. Van Citters and D. Morton. Model atmospheres for B-type stars with blanketing by ultraviolet lines. *Astrophys. J.*, 161, 695, 1970.

[239] G. Clementini, R. Gratton, E. Carretta, and C. Sneden. Homogeneous photometry and metal abundances for a large sample of HIPPARCOS metal-poor stars. *Mon. Not. Roy. Astr. Soc.*, 302, 22, 1999.

[240] A. Code. Stellar energy distribution. In Greenstein [397], chapter 2, page 50.

[241] A. Code. The role of space observations in the calibration of fundamental stellar quantities. In Hayes et al. [464], page 209.

[242] A. Code, R. Bless, J. Davis, and R. Brown. Empirical effective temperatures and bolometric corrections for early-type stars. *Astrophys. J.*, 203, 417, 1976.

[243] C. Cohen-Tannoudji, J. Dupont-Roc, and G. Grynberg. *Photons and Atoms: Introduction to Quantum Electrodynamics*. (New York: John Wiley & Sons), 1989.

[244] C. Cohen-Tannoudji, J. Dupont-Roc, and G. Grynberg. *Atom-Photon Interactions: Basic Processes and Applications*. (New York: John Wiley & Sons), 1992.

[245] L. Colina and R. Bohlin. Absolute flux distributions of solar analogs from the UV to the near-IR. *Astr. J.*, 113, 1138, 1997.

[246] G. Collins. *The Fundamentals of Stellar Astrophysics*. (New York: W. H. Freeman & Company), 1989.

[247] A. Compton. A quantum theory of the scattering of X-rays by light elements. *Phys. Rev.*, 21, 483, 1923.

[248] E. Condon and G. Shortley. *Theory of Atomic Spectra*. (Cambridge: Cambridge University Press), 1963.

[249] J. Cooper. Broadening of isolated lines in the impact approximation using a density matrix formulation. *Rev. Mod. Phys.*, 39, 167, 1967.

[250] J. Cooper, R. Ballagh, K. Burnett, and D. Hummer. On redistribution and the equations for radiative transfer. *Astrophys. J.*, 260, 299, 1982.

[251] J. Cooper, R. Ballagh, and I. Hubeny. Approximate formulation of redistribution in the Lyα, Lyβ, Hα system. *Astrophys. J.*, 344, 949, 1989.

[252] J. Cooper, I. Hubeny, and J. Oxenius. On the line-profile coefficient for stimulated emission. *Astr. Astrophys.*, 127, 224, 1983.

[253] J. Cooper and G. Oertel. Electron-impact broadening of isolated lines in neutral atoms in a plasma. I. *Phys. Rev.*, 180, 286, 1969.

[254] D. Cope, R. Khoury, and R. Lovett. Efficient calculation of general Voigt profiles. *J. Quantit. Spectrosc. Radiat. Transf.*, 39, 163, 1988.

[255] J. Coste and J. Peyraud. Kinetics of the Compton scattering and the Bose condensation of a photon gas. *Phys. Rev. A*, 12, 2144, 1975.

[256] M. Costes and C. Naulin. Dissociation energies and partition functions of small molecules. In Jørgensen [602], page 250.

[257] R. Courant and D. Hilbert. *Methods of Mathematical Physics*. Two volumes. (New York: Wiley-Interscience), 1989.

[258] J. Cowan, D. Burris, C. Sneden, A. McWilliam, and G. Preston. Evidence of heavy element nucleosynthesis early in the history of the Galaxy: The ultra-metal-poor star CS 22892-052. *Astrophys. J.*, 439, 51, 1995.

[259] R. Cowan. *The Theory Of Atomic Structure and Spectra*. University of California Press, 1981.

[260] C. Cowley. An approximate Stark broadening formula for use in spectrum synthesis. *Observatory*, 91, 139, 1971.

[261] A. Cox, editor. *Allen's Astrophysical Quantities*. (New York: Springer-Verlag), 4th edition, 2000.

[262] J. Cox and R. Giuli. *Principles of Stellar Structure*. Two volumes. (New York: Gordon and Breach), 1968.

[263] A. Cox, W. Livingston, and M. Matthews, editors. *Solar Interior and Atmosphere*. (Tucson: University of Arizona Press), 1991.

[264] L. Crivellari, I. Hubeny, and D. Hummer, editors. *Stellar Atmospheres: Beyond Classical Models*. NATO Advanced Science Institute. (Dordrecht: Kluwer), 1991.

[265] P. Crowther, L. Smith, D. Hillier, and W. Schmutz. Fundamental parameters of Wolf-Rayet stars. III. The evolutionary status of WNL stars. *Astr. Astrophys.*, 293, 427, 1995.

[266] A. Dalgarno. Rayleigh scattering near an absorption line. *J. Optical Soc. America*, 53, 1223, 1963.

[267] A. Dalgarno and A. Kingston. Refractive indices and Verdet constants of the inert gases. *Proc. Roy. Soc. London*, A259, 424, 1960.

[268] A. Dalgarno and N. Lane. Free-free transitions of electrons in gases. *Astrophys. J.*, 145, 623, 1966.

[269] A. Dalgarno and D. Williams. Rayleigh scattering by molecular hydrogen. *Astrophys. J.*, 136, 690, 1962.

[270] J. Davis and R. Shobbrook. On *uvby* indices and empirical effective temperatures and bolometric corrections for B stars. *Mon. Not. Roy. Astr. Soc.*, 178, 651, 1977.

[271] J. Davis, W. Tango, A. Booth, E. Thorvaldson, and J. Giovannis. The Sydney University Stellar Interferometer. II. Commissioning observations and results. *Mon. Not. Roy. Astr. Soc.*, 303, 783, 1999.

[272] J. De Greve, R. Blomme, and H. Hensberge, editors. *Stellar Atmospheres: Theory and Observations*. Lecture Notes in Physics. (Berlin: Springer-Verlag), 1997.

[273] E. Depagne, V. Hill, M. Spite, F. Spite, B. Plez, T. Beers, B. Barbuy, R. Cayrel, J. Andersen, P. Bonifacio, P. François, B. Nordström, and F. Primas. First stars. II. Elemental abundances in the extremely metal-poor star CS 22949–037. A diagnostic of early massive supernovae. *Astr. Astrophys.*, 390, 187, 2002.

[274] G. Deridder and W. van Rensbergen. Tables of damping constants of spectral lines broadened by H and He. *Astr. Astrophys. Suppl.*, 23, 147, 1976.

[275] R. Dicke and J. Wittke. *Introduction to Quantum Mechanics*. (Reading: Addison-Wesley Publishing Company), 1960.

[276] M. Dimitrijević and S. Sahal-Bréchot. Stark broadening of neutral helium lines. *J. Quantit. Spectrosc. Radiat. Transf.*, 31, 301, 1984.

[277] M. Dimitrijević and S. Sahal-Bréchot. Stark broadening of neutral He lines of astrophysical interest. Regularities within spectral series. *Astr. Astrophys.*, 136, 289, 1984.

[278] M. Dimitrijević and S. Sahal-Bréchot. Comparison of measured and calculated Stark broadening for neutral He lines. *Phys. Rev. A*, 31, 316, 1985.

[279] M. Dimitrijević and S. Sahal-Bréchot. Stark broadening of He I lines. *Astr. Astrophys. Suppl.*, 82, 519, 1990.

[280] P. Dirac. The quantum theory of the emission and absorption of radiation. *Proc. Roy. Soc. London*, A114, 243, 1927.

[281] P. Dirac. The quantum theory of dispersion. *Proc. Roy. Soc. London*, A114, 710, 1927.

[282] P. Dirac. The quantum theory of the electron. *Proc. Roy. Soc. London*, A117, 610, 1928.

[283] P. Dirac. The quantum theory of the electron. Part II. *Proc. Roy. Soc. London*, A118, 351, 1928.

[284] P. Dirac. Quantum mechanics of many-electron systems. *Proc. Roy. Soc. London*, A123, 714, 1929.

[285] P. Dirac. The basis of statistical quantum mechanics. *Proc. Cambridge Phil. Soc.*, 25, 62, 1929.

[286] P. Dirac. *The Principles of Quantum Mechanics*. (Oxford: Clarendon Press), 4th edition, 1958.

[287] E. Dorfi and L. Drury. Simple adaptive grids for 1-D initial value problems. *J. Comp. Phys.*, 69, 175, 1987.

[288] N. Doughty and P. Fraser. The free-free absorption coefficient of the negative hydrogen ion. *Mon. Not. Roy. Astr. Soc.*, 132, 267, 1966.

[289] H. Drawin. Influence of atom-atom collisions on the collisional-radiative ionization and recombination coefficients of hydrogen plasmas. *Z. für Phys.*, 225, 483, 1969.

[290] S. Dreizler. Spectral analysis of extremely He-rich subdwarf O-stars. *Astr. Astrophys.*, 273, 212, 1993.

[291] S. Dreizler. Temperature correction schemes. In Hubeny et al. [536], page 69.

[292] S. Dreizler and K. Werner. Line blanketing by Fe group elements in non-LTE model atmospheres for hot stars. *Astr. Astrophys.*, 278, 199, 1993.

[293] P. Dufton. An early-type metal-rich star, HD 135485. *Astr. Astrophys.*, 28, 267, 1973.

[294] H. Dwight. *Tables of Integrals and Other Mathematical Data*. (New York: Macmillan Publishing Co.), 4th edition, 1961.

[295] M. Dworetsky, F. Castelli, and R. Faraggizna, editors. *Peculiar Versus Normal Phenomena in A-type and Related Stars*. (San Francisco: Astronomical Society of the Pacific), 1993.

[296] P. Dykema, R. Klein, and J. Castor. A new scheme for multidimensional line transfer. III. A two-dimensional Lagrangian variable tensor method with discontinuous finite-element S_n transport. *Astrophys. J.*, 457, 892, 1996.

[297] G. Ecker. Zur statistischen Beschreibung von gesamtheiten mit kollecktiver Wechselwirkung. I. Grundlagen und Grenzen kollecktiver Beschreibung. *Z. für Phys.*, 140, 274, 1955.

[298] G. Ecker. Zur statistischen Beschreibung von gesamtheiten mit kollecktiver Wechselwirkung. III. Die verschiedenen Formulierungen der Tragerkinetik. *Z. für Phys.*, 141, 294, 1955.

[299] G. Ecker. Das Mikrofeld in gesamtheiten mit Coulombscher Wechselwirkung. *Z. für Phys.*, 148, 593, 1957.

[300] G. Ecker. Abweichungen von dem Holtsmark-Profil der Balmer-Linen im Plasma. *Z. für Phys.*, 149, 254, 1957.

[301] A. Eddington. The formation of absorption lines. *Mon. Not. Roy. Astr. Soc.*, 89, 620, 1929.

[302] A. Eddington. *The Internal Constitution of the Stars*. (New York: Dover Publications), 1959.

[303] A. Edmonds. *Angular Momentum in Quantum Mechanics*. (Princeton: Princeton University Press), 1957.

[304] O. Eggen. On the existence of subdwarfs in the (M_{bol}, log T_{eff}) plane. III. *Astrophys. J.*, 182, 821, 1973.

[305] O. Eggen, D. Lynden-Bell, and A. Sandage. Formation of the Milky Way. *Astrophys. J.*, 136, 748, 1962.

[306] O. Eggen and A. Sandage. On the existence of subdwarfs in the (M_{bol}, log T_{eff}) plane. II. *Astrophys. J.*, 136, 735, 1662.

[307] A. Einstein. Über einen die Erzeugung und Verwandlung des Lichtes betreffenden heuristischen Gesichtspunkt. *Ann. Physik*, 17, 132, 1905. English translation in Beck and Havas [99].

[308] A. Einstein. Beiträge zur Quantentheorie. *Deutsche Phys. Gesells. Verhandlungen*, 16, 820, 1914. English translation in Engel and Schucking [319].

[309] A. Einstein. Strahlungs-Emission und -Absorption nach der Quantentheorie. *Deutsche Phys. Gesells. Verhandlungen*, 18, 318, 1916. English translation in Engel and Schucking [319].

[310] A. Einstein. Zur Quantentheorie des Strahlung. *Phys. Gessels. Zürich Mitteilungen*, 18, 47, 1916. English translation in Engel and Schucking [319].

[311] A. Einstein. Zur Quantentheorie des Strahlung. *Z. für Phys.*, 18, 121, 1917.

[312] A. Einstein. Quantentheorie des einatomigen idealen Gases. *Berliner Berichte*, page 261, 1924.

[313] A. Einstein. Quantentheorie des einatomigen idealen Gases. *Berliner Berichte*, page 3, 1925.

[314] A. Einstein. *The Meaning of Relativity*. (Princeton: Princeton University Press), 1955.

[315] S. Eisenstadt, H. Eleman, and M. Schultz. Variational iterative methods for nonsymmetric systems of linear equations. *SIAM J. Num. Anal.*, 20, 345, 1983.

[316] E. Eissner, M. Jones, and H. Nussbaumer. Techniques for the calculation of atomic structures and radiative data including relativistic corrections. *Comp. Phys. Commun.*, 8, 270, 1974.

[317] J. Elias, R. Bell, K. Matthews, and G. Neugebauer. Infrared measurements of metal-poor subdwarfs and a comparison with model atmospheres. *Pub. Astr. Soc. Pacific*, 101, 1121, 1989.

[318] R. Emden. *Gaskugeln*. (Leipzig), 1907.

[319] A. Engel and E. Schucking, editors. *Collected Papers of Albert Einstein, Volume 6. The Berlin Years: 1914–1917*. (Princeton: Princeton University Press), 1997.

[320] P. Epstein. Zur Theorie des Starkeffektes. *Ann. Physik*, 355, 489, 1916.

[321] U. Fano. Description of states in quantum mechanics by density matrix and operator techniques. *Rev. Mod. Phys.*, 29, 74, 1957.

[322] N. Feautrier, N. Tran-Minh and H. Van Regemorter. The quantum unified theory applied to the Lyman-alpha profile calculation. *J. Phys. B*, 9, 1871, 1976.

[323] P. Feautrier. Sur la résolution numerique de l'équations de transfert. *C. R. Acad. Sci. Paris, Ser. B*, 258, 3189, 1964.

[324] P. Feautrier. Théorie des classifications stellaires. I. Construction de modèles en équilibre thermodynamique local. *Ann. d'Astrophys.*, 30, 125, 1967.

[325] P. Feautrier. Théorie des classifications stellaires. II. Étude des écarts à l'équilibre thermodynamique local dans le continu. *Ann. d'Astrophys.*, 31, 257, 1968.

[326] E. Feenberg and G. Pake. *Notes on the Quantum Theory of Angular Momentum*. (Stanford: Stanford University Press), 1959.

[327] S. Feltzing and B. Gustafsson. Abundances in metal-rich stars: Detailed abundance analysis of 47 G and K dwarf stars with $[M/H] > 0.10$ dex. *Astr. Astrophys. Suppl.*, 129, 237, 1998.

[328] E. Fermi. Quantization of the monatomic perfect gas. *Atti della R. Accad. Naz. dei Lincei*, 3, 145, 1926.

[329] E. Fermi. Eine statistische Methode zur Bestimmung einiger Eigenschaften des Atomes und ihre Anwendung auf die Theorie des Periodische Systems der Elemente. *Z. für Phys.*, 48, 73, 1928.

[330] R. Feynman. Forces in molecules. *Phys. Rev.*, 56, 340, 1939.

[331] G. Finn. Frequency redistribution on scattering. *Astrophys. J.*, 147, 1085, 1967.

[332] G. Finn. Probability distribution for photon exit. *J. Quantit. Spectrosc. Radiat. Transf.*, 12, 35, 1972.

[333] G. Finn. Statistical functions in radiative transfer. *J. Quantit. Spectrosc. Radiat. Transf.*, 12, 149, 1972.

[334] G. Finn. Probability distributions for photon exit: Photons formed with specified frequency. *J. Quantit. Spectrosc. Radiat. Transf.*, 12, 1217, 1972.

[335] G. Finn and D. Mugglestone. Tables of the line broadening function $H(a, v)$. *Mon. Not. Roy. Astr. Soc.*, 129, 221, 1965.

[336] J. Fiutak, and J. Van Kranendonk. Impact theory of Raman line broadening. *Canadian J. Phys.*, 40, 1085, 1062.

[337] P. Flower. Bolometric corrections for late-type giants and supergiants. *Astr. Astrophys.*, 41, 391, 1975.

[338] V. Fock. Näherungsmethode zur Lösung des quantenmechanischen Mehrkörperproblems. *Z. für Phys.*, 61, 126, 1930.

[339] H. Foley. The pressure broadening of spectral lines. *Phys. Rev.*, 69, 616, 1946.

[340] R. Fowler. Statistical equilibrium with special reference to the mechanism of ionization by electronic impact. *Phil. Mag.*, 47, 257, 1924.

[341] R. Fowler. On dense stars. *Mon. Not. Roy. Astr. Soc.*, 87, 114, 1926.

[342] R. Fowler and E. Guggenheim. *Statistical Thermodynamics*. (Cambridge: Cambridge University Press), 1956.

[343] R. Fowler and E. Milne. The intensities of absorption lines in stellar spectra, and the temperatures and pressures in the reversing layers of stars. *Mon. Not. Roy. Astr. Soc.*, 84, 403, 1923.

[344] R. Fowler and E. Milne. The maxima of absorption lines in stellar spectra. II. *Mon. Not. Roy. Astr. Soc.*, 84, 499, 1924.

[345] L. Fox, editor. *Numerical Solution of Ordinary and Partial Differential Equations*. (Oxford: Pergamon Press), 1962.

[346] L. Fox. *An Introduction to Numerical Linear Algebra*. (New York: Oxford University Press), 1965.

[347] P. François, E. Depagne, V. Hill, M. Spite, F. Spite, B. Plez, T. Beers, B. Barbuy, R. Cayrel, J. Andersen, P. Bonifacio, P. Molaro, B. Nordström, and F. Primas. First stars. III. A detailed elemental abundance study of four extremely metal-poor giant stars. *Astr. Astrophys.*, 1105, 403, 2003.

[348] D. Friend and D. Abbott. The theory of radiatively driven stellar winds. III. Wind models with finite disk correction and rotation. *Astrophys. J.*, 311, 701, 1986.

[349] H. Frisch. Non-LTE transfer V. Asymptotics of partial redistribution. *Astr. Astrophys.*, 83, 166, 1980.

[350] U. Frisch and H. Frisch. Non-LTE transfer. $\sqrt{\epsilon}$ revisited. *Mon. Not. Roy. Astr. Soc.*, 173, 167, 1975.

[351] C. Froese-Fischer. The MCHF atomic structure package. *Comp. Phys. Commun.*, 64, 369, 1991.

[352] C. Fröhlich. Observations of irradiance variability. *Space Sci. Rev.*, 94, 15, 2000.

[353] L. Frommhold. *Collision-Induced Absorption in Gases*. (New York: Cambridge University Press), 1993.

[354] R. Gabler, A. Gabler, R.-P. Kudritzki, and R. Méndez. Unified NLTE model atmospheres including spherical extension and stellar winds. III. The EUV fluxes of hot massive stars and He II emission in extragalactic giant H II regions. *Astr. Astrophys.*, 265, 656, 1992.

[355] R. Gabler, A. Gabler, R.-P. Kudritzki, J. Puls, and A. Pauldrach. Unified NLTE model atmospheres including spherical extension and stellar winds: Method and first results. *Astr. Astrophys.*, 226, 162, 1989.

[356] R. Gabler, A. Gabler, R.-P. Kudritzki, J. Puls, and A. Pauldrach. Unified NLTE model atmospheres including spherical extension and stellar winds. In Garmany [362], page 64.

[357] A. Gabler, R. Gabler, A. Pauldrach, J. Puls, and R.-P. Kudritzki. On the use of Hα and He II λ 4686 as mass-loss and luminosity indicators for hot stars. In Garmany [362], page 218.

[358] R. Gabler, R.-P. Kudritzki, and R. Méndez. Unified NLTE model atmospheres including spherical extension and stellar winds. II. EUV fluxes and the He II Zanstra discrepancy in central stars of planetary nebulae. *Astr. Astrophys.*, 245, 587, 1991.

[359] J. Gallagher, editor. *Luminous Blue Variable Stars*. (San Francisco: Astronomical Society of the Pacific), 1997.

[360] G. Gamow. Zur Quantentheorie des Atomkernes. *Z. für Phys.*, 51, 204, 1928.

[361] G. Gamow. Zur Quantentheorie der Atomzertrümmerung. *Z. für Phys.*, 52, 510, 1928.

[362] C. Garmany, editor. *Properties of Hot Luminous Stars: Boulder-Munich Workshop on Stellar Atmospheres*. (San Francisco: Astronomical Society of the Pacific), 1990.

[363] R. Garrison, editor. *The MK Process and Stellar Classification*. (Toronto: University of Toronto), 1984.

[364] R. Garrison. The use and abuse of standard stars. In Hayes et al. [464], page 17.

[365] J. Gaunt. Continuous absorption. *Phil. Trans. Roy. Soc. London*, A229, 163, 1930.

[366] J. Gaunt. Radiation of free electrons in a Coulombian field. *Z. für Phys.*, 59, 508, 1930.

[367] M. Gavrila. Elastic scattering of photons by a hydrogen atom. *Phys. Rev.*, 163, 147, 1967.

[368] K. Gayley. An improved line-strength parameterization in hot-star winds. *Astrophys. J.*, 454, 410, 1995.

[369] K. Gayley and S. Owocki. Acceleration efficiency in line-driven flows. *Astrophys. J.*, 434, 684, 1994.

[370] K. Gayley, S. Owocki, and S. Cranmer. Momentum deposition in Wolf-Rayet winds: Nonisotropioc diffusion with effective gray opacity. *Astrophys. J.*, 442, 296, 1995.

[371] S. Geltman. Free-free radiation in electron-neutral atom collisions. *J. Quantit. Spectrosc. Radiat. Transf.*, 13, 601, 1973.

[372] B. Gerasimovic. Note on the deviation of stellar atmospheres from thermodynamic equilibrium. *Mon. Not. Roy. Astr. Soc.*, 89, 272, 1929.

[373] B. Gerasimovic. The contours of emission lines in expanding nebular envelopes. *Z. für Astrophys.*, 7, 335, 1933.

[374] M. Gerbaldi, R. Faraggiana, R. Burnage, F. Delmas, A. Gómez, and S. Grenier. Search for reference A0 dwarf stars: Masses and luminosities revisited with HIPPARCOS parallaxes. *Astr. Astrophys. Suppl.*, 137, 273, 1999.

[375] D. Gezari, A. Labeyrie, and R. Stachnik. Speckle interferometry: Diffraction-limited measurements of nine stars with the 200-inch telescope. *Astrophys. J.*, 173, L1, 1972.

[376] H. Gieske and H. Griem. Calculated electron and ion Stark broadening of the allowed and forbidden $2P - nL$ $(n \geq 5, L = 1, 2, \ldots, n-1)$ triplet and singlet transitions in neutral He. *Astrophys. J.*, 157, 963, 1969.

[377] R. Glass and A. Hibbert. Relativistic effects in many-electron atoms. *Comp. Phys. Commun.*, 16, 19, 1978.

[378] I. Glushneva. Synthetic color indices and energy distribution in the spectra of solar type stars. *Astr. Reports*, 38, 578, 1994.

[379] L. Goldberg, E. Müller, and L. Aller. The abundances of the elements in the solar atmosphere. *Astrophys. J. Suppl.*, 5, 1, 1960.

[380] H. Goldstein. *Classical Mechanics*. (Reading: Addison-Wesley Publishing Company), 1950.

[381] G. Gontcharov and O. Kiyaeva. Astrometric orbits from a direct combination of ground-based catalogs with the HIPPARCOS catalog. *Astr. Let.*, 28, 261, 2002.

[382] W. Gordon. Zur Berecnung der Matrizen beim Wasserstoffatom. *Ann. Physik*, 2, 1031, 1929.

[383] S. Goudsmit and G. Uhlenbeck. Die Kopplungsmöglichkeiten der Quantenvektoren im Atom. *Z. für Phys.*, 35, 618, 1926.

[384] R. Gould. Boltzmann equation for a photon gas interacting with a plasma. *Annals Phys.*, 69, 321, 1972.

[385] P. Gouttebroze, P. Lemaire, J.-C. Vial, and G. Artzner. *Astrophys. J.*, 225, 655, 1978.

[386] I. Gradshyteyn and I. Ryzhik. *Table of Integrals, Series, and Products*. (San Diego: Academic Press), 5th edition, 1994.

[387] G. Gräfener and W.-R. Hamann. Hydrodynamic model atmospheres for WR stars. Self-consistent modeling of a WC star wind. *Astr. Astrophys.*, 432, 633, 2005.

[388] G. Gräfener, L. Koesterke, and W.-R. Hamann. Line-blanketed model atmospheres for WR stars. *Astr. Astrophys.*, 387, 244, 2002.

[389] I. Grant. Calculation of Gaunt factors for free-free transitions near positive ions. *Mon. Not. Roy. Astr. Soc.*, 118, 241, 1958.

[390] R. Gratton, F. Fusi-Pecci, E. Carretta, E. Clementini, C. Corsi, and M. Lattanzi. Ages of globular clusters from HIPPARCOS parallaxes of local subdwarfs. *Astrophys. J.*, 491, 749, 1997.

[391] D. Gray. *The Observation and Analysis of Stellar Photospheres*. (Cambridge: Cambridge University Press), 3rd edition, 2008.

[392] L. Green, P. Rush, and C. Chandler. Oscillator strengths and matrix elements for the electric dipole moment for hydrogen. *Astrophys. J. Suppl.*, 3, 37, 1957.

[393] R. Green. *Spherical Astronomy*. (Cambridge: Cambridge University Press), 1985.

[394] R. Greene and J. Cooper. A unified theory of Stark broadening for hydrogenic ions. II. Line wings. *J. Quantit. Spectrosc. Radiat. Transf.*, 15, 1037, 1975.

[395] R. Greene and J. Cooper. A unified theory of Stark broadening for hydrogenic ions. III. Results for He II Lyα. *J. Quantit. Spectrosc. Radiat. Transf.*, 15, 1045, 1975.

[396] R. Greene, J. Cooper and E. Smith A unified theory of Stark broadening for hydrogenic ions. I. A general theory (including time ordering). *J. Quantit. Spectrosc. Radiat. Transf.*, 15, 1025, 1975.

[397] J. Greenstein, editor. *Stellar Atmospheres*. (Chicago: University of Chicago Press), 1960.

[398] W. Greiner and J. Reinhardt. *Quantum Electrodynamics*. (Berlin: Springer-Verlag), 1994.

[399] H. Griem. Stark broadening of higher H and H-like lines by electrons and ions. *Astrophys. J.*, 132, 883, 1960.

[400] H. Griem. *Plasma Spectroscopy*. (New York: McGraw-Hill Book Company), 1964.

[401] H. Griem. Calculated electron and ion Stark broadening of the allowed and forbidden $2\,^3P - 4\,^3P, ^3D, ^3F$ transitions in neutral He. *Astrophys. J.*, 154, 1111, 1968.

[402] H. Griem. Semi-empirical formulas for the electron-impact widths and shifts of isolated ion lines in plasmas. *Phys. Rev.*, 165, 258, 1968.

[403] H. Griem. *Spectral Line Broadening by Plasmas*. (New York: Academic Press), 1974.

[404] H. Griem. *Principles of Plasma Spectroscopy*. (Cambridge: Cambridge University Press), 1997.

[405] H. Griem, M. Baranger, A. Kolb, and G. Oertel. Stark broadening of neutral He lines in a plasma. *Phys. Rev.*, 125, 177, 1962.

[406] R. Griffin. *A Photometric Atlas of the Spectrum of Arcturus λλ 3600–8825*. (Cambridge: Cambridge Philosophical Society), 1968.

[407] R. Griffin and R. Griffin. *A Photometric Atlas of the Spectrum of Procyon λλ 3140–7470Å*. (Cambridge: Institute of Astronomy), 1979.

[408] H.-G. Groth and P. Wellmann, editors. *Spectrum Formation in Stars With Steady-State Extended Atmospheres*. NBS Special Publication No. 332. (Washington, DC: U.S. Government Printing Office), 1970.

[409] F. Grupp. MAFAGS-OS: New opacity sampling model atmospheres for A, F, and G stars I. The model and the solar flux. *Astr. Astrophys.*, 420, 289, 2004.

[410] F. Grupp. MAFAGS-OS: New opacity sampling model atmospheres for A, F, and G stars II. Temperature determination and three "standard" stars. *Astr. Astrophys.*, 426, 309, 2004.

[411] J. Gruschinske and R.-P. Kudritzki. Spherical extended non-LTE model atmospheres of low gravity subluminous O-stars. *Astr. Astrophys.*, 77, 341, 1979.

[412] T. Gull, S. Johannson, and K. Davidson, editors. *Eta Carinae and Other Mysterious Stars.* (San Francisco: Astronomical Society of the Pacific), 2001.

[413] R. Gurner and E. Condon. Wave mechanics and radioactive disintegration. *Nature*, 122, 439, 1928.

[414] B. Gustafsson. A Feautrier-type method for model atmospheres including convection. *Astr. Astrophys.*, 10, 187, 1971.

[415] B. Gustafsson. Fundamental parameters and models of stellar atmospheres. In Hayes et al. [464], page 303.

[416] B. Gustafsson. Chemical analyses of cool stars. *Ann. Rev. Astr. Astrophys.*, 27, 701, 1989.

[417] B. Gustafsson. Opacity incompleteness and atmospheres of cool stars. In Adelman and Wiese [10], page 347.

[418] B. Gustafsson. Molecules in stellar atmospheres. *Astrophys. Space Sci.*, 255, 241, 1997.

[419] B. Gustafsson. The current status in the modeling of stellar atmospheres. In Piskunov et al. [881], page 3.

[420] B. Gustafsson and R. Bell. The colors of G and K type giants. I. *Astr. Astrophys.*, 74, 313, 1979.

[421] B. Gustafsson, R. Bell, K. Eriksson, and Å. Nordlund. A grid of model atmospheres for metal-deficient giant stars. I. *Astr. Astrophys.*, 42, 407, 1975.

[422] B. Gustafsson, B. Edvardsson, K. Eriksson, U. Jørgensen, Å. Nordlund, and B. Plez. A grid of MARCS model atmospheres for late-type stars I. Methods and general properties. *Astr. Astrophys.*, 486, 951, 2008.

[423] B. Gustafsson, K. Fredga, G. Gahm, and R. Bell. The ultraviolet flux of HD 122563. *Astr. Astrophys.*, 89, 255, 1980.

[424] B. Gustafsson and U. Jørgensen. Models of late-type stellar photospheres. *Astr. Astrophys. Rev.*, 6, 19, 1994.

[425] B. Gustafsson, T. Karlsson, E. Olsson, B. Edvardsson, and N. Ryde. The origin of carbon, investigated by spectral analysis of solar-type stars in the galactic disk. *Astr. Astrophys.*, 342, 426, 1999.

[426] B. Gustafsson and P. Nissen. The metal-to-H ratio in F1–F5 stars, as determined by a model-atmosphere analysis of photoelectric observations of a group of weak metal lines. *Astr. Astrophys.*, 19, 261, 1972.

[427] S. Haas, S. Dreizler, U. Heber, S. Jeffery, and K. Werner. Iron and nickel abundances of subluminous O-stars. I. NLTE model atmospheres with line blanketing by Fe group elements. *Astr. Astrophys.*, 311, 669, 1996.

[428] H. Haken and H. Wolf. *The Physics of Atoms and Quanta.* (Berlin: Springer-Verlag), 5th edition, 1996.

[429] W.-R. Hamann. Line formation in expanding atmospheres: On the validity of the Sobolev approximation. *Astr. Astrophys.*, 93, 353, 1981.

[430] W.-R. Hamann. Line formation in expanding atmospheres: Accurate solution using approximate lambda operators. *Astr. Astrophys.*, 148, 364, 1985.

[431] W.-R. Hamann. Line formation in expanding atmospheres: Multi-level calculations using approximate lambda operators. *Astr. Astrophys.*, 160, 347, 1986.

[432] W.-R. Hamann. Basic ALI in moving atmospheres. In Hubeny et al. [536], page 171.

[433] W.-R. Hamann, L. Koesterke, and U. Wessolowski. Iteration with approximate lambda operators, and its application to the expanding atmospheres of WR stars. In Crivellari et al. [264], page 69.

[434] W.-R. Hamann, L. Koesterke, and U. Wessolowski. Spectral atlas of the galactic Wolf-Rayet stars (WN sequence). *Astr. Astrophys. Suppl.*, 113, 459, 1995.

[435] W.-R. Hamann, U. Wessolowski, W. Schmutz, E. Schwarz, G. Dünnebeil, L. Koesterke, E. Baum, and U. Leuenhagen. Analyses of Wolf-Rayet stars. *Rev. Mod. Astr.*, 3, 174, 1990.

[436] W.-R. Hamann, U. Wessolowski, E. Schwarz, G. Dünnebeil, and W. Schmutz. High-density winds: Wolf-Rayet stars – A progress report about quantitative spectral analyses. In Garmany [362], page 259.

[437] C. Hansen, S. Kawaler, and V. Trimble. *Stellar Interiors: Physical Principles, Structure, and Evolution*. (New York: Springer-Verlag), 2nd edition, 2004.

[438] R. Härm and M. Schwarzschild. Red giants of Population II. III. *Astrophys. J.*, 139, 594, 1964.

[439] R. Härm and M. Schwarzschild. Red giants of Population II. IV. *Astrophys. J.*, 145, 496, 1966.

[440] R. Härm and M. Schwarzschild. Oscillatory thermal instabilities at the onset of helium shell burning. *Astrophys. J.*, 172, 403, 1972.

[441] R. Härm and M. Schwarzschild. Transition from a red giant to a blue nucleus after ejection of a planetary nebula. *Astrophys. J.*, 200, 324, 1975.

[442] J. Harrington. The scattering for resonance-line radiation in the limit of large optical depth. *Mon. Not. Roy. Astr. Soc.*, 162, 43, 1973.

[443] D. Harris. On the line-absorption coefficient due to Doppler effect and damping. *Astrophys. J.*, 108, 112, 1948.

[444] E. Harris. *A Pedestrian Approach to Quantum Field Theory*. (New York: Wiley-Interscience), 1972.

[445] D. Hartree. Wave mechanics of an atom with a non-Coulomb central field. Part I. *Math. Proc. Cambridge Philosophical Soc.*, 24, 89, 1928.

[446] D. Hartree. Wave mechanics of an atom with a non-Coulomb central field. Part II. *Math. Proc. Cambridge Philosophical Soc.*, 24, 111, 1928.

[447] A. Hauer and A. Merts, editors. *Atomic Processes in Plasmas*. (New York: American Institute of Physics), 1988.

[448] P. Hauschildt. Radiative equilibrium in rapidly expanding shells. *Astrophys. J.*, 398, 224, 1992.

[449] P. Hauschildt. A fast operator perturbation method for the solution of the special relativistic equation of transfer in spherical geometry. *J. Quantit. Spectrosc. Radiat. Transf.*, 47, 433, 1992.

[450] P. Hauschildt. Multi-level non-LTE radiative transfer in expanding shells. *J. Quantit. Spectrosc. Radiat. Transf.*, 50, 301, 1993.

[451] P. Hauschildt, F. Allard, T. Barman, A. Schweitzer, E. Baron, and S. Leggett. Cool stellar atmospheres. In Woodward et al. [1172], page 427.

[452] P. Hauschildt, F. Allard, and E. Baron. The NextGen model atmosphere grid for $3000 \leq T_{\rm eff} \leq 10{,}000$ K. *Astrophys. J.*, 512, 377, 1999.

[453] P. Hauschildt, F. Allard, J. Ferguson, E. Baron, and D. Alexander. The NextGen model atmosphere grid. II. Spherically symmetric model atmospheres for giant stars with effective temperatures between 3000 and 6000 K. *Astrophys. J.*, 525, 871, 1999.

[454] P. Hauschildt, F. Allard, A. Schweitzer, and E. Baron. Cool stellar atmospheres. In Lejeune and Fernandes [691], page 95.

[455] P. Hauschildt and E. Baron. Numerical solution of the expanding stellar atmosphere problem. *J. Comp. Appl. Math.*, 109, 41, 1999.

[456] P. Hauschildt, E. Baron, and F. Allard. Parallel implementation of the PHOENIX generalized stellar atmosphere program. *Astrophys. J.*, 483, 390, 1997.

[457] P. Hauschildt, D. Lowenthal, and E. Baron. Parallel implementation of the PHOENIX generalized stellar atmosphere program. III. A parallel algorithm for direct opacity sampling. *Astrophys. J. Suppl.*, 134, 323, 2001.

[458] P. Hauschildt, H. Störzer, and E. Baron. Convergence properties of the accelerated Λ-iteration method for the solution of radiative transfer problems. *J. Quantit. Spectrosc. Radiat. Transf.*, 51, 875, 1994.

[459] P. Hauschildt and R. Wehrse. Solution of the special relativistic equation of radiative transfer in rapidly expanding spherical shells. *J. Quantit. Spectrosc. Radiat. Transf.*, 46, 81, 1991.

[460] D. Hayes. Unpublished Ph.D. thesis, UCLA, 1968.

[461] D. Hayes. Stellar absolute fluxes and energy distributions from 0.32 to 4.0 μ. In Hayes et al. [464], page 225.

[462] D. Hayes and D. Latham. A rediscussion of the atmospheric extinction and the absolute spectral-energy distribution of Vega. *Astrophys. J.*, 197, 593, 1975.

[463] D. Hayes, D. Latham, and S. Hayes. Measurements of the monochromatic flux from Vega in the near infrared. *Astrophys. J.*, 197, 587, 1975.

[464] D. Hayes, L. Pasinetti, and A. Philip, editors. *Calibration of Fundamental Stellar Quantities*. IAU Symposium No. 111. (Dordrecht: Reidel), 1985.

[465] G. Hébrard, N. Allard, I. Hubeny, S. Lacour, R. Ferlet, and A. Vidal-Madjar. Quasi-molecular lines in Lyman wings of cool DA white dwarfs. Application to *FUSE* observations of G 231–40. *Astr. Astrophys.*, 394, 647, 2002.

[466] G. Hébrard, N. Allard, J. Kielkopf, P. Chayer, J. Dupuis, J. Kruk, and I. Hubeny. Modeling of the Lyγ satellites in FUSE spectra of DA white dwarfs. *Astr. Astrophys.*, 405, 1153, 2003.

[467] U. Heber and C. Jeffery, editors. *The Atmospheres of Early-Type Stars*. Lecture Notes in Physics (Heidelberg: Springer-Verlag), 1992.

[468] P. Heinzel. Non-coherent scattering in subordinate lines: A unified approach to redistribution functions. *J. Quantit. Spectrosc. Radiat. Transf.*, 25, 483, 1981.

[469] P. Heinzel and I. Hubeny. Non-coherent scattering in subordinate lines. II. Collisional redistribution. *J. Quantit. Spectrosc. Radiat. Transf.*, 27, 1, 1982.

[470] P. Heinzel and I. Hubeny. Non-coherent scattering in subordinate lines. IV. Angle-averaged redistribution functions. *J. Quantit. Spectrosc. Radiat. Transf.*, 30, 77, 1983.

[471] W. Heitler. *The Quantum Theory of Radiation*. (New York: Dover Publications), 3rd edition, 1984.

[472] H. Helfer, G. Wallerstein, and J. Greenstein. Abundances in some Population II K giants. *Astrophys. J.*, 129, 700, 1959.

[473] H. Helfer, G. Wallerstein, and J. Greenstein. Abundances in G dwarf stars. III. Stars in moving clusters. *Astrophys. J.*, 132, 335, 1960.

[474] H. Helfer and G. Wallerstein. Abundances in K giant stars. I. A determination of the abundances in the Hyades K giants. *Astrophys. J. Suppl.*, 9, 81, 1964.

[475] H. Helfer and G. Wallerstein. Abundances in K stars. II. A survey of field stars. *Astrophys. J. Suppl.*, 16, 1, 1968.

[476] H. Helfer, G. Wallerstein, and J. Greenstein. Metal abundances in the subgiant ζ Herculis and three other dG stars. *Astrophys. J.*, 138, 97, 1963.

[477] C. Helling and U. Jørgensen. Optimizing the opacity sampling method. *Astr. Astrophys.*, 337, 477, 1998.

[478] K. Hempe and K. Schönberg. Line formation in the comoving frame: Accurate solution using an approximate Newton-Raphson operator. *Astr. Astrophys.*, 160, 141, 1986.

[479] L. Henyey. Near thermodynamic radiative equilibrium. *Astrophys. J.*, 103, 332, 1946.

[480] L. Henyey, J. Forbes, and N. Gould. A new method for automatic computation of stellar evolution. *Astrophys. J.*, 139, 306, 1964.

[481] L. Henyey, L. Wilets, K.-H. Böhm, R. Lelevier, and R. Levee. A method for automatic computation of stellar evolution. *Astrophys. J.*, 129, 628, 1959.

[482] A. Heras, R. Shipman, S. Price, T. de Graauw, H. Walker, M. Jourdain de Muizon, M. Kessler, T. Prusti, L. Decin, B. Vandenbussche, and L. Waters. Infrared spectral classification of normal stars. *Astr. Astrophys.*, 394, 539, 2002.

[483] F. Herwig. Evolution of asymptotic giant branch stars. *Ann. Rev. Astr. Astrophys.*, 43, 435, 2005.

[484] G. Herzberg. *Infrared and Raman Spectra of Polyatomic Molecules*. (Princeton: Van Nostrand), 1945.

[485] G. Herzberg. *Molecular Spectra and Molecular Structure. I. Spectra of Diatomic Molecules*. (New York: Van Nostrand Reinhold Company), 2nd edition, 1950.

[486] G. Herzberg. *Electronic Spectra and Electronic Structure of Polyatomic Molecules*. (Princeton: Van Nostrand), 1966.

[487] G. Herzberg and L. Howe. The Lyman bands of molecular H. *Canadian J. Phys.*, 37, 636, 1959.

[488] E. Hertzsprung. Über die Sterne der Unterabteilungen *c* und *ab* nach der Spektralklassifikation von Antonia C. Maury. *Astr. Nachrichten*, 19, 91, 1909.

[489] D. Hestroffer and C. Magnan. Wavelength dependency of the solar limb darkening. *Astr. Astrophys.*, 333, 338, 1998.

[490] M. Heydari-Malayeri, P. Stee, and J.-P. Zahn, editors. *Evolution of Massive Stars, Mass Loss and Winds*. EAS Publ. Ser. 13, 2004.

[491] A. Hibbert. `CIV3` – A general program to calculate configuration interaction wave functions and electric-dipole oscillator strengths. *Comp. Phys. Commun.*, 9, 141, 1975.

[492] D. Hillier. An iterative method for the solution of the statistical and radiative equilibrium equations in expanding atmospheres. *Astr. Astrophys.*, 231, 116, 1990.

[493] D. Hillier and D. Miller. The treatment of non-LTE line blanketing in spherically expanding outflows. *Astrophys. J.*, 496, 407, 1998.

[494] W. Hiltner, editor. *Astronomical Techniques*. (Chicago: University of Chicago Press), 1962.

[495] W. Hiltner and H. Johnson. The law of interstellar absorption and reddening. *Astrophys. J.*, 124, 367, 1956.

[496] W. Hindmarsh. Collision broadening and shift in the resonance line of Ca. *Mon. Not. Roy. Astr. Soc.*, 119, 11, 1959.

[497] W. Hindmarsh. Collision broadening and shift in the λ 6573 line of Ca. *Mon. Not. Roy. Astr. Soc.*, 121, 48, 1960.

[498] W. Hindmarsh, A. Petford and G. Smith. Interpretation of collision broadening and shift in atomic spectra. *Proc. Roy. Soc. London A*, 297, 296, 1967.

[499] J. Hirschfelder, C. Curtiss, and R. Bird. *Molecular Theory of Gases and Liquids*. (New York: John Wiley & Sons), 1964.

[500] F. Hjerting. Tables facilitating the calculation of line absorption coefficients. *Astrophys. J.*, 88, 508, 1938.

[501] J. Holberg. Using DA white dwarfs to calibrate synthetic photometry. In Sterken [1052], page 533.

[502] T. Holstein. Pressure broadening of spectral lines. *Phys. Rev.*, 79, 744, 1950.

[503] J. Holtsmark. Über die Verbreiterung von Spektrallinien. *Ann. Physik*, 58, 577, 1919.

[504] C. Hooper. Electric microfield distributions in plasmas. *Phys. Rev.*, 149, 77, 1966.

[505] C. Hooper. Low-frequency component electric microfield distributions in plasmas. *Phys. Rev.*, 165, 215, 1968.

[506] C. Hooper. Asymptotic electric microfield distributions in low-frequency component plasmas. *Phys. Rev.*, 169, 193, 1968.

[507] E. Hopf. Zum Problem des Strahlungsgleichgewichts in den äusseren Schichten der Sterne. Strenge Lösung der singulären Integralgleichung von Milne. *Z. für Phys.*, 46, 374, 1927.

[508] E. Hopf. Über Strahlungsgleichgewicht in den äusseren Schichten der Sterne. II. Eindeutingkeit beweis für die Lösung des Problem von Milne. *Z. für Phys.*, 49, 155, 1928.

[509] E. Hopf. Remarks on the Schwarzschild-Milne model of the outer layers of a star. *Mon. Not. Roy. Astr. Soc.*, 90, 287, 1930.

[510] E. Hopf. Remarks on the Schwarzschild-Milne model of the outer layers of a star. II. *Mon. Not. Roy. Astr. Soc.*, 92, 863, 1932.

[511] E. Hopf. *Mathematical Problems of Radiative Equilibrium.* Cambridge Tracts in Mathematics and Mathematical Physics No. 31. (Cambridge: Cambridge University Press), 1934.

[512] N. Houk, H. Irvine, and D. Rosenbush. *An Atlas of Objective-Prism Spectra.* (University of Michigan), 1974.

[513] N. Houk and M. Newberry. *A Second Atlas of Objective-Prism Spectra.* (University of Michigan), 1984.

[514] I. Howarth, editor. *Boulder-Munich II: Properties of Hot, Luminous Stars.* (San Francisco: Astronomical Society of the Pacific), 1998.

[515] K. Huang. *Statistical Mechanics.* (New York: John Wiley & Sons), 1963.

[516] I. Hubeny. Non-coherent scattering in subordinate lines. III. Generalized redistribution functions. *J. Quantit. Spectrosc. Radiat. Transf.*, 27, 593, 1982.

[517] I. Hubeny. A modified Rybicki method and the partial coherent scattering approximation. *Astr. Astrophys.*, 145, 461, 1985.

[518] I. Hubeny. Non-LTE line transfer with partial redistribution. II. An equivalent-two-level-atom approach. *Bull. Astr. Inst. Czech.*, 36, 1, 1985.

[519] I. Hubeny. General aspects of partial redistribution and its astrophysical importance. In Beckman and Crivellari [100], page 27.

[520] I. Hubeny. Probabilistic interpretation of radiative transfer I. The $\sqrt{\epsilon}$-law. *Astr. Astrophys.*, 185, 332, 1987.

[521] I. Hubeny. A computer program for calculating non-LTE model stellar atmospheres. *Comp. Phys. Commun.*, 52, 103, 1988.

[522] I. Hubeny. Accelerated lambda iteration. In Heber and Jeffery [467], page 377.

[523] I. Hubeny. Accelerated lambda iteration: An overview. In Hubeny et al. [536], page 17.

[524] I. Hubeny. From complete linearization to ALI and beyond. In Hubeny et al. [541], page 3.

[525] I. Hubeny, O. Blaes, J. Krolik, and E. Agol. Non-LTE models and theoretical spectra of accretion disks in active galactic nuclei. IV. Effects of Compton scattering and metal opacities. *Astrophys. J.*, 559, 680, 2001.

[526] I. Hubeny and A. Burrows. A new algorithm for two-dimensional transport for astrophysical simulations. I. General formulation and tests for one-dimensional spherical case. *Astrophys. J.*, 659, 1458, 2007.

[527] I. Hubeny, A. Burrows, and D. Sudarsky. A possible bifurcation in atmospheres of strongly irradiated stars and planets. *Astrophys. J.*, 594, 1011, 2003.

[528] I. Hubeny and J. Cooper. Redistribution of radiation in the presence of velocity-changing collisions. *Astrophys. J.*, 305, 852, 1986.

[529] I. Hubeny, S. Heap, and T. Lanz. Non-LTE line-blanketed model atmospheres of O stars. In Howarth [514], page 108.

[530] I. Hubeny and P. Heinzel. Non-coherent scattering in subordinate lines. V. Solutions of the transfer problem. *J. Quantit. Spectrosc. Radiat. Transf.*, 32, 159, 1984.

[531] I. Hubeny, D. Hummer, and T. Lanz. NLTE model stellar atmospheres with line blanketing near the series limits. *Astr. Astrophys.*, 282, 151, 1994.

[532] I. Hubeny and T. Lanz. Accelerated complete-linearization method for calculating NLTE model stellar atmospheres. *Astr. Astrophys.*, 262, 501, 1992.

[533] I. Hubeny and T. Lanz. Non-LTE line-blanketed model atmospheres of hot stars. I. Hybrid complete linearization/accelerated lambda iteration method. *Astrophys. J.*, 439, 875, 1995.

[534] I. Hubeny and T. Lanz. Model photospheres with accelerated lambda iteration. In Hubeny et al. [536], page 51.

[535] I. Hubeny and B. Lites. Partial redistribution in multilevel atoms. I. Method and application to the solar H-line formation. *Astrophys. J.*, 455, 376, 1995.

[536] I. Hubeny, D. Mihalas, and K. Werner, editors. *Stellar Atmosphere Modeling*. (San Francisco: Astronomical Society of the Pacific), 2003.

[537] I. Hubeny and J. Oxenius. Absorption and emission line-profile coefficients of multilevel atoms. III. Generalized atomic redistribution functions for three-photon processes. *J. Quantit. Spectrosc. Radiat. Transf.*, 37, 65, 1987.

[538] I. Hubeny and J. Oxenius. Absorption and emission line-profile coefficients of multilevel atoms. IV. Velocity-averaged generalized redistribution functions for three-photon processes. *J. Quantit. Spectrosc. Radiat. Transf.*, 37, 397, 1987.

[539] I. Hubeny, J. Oxenius, and E. Simonneau. Absorption and emission line profile coefficients of multilevel atoms. I. Atomic profile coefficients. *J. Quantit. Spectrosc. Radiat. Transf.*, 29, 477, 1983.

[540] I. Hubeny, J. Oxenius, and E. Simonneau. Absorption and emission line profile coefficients of multilevel atoms. II. Velocity-averaged profile coefficients. *J. Quantit. Spectrosc. Radiat. Transf.*, 29, 495, 1983.

[541] I. Hubeny, J. Stone, K. MacGregor, and K. Werner, editors. *Recent Directions in Astrophysical Quantitative Spectroscopy and Radiation Hydrodynamics*. (New York: American Institute of Physics), 2009.

[542] K. Huber and G. Herzberg. *Molecular Spectra and Molecular Structure IV. Constants of Diatomic Molecules*. (New York: Van Nostrand Reinhold Co.), 1979.

[543] J. Humlíček. Optimized computation of the Voigt and complex probability functions. *J. Quantit. Spectrosc. Radiat. Transf.*, 27, 437, 1982.

[544] D. Hummer. Non-coherent scattering. I. The redistribution functions with Doppler broadening. *Mon. Not. Roy. Astr. Soc.*, 125, 21, 1962.

[545] D. Hummer. The Voigt function. An eight significant-figure table and generating procedure. *Mem. Roy. Astr. Soc.*, 70, 1, 1965.

[546] D. Hummer. Non-coherent scattering. III. The effect of continuous absorption on the formation of spectral lines. *Mon. Not. Roy. Astr. Soc.*, 138, 73, 1968.

[547] D. Hummer. Non-coherent scattering. VI. Solutions of the transfer problem with a frequency-dependent source function. *Mon. Not. Roy. Astr. Soc.*, 145, 95, 1969.

[548] D. Hummer. Line formation in expanding atmospheres. In Slettebak [1014], page 281.

[549] D. Hummer. A fast and accurate method for evaluating the nonrelativistic free-free Gaunt factor for hydrogenic ions. *Astrophys. J.*, 327, 477, 1988.

[550] D. Hummer, C. Kunasz, and P. Kunasz. Numerical evaluation of the formal solution of radiative transfer problems in spherical geometries. *Comp. Phys. Commun.*, 6, 38, 1973.

[551] D. Hummer and G. Rybicki. Non-coherent scattering. VII. Frequency dependent thermalization lengths and scattering with continuous absorption. *Mon. Not. Roy. Astr. Soc.*, 150, 419, 1970.

[552] D. Hummer and G. Rybicki. Radiative transfer in spherically symmetric systems. The conservative grey case. *Mon. Not. Roy. Astr. Soc.*, 152, 1, 1971.

[553] D. Hummer and G. Rybicki. The Sobolev approximation for line formation with continuous opacity. *Astrophys. J.*, 293, 258, 1985.

[554] A. Hundhausen. *Coronal Expansion and Solar Wind.* (New York: Springer-Verlag), 1972.

[555] G. Hunt, I. Grant, J. Houghton, A. Underhill, and M. Williams, editors. *Transport Theory.* Atlas Symposium No. 3. *J. Quantit. Spectrosc. & Radiat. Transf.*, 11, 511, 1971.

[556] G. Hunter, A. Yau, and H. Pritchard. Rotation-vibration level energies of the hydrogen and deuterium ion molecules. *Atomic Data and Nuclear Data Tables*, 14, 11, 1974.

[557] I. Iben. Stellar evolution within and off the main sequence. *Ann. Rev. Astr. Astrophys.*, 5, 571, 1967.

[558] I. Iben. Post main sequence evolution of single stars. *Ann. Rev. Astr. Astrophys.*, 12, 215, 1976.

[559] I. Iben and A. Renzini. Asymptotic giant branch evolution and beyond. *Ann. Rev. Astr. Astrophys.*, 21, 271, 1983.

[560] A. Illarionov and R. Syunyaev. Comptonization, characteristic radiation spectra, and thermal balance of low-density plasma. *Sov. Astr.*, 18, 413, 1975.

[561] A. Illarionov and R. Syunyaev. Comptonization, the background-radiation spectrum, and the thermal history of the universe. *Sov. Astr.*, 18, 691, 1975.

[562] F. Irons. On the equality of the mean escape probability and mean net radiative bracket for line photons. *Mon. Not. Roy. Astr. Soc.*, 182, 705, 1978.

[563] F. Irons. The escape factor in plasma spectroscopy. I. The escape factor defined and evaluated. *J. Quantit. Spectrosc. Radiat. Transf.*, 22, 1, 1979.

[564] F. Irons. The escape factor in plasma spectroscopy. II. The case of radiative decay. *J. Quantit. Spectrosc. Radiat. Transf.*, 22, 21, 1979.

[565] J. Isern, M. Hernanz, and E. García-Berro, editors. *White Dwarfs.* (Dordrecht: Reidel), 1997.

[566] Y. Itakawa, editor. *Photon and Electron Interactions with Atoms, Molecules, and Ions.* (New York: Springer-Verlag), 2001.

[567] N. Itoh. Relativistic free-free Gaunt factors for high-temperature stellar plasmas. *Rev. Mex. Astr. Astrof.*, 23, 911, 1992.

[568] N. Itoh, K. Kojo, and M. Nakagawa. Relativistic free-free Gaunt factor of the dense high-temperature stellar plasma II. Carbon and oxygen plasmas. *Astrophys. J. Suppl.*, 74, 291, 1990.

[569] N. Itoh, F. Kuwashima, K. Ichihashi, and H. Mutoh. The Rosseland mean free-free Gaunt factor of the dense high-temperature stellar plasma. *Astrophys. J.*, 382, 636, 1991.

[570] N. Itoh, M. Nakagawa, and Y. Kohyama. Relativistic free-free opacity for a high-temperature stellar plasma. *Astrophys. J.*, 294, 17, 1985.

[571] V. Ivanov. Diffusion of resonance radiation in stellar atmospheres and nebulae. I. Semi-infinite medium. *Sov. Astr.*, 6, 793, 1963.

[572] V. Ivanov. Diffusion of resonance radiation in stellar atmospheres and nebulae. II. A layer of finite thickness. *Sov. Astr.*, 7, 199, 1963.

[573] V. Ivanov. *Transfer of Radiation in Spectral Lines*. NBS Special Publication No. 385. (Washington, DC: U.S. Government Printing Office), 1973.

[574] V. Ivanov. Analytical methods of line formation theory: are they still alive? In Crivellari et al. [264], page 583.

[575] J. Jackson. *Classical Electrodynamics*. (New York: John Wiley & Sons), 2nd edition, 1975.

[576] M. Jammer. *The Conceptual Development of Quantum Mechanics*. (New York: McGraw-Hill Book Company), 1966.

[577] J. Jefferies. The source function in a non-equilibrium atmosphere. VII. The interlocking problem. *Astrophys. J.*, 132, 775, 1960.

[578] J. Jefferies. *Spectral Line Formation*. (Waltham: Blaisdell), 1968.

[579] J. Jefferies and R. Thomas. The source function in a non-equilibrium atmosphere. II. Depth dependence of source function for resonance and strong subordinate lines. *Astrophys. J.*, 127, 667, 1958.

[580] J. Jefferies and R. Thomas. The source function in a non-equilibrium atmosphere. III. The influence of a chromosphere. *Astrophys. J.*, 129, 401, 1959.

[581] J. Jefferies and R. Thomas. The source function in a non-equilibrium atmosphere. V. Character of the self-reversed emission core of Ca^+ H and K. *Astrophys. J.*, 131, 695, 1960.

[582] J. Jefferies and O. White. The source function in a non-equilibrium atmosphere. VI. The frequency dependence of the source function for resonance lines. *Astrophys. J.*, 132, 767, 1960.

[583] T. John. The free-free transitions of the negative hydrogen ion in the exchange approximation. *Mon. Not. Roy. Astr. Soc.*, 128, 93, 1964.

[584] T. John. The free-free transitions of He^- at threshold. *Astrophys. J.*, 149, 449, 1967.

[585] T. John. The continuous absorption coefficient of alkali metal negative ions. *J. Phys. B.*, 5, 1662, 1972.

[586] T. John. The free-free transitions of atomic and molecular negative ions in the infrared. *Mon. Not. Roy. Astr. Soc.*, 170, 5, 1975.

[587] T. John. The continuous absorption coefficient of atomic and molecular negative ions. *Mon. Not. Roy. Astr. Soc.*, 172, 305, 1975.

[588] T. John. The reliability of H^- free-free absorption coefficients. *Astr. Astrophys.*, 282, 890, 1994.

[589] T. John. The free-free absorption coefficient of the negative ion of molecular hydrogen in the far-IR spectrum. *Mon. Not. Roy. Astr. Soc.*, 269, 865, 1994.

[590] T. John. The free-free absorption coefficient of the negative He ion. *Mon. Not. Roy. Astr. Soc.*, 269, 871, 1994.

[591] T. John and A. Williams. The free-free transitions of neon and argon negative ions. *J. Phys. B*, 6, L384, 1973.

[592] H. Johnson. Infrared stellar photometry. *Astrophys. J.*, 135, 69, 1962.

[593] H. Johnson. Interstellar extinction in the Galaxy. *Astrophys. J.*, 141, 923, 1965.

[594] H. Johnson. Astronomical measurements in the infrared. *Ann. Rev. Astr. Astrophys.*, 4, 193, 1966.

[595] H. Johnson and B. Krupp. Treatment of atomic and molecular line-blanketing by opacity sampling. *Astrophys. J.*, 206, 201, 1976.

[596] H. Johnson and R. Mitchell. The spectral-energy curves of subdwarfs. II. *Astrophys. J.*, 153, 213, 1968.

[597] H. Johnson and W. Morgan. Fundamental stellar photometry for standards of spectral on the revised system of the Yerkes Spectral Atlas. *Astrophys. J.*, 117, 313, 1953.

[598] H. Johnson and A. Sandage. The galactic cluster M 67 and its significance for stellar evolution. *Astrophys. J.*, 121, 616, 1955.

[599] H. Johnson and A. Sandage. Three-color photometry in the globular cluster M 3. *Astrophys. J.*, 124, 379, 1956.

[600] K. Johnston and C. Vegt. Reference frames in astronomy. *Ann. Rev. Astr. Astrophys.*, 37, 97, 1999.

[601] U. Jørgensen. Molecules in stellar and star-like atmospheres. In Hubeny et al. [536], page 303.

[602] U. Jørgensen, editor. *Molecules in the Stellar Environment.* (Berlin: Springer-Verlag), 1994.

[603] U. Jørgensen. Dominating molecules in the photospheres of cool stars. In *Molecules in the Stellar Environment* [602], page 29.

[604] P. Joss, E. Salpeter, and J. Ostriker. On the "critical luminosity" in stellar interiors and stellar surface boundary conditions. *Astrophys. J.*, 181, 429, 1973.

[605] F. Jüttner. Das Maxwellische Gesetz der Geschwindigkeitverteilung in der Relativtheorie. *Ann. Physik*, 34, 856, 1911.

[606] W. Kalkofen, editor. *Methods in Radiative Transfer.* (Cambridge: Cambridge University Press), 1984.

[607] W. Kalkofen. Operator perturbation methods: A synthesis. In *Methods in Radiative Transfer* [606], page 427.

[608] W. Kalkofen. Numerical methods in radiative transfer. In Beckman and Crivellari [100], page 153.

[609] W. Kalkofen, editor. *Numerical Radiative Transfer.* (Cambridge: Cambridge University Press), 1987.

[610] L. Kantorovich and G. Akilov. *Functional Analysis in Normed Spaces.* (New York: Pergamon Press), 1964.

[611] W. Karzas and R. Latter. Electron radiative transitions in a Coulomb field. *Astrophys. J. Suppl.*, 6, 167, 1961.

[612] P. Keenan. The MK classification and its calibration. In Hayes et al. [464], page 121.

[613] P. Keenan and C. Barnbaum. Revision and calibration of MK luminosity classes for cool giants by HIPPARCOS parallaxes. *Astrophys. J.*, 518, 859, 1999.

[614] P. Kepple. Improved Stark profile calculations for the He II lines at 256, 304, 1085, 1216, 3202, and 4686Å. *Phys. Rev.*, A6, 1, 1972.

[615] D. Kershaw. A fast method for computing the integrals of the relativistic Compton scattering kernel for radiative transfer. *J. Quantit. Spectrosc. Radiat. Transf.*, 38, 347, 1987.

[616] J. Kielkopf and N. Allard. Satellites of Lyman alpha due to H-H and H-H^+ collisions. *Astrophys. J.*, 450, L75, 1995.

[617] H. Kienle, D. Chalonge, and D. Barbier. Comparaison de sources étalons utilisées en spectrophotométrie stellaire à l'Observatoire de Göttingen et au Laboratoire d'Astrophysique de Paris. *Ann. d'Astrohys.*, 1, 396, 1938.

[618] J. Kingdon and G. Ferland. Rate coefficients for charge transfer between hydrogen and the first 30 elements. *Astrophys. J. Suppl.*, 106, 205, 1996.

[619] R. Kippenhahn and A. Weigert. *Stellar Structure and Evolution*. (Berlin: Springer-Verlag), 1990.

[620] J. Kirkpatrick, D. Kelly, G. Rieke, J. Liebert, F. Allard, and R. Wehrse. M dwarf spectra from 0.6 to 1.5 micron: A spectral sequence, model atmosphere fitting, and the temperature scale. *Astrophys. J.*, 402, 643, 1993.

[621] O. Klein and Y. Nishina. Über die Streuung von Strahlung durch freie Elektronen nach der neuen relativistischen Quantendynamik von Dirac. *Z. für Phys.*, 52, 853, 1929.

[622] R. Klein and J. Castor. H and He II spectra of Of stars. *Astrophys. J.*, 220, 902, 1978.

[623] R. Klein, J. Castor, A. Greenbaum, D. Taylor, and P. Dykema. A new scheme for multidimensional line transfer. I. Formulation and 1D results. *J. Quantit. Spectrosc. Radiat. Transf.*, 41, 199, 1989.

[624] F. Kneer. Comments on the redistribution function of Jefferies and White. *Astrophys. J.*, 200, 367, 1975.

[625] H. Kobayashi. Line-by-line calculation using Fourier-transformed Voigt function. *J. Quantit. Spectrosc. Radiat. Transf.*, 62, 477, 1999.

[626] D. Koester, U. Sperhake, N. Allard, D. Finley, and S. Jordan. Quasi-molecular satellites of Lyβ in ORFEUS observations of DA white dwarfs. *Astr. Astrophys.*, 336, 276, 1998.

[627] L. Koesterke, W.-R. Hamann, and G. Gräfener. Expanding atmospheres in non-LTE: Radiation transfer using short characteristics. *Astr. Astrophys.*, 384, 562, 2002.

[628] L. Koesterke, W.-R. Hamann, and P. Kosmol. Upgrading the accelerated lambda-iteration technique by means of "least-change secant methods." *Astr. Astrophys.*, 255, 490, 1992.

[629] A. Kompaneets. The establishment of thermal equilibrium between quanta and electrons. *Sov. Phys. JETP*, 4, 730, 1957.

[630] N. Konjević. Plasma broadening and shifting of non-hydrogenic spectral lines: Present status and applications. *Physics Reports*, 316, 339, 1999.

[631] N. Konjević and W. Wiese. Experimental Stark widths and shifts for spectral lines of neutral and ionized atoms. *J. Phys. Chem. Ref. Data*, 19, 1307, 1990.

[632] N. Kosirev. Radiative equilibrium of the extended photosphere. *Mon. Not. Roy. Astr. Soc.*, 94, 430, 1934.

[633] V. Kourganoff. *Basic Methods in Transfer Problems*. (New York: Dover Publications), 1963.

[634] J. Kovalevsky. First results from HIPPARCOS. *Ann. Rev. Astr. Astrophys.*, 36, 99, 1998.

[635] R. Kraft. Helfer, Wallerstein, & Greenstein's abundances in Population II giants. *Astrophys. J.*, 525, 896, 1999.

[636] H. Kramers. Theory of X-ray absorption and of the continuous X-ray spectrum. *Phil. Mag.*, 46, 836, 1923.

[637] H. Kramers and W. Heisenberg. Über die Streuung von Strahlung durch Atome. *Z. für Phys.*, 31, 661, 1925.

[638] K. Krishna Swamy. Adiabatic gradient for ionizable matter and radiation *Astrophys. J.*, 134, 1017, 1961.

[639] P. Kuan and L. Kuhi. P Cygni stars and mass loss. *Astrophys. J.*, 199, 148, 1975.

[640] J. Kubát. Spherically symmetric model atmospheres using approximate lambda operators. I. First results for static NLTE atmospheres. *Astr. Astrophys.*, 287, 179, 1994.

[641] J. Kubát, J. Puls, and A. Pauldrach. Thermal balance of electrons in calculations of model stellar atmospheres. *Astr. Astrophys.*, 341, 587, 1999.

[642] R.-P. Kudritzki. Spectroscopic constraints on the evolution of massive stars. In Lambert [667], page 97.

[643] R.-P. Kudritzki, A. Pauldrach, and J. Puls. Winds of hot stars as a diagnostic tool of stellar evolution. In Nomoto [799], page 114.

[644] R.-P. Kudritzki, A. Pauldrach, J. Puls, and D. Abbott. Radiation-driven winds of hot stars. VI. Analytical solutions for wind models including the finite cone angle effect. *Astr. Astrophys.*, 219, 205, 1989.

[645] R.-P. Kudritzki and J. Puls. Winds from hot stars. *Ann. Rev. Astr. Astrophys.*, 38, 613, 2000.

[646] R. Kulsrud. The Gaunt factor for free-free transitions in pure hydrogen. *Astrophys. J.*, 119, 386, 1954.

[647] P. Kunasz and L. Auer. Short characteristic integration of radiative transfer problems: Formal solution in two-dimensional slabs. *J. Quantit. Spectrosc. Radiat. Transf.*, 39, 67, 1988.

[648] P. Kunasz and D. Hummer. Radiative transfer in spherically symmetric systems. III. Fundamentals of line formation. *Mon. Not. Roy. Astr. Soc.*, 166, 19, 1974.

[649] P. Kunasz and D. Hummer. Radiative transfer in spherically symmetric systems. IV. Solution of the line transfer problem with radial velocity fields. *Mon. Not. Roy. Astr. Soc.*, 166, 57, 1974.

[650] P. Kunasz and G. Olson. Short characteristic solution of the non-LTE line transfer problem by operator perturbation. II. The two-dimensional planar slab. *J. Quantit. Spectrosc. Radiat. Transf.*, 39, 1, 1988.

[651] M. Kuntz. A new implementation of the Humlíček algorithm for the calculation of the Voigt profile function. *J. Quantit. Spectrosc. Radiat. Transf.*, 57, 819, 1997.

[652] M. Künzli, P. North, R. Kurucz, and B. Nicolet. A calibration of Geneva photometry for B to G stars in terms of $T_{\rm eff}$, log g, and [M/H]. *Astr. Astrophys. Suppl.*, 122, 51, 1997.

[653] R. Kurth. Zur Schwarzschildschen Integralgleichung. *Z. für Astrophys.*, 31, 115, 1952.

[654] R. Kurucz. ATLAS: A computer program for calculating model stellar atmospheres. Technical Report 309, Smithsonian Astrophysical Observatory, 1970.

[655] R. Kurucz. Stellar spectral synthesis in the ultraviolet. *Astrophys. J.*, 188, 21, 1974.

[656] R. Kurucz. A preliminary theoretical line-blanketed model solar photosphere. *Solar Phys.*, 34, 17, 1974.

[657] R. Kurucz. Model atmospheres for G, F, A, B, and O stars. *Astrophys. J. Suppl.*, 40, 1, 1979.

[658] R. Kurucz. New opacity calculations. In Crivellari et al. [264], page 441.

[659] R. Kurucz. The solar spectrum. In Cox et al. [263], page 663.

[660] R. Kurucz. A new opacity-sampling model atmosphere program for arbitrary abundances. In Dworetsky et al. [295], page 87.

[661] R. Kurucz. The Kurucz-Smithsonian atomic and molecular database. In Adelman and Wiese [10], page 205.

[662] R. Kurucz. Status of the ATLAS 12 opacity-sampling program and of new programs for Rosseland and distribution-function opacity. In Adelman et al. [9], page 106.

[663] R. Kurucz, E. Peytremann, and E. Avrett. *Blanketed Model Atmospheres of Early Type Stars*. Smithsonian Institution. (Washington, DC: U.S. Government Printing Office), 1974.

[664] H. Kusch. Experimentelle Untersuchung der Druckverbreiterung von Eisenlinien durch neutrale Wasserstoffatome und Wasserstoffmolekle. *Z. für Astrophys.*, 45, 1, 1958.

[665] A. Labeyrie. Attainment of diffraction limited resolution in large telescopes by Fourier analyzing speckle patterns in star images. *Astr. Astrophys.*, 6, 85, 1970.

[666] A. Labeyrie. Stellar interferometry methods. *Ann. Rev. Astr. Astrophys.*, 16, 77, 1978.

[667] D. Lambert, editor. *Frontiers of Stellar Evolution*. (San Francisco: Astronomical Society of the Pacific), 1991.

[668] H. Lamers. Stellar wind theories. In de Grève et al. [272], page 69.

[669] H. Lamers and J. Cassinelli. *Introduction to Stellar Winds*. (Cambridge: Cambridge University Press), 1999.

[670] H. Lamers, M. Cerruti-Sola, and M. Perinotto. The "SEI" method for accurate and efficient calculations of line profiles in spherically symmetric stellar winds. *Astrophys. J.*, 314, 726, 1987.

[671] H. Lamers and D. Morton. Mass ejection from the O4f star ζ Puppis. *Astrophys. J. Suppl.*, 32, 715, 1976.

[672] A. Lançon and P. Wood. A library of 0.5μ to 2.5μ spectra of luminous cool stars. *Astr. Astrophys. Suppl.*, 146, 217, 2000.

[673] L. Landau and E. Lifschitz. *The Classical Theory of Fields*. (Oxford: Pergamon Press), 4th edition, 1975.

[674] L. Landau and E. Lifschitz. *Fluid Mechanics*. (Oxford: Pergamon Press), 2nd edition, 1987.

[675] A. Landé. Termstruktur und Zeemaneffekt der Multipletts. *Z. für Phys.*, 15, 189, 1923.

[676] A. Landé. Termstruktur und Zeemaneffekt der Multipletts. *Z. für Phys.*, 19, 112, 1923.

[677] K. Lang. *Astrophysical Data*. (New York: Springer-Verlag), 1992.

[678] K. Lang and O. Gingerich. *A Source Book in Astronomy and Astrophysics, 1900–1975*. (Cambridge: Harvard University Press), 1979.

[679] T. Lanz, M. Barstow, I. Hubeny, and J. Holberg. A self-consistent optical, ultraviolet, and extreme-ultraviolet model for the spectrum of the hot white dwarf G191-B2B. *Astrophys. J.*, 473, 1089, 1996.

[680] T. Lanz, A. de Koter, I. Hubeny, and S. Heap. Toward resolving the "mass discrepancy" in O-type stars. *Astrophys. J.*, 465, 359, 1996.

[681] T. Lanz and I. Hubeny. Atomic data in non-LTE model stellar atmospheres. In Hubeny et al. [536], page 117.

[682] T. Lanz and I. Hubeny. A grid of non-LTE line-blanketed model atmospheres of O-type stars. *Astrophys. J. Suppl.*, 146, 417, 2003.

[683] T. Lanz and I. Hubeny. A grid of NLTE line-blanketed model atmospheres of early B-Type stars. *Astrophys. J. Suppl.*, 169, 83, 2007.

[684] T. Lanz, I. Hubeny, and S. Heap. Non-LTE line-blanketed model atmospheres of hot stars. III. Hot subdwarfs: The sdO star BD +75°325. *Astrophys. J.*, 485, 843, 1997.

[685] R. Larson. The emitted spectrum of a protostar. *Mon. Not. Roy. Astr. Soc.*, 145, 297, 1969.

[686] Y. Lebreton. Stellar structure and evolution: Deductions from HIPPARCOS. *Ann. Rev. Astr. Astrophys.*, 38, 35, 2001.

[687] H.-W. Lee. Raman-scattering wings of Hα in symbiotic stars. *Astrophys. J.*, 541, L25, 2000.

[688] H.-W. Lee. Exact low-energy expansion of the Rayleigh scattering cross-section by atomic H. *Mon. Not. Roy. Astr. Soc.*, 358, 1472, 2005.

[689] H.-W. Lee and H. Kim. Rayleigh scattering cross-section redward of Lyα by atomic hydrogen. *Mon. Not. Roy. Astr. Soc.*, 347, 802, 2004.

[690] R. Leighton. *Principles of Modern Physics*. (New York: McGraw-Hill Book Company), 1959.

[691] T. Lejeune and J. Fernandes, editors. *Observed HR Diagrams and Stellar Evolution*. (San Francisco: Astronomical Society of the Pacific), 2002.

[692] M. Lemke. Extended VCS Stark broadening tables of hydrogen Lyman to Brackett series. *Astr. Astrophys. Suppl.*, 122, 285, 1997.

[693] P. Léna. Continuum infrared radiation of the solar photosphere. *Astr. Astrophys.*, 4, 202, 1970.

[694] J. Lester, R. Gray, and R. Kurucz. Theoretical $uvby\beta$ indices. *Astrophys. J. Suppl.*, 61, 509, 1986.

[695] H. Li, J. Wu, B. Zhou, J. Zhu, and Z. Yan. Calculations of energies of the hydrogen molecular ion. *Phys. Rev. A*, 75, 012504, 2007.

[696] R. Liboff. *Kinetic Theory: Classical, Quantum, and Relativistic Descriptions*. (New York: John Wiley & Sons), 2nd edition, 1998.

[697] E. Lindholm. Zur theorie der verbreiterung von spektrallinien. *Arkiv for Math., Astr., och Fysik*, 28B, No. 3, 1941.

[698] E. Lindholm. Pressure broadening of spectral lines. *Arkiv for Math., Astr., och Fysik*, 32A, No. 17, 1945.

[699] G. Lombardi, D. Kelleher, and J. Cooper. Redistribution of radiation in the absence of collisions. *Astrophys. J.*, 288, 820, 1985.

[700] C. de Loore, J. de Grève, and D. Vanbeveren. Parameters of massive stars and main-sequence evolution without and with stellar wind. *Astr. Astrophys. Suppl.*, 34, 363, 1978.

[701] W. Lotz. Electron-impact ionization cross-sections and ionization rate coefficients for atoms and ions from H to Ca. *Z. für Phys.*, 216, 241, 1968.

[702] R. Loudon. *The Quantum Theory of Light*. (Oxford: University of Oxford Press), 1973.

[703] L. Lucy. A temperature correction procedure. In Avrett et al. [64], page 93.

[704] L. Lucy. The asymmetry of resonance line scattering in a velocity gradient. *Astr. Astrophys.*, 140, 210, 1984.

[705] L. Lucy and P. Solomon. Mass loss by hot stars. *Astrophys. J.*, 159, 879, 1970.

[706] K. MacGregor, L. Hartmann, and J. Raymond. Radiative amplification of sound waves in the winds of O and B stars. *Astrophys. J.*, 231, 514, 1979.

[707] J. MacFarlane. Non-LTE radiative transfer with lambda acceleration: convergence properties using exact full and diagonal lambda operators. *Astr. Astrophys.*, 264, 153, 1992.

[708] P. Magain. Spectroscopic analysis of extreme metal-poor 'dwarfs'. II. Improved model atmospheres and detailed abundances. *Astr. Astrophys.*, 146, 95, 1985.

[709] C. Magnan. Radiative transfer in a moving medium. *J. Quantit. Spectrosc. Radiat. Transf.*, 14, 123, 1974.

[710] E. Makarova. A photometric investigation of the energy distribution in the continuous solar spectrum in absolute units. *Sov. Astr.*, 1, 531, 1957.

[711] C. Mark. The neutron density near a plane surface. *Phys. Rev.*, 72, 558, 1947.

[712] W. Martin. Energy levels of neutral He. *J. Phys. Chem. Ref. Data*, 2, 257, 1973.

[713] L. Martins, R. González-Delgado, C. Leitherer, M. Cerviño, and P. Hauschildt. A high-resolution stellar library for evolutionary population sysnthesis. *Mon. Not. Roy. Astr. Soc.*, 358, 49, 2005.

[714] F. Martins, D. Schaerer, and D. Hillier. A new calibration of stellar parameters of galactic O stars. *Astr. Astrophys.*, 436, 1049, 2005.

[715] T. Matthews and A. Sandage. Optical identification of 3C 48, 3C 196, and 3C 286 with stellar objects. *Astrophys. J.*, 138, 30, 1963.

[716] H. McAlister. High angular resolution measurements of stellar properties. *Ann. Rev. Astr. Astrophys.*, 23, 59, 1985.

[717] H. McAlister. The calibration of interferometrically determined properties of binary stars. In Hayes et al. [464], page 97.

[718] W. McCrea. A note on Dr. P. A. Taylor's paper "The equlibrium of the calcium chromosphere. *Mon. Not. Roy. Astr. Soc.*, 88, 729, 1928.

[719] W. McCrea. Model stellar atmospheres. *Mon. Not. Roy. Astr. Soc.*, 91, 836, 1931.

[720] D. Menzel, editor. *Selected Papers on Physical Processes in Ionized Plasmas*. (New York: Dover Publications), 1962.

[721] D. Menzel, editor. *Selected Papers on the Transfer of Radiation*. (New York: Dover Publications), 1966.

[722] D. Menzel and C. Pekeris. Absorption coefficients and H-line intensities. *Mon. Not. Roy. Astr. Soc.*, 96, 77, 1935.

[723] E. Merzbacher. *Quantum Mechanics*. (New York: John Wiley & Sons), 3rd edition, 1998.

[724] M. Meyer, S. Edwards, K. Hinkle, and S. Strom. Near-infrared classification spectroscopy: H-band spectra of fundamental MK standards. *Astrophys. J.*, 508, 397, 1998.

[725] G. Meynet, A. Maeder, G. Schaller, D. Schaerer, and C. Charbonnel. Grids of massive stars with high mass-loss rates. V. From 12 to 120 $M_\odot$ at $(Z = 0.001, 0.004, 0.008, 0.020)$. *Astr. Astrophys. Suppl.*, 103, 97, 1996.

[726] G. Michaud. Diffusion processes in peculiar A stars. *Astrophys. J.*, 160, 641, 1970.

[727] A. Michelson and F. Pease. Measurement of the diameter of α Orionis with the interferometer. *Astrophys. J.*, 53, 249, 1921.

[728] D. Mihalas. Balmer-line-blanketed model atmospheres for A-type stars. *Astrophys. J. Suppl.*, 13, 1, 1966.

[729] D. Mihalas. *Stellar Atmospheres*. (San Francisco: W. H. Freeman & Company), 1st edition, 1970.

[730] D. Mihalas. Non-LTE model atmospheres for B and O stars. Technical Report NCAR-TN/STR-76, National Center for Atmospheric Research, Boulder, Colorado, 1972.

[731] D. Mihalas. *Stellar Atmospheres*. (San Francisco: W. H. Freeman & Company), 2nd edition, 1978.

[732] D. Mihalas. Solution for the comoving-frame equation of transfer in spherically symmetric flows. VI. Relativistic flows. *Astrophys. J.*, 237, 574, 1980.

[733] D. Mihalas. Solution for the comoving-frame equation of transfer in spherically symmetric flows. VII. Angle-dependent partial redistribution. *Astrophys. J.*, 238, 1034, 1980.

[734] D. Mihalas and L. Auer. Non-LTE model atmospheres. V. Multi-line H-He models for O and early B stars. *Astrophys. J.*, 160, 1161, 1970.

[735] D. Mihalas, A. Barnard, J. Cooper, and E. Smith. He I λ 4471 profiles in B stars: Calculations with an improved line broadening theory. *Astrophys. J.*, 190, 315, 1974.

[736] D. Mihalas, A. Barnard, J. Cooper, and E. Smith. He I λ 4922 profiles in B stars: Calculations with an improved line broadening theory. *Astrophys. J.*, 197, 139, 1975.

[737] D. Mihalas, J. Heasley, and L. Auer. A non-LTE model stellar atmosphere computer program. Technical Report NCAR-TN/STR-104, National Center for Atmospheric Research, Boulder, Colorado, 1975.

[738] D. Mihalas and D. Hummer. Theory of extended stellar atmospheres. I. Computational method and first results for static spherical models. *Astrophys. J. Suppl.*, 28, 343, 1974.

[739] D. Mihalas and R. Klein. On the solution of the time-dependent inertial-frame equation of radiative transfer in moving media to $O(v/c)$. *J. Comp. Phys.*, 46, 97, 1982.

[740] D. Mihalas and P. Kunasz. Solution for the comoving-frame equation of transfer in spherically symmetric flows. V. Multilevel atoms. *Astrophys. J.*, 219, 635, 1978.

[741] D. Mihalas and P. Kunasz. The computation of radiation transport using Feautrier variables. II. Spectrum line formation in moving media. *J. Comp. Phys.*, 64, 1, 1986.

[742] D. Mihalas, P. Kunasz, and D. Hummer. Solution for the comoving-frame equation of transfer in spherically symmetric flows. I. Computational method for equivalent-two-level-atom source functions. *Astrophys. J.*, 202, 465, 1975.

[743] D. Mihalas, P. Kunasz, and D. Hummer. Solution for the comoving-frame equation of transfer in spherically symmetric flows. II. Picket-fence models. *Astrophys. J.*, 203, 647, 1976.

[744] D. Mihalas, P. Kunasz, and D. Hummer. Solution for the comoving-frame equation of transfer in spherically symmetric flows. III. Effect of aberration and advection terms. *Astrophys. J.*, 206, 515, 1976.

[745] D. Mihalas, P. Kunasz, and D. Hummer. Solution for the comoving-frame equation of transfer in spherically symmetric flows. IV. Frequency-dependent source functions for scattering by atoms and electrons. *Astrophys. J.*, 210, 419, 1976.

[746] D. Mihalas and B. Mihalas. *Foundations of Radiation Hydrodynamics*. (New York: Dover Publications), 1999.

[747] D. Mihalas and D. Morton. A model for a B1 V star with line blanketing. *Astrophys. J.*, 142, 253, 1965.

[748] D. Mihalas, B. Pagel, and P. Souffrin. *Theorie des Atmospheres Stellaires*. (Geneva: Observatoire de Genève), 1971.

[749] R. Milkey and D. Mihalas. Calculation of the solar chromospheric Lyα allowing for partial redistribution effects. *Solar Phys.*, 32, 361, 1973.

[750] R. Milkey and D. Mihalas. Resonance-line transfer with partial redistribution: A preliminary study of Lyα in the solar chromosphere. *Astrophys. J.*, 185, 709, 1973.

[751] R. Milkey and D. Mihalas. Resonance-line transfer with partial redistribution. II. The solar Mg II lines. *Astrophys. J.*, 192, 769, 1974.

[752] R. Milkey, R. Shine, and D. Mihalas. Resonance-line transfer with partial redistribution. IV. A generalized formulation for lines with common upper states. *Astrophys. J.*, 199, 718, 1975.

[753] E. Milne. Radiative equilibrium in the outer layers of a star. *Mon. Not. Roy. Astr. Soc.*, 81, 361, 1921.

[754] E. Milne. The temperature in the outer atmosphere of a star. *Mon. Not. Roy. Astr. Soc.*, 82, 368, 1922.

[755] E. Milne. Statistical equilibrium in relation to the photoelectric effect, and its application to the determination of absorption coefficients. *Phil. Mag.*, 47, 209, 1924.

[756] E. Milne. The theoretical contours of absorption lines in stellar atmospheres. *Mon. Not. Roy. Astr. Soc.*, 89, 3, 1928.

[757] E. Milne. Ionization in stellar atmospheres. I. Generalized Saha formulae, maximum intensities, and the determination of the coefficient of opacity. *Mon. Not. Roy. Astr. Soc.*, 89, 17, 1928.

[758] E. Milne. Thermodynamics of the stars. In Menzel [721], page 77.

[759] E. Milne. The radiative equilibrium of a planetary nebula. *Z. für Astrophys.*, 1, 98, 1930.

[760] E. Milne and S. Chandrasekhar. Ionization in stellar atmospheres. III. *Mon. Not. Roy. Astr. Soc.*, 92, 150, 1932.

[761] A. Mitchell. *Computational Methods in Partial Differential Equations.* (London: John Wiley & Sons), 1969.

[762] W. Mitchell. The absorption coefficient of the negative hydrogen ion. *Astrophys. J.*, 130, 872, 1959.

[763] W. Mitchell. Limb darkening in the solar ultraviolet. *Solar Phys.*, 69, 39, 1981.

[764] O. Moe and E. Milone. Limb darkening 1945–3245 Å for the quiet Sun from SKYLAB data. *Astrophys. J.*, 226, 301, 1978.

[765] C. Møller. *The Theory of Relativity.* (Oxford: Oxford University Press), 2nd edition, 1972.

[766] W. Morgan, H. Abt, and J. Tapscott. *Revised MK Spectral Atlas for Stars Earlier Than the Sun.* (Yerkes Observatory, University of Chicago, and Kitt Peak National Observatory), 1978.

[767] W. Morgan, D. Harris, and H. Johnson. Some characteristics of color systems. *Astrophys. J.*, 118, 92, 1953.

[768] W. Morgan and P. Keenan. Spectral Classification. *Ann. Rev. Astr. Astrophys.*, 11, 29, 1973.

[769] W. Morgan, P. Keenan, and E. Kellman. *An Atlas of Stellar Spectra.* (Chicago: University of Chicago Press), 1943.

[770] C. Morossi, M. Franchini, M. Malagnini, R. Kurucz, and R. Buser. Synthetic DDO colors. *Astr. Astrophys.*, 295, 471, 1995.

[771] P. Morse. *Thermal Physics*. (Reading: Benjamin/Cummings Publishing Company), 1969.

[772] D. Morton and T. Adams. Effective temperatures and bolometric corrections of early-type stars. *Astrophys. J.*, 151, 611, 1968.

[773] N. Mowlavi, D. Schaerer, G. Meynet, P. Bernasconi, C. Charbonnel, and A. Maeder. Grids of stellar models. VII. From 0.8 to 60 $M_\odot$ at $Z = 0.10$. *Astr. Astrophys. Suppl.*, 128, 471, 1998.

[774] B. Mozer and M. Baranger. Electric field distributions in an ionized gas. II. *Phys. Rev.*, 118, 626, 1960.

[775] G. Münch. A theoretical discussion of the continuous spectrum of the Sun. *Astrophys. J.*, 102, 385, 1945.

[776] G. Münch. The effect of the absorption lines on the temperature distribution of the solar atmosphere. *Astrophys. J.*, 104, 87, 1946.

[777] G. Münch. Model solar atmospheres. *Astrophys. J.*, 106, 217, 1947.

[778] G. Münch. The effect of nongrayness on the temperature distribution of the solar atmosphere. *Astrophys. J.*, 107, 265, 1948.

[779] G. Münch. The effect of electron scattering on the line spectrum of high-temperature stars. *Astrophys. J.*, 108, 116, 1948.

[780] G. Münch. On the formation of absorption lines by noncoherent scattering. *Astrophys. J.*, 109, 275, 1949.

[781] G. Münch. Study of the H and He II lines in a high-temperature subdwarf. *Astrophys. J.*, 217, 642, 1958.

[782] G. Münch. The theory of model stellar atmospheres. In Greenstein [397], chapter 1, page 1.

[783] G. Münch and A. Slettebak. A new O-type subdwarf. *Astrophys. J.*, 129, 852, 1959.

[784] A. Munier and R. Weaver. Radiation transfer in the fluid frame: A covariant formulation. I. Radiation hydrodynamics. *Comp. Phys. Reports*, 3, 125, 1986.

[785] A. Munier and R. Weaver. Radiation transfer in the fluid frame: A covariant formulation. II. The radiation transfer equation. *Comp. Phys. Reports*, 3, 165, 1986.

[786] T. Murai. Electronic energies of the hydrogen molecular ion H_2^+. *Science of Light*, 23, 83, 1974.

[787] C. Murray. *Vectorial Astronomy*. (Bristol: Adam Hilger Ltd.), 1983.

[788] T. Nagel, S. Dreizler, T. Rauch, and K. Werner. AcDc - A new code for the NLTE spectral analysis of accretion disks: application to the helium CV AM CVn. *Astr. Astrophys.*, 428,109, 2004.

[789] K. Nagendra. Numerical solutions of polarized line transfer equations. In Hubeny et al. [536], page 583.

[790] M. Nakagawa, Y. Kohyama, and N. Itoh. Relativistic Gaunt factor of the dense high-temperature stellar plasma. *Astrophys. J. Suppl.*, 63, 661, 1987.

[791] D. Nagirner and Yu. Poutanen. Compton scattering by Maxwellian electrons: Frequency and directional redistribution of radiation. *Astr. Let.*, 19, 262, 1993.

[792] A. Nayfonov, W. Däppen, D. Hummer, and D. Mihalas. The MHD equation of state with post-Holtsmark microfield distributions. *Astrophys. J.*, 526, 451, 1999.

[793] J. von Neumann. Wahrscheinlichkeitstheoretischer Aufbau der Quantenmechanik. *Göttinger Nachrichten*, 1, 245, 1927.

[794] K. Ng. Hypernetted chain solutions for the classical one-component plasma up to $\Gamma = 7000$. *J. Chem. Phys.*, 61, 2680, 1974.

[795] G. Nienhuis, and F. Schuller. Collisional redistribution of fluorescence radiation. *Physica*, 92C, 397, 1977.

[796] Y. Nishina. Die Polarisation der Comptonstreuung nach der Diracschen Theorie des Elektrons. *Z. für Phys.*, 52, 869, 1929.

[797] Å. Nordlund. On convection in stellar atmospheres. *Astr. Astrophys.*, 32, 407, 1974.

[798] Å. Nordlund. A two-component representation of stellar atmospheres with convection. *Astr. Astrophys.*, 50, 23, 1976.

[799] K. Nomoto, editor. *Atmospheric Diagnostics of Stellar Evolution*. (New York: Springer-Verlag), 1988.

[800] C. Norman, A. Renzini, and M. Tosi, editors. *Stellar Populations*. (Cambridge: Cambridge University Press), 1987.

[801] M. van Noort, I. Hubeny, and T. Lanz. Multidimensional non-LTE radiative transfer. I. A universal two-dimensional short-characteristics scheme for Cartesian, spherical and cylindrical coordinate systems. *Astrophys. J.*, 568, 1066, 2002.

[802] J. Norris, T. Beers, and S. Ryan. Extremely metal-poor stars. VII. The most metal-poor dwarf, CS 22876–032. *Astrophys. J.*, 540, 456, 2000.

[803] A. Nota and H. Lamers, editors. *Luminous Blue Variables: Massive Stars in Transition*. (San Francisco: Astronomical Society of the Pacific), 1997.

[804] H. Nussbaumer, H. Schmid, and M. Vogel. Raman scattering as a diagnostic possibility in astrophysics. *Astr. Astrophys.*, 211, L27, 1989.

[805] D. O'Connell, editor. *Stellar Populations*. (Amsterdam: North Holland), 1958.

[806] B. Odom, D. Hanneke, B. D'Urso, and G. Gabrielse. New measurement of the electron magnetic moment using a one-electron quantum cyclotron. *Phys. Rev. Let.*, 97, 030801, 2006.

[807] T. Ohmura. Evaluation of free-free absorption coefficient of the negative hydrogen ion. *Astrophys. J.*, 140, 282, 1964.

[808] T. Ohmura and H. Ohmura. Continuous absorption due to free-free transition in hydrogen. *Phys. Rev.*, 121, 513, 1961.

[809] J. Oke. Absolute spectral energy distributions in stars. *Ann. Rev. Astr. Astrophys.*, 3, 23, 1965.

[810] J. Oke. Absolute spectral energy distributions for white dwarfs. *Astrophys. J. Suppl.*, 27, 21, 1974.

[811] J. Oke. Faint spectrophotometric standard stars. *Astr. J.*, 99, 1621, 1990.

[812] J. Oke and J. Gunn. Secondary standard stars for absolute spectrophotometry. *Astrophys. J.*, 266, 713, 1983.

[813] J. Oke and R. Schild. The absolute spectral energy distribution of α Lyrae. *Astrophys. J.*, 161, 1015, 1970.

[814] E. Olson. The calibration of *uvby* photometry. *Pub. Astr. Soc. Pacific*, 86, 80, 1974.

[815] G. Olson, L. Auer, and J. Buchler. A rapidly convergent iterative solution of the non-LTE radiation transfer problem. *J. Quantit. Spectrosc. Radiat. Transf.*, 35, 431, 1986.

[816] G. Olson and P. Kunasz. Short characteristic solution of the non-LTE transfer problem by operator perturbation. I. The one-dimensional planar slab. *J. Quantit. Spectrosc. Radiat. Transf.*, 38, 325, 1987.

[817] A. Omont, E. Smith, and J. Cooper. Redistribution of resonance radiation. I. The effect of collisions. *Astrophys. J.*, 175, 185, 1972.

[818] E. Öpik. Stellar structure, source of energy, and evolution. *Pub. de l'Observatoire Astr. de l'Universite de Tartu*, No. 3, 30, 1, 1938.

[819] J. Oppenheimer. Note on the theory of the interaction of field and matter. *Phys. Rev.*, 35, 461, 1930.

[820] D. Osterbrock. The escape of resonance-line radiation from an optically thick medium. *Astrophys. J.*, 135, 195, 1962.

[821] D. Osterbrock. *Astrophysics of Gaseous Nebulae and Active Galactic Nuclei.* (Mill Valley: University Science Books), 1989.

[822] R. Oudmaijer, M. Groenewegen, and H. Schrijver. The absolute magnitude of K0 V stars from BT HIPPARCOS parallaxes. *Astr. Astrophys.*, 341, L55, 1999.

[823] S. Owocki. Winds from hot stars. *Reviews in Modern Astronomy*, 3, 98, 1990.

[824] S. Owocki. Instabilities in hot-star winds: Basic physics and recent developments. In Heber and Jeffery [467], page 393.

[825] S. Owocki. Stellar wind mechanisms and instabilities. In Heydari-Malayeri et al. [490], page 163.

[826] S. Owocki. Radiation hydrodynamics of line-driven winds. In Hubeny et al. [541], page 173.

[827] S. Owocki, J. Castor, and G. Rybicki. Time-dependent models of radiatively driven stellar winds. I. Nonlinear evolution of instabilities for a pure absorption model. *Astrophys. J.*, 335, 914, 1988.

[828] S. Owocki, S. Cranmer, and K. Gayley. Inhibition of wind compressed disk formation by nonradial line-forces in rotating hot-star winds. *Astrophys. J.*, 472, L115, 1996.

[829] S. Owocki and K. Gayley. In Nota and Lamers [803], page 121.

[830] S. Owocki, K. Gayley. and N. Shaviv A porosity-length formalism for photon-tiring-limited mass loss from stars above Eddington limit. *Astrophys. J.*, 616, 525, 2004.

[831] S. Owocki and J. Puls. Nonlocal escape-integral approximations for the line force in structured line-driven stellar winds. *Astrophys. J.*, 462, 894, 1996.

[832] S. Owocki and G. Rybicki. Instabilities in line-driven stellar winds. I. Dependence on perturbation wavelength. *Astrophys. J.*, 284, 337, 1984.

[833] S. Owocki and G. Rybicki. Instabilities in line-driven stellar winds. II. Effect of scattering. *Astrophys. J.*, 299, 265, 1985.

[834] J. Oxenius. *Kinetic Theory of Particles and Photons.* (Berlin: Springer-Verlag), 1986.

[835] F. Paletou and L. Auer. A new approximate operator method for partial frequency redistribution problems. *Astr. Astrophys.*, 297, 771, 1995.

[836] A. Pannekoek. Ionization in stellar atmospheres. *Bull. Astr. Inst. Neth.*, 1, 107, 1922.

[837] A. Pannekoek. The theoretical contours of absorption lines. I. *Mon. Not. Roy. Astr. Soc.*, 91, 139, 1930.

[838] W. Panofsky and M. Philips. *Classical Electricity and Magnetism.* (Reading: Addison-Wesley Publishing Company), 2nd edition, 1962.

[839] E. Parker. Dynamics of the interplanetary gas and magnetic fields. *Astrophys. J.*, 128, 664, 1958.

[840] E. Parker. The hydrodynamic theory of solar corpuscular radiation and stellar wind. *Astrophys. J.*, 132, 821, 1960.

[841] E. Parker. *Cosmical Magnetic Fields: Their Origin and Their Activity.* (New York: Oxford University Press), 1979.

[842] A. Pauldrach and A. Herrero. Multi-level non-LTE calculations for very optically thick winds and photospheres under extreme NLTE conditions. *Astr. Astrophys.*, 199, 262, 1988.

[843] A. Pauldrach, J. Puls, and R.-P. Kudritzki. Radiation-driven winds of hot luminous stars. Improvements of the theory and first results. *Astr. Astrophys.*, 164, 86, 1986.

[844] W. Pauli. Über das thermische Gleichgewicht zwischen Strahlung und freien Elektronen. *Z. für Phys.*, 18, 272, 1923.

[845] W. Pauli. Über den Zusammenhang des Abschlusses der Electrongrupen im Atom mit der Komplexstruktur des Spektren. *Z. für Phys.*, 31, 765, 1925.

[846] W. Pauli. Zur Quantenmechanik des magnetischen Elektrons. *Z. für Phys.*, 43, 601, 1927.

[847] W. Pauli. The connection between spin and statistics. *Phys. Rev.*, 58, 716, 1940.

[848] L. Pauling and E. Wilson. *Introduction to Quantum Mechanics.* (New York: McGraw-Hill Book Company), 1935.

[849] C. Payne. *Stellar Atmospheres.* (Cambridge: Harvard College Observatory), 1925.

[850] G. Peach. Continuous absorption coefficients for non-hydrogenic atoms. *Mon. Not. Roy. Astr. Soc.*, 124, 371, 1962.

[851] G. Peach. A general formula for the calculation of absorption cross-sections for free-free transitions in the field of positive ions. *Mon. Not. Roy. Astr. Soc.*, 130, 361, 1965.

[852] G. Peach. Total continuous absorption coefficients for complex atoms. *Mem. Roy. Astr. Soc.*, 70, 29, 1965.

[853] G. Peach. Total continuous absorption coefficient for complex atoms. *Astr. J.*, 71, 174, 1966.

[854] G. Peach. Free-free absorption coefficients for non-hydrogenic atoms. *Mem. R. Astr. Soc.*, 71, 1, 1967.

[855] G. Peach. A revised general formula for the calculation of atomic photoionization cross sections. *Mem. Roy. Astr. Soc.*, 71, 13, 1967.

[856] G. Peach. Continuous absorption coefficients for non-hydrogenic atoms. *Mem. Roy. Astr. Soc.*, 73, 1, 1970.

[857] F. Pease. The angular diameter of α Bootis by the interferometer. *Pub. Astr. Soc. Pacific*, 33, 171, 1921.

[858] F. Pease. Interferometric observations of star diameters. *Pub. Astr. Soc. Pacific*, 34, 183, 1922.

[859] J.-C. Pecker and E. Schatzman. *Astrophysique Générale*. (Paris: Masson et Cie.), 1959.

[860] A. Peraiah. *An Introduction to Radiative Transfer*. (Cambridge: Cambridge University Press), 2001.

[861] D. Peterson. The Balmer lines in early type stars. *S.A.O. Special Report No. 293*. (Cambridge: Smithsonian Astrophysical Observatory), 1968.

[862] E. Peytremann. Line blanketing and model stellar atmospheres. III. Tables of models and broad-band colors. *Astr. Astrophys. Suppl.*, 18, 81, 1974.

[863] H. Pfennig and E. Trefftz. Zur quasistatischen Druckverbreiterung der diffusen Heliumlinien. *Z. für Phys.*, 190, 253, 1966.

[864] A. Philip and D. Hayes, editors. *Multicolor Photometry and the Theoretical HR Diagram*. Dudley Observatory Report No. 9. (Albany: State University of New York), 1975.

[865] A. Phillips. LTE line blanketing of an early B star atmosphere by ultraviolet lines using accurate damping constants. *Mon. Not. Roy. Astr. Soc.*, 181, 777, 1977.

[866] A. Phillips and S. Wright. A fully blanketed early B star LTE model atmosphere using an opacity sampling technique. *Mon. Not. Roy. Astr. Soc.*, 192, 197, 1980.

[867] A. Pierce. Relative solar energy distribution in the spectral region 10,000–25,000Å. *Astrophys. J.*, 119, 312, 1954.

[868] A. Pierce and C. Slaughter. Solar limb darkening. I. At wavelength of 3033–7297Å. *Solar Phys.*, 51, 52, 1977.

[869] M. Pinsonneault, D. Terndrup, R. Hanson, and J. Stauffer. Distances to open clusters from main-sequence fitting. I. New models and comparison with the properties of the Hyades eclipsing binary VB 22. *Astrophys. J.*, 598, 588, 2003.

[870] M. Pinsonneault, D. Terndrup, R. Hanson, and J. Stauffer. Distances to open clusters as derived from main-sequence fitting. II. Construction of empirically calibrated isochrones. *Astrophys. J.*, 600, 946, 2004.

[871] N. Piskunov, W. Weiss, and D. Gray, editors. *Modeling of Stellar Atmospheres*. (San Francisco: Astronomical Society of the Pacific), 2003.

[872] G. Placzek. The angular distribution of neutrons emerging from a plane surface. *Phys. Rev.*, 72, 556, 1947.

[873] G. Placzek and W. Seidel. Milne's problem in transport theory. *Phys. Rev.*, 72, 550, 1947.

[874] M. Planck. Über das Gesetz der Energieverteilung im Normalspektrum. *Ann. Physik*, 4, 553, 1901.

[875] M. Planck. *The Theory of Heat Radiation*. English translation of *Vorlesungen über die Theorie der Wärmestrahlung*, 1913. (New York: Dover Publications), 1991.

[876] B. Podolsky. Dispersion by H-like atoms in undulatory mechanics. *Proc. Nat. Acad. Sci.*, 14, 253, 1928.

[877] G. Pomraning. *The Equations of Radiation Hydrodynamics*. (Oxford: Pergamon Press), 1973.

[878] F. Pont, M. Mayor, C. Turon, and D. Vandenberg. HIPPARCOS subdwarf and globular cluster ages: The distance and age of M 92. *Astr. Astrophys.*, 329, 87, 1998.

[879] D. Popper. Masses of hot main-sequence stars. *Astrophys. J.*, 220, 11, 1978.

[880] D. Popper. HIPPARCOS parallaxes of eclipsing binaries and the radiative flux scale. *Pub. Astr. Soc. Pacific*, 110, 919, 1998.

[881] A. Pradhan. Radiative recombination of the ground state of lithium-like ions. *Astrophys. J.*, 270, 339, 1983.

[882] A. Pradhan and S. Nahar. *Atomic Astrophysics and Spectroscopy*. (Cambridge: Cambridge University Press), 2011.

[883] A. Pradhan and H.-L. Zhang. Electron collisions with atomic ions. In Itikawa [566], page 1.

[884] L. Prandtl. Bericht über Untersuchungen zur ausgebildeten Turbulenz. *Z. Angew. Math. Mech.*, 5, 136, 1925.

[885] L. Prandtl and O. Tietjens. *Fundamentals of Hydro and Aeromechanics*. (New York: Dover Publications), 1957.

[886] M. Prasad, D. Kershaw, and J. Beason. A simple method for computing relativistic Compton scattering kernel for radiative transfer. *Appl. Phys. Let.*, 48, 1193, 1986.

[887] G. Preston. The chemically peculiar stars of the upper main sequence. *Ann. Rev. Astr. Astrophys.*, 12, 257, 1974.

[888] E. Priest. *Solar Magnetohydrodynamics*. (Dordrecht: Reidel), 1982.

[889] J. Puls. Approximate lambda-operators working at optimum convergence rate. II. Line transfer in expanding atmospheres. *Astr. Astrophys.*, 248, 581, 1991.

[890] J. Puls and D. Hummer. The Sobolev approximation for the line force and line source function in a spherically-symmetrical stellar wind with continuum opacity. *Astr. Astrophys.*, 191, 87, 1988.

[891] J. Puls, R.-P. Kudritzki, A. Herrero, A. Pauldrach, S. Haser, D. Lennon, R. Gabler, S. Voels, J. Vilchez, S. Wachter, and A. Feldmeier. O star mass-loss and wind momentum rates in the Galaxy and the Magellanic Clouds: Observations and theoretical predictions. *Astr. Astrophys.*, 305, 171, 1996.

[892] A. Ralston and P. Rabinowitz. *A First Course in Numerical Analysis*. (New York: Dover Publications), 2nd edition, 2001.

[893] J. Rast, F. Cartier, F. Kneubühl, D. Huguenin, and E. Müller. Measurement of the absolute solar brightness temperature in the far-infrared with a balloon-borne interferometer. *Astr. Astrophys.*, 83, 199, 1980.

[894] T. Rauch and J. Deetjen. Handling of atomic data. In Hubeny et al. [536], page 103.

[895] T. Rauch, S. Dreizler, and K. Werner. New spectral analyses of pre-white dwarfs. In Isern et al. [565], page 221.

[896] T. Rauch, J. Köppen, and K. Werner. Spectral analysis of the planetary nebula K1–27 and its very hot H-deficient central star. *Astr. Astrophys.*, 286, 543, 1994.

[897] T. Rauch and K. Werner. NLTE model atmospheres for super-soft X-ray sources. In Hubeny et al. [541], page 85.

[898] A. Reichel and I. Vardavas. A simple quadrature method for the evaluation of the redistribution functions $R_{\rm III,IV}(x', \mathbf{n}'; x, \mathbf{n})$. *J. Quantit. Spectrosc. Radiat. Transf.*, 15, 929, 1975.

[899] I. Reid. Younger and brighter – New distances to globular clusters based on HIPPARCOS parallax measurements of local subdwarfs. *Astr. J.*, 114, 161, 1997.

[900] I. Reid. The HR diagram and the galactic distance scale after HIPPARCOS. *Ann. Rev. Astr. Astrophys.*, 37, 191, 1999.

[901] I. Reid, F. van Wyk, F. Marang, G. Roberts, D. Kilkenny, and S. Mahoney. A search for previously unrecognized metal-poor subdwarfs in the HIPPARCOS astrometric catalog. *Mon. Not. Roy. Astr. Soc.*, 325, 931, 2001.

[902] A. Reiz. The structure of stars with negligible content of heavy metals. *Astrophys. J.*, 120, 342, 1954.

[903] L. Relyea and R. Kurucz. A theoretical analysis of *uvby* photometry. *Astrophys. J. Suppl.*, 37, 45, 1978.

[904] R. Richtmyer and K. Morton. *Difference Methods for Initial-Value Problems*. (New York: Interscience Publishers), 2nd edition, 1967.

[905] W. Rindler. *Introduction to Special Relativity*. (Oxford: Oxford University Press), 1982.

[906] P. Rosenzweig and L. Anderson. A determination of the basic atmospheric parameters of ϕ Cassiopeiae. *Astrophys. J.*, 411, 207, 1993.

[907] S. Rosseland. Note on the absorption of radiation within a star. *Mon. Not. Roy. Astr. Soc.*, 84, 525, 1924.

[908] S. Rosseland. *Theoretical Astrophysics*. (Oxford: Oxford University Press), 1936.

[909] L. Rossi. Theoretical $B - V$ color indices and bolometric corrections for hot horizontal branch stars. *Astr. Astrophys.*, 74, 195, 1979.

[910] E. Rouef. Broadening of some solar Na I lines by atomic H. *Astr. Astrophys.*, 38, 41, 1975.

[911] E. Roueff and H. van Regemorter. Spectral line broadening due to atomic collisions. *Astr. Astrophys.*, 1, 69, 1969.

[912] J. Rountree and G. Sonneborn. *Spectral Classification with the International Ultraviolet Explorer: An Atlas of B-type Spectra*. NASA Ref. Pub. 1312. (Washington, DC: U.S. Government Printing Office), 1993.

[913] D. Roussel-Dupre. H I Lyα in the Sun: The effects of partial redistribution in the line wings. *Astrophys. J.*, 272, 723, 1983.

[914] F. Ruland, R. Griffin, R. Griffin, D. Biehl, and H. Holweger. Line blocking and equivalent widths in the spectrum of Pollux. *Astr. Astrophys. Suppl.*, 42, 391, 1980.

[915] H. Russell. Relations between the spectra and other characteristics of stars. *Popular Astronomy*, 22, 275, 1914.

[916] H. Russell and F. Saunders. New regularities in the spectra of the alkaline earths. *Astrophys. J.*, 61, 38, 1925.

[917] G. Rybicki. A modified Feautrier method. In Hunt [555], page 589.

[918] G. Rybicki. A novel approach to the solution of multilevel transfer problems. In Athay et al. [40], page 145.

[919] G. Rybicki. Escape probability methods. In Kalkofen [606], page 21.

[920] G. Rybicki. Recent advances in computational methods. In Crivellari et al. [264], page 1.

[921] G. Rybicki. A new kinetic equation for Compton scattering. *Astrophys. J.*, 584, 528, 2003.

[922] G. Rybicki. Improved Fokker-Planck equation for resonance line scattering. *Astrophys. J.*, 647, 709, 2006.

[923] G. Rybicki and D. Hummer. Non-coherent scattering. V. Thermalization distances and their distribution function. *Mon. Not. Roy. Astr. Soc.*, 144, 313, 1969.

[924] G Rybicki and D. Hummer. A generalization of the Sobolev method for flows with nonlocal radiative coupling. *Astrophys. J.*, 219, 654, 1978.

[925] G. Rybicki and D. Hummer. An accelerated lambda iteration method for multilevel radiative transfer. I. Non-overlapping lines with background continuum. *Astr. Astrophys.*, 245, 171, 1991.

[926] G. Rybicki and D. Hummer. An accelerated lambda iteration method for multilevel radiative transfer. II. Overlapping transitions with full continuum. *Astr. Astrophys.*, 262, 209, 1992.

[927] G. Rybicki and A. Lightman. *Radiative Process in Astrophysics*. (New York: John Wiley & Sons), 1979.

[928] Y. Saad. *Iterative Methods for Sparse Linear Systems*. (Philadelphia, SIAM), 2nd edition, 2003.

[929] Y. Saad and M. Schultz. Conjugate gradient-like algorithms for solving nonsymmetric linear systems. *Math. Comp.*, 44, 417, 1985.

[930] Y. Saad and M. Schultz. GMRES: A generalized minimal residual algorithm for solving nonsymmetric linear systems. *SIAM J. Sci. Statis. Comp.*, 7, 178, 1986.

[931] M. Saha. Ionization in the solar chromosphere. *Phil. Mag.*, 40, 472, 1920.

[932] M. Saha. On a physical theory of stellar spectra. *Proc. Roy. Soc*; 99A, 135, 1921.

[933] S. Sahal-Bréchot. Impact theory of the broadening and shift of spectral lines due to electrons and ions in a plasma. *Astr. Astrophys.*, 2, 322, 1969.

[934] S. Sahal-Bréchot. Stark broadening and isolated lines in the impact approximation. *Astr. Astrophys.*, 35, 319, 1974.

[935] S. Sahal-Bréchot, M. Dimitrijevic and N. Moreau. Virtual Laboratory Astrophysics: the STARK-B database for spectral line broadening by collisions with charged particles and its link to the European project VAMDC. *J. Phys. Conf. Ser.*, 397, 1, 2012.

[936] S. Sahal-Bréchot and E. Segré. Semi-classical calculations of electron and ion broadening of the strongest UV ion lines of astrophysical interest. *Astr. Astrophys.*, 13, 161, 1971.

[937] E. Salpeter. Nuclear reactions in stars. Buildup from He. *Phys. Rev.*, 107, 515, 1957.

[938] A. Sandage. The population concept, globular clusters, subdwarfs, ages, and the collapse of the Galaxy. *Ann. Rev. Astr. Astrophys.*, 24, 421, 1986.

[939] A. Sandage and O. Eggen. On the existence of subdwarfs in the $(M_{\rm bol}, \log T_{\rm eff})$ diagram. *Mon. Not. Roy. Astr. Soc.*, 119, 278, 1959.

[940] A. Sandage and M. Schwarzschild. Inhomogeneous stellar models. II. Models with exhausted cores in gravitational contraction. *Astrophys. J.*, 116, 463, 1952.

[941] A. Sandage and G. Wallerstein. Color-magnitude diagram for the disk globular cluster NGC 6356 compared with halo clusters. *Astrophys. J.*, 131, 598, 1960.

[942] A. Santolaya-Rey, J. Puls, and A. Herrero. Atmospheric NLTE-models for the spectroscopic analysis of luminous blue stars with winds. *Astr. Astrophys.*, 323, 488, 1997.

[943] A. Sarmiento and J. Canto. A fast and reliable method for computing free-bound emission coefficients for hydrogenic ions. *Rev. Mex. Astr. Astrof.*, 11, 61, 1985.

[944] W. Saslow and D. Mills. Raman scattering by hydrogenic systems. *Phys. Rev.*, 187, 1025, 1969.

[945] A. Sauval and J. Tatum. A set of partition functions and equilibrium constants for 300 diatomic molecules of astrophysical interest. *Astrophys. J. Suppl.*, 56, 193, 1984.

[946] D. Saxon. *Elementary Quantum Mechanics*. (San Francisco: Holden-Day), 1968.

[947] D. Schaerer. Combined stellar structure and atmosphere models for massive stars. Wolf-Rayet models with spherically outflowing envelopes. *Astr. Astrophys.*, 309, 129, 1996.

[948] D. Schaerer. Combined stellar structure and atmosphere models for massive stars: New ionizing fluxes and their impact on H II regions. In Gallagher [359], page 71.

[949] D. Schaerer, C. Charbonnel, G. Meynet, A. Maeder, and G. Schaller. Grids of stellar models. IV. From 0.8 to 120 $M_\odot$ at Z = 0.040. *Astr. Astrophys. Suppl.*, 102, 339, 1993.

[950] D. Schaerer, A. de Koter, W. Schmutz, and A. Maeder. Combined stellar structure and atmosphere models for massive stars. I. Interior evolution and wind properties on the main sequence. *Astr. Astrophys.*, 310, 837, 1996.

[951] D. Schaerer, A. de Koter, W. Schmutz, and A. Maeder. Combined stellar structure and atmosphere models for massive stars. II. Spectral evolution on the main sequence. *Astr. Astrophys.*, 312, 475, 1996.

[952] D. Schaerer, G. Meynet, A. Maeder, and G. Schaller. Grids of stellar models. II. From 0.8 to 120 $M_\odot$ at $Z = 0.008$. *Astr. Astrophys. Suppl.*, 98, 523, 1993.

[953] D. Schaerer and W. Schmutz. Hydrodynamic atmosphere models for hot luminous stars. *Astr. Astrophys.*, 288, 231, 1994.

[954] D. Schaerer and W. Schmutz. Hydrodynamic atmosphere models for hot luminous stars. II. Method and improvements over unified models. *Space Sci. Rev.*, 66, 177, 1994.

[955] G. Scharmer. Solutions to radiative transfer problems using approximate lambda operators. *Astrophys. J.*, 249, 720, 1981.

[956] G. Scharmer. A linearization method for solving partial redistribution problems. *Astr. Astrophys.*, 117, 83, 1983.

[957] G. Scharmer. Accurate solutions to non-LTE problems using approximate lambda operators. In Kalkofen [606], page 173.

[958] G. Scharmer and M. Carlsson. A new approach to multi-level non-LTE radiative transfer problems. *J. Comp. Phys.*, 59, 56, 1985.

[959] G. Scharmer and M. Carlsson. A new method for solving multi-level non-LTE problems. In Beckman and Crivellari [100], page 189.

[960] G. Scharmer and Å. Nordlund. `DQPT`: A computer program for solving non-LTE problems for two-level atoms in one-dimensional semi-infinite media with velocity fields. Technical Report 19, Stockholms Observatorium, Saltsjobaden, Sweden.

[961] L. Schiff. *Quantum Mechanics*. (New York: McGraw-Hill Book Company), 3rd edition, 1968.

[962] J. Schmid-Burgk. Radiative transfer through spherically-symmetric atmospheres and shells. *Astr. Astrophys.*, 40, 249, 1975.

[963] H. Schmid. Identification of the emission bands at $\lambda\lambda$ 6830, 7088. *Astr. Astrophys.*, 211, L31, 1989.

[964] H. Schmid. Raman scattering and symbiotic stars. In Gull et al. [412], page 347.

[965] J. Schmid-Burgk and M. Scholz. Transfer in spherical media using integral equations. In Kalkofen [606], page 381.

[966] M. Scholz and Y. Takeda. Model study of wavelength-dependent limb darkening and radii of M-type giants and supergiants. *Astr. Astrophys.*, 186, 200, 1987.

[967] K. Schönberg and K. Hempe. Multilevel line formation in the comoving frame: Accurate solution using an approximate Newton-Raphson operator. *Astr. Astrophys.*, 163, 151, 1986.

[968] M. Schönberg and S. Chandrasekhar. On the evolution of main-sequence stars. *Astrophys. J.*, 96, 161, 1942.

[969] D. Schönberner. Late stages of stellar evolution. II. Mass loss and the transition of asymptotic giant branch stars into hot remnants. *Astrophys. J.*, 272, 708, 1983.

[970] T. Schöning and K. Butler. Stark broadening of He II lines. *Astr. Astrophys. Suppl.*, 78, 51, 1989.

[971] T. Schöning and K. Butler. Stark broadening of He II lines and new results in astrophysical spectroscopy. *Astr. Astrophys.*, 219, 326, 1989.

[972] E. Schrödinger. Quantization as a problem of eigenvalues. *Ann. Physik*, 79, 361, 1926. English translation in Shearer and Deans [995], page 1.

[973] E. Schrödinger. Quantization as a problem of eigenvalues. II. *Ann. Physik*, 79, 489, 1926. English translation in Shearer and Deans [995], page 13.

[974] E. Schrödinger. The relation between the quantum mechanics of Heisenberg, Born, and Jordan, and that of Schrödinger. *Ann. Physik*, 79, 734, 1926. English translation in Shearer and Deans [995], page 45.

[975] E. Schrödinger. Quantization as a problem of eigenvalues. III. Perturbation theory, with application to the Stark effect of the Balmer lines. *Ann. Physik*, 80, 437, 1926. English translation in Shearer and Deans [995], page 62.

[976] E. Schrödinger. Quantization as a problem of eigenvalues. IV. *Ann. Physik*, 81, 109, 1926. English translation in Shearer and Deans [995], page 102.

[977] L. Schubert. Modification of a quasi-Newton method for nonlinear equations with a sparse Jacobian. *Math. Comput.*, 24, 27, 1970.

[978] R. Schulte-Ladbeck, D. Hillier, and J. Herald. The Hopkins Ultraviolet Telescope far-ultraviolet spectral atlas of Wolf-Rayet stars. *Astrophys. J.*, 454, L51, 1995.

[979] A. Schuster. Radiation through a foggy atmosphere. *Astrophys. J.*, 21, 1, 1905.

[980] K. Schwarzschild. On the equilibrium of the Sun's atmosphere. English translation in Menzel [721], page 25.

[981] K. Schwarzschild. Diffusion and absorption in the Sun's atmosphere. English translation in Menzel [721], page 35.

[982] M. Schwarzschild. *Structure and Evolution of the Stars*. (Princeton: Princeton University Press), 1958.

[983] M. Schwarzschild and R. Härm. Red giants of Population II. II. *Astrophys. J.*, 136, 158, 1962.

[984] M. Schwarzschild and R. Härm. Thermal instability in non-degenerate stars. *Astrophys. J.*, 142, 855, 1965.

[985] M. Schwarzschild and R. Härm. Hydrogen mixing by helium-shell flashes. *Astrophys. J.*, 150, 961, 1967.

[986] M. Schwarzschild, L. Searle, and R. Howard. On the colors of subdwarfs. *Astrophys. J.*, 122, 353, 1955.

[987] M. Schwarzschild and H. Selberg. Red giants of Population II. *Astrophys. J.*, 136, 150, 1962.

[988] N. Scott and P. Burke. Electron scattering by atoms and ions using the Breit-Pauli Hamiltonian: An *R*-matrix approach. *J. Phys. B*, 13, 4299, 1980.

[989] R. Sears and A. Whitford. Six-color photometry of stars. XII. Colors of Hyades and subdwarf stars. *Astrophys. J.*, 155, 899, 1969.

[990] M. Seaton. The quantum defect method. *Mon. Not. Roy. Astr. Soc.*, 118, 504, 1958.

[991] M. Seaton. Outer-region contributions to radiative transition probabilities. *J. Phys. B*, 19, 2601, 1986.

[992] M. Seaton. Wing formulae for plasma-broadened spectral lines of hydrogenic ions. *J. Phys. B*, 28, 565, 1995.

[993] M. Selby, D. Blackwell, A. Petford, and M. Shallis. Measurement of the absolute flux from Vega in the K-band, 2.2μ. *Mon. Not. Roy. Astr. Soc.*, 193, 111, 1980.

[994] M. Selby, C. Mountain, D. Blackwell, A. Petford, and S. Leggett. Measurement of the absolute monochromatic flux from Vega at λ 2.20 and λ 3.80 μ by comparison with a furnace. *Mon. Not. Roy. Astr. Soc.*, 203, 795, 1983.

[995] J. Shearer and W. Deans, editors. *Collected Papers on Wave Mechanics by E. Schrödinger*. (New York: Chelsea Publishing Company), 1978.

[996] A. Shestakov, D. Kershaw, and M. Prasad. Evaluation of integrals of the Compton scattering cross-section. *J. Quantit. Spectrosc. Radiat. Transf.*, 40, 377, 1988.

[997] R. Shine, R. Milkey, and D. Mihalas. Resonance line transfer with partial redistribution. V. The solar Ca II lines. *Astrophys. J.*, 199, 724, 1975.

[998] R. Shine, R. Milkey, and D. Mihalas. Resonance line transfer with partial redistribution. VI. The Ca II K-line in solar-type stars. *Astrophys. J.*, 201, 222, 1975.

[999] Z. Shippony and W. Read. A highly accurate Voigt function algorithm. *J. Quantit. Spectrosc. Radiat. Transf.*, 50, 635, 1993.

[1000] B. Shore and D. Menzel. *Principles of Atomic Spectra*. (New York: John Wiley & Sons), 1968.

[1001] S. Shore. *An Introduction to Astrophysical Hydrodynamics*. (San Diego: Academic Press), 1992.

[1002] S. Shore. Blue sky and hot piles: The evolution of radiative transfer theory from atmospheres to nuclear reactors. *Historica Mathematica*, 29, 463, 2002.

[1003] S. Shore. *The Tapestry of Modern Astrophysics*. (Hoboken: Wiley-Interscience), 2003.

[1004] C. Short and P. Hauschildt. Massive multispecies, multilevel non-LTE model atmospheres for novae in outburst. *Astrophys. J.*, 525, 375, 1999.

[1005] C. Short and P. Hauschildt. Atmospheric models of red giants with massive-scale non-local thermodynamic equilibrium. *Astrophys. J.*, 596, 501, 2003.

[1006] F. Shu. *The Physics of Astrophysics*. Volume 1: Radiation. (Mill Valley: University Science Books), 1991.

[1007] F. Shu. *The Physics of Astrophysics*. Volume 2: Gas Dynamics. (Mill Valley: University Science Books), 1992.

[1008] A. Sidi, W. Ford, and D. Smith. Acceleration of convergence of vector sequences. *SIAM J. Num. Anal.*, 23, 178, 1986.

[1009] J. Slater. Radiation and absorption on Schrödinger's theory. *Proc. Nat. Acad. Sci. U.S.*, 13, 7, 1927.

[1010] J. Slater. The theory of complex spectra. *Phys. Rev.*, 34, 1293, 1929.

[1011] J. Slater. *Quantum Theory of Atomic Structure*. Volume I. (New York: McGraw-Hill Publishing Company), 1960.

[1012] J. Slater. *Quantum Theory of Atomic Structure*. Volume II. (New York: McGraw-Hill Publishing Company), 1960.

[1013] J. Slater. *Quantum Theory of Molecules and Solids*. Volume I. (New York: McGraw-Hill Publishing Company), 1963.

[1014] A. Slettebak, editor. *Be and Shell Stars*. IAU Symposium No. 70. (Dordrecht: Reidel), 1976.

[1015] W. Smart. *Spherical Astronomy*. (Cambridge: Cambridge University Press), 1956.

[1016] B. Smirnov. Excitation transfer in atomic collisions. *Soviet Phys. J.E.T.P.*, 24, 314, 1967.

[1017] D. Smith, W. Ford, and A. Sidi. Extrapolation methods for vector sequences. *SIAM Rev.*, 29, 199, 1987.

[1018] E. Smith, J. Cooper, and C. Vidal. Unified classical-path treatment of Stark broadening in plasmas. *Phys. Rev.*, 185, 140, 1969.

[1019] G. Smith. *Numerical Solution of Partial Differential Equations*. (New York: Oxford University Press), 1965.

[1020] G. Smith and M. Ruck. A new abundance analysis of the super-metal-rich K2 giant μ Leonis. *Astr. Astrophys.*, 356, 570, 2000.

[1021] P. Smith and W. Wiese, editors. *Atomic and Molecular Data for Space Astronomy*. (Berlin: Springer-Verlag), 1992.

[1022] C. Sneden, J. Cowan, J. Lawler, I. Ivans, S. Burles, T. Beers, F. Primas, V. Hill, J. Truran, G. Fuller, B. Pfeiffer, and K. Kratz. The extremely metal-poor, neutron capture-rich star CS 22892–052: A comprehensive abundance analysis. *Astrophys. J.*, 591, 936, 2003.

[1023] C. Sneden, H. Johnson, and B. Krupp. A statistical method for treating molecular line opacities. *Astrophys. J.*, 204, 281, 1976.

[1024] C. Sneden, G. Preston, A. McWilliam, and L. Searle. Ultra-metal-poor halo stars: The remarkable spectrum of CS 22892-052. *Astrophys. J.*, 431, 27, 1994.

[1025] I. Sobelman. *Atomic Spectra and Radiative Transitions*. (Berlin: Springer-Verlag), 1979.

[1026] I. Sobelman, L. Vainshtein, and E. Yukov. *Excitation of Atoms and Broadening of Spectral Lines*. (Berlin: Springer–Verlag), 1981.

[1027] V. Sobolev. The diffusion of Lα radiation in nebulae and stellar envelopes. *Sov. Astr.*, 1, 678, 1957.

[1028] V. Sobolev. Stars with bright spectral lines. In Ambartsumyan [27], chapter 28–29, pages 478–519.

[1029] V. Sobolev. *Moving Envelopes of Stars*. (Cambridge: Harvard University Press), 1960.

[1030] V. Sobolev. *A Treatise on Radiative Transfer*. (Princeton: D. Van Nostrand Company), 1963.

[1031] H. Socas–Navarro and J. Trujillo Bueno. Linearization versus preconditioning: Which approach is best for solving multilevel transfer problems? *Astrophys. J.*, 490, 383, 1997.

[1032] W. Somerville. The continuous absorption coefficient of the negative hydrogen molecular ion. *Astrophys. J.*, 139, 192, 1964.

[1033] G. Sonneborn, H. Moos, and B.-G. Andersson, editors. *Astrophysics in the Far Ultraviolet: Five Years of Discovery With FUSE*. (San Francisco: Astronomical Society of the Pacific), 2006.

[1034] E. Spiegel. Convection in stars I. Basic Boussinesq convection. *Ann. Rev. Astr. Astrophys.*, 9, 323, 1971.

[1035] E. Spiegel. Convection in stars II. Special effects. *Ann. Rev. Astr. Astrophys.*, 10, 261, 1972.

[1036] M. Spite, F. Spite, R. Cayrel, V. Hill, B. Nordström, B. Barbuy, T. Beers, and P. Nissen. Abundances in very metal–poor stars. *Astrophys. Space Sci.*, 265, 141, 1969.

[1037] L. Spitzer. *Physics of Fully Ionized Gases*. (New York: John Wiley & Co.), 1956.

[1038] L. Spitzer and J. Greenstein. Continuous emission from planetary nebulae. *Astrophys. J.*, 114, 407, 1951.

[1039] G. Stasinska and D. Schaerer. Combined stellar structure and atmosphere models for massive stars. IV. The impact on the ionization structure of single star H II regions. *Astr. Astrophys.*, 322, 615, 1997.

[1040] J. Stebbins and G. Kron. Six color photometry of stars. X. The stellar magnitude and color index of the sun. *Astrophys. J.*, 126, 266, 1957.

[1041] W. Steenbock and H. Holweger. Statistical equilibrium of lithium in cool stars of different metallicity. *Astr. Astrophys.*, 130, 319, 1984.

[1042] C. Stehlé. Stark broadening of the hydrogen Lyman-α line from the center to the near line wings for low-density plasmas. *Phys. Rev. A*, 34, 4153, 1986.

[1043] C. Stehlé. Stark broadening of hydrogen Lyman and Balmer in the conditions of stellar envelopes. *Astr. Astrophys. Suppl.*, 104, 509, 1994.

[1044] C. Stehlé. Stark profiles of He^+. *Astr. Astrophys.*, 292, 699, 1994.

[1045] C. Stehlé and R. Hutcheon. Extensive tabulations of Stark broadened H–line profiles. *Astr. Astrophys. Suppl.*, 140, 93, 1999.

[1046] R. Stein and Å. Nordlund. Simulations of solar granulation. I. General properties. *Astrophys. J.*, 499, 914, 1998.

[1047] R. Stein and Å. Nordlund. Realistic solar convection simulations. *Solar Phys.*, 192, 91, 2000.

[1048] R. Stein and Å. Nordlund. Radiative transfer in 3D numerical simulations. In Hubeny et al. [536], page 519.

[1049] O. Steiner. A rapidly converging temperature correction procedure using operator perturbation. *Astr. Astrophys.*, 231, 278, 1990.

[1050] O. Steiner. Fast solution of radiative transfer problems using a method of multiple grids. *Astr. Astrophys.*, 242, 290, 1991.

[1051] M. Steffen and D. Schönberner. Long–term evolution of AGB wind envelopes: Insights from hydrodynamical models. In Le Bertre et al. [113], page 379.

[1052] C. Sterken, editor. *The Future of Photometric, Spectrophotometric and Polarimetric Standardization*. (San Francisco: Astronomical Society of the Pacific), 2007.

[1053] J. Stilley and J. Callaway. Free–free absorption coefficient of the negative hydrogen ion. *Astrophys. J.*, 160, 245, 1970.

[1054] M. Stix. *The Sun: An Introduction*. (Berlin: Springer–Verlag), 2nd edition, 2002.

[1055] E. Stoner. The limiting density in white dwarf stars. *Phil. Mag.*, 7, 63, 1929.

[1056] E. Stoner. The equilibrium of dense stars. *Phil. Mag.*, 9, 944, 1930.

[1057] S. Strom and E. Avrett. The temperature structure of early–type model stellar atmospheres. II. A grid of stellar models. *Astrophys. J. Suppl.*, 12, 1, 1965.

[1058] S. Strom, O. Gingerich, and K. Strom. Studies in non–grey stellar atmospheres. III. The metal abundances of Sirius and Vega. *Astrophys. J.*, 146, 880, 1966.

[1059] S. Strom and R. Kurucz. A statistical procedure for computing line–blanketed model stellar atmospheres with applications to the F5 IV star Procyon. *J. Quantit. Spectrosc. Radiat. Transf.*, 6, 591, 1966.

[1060] B. Strömgren. The opacity of stellar matter and the H content of the stars. *Z. für Astrophys.*, 4, 118, 1932.

[1061] B. Strömgren. The boundary–value problem of the theory of stellar absorption lines. *Astrophys. J.*, 86, 1, 1937.

[1062] O. Struve. The Stark effect in stellar spectra. *Astrophys. J.*, 69, 173, 1929.

[1063] O. Struve. Pressure effects in stellar spectra. *Astrophys. J.*, 70, 85, 1929.

[1064] O. Struve and C. Elvey. The intensities of stellar absorption lines. *Astrophys. J.*, 79, 406, 1934.

[1065] J. Surdej. Formation of resonance doublet profiles in rapidly expanding envelopes. *Astrophys. Space Sci.*, 73, 101, 1980.

[1066] R. Sutherland. Accurate free–free Gaunt factors for astrophysical plasmas. *Mon. Not. Roy. Astr. Soc.*, 300, 321, 1998.

[1067] J. Sykes. Approximate integration of the equation of transfer. *Mon. Not. Roy. Astr. Soc.*, 111, 377, 1951.

[1068] J. Synge and A. Schild. *Tensor Calculus*. (New York: Dover Publications), 1978.

[1069] J. Tatum. Accurate partition functions and dissociation equilibrium constants of diatomic molecules of astrophysical interest. *Pub. Domin. Astrophys. Obs.*, 13, 1, 1966.

[1070] B. Taylor. Catalogs of temperatures and [Fe/H] averages for G and K stars. *Astr. Astrophys. Suppl.*, 134, 121, 1999.

[1071] P. Ten Bruggencate. Die Entstehung der Fraunhoferschen Linien in der Sonnenatmosphäre. *Z. für Astrophys.*, 4, 159, 1932.

[1072] P. Ten Bruggencate, H. Gollnow, S. Günther, and W. Strohmeier. Die Mitte–Rand–Variation der Balmerlinien Hα auf der Sonnenscheibe. *Z. für Astrophys.*, 26, 51, 1949.

[1073] L. Thomas. The kinematics of an electron with an axis. *Phil. Mag.*, 3, 1, 1927.

[1074] L. Thomas. The radiation field in a fluid in motion. *Quart. J. Math.* (Oxford), 1, 239, 1930.

[1075] R. Thomas. The source function in a non–equilibrium atmosphere. I. The resonance lines. *Astrophys. J.*, 125, 260, 1957.

[1076] R. Thomas. Comment on the use of net rate processes and the equivalent 2-level atom in non–LTE computations. *Ann. d'Astrophys.*, 23, 871, 1960.

[1077] R. Thomas. The source function in a non–equilibrium atmosphere. IV. Evaluation and application of the net radiative bracket. *Astrophys. J.*, 131, 429, 1960.

[1078] R. Thomas. Comment on the difference between a non–LTE and a pure–absorption model for the line–blanketing effect. *Astrophys. J.*, 141, 333, 1965.

[1079] R. Trampedach, W. Däppen, and V. Baturn. A synoptic comparison of the Mihalas–Hummer–Däppen and OPAL equations of state. *Astrophys. J.*, 646, 560, 2006.

[1080] N. Trams, L. Waters, C. Waelkens, H. Lamers, and W. van der Veen. The effect of mass loss on the evolution of low–mass post–AGB stars. *Astr. Astrophys.*, 218, L1, 1989.

[1081] N. Tran-Minh, N. Feautrier and H. Van Regemorter. On simple quantum theories of pressure broadening: the exact resonance approximation applied to Lyman alpha. *J. Quantit. Spectrosc. Radiat. Transf.*, 16, 849, 1976.

[1082] N. Tran-Minh and H. Van Regemorter. Quantum theory of Stark broadening by electrons. *J. Phys. B*, 5, 903, 1972.

[1083] G. Traving. *Über die Theorie der Druckverbreiterung von Spektrallinien*. (Karlsruhe: Verlag G. Braun), 1960.

[1084] P. Tremblay and P. Bergeron. Spectroscopic analysis of DA white dwarfs: Stark broadening of hydrogen lines including nonideal effects. *Astrophys. J.*, 696, 1755, 2009.

[1085] J. Trujillo–Bueno. The generation and transfer of polarized radiation in stellar atmospheres. In Hubeny et al. [536], page 551.

[1086] J. Trujillo Bueno and P. Fabiani Bendicho. A novel iterative scheme for the very fast and accurate solution of non–LTE radiative transfer problems. *Astrophys. J.*, 455, 646, 1995.

[1087] D. Turnshek, R. Bohlin, R. Williamson, O. Lupie, J. Koorneef, and D. Morgan. An atlas of Hubble Space Telescope photometric, spectrophotometric, and polarimetric calibration objects. *Astr. J.*, 99, 1243, 1990.

[1088] D. E. Turnshek, D. A. Turnshek, E. Craine, and P. Boeshaar. *An Atlas of Digital Spectra of Cool Stars: Types G, K, M, S and C* (Tucson: Western Research Company), 1985.

[1089] J. Twitty, P. Rarig, and R. Thompson. A comparison of fast codes for the evaluation of the Voigt profile. *J. Quantit. Spectrosc. Radiat. Transf.*, 24, 529, 1980.

[1090] G. Uhlenbeck and S. Goudsmit. Spinning electrons and the structure of spectra. *Nature*, 117, 264, 1926.

[1091] H. Uitenbroek. Multilevel radiative transfer with partial frequency redistribution. *Astrophys. J.*, 557, 389, 2001.

[1092] H. Uitenbroek. The effect of coherent scattering on radiative losses in the solar Ca II K Line. *Astrophys. J.*, 565, 1312, 2002.

[1093] H. Uitenbroek. Multi–level accelerated lambda iteration with partial redistribution. In Hubeny et al. [536], page 597.

[1094] A. Underhill and J. Waddell. *Stark Broadening Functions for the H–Lines*. NBS Circular No. 603. (Washington, DC: U.S. Government Printing Office), 1959.

[1095] W. Unno. On the radiation pressure in a planetary nebula. I. *Pub. Astr. Soc. Japan*, 2, 53, 1950.

[1096] W. Unno. On the radiation pressure in a planetary nebula. II. *Pub. Astr. Soc. Japan*, 3, 158, 1952.

[1097] W. Unno. On the radiation pressure in a planetary nebula. III. The problem of He II. *Pub. Astr. Soc. Japan*, 3, 178, 1952.

[1098] W. Unno. Note on the Zanstra redistribution in planetary nebula. *Pub. Astr. Soc. Japan*, 4, 100, 1952.

[1099] A. Unsöld. Wasserstoff und Helium in Sternatmosphären. *Z. für Astrophys.*, 3, 81, 1931.

[1100] A. Unsöld. Quantitative Analyse des B0–Sternes τ Scorpii. II. Teil. Deutung des Linienspektrums, Kosmische Häufigkeit der leichten Element, Elektronendruck p_e, Temperatur T und Schwerebeschleunigung g in der Atmosphäre von τ Scorpii. *Z. für Astrophys.*, 21, 22, 1942.

[1101] A. Unsöld. Über den Einfluss der Fraunhofer–Linien auf die Temperaturschichtung der Sternatmosphären. *Z. für Astrophys.*, 22, 356, 1943.

[1102] A. Unsöld. *Physik der Sternatmosphären*. (Berlin: Springer–Verlag), 2nd edition, 1955.

[1103] A. Unsöld. Zur Deutung der kontinuierlichen Sternspektren. *Z. für Astrophys.*, 49, 1, 1960.

[1104] D. Vanbeveren, W. Van Rensbergen, and C. de Loore, editors. *Evolution of Massive Stars: A Confrontation Between Theory and Observation*. (Dordrecht: Kluwer), 1994.

[1105] D. Van Blerkom. Theory of Wolf–Rayet spectra. In Bappu and Sahade [79], page 165.

[1106] D. Van Blerkom and D. Hummer. The normalized on-the-spot approximation for line transfer problems. *J. Quantit. Spectrosc. Radiat. Transf.*, 9, 1567, 1969.

[1107] H. van Regemorter. Rate of collisional excitation in stellar atmospheres. *Astrophys. J.*, 136, 906, 1962.

[1108] M. Vardya. Pressure dissociation. *Astrophys. J.*, 132, 905, 1960.

[1109] M. Vardya. Hydrogen–helium adiabats for late–type stars. *Astrophys. J. Suppl.*, 4, 281, 1960.

[1110] M. Vardya. Physical atmospheric parameters for late–type stars. *Astrophys. J.*, 133, 107, 1961.

[1111] M. Vardya. Planck and Rosseland mean of Rayleigh scattering by H_2. *Astrophys. J.*, 135, 303, 1962.

[1112] M. Vardya. Opacity at $\lambda 1.65\mu$ in the late–type stars . In Avrett et al. [64], page 50.

[1113] M. Vardya. Thermodynamics of a solar composition gaseous mixture. *Mon. Not. Roy. Astr. Soc.*, 129, 205, 1965.

[1114] M. Vardya. Role of negative ions in late–type stars. *Mem. Roy. Astr. Soc.*, 70, 249, 1967.

[1115] M. Vardya. Atmospheres of very late–type stars. *Ann. Rev. Astr. Astrophys.*, 8, 87, 1970.

[1116] M. Vardya and R. Wildt. Molecules and late–type stellar models. *Astrophys. J.*, 131, 448, 1960.

[1117] C. van't Veer-Menneret, D. Katz, R. Cayrel, and C. Soubrian. A grid of metal–poor model stellar atmospheres for stars born in the early Galaxy. *Astrophys. Space Sci.*, 265, 257, 1969.

[1118] J. Vernazza, E. Avrett, and R. Loeser. Structure of the solar chromosphere. Basic computations and summary of the results. *Astrophys. J.*, 184, 605, 1973.

[1119] J. Vernazza, E. Avrett, and R. Loeser. Structure of the solar chromosphere. III. Models of the EUV brightness components of the quiet Sun. *Astrophys. J. Suppl.*, 45, 635, 1981.

[1120] C. Vidal, J. Cooper, and E. Smith. Hydrogen Stark broadening calculations with the unified classical path theory. *J. Quantit. Spectrosc. Radiat. Transf.*, 10, 1011, 1970.

[1121] C. Vidal, J. Cooper, and E. Smith. *Unified Theory Calculations of Stark Broadened H–Lines Including Lower State Interactions*. NBS Monograph No. 120. (Washington, DC: U.S. Government Printing Office), 1971.

[1122] C. Vidal, J. Cooper, and E. Smith. Hydrogen Stark–broadening tables. *Astrophys. J. Suppl.*, 25, 37, 1973.

[1123] T. Viik. An efficient method to calculate Ambarzumian's, Chandrasekhar's, and Hopf's functions. *Astrophys. Space Sci.*, 127, 285, 1986.

[1124] W. Vincenti and C. Kruger. *Introduction to Physical Gas Dynamics*. (New York: John Wiley & Sons), 1965.

[1125] E. Vitense. Die Wasserstoffkonvektionszone der Sonne. *Z. für Astrophys.*, 32, 135, 1953.

[1126] J. Waddell. Solar center–limb variations of the Na D lines from photoelectric observations. *Ann. d'Astrophys.*, 23, 917, 1960.

[1127] N. Walborn. Some spectroscopic characteristics of the OB stars: An investigation of the space distribution of certain OB stars and the reference frame of the classification. *Astrophys. J. Suppl.*, 23, 257, 1971.

[1128] N. Walborn, J. Nichols–Bohlin, and R. Panek. *International Ultraviolet Explorer Atlas of O–Type Spectra from 1200 to 1900Å*. NASA Ref. Pub. 1155. (Washington, DC: U.S. Government Printing Office), 1985.

[1129] G. Wallerstein. Abundances in G dwarfs. VI. A survey of field stars. *Astrophys. J. Suppl.*, 6, 407, 1962.

[1130] G. Wallerstein, T. Greene, and L. Tomley. Abundances of the light elements in the H–poor star HD 30353. *Astrophys. J.*, 150, 245, 1967.

[1131] G. Wallerstein, J. Greenstein, R. Parker, H. Helfer, and L. Aller. Red giants with extreme metal deficiencies. *Astrophys. J.*, 137, 280, 1963.

[1132] G. Wallerstein and H. Helfer. Abundances in G dwarf stars. I. A comparison of two stars in the Hyades with the Sun. *Astrophys. J.*, 129, 347, 1959.

[1133] G. Wallerstein and H. Helfer. Abundances in G dwarf stars. II. The high–velocity star 85 Pegasi. *Astrophys. J.*, 129, 720, 1959.

[1134] G. Wallerstein and H. Helfer. Abundances in G dwarfs. V. The metal–rich star 20 Leo Minoris and two comparison stars. *Astrophys. J.*, 133, 562, 1961.

[1135] G. Wallerstein, H. Helfer, L. Aller, R. Parker, and J. Greenstein. Red giants with extreme metal deficiencies. *Astrophys. J.*, 137, 280, 1963.

[1136] G. Wallerstein and W. Hunziker. Abundances in high–velocity A stars. II. The metal–poor star HD 109995. *Astrophys. J.*, 140, 214, 1964.

[1137] G. Wallerstein, I. Iben, P. Parker, A. Boesgaard, G. Hale, A. Champagne, C. Barnes, F. Käppler, V. Smith, R. Hoffman, F. Timmes, C. Sneden, R. Boyd, B. Meyer, and D. Lambert. Synthesis of the elements in stars: Forty years of progress. *Rev. Mod. Phys.*, 69, 995, 1997.

[1138] W. Weise and J. Fuhr. Atomic transition probabilities for hydrogen, helium, and lithium. *J. Phys. Chem. Ref. Data*, 38, 565, 2009.

[1139] W. Weiss and A. Baglin, editors. *Inside the Stars.* (San Francisco: Astronomical Society of the Pacific), 1993.

[1140] V. Weisskopf. Zur Theorie der Resonanzfluoreszenz. *Ann. Physik*, 401, 23, 1931.

[1141] V. Weisskopf. Zur Theorie der Kopplungsbreite und der Stoss–dämpfung. *Z. für Phys.*, 75, 287, 1932.

[1142] V. Weisskopf. Zur Theorie der Kopplungsbreite. *Z. für Phys.*, 77, 398, 1933.

[1143] V. Weisskopf. Die Streuung des Lichtes an angeregten Atomen. *Z. für Phys.*, 85, 451, 1933.

[1144] V. Weisskopf and E. Wigner. Berechnung der natürlichen Linienbreite auf Grund der Diracschen Lichttheorie. *Z. für Phys.*, 63, 54, 1930.

[1145] V. Weisskopf and E. Wigner. Über die natürliche Linienbreite in der Strahlung des harmonischen Oszillators. *Z. für Phys.*, 65, 18, 1930.

[1146] C. von Weizsäcker. Element transformation inside stars. II. *Phy. Z.*, 39, 633, 1938.

[1147] R. Wells. Rapid approximation to the Voigt/Faddeeva function and its derivatives. *J. Quantit. Spectrosc. Radiat. Transf.*, 62, 29, 1999.

[1148] K. Werner. Construction of non–LTE model atmospheres using approximate lambda operators. *Astr. Astrophys.*, 161, 177, 1986.

[1149] K. Werner. Non–LTE model atmosphere calculations with approximate lambda operators: Application of tridiagonal operators. *Astr. Astrophys.*, 226, 265, 1989.

[1150] K. Werner, J. Deetjen, S. Dreizler, T. Nagel, T. Rauch, and S. Schuh. Model photospheres with accelerated lambda iteration. In Hubeny et al. [536], page 31.

[1151] K. Werner and S. Dreizler. The classical stellar atmosphere problem. *J. of Comp. and Appl. Math.*, 109, 65, 1999.

[1152] K. Werner, S. Dreizler, U. Heber, N. Kappelmann, J. Kruk, T. Rauch, and B. Wolff. Ultraviolet spectroscopy of hot compact stars. *Rev. Mod. Astr.*, 10, 219, 1997.

[1153] K. Werner and F. Herwig. The elemental abundances in bare planetary nebula central stars and the shell burning in AGB stars. *Pub. Astr. Soc. Pacific*, 118, 183, 2006.

[1154] K. Werner and D. Husfeld. Multi–level non–LTE line formation calculations using approximate Λ–operators. *Astr. Astrophys.*, 148, 417, 1985.

[1155] K. Werner, T. Rauch, M. Barstow, and J. Kruk. Chandra and FUSE spectroscopy of the hot bare stellar core H 1504+65. *Astr. Astrophys.*, 421, 1169, 2004.

[1156] K. Werner, T. Rauch, and J. Kruk. Quantitative spectral analysis of evolved low-mass stars. In Hubeny et al. [541], page 15.

[1157] K. Werner and B. Wolff. The EUV spectrum of the unique bare stellar core H 1504+65. *Astr. Astrophys.*, 347, L9, 1999.

[1158] A. Weselink. The color index of a black body with infinite temperature. *Bull. Astr. Inst. Neth.*, 10, 99, 1948.

[1159] J. Wheeler and R. Wildt. The absorption coefficient of the free–free transitions of the negative hydrogen ion. *Astrophys. J.*, 95, 281, 1942.

[1160] O. White. Limb–darkening observations of Hα, Hβ, and Hγ. *Astrophys. J. Suppl.*, 7, 333, 1962.

[1161] G. Wick. Über ebene Diffusionsprobleme. *Z. für Phys.*, 121, 702, 1943.

[1162] N. Wiener and E. Hopf. Über eine Klasse singulären Integralgleichungen. *Sitz. Der Preussischen Akad. Der Wiss., Math. Phys. Klasse*, p. 696, 1931.

[1163] B. Wienke. Relativistic invariance and photon–electron scattering kernels in transport theory. *Nucl. Sci. Engin.*, 52, 247, 1973.

[1164] R. Wildey, E. Burbidge, A. Sandage, and G. Burbidge. On the effects of Fraunhofer lines on *UBV* measurements. *Astrophys. J.*, 135, 94, 1962.

[1165] R. Wildt. Electron affinity in astrophysics. *Astrophys. J.*, 89, 295, 1939.

[1166] J. Wilkinson. *The Algebraic Eigenvalue Problem.* (Oxford: Oxford University Press), 1965.

[1167] R. Williams. Spectrophotometric determination of stellar temperatures. *Pub. Obs. Michigan*, 7, 93, 1939.

[1168] A. Willis, K. van der Hucht, P. Conti, and C. Garmany. An atlas of high resolution IUE ultraviolet spectra of 14 Wolf–Rayet stars. *Astr. Astrophys. Suppl.*, 63, 417, 1986.

[1169] O. Wilson. The structure of the atmosphere of the K–type component of ζ Aurigae. *Astrophys. J.*, 107, 126, 1948.

[1170] O. Wilson. Eclipses by extended atmospheres. In Greenstein [397], chapter 11, page 436.

[1171] O. Wilson and H. Abt. Chromospheric structure of the K–type component of ζ Aurigae. *Astrophys. J. Suppl.*, 1, 1, 1954.

[1172] C. Woodward, M. Bicay, and J. Shull, editors. *Tetons 4: Galactic Structure, Stars and the Interstellar Medium.* (San Francisco: Astronomical Society of the Pacific), 2001.

[1173] R. Woolley. Interlocking of lines in absorption spectra. *Mon. Not. Roy. Astr. Soc.*, 90, 779, 1930.

[1174] R. Woolley. Interlocking of triplets of absorption lines. *Mon. Not. Roy. Astr. Soc.*, 91, 864, 1931.

[1175] R. Woolley. The analysis of line intensities in stellar spectra. *Mon. Not. Roy. Astr. Soc.*, 92, 482, 1932.

[1176] R. Woolley. Fluoresence in Hα and Hβ. *Mon. Not. Roy. Astr. Soc.*, 94, 631, 1934.

[1177] R. Woolley. Non–coherent formation of absorption lines. *Mon. Not. Roy. Astr. Soc.*, 98, 624, 1938.

[1178] R. Woolley and W. Stibbs. *The Outer Layers of a Star*. (Oxford: Clarendon Press), 1953.

[1179] K. Wright. A study of line intensities in the spectra of four solar–type stars. *Pub. Domin. Astrophys. Obs.*, 8, 1, 1948.

[1180] M. Wrubel. Exact curves of growth for the formation of absorption lines according to the Milne–Eddington model. I. Total flux. *Astrophys. J.*, 109, 66, 1949.

[1181] M. Wrubel. Exact curves of growth for the formation of absorption lines according to the Milne–Eddington model. II. Center of the disk. *Astrophys. J.*, 111, 157, 1950.

[1182] M. Wrubel. Exact curves of growth. III. Schuster–Schwarzschild model. *Astrophys. J.*, 119, 51, 1954.

[1183] B. Yada. The effect of two–photon emission on the radiation field of planetary nebulae. I. *Pub. Astr. Soc. Japan*, 5, 128, 1953.

[1184] B. Yada. The effect of two–photon emission on the radiation field of planetary nebulae. II. *Pub. Astr. Soc. Japan*, 6, 76, 1954.

[1185] Y. Yamashita, K. Nariai, and T. Norimoto. *An Atlas of Representative Stellar Spectra*. (Tokyo: University of Tokyo Press), 1978.

[1186] J. Yelnik, K. Burnett, J. Cooper, R. Ballagh, and D. Voslamber. Redistribution of radiation for the wings of Lyα. *Astrophys. J.*, 248, 705, 1981.

[1187] G.–J. van Zadelhoff, C. Dullemond, F. van der Tak, J. Yates, S. Doty, V. Ossenkopf, M. Hogerheijde, M. Juvela, H. Wiesemeyer, and F. Schöier. Numerical methods for non–LTE line radiative transfer: Performance and convergence characteristics. *Astr. Astrophys.*, 395, 373, 2002.

[1188] Ya. Zel'dovich, V. Kurt, and R. Syunyaev. Recombination of hydrogen in the hot model of the universe. *Sov. Phys. JETP*, 28, 146, 1969.

[1189] Ya. Zel'dovich, E. Levich, and R. Syunyaev. Stimulated Compton interaction between Maxwellian electrons and spectrally narrow radiation. *Sov. Phys. JETP*, 35, 733, 1972.

[1190] Y. Zhang, C. Cheng, J. Kim, J. Stanojevic, and E. Eyler. Dissociation energies of molecular and the hydrogen molecular ion. *Phys. Rev. Let.*, 92, 203003, 2004.

[1191] O. Zienkiewicz. *Finite Element Methods in Engineering Science*. (London: McGraw–Hill Book Company), 1971.

OPACITY PROJECT

[1192] D. Hummer and D. Mihalas. The equation of state for stellar envelopes. I. An occupation probability formalism for the truncation of internal partition functions. *Astrophys. J.*, 331, 794, 1988.

[1193] D. Mihalas, W. Däppen, and D. Hummer. The equation of state for stellar envelopes. II. Algorithm and selected results. *Astrophys. J.*, 331, 815, 1988.

[1194] W. Däppen, D. Mihalas, D. Hummer, and B. Mihalas. The equation of state for stellar envelopes. III. Thermodynamic quantities. *Astrophys. J.*, 332, 261, 1988.

[1195] D. Mihalas, D. Hummer, B. Mihalas, and W. Däppen. The equation of state for stellar envelopes. IV. Thermodynamic quantities and selected ionization fractions for six elemental mixes. *Astrophys. J.*, 350, 300, 1990.

[1196] D. Hummer. The atomic internal partition function. In Hauer and Merts [447], page 1.

[1197] W. Däppen, L. Anderson, and D. Mihalas. Statistical mechanics of partially ionized plasmas. The Planck–Larkin partition function, polarization shifts, and simulations of optical spectra. *Astrophys. J.*, 319, 195, 1987.

[1198] M. Seaton. Atomic data for opacity calculations. I. General description. *J. Phys. B*, 20, 6363, 1987.

[1199] K. Berrington, P. Burke, K. Butler, M. Seaton, P. Storey, K. Taylor, and Y. Yu. Atomic data for opacity calculations: II. Computational methods. *J. Phys. B*, 20, 6379, 1987.

[1200] Y. Yu, K. Taylor, and M. Seaton. Atomic data for opacity calculations: III. Oscillator strengths for C II. *J. Phys. B*, 20, 6399, 1987.

[1201] Y. Yu and M. Seaton. Atomic data for opacity calculations: IV. Photoionization cross sections for C II. *J. Phys. B*, 20, 6409, 1987.

[1202] M. Seaton. Atomic data for opacity calculations: V. Electron impact broadening of some C II lines. *J. Phys. B*, 20, 6431, 1987.

[1203] J. Fernley, K. Taylor, and M. Seaton. Atomic data for opacity calculations: VII. Energy levels, f–values, and photoionization cross sections for He–like ions. *J. Phys. B*, 20, 6457, 1987.

[1204] M. Seaton. Atomic data for opacity calculations: VIII. Line–profile parameters for 42 transitions in Li–like and Be–like ions. *J. Phys. B*, 21, 3033, 1988.

[1205] G. Peach, H. Saraph, and M. Seaton. Atomic data for opacity calculations: IX. The lithium isoelectronic sequence. *J. Phys. B*, 21, 3669, 1988.

[1206] D. Luo, A. Pradhan, H. Saraph, P. Storey, and Y. Yu. Atomic data for opacity calculations: X. Oscillator strengths and photoionization cross sections for O III. *J. Phys. B*, 22, 389, 1989.

[1207] D. Luo and A. Pradhan. Atomic data for opacity calculations: XI. The carbon isoelectronic sequence. *J. Phys. B*, 22, 3377, 1989.

[1208] M. Seaton. Atomic data for opacity calculations: XII. Line–profile parameters for neutral atoms of He, C, N, and O. *J. Phys. B*, 22, 3603, 1989.

[1209] M. Seaton. Atomic data for opacity calculations: XIII. Line profiles for transitions in hydrogenic ions. *J. Phys. B*, 23, 3255, 1990.

[1210] J. Tully, M. Seaton, and K. Berrington. Atomic data for opacity calculations: XIV. The beryllium sequence. *J. Phys. B*, 23, 3811, 1990.

[1211] P. Sawey and K. Berrington. Atomic data for opacity calculations: XV. Fe I – IV. *J. Phys. B*, 25, 1451, 1992.

[1212] V. Burke. Atomic data for opacity calculations: XVII. Calculation of line broadening parameters and collision strengths between n = 2, 3, and 4 states of C IV. *J. Phys. B*, 25, 4917, 1992.

[1213] S. Nahar and A. Pradhan. Atomic data for opacity calculations: XVIII. Photoionization and oscillator strengths of Si–like ions Si^0, S^{2+}, Ar^{4+}, Ca^{6+}. *J. Phys. B*, 26, 1109, 1993.

[1214] K. Butler, C. Mendoza, and C. Zeippen. Atomic data for opacity calculations: XIX. The Mg isoelectronic sequence. *J. Phys. B*, 26, 4409, 1993.

[1215] S. Nahar and A. Pradhan. Atomic data for opacity calculations: XX. Photoionization cross sections and oscillator strengths for Fe II. *J. Phys. B*, 27, 429, 1994.

[1216] A. Hibbert and M. Scott. Atomic data for opacity calculations: XXI. The Ne sequence. *J. Phys. B*, 27, 1315, 1994.

[1217] M. Le Dourneuf, S. Nahar, and A. Pradhan. Photoionization of Fe^+. *J. Phys. B*, 26, L1, 1993.

[1218] M. Bautista and A. Pradhan. Photoionization of neutral iron. *J. Phys. B*, 28, L1, 1995.

[1219] S. Nahar. Photoionization cross sections and oscillator strengths for Fe III. *Phys. Rev. A, 53, 38, 1996.*

[1220] M. Seaton, C. Zeippen, J. Tully, A. Pradhan, C. Mendoza, A. Hibbert, and K. Berrington. The Opacity Project – Computation of atomic data. *Rev. Mexicana Astr. Astrof.*, 23, 19, 1992.

[1221] M. Seaton. The Opacity Project – A postscript. *Rev. Mexicana Astr. Astrof.*, 23, 180, 1992.

[1222] M. Seaton. Radiative opacities. In Weiss and Baglin [1139], page 222.

[1223] W. Cunto, C. Mendoza, F. Ochsenbein, and C. Zeippen. TOPbase at the CDS. *Astr. Astrophys.*, 275, L5, 1993.

[1224] M. Seaton. Fitting and smoothing of opacity data. *Mon. Not. Roy. Astr. Soc.*, 265, L25, 1993.

[1225] M. Seaton, Y. Yu, D. Mihalas, and A. Pradhan. Opacities for stellar envelopes. *Mon. Not. Roy. Astr. Soc.*, 266, 805, 1994.

[1226] M. Seaton. New atomic data for astronomy: An introductory review. In Adelman and Wiese [10], page 1.

[1227] M. Seaton. Interpolations of Rosseland–mean opacities for variable X and Z. *Mon. Not. Roy. Astr. Soc.*, 279, 95, 1996.

[1228] M. Seaton, editor. *The Opacity Project. Volume 1.* (Bristol: Institute of Physics Publishing), 1995.

[1229] K. Berrington, editor. *The Opacity Project. Volume 2.* (Bristol: Institute of Physics Publishing), 1997.

[1230] M. Seaton. Free–free transitions and spectral line broadening. *J. Phys. B.*, 33, 2677, 2000.

[1231] M. Seaton and N. Badnell. A comparison of Rosseland–mean opacities from OP and OPAL. *Mon. Not. Roy. Astr. Soc.*, 354, 457, 2004.

[1232] N. Badnell, M. Bautista, K. Butler, F. Delahaye, C. Mendoza, P. Palmeri, C. Zeippen, and M. Seaton. Updated opacities from the Opacity Project. *Mon. Not. Roy. Astr. Soc.*, 360, 458, 2005.

IRON PROJECT

[1233] K. Berrington. Summary of the IRON and Opacity Projects. In Adelman and Wiese [10], page 19.

[1234] D. Hummer, K. Berrington, W. Eissner, A. Pradhan, H. Saraph, and J. Tully. Atomic data from the IRON Project. I. Goals and methods. *Astr. Astrophys.*, 279, 298, 1993.

[1235] D. Lennon and V. Burke. Atomic data from the IRON Project. II. Effective collision strengths for infrared transitions in carbon–like ions. *Astr. Astrophys. Suppl.*, 103, 273, 1994.

[1236] K. Butler and C. Zeippen. Atomic data from the IRON Project. V. Effective collision strengths for transitions in the ground configuration of oxygen–like ions. *Astr. Astrophys. Suppl.*, 108, 1, 1994.

[1237] H.–L. Zhang and A. Pradhan. Atomic data from the IRON Project. VI. Collision strengths and rate coefficients for Fe II. *Astr. Astrophys.*, 293, 953, 1995.

[1238] S. Nahar. Atomic data from the IRON Project. VII. Radiative dipole transition probabilities for Fe II. *Astr. Astrophys.*, 293, 967, 1995.

[1239] M. Galavis, C. Mendoza, and C. Zeippen. Atomic data from the IRON Project. X. Effective collision strengths for infrared transitions in Si– and S–like ions. *Astr. Astrophys. Suppl.*, 111, 347, 1995.

[1240] R. Kisielius, K. Berrington, and P. Norrington. Atomic data from the IRON Project. XV. Electron excitation of the fine–structure transitions in H–like ions He II and Fe XXVI. *Astr. Astrophys. Suppl.*, 118, 157, 1996.

[1241] M. Bautista. Atomic data from the IRON Project. XVI. Photoionization cross sections and oscillator strengths for Fe V. *Astr. Astrophys. Suppl.*, 119, 105, 1996.

[1242] S. Nahar and A. Pradhan. Atomic data from the IRON Project. XVII. Radiative transition probabilities for dipole allowed and forbidden transitions in Fe III. *Astr. Astrophys. Suppl.*, 119, 509, 1996.

[1243] H.–L. Zhang. Atomic data from the IRON Project. XVIII. Electron impact excitation collision strengths and rate coefficients for Fe III. *Astr. Astrophys. Suppl.*, 119, 523, 1996.

[1244] M. Bautista. Atomic data from the IRON Project. XX. Photoionization cross sections and oscillator strengths for Fe I. *Astr. Astrophys. Suppl.*, 122, 167, 1997.

[1245] M. Bautista and A. Pradhan. Atomic data from the IRON Project. XXVI. Photoionization cross sections and oscillator strengths for Fe IV. *Astr. Astrophys. Suppl.*, 126, 365, 1997.

[1246] H.–L. Zhang and A. Pradhan. Atomic data from the IRON Project. XXVII. Electron impact excitation collision strengths and rate coefficients for Fe IV. *Astr. Astrophys. Suppl.*, 126, 373, 1997.

[1247] M. Galavis, C. Mendoza, and C. Zeippen. Atomic data from the IRON Project. XXIX. Radiative rates for transitions within the $n = 2$ complex in ions of the boron isoelectronic sequence. *Astr. Astrophys. Suppl.*, 131, 119, 1998.

[1248] G.–X. Chen and A. Pradhan. Atomic data from the IRON Project. XXXVII. Electron impact excitation collision strengths and rate coefficients for Fe VI. *Astr. Astrophys. Suppl.*, 136, 395, 1999.

[1249] P. Storey, H. Mason, and P. Young. Atomic data from the IRON Project. XL. Electron excitation of the Fe V EUV transitions. *Astr. Astrophys. Suppl.*, 141, 285, 2000.

[1250] S. Nahar, F. Delahaye, A. Pradhan, and C. Zeippen. Atomic data from the IRON Project. XLIII. Transition Probabilities for Fe V. *Astr. Astrophys. Suppl.*, 144, 141, 2000.

[1251] G.–X. Chen and K. Pradhan. Atomic data from the IRON Project. XLIV. Transition probabilities and line ratios for Fe VI with fluorescent excitation in planetary nebulae. *Astr. Astrophys. Suppl.*, 147, 111, 2000.

[1252] S. Nahar and A. Pradhan. Atomic data from the IRON Project. LIX. New radiative transition probabilities for Fe IV including fine structure. *Astr. Astrophys.*, 437, 345, 2005.

[1253] S. Nahar. Atomic data from the IRON Project. LXI. Radiative E1, E2, E3, and M1 transition probabilities for Fe IV. *Astr. Astrophys.*, 448, 779, 2006.

OPAL

[1254] F. Rogers. Statistical mechanics of Coulomb gases of arbitrary charge. *Phys. Rev. A*, 10, 2441, 1974.

[1255] F. Rogers. Formation of composites in equilibrium plasmas. *Phys. Rev. A*, 19, 375, 1979.

[1256] F. Rogers. Occupation numbers for reacting plasmas. The role of the Planck–Larkin partition function. *Astrophys. J.*, 310, 723, 1986.

[1257] C. Iglesias, F. Rogers, and B. Wilson. Reexamination of the metal contribution to astrophysical opacity. *Astrophys. J.*, 322, L45, 1987.

[1258] F. Rogers and C. Iglesias. Parametric potential method of generating atomic data. *Phys. Rev. A*, 38, 5007, 1988.

[1259] C. Iglesias, F. Rogers, and B. Wilson. Opacities for classical Cepheid models. *Astrophys. J.*, 360, 221, 1990.

[1260] C. Iglesias and F. Rogers. Opacity tables for Cepheid variables. *Astrophys. J.*, 371, L73, 1991.

[1261] C. Iglesias and F. Rogers. Opacities for the solar radiative interior. *Astrophys. J.*, 371, 408, 1991.

[1262] F. Rogers and C. Iglesias. Radiative atomic Rosseland mean opacity tables. *Astrophys. J. Suppl.*, 79, 507, 1992.

[1263] C. Iglesias, F. Rogers, and B. Wilson. Spin–orbit interaction effects on the Rosseland mean opacity. *Astrophys. J.*, 397, 717, 1992.

[1264] F. Rogers and C. Iglesias. Rosseland mean opacities for variable compositions. *Astrophys. J.*, 401, 361, 1992.

[1265] C. Iglesias and B. Wilson. Statistical simulation of atomic data in opacity calculations. *J. Quantit. Spectrosc. Radiat. Transf.*, 52, 127, 1994.

[1266] C. Iglesias and F. Rogers. Discrepancies between OPAL and OP opacities at high densities and temperatures. *Astrophys. J.*, 443, 460, 1995.

[1267] C. Iglesias, B. Wilson, F. Rogers, W. Goldstein, A. Bar–Shalom, and J. Oreg. Effects of heavy metals on astrophysical opacities. *Astrophys. J.*, 445, 855, 1995.

[1268] F. Rogers and C. Iglesias. The OPAL opacity code: New results. In Adelman and Wiese [10], page 31.

[1269] C. Iglesias and F. Rogers. Updated OPAL opacities. *Astrophys. J.*, 464, 943, 1996.

[1270] F. Rogers and C. Iglesias. Opacity of stellar matter. *Space Sci. Rev.*, 85, 61, 1998.

[1271] C. Iglesias, F. Rogers, and D. Saumon. Density effects on the opacity of cool He white dwarf atmospheres. *Astrophys. J.*, 569, 111, 2002.

[1272] **WEB SITES**

opacities

Opacity Project
http://cdsweb.u-strasbg.fr/topbase/op.html
Iron Project
http://www.usm.lmu.de/people/ip/papers/papers.html
Opserver
http://opacities.osc.edu
Opacity Tables
http://cdsweb.u-strasbg.fr/topbase/OpacityTables.html
Topbase
http://cdsweb.u-strasbg.fr/topbase/topbase.html
Tipbase
http://cdsweb.u-strasbg.fr/tipbase/home.html
Tiptopbase
http://cdsweb.u-strasbg.fr/OP.htx
Radiative Data
http://www.astronomy.ohio-state.edu/~nahar

Index

Damping constant (width), 235, 237, 239, 240, 256